ACS SYMPOSIUM SERIES 285

Applied Polymer Science

Second Edition

Roy W. Tess, EDITOR

Gary W. Poehlein, EDITOR

American Chemical Society, Washington, D.C. 1985

Library of Congress Cataloging in Publication Data

Applied polymer science.
  (ACS symposium series, ISSN 0097-6156; 285)

  Includes bibliographies and indexes.

  1. Polymers and polymerization.

  I. Tess, Roy W. (Roy William), 1915–
II. Poehlein, Gary W. III. Series.

QD381.A66   1985   547.7   85-15766
ISBN 0-8412-0891-3

ACS Symposium Series

M. Joan Comstock, *Series Editor*

FOREWORD

The ACS SYMPOSIUM SERIES was founded in 1974 to provide a medium for publishing symposia quickly in book form. The format of the Series parallels that of the continuing ADVANCES IN CHEMISTRY SERIES except that, in order to save time, the papers are not typeset but are reproduced as they are submitted by the authors in camera-ready form. Papers are reviewed under the supervision of the Editors with the assistance of the Series Advisory Board and are selected to maintain the integrity of the symposia; however, verbatim reproductions of previously published papers are not accepted. Both reviews and reports of research are acceptable, because symposia may embrace both types of presentation.

CONTENTS

vi

PREFACE

SURFACE COATINGS, PLASTICS, POLYMER CHEMISTRY, AND RELATED TOPICS were covered in the 1975 edition of *Applied Polymer Science*. The favorable reception of the original edition, as well as specific requests for a new edition because of the many new developments in the field, has encouraged us to undertake the preparation of this new volume. Some entirely new chapters on new topics include the following: Transport Properties of Polymers, Fracture Mechanics of Polymers, Flammability of Polymers, Foamed Plastics, Polymers and the Technology of Electrical Insulation, Medical Applications of Polymers, Resins for Aerospace, Polymer Processing, Corrosion Control by Organic Coatings, Appliance Coatings, Polymer Coatings for Optical Fibers, and Paint Manufacture. A chapter entitled Introduction to Polymer Science and Technology has been included for the first time to provide some definitions of polymer terms, to cover certain omissions in the original and current editions, and to ease the way for the general reader or student. In this chapter, certain topics pertinent to an introduction have been covered only briefly or occasionally not at all because the topics are covered thoroughly in subsequent chapters. On the other hand, limited repetition has been tolerated among different chapters when it seemed desirable to provide continuity of thought within a chapter.

Several chapters in the original edition have been deleted in this volume for the following reasons: in two cases because of their strictly historical nature; the lack of substantive new developments in the topic; and the nonavailability of the original authors. Quite a few new authors were recruited to prepare entirely new or revised versions of former chapters. Unfortunately, some promised chapters never were forthcoming and had to be abandoned. Although all topics of potential interest simply could not be included in the present volume, nevertheless a very broad range of topics has been covered. The original edition should continue to be useful because it contains some chapters, including those of historical importance, that are not repeated in this new edition.

The subject matter of this book represents a major segment of chemical industry and a growing segment of the chemical curriculum in academia. Reflecting the interest of both industry and academia in the topics, the authors and editors of this volume are affiliated with industry, academia, or both. It is hoped that this book will find use by students and teachers in coursework and research as well as by scientific and technical personnel in industrial and governmental work.

Acknowledgments

We are indebted to the capable and conscientious contributing authors who are among the foremost authorities in their fields. We also thank the Executive Committee of the Division of Polymeric Materials: Science and Engineering for its endorsement of this book, and the staff of the American Chemical Society Books Department for their help and advice. Useful comments were made by an advisory committee consisting of J. K. Craver, R. R. Myers, J. L. Gardon, D. R. Paul, R. D. Deanin, R. B. Seymour, J. K. Gillham, C. E. Carraher, J. C. Weaver, K. L. DeVries, and L. F. Thompson.

Many experts have been helpful in providing outside reviews of chapters. We express our appreciation to the following: D. R. Paul, Aldo DeBenedictis, F. M. McMillan, P. E. Pierce, C. W. Schroeder, Andrew Klein, J. W. Gooch, Charles Aloisio, John Muzzy, Alan Berens, F. J. Schork, A. P. Yoganathan, Albin Turbak, Miroslav Marek, T. E. Futern, Robert Samuels, Wayne Tincher, J. D. Willons, George Fowles, S. S. Labana, T. J. Miranda, J. E. Carey, J. P. McGuigan, A. L. Rocklin, J. K. Gillham, G. D. Edwards, J. M. Klarquist, G. G. Velten, G. A. Short, T. K. Rehfeldt, E. W. Starke, Jr., D. C. Rich, A. G. Rook, and S. L. Davidson. The generosity of several publishers in permitting reproduction of text and figures is gratefully acknowledged and mentioned specifically at various places in the text.

Roy W. Tess
Consultant
P.O. Box 577
Fallbrook, CA 92028

Gary W. Poehlein
School of Chemical Engineering
Georgia Institute of Technology
Atlanta, GA 30332–0100

April 1985

INTRODUCTION

A Century of Polymer Science and Technology

HERMAN F. MARK

Polytechnic Institute of New York, Brooklyn, NY 11201

Since the beginning of history, natural polymers such as fur, wood, hide, wool, horn, cotton, flax, resins, and gum, together with stone and a few metals, were the backbone of all civilization and art.

We would have no Bible, no Greek epics and tragedies, and no Roman history without parchment and papyrus. There would be no paintings of Leonardo, Raphael, and Rembrandt without canvas and polymerizing oils. And were would be no music of Corelli, Beethoven, and Tchaikovsky without string instruments, all of which consist entirely of natural organic polymers such as wood, resins, and lacquers.

All naval battles until 100 years ago were fought with wooden ships that were kept afloat and moving by rosin, ropes, and sails. Hardened wood and strongly tanned leather were the first offensive and defensive weapons, and later, catapults and artillery were placed in position on wooden carriages drawn by horses or men using cellulosic ropes. Even today, the common propellants for all firearms are based on cellulose nitrate or on equivalent organic polymers.

Most of all, in daily life, shelter, clothing, food, education, and recreation depended, and still depend, essentially on the use of natural polymers--wood, cotton, fur, wool, silk, starch, leather, paper, rubber, and a variety of resins, glues, and coatings. Around each of these materials a highly sophisticated art developed--entirely empirical and without any basic knowledge and, in fact, in most cases, without any concern about the material's composition and structure.

No wonder then that leading philosophers and scientists always have been strongly attracted by the exceptional properties and the outstanding capabilities of these materials and have studied them with whatever methods they had available. Such fascinating phenomena as the spinning of strong, tough, glossy, and extremely durable threads by spiders and silkworms are said to have caused early speculation in China about making artificial silk long before Robert Hooke suggested it in his "Micrographia" in 1664.

Evidently, the idea was there but the material was lacking then to perform successfully the process of fiber spinning. But when Henri Braconnot in 1832 and Christian Friedrich Schoenbein in 1846 discovered how to make cellulose nitrate, the time for the "spark" had arrived. British Patent 283, issued in 1855, disclosed the treating of bast fibers from mulberry twigs with nitric acid, dissolving the product in a mixture of alcohol and ether together with rubber, and from this viscous mass drawing fibers with a steel needle; after these fibers solidified in air, they were wound on a spool.

The first man-made fiber was prepared by the manipulation of a natural polymer, cellulose, which was made soluble through nitration. These fibers, which were highly flammable, represented an impractical but pioneering step in a promising direction.

However, after Ozanam in 1862 constructed the first spinning jets, and Joseph W. Swan in 1883 found a method to "denitrate" the filaments and convert them into cellulose hydrate, the way was open for Count Hilaire de Chardonnet to obtain a patent in 1885 and to bridge the gap from the laboratory to the plant by simplifying and coordinating the four essential steps: nitration, dissolution, spinning, and regeneration. His invention and enterprise initiated a new era in the textile business, which now—only 90 years later—has grown into a multifaceted industry whose output has a value of several billion dollars per year. Chardonnet's procedure made it clear that to form fibers the cellulose (or some other natural polymer) must first be made soluble, then the solution must be extruded, and finally the cellulose (or the original polymer) must be regenerated in the form of a fine fiber.

Soon after 1890, additional methods were found to solubilize cellulose by acetylation, xanthation, and cuproxyammoniation; to spin the resulting solutions by coagulating them into the form of a filament; and to use the resulting fiber as it is or to regenerate it into cellulose. These artificial products were started with an already existing natural polymer, generally cellulose, and modified chemically and brought into fiber form by coagulation, stretching, and drying. The resulting rayons dominated the field of man-made fibers until the mid-1930s.

Rubber was another natural polymer whose exceptionally useful properties aroused the interest of many prominent scientists. In 1826, Michael Faraday performed elemental analysis of rubber and established its correct empirical formula as C_5H_8. This formula was confirmed by Jean Baptiste Andre Dumas in 1838. Destructive distillation then was used to explore the structure of complex materials because this procedure has capacity to decompose large molecules into simpler structural units. Justus von Liebig, John Dalton, and others used this method and obtained several low-boiling liquids from rubber. In 1860, C. Greville Williams isolated the most preponderant species, which had the formula C_5H_8, and called it isoprene. In the best tradition of classical organic chemistry, he proceeded from analysis to synthesis and found that a white spongy elastic mass could be obtained from isoprene through the action of oxygen. But it remained for Gustave Bouchardat in 1879 to take isoprene obtained from natural rubber by dry distillation and convert it by treatment with hydrochloric acid into an elastic, rubberlike solid.

That same year the first laboratory sample of synthetic rubber was developed. Otto Wallach in 1887 and William A. Tilden in 1892 confirmed this synthesis and showed that this synthetic elastomer reacted with sulfur in the same way as ordinary rubber to form a tough, elastic, insoluble product.

At the end of the 19th century, rubber, with gutta-percha, was used mainly as an electrical insulator on wires and cables. Demand was limited, and the supply of natural rubber at a reasonable price (about $1.00/1b in 1900) was ensured. Some work was done during these years on practical syntheses of isoprene and on the replacement of isoprene by its simpler homolog, butadiene, which had been known since 1863. However, advent of the automobile and accelerated use of electric power rapidly increased the demand for rubber, thus raising its price to about $3.00/1b in 1911. These circumstances focused new attention on the production of a synthetic rubber. S. B. Lebedev polymerized butadiene in 1910, and Carl Dietrich Harries, between 1900 and 1910 established qualitatively the structure of rubber as a 1,4-polyisoprene and synthesized larger quantities of rubberlike materials from isoprene and other dienes.

Little was known, however, about the exact configuration of the rubber molecule or the molecular mechanism of rubber elasticity. Two world wars and the mushrooming development of the automobile and the airplane raised the demand for elastomeric materials to a level that, by the 1950s, dozens of large industrial organizations produced some 2 million tons of synthetic rubber per year. Most of the production steps are now fundamentally well known, and most of the basic properties of raw and cured elastomers are now reasonably well understood.

1900-1910

At the beginning of this century the first fully synthetic polymer was made, that is, a material that was prepared by the interaction of small, ordinary organic molecules and represented a system of very high molecular weight. This synthesis was not only step of scientific importance but also the beginning of a new technology.

The development of Bakelite by Leo H. Baekeland was actually an outgrowth of his search for a synthetic substitute for shellac. Such a material, he believed, might offer properties superior to those of natural shellac.

Baekeland decided to make the synthetic material by reacting phenol with formaldehyde to form a hard resin and then dissolving the resin in a suitable solvent. He had no difficult forming various resins, but to his dismay he could find no satisfactory solvent. However, he realized that some of the hard, solvent-resistant resins he had produced in the laboratory might have great commercial value in themselves. One advantage they had was an outstanding ability to maintain their shape. In addition, they were good electrical insulators, could be machined easily, and were resistant to heat and many chemicals.

In 1909, at a meeting of the New York Section of the American Chemical Society, Baekeland announced his development of Bakelite.

The thermosetting plastic was first made commercially in 1910 by
General Bakelite Company (later called Bakelite Corporation), which
Union Carbide acquired in 1939.

During the first decade extensive descriptive studies were
carried out on natural polymers of all kinds: proteins (wool,
silk, and leather), carbohydrates (cellulose, starch, and gums),
and other resinous products (shellac, rubber, and gutta-percha).
Three large domains of scientific and technical interest came into
being:

1. The field of proteinic materials with special institutes for
   research on leather, wool, silk, with corresponding textbooks,
   journals, and societies serving and advancing large industries:
   shoe, luggage, textiles, and others.
2. The discipline of cellulosics having its own special textbooks,
   for example, "Wood Chemistry and Technology"; special journals,
   for example, Paper Trade Journal; numerous societies, for
   example, the Paper and Textile Society.
3. The domain of rubber and resin science and industry with its
   own largely empirical know-how and its own highly sophisticated
   technologies.

Rubber chemists, fiber chemists, and resin chemists pursued
their eminently practical goals with admirable empirical skill and
success without being too much concerned about basic structural
problems. For them, in general, these three disciplines were
different worlds as were Jupiter and Saturn before Copernicus.

During this period, two events clearly foreshadowed the existence of very large chainlike molecules.

In 1900, E. Bamberger and F. Tschirner reacted diazomethane
with β-arylhydroxylamines to methylate the hydroxyl groups of the
substituted hydroxylamine. Instead they obtained a product in
which two phenylhydroxylamines were connected by the methylene
bridge:

$$CH_2 = N \equiv N$$

They concluded that diazomethane dissociates into N_2 and CH_2, which
in turn can either react with the amine or polymerize to form
polymethylene. This material $(CH_2)_x$ was found to be a white,
chalklike, fluffy powder, apparently amorphous, with a melting
point of 128 °C.

This work was the first correct formulation and description of
a polyhydrocarbon, linear polymethylene, which in its structure and
properties is identical with linear 1,2-polyethylene. However,
because interest in the amorphous byproducts of organic chemical
syntheses was lacking, and evidently also because diazomethane was
limited in availability, no impression was made on the chemists of
1900.

In 1906, Hermann Leuchs, one of Emil Fischer's most distinguished associates, made an ingenious step in the direction of

forming long linear polypeptide chains. He prepared α-amino acid N-carboxylic anhydrides of the type

$$
\begin{array}{c}
\text{O} \\
\| \\
\text{C} \!-\!\!-\!\!-\! \text{O} \\
\diagup \qquad\qquad \diagdown \\
\text{HN} \qquad\qquad\qquad \text{C}\!=\!\text{O} \\
\diagdown \qquad\qquad \diagup \\
\text{CHR}
\end{array}
$$

and found that they split off CO_2 at elevated temperatures and in the presence of moisture to produce solid bodies which he considered to be linear polymers:

$$
\left[\begin{array}{c}
\text{HN}\!-\!\text{CHR}\!-\!\text{C}\!=\!\text{O} \\
\ | \qquad\qquad\quad\ | \\
\end{array}\right]_x
$$

Leuchs described the synthesis of several representative anhydrides of type 1 having different substituents. Although the discovery of Hermann Leuchs did not attract much attention at first, it has been expanded into one of the most important synthetic tools in protein research. Emil Fischer, whose institute in Berlin occupied a leading role in the research on sugars and proteins for 30 years, expressed at several occasions the opinion that his own synthetic polypeptides, and by extrapolation natural proteins, would be represented by long chains with the –CO–NH– unit as a recurring bond.

1910–1920

Rapidly increasing commercial use of cellulose and its derivatives in the industries of paper, textiles, films, and coatings during this period--partly in connection with World War I--resulted in a strong intensification and establishment of new methods to analyze and characterize natural polymers of all kinds. During this period large and important industries began to get involved with the chemical reactivity of cellulosic materials. Nitrocellulose and cellulose acetate emerged as the first important commercial thermoplastics; casein, phenol, urea- and melamine-formaldehyde systems began to represent the family of the thermosettings. Cellulose xanthate and cuprammonia cellulose were used in mounting scale in the manufacturing processes for manmade fibers (rayon) and films (cellophane).
Strong interest in process and product control led to the use of new methods for the characterization and testing of polymers. E. Berl carried out extensive studies on the viscosity of cellulose and cellulose derivatives in solution. One of his results was that less viscous solutions always gave fibers or films of decreased strength, which he ascribes to a chemical degradation of the material. A. Samec made corresponding observations on the degradation of starch.

In addition to the accumulation of empirical data in relation
to the properties of many natural polymers, the literature of this
period also contains numerous speculations on the structural
formula of these materials, particularly of rubber and cellulose.
On the basis of the fact that these materials consist of specific
basic units such as isoprene and glucose, a variety of proposals
were offered as to how these units are arranged and linked together
in the molecule. Chains were preferred for rubber and cellulose by
some authors, cyclic structures by others. No absolutely
convincing arguments were available at that time for any of the
numerous suggestions. However, in 1920, Herman Staudinger postulated in a basic article on polymerization that rubber and two of
the then known synthetic polymers are formed of long chains in
which the basic units are held together by normal chemical
valences. He constructed formulas for polystyrene, polyoxymethylene, and rubber which are still used today:

Polystyrene

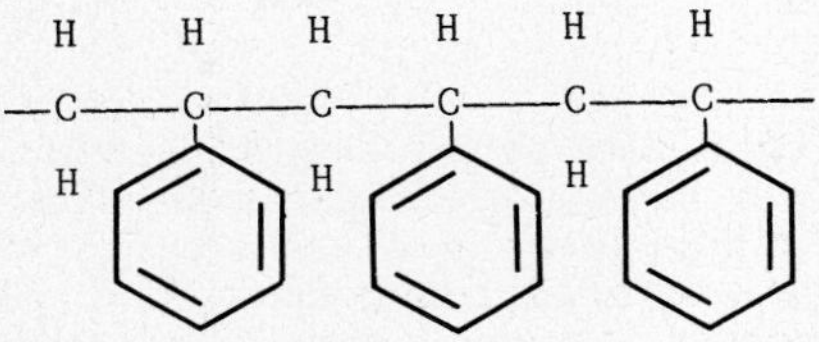

Polyoxymethylene

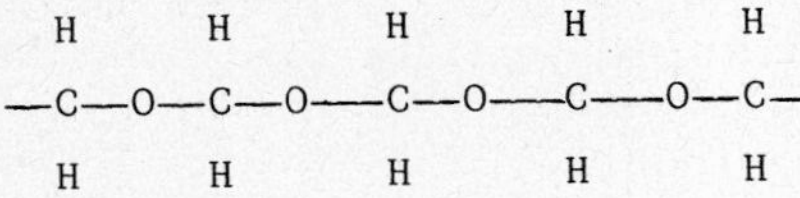

Rubber

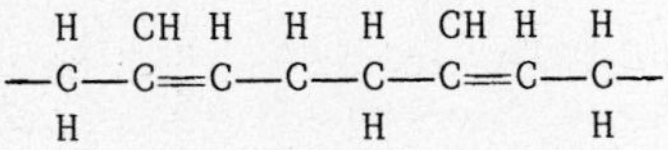

1920–1930

This heroic decade of polymer science and engineering is characterized by the introduction of several novel physical chemical
methods for the study of polymers in solution and in the solid
state. Precision osmometry, ultracentrifugation, and electrophoresis provided important new data on polymers--soon to be termed
macromolecules--in solution, whereas X-ray diffraction and IR
spectroscopy made decisive contributions to our knowledge of these
materials in the solid state such as membranes, fibers, or gels.
    For many eminent chemists of those days, for example,
P. Karrer, K. Hess, R. O. Herzog, M. Bergmann and H. Pringsheim,
the existence of organic substances with molecular weights of
several hundred thousand seemed unlikely, and Staudinger's arguments in favor of their existence appeared entirely insufficient.
These scientists and many others preferred to think of these
substances as consisting of small building units that are held
together by exceptionally strong forces of aggregation or associa-
tion--forces that, at that time, were still of unknown origin.
    The fact that proteins, cellulose, and rubber are the products
of living organisms (plants or animals) was an attractive and,

probably, legitimate argument in favor of something new, something that still had to be learned and clarified to understand their structure and properties.

However, as is often the case in science and history, this somewhat romantic approach had to fade away gradually under the influence of more and better experimental evidence pointing toward the macromolecular concept. This change did not happen without contradiction and opposition; on the contrary, during many meetings, symposia, and seminars, opposing views were presented and defended with great emphasis and insistence. Strangely enough, even the champions of the long-chain aspect—Freudenberg, Meyer, and Staudinger—did not agree with each other, as they easily could have done. Instead of concentrating on the essential principle, they disagreed on specific details, and on certain occasions they argued with each other more vigorously than with the defenders of the association theory. Of course, none of them was, at that time, completely correct in all the details of his approach. But they all were thinking and working in the right direction, and in the end they emerged as the natural leaders for future developments.

Many factors eventually tipped the scales in favor of the concept of very long, chainlike molecules. One factor was the rapid improvement and refinement of the X-ray diffraction method that, properly and precisely applied, not only gave answers in favor of long chains but permitted, and still permits, a progressively detailed description of their microstructure. Regarding proteins, it was becoming increasingly clear that each turn and wiggle of the chains represents a significant design and has far-reaching consequences for the actions of the substance in the living organism.

Another important factor in favor of long chains was the introduction of the ultracentrifuge by Th. Svedberg. The ultracentrifuge played a decisive role because it was the first method that permitted direct and reproducible measurements of molecular weights in the range above 40,000. In addition, improved osmometers added significance and reliability to these data.

Many European scientists made valuable contributions to the long-chain concept, but at the same time in the United States W. H. Carothers and his associates provided additional important evidence for the existence of very long-chain molecules by the quantitative analytical determination of their end groups. Right from the beginning, Carothers focused his attention on purity. He realized that purities of 99% or even 99.5% would not open the door to the realm of true macromolecules, even though in ordinary organic chemistry materials of this specification are quite normal and useful. An overall classification as addition and condensation polymers, an idea not clearly formulated earlier by anybody, introduced immediately the important element of order and facilitated the planning and the tracking of work in progress.

Both classes of polymers were attacked simultaneously, so that free-radical-initiated, self-propagating chain reactions and slow, endothermic step reactions were studied side by side. After the first results were attained, a grand strategy for practical applications developed quite naturally; the vinyl- and diene-type addition polymers were pursued with the ultimate aim being the production of a synthetic rubber. The signals coming from the

polycondensation front, meanwhile, strongly indicated the existence
of superior fiber and film formers.

1930-1940

With the basic structure of polymers of macromolecules clarified,
scientists now searched for a quantitative understanding of the
various polymerization processes, the action of specific catalysts,
and initiation and inhibitors. In addition, they strived to
develop methods to study the microstructure of long-chain compounds
and to establish preliminary relations between these structures and
the resulting properties. In this period also falls the origin of
the kinetic theory of rubber elasticity and the origin of the
thermodynamics and hydrodynamics of polymer solutions. Industrially polystyrene, poly(vinyl chloride), synthetic rubber, and
nylon appeared on the scene as products of immense value and utility. One particularly gratifying, unexpected event was the
polymerization of ethylene at very high pressures.

1940-1950

World War II accelerated the movement of many materials from
laboratory scale to production level, particularly in the domains
of synthetic rubbers, fibers, films, and coatings. In the synthetic rubber field, the techniques of emulsion and suspension
polymerization were put to work in a surprisingly short time and
removed the threat of a rubber shortage. New fiber formers were
developed on commercial scale and three types--polyamide, polyester, and acrylics--were firmly established for further improvement
and expansion. The technology of films and plastics was enlivened
by the existence of these materials in large scale and by the
availability of others such as polyethylene, fluoropolymers, and
silicones. Polyepoxides, polycarbonates, and polyurethanes
appeared on the scene. Synthetic rubbers did not only expand in
quantity but even more so in diversity of composition, structure,
and application. At the same time all existing methods for the
characterization of polymers were improved, and new ones were
added: gas chromatography, differential thermal analysis,
polarized-IR spectroscopy, and small-angle X-ray diffraction
spectroscopy.

1950-1960

Several unexpected and eminently important events occurred during
this period: A. Keller's discovery of chain folding, which led to
a completely new and very fruitful concept regarding the
supermolecular structure of polymeric systems; the discovery of
coordination of complex catalysts by K. Ziegler; and the subsequent
synthesis of stereoregulated polymers by G. Natta; and last but
certainly not least, the firm concept of the alpha helix by L.
Pauling and with it the beginning of the systematic, step-by-step
analysis of protein molecules, supported and followed by the
completely controlled build up of synthetic polypeptides. Simultaneously the double helix of Crick and Watson led to the first
basic understanding of the genetic code and to a galaxy of

important research projects that are still in progress. At the same time the understanding of and experience in preparing and using synthetics reached such a level that specially designed macromolecules could be made for specific demands without any trial and error type of random laboratory work.

1960-1970

By the beginning of this decade, so many new materials and so many unexpected ways to make and use them had been established that a shakedown period was indicated during which the fundamental laboratory work and the pilot plant efforts were transformed into profitable large-scale operations. From the tailoring of a rubber or a fiber in the laboratory, one had to advance to its tailoring on a large commercial scale smoothly, uniformly, and with profit. Engineering was written on the flag of the emerging generation of polymer chemists and physicists. At the same time the thoughts and ideas went further and, besides satisfying existing demands, progressive pioneers started to imagine newer vistas and to create new dimensions in living, transportation, communications, education, and relaxation. All these efforts had to be geared in such a way that they would meet such societal needs as safety, purity of the environment, and health. These effort also had to take into account the necessity of saving energy, first through improved design of planes, cars, ships, and houses, and later through recycling and waste utilization.

1970-1980

On the basis of strong demands to save energy in the construction of vehicles of all kinds, the large family of engineering plastics began to appear on the scene and gained more and more momentum.

The engineering thermoplastics possess high crystalline melting temperature (T_m) and high glass transition temperature (T_g) but are still extendable and moldable. Preferably they are used as composites, the matrices being aromatic polyesters, polyamides, polysulfones, polysulfides, polyethers, and polyimides, and the fillers being carbon black, silica, glass fibers, and carbon fibers. The resulting sheet-molding compounds (SMC) are fabricated into integral parts of cars, trucks, buses, and planes. Their thermosetting counterparts are the short fiber-filled reaction injection moldings (RIM) where the reinforced matrix of polyepoxide, polyurethane, or unsaturated polyesters sets during formation to give a hard, tough, insoluble, and infusible object.

Other areas of research and development that are now taking shape are studies of organic polymers that are photo- and electroresponsive, and studies of systems that are biocompatible and promise to find extensive use in biology and medicine.

Literature Cited

1. Hoesch, K. "Emil Fischer"; DCG: Berlin, 1921.
2. Flory, P. J. "Principles of Polymer Chemistry"; Cornell University Press: Ithaca, NY, 1953; p. 6.
3. Mark, H. "Polymers, Past, Present, and Future"; Welch Foundation: Houston, 1967.
4. Olby, R. "Double Helix"; University of Washington Press: Seattle, 1974.
5. Stahl, G. A. in "A Short History of Polymer Science," Stahl, G. A., Ed.; ACS SYMPOSIUM SERIES 175, American Chemical Society, Washington, D.C., 1981; p. 26.

Introduction to Polymer Science and Technology

CHARLES E. CARRAHER, JR.[1], and RAYMOND B. SEYMOUR[2]

[1] Department of Chemistry, Wright State University, Dayton, OH 45435
[2] Polymer Science Department, University of Southern Mississippi, Hattiesburg, MI 39406–0076

Structure of Polymers
Kinetics of Polymerization
Property–Molecular Weight Relationships
Interchain and Intrachain Forces
Crystalline–Amorphous Structures
Transitions
End Uses of Polymers as Related to Structure
Physical Characterization and Testing
Educational Aspects
Nomenclature

Polymer science and technology are interdisciplinary in that they borrow and contribute to other fields of science. They borrow in the sense that the laws that serve as the basis of chemistry, physics, and engineering are equally applicable to macromolecules. They contribute in a similar manner, that is, basic principles formulated within the framework of polymer science and technology are applicable to chemistry and other disciplines. Thus, the technological principles applicable to the processing of metals are applicable to the processing of polymers and vice versa.

Briefly, polymer science is the science that deals with large molecules wherein the chemical bonds are largely covalent. Polymer technology is the practical application of polymer science. The word "polymer" is derived from the Greek "poly" (many) and "meros" (parts). The word "macromolecule" ("macro" meaning large) is often used synonymously for polymer, and vice versa. Some scientists tend to differentiate between the two terms with macromolecule being used to describe large molecules such as DNA and proteins that cannot be depicted as being (exactly) derived from a single, simple (monomeric) unit, and polymer is used to describe larger molecules such as polystyrene that can be depicted as being composed of styrene units. This differentiation is not always observed and will not be in this text. The process of forming a polymer is called polymerization.

0097–6156/85/0285–0013$09.75/0

The degree of polymerization (DP) or average degree of polymerization ($\overline{DP}$) is the number (or average number) of units (mers) composing a chain(s). The term "chain length" is used as a synonym for DP. The DP of a dimer is 2, that of a trimer is 3, that of a tetramer is 4, etc. Chains with DP values below 10 to 20 are referred to as oligomers (small units) or telomers. Many polymer properties are dependent on chain length, but for most commercial polymers the change in polymer property with change in DP is small when the DP is greater than 100.

Structure of Polymers

Two terms, configuration and conformation, are often confused. Configuration refers to arrangements fixed by chemical bonding that cannot be altered except by primary bond breakage. Terms such as head-to-tail, d- and l-, cis, and trans refer to the configuration of a chemical species. Conformation refers to arrangements around single primary bonds. Polymers in solution or in melts continuously undergo conformational changes.

Monomer units in a growing chain usually form what is referred to as a head-to-tail arrangement where the repeating polymer chain for a vinyl monomer $H_2C = CHX$ can be described by

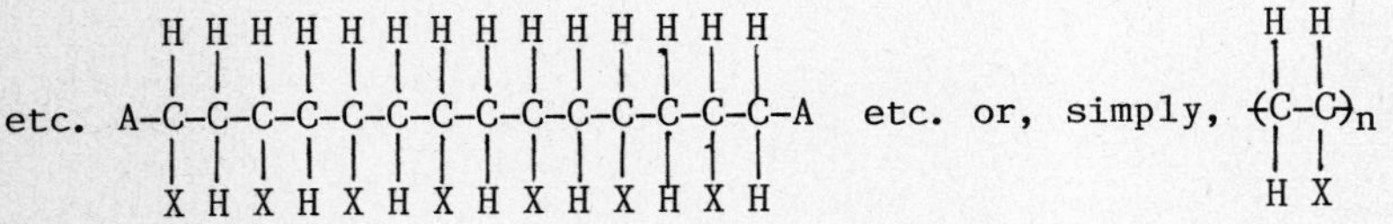

```
       H H H H H H H H H H H H H H                          H H
       | | | | | | | | | | | | | |                         | |
etc. A-C-C-C-C-C-C-C-C-C-C-C-C-C-C-A   etc. or, simply, (C-C)n
       | | | | | | | | | | | | | |                         | |
       X H X H X H X H X H X H X H                          H X
```

A head-to-head configuration would be represented by $(CH_2CHXCHXCH_2)_n$.

Even with a head-to-tail configuration, a variety of possible structures exists. For illustrative purposes, we will consider possible combinations derived from the homopolymerization of A and the copolymerization of A with B. The following are types of polymers that can be prepared. Homopolymerization (involves one monomeric unit in the chain):

Linear

    A ---> -A-A-A-A-A-

Branched

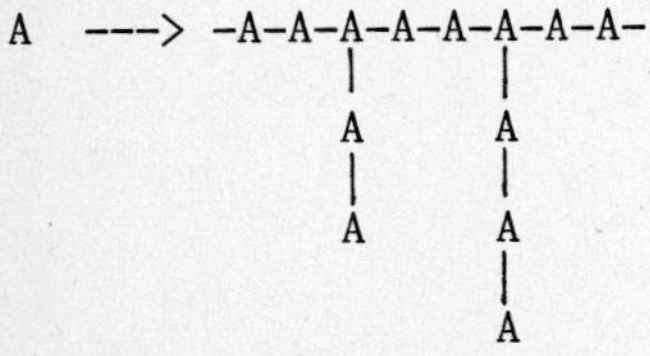

    A ---> -A-A-A-A-A-A-A-A-
                | |
                A   A
                | |
                A   A
                        |
                        A

Cross-linked

```
                          -A-A-A-A-              -A-A-A-A-
   A ─────────>               |        and/or        |
                          -A-A-A-A-                   C
                                                      |
                                                  -A-A-A-A-
```

where C = a cross-linking agent

Copolymerization (involves more than one monomeric unit in the chain):

Linear-Random

     -A-A-B-A-B-B-A-A-B-A-B-

Linear-Alternating

     -A-B-A-B-A-B-A-B-

Linear-Block

     -A-A-A-A-B-B-B-B-B-B-A-A-A-A-B-B-

Graft

```
   -A-A-A-A-A-A-A-A-A-A
          |         |
          B         B
          |         |
          B         B
          |         |
          B         B
                    |
                    B
                    |
                    B
```

Cross-linked or Network (three-dimensional; wide variation in possible structures

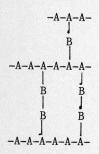

```
          -A-A-A-
              |
              B
              |
  -A-A-A-A-A-A-A-
      |       |
      B       B
      |       |
      B       B
      |       |
  -A-A-A-A-A-A-A-
```

Polymers with the above structures can be tailored to exhibit desired properties by using combinations of many of the common monomers.

Configuration also refers to structural regularity with respect to the substituted carbon within the polymer chain. For linear homopolymers derived from monomers such as styrene and vinyl chloride of the form $H_2C = CHX$, configuration from monomeric unit to monomeric unit can vary somewhat randomly (atactic) with respect to the carbon to which the X is attached, or can vary alternately (syndiotactic), or can be alike such that all of the X groups can be placed on the same side of a backbone plane (isotactic).

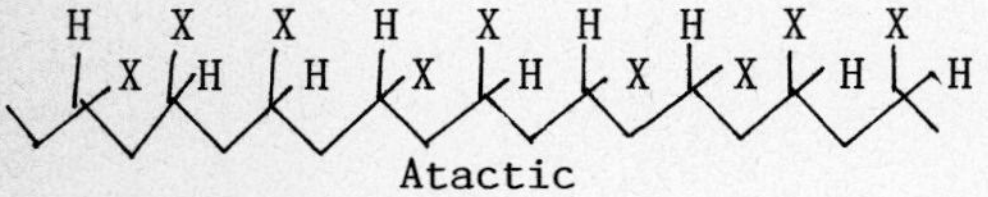

Atactic 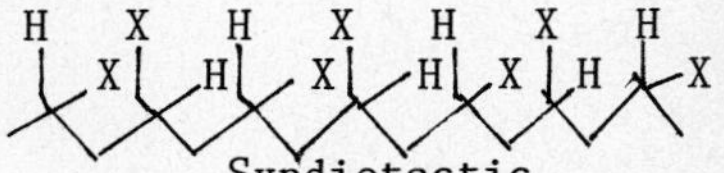   Syndiotactic

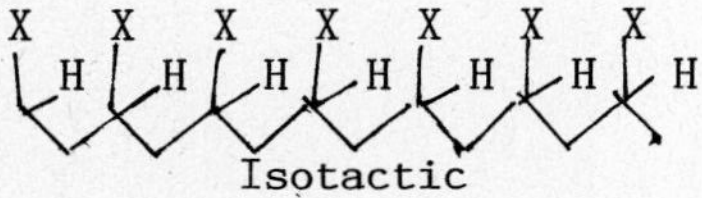

Isotactic

Polymerization of 1,3-dienes such as 1,3-butadiene can occur by 1,2-polymerization or 1,4-polymerization as follows:

$$n \; H_2C=CH-CH=CH_2 \longrightarrow (CH_2-CH)_n$$
$$\text{1,2-polymerization} \qquad \qquad CH=CH_2$$

$$n \; H_2C=CH-CH=CH_2 \longrightarrow (CH_2-CH=CH-CH_2)_n$$

1,4-polymerization

The 1,2-products can exist in the stereoregular isotactic or syndiotactic forms and the irregular atactic forms. The stereo-regular forms are rigid, crystalline materials, and the atactic forms are soft elastomers.

In the case of 1,4-polymers, rotation is restricted because of the double bond in the chain. The *cis* isomer of 1,4-polybutadiene is a soft elastomer with a T_g of −108 °C, but the trans isomer is harder with a T_g of −83 °C.

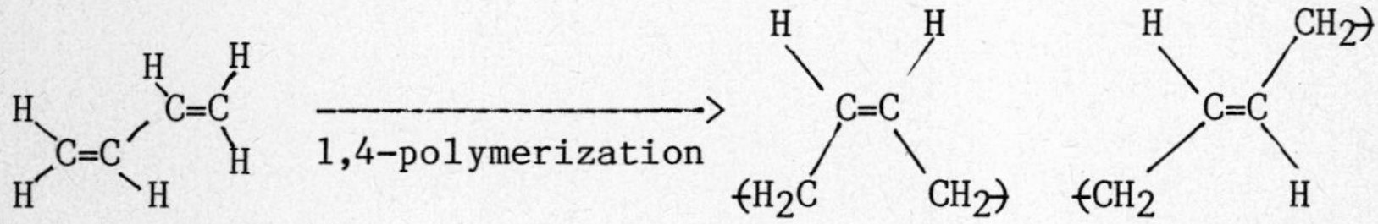

Butadiene   1,4-polymerization   1,4-*cis*   1,4-*trans*
                                             polymer   polymer

Kinetics of Polymerization

General Considerations. The terms addition and condensation polymers were first used by Carothers and are based on whether the repeating unit, mer, of a polymer chain contains the same atoms as the monomer (1-3). Addition polymers have the same atoms as the monomer in the repeat unit, with the atoms in the backbone typically being only carbon. Condensation polymers typically contain fewer atoms within the repeat unit than the reactants because of the formation of byproducts during the polymerization process, and the polymer backbone typically contains atoms of more than one element. Polystyrene, poly(vinyl chloride), polyethylene, and poly(vinyl alcohol) are illustrative of addition polymers, and polyesters and polyamides (nylons) are illustrative of condensation polymers. The corresponding polymerizations are then called addition and condensation polymerizations.

Stepwise, or step-growth, kinetics refers to polymerizations in which polymer molecular weight increases in a slow, stepwise manner as reaction time increases. Thus, for a polyamide, polymerization begins with one dicarboxylic acid molecule reacting with one diamine and the resulting formation of an amide linkage. This amine-acid unit can now react with either another acid or amine to produce a chain capped with either two acid or two amine groups. This process continues throughout the reaction system wherever molecules with the correct functionality, necessary activation energy, and correct geometry collide. The activation energy for each step-growth reaction is about 20 to 50 kcal/mol and is approximately constant throughout the reaction.

Chain growth through chain kinetics requires initiation to begin the growth. For free-radical processes, the initiation step produces a free radical derived from light or peroxides. For cationic polymerizations, the initiation step produces a cation typically derived from a catalyst-cocatalyst complex such as $H^+[BF_3OH^-]$. Anionic polymerizations begin through initiation with metal alkyls, alkali amines, etc. that form carbanions.

Polymerization rapidly occurs only with chains possessing a free radical, cation, or anion (referred to as active chains), with rapid addition of units and subsequent chain growth. This process results in a reaction mixture largely composed of polymer and monomer throughout the entirety of the polymerization process. Polymerization occurs until the reactive end is terminated by chemical or physical means.

The activation energy for each chain-growth reaction is only approximately 0 to 5 kcal/mol. Thus, the driving force for differently observed kinetic processes is directly related to the ease of addition of subsequent units during polymerization.

Most addition polymers are formed from polymerizations exhibiting chain-growth kinetics. Such processes include the typical polymerizations of the vast majority of vinyl monomers such as ethylene, styrene, vinyl chloride, propylene, methyl acrylate, and vinyl acetate. Furthermore, most condensation polymers are formed from systems exhibiting stepwise kinetics. Industrially, such systems include those used for the formation .pa of polyesters and polyamides. Thus, a large overlap exists between the terms addition polymers and chain-growth kinetics and the terms condensation polymers and stepwise kinetics.

Although the overlap of terms is great, many exceptions exist. For example, the formation of polyurethanes typically occurs through stepwise kinetics. The polymers are classified as condensation polymers, and the backbone is heteratomed, yet no byproduct is released when the isocyanate and diol are condensed. The formation of nylon 6, a condensation polymer, from the corresponding internal lactam occurs through chain-growth kinetics.

$$OCN-R-NCO + HO-R'-OH \longrightarrow$$

$$\{ \overset{O}{\overset{\|}{C}}-\overset{H}{\overset{|}{N}}-R-\overset{H}{\overset{|}{N}}-\overset{O}{\overset{\|}{C}}-R'-O \}_n$$

Polyurethane

$$\overset{H^+}{\longrightarrow} \{ \overset{O}{\overset{\|}{C}}(CH_2)_5 \overset{H}{\overset{|}{N}} \}_n$$

Nylon-6

Stepwise Polymerization. Although condensation polymers account for only about one-fourth of synthetic polymers (bulkwise), most natural polymers are of the condensation type. As shown by Carothers in the 1930s (2, 3), the chemistry of condensation polymerizations is essentially the same as classic condensation reactions that result in the synthesis of monomeric amides, urethanes, esters, etc.; the principle difference is that the reactants employed for polymer formation are bifunctional (or higher) instead of monofunctional. Although more complicated situations can occur, we will consider only the kinetics of simple polyesterification. The kinetics of most other common condensations follow an analogous pathway.

For uncatalyzed reactions in which the dicarboxylic acid and diol are present in equimolar amounts, one diacid molecule is experimentally found to act as a catalyst and leads to the following kinetic expression.

$$\text{Rate of polycondensation} = -d[A]/dt = k[A]^2[D] = k[A]^3 \qquad (1)$$

where [A] is the dicarboxylic acid concentration and [D] is the diol concentration. When [A] = [D], rearrangement gives

$$-d[A]/[A]^3 = kdt \qquad (2)$$

Integration over the limits of $A = A_0$ to $A = A_t$ and $t = 0$ to $t = t$ gives

$$2kt = 1/[A_t]^2 - 1/[A_0]^2 = 1/[A_t]^2 + \text{constant} \qquad (3)$$

It is convenient to express Equation 3 in terms of extent of reaction, p, where p is the fraction of functional groups that have reacted at time t. Thus 1-p is the fraction of unreacted groups and

$$A_t = A_0(1-p) \qquad (4)$$

Substitution of the expression for A_t from Equation 4 into Equation 3 and rearrangement yields

$$2A_0^2 kt = 1/(1-p)^2 + \text{constant} \tag{5}$$

A plot of $1/(1-p)^2$ as a function of time should be linear with a slope of $2A_0^2 k$ from which k is determinable. Determination of k as a function of temperature enables the calculation of activation energy.

The number average degree of polymerization, $\overline{DP}_N$, can be expressed as

$\overline{DP}_N$ = number of original molecules/number of molecules at a specific time t is given by

$$\overline{DP}_N = N_0/N = A_0 A_t = A_0/A_0(1 - p) = 1/1 - p \tag{6}$$

This relationship, Equation 6, is called the Carothers equation because it was first found by Carothers while working with the synthesis of nylon 66. (1, 3).

Because the value of k at any temperature can be determined from the slope $(2A_0^2 k)$ of the line when $1/(1-p)^2$ is plotted against t, DP can be determined at any time t from the expression

$$DP^2 = 2kt[A_0]^2 + \text{constant} \tag{7}$$

<u>Free-Radical Chain Polymerization</u>. In contrast to the typically slow stepwise polymerizations, chain reaction polymerizations are usually rapid with the initiated species rapidly propagating until termination. A kinetic chain reaction usually consists of at least three steps, namely, initiation, propagation, and termination. The initiator may be an anion, cation, free radical, or coordination catalyst.

Because most synthetic plastics, elastomers, and fibers are prepared by free-radical chain polymerizations, this method will be discussed here. Initiation can occur through decomposition of an initiator such as azobisisobutyronitrile (AIBN), light, heat, sonics, or other technique to form active free radicals. Here initiation will be considered as occurring by decomposition of an initiator, I, and is described as follows.

$$I \xrightarrow{\;k_d\;} 2R^{\cdot} \tag{8}$$

$$\text{Rate of initiator decomposition} = R_d = -d[I]/dt = k_d[I] \tag{9}$$

Initiation of polymerization occurs by addition of the generated initiator free radical $R^{\cdot}$ to a vinyl molecule, M.

$$R^{\cdot} + M \xrightarrow{\;k_i\;} RM^{\cdot} \tag{10}$$

$$R_i = -d[M]/dt = d[RM^{\cdot}]/dt = k_i[R^{\cdot}] \, M = 2k_d f[I] \tag{11}$$

where f is an initiator efficiency factor, essentially the fraction of decomposed initiator fragments, $R^{\cdot}$, that successfully start chain growth.

Propagation is a bimolecular reaction in which the radical $RM^{\cdot}$ adds to another monomer molecule. Although slight changes exist in the propagation rate constant, k_p, in the first few steps of chain growth, the rate constant is generally considered to be independent of chain length. Thus, all of the propagation steps can be described by a single specific rate constant, k_p.

$$M\cdot + M \xrightarrow{k_p} M\cdot \tag{12}$$

The rate of decrease of monomer with time is described by

$$-d[M]/dt = k_p[M\cdot][M] + k_i[R\cdot][M] \tag{13}$$

For long chains the amount of monomer consumption by the initiation step is negligible and permits Equation 13 to be rewritten as

$$R_p = -d[M]/dt = k_p[M\cdot][M] \tag{14}$$

Termination typically occurs by coupling of two macroradicals (Equations 15 and 16) or through disproportionation (Equations 17 and 18)

$$M\cdot + M\cdot \longrightarrow M-M \tag{15}$$

$$R_t = -d[M\cdot]/dt = 2k_t[M\cdot]^2 \tag{16}$$

$$M\cdot + M\cdot \longrightarrow 2M \tag{17}$$

as $2M-CH_2-\underset{\underset{R}{|}}{C}H\cdot \longrightarrow M-CH = \underset{\underset{R}{|}}{C}H + M-CH_2-\underset{\underset{R}{|}}{C}H_2$

$$R_t = -d[M\cdot]/dt = 2k_t[M\cdot]^2 \tag{18}$$

Although Equations 14, 16, and 18 are theoretically important, they contain [M·], which is difficult to experimentally determine. The following approach is used to render a more usable form of these equations.

The rate of monomer change is described by

$$d[M\cdot]/dt = k_i[R\cdot][M] - 2k_t[M\cdot]^2 \tag{19}$$

Experimentally the number of growing chains is found to be approximately constant over most of the reaction. This situation is called a "steady state" and results in $d[M\cdot]/dt = 0$ and

$$k_i[R\cdot][M] = 2k_t[M\cdot]^2 \tag{20}$$

Furthermore, experiments indicate that if the temperature, amount of light, etc. are constant, then the generation of R· and number of growing chains is constant and leads to a steady state for R· with

$$d[R\cdot]/dt = 2k_d f[I] - k_i[R\cdot][M] = 0 \tag{21}$$

Solving for [M·] from Equation 20 and [R·] from Equation 21 and substituting into Equation 20 the expression for [R·] from Equation 21 gives an expression for [M·] that contains readily determinable terms.

$$[M\cdot] = (k_d f[I]/k_t)^{1/2} \tag{22}$$

Useful rate and kinetic chain length expressions are then obtainable from Equations 14, 16, 18, and 20.

$$R_p = k_p[M][M\cdot] = k_p[M](k_d f[I]/k_t)^{1/2} = k'[M][I]^{1/2} \tag{23}$$

where $k' = (k_p^2 k_d f/k_t)^{1/2}$.

$$R_t = 2k_t[M\cdot]^2 = 2k_d f[I] \tag{24}$$

$$\overline{DP} = R_p/R_i = R_p/R_t = k_p[M](k_d f[I]/k_t)^{1/2}/2k_d f[I] = k''[M]/[I]^{1/2} \tag{25}$$

where $k'' = k_p/(2k_d k_t f)^{1/2}$.

Because of the great industrial importance of free-radical polymerizations, these reactions are the most studied reactions in all of chemistry.

Free-Radical Copolymerization. Although the mechanism of copolymerization is similar to that described for homopolymerizations, the reactivities of monomers may differ when more than one monomer is present in the feed and lead to polymer chains with varying amounts of the reactant monomers. The difference in the reactivity of monomers can be expressed with reactivity ratios, r.

The copolymer equation, which expresses the composition of growing chains at any reaction time t, was developed in the 1930s by a group of investigators including Wall, Mayo, Simha, Alfrey, Dorstal, and Lewis (for instance, 1 and 4-6).

Four chain extension reactions are possible when monomers M_1 and M_2 are present in a polymerization reaction mixture. Two of these steps are self-propagating steps (Equations 26 and 28) and two are cross-propagating steps (Equations 27 and 29). The difference in the reactivity of the monomers can be expressed in terms of reactivity ratios that are ratios of the propagating rate constant where $r_1 = k_{11}/k_{12}$ and $r_2 = k_{22}/K_{21}$.

$$M_1\cdot + M_1 \xrightarrow{k_{11}} M_1M_1\cdot \tag{26}$$

$$M_1\cdot + M_2 \xrightarrow{k_{12}} M_1M_2\cdot \tag{27}$$

$$M_2\cdot + M_2 \xrightarrow{k_{22}} M_2M_2\cdot \tag{28}$$

$$M_2\cdot + M_1 \xrightarrow{k_{21}} M_2M_1\cdot \tag{29}$$

Experimentally, as in the case of chain polymerizations, the specific rate constants are found to be essentially independent of chain length, with monomer addition primarily dependent only on the adding monomer unit and the growing end. Thus, Equations 26–29 are sufficient to describe the polymerization.

The rate of monomer consumption can be described by the following:

Rate of consumption of M_1 = $-d[M_1]/dt = k_{11}[M_1^\cdot][M_1] + k_{21}[M_2^\cdot][M_1]$ (30)

Rate of consumption of M_2 = $-d[M_2]/dt = k_{22}[M_2^\cdot][M_2] + k_{12}[M_1^\cdot][M_2]$ (31)

Again, experiments indicate that the number of growing chains remains essentially constant throughout much of the copolymerization reaction and gives a steady state concentration of M·. The concentration of $M_1^\cdot$ and $M_2^\cdot$ can then be described and inserted into Equation 32 which describes the ratio of monomer uptake in growing chains, $d[M_1]/d[M_2]$, and leads to Equation 33.

$$\frac{d[M_1]}{d[M_2]} = \frac{k_{11}[M_1^\cdot][M_1] + k_{21}[M_2^\cdot][M_1]}{k_{12}[M_1^\cdot][M_2] + k_{22}[M_2^\cdot][M_2]} = n \qquad (32)$$

$$n = \frac{[M_1]}{[M_2]} \frac{r_1[M_1]+[M_2]}{[M_1]+r_2[M_2]} = \frac{r_1([M_1]/[M_2])+1}{r_2([M_2]/[M_1])+1} = \frac{r_1 X+1}{(r_2/X)+1} \qquad (33)$$

where $X = [M_1]/[M_2]$ = molar ratio of the feed composition, and n is the molar ratio of monomers incorporated into growing polymer chains, that is, the composition of copolymer chains formed.

Values of r have been determined for most of the industrially important vinyl monomers and allow a quick prediction of polymer composition. Thus, values of r_1 greater than one signify that radical $M_1^\cdot$ will tend to add monomer M_1 rather than M_2, etc. When both r_1 and r_2 are near 1, a random (not alternating) chain will result. An alternating copolymer is produced when both r_1 and r_2 are near zero. The presence of r values in excess of one indicates the formation of block copolymers and/or mixtures.

The copolymerization equation (Equation 33), allows one to adjust the monomer feed to produce a copolymer of desired composition. The copolymer equation is found to be valid except when strong steric or polar restrictions are present.

Property–Molecular Weight Relationships

Polymerization reactions, both synthetic and natural, lead to polymers with a heterogeneous molecular weight, that is, polymer chains with a different number of units. Molecular weight distributions may be relatively broad as is the case for most synthetic polymers and many naturally occurring polymers, relatively narrow as occurs for certain natural polymers (because of the imposed steric and electronic constraints), or may be mono-, bi-, tri-, or polymodal. A bimodal curve often characterizes a polymerization occurring under two distinct pathways or environments. Thus, most synthetic polymers and many naturally occurring polymers consist of molecules with different molecular weights and are said to be polydisperse. In contrast, specific proteins and nucleic acids consist of molecules with a specific molecular weight and are said to be monodisperse.

Because typical small molecules and large molecules with molecular weights less than a critical value required for chain entanglement are weak and are readily attacked by appropriate reactants, these properties are related to molecular weight. Thus, melt viscosity and chemical resistance of amorphous polymers is dependent on the molecular weight distribution. In contrast, density, specific heat capacity, and refractive index are essentially independent of the molecular weight at molecular weight values above the critical molecular weight, typically above chain lengths of 100 units.

The melt viscosity is usually proportional to the 3.4 power of the average molecular weight at values above the critical molecular weight required for chain entanglement, that is, $\eta = \overline{M}^{3.4}$. Thus, the melt viscosity increases rapidly as the molecular weight increases, and more energy is required for the processing and fabrication of these large molecules. However, as shown in Figure 1, the strength of polymers increases as the molecular weight increases and then tends to level off.

Thus, although a value above the threshold molecular weight value (TMWV) is essential for most practical applications, the additional cost of energy required for processing extremely high molecular weight polymers is seldom justified. Accordingly, a commercial polymer range is customarily established above the TMWV but below the extremely high molecular weight range. However, because toughness increases with molecular weight, extremely high molecular weight polymers, such as ultrahigh molecular weight polyethylene, are used for the production of tough articles including films employed as trash bags.

Oligomers and other low molecular weight polymers are not useful for applications for which high strength is required. The value of TMWV will be dependent on the glass transition temperature, T_g, the cohesive energy density (CED) of amorphous polymers, the extent of crystallinity in crystalline polymers, and the effect of reinforcements in polymeric composites. Thus, although a low molecular weight amorphous polymer may be satisfactory for use as a coating or adhesive, a polymer with a DP value of at least 1000 may be required if it is used as an elastomer or plastic. With the exception of polymers with highly regular structures, such as isotactic polypropylene, polymers with strong hydrogen intermolecular bonds are required for fibers. Because of their higher CED values, polar polymers having DP values lower than 1000 are satisfactory for use as fibers. In a later section of this chapter, molecular weight and its determination will be discussed in some detail.

Interchain and Intrachain Forces

The forces present in molecules are often divided into primary forces (typically greater than 50 kcal/mol of interaction) and secondary forces (typically less than 10 kcal/mol of interaction). Primary bonding forces can be subdivided into ionic bonds (not typically present in polymer backbones and characterized by a lack of directional bonding); metallic bonds (often considered as charged

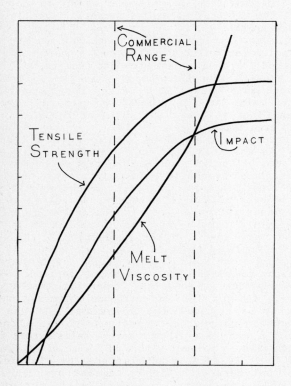

Figure 1. Relationship of polymer properties to molecular weight.
(Reproduced with permission from McGraw-Hill. Copyright 1971.)

atoms surrounded by a potentially fluid sea of electrons, not found in polymers); and covalent bonds (including coordinate and dative). Covalent bonds are directional and the major means of bonding within polymers. The bonding lengths of primary bonds are usually 9 to 20 nm, and the carbon-carbon bond length is approximately 15 to 16 nm.

Secondary forces, frequently called van der Waals forces because they are the forces responsible for the van der Waals correction to the ideal gas relationships, interact over longer distances and generally exhibit significant interaction between 25 and 50 nm. The force of these interactions is inversely proportional to some power of r, generally 2 or greater [force 1/(distance)^r], and thus is dependent on the distance between the interacting molecules. Therefore, many physical properties of polymers are dependent on both the conformation (arrangements related to rotation about single bonds) and configuration (arrangements related to the actual chemical bonding about a given atom) because both affect the proximity of one chain relative to another.

Atoms in individual polymer molecules are joined to each other by relatively strong covalent bonds. The bond energies of the carbon-carbon bonds are approximately 80 to 90 kcal/mol. Polymer molecules, like all other molecules, are attracted to each other (and in long-chain polymer chains they are attracted to each other even between segments of the same chain) by intermolecular, secondary forces.

Intermolecular forces are also partly responsible for the increase in boiling points within a homologous series such as the alkanes, for the higher than expected boiling points of polar organic molecules such as alkyl chlorides, and for the abnormally high boiling points of alcohols, amines, and amides. Although the forces responsible for these increases in boiling points are all called van der Waals forces, these forces are subclassified in accordance with their source and intensity. Secondary, intermolecular forces include London dispersion forces, induced permanent forces, and dipolar forces, including hydrogen bonding.

Nonpolar molecules such as ethane $H(CH_2)_2H$ and polyethylene are attracted to each other by weak London or dispersion forces resulting from induced dipole-dipole interaction. The temporary or transient dipoles in ethane or along the polyethylene chain are due to instantaneous fluctuations in the density of the electron clouds. The energy range of these forces is about 2 kcal per unit in nonpolar and polar polymers alike, and this force is independent of temperature. These dispersion forces are the major forces present between chains in elastomers and soft plastics.

Methane, ethane, and ethene are all gases; hexane, octane, and nonane are all liquids (under standard conditions); and polyethylene is a waxy solid. This trend is primarily due to both an increase in mass per molecule and to an increase in the London forces per molecule as the chain length increases. With the assumption that the attraction between methylene or methyl units is 2 kcal/mol or 3×10^{-24} kcal/molecular interaction, the interaction per molecule can be calculated as 3×10^{-24} kcal/molecule for methane, 2×10^{-23} kcal/molecule for hexane, and 6×10^{-21} kcal/molecule for a polyethylene chain of 2000 repeating units.

Polar molecules such as ethyl chloride, H_3C-CH_2Cl, and poly(vinyl chloride) (PVC), $+CH_2-CHCl+_n$, are attracted to each other by dipole–dipole interactions resulting from the electrostatic attraction of a chlorine atom in one molecule to a hydrogen atom in another molecule. Because this dipole–dipole interaction, which ranges from 2 to 6 kcal/mol repeat unit in the molecule, is temperature dependent, these forces are reduced as the temperature is increased in the processing of .pa polymers. Although the dispersion forces are typically weaker than the dipole–dipole forces, they are also present in polar compounds such as ethyl chloride and PVC.

Strongly polar molecules such as ethanol, poly(vinyl alcohol), and cellulose are attracted to each other by a special type of dipole–dipole interaction called hydrogen bonding in which the oxygen atoms in one molecule are attracted to the hydrogen atoms in another molecule. These attractions are the strongest of the intermolecular forces and may have energies as high as 10 kcal/mol repeat unit (the H–F hydrogen bond is higher). Intermolecular hydrogen bonds are usually present in fibers such as cotton, wool, silk, nylon, polyacrylonitrile, polyesters, and polyurethanes. Intramolecular hydrogen bonds are responsible for the helices observed in starch and globular proteins.

The high melting point of nylon 66 (265 °C) is the result of a combination of dispersion, dipole–dipole, and hydrogen bonding forces between the polyamide chains. The hydrogen bonds are decreased when the hydrogen atoms in the amide groups in nylon are replaced by methyl groups and when the hydroxyl groups in cellulose are esterified or etherified.

In addition to the contribution of intermolecular forces, chain entanglement is also an important contributory factor to the physical properties of polymers. Although paraffin wax and high density polyethylene (HDPE) are homologs with relatively high molecular weights, the chain length of paraffin is too short to permit entanglement, and hence it lacks the strength and other characteristic properties of HDPE.

Crystalline–Amorphous Structures

General Considerations.

There are numerous theories associated with crystallization tendencies and the form(s) and mix(es) of crystalline–amorphous regions within polymers. Here we will only briefly consider a few contributing factors.

A three-dimensional crystalline polymer often can be described as a fringed micelle (chains packed as a sheaf of grain) or as a folded chain. Regions where the polymer chains exist in an ordered array are called crystalline domains. These crystalline domains in polymers are typically smaller than crystalline portions of smaller molecules. Furthermore, imperfections in polymer crystalline domains are more frequent, and one polymer chain may reside both within a crystalline domain and within amorphous regions. These connective chains are responsible for the toughness of a polymer. Sharp boundaries between the ordered (crystalline) and disordered (amorphous) portions are the exception but do occur in some instances such as with certain proteins, poly(vinyl alcohol), and certain cellulosic materials. Highly crystalline polymers exhibit high melting points and high densities, resist dissolution and swelling in solvents, and have high moduli of rigidity (are stiff) relative to the polymers with less crystallinity.

The amount and kind of crystallinity depends on both the polymer structure and on its treatment. The latter point is illustrated by noting that the proportion of crystallinity can be effectively regulated for many common polymers by controlling the rate of formation of crystalline segments. Thus, polypropylene can be heated above its melting range and cooled quickly (quenched) to produce a product with only a moderate amount of crystalline domains. Yet, if it is cooled at a slower rate (such as 1 °C/10 min), the resulting polypropylene will be largely crystalline.

Polymer properties are directly dependent on both the inherent shape of the polymer and on its treatment. Contributions of polymer shape to polymer properties are often complex and interrelated, but can be broadly divided into terms dealing with chain flexibility, chain regularity, interchain forces, and steric effects.

Chain Flexibility. Flexibility is related to the activation energies required to initiate rotational and vibrational segmental chain motions. For some polymers, as flexibility is increased, the tendency toward crystallinity increases. Polymers containing regularly spaced single C-C, C-N, and C-O bonds allow rapid conformational changes that contribute to the flexibility of a polymer chain and to the tendency toward crystalline formation. Yet, chain stiffness may also enhance crystalline formation by permitting or encouraging only certain "well-ordered" conformations to occur within the polymer chain. Thus p-polyphenylene is a linear chain that cannot fold over at high temperatures. Such species are highly crystalline, high melting, rigid, and insoluble.

Intermolecular Forces. Crystallization is favored by the presence of regularly spaced units that permit strong intermolecular interchain associations. The presence of moieties that carry dipoles or that are highly polarizable encourages strong interchain attractions. This tendency is particularly true for situations in which interchain hydrogen bonds are formed. Thus, the presence of regularly spaced carbonyl, amine, amide, and alcohol moities encourages crystallization tendencies.

Structural Regularity. Structural regularity enhances the tendency for crystallization. Thus, linear polyethylene is difficult to obtain in any form other than a highly crystalline one. Low density, branched polyethylene is typically largely amorphous. The linear polyethylene chains are nonpolar, and the crystallization tendency is mainly based on the flexibility of the chains to achieve a regular, tightly packed conformation that takes advantage of the special restrictions inherent in dispersion forces.

Monosubstituted vinyl monomers can produce polymers with different configurations--two regular structures (isotactic and syndiotactic) and a random (atactic) form. For polymers with the same chemical makeup, those forms derived from regular structures exhibit greater rigidity, are higher melting, and are less soluble relative to the atactic form.

Extensive work with condensation polymers and copolymers fully confirms the importance of structural regularity on crystallization tendency, and consequently on associated properties. Thus, copolymers containing regular alteration of each copolymer unit, either ABABAB type or block type, show a distinct tendency to crystallize, and corresponding copolymers with random distributions of the two are intrinsically amorphous, less rigid, lower melting, and more soluble.

Steric Effects. The effect of substituents on polymer properties depends on a number of items including location, size, shape, and mutual interactions. Although methyl and phenyl substituents tend to lower chain mobility, they do permit good packing of chains, and their presence produces dipoles that further contribute to the crystallization tendency. The presence of aromatic substituents, even comparatively large substituents such as anthracene, further contributes to intrachain and interchain attraction tendencies through the mutual interactions of the aromatic substituents. Although their bulky size discourages crystallization by increasing the interchain distances, they do encourage rigidity. Thus, polymers containing bulky aromatic substituents tend to be rigid, high melting, less soluble, yet fairly amorphous.

Ethyl to hexyl substituents tend to lower the tendency for crystallization because their major contribution is to increase the average distance between chains and thus decrease the contributions of secondary bonding forces. If the substituents become longer (from 12 to 18 carbon atoms) and remain linear, a new phenomenon occurs—the tendency of the side chains to form crystalline domains of their own.

Transitions

Polymers can exhibit a number of different conformational changes with each change accompanied by differences in polymer properties. Two major transitions occur at T_g, which is associated with local, segmental chain mobility in the amorphous regions of a polymer, and the melting point (T_m), which is associated with whole chain mobility. The T_m is called a first-order transition temperature, and T_g is often referred to as a second-order transition temperature. The values for T_m are usually 33 to 100% greater than for T_g, and T_g values are typically low for elastomers and flexible polymers and higher for hard amorphous plastics.

Viscosity is a measure of the resistance to flow. Flow, which is the result of cooperative movement of the polymer segments from hole to hole in a melt or solution, is impeded by chain entanglement, intermolecular forces, cross-links, and the presence of reinforcing agents.

The flexibility of amorphous polymers above the glassy state, which is governed by the same forces as melt viscosity, is dependent on a wriggling type of segmental motion in the polymer chains. This flexibility is increased when many methylene groups $-CH_2-$ are present between stiffening groups in the chain and when oxygen atoms are present in the chain. Thus, the flexibility of aliphatic polyesters usually increases as the number of methylene groups is increased. In contrast, the flexibility of amorphous polymers above the glassy state is decreased when stiffening groups such as phenylene, sulfone, and amide are present in the backbone.

The flexibility of amorphous polymers is reduced drastically when they are cooled below T_g. At temperatures below T_g, no segmental motion exists, and any dimensional changes in the polymer chain are the result of temporary distortions of the primary valence bonds. Amorphous plastics perform best below T_g, but elastomers must be used above the brittle point, or T_g.

The T_g value of isotactic polypropylene is approximately -10 °C, yet because of its high degree of crystallinity, it does not readily flow below its T_m of approximately 150 °C. Thus, physical flow tendencies are related to both the T_g and T_m values and to the real physical nature of the product (proportion and type of crystallinity).

Because the specific volume of polymers increases at T_g in order to accommodate the increased segmental chain motion, T_g values may be estimated from plots of the change in specific volume with temperature. Other properties such as stiffness (modulus), refractive index, dielectric properties, gas permeability, X-ray adsorption, and heat capacity all change at T_g. Thus, T_g may be estimated by noting the change in any of these values such as the increase in gas permeability.

End Uses of Polymers as Related to Structure

General Discussion. The usefulness of polymers depends not only on their properties but also on their abundance and reasonable cost. Polymer properties are related not only to the chemical nature of the polymer, but also to such factors as extent and distribution of crystallinity and distribution of polymer chain lengths. These factors influence properties such as hardness, comfort, chemical resistance, biological response, weather resistance, tear strength, dyeability, flex life, stiffness, electrical properties, and flammability.

Elastomers. Elastomers are characterized by the ability to elongate upon application of stress and to return quickly to the original length upon release of stress. They are high polymers that possess chemical cross-links. For industrial application they must be used above T_g to allow for full chain mobility. The normal unextended state of elastomers must be amorphous. The restoring force, after elongation, is largely due to entropy effects. As the material is elongated, the random chains are forced to occupy more ordered positions. Upon release of the applied force the chains tend to return to a more random state.

Gross, actual mobility of chains must be low. The cohesive energy forces between chains should be low and permit rapid, easy expansion. In its extended state a chain should exhibit a high tensile strength, whereas at low extensions it should have a low tensile strength. Polymers with low cross-linked density usually meet the desired property requirements. The material after deformation should return to its original shape because of the cross-linking. This property is often referred to as rubber "memory."

Fibers. Fiber properties include high tensile strength and high modulus (high stress for small strains, i.e., stiffness). These properties can be obtained from high molecular symmetry and high cohesive energies between chains, both requiring a fairly high degree of polymer crystallinity. Fibers are normally linear and drawn (oriented) in one direction to produce high mechanical properties in that direction.

Typical condensation polymers, such as polyester and nylon, often exhibit these properties. If the fiber is to be ironed, its T_g should be above 200 °C; if it is to be drawn from the melt, its T_g should be below 300 °C. Branching and cross-linking are undesirable because they disrupt crystalline formation even though a small amount of cross-linking may increase some physical properties if effected after the material is suitably drawn and processed.

Plastics. Materials with properties intermediate between elastomers and fibers are grouped together under the term "plastics." Thus, plastics exhibit some flexibility with hardness with varying degrees of crystallinity. The molecular requirements for a plastic are the following: if it is linear or branched, with little or no cross-linking, then it should be below its T_g if amorphous and/or below its melting point if crystallizible when it is used; or if it is cross-linked, the cross-linking must be sufficient to severely restrict molecular motion.

Adhesives. Adhesives can be considered as coatings between two surfaces. The classic adhesives were water-susceptible animal and vegetable glues obtained from hides, blood, and starch. Adhesion may be defined as the process that occurs when a solid and movable material (usually in a liquid or solid form) are brought together to form an interface, and the surface energies of the two substances are transformed into the energy of the interface.

A unified science of adhesion is still being developed. Adhesion can result from mechanical bonding between the adhesive and adherend and/or primary and/or secondary chemical forces. Contributions through chemical forces are often more important and illustrate why nonpolar polymeric materials such as polyethylene are difficult to bond, although polycyanoacrylates are excellent adhesives. Numerous types of adhesives are available such as solvent-based, latex, pressure-sensitive, reactive, and hot-melt adhesives.

The combination of an adhesive and adherend is a laminate. Commercial laminates are produced on a large scale with wood as the adherend and phenolic, urea, epoxy, resorcinol, or polyester resins as the adhesives. Many wood laminates are called plywood. Laminates of paper or textile include items under the trade names of Formica and Micarta. Laminates of phenolic, nylon, or silicone resins with cotton, asbestos, paper, or glass textile are used as mechanical, electrical, and general purpose structural materials. Composites of fibrous glass, mat or sheet, and epoxy or polyester resins are widely employed as reinforced plastic (FRP) structures.

Coatings. The traditional uses of coatings for decorative and protective purposes are expanding into future concepts of coatings as energy collective devices, burglar alarm systems, and other novel end uses. Even so, such properties as adhesion, corrosion protection, weatherability, color stability, water and chemical resistance, toughness, hardness, and application properties continue to be major requirements. Recent emphasis is on coatings that can be used with low quantities of solvent or can be diluted with water. Acrylates, epoxies, and urethanes are important types of resins used, although alkyds are still popular.

Polyblends and Composites. Polyblends are made by mixing components together in extruders or mixers, on mill rolls, etc. Most are heterogeneous systems consisting of a polymeric matrix in which another polymer is imbedded. Although the units of copolymers are connected through primary bonds, the components of polyblends adhere only through secondary bonding forces. In contrast to polyblends, composites consist of a polymeric matrix in which a foreign material is dispersed. Composites typically contain fillers such as carbon black, wood flour, talc, or reinforcing materials such as glass fibers, hollow spheres, and glass mats.

Physical Characterization and Testing

Testing Societies. Public acceptance of polymers is usually associated with an assurance of quality based on a knowledge of successful long-term and reliable tests. In contrast, much of the dissatisfaction with synthetic polymers is related to failures that possibly could have been prevented by proper testing, design, and quality control. The American Society for Testing and Materials (ASTM), through its committees D-1 on paint and D-20 on plastics, has developed many standard tests that should be referred to by all producers and consumers of finished polymeric materials. Cooperating groups in many other technical societies also exist: the American National Standards Institute, the International Standards Organization, and standards societies such as the British Standards Institution in England, the Deutsche Normenausschuss in Germany, and comparable groups in every developed nation throughout the entire world.

Thus, standard tests account for much of the testing done in industry and are used to ensure product specifications. These tests measure stress-strain relationships, flex life, tensile strength, abrasion resistance, moisture retention, dielectric constant, hardness, thermal conductivity, etc. New tests are continually being developed and submitted to ASTM, and after adequate verification through "round robin" testing are finally accepted as standard tests.

Some tests are developed within a given company to measure a certain property peculiar to that company, and these tests may or may not be submitted to ASTM for verification and acceptance. Data obtained from such tests may be quite valuable to that company, but such data should not be used for comparative results with other tests until adequate precautions are taken to ensure that conditions of the two sets of tests are the same. Furthermore, it is not always clear what particular property or other property-structure relationship is being tested with many standard and nonstandard tests because the tests are more often use oriented. Even so, such

tests form the basis of product reliability and reproducibility and
are benchmarks of industry.

Each standardized test is specified by a unique combination of
letters and numbers and contains specifications regarding data
gathering, instrument design, and test conditions. These
specifications make it possible for laboratories across the country
to compare data with some confidence. Thus, the Izod test, a
popular impact test, is given the ASTM number D256-56 (1961), the
latter number being the year it was first accepted. The ASTM
instructions for the Izod test specify test material shape and size,
test equipment, test procedure, and result reporting. A list of
available tests along with brief descriptions of the use of each
test is contained in "Compilation of ASTM Standard Definitions" (7).

Molecular Weight. A polymer is a polymer because of its large size,
and it is necessary to obtain physical measurements of this size.
The physical parameter typically used to describe polymer size is
molecular weight. With the exception of a few natural occurring
polymers, polymer samples consist of chains having varying length
and varying molecular weights as depicted in Figure 2. Several
mathematical moments can be described by using this differential or
frequency distribution curve, and these moments can be described by
equations and determined physically by using various techniques.

The first moment is called the number-average molecular weight,
$\bar{M}_n$. Any measurement that leads to the number of molecules,
functional groups, or particles that are present in a given weight
of sample allows the calculation of $\bar{M}_n$. Most thermodynamic
properties are related to the number of particles present and thus
are dependent on $\bar{M}_n$.

Colligative properties dependent on the number of particles
present are obviously related to $\bar{M}_n$. $\bar{M}_n$ values are independent of
molecular size and are highly sensitive to the presence of small
molecules in the mixture. Values for $\bar{M}_n$ are determined by Raoult's
techniques and are dependent on colligative properties such as
ebulliometry (boiling point elevation), cryometry (freezing point
depression), osmometry, and end-group analysis.

The number-average molecular weight, $\bar{M}_n$, is calculated like any
other numerical average by dividing the sum of the individual
molecular weight values, $M_i N_i$, by the number of molecules, N_i.

$$\bar{M}_n = \frac{\text{total weight of sample}}{\text{no. of molecules of } N_i} = \frac{W}{\sum_{i=1}^{\infty} N_i} = \frac{\sum_{i=1}^{\infty} M_i N_i}{\sum_{i=1}^{\infty} N_i} \tag{34}$$

Weight-average molecular weight, $\bar{M}_w$, is determined from
experiments in which each molecule or chain makes a contribution to
the measured result. This average is more dependent on the number
of heavier molecules than is the number-average molecule weight,
which is dependent simply on the total number of particles.

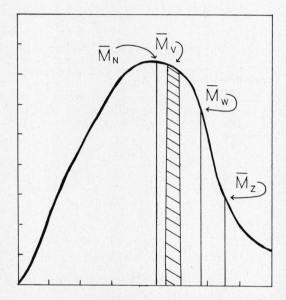

Figure 2.   Moelcular weight distributions. (Reproduced with permission from McGraw—Hill. Copyright 1971.)

The weight–average molecular weight, $\overline{M}_w$, is the second moment or second power average as shown mathematically:

$$\overline{M}_w = \frac{\sum\limits_{i=1}^{\infty} M_i^2 N_i}{\sum\limits_{i=1}^{\infty} M_i N_i} \tag{35}$$

Bulk properties associated with large deformations, such as viscosity and toughness, are particularly related to $\overline{M}_w$ values. $\overline{M}_w$ is determined by light scattering and ultracentrifugation techniques.

Melt elasticity is more closely dependent on $\overline{M}_z$ (the z–average molecular weight) which can also be obtained by ultracentrifugation techniques. $\overline{M}_z$ is the third moment or third power average and is shown mathematically as

$$\overline{M}_z = \frac{\sum\limits_{i=1}^{\infty} M_i^3 N_i}{\sum\limits_{i=1}^{\infty} M_i^2 N_i} \tag{36}$$

Although z + 1 and higher average molecular weights may be calculated, the major interests are in $\overline{M}_n$, $\overline{M}_w$, and $\overline{M}_z$, which are listed in order of increasing size in Figure 2. Because $\overline{M}_w$ is always greater than $\overline{M}_n$ except in monodisperse systems, the ratio $\overline{M}_w/\overline{M}_n$ is a measure of polydispersity and is called the polydispersity index. The most probable distribution for polydisperse polymers produced by condensation techniques is a polydispersity index of 2.0. Thus, for a polymer mixture that is heterogeneous with respect to molecular weight, $\overline{M}_z > \overline{M}_w > \overline{M}_n$. As the heterogeneity decreases, the various molecular weight values converge until for homogeneous mixtures $\overline{M}_z = \overline{M}_w = \overline{M}_n$. The ratios of such molecular weight values are often used to describe the molecular weight heterogeneity of polymer samples.

Viscometry is the most widely used method for the characterization of polymer molecular weight because it provides the easiest and most rapid means of obtaining molecular weight related data and requires a minimum amount of instrumentation. A most obvious characteristic of polymer solutions is their high viscosity, even when the amount of added polymer is small. Even so, because viscometry does not yield absolute values of $\overline{M}$, one must calibrate the viscometry results with values obtained for the same polymer and solvent by using an absolute technique such as light scattering photometry.

All classic molecular weight determination methods require the polymer to be in solution. To minimize polymer–polymer interactions, solutions equal to and less than 1 g of polymer per 100 ml of solution are used. To further minimize solute interactions, extrapolation of the measurements to infinite dilution is normally practiced.

Molecular Weight Distribution. Polymer properties are dependent on both average chain length and distribution of chain lengths. Curves such as Figure 2 are now usually determined by gel permeation chromatography (GPC). Prior to the introduction of GPC, polydisperse polymers were fractionated by the addition of a nonsolvent to a polymer solution, cooling a solution of polymer, solvent evaporation, zone melting, extraction, diffusion, or centrifugation. The molecular weight of the fractions may be determined by any of the classical techniques mentioned earlier.

The least sophisticated but most convenient technique illustrating polymer fractionation is fractional precipitation, which is dependent on the slight change in the solubility parameter with molecular weight. Thus, when a small amount of miscible nonsolvent is added to a polymer solution at a constant temperature, the product with the highest molecular weight precipitates. This procedure may be repeated after the precipitate is removed. These fractions may also be redissolved and fractionally precipitated. The shape of the distribution curve is then constructed from the fractional amount of each sample after chain length determination.

Rheology. The branch of science related to the study of deformation and flow of materials was given the name rheology by Bingham, who has been called the father of modern rheology (1). The prefix "rheo" is derived from the Greek term "rheos," meaning current or flow. The study of rheology includes two vastly different branches of mechanics called fluid and solid mechanics. The polymer chemist is usually concerned with viscoelastic materials that act as both solids and fluids.

The elastic component is dominant in solids, hence their mechanical properties may be described by Hooke's law (Equation 37) which states that the applied stress (S) is proportional to the resultant strain (γ) but is independent of the rate of this strain ($d\gamma/dt$).

$$S = E\gamma \tag{37}$$

Stress is equal to the force per unit area, and strain or elongation is the extension per unit length. For an isotropic solid, that is, one having the same properties independent of direction, the strain is defined by Poisson's ratio, $V = \gamma_l/\gamma_w$, which is the percentage change in longitudinal strain, γ_l, per percentage change in lateral strain, γ_w.

When there is no volume change, as when an elastomer is stretched, Poissons's ratio is 0.5. This value decreases as the T_g of the substance increases and approaches 0.3 for rigid PVC and ebonite. For simplicity, the polymers can be considered to be isotropic viscoelastic solids with a Poisson ratio of 0.5, and only deformations in tension and shear will be considered. Thus, a shear modulus (G) will usually be used in place of Young's modulus of elasticity (E). Hooke's law for shear is given in Equation 38. E is approximately 2.6 G at temperatures below T_g.

$$S = G\gamma \tag{38}$$

The viscous component is dominant in liquids, thus their flow properties can be described by Newton's law (Equation 39) which states that the applied stress S is proportional to the rate of strain $d\gamma/dt$, but is independent of the strain γ or applied velocity gradient.

$$S = \eta(d\gamma/dt) \tag{39}$$

Both Hooke's and Newton's laws are valid for small changes in strain or rate of strain, and both are useful in describing the effect of stress on viscoelastic materials. The initial elongation of a stressed polymer below T_g is the reversible elongation due to the stretching of covalent bonds and distortion of bond angles. Some of the early stages of elongation by disentanglement of polymer chains may also be reversible.

However, the rate of flow, which is related to slow disentanglement and slippage of polymer chains past one another, is irreversible and increases as the temperature increases in accordance with the Arrhenium relationship

$$\eta = Ae^{E/RT} \tag{40}$$

A single weightless Hookean, or ideal, elastic spring with a modulus of G and a simple Newtonian (fluid) dash pot or shock absorber having a liquid with a viscosity are convenient to use as models illustrating the deformation of an elastic solid and an ideal liquid. Because polymers are often viscoelastic solids, combinations of these models are used to demonstrate deformations resulting from the application of stress to an isotropic solid polymer.

Physical Tests. Numerous physical tests are routinely employed to predict the performance of polymers. Many of these can be termed "use tests," to indicate that some relationship exists between the test results and some performance operation. The following are descriptions of several of the more routinely performed physical tests.

Tensile strength, which is a measure of the ability of a polymer to withstand pulling stresses, is usually determined by pulling a dumbbell-shaped specimen (ASTM-D638-72). These test specimens, like all others, must be conditioned under standard conditions of humidity (50%) and temperature (23 °C) before testing. The ultimate tensile strength is equal to the load that caused failure divided by the minimum cross-sectional area.

Flexural strength, or cross-breaking strength, is a measure of the bonding strength or stiffness of a bar test specimen used as a simple beam (ASTM-D790-71). The flexural strength is based on the load required to rupture a simple beam before its deflection is 5%. Compressive strength, or the ability of a specimen to resist crushing forces, is measured by crushing a cylindrical specimen (ASTM-D695-69). The ultimate compression strength is equal to the load that caused failure divided by the minimum cross-sectional area.

Impact strength is a measure of toughness or the ability of a specimen to withstand a sharp blow, such as the ability to withstand a given object being dropped from a specific height. Impact resistance can be determined by measuring the energy of a pendulum necessary to break a plastic specimen. An unnotched specimen is used in the Charpy test (ASTM-D256-73), and a notched specimen is used in the Izod impact test. Because the results of these impact tests are controversial, other tests simulating actual use have been developed, such as dropping specimens from specific heights.

Shear strength is a measure of the load required to cause failure in the area of the sheared specimen in accordance with ASTM-D732-46. The shear strength is equal to the load divided by the area.

Hardness is a general term that can describe a combination of properties including resistance to penetration, abrasion, and scratching. Indentation hardness of thermosets can be measured by a Barcol Impressor [ASTM-E-2583-67(1972)].

Rockwell hardness tests [ASTM-D785-65 (1970)] measure hardness in progressive numbers on different scales corresponding to the size of the ball indentor used.

Scratch hardness may be measured on Mohs scale, which ranges from 1 for talc to 10 for diamonds, or by scratching with pencils of specified hardness (ASTM-D-3363). Hardness may also be measured by the number of bounces of a ball or the amount of rocking by a Sward hardness rocker. Abrasion resistance may be measured by the loss in weight caused by the rubbing of the wheels of a Taber abraser (ASTM-D-1044).

The tests for change in dimensions of a polymer under long-term stress, called creep or cold flow (ASTM-D74-56), are no longer recommended by ASTM. ASTM-D-671 describes suggested tests for fatigue or endurance of plastics under repeated flexure.

Electrical Measurements. The electrical properties of polymers have much in common with mechanical properties. They can be divided into static properties equivalent to direct current properties and dynamic properties resulting from alternating current measurements. The most used parameter is the volume or bulk resistivity (ASTM-D257-75b) which is the resistance in ohms of a material 1 cm thick and 1 cm^2 in area. Bulk resistivity is one of only a few properties that vary nearly 10^{25} in typical use (materials with values above 10^{17} ohm-cm for polystyrene to 10^{-5} ohm-cm for copper).

Most polymeric applications call for materials with high resistivities that act as insulators of electrical wire, etc. More recently a number of polymeric semiconductors and conductors have been developed.

Conductivity is the reciprocal of resistivity. The electrical properties can be altered greatly by addition of impurities and fillers. Thus, the bulk resistivity of ABS (terpolymer of acrylonitrile, butadiene, and styrene) can be decreased from 10^{14} to 10^{-1} ohm-cm by addition of silver powder.

Several tests are essential for the evaluation of plastics in electrical applications. These tests include dielectric constant (permittivity; ASTM-D150-74), which is the ratio of the capacitance of the polymer compared to air, dielectric strength, and dielectric breakdown voltage (ASTM-D149-75). Dielectric breakdown voltage is the maximum applied voltage a polymer can withstand for 1 min divided by the thickness of the sample in mils (10^{-3} in.).

The power factor is the energy required for the rotation of the dipoles of a polymer in an applied electrostatic field of increasing frequency. These values, which typically range from 1.5 x 10^4 for polystyrene to 5 x 10^{-2} for plasticized cellulose acetate, increase at T_g because of increased chain mobility. The loss factor is the product of the power factor and the dielectric constant. The arc resistance, or resistance to tracking, is considered the minimum time required for a high-voltage discharge to find a conducting carbonized path across the surface of a polymer as evidenced by the disappearance of the arc into the test specimen (ASTM-D495-73).

Thermal Analysis. Thermal and related properties of polymers can be determined by various procedures including thermal gravimetric analysis (TGA), differential scanning calorimetry (DSC), differential thermal analysis (DTA), torsional braid analysis (TBA), thermal mechanical analysis (TMA), and pyrolysis gas chromatography (PGC).

One of the simplest techniques is PGC in which the gases resulting from the pyrolysis of a polymer are analyzed by gas chromatography. This technique may be used for qualitative and quantitative analysis. Quantitative analysis requires calibration with known amounts of standard polymer pyrolyzed under the same conditions as the unknown.

Several different modes of thermal analysis are described as DSC. DSC is a technique of nonequilibrium calorimetry in which the heat flow into or away from the polymer is measured as a function of temperature or time. This technique is different than DTA in which the temperature difference between a reference and a sample is measured as a function of temperature or time. Currently available DSC equipment measures the heat flow by maintaining a thermal balance between the reference and sample by changing a current passing through the heaters under the two chambers. For example, the heating of a sample and reference proceeds at a predetermined rate until heat is emitted or consumed by the sample. If an endothermic occurrence takes place, the temperature of the sample will be less than that of the reference. The circuitry is programmed to maintain the reference and the sample compartments at the same temperature by raising the temperature of the sample to that of the reference. The current necessary to maintain the temperature of the sample at that of the reference is recorded. The area under the resulting curve is a direct measure of the heat of transition.

Possible determinations from DSC and DTA measurements include heat of transition; heat of reaction; sample purity; phase diagram; specific heat; sample identification; rate of crystallization, melting, or reaction; and activation energy.

In TGA, a sensitive balance is used to follow the weight change of a polymer as a function of time or temperature. In making both TGA and thermocalorimetric measurements, the same heating rate and flow of gas should be employed to give the most comparable thermograms. TGA can allow determination of the following: sample purity, material identification, solvent retention, reaction rate, activation energy, heat of reaction, and polymer thermal stability.

TMA measures the mechanical response of a polymer as a function of temperature. Typical measurements as a function of temperature include the following: expansion properties, that is, expansion of a material to calculate the linear expansion coefficient; tension properties, that is, the measurement of shrinkage and expansion of a material under tensile stress, for example, elastic modulus; dilatometry, that is, volumetric expansion within a confining medium, for example, specific volume; single fiber properties, that is, tensile response of single fibers under a specific load, for example, single-fiber modulus; and compression properties, such as measuring softening, or penetration under load.

Compressive, tensile, and single-fiber properties are usually measured under some load and yield information about softening points, modulus changes, phase transitions, and creep properties. For compressive measurements, a probe is positioned on the sample and loaded with a given stress. A record of the penetration of the probe into the polymer is obtained as a function of temperature. Tensile properties can be measured by attaching the fiber to two fused quartz hooks. One hook is loaded with a given stress. Elastic modulus changes are recorded by monitoring a probe displacement.

In TBA the changes in tensile strength as the polymer undergoes thermal transition is measured as a function of temperature and sometimes also as a function of the applied frequency of vibration of the sample. As thermal transitions are measured, irreversible changes such as thermal decomposition of cross-linking are observed, if present. In general, a change in T_g or change in the shape of the curve (shear modulus versus temperature) during repeated sweeps through the region, such as a region containing the T_g, is evidence of irreversible change. The name TBA is derived from the fact that measurements are made on fibers that are "braided" together to give test samples connected between or onto vicelike attachments or hooks.

DSC, DTA, TMA, and TBA analyses are all interrelated and signal changes in thermal behavior as a function of heating rate or time. TGA is also related to other analyses in the assignment of phase changes associated with weight changes.

The polymer softening range, although not a specific thermo-dynamic property, is a valuable "use" property and is normally a simple and readily obtainable property. Softening ranges generally lie between T_g and T_m. Some polymers do not exhibit a softening range but rather undergo a solid state decomposition before softening.

Softening ranges depend on the technique and procedure used to determine them. Thus, listings of softening ranges should be accompanied by the specific technique and procedure employed for the determination. The following are techniques often used for the determination of polymer softening ranges.

The capillary technique is analogous to the technique employed to determine melting points of typical organic compounds. The sample is placed in a capillary tube and heated, and the temperature is recorded from beginning to end of melting. Control of the heating rate gives more significance to the measurements. Instruments such as the Fisher-Johns melting point apparatus are useful in this respect.

Another technique requires a plug or film (or other suitable form) of the polymer to be stroked along a heated surface for which the temperature is increased until the polymer sticks to the surface. A modification of this uses a heated surface containing a temperature gradient between the ends of the surface.

The Vicat needle method consists of determining the temperature at which a 1-mm penetration of a needle (having a point with an area of 1 mm) occurs on a standard sample (0.32 cm thick with a minimum width of 1.8 cm) at a specified heating rate (often 50 °C/h) under specific stress (generally less than 1 kg). This determination is related to the heat deflection point.

In the ring and ball method the softening range of a sample is determined by noting the temperature at which the sample, held within a horizontal ring, is forced downward by the weight of a standard steel ball supported by the sample. The ball and ring are generally heated by inserting them in a bath.

Softening range data are useful in selecting proper temperatures for melt fabrication, such as melt pressing, melt extruding, and molding. They also indicate the thermal stability of products.

Other Tests. Flammability tests for polymers include tests for ignition (ASTM-D-1929-68), rate of burning of cellular plastics (ASTM-D-1692-74), many flammability tests such as ASTM-D-635-74, measurements of smoke density (ASTM-D-2843-70), and the oxygen index test (ASTM-D-2863-74). The oxygen index is the minimum concentration of oxygen in an oxygen-nitrogen mixture that will support candlelike combustion. Flammability tests are useful for comparative purposes, but because of the presence of many variables in actual fires, they are not reliable for assuring lack of flammability in large-scale fires.

Because many polymers are resistant to attack by corrosives, tests for chemical resistance of polymers are particularly important. ASTM-D-543-67 (1977) measures weight and dimensional changes of test samples immersed for 7 days in many different test solutions. These tests may be coupled with tensile tests. Other ASTM tests include those under accelerated service conditions [ASTM-D756-76 (1971)], water absorption [ASTM-D570-63 (1972)], and environmental stress cracking of ethylene plastics [(ASTM-D1693-70)].

Other tested properties include those associated with optical properties, acoustical properties, colorability, and degradation.

Spectroscopy. The usual spectral instrumentation applicable to smaller molecules are equally applicable to macromolecules with little modification. For example, the old axiom that good IR spectra cannot be obtained for polymers, although untrue (because polystyrene film is used to standardize spectra), does have some factual basis. Spectra of amorphous polymers, or polymers containing an abundance of amorphous regions, are typically less sharp relative to spectra of solid, well-ordered organic crystalline compounds. Thus, "background" contributions and lack of band sharpness are responsible for "hiding" many pertinent vibrational bands within polymers. This phenomena is readily overcome by using Fourier transform IR spectrophotometry. Thus, although bands, etc.

tend to be less sharp for many polymers, this lack of sharpness can often be overcome with proper instrumentation.

Spectroscopic instrumentation that has been widely and successfully applied to polymers includes IR, NMR, electron spin resonance, UV, X-ray, near IR, SIMS (secondary ion mass spectrometry), MS (mass spectrometry), photoacoustic, Raman, and microwave spectroscopy, and electron spectroscopy for chemical analysis.

Educational Aspects

Because the majority of scientists and engineers are employed in some aspect of the polymer industry, a number of societies have been active in the education areas. These groups include the Society of Plastics Engineers, Plastics Institute of America, the Society of the Plastics Industry, and the American Chemical Society (ACS). With the exception of the ACS, these groups have focused on continuing education by offering short courses under a wide variety of formats.

Much of the activity within the ACS has focused on the Joint Polymer Education Committees largely composed of members from the Divisions of Polymer Chemistry and Polymeric Materials: Science and Engineering (formerly Organic Coatings and Plastics Chemistry). The first standardized ACS examination in polymer chemistry was developed in 1978. A model syllabus was generated in 1980.

Probably the most significant single event in polymer education occurred in 1978. The latest edition of "Undergraduate Professional Education in Chemistry: Criteria and Evaluation Procedures" by the ACS Committee on Professional Training states, "In view of the current importance of inorganic chemistry, biochemistry, and polymer chemistry, advanced courses in these areas are especially recommended and students should be strongly encouraged to take one or more of them. Furthermore, the basic aspects of these three important areas should be included at someplace in the core materials." After almost 30 years as a stalwart of the sciences, polymer chemistry has been recognized as essential core material in the training of all ACS accredited undergraduate majors.

The full impact of these new provisions is yet to be fully recognized. Educators are advocating that polymer chemistry be recommended as advanced work, and possibly of greater importance that "basic aspects" of polymer chemistry be included in the core material. The education committees of a number of divisions and societies associated with polymer science are working toward adopting recommendations involved with these two major related points.

Nomenclature

The International Union of Pure and Applied Chemistry (IUPAC) formed a Subcommission on Nomenclature of Macromolecules in 1952 and has proceeded to study various topics related to cyclic polymers, blends, composites, cross-linked polymers, block copolymers, etc. IUPAC periodically reports its decisions regarding nomenclature (1, 7, and 8). Even so, these rules have not been generally accepted for common polymers by the majority of those in polymer science,

Table I. Sample Names of Some Common Polymers

Industrial	Common	IUPAC
Polyethylene	Polyethylene	Poly(methylene)
Polystyrene	Polystyrene	Poly(1-phenylethylene)
Polyhexamethylene adipamide	Poly(hexamethylene adipamide)	Poly(iminohexamethylene iminoadepoly)
Polyphenylene oxide	Poly(phenylene oxide)	Poly(oxy-1,4-phenylene)
Polyethylene terephthalate	Poly(ethylene terephthalate)	Poly(oxyethyleneoxytere-phthaloyl
Polyvinyl chloride	Poly(vinyl chloride)	Poly(1-chloroethylene)
Polyvinyl butyral	Poly(vinyl butyral)	Poly[(2-propyl-1,3-dioxene-4,6-diyl)methylene]
Polymethyl methacrylate	Poly(methyl metha-acrylate	Poly[1-(methoxycarbonyl)-crylate)1-methylethylene]

such as textbook and monograph authors, although many journals require conformity to IUPAC rules.

Although a wide diversity exists in the practice of naming polymers, three approaches represent the most used systems. Table I gives the names of some common polymers to illustrate the three systems. The only formal system is the IUPAC system (1, 7, and 8). The second system is referred to as simply the industrial system because it is used by a number of industrial societies for their publications. The third system is referred to as the common system because of past historical use. The latter two systems are informal (semisystematic) and are only useful for the more common, simple polymers and typically differ from one another only by the absence or presence of parentheses.

The majority of undergraduate texts use the industrial system, and a few of the polymer texts have adopted the IUPAC system for common polymers. An IUPAC report (9) states "The Commission recognized that a number of common polymers have semisystematic or trivial names that are well established by usage; it is not intended that they be immediately supplanted by the structure-based names. Nonetheless, it is hoped that for scientific communication the use of semisystematic or trivial names for polymers will be kept to a minimum." Nevertheless, the trend is toward usage of the industrial system.

In summary, currently a diversity of polymer nomenclature exists with regard to common polymers with none being universally accepted as correct.

Literature Cited

1. Seymour, R.; Carraher, C. "Polymer Chemistry: An Introduction"; Plenum: New York, 1981.
2. Marvel, C.; Carraher, C. CHEMTECH 1985, 716-20.
3. Carothers, W. Chem. Rev. 1931, 8, 353-426.
4. Alfrey, T.; Bohrer, J.; Mark, H. "Copolymerization"; Interscience: New York, 1952.
5. Wall, F. J. Am. Chem. Soc. 1944, 66, 2050-7.
6. Dostal, H. Monatsh. Chem. 1936, 69, 424-6.
7. "Compilation of ASTM Standard Definitions"; American Society for Testing and Materials: Philadelphia.
8. Elias, H. G.; Pethrick, R. "Polymer Yearbook"; Harwood Academic: New York, 1984; Chap. 1. Hall, C. "Polymer Materials"; Macmillan: New York, 1981; pp. 184-7.
9. Macromolecules 1973, 6(2), 149-54.

Bibliography

Aggarwal, S. L. "Block Copolymers"; Plenum: New York, 1970.
Albright, L. F. "Processes for Major-Addition-Type Plastics and Their Monomers"; McGraw-Hill: New York, 1974.
Alfrey, T. "Mechanical Behavior of High Polymers"; Wiley-Interscience: New York, 1948.
Allcock, H. R.; Lampe, F. W. "Contemporary Polymer Chemistry"; Prentice Hall: Englewood Cliffs, N.J., 1981.
Allen, P. W.; "Techniques of Polymer Characterization"; Butterworths: London, 1959.

Allpert, D. C.; Jones, W. H. "Block Copolymers"; Halsted: New York, 1973.

Anderson, J. C.; Leaver, K. D.; Alexander, J. M.; Rawlins, R. D. "Material Science"; Halsted: New York, 1975.

Bailey, W. J. "Macromolecular Synthesis"; Wiley: New York, 1972.

Barrett, K. E. G. "Dispersion Polymerization in Organic Media"; Wiley: New York, 1975.

Baun, C. E. H. "Macromolecular Science"; Wiley: New York, 1972.

Bikales, N. M. "Mechanical Properties of Polymers"; Wiley-Interscience: New York, 1971.

Billmeyer, F. W. "Textbook of Polymer Science"; Wiley-Interscience: New York, 1971.

Blackley, D. C. "Emulsion Polymerization"; Halsted: New York, 1975.

Bloch, B.; Hastings, G. W. "Plastic Materials in Surgery"; Charles C. Thomas: Springfield, Ill., 1972.

Boenig, H. V. "Structure and Properties of Polymers"; Halsted: New York, 1973.

Bolker, H. I. "Natural and Synthetic Polymers"; Dekker: New York, 1974.

Braun, D.; Cherdron, H.; Keru, W. "Techniques of Polymer Synthesis and Characterization"; Wiley-Interscience: New York, 1972.

Carraher, C. E.; Gebelein, C. "Biological Activities of Polymers"; American Chemical Society: Washington, D.C., 1982.

Carraher, C. E.; Gebelein, C. "Bioactive Polymeric Systems"; Plenum: New York, 1985

Carraher, C. E.; Moore, J. "Chemical Modification of Polymers"; Plenum: New York, 1983.

Carraher, C. E.; Preston, J. "Advances in Interfacial Synthesis"; Dekker: New York, 1982.

Carraher, C. E.; Sheats, J.; Pittman, C. U. "Organometallic Polymers"; Academic: New York, 1978.

Carraher, C. E.; Sheats, J.; Pittman, C. U. "Metallo-Organic Polymers"; MER: Moscow, Russia, 1981.

Carraher, C. E.; Sheats, J.; Pittman, C. U. "Advances in Organometallic Polymers"; Dekker: New York, 1982.

Carraher, C. E.; Sperling, L. "Renewable Resources for Polymer Applications"; Plenum: New York, 1982.

Carraher, C. E.; Tsuda, M. "Modification of Polymers"; American Chemical Society: Washington, D.C., 1980.

Chiu, J. "Polymer Characterization by Thermal Methods of Analysis"; Dekker: New York, 1974.

Chompff, A. J.; Newman, S. "Polymer Networks"; Plenum: New York, 1973.

Collins, E. A.; Bares, J.; Billmeyer, F. W. "Experiments in Polymer Science"; Wiley-Interscience: New York, 1973.

Conley, R. T. "Infrared Spectroscopy"; Allyn and Bacon: Boston, 1966.

Cowie, J. M. G. "Polymers: Chemistry and Physics of Modern Materials"; Intext Educational: New York, 1974.

Craver, C. D. "Infrared Spectra of Plasticizers and Other Additives"; Coblens Society: Kirkwood, Mass., 1980.

Craver, J. K.; Tess, R. W. "Applied Polymer Science"; American Chemical Society: Washington, D.C., 1975.

Dack, M. R. J. "Solutions and Solubilities Techniques of Chemistry"; Wiley: New York, 1975.

Deanin, R. D. "Polymer Structure, Properties and Applications"; Cahners Books: Boston, 1972.

Deanin, R. D. "New Industrial Polymers"; American Chemical Society: Washington, D.C., 1974.

Dole, M. "The Radiation Chemistry of Macromolecules"; Academic: New York, 1972.

Dubois, J. H.; John, F. W. "Plastics"; Van Nostrand-Reinhold: New York, 1974.

Economy, J. "New and Specialty Fibers"; Wiley: New York, 1976.

Eisenman, G. "Membranes"; Dekker: New York, 1975.

Elias, H. G. "Macromolecules, Structure and Properties"; Plenum: New York, 1975.

Flory, P. J. "Principles of Polymer Chemistry"; Cornell University Press: Ithaca, N.Y., 1953; Chap. 1.

Frisch, K. "Electrical Properties of Polymers"; Technomic: Westport, Conn., 1972.

Furukawa, J.; Vogl, O. "Ionic Polymerization"; Dekker: New York, 1976.

Gaylord, N. W. "Reinforced Plastics"; Cahners Books: Boston, 1974.

Ham, G. E. "Copolymerization"; Interscience: New York, 1964.

Ham, G. E. "Vinyl Polymerization"; Interscience: New York, 1967.

Harris, F.; Seymour, R. B. "Solubility Property Relationships in Polymer"; Academic: New York, 1977.

Harward, R. N. "The Physics of Glassy Polymers"; Halsted: New York, 1973.

Hay, J. M. "Reactive Free Radicals"; Academic: New York, 1974.

Herman, B. S. "Adhesives"; Noyes Data Corp.: Park Ridge, N. J., 1976.

Hixon, H. F.; Goldberg, E. P. "Polymer Grafts in Biochemistry"; Dekker: New York, 1976.

Hopfinger, H. J. "Conformational Properties of Macromolecules"; Academic: New York, 1974.

Jenkins, A. D. "Polymer Science"; American Elsevier: New York, 1972.

Kambe, K.; Garn, P. D. "Thermal Analysis"; Halsted: New York, 1975.

Kennedy, J. P. "Cationic Polymerization of Olefins"; Wiley-Interscience: New York, 1975.

Kirshenbaum, G. S. "Polymer Science Study Guide"; Gordon and Breach: New York, 1973.

Kromenthal, R. L.; Oser, Z.; Martin, E. "Polymer in Medicine and Surgery"; Plenum: New York, 1975.

Ledwith, A.; North, A. M. "Molecular Behavior and Development of Polymeric Materials"; Halsted: New York, 1975.

Lenz, R. W. "Organic Chemistry of High Polymers"; Wiley-Interscience: New York, 1967.

Lenz, R. W. "Coordination Polymerization"; Academic: New York, 1975.

Mandelkern, L. L. "An Introduction to Macromolecules"; Springer-Verlag: New York, 1972.

Mark, H.F.; Gaylord, N. G.; Bikales, N. M. "Encyclopedia of Polymer Science and Technology"; Wiley-Interscience: New York, 1964-70 (14 volumes plus supplement).

Mathias, L.; Carraher, C. E. "Crown Ethers and Phase Transfer Agents for Polymer Applications"; Plenum: New York, 1984.

McCaffery, E. M. "Laboratory Preparation for Macromolecular Chemistry"; McGraw-Hill: New York, 1970.

Milby, R. V. "Plastics Technology"; McGraw-Hill: New York, 1973.

Millich, F.; Carraher, C. E. "Interfacial Synthesis"; Dekker: New York, 1977.

Moncreif, R. W. "Manmade Fibers"; Halsted: New York, 1975.

Morawetz, H. "Macromolecules in Solution"; Wiley-Interscience: New York, 1975.

Morgan, P. W. "Condensation Polymers by Interfacial and Solution Methods"; Wiley-Interscience: New York, 1965.

Morton, M. "Rubber Technology"; Van Nostrand-Reinhold: New York, 1973.

Nielsen, L. E. "Mechanical Properties of Polymers"; Dekker: New York, 1974.

Odian, G. "Principles of Polymerization"; Wiley: New York, 1981.

Ogorkiewiz, R. M. "Thermoplastic Properties and Design"; Wiley: New York, 1974.

Patten, W. J. "Plastics Technology"; Reston: Reston, Va., 1976.

Plueddemann, E. P. "Composite Materials"; Academic: New York, 1974.

Renner, E.; Samsonov, G. V. "Protective Coatings on Metals"; Plenum: New York, 1973.

Rodriguez, F. "Principles of Polymer Systems"; McGraw-Hill: New York, 1971.

Rosen, S. L. "Fundamental Principles of Polymeric Materials for Practicing Engineers"; Cahners Books: Boston, 1971.

Schlenker, B. R. "Introduction to Material Science"; Wiley: New York, 1975.

Schultz, J. M. "Polymer Materials Science"; Prentice-Hall: Englewood Cliffs, N. J., 1973.

Seymour, R. B. "Introduction to Polymer Chemistry"; McGraw-Hill: New York, 1971.

Seymour, R. B. "Modern Plastics Technology"; Reston: Reston, Va., 1975.

Seymour, R. B.; Carraher, C. E. "Polymer Chemistry: An Introduction"; Dekker: New York, 1981.

Seymour, R. B. "Plastics vs. Corrosives"; Wiley: New York, 1982.

Seymour, R. B., Carraher, C. E. "Structure-Property Relationships in Polymers"; Plenum: New York, 1984.

Seymour, R. B.; Stahl, G. A. "Macromolecular Solutions: Structure vs. Properties"; Pergamon: Elmsford, N.Y., 1982.

Slade, P. E. "Polymer Molecular Weights"; Dekker: New York, 1975.

Slade, P. E.; Jenkins, P. E. "Thermal Characterization Techniques"; Dekker: New York, 1974.

Small, P. A. "Long Chain Branching Polymers"; Springer-Verlag: New York, 1975.

Soloman, D. H. "Step Growth Polymerizations"; Dekker: New York, 1974.

Sorenson, W. R.; Campbell, T. W. "Preparative Methods of Polymer Chemistry," 2nd ed.; Wiley-Interscience: New York, 1968.

Starks, C. "Free Radical Telomerization"; Academic: New York, 1975.

Steven, M. P. "Polymer Chemistry"; Addison-Wesley: Reading, Mass., 1975.

Stille, J. K.; Campbell, T. W. "Condensation Monomers"; Wiley-Interscience: New York, 1972.

Tanford, C. "Physical Chemistry of Macromolecules"; Wiley: New York, 1961.

Treloar, L. R. G. "Introduction to Polymer Science"; Wykeham: England, 1975.

Van Krevelen, D. W. "Properties of Polymers"; American Elsevier: New York, 1972.

Ward, I. M. "Structure and Properties of Oriented Polymers"; Wiley: New York, 1975.

Whitby, B. "Synthetic Rubber"; Wiley: New York, 1973.

Williams, H. T. "Polymer Engineering"; Elsevier Scientific: New York, 1975.

Williams, J. G. "Stress Analysis of Polymers"; Halsted: New York, 1972.

Yokum, R. H.; Nyquist, E. G. "Functional Monomers: Their Preparation, Polymerization and Application"; Dekker: New York, 1974.

POLYMERIZATION AND POLYMERIZATION MECHANISMS

Anionic Polymerization

Maurice Morton

Institute of Polymer Science, The University of Akron, Akron, OH 44325

Historical Review
Special Features of the Anionic Mechanism
  Absence of Termination Processes
  Effect of Counterion (Initiator)
  Effect of Solvents and Reaction Conditions
Synthesis Capabilities
  Block Copolymers
  Functional End-Group Polymers
Initiation Processes in Anionic Polymerization
  Initiation by Electron Transfer
  Initiation by Nucleophilic Attack
Mechanism and Kinetics of Homogeneous Anionic Polymerization
  Polar Media
  Nonpolar Media

The term "ionic polymerization" basically involves the chemistry of heterolytic cleavage of chemical bonds, as opposed to the homolytic reactions that characterize the well-known free-radical polymerization mechanism. Hence, essential and profound differences exist between these two mechanisms of polymerization. Although these differences are also found between radical and ionic mechanisms in ordinary reactions, they exert a much more drastic influence on the result, that is, the growth of a long chain molecule to macro dimensions. Thus, one would expect that the two mechanisms could lead to quite different results in most simple reactions, in terms of rate, yield, or mode of the reaction. In the case of polymerization, however, such differences, can, in fact, decide whether any high polymer is obtained at all.

The differences between the homolytic and heterolytic mechanisms of polymerization are reflected in the influence of the following factors on the course of the reaction:

0097–6156/85/0285–0051$06.00/0
© 1985 American Chemical Society

1. The effect of monomer structure on reactivity (polymeriza-
   bility) and on presence of "side reactions," that is, those
   reactions that interfere with chain growth, for example,
   termination, transfer, and rearrangement. This factor is
   relevant to both mechanisms.
2. Type of initiator. This factor does not influence the propagation reaction in the free-radical mechanism, but can have a
   profound effect on ionic propagation.
3. Nature of the medium. Again, the free-radical mechanism
   generally exhibits little or no dependence of the propagation
   reaction on the type of solvent or medium (1), whereas ionic
   systems can show very large effects, not only on the kinetics
   but on the actual chain structure as well.

Items 2 and 3 arise from the fact that both the "counterion"
and the medium itself can markedly affect the nature of the growing
chain end. Thus, the growing chain end may assume various forms
that depend on the extent of electrical charge separation and range
all the way from a polarized covalent (sigma) bond to a completely
dissociated state of free ions. This characteristic presents the
greatest distinction between the mechanisms of free-radical and
ionic polymerization.

Hence, in contrast to free-radical polymerization, the nature
of the active species in ionic polymerization is still surrounded
by a great deal of mystery and ignorance, especially because it may
vary from one system to another. Furthermore, the elucidation of
the active species in such ionic systems is made even more
difficult by the known sensitivity of these systems to traces of
impurities.

Ionic polymerization has been subdivided into two broad categories; cationic and anionic, on the basis of the nature of the
"charge" on the tip of the growing chain. The above classification
does not necessarily refer to the presence of free cations or
anions. In fact, it would seem that, in the majority of cases, the
main species involved are "ion-pairs" in which a counterion is
associated with the cation or anion. These ionic systems are quite
complex and require intensive study and a certain degree of
specialization; therefore, they have generally been considered as
separate fields of investigation. The "Ziegler-Natta" systems
(which may be heterogeneous or homogeneous) are not yet understood
enough to be classified into a particular ionic system, although in
the majority of cases they are considered as coordinated anionic
polymerizations. For the sake of clarity and simplicity, this
discussion will be limited to the simpler homogeneous anionic
polymerization systems.

Finally, this chapter is not a comprehensive review of the work
done in this field, but rather a concise discussion of the present
state of understanding. Further exploration of anionic
polymerization can be found in two recent books (2, 3), the former
being an assembly of topical papers presented at a recent
symposium, and the latter being a more comprehensive treatment of
this subject.

Historical Review

Because the mechanism of an addition polymerization reaction is defined by the character of the growing chain end, and because most polymerizations involve vinyl and allied compounds, we are concerned with the presence of carbon atoms that are free radicals, carbenium ions, or carbanions. Such definitions made it difficult to understand and classify the earliest instances of anionic polymerization, that is, the polymerization of butadiene and isoprene by sodium, probably first reported by Harries (4) and Matthews and Strange (5). Although this polymerization soon became a commercial process for synthetic rubber of the polybutadiene type and was studied intensively by Ziegler for many years, its significance as a carbanionic chain reaction was not recognized for a long time because of the heterogeneous nature of the reaction components. Thus, it was known that metallic sodium is capable of adding two atoms either 1,4 or 1,2 across a conjugated 1,3-diene, and that such di-adducts can further add Na and RNa groups across additional molecules of the diene, but this reaction was considered a "step-wise" polymerization without regard to the prevailing mechanism. Eventually scientists recognized (ca. 1950) that, because the terminal carbon atom in these diene "adducts" is attached to an alkali metal, it must have the nature of a carbanion.

Actually, the earliest examples of anionic polymerization studied were the base-catalyzed polymerizations of ethylene oxide involving a ring-opening chain reaction:

$$RONa + CH_2\text{-}CH_2 \xrightarrow{\quad} ROCH_2CH_2ONa \xrightarrow{\quad} \cdots\cdots \xrightarrow{\quad} \tag{1}$$

Because the monomer was not a vinyl compound and the active chain end was an alkoxide, this reaction was not considered an important case of anionic polymerization. Ironically, this reaction actually is a very good example of the anionic mechanism and can be satisfactorily studied because it is a homogeneous reaction. In fact, it was Flory (6) who first pointed out the unique consequences that arise from such a polymerization in which presumably the alkoxide chain end does not undergo any "side reactions," that is, termination. Flory remarked that in such a situation in which all the growing chains have equal access to the monomer, the chains will tend to reach similar lengths, that is, the molecular weight of the polymer will have the very narrow Poisson distribution:

$$P_w/P_n = 1 + 1/P_n \tag{2}$$

where P_w is the weight-average number of units per chain and P_n is the number-average number of units per chain. This example was apparently the first case of a "living polymer" that did not involve a vinyl monomer.

A similar case was later found in the base-catalyzed ring-opening polymerization of cyclic siloxanes. In this case, on the basis of kinetic measurements, it was shown (7, 8) that the active species is the free silanolate anion ($-Si\text{-}O^-$) and no true termination step exists whereby chains cease growing. However, although it was proven conclusively (8) that each initiator molecule produced a polymer chain, the Poisson distribution was not attained because of the prevalence of siloxane bond interchange between the polymer and the growing silanolate ion. More recently, a narrow distribution of chain lengths was attained (9) in these systems if the base-catalyzed siloxane bond interchange was suppressed (by using the much more reactive cyclic trimer and a weaker base).

Although the unique features of the above polymerizations were recognized, the same was difficult to do for vinyl and diene monomers in which the much more reactive anion, that is, a carbanion, was involved. Thus, although some of the earlier studies, such as those of Robertson and Marion (10) on sodium polymerization of butadiene in toluene, and Higginson and Wooding (11) on styrene polymerization by potassium amide in liquid ammonia, demonstrated the presence of an anionic mechanism, the absence of any true termination step in these investigations was not recognized because of the presence of many transfer reactions.

Much more was learned about the true nature of the carbanionic mechanism after more intensive investigations that followed two special developments. The first was the discovery of the stereospecific polymerization of isoprene to the high *cis*-1,4 polymer (12) by lithium or its organic compounds. This polymerization virtually synthesized the structure of natural rubber. The other development was the discovery of the nonterminating ("living") nature of the polymerization of styrene by sodium naphthalene (13). Because both of these systems involve homogeneous reactions, it was possible to subject them to investigation with regard to kinetics and stoichiometry, and thus determine the main characteristics of the anionic mechanism. These developments and possibilities aroused an increased interest in the field.

Special Features of the Anionic Mechanism

The special features of the anionic mechanism that distinguish it from the other mechanisms can be classified as follows:

Absence of Termination Processes. The possibility of having carbanionic species that show a negligible rate of termination is now realized. In other words, just as the growing chain end in the alkoxide polymerization of ethylene oxide represents a "stable" salt of an alkali metal and an alcohol, the styryl sodium chain end, in the polymerization of styrene by sodium naphthalene, represents a "stable" salt of sodium and a hydrocarbon. This relationship was first noted in the particular case of the sodium naphthalene systems in which the organometallic species is stabilized by a high degree of solvation by an ether, such as tetrahydrofuran, so that no observable side reactions exist, that is, termination of chains, at least not within the time scale of the polymerization reaction.

Such "living" polymer systems, however, are not limited to polymerizations in solvating media, such as ethers. Thus, the lithium-catalyzed polymerizations, which can lead to the synthesis of *cis*-1,4-polyisoprene, also demonstrate the virtual absence of

termination (14). Because lithium polymerizations can be carried
out in both ether and hydrocarbon solvents, they can be used to
demonstrate the role of the solvent in these reactions. In all of
these cases, the absence of any noticeable termination process can
lead to a very narrow distribution of chain lengths, in accordance
with Equation 2. However, although necessary, this condition is
not sufficient to guarantee such a result because the rate of
initiation of chains also is a factor. Hence, the question of the
molecular weight and its distribution will be discussed together
with the question of initiation processes.

Effect of Counterion (Initiator). As stated earlier, in all ionic
polymerizations a possible influence exerted by the counterion
exists and originates from the initiator. In other words, unlike
the case of free-radical polymerization, the type of initiator used
may actually affect the nature of the chain growth process. This
relationship is strikingly demonstrated by the organoalkali
polymerization of dienes in which the proportion of *cis-* and *trans-*
1,4 as well as 1,2 units in the chain is governed by the alkali
metal counterion. This result was, of course, the basis for the
discovery of the stereospecific polymerization of isoprene by
lithium initiators (12). Foster and Binder (15) showed the effect
on the microstructure of both polybutadiene and polyisoprene
prepared by means of the five well-known alkali metals, and
demonstrated that the 1,4 type of addition decreases with
increasing electropositivity of the metal. Hence, the
microstructure of polyisoprene varies all the way from a very high
1,4 content with lithium to a very high 3,4 content with rubidium
or cesium. Foster and Binder also showed the a similar effect on
the structure of polybutadiene with regard to 1,4 versus 1,2 units,
but the proportion of the *cis*-1,4 isomer was, of course, much
lower.

Effect of Solvents and Reaction Conditions. The term "solvent" is
customarily used rather loosely in polymerization reactions because
such "solvents" may refer either to the actual medium in which the
reaction is carried out, or to trace materials present in the
medium. Hence, the term really encompasses any component other
than monomer and initiator. Thus, in free-radical polymerization,
the role of the solvent is limited to "interfering" with the normal
propagation reaction, either through chain transfer or even by
termination (inhibition or retardation). Either of these events
can affect only the chain length or the overall rate, or both.
    In contrast, in anionic systems in which the solvent may not
actually interrupt the propagation process, it may play an active
role in controlling both the rate and mode of the chain growth
step. This control is perhaps most dramatically illustrated in the
case of the organolithium polymerizations in connection with two
specific aspects: chain microstructure of polydienes and
copolymerization of dienes and styrene.
    The dramatic effect of even traces of ethers on the micro-
structure of polybutadiene and polyisoprene in organolithium
polymerizations in hydrocarbon media was demonstrated very effectively by Tobolsky et al. (16, 17) They showed that highly
solvating ethers, such as H_4-furan, when present in approximately

stoichiometric proportion to the amount of lithium, can cause a
sharp decrease in the 1,4 addition reaction in favor of the 1,2 (or
3,4) reaction. They suggested that the ethers, by coordinating
with (solvating) the lithium cation, cause a greater charge
separation and thereby mimick the behavior of the more electropositive, larger cations of the alkali metal series, which are also
known to lead to high side-vinyl structures in the polymer. The
effect of such solvents on the microstructure is illustrated in
Table I on the basis of NMR analysis (18, 19), which gives more
accurate results than IR spectroscopy methods. However, the
results are, in general, similar to those published previously (16,
17).

Table I also shows that the initiator and monomer concentration
can also markedly affect the chain structure in nonpolar media
(these factors apparently have no effect in the presence of polar
solvents). Thus, the highest *cis*-1,4 content in the case of both
polybutadiene and polyisoprene is obtained in the absence of any
solvents and at very low initiator levels.

The equally dramatic effect of polar solvents on the copolymerization behavior of dienes with styrene is illustrated in Table
II.

The r values show that the relative rate of entry of the diene
and styrene monomers is apparently controlled very closely by the
nature of the carbon-lithium bond. Thus, in hydrocarbons the
preference is very strong for the dienes, whereas, in the presence
of a highly solvating medium such as H_4-furan, the exact reverse is
true. Solvents of intermediate polarity show a lesser effect.
Apparently, the effect of the solvent in influencing the charge
separation at the carbon-lithium bond profoundly influences the
kinetics of the copolymerization.

Synthesis Capabilities

The "living" nature of polymer chains in homogeneous anionic
polymerization leads to novel possibilities of chemical synthesis
that are not available in polymerization systems in which the
growing chain has a short lifetime. The three possibilities
offered by the "living" ends of these polymer chains are A.
attainment of a very narrow molecular weight distribution, much
narrower than previously accomplished by fractionation; B. formation of block copolymers by sequential addition of different
monomers; and C. formation of polymers having functional end groups
by addition of an appropriate reagent at the conclusion of the
polymerization.

In regard to Item A, the factors controlling the molecular
weight distribution are discussed more fully in a later section
under the topic of initiation reactions. Furthermore, a comprehensive review of accomplishments in this area can be found in a
very recent book (3). Hence, this discussion will be limited to
Items B and C.

Block Copolymers. In principle, a block copolymer having any
number of blocks can be prepared by sequential addition of different monomers. However, in practice, this preparation may have
some limitations imposed by the inclusion, with the added monomers,

Table I. Chain Microstructure of Organolithium Polydienes

| Solvent | Initiator Conc. (M) | Microstructure (mole %) | | | |
		cis-1,4	*trans*-1,4	3,4	1,2
Polyisoprene ([18], [19])					
Benzene	3×10^{-4}	69	25	6	--
n-Hexane	1×10^{-2}	70	25	5	--
n-Hexane	1×10^{-5}	86	11	3	--
None	3×10^{-3}	77	18	5	--
None	8×10^{-6}	96	0	4	--
H_4-furan ([20])	3×10^{-4}	--	26	66	9
Polybutadiene ([18], [19], [21])					
Benzene	8×10^{-6}	52	36	--	12
Cyclohexane	1×10^{-5}	68	28	--	4
n-Hexane	2×10^{-5}	56	37	--	7
n-Hexane	3×10^{-2}	36	62	--	8
None	7×10^{-6}	86	9	--	5
None	3×10^{-3}	39	52	--	9
H_4-furan	2×10^{-4}	7	10	--	85

Table II. Copolymerization of Styrene and Dienes in Organolithium Systems (25 °C)

Monomer (M_2)	Solvent	r_1	r_2
Butadiene ([22])	None	0.04	11.2
	Benzene	0.04	10.8
	Triethylamine	0.5	3.5
	Diethyl ether	0.4	1.7
	H_4-furan	4.0	0.3
Isoprene	Benzene ([23])	0.26	10.6
	Triethylamine ([24])	1.0	0.8
	H_4-furan ([25])	9.0	0.1

of adventitious impurities capable of terminating some of the active chain ends. These anionic systems, therefore, offer unique opportunities for the synthesis of block copolymers of known architecture, in regard to both composition and chain length. The special feature of such polymers is that each block can, in principle, have the Poisson distribution (Equation 2) of chain lengths, that is, approach virtual monodispersity and thus add a new dimension to the synthesis of "tailor-made" polymers.

A striking example of the unusual properties of such well-characterized block copolymers is offered by the "ABA" types of polymers, known as "thermoplastic elastomers." In these elastomers, the A blocks consist of polystyrene (molecular weight approximately 10,000–15,000), and the B block is a polydiene, such as polybutadiene or polyisoprene (molecular weight approximately 50,000–75,000). These elastomers can be synthesized by the sequential addition of the two monomers to an organolithium initiator (26, 27) in hydrocarbon solvents that yield rubbery polydienes of high 1,4 content. Because these elastomers are "pure" block copolymers, the homopolymer blocks are incompatible and tend to separate into two phases. The net effect is that the polystyrene, which is the lesser component, separates as a dispersed phase within the polydiene medium. In view of the restrictions imposed by the chemical bonds between the blocks, the two phases do not separate on a macro scale, but form a very finely divided (approximately 200 Å) dispersion of polystyrene in the polydiene. The regularity of such a dispersion of polystyrene spheres in a polyisoprene matrix is shown in the transmission electron microphotograph of Figure 1. All of these phase separations can only occur from solutions or melts. At ambient temperatures, then, the polystyrene dispersion is in a glassy state in which each polystyrene particle acts as a virtually rigid juncture for a large number (approximately 200) of polydiene chain ends and thus creates an elastic network. However, because the network junctions are in the form of thermoplastic polystyrene particles, the network is reversibly destroyed and reformed on heating and cooling.

These ABA block copolymers thus owe their properties to the incompatibility of the individual blocks and to the unusual morphology of the resulting material. They are an excellent example of the special features that arise from block and graft polymers in which different homopolymer chains are linked together chemically. Their mechanical properties can also be quite unusual. Figure 2 shows typical stress-strain curves of a series of styrene-isoprene-styrene block copolymers. The effect of the amount of dispersed polystyrene is obvious, as is the unusually high tensile strength of these elastomers, especially in view of their amorphous nature. These high values of tensile strength have been ascribed (28, 29) to the "filler" effect of the polystyrene particles and to their ability to yield at high stress and absorb the strain energy that would otherwise cause rupture.

0.2μ

ULTRA-THIN SECTION—LINEAR SIS-1

Figure 1. Transmission electron micrograph of a styrene–isoprene–styrene triblock copolymer.

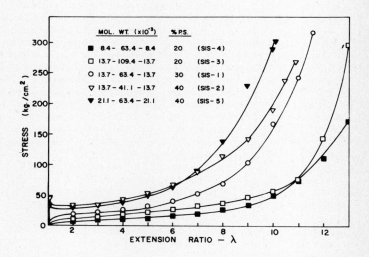

Figure 2. Tensile properties of styrene–isoprene–styrene triblock copolymers.

Functional End-Group Polymers. The formation of various functional groups on the "living" ends of anionic polymer chains is based on the well-known reaction of organometallic compounds with various reagents. Thus, organoalkali compounds can react as illustrated in the following typical reactions:

$$RM + CO_2 \longrightarrow RCOOM \tag{3}$$

$$RM + \underset{\displaystyle \underset{O}{\diagdown\diagup}}{CH_2-CH_2} \longrightarrow R-CH_2-CH_2OM \tag{4}$$

$$RM + R'CHO \longrightarrow R-\underset{\displaystyle \underset{H}{|}}{\overset{\displaystyle \overset{R'}{|}}{C}}-OM \tag{5}$$

where M represents an alkali metal. The resulting products, on hydrolysis, yield carboxylic acids, alcohols, etc.

The interest in such polymers with functional end groups centers mainly on the difunctional variety because this variety can offer interesting possibilities of various linking reactions, such as chain extension or network formation, which are important in liquid polymer technology. To synthesize such difunctional polymers it is necessary first to have a di-anionic polymer chain, such as is formed from initiation reactions involving an electron transfer process as will be discussed later. However, di-anionic species are generally insoluble in inert solvents such as hydrocarbons because of the tendency of organometallics to associate in such solvents. Hence, solvating solvents such as ethers are generally required to obtain homogeneous systems, and the presence of such ethers, as mentioned earlier, is a serious obstacle to the synthesis of rubbery, high 1,4-polydienes. This situation has posed a serious problem in the development of liquid polymer technology for the polydienes. Recently, however, aromatic ethers and tertiary amines were used as solvents to form homogeneous solutions of organodilithium initiators and polymers without exerting any major effect on the microstructure of polydienes (30-32). Hence, these ethers and amines can be used in the synthesis of rubbery, difunctional polydienes having a very narrow molecular weight distribution (33). A recent publication (34) describes how difunctional polyisoprenes made in this way were end-linked to form relatively uniform elastic networks that exhibited superior properties.

Initiation Processes in Anionic Polymerization

The anionic polymerization systems based on the alkali metals involve only two processes, initiation and propagation, because termination can be avoided, if desired. The initiation processes fall into two broad classes: electron transfer and nucleophilic attack.

Initiation by Electron Transfer. This mechanism of initiation operates in polymerizations by alkali metals or their complexes and was best elucidated in the case of the sodium naphthalene complex (13), which was used to form the well-known "living" polymers. In these complexes, the naphthalene is a radical anion (35) formed by transfer of an electron from sodium in the presence of a highly solvating solvent such as H_4-furan:

$$Na + \text{[naphthalene]} + H_4\text{-furan} \rightleftharpoons \left[\text{naphthalene}\right]^{\cdot-} Na^+ \ (H_4\text{-furan}) \quad (6)$$

Greenish blue

This process thus has the effect of dissolving sodium metal to form a bright greenish-blue solution. This solution can actually be considered as a solution of sodium in its metallic state. In the presence of a monomer, such as styrene, a rapid transfer of electrons occurs from the naphthalene:

$$\left[\text{naphthalene}\right]^{\cdot-} Na^+ \ (H_2\text{-furan}) + \underset{\text{CH}_2=\text{CH}-\text{C}_6\text{H}_5}{} \rightleftharpoons$$

Greenish blue

$$\left[\begin{array}{c} CH_2=CH \\ C_6H_5 \end{array}\right]^{\cdot-} Na^+ \ (H_4\text{-furan}) + \text{naphthalene} \quad (7)$$

Red

followed by coupling of the styryl radical-anions, thus:

$$2\left[\begin{array}{c} CH_2=CH \\ C_6H_5 \end{array}\right]^{\cdot-} Na^+ \longrightarrow Na^+ \ \overset{=}{C}H-CH_2-CH_2-\overset{=}{C}H \ Na^+ \xrightarrow{\text{pzn.}} \quad (8)$$

Hence, the final result is the initiation of a di-anionic growing chain.

This initiation process is quite fast, relative to the propagation process, so that all the growing chains are formed during a short-time interval and have equal opportunity to grow. This condition produces a narrow molecular weight distribution. These reactions with styrene are so rapid that the experimental technique involves dropwise addition of the monomer to the sodium naphthalene solution at low temperature (−78 °C). Hence, the first few drops convert all the sodium naphthalene into styryl disodium. Further additions then result in chain growth that is governed mainly by mixing conditions.

Reactions 7 and 8 are somewhat oversimplified presentations. The transfer of a second electron from the naphthalene to the styrene is also possible and produces a monostyrene di-anion rather than the distyrene di-anion shown in Reaction 8. In addition, the possibility of coupling of naphthalene and styrene radical-anions exists. However, the only species ever isolated (e.g., by hydrolysis to 1,4-diphenylbutane) from this type of initiation process has been related to the distyrene species of Reaction 8.

This result can probably be ascribed to the fact that the styrene radical anions are the most reactive (least stable) of the species.

As stated earlier, the electron-transfer processes in these sodium naphthalene systems are actually the homogeneous analogs of the reactions of monomers with the metals. Thus, the initiation of butadiene by sodium may be written as

$$Na + CH_2=CH-CH=CH_2 \longrightarrow \overset{\cdot}{C}H_2-CH=CH-\overset{\bar{\bar{}}}{C}H_2 Na^+ \tag{9}$$

$$\overset{\cdot}{C}H_2-CH=CH-\overset{\bar{\bar{}}}{C}H_2 Na^+ \overset{\displaystyle\nearrow Na^+ \ \overset{\bar{\bar{}}}{C}H_2-CH=CH-CH_2-CH_2-CH=CH-\overset{\bar{\bar{}}}{C}H_2 \ Na^+}{\underset{\displaystyle Na^+ \ \overset{\bar{\bar{}}}{C}H_2-CH=CH-\overset{\bar{\bar{}}}{C}H_2 \ Na^+}{\overset{\displaystyle Na}{\longrightarrow}}} \tag{10}$$

Reaction 10 represents coupling of the radical anions, and Reaction 11 represents a second electron-transfer step. Reaction 11 does occur in the case of sodium and butadiene (10). This result is not surprising because an electron transfer step such as Reaction 9 can only occur at the metal surface and cannot give rise to a high concentration of radical anions in the homogeneous phase (as in the case of the naphthalene complex). Hence, these radical anions are quite likely to remove a second electron from the sodium when they are in the vicinity of the metal surface. The above reactions of butadiene are all shown as 1,4 for the sake of simplicity, although a substantial proportion of 1,2 additions could occur as well. These additions depend on the nature of the medium.

A heterogeneous initiation reaction, such as shown in Reactions 9-11, would be expected to be quite slow relative to the homogeneous propagation steps that follow. Hence, this situation would not lead to the conditions necessary for the Poisson distribution of chain lengths. This aspect of the heterogeneous initiation probably obscured the absence of chain termination in these systems that were the basis of the well-known sodium polybutadiene synthetic rubbers. The phenomenon of the "living" chain end in these polymerizations was recognized only with the advent of the homogeneous initiation systems of the sodium naphthalene type.

<u>Initiation by Nucleophilic Attack</u>. In contrast to the alkali metals, the alkali organometallics can initiate polymerization by a direct nucleophilic attack on the π electrons of the monomer. This polymerization can happen with both soluble and insoluble initiators, but the latter will lead to a very slow reaction. The homogeneous initiation systems of this type are best exemplified by the organolithium compounds, largely alkyllithium, which are soluble in a variety of solvents, including the hydrocarbons. However, even in these homogeneous systems, the rate of the initiation reaction is greatly affected by the nature of the medium and is much more rapid in the presence of solvating media, for example, ethers, than in hydrocarbons. The kinetics of these initiation reactions are also influenced by the structure of the initiator. For example, *sec*-butyl lithium is a much more active initiator than *n*-butyl lithium in hydrocarbon solvents and can lead

to polymers of narrow molecular weight distribution. When the initiation step is not sufficiently fast, however, then the molecular weight distribution may be considerably broadened. This is often the case in hydrocarbon media (14) despite the complete absence of any termination. Some of the aspects of the initiation reaction will be further explored in the following discussion of the kinetics and mechanisms of these polymerizations.

Mechanism and Kinetics of Homogeneous Anionic Polymerization

The overall polymerization reaction in a homogeneous medium involving, for example, an organolithium initiator, may be written as follows:

$$RLi \ + M \ \longrightarrow \ RMLi \tag{12}$$

$$RM_jLi + M \ \longrightarrow \ RM_{j+1}Li \tag{13}$$

where M represents the monomer. Reaction 12 represents initiation and Reaction 13 represents the propagation step. The determination of the individual rate constants of the two reactions from polymerization rate measurements poses a rather intractable mathematical problem (36). Fortunately, however, it is possible to measure these rate constants separately. The rate constant of the initiation step can be measured from the initial rate of disappearance of initiator, and that of the propagation step can be measured from the rate of disappearance of monomer after all the initiator has reacted. These techniques have been applied to a number of such anionic polymerizations. As might be expected, most attention has been given to the propagation reaction because it is the actual chain-building process, but some initiation studies have also been carried out. Here again the nature of the medium has been found to exert a profound effect on the kinetics and mechanism; therefore, the two classes of solvents, solvating and nonsolvating, should be treated separately.

Polar Media. The polar solvents, which generally act as coordinating ligands to solvate the metal counterion, encompass such compounds as ethers and amines. A number of kinetic studies of the propagation reaction have been carried out (37-40) for several monomers and initiators in this type of solvent, for example, H_4-furan and dimethoxyethane. In all cases, the reaction rate was found to be first order with respect to monomer, in accordance with Reaction 13. However, the rate dependence on the number of growing chains (RM_jLi) was found to vary anywhere from half order to first order. Hence, Reaction 13 may not describe the actual mechanism in this respect.

On the basis of kinetic and electrolytic measurements (35-38), the actual active species in these solvating media may be twofold, that is, ion pairs and dissociated anions. In fact, more than one type of ion pair is possible (for example, contact, solvent separated, etc.). However, for the sake of simplicity, the actual propagation steps are best represented as follows:

$$RM_j^-Li^+ \xrightleftharpoons{K_e} RM_j^- + Li^+$$

$$M \quad k_p^{\pm} \qquad M \quad k_p^- \qquad\qquad\qquad (14)$$

$$RM_{j+1}^-Li^+ \xrightleftharpoons{K_e} RM_{j+1}^- + Li^+$$

where $RM_j^-Li^+$ represents ion-pair species, RM_j^- represents free anions, and $k_p^{\pm}$ and k_p^- are the respective propagation rate constants. Thus, the overall rate can be expressed as

$$R_p = k_p^{\pm}[RM_j^-Li^+] \, [M] + k_p^-[RM_j^-] \, [M] \qquad\qquad (15)$$

Taking into account the equilibrium constant K_e, and assuming that it is small (little dissociation), we obtain

$$R_p/[M] = k_p^{\pm}[RM_jLi] + k_p^- \, K_e^{\frac{1}{2}} \, [RM_jLi]^{\frac{1}{2}} \qquad\qquad (16)$$

where $[RM_jLi]$ represents the total concentration of both types of growing chains (total initiator). Equation 16 can then be rearranged to a convenient linear form:

$$\frac{R_p}{[M] \, [RM_jLi]^{\frac{1}{2}}} = k_p^{\pm}[RM_jLi]^{\frac{1}{2}} + k_p^- K_e \qquad\qquad (17)$$

Equation 17 can then be plotted as a linear function of rate versus initiator concentration to yield k_p^- and $k_p^- K_e$. Furthermore, if K_e is determined, for example, from conductivity measurements, then the absolute value of k_p^- is also available. A number of such measurements have been taken and yield rate constant values for various monomers (mainly styrenes and dienes) and various counterions and solvents (3). In general these data indicate that, although the free anions are only present in very small proportion ($K_e \sim 10^{-6}$), they are responsible for most of the chain propagation because their rate constants ($k_p \sim 10^4-10^5 M^{-1}sec^{-1}$) are several orders greater than those of the ion pairs ($k_p^{\pm} \sim 10^2 M^{-1}sec^{-1}$). Hence, Reaction 14 seems to represent an adequate picture of the anionic mechanism in these systems.

Nonpolar Media. Because organolithium initiators are soluble in hydrocarbons, the kinetics of these polymerizations have also been studied in these nonsolvating media. A large number of such studies have been carried out (3, 41) mainly on styrene and the dienes. Again the propagation rate is first order with respect to monomer, in accordance with Reaction 13. However, the rate dependence on growing chain concentration has been found to show marked variation from one system to another with the orders varying from one half to much lower values (3, 41). These systems pose increased difficulties for kinetic measurements, because the rates are much slower in hydrocarbons and hence quite sensitive to traces of polar impurities.

By analogy with the mechanism proposed for the solvated anions, these fractional kinetic orders can be ascribed to association-dissociation phenomena involving the growing chains. Because ionic dissociation is not a viable assumption for such low dielectric media, ion-pair association is assumed. Thus,

$$(RM_j^-Li^+)_x \underset{}{\overset{K_e}{\rightleftarrows}} xRM_j^-Li^+$$

$$M \downarrow k_p \qquad\qquad (18)$$

$$xRM_{j+1}^-Li^+$$

In this case, if the associated form is assumed to be incapable of propagating, then the reaction order will correspond with the value of $1/x$. To prove such a mechanism, however, requires an independent measurement of x. Fortunately, such measurements have been made (3, 37, 41-43) and show quite unequivocally that the value of x is consistently 2, that is, the growing polymer chains are associated in pairs in the case of all systems for which kinetic data are available, that is, styrene, butadiene, and isoprene. Hence, the kinetic scheme shown in Reaction 18 could be invoked only to explain half-order rate dependencies but not the lower orders (1/3 to 1/6), such as found for the dienes (3, 41). Therefore, this scheme does not offer a general mechanism for these reactions.

Thus, the detailed mechanism of the chain-growth reaction in hydrocarbon media still remains to be elucidated. It is undoubtedly complicated by the fact that these polymer-lithium species are in an associated state in these solutions, presumably because of interaction between the lithium-bonded chain ends similar to the behavior of most organolithium compounds. Furthermore, studies (3, 41, 44-47) of the organolithium chain-end protons by NMR spectroscopy have indicated that the carbon-lithium bond is essentially σ in character in hydrocarbon media, but that it is in equilibrium with a small contribution of π-bonding, which is more reactive and gives rise to the minor proportion of 1,2 microstructure. Because of these complexities in the structure of the growing chain end in hydrocarbon media, the propagation reaction does not show simple kinetic behavior in these media. The most probable mechanism for the propagation reaction undoubtedly involves a direct interaction between the monomer and the associated chain ends, and the puzzlingly low kinetic orders probably reflect changes in structure and reactivity of the associated complex with changes in concentration.

In conclusion, although a number of anionic polymerization systems involving alkali metal compounds in ionizing solvents have lent themselves to a satisfactory elucidation, much remains to be learned about such systems in nonpolar media such as hydrocarbons. Furthermore, the systems discussed herein have involved only the alkali metals and a very small group of hydrocarbon monomers, for example, styrene, butadiene, and isoprene. In principle, the anionic mechanism (nucleophilic attack) should apply to a wide variety of monomers having electronegative substituents, for example, acrylonitrile, acrylates, and vinyl chloride. In fact, such monomers do not give "clean" systems with the organoalkali initiators because of frequent side reactions with reactive groups on the monomer (H, Cl, etc.). Hence, the "living" polymer systems are largely restricted to the above mentioned hydrocarbon monomers.

In addition, the organoalkali initiators only work effectively with the conjugated monomers. They are ineffective with the olefins, or even with ethylene. (Some success has been reported (48) in the polymerization of ethylene to a reasonably high molecular weight in highly chelated organolithium systems. However, these polymerizations required relatively higher temperatures and showed much evidence of termination reactions.)

One suspects that the π–allyl structure that is available in these conjugated systems is required for the necessary reactivity. Such is not the case with transition metal compounds that can coordinate with the π electrons of the isolated double bond. More elucidation is needed of the actual bond structure of the growing chain and might be obtained from NMR spectroscopy (44–47) or similar techniques.

Literature Cited

1. Lewis, F. M.; Walling, C.; Cummings, W.; Briggs, E. R.; Mayo, F. R. J. Am. Chem. Soc. 1948, 70, 1519.
2. "Anionic Polymerization"; McGrath, J. E., Ed.; ACS SYMPOSIUM SERIES No. 166, American Chemical Society: Washington, D.C., 1981; p. 19.
3. Morton, Maurice, "Anionic Polymerization: Principles and Practice"; Academic Press: New York, 1983.
4. Harries, C Ann. 1911, 383, 213.
5. Matthews, F.; Strange, E. Brit. Pat. 24 790, 1910.
6. Flory, P.J. J. Am. Chem. Soc. 1940, 62, 1561.
7. Grubb, W. T.; Osthoff, R. C. J. Am. Chem. Soc. 1955, 77, 1405.
8. Morton, M.; Deisz, M. A.; Bostick, E. E. J. Polym. Sci. 1964, A2, 513.
9. Saam, J. G.; Gordon, D. J.; Lindsey, S. Macromolecules 1970, 3, 1.
10. Robertson, R. E.; Marion, L. Can. J. Res. 1948, 26B, 657.
11. Higginson, W. C. E.; Wooding, N. S. J. Chem. Soc. 1952, 760.
12. Stavely, F. W., et al. Ind. Eng. Chem. 1956, 48, 778.
13. Szwarc, M.; Levy, M.; Milkovich, R. J. Am. Chem. Soc. 1956, 78, 2656.
14. Morton, M.; Rembaum, A. A.; Hall, J. L. 132nd Meeting, American Chemical Society, New York, 1957; J. Polym. Sci. 1963, A1, 461.
15. Foster, F. C.; Binder, J. R. Adv. Chem. Series 1957, 19, 26.
16. Morita, H.; Tobolsky, A. V. J. Am. Chem. Soc. 1957, 79, 5853.
17. Tobolsky, A. V.; Rogers, C. E. J. Polym. Sci. 1959, 40, 73.
18. Morton, M.; Fetters, L. J. Rubber Chem. Technol. 1975, 48, 359.
19. Morton, M.; Rupert, J. ACS SYMPOSIUM SERIES No. 212, American Chemical Society: Washington, D.C., 1983.

20. Worsfold, D. J.; Bywater, S. Can. J. Chem. 1964, 42, 2884.
21. Santee, E. R., Jr.; Malotky, L. O.; Morton, M. Rubber Chem. Technol. 1973, 46, 1156.
22. Ref. 3, p. 142; Huang, L. K., Ph. D.; Dissertation University of Akron, 1980.
23. Morton, M.; Ells, F. R. J. Polym. Sci. 1962, 61, 25. Ells, R. F., Ph. D. Dissertation, University of Akron, 1963.
24. Spirin, Yu. L.; Polyakov, D. R.; Gantmakher, A. R.; Medvedev, S. S. J. Polym. Sci. 1961, 53, 233.
25. Spirin, Yu. L.; Arest-Yakubovich, A. A.; Polyakov, D. K.; Gantmakher, A. R.; Medvodev, S. S. J. Polym. Sci. 1962, 58, 1161.
26. Holden, G.; Milkovich, R. Belg. Pat. 627 652, 1963; Chem. Abstr.1964, 60, 14174.
27. Fetters, L. J. J. Polym. Sci., Part C 1969, C, No. 26, 1.
28. Smith, T. L.; Dickie, R. A. ibid., 163.
29. Morton, M.; McGrath, J. E.; Juliano, P. C. ibid., 99.
30. Fetters, L. J.; Morton, M. Macromolecules 1969, 2, 453.
31. Morton, M.; Fetters, L. J. U. S. Patent 3 663 634, 1972.
32. Fetters, L. J. U. S. Patent 3 848 008, 1974.
33. Morton, M.; Fetters, L. J.; Inomata, J.; Rubio, D. C.; Young, R. N. Rubber Chem. Technol. 1976, 49, 303.
34. Morton, M.; Rubio, D. C. Plast. Rubber: Mater. Appl. 1978, 3, 139.
35. Paul, D. E.; Lipkin, D.; Weissman, S. I. J. Am. Chem. Soc. 1956, 78, 116.
36. Bauer, E.; Magat, M. J. Chem. Phys. 1950, 47, 841.
37. Morton, M.; Bostick, E. E.; Livigni, R. A. Rubber Plastics Age 1961, 42, 397.
38. Morton, M.; Bostick, E. E.; Livigni, R. A.; Fetters, L. J. J. Polym. Sci. 1963, A1, 1735.
39. Szwarc, M. Adv. Chem. Ser. 1962, 34, 96.
40. Bhattacharya, D. N.; Lee, C. L.; Smid, J.; Szwarc, M. Polymer 1964, 5, 54.
41. Morton, M.; Fetters, L. J. Rubber Chem. Technol. 1975, 48, 359.
42. Morton, M.; Fetters, L. J.; Pett, R. A.; Meier, J. F. Macromolecules 1970, 3, 327.
43. Fetters, L. J.; Morton, M. Macromolecules 1974, 7, 552.
44. Morton, M.; Sanderson, R. D.; Sakata, R. J. Polym. Sci. 1971, B9, 61.
45. Morton, M.; Sanderson, R. D.; Sakata, R. Macromolecules 1973, 6, 181.
46. Morton, M.; Sanderson, R. D.; Sakata, R.; Falvo, L. A. ibid., 186.
47. Morton, M.; Falvo, L. A. ibid., 190.
48. Langer, A. W. Trans. N. Y. Acad. Sci. 1961, 27, 741.

Coordinated Anionic Polymerization and Polymerization Mechanisms

FREDERICK J. KAROL

UNIPOL Systems Department, Union Carbide Corporation, Bound Brook, NJ 08805

Early Developments (1950–1965)
Developments Since 1965
Catalyst Systems and Chemistry
Types of Olefin Monomers and Polymers
Polymerization Mechanisms
Origin of Stereoregulation
Chemically Anchored, Supported Catalysts for Olefin
    Polymerization
General Features of Polymerization Catalysts,
Future Research and Related Areas,

Early Developments (1950–1965)

Discoveries in the laboratories of Ziegler and Natta (1–5) in the early 1950s caused a revolution in polymer and organometallic chemistry. The ability to polymerize ethylene at atmospheric pressure and room temperature was a result of extensive studies by Ziegler over many years in the field of reactions of organometallic compounds with olefins (6). Prior to this discovery extremely high pressure (>20,000 lb/in^2) and temperatures (approximately 250 °C) were required to convert ethylene to solid polyethylene. Direct polymerization of ethylene by this high pressure route had been achieved in the 1930s. This free-radical process normally produces branched polyethylenes of the low-density type. Ziegler claimed the discovery of a new process for polyethylene, but acknowledged he had not discovered a new product. He recognized its identity with polymethylene made from catalyzed decompositions of diazomethane.

Ziegler and Gellert (6) in 1949 showed that aluminum hydride reacts with ethylene at 60–80 °C to yield triethylaluminum. At 100–120 °C reaction with additional ethylene leads to formation of higher alkyls of aluminum (Reaction 1). At temperatures above 120 C higher aluminum alkyls react with ethylene through a displacement reaction to give olefins and triethyl aluminum (Reaction 2). These reactions represent a catalytic process for the conversion of ethylene into higher α-olefins.

$$al-C_2H_5 + nCH_2=CH_2 \xrightarrow{\text{100-120 °C}} al -(CH_2-CH_2)_n-C_2H_5 \quad (1)$$

$$al-(CH_2-CH_2)_n--C_2H_5 + CH_2=CH_2 \xrightarrow{\text{120-250 °C}} \quad (2)$$

$$CH_2=CH-(CH_2-CH_2)_{n-1}--- C_2H_5 + al-C_2H_5$$

$$\text{where al} = 1/3 \text{ Al}$$

In the course of these investigations, an experiment was carried out to prepare hexyl and octyl derivatives of aluminum by reaction of triethylaluminum with ethylene. Instead of the anticipated aluminum alkyls, an almost quantitative yield of 1-butene was obtained. After a strenuous investigation, Ziegler and his coworkers found that an extremely small trace of metallic nickel caused this change in the course of the reaction. The nickel, present from a previous hydrogenation experiment, catalyzed the displacement reaction (Reaction 2) of 1-butene from butylaluminum (Reaction 3).

$$al-C_2H_5 + C_2H_4 \xrightarrow[\text{Ni}]{\text{100-120 °C}} al-CH_2-CH_2-C_2H_5$$

$$\downarrow C_2H_4 \quad (3)$$

$$al-C_2H_5 + CH_3CH_2CH=CH_2$$

At this point Ziegler and his coworkers carried out experiments on the effects of adding various other metal compounds to triethylaluminum. In one of these experiments with zirconium acetylacetonate, ethylene, and triethylaluminum, they found, to their surprise, an autoclave filled with a solid cake of snow-white polyethylene (1, 6). Further work revealed that aluminum alkyls in conjunction with certain transition metal compounds of Groups IV-VI, as well as uranium and thorium, were active ethylene polymerization catalysts. Ultimately, Ziegler catalysts were described to be the product of reaction of metal alkyls, aryls, or hydrides of Groups I-IV and certain transition metal compounds of Groups IV-VIII (Reaction 4). The choice of a particular catalyst and experimental conditions is dictated by the structure of the monomer to be polymerized,

metal alkyl, aryl, + transition metal ⟶ active Ziegler
   or hydride   compound   catalyst
   (Group I-IV)   (Group IV-VIII)   (4)

Ziegler and his coworkers were primarily interested in ethylene polymerization and copolymerization with α-olefins. After Ziegler revealed details of his work to Montecatini, Natta, working with combinations of Ziegler-type catalysts, discovered stereoregular polymers of propylene, 1-butene, styrene, etc. (2, 3). Ziegler catalysts containing highly ordered (crystalline) transition metal salts in a lower valence state, for example, $TiCl_3$ and VCl_3, polymerize α-olefins to crystalline stereoisomeric polymers. Under the direction of Natta, basic principles of controlling stereoregularity were established (7-9). For contributions in this area, Ziegler and Natta were awarded the 1963 Nobel Prize for chemistry.

Independent catalyst research, carried out by several U. S. oil companies in the early 1950s with transition metal oxides supported on refractory metal oxides, led to the discovery of some of the earliest low-pressure catalysts for olefin polymerization (10-13). These catalysts generally consist of oxides of transition metal elements from Groups V-VII of the periodic table. For catalytic activity the transition metal oxides are supported on high-surface-area solids such as silica, alumina, silica-alumina, and clay. Silica-supported chromium trioxide (CrO_3/SiO_2) catalyst is the most important transition metal oxide catalyst for ethylene polymerization (12). Ethylene homopolymers made with these catalysts are predominantly linear, high-density products. With propylene and higher linear and branched α-olefins, polymerization rates, polymer yields, and degree of crystallinity are much lower than for the polymerization of ethylene.

Developments Since 1965

Developments toward higher activity ($\geq$200 kg polymer/g Ti vs. 1-5 kg polymer/g Ti) Ziegler-Natta catalysts during the last 15 years have, to a considerable extent, been based on reaction of specific magnesium, titanium, and aluminum compounds (14-19). Catalysts, chemically anchored on Mg(OH)Cl-type supports, provided some of the early impetus in the area of high activity systems (20, 21). Other studies concentrated on the use of $MgCl_2$ as a substrate. Grinding of $MgCl_2$ and treatment with $TiCl_4$ provided one route to a higher surface area substrate of magnesium and titanium (22-24). Some developments focused on reaction products of magnesium alkyls and titanium compounds (25-27). Other workers described the advantages of preparing trimetallic sponges by the addition of certain aluminum compounds to a magnesium substrate that had been treated with a titanium compound (28). Catalysts based on reaction products of magnesium alkoxides with transition metal compounds have also received attention (29). During catalyst preparation the original structure of the alkoxide is destroyed, and a new crystalline species of higher surface area is formed.

The discovery of ether-treated $TiCl_3$-based catalysts of high activity and stereospecificity, particularly for the polymerization of propylene, has been of considerable importance (30, 31). These catalysts are prepared by reduction of $TiCl_4$ with aluminum alkyls and subsequent treatment with Lewis bases such as ethers. The treated $TiCl_3$ product can be transformed to a highly active, stereospecific catalyst by treatment with $TiCl_4$.

Attractive, high-activity catalysts for propylene polymerization have also been described (32, 33). Montedison/Mitsui catalysts comprise an aluminum alkyl complexed with a electron donor such as ethyl benzoate, and a solid matrix containing the reaction products of halogenated magnesium compounds with a Ti(IV) compound and an electron donor. The specific surface area of the solid matrix after treatment is in the range of 100–200 m^2/g. High catalyst productivities based on titanium have been reported for polymerizations with these catalysts.

The direct use of organometallic compounds of transition metals for the preparation of solid catalysts for olefin polymerization, particularly ethylene polymerization, developed in the 1960s. Catalysts obtained by supporting π–cyclopentadienyl, π–allyl, and σ–organometallic compounds of transition metals, such as titanium, zirconium, and chromium, proved to be highly active for ethylene polymerization (34). A chromocene catalyst, $(C_5H_5)_2Cr/SiO_2$, has been described in some detail (35). Catalysis by supported complexes of transition metals has recently been well documented (19), and much active research continues in this area.

Catalyst Systems and Chemistry

Before the 1970s, Ziegler–Natta catalysts for α–olefin production were normally prepared from certain compounds of transition metals of Groups IV–VI of the periodic table (Ti, V, Cr, etc.) in combination with an organometallic alkyl or aryl (Table I). Practically all subhalides of transition metals have been claimed as catalysts in stereoregular polymerization. Only those elements with a first work function <4 eV and a first ionization potential <7 V yield sufficiently active halides, that is, titanium, vanadium, chromium, and zirconium (7, 8). Only titanium chlorides have gained widespread acceptance in crystalline polyolefin production.

The reaction of an alkylaluminum compound with titanium tetrachloride produces at room temperatures an almost instantaneous precipitation of lower valence titanium chloride (β–TiCl$_3$) and evolution of gaseous hydrocarbons (36). This process (Reactions 5 and 6) represents a ligand exchange reaction of an alkyl and a chlorine group followed by decomposition of the unstable σ–bonded alkyl titanium trichloride. These reactions are typical for numerous Ziegler-type catalysts:

$$R_3Al + TiCl_4 \longrightarrow RTiCl_3 + R_2AlCl \qquad (5)$$

$$RTiCl_3 \longrightarrow TiCl_3 + [R\bullet] \qquad (6)$$

The fate of the alkyl fragment [R•] remains uncertain. There is still controversy concerning the precise extent of reduction that is reached with trialkylaluminum compounds at different ratios of aluminum to titanium. However, scientists generally agree that the more active organometallic compounds such as trialkylaluminum provide more extensive alkylation, and reduction of the intermediate β–form of TiCl$_3$ may occur (Reactions 7–10). This reduction process leads to lower valence states presumable by reactions of the following type:

$$RTiCl_3 + R_3Al \longrightarrow R_2TiCl_2 + R_2AlCl \tag{7}$$

$$R_2TiCl_2 \longrightarrow RTiCl_2 + [R\cdot] \tag{8}$$

$$\beta\text{-}TiCl_3 + R_3Al \longrightarrow RTiCl_2 + R_2AlCl \tag{9}$$

$$RTiCl_2 \longrightarrow TiCl_2 + [R\cdot] \tag{10}$$

Intermediate alkyltitanium halides and the titanium subhalides remain, in most cases, tightly complexed with the product organoaluminum compounds. A large number of compounds and complexes are possible, of which only a few are catalytically active. Because low-valent transition metal compounds are electron-deficient molecules, they will attempt to expand their coordination number by sharing ligands between two metal centers with the formation of bimetallic complexes. One cannot be certain whether the compounds and complexes that are isolated are the true catalysts or are merely precursors of other compounds that are the true catalysts. By analogy with these reactions, the solid crystalline surface of titanium dichloride or trichloride and organoaluminum compounds in solution might be expected to undergo similar reactions. Bimetallic complexes have been suggested to be formed primarily at sites of structural defects such as edges, step faults, and chlorine vacancies where the hexacoordination ability of exposed ions would be incompletely satisfied (37).

For the higher activity Ziegler-Natta catalysts (Table II) based on reaction products of specific magnesium, titanium, and aluminum compounds, the similarity in size, coordination preference, electronic structure, and electronegativity of Ti(IV), Mg(II), and Al(III) ions is reflected in structural parameters and chemical properties (38) (Table III). The similarity in size between Mg(II) and Ti(IV) probably permits an easy substitution between ions in a catalyst framework.

The role of magnesium ions in high activity Ziegler-Natta catalysts has received some recent attention with particular emphasis on four points (39):

1. Titanium centers may be diluted by magnesium ions that influence the number of active centers. This dilution effect increases the number of active centers that tend to be isolated. The active centers are at least an order of magnitude higher than the earlier Ziegler-Natta catalysts.
2. The presence of magnesium ions stabilizes active titanium centers from deactivation processes relative to soluble systems.
3. The presence of magnesium ions enhances chain-transfer processes because the number-average molecular weight decreases when the Mg/Ti ratio increases.
4. The presence of magnesium ions leads to catalysts that provide polyethylenes with a narrow molecular weight distribution ($\overline{M}w/\overline{M}n$ ca.3-5).

Several reviewers have attempted to summarize existing data on the determination of propagation rate constants (k_p) and the number of active centers (C*) in olefin polymerization (40-43). Although

Table I. Selected First-Generation Ziegler-Natta Catalysts

Transition Metal Compound	Metal Alkyl	Polymer
$TiCl_4$	$(C_2H_5)_3Al$	polyethylene
$TiCl_3$	$(C_2H_5)_2AlCl$	isotactic polypropylene
VCl_4	$(C_2H_5)_2AlCl$	syndiotactic polypropylene
VCl_4	$(i-C_4H_9)_3Al$	poly-4-methyl-1-pentene
Soluble cobalt salt	$(C_2H_5)_2AlCl$	cis-1,4-polybutadiene

Table II. High-Activity Ziegler-Natta Catalysts

Titanium/Magnesium Composition	Metal Alkyl	Polymer
$TiCl_4/MgC_8H_{17}Br$	$(C_2H_5)_3Al$	polyethylene
$TiCl_4/Mg(OC_2H_5)_2$	$(C_2H_5)_3Al$	polyethylene
$TiCl_4/MgCl_2$ (activated)	$(C_2H_5)_3Al$	polyethylene
$TiCl_4/MgCl_2$/electron donor	$(C_2H_5)_3Al$	polyethylene
$\beta TiCl_3/AlCl_3$/ether	$(C_2H_5)_2AlCl$	isotactic polypropylene
$TiCl_4/MgCl_2$/ ethyl-p-toluate	$(C_2H_5)_3Al\cdot$ ethyl-p-toluate	isotactic polypropylene

Table III. Geometric and Electronic Properties of Ions of Catalyst Components

Ion	Radius (Å)	Size/ Charge	Electro- negativity	Electronic Structure	Coordination Number	Geometry
Ti(IV)	0.68	0.17	1.5	$3s^2 3p^6 3d^0$	4	tetrahedral
					5	trigonal bypyrimidal
					6	octahedral
Mg(II)	0.65	0.33	1.3	$2s^2 2p^6$	4	tetrahedral
					6	octahedral
Al(III)	0.50	0.17	1.6	$2s^2 2p^6$	4	tetrahedral
					6	octahedral
Cl(−I)	1.81	1.81	3.2	$3s^2 3p^6$	1	--
					2	bent-bridging

the absolute values for the rate constants and active centers
sometimes differ from one reviewer to another, some conclusions
appear to be generally accepted. Results of experiments to
determine the number of active centers have indicated that for
first-generation Ziegler-Natta catalysts only a small fraction of
the total amount of transition metal compound is catalytically
active at any specific time. Generally for titanium catalysts of
low productivity, C* values range from 10^{-2}-10^{-4} mol/mol titanium
compound. The active site concentration in $TiCl_3$-based Ziegler-
Natta catalysts has been calculated to be 10^{-2}-10^{-3} mol/mol $TiCl_3$
(44, 45). Similarly, an active site concentration (5 x 10^{-2}
mol/mol $TiCl_4$) was calculated for an alkylaluminum-titanium
tetrachloride catalyst (46). If the active sites are assumed to be
titanium centers, only a very small proportion of these centers in
the $TiCl_3$ solid is active as polymerization sites. Some
potentially active sites may not be used because they are
inaccessible or deactivated by impurities in the system. Titanium
catalysts of higher productivity (second/third generation
catalysts) frequently show higher C* values of 7 x 10^{-1}-10^{-2}
mol/mol titanium compound. Some uncertainty continues to exist
about the value of k_p in high activity catalysts because of the
decay of catalytic activity with time (47, 48).

Types of Olefin Monomers and Polymers

Ziegler-Natta catalysts can polymerize a variety of structurally
different monomers. Examples of stereoregular homopolymers (Table
IV), elastomeric or crystalline copolymers, as well as block
copolymers may be found in the patent and open literature (4, 49-
51). Ethylene polymerizes easily with many soluble and
heterogeneous Ziegler catalysts. Some ethylene-active catalysts,
for example, Cp_2TiCl_2 + aluminum alkyl (52), are not active for α-
olefin polymerizations. However, all known Ziegler catalysts that
polymerize propylene are also active in ethylene polymerization.

Many α-olefins, in addition to propylene, have been polymerized
to isotactic polymers. The reactivity of the olefin diminishes as
the size of the olefin increases, for example, ethylene >
propylene > 1-butene > 4-methyl-1-pentene. Reactivity also
diminishes as branching comes closer to the double bond, for
example, 4-methyl-1-pentene > 3-methyl-1-pentene. The lower
polymerization activity of higher α-olefins has been ascribed to
difficulty in approach or coordination to the active site. Reports
of polymerization of isobutylene (2-methylpropene) with $TiCl_4$-
Ziegler catalyst are now attributed to a cationic polymerization
initiated by residual $TiCl_4$ present in the particular Ziegler
catalyst. Attempts to polymerize internal, noncyclic olefins such
as cis- and trans-2-butene have not been successful (53). However,
under special conditions Natta and coworkers (54) were able to
copolymerize ethylene with cis- and trans-2-butene, cyclopentene,
cyclohexene, cycloheptene, and cyclooctene. Homopolymerization of
many cycloolefins has been reported (55). Polymerization can occur
with these olefins by 1,2-addition to the double bond or by various
ring opening processes. Choice of catalyst components and
polymerization conditions determine the mode of polymerization.

Table IV. Polymer Nomenclature and Structure

Nomenclature	Structure
Isotactic Polypropylene	$\sim\!\!\sim\!\!\sim CH_2\text{-}CH\text{-}CH_2\text{-}CH\text{-}CH_2\text{-}CH\sim\!\!\sim\!\!\sim$ with CH_3 substituents
Syndiotactic Polypropylene	$\sim\!\!\sim\!\!\sim CH_2\text{-}CH\text{-}CH_2\text{-}CH\text{-}CH_2\text{-}CH\sim\!\!\sim\!\!\sim$ with CH_3 substituents
cis-1,4-Polybutadiene	cis double bond structure
trans-1,4-Polybutadiene	trans double bond structure
Isotactic 1,2-Polybutadiene	$\sim\!\!\sim\!\!\sim CH_2\text{-}CH\text{-}CH_2\text{-}CH\text{-}CH_2\text{-}CH\sim\!\!\sim\!\!\sim$ with $CH=CH_2$ substituents
Syndiotactic 1,2-Polybutadiene	$\sim\!\!\sim\!\!\sim CH_2\text{-}CH\text{-}CH_2\text{-}CH\text{-}CH_2\text{-}CH\sim\!\!\sim\!\!\sim$ with $CH=CH_2$ substituents

Some Ziegler-type catalysts have been used to polymerize terminal acetylenes (56, 57). Highly dispersed or soluble catalysts based on $TiCl_4$, $Ti(OR)_4$, metal chelates (Co, Ni, V, Fe) plus $AlEt_3$ have been most successful. Acetylenes, unlike α-olefins, polymerize with Group VIII transition metal compounds. With some acetylenes, trimerization reactions leading to the corresponding substituted benzene derivatives take place.

Nonconjugated dienes of the type $H_2C=CH-(CH_2)_n-CH=CH_2$ have been polymerized by 1,2- and cycloaddition routes (58). Conjugated dienes are readily polymerized by Ziegler catalysts (59). By proper selection of catalyst it is feasible to prepare polymers having any desired structure (Table V). It is possible in some cases to change the type of structural units in the polymer by merely altering the ratio of catalyst components. Factors that determine stereoregulation in these polymers include the types of metal ligands, crystal structure of the transition metal salt, specific transition metal, relative concentrations of catalyst components, and experimental conditions.

Polymerization studies with polar monomers indicate that some of these monomers can be polymerized at Ziegler-type sites (4, 5). Frequently secondary reactions often prevent propagation from occurring. The polar monomer may complex or react irreversibly with one or both of the catalyst components, or else one of the catalyst components may serve as a radical or cationic initiator for polymerization of the monomer. Prevention of these side reactions permits a more favorable Ziegler-type polymerization.

Polymerization Mechanisms

The chemistry of catalyst formation and the nature of active sites have been extensively debated since the initial discoveries of Ziegler and Natta (4, 5, 36, 60). The kinetics and mechanism have been the subject of numerous studies. Nearly all possibilities in assigning the nature of the active site have been exhausted by proposals of different theories. A considerable amount of experimental evidence has been reported frequently to claim proof that a particular mechanism is operative. One may classify these proposed mechanisms according to the charge distribution in the transition state of the propagation reaction, viz, coordinated anionic, coordinated cationic, and coordinated radical, in accordance with whether the propagating polymer chain is considered a carbanion, a carbonium ion, or a radical species. "Coordinate" indicates the common feature of complexation of the olefin before introduction into the growing chain (18). In the case of α-olefins, proposals that have received most attention are those of the coordinate-anionic type in which the coordinated monomer enters the chain through a catalyst-polymer bond polarized in the sense $M^{\sigma+}-R^{\sigma-}$. Proposals for coordinate anionic polymerization may be further distinguished in accordance with whether the transition metal, or the base metal center, or a bimetallic complex involving both centers is considered the site of chain growth. Propagation at one metal center of a bimetallic complex would be classified in the monometallic category. Bimetallic mechanisms that employ two different metals have been proposed by different workers (37, 61, 62). One proposal for an active site model for the bimetallic mechanism would be

$$\text{Cl} \diagdown \atop \text{Cl} \diagup \text{Ti} \diagup \substack{\cdots \text{Cl} \cdots \\ \cdots \text{Pn} \cdots} \text{Al} \substack{\diagup \text{R} \\ \diagdown \text{R}}$$

where Pn is the growing polymer chain. Propagation occurs by coordination of the olefin to the titanium center with cleavage of the titanium—polymer partial bond. The polymer chain in this mechanism is always bound, at least partially, to aluminum (Reaction 11).

$$(11)$$

where R = alkyl and P = polymer chain.

The basic feature of proposals for the monometallic mechanism is that propagation occurs entirely at one metal center. A monometallic mechanism involving titanium in a lower valence state, for example, RTiCl, has been proposed (63) to be an active site for ethylene polymerization with propagation occurring by coordination and insertion into the titanium—carbon bond (Reaction 12).

$$Cl-Ti-(CH_2-CH_2)_{\overline{n}}-R + CH_2=CH_2 \longrightarrow Cl-Ti-(CH_2-CH_2)_{\overline{n}} \to R \qquad (12)$$
$$\uparrow$$
$$CH_2=CH_2$$

$$Cl-Ti-(CH_2-CH_2)_{\overline{n+1}}-R, \text{ etc.}$$

The rate of polymerization of ethylene with $Et_3Al-TiCl_4$ or $(i-Bu)_3Al-TiCl_4$ was maximum when the average valence of titanium was two.

The view of chain growth at the transition metal center finds strong support from the results of copolymerization experiments

with ethylene and propylene (64). Ethylene and propylene show
different relative activities that depend on the transition metal
compound structure but not on the metal alkyl structure (Table VI).
Propylene relative reactivities increased in the series $HfCl_4$ <
$ZrCl_4$ < $TiCl_4$ < $VOCl_3$ < VCl_4, with the increase corresponding to
increasing electronegativity of the transition metal. No
significant differences in relative activities of ethylene and
propylene were observed when different metal alkyls were combined
with VCl_4. Because a change in the reducing agent did not affect
the active site but a change in the transition metal compound
produced a difference active site in every case, it was concluded
that chain growth occurred at the transition metal center.
Propagation occurs by coordination to the transition metal center
and subsequent insertion into the carbon–metal bond (Reaction 12).

Olefin polymerization can take place with some transition metal
compound-based catalysts in the absence of added metal alkyl (5,
36, 65, 66). Crystalline $TiCl_2$, free of any organometallic
cocatalyst, can be transformed into a very active catalyst by
mechanical activation (ball milling) (66). Ball milling caused not
only the expected increase in surface area, but also
disproportionation of Ti(II) to Ti(III) and Ti. Catalyst activity
was proportional to surface area and a direct function of Ti(II)
content of the catalyst. This catalyst was also effective for
copolymerization of ethylene with propylene and 1-butene (67).
Finally, this catalyst is capable of stereospecifically
polymerizing propylene at a very low rate (68). Moreover, the
fraction insoluble in boiling heptane (approximately 30%) had a
partially crystalline isotactic structure as shown by X-ray
analysis. The initiation reaction with this catalyst has not been
explained. These studies suggest that all the essential features
of the stereospecific polymerization are present in the surface of
the titanium halide.

Similar conclusions have been reached on the basis of studies
with $TiCl_3$ and amines (69). These catalysts polymerize propylene
at a slow rate to highly isotactic polymers. The active site was
long lived because polypropylene molecular weight increased with
polymerization time over a long period. Mechanism studies (70)
with the $TiCl_3$-Et_3N and Ziegler catalysts indicate a close
resemblance and support the view that both catalyst types operate
by propagation at a transition metal–carbon bond.

Dienes have also been polymerized with transition metal
compounds with no added metal alkyl. Allyl nickel bromide polymerized butadiene to a polymer having 95% trans-1,4 structure when
benzene was used as a solvent and 75% cis-1,4 when diethyl ether
was used (71).

Although numerous mechanism proposals have been offered for the
coordinated anionic polymerization of olefins, no direct proof
exists to confirm the proposed structures. Although the structures
have value in stimulating future research to verify certain
features of the mechanism, noncritical acceptance of the proposal
as fact could hinder progress.

Table V. Catalysts for Stereospecific Polymerization of Butadiene ($\underline{4}$)

Metal Alkyl	Transition Metal Compound	Conditions	Structure of Polymer (%)		
			1,2	*cis*-1,4	*trans*-1,4
Et_2AlCl, $EtAlCl_2$	cobalt compounds	Al/Co $\geq$100	--	95-98	--
Et_3Al	VCl_4, $VOCl_3$	Al/V = 2	--	--	95-100
Et_3Al	TiI_4	Al/Ti = 5	5	92	4
Et_3Al	chromium acetylacetonate	Al/Cr = 10	isotactic	--	--
Et_3Al	chromium acetylacetonate	Al/Cr = 3	syndiotactic	--	--
$EtAlCl_2$	β-$TiCl_3$	Al/Ti = 1-5	--	--	100
Et_3Al	$Ti(OBu)_4$	Al/Ti = 3	90	10	--
Et_3Al	$TiCl_4$	Al/Ti = 1	2	49	49

Table VI. Reactivity Ratios for Ethylene-Propylene Copolymerization with Organometallic Mixed Catalysts

Catalysts	$r_{C_2H_4}$
$Al(i-Bu)_3$ + VCl_4	16
$Al(i-Bu)_3$ + $VOCl_3$	28
$Al(i-Bu)_3$ + $TiCl_4$	37
$Al(i-Bu)_3$ + $ZrCl_4$	61
$Al(i-Bu)_3$ + $HfCl_4$	76

Note: Substitution of $Al(i-Bu)_3$ with $Zn(n-Bu)_2$, $Zn(C_6H_5)_2$, $Sn(C_6H_5)_4$ + $AlBr_3$, and CH_3TiCl_3 led to same $r_{C_2H_4}$ value.

Origin of Stereoregulation

Experimental data provide support for the proposal originally presented by Natta and coworkers that a distribution of localized active sites exists on the crystal surface. These sites are located primarily at the boundaries of the elemental crystal layers and exhibit various specificities. The steric environment at these active sites is different. Less stereospecific synthesis occurs at the most exposed, accessible sites. More stereospecific polymerization takes place at the most sterically hindered centers. Detailed models for stereoregulation must be compatible with a considerable body of experimental data. Insertion of the α-olefin into the growing polymer chain occurs by a cis opening of the double bond. The α-olefin adds in a head-to-tail fashion with the unsubstituted carbon of the α-olefin attached to the metal center. Highly isotactic polymers have only been formed by using a variety of well-formed crystalline transition metal compounds. Isotactic polypropylene of high steric purity has been prepared using α, γ, δTiCl$_3$, VCl$_3$, TiCl$_2$, and various mixed crystals of TiCl$_3$ with AlCl$_3$ or other compounds. Isotactic polymerization can take place in the absence of metal alkyl. The stereoregulating capacity and activity of the catalyst depend upon the particular combination of components that form the catalyst. Highly crystalline syndiotactic polypropylene is produced with soluble catalysts. Stereochemistry of polymers produced from conjugated dienes is determined by the particular soluble or heterogeneous catalyst.

Proposals by Cossee and Arlman (72-75) have received considerable attention in Ziegler-Natta catalysis. These workers attributed stereoregularity to the special geometry arising from active sites that are adjacent to crystal defects in a solid substrate. The layer-lattice structure of TiCl$_3$ consists of titanium ions interspersed with layers of chloride ions. For a stoichiometric crystal of TiCl$_3$ to achieve electroneutrality, the crystal should contain a number of chlorine vacancies on the surface. These vacancies should be found on the edges of the elementary sheets and not in the basal planes of the crystals. Microscopic observations showing that the polymer grows from the edges support this conclusion (76). Figure 1 shows one of the surface arrangements of α-TiCl$_3$ (73). Each Ti(III) in the surface is only five coordinated. Of the five chloride ions around each titanium ion and arranged at five of the six corners of an octahedron, three are deeply buried in the interior of the crystal. Of the two remaining chloride ions one is attached to a second titanium ion. The initiation reaction consists of alkylation of a pentacoordinate Ti(III) ion (Reaction 13).

$$
\begin{array}{c}
\square \\
\vert \\
Cl \\
\diagup \\
Cl-Ti-Cl + AlEt_3 \longrightarrow \\
\vert \quad \vert \\
Cl \quad Cl
\end{array}
\left[
\begin{array}{c}
Et \\
Et_\diagdown \ \vert \diagup Et \\
\cdot \cdot Al \\
\vert \ Cl \\
Cl-Ti--Cl \\
\vert \\
Cl \\
\vert \\
Cl
\end{array}
\right]
\longrightarrow
\begin{array}{c}
Et \\
\vert \quad Cl \\
\diagup \\
Cl-Ti---\square \\
\diagup \ \vert \\
Cl \quad \vert \\
Cl
\end{array}
\qquad (13)
$$

pentacoordinate exposed   transition state   + Et_2AlCl
Ti site   active site

$\square$ = octahedral vacancy

Chain propagation occurs by coordination of the olefin at the vacant octahedral position and migration of the alkyl group to regenerate an octahedral vacancy and a new, extended alkyltitanium bond (Reaction 14).

$$
\begin{array}{c}
R \\
\vert \ Cl \\
Cl-Ti---\square \\
Cl \ \vert \\
Cl
\end{array}
\longrightarrow
\begin{array}{c}
R \quad CH-CH_3 \\
\vert \ Cl \ \Vert \\
Cl-Ti \quad CH_2 \\
Cl \ \vert \\
Cl
\end{array}
\longrightarrow
\begin{array}{c}
\ \ \diagup R \diagdown \ CH_3 \\
Cl-Ti-Cl \ \diagdown CH \\
Cl \ \vert \ \diagdown CH_2 \\
Cl
\end{array}
\longrightarrow
$$

$$
\begin{array}{c}
\square \\
\vert \ Cl \quad R \\
\diagup \quad \vert \\
Cl-Ti \quad CH-CH_3 \\
Cl \ \vert \diagdown CH_2 \\
Cl
\end{array}
\qquad (14)
$$

In the olefin complex the carbon–carbon double bond is believed to be perpendicular to one of the valency directions of the transition metal. The theoretical treatment of π–bonding by Chatt and Duncanson (77) furnishes the basic idea underlying the molecular orbital treatment for a variety of complexes between transition elements and unsaturated hydrocarbons.

For discussion of stereospecificity the placement of the active center ($EtTiCl_4 \ \square$) in the solid lattice must be considered. Figures 2 and 3 illustrate that this center is asymmetrically placed because the plane through the alkyl group, titanium, Cl_3, and Cl_4 is not a plane of symmetry and the positions of alkyl group and vacancy are not equivalent. For the olefin complex the only possibility for a propylene molecule is to be placed with its $-CH_2-$ end downward. Because of the asymmetry of the center, one position of the methyl group will be favored. Sterically the orientation with the methyl group on the side of Cl_1 (Figure 2) will be favored. After rearrangement the polymer alkyl group and the vacancy will have changed position (Figure 3). Two possibilities now exist. If monomer adds to the rearrangement complex, the polymer will be syndiotactic. If, however, the vacancy and polymer alkyl shift position and then monomer addition occurs, the polymer

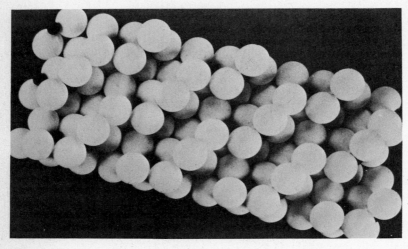

Figure 1.   One of the four possible regular structures of face
(10$\bar{1}$0) of a α–TiCl$_3$.   Small black sphere is a titanium.
(Reproduced with permission from Ref. 75.   Copyright
1967 Dekker.)

Figure 2.   Position of octahedral active center in surface of α–
TiCl$_3$ (10$\bar{1}$0).   Large black sphere is a polymer chain and
small black sphere is a titanium.   (Reproduced with
permission from Ref. 75.   Copyright 1967 Dekker.)

will be isotactic. Apparently, under normal conditions of polymerization the growing alkyl shifts to the most favorable position before the next monomer is inserted; hence, isotactic polymer is formed. Actually only at very low temperatures (<-40 °C) has highly syndiotactic polypropylene been produced (78).

An alternate model involving direct insertion, without prior coordination of the olefin to the transition metal, has been proposed. This model requires that the metal—carbon bond be highly polarized. At this time, considerably more experimental evidence must be gathered before a choice between the different proposals can be rigorously established.

More recent views (17) of the origin of stereospecificity in the synthesis of isotactic polymers connect this origin with the ability of the catalyst—growing polymer chain system, that is, the catalytic center, to discriminate between the two prochiral faces of the α—olefin. The catalytic system must possess one or more chirality centers (Figure 4).

If one considers the simplest model of a monometallic catalytic center M in Figure 4, a chiral carbon atom exists in the growing polymer chain in the β—position with respect to the metal atom. The metal atom itself can be a chirality center, which, being bound to a solid surface, could maintain its absolute configuration during the insertion reaction.

Different types of calculations, including calculations of nonbonded interactions at the active site, have been attempted by different authors (79-83). These attempts include the initial work by Cossee and Arlman about 20 years ago (74, 75). The weak point of this approach is the choice of the model on which calculations are made. Because knowledge of the structure of catalytic centers is not large, calculations are seldom helpful in providing a better understanding of experimental results and designing new experiments. Calculations will become much more fruitful when more reliable models based on greater knowledge of the structure of the catalytic centers become available for these calculations.

Certain soluble catalysts polymerize propylene to highly syndiotactic polymers that are free of the isotactic form. Natta, Pasquon, and Zambelli (78) showed that VCl_4 (or vanadium triacetylacetonate) in combination with AlR_2X-type metal alkyls and anisole, polymerize propylene to highly syndiotactic polypropylene. These apparently homogeneous catalysts were used at low temperatures (-40 to -78 °C). Stereochemistry was initially explained by a mechanism involving repulsion between the methyl groups on the last added and new monomer unit to achieve orientation (83, 84).

This syndiotactic model resembles the isotactic one in that it also is based on an octahedral complex that has an alkyl ligand as the growth site, and a vacant octahedral position through which propylene is complexed. In this homogeneous model for polymerization, methyl—methyl interactions force the propylene to be complexed in opposite configuration at each consecutive growth step. Copolymerization studies (85) with soluble syndiotactic-specific catalysts support this view.

Recent studies (86) on the mechanism of syndiotactic poly-merization of propylene showed that in the presence of homogeneous

Figure 3.   The same active center as shown in Figure 2.   Vacancy
and alkyl group have interchanged.   (Reproduced with
permission from Ref. 75.   Copyright 1967 Dekker.)

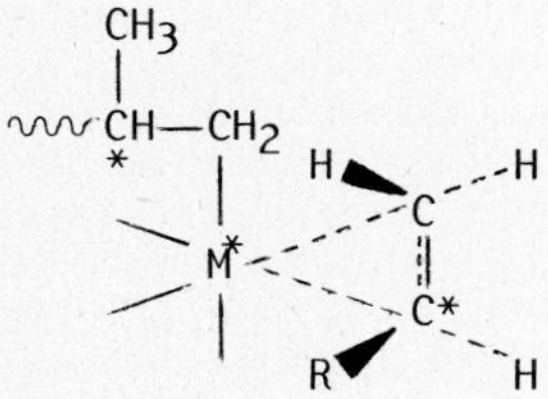

Figure 4.   Schematic representation of monometallic catalytic
centers for isotactic polymerization.   An asterisk
indicates a chirality center.

catalytic systems the syndiotactic chain propagation occurred via secondary insertion steps. Pentacoordinated V(III) catalytic complexes were proposed. Polymerization was considered to occur via equatorial coordination of the monomer on the vanadium atom, rotation of the coordinated olefin, and insertion on the reactive V–C bond (Reactions 15 and 16).

$$
\begin{array}{c}
\text{Cl}\quad\text{CH}_3 \\
|\qquad| \\
\text{Cl—V—C—CH}_2\text{—R} \quad + \quad \text{CH}_3\text{CH}=\text{CH}_2 \\
\qquad| \\
\square\ \ \text{Cl}\ \ \text{H}
\end{array}
\qquad\qquad\downarrow\ \text{Coordination} \qquad (15)
$$

$$
\begin{array}{c}
\text{Cl}\quad\text{CH}_3 \\
|\qquad| \\
\text{Cl—V—C—CH}_2\text{—R} \\
\text{CH}_3\!-\!\text{C}\ \overset{H}{\underset{C}{\diagup}}\ \text{Cl}\ \text{H} \\
H\!-\!\overset{H}{\underset{}{C}}
\end{array}
\qquad\qquad\downarrow\ \text{Insertion} \qquad (16)
$$

$$
\begin{array}{c}
\text{Cl}\quad\text{H}\qquad\qquad\text{CH}_3 \\
|\qquad|\qquad\qquad| \\
\text{Cl—V—C—CH}_2\text{—C—CH}_2\text{—R} \\
|\qquad\qquad| \\
\square\ \ \text{Cl}\ \ \text{CH}_3\qquad\qquad\text{H}
\end{array}
$$

28

The stereochemical features of the polymerization were satisfactorily accounted for by considering the nonbonded interactions between the monomer molecule undergoing insertion and the ligands on the vanadium atom. Coordination number five for vanadium was chosen to justify the more relevant data concerning the propylene polymerization and ethylene, 1-butene, and *cis*-2-butene copolymerization.

In diene polymerizations many different origins of stereochemistry have been suggested (87, 88). Different proposals for stereochemistry have suggested the importance of the complexing step, the formation of the transition state, and the importance of the diene after insertion and while the next added diene is being inserted. Attempts to correlate microstucture of polydienes with the nature of a particular Ziegler-type catalyst have proven extremely difficult. This difficulty has been due to the larger number of stereoregulating paths that are feasible and the sensitivity of stereoregulation to the conditions of polymerization.

Chemically Anchored, Supported Catalysts for Olefin Polymerization

Chemically anchored, supported transition metal catalysts have been used extensively for production of high-density, linear polyethylene and ethylene-α-olefin copolymers. These supported catalysts continue to attract considerable attention and appear to be an object of study in many industrial and academic laboratories throughout the world. These catalysts for olefin polymerization may be divided into three classes: metal oxide, particularly CrO_3/SiO_2 ([11], [12]); Ziegler, particularly $R_3Al + TiCl_4/Mg(OH)Cl$ ([20], [21]); and organotransition metal compounds, particularly $(C_5H_5)_2Cr/SiO_2$ ([35], [89]).

The significant improvement in polymerization activity probably represents the single most important advantage of a chemically anchored catalyst. These catalyst systems show high efficiency, presumably because the active transition metal compound resides only on the support surface and thus permits availability of a larger concentration of active sites for polymerization. Reaction of a transition metal compound with a support surface provides an anchoring device preventing destruction of potential sites by mutual interaction.

Studies ([90], [91]) with CrO_3/SiO_2 catalyst have shown that formation of a surface chromate takes place by reaction of CrO_3 and surface silanol groups on silica (Reaction 17). Reaction of this chemisorbed chromate with ethylene results in an oxidation-reduction reaction ([90]-[95]) with formation of a low-valent chromium center (Reaction 18). Proposals for Cr(II) as the active site are based on studies of the catalyst after reduction by ethylene, carbon monoxide, or hydrogen. One study ([93], [94]) showed that the polymerization rate increased with the fraction of Cr(II) in the catalyst. Another study ([92]) showed by polarography that the chromium is reduced to a divalent state by ethylene.

The primary initiation reaction remains unclear. A similar situation exists with other metal-alkyl free catalysts ([66], [67]). Polymer chain growth may be pictured to occur by a coordinated anionic mechanism (Reaction 19).

Chemisorption

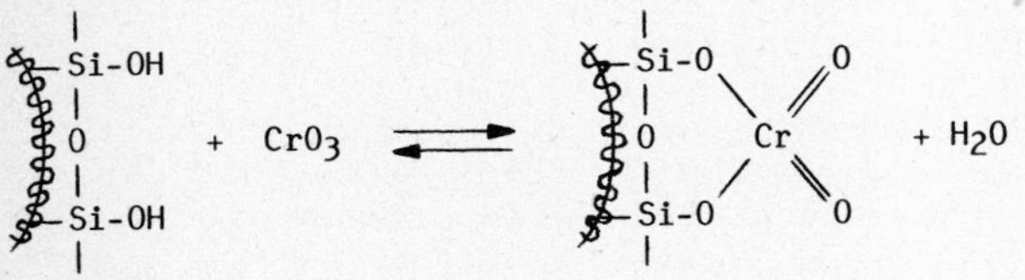

$$(17)$$

Reduction

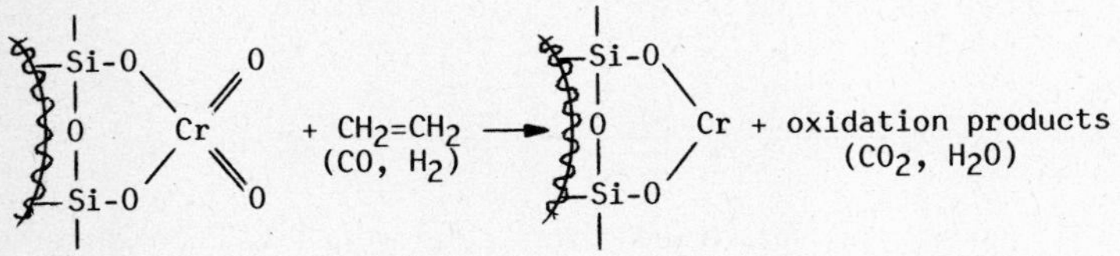

$$(18)$$

Initiation

Reaction uncertain, but believed to result in formation of divalent chromium hydride or alkyl.

Propagation

$$Cr-R + nCH_2=CH_2 \longrightarrow Cr-(CH_2-CH_2)_n-R, \text{ etc.} \qquad (19)$$

R = hydride or alkyl

The rate and extent to which silica-based catalysts fracture during the polymerization of ethylene have been discussed (96). Fragmentation of the catalyst was complete within the first minute or two of polymerization, whereas the rate of reaction continued to increase for more than an hour.

With the Ziegler catalyst, $R_3Al + TiCl_4$ supported on magnesium hydroxychloride, chemisorption can be represented by Reaction 20.

$$TiCl_4 + Mg(OH)Cl \longrightarrow Cl_3TiOMgCl + HCl \qquad (20)$$

Reduction and chain propagation probably occur in a manner described earlier for unsupported Ziegler catalysts.

Chromocene deposited on silica supports forms a highly active catalyst for polymerization of ethylene (35, 89). The catalyst formation step liberates cyclopentadiene and leads to a new divalent chromium species containing a cyclopentadienyl ligand. Polymerization is believed to occur by a coordinated anionic mechanism (Reaction 21) outlined earlier.

$$Cr-R + nCH_2=CH_2 \longrightarrow Cr-(CH_2-CH_2)_n-R, \text{ etc.} \qquad (21)$$

The presence of the cyclopentadienyl ligand at the chromium center provides a catalyst with a unique high response to hydrogen as a chain transfer agent (97). A number of π- and σ-bonded transition metal compounds in solution or supported (34) have been described as polymerization catalysts. Unsupported and supported transition metal-allyl compounds have been proposed to initiate polymerization by reaction with monomer in a manner illustrated by Reaction 22 for $(allyl)_3ZrBr/SiO_2$ and ethylene.

$$(22)$$

General Features of Polymerization Catalysts

The current statue of olefin and diene polymerizations catalyzed by transition metal compounds that function by a coordinated anionic mechanism suggests a number of general conclusions.

1. Many distinct catalyst types are possible. The voluminous patent literature and scientific publications indicate the vast scope of transition metal catalyzed polymerization of monomers.
2. Propagation occurs by monomer coordination and insertion into a transition metal–carbon bond. The total experimental evidence strongly supports the proposal that the active site in Ziegler catalysts is a transition metal–carbon bond, and the propagation reaction consists of repeated insertion of the olefin into this bond.
3. Several routes to transition metal–carbon bond exist. Transition metal carbon bonds may be generated by alkylation of a transition metal compound with a metal alkyl. Low-valent transition metal compounds, per se, that is, $TiCl_2$, may function as catalysts. Reduction by the olefin that occurs with the CrO_3/SiO_2 catalyst may also provide sites for polymerization. Finally, transition metal compounds in solution (34) or supported may function as polymerization sites.
4. Ligand environment at active sites plays a significant role in polymerization behavior. Ligand effects in diene polymerization (87, 88) and work with supported chromocene catalysts (98) dramatically illustrate this point.
5. Isotactic placements originate from catalyst–monomer interactions. These placements do not require the participation of metal alkyl in the active site.
6. Syndiotactic placements originate from nonbonded interactions between the monomer molecule undergoing insertion and ligands on the vanadium atom.
7. Catalyst supports can lead to or improve polymerization activity by generating or increasing active site concentration.
8. Magnesium compounds or ions in high activity catalysts play several roles by increasing the number and stability of active transition metal centers. The presence of magnesium enhances chain transfer processes and can provide polyethylenes of narrow molecular weight distribution.

Future Research and Related Areas

After 30 years, olefin polymerization by a coordinated anionic mechanism continues to receive worldwide attention as evidenced by a voluminous patent and journal literature. Much attention has been directed to catalyst and process optimization and understanding of key reaction variables. The development of high-activity Ziegler-Natta catalysts has spurred a renewed interest in simplified processes requiring no post-treatment of the polymers. Recent announcements by Union Carbide of a low-pressure, fluid bed process to produce granular, low-density polyethylenes have caused a revolution in the polyethylene field (99, 100). A whole new generation of low-density polyethylenes, described as linear low-density polyethylenes, has appeared (101, 102). Catalyst research in this area continues to be intense with considerable emphasis on copolymerization kinetics, and also control of catalyst morphology to regulate polymer morphology. Announcements by Montedison and Mitsui Petrochemicals and by others in regard to high-mileage catalysts for isotactic polypropylene will also continue to receive worldwide attention (32).

The advent of the energy crisis has caused us to examine traditional views of the relative costs of different monomers and to consider the potential of less costly monomers for polymerization. One can expect that catalysis of the coordinated anionic type will play a major role in any new developments in olefin and diene polymerizations. Finally, one should recall that Ziegler catalysts have found many uses in other areas of chemistry such as metathesis of olefins, oligomerization, isomerization, hydrogenation, and alkylation. The vast scope of these catalysts will almost certainly achieve a wider range as these types of studies continue in the future.

Literature Cited

1. Ziegler, K.; Holzkamp, E.; Breil, H.; Martin, H. Angew. Chem. 1955, 67, 541.
2. Natta, G. J. Polym. Sci. 1955, 16, 143.
3. Natta, G.; Pino, P.; Corradini, P; Danusso, F.; Mantica, E.; Mazzanti, G.; Moraglio, G. J. Am. Chem. Soc. 1955, 77, 1708.
4. Boor, J., Jr. In "Macromolecular Reviews"; Peterlin, A., et al., Eds.; Interscience: New York, 1967; Vol. 2, pp. 115-268.
5. Boor, J., Jr. "Ziegler-Natta Catalysts and Polymerizations"; Academic: New York-San Francisco-London, 1979.
6. Ziegler, K. In "Advances in Organometallic Chemistry"; Stone, F. G. A.; West, R, Eds.; Academic: New York-San Francisco-London, 1979.
7. Natta, G. Angew. Chem. 1956, 68, 393.
8. Natta, G. Mod. Plastics 1956, 34(4), 1969.
9. Natta, G. Angew. Chem. 1964, 76, 553.
10. Peters, E. F.; Zletz, A.; Evering, B. L Ind. Eng. Chem. 1957, 49, 1879.
11. Clark, A.; Hogan, J. P.; Banks, R. L.; Lanning, W. C. Ind. Eng. Chem. 1956, 48, 1152.
12. Hogan, J. P.; Banks, R. L. U. S. Patent 2 825 721, 1958. Sailors, H. R.; Hogan, J. P. Macromol. Sci. Chem. 1981, A(7), 1377.
13. Friedlander, H. N. In "High Polymers"; Raff, R. A. V.; Doak, K. W., Eds.; Interscience: New York, 1965; Vol. XX, pp. 215-66.
14. Karol, F. J. In "Encyclopedia of Polymer Science and Technology, Supplement I"; Mark, H. F.; Bikales, N. M., Eds.; Interscience: New York, 1976; pp. 120-46.
15. Weissermel, K.: Cherdron, H.; Berthhold, J.; Diedrich, B.; Keil, K. D.; Rust, K.; Strametz, H.; Toth, T. J. Polym. Sci. Symp. 1975, 51, 187 and references therein.

16. Diedrich, B. Appl. Polym. Symp. 1975, 26, 1 and references therein.
17. Pino, P.; Mulhaupt, R. Angew. Chem. Int. Ed. Engl. 1980, 19, 857.
18. Galli, P.; Luciani, L.; Cecchin, G. Angew. Makrol. Chemie 1981, 94, 63.
19. Yermakov, Y. I.; Kuznetsov, B. N.; Zakharov, V. A. "Studies in Surface Science and Catalysis"; Elsevier: Amsterdam-Oxford-New York, 1981; Vol. 8, Chap. 5.
20. Dassesse, P.; Dechenne, R. U. S. Patent 3 400 110, 1968.
21. Stevens, J. Hydrocarbon Process 1970, 179–82.
22. Hewett, W. A.; Shokal, E. C. British Patent 904 510, 1960.
23. Longi, P.; Giannini, U.; Cassata, A. British Patent 1 335 887, 1970.
24. Giannini, U.; Cassata, A.; Longi, P.; Mazzochi, R. British Patent 1 387 890, 1971.
25. Haward, R. N.; Rober, A. N.; Fletcher, K. L. Polymer 1973, 14, 365.
26. Lassalle, D.; Vidal, J. L.; Roustant, J. C.; Mangin, P. 5th Conf. European Plastics Caoutch. Soc. Chim. Ind.; Paris, France, 1978.
27. Boucher, D. C.; Parsons, I. W.; Haward, R. N. Makrol. Chem. 1971, 175, 3461.
28. Berger, E.; Derroitte, J. L. U.S. Patent 3 901 863, 1975.
29. Delbouille, A.; Derroitte, J. L.; Berger, E.; Gerard, P. British Patent 1 275 641, 1972.
30. Hermans, J. P.; Henrioulle, P. U.S. Patent 3 769 233, 1973.
31. Hermans, J. P.; Henrioulle P. U.S. Patent 4 210 738, 1980.
32a. Giannini, U.; Albizzati, E.; Parodi, S. U.S. Patent 4 149 990, 1979.
  b. Luciani, L.; Kashiwa, N.; Barbé, P. C.; Toyota, A. U.S. Patent 4 226 741, 1980.
  c. Crespi, G. Paper presented at symposium on "Frontiers in Organic Chemical Technology"; Baroda, India, March 28–29, 1979; pp. 333–42.
  d. Crespi, G. Pet. Chem. Ind., Dev. Ann. 1979, 121.
  e. DiDrusco, G.; Rivaldi, R. Hydrocarbon Proc. 1981, 153–55.
33a. Goodall, B. L. Paper presented at Midland Macromolecular Symposium; Midland, Mich., August 17–21, 1981.
  b. Goodall, B. L.; van der Sar, J. C.; Keuzenkamp, A. European Patent Appl. 0 029 623, 1981.
34. Ballard, D. G. H. Advan. Catal. 1973, 23, 263.
35. Karol, F. J.; Karapinka, G. L.; Wu, C.; Dow, A. W.; Johnson, R. N.; Carrick, W. L. J. Polym. Sci., A-1 1972, 10, 2621.
36. Hoeg, D. F. In "The Stereochemistry of Macromolecules"; Ketley, A. D., Ed.; Dekker: New York, 1967; Chap. 2.
37. Natta, G.; Mazzanti, G. Tetrahedron 1960, 8, 86.
38. Cotton, F. A.; Wilkinson, G. "Advanced Inorganic Chemistry," 4th ed.; Interscience: New York, 1980.
39. Greco, A.; Bertolini, G.; Cesca, S. J. Appl. Polym. Sci. 1980, 25, 2045.
40. Tait, P. J. T. In "Developments in Polymerization"; Haward, R. M., Ed.; Appl. Sci. Publ.: London, 1979; Vol. 2.

41. Tait, P. J. T. "Studies on Active Center Determination in Ziegler-Natta Polymerization"; Paper presented at Midland Macromolecular Symposium; Midland, Mich., August 17-21, 1981.
42. Caunt, A. D. "Specialist Periodical Report, Catalysis"; Chemical Society: London, 1977; Vol. 1.
43. See reference 19, Chapter 3.
44. Natta, G.; Pasquon, I. Advan. Catal. 1959, 11, 1.
45. Grieveson, B. M. Makromol. Chem. 1965, 84, 93.
46. Feldman, C. F.; Perry, E. J. Polym. Sci. 1960, 46, 217.
47. Chien, J. C. W. In "Preparation and Properties of Stereoregular Polymers"; Lenz, R. W.; Ciardelli, F., Eds.; D. Reidel Publ.: Holland-Boston-London, 1980.
48. Giannini, U. Makrol. Chem. Suppl. 1981, 5, 216.
49. See reference 5, Chapters 19-21.
50. Breuer, F. W.; Geipel, L. E.; Loebel, A. B. In "High Polymers"; Raff, R. A. V.; Doak, K. W., Eds.; Interscience: New York, 1965; Vol. XX.
51. Lukach, C. A.; Spurlin, H. M. In "High Polymers"; Ham, G. E., Ed.; Interscience: New York, 1964; Vol. XVIII.
52. Breslow, D. S.; Newburg, N. R. J. Am. Chem. Soc. 1959, 81, 81.
53. Natta, G. Experientia 1963, 19, 609.
54. Natta, G.; Dall'Asta, G.; Mazzanti, G. Angew. Chem. 1964, 76, 765.
55. See reference 4, pp. 241-43.
56. Reikhsfeld, V. O.; Makovetskii, K. L. Dokl. Akad. Nauk. SSSR 1964, 155, 414.
57. Nicolescu, I. V.; Angelescu, E. M. J. Polym. Sci. 1965, 3, 1227.
58. Marvel, C. S.; Gall, E. J. J. Org. Chem. 1960, 25, 1784.
59. Marconi, W. In "The Stereochemistry of Macromolecules"; Ketley, A. D, Ed.; Dekker: New York, 1967; Chap. 5.
60. Schindler, A. In "High Polymers"; Raff, R. A. V.; Doak, K. W., Eds.; Interscience: New York, 1965; Vol. XX, Chap. 5.
61. Patat, F.; Sinn, H. Angew. Chem. 1958, 70, 496.
62. Uelzmann, H. J. Polym. Sci. 1958, 32, 457.
63. Ludlum. B. D.; Anderson, A. W.; Ashby, C. E. J. Am. Chem. Soc. 1958, 80, 1380.
64. Karol, F. J.; Carrick, W. L. Ibid. 1961, 83, 2654.
65. Matlack, A. S.; Breslow, D. S. J. Polym. Sci., Part A 1965, 2853.
66. Werber, F. X.; Benning, C. J.; Wszolek, W. R.; Ashby, G. E. J. Polym. Sci., Part A-1 1968, A-1, 743.

67. Benning, C. J.; Wszolek, W. R.; Werber, F. X. J. Polym. Sci. Part A-1 1968, 755.
68. Hoeg, D. F.; Liebman, S. Ind. Eng. Chem. Process Design Develop. 1962, 1, 120.
69. Boor, J., Jr. J. Polym. Sci., Part A-2 1964, 265.
70. Boor, J., Jr. Polym. Preprints 1965, 6, 890.
71. Porri, L.; Natta, G.; Gallazzi, M. C. Chim. Ind. 1964, 46, 428.
72. Cossee, P. J. Catal. 1964, 3, 80.
73. Arlman, E. J. Ibid., 1964, 3, 89.
74. Arlman, E. G. Ibid., 1964, 3, 99.

75. Cossee, P. In "The Stereochemistry of Macromolecules";
    Ketley, A. D., Ed.; Dekker: New York, 1967; Chap. 3.
76. Hargitay, B.; Rodriguez, L.; Miotto, M. J. Polym. Sci. 1959,
    35, 559.
77. Chatt, J.; Duncanson, L. A. J. Chem. Soc. 1953, 2939.
78. Natta, G.; Pasquon, I.; Zambelli, A. J. Am. Chem. Soc. 1962,
    84, 1488.
79. Allegra, G. Makromol. Chem. 1971, 145, 235.
80. Corradini, P.; Barone, V.; Fusco, R.; Guerra, G.;
    Eur. Polym. J. 1979, 15, 1133.
81. Armstrong, D. R.; Perkins, P. G.; Steward, J. J. P J. Chem.
    Soc. Dalton Trans. 1972, 1, 1972.
82. Cassoux, P.; Crasnier, F.; Labarre, J. F. J. Organomet. Chem.
    1979, 165, 303.
83. Boor, J., Jr.; Youngman, E. A. J. Polym. Sci., Part A-1 1966,
    4, 1861.
84. Youngman, E. A.; Boor, J., Jr. In "Macromolecular Reviews";
    Peterlin, A., et al., Eds.; Interscience: New York, 1967;
    Vol. 2, pp. 33-69.
85. Zambelli, A.; Léty, A.; Tosi, C.; Pasquon, I. Makromol. Chem.
    1967, 115, 73.
86. Zambelli, A.; Allegra, G. Macromolecules 1980, 13, 42.
87. Marconi, W. In "The Stereochemistry of Macromolecules";
    Ketley, A. D., Ed.; Dekker: New York, 1967; Chap. 5.
88. Cooper, W. Ind. and Eng. Chem. Prod. Res. and Dev. 1970, 9,
    457.
89. Karapinka, G. L. U.S. Patent 3 709 853, 1973.
90. Hogan, J. P. J. Polym. Sci., Part A-1 1970, 8, 2637.
91. McDaniel, M. P. Paper presented at Midland Macromolecular
    Symposium; Midland, Mich., August 17-21, 1981.
92. Baker, L. M.; Carrick, W. L. J. Org. Chem. 1968, 33, 616.
93. Krauss, H. L.; Stach, H. Inorg. Nucl. Chem. Lett. 1968, 4,
    393.
94. Krauss, H. L.; Stach, H. Z. Anorg. Allg. Chem. 1969, 366, 380.
95. Henrici-Olive, G.; Olive, S. Angew. Chem. Int. 1971, 10, 766.
96. McDaniel, M. P. J. Polym. Sci. Chem. Ed. 1981, 19, 1967.
97. Karol, F. J.; Brown, G. L.; Davison, J. M. J. Polym. Sci.,
    Part A-1 1973, 11, 413.
98. Karol, F. J.; Wu, C. Ibid., 1974, A-1, 12, 1549.
99. Modern Plastics, October 1979, pp. 42-43.
100. Chem. Eng., 1979 Kirkpatrick Chemical Engineering Achievement
    Awards, December 3, 1979, pp. 80-85.
101. Modern Plastics, March 1980, pp. 59-61; July 1980, pp. 42-45.
102. Plastics World, December 1981, pp. 69-72.

Recent Developments in Cationic Polymerization

VIRGIL PERCEC

Department of Macromolecular Science, Case Western Reserve University, Cleveland, OH 44106

Ring Opening Polymerization
  General Considerations
  Initiation
  Propagation
  Termination and Transfer Processes
  Living Cationic Ring-Opening Polymerization
  New Polymers by Cationic Ring-Opening Polymerization
  Graft Copolymers
  Block Copolymers
Cationic Polymerization of Olefins
  General Considerations
  Graft Copolymers
  Inifer Technique
  Quasi-living Carbocationic Polymerization
  Proton Traps
  Block Copolymers
  Heterogeneous Graft Copolymerization
Polymers with Functional End Groups
  Polymers with Two Functional End Groups: Telechelics
  Polymers with One Functional End Group: Macromonomers

Cationic polymerization of heterocyclic and vinylic monomers is currently one of the most active areas of polymer chemistry. In the past two years four monographs dedicated to different topics in this field were published (1-4). Cationic polymerization is also one of the few areas of polymer science that has its own scientific meeting. The fifth meeting was organized in Kyoto (1979) (5); the previous meetings were in Akron (1976) (6); Rouen (1973) (7); Keele (1952) (8); and Dublin (1949) (9). At the 26th International Union of Pure and Applied Chemistry (IUPAC) International Symposium on

0097–6156/85/0285–0095$09.75/0

Macromolecules in Mainz (1979), four of the 33 main lectures were
dedicated entirely to the field of cationic polymerization (10-13).

A short history of cationic polymerization was published in
1975 (14, 15). An excellent collection of classic papers was
published in 1963 (16). The first recorded cationic polymerization
was described in 1789 (17), and the first industrial polymer
prepared by cationic polymerization, butyl rubber, appeared on the
market in 1943 (14).

In spite of these achievements and many others, our basic
knowledge about cationic polymerization started to develop only
very recently. The main reason is that the electrophilic species
through which cationic polymerization takes place (carbenium,
oxonium, sulfonium, phosphonium, and ammonium ions) are very
reactive. Consequently, in addition to propagation reactions,
chain transfer, termination, and reactions with traces of nucleophilic impurities take place. Only highly sophisticated techniques
such as high-vaccuum reaction conditions, adiabatic calorimetry,
and Fourier transform NMR measurements made progress possible in
this area of chemistry. Even so, a large difference exists between
our knowledge of ring-opening polymerization and vinylic
polymerization. Cationic polymerization of vinylic monomers takes
place with carbenium ions, which are more reactive than oxonium,
sulfonium, phosphonium, or ammonium ions. Chain transfer to
monomer can be decreased only at very low   polymerization
temperature. Consequently, polymers with high molecular weights
can be obtained by reaction conditions that are not of interest to
industry.

Ring-opening polymerization can be followed by NMR techniques;
therefore, direct evidence for the polymerization mechanism could
be obtained (1, 4, 17). Our knowledge about vinylic polymerization
mechanisms is obtained mainly from indirect evidence.

The present state of both ring-opening (1, 4, 11, 17-20) and
vinylic polymerization (2, 3, 14) was recently reviewed. The
present review will present the most recent developments in both
areas mainly in regard to the preparative power of cationic
polymerization. Only a few basic achievements of the mechanistic
aspects will be considered.

Ring-Opening Polymerization

General Considerations. The reactivity of heterocyclic monomers is
governed by the size of the ring, nature of the heteroatom and its
electronegativity and bond strength with the carbon atom, and
steric factors. A detailed discussion of all these factors is
presented by Penczek et al. (1). Two basic principles will be
outlined here. They will refer to the most simple heterocyclic
monomers only.

The size of the ring, that is, the number of atoms in the ring,
controls the ring strain by two factors. The first factor refers
to the difference between the bond angles that result from normal
orbital overlap and the bond angles that are a function of the
number of atoms in the ring, that is, the angle strain. The second
factor is the consequence of the interactions of the nonbonded
atoms. For a certain geometry of the molecule, nonbonded atoms are
situated in a close proximity that gives rise to this kind of
interaction. Both bond angles and nonbonding interactions are

responsible for the heterocycle ring strain. Consequently, with the exception of the three-membered rings, all rings exhibit a noncoplanar conformation of minimum energy. As a function of the number of atoms in the ring, the ring strain is dominated by angle strain or by nonbonded atom interactions. Table I summarizes the ring strain values for the most conventional rings.

According to the ring strain energies presented in Table I, the heterocyclic derivatives containing six atoms in the ring are generally unpolymerizable. An exception is <u>sym</u>-trioxane which polymerizes under conditions in which the polymer can crystallize.

The reactivity of the ring opening toward a cationic mechanism is mainly dictated by the nucleophilicity or the basicity of the monomer. The nucleophilicity of a monomer, that is, its ability to combine with electrophilic species, is determined by kinetically controlled conditions, and unfortunately, no general order of nucleophilicity is known. The monomer basicity, that is, its ability to interact with a proton, can be measured from thermodynamically controlled conditions. The most common method used to determine the basicity is based on the proportion of hydrogen-bonded compound measured at equilibrium. The basicity decreases in the following order: $R_3N > R_3P > R_2O > R_2S$, although R_2S is more basic than R_2O when the ability of bonding with softer acids is measured. In the case of cyclic ethers, the basicity order is as follows: $O(CH_2)_3 > O(CH_2)_4 > O(CH_2)_5 > O(CH_2)_4 > O(CH_2)_2$. Usually the basicity of a heterocyclic compound is affected in the expected order by the inductive effects, conjugation, steric effects, and ring size (1). The order of basicities is the only estimation of the monomer nucleophilicities, and it reflects fairly well the overall reactivity observed in ring-opening cationic polymerization.

<u>Initiation</u>. The most recent classification of initiators for cationic ring-opening polymerization was presented and discussed by Penczek et al. (1). Only a few classes of initiators that are very useful both for mechanistic studies as well as for synthesis of well-defined polymers will be presented here.

Protonic Acids. The simplest way of initiation and polymerization by a protonic acid is the following:

$$HA + Z\bigcirc \xrightarrow{k_i} H-^+Z\bigcirc \ A^-$$

$$H-^+Z\bigcirc A^- + n\ Z \xrightarrow{k_p} H-(Z\rightsquigarrow)_n-^+Z\bigcirc A^-$$

$$H-(Z\rightsquigarrow)_n-^+Z\bigcirc A^- \xrightarrow{k_t} H-(Z\rightsquigarrow)_{n+1}-A$$

When A^- is a noncomplex anion, that is, Cl^-, FSO_3^-, CF_3COO^-, ClO_4^-, or $CF_3SO_3^-$, competition always exists between the propagation (k_p) and the recombination of the growing macrocation with the counteranion (k_t). The ratio k_p/k_t will control the polymerization degree of the obtained polymer. The k_p/k_t is determined mainly by the ratio of the monomer nucleophilicity to that of the counteranion. Fluorosulfonic acid (FSO_3H and trifluoromethanesulfonic acid (CF_3SO_3H) are the most conventional initiators used both for kinetic studies as well as for new monomer reactivity testing studies.   They already have replaced the conventional Lewis acids such as $AlCl_3$ or BF_3 for two reasons: Their anions are weak nucleophiles, and in the case of heterocyclic monomers the initiation takes place by direct and quantitative protonation.

At the other extreme of this simple initiator class is HCl. Its anion is a strong nucleophile, and only highly nucleophilic N-substituted amines can be polymerized by HCl.   Simple addition of the initiator to the first monomer molecule takes place in other cases:

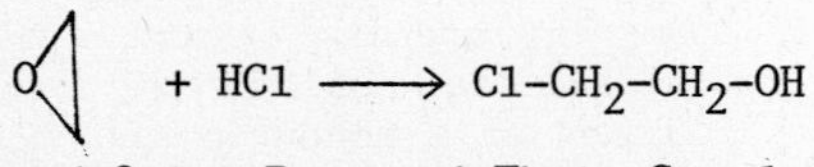

$$O\!\!\diagdown\!\!\diagup \quad + \ HCl \ \longrightarrow \ Cl\text{-}CH_2\text{-}CH_2\text{-}OH$$

Stable Carbenion and Onium Ions and Their Covalent Precursors.   The most representative initiators from this class are the following:

1. Carbenium ions: R_3C^+, $[(C_6H_5)_3C^+A^-]$
2. Alkoxycarbenium ions: $ROCH_2^+$, $(CH_3OCH_2^+A^-)$
3. Oxocarbenium ions: $R\text{-}C{\equiv}O^+$, $(C_6H_5C{\equiv}O^+A^-)$
4. Onium ions: R_nX^+, $[(C_2H_5)_3O^+A^-]$
5. Covalent initiators: RA, $[CH_3OSO_2CF_3$, $(CF_3SO_2)_2O$, CH_3I, etc.]

Carbenium Ions.   Stable carbenium ions (triphenylmethyl and tropylium salts) were developed by Ledwith ($\underline{21}$-$\underline{23}$).   Their merit is that they can initiate the polymerization of certain olefins by direct addition.   These initiators are very useful in kinetic studies, especially when weak nucleophiles such as SbF_6^- or $A_5F_6^-$ are used as counteranions.   Stable trityl salts, which might lead to systems devoid of side reactions, do not, however, initiate the polymerization of the majority of heterocyclic monomers by direct addition ($\underline{24}$, $\underline{25}$).   On the other hand these initiators can be produced in situ by the reaction of a suitable organic halide with a silver salt:

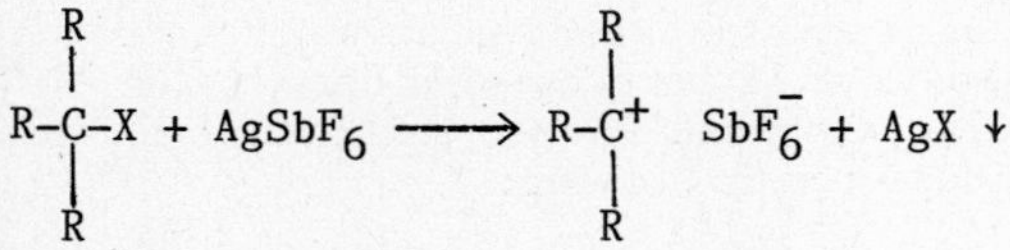

$$\underset{\underset{R}{|}}{\overset{\overset{R}{|}}{R\text{-}C\text{-}X}} + AgSbF_6 \ \longrightarrow \ \underset{\underset{R}{|}}{\overset{\overset{R}{|}}{R\text{-}C^+}} \ SbF_6^- + AgX \!\downarrow$$

Richards et al. ($\underline{26}$-$\underline{29}$) investigated the initiation of tetrahydrofuran (THF) polymerization induced by a large variety of organic halides in conjunction with $AgPF_6$.   The reactivity of the saturated alkyl halides are in the anticipated sequence: iodide > bromide > chloride > fluoride.   Cynnamyl bromide and p-methylbenzyl bromide are the most useful initiators ($\underline{28}$, $\underline{29}$), and 1,4-dibromo-2-butene and α,α'-dibromoxylene are excellent difunctional initiators.

Franta et al. ($\underline{30}$) studied the efficiency and the mechanism of initiation of the polymerization induced by several alkyl halides and $AgSbF_6$.   The initiation occurs by addition, proton elimination, and/or hydride abstraction.

Addition:

$$-\overset{|}{\underset{|}{C}}{}^{+} + O\big\langle\underset{\quad}{\overset{\quad}{\big]}} \xrightarrow{\ a\ } -\overset{|}{\underset{|}{C}}-\overset{+}{O}\big\langle\underset{\quad}{\overset{\quad}{\big]}}\ SbF_6^{-}$$

Proton elimination:

$$CH_3-\overset{|}{\underset{|}{C}}{}^{+} + O\big\langle\underset{\quad}{\overset{\quad}{\big]}} \xrightarrow{\ H^{+}\ } CH_2 = \overset{|}{\underset{|}{C}} + H-\overset{+}{O}\big\langle\underset{\quad}{\overset{\quad}{\big]}}\ SbF_6^{-}$$

Hydride abstraction:

$$-\overset{|}{\underset{|}{C}}{}^{+} + O\big\langle\underset{\quad}{\overset{\quad}{\big]}} \xrightarrow{\ H^{-}\ } -\overset{|}{\underset{|}{C}}H + \overset{+}{O}\big\langle\underset{\quad}{\overset{\quad}{\big]}}\ S_6F_6^{-}$$

Table II summarizes the data obtained from the reactions outlined above. The importance of these initiators in the synthesis of sequential copolymers will be discussed later.

Alkoxycarbenium Ions. Methoxycarbenium hexachloroantimonate prepared by a reaction developed by Olah (31) was exploited by Penczek et al. (32) to solve one of the most disputed controversies concerning the mechanism of 1,3-dioxolane polymerization.

$$CH_3OCH_2Cl + HF \cdot SbF_6 \longrightarrow CH_3O - \overset{+}{C}H_2SbF_6^{-} + HCl$$

$$CH_3)CH_2^{+} + \Big\langle \underset{O}{\overset{O}{}} \Big\rangle \rightleftharpoons CH_3OCH_2-{}^{+}\Big\langle \underset{O}{\overset{O}{}} \Big\rangle \rightleftharpoons CH_3-{}^{+}O\Big\langle \underset{O}{\overset{O}{}} \Big\rangle$$

A more detailed discussion will be presented in another chapter.

1,3-Dioxolan-2-ylium (Dioxolenium) Salts. Triphenylmethylium salts react with dioxolane by hydride transfer to form the corresponding dioxolenium salts (33), which react with nucleophiles exclusively by addition.

$$Ph-\overset{Ph}{\underset{Ph}{\overset{|}{\underset{|}{C}}}}{}^{+}\ SbF_6^{-} + \big\langle\underset{O}{\overset{O}{\big]}} \longrightarrow Ph-\overset{Ph}{\underset{Ph}{\overset{|}{\underset{|}{C}}}}H + \big\langle\underset{O}{\overset{O}{\big]}}{}^{+}\ SbF_6^{-}$$

$$\big\langle\underset{O}{\overset{O}{\big]}}{}^{+}\ SbF_6^{-} + \Big\langle\underset{O}{}\Big\rangle \rightleftharpoons \overset{H}{\underset{O}{\rangle}}C-O-CH_2-CH_2-\overset{+}{O}\big\langle\underset{\quad}{\overset{\quad}{\big]}}SbF_6^{-}$$

Dioxolenium salts can be considered derivatives of dialkoxycarbenium ions and are excellent initiators for ring-opening polymerization of THF.

Bisdioxolenium salts were used by Yamashita et al. (37) to produce dicationically growing poly(THF).

$$ClCH_2-CH_2-O-\overset{O}{\overset{\|}{C}}-(CH_2)_8-\overset{O}{\overset{\|}{C}}-O-CH_2CH_2Cl + 2\ AgClO_4 \longrightarrow$$

$$\longrightarrow ClO_4^{-}\ \Big[\overset{O}{\underset{O}{\big\rangle}}{}^{+}\Big]-(CH_2)_8-\Big\langle\underset{O}{\overset{O}{}}\Big]{}^{+}\ ClO_4^{-} + 2AgCl \downarrow$$

Table I. Ring Strain (kcal/mol) in Heterocyclic Monomers ([1])

n	$\left(\begin{array}{c}CH_2 \\ (CH_2)_{n-1}\end{array}\right)$	$\left(\begin{array}{c}O \\ (CH_2)_{n-1}\end{array}\right)$	$\left(\begin{array}{c}CH_2 \\ O \quad (CH_2)_{n-3} \quad O\end{array}\right)$	$\left(\begin{array}{c}NH \\ (CH_2)_{n-1}\end{array}\right)$	$\left(\begin{array}{c}S \\ (CH_2)_{n-1}\end{array}\right)$
3	27.4	27.3	--	26.9	19.8
4	26.0	25.5	--	--	19.7
5	6.0	5.6	6.2	5.8	1.97
6	-0.02	1.2	0	-0.15	-0.3
7	5.1	8.0	4.7	--	3.5
8	8.2	10.0	12.8	--	--

Table II. Reaction Path in the System RX + THF + $AgSbF_6$
           (a=addition, H^+=proton elimination, and H^-=
           hydride abstraction)

Alkyl Halide (corresponding cation)	Mechanism of Initiation		
	(X=I)	(X=Br)	(X=Cl)
$(C_6H_5)_3C^+$	H^-	H^-	H^-
$(C_6H_5)_2C^+H$			a
$C_6H_5CH_2^+$		a	H^+
$p\text{-}CH_3C_6H_4CH_2^+$		a([29])[a]	H^+
$(CH_3)_3C^+$	H^+	H^+	
$(CH_3)_2CH^+$	a	H^+	
$CH_2{=}CH{-}CH_2^+$		a/H^+	

[a] $AgPF_6$ salt

Onium Ions. Trialkyloxonium ions ($R_3O^+A^-$) became the conventional initiators for the cationic ring-opening polymerization of all classes of heterocycles (cyclic acetals, ethers, sulfides, lactones, phosphates, and amines). They are prepared by two methods developed by Meerwin (38) and Olah (39). Another more general and convenient synthesis method was recently developed by Penczek et al. (40):

$$R-C\underset{X}{\overset{O}{\diagdown}} + O\underset{R'}{\overset{R'}{\diagup}} + MtX_n \rightarrow R-C\overset{O}{\underset{R'}{\diagdown}}O\underset{R'}{\overset{R'}{\diagup}} MtX_{n+1}^-$$

$$R-\overset{O}{\underset{R'}{\overset{||}{C}}}O\underset{R'}{\overset{+}{\diagup}} MtX_{n-1}^- + O\underset{R}{\overset{R'}{\diagup}} \rightarrow R-\overset{O}{\overset{||}{C}}-OR' + R'_3O^+MtX_{n+1}^-$$

Trialkyloxonium ions are strong alkylating agents and initiate the polymerization of heterocyclic monomers by simple alkylation of the most nucleophilic site of the monomer. The initiation occurs quantitatively and without side reactions when stable anions are used. Consequently, these initiators are very useful for kinetic measurements. The initiation was followed directly by NMR spectroscopy in the case of THF (41), cyclic sulfides (42), and cyclic esters of phosphonic acid such as 2-methoxy-2-oxo-1,3,2-dioxaphosphorinane (43).

$$(C_2H_5)_3O^+BF_4^- + O\langle \rangle \rightarrow C_2H_5-\overset{+}{O}\langle \rangle BF_4^- + (C_2H_5)_2O$$

$$(C_2H_5)_3O^+BF_4^- + S\langle\overset{CH_3}{\underset{CH_3}{}}\rangle \rightarrow C_2H_5-\overset{+}{S}\langle\overset{CH_3}{\underset{CH_3}{}}\rangle BF_4^- + (C_2H_5)_2O$$

$$(C_2H_5)_3O^+ SbF_6^- + \underset{O\overset{||}{P}\diagdown OCH_3}{O\langle\rangle O} \rightarrow \underset{C_2H_5O\overset{|}{P}\diagdown OCH_3}{O\langle\overset{+}{}\rangle O} SbF_6^- + (C_2H_5)_2O$$

Covalent Initiators. The initiation with alkylating compounds depends both on the ability of the initiator to form a cation and on the monomer nucleophilicity. Strong alkylating agents such as esters of superacids (CF_3SO_3R, FSO_3R, and $ClSO_3R$) are able to initiate directly without sides reactions both strong and weak nucleophilic monomers (44, 45).

$$C_2H_5OSO_2CF_3 + O\langle \rangle \rightleftharpoons C_2H_5-\overset{+}{O}\langle \rangle CF_3SO_3^-$$

Weak cationating agents such as alkyl iodine, benzyl halides, and methyl-p-toluenesulfonate are able to initiate the polymerization of strong nucleophilic monomers such as cyclic amines (46, 47) and cyclic imino ethers (48-52).

Competition always exists between the nucleophilicity of the counteranion and that of the monomer when these types of initiators are used. In the case of THF polymerization with superacid ester initiators, macroions and macroesters are in equilibrium (44, 45). The polymerization of cyclic imino ethers takes place with either macroions or covalent species. The classic example is 2–methyl–2–oxazoline, which polymerizes exclusively via covalently bonded alkyl chloride species (benzyl chloride initiator) or via oxazolinium species (benzyl bromide initiator) (51).

$$C_6H_5CH_2X + \underset{\underset{CH_3}{\overset{|}{C}}}{\overset{|}{N}} \quad O \rightleftharpoons C_6H_5-CH_2-\underset{\underset{CH_3}{\overset{|}{C}}}{\overset{|}{N}}\overset{+}{,}O \quad X^- \rightleftharpoons C_6H_5CH_2-\underset{\underset{\underset{CH_3}{\overset{|}{C=O}}}{\overset{|}{N}}}{}-CH_2-CH_2-X$$

$$X = Cl, Br \qquad\qquad X = Cl$$

The mechanistic difference is explained by the different nucleophilicities of the counteranions Cl^- and Br^-. Dicationically terminated macromolecules can be obtained by the initiation with anhydrides of strong protonic acids such as trifluoromethanesulfonic anhydride (53, 54).

$$\overset{CF_3SO_2}{\underset{CF_3SO_2}{}} O + O\underset{}{\triangleleft} \overset{k_1}{\longrightarrow} CF_3SO_2-\overset{+}{O}\underset{}{\triangleright} CF_3SO_3^- \xrightarrow[THF]{k_{21}}$$

$$\xrightarrow[THF]{k_{21}} CF_3-SO_2-O-(CH_2)_4-\overset{+}{O}\underset{}{\triangleleft} CF_3SO_3^-$$

Photoinitiators for Cationic Polymerization.   Recently a class of photoinitiators for cationic polymerization was discovered by Crivello et al. (55).   This class includes diaryliodonium (Structure I) (56, 57), triarylsulfonium (Structure II) (58–62), dialkylphenacylsulfonium (Structure III) (63), dialkyl–4–hydroxy–phenylsulfonium salts (Structure IV) (64), and triarylselenonium salts (Structure V) (65).

$$\left[\overset{Ar}{\underset{Ar}{}}\overset{+}{I}\right] X^- \qquad \left[\overset{Ar}{\underset{Ar}{\overset{|}{S}^+}}\right] X^- \qquad \left[\overset{R}{\underset{R'}{ArC\overset{\|}{C}H_2-\overset{|}{S}^+}}\right] X^- \qquad HO-\left\langle\overset{R_1\ R_2}{\underset{R_3\ R_4}{}}\right\rangle-\overset{R_5}{\underset{R_6}{\overset{|}{S}^+}} X^-$$

$$\text{I} \qquad\qquad \text{II} \qquad\qquad \text{III} \qquad\qquad \text{IV}$$

$$\text{V} \quad \left[Ar-\overset{Ar}{\underset{Ar}{\overset{|}{Se}^+}}\right] X^- \qquad \text{where:}\quad X^- = BF_4^-, \; AsF_6^-, \; PF_6^-, \; SbF_6^-$$

In the absence of light these salts are stable even at high temperatures and do not exhibit catalytic activity.   On irradiation, for example in the case of diaryliodonium salts, the major photochemical process that occurs involves the homolytic cleavage of a carbon–iodine bond to produce a strong acid HX (i.e., HBF_4, $HAsF_6$, HPF_6, or $HSbF_6$).   These acids are among the strongest known and are excellent initiators for cationic polymerization. This mechanism is outlined as follows:

Major $ArI^+X^- \xrightarrow{h\nu} [Ar_2I^+X^-]^*$

$[Ar_2I^+X^-]^* \longrightarrow Ar-I^{\overset{+}{\bullet}} + Ar^{\bullet} + X^-$

$Ar-I^{\overset{+}{\bullet}} + Y-H \longrightarrow Ar-I^+-H+Y^{\bullet}$

where Y is a solvent or a monomer.

$Ar-I^+-H \longrightarrow Ar-I + H^+$

Minor $[ArI^+X^-]^* + Y-H \longrightarrow [Ar-Y-H]^+ + ArI + X^-$

$[Ar-Y-H]^+ \longrightarrow ArY + H^+$

Both vinylic (styrene, α–methylstyrene, and vinyl ethers) and heterocyclic monomers (cyclic ethers or epoxide, oxetane, THF, and trioxane; cyclic sulfides or propylene sulfide and thietane; and lactones and spiro bicyclic orthoesters) were polymerized at room temperature by photoinitiation at wavelengths shorter than 360 nm. The photodecomposition of diaryliodonium salts can be sensitized at wavelengths longer than 360 nm by the use of dyes such as Acridine orange, Acridine yellow, Phosphine R, Benzoflavin, and Setoflavin T (66). Triarylsulfonium, dialkylphenacylsulfonium, and dialkyl (4–hydroxyphenyl) sulfonium salts were sensitized by perylene and other polynuclear hydrocarbons (65, 67). Under these conditions, photoinitiated cationic polymerization can be performed by incandescent light sources or even ambient sunlight.

The polymerization rate depends on both the monomer reactivity and the nucleophilicity of the counteranion of the initiator salt. The order of reactivity in photoinduced polymerization correlates well with the known relative nucleophilicities of the anions, that is, $SbF_6^- > AsF_6^- > PF_6^- > BF_4^-$.

During the photodecomposition, free-radical species (Ar• and Y•) are also produced as transient intermediates. Therefore, the photolysis of the sulfonium salts should also initiate the free-radical polymerization (59). The amphifunctional character of sulfonium salts was demonstrated by the following series of experiments. Irradiation of an equimolar mixture of 1,4cyclohexene oxide and methyl methacrylate with $Ph_3S^+SbF_6^-$ as the photoinitiator gave a mixture of two homopolymers. Thus, both cationic (cyclohexene oxide) and free-radical (methyl methacrylate) polymerizations took place independently. The same system containing 2,6–di–tert–butyl–4–methylphenol (radical inhibitor) gave only poly(cyclohexene oxide). Alternatively the system with triethylamine (poison for cationic species) yielded only poly(methyl methacrylate). Monomers such as glycidyl acrylate and glycidyl methacrylate that contain functional groups capable of cationic and free-radical polymerizations are converted into a cross-linked insoluble polymer. The use of these hybride photo-initiators is very interesting in the synthesis of interpenetrating network structures.

Irradiation of dialkylphenacylsulfonium salts also produces strong protonic acids (63).

$$Ar{-}\overset{\overset{O}{\|}}{C}{-}CH_2{-}\overset{+}{S}\overset{R}{\diagdown}_R \; X^- \; \underset{}{\overset{h\nu}{\rightleftharpoons}} \; Ar{-}\overset{\overset{O}{\|}}{C}{-}CH{=}S\overset{R}{\diagdown}_R \; + \; HX$$

However, unlike the triarylsulfonium salts, these compounds undergo reversible photoinduced ylid formation rather than homolytic carbon–sulfur bond cleavage. Because the rate of the thermal back reaction is appreciable at room temperature, only those monomers that are more nucleophilic than the ylid will polymerize. Epoxides, vinyl ethers, and cyclic acetals undergo facile cationic polymerization when irradiated in the presence of dialkylphenacylsulfonium salts as photoinitiators.

The photodecomposition of dialkyl-4-hydroxyphenylsulfonium salts (64) gives rise to a resonance-stabilized ylid and an acid HX. Styrene oxide, 1,4-cyclohexene oxide, trioxane, and vinyl

ether were polymerized with satisfactory rates. However, THF, ε–caprolactone, and α–methylstyrene could not be polymerized (64).

Recently, Ledwith (68) continuing his interest in the chemistry of cation radicals (69, 70) demonstrated that the photoinitiation by triarylaminium, sulfonium, and iodonium salts occurs by a mechanism that is different from that proposed by Crivello.

Both cation radicals and protons are responsible for the photoinitiation with these initiators. Stable cation radicals based on phenothiazine or its derivatives, and triarylpyrylium and thiopyrylium salts are excellent photoinitiators for different heterocyclic monomers (68).

Photoinitiated cationic polymerizations are widely used for photocurable coatings for coatings of metal containers, wood, paper, and floor tiles, and also have considerable promise in applications involving photoimaging. Epoxy-based photoresists with high resolution have been developed, and the use of these materials in photography and plastic flexographic printing plates has been demonstrated.

Initiation of Cationic Polymerization by Free-Radical Initiators. A new procedure for the initiation of cationic polymerization was developed by Ledwith (13, 23, 71, 72). This procedure consists of the oxidation of the electron-donor radicals by aryldiazonium, diaryliodonium, and triarylsulfonium salts containing a stable counteranion. The parent radical can be obtained by thermal or photochemical decomposition.

$$AIBN \xrightarrow[\text{or } h\nu]{\Delta} R^{\bullet}$$

$$R^{\bullet} + THF \longrightarrow RH + O\langle \cdot \rangle \xrightarrow[-e]{Ar_2I^+PF_6^-} O\langle + \rangle \ PF_6^- + ArI + A\dot{r}$$

$$R^{\bullet} + CH_2{=}CH{-}OR' \longrightarrow RCH_2{-}\overset{\cdot}{C}H{-}OR' \xrightarrow[-e]{Ar_2I^+PF_6^-} RCH_2{-}\overset{+}{C}H{-}OR' \ PF_6^- + ArI + A\dot{r}$$

2,2'-Azobis(2-methylpropionitrile) (AIBN), benzoyl peroxide, phenylazotriphenylmethane, and benzpinacol were used as thermal radical initiators. Phenylazotriphenylmethane is especially an interesting radical initiator because by its radical oxidation, a well-known stable carbenium salt is obtained.

$$PhN = NCPh_3 \xrightarrow[\text{or } \Delta]{h\nu} Ph^{\bullet} + N_2 + {}^{\bullet}CPh_3$$

$$Ph_3C^{\bullet} + Ph_3S^+PF_6^- \longrightarrow Ph_3C^+PF_6^- + Ph_2S + Ph^{\bullet}$$

2,2-Dimethoxy-2-phenylacetophenone, benzophenone, benzil, and many other radical photoinitiators were used to induce the cationic polymerization in the presence of different oxidants.

Transformation of Anionic Polymerization into Cationic Polymerization. Richards et al. (26, 27, 73-75) proposed several methods for the transformation of a living anionic polymeric chain end into a cationic one. Such a process requires three distinct stages: polymerization of a monomer I by an anionic mechanism, and capping of the propagating end with a suitable but potentially reactive functional group; isolation of polymer I, dissolution in a solvent suitable for mechanism (2), and addition of monomer II; and reaction, or change of conditions, to transform the functionalized end into propagating species II that will polymerize monomer II by a cationic mechanism (73).

Two simple ways are the reaction of polystyryllithium with excess bromine or α,α'-dibromoxylene. The cationic initiation can be carried out by reacting the labile halide end group with a silver salt containing a weak nucleophilic anion.

$$\sim\sim\sim M^- Li^+ + Br_2 \longrightarrow \sim MBr + LiBr \downarrow$$

$$\sim\sim\sim MBr + AgX \longrightarrow \sim M^+ X^- + AgBr \downarrow$$

or

$$\sim\sim M^- Li^+ + BrCH_2 - \langle\!\langle \bigcirc \rangle\!\rangle - CH_2 Br \longrightarrow \sim\sim MCH_2 - \langle\!\langle \bigcirc \rangle\!\rangle - CH_2 Br + LiBr$$

$$\sim\sim MCH_2 - \langle\!\langle \bigcirc \rangle\!\rangle - CH_2 Br + AgX \longrightarrow \sim\sim MCH_2 - \langle\!\langle \bigcirc \rangle\!\rangle - CH_2^+ X^- + AgBr \downarrow$$

In both cases Wurtz condensation reactions do not allow high chain end functionality to be obtained.

$$\sim\sim\sim M^- Li^+ + \sim\sim MBr \dashrightarrow \sim\sim M-M \sim\sim + LiBr$$

The transformation of anionic living chain ends into a Grignard less reactive chain end allows the Wurtz condensation reaction to be completely eliminated (75). Yields as high as 95% were obtained by using this procedure.

$$\sim\sim M^- Li^+ + MgBr \dashrightarrow \sim\sim\sim MMgBr + LiBr$$

Even so, for polystyrene, benzyl bromide chain ends were preferred over 1-bromoethylbenzene chain ends for cationic initiation. 1-Bromoethylbenzene functional groups possess hydrogens on the β-carbon atom and do not initiate the polymerization entirely by an additive process. The principal side reaction is one of β-hydrogen elimination (28).

$$CH_3 - CHBr + AgPF_6 \longrightarrow AgBr\downarrow + CH_3 - CH^+ PF_6^- \rightleftharpoons CH_2 = CH + HPF_6$$

Propagation. The structure of the growing species (tertiary oxonium ions) in ring-opening polymerization of several monomers was already characterized by NMR spectroscopy (Table III). Carbenium-oxonium equilibria were also evidenced and measured.

$$\sim\sim OCH_2CH_2CH_2CH_2^+ + O\bigcirc \xrightarrow[\text{slow}]{\text{fast}} \sim\sim OCH_2CH_2CH_2CH_2 - O^+\bigcirc$$

The S_N2 mechanism of propagation in polymerization of heterocyclic monomers was generally accepted.

$$\sim\sim(O-CH_2CH_2CH_2CH_2)_n - O^+\bigcirc + :O\bigcirc \xrightarrow[kd]{kp} \sim\sim(OCH_2CH_2CH_2CH_2)_{n+1}^+ O\bigcirc$$

One of the most important achievements was the demonstration that the rate constant of propagation by free ions does not differ from that by ion pairs. Also the rate constant of propagation is not affected by the counteranion nature (45).

Table III. Growing species in Cationic Polymerization of Heterocyclic Monomers Observed Directly by ^{1}H–NMR Spectroscopy

Monomer	Structure of growing species (anions omitted)	Reference
(tetrahydrofuran)	$\sim\sim\sim CH_2-\overset{+}{O}\big\langle \begin{smallmatrix} CH_2 & CH_2 \\ & CH_2 \\ CH_2 \end{smallmatrix}$	76
(oxepane)	$\sim\sim\sim CH_2-\overset{+}{O}\big\langle \begin{smallmatrix} CH_2 CH_2 CH_2 \\ \quad \| \\ CH_2 CH_2 CH_2 \end{smallmatrix}$	77
(1,3-dioxolane)	$\sim\sim\sim CH_2-\overset{+}{O}\big\langle \begin{smallmatrix} CH_2-O-CH_2 \\ \quad \| \\ CH_2-O-CH_2 \end{smallmatrix}$	32
(cyclic phosphate, OCH$_3$)	$\cdots -CH_2-O-\overset{+}{P}\big\langle \begin{smallmatrix} O-CH_2 \\ \quad CH_2 \\ O-CH_2 \end{smallmatrix} \\ \overset{|}{O}CH_3$	78
(dimethyl thietane, H$_3$C CH$_3$, S)	$\cdots -CH_2-\overset{+}{S}\big\langle \begin{smallmatrix} CH_2 & CH_3 \\ & C \\ CH_2 & CH_3 \end{smallmatrix}$	79
(dimethyl azetidine, CH$_3$ CH$_3$, N-CH$_3$)	$\cdots -CH_2-\overset{+}{N}\big\langle \begin{smallmatrix} CH_2 & CH_3 \\ & C \\ CH_2 & CH_3 \end{smallmatrix} \\ \overset{|}{C}H_3$	80
(2-methyl oxazoline, N O, CH$_3$)	$\cdots -CH_2-\overset{+}{N}\big\langle \begin{smallmatrix} CH_2-CH_2 \\ \quad\quad\quad O \\ C \\ CH_3 \end{smallmatrix}$	51

The long dispute of the growing species structure in the polymerization of cyclic acetals seems to be at its end. Penczek et al. (32) showed clearly that propagation proceeds on linear growing species. Growth by ring expansion seems to be unlikely on the basis of the existing experimental evidence. The presence of the end groups was clearly demonstrated by ^{1}H-, ^{31}P[^{1}H]-NMR, and UV. DP_n calculated from the end groups agrees well with the DP_n determined osmometrically. All equilibrium constants were measured recently for this propagation scheme:

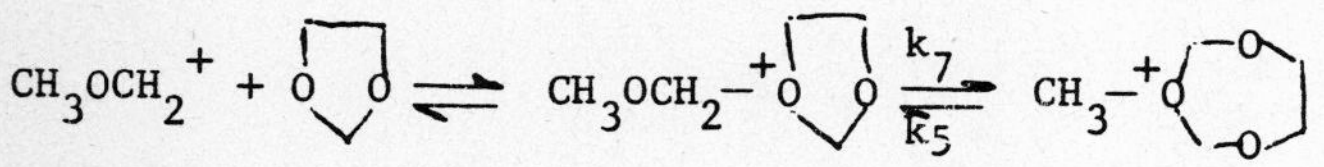

$k_7/k_5 = 3 \cdot 10^2$, and explains the tendency of isomerization of less stable (more strained) five-membered rings into less strained seven-membered ones. A more detailed discussion on this topic can be found in a recent review (1).

<u>Termination and Transfer Processes</u>. Clear evidence about the mechanism of termination and chain transfer processes can be obtained from the polymer chain end structure. One polymer chain end is controlled by the initiation mechanism, and the second one is controlled by termination and/or chain transfer. The chemical structure of the end groups has been studied in a few cases only (1). Several peculiarities of these reactions will be outlined here.

Temporary Termination: Reversible Recombination with Noncomplex Anions. Temporary termination was evidenced for the first time in the polymerization of cyclic imino ethers (51).

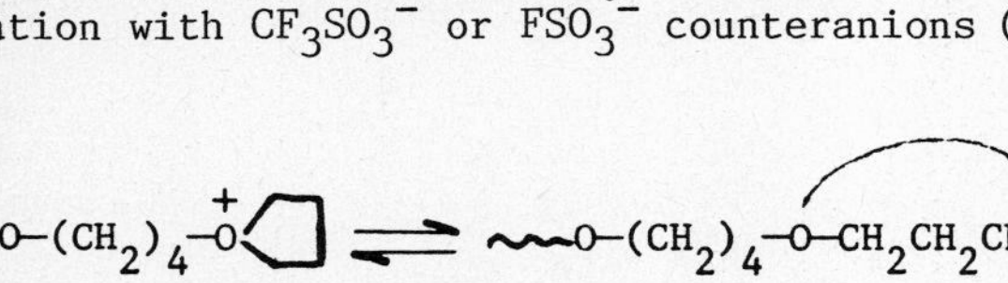

The same reaction was recently evidenced in the case of THF polymerization with $CF_3SO_3^-$ or FSO_3^- counteranions (1):

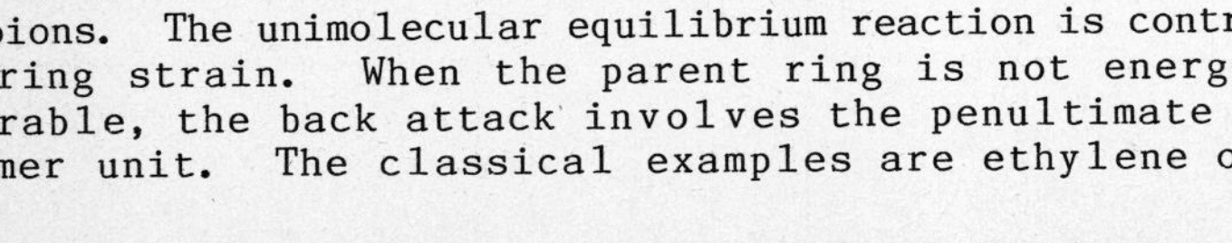

The propagation can take place with covalent or macroester species, but in these cases the rate constants are lower than with macroions. The unimolecular equilibrium reaction is controlled by the ring strain. When the parent ring is not energetically favorable, the back attack involves the penultimate or other monomer unit. The classical examples are ethylene oxide and aziridine polymerization. In both cases a six-membered ring is formed (k_6/k_3 = 10/1) (81).

The collapse of the macroion pair can be irreversible if the counteranion is a strong nucleophile and the monomer nucleophilicity is low. For monomers with high nucleophilicity, the termination involving anions becomes less important.

The macroions concentration can be measured by two end-capping methods that were developed by Saegusa (82) and Penczek (83).

Transfer and Termination Involving the Polymer Backbone. The polymer chain contains the same heteroatom as the parent monomer. Therefore, the nucleophilic attack of a monomer molecule on the strained onium ion can be replaced by the attack of a polymer chain. The polymer chain can react with active species of a foreign macromolecule or of its own macromolecule (back biting).

Both intermolecular and intramolecular reactions can be either reversible or irreversible (termination). In reversible reactions true chain transfer takes place when the rate constant of the backward reaction (k_{ri}) becomes comparable with the rate constant of propagation. This is valid in the case of cyclic acetal polymerization in which the product of chain transfer is equally active in propagation.

Transfer or termination reactions that involve the reaction of the growing center with its own backbone are inherent features of the cationic polymerization of heterocyclic monomers.

Evidence for the back-biting reaction is the formation of cyclic oligomers during the ring-opening polymerization of some heterocyclic monomers (81, 84-87). The classic example is dioxane formation during polymerization of ethylene oxide.

The intermolecular chain transfer (termination) reaction was demonstrated by polymerizing a given monomer in the presence of another preformed polymer. A copolymer is obtained in this case (88). Mechanisms and kinetics of the reactions have been recently discussed (1).

<u>Living Cationic Ring-Opening Polymerization</u>. When initiation is complete and rapid, compared to propagation, termination reactions lead to incomplete conversions even after "infinite" time. It was mentioned earlier that termination or transfer reactions are caused by the reaction of the growing species with counteranions and the formed polymer. Termination by reaction with counteranions can be controlled by using an initiator that produces a stable counteranion. Termination by reaction with the formed polymer occurs because the polymer contains the same nucleophilic heteroatoms as the monomer and so is able to react with the active chain end. As a consequence, the strained onium ion that provided the driving force for the propagation is transformed into a nonstrained and, therefore, unreactive onium ion. This termination can be irreversible and thus cause a "suicidal" polymerization, or may be temporary and lead to a "dormant" polymer.

When the termination is irreversible, and in the absence of transfer reactions, the living character of the polymerization can in a first approximation be defined as the ratio k_p/k_t, where k_p and k_t are the rate constants for propagation and for termination, respectively.

From a series of kinetic studies with cyclic sulfides, amines (47, 81, 89), and cyclic esters of phosphonic acid (90), it was shown that in a given series of monomers with the same parent heterocycle, the ratio k_p/k_t increases with the bulkiness and the number of substituents on the monomer. Although the absolute rate of polymerization decreases also, the presence of substituents apparently retards the termination more than the propagation. A few examples are presented below.

$R = -C(CH_3)_3$)	$R = -C_2H_5$	$R = -C(CH_3)_3$	$R = -CH_3$
$k_p/k_t \simeq \infty$	$k_p/k_t \simeq 360$	$k_p/k_t \simeq \infty$	$k_p/k_t \simeq \infty$
$R = -CH_3$	$R = -H$	$R = -C_2H_5$	$R = -H$
$k_p/k_t = 286$	$k_p/k_t = 4$	$k_p/k_t = 6$	$k_p/k_t = 250$

The larger decrease of k_t with the size of the exocyclic group is mostly due to the increasingly negative entropy of activation (90).

Several oxazines and oxazolines also polymerize by a living

mechanism (48–52, 91). Chain transfer to polymer does not occur in this case because the nucleophilicity of the heteroatoms in the polymer chain is lower than that in the monomer molecules

(because of the presence of R–C– substituents on the polymer heteroatoms).

Processes that do not terminate were described for the polymerization of five- and seven-membered cyclic ethers: THF (4, 91) and oxepane (93). Under proper working conditions both cyclic acetals, namely 1,3-dioxolane and 1,3-dioxepane (94, 95), as well as bicyclic acetals or 1,6-anhydro-1,3,4-tri-O-benzyl-α-D-glucopyranose (96) can polymerize under living conditions.

New Polymers by Cationic Ring-Opening Polymerization. Models for two of the most important classes of biopolymers, that is, poly-saccharides and polyphosphates (nucleic and teichoic acids) can be obtained by cationic ring-opening polymerization.

High molecular weight stereoregular polysaccharides were prepared by the cationic polymerization of bicyclic acetals. Schuerch developed and reviewed this field comprehensively several times and described the mechanism of polymerization, synthetic scope of the method, and applications of these synthetic polysaccharides (97–100).

The synthesis of polyphosphates as a model for nucleic and teichoic acids by ring-opening polymerization of cyclic esters of phosphoric acid is studied by Penczek's group and was reviewed recently (43, 101, 102).

Sumimoto and Okada reviewed the reaction mechanism of the ring-opening polymerization of bicyclic acetals, oxalactones, oxalactames, and related heterocyclic compounds (103).

The ring-opening polymerization of atom-bridged and bond-bridged bicyclic ethers, acetals, and orthoesters was reviewed by Yokoyama and Hall (104).

The polymerization of macrocyclic acetals is a very active field for Schulz (105–107). The polymerization of 4H, 7H–1,3-dioxepin (106) is of particular interest in the synthesis of functional polyethers.

Ring-opening polymerization of bicyclic ethers is intensively studied by Kops et al. (108).

Bailey continues his series of papers in the area of ring-opening polymerization with expansion in volume (109, 110). Recently this field has attracted the interest of another group of researchers (111).

Vogl (112, 113) is developing an ingenious preparative method for ionomers synthesis by copolymerization of ethyl glycidate with other cyclic ethers and then hydrolysis of the pendant ester groups.

Graft Copolymers. Graft copolymers can be synthesized in two ways: "grafting from" and "grafting onto." Grafting from makes use of knowledge of initiating mechanisms. Grafting onto is based on the coupling reaction (termination) of a living polymer with a suitable functional group from the polymer main chain. Both methods have been used for the synthesis of graft copolymers.

Grafting from is based on the use of a macromolecular initiator, that is, a polymer backbone containing dioxolenium salts as precursors of carbenium and oxocarbenium ions (i.e., labile halides or acylhalides). The last two functional groups can be transformed into a carbenium ion by reaction with a silver salt containing a stable anion. When the monomer used is a strong nucleophile (such a cyclic imino ethers) the labile halide itself can initiate graft copolymerization.

Okada et al. (114) initiated the graft copolymerization of THF from a polystyrene containing dioxolenium pendant groups.

Dreyfuss et al. (115-118) investigated the initiation of THF, 7-oxabicyclo[2.2.1] heptane, oxetane, propylene oxide, styrene oxide, dioxolane, trioxane, ε-caprolactame, and thietane from a variety of polymers containing labile halides. Graft copolymers have been prepared from several backbones: poly(vinyl chlorides), polychloroperene, chlorinated ethylene-propylene-diene rubber (EPDM), chlorobutyl rubber, bromobutyl rubber, chlorinated polybutadiene, and chlorinated butadiene-styrene copolymer. Soluble silver salts with anions such as $CF_3SO_3^-$, BF_4^-, PF_6^-, AsF_6^-, SbF_6^-, and ClO_4^- were used to produce carbenium ions in situ.

$$\text{Polymer } -X + AgY + nZ \ ---- \ \text{Polymer} - (Z\sim)_n^{\sim} + AgX\downarrow$$

Because silver salts are very expensive and the byproduct, AgX, darkens on exposure to light and gives black polymers, Dreyfuss et al. (119, 120) investigated a large number of other metal salts for these initiation processes. The same authors (121, 122) used for the first time nitrosyl and nitryl hexafluorophosphates for the initiation of graft copolymerization of THF. The merit of these initiators is that the byproduct obtained is a volatile compound and can be easily removed from the reaction medium.

$$\text{THF} + NOPF_6 + RX \ ---- > \ R - PTHF^+PF_6^- + NOCl$$

Recently Dreyfuss et al. (113, 114) developed a new method for the determination of the number of poly(THF) branches in a graft copolymer. The method is based on the termination of the living cationic chain ends with NH_4OH-NH_4Cl buffer and reaction with fluorescamine. The chain ends concentration is determined by fluorescent spectroscopy.

Franta et al. (115) synthesized graft copolymers by initiation of THF polymerization from chloromethylated polystyrene, partially brominated 1,4-polybutadiene, and a random copolymer of styrene and methacryloyl chloride in the presence of $AgSbF_6$.

A large number of graft copolymers were obtained by Dreyfuss and Kennedy (126–128) by grafting of pendant epoxy groups of a variety of polymer backbones. This grafting mechanism is based on Saegusa's findings that several ring compounds are able, in conjunction with Lewis acids, to generate tertiary oxonium ions, the true initiating species (129).

Chloromethylated cross-linked polystyrene was used to initiate the graft copolymerization of several 2-substituted-2-oxazoline derivatives (130, 131). Allylic chloride from 1-chloro-1,3-butadiene–butadiene copolymer and from poly(vinyl chloride) was used to initiate the graft copolymerization of 2-methyl-2-oxazoline (132, 133).

The grafting onto method was used to prepare graft copolymers by deactivation reaction onto a backbone fitted with nucleophilic sites. Franta et al. (134) used this technique to synthesize graft copolymers of poly(THF) with nucleophilic backbones poly(p-dimethylaminostyrene) and poly(2-vinylpyridine).

Richards et al. (135) succeeded in quantitative graft copolymerization of poly(THF) onto poly(4-vinylpyridine). Goethals et al. (89) also used the grafting onto method to prepare graft copolymers in a quantitative yield by reacting a living poly(N-*tert*-butylaziridine) with poly(2-vinylpyridine) as a "deactivating" polymer.

Block Copolymers. Several methods have already been used for the synthesis of block copolymers. The most conventional method, that is, the addition of a second monomer to a living polymer, does not produce the same spectacular results as in anionic polymerization. Chain transfer to polymer limits the utility of this method. A recent example was afforded by Penczek et al. (136). The addition of the 1,3-dioxolane to the living bifunctional poly(1,3-dioxepane) leads to the formation of a block copolymer, but before the second monomer polymerizes completely, the transacetalization process (transfer to polymer) leads to the conversion of the internal homoblock to a more or less (depending on time) statistical copolymer. Thus, competition of homopropagation and transacetalization is not in favor of formation of the block copolymers with pure homoblocks, at least when the second block, being built on the already existing homoblock, is formed more slowly than the parent homoblock that is reshuffled by transacetalization.

In systems devoid of chain transfer to polymer, for example, cyclic imino ethers, this method gives rise to pure block copolymers (137, 138).

A related method was developed by Pepper and Goethals (139). Taking advantage of the dormant character of polystyrene prepared at low temperature with perchloric acid (in the form of a macroester) they prepared well-characterized block copolymers of polystyrene with cyclic amines.

ABA and AB block copolymers were synthesized by the initiation of 2-oxazoline polymerization by a polymer containing tosylate (140–143) or alkyl halide end groups (144–147).

Initiation of THF polymerization from polystyrene or polybutadiene containing labile halides as chain ends in conjunction with silver salts was used by Richards et al. (26, 27, 74) to initiate the block copolymerization of THF. Richards et al. (148) developed a new route for preparing block copolymers by a macrocation to macroanion transformation. This process consists in reacting living poly(THF) with the lithium salt of cinnamyl alcohol to prepare a polymer possessing a styryl terminal group. This reaction is quantitative. The second stage involves the reaction of this product with n-butyl lithium in benzene to form an adduct to which a monomer such as styrene or isoprene is added to prepare a block copolymer anionically. This last stage unfortunately operates with only 20% efficiency.

The coupling of an anionic living polymer with a cationic living polymer gives rise to AB or $(AB)_n$ block copolymers. In the case of polystyrene with poly(THF) the coupling efficiency seems to depend on the nucleophilicity of the counteranion and of the anionic chain ends. For example, the grafting yield is very low when poly(THF) with BF_4^- counteranion is used (188), and it increases in the case of FSO_3^- counteranion (179). The yield is quantitative when carboxylating polystyrene anions are used (37, 150). Multiblock copolymers $(AB)_n$ were obtained by coupling of dianionic polystyrene with dicationic poly(THF) (151, 152). The coupling of living anionic poly(α-methylstyrene) with cationic living poly(THF) occurs with only 20% efficiency. Proton transfer and hydride transfer gives rise to poly(THF) and poly(α-methylstyrene) with vinylic end groups as byproducts (153).

Attempts to produce block copolymers by coupling of living cationic poly(N-tert-butylaziridine) with living anionic polystyrene failed because the result was a mixture of the two homopolymers. However, when the carbanion of the anionic polystyrene was first converted into a thiolate anion by reaction with propylene sulfide, coupling with living poly(N-tert-butylaziridine) was successful in producing a block copolymer (89). ABA-type block copolymers were synthesized by reaction of a living poly(N-tert-butylaziridine) with telechelic amino- or carboxy-terminated polymers having polybutadiene or polybutadiene-co-acrylonitrile as backbones (89).

Chain transfer to a second polymer can be exploited as a possible avenue for block copolymer synthesis. The homopolymerization of a given monomer A in the presence of a preformed polymer B or the interaction between two homopolymers in the presence of a cationic initiator (1, 153, 154) produce in the first step of the reaction block copolymers. Synthesis of other block copolymers was reviewed by Yamashita et al. (152). Trimethylcellulose-[b-poly(THF)]-star block copolymers were synthesized by Feger and Cantow (156, 157). Trimethylcellulose containing a labile chlorine end group was used to initiate the living polymerization of THF in the presence of $AgSbF_6$. The living chain end of this AB block copolymer was reacted with poly(4-vinylpyridine) oligomers to form star-shaped block copolymers.

Cationic Polymerization of Olefins

General Considerations. This field was presented in a few recent monographs (2, 3, 14, 158).

Carbenium species are more active than oxonium, sulfonium, or ammonium. Side reactions, or transfer to monomer and nucleophilic attack of the aromatic ring in the case of styrene polymerization (to produce a 3-phenylindane type end groups) lead to low molecular weight polymers. The only way to avoid these side reactions is to decrease the polymerization temperature. The first carbenium ion was observed by ^{13}C-NMR spectroscopy in 1979 (159). In these conditions all the mechanistic approaches are supported by indirect evidence only. In the case of ring-opening polymerization, sequential and functional polymers are synthesized by using our knowledge of polymerization mechanisms. In the case of olefin polymerization, sequential copolymers are evidence for the suggested mechanism of polymerization.

Information about the propagation rate constants with free ions are obtained, as in the case of ring-opening polymerization, by using stable carbenium ions as initiators (23). Unfortunately these initiators can be used only for the polymerization studies of very active cationic monomers (i.e., strong bases such as vinyl ethers or vinyl derivatives containing strong electron donor pendant groups) (23). Another way to determine propagation rate constants with free ions is to use radiation-induced ionic polymerization techniques (160).

A new method for the study of nonstationary polymerization is the flow and stopped-flow spectroscopy developed by Sawamoto and Higashimura (161, 162). Although these methods offer the only available data about polymerization kinetics through known species, their preparative applications are very limited. A number of useful discoveries are coming from Kennedy's laboratory (163-175). Part of these discoveries will be presented later. Kennedy's research philosophy consists in understanding the mechanisms of polymerization of conventional monomers, and its use in the design of new polymeric materials (163) has proven very productive.

Graft Copolymers. As in the case of ring-opening polymerization, labile halides can be used in conjunction with a Lewis acid in this case to produce carbenium species. If the initiation takes place by addition, graft copolymers can be obtained by the grafting from technique when the labile halide is part of a polymer chain (10, 163-165).

$$P-Cl + Et_2AlCl \longrightarrow [P^+Et_2AlCl_2^-] \xrightarrow{M} P - poly(M) + EtAlCl_2$$

A large variety of graft copolymers was prepared by this technique and some are presented in the box (159). Under suitable reaction conditions, graft copolymers free of homopolymers could be prepared.

Initiation of graft copolymerization by radiation-induced cationic mechanism was recently reviewed by Stannett (160). This method is especially useful for cationic graft copolymerization from inert polymer supports.

Inifer Technique.   Inifers are bifunctional initiator-chain transfer agents that have been used for the preparation of α,ω-difunctional polyisobutylene carrying $\text{---CH}_2\text{C(CH}_3)_2\text{Cl}$ end groups (10, 164, 166-168).

   The mechanism of the inifer system based on dicumyl chloride-BCl_3 and isobutylene is outlined below.

Representative Graft Copolymers Prepared by Carbocationic Techniques

---

I. Elastomeric backbones

    A. Elastomeric branches

        Poly(butadiene-g-isobutylene)[a]
        Poly[chloroprene-g-(isobutylene-co-isoprene)]
        Poly[isobutylene-co-isoprene)-g-chloroprene][a]

    B. Glassy branches

        Poly[(ethylene-co-propylene)-g-styrene][a]
        Poly[(isobutylene-co-isoprene)-g-styrene][a,b]
        Poly(butadiene-g-α-methylstyrene)[a]

    C. Two branches (bigrafts)

        1. A glassy and an elastomeric branch

            Poly[ethylene-co-propylene-co-1,4-hexadiene)
            -g-styrene-g-isobutylene][a,b]

        2. Two glassy branches

            Poly[(ethylene-co-propylene-co-1,4-hexadiene)
            -g-styrene-g-α-methylstyrene][a,b]

II. Glassy backbones

    A. Elastomeric branches

        Poly(vinyl chloride-g-isobutylene)
        Chloromethylated polystyrene-g-polyisobutylene

    B. Glassy branches

        Poly(vinyl chloride-g-styrene)

---

[a]Lightly chlorinated backbone used.

[b]Lightly brominated backbone used.

The functionality of the obtained telechelic polyisobutylenes is affected mainly by the intramolecular cycloalkylation of the initiator.

Cycloindane formation can be avoided by working under proper temperature and solvent mixture conditions (169, 170). By using tricumyl chloride-BCl_3 trinifer system, three-arm star telechelic polyisobutylenes carrying exactly three $-C(CH_3)_3Cl$ end groups have been synthesized by Kennedy et al. (171, 172).

Quasi-living Carbocationic Polymerization. Recently, Kennedy et al. (165, 173) developed polymerization systems in which under well-defined conditions (a special manner of continuous mixing of monomer with initiating systems), chain termination and chain transfer to monomer are reversible or avoidable, and for all practical purposes the system behaves as if R_t and R_{trM} are equal to zero. Fast R_i was achieved by premixing the ingredients of the initiating systems.

Kinetic equations have been derived according to which the molecular weight of the polymer can be controlled by the cumulative amount of monomer added and initial concentrations of initiator: $DP_n = M_{total}/[I]_0$. This equation is very similar to that defining living conditions: $DP_n = [M]/[I]_0$. A few studied systems that polymerize under quasi-living conditions are H_2O-BCl_3-α-methylstyrene, $C_6H_5C(CH_3)_2Cl-BCl_3$-α-methylstyrene, tert-BuCl-$TiCl_4$-isobutylene.

Proton Traps. 2,6-Di-tert-butyl-4-methylpyridine (DBMP) and 2,6-di-tert-butylpyridine (DtBP) are hindered bases incapable of reacting with electrophiles other than protonic acids. Consequently, they can be successfully used for the trapping of protons during their transfer to monomer (174, 175). At the same time they can disseminate between the two major initiation mechanisms encountered in cationic polymerization, that is, protonic initiation or carbenium initiation (176).

Block Copolymers. Two conventional techniques were applied for the synthesis of block copolymers: initiation of monomer polymerization from a preformed polymer containing an initiator as chain end or ends, and the addition of a second monomer to a living polymerization of the first one.

Poly(α-methylstyrene-<u>b</u>-isobutylene-<u>b</u>-α-methylstyrene) was prepared by the initiation of α-methylstyrene polymerization from a ditelechelic polyisobutylene containing *tert*-chlorine end groups. AlEt$_2$Cl was used as coinitiator (<u>177</u>). Poly(isobutylene-<u>b</u>-styrene) and poly(isobutylene-<u>b</u>-α-methylstyrene) were prepared by the initiation of styrene and α-methylstyrene polymerization from an asymmetric telechelic polyisobutylene, that is, $(CH_3)_2$-C=Ch-CH$_2$~ PiB~CH$_2$-C(CH$_3$)$_2$Cl and AlEt$_2$Cl (<u>178</u>).

Addition of a second monomer to a living polymer chain was used to produce block copolymers from <u>N</u>-vinylcarbazole and vinyl ethers by using stable carbenium salts as initiators (<u>179</u>). The same avenue was used by Higashimura et al. (<u>180</u>) to produce a block copolymer by initiation of the polymerization of isobutyl vinyl ether from a long-lived poly-<u>p</u>-methoxystyrene. Living polymerization of <u>p</u>-methoxystyrene and <u>N</u>-vinylcarbazole has successfully been achieved by using iodine as initiator (<u>180</u>, <u>181</u>). By using a programmed successive addition of monomers, Giusti (<u>182</u>) successfully prepared block copolymers from styrene and isobutylene.

A detailed description of the sequential copolymer synthesis by carbocationic polymerization is presented in a recent book (<u>3</u>).

<u>Heterogeneous Graft Copolymerization</u>. Poly(vinyl chloride) films and powders (<u>183</u>) and chlorinated polypropylene (<u>184</u>) fibers were grafted with styrene, isobutylene, and styrene, respectively. Grafting from techniques were used. By using the same technique a silica surface first treated with chlorosilyl functional groups was grafted with polyisobutylene and butyl rubber (<u>185</u>, <u>186</u>):

$$-Si-OH + Cl-\underset{\underset{CH_3}{|}}{\overset{\overset{CH_3}{|}}{Si}}-CH_2CH_2\phi CH_2Cl_2 \xrightarrow{-HCl} -SiO-\underset{\underset{CH_3}{|}}{\overset{\overset{CH_3}{|}}{Si}}CH_2CH_2\phi CH_2Cl + AlEt_2Cl \rightarrow$$

$$\underset{\xrightarrow{\hspace{2cm}}}{isobutylene} \quad -Si-\underline{g}-polyisobutylene$$

Grafting on technique was used to produce a graft copolymer silica-<u>g</u>-poly(<u>N</u>-*tert*-butylaziridine) (<u>187</u>). The technique used was the coupling of a silica-containing amine group with a temporarily living poly(*tert*-butylaziridine).

$$\begin{array}{l} -OH \\ -OH \\ -OH \end{array} + (Et_3O)_3Si-(CH_2)_3NH_2 \longrightarrow \overset{-O}{\underset{-O}{\overset{\diagdown}{\diagup}}}Si(CH_2)_3NH_2 \quad +$$

$$poly\left(TBA\right)^+_{\underset{+}{N}}\diagdown \; CF_3SO_3^- \longrightarrow \overset{-O}{\underset{-O}{\overset{\diagdown}{\diagup}}}Si(CH_2)_3\underset{+}{N} \, poly\left(TBA\right)$$

or

$$poly\left(TBA\right)^+_{\underset{+}{N}}\diagdown + NH_2(CH_2)_3Si(OEt)_3 \longrightarrow poly\left(TBA\right)\underset{+}{N}-poly\left(TBA\right)NH(CH_2)_3Si(OEt)_3$$

These heterophase methods are very interesting from both an academic as well as technological point of view.

Polymers with Functional End Groups

Polymers with Two Functional End Groups: Telechelics. In accordance with their historical appearance, polymers with two functional end groups will be considered first. A very large range of telechelic polyisobutylenes (PIB) were synthesized and characterized by Kennedy and his coworkers (165). These data were already reviewed several times (3, 165, 188). A few avenues for PIB telechelics preparation based on the inifer technique will be presented. The dehydrochlorination of α,ω-di(tert-chloro)-polyisobutylene led to α,ω-di(isopropenyl)polyisobutylene (189) in the quantitative yield. The regioselective hydroboration of α,ω-di(isopropenyl)polyisobutylene followed by alkaline hydrogen peroxide oxidation led to a new telechelic polyisobutylene diol carrying two terminal primary hydroxyl end groups (190), that is, α,ω-di(hydroxy)polyisobutylene.

$$Cl-\overset{\overset{\displaystyle CH_3}{|}}{\underset{\underset{\displaystyle CH_3}{|}}{C}}-CH_2 \sim PIB \sim \overset{\overset{\displaystyle CH_3}{|}}{\underset{\underset{\displaystyle CH_3}{|}}{C}}-C_6H_4-\overset{\overset{\displaystyle CH_3}{|}}{\underset{\underset{\displaystyle CH_3}{|}}{C}} \sim PIB \sim CH_2-\overset{\overset{\displaystyle CH_3}{|}}{\underset{\underset{\displaystyle CH_3}{|}}{C}}-Cl$$

$$\downarrow \quad \begin{array}{l} tBuOK \\ THF \end{array}$$

$$CH_2{=}\overset{\overset{\displaystyle }{}}{\underset{\underset{\displaystyle CH_3}{|}}{C}}-CH_2 \sim PIB \sim \overset{\overset{\displaystyle CH_3}{|}}{\underset{\underset{\displaystyle CH_3}{|}}{C}}-C_6H_4-\overset{\overset{\displaystyle CH_3}{|}}{\underset{\underset{\displaystyle CH_3}{|}}{C}} \sim PIB \sim CH_2-\overset{\overset{\displaystyle CH_3}{|}}{\underset{\underset{\displaystyle }{}}{C}}{=}CH_2$$

$$\left| \begin{array}{l} 1) \ BH_3/THF/0°C \\ 2) \ NaOH/H_2O_2/30{-}45°C \end{array} \right.$$

$$HO{-}CH_2-\underset{\underset{\displaystyle CH_3}{|}}{CH}-CH_2 \sim PIB \sim \overset{\overset{\displaystyle CH_3}{|}}{\underset{\underset{\displaystyle CH_3}{|}}{C}}-C_6H_4-\overset{\overset{\displaystyle CH_3}{|}}{\underset{\underset{\displaystyle CH_3}{|}}{C}} \sim PIB \sim CH_2-\underset{\underset{\displaystyle CH_3}{|}}{CH}-CH_2{-}OH$$

The same chemistry was used for the three-arm star telechelic polyisobutylene synthesis (171). These two functional groups (that is, propenyl and hydroxyl) afford the possibility of almost any kind of functional groups to be introduced at the chain ends of polyisobutylene. By derivation of hydroxyl and propenyl terminated polyisobutylene, telechelics containing carboxylic (191, 192), −SiCl and H−SiH (193, 194), acryloyl and methacryloyl (195), oxycarbonyl (196), amine, cyanato (197), ethynyl, nitrile (198), phenol, and epoxy (199) groups have been synthesized. These materials are useful for a large variety of applications such as chain extension, networks, and block copolymers.

α,ω-Di(acryloyl)poly(THF) and α,ω-di(methacryloyl)poly(THF) were prepared by the polymerization of THF initiated by protonic acids (HSbF$_6$ or CF$_3$SO$_3$H) in the presence of acrylic or methacrylic anhydride as a transfer agent (200). On the basis of the fact that traces of acrylic or methacrylic acids accelerate the polymerization, the following mechanism of reaction was proposed:

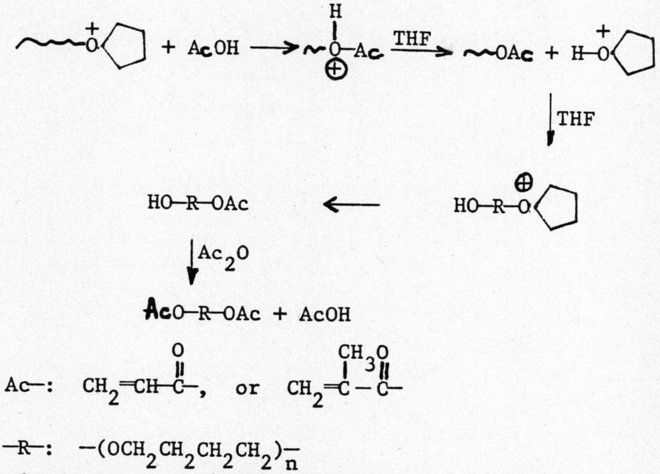

Ac—: $CH_2{=}CH{-}\overset{O}{\underset{\|}{C}}{-}$,   or $CH_2{=}\overset{CH_3}{\underset{|}{C}}{-}\overset{O}{\underset{\|}{C}}{-}$

—R—: $-(OCH_2CH_2CH_2CH_2)_{\overline{n}}$

A telechelic poly(THF) containing azetidinium salts as chain ends was synthesized recently (47). The method used is outlined below:

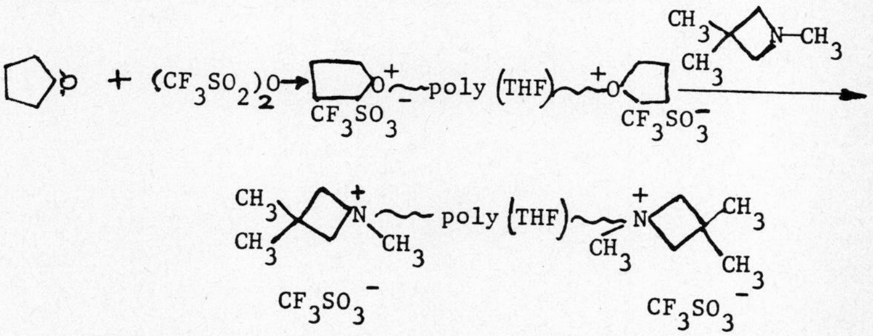

1,3,3-Trimethylazetidine was polymerized by this macroinitiator, and ABA block copolymers were obtained. Reaction with a polyfunctional nucleophile such as diethylenetriamine leads to high molecular weight polymers and eventually cross-linking. The coupling reaction with amino-terminated polybutadiene polymers led to polyblock copolymers (47).

<u>Polymers with One Functional End Group: Macromonomers.</u> The term Macromer was used by Milkovich (201) to describe a type of oligomeric material having a polymerizable group at one end of the molecule. Typical reactive groups include epoxy, acrylate, olefin, and glycol groups. By homopolymerization or copolymerization with a second reactive monomer, they give rise to comblike polymers or graft copolymers.

Interest in this area was aroused after 1980. Three recent reviews summarize the research in this field (202-204). They also present the advantages of this synthetic method in comparison with other techniques for the synthesis of graft copolymers. Among radical, anionic, and cationic types of polymerization, cationic polymerization seems to be the most successful for the synthesis of polymerizable oligomers.

Kennedy and Frisch (205) proposed the synthesis of polymerizable oligomers by cationic polymerization of isobutylene:

Copolymers with butyl acrylate and methyl methacrylate were prepared.

Rempp et al. (203) synthesized polymerizable oligomers of poly(THF) by initiation from an initiator containing a polymerizable double bond (206), as well as by coupling a living poly(THF) with a nucleophile containing a polymerizable group (202, 203).

or

Copolymers with styrene and alkyl methacrylates were synthesized also. The coupling method was initially developed by Asami et al. (200, 210) who succeeded in preparing polymerizable oligomers from living poly(THF) by coupling with vinyl phenolate ($CH_2=CH-C_6H_4-O^-$) or with vinylbenzyl alcoholate ($CH_2=CH-C_6H_4-CH_2O^-$). The homopolymerization and copolymerization of these Macromers were studied (209).

Another Macromer was designed by Goethals and Vlegels (211). The deactivation of the active species (aziridinium ions) of living poly(1-*tert*-butylaziridine) by methacrylic acid led to the corresponding polyamine-methacrylate ester Macromer.

Additional Research Reports

The 6th International Symposium on Cationic Polymerization and Related Processes was held in Ghent, Belgium (August 30 - September 2, 1983). Its Proceedings were published as a book (212).

Recent review articles on the following topics were published: the controversy concerning the cationic ring-opening polymerization of cyclic acetals (213), photoinitiators for cationic polymerization (214), living polymerization and selective dimerization (215), macromonomers (216), and functional polymers and sequential copolymers by carbocationic polymerization (217).

Two special issues containing Kennedy's work on the use of sterically hindered amines in carbocationic polymerization (218) and on quasi-living carbocationic polymerization (219) were also published.

Acknowledgment

The author wishes to express his gratitude to S. Penczek for his careful reading and criticism of this manuscript and to J. P. Kennedy for many discussions. The financial support of the National Science Foundation is gratefully acknowledged. (Grant DMR: 82-13895)

Literature Cited

1.   Penczek, S.; Kubisa, P.; Matyjaszewski, K. Adv. Polym. Sci. 1980, 37, 1. Kubisa, P.; Penczek, S. "Encycl. Polym. Sci. Technol. Suppl."; Wiley: New York, 1977; Vol. 2, p. 161.
2.   Gandini, A.; Cheradame, H. Adv. Polym Sci. 1980, 34/35. 1.
3.   Kennedy, J. P.; Marechal, E. "Carbocationic Polymerization"; Wiley: New York, 1982.
4.   Dreyfuss, P. "Poly(tetrahydrofuran)"; Gordon and Breach: New York, 1982.
5.   Polymer J. 1980, 12, 9.
6.   Kennedy, J. P.; J. Polym. Sci. Polym. Symp. Ed. 1977, 56.
7.   Kern, W.; Sigwalt, P. Makromol. Chem. 1974, 175, 1017.
8.   "Cationic Polymerization and Related Complexes"; Plesch, P. H., Ed.; W. Heffer and Sons: Cambridge, 1973.
9.   Pepper, D. C. Sci. Proc. Roy. Dublin Soc. 1950, 25, 131.

10. Kennedy, J. P. Makromol. Chem., Suppl. 1979, 3, 1.
11. Penczek, S. Makromol. Chem., Suppl. 1979, 3, 17.
12. Szwarc, M. Makromol. Chem., Suppl. 1979, 3, 327.
13. Ledwith, A. Makromol. Chem., Suppl. 1979, 3, 348.
14. Kennedy, J. P. "Cationic Polymerization of Olefins: A Critical Inventory"; Wiley: New York, 1975.
15. Kennedy, J. P. In "Applied Polymer Science"; Craver, J. K.; Tess, R. W., Eds.; American Chemical Society: Washington, D.C., 1975; p. 195.
16. "The Chemistry of Cationic Polymerization"; Plesch, P. H., Ed.; Pergamon Press, 1963.
17. Watson, Bishop "Chemical Essays, 5th Ed."; J. Evans: London, 1789; Vol. III.
18. Dunn, D. J. "Developments in Polymerization"; Howard, R. N., Ed.; Appl. Sci. Publ.: London, 1979; Vol. 1, p. 45.
19. "Ring-Opening Polymerization"; Saegusa, T.; Goethals, E. J., Eds.; ACS SYMPOSIUM SERIES No. 59, American Chemical Society: Washington, D.C., 1977.
20. "Polymerization of Heterocycles"; Penczek, S., Ed.; Pergamon Press: Oxford, 1977.
21. Billingham, N. C. "Developments in Polymerization"; Howard, R. N., Ed.; Appl. Sci. Publ.: London, 1979; p. 47.

22. Ledwith, A.; Sherington, D. C. Adv. Polym. Sci. 1974, 19, 1.
23. Ledwith, A. Pure Appl. Chem. 1979, 51, 159.
24. Dreyfuss, M. P.; Westfahl, J. C.; Dreyfuss, P. Macromolecules 1968, 1, 437.
25. Penczek, S. Makromol. Chem. 1974, 175, 1217.
26. Burgess, F. J.; Cunliffe, A. V.; MacCallum, D. R.; Richards, D. H. Polymer 1977, 18, 719.
27. Ibid., 726.
28. Burgess, F. J.; Cunliffe, A. V.; Richards, D. H.; Thompson, T. Polymer 1978, 19, 334.
29. Richards, D. H.; Thompson, T. Polymer 1979, 20, 1439.
30. Zilliox, J. G.; Reibel, L.; Scheer, M.; Schweickert, J. C.; Franta, E. IUPAC, Int. Symp. Macromol., Mainz, 1979; Preprints, Vol. I, p. 56.
31. Olah, G. A.; Svoboda, J. J. Synthesis 1973, 52.
32. Kubisa, P.; Szymanski, R.; Penczek, IUPAC Int. Symp. Macromol., Strasbourg, 1981; Preprints, Vol. I, p. 256.
33. Stolarczyk, A.; Kubisa, P.; Penczek, S. J. Macromol. Sci., Chem. 1977, A11, 2047.
34. Olah, G. A.; Kuhn, S. J.; Toglyesi, W. S.; Baker, E. B. J. Am. Chem. Soc. 1962, 84, 2733.
35. Franta, E.; Reibel, L.; Lehmann, J.; Penczek,S. J. Polym. Sci. Polym. Symp. 1976, 56, 139.
36. Kubisa, P.; Penczek, S. Makromol. Chem 1978, 179, 445.
37. Yamashita, Y.; Hirota,M.; Nobutoki, K.; Nakamura, Y.; Nirao, A.; Kozawa, S.; Chiba, K.; Matsui, H.; Hattori, G.; Okada, M. J. Polym. Sci., Part B 1970, 8, 481.
38. Meerwin, H. In "Houben-Weyl Methoden der Organikhen Chemie"; Muller, E., Ed.; 4th ed., Vol. VI/3, Stuttgart, George Thieme Verlag, 1965; p. 325.
39. Olah, G. A.; Olah, J. A.; Suoboda, J. J. Synthesis 1973, 490.

40. Szymanski, R.; Wieczorek, H.; Kubisa, P.; Penczek, S. Chem. Commun. 1976, 33.
41. Saegusa, T.; Kimura, Y.; Fujii, H.; Kobayashi, S. Macromolecules 1973, 6, 657.
42. Drijvers, W.; Goethals, E. J. Makromol. Chem. 1971, 148, 311.
43. Penczek, S. Pure Appl. Chem. 1976, 48, 363. Penczek, S. J. Polym. Sci. Polym. Symp. 1980, 67, 149.
44. Saegusa, T.; Kobayashi, S. J. Polym. Sci. Polym. Symp. 1976, 56, 241.
45. Penczek, S.; Matyjaszewski, K. J. Polym. Sci. Polym. Symp. 1976, 56, 255.
46. Nekrasov, A. V.; Pushchaeva, L. M.; Morozova, I. S.; Markevich, Al. Berlin, M. A.; Ponomarenko, A. T.; Enikolopyan, N. S. J. Macromol. Sci. Chem. 1974, A8, 241.
47. Goethals, E. J.; Schacht, E. H.; Bogaert, K. E.; Ali, S. I.; Tezuka, Y. Polym. J. 1980, 12, 571.
48. Saegusa, T. Pure Appl. Chem. 1974, 39, 81.
49. Saegusa, T.; Kobayashi, S. "M. T. P. Int. Rev. Sci.: Phys. Chem. Ser. Two"; Bawn, C. E. H., Ed.; Butterworths, 8, 153 (1975).
50. Saegusa, T. Makromol. Chem. 1974, 175, 1199.
51. Saegusa, T.; Kobayshi, S.; Yamada, A. Makromol. Chem. 1976, 177, 2271.
52. Saegusa, T.; Kobayashi, S. "Encycl. Polym. Sci. Technol., Suppl."; Mark, H. F., Bikales, N. M., Eds.; Wiley: 1976; Vol. 1, p. 220.
53. Smith, S.; Hubin, A. J.; J. Macromol. Sci.-Chem. 1973, A-7, 1399.
54. Smith, S.; Schultz, W. J.; Newmark, R. A. In "Ring-Opening Polymerization"; Saegusa, T.; Goethals, E. J., Eds.; ACS SYMPOSIUM SERIES No. 59, American Chemical Society: Washington, D.C., 1977; p. 13.
55. Crivello, J. V. CHEMTECH 1980, 624.
56. Crivello, J. V.; Lam, H. J. W. Macromolecules 1977, 10, 1307.
57. Crivello, J. V.; Lam, H. J. W. J. Polym. Sci. Symp. 1976, 56, 383.
58. Crivello, J. V.; Lam, H. J. W. J. Polym. Sci. Polym. Chem. Ed. 1979, 17, 977.
59. Crivello, J. V.; Lam, H. J. W. J. Polym. Sci. Polym. Lett. Ed. 1979, 17, 759.
60. Crivello, J. V.; Lam, H. J. W. J. Polym. Sci. Polym. Chem. Ed. 1980, 18, 2677.
61. Ibid., 2697.
62. Ibid., 1047.
63. Ibid., 2877.
64. Ibid., 1021.
65. Ibid., 1059.
66. Ibid., 2441.
67. Crivello, J. V.; Lam. H. J. W. Macromolecules 1981, 14, 1141.
68. Ledwith, A. Polym. Prepr. 1982, 23(1), 323.
69. Ledwith, A. Ann. N.Y. Acad. Sci. 1969, 155(2), 482.
70. Ledwith, A. Acc. Chem. Res. 1972, 5, 133.
71. Ledwith, A. Polymer 1978, 19, 1217.
72. Abdul-Rasoul, F. A. M.; Ledwith, A.; Yagci, Y. Polymer 1978, 19, 1219.

73. Burgess, F. J.; Cunliffe, A. V.; Richards, D. H.; Sherrington,
     D. C. J. Polym. Sci. Polym. Lett. Ed. 1976, 14, 471.
74. Burgess, F. J.; Cunliffe, A. V.; Dawkins, J. V.; Richards, D.
     H. Polymer 1977, 18, 733.
75. Burgess, F. J.; Richards, D. H. Polymer 1976, 17, 1020.
76. Matyjaszewski, K.; Penczek, S. J. Polym. Sci. Polym. Chem. Ed.
     1974, 12, 1905.
77. Matyjaszewski, K.; Brzezinska, K.; Penczek, S. IUPAC Int.
     Symp. Macromol., Strasbourg; Preprints, Vol. I, p. 260, 1981.
78. Lapienis, G.; Penczek, S. Macromolecules 1974, 7, 166.
79. Goethals, E. J.; Drijvers, W. Makromol. Chem. 1973, 165, 329.
80. Goethals, E. J.; Schacht, E. H. J. Polym. Sci. Polym. Lett.
     Ed. 1973, 11, 497.
81. Goethals, E. J. J. Polym. Sci. Polym. Symp. 1976, 56, 271.
82. Saegusa, T.; Matsumoto, S. J. Polym. Sci. Part A-1 1968, 6,
     1559.
83. Brzezinska, K.; Chwialkowska, W.; Kubisa, P.; Matyjaszewski,
     K.; Penczek, S. Makromol. Chem. 1977, 178, 2491.
84. Dreyfuss, P.; Dreyfuss, P. M. Polym. J. 1976, 8, 81.
85. Black, P. E.; Worfold, D. J. Can. J. Chem. 1976, 54, 3325.
86. Bucquoye, M.; Goethals, E. J. Makromol. Chem. 1978, 179, 1681.
87. Goethals, E. J. Adv. Polym. Sci. 1977, 23, 103.
88. Jaacks, V. ADVANCES IN CHEMISTRY SERIES No. 91, American
     Chemical Society: Washington, D.C., 1969; p. 371.
89. Goethals, E. J.; Munir, A.; Bossaer, P. Pure Appl. Chem.
     1981, 53, 1753.
90. Lapienis, G.; Penczek, S. Macromolecules 1977, 10, 1301.
91. Percec, V. Polym. Bull. 1981, 5, 651.
92. Dreyfuss, P.; Dreyfuss, P. M. Adv. Polym. Sci. 1967, 4, 528.
93. Brzezinska, K.; Matyjaszewski, K.; Penczek, S.
     Makromol. Chem. 1978, 179, 2387.
94. Penczek, S.; Kubisa, P. ACS SYMPOSIUM SERIES No. 59, American
     Chemical Society: Washington, D.C., 1977; p. 60.
95. Kubisa, P.; Penczek, S. Makromol. Chem. 1978, 179, 445.
96. Uryn, T.; Ito, K.; Kobayashi, K. I.; Matsuzaki, K. Makromol.
     Chem. 1979, 180, 1509.
97. Schuerch, C. Adv. Polym. Sci. 1972, 10, 173.
98. Schuerch, C. Acc. Chem. Res. 1973, 6, 184.
99. Schuerch, C. "Encycl. Polym. Sci. Technol., Suppl."; Vol. 1,
     1976, p. 510.
100. Schuerch, C. Adv. Carbohydr. Chem. Biochem. 1981, 39, 157.
101. Kosinski, P.; Penczek, S. IUPAC Intern. Symp. Macromol.,
     Strasbourg, 1981, Vol. I, p. 279.
102. Penczek, S. In "Phosphorus Chemistry Directed Towards
     Biology"; Stec, W. J., Ed.; Pergamon Press: Oxford and New
     York, 1980; p. 133.
103. Sumitomo, H.; Okada, M Adv. Polym. Sci. 1978, 28, 47.
104. Yokoyama, Y.; Hall, H. K., Jr.; Adv. Polym. Sci. 1982, 42,
     107.
105. Schulz, R. C.; Albrecht, K.; Thi, Q. V. T.; Nienburg, J.;
     Engel, D. Polym. J. 1980, 12, 639.
106. Schulz, R. C.; Albrecht, K.; Hellermann, W.; Kane, A.; Thi,
     Q. V. T. Pure Appl. Chem. 1981, 53, 1763.
107. Hellermann, W.; Schulz, R. C. Makromol. Chem., Rapid Commun.
     1981, 2, 585.

108. Kops, J.; Hvilsted, S.; Spanggaard, H. Pure Appl. Chem. 1981, 53, 1777.
109. Bailey, W. J.; Sun, R. L.; Katsuki, H.; Endo, T.; Iwama, H.; Tsushima, R.; Saigou, K.; Bitritto, M. M. ACS SYMPOSIUM SERIES No. 59, American Chemical Society: Washington, D.C., 1977; p. 38.
110. Endo, T.; Okawara, M.; Bailey, W. J. Polym. J. 1981, 13, 715.
111. Trathnigy, B.; Hippmann, G. Angew. Makromol. Chem. 1982, 105, 9.
112. Saegusa, T; Kobayashi, T.; Kobayashi, S.; Couchman, S. L.; Vogl, O. Polym. J. 1979, 11, 463.
113. Vogl, O.; Muggel, J.; Bansleben, D. Polym. J. 1980, 12, 677.
114. Okada, M.; Sumimoto, H.; Kakezawa, T. Makromol. Chem. 1972, 162, 285.
115. Dreyfuss, P.; Kennedy, J. P. J. Polym. Sci. Polym. Lett. Ed. 1976, 14, 135. Ibid., 139.
116. Dreyfuss, P.; Kennedy, J. P. J. Polym. Sci. Symp. 1976, 56, 129.
117. Lee, K. I.; Dreyfuss, P. ACS SYMPOSIUM SERIES No. 59, American Chemical Society: Washington, D.C., 1977; p. 24.
118. Adaway, P. D. T; Kennedy, J. P. J. Appl. Polym. Sci., Polym. Symp. 1977, 30, 183.
119. Eckstein, Y.; Dreyfuss, P. J. Polym. Sci. Polym. Chem. Ed. 1980, 18, 1799.
120. Eckstein, Y.; Dreyfuss, P. J. Inorg. Nucl. Chem. 1981, 43, 23.
121. Eckstein, Y.; Dreyfuss, P. J. Polym. Sci. Polym. Chem. Ed. 1979, 17, 4115.
122. Seung, N.; Fetters, L. J.; Dreyfuss, P. IUPAC Int. Symp. Macromol., Strasbourg, 1981; Vol. I, p. 245.
123. Lee, D. P.; Dreyfuss, P. J. Polym. Sci. Polym. Chem. Ed. 1980, 18, 1627.
124. Eckstein, Y.; Lee, D. P.; Quirk, R. P.; Dreyfuss, P. J. Polym. Sci. Polym. Chem. Ed. 1980, 18, 2021.
125. Franta, E.; Taromi, F. A.; Rempp, P. Makromol. Chem. 1976, 177, 2191.
126. Dreyfuss, P.; Kennedy, J. P. J. Appl. Polym. Sci. Symp. 1977, 30, 153.
127. Ibid., 165.
128. Ibid., 179.
129. Saegusa, T.; Matsumoto, S.; Hashimot, Y. Polym. J. 1970, 1, 31.
130. Saegusa, T.; Kobayashi, S.; Yamada, A. Macromolecules 1975, 8, 390.

131. Simionescu, C. I.; Denes, F.; Percec, V.; Totolin, M.; Kennedy, J. P. IUPAC Int. Symp. Macromol., Strasbourg, 1981; Vol. I., p. 298.
132. Saegusa, T.; Yamada, A.; Kobayashi, S. Polym. J. 1978, 11, 53.
133. Trivedi, P. D.; Schulz, D. N. Polym. Bull. 1980, 3, 37.
134. Dondos, A.; Lutz, P.; Reibel, L.; Rempp, P.; Franta, E. Makromol. Chem. 1978, 179, 2549.
135. Cunliffe, A. V.; Hartley, D. B.; Kingston, S. B.; Richards, D. H.; Thompson, D. Polymer 1981, 22, 101.

136. Chwialkowska, W.; Kubisa, P.; Penczek, S. Makromol. Chem. 1982, 183, 753.
137. Litt, M.; Herz, J. Polym. Prepr. 1969, 20(2), 905.
138. Litt, M.; Matsuda, T ADVANCED CHEMISTRY SERIES No. 142; Platzer, N. A. J., Ed.; American Chemical Society: Washington, D.C., 1975; p. 321.
139. Bossar, P. K.; Goethals, E. J.; Hackett, P. J.; Pepper, D. C. Eur. Polym. J. 1977, 13, 489.
140. Saegusa, T.; Ikeda, H. Macromolecules 1973, 6, 805.
141. Percec, V. Polym. Bull. 1981, 5, 643.
142. Percec, V. Polym. Prepr. 1982, 23(1), 301.
143. Percec, V.; Guhaniyogi, S. C.; Kennedy, J. P.; Ivan, B. Polym. Bull. 1982, 8, 25.
144. Seung, S. L. N.; Young, R. N. J. Polym. Sci. Polym. Lett. Ed. 1979, 17, 233.
145. Seung, S. L. N.; Young, R. N. Polym. Bull. 1979, 1, 481.
146. Seung, S. L. N.; Young, R. N. J. Polym. Sci. Polym. Lett. Ed. 1980, 18, 89.
147. Morishima, Y.; Tanaka, T.; Nozakura, S. Polym. Bull. 1981, 5, 19.
148. Abadie, M. J. M.; Schue, F.; Souel, T.; Hartley, D. B.; Richards, D. H. Polymer 1982, 23, 445.
149. Simonds, R. P.; Goethals, E. J.; Spassky, N Makromol. Chem. 1978, 179, 1851.
150. Takahashi, A.; Yamashita, Y. Polym. Prepr. 1974, 15(1), 184.
151. Yamashita, Y.; Nobutoki, K.; Nakamura, Y., Hirota, H. Macromolecules 1971, 4, 548.
152. Richards, D. H.; Kingston, S. B.; Sonel, T. Polymer 1978, 19, 68.
153. Ibid., 806.
154. Berger, G.; Levy, M.; Vofsi, D. J. Polym. Sci., Part B 1966, 4, 183.
155. Yamashita, Y. Adv. Polym. Sci. 1978, 28, 1.
156. Feger, C.; Cantow, H. J. Polym. Bull. 1980, 3, 407.
157. Ibid., 1982, 6, 321.
158. Kennedy, J. P.; Marechal, E. J. Polym. Sci. Macromol. Rev. 1981, 16, 123.
159. Lyerla, J. R.; Yannoni, C. S.; Bruck, D.; Fyfe, C. A. J. Am. Chem. Soc. 1979, 101, 4770.
160. Stannett, V. T. Pure Appl. Chem. 1981, 53, 673.
161. Sawamoto, M.; Higashimura, T. Polym. Prepr. 1979, 20, 727.
162. Sawamoto, M.; Higashimura, T. Macromolecules 1978, 11, 328.
163. Kennedy, J. P. J. Polym. Sci. Symp. 1976, 56, 1.
164. Kennedy, J. P. J. Appl. Polym. Sci. Symp. 1977, 30.
165. Kennedy, J. P. Polym. J. 1980, 12, 609.
166. Kennedy, J. P.; Smith, R. A. Polym. Prepr. 1979, 20, 316.
167. Kennedy, J. P.; Smith, R. A. J. Polym. Sci. Polym. Chem. Ed. 1980, 18, 1523.
168. Wondraczek, R. H.; Kennedy, J. P.; Storey, R. F. J. Polym. Sci. Polym. Chem. Ed. 1982, 20, 43.
169. Chang, V. S. C.; Kennedy, J. P.; Ivan, B. Polym. Bull. 1980, 3, 339.
170. Chang, V. S. C.; Kennedy, J. P. Polym. Bull. 1980, 4, 513.
171. Kennedy, J. P.; Ross, L. R.; Lackey, J. E.; Nuyken, O. Polym. Bull. 1981, 4, 67.

172. Kennedy, J. P.; Ross, L. R.; Nuyken, O. Polym. Bull. 1981, 5, 5.

173. Sawamoto, M.; Kennedy, J. P. Polym. Prepr. 1981, 22(2), 140.

174. Kennedy, J. P.; Chou, R. T. Polym. Prepr. 1979, 20, 306.

175. Guhaniyogi, S. C.; Kennedy, J. P. Polym. Bull. 1981, 4, 267.

176. Manlis, J. M.; Collomb, J.; Gandini, A.; Cheradame, H. Polym. Bull. 1980, 3, 197.

177. Kennedy, J. P.; Smith, R. A. J. Polym. Sci. Polym. Chem. Ed. 1980, 18, 1539.

178. Kennedy, J. P.; Huang, S. Y.; Smith, R. A. J. Macromol. Sci. Chem. 1980, A17, 1085.

179. Rooney, J. M.; Squire, D. R.; Stannett, V. T. J. Polym. Sci. Polym. Chem. Ed. 1976, 14, 1877.

180. Higashimura, T.; Mitsuhashi, M.; Sawamoto, M. Macromolecules 1979, 12, 178.

181. Higashimura, T.; Teranishi, H.; Sawamoto, M. Macromolecules 1980, 12, 393.

182. Giusti, P. Polym. J. 1980, 12, 555.

183. Vidal, A.; Donnet, J. B.; Kennedy, J. P. J. Polym. Sci. Polym. Lett. Ed. 1977, 15, 585.

184. Denes, F.; Percec, V.; Totolin, M.; Kennedy, J. P. Polym. Bull. 1980, 2, 499.

185. Vidal, A.; Guyot, A.; Kennedy, J. P. Polym. Bull. 1980, 2, 315.

186. Vidal, A.; Guyot, A.; Kennedy, J. P. IUPAC Intern. Symp. Macromol., Strasbourg, 1981; Prepr. Vol. I, p. 303.

187. Munir, A.; Goethals, E. J. Macromol. Chem., Rapid Commun. 1981, 2, 693. Munir, A.; Goethals, E. J. IUPAC Int. Symp. Macromol., Strasbourg, 1981; Prepr. Vol. I. 299.

188. Kennedy, J. P. J. Macromol. Sci. Chem. 1979, A13, 695.

189. Kennedy, J. P.; Chang, V. S. C.; Smith, R. A.; Ivan, B. Polym. Bull. 1979, 1, 575.

190. Ivan, B.; Kennedy, J. P.; Chang, V. S. C. J. Polym. Sci. Polym. Chem. Ed. 1980, 18, 3177.

191. Liao, T. P.; Kennedy, J. P. Polym. Bull 1981, 5, 11.

192. Percec, V.; Guhaniyogi, S. C.; Kennedy, J. P. Polym. Bull. 1982, 8, 319.

193. Chang, V. S. C.; Kennedy, J. P. Polym. Bull. 1981, 5, 379.

194. Kennedy, J. P.; Chang, V. S. C.; Guyot, A. Adv. Polym. Sci. 1982, 43, 1.

195. Liao, T. P.; Kennedy, J. P. Polym. Bull. 1981, 6, 135.

196. Wondraczek, R. H.; Kennedy, J. P. Polym. Bull. 1981, 4, 445.

197. Percec, V.; Guhaniyogi, S. C.; Kennedy, J. P. Polym. Bull. 1983, 9, 27.

198. Percec, V.; Kennedy, J. P. Polym. Bull. 1983, 9, 570. Ibid., 10, 31.

199. Kennedy, J. P.; Guhaniyogi, S. C.; Percec, V. Polym. Bull. 1982, 8, 563; ibidem, Polym. Bull. 1982, 8, 571.

200. Kress, H. J.; Heitz, W. Makromol. Chem., Rapid Commun. 1981, 2, 427.

201. Milkovich, R. Polym. Prepr. 1980, 21(1), 40.

202. Yamashita, Y. J. Appl. Polym. Sci., Symp. 1981, 36, 193.

203. Rempp, P.; Masson, P.; Vargas, J. S.; Franta, E. Plaste Kautsch. 1981, 28, 365.

204. Saegusa, T. Topics Curr. Chem. 1982, 100, 75.

205. Kennedy, J. P.; Frisch, K. C., Jr.; IUPAC Int. Symp. Macromol., Florence, 1980; Preprints, Vol. 2, p. 162.
206. Sierra-Vargas, J.; Zilliox, J. G.; Rempp, P.; Franta, E. Polym. Bull. 1980, 3, 83.
207. Sierra-Vargas, J.; Franta, E.; Rempp, P. Makromol. Chem. 1981, 182, 2603.
208. Masson, R.; Sierra-Vargas, J.; Franta, E.; Rempp, P. IUPAC Int. Symp. Macromol., Strasbourg, 1981; Preprints, Vol. I, p. 235.
209. Asami, R.; Takaki, T.; Kita, K.; Asakura, E. Polym. Bull. 1980, 2, 713.
210. Asami, R.; Takaki, M. IUPAC Internat. Symp. Macromol., Strasbourg, 1981; Preprints, Vol. I, p. 240.
211. Goethals, E. J. Vlegels, M. A. Polym. Bull. 1981, 4, 521.
212. "Cationic Polymerization and Related Processes"; Goethals, E. J., Ed.; Academic Press: London, 1984.
213. Szymanski, R.; Kubisa, P.; Penczek, S. Macromolecules 1983, 16, 1000.
214. Crivello, J. V. Adv. Polym. Sci. 1984, 62, 1.
215. Higashimura, T.; Sawamoto, M. Adv. Polym. Sci. 1984, 62, 49.
216. Rempp, P.; Franta, E. Adv. Polym. Sci. 1984, 58, 1.
217. Kennedy, J. P. Makromol. Chem., Suppl. 1984, 7, 171.
218. Kennedy, J. P., et al. J. Macromol. Sci. Chem. 1982, A18(1), 3.
219. Kennedy, J. P., et al. J. Macromol. Sci. Chem. 1982-83, A18(9), 1185.

Emulsion Polymerization

GARY W. POEHLEIN

School of Chemical Engineering, Georgia Institute of Technology, Atlanta, GA 30332

Emulsion polymerization is the process of choice for the commercial production of many polymers used for coating and adhesive applications, especially for those products that can be used in latex form. Emulsion polymerization uses free-radical polymerization mechanisms with unsaturated monomers. The heterogeneous nature of the reaction mixture, however, has a significant influence on the chemical and physical reaction mechanisms and on the nature of the final product.

A simple recipe for emulsion polymerization would be comprised of hydrophobic monomers (40 to 60 volume percent), a continuous aqueous phase (40 to 60 volume percent), a water-soluble initiator, and an emulsifier or stabilizer. Other minor ingredients such as chain transfer agents, inhibitors or retarders, and buffers may also be present. Emulsion polymerization is characterized by a large number of reaction sites (the polymer particles) that contain a small number of free radicals. These free radicals are isolated because of the water phase between the particles. Typical polymer particle diameters would be in the range of 50 to 500 nm, although new technology involving high-swelling particles can produce particles as large as 50 μm (50,000 nm).

A considerable amount of work has been published during the past 20 years on a wide variety of emulsion polymerization and latex problems. A list of 11, mostly recent, general reference books is included at the end of this chapter. Areas in which significant advances have been reported include reaction mechanisms and kinetics, latex characterization and analysis, copolymerization and particle morphology control, reactor mathematical modeling, control of adsorbed and bound surface groups, particle size control reactor parameters. Readers who are interested in a more in-depth study of emulsion polymerization will find extensive literature sources.

Reaction Ingredients and Mechanisms

The colloidal nature of the reaction media has a significant influence on the course of an emulsion polymerization reaction. A number of distinct phases exist during different intervals of a batch reaction. Chemical and physical phenomena within these phases and at the interfaces can be important in determining reaction kinetics and the properties of the latex product.

At the beginning of a batch reaction the continuous aqueous phase contains the water-soluble initiator, emulsifiers, and buffers. Common ionic emulsifiers will be present as molecularly dissolved electrolytes, as surface active agents at the various interfaces, and as molecular clusters called micelles. The monomer will be in three different locations. A small amount will be dissolved in the water phase. Some will be solublized within the emulsifier micelles. The bulk of the monomer, however, will exist in the relatively large (ca. 5 μm) monomer droplets. Any oil-soluble components such as chain transfer agents will be distributed with the monomer if the water solubility is sufficient to permit transport from the droplets.

The polymer particles that are formed after the reaction begins represent another distinct phase. These particles will be swollen by monomer and other oil-soluble ingredients. When relatively water-insoluble monomers are used, the particle formation period, called Interval 1, extends from the beginning of the polymerization to the point at which the system is not capable of stabilizing any new particles. At this point, the free emulsifier is completely adsorbed on the surface of the polymer particles, and any micelles initially present will have disappeared. The polymer particles are swollen with monomer, but in most cases the bulk of the monomer still remains in the droplets. Homogeneous nucleation of polymer particles can be significant throughout the conversion range with monomers that are more water soluble such as vinyl acetate. These new particles may not be stable, and they can flocculate onto the larger particles that were formed earlier.

The polymer particles continue to grow during Interval 2 with monomer being supplied by diffusion from the droplets through the aqueous phase. Interval 3 begins when the monomer droplets disappear. The monomer in the polymer particles continues to polymerize during Interval 3, and the particle interior becomes more viscous with some of the reactions becoming diffusion controlled.

Most of the important chemical and physical phenomena that occur in emulsion polymerization are listed and discussed briefly in the remainder of this section.

Aqueous Phase Phenomena. Although the continuous phase is not normally a locus for significant conversion of monomer to polymer, a number of important reaction phenomena can take place in this phase. These reaction components are the following:

1. Free-radical generation. Water-soluble initiators such as the persulfates and various redox systems are used to generate free radicals.
2. Radical propagation. The free radicals formed in the aqueous phase are hydrophilic and usually charged. Thus, they are very likely to react with monomer that is dissolved in the aqueous phase before entering particles, micelles, or droplets.
3. Particle nucleation. Polymer particles can be formed by several mechanisms. Homogeneous nucleation, a term used by Fitch and coworkers (1), can occur in the water phase by precipitation of the growing oligomeric radicals.
4. Radical termination. Normal termination reactions would be expected to take place in the continuous phase. These reactions will not be the dominant method of radical termination, but water-phase termination can be quite important if water-soluble monomers are used. Water-soluble polymer can be formed with such monomers even though most of the monomer will copolymerize with the hydrophobic monomer in the particles.

Monomer Droplets. The monomer droplets serve primarily as reservoirs that supply monomer to the reaction sites in the polymer particles. These droplets can also contain a variety of other oil-soluble ingredients including dissolved polymer, chain transfer agents, and in unusual cases oil-soluble initiator. The monomer and other ingredients, if they have the requisite water solubility, are transported to the primary polymerization locus in the polymer particles. Reaction phenomena that can occur in the monomer droplets include the following:

1. Polymerization (Propagation and Termination). The number of particles formed from standard recipes is considerably greater (usually by two to four orders of magnitude) than the number of monomer droplets present at the beginning of the reaction. Thus, polymerization within monomer droplets is usually not considered to be significant. Ugelstad, ElAasser, and Vanderhoff (2), however, demonstrated that polymerization in monomer droplets can be significant, even dominant, if the droplets can be made small.
2. Initiation. Water-soluble initiators are normally used in emulsion polymerization, and droplet initiation can only take place when a waterborne oligomer diffuses into the monomer droplet. Although such diffusion does take place, in most emulsion polymerization systems the bulk of the initiation and propagation occurs in the particles. Oil-soluble initiators

can be used in systems that generate a large number of small monomer drops (2). In such cases the dominant initiation reaction can take place in the droplets.

3. Transport size reduction. As mentioned earlier, the monomer droplets serve as a reservoir for supplying oil-soluble components to the reaction site in the particles. Thermodynamic driving forces will cause diffusion of such components from the droplets to the particles. Such transport will take place even if the droplets contain polymer or other water-insoluble components. In such cases, however, the diffusion transport will stop before the droplets disappear, and the droplets, greatly reduced in size, will be a part of the final particle population.

Polymer Particles. The polymer particles are almost always the dominant site for polymerization. The following phenomena play important roles:

1. Free-radical and reagent absorption. Free radicals, monomers, and other reagents are transported into the polymer particles. The free radicals are likely to be oligomers because a hydrophilic ion-radical would remain in the aqueous phase. Monomers and other reagents can diffuse from the monomer droplets to the particles if they possess adequate water solubility. Dissolved polymer and other water-insoluble ingredients would remain in the droplets.

2. Radical propagation. Free radicals within the polymer particles will react with monomer until the propagation reaction is stopped by transfer or termination reactions or until the monomer supply is exhausted. Free radicals that contain a hydrophilic end group may have a somewhat reduced mobility because the end group will tend to remain at the particle surface.

3. Radical transfer. Free-radical transfer reactions with monomer, polymer, and added transfer agents can take place in the particles. The polymer transfer reaction will be more important in emulsion polymerization because of the high concentration of polymer in the particles.

4. Radical desorption. Data for a number of experimental studies have been modeled by a kinetic scheme that includes desorption of free radicals. Presumably, radical desorption follows a radical transfer reaction. The mobile free radical could possibly cross the particle-water interface into the water phase. Nomura (3) has published a recent review paper on radical desorption.

5. Radical termination. Free-radical termination reactions are very fast reactions. The combination of reaction speed and the small reactor volume (i.e., the polymer particle) alters the kinetic model in some cases. The simplest model (Smith-Ewart Case 2) is based on the assumption that instant termination occurs when a free radical enters a particle that already contains an active radical. As the particles become larger and/or the radical mobility decreases because of the gel effect, the termination rate becomes slower.

Monomer-Swollen Emulsifier Micelles. The role of micelles in emulsion polymerization has been extensively discussed. The original work of Harkins (4) and Smith and Ewart (5) treated micelles as sites for particle nucleation. They proposed that particle nucleation occurs when a free radical enters a monomer-swollen micelle and begins polymerization. As the particles grow the micelles disband to provide emulsifier for the new organic surface.

Other workers suggest that the micelles simply exist as reservoirs of emulsifier to stabilize particles nucleated in the aqueous phase. Roe (7) demonstrated that the Smith-Ewart equations can be derived without evoking the concept of a micelle. Perhaps each of these concepts is valid under the proper reaction conditions.

Summary. All of the phases and the physical and chemical mechanisms discussed in this section are important during the course of an emulsion polymerization reaction. They influence the reaction kinetics and the properties of the latex produced. Not all of the phenomena that can occur are understood in a quantitative manner. Nevertheless, considerable advances have been made in the fundamental understanding and the commercial exploitation of emulsion polymerization processes. The remainder of this chapter will focus on reactor types and reaction kinetics.

Types of Reactor Processes

Batch Reactors. Polymer latexes are produced in a wide variety of reactors. The bottle polymerizer was employed for early product development studies, and, in fact, such equipment is still used widely today. Bottles are partially filled with the recipe ingredients, attached to a rotating shaft that is immersed in a temperature-controlled bath, and allowed to react for a fixed time. The latex is then removed for evaluation. The advantage of a bottle polymerizer is that a large number of recipe variations can be run simultaneously. Some polymerizers will hold more than 100 bottles.

Some early batch polymerization reactors were built on rotating shafts to copy the action of bottle polymerizers. These reactors were expensive and difficult to maintain. They were replaced by standard stirred vessels which are commonly used today. Typical batch reactors contain an agitator that is mounted in the center of the reactor top. The reactors are often glass lined and contain one or more baffles to enhance mixing. Heat removal is accomplished by circulating a coolant through the reactor jacket.

Numerous reactor design variations have been employed in commercial processes. Polished stainless steel reactors have been used in place of glass-line tanks for some systems. The smooth surfaces (glass or polished metal) are desirable to minimize surface fouling. The use of stainless steel increases heat transfer rates and reduces maintenance costs. The stainless surface may, however, be more prone to fouling.

The heat of polymerization can be removed by a number of techniques. Cooled reactor jackets are most common, but internal cooling surfaces in the form of coils or pipe baffles are also

used. Shell-and-tube heat exchangers can be used in a circulation
loop external to the reactor. Reflux condensers are another
alternative. These forms of cooling, however, add heat transfer
surface that may become fouled and require frequent cleaning.

When stirred-tank reactors are operated in the batch mode, all
ingredients are added at or near the beginning of the reaction
cycle, the reaction is allowed to proceed to a desired end point,
and the product latex is removed for further processing. Strict
batch operation has a number of disadvantages. First, the heat
load on the cooling system can be very nonuniform. The production
rates from such reactors can be limited by the capability of the
heat removal system during the peak in the exotherm. The use of
mixed initiator systems (fast and slow) and the continuous addition
of a fast initiator are two ways of trying to deal with this
problem.

A second possible problem with batch reactors is composition
drift of copolymer systems. As with bulk, solution, and suspension
systems, the more reactive monomer polymerizes first, and the least
reactive polymerizes last. Two additional factors must be
considered in emulsion polymerization. First, the water solubilities of the monomers can influence the course of the
polymerization because of reaction in the water phase to produce
copolymer oligomers or even water-soluble polymer. These molecules
can be rich in the water-soluble monomer even if its reactivity is
relatively low. Second, the high degree of subdivision achieved by
producing small polymer particles can lead to phase domains that
are smaller than those in copolymer produced by other processes.

Latex produced from the more water-insoluble monomers in a
batch reactor normally would have a relatively narrow particle size
distribution (PSD). Interval 1, the particle nucleation part of
the reaction, is usually completed early in the polymerization
cycle, and thus all particles in the final latex would have about
the same age and the same size. Several factors can counter this
normal trend, however. A low initiation rate will extend Interval
1 and broaden the PSD. Particle nucleation later in the reaction
can also generate broader size distributions. Several factors can
lead to a second nucleation. Water-soluble initiators generate
surface active oligomers that add to the stabilizing capability of
the system. Any limited flocculation within the system would
reduce surface area and possibly free emulsifier for stabilizing
new particles. In addition, the total interfacial area decreases
during Interval 3 because the density of polymer is greater than
monomer. A second-stage nucleation can be caused by any one or a
combination of the above factors.

<u>Semibatch Reactors</u>. Part of the recipe ingredients are withheld
from the initial charge in semibatch (sometimes called semicontinuous) operation. These ingredients are added later in a
programmed manner to control the course of the reaction and to
produce a desired product. One reason for using semibatch
operation is to control heat release and/or the rate of polymerization. This control is most commonly accomplished by withholding
part of the monomer and feeding it at a controlled rate later in
the reaction cycle. Such reactions operate in a monomer-starved
condition, and branching (polymer transfer) mechanisms can be more

significant. Likewise, the high polymer concentration in the particles can cause the termination reaction to be diffusion controlled. If heat removal is a problem the monomer feed rate can control the reaction so that desired temperatures can be maintained. If heat removal is not a problem one may still choose to operate the monomer-starved regime because the diffusion-limited termination step can cause higher reaction rates and hence increased reactor productivity. A second, but less common, method of reaction rate control is by the programmed addition of a highly reactive initiation system.

A second motivation for using a monomer-feed semibatch procedure is to control copolymer composition and/or particle morphology. Delayed feed of part of the more reactive monomer can be used to eliminate or reduce the composition drift of the copolymer. The delayed feed of a comonomer mixture when the reactor is operated in the monomer-starved regime can also be used to prevent copolymer composition drift. Such operations will produce polymer particles with more uniform morphology. Different monomer addition schemes can be employed to control nonuniform particle morphology (see papers by Bassett et al. in General References 7 and 9).

A third motivation for delayed monomer addition is to produce high-solids latexes. The use of semibatch operation permits one to pass through the flocculation-sensitive parts of the reaction cycle and then to build the solids level near the end of the polymerization. Sometimes the addition of more emulsifier is used in the production of high-solids latexes.

The nature of the PSD can also be controlled by semibatch operation. If a narrow PSD is desired the emulsifier and initiator components are charged to yield a very short particle-nucleation period. Narrowly distributed seed latexes can also be used for this purpose. In such systems the age distribution is narrow and the size distribution follows.

If the emulsifier feed and initiation system are formulated to yield a long nucleation period, a relatively broad PSD latex can be produced. In extreme cases of delayed-emulsifier feed, multiple nucleation periods will result, and even broader PSDs can be produced. Broad distributions can be an asset when high-solids, low-visibility products are desired.

A semibatch system will be influenced differently by the presence of inhibitor. If inhibitor is present in the recipe ingredients of a batch reactor it will delay the start of polymerization, after which the reaction will proceed in a normal manner. Inhibitor in the delayed feed stream(s) to a semibatch system will reduce the effective rate of initiation. This reduction may require the use of more initiator, and because the inhibitor reacts rapidly, the polymerization rate may increase dramatically when the delayed-feed part of the cycle is finished.

Continuous Reactors. A variety of continuous reactor systems are used commercially, but the most common are comprised of a number of stirred-tank reactors (CSTR) connected in series. Operation normally involves pumping all ingredients into the first CSTR and removing the partially converted latex from the final reactor. Intermediate feed streams can also be employed. Detailed reviews of emulsion polymerization in continuous systems are given by Poehlein et al. (8, General References 2, 8, and 9). The most common differences between continuous and batch systems will be discussed in the remainder of this section.

The differences between a single CSTR and a batch reactor are similar to those between semibatch and batch reactors, except that they are usually more pronounced. The addition of more reactors to a series system tends to reduce some of the observed performance differences. A typical example of different behavior is the heat release profile. An advantage often cited for continuous reactor systems is a constant heat load with fully used reactor volume. Batch reactors are not usually operated full, and the heat load is nonuniform. In addition, portions of the batch reaction cycle are devoted to charging and emptying the reactor and sometimes for heating the reagents to polymerization temperature. Thus, the production rate per unit volume can be higher in a continuous system.

Uniform product quality is also an advantage claimed for continuous reactors. If a continuous reactor can be controlled at a desired steady state, product quality transients should be considerably less than typical batch-to-batch variations. Start-up transients and shut-down procedures can produce off-spec product; hence, frequent start-ups and/or product changes represent a real problem for continuous systems.

A single CSTR is a valuable tool for the study of polymerization kinetics but not for commercial production. As mentioned earlier, systems comprised of a series of CSTRs are most common. Some of the early synthetic rubber processes contain as many as 15 CSTRs. More recent systems are comprised of three to five reactors. Not all reactors need to be the same size. In fact there are substantial reasons for using reactors of different sizes.

Nearly all particles are likely to be produced in the first reactor. The size of this reactor will influence the number of particles formed. If a maximum number of particles is desired this first reactor will be operated at a relatively small mean residence time and thus will be smaller than the other reactors in the system.

Another reason for using different reactor sizes along the CSTR train is the variation of polymerization rate with monomer conversion. This factor is not a major consideration if the final conversion is modest as in the case of styrene–butadiene rubber (SBR) processes. Normal exit conversions are 55 to 65% in such systems, and the residual monomer is recovered and recycled. If a very high conversion is desired one must deal with the problem that the polymerization rate is low at high conversions. The final reactor in the series needs to be very large if the desired conversion approaches 100%. Likewise, batch reaction cycle times become large if high conversions are desired.

High-conversion continuous processes will require large reactors near the end of the CSTR series. In fact it may be advantageous to permit the reaction to continue in the product storage tanks in high-conversion processes. This condition will reduce the level of residual monomer in the end products or in the downstream processing steps.

Continuous processes involving tubular reactors have been reported in the literature (General References 1, 2, 7, and 9). Continuous tubular reactors have been used in three ways. Gonzalez (9) used a tubular prereactor to feed a CSTR system. The tubular prereactor served as a particle nucleation system and thus solved the problem of conversion oscillations often observed in CSTR systems. A tubular prereactor also can be used to generate a higher particle concentration than would be produced with the same recipe in a CSTR system.

Emulsion polymerization reactions have also been studied in reactors consisting only of tubes. Such reactors offer the potential advantage of a large area for heat transfer per unit volume and hence a high polymerization rate. One potential problem with tubular reactors, namely plugging, has discouraged commercial use. A number of studies have been reported on once-through continuous tubular reactors but commercial reactors of this type have not been publicized.

A continuous tubular-loop process has been patented (10) and used for relatively small-scale production. The loop process consists of a tube-pump system in which the rate of latex circulation around the tube loop is considerably greater than the throughput rate. Thus, the distribution of residence times should be nearly the same as that of a single CSTR.

The PSD of a latex is strongly related to the particle age distribution. Thus, one would expect latexes produced in continuous systems to have size characteristics different from batch products. The most extreme differences are seen for a single CSTR. The particle age distribution in the latex product from a CSTR is given by Equation 1.

$$f(t) = \frac{1}{\Theta} e^{-t/\Theta}$$

(1)

where t is particle age and Θ is the reactor mean residence time. This distribution is very broad in comparison to the narrow age distribution of particles produced in a batch reactor. The distribution of particle ages and the corresponding PSDs become narrower as more CSTRs are connected in series. In fact rather narrow-distribution SBR latex can be produced in the commercial systems containing 12–15 reactors in series.

The response of a CSTR system to inhibitors in the feed streams is, in some respects, similar to a semibatch system. Because inhibitor enters with the feed stream, the rate of initiation is reduced in proportion to the inhibitor flow. In extreme cases, the flow of inhibitor may be sufficient to prevent any initiation in the first reactor. When this happens the particle nucleation phenomena is shifted to the second tank, and serious control problems can be experienced. Increased initiator concentrations can be used to overcome high inhibitor concentrations, but such a course of action can produce initiation rates that are considerably higher in the downstream reactors.

Continuous reactors comprised of a CSTR train are often operated with a single feed location in the first reactor. The use of intermediate feed locations can be advantageous for several reasons. First, if the conversion in the first reactor is low or

modest, as is often the case, feeding the entire monomer charge
through the first reactor is an inefficient use of reactor volume.
Feeding part of the monomer downstream can increase process
productivity and perhaps even reduce wall fouling.

A second reason for intermediate monomer feed locations is to
control copolymer composition and particle morphology. Unlike the
batch reactor in which the more reactive monomer reacts first, the
copolymer produced in a CSTR should have uniform composition. If
all monomers are fed to the first reactor, however, the polymer
formed in the downstream reactors will contain less of the most
reactive monomer. Thus, the use of intermediate feed locations in
a CSTR system is analogous to the programmed addition of reactive
monomers in semibatch reactors. The location and rates of the
various monomer streams will influence copolymer composition and
particle morphology.

A third reason for feeding some of the monomer to downstream
reactors is to build the solids level of the product. This
increase in solids can occur without the flocculation and rheology
problems that might exist if the organic phase concentration were
high in the early reactors.

Other recipe ingredients such as emulsifiers and chain transfer
agents could be subdivided and introduced to the reactor system in
several locations. Emulsifiers and/or stabilizers may be necessary
to produce a stable effluent. Chain transfer agents that are used
to limit branching reactions may be of particular importance in the
high-conversion end of the reaction system. Likewise, initiator
systems that do not have a long half-life will tend to decompose in
the first few reactors, and downstream additions may be necessary
to achieve reasonable polymerization rates.

The methods used for introducing feed streams into continuous
reactors can be quite important. All ingredients are charged and
mixed before the latex is formed in most batch reactor processes.
The major purposes of mixing after the reaction begins are to
facilitate heat removal through the cooling surface and to maintain
mass transfer from the monomer phase to the polymer particles.
With a CSTR reaction system, however, the feed streams are added to
partially converted latexes, and other factors need to be
considered.

Initiator streams are normally electrolyte solutions that can
cause flocculation. These streams should be as dilute as feasible
and added at a location where rapid mixing takes place. Emulsifier
feed streams may cause local nucleation in the reactor if they are
not mixed properly. Monomer additions can also be a problem if the
dispersion is not adequate to provide sufficient mass transfer
rates. This situation can be especially important in systems
containing gaseous monomers.

In summary, continuous reactions will be used with increased
frequency as production requirements grow and as design procedures
improve. If continuous systems are to be employed in a commercial
process, product development and pilot plant studies should use
small-scale continuous systems early in the development process.
Such studies will substantially increase the probability of a
successful commercial process.

Emulsion Polymerization Kinetics

The kinetics of emulsion polymerization reactions are complex because of the numerous chemical and physical phenomena that can occur in the multicomponent, multiphase mixture. A large amount of literature exists on kinetics problems. The general references listed at the end of this chapter contain many important papers. The review paper by Ugelstad and Hansen (11) is a comprehensive treatment of batch kinetics. The purpose of the remainder of this chapter is to present the general kinetics problems and some of the published results. The reader should use the references cited earlier for a more detailed study.

Kinetics models are useful for designing commercial reactors and for studying the fundamental mechanisms of the important reactions. The free-radical polymerization that takes place in emulsion systems is characterized by three main reactions: initiation, propagation, and termination. Various radical transfer reactions can also be important. The rate of polymerization for bulk, solution, and suspension processes can be expressed as shown by Equation 2:

$$R_p = k_p \, [M][R\cdot] \tag{2}$$

where k_p is the propagation rate constant, $[M]$ is the monomer concentration, and $[R\cdot]$ is the free-radical concentration.

In bulk, solution, and suspension polymerization, the problem of determining $[R\cdot]$ is handled by assuming a steady-state (actually slowly changing) radical concentration and thus equating the rates of initiation and termination.

$$R_i = k_t[R\cdot]^2 \tag{3}$$

where R_i is the rate of initiation, and k_t is the termination rate constant. By combining Equations 2 and 3 one obtains

$$R_p = \frac{k_p}{k_t^{\frac{1}{2}}} \, R_i^{\frac{1}{2}} \, [M] \tag{4}$$

The major problem associated with the use of Equation 4 for reactor design calculations stems from the fact that $(k_p/k_t^{\frac{1}{2}})$ varies with monomer conversion, sometimes by several orders of magnitude.

The free radicals in emulsion polymerization are isolated in the polymer particles, and $[R\cdot]$ is expressed by Equation 5:

$$[R\cdot] = \langle n \rangle \, N/N_A \tag{5}$$

where $\langle n \rangle$ is the average number of free radicals per particle, N is the particle concentration (usually expressed as the number of particles per liter of aqueous phase), and N_A is the Avogadro number. Note that $\bar{n}$ is the average number of free radicals per particle for a monodisperse system. $\langle n \rangle$ represents the average of $\bar{n}$ over the latex PSD.

Combining Equations 2 and 5 gives the following relationship for the rate of emulsion polymerization:

$$R_p = k_p \ [M]_p \left(\frac{\langle \bar{n} \rangle N}{N_A} \right) \tag{6}$$

where the subscript "p" on the monomer concentration indicates the concentration of monomer in the polymer particles that are considered the dominant site for conversion of monomer to polymer. Note that R_p in Equation 6 will also be based on a unit volume of aqueous phase.

Values for the propagation rate constant can be determined from bulk or solution experiments. Values of k_p have been published for a wide variety of monomers as a function of temperature. With standard emulsion polymerization recipes the value of $[M]_p$ is determined from equilibrium swelling measurements if a free monomer phase is present and by a mass balance if all the monomer is in the polymer particles. One normally assumes that $[M]_p$ is not dependent on particle size in latexes comprised of different-sized particles. This assumption will be questionable in some systems, especially those involving high-swelling particles.

By assuming the $[M]_p$ can be determined by published methods, the problem of computing R_p is reduced to the prediction of $\langle \bar{n} \rangle$ and N. The value of N can be known if a latex seed is used in the reactor. Seeds are used in many scientific studies because the problem of predicting N is eliminated. If a seed is not employed N must be predicted by published correlations or measured. Smith and Ewart (6) published an early theory on particle nucleation that resulted in the relationship shown in Equation 7:

$$N = k \ R_i^{\ 0.4} \ [S]^{0.6} \tag{7}$$

where [S] is the emulsifier concentration, and k is a constant that depends on the assumption used in the model. Equation 7 gives a reasonable prediction of N for styrene polymerizations with standard emulsifiers, but is not adequate for many systems.

More detailed and complete theories on particle nucleation have been published by Fitch et al. (General References 3 and 4) and by Hansen and Ugelstad (12). These publications consider several mechanisms for particle nucleation, and they present mathematical models that account for these various mechanisms. The present state of the art, however, will not permit one to compute N from a knowledge of the recipe ingredients and reaction conditions, except for special cases. Thus, most product and process development work should probably include the measurement of N as a function of the important variables.

One interesting comparison between a CSTR and a batch reactor is, however, in the prediction of N. If the Smith–Ewart concepts are applied to a single CSTR, the particle concentration prediction is given by Equation 8:

$$N = k' \ [S]^{1.0} \ \Theta^{-0.67} \tag{8}$$

where k' is a constant, and Θ is the mean residence time of the reactor. The equation is quite different from Equation 7. Hence, one should expect particle nucleation models for batch and CSTR systems to be quite different, even for identical recipes and reaction mechanisms. Such differences are one reason for using continuous research reactors early in the product-process development program. The use of batch data to design a continuous process is a high-risk venture.

The second part of the R_p problem, the determination of $\langle \bar{n} \rangle$ is the subject of numerous papers (5,10,12,13). Most of these papers are concerned with obtaining solutions to the Smith-Ewart recursion relationship given by Equation 9:

$$\frac{dN_n}{dt} = \left(\frac{\rho_A}{N}\right)(N_{n-1}-N_n) + k_d \left\{(n+1)N_{n+1}-nN_n\right\}$$

$$+\left(\frac{k_t}{v}\right)\left\{(n+2)\ (n+1)N_{n+2} -n(n-1)N_n\right\} = 0 \qquad (9)$$

where ρ_A is the total rate of radical absorption by the particles, N_n is the number of particles that contain "n" free radicals, k_d is a rate constant for desorption of radicals from the particles, and v is the volume of the monomer-swollen particles.

Smith and Ewart (5) obtained solutions to Equation 9 for three special cases: $\bar{n} \ll 1.0$, $\bar{n} = 0.5$, and $\bar{n} \gg 1.0$. Case 2 ($\bar{n} = 0.5$) has been the most widely publicized solution. Data for a broad range of styrene emulsion polymerization experiments are consistent with this model. More general solutions for Equation 9 were published by Stockmayer (13), O'Toole (14), and Ugelstad et al. (15). A relationship for the average number of radicals per particle in a monodisperse system is given by Equation 10:

$$\bar{n} = \frac{a}{4}\ \frac{I_m(a)}{I_{m-1}(a)} \qquad (10)$$

where the I's are modified Bessel functions of the first kind, m = $k_d v/k_t$, a = $\sqrt{8\alpha}$, and $\alpha = \rho_A v/Nk_t$.

The parameter α cannot be directly evaluated because ρ_A, the radical absorption rate, is dependent on initiation rate, radical desorption from the particles, and water-phase termination. Ugelstad et al. (15) derived the following equation relating these phenomena:

$$\alpha = \alpha' + \bar{mn} - Y\alpha^2 \qquad (11)$$

where $\alpha' = \rho_i v/Nk_t$, ρ_i is the rate of initiation, Y = $2Nk_t k_{tw}/k_a^2 v$, k_{tw} is the water-phase termination constant, and k_a is a radical absorption coefficient. Numerical computation schemes were used to solve Equations 10 and 11 to yield results such as those shown in Figures 1 and 2.

The various Smith-Ewart cases can be identified on these graphs. The horizontal line at $\bar{n} = 0.5$ with m = 0 would be Case 2. The lower left area where $\bar{n}$ is small would correspond to Case 1, and the curves in the upper right side ($\bar{n} > 0.5$) would approach Case 3 or bulk kinetics.

If all the relevant parameters are known (α', m, Y) one can use graphs such as Figures 1 and 2 to determine $\bar{n}$. However, two limitations must be considered. First, as mentioned earlier, the solution is for a monodisperse latex. If a broad size distribution exists, one must account for differences in free-radical transport among the particles. If these differences are accounted for, one obtains $\langle\bar{n}\rangle$ that is an average of $\bar{n}$ over the distribution.

The second limitation stems from the fact that $\bar{n}$ can change with time during the course of a polymerization reaction because α', m, or Y can change. Thus, the course of a reaction will trace a curve for n in the space [α', m, Y].

Napper, Gilbert et al. (16-19) have obtained solutions for the system of equations represented by Equation 9 without assuming that $dN_n/dt = 0.0$. They have, however, neglected re-entry of free radicals that have desorbed from the particles. Among other things, they conclude that water-phase termination can be important, that is, Y > 0.0.

Dubner, Poehlein, and Lee (20) developed a steady-state kinetics model for a single, seed-fed CSTR on the basis of the O'Toole-Ugelstad concepts. Dimensionless groups analogous to α', m, and Y are α'_c, γ, and Y_c. These groups are defined in terms of v_s, the size of the seed latex particles, instead of the size of the polymerizing particles. An additional dimensionless group, $\beta = v_s/\Theta K_1[M]_p$, is needed to include the new parameter Θ, the CSTR mean residence time. K_1 is a particle growth parameter.

Figure 3 is a typical result of Dubner's work. The plot of $\langle\bar{n}\rangle$ versus α'_c includes consideration for the variation of $\bar{n}$ with particle size in the reacting mixture. Figure 3 is analogous to the single curve for m = 0 in Figure 1. The parametric curves result because of variations in β. As β decreases (higher Θ values) deviations from S-E Case 2 behavior occurs at smaller values of α'_c.

Figure 4 shows the influence of radical desorption from particles (the parameter γ) on $\langle\bar{n}\rangle$. The shape of the curves are, as expected, quite similar to those reported by Ugelstad et al. In this case two parameters, β and Y_c, are fixed. Dubner et al. also report that the radical desorption rate (as accounted for by γ) could have a very substantial influence on the PSD of the CSTR effluent latex. Figure 5 illustrates this effect. The distributions are plotted in terms of a dimensionless diameter, that is, particle diameter divided by seed-particle diameter.

The curve for $\gamma = 0.0$ is single peaked, similar to published results for styrene. As γ increases the PSD has two peaks; one at the seed size and one larger. If γ is increased even further, the distribution becomes single peaked again, but skewed to the size of the seed particles. Hence, the measurement of PSDs from a seed-fed CSTR would be one way to study radical desorption mechanisms and rates.

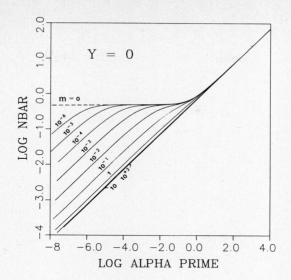

Figure 1

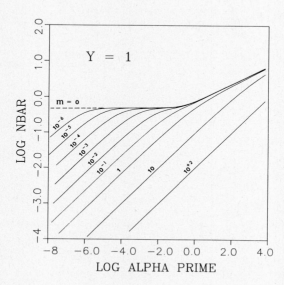

Figure 2

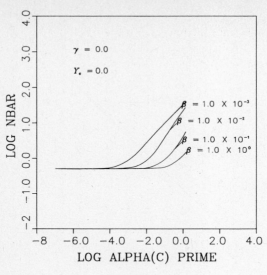

Figure 3

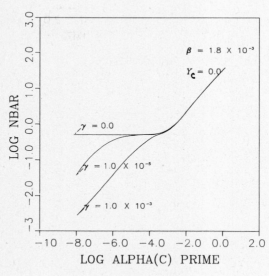

Figure 4

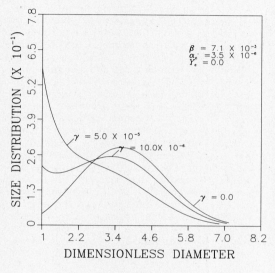

Figure 5

Other Preparation Methods

Synthetic polymeric latexes can be produced by processes that are
different from the standard emulsion polymerization methods
described in this chapter. Two such processes, inverse emulsion
polymerization and direct emulsification, are described briefly in
order to make this paper more complete. The literature on these
processes is less extensive, but interest in such processes has
recently increased.

Inverse Emulsion Polymerization. Water-soluble monomers can be
polymerized by emulsifying water solutions of these monomers in an
organic continuous phase. This process, called inverse emulsion
polymerization, yields a product comprised of a colloidal
suspension of droplets of aqueous polymer solution. The original
study of an inverse system by Vanderhoff et al. (20) involved the
monomer sodium p-vinylbenzene sulfonate, an organic phase of o-
xylene, Span 60 as the emulsifier, and either benzoyl peroxide or
potassium persulfate initiator. Later work by Kurenkov et al. (21)
involved acrylamide in a toluene continuous phase, potassium
persulfate, and Sentamid-5 (emulsifier). DiStefano (22) examined
three monomers: acrylamide, dimethylaminoethyacrylate
hydrochloride, and methacrylamide. The analogies between standard
and inverse emulsion polymerization are obvious, but not complete.
High polymerization rates coupled with high molecular weights are
similar, but the detailed mechanisms and kinetics appear to be
quite complex. Evidence is presented for multiple emulsion
droplets (oil-in-water-in-oil), a strong gel effect causing an
inverse relationship between molecular weight and emulsifier
concentration, and sensitivity to initiator and salt concentrations
(22).
    High molecular weights are important for many applications of
water-soluble polymers, and thus inverse emulsion polymerization
processes are becoming more important. Work presently in progress
should help to generate a better understanding of the chemical and
physical mechanisms involved.

Direct Emulsification. Polymer colloids called "artificial
latexes" can be prepared by dispersion of bulk polymers or polymer
solutions into an aqueous medium. Direct emulsification processes
are reviewed by ElAasser (23). The preparation procedures involve
mechanical dispersion that may be followed by removal of solvent.
According to ElAasser "the efficiency of emulsification," and hence
the particle size characteristics of the latex, "is determined by
the efficiency of formation of fine droplets and the efficiency of
stabilization of the formed droplets." Important parameters in the
process include the source of energy or agitation, its intensity,
and duration; type and concentration of emulsifiers; mode of
addition of emulsifier and the two phases; density ratio of the two
phases; temperature; and the rheology of the two phases.
    Direct emulsification can be used to produce latexes from
polymers that cannot be polymerized by free-radical mechanisms and
from natural polymers or their derivatives. Three methods or a
combination of methods can be applied in direct emulsification

processes: solution emulsification, phase inversion, and self-emulsification.   The first method, as the name implies, involves emulsification of a polymer solution followed by solvent stripping. Phase inversion occurs when a dilute aqueous alkali is added to a polymer that has been compounded with a long-chain fatty acid. Self-emulsification can be used with polymers that are modified with functional groups, such as amino or quaternary ammonium groups, so that they can be dispersed in water or acids without emulsifiers.

Acknowledgment

This material is based, in part, upon work supported by the National Science Foundation under Grant No. CPE-801445.

Literature Cited

1. Fitch, R. M.; Tsai, C. H. In "Polymer Colloids"; Fitch, R. M., Ed.; Plenum: New York, 1971; Chap. 5 and 6.
2. Ugelstad, J.; ElAasser, M.; Vanderhoff, J. Poly. Letter II 1973, 503.
3. Nomura, M. In "Emulsion Polymerization"; Piirma, I., Ed.,; Academic Press: New York, 1982; Chap. 5, p. 191-219.
4. Harkins, W. D. J. Am. Chem. Soc. 1947, 69, 1429.
5. Smith, W. V.; Ewart, R. H. J. Chem. Phys. 1948, 16, 592.
6. Roe, C. P. Ind. Eng. Chem. 1968, 60, 20.
7. Poehlein, G. W.; Dougherty, D. J. Rubber Chem. Technol. 1977, 50(3), 601.
8. Gonzalez, R. A. M.S. Thesis, Lehigh University, Bethlehem, Pa., 1974.
9. Gulf Oil Canada Ltd., U.S. Patent 3 551 396, 1970.
10. Ugelstad, J.; Hansen, F. K. Rubber Chem. Technol. 1976, 49(3), 536.
11. Hansen, F. K.; Ugelstad, J. J. Polym. Sci., Polym. Chem. Ed. 1978, 16, 1953-79; 1979, 17, 3033-45; 1979, 17, 3047-67.
12. Stockmayer, W. H. J. Polym. Sci. 1957, 24, 314.
13. O'Toole, J. T. J. Appl. Polym. Sci. 1965, 9, 1291.
14. Ugelstad, J.; Mørk, P. C.; Aasen, J. O. J. Polym. Sci. A-1 1967, 5, 2281.
15. Gilbert, R. G.; Napper, D. H. J. Chem. Soc. Faraday Trans. 1 1974, 70, 391.
16. Hawkett, B. S.; Napper, D. H.; Gilbert, R. G. J. Chem. Soc. Faraday Trans. 1 1975, 71, 2288; 1977, 73, 690; 1980, 76, 1323.
17. Lansdowne, S. W.; Gilbert, R. G.; Napper, D. H.; Sangster, D. F. J. Chem. Soc. Faraday Trans. 1 1980, 76, 1344.
18. Lichti, G.; Gilbert, R. G.; Napper, D. H. J. Polym. Sci., A-1 1980, 18, 1297.
19. Poehlein, G. W.; Dubner, W.; Lee, H. C. Brit. Polym. J. 1982, 14.
20. Vanderhoff, J. W.; Bradford, E. B.; Tarkowski, H. L; Shaffer, J. B.; Wiley, R. M. In "Polymerization and Polycondensation Processes"; ADVANCES IN CHEMISTRY SERIES No. 34, American Chemical Society: Washington, D.C., 1962; p. 32.

21. Kurenkov, V. K.; Osipova, T. M.; Kuznetsov, E. V.;
     Myagchenkov, V. A. Vysokonol. Soldin., Ser. B20 1978, 647.
22. DiStefano, F. M.S. Thesis, Lehigh University, Bethlehem, Pa.,
     1981.
23. ElAasser, M.S. Paper 15, short course notes, "Advances in
     Emulsion Polymerization and Latex Technology"; Lehigh
     University, Bethlehem, Pa., June 1982.

General References

1.   "Emulsion Polymerization"; Piirma, I.; Gardon, J. L., Eds.;
     ACS SYMPOSIUM SERIES No. 24, American Chemical Society:
     Washington, D.C., 1976.
2.   "Polymerization Reactors and Processes"; Henderson, J. N.;
     Bouton, T. C.; ACS SYMPOSIUM SERIES No. 104, American
     Chemical Society: Washington, D.C., 1979.
3.   "Polymer Colloids"; Fitch, R. M., Ed.; Plenum: New York,
     1971.
4.   "Polymer Colloids II"; Fitch, R. M., Ed.; Plenum: New York,
     1980.
5.   "Emulsion Polymerization and Its Applications in Industry";
     Eliseeva, V. I.; Ivanchev, S. S.; Kuchanov, S. I.; Lebedev,
     A. V.; translated from Russian by Teaque, S. J.; Plenum: New
     York, 1981.
6.   "Emulsion Polymerization of Vinyl Acetate"; ElAasser, M. S.;
     Vanderhoff, J. W., Eds.; Applied Science Publ: Englewood,
     N.J., 1981.
7.   "Emulsion Polymers and Emulsion Polymerization"; Bassett, D.
     R.; Hamielec, A. E., Eds.; ACS SYMPOSIUM SERIES No. 165,
     American Chemical Society: Washington, D.C., 1981.
8.   "Emulsion Polymerization"; Piirma, I., Ed.; Academic Press:
     New York, 1982.
9.   "Science and Technology of Polymer Colloids"; Poehlein, G.
     W.; Goodwin, J. W.; Ottewill, R. H., Eds.; Martinus Nijhoff
     Publ.: The Hague, Netherlands, 1983.
10. "New Concepts in Emulsion Polymers"; Hwa, J.; Vanderhoff, J.
     W., Eds.; J. Polym. Sci. 1969, C27.
11. Blackley, D. C. "Emulsion Polymerization: Theory and
     Practice"; Applied Science Publ.: London, 1975.

New Developments and Trends in Free-Radical Polymerization

GEORGE E. HAM

G. E. Ham Associates, 284 Pine Road, Briarcliff Manor, NY 10510

Catalysts
Telechelic Polymers
Photopolymers
Polyolefins and Modified Polyolefins
Weather-Resistant Grafts of Vinyl Monomers to EPDM and
    Polyacrylates
Poly(phenylene oxide) Grafts and Blends
Styrene-Acrylonitrile-Maleic Anhydride Terpolymers
Modifiers for Poly(vinyl chloride)
Nitrile Barrier Resins
Polymerization in Air
Emulsion Polymerization
Powders
Diallyldimethylammonium Chloride
Charge-Transfer Polymerization
Template Polymerization
Incorporation of Polymer Additives in Chains
Biologically Active Polymers

Polymerization of vinyl monomers by free-radical mechanisms is perhaps the most widely encountered and best understood mode of vinyl polymerization. The popularity of free-radical polymerization is due in substantial part to the many advantages that this route to polymers offers to industry. The polymerization process is noteworthy for its ease, convenience, and relative insensitivity to impurities, such as water and oxygen, that plague ionic polymerizations. Indeed, it is common to carry out free-radical polymerizations in water as a suspending medium, as in emulsion and suspension polymerization. Another advantage of free-radical polymerization is that it offers a convenient approach toward the design and synthesis of myriad specialty polymers for use in almost every area.

0097-6156/85/0285-0151$06.00/0

Catalysts

A great impetus has been furnished toward the syntheses of new specialty polymers as well as toward improved processes for the synthesis of commodity polymers by the increased availability of novel low- and high-temperature and more versatile initiators. These catalysts belong to the classes of peroxides, percarbonates, peroxyphosphates, and others. These new and improved initiators allow tailoring of the catalyst system to the specific requirements of the monomer and polymer system. For example, the polymerization of ethylene at high pressures has been revolutionized by the availability of selected hydrocarbon-soluble catalysts such as *tert*-butyl perpivalate and bis(2-ethylhexyl) percarbonate. In addition, a new class of asymmetrical azo catalysts now offered is characterized by an α-cyanoisoalkyl group on one side and a *tert*-butyl group on the other side. Such components avoid the deleterious oxidative characteristics of peroxides and offer improved solubility. They can be tailored to decompose in the appropriate ranges for selected vinyl monomers. Another novel catalyst class based on labile carbon compounds may allow the avoidance of some of the problems of conventional free-radical initiators, such as degradative chain transfer and loss of unused catalyst.

Telechelic Polymers

Interest in free-radical catalysts containing organic functionality, such as hydroxyl groups, carboxylic acid groups, and amide groups is increasing. For example, butadiene has been polymerized in the presence of 4,4'-azobis (β-hydroxyisobutyronitrile) to produce a so-called telechelic polybutadiene containing terminal hydroxyl groups. Of course, in free-radical polymerization one encounters not only bimolecular coupling but disproportionation as well. Accordingly, under normal circumstances, it is impossible to produce 100% polymer with the desired functionality on both ends. In the case of butadiene polymerization, sufficient chain transfer with the polymer takes place so that almost every polymer molecule will have at least dual functionality. When the polymer has several branches, some chain ends may not have the desired functionality. However, for many purposes it is sufficient that each molecule have merely, two functional groups.

Appropriate telechelic polymers produced from a variety of monomers may be incorporated as reactive components in a variety of applications such as sealants, elastomers, foams, and fibers. Such telechelic polymers may impart almost any desired characteristic such as hydrophilic properties, elastomeric properties, dyeability, and solvent resistance. A dihydroxy telechelic polymer may be reacted with a telechelic polymer containing two carboxylic acid groups to produce a condensed polyester. Accordingly, from these telechelic entities one may produce block copolymers that will have alternating addition polymer residues of like or unlike repeat units. In addition, block copolymers can be produced by modifications of this procedure in which addition polymers are alternated with condensation polymer units.

Photopolymers

These materials, although known and used for many years in small
quantities in demanding applications, have only recently burgeoned
into a major outlet for specialty monomers and polymers. Two
important modes of use of these materials should be noted:
photopolymerization and photocross-linking and chain extension. In
photopolymerization, light-sensitive acrylic monomers, acrylated
polyurethanes, polyethers, polyesters, epoxies, and other compounds
are exposed to light in a controlled manner for rapid curing to
products of industrial importance. In general, the materials are
converted on curing from sticky or low-viscosity materials to tough,
durable finishes. In photocross-linking and chain extension, pre-
existing oligomers or polymers, generally in film form, are
photocross-linked in the presence of photosensitizers. These
products include ene-thiol-type products, epoxies, unsaturated
polyesters, or unmodified polymers exposed to high radiation.
     Areas of application of photopolymerized or photocross-linked
products include asbestos-vinyl floor-tile coatings, negative
photoresists, tough protective coatings, printing inks, lithography,
printing plates, "plaster" casts, can coatings, coatings for plastic
containers, coatings for wood, paper, plastics, microelectronics,
communications, adhesives, chemical specialties, folding cartons,
graphic arts, medical and dental uses, and pigments. Suitable
photosensitizers for the conversion of the photopolymers include
benzoin ethers, acetophenone, halogenated ketones, and halogenated
aromatics. The market for radiation-curable materials, currently at
a level of about 100 million lb and valued at $450 million, is
expected to grow to over 200 million lb by 1990.

Polyolefins and Modified Polyolefins

In the industrial area, polyethylene is now being produced at entry
pressures as high as 45,000 to 47,000 $lb/in.^2$. One of the important
consequences of the use of higher pressures and temperatures is to
gain homogeneous (one phase) reaction conditions. In this way
deleterious branching and poor rheological characteristics in
fabricated products from polyethylenes may be avoided. In addition,
improved film resin clarity may result.
     The use of homogeneous reaction conditions allows one to carry
out the polymerization of ethylene in a way to limit the important
grafting reaction that leads to long chain branching in high-
pressure polyethylene. This grafting reaction is especially
pronounced on polymers that are swollen rather than dissolved in the
reaction phase. In a related development, modification of
polyethylene with small concentrations of vinyl acetate has become
very popular. The presence of vinyl acetate monomer and vinyl
acetate units in polyethylene also increases the solubility of
polyethylene in the reaction phase with the desired reduction in
deleterious grafting. Such ethylene copolymers are important in
their own right, however, because of improved toughness and clarity
in comparison with straight polyethylene. Indeed, the next 10 years
should see important extension of commercial ethylene copolymers to
include not only the widely used ethylene-vinyl acetate copolymers

but ethylene–acrylate copolymers, ethylene–acrylic acid copolymers, ethylene–carbon monoxide copolymers, and perhaps ethylene–acrylonitrile copolymers.

For many years, polymer chemists have believed that the availability of amorphous polyethylene that would not crystallize under ordinary use conditions would be an important addition to the spectrum of polymeric materials. Indeed ethylene copolymers accomplish this result by preventing the crystallization of polyethylene. Improved toughness, clarity, and low temperature properties can be effected.

Another benefit of copolyermization of ethylene is incorporation of appropriate polar groups such as carboxylic acid groups or salt groups in the ionomers. Du Pont is still the only producer of the Surlyn–type of polyethylene ionomers. However, several companies produce copolymers of ethylene with acrylic acid (EAA). Such copolymers offer improved thermal properties with retention of clarity, improved heat–seal strength, and impact strength, as well as improved adhesion to a variety of substances, such as aluminum foil, glass fibers, and mineral fibers.

Weather–Resistant Grafts of Vinyl Monomers to Ethylene–Propylene–Diene–Monomer (EPDM) and Polyacrylates

Because of increasing sales resistance to conventional acrylonitrile–butadiene–styrene (ABS) materials, manufacturers are looking for improved alternatives. Important and promising materials are the grafts of styrene and/or styrene–acrylonitrile onto weather–resistant elastomers, such as ethylene propylene–diene terpolymers (EPDM) and polyacrylates. These products are produced by free–radical polymerization of styrene and acrylonitrile in the presence of the elastomers in mass, solution, or emulsion processes. Care must be taken to employ elastomers containing suitable grafting sites such as, in the case of EPDM, 1,4–hexadiene or ethylidene norbornene. Appropriate conditions of grafting, including suitable peroxide initiators, temperature and solvents, must be selected. The resulting products possess far superior weather resistance than that of conventional ABS materials. Products have been introduced to the market both in the United States and in Europe. These products include not only the EPDM graft materials but, in addition, grafts of styrene and methacrylates to elastomeric polyacrylates.

Poly(phenylene oxide) Grafts and Blends

Poly(phenylene oxide) (PPO)/polystyrene blends are widely known and used as engineering molding resins. What is not generally known is that certain types of these materials are grafts of polystyrene and copolymers onto PPO. This practice assures compatibility and homogeneity during manufacture and use of the product and is one avenue toward improved blend compatibility between PPO and polystyrene. In general, grafting techniques are similar to those described earlier. Bulk, solution, and emulsion methods may be used for the grafting of styrene to PPO. Appropriate attention must be given to the use of suitable peroxides and control of temperature and other reaction conditions.

Styrene-Acrylonitrile-Maleic Anhydride Terpolymers

Recently, products composed of styrene, acrylonitrile, and maleic anhydride have been introduced as improvements over the older ABS materials. These products exhibit superior heat resistance and thermal stability in comparison to conventional ABS materials. In addition, the impact resistance of styrene-acrylonitrile-maleic anhydride terpolymers can be improved by the addition of or by grafting onto appropriate elastomers. The product also exhibits desirable rheological behavior due to polar interaction of anhydride groups during processing and drawing.

Modifiers for Poly(vinyl chloride)

In a quite different but very important industrial area, free-radical polymerizations have made great inroads in the optimization of the desired commercial properties of impact-modified poly(vinyl chloride) (PVC). In a most sophisticated variation, grafted impact modifiers based on the quaterpolymerization of acrylic esters, butadiene, styrene, and acrylonitrile have been produced and almost precisely match the refractive index of PVC. The blending of the impact modifier with PVC yields a completely clear polymer suitable for shampoo bottles and food containers. In addition to excellent clarity these polymers have extremely good impact strength combined with improved fabricability by flow molding equipment.

Nitrile Barrier Resins

Another potentially important development in the free-radical polymerization area is the nitrile barrier resins. These products generally are graft polymers containing a glassy phase comprised of 80% acrylonitrile and 20% styrene or other vinyl monomer grafted onto a rubber substrate. The commercialization of these materials has been interrupted by an adverse ruling by the Food and Drug Administration (FDA) in connection with extraction of traces of acrylonitrile in food-related applications. The amounts involved are very small. A new attitude under development in the FDA may lead to the eventual clearance of these materials.

These resins may be blow molded to yield excellent containers for carbonated beverages. In addition, they show a desirable low level of transmission of carbon dioxide, oxygen, and water vapor in comparison with other plastics available today.

Polymerization in Air

In another commercial application of free-radical polymerization, polymerizations may be carried out in industrial coatings in the presence of air to yield a variety of coatings and structures of commercial import. This development is possible, in part, because certain vinyl monomers, particularly the acrylates, are less sensitive to retardation by oxygen compared with other monomers. It is therefore possible to produce radiation-cured coatings. UV-cured printing inks and the photopolymers are important in imaging for printing, photoresist, and related applications.

Emulsion Polymerization

Perhaps one of the most powerful and yet oldest polymerization techniques for vinyl monomers is emulsion polymerization. The technique is highly versatile and readily applicable to the manufacture of copolymers containing two to five monomer constituents. In this way, two or three components may be used for the optimization of the glass transition temperature or thermal stability of the resulting product, and the remaining one or two components can impart desirable chemical properties such as adhesion, cross-linkability, photosensitivity, and compatibility. Emulsion polymerization may be employed for the manufacture of products to be isolated and used as molding pellets, or for end-use polymer emulsions, which can be applied directly or with suitable formulation as surface coatings, adhesives, pigment vehicles, lacquers, paints, etc. Emulsion polymerization also offers the versatility, by gradual monomer addition during polymerization or selective addition of monomers at different rates, of producing uniform copolymers. For the manufacture of uniform copolymers by other methods, such as mass and solution polymerization, the manufacturing process must be interrupted at conversions of only 10 or 20% and this results in higher costs of manufacture. In the case of emulsion polymerization, uniform product may be produced at conversions of 85-95+%.

Powders

Another related development of importance is powder coatings, which are polymer concentrates in monomers of low volatility and may be sprayed as powder onto a surface and cured by subsequent heating. In a sense, these are nonsolvent-based coatings in that no solvent must be subsequently removed from the system.

Diallyldimethylammonium Chloride

Another free-radical polymer of importance is diallyldimethylammonium chloride, which is employed as a coating on paper in electrographic reproduction for the purpose of improving moisture retention of paper. The polymer allows better dissipation of static electricity, an essential step in the imaging process. The polymer is also used in water purification as a flocculant.

Charge-Transfer Polymerization

Charge-transfer polymerization has been regarded by some scientists as a nonradical type of polymerization or, at least, as an unconventional polymerization. Some viewpoints hold that in this polymerization the charge-transfer complex involving, for example, styrene and acrylonitrile complexed with zinc chloride undergoes decomposition with monomers adding to both ends of the nascent chain in a matrix-type polymerization. The evidence of nonradical character of these polymerizations includes, for example, insensitivity to charge-transfer agents in reducing molecular weight, nonessentiality of free-radical catalysts, and noninhibition by the usual free-radical inhibitors.

By now most of these points have been successfully overcome. Possibly, the last remaining questionable issue surrounds the nonefficacy of chain-transfer agents in limiting the molecular weight of the charge-transfer polymers. Recent work in Great Britain suggests that the use of more effective chain-transfer agents, such as aromatic amines, leads to the expected molecular weight reduction.

Charge-transfer polymerization is characterized by a highly efficient propagation step that tends to exclude other free-radical reactions, except those that can compete in an efficient manner.

Template Polymerization

Still another development of increasing interest involves attempts to control free-radical polymerization by offering during the propagation step a suitable surface or template for polymerization. In one example, chloroprene has been successfully polymerized to give exclusively head-to-tail *trans* polymer by using frozen monomer. In addition, polyacrylonitrile, when produced in the presence of pre-existing polymer, has an ordered character suggestive of template polymerization. Vinyl chloride and vinylidene chloride also show similar behavior.

Studies by Wegner of Mainz on conjugated diacetylenes confirm that one may polymerize crystalline monomer and retain much of the crystalline character in the resulting polymer. It is now settled as to the mechanism of this particular polymerization, but it seems probable that free-radical as well as ionic polymerizations may display such effects in appropriate cases.

Such ordered polymerizations in any of the categories discussed offer the potential for polymerization "without end" by minimizing polymer radical-transfer possibilities. The amount of external catalyst required for such polymerizations would be very small for similar reasons.

Incorporation of Polymer Additives in Chains

Interesting possibilities for replacing polymer additives exist. For example, polymerizable vinyl monomers containing light-stabilizing moieties, such as benzophenone or benzotriazole groups, have been produced. Incorporation of such monomers at a level of 1 or 2% into a parent polymer protects the polymer against the deleterious effects of sunlight.

Biologically Active Polymers

Finally, there should be mention of a development of Ringsdorff (Mainz) on the preparation of biologically active polymers (mostly of free-radical origin) to revolutionize the whole area of chemotherapy. Ringsdorff has pointed out the important differences in drug treatment between the step of taking the drug to the desired site of application and the release of a suitable pharmacological or biological function at the site. In his research, polymers with a molecular weight of 25,000 were employed because of limitations on diffusivity through cell walls. Attached to these polymers were a variety of biologically active functions held to the backbone by labile segments that undergo cleavage in the appropriate environments. This approach offers promise of producing anticarcinogenic, antiviral, antiradiative, and birth control agents, as well as numerous other drugs or biologically active agents of importance.

Condensation Polymerization and Polymerization Mechanisms

I. K. MILLER and J. ZIMMERMAN

Experimental Station, E. I. du Pont de Nemours & Company, Inc., Wilmington, DE 19898

Preparation Methods for Condensation Polymers
    Polyamides
    Polyesters
    Polyimides
    Polyurethanes
    Polyureas
    Polycarbonates
    Polyanhydrides
Molecular-Weight Distribution of Condensation Polymers
Kinetics and Mechanism of Nylon 66 Polyamidation
    Amidation
    Effect of End-Group Imbalance on Molecular Weight
    Effect of Monofunctional Monomers on Molecular Weight
    Kinetics
    Thermal Degradation
    Branching and Cross-linking

The term "condensation polymers" was introduced by W. H. Carothers in his early work on the preparation of polyesters and polyamides to distinguish this class of polymers from vinyl polymers made by addition reactions. Condensation polymers were defined as polymeric molecules that may be converted by hydrolysis, or its equivalent, into monomers that differ from the structural units by one molecule of H_2O, HCl, NH_3, etc. This broad definition includes many polymers made by ring-opening or addition reactions, for example, lactone and lactam polymers, polyurethanes, polyureas, polyimides, and polyhydrazides, as well as polymers made by true condensation reactions.

   Although the formation of a low molecular weight polyester by condensation of ethylene glycol and succinic acid was reported as early as 1863 (1) and low molecular weight ε-aminocaproic acid

0097–6156/85/0285–0159$06.00/0

polymers were prepared in 1899 ($\underline{2}$), systematic work on polycondensation reactions leading to linear high molecular weight polymers of practical utility derives from the work begun in 1929 by Carothers and coworkers ($\underline{3}$). In initial work, polyesters with molecular weights ranging from 2500 to 5000 were prepared by the condensation of dibasic acids and glycols. As techniques were refined, molecular weights in excess of 10,000 were obtained. Structure–property relationships developed in early work soon showed that solubility characteristics and low melting points made linear aliphatic polyesters unsuitable for fiber applications. Similar studies of polyamidation showed that the desired properties could be obtained with polyamides. Many polyamides were prepared both from amino acids and from diamines and diacids. The polyamide of hexamethylene diamine and adipic acid was selected for the first commercial development of a textile fiber because of an optimum balance of properties and the prospective manufacturing cost.

The major research contributions leading to practical technology for production of high molecular weight polymers by the relatively simple condensation reactions were in the areas of understanding both the mechanism of molecular weight increase and the requirements for attainment of high molecular weight. The dependence of molecular weight on the extent of the reaction was shown by W. H. Carothers, and the concept of nondependence of functional–group reactivity on molecular size and derivation of the molecular–weight distribution for condensation polymers were shown by P. J. Flory. These contributions, as well as the development of analytical procedures for polymer end groups were essential. In the following discussion of the mechanism of condensation polymerization, the general requirements for production of high molecular weight polymers by condensation reactions rather than the details of the functional group reactions will be emphasized. A discussion of polyamidation will be used to illustrate the principles which apply generally to polymers of this class.

Preparation Methods for Condensation Polymers

Polyamides. Many methods have been reported for laboratory synthesis of polyamides ($\underline{4}$). Commercial methods include melt polycondensation, ring opening of lactams, and low–temperature solution polymerization.

Direct Amidation. Formation of polyamides by condensation of amine and carboxyl groups is illustrated by

$$H_2NRCO_2H \rightleftharpoons H(NHRCO)_n\text{------}OH + H_2O \qquad (1)$$

In this reaction, an amino acid may be polymerized or a diamine may be polymerized with a diacid. Similar condensation reactions occur with derivatives of the acid moiety. Examples include ester aminolysis (Reaction 2) in which phenyl or higher alkyl esters are preferred, amide aminolysis (Reaction 3), and acidolysis of acyl derivatives of diamines (Reaction 4).

$$H_2NR_1NH_2 + R_2O_2CR_3CO_2R_2 \longrightarrow H(HNR_1NHCOR_3CO)_n OR_2 + R_2OH \qquad (2)$$

$$H_2NR_1NH_2 + H_2NOCR_2CONH_2 \longrightarrow H(HNR_1NHCOR_2CO)_n NH_2 + NH_3 \qquad (3)$$

$$R_1CONHR_2NHCOR_1 + HO_2CR_3CO_2H \longrightarrow$$
$$R_1CO(HNR_2NHCOR_3CO)_n OH + R_1CO_2H \qquad (4)$$

Reaction of Acid Chlorides. Low-temperature polycondensation of
diamines and diacid chlorides is an important route for preparing
high-melting polyamides such as aromatic polyamides, which decompose
or cross-link if prepared by high-temperature melt routes. The
reaction may involve an interfacial reaction between the diacid
chloride in a water-immiscible solvent with an aqueous diamine
solution, or the reaction may be carried out in a homogeneous
solution. The presence of a base is usually needed to remove HCl so
that polymerization is complete. With weakly basic aromatic
diamines, an acid acceptor is not always needed because HCl can be
evaporated from the reaction mixture. The general reaction is given
by

$$H_2NR_1NH_2 + ClCOR_2COCl \xrightarrow{\text{base}} H(HNR_1NHCOR_2CO)_n Cl + HCl \qquad (5)$$

Polysulfonamides (Reaction 6) and polyphosphoramides (Reaction
7) can be made by similar reactions.

$$H_2NR_1NH_2 + ClSO_2R_2SO_2Cl \longrightarrow (HNR_1NHSO_2R_2SO_2)_n + HCl \qquad (6)$$

$$H_2NR_1NH_2 + R_2POCl_2 \longrightarrow (HNR_1NHR_2PO)_n + HCl \qquad (7)$$

Other low-temperature polyamide syntheses include reaction of N,N'-
(alkanedioyl) lactams (Reaction 8) or N,N'-(alkanedioyl) diimides
(Reaction 9) with diamines (5).

Ring-Opening Polymerization. Polyamides can be formed by ring-opening reactions as in the polymerization of caprolactam. The polymerization may be carried out at high temperature with water, amino acid, or amine carboxylate salt as initiator. The predominate ring-opening mechanism with these initiators is carboxyl-catalyzed aminolysis. In the case of water initiation, hydrolysis occurs (Reaction 10) to form small amounts of the amino acid. The amino acid then reacts with lactam amide groups (Reaction 11).

$$\text{(ring)} \quad + \quad H_2O \longrightarrow H_2N(CH_2)_5CO_2H \tag{10}$$

$$n\,\text{(ring)} \quad + \quad H_2N(CH_2)_5CO_2H \longrightarrow H\text{-}(NH(CH_2)_5CO)_{n+1}\text{---}OH \tag{11}$$

Low-temperature polymerization of caprolactam can be accomplished by an anionic mechanism in which ring opening is effected by a strong base, usually with the addition of an acylating cocatalyst (e.g., acetic anhydride) as illustrated by Reaction 12.

$$n\,\text{(ring)} \xrightarrow[\text{Ac}_2\text{O}]{\text{NaH}} \quad \text{N---}(CO(CH_2)_5NH)_{n-1}\text{---}H \tag{12}$$

Whether lactams can be polymerized to form high molecular weight linear polymers depends on ring size and other structural factors. Ring substituents, especially on the nitrogen atom, inhibit ring opening. Linear polymers are the favored form at equilibrium only when the ring-opening rate is much greater than the cyclization rate. (This subject is reviewed in References 6 and 7.)

Miscellaneous Routes. Polyamides have been prepared by other reactions, including addition of amines to activated double bonds, polymerization of isocyanates, reaction of formaldehyde with dinitriles, reaction of dicarboxylic acids with diisocyanates, reaction of carbon suboxide with diamines, and reaction of diazlactones with diamines. These reactions are reviewed in Reference 4.

Polyesters. Direct Esterification. Polyesters can be made from hydroxy acids and from diacids and diols by direct condensation:

$$HORCOOH \rightleftharpoons H(ORCO)_n\text{----}OH + H_2O \tag{13}$$

This reaction is best suited for the preparation of low molecular weight polyesters (degree of polymerization, 5-50) of aliphatic hydroxy acids or diacids reacted with glycols having either primary or secondary hydroxyl groups.

High molecular weight polyesters can be obtained only with special techniques because of the difficulty of obtaining complete water removal without loss of monomers. The reaction is self-catalyzed by carboxyl groups and can be catalyzed by other acids, for example, p-toluenesulfonic acid and by compounds such as titanium alkoxides, dialkyltin oxides, and antimony pentafluoride.

Reaction of Acid Chlorides. Polyesters can be formed by the reaction of diacid chlorides with diols:

$$HOR_1OH + ClCOR_2COCl \longrightarrow H(OR_1OCOR_2CO)_n Cl + HCl \qquad (14)$$

Polyesters of aliphatic diols can be made by heating the reactants to remove HCl or by refluxing the reactants in an inert solvent. With dihydric phenols, in the presence of a strong base, polymerization can be carried out at or near room temperature either in homogeneous solution or by interfacial reaction using aqueous NaOH as the base.

Ester Exchange. The most practical methods for the preparation of high molecular weight polyesters involve ester exchange reactions. The simplest exchange method involves a catalyzed alcoholysis reaction between a diol and a dicarboxylic acid ester with elimination of the alcohol:

$$HOR_1OH + CH_3OCOR_2CO_2CH_3 \longrightarrow H(OR_1OCOR_2COOR_1O)_n \!\!-\!\!-\!\!-\!\!H + CH_3OH \qquad (15)$$

An excess of glycol is used to cause complete alcohol removal and form chains with hydroxyl end groups. High molecular weights are accomplished by self-alcoholysis of these molecules to remove the excess glycol and approach a glycol/diacid ratio of 1 in the final polymer.

Polyesterification by acidolysis ester exchange is also a useful synthetic method, for example, reaction of the diacetate ester of a dihydric phenol with a dicarboxylic acid in the presence of a catalyst to eliminate acetic acid.

Ring-Opening Polymerization. Many lactones can be converted into linear polymers by ring-opening reactions as illustrated by Reaction 16. The reaction usually requires a catalyst and can be carried out in bulk or in an inert solvent. As in the case of lactam polymerization, ring size and other structural factors are important in determining whether high-molecular-weight polymers can be obtained. A review of these factors, as well as a discussion of other reactions for polyester formation, is given in Reference 8.

Polyimides (8). Polyimides are prepared by reaction of a diamine with a tetracarboxylic acid or its derivative. Polyimides that melt without decomposition can be synthesized by heating a mixture of the diamine and tetracarboxylic acid to eliminate water (Reaction 17) or by a similar reaction using a diacid–diester to eliminate water and an alcohol (Reaction 18).

$$\text{HOOC} \underset{\text{HOOC}}{\overset{\text{COOH}}{\bigcirc}} \text{COOH} + H_2NRNH_2 \longrightarrow -R-N \underset{}{\bigcirc} N-R- + H_2O \qquad (17)$$

$$\underset{\text{HOOC}}{\overset{CH_3OOC}{\bigcirc}} \underset{\text{COOCH}_3}{\overset{\text{COOH}}{\bigcirc}} + H_2NRNH_2 \longrightarrow -RN \underset{}{\bigcirc} NR- + H_2O + CH_3OH \qquad (18)$$

A more general method involves the reaction of a diamine with a dianhydride in a solvent at or near room temperature to form a poly(amide–acid) that is subsequently dehydrated to form the polyimide (Reaction 19).

$$\bigcirc + H_2NRNH_2 \longrightarrow -NC \underset{\text{HOOC}}{\overset{\text{O}}{\bigcirc}} \underset{}{\overset{\text{COOH}}{\bigcirc}} \longrightarrow$$

$$-N \underset{}{\bigcirc} N- + H_2O \qquad (19)$$

Polyurethanes (8). Polyurethanes are prepared by the reaction of a diol with a diisocyanate:

$$HOR_1OH + OCNR_2NCO --- \overset{O}{\underset{}{}} \{OR_1OCNR_2NC\}_n \qquad (20)$$

Catalysts, including bases, metal complexes, and organometallic compounds, are usually used.

Polyureas (8). Many synthetic methods for preparation of polyureas are known. Examples of reactions that can be used for the preparation of a variety of polyurea structures include reaction of diamines with phosgene (Reaction 21), with urea (Reaction 22) or with diisocyanates (Reaction 23).

$$H_2NRNH_2 + ClCOCl \xrightarrow{\text{NaOH}} \{NHRNHCO\}_n + NaCl \tag{21}$$

$$H_2NRNH_2 + H_2NCONH_2 \longrightarrow \{NHRNHCO\}_n + NH_3 \tag{22}$$

$$H_2NR_1NH_2 + OCNR_2NCO \longrightarrow \{NHR_1NHCONHR_2NHCO\}_n \tag{23}$$

Polycarbonates (4). Polycarbonates may be prepared by a variety of synthetic methods. Procedures most commonly used include reaction of diols with phosgene (Reaction 24), transesterification of diesters of carbonic acid with dihydroxy compounds (Reaction 25), and polymerization of cyclic carbonates (Reaction 26).

$$HOROH + COCl_2 \longrightarrow H \{ORO\overset{\overset{\displaystyle O}{\|}}{C}\}_n OROH + HCl \tag{24}$$

$$HOR_1OH + R_2OCOR_2 \longrightarrow H\{OR_1\overset{\overset{\displaystyle O}{\|}}{C}\}_n OR_2 + R_2OH \tag{25}$$

$$R \overset{\displaystyle O}{\underset{\displaystyle O}{\diamond}} C = O \longrightarrow \{ORO - \overset{\overset{\displaystyle O}{\|}}{C}\}_n \tag{26}$$

Polyanhydrides (4). Elimination of water between carboxyl groups of a dicarboxylic acid can lead to the formation of polyanhydrides. The polymerization is best accomplished by reacting the dicarboxylic acid with acetic anhydride to form a mixed anhydride that is then heated under vacuum to eliminate acetic anhydride. The overall reaction is represented by Reaction 27.

$$HO\overset{\overset{\displaystyle O}{\|}}{C}R\overset{\overset{\displaystyle O}{\|}}{C}OH \xrightarrow{(CH_2CO)_2O} \{R\overset{\overset{\displaystyle O}{\|}}{C}O\overset{\overset{\displaystyle O}{\|}}{C}\}_n \tag{27}$$

Molecular-Weight Distribution of Condensation Polymers

Polymers made by stepwise intermolecular reaction between bifunctional monomers have a characteristic molecular-weight distribution. A quantitative expression of this molecular-weight distribution, first derived by P. J. Flory (9), is given by Equations 28–30. The extent of the reaction p is equal to the fraction of the original functional groups that have reacted at any point in the polymerization and can be calculated by:

$$p = (T_o - T)/T_o \tag{28}$$

In Equation 28, T_o is the initial concentration of functional groups, and T is the concentration of unreacted functional groups.

The mole or number fraction of chains containing x repeat units is given by Equation 29; the weight fraction is given by Equation 30. The curves for these distributions are given in Figure 1.

$$n_x = p^{x-1} (1-p) \tag{29}$$

$$w_x = xp^{x-1} (1-p)^2 \tag{30}$$

The number average degree of polymerization for a polymer with a most probable molecular-weight distribution is given by the value of x at the maximum in the weight-fraction curve. For values of p approaching unity, the number average degree of polymerization (DP) is given by Equation 31.

$$\overline{DP}_n = 1/(1-p) \tag{31}$$

Although the mechanism of condensation reactions is generally simple, the preparation of condensation polymers with useful molecular weight can be very difficult and can require very precisely controlled reaction conditions because of the very high reaction yields needed, as shown by Equation 31.

Table I shows the effect of the extent of reaction p on the molecular weight of nylon 66.

A degree of conversion of nearly 99% is needed to obtain nylon 66 with a number-average molecular weight of 10,000, which is the minimum molecular weight for linear polymers that is the most useful for plastics, film, and fiber uses. To achieve a molecular weight in excess of 20,000 a degree of conversion of about 99.4% is needed. Even at this high reaction yield the polymer molecular weight is low compared to molecular weights attainable, for example, by free-radical addition reactions. The need for a high reaction yield introduces other requirements for monomer purity, monomer balance, reaction rate, and control of side reactions that must be met for preparation of condensation polymers of high molecular weight. These requirements are illustrated by a discussion of the mechanism and kinetics of polymerization of the polyamide of hexamethylene diamine and adipic acid (nylon 66).

Kinetics and Mechanism of Nylon 66 Polyamidation

Amidation. Synthesis of nylon 66 involves condensation of amine with carboxyl groups in a melt polymerization. The reaction is represented by Reaction 32.

$$--NH_2 + --COOH \rightleftharpoons \overset{\overset{O}{\underset{|}{H||}}}{-NC-} + H_2O \tag{32}$$

The equilibrium constant for this reaction is given by Equation 33.

$$\frac{[CONH] [H_2O]}{[COOH] [NH_2]} = K \tag{33}$$

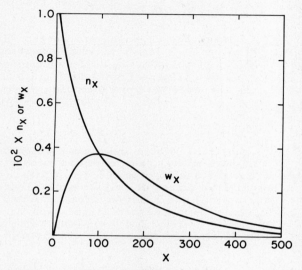

Figure 1. Most probable molecular weight distribution for condensation polymers (p = 0.99).

TABLE I.   Extent of Reaction
vs. the Number-Average
Molecular Wt. of Nylon 66

p	$\overline{M}_n$
0.995	22,600
0.993	16,142
0.990	11,300
0.980	5,650

If nylon 66 is equilibrated at 280 °C under steam at atmospheric pressure, the water content of the polymer melt is about 0.16% or 89 mol per 10^6 g of polymer, and a value of 3000 (eq/10^6 g)^2 is found for the product of the concentrations of uncondensed amine and carboxyl groups. These results yield a value of 260 for the equilibrium constant (Equation 33) because the amide group concentration is about 8800 eq/10^6 g.

Effect of End-Group Imbalance on Molecular Weight. If T is the concentration of end groups in a linear polymer, expressed in units of equivalents per 10^6 g of polymer, the number of chains per 10^6 g is T/2 and the number of chains per gram is T/ (2 x 10^6). The reciprocal of this is the number-average molecular weight (Equation 34).

$$\overline{M}_n = (2 \times 10^6)/T \tag{34}$$

If nylon 66 is equilibrated at a known water vapor pressure to achieve a fixed end group product, the amine and carboxyl end-group concentrations are determined by the end-group imbalance. This imbalance can result from a lack of initial equivalence of hexanedioic acid (adipic acid) and diamine, from diamine loss, from monomer impurities, or from polymer degradation by adipamide cyclization. If $P = [COOH] \cdot [NH_2]$ and $D = [COOH] - [NH_2]$, the end-group concentrations in equilibrated polymer are given by Equations 35 and 36.

$$[NH_2] = \frac{-D + (D^2 + 4P)^{1/2}}{2} \tag{35}$$

$$[COOH] = \frac{D + (D^2 + 4P)^{1/2}}{2} \tag{36}$$

For polymers containing no monofunctional monomers, these end-group concentrations determine the number-average molecular weight (Equation 34), as indicated in Equation 37.

$$\overline{M}_n = \frac{2 \times 10^6}{(D^2 + 4P)^{1/2}} \tag{37}$$

Figure 2 shows the effect of end-group imbalance D on the molecular weight of nylon 66 equilibrated with steam at atmospheric pressure at 280 °C to yield a product of end-group concentrations P = 3000 (eq/10^6 g)^2. From Figure 2, the needed degree of precise control of monomer balance can be deduced for attainment of high molecular weight polyamide and for precise molecular-weight control. Monomer balance is important also in the preparation of condensation polymers by nonequilibrium reactions, for example, polyamides from diacid chlorides and diamines, polyesters from diacid chlorides and diols, or polyure.pa thanes from diamines and diisocyanates.   To obtain a conversion of monomer functional groups of 99.5%, monomer imbalance cannot exceed 0.5%.

Precise control of initial monomer balance with many polyamide systems can be accomplished by measuring the pH of an aqueous solution of the diamine and diacid. With reaction systems not amenable to this type of measurement, very precise metering techniques for monomer addition must be used.

Effect of Monofunctional Monomers on Molecular Weight. The molecular weight of a polyamide (Equation 34) is determined by the total concentration of polymer end groups. These end groups, in addition to unreacted carboxyl groups, C, and amine groups, A, may include nonfunctional or "stabilized" end groups, E_S, due to the presence of monofunctional impurities in monomers or to the addition of monofunctional acids or amines for molecular-weight control. The molecular weight of a polyamide containing nonfunctional end groups is given by Equation 38. End-group concentrations are expressed in units of equivalents per 10^6 g of polymer.

$$\overline{M}_n = \frac{2 \times 10^6}{A + C + E_S} \tag{38}$$

Figure 3, which shows the effect of nonfunctional end groups on the molecular weight of nylon 66 with equal amine and carboxyl end-group concentrations at 50 eq/10^6 g, illustrates the necessity for the absence of monofunctional impurities in condensation polymer intermediates if maximum molecular weight is to be obtained.

Kinetics. In aqueous solution, second-order kinetics (Equation 39) are found for condensation of simple carboxylic acids and amines (10). Studies of polyamidation when a 90% conversion level is reached indicate second-order kinetics for this reaction (11). At conversions above 90%, evidence suggests that a carboxyl-catalyzed third-order reaction assumes increasing importance and becomes predominant.

$$\frac{-d\,[COOH]}{dt} = \frac{-d\,[NH_2]}{dt} = k_2\,[COOH]\,[NH_2] \tag{39}$$

With an anhydrous melt at conversions above 90%, Equation 40 applies. Although various values have been reported for the activation energy of polyamidation, a value of about 20 kcal/mol appears to be justified (12). Values of 2.6×10^{-4}/(eq/10^6 g) • min (280 °C) and 3.5×10^{-4}/(eq/10^6 g) • min (290 °C) have been reported for the second-order rate constant for melt amidation (k_2 in Equation 38). Values of 5×10^{-6}/(eq/10^6 g)^2• min (280 °C) and 9×10^{-6}/(eq/10^6 g)^2• min (290 °C) have been obtained for the third-order rate constant (k_3 in Equation 40). From these rate constants, the time required for the very high conversions needed for polyamidation can be approximated.

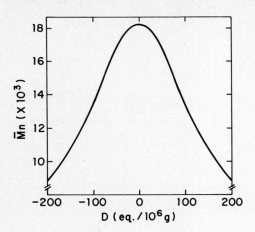

Figure 2. Effect of end group imbalance on number average molecular weight.

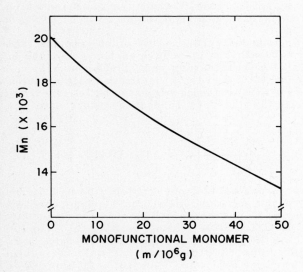

Figure 3. Effect of monofunctional monomer concentration on number average molecular weight of condensation polymer.

$$\frac{-d \text{ [COOH]}}{dt} = k_3 \text{ [COOH]}^2 \text{ [NH}_2\text{]} \tag{40}$$

For a simplified case in which the entire polymerization reaction is assumed to occur at a constant temperature under conditions that eliminate the reverse reaction (see Reaction 32), the time needed to polymerize nylon 66 to 99% conversion is given by Equation 41. This equation yields a reaction time of 17 min for polymerization at 280 °C of nylon 66 to 99% conversion (DP = 100). This calculation emphasizes the need for condensation reactions with a fast reaction rate in view of the high reaction yield needed. Reactions with rate constants significantly smaller than the constants for aliphatic amidation will require long reaction times and, therefore, will be impractical for polymer formation, particularly if monomer side reactions or polymer degradation can occur under conditions of the polymerization.

$$t99 = \frac{9}{k_2(8,800)} + \frac{99}{2k_3(880)} 2 \tag{41}$$

Thermal Degradation. Aliphatic polyamides undergo chain cleavage reactions at polymerization and processing temperatures. Degradation of many polyamides in the absence of oxygen at temperatures above 300 °C can occur by cleavage of the C–N bond to form a double bond and primary amide group as shown by Reaction 42.

$$\underset{\text{---CNCH}_2\text{CH}_2\text{---}}{\overset{\text{OH}}{}} \quad \text{----->} \quad \text{---CNH}_2 + H_2C = CH\text{---} \tag{42}$$

Polyadipamides are less thermally stable than other aliphatic polyamides because they are subject to a chain-cleavage reaction that occurs at temperatures below 300 °C. This reaction involves adipamide cyclization (Reaction 43).

$$R-NHCO(CH_2)_4CONHR \quad \text{----->} \quad RNHCO-\underset{\underset{CH_2CH_2}{|\quad|}}{\overset{CO}{\overset{/\backslash}{CHCH_2}}} + RNH_2 \tag{43}$$

For an aliphatic polyamide that degrades by amide cleavage, the net rate of polymerization at high conversion under vacuum (that is, where the water concentration approaches zero) is given by Equation 44 where k_D is the rate constant for chain cleavage, and $[A_m]$ is the amide group concentration.

$$\frac{d[A_m]}{dt} = k_3 \text{ [NH}_2\text{] [COOH]}^2 - k_D[\text{Am}] \tag{44}$$

A limiting molecular weight is obtained when the amidation and degradation terms of the equation are equal. To obtain the highest possible molecular weight with a condensation polymerization the rate constant for polymerization must be very large compared to the rate constant for any chain-cleavage reactions.

Compounds that catalyze polyamidation but not degradation will increase the attainable molecular weight. Catalysts for polyamidation include boric acid, hypophosphorous acid and its salts, and phosphoric acid (13).

Branching and Cross-linking. Branching or cross-linking may occur in condensation polymers either from the presence of polyfunctional monomers or in some cases from side reactions that occur during polymer preparation or processing. The most important effect of polyfunctionality is a broadening of the molecular-weight distribution. Equations 45 and 46 express the weight-average molecular weight for condensation polymers containing trifunctional monomers of concentration b or tetrafunctional monomers (equivalent to cross-links) of concentration ℓ. The quantity T is the total polymer end-group concentration. Units for T, b, and ℓ are eq/10^6g.

$$\overline{M}_w = \frac{4 \times 10^6}{T - 3b} \tag{45}$$

$$\overline{M}_w = \frac{4 \times 10^6}{T - 8\ell} \tag{46}$$

At high molecular weight where T is small, very small concentrations of branches or cross-links have a large effect on the weight-average molecular weight. A branch concentration equal to T/3 or a cross-link concentration equal to T/8 will produce an infinite polymer network.

Polymer melt viscosity is very sensitive to branching or cross-linking because melt viscosity is an exponential function of weight-average molecular weight. For this reason, conditions that cause branching or cross-linking must be carefully avoided in cases where high-molecular-weight polymers are to be melt polymerized or melt processed.

Literature Cited

1. Lourenco, A. V. Ann. Chim. Phys. 1863, 3, 67, 293.
2. Gabriel, S.; Maas, T. A. Ber. 1899, 32, 1266.
3. Carothers, W. H. in "High Polymers Series"; Vol. 1; Interscience: New York, 1940.
4. "Encyclopedia of Polymer Science and Technology," Vol. 10, Wiley: New York, 1969.
5. Ogata, N.; Sanui, K.; Nohmi, T. J. Polym. Sci. 1974, 12, 1327.
6. Small, P. A. Trans. Far. Soc. 1955, 51, 1717. Dainton, F. S.; Ivin, R. J. Quart. Rev. 1958, 12, 82.
7. Hall, H. K. J. Am. Chem. Soc. 1958, 80, 6404. Ibid., 1960, 82, 1209.
8. "Encyclopedia of Polymer Science and Technology," Vol. 11, Wiley: New York, 1969.
9. Flory, P. J. J. Am. Chem. Soc. 1936, 58, 1877.

10. Morawetz, H.; Otaki, P. S. J. Am. Chem. Soc. 1963, 85, 463.
11. Flory, P. J. Chem. Rev. 1946, 39, 137.
12. "Kinetics and Mechanisms of Polymerization," Vol. 3, Solomon, D. H., Ed.; Dekker: New York, 1972.
13. Charles, J.; Colonge, J.; Descotes, G. Compt. Rend. 1963, 265, 3107. Wyness, K. G. J. Chem. Soc. 1958, 2934. Flory, P. J. (to Du Pont), U. S. Patent 2 244 192 (1941). Sum, W. M. (to Du Pont), U. S. Patent 3 173 898 (1965).

Block Polymers and Related Materials

N. R. LEGGE[1], S. DAVISON, H. E. DE LA MARE, G. HOLDEN, and M. K. MARTIN[2]

Shell Development Company, Houston, TX 77001

General Discussion

The science and technology of block polymers have grown enormously since 1960. Concrete evidence of this may be seen in the several excellent texts (1–15) and proceedings of symposia (16–18) that have appeared in the past few years. Our intention is to present here a comprehensive yet concise view; namely, to show the past, present, and future of block polymers in perspective.

[1]Current address: 19 Barkentine Road, Rancho Palos Verde, CA 90274
[2]Current address: Specialty Chemical Division, 3M Center, St. Paul, MN 55144

<u>Definition and Notation</u>. In this discussion, we have elected to
define block polymers as polymers that contain chemically different
polymeric segments (blocks) attached at one or more junction points
and show definite phase separation. This definition has two
advantages. First, "block polymer" may be used in its simplest
sense, namely, a polymer that contains "blocks" or mer units.
Second, by imposing the phase segregation restriction, the
difficulties in defining the minimum segment length constituting a
"block" in different systems may be avoided. We point out that this
definition is independent of the type of synthesis used and includes
many graft copolymers. Two important exceptions should be
mentioned. 1. Blocks, but no phase segregation. This situation
occurs when the segment length is sufficiently long to constitute a
"block" but because of compatibility of the blocks, phase
segregation does not occur. An example of such a material is
poly(α-methylstyrene)polystyrene (<u>19</u>). 2. Phase separation, but no
blocks. This possibility occurs with the so-called "ionomer"
polymers. Here it is surmised that agglomeration of ionic groups
apparently occurs to an extent sufficient to impart characteristic
properties to the polymers.

There are many notations for representing block polymers. We
have followed the most common in using A to represent a block of A
mer units, with similar representation for B, C, etc., as shown in
Table I. Here A-B represents a two block polymer, A-B-C a three
block polymer, and A-B-A represents a three block polymer with A
terminal blocks and B center blocks. $(A-B)_n$ represents a linear
multiblock polymer of alternating A and B blocks.

For simplicity in this discussion we view nearly all block
polymers as falling into the six fundamental A-B or A-B-C types
shown in the box. $(A-B)_n-X$ represents a coupled polymer where the
coupler radical, X, may either be a significant structural feature,
or it may be indicated only to denote the "coupling" process; $\underline{n}$ is
normally two, three, or four.

Table I. Block Polymer Notation

Notation	Polymer Structure
A-B	A-A-A-A$\cdots$A-B-B-B-B$\cdots$B
A-B-A	A-A-A-A$\cdots$A-B-B-B-B$\cdots$B-A-A-A-A$\cdots$A
A-B-C	A-A-A-A$\cdots$A-B-B-B-B$\cdots$B-C-C-C-C$\cdots$C
$(A-B)_n$	A-A-A$\cdots$A-B-B-B$\cdots$B-A-A-A$\cdots$A-B-B-B$\cdots$B-A-A$\cdots$A

```
┌─────────────────────────────────────────────────────────────┐
│                                                               │
│     Important Types of Block Polymers                         │
│                                                               │
│   A-B              A-B-A              A-B-C                    │
│                                                               │
│   (A-B)ₙ           (A-B)ₙ-X           (X = coupler            │
│                                        n = 2,3,4)             │
│                                                               │
│   B···B-B-B┌B┐B-B-B···B                                       │
│            │A│ₙ          (n random grafts                     │
│            └ ┘           of A-block on B-block)               │
│                                                               │
└─────────────────────────────────────────────────────────────┘
```

In this notation A and B can represent different stereo configurations, or microstructures of the same mer. Likewise, either A and/or B may be random or tapered copolymer blocks rather than homopolymeric blocks. A tapered A-B block is one that begins with a section of pure A mers and gradually incorporates B mers until at the end it becomes a section of pure B mers. The A-B-C designation represents the incorporation of a third monomer segment and would most probably comprise a three-phase system. An A-B-A block polymer based on styrene and butadiene would mean in this notation a polystyrene block-polybutadiene block-polystyrene block structure, that is, S-B-S.

S-B-S and S-I-S. Much of our discussion will refer directly to data for S-B-S and S-I-S block polymers. We justify this on several counts. These block polymers can be clearly defined as to structures, molecular weights, and compositions. They have served as model systems for much of the recent work in block polymers. They also comprise the largest volume of commercial block polymers. Finally, we believe that the discovery (20) of the S-B-S and S-I-S thermoplastic rubbers which are strong, resilient rubbers without vulcanization, and the concomitant, readily understood theory (21), provided a paradigm (terminology of T. S. Kuhn (22)) that significantly accelerated the scientific work on these polymers in recent years.

Phase Separation. Except in rare instances, different polymers do not mix at the molecular level. This behavior is a natural consequence of the very high molecular weight of polymers which results in a very small entropy of mixing, of insufficient magnitude to offset the usual small but positive heat of mixing. As a result, at equilibrium the dissimilar segments in block polymers undergo separation into different phases in accordance with the phase rule. Reduction of the positive heat of mixing is sometimes achieved by introduction of interacting groups into the polymer pair. As a matter of fact, in some of the few existing cases of compatible polymer pairs, miscibility is achieved because of the presence of interacting polar groups, rather than because of structural similarity.

In block polymers, because of the constraints imposed by the chemical bonding between segments, the individual phases are highly interspersed with one another, usually on a very small scale. The phenomenon has been referred to as a "microphase" separation. However, the phase dimensions are frequently submicroscopic on an optical instrument, and separation can only be observed by electron microscope. The phase topology is influenced by a number of factors that determine whether the various phases exist in the form of dispersed spheroids, cylinders, layers, or as a continuous matrix. Submicroscopic dispersed phase entities have been referred to as domains.

Properties. Physical and mechanical properties of block polymers are significantly influenced by the morphology of the multiphase structure and thus differ greatly from properties predicted from a linear combination of the polymer components. For example, a hard phase dispersed in a soft phase may introduce virtual cross-links or reinforcement to increase the strength or elastic response of the polymer. On the other hand, each phase can still behave characteristically, independent of the other phases, showing transition temperatures, dynamic loss responses, and individual segment solubility behavior, characteristic of the corresponding homopolymers. As a result, opportunities are presented for the design of novel materials for specific applications requiring composite materials. Here the advantages are high uniformity, control of size and shape of phases, and assured wetting between phases.

Historical Review--A Pathway to Discovery

Kuhn (22) suggests that progress in science proceeds via a succession of paradigms that can be accepted as models by scientific groups. These paradigms are continuously subject to further redefinition and articulation by additional scientific research. As new technological areas are entered, there is a period of exploration and groping. At some point a paradigm appears and is accepted by a large enough group to generate a burst of activity. Then, in due course of time, new paradigms appear, and the field expands.

We suggest that the S-B-S thermoplastic rubbers and the domain theory have produced a paradigm upon which the block polymer field advanced significantly. In reviewing the technological discovery of the S-B-S thermoplastic rubbers and the virtually simultaneous formation of a theory that enabled us to move rapidly toward commercialization, it was of interest to us to trace a pathway to discovery for the relatively small group concerned with it. We have set forth in Table II some selected events that we believe enabled us to very rapidly understand the physical phenomena and proceed. Table II is, then, not intended in any way to be a comprehensive illustration of the history of block polymers but is rather a discrete list of events leading to the discovery of S-B-S and S-I-S thermoplastic rubbers in our laboratory.

In the early part of the 50 years commemorated by this book, there was very little block polymer science or technology to record. Probably the first block polymer reported was in 1938 by Bolland and Melville (23), who found that a film of poly(methyl methacrylate) deposited on the walls of an evacuated tube could initiate the polymerization of chloroprene. Later Melville concluded (24) that

Table II. Events of Significance in the Pathway to Discovery

Year	Event
1938	Bolland and Melville--Free radicals trapped in poly(methyl methacrylate) initiate chloroprene polymerization to give graft polymers.
1946	Baxendale, Evans, and Parks--Styrene and methyl acrylate prepolymers linked with diisocyanate to give (A-B)n block polymers.
1951	Lundsted (Wyandotte)--Pluronic (A-B-A block polymer of ethylene oxide and propylene oxide).
1953	Mark--Discussion of A-B and (A-B)n block polymers and their surface-active character.
1955	Horne (Goodrich)--Ziegler-type synthesis of high cis-polyisoprene.
1955	Stavely (Firestone)--Lithium metal-initiated anionic high cis-polyisoprene.
1956	Szwarc--Sodium naphthalene anionic diinitiators. Living polymers. A-B-A isoprene-styrene-isoprene block polymers. A-B and A-B-C discussed.
1957	Bateman (BRPRA)--Methyl methacrylate grafts on natural rubber.
1957	Porter (Shell)--Polydiene-styrene block polymers.
1958	Schollenberger (Goodrich)--Virtually cross-linked polyurethane elastomers.
1959	Shell team--First commercial production of high cis-polyisoprene by anionic polymerization.
1961	Crouch (Phillips)--Commercial production of S-B block polymers for vulcanized applications.
1965	Shell team--Thermoplastic elastomers with high strength and resilience through physical cross-linking
1966	Cooper and Tobolsky--Correlated behavior of polyester-polyurethane (A-B)n block polymers with that of S-B-S block polymer.
1967	Shell team--Formal domain theory. S-B-S and S-I-S physical chemistry and rheology.
1970	Allied Chemical--ET polymers (graft of butyl rubber on polyethylene).
1971	Uniroyal--TPR thermoplastic rubber.
1972	Du Pont--Hytrel (polyether-crystalline polyester thermoplastic rubber).
1972	Shell team--Second generation block polymers Kraton G thermoplastic rubber (S-EB-S)

the second polymerization was initiated by trapped free radicals in the film.

In 1946, Baxendale, Evans, and Parks reported (25) on a step-wise formation of an $(A-B)_n$ block polymer in which low molecular weight polystyrene and methyl acrylate prepolymers were linked with diisocyanate to form a high molecular weight, multiblock polymer.

An early industrial A-B-A polymer was introduced in 1951 by Wyandotte Chemical (26). This polymer was the nonionic detergent, Pluronic, a three block polymer with poly(ethylene oxide) end blocks and a poly(propylene oxide) center block. The degree of polymerization of the blocks was between 20 and 80.

A stimulating paper on the synthesis and application of "block" and "graft" polymers by H. F. Mark was published in 1953 (27). "Multiblock" and "multigraft" polymers of A and B where, in general, the A or B segments were 50 Å long, were discussed principally from the viewpoint of detergency.

Commencing in 1955, three major breakthroughs in polymerization profoundly affected future work in polymer synthesis. Two of these were presented at the same meeting of the American Chemical Society (ACS) Rubber Division in 1955: the work of Horne and coworkers of Goodrich on Ziegler-catalyzed high *cis*-polyisoprene (28), and that of Stavely and coworkers of Firestone on lithium metal anionic polymerization of high *cis*-polyisoprene (29). With respect to block polymers, the Firestone anionic polymerization was a key advance. These solution polymerization systems were very sensitive to hydrocarbon and polar impurities which affected the purity of polymer structure. Fortunately, the gas-liquid chromatographic (GLC) techniques advanced rapidly enough to permit commercial development of these systems. It is probable that by early 1956 most major synthetic rubber research groups were examining these new polymerization systems. Over the next few years, high *cis*-polybutadiene was also under study.

The third major breakthrough was by Szwarc and coworkers who reported in 1956 (30, 31) that certain anionic polymerization systems gave "living polymers" to which a second monomer could be added without termination, so that block polymers could be formed. Interest of research groups in this system was dissipated somewhat by the finding that the sodium naphthalene diinitiator used by Szwarc resulted in undesirable 1,2- or 3,4-addition structures in diene polymers.

It is likely that other groups that studied the Firestone lithium metal initiator came to the same conclusion as Firestone (32), namely, that lithium alkyls would be more suitable initiators than lithium metal. Also, the polymerization was determined to be anionic, and thus the Szwarc living polymer systems and the alkyl lithium initiator systems were easily connected. In time, many published papers confirmed these points. Workers at Shell (40) and other research groups concluded that the lithium alkyl-initiated anionic polymerizations of isoprene, butadiene, styrene, etc. were living polymer systems capable of producing block polymers. The group at the University of Akron, under the   direction of Morton, contributed very significantly to the study of anionic polymerization and block polymers based on this system.

Ceresa (33, 34) reviewed "block" and "graft" polymerization of natural rubber extensively and pointed out (34) that this work reached its peak in 1959. For many in synthetic rubber research, the comprehensive work of Bateman and coworkers at the British Rubber Producers Research Association was the most familiar. In 1957 Bateman (35) and Merrett (36) showed that graft polymers of methyl methacrylate (PMMA) on natural rubber (NR) could exist in two physical forms depending on the method of precipitating the polymer. If precipitated so that the NR chains were collapsed and the PMMA chains were extended, the material was hard, stiff, and nontacky. If precipitated so that the NR chains were extended and the PMMA chains were collapsed, it was soft and flabby. Both of these forms were stable under heavy milling. They were evaluated as vulcanized systems and compared with vulcanized NR. It was found that the physical form of the graft affected the vulcanizate properties (35). Merrett (36) discussed the phenomenon in terms of microseparation of phases. He pointed out that microseparation to give either block collapsed will represent negative free energies of interaction and once formed, any attempt to approach homogeneity would be improbable. He also postulated that either form of the dry polymer would consist of domains of collapsed chains as a discrete phase in a continuous phase of the extended chains.

From the viewpoint of history, Bateman's introductory comments (35) in 1957, characterizing the state of the work on NR graft polymers, are worth noting. He described the current (1957) state of knowledge as rudimentary and pointed out that before that time no evidence was available to show how, or whether, the NR-PMMA grafts behaved differently from the corresponding mixtures of homopolymers.

In 1958 Schollenberger, Scott, and Moore (37) of Goodrich reported on a polyurethane elastomer, Estane, which was a linear polyurethane described as being "virtually cross-linked" by secondary rather than by primary valence forces.

At Shell, research on anionic solution polymerization systems commenced in mid-1955 and covered a broad area. One of the outgrowths of this work was the first commercial production of high *cis*-polyisoprene (IR) in 1959 at the Torrance, California, plant. In this case, our confidence in this polymerization system was such that we went directly from bench scale to commercial scale. The very strong analytical background, particularly in GLC, developed at the Emeryville Research Center, was a great strength in all of this anionic research.

In 1961 Crouch and Short (38) discussed the use of S-B block polymers, although not identified as such, in vulcanized rubber applications. Commercial production of the S-B block polymer commenced late in 1962. Identification of the S-B block structure was presented by Railsback, Beard, and Haws (39) in 1964.

We can also assess the state of knowledge on block polymers in 1962. Ceresa (33), in his book "Block and Graft Copolymers" included the syntheses of 1400 block copolymers. He pointed out that less than 5% of the block and graft polymers described had been isolated with a reasonable purity and that fewer than 20 species had been analyzed and characterized fully. He attributed this situation to the fact that linear or graft block polymers are contaminated with homopolymers. In many cases the block polymer was present as only a small percentage of the final mixture.

In the first Shell work in block polymers, two- and three-block polymers of styrene and butadiene and of isoprene and styrene were successfully made very early (40). Although these early polymers were not optimal compositions, they contributed most importantly to our background of knowledge.

During this period Shell also had active programs on high *cis*-polybutadiene rubber (*cis*-BR) and ethylene/propylene rubber (EPR). We, as well as others, found that the solution-polymerized BRs and IRs had cold flow and processability problems that we had not seen before in years of research in synthetic rubber emulsion systems. Processability was a major problem in the *cis*-BR project, and we gathered together a team that spent considerable time in studying rheology and processing of these new polymer systems.   It was apparent that the copolymerization of styrene with the butadiene would improve processability significantly, as in emulsion polymers. However, we were aiming with great intensity at high *cis* structures and at high resilience so that we generally discarded the idea of styrene copolymers, which had lower resilience than natural rubber (NR), attributed to the bulky styrene groups on the chain.   Somewhat later, work at the Torrance Research Center showed that alkyl lithium-initiated BR, although an attractive blending polymer, had too low a viscosity in the raw state to be handled in the existing drying process.   The alkyl lithium-initiated BR is not a structurally pure polymer but contains about equal amounts of *cis* and *trans* structures with 10-15% of 1,2-addition.   In the blending case we were not so concerned about resilience and looked upon the lower *cis*-BR as contributing high abrasion resistance.

The next step to discovery was the suggestion that we might lessen the cold flow of BR by adding very short end blocks of polystyrene.   Because the chain ends do not contribute appreciably to the vulcanized state, we hoped that the abrasion resistance and resilience of polybutadiene would be little affected by the very low molecular weight end blocks of polystyrene.   Processability problems with the polybutadiene led us to use a low molecular weight in the polybutadiene portion.   Samples of this type were made for evaluation.

Fortunately for our project and for the development of block polymers, we had an acutely observant team member, Hendricks, who prepared unvulcanized sheets of the three-block polymer for rheological measurements.   He immediately reported that the "unvulcanized" sheets were unusually "snappy" and appeared to be "scorched."   Subsequent stress-strain measurements showed high tensile strength and high ultimate elongation.   However, solution data showed complete solubility, with no gel or microgel present.

At this point, we had the first four of the seven characteristic features of A-B-A thermoplastic elastomers, as shown in the box. That is, we were completely confident that we had a three-block polymer, rubbery behavior with high tensile strength in the unvulcanized state, and also complete solubility. We concluded from these properties that these polymers were two-phase systems.   We then generated the essentials of the two-phase, domain theory and visualized the physical structure illustrated schematically in Figure 1.   We also visualized applications in footwear, in injection-molded items, and in solution-based adhesives.   Positive confirmation of the two-phase structure quickly followed, by detection of two separate glass transition temperatures, as well as observation of the thermoplasticlike reversibility of bulk- and

melt-state properties with temperature. These events established
the reality of the last three items shown in the box.

Characteristic Features of A-B-A Thermoplastic Elastomers

Three-Block Polymer

High-Strength Rubber

No Vulcanization Required

Completely Soluble

Reversible Melt-Bulk Properties

Two Glass Transition Temperatures

Two Phases

In retrospect, the pathway to discovery was not direct. We were
looking for improved vulcanizable rubbers. Instead we discovered
rubbers that did not require vulcanization (Figure 2). However, the
events listed in Table II suggest that one could have found a more
direct pathway to the discovery of the importance of two
interconnecting phases in polymers.

The immediate accessibility of thermoplastics processing
equipment and plastics applications knowledge enabled us to move
rapidly in product development. The availability of readily
modified plant equipment at Torrance and a receptive, knowledgeable
manufacturing organization accelerated our progress.

This discovery culminated in the commercial production and the
announcement (41) in 1965 of thermoplastic elastomers from block
polymers of styrene and butadiene (S-B-S) and of styrene and
isoprene (S-I-S). To rubber scientists and technologists the most
outstanding property of S-B-S and S-I-S was the unvulcanized tensile
strength compared to that of vulcanized NR and vulcanized SBR carbon
black stocks. Stress-strain curves, to break, of these latter
materials are compared to that of S-B-S in Figure 2. It was pointed
out that the high strength of S-B-S must be due to physical cross-
links.

Although we did not publish our domain theory in 1965, it was
fairly evident to those acquainted with the historical developments
in Table II that the new products were three-block polymers.
Researchers in the field very rapidly commenced to use these
polymers as models for comparison and as subject for physicochemical
studies. Evidence for this may be seen in the work reported by
Cooper and Tobolsky (42) in 1966, in which they correlated the
behavior of polyester-polyurethane thermoplastic rubbers (Estane
products) (37) with those of a Shell S-B-S polymer. They concluded
that the presence of segregated hard and soft phases in the Estane

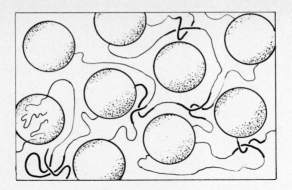

Figure 1.   Phase arrangement in A–B–A block polymers. (Reproduced
            with permission from Ref. 100.   Copyright 1969 Wiley.)

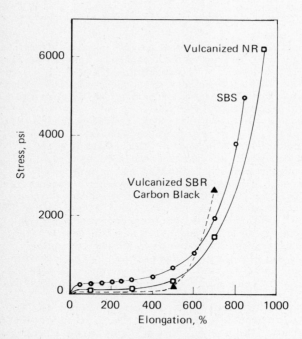

Figure 2.   Comparison of stress-strain responses of S–B–S and two
            conventional vulcanizates.

polymers, not hydrogen bonding, was the source of the thermoplastic elastomer behavior.

Structural and thermodynamical aspects of the phase separation, domain theory, were presented in 1967 to the International Rubber Conference (21) in the United Kingdom and to a symposium (18) at the California Institute of Technology later in the year.   Network properties of the A–B–A polymers were also described (18). Following the commercial announcement in 1965 and the meetings in 1967, there was an explosion of requests for samples of S–B–S and S–I–S products. We then prepared laboratory samples and offered these along with detailed molecular characterization data to outside research workers for study.

The impetus given to the block polymer field by discovery of the diene–based thermoplastic rubbers, S–B–S and S–I–S, is suggested by the data of Figure 3. Here we show a histogram of the number of U.S. patents on block polymers generally related to the thermoplastic rubbers, over the period of 1961–80. By taking into account the lag in patent issuance, the sharp increase of patents issued from 1966–1970 and subsequently confirms the upsurge of interest.

For these reasons, we feel that the discovery of A–B–A thermoplastic rubbers and the domain theory gave the block polymer field a paradigm (22) that greatly accelerated research on a worldwide basis. The content as well as the volume of literature support this view (Figure 3). The usefulness of the paradigm was based on the well–characterized block polymers that   can be produced from the anionic polymerization system as shown in the box and on the readily understood basics of the domain theory.

Anionically Polymerized A–B–A Model Polymers

Predictable Molecular Weights

Narrow Molecular Weight Distributions

High–Purity Block Compositions

Facile Coupling or End–Group Functionalization

In the last years of the history shown in Table II we see the announcements of new commercial thermoplastic rubbers. Uniroyal TPR appeared in 1971. (This thermoplastic rubber many not be a block polymer. Presumably it is a blend that achieves its properties by virtue of interpenetrating networks between the plastic and rubber constituents. The exact structure has not been disclosed.) Du Pont's Hytrel, an $(A-B)_n$ polyetherpolyester thermoplastic rubber, came out in 1972. Also in 1972, Shell announced a second generation block polymer, Kraton–G, which is a three–block S–EB–S thermoplastic rubber (EB represents an ethylene–butylene rubbery midblock).

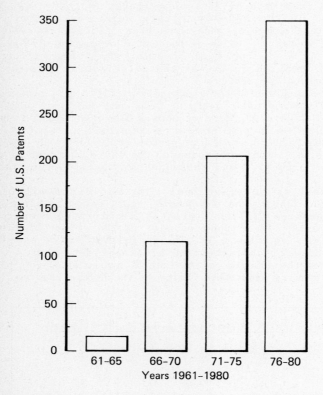

Figure 3.   U.S. patents on block copolymers issued between 1960 and
1980.

Synthesis--A Survey of Methods

<u>Nongrafted Types</u>.   (See Glossary at end of chapter for the meaning of the various letter abbreviations for monomers and polymers.)

Anionic Methods. Of the various methods used for nonpolar block polymer synthesis, the anionic approach is by far the most useful and most precise in producing a given segmented chain design [A-B-A, $(A-B)_n$, A-B-C, etc.]. The linear A-B-A block polymer may be synthesized by two basic routes, sequential polymerization of all three blocks or sequential polymerization of two blocks, followed by coupling. The "sequential" method is illustrated in Reaction la for the preparation of S-B-S; the "coupling" process is depicted in Reaction lb.

$$s\text{-BuLi} \xrightarrow{\text{Styrene}} \text{S-Li} \xrightarrow{\text{Butadiene}} \text{S-B-Li} \xrightarrow{\text{Styrene}} \text{S-B-S} \qquad (1a)$$
$$\phantom{s\text{-BuLi} \xrightarrow{\text{Styrene}} \text{S-Li} \xrightarrow{\text{Butadiene}} \text{S-B-Li}} \xrightarrow{\text{Coupler}} \text{S-B-B-S} \qquad (1b)$$

In either method alkyllithiums are the standard initiators employed. *Sec*-Butyllithium (*s*-BuLi) is frequently used because its rate of addition to styrene or conjugated diolefins is very fast. In Reaction la there are essentially three initiation steps, or more precisely one initiation and two crossover steps.

Reaction	Rate Constants
s-BuLi + Styrene ⟶ S-Li	$k_{s\text{-BuLi-S}}$
S-Li + Butadiene ⟶ S-B-Li	$k_{S\text{-B}}$
S-B-Li + Styrene ⟶ S-B-S-Li	$k_{B\text{-S}}$

Narrow molecular weight distribution and successful block construction depend upon favorable ratios of the rate constants of each step to their respective propagation rate constants, that is, $k_{s\text{-BuLiS}}/k_{SS}$, k_{SB}/k_{BB}, and $k_{B\text{-S}}/k_{SS}$. In an extreme case for which the ratio for any given stage is much less than one ($k_{propagation} \gg k_{crossover}$) then only a fraction of the chains would be fully "blocked," and a mixture of block polymer species would be obtained. For example, a mixture of A-B and A-B-A would result if $k_{BS}/k_{SS} \ll$ 1. Favorable crossover and initiation kinetics are often obtained by addition of RLi-complexing agents (ethers, amines, etc.) to hydrocarbon solvents. Microstructure of the diene block can also be regulated by the amount of such complexers added to the hydrocarbon solvent. If high 1,4-polydiene microstructure is desired, one works in an all-hydrocarbon solvent or in a hydrocarbon solvent with low levels of carefully selected complexing agents, for example, anisole (<u>43</u>). If high 1,2- or 3,4-diene microstructure is desired, then one works with larger amounts of complexing agents or in an all-ether or an all-tetrahydrofuran solvent.

As shown in Reaction lb, another common route in anionic polymerization of A-B-A polymers is that of coupling living A-B polymer chains to give A-B-B-A (equivalent to A-B-A) polymers. A particular advantage of this method is that it avoids any skewing of the last block because of unfavorable crossover kinetics, and therefore the molecule is completely symmetrical. Reactions of living polymers with pure coupling reagents such as the following (Reactions 2-5) are surprisingly fast and efficient and give 80 to 95% coupling at 25 to 75 °C. Coupling agents may be incorporated into the chain; however, some function without incorporation, for example, Reaction 4.

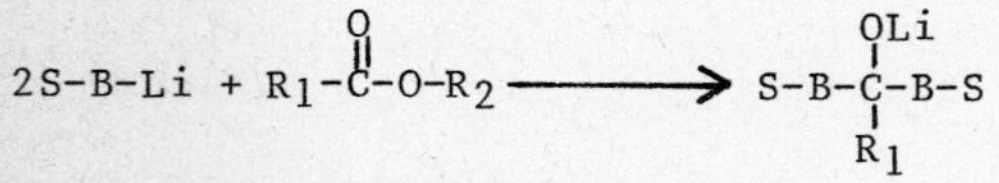
2S-B-Li + R1-C-O-R2
S-B-C-B-S
OLi
R1

(2)

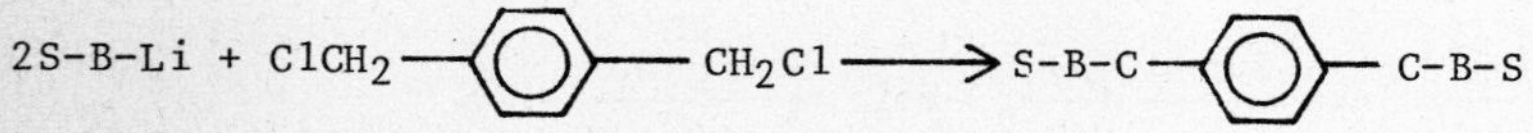
2S-B-Li + ClCH2
CH2Cl
S-B-C
C-B-S

(3)

$$2S\text{-}B\text{-}Li + I_2 \longrightarrow S\text{-}B\text{-}B\text{-}S \qquad (4)$$

$$4S\text{-}B\text{-}Li + SiCl_4 \longrightarrow (S\text{-}B)_4Si \text{ and } (S\text{-}B)_3SiCl \qquad (5a)$$

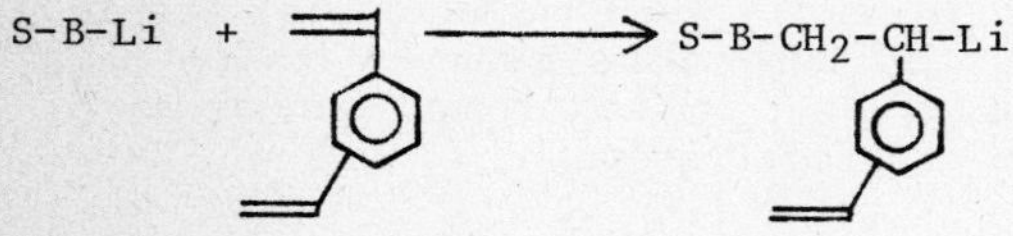
S-B-Li +
S-B-CH2-CH-Li

(5b)

(I)

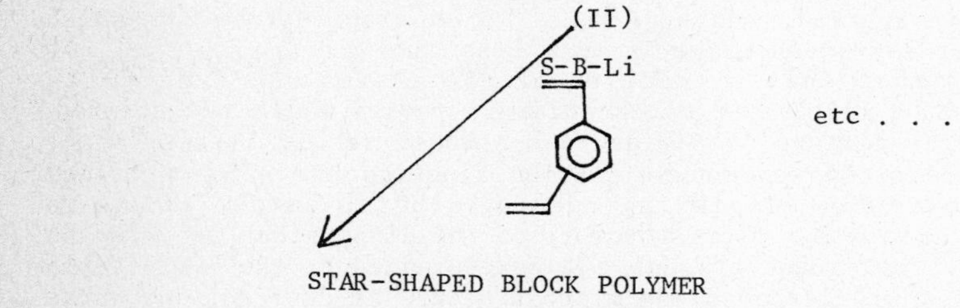
(I) + n
+ S-B-Li
S-B(CH2-CH-CH2-CH)n-CH2-CHLi
CHLi
S-B-CH2
(II)
S-B-Li
etc . . .

(5c)

STAR-SHAPED BLOCK POLYMER

(III)

Coupling Reaction 5b involves reacting living A-B polymer chains with divinylbenzene (DVB). The polymerization of DVB then results in a hub containing pendent vinyl groups that serve as branch points for the star-shaped polymer. This procedure has led to the formation of star-branched polyisoprenes containing up to 56 weight-average number of arms (44). Recent investigations (45–47) have further elucidated the kinetics of the DVB coupling process.

Di-initiation of anionic block polymers is another useful way to begin block polymer syntheses. However, di-initiation has not yet achieved the prominent position that simple monofunctional initiation has in commercial production. For one reason, di-initiators are difficult to obtain in a stable soluble form without some ether, amine, or other solvating agent present, and thus some alteration of microstructure is expected. However, recent investigations claim there are difunctional initiators that do not significantly alter the polydiene microstructure (48–50). Initiator agglomeration was prevented by capping each lithio site with isoprene to form polyisoprenes of short chain length (i.e., DP=5). Small quantities (0.1 equivalents) of triethylamine were added to further improve the initiator solubility in hydrocarbon solvents:

$$\left[Et_3\text{-N}\right]_{0.1} \text{Li}-(\text{PI})_5 - \overset{\overset{R}{\overset{|}{C}}}{\underset{CH_3}{\overset{|}{C}}} - \underset{}{\bigcirc} - \overset{\overset{R}{\overset{|}{C}}}{\underset{CH_3}{\overset{|}{C}}} - (\text{PI})_5\text{-Li}\left[\text{NEt}_3\right]_{0.1}$$

The microstructure of a polyisoprene prepared by this initiator was reported to contain 92% 1,4-units and 8% 3,4-units.

Di-initiators are indispensable to the preparation of such polymers as 2VP-I-2VP (Reaction 6) and EO-B-EO (Reaction 7). As shown in these reactions (51), crossover from I-Li to B-K to 2VP or EO monomer, respectively, is possible, but the reverse steps, that is, 2VP-Li to isoprene, or EO-K to butadiene, are not possible; the rate constants are 0.

$$\text{Li-}(\phi)_2\text{C-C-C-C}(\phi)_2\text{-Li} \xrightarrow{\text{Isoprene}} \text{Li-I-Li} \xrightarrow{\text{2-Vinylpyridine}} \text{2VP-I-2VP}$$

$$\quad\quad\quad C \quad\quad\quad\quad C \tag{6}$$

$$\text{K-}(\phi)\text{C-C}(\alpha\text{MS})_2\text{-C-C}(\phi)\text{-K} \xrightarrow{\text{Butadiene}} \text{K-B-K} \xrightarrow{\text{Ethylene Oxide}} \text{EO-B-EO} \tag{7}$$

Thus, mono-initiators are useful for making A-B-C where C is polar, but reverse crossing from C back to B or A is not possible (k_{CB}, k_{CA} = 0). Therefore, block construction such as C-B-A-B-C, C-A-C, or simply C-B-C where C is a polar block, are of necessity, prepared dianionically.

In a practical sense the hydrocarbon monomers that work best in anionic systems are styrene, α-methylstyrene, p-(tert-butyl)styrene, butadiene, isoprene, 2,3-dimethylbutadiene, piperylene, stilbene, and 1,1-diphenylethylene. The latter two monomers give rise to alternating copolymers with other dienes but do not homopolymerize. Among the polar monomers (C) that can be polymerized are such monomers as 2-vinylpyridine, pivalolactone, methacrylonitrile, methyl-methacrylate, ethylene oxide (not with Li-counterion), ethylene sulfide, and propylene sulfide. However, polymerization of many of these polar monomers suffers from side reactions and complicating termination or transfer reactions not present in the well-behaved hydrocarbon system. In some cases the Li counterion must be replaced by, for example, a K counterion for successful polymerization. In other cases, the Li-C bond must be first converted to a Li-O bond, for example, with CO_2 or $\underset{O}{C}-C$.

Yamashita has reported the anionic synthesis of a poly-α-methylstyrene-poly-ε-caprolactone (α-MeS-ε-C, Mn ∿4 x 10^3 - 5.5 x 10^3), but the second block formation is complicated by a depolymerization equilibrium to give cyclic oligomers (52).

Richards has recently reported and speculated on some techniques using anionic and cationic living chains to produce polymers and copolymers possessing ionic groups, that is, quaternary ammonium bonds, at specified points along the backbone (53).

Recent developments have also been reviewed for the synthesis of telechelic (functional groups at both ends) and semitelechelic (functional group at one end) polymers via anionic methods (54). The use of two basic procedures is reported: 1. termination of living anionic chains with suitable electrophiles, and 2. the use of functionally substituted anionic initiators. Two of these latter initiators are acetals that give good molecular weight control and monodispersity:

$$\underset{\underset{OC_2H_5}{|}}{Li-R-O-CHCH_3} \qquad Li-R-O \overset{S}{\underset{O}{\bigcirc}}$$

Hydroxyl functionality can thus be introduced at the chain terminus by hydrolysis of the acetal function.

Anionic systems generally have the following advantages when stringent purification and polymerization techniques are followed:

1. Predictable molecular weight. MW = grams of monomer/moles of initiator or MW = grams of monomer x functionality of initiator/equivalents of C-Li. These simple equations hold for either mono- or di-initiators.
2. Narrow molecular weight distribution with high-purity block compositions. (M_w/M_n) values may approach 1.0 when $k_{initiation}/k_{propagation}$ values are very favorable.
3. Facile and efficient paths for "coupling" of segments into linear, branched, or star-shaped species.
4. Efficient "end-of-chain functional capping" with reagents such as $\underset{O}{C}-C$, CO_2, and $\underset{O}{C-C-C}-SO_2$.

Cationic Methods. Cationic methods have also found utility in preparation of block polymers; however, generally speaking these methods suffer from a variety of synthesis problems including monomer-chain end equilibria, facile chain end termination or transfer with certain counteranions, and difficult molecular weight control. Most examples of living cationic systems include the use of an oxonium cation with a carefully selected counterion (Reaction 8).

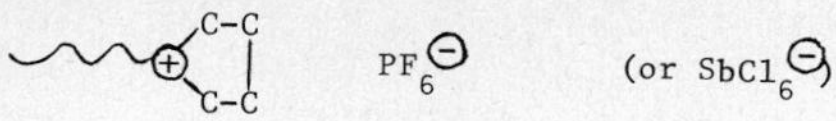

$$\sim\!\!\!\bigoplus\!\!\begin{array}{c} C-C \\ | \\ C-C \end{array} \quad PF_6^{\ominus} \qquad (\text{or } SbCl_6^{\ominus}) \tag{8}$$

Some A–B block copolymers have been reported (55) by using the approach of the living oxonium ion system and procedure (Reaction 9):

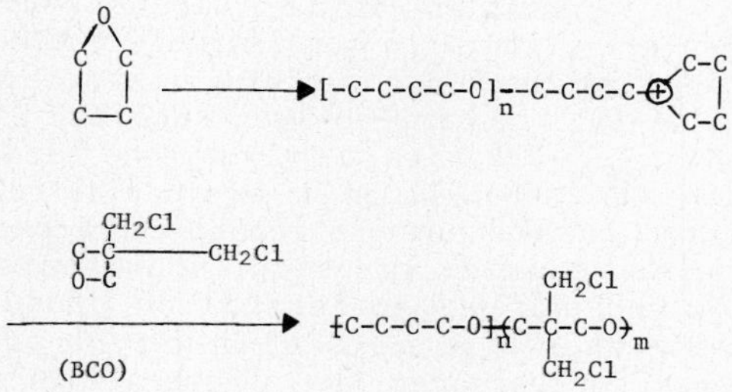

$$\begin{array}{c} O \\ C \quad C \\ | \quad | \\ C-C \end{array} \longrightarrow [-C-C-C-C-O]_n\!-\!-C-C-C-C\!\!\bigoplus\!\!\begin{array}{c} C-C \\ | \\ C-C \end{array}$$

$$\begin{array}{c} CH_2Cl \\ | \\ C-C\!-\!\!-\!\!-CH_2Cl \\ | \quad | \\ O-C \end{array} \longrightarrow \{C-C-C-C-O\}_n\{C-\!\!\!\!\begin{array}{c} CH_2Cl \\ | \\ C \\ | \\ CH_2Cl \end{array}\!\!\!\!-C-O\}_m \tag{9}$$

(BCO)

THF–BCO

However, identification of cationic block polymers has not always been definitive because of the complexities of the reaction systems and a failure to do complete characterization studies on the polymers.

A novel two-stage initiator (56) for cationic polymerization has been described (Reactions 10 and 11):

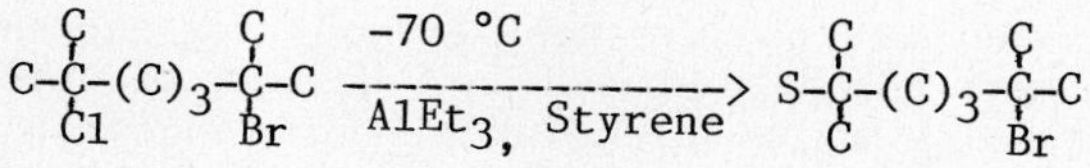

$$\begin{array}{ccc} C & & C \\ | & & | \\ C-C-(C)_3-C-C \\ | & & | \\ Cl & & Br \end{array} \xrightarrow[\text{AlEt}_3,\ \text{Styrene}]{-70\ °C} \begin{array}{ccc} C & & C \\ | & & | \\ S-C-(C)_3-C-C \\ | & & | \\ C & & Br \end{array} \tag{10}$$

$$\xrightarrow[\text{Isobutylene}]{\text{AlEt}_3} \begin{array}{ccccc} C & C & C \\ | & | & | \\ S-C-(C)_3-C-(C-C)_n-C-C=C \\ | & | & | \\ C & C & C \end{array} \tag{11}$$

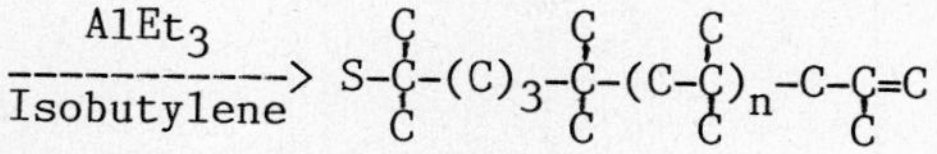

S–BU

Reaction 10 depends on the selective removal of the *tert*-chloride ion and cationic polymerization of styrene in the first step. Subsequently and at higher temperatures the *tert*-bromide function is used to initiate the cationic polymerization of isobutylene (Reaction 11).

Anionic–Cationic (Ion Coupling) Methods. Anionic–cationic termination reactions have been reported (57) to be useful in the synthesis of block polymers of the type THF–S and THF–S–THF (Reaction 13):

$$\ominus\!\sim\!\!\!\sim\!\! S \sim\!\!\!\sim\!\ominus \quad + \; 2\sim\!\! \overset{\oplus}{O}\!\!\begin{array}{c} C\text{--}C \\ | \\ C\text{--}C \end{array} \quad \text{------->} \; THF\text{--}S\text{--}THF \qquad (13)$$

THF

Although the process looks straightforward on paper, it is difficult to carry out in practice because of facile elimination and other side reactions. Consequently, there is considerable doubt about the utility of the ion coupling method for block polymer synthesis. Multiblock polymers, $(A\text{--}B)_n$, have also been reported (58) by ion coupling of a dicationic poly–THF with a polystyrene dicarboxylate. The polystyrene dianion was carboxylated to avoid side reactions in the ion coupling process, and the authors report the formation of a 60–segment polymer. The reaction was slow, accompanied by some homopolymer formation and was not successful in producing high molecular weight $(A\text{--}B)_n$ when high molecular weight presegments (>10,000) were employed. Furthermore, the stress–strain properties of the product were not determined; therefore, the quality of the $(A\text{--}B)_{30}$ was not fully defined. Although the ion–coupling reaction may have some laboratory utility, it is not currently a practical route to block polymers.

Polycondensation or Step–Growth Methods. Polycondensation routes offer a variety of block polymers, most notably the elastomeric polyurethanes. The latter polymers contain flexible or "soft" segments alternating with rigid or "hard" segments. (Soft and hard phases used in this sense refer to polymers that are above and below their respective glass transition temperature (or crystalline melting point) at the use temperature.) The soft segments are usually rubbery or amorphous polyether or polyester segments, and the hard segments are polyurethane or polyurea segments. These $(A\text{--}B)_n$ block polymers can be prepared in a variety of ways, but they are often made by using a low molecular weight prepolymer (polyester or polyether with a hydroxyl difunctionality) that has been terminated with excess diisocyanate (Reaction 14):

HO–(Eth)–OH+ excess OCN–R_1–NCO ------>

$$OCN\text{--}R_1\text{--}N\text{--}\overset{\overset{O}{\|}}{C}\text{--}O\text{--}(Eth)\text{--}O\text{--}\overset{\overset{O}{\|}}{C}\text{--}N\text{--}R_1\text{--}NCO \qquad (14)$$

I (Soft Segment)

The soft prepolymer segment from Reaction 14 is in turn condensed with a low molecular weight chain extender (HO–R_2OH) and excess diisocyanate (Reaction 15) to give the block polymer $(A\text{--}B)_n$ which possesses soft polyether segments and hard crystalline (high MP) polyurethane segments:

$$I + xHO-R_2-OH + (X-1)OCN-R_1-NCO \longrightarrow$$

$$\underset{\text{Soft}}{\overset{O\ H}{\underset{}{\text{[C-N-R}_1}\text{-N-C-O-(Eth)-O-}}}\underset{\text{Hard}}{\overset{H\ O}{\underset{}{(\text{C-N-R}_1}\text{-N-C-O-R}_2\text{-O)}_x]_n}} \tag{15}$$

A range of properties is obtainable and depends on the amount of diisocyanate used in excess of the molar amount of prepolymer diisocyanate (59).

Hytrel is a random segmented polyester made (60, 61) by the equilibrium melt transesterification of dimethyl terephthalate (DMT), 1,4-butanediol, and polytetramethyleneglycol (PTMEG) (Reaction 16):

$$[-O-PTMEG-O-\overset{O}{\overset{\|}{C}}-\underset{\text{Soft}\atop\text{Rubbery}}{\bigcirc}-\overset{O}{\overset{\|}{C}}]_x[O-C-C-C-C-O-\overset{O}{\overset{\|}{C}}-\underset{\text{Hard Crystalline}}{\bigcirc}-\overset{O}{\overset{\|}{C}}]_y \tag{16}$$

The process involves transesterification (catalyst, 200 °C) followed by polycondensation (250 °C, <1mm) in a second stage. Morphological studies show the presence of crystalline (mp 190–200 °C) polyester lamellae in a continuous amorphous phase. In contrast to the A–B–A thermoplastic elastomers where the domains are formed from amorphous polystyrene segments, the domains here are formed from crystalline hard segments containing the 1,4-glycol polyester moiety.

A variety of block polymers, $(A-B)_n$, of variable "hard" and "soft" segments have been made (62, 63) by condensing rubbery α,ω-bis(dimethylamino)-polydimethylsiloxane (DMS) oligomers with hard oligomers containing *bis*-phenolic or hydroxyl end functions (Reaction 17):

$$HO-C-C-\alpha MS-C-C-OH \quad + \quad \underset{C_2H_5}{\overset{C_2H_5}{N-DMS-N}}\overset{C_2H_5}{\underset{C_2H_5}{}} \xrightarrow[\text{(8 hrs)}]{180°C} \underset{\text{Hard}\quad\text{Soft}}{[(O-\alpha MS-O)-(DMS)]_n} \tag{17}$$

By the process shown in Reaction 17, block copolymers have been constructed with hard blocks derived from polysulfones, polycarbonates, poly(BPA-terephthalates), etc.

Free–Radical Methods. Potentially there are several free-radical routes to block polymers; however, most of them suffer from simultaneous homopolymer formation, chain transfer, and other side reactions common to radical systems. Nevertheless, on a laboratory scale, some block polymers continue to be made by this route. Bamford and others (64) have used photochemical techniques to generate A–B–A and $(A-B)_n$ block polymers (Reaction 18):

$$S-CBr_3 \xrightarrow[\text{Mn}_2(\text{CO})_{10}]{h\nu} S-CBr_2\cdot \xrightarrow{\text{Monomer M}} S-(M)_n\cdot \tag{18}$$

$$S-(M)_n\cdot \xrightarrow{\text{Combination}} S-(M)_{2n}-S$$

Other approaches to free-radical polymers include capping with a peroxy end group or using a dual path initiator. The former approach was used (65) to convert α,ω-hydroxy polyethers (Eth) to esters to peroxycarbamate end groups (Reaction 19):

$$\text{HO-Eth-OH} \xrightarrow[\text{b) t-BuO}_2\text{H}]{\text{a) 2,4 TDI}} \left(\sim\sim\text{Eth-O-}\overset{\text{O}}{\overset{\|}{\text{C}}}\text{-N-}\bigcirc \right)_2 \xrightarrow[\text{M}]{\Delta}$$

with N-C-O-O-t-Bu group and CH$_3$

$$\longrightarrow \text{M-Eth-M} \quad + \quad \text{Homo M (from R-O)} \tag{19}$$

(M = polymer from vinyl monomer)

Reaction 19 suffers from several problems including homopolymer formation and, as carried out, would be expected to give a mixture of polymers.

The dual temperature initiators (shown schematically in Reaction 20) depend on initiation of one block at a low temperature, T_1 (using metal^{n+} + ROOH, or low temperature azo compound), followed by a second block at a higher temperature, T_2, where the polymer-containing radical is produced by thermal cleavage of the more stable R-O-O-R. An example of this approach is shown schematically in (20) and (21); it too suffers from the fact that a mixture of block and homopolymers is produced.

$$\tag{20}$$

$$\text{H-O-O-}\overset{\text{C}}{\underset{\text{C}}{\text{C}}}-\bigcirc-\overset{\text{C}}{\underset{\text{C}}{\text{C}}}\text{-O-O-}\overset{\text{C}}{\underset{\text{C}}{\text{C}}}-\bigcirc-\overset{\text{C}}{\underset{\text{C}}{\text{C}}}\text{-OOH} \xrightarrow[T_1 \text{ (low)}]{\substack{\text{MMA Monomer} \\ \text{Osmic Acid}}} \text{MMA-RO-OR-MMA}$$

with R below

$$\text{MMA-R-O-O-R-MMA} \xrightarrow[T_2 \text{ (high)}]{\text{Styrene}} \text{MMA-S}\cdot \xrightarrow{\text{Combination}} \text{MMA-S-S-MMA} \tag{21}$$

Ziegler-Natta. The tremendous impact that Ziegler-Natta catalysts have had on homopolyolefin polymerizations has also carried over to some extent into polyolefin block copolymers. The literature, particularly the patent literature (66) abounds with accounts of polyolefin block polymers that have been prepared by sequential introduction of monomers into Ziegler-Natta catalyst systems (usually heterogeneous Ti-based) capable of maintaining significant chain lifetimes. For example, it is claimed that the impact properties of polypropylene may be modified by tailblocking a copolymer segment on the propylene block:

$$P \xrightarrow[\text{Ti System}]{C_2/C_3} P\text{-EP} \tag{22}$$

Similarly, a variety of two, three, and multiblock olefin polymers [(P-E)$_n$, P-EP-P, P-E-P, E-EP] have been reported. However, in most cases the nature of Ziegler-Natta catalysis (low chain lifetimes, chain transfer from Ti to Al or Zn, thermal dieout, different reactive sites, etc.) is such that the polymer produced is often a mixture of homopolymer, unattached copolymers, and two or three segment block polymers. It is clear that the larger the number of blocks desired, the more difficult it is to obtain a

selected polyolefin block polymer. The precise segmented structure
that can be obtained with the long-lived anionic systems is simply
not possible with these Ziegler-Natta catalysts. Nevertheless,
block polymers have been made with the application of careful
experimental techniques, and in some cases the mixtures obtained may
have useful properties. Alternatively, solvent extraction
techniques may be employed to yield the block polymers in a purer
state.

In using Ziegler-Natta catalysis to make copolymers of EP, one
often finds that the product is not a random copolymer but a
"blocky" copolymer in which segments of polyethylene or polypropylene are produced in lengths that are long enough to exhibit
crystallinity of the respective polyolefin. Still another type of
block polymer that is possible with Ziegler-Natta catalysis is the
"stereoblock" polymer that can be made with propylene or higher α-
olefins where, for example, a mixture of isotactic and atactic
blocks of the same monomer may occur along a given polymer chain.

Finally, there are interesting reports of combinations of
Ziegler-Natta catalysis with free-radical methods for synthesizing
polyolefin-poly(polar monomer) block polymers. A typical example
(67) is that in which zinc diethyl was used as a transfer agent to
increase the number of reactive metal-polymer bonds:

$$C_2H_4 \xrightarrow[\substack{ZnEt_2 \\ (45\ °C)}]{Ti/Al} E\text{-}Zn \xrightarrow[\text{Cumyl Hydroperoxide} \\ (60\ °C)]{\text{MMA Monomer}} E + E\text{-}MMA + MMA \qquad (23)$$

Thus, again a mixture of polymers was produced, but extraction
experiments suggested that a significant fraction ($\leq$ 20%) of the
polar monomer was attached to an E block. Presumably a radical is
produced at the end of an E chain, perhaps by a displacement of an
R-O$\cdot$ radical on a Zn-E bond (Reaction 23).

Grafted Types. Graft polymers have been synthesized by a variety of
approaches including all of those mentioned earlier (anionic,
cationic, radical, etc.). Four of these approaches will be
described in this chapter. However, the reader should recognize
that many chemical paths to grafting are available, albeit many of
these suffer from the complexity of products produced and overall
inefficiency of the grafting method.

Cationic Methods. Cationic grafting has been employed by Kennedy
(68) using chlorinated butyl rubber (BU) as a backbone:

$$\sim\!\!\sim BU \sim\!\!\sim C\!=\!\overset{\overset{\textstyle C}{|}}{\underset{\underset{\textstyle Cl}{|}}{C}}\!-\!C\!-\!C \sim\!\!\sim \xrightarrow[\text{Styrene}]{AlR_3} \sim\!\!\sim \left(\overset{\textstyle C}{\underset{\textstyle S}{|}}\right)_{\!n}\!\!\sim\!\!BU\sim\!\!\sim \qquad (24)$$

Polystyrene (S) was grafted on the butyl backbone by converting the
allylic chloride sites to initiating carbonium ions (R^+ AlR_3Cl^-).
The product has thermoplastic elastomer properties. Most recently
Kennedy and coworkers have developed techniques for grafting

polyisobutylene onto a polychloroprene backbone. These grafts are unusual in that they carry *tert*-chloride branch termini (68).

Metalation Methods. It is possible to metalate a hydrocarbon polymer chain and thereby introduce a well-controlled number of random graft sites by appropriate control of the metalating agent (Reaction 25). Graft polymers are then easily made (69) by introduction of styrene or other monomers that can be anionically polymerized (Reaction 26).

$$\sim\sim B \sim\sim C-C=C-C \xrightarrow{n(s-RLi/TMED)} \sim\sim B-C-C=C \left(\underset{Li}{\overset{C}{|}} \right)_n \sim\sim \qquad (25)$$

$$\text{Metalated B} \xrightarrow{\text{Styrene}} \sim\sim B \sim\sim \left(\underset{S}{\overset{C}{|}} \right)_n \sim\sim \qquad (26)$$

Clearly, metalation can be used to make a variety of block polymers and is advantageous because of precise control of molecular weight of the graft. It is, of course, limited primarily to hydrocarbon blocks and a few select polar blocks. It is of interest that the graft polymer requires 6 to 10 grafts to give mechanical properties equivalent to an A-B-A where the A block molecular weight is equal to that of the graft. Thus, the metalation method is less attractive because of the greater consumption of alkyl lithium.

Anionic-Radical Combinations. Radical grafting of one monomer on the backbone of another polymer is well known and is the basis of an important commercial process for making high impact polystyrene. Styrene is thermally bulk polymerized in the presence of 5 to 10% (by weight) polybutadiene, the polymerization proceeding by a free-radical grafting path (70).

A patent (71) describes a combination of anionic and radical techniques to produce a variety of graft polymers. The synthesis proceeds via a capped anionic block polymer, the capping producing a polymerizable vinyl end unit. The capped polymer is then free-radical polymerized with a selected vinyl monomer to produce the graft polymer:

$$\text{Styrene} \xrightarrow[\text{b)C-C}]{\text{a)RLi}} \text{S-C-C-O-Li} \xrightarrow[\text{C=C-C-Cl}]{\text{Capping}} \overset{O}{\overset{\parallel}{\text{S-O-C-C=C}}} \qquad (27)$$

$$\overset{O}{\overset{\parallel}{\text{S-O-C-C=C}}} + \overset{O}{\overset{\parallel}{\text{C-C-O-C-C=C}}} \xrightarrow{R^\bullet} \sim\sim \text{EA} \sim \left(\underset{S}{\overset{C}{|}} \right)_n \sim\sim$$

This system offers the advantage of being able to synthesize a variety of hydrocarbon, polar, block polymers. It is, however, limited in the first stage to those monomers that are capable of good "living" anionic character.

These "polymeric monomers" were studied by Milkovich and coworkers at CPC International and assigned the trademark Macromer. Several kinds of Macromers have been synthesized and offer functionalized macromolecules that potentially can be polymerized by a variety of polymerization mechanisms. Such anionically prepared Macromers can be used to attach controlled size grafts to the backbone of another type of polymer (15).

Anionic Coupling or Attachment. Graft polymers have also been made by the reaction of living polymeric lithium salts at pendant functional groups on a polymer backbone, for example:

$$\sim\!\!\sim\!\!S\!\!\sim\!\!\sim\!\!(C\text{--}C)_x\!\!\sim\!\!\sim \xrightarrow{\ B\text{--}Li\ } \sim\!\!\sim\!\!S\!\!\sim\!\!\sim\!\!(C\text{--}C)_x\!\!\sim\!\!\sim \tag{28}$$

with pendant groups CH_2Cl converting to $CH_2\text{--}B$.

$$\sim\!\!\sim\!\!MMA\!\!\sim\!\!\sim\!\!(C\text{--}C)_{x+y}\!\!\sim\!\!\sim \xrightarrow{\ S\text{--}Li\ } \sim\!\!\sim\!\!MMA\!\!\sim\!\!\sim\!\!(C\text{--}C)_x\!\!\text{---}(C\text{--}C)_y\!\!\sim\!\!\sim \tag{29}$$

Any protic impurities in the parent polymer and any side reactions lead to loss of B–Li and thus to inefficient grafting. Furthermore, because the grafts are randomly placed along the backbone, the product is a statistical array of species, each varying in the number of grafts/chains.

In general, graft polymers have the preparative advantage of combining unlike polymeric segments that otherwise cannot be easily combined by direct synthesis. However, they are often produced as mixtures with homopolymers arising from the parent polymer and the grafting species.

Derivatization

The performance properties and processability of a block polymer are dependent upon the degree of incompatibility between the phases, and the glass transition temperatures of the hard and soft phases. A convenient approach to alter these parameters is to react one or more of the different segments chemically to give partial or complete functionalization. These chemical changes can also be used to change significantly hydrophobicity, weatherability, solubility, water absorptive capacity, flammability, etc., of a block polymer. We have listed a number of methods for obtaining such derivatives in the following sections.

Hydrogenation Methods. One way to significantly alter the character of a block polymer, particularly to improve weatherability, is by hydrogenation. Techniques (e.g., see Ref. 72) exist for hydrogenating A–B–A or graft polymers selectively or completely as illustrated in the following reactions:

$$S-B_{1,4}-S \xrightarrow[\text{Mild}]{H_2, \text{ Catalyst}} S-E-S \tag{30}$$

$$S-B_{1,4}/B_{1,2}-S \xrightarrow{\hspace{2cm}} S-EB-S$$

$$S-I-S \xrightarrow{\hspace{2cm}} S-EP-S$$

$$B_{1,4}/B_{1,2} \sim \left(\begin{matrix} C \\ | \\ S \end{matrix}\right)_n \sim \xrightarrow{\hspace{2cm}} \sim EB \sim \left(\begin{matrix} C \\ | \\ S \end{matrix}\right)_n \sim$$

$$B_{1,4}-I-B_{1,4} \xrightarrow{\hspace{2cm}} E-EP-E$$

$$S-I-S \xrightarrow[\text{Rigorous}]{H_2, \text{ Catalyst}} VCH-EP-VCH$$

Hydrogenation is also a useful process for converting block polymers with only soft segments (e.g., $B_{1,4}-I-B_{1,4}$) into one containing a hard crystalline segment (E) and a soft center segment (EP). Thus, a polymer with no self-vulcanizing character can be converted into a thermoplastic elastomer. A variety of polymers with intermediate properties can be achieved by partial hydrogenation in contrast to complete hydrogenation.

A recent review article discusses in some detail the use of anionic polymerization to make a number of block diene structures with a variety of microstructures (73). These unsaturated polymers can be hydrogenated to give block thermoplastic elastomers with self-vulcanizing properties. For example, anionic polymerization has been used to synthesize diblock and multiblock copolymers of polybutadiene–polyisoprene. The unsaturated polymers were then hydrogenated to give elastomeric poly(ethylene–ethylene/propylene) materials of the E–EP, E–EP–E, or higher block content (total M_n ca. 100,000). Even the E–EP diblock showed a tensile strength of approximately 11.4 MPa (1650 psi) at break. The author has suggested that the soft rubbery EP chain is capable of interpenetrating the hard crystallizable polyethylene chain via chain folding. Another possible explanation is that an interpenetrating network exists with a continuous crystalline polyethylene phase and an amorphous polyethylene phase that is interpenetrated by the rubbery amorphous ethylene-propylene-copolymer.

Ionic Substitution Methods. Ionic derivatives have been prepared from S–B–S and S–I–S, for which it has been possible to selectively derivatize the center blocks (Reactions 31–33). Such ionic polymers have been examined as membranes for piezo-dialysis and reverse osmosis (74).

$$\text{S-I-S} \xrightarrow[\text{2) } H_2O, \text{ NaOH}]{\text{1) } ClSO_3H} \text{S-I-S} \atop \underset{SO_3^-Na^+}{|} \qquad (31)$$

$$\text{S-B-S} \xrightarrow[\text{2) } (CH_3)_3N]{\text{1) N-BrSuccinimide}} \text{S-B-S} \atop \underset{+N(CH_3)_3Br^-}{|} \qquad (32)$$

$$\text{S-EP-S} \xrightarrow[\text{2) } H_2O, \text{ NaOH}]{\text{1) } SO_2CL_2, \text{ Perox.}} \text{S-EP-S} \atop \underset{SO_3^-Na^+}{|} \qquad (33)$$

S-B-S and S-I-S may also be completely or partially hydroxylated (75) as shown in Reactions 34 and 35 or carboxylated as shown schematically in Reactions 36 and 37.

$$\text{S-I-S} \xrightarrow[\text{2) } H_2O_2, \text{ OH}^-]{\text{1) } BH_3} \text{S-(C-}\overset{\text{C}}{\underset{OH}{\text{C}}}\text{-C-C)}_x\text{-S} \qquad (34)$$

$$\text{S-I-S} \xrightarrow[\text{2) } H_2O, \text{ H}^+]{\text{1) } CH_3\overset{O}{\overset{\|}{C}}\text{-OOH}} \text{S-I-(C-}\overset{\text{C}}{\underset{OHOH}{\text{C}}}\text{-C-C)}_x\text{-S} \qquad (35)$$

$$\text{S-I-S} \xrightarrow[\text{2) } CO_2]{\text{1) RLi, TMED}} \underset{(CO_2H)_x \quad (CO_2H)_x \quad (CO_2H)_x}{\text{S---------I---------S}} \qquad (36)$$

$$\text{S-I-S} \xrightarrow[\substack{\text{2) Maleic An.} \\ \text{3) } H_2O, \text{ OH}^-}]{\text{1) R}\cdot} \text{S-I-(C-}\overset{\text{C}}{\underset{\underset{NaO-\overset{\|}{C} \quad \overset{\|}{C}-ONa}{\overset{C--C}{|}}}{\text{C}}}\text{=C-C)-------)}_x\text{-S} \qquad (37)$$

Of the four reactions shown, only Reaction 36 is nonselective and carboxylates the end blocks (S) at the benzylic or ring positions and the mid blocks (I) at the allylic carbon.

Block polymers can also be chlorosulfonated (accompanied by some chlorination) and in turn hydrolyzed to the sodium salt (Reactions 38 and 39) (76).

$$\text{S-EP-S} \xrightarrow[\text{Peroxide}]{SO_2Cl_2} \underset{(\overset{|}{C}l)_x(SO_2Cl)_x(\overset{|}{C}l)_x}{\text{S----EP------S}} \xrightarrow{H_2O/OH-} \qquad (38, 39)$$

$$\underset{(\overset{|}{C}l)_x \ (SO_3^-Na^+) \ (\overset{|}{C}l)_x}{\text{S------EP-------S}}$$

Most of the chlorosulfonyl groups are found in the EP segment, whereas the bulk of the bound chlorine is found in the styrene

segment. Some chlorosulfonyl units arise from a small amount of
unsaturation left in the EP segment. Selective sulfonation of S-EP-
S may also be carried out as shown in Reaction 40 by sulfonating
with an SO_3 complex (77).

$$
\text{S-EP-S} \xrightarrow[\text{2) } H_2O]{\text{1) } SO_3 \cdot (RO)_3P=O} \underset{(SO_3H)_x}{\text{S------EP------S}} \underset{(SO_3H)_x}{} \tag{40a}
$$

$$
\text{-EPDM-} \xrightarrow[\substack{\text{2) } H_2O \\ \text{3) } Zn(OAc)_2}]{\text{1) Acetyl Sulfate}} \text{-EP-DM-} \tag{40b}
$$
$$
\underset{(SO_3Zn)}{|}
$$

The sulfonation of EPDM as shown in Reaction 41b has recently
been announced to be under commercial development by Uniroyal (78).
Sulfonated EPDM can be classified as an ionic thermoplastic
elastomer, whereby ionic interactions between chains act as
thermally reversible cross-links. Above certain temperatures these
ionic interactions break down, and plastic flow can occur. Upon
cooling, these ionic cross-links reform to give the desired property
characteristics. To enhance the processability of sulfonated EPDM
an ionic plasticizer "ionolyzer" (i.e., zinc stearate) is added to
about 20 phr (parts per 100 rubber). Care must be taken to ensure
complete neutralization, because it was observed that residual acid
can result in covalent cross-linking during fabrication.

Clearly a variety of other chemical reactions may be carried out
on existing block polymers. Complete derivatization is, however,
often subject to the usual problems of polymer reactions such as
steric hindrance, neighboring group effects, solvent problems, and
gelation. However, judicious choice of reagent, conditions, and
moderate conversions can produce an attractive array of block
polymers that can be synthesized in no other way.

Structure and Properties

Motivation for investigating the relationship between the composite
structure and the unique properties of block polymers was especially
strong during the past several years. There were the usual powerful
incentives for developing new polymers and modifying existing ones,
accentuated by the rich variety of combinations and variations
possible. Also, unusual opportunities for theoretical and
experimental characterization were presented by the use of
relatively simple model systems that possessed narrow molecular
weight distribution and high-purity block composition. These were
typified by the A-B-A block polymers made by anionic polymerization.

A definitive picture of the phase structure emerged and was made
possible by the simultaneous development of powerful investigative
techniques such as phase staining/electron micrography, low-angle X-
ray diffraction, and GPC which were combined with conventional
physicochemical characterizations of bulk, melt, and solution states
of the block polymers.

The following summarizes the important characteristic features
of block polymers that have proved immensely useful for
understanding their special behavior.

Phase Structure of Block Polymers. Block polymers have heterophase structure in which the phases are highly interspersed. Two or more phases may be present, and each of these may be continuous or discrete depending on the relative amounts of the various polymeric species present and the conditions of preparation. The discrete phases may have various shapes usually referred to as "domains." Spheres, rods, and lamellae have been experimentally observed and theoretically justified. Molau (79) symbolizes the change in A-B-A structure as a function of A or B content, as shown in Figure 4.

Dimensions of the fine phase structure are usually less than 0.1 µm. Thus, details of the structure cannot be resolved except with an electron microscope used in combination with a phase discriminating technique typified by the OsO_4 staining of unsaturated blocks (80). The individual phases may be crystalline, glassy or rubbery, depending on chemical composition and temperature. Whether crystalline or amorphous, the structure is almost always highly ordered so that measurements of dispersed phase size and phase spacing may be made by low angle X-ray diffraction. The highly ordered structure may persist even in solution as is evidenced occasionally in the irridescent appearance of concentrated solutions of block polymers (81)

The interfacial regions between phases may be of significant dimensions so that they may exert some behavior characteristic of a bulk phase. They can be considered to be limited regions of polymer-polymer mixing, and their size depends upon the interacting polymers and the specimen preparation conditions (82). Because the blocks are covalently bound to one another, the interfacial regions have high strength assuring close and intimate contact between phases, that is, good "wetting" of the dispersed phase by the matrix in the descriptive terms often used for physical blends.

Influence of Structure on Properties. The characteristic behavior of block polymers may be thought of as a combination of properties arising from three distinct aspects of the phase structure: individual phases, heterophase structure, and interfacial regions.

Individual Phases. Certain properties of the polymer species constituting the individual phases may be exerted almost unchanged in the block polymer, such as thermal and mechanical properties as well as segmental solubility in various solvents. For example, measurements of melting points, glass transition temperatures, and dynamic mechanical absorption peaks at phase changes will be characteristic of individual phases. Also an individual phase can be dissolved by the same solvents for the corresponding homopolymer. The overall physical behavior of the block polymer is influenced by the contributions of each individual phase. The influence of the phase in the overall behavior of the block polymer depends strongly on whether the phase is dispersed or is present as a continuous matrix. A continuous phase will contribute various properties to the block polymer almost equivalent to those of the corresponding homopolymer, for example, adhesion, wettability, frictional coefficient, static build-up, paintability, etc. By contrast, contribution of the dispersed phase to these properties is considerably less.

Heterophase Structure.   The heterophase structure of block polymers
imparts features that are reminiscent of certain valuable composite
materials made by physically mixing components. Improvements in
impact and tensile strengths and increase or decrease of hardness
and flexibility may be attained. The uniformity, high degree of
dispersion, and perfect wetting possible with block polymers extend
the versatility that can be achieved with physically mixed
composites. The strong chemical bonding between dispersed phase and
matrix can result in effective cross-linking when the dispersed
phase is hard, so that the dispersed phase particles can behave as
cross-link junctions of very high functionality. These junctions
can also "entrap" entanglements of polymeric chains in the matrix so
that these entanglements can act as cross-links themselves. As a
matter of fact, in the A-B-A thermoplastic elastomers, the effective
molecular weight of chains between cross-links (M_c) turns out to be
closer to the average chain length between entanglements rather than
to the length of the B blocks. This result has been shown by
measurements of the swelling volume (83) and the modulus (21). The
strong connections between dispersed phase and matrix can also
transmit stresses exerted on the matrix to the dispersed phase so
that deformation of the bulk material results in deformation of the
dispersed phase. This condition may occur even though the dispersed
phase may be in the glassy state. Distortion of the dispersed phase
in the A-B-A thermoplastic elastomers has been proposed as an energy
absorbing mechanism (84) accounting for the "reinforcing" action of
the polystyrene domains. In this way the domains assume the role
exerted by reinforcing fillers, such as carbon black, although the
mechanism of reinforcement in the two cases may be different. The
characteristic "stress-softening" or "Mullin's effect" shown by
filler-containing rubber vulcanizates is also demonstrated by A-B-A
thermoplastic elastomers (85). This result also supports the
concept of the domains behaving as filler particles.

Interfacial Regions. Properties arising from the interfacial
regions between phases of block polymers are not as readily
identifiable as those arising from the individual phases of the
heterophase structure. We can plausibly identify the broadening or
diffuseness that appears in phase transition measurements of some
block polymers as arising from these regions. Modifications in melt
and solution viscosities as well as mechanical properties are
expected to appear and depend on the volume fraction occupied by
interfacial regions. These changes will roughly be in a direction
expected from a linear combination of the homopolymer properties; in
certain cases degradation of those properties unique to the
heterophase structure will result.

Bulk Properties. Block polymers can show mechanical properties in
the bulk state that are superior to those that can be achieved with
the corresponding homopolymers or random copolymers. This
improvement in behavior is made possible by the segregated phase
structure in block polymers. Of primary interest have been
structures possessing both soft and hard phases. These polymers may
range from the thermoplastic elastomers in which the hard phase is
dispersed in a soft phase matrix, to toughened (i.e., high impact
strength) thermoplastics in which the soft phase is dispersed in a

hard phase matrix. At processing temperatures, the hard phase is raised above its glass transition or crystalline melting temperature so that flow may take place.

The A-B-A thermoplastic elastomers below the glass transition temperature of the hard phase at service temperatures possess high tensile strength and rapid return from high extension and thus behave like reinforced rubber vulcanizates. Unlike vulcanizates at higher temperatures, they can undergo a transition to a melt and be processed like thermoplastic materials. The transition between the meltlike and vulcanizatelike behavior is reversible and can be repeated many times. In these polymers A and B are chosen so that A is hard and B is soft at use temperatures, that is, in use, A is a thermoplastic polymer well below its glass transition temperature. Molecular weights of A and B are selected so that a favorable balance between mechanical performance and processing behavior is achieved (see Figure 5).

The distinguishing feature of the A-B-A thermoplastic elastomer structure is that a three-dimensional network is established by the dispersed domains serving as cross-link junctions of high functionality (see Figure 1). With increase of the proportion of A, the stress-strain response changes and successively approximates that of a nonreinforced vulcanizate, a reinforced vulcanizate, a flexible thermoplastic, and a toughened thermoplastic.

If the hard phase is the midblock, that is, B-A-B, no three-dimensional network can be established by phase separation of the end blocks, and mechanical properties are those of a nonvulcanized rubber (21). This condition also holds true for the two-block polymer A-B.

Thermal behavior of the bulk polymer is critically determined by the thermal response of each individual phase. The block polymer will lose strength rapidly as temperature is raised to approach the glass transition temperature of the dispersed hard phase. On the other hand, modulus and hardness will increase rapidly as the temperature is lowered to approach the glass transition temperature of the soft phase matrix. At temperatures between these two extremes, stress-strain behavior is comparable to that of the homopolymer constituting the continuous phase, constrained by cross-linking features and modified by the fillerlike effect of any dispersed phase. The block polymer is also subject to energy dissipating mechanisms such as hysteresis and stress softening effects displayed by various filled and unfilled rubber systems.

<u>Melt Properties</u>. Block polymers may display thermoplasticlike processability in the melt state as exemplified by the A-B-A thermoplastic elastomers. However, various characteristic features distinguish the melt behavior of block polymers from that of conventional thermoplastic polymers (21, 86). These can be summarized as follows:

1. Limiting melt viscosity shifted to low shear stress, or absent entirely. Whereas conventional polymers approach a Newtonian (constant) viscosity with decreasing shear stress, a block polymer of comparable molecular weight will approach it much more slowly, or in some cases, the viscosity may increase

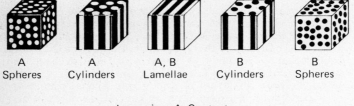

A   A   A, B   B   B
Spheres   Cylinders   Lamellae   Cylinders   Spheres

Increasing A-Content
Decreasing B-Content

Figure 4. Changes in A-B-A structure as a function of composition.
(Reproduced with permission from Ref. 4. Copyright 1971
Plenum.)

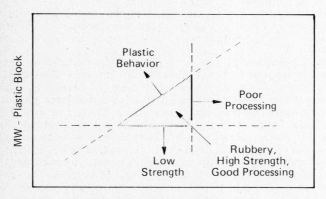

Figure 5. Effect of block molecular weight on properties of A-B-A
thermoplastic elastomers. (Reproduced with permission
from Ref. 18. Copyright 1969 Wiley.)

without limit.   The behavior in the latter case has been compared to that of a rubbery polymer exerting a yield stress.

2. Anomalous relationship of melt viscosity to shear stress. Compared to a homopolymer of the same molecular weight, the melt viscosity of a block polymer frequently is anomalously high at low or moderate shear stresses.   At high shear stresses, however, the melt viscosity may approach that of the homopolymer.

3. Anomalous relationship of melt viscosity to molecular weight. The melt viscosity of a block polymer is significantly higher than that of the corresponding random copolymer of the same molecular weight, and this deviation itself increases with molecular weight of the blocks.

4. Melt viscosity is critically affected by the difference in segmental solubility parameters, that is, by the interaction parameter between blocks.   The deviation from normal behavior is dependent on the composition and block structure of the particular block polymer examined.   It is most dramatic when the block polymer is capable of forming a three-dimensional network in the bulk state and the different polymer segments have high mutual incompatibility as evidenced by diverse solubility parameters.

The anomalous melt behavior of block polymers has been attributed to the phase structure that persists in the melt. Although all polymeric segments are at a temperature sufficiently high to place them in the melt state, this increase in temperature is not sufficiently great to eliminate phase segregation.   (An increase of temperature increases the negative contribution of the entropy of mixing term to the free energy of mixing.   However, in the melt temperature region this increase is ordinarily not sufficient to offset the positive heat of mixing, except in rare cases in which phase separation already may be borderline in the bulk state.)   For A-B-A polymers the higher viscosity due to the phase structure in the melt may be pictured as the so-called "switchboard" effect.   Because the A domains persist in the melt, a three-dimensional network exists, and flow can only occur by A segments being removed from the domains, then being pulled through the incompatible B matrix, and finally being taken up in domains again.   The extra energy required for the process increases the total energy required for flow, that is, increases the melt viscosity of the block polymer.   The higher the molecular weight of the blocks, the more energy that will be required for this transfer process.   Higher viscosities also result when large differences in solubility parameters exist between the blocks because of greater dissipation of energy in the forced mixing/demixing process.

To obtain adequate processability of block polymers while retaining good bulk properties, adjustment of molecular weights of the individual blocks is the most critical parameter.   Additives that involve plasticization of either the hard or soft phases such as oils, resins, and other polymers can also be used.   Additives that melt at temperatures in the melt region of the block polymer are especially effective.

Solution Properties.   Block polymers when dissolved in solvents display behavior that is unusual when compared to that of the

corresponding homopolymer or random copolymer. The most important
feature of the block polymer in solution is independent phase
solubility. That is, each phase will tend to be solvated by those
solvents that dissolve the corresponding homopolymers. Thus, it is
possible that a block polymer may disperse on a molecular scale in a
solvent or solvent blend so that certain blocks are solvated whereas
other blocks remain in a collapsed state. The solubility parameter
is useful for predicting this behavior. Generally a block will
dissolve in a solvent that has a solubility parameter close to its
own. The usual contributing features of hydrogen bonding and
polarity must be taken into account. Also some complications in
solubility behavior are introduced by the constraints imposed by the
chemical bonding between blocks.

In dilute solutions using a solvent system good for all blocks,
there is little intermolecular interaction, and the behavior of
block polymers approaches that of homopolymers or random copolymers.
Conventional dilute polymer solution theory may be applied with
moderate success, as in osmometry and viscometry, provided that the
block composition is taken into account. Some complicating features
may arise because of the possibility of intramolecular block
separation, for example, prevention of a random coil configuration,
which is usually assumed in some conventional treatments (87).

In concentrated solutions, discrepancies between block polymers
and the corresponding homopolymers or random copolymers are more
dramatic. This condition is because interaction between block
polymer molecules may lead to phase separation of a given block if
the solvent is poor for that particular block. In the case of A–B–A
block polymers, use of a solvent system poor for A but good for B
will lead to incipient cross–linking of the system by the A phase.
The structure then behaves like a highly swelled, three–dimensional
network (gel). Large anomalous increases in solution viscosity may
result and depend on the particular solvent system and the polymer
concentration. However, if the solvent is good for A but poor for
B, clusters of B will form and be held suspended in solution by the
solvated A blocks. These may be of colloidal size and show
properties of colloidal suspensions. Thus, the solution viscosity
of these suspensions may be lower by orders of magnitude than that
of the corresponding solution in which all blocks are solvated.

Even if all blocks of a block polymer are solvated in solution,
that is, by using a nonselective solvent, phase segregation of the
solvated blocks may occur at high concentration. The result is a
supermolecular ordering in solution comparable to that in liquid
crystals. In certain cases the aggregates are sufficiently large to
diffract visible light, and, as has been mentioned earlier, block
polymer solutions that are irridescent above a critical
concentration have been observed (81).

Because the individual blocks in a block polymer exhibit
selective solution properties, a block polymer may act as a surface
active agent. It can be accommodated at an interface between two
phases of other materials if it contains blocks compatible with each
phase, respectively. Thus, it can act as an emulsifier between two
incompatible solvents, or other liquids. In the same way it can act
as a compatibilizer or dispersing agent between two incompatible
resins or polymers if these resins or polymers are compatible with
the respective blocks.

Commercial Block Polymers

Various commercial thermoplastic rubbers are listed in Table III. Also listed in Table III are the structural type and the composition of the hard and soft segments. It is not within the scope of this paper to discuss the differences between these products or their exact composition. However, we can comment briefly on preferred areas of application as we understand them.

Polyurethane thermoplastic rubbers are generally in the higher price range (over \$1.25 per lb) and are used in automotive, mechanical goods, and recreational vehicle applications.

Du Pont's Hytrel copolyester thermoplastic elastomers are slightly higher in price than the polyurethane types and have grown rapidly. They have a wide range of service temperatures and have found major uses in recreational vehicles, belting, hydraulic hose, and mechanical goods.

Uniroyal's TPR grades are thought to be physical blends of polypropylene (or less frequently polyethylene) with EPDM rubber, the latter probably slightly cross-linked. Properties of TPR fit into the class of a thermoplastic elastomer, although the TPR structure itself does not readily fit into our definition of a block polymer. Applications are in automotive, wire and cable, mechanical and sporting goods. Monsanto has recently published details of a whole series of thermoplastic rubbers based on the vulcanization of a rubbery component while being sheared in a mixture with a hard thermoplastic. Some of these products based on EPDM and polypropylene are being produced under the trade name of Santoprene (88-91). Similar materials are made from blends of uncross-linked EPDM with polypropylene by other manufacturers.

The Shell Kraton thermoplastic rubbers are based on S-B-S, S-I-S, and S-EB-S. Applications include adhesives, footwear, mechanical goods, and automotive applications. Shell also manufactures S-EP block copolymers under the Shellvis trade name. These are used as viscosity index improvers in lubricating oils.

Phillips' Solprene thermoplastic rubbers are $(S-B)_n-X$ and $(S-I)_n-X$ types. Phillips also manufactures S-B two-block polymers that are not thermoplastic rubbers. According to a recent report (92), production of these polymers is being discontinued. Other companies including ANIC (Italy), Firestone (U.S.A), and ASAMI (Japan) market block copolymers similar to the Shell and Phillips thermoplastic rubbers.

Applications of Block Polymers

One noteworthy feature of the block copolymers with polystyrene end segments is that they have no commercial applications as pure materials. Instead, they are mixed (or compounded, to use the common term) with other polymers, fillers, oils, resins, etc. to give final products designed for specific end uses. In contrast, the polyurethanes and polyesters are used in their undiluted state. The TPR-type materials are blends to begin with and often contain fillers. Other polymers, oils, and resins are not added to these materials in significant amount.

Table III. Classification of Thermoplastic Rubbers

Trade Name (mfgr)	Type	Hard Segment(A)	Soft Segment (B)	Notes
Kraton D (Shell)	3-block (S-B-S or S-I-S)	S	B or I	general purpose, soluble
Solprene 400 (Phillips)	branched (S-B)n (S-I)n			
Stereon (Firestone)	3-block (S-B-S)	S	B	general purpose, soluble
Tufprene (Asahi)	3-block (S-B-S)			
Europrene Solt (ANIC)	3-block (S-B-S)			
Kraton G (Shell)	3-block (S-EB-S)	S	EB	improved stability, soluble, uncompounded
Elexar (Shell)	3-block (S-EB-S and S-B-S	S	EB	wire and cable
Estane (B.F. Goodrich) Roylar (Uniroyal) Texin (Mobay) Pellethane (Upjohn)	multiblock	poly-urethane	ether or ester	hard, abrasion and oil resistant high cost
Hytrel (Du Pont)	multiblock	ester	ester (amorphous)	similar to polyurethane and better low temperature
TPR (Uniroyal)	blend	P (or E)	EPDM (cross-linked)	
Santoprene	blend	P	EPDM (cross-linked)	
Ren-Flex (Ciba-Geigy)	blend	P	EPDM	low density, hard, not highly filled
Polytrope (Schulman)	blend	P	EPDM	
Somel (Du Pont)a	blend	P	EPDM	
Telcar (B.F. Goodrich)b	blend	P	EPDM	
Vistaflex (Exxon)c	blend	P	EPDM	

a Now sold by Colonial
b Now sold by Teknor-Spex
c Now sold by Cook-Reichhold

Mechanical Goods. Thermoplastic elastomers can be formed into useful rubber articles by conventional plastics processing methods such as extrusion, blow molding, and injection molding. Vulcanizing agents or cure cycles are not required. There are no vulcanization residues, and scrap is reusable. Compounds based on S-B-S, S-I-S, $(S-B)_n-X$, or S-EB-S can have hardnesses as low as 40 Shore A. They can be compounded with various additives to give materials designed for specific applications. Table IV indicates the effects of additives on various properties. It has been found that polystyrene helps the processability of the product and makes it harder. Oils also help processability but lower the hardness. Oils with low aromatic content must be used because other oils plasticize the polystyrene domains and weaken the product. Large amounts of oil and polystyrene can be added to these block polymers, up to the weight of the polymer in some cases.

In addition to, or instead of, polystyrene and oils, polymers such as polypropylene, polyethylene, or ethylene-vinyl acetate copolymer can be blended with these block copolymers. Blends with S-B-S or $(S-B)_n-X$ block polymers usually show greatly improved ozone resistance (S-EB-S already has excellent ozone resistance). In addition, these blends have some solvent resistance. In certain cases, some oils that are stable to UV radiation reduce the stability of the blends; however, the effects can be minimized by the use of UV stabilizers and absorptive or reflective pigments (e.g., carbon black or titanium dioxide).

Large amounts of inert fillers such as whiting, talc, and clays can be used in these compounds. Reinforcing fillers such as carbon black are not required, and, in fact, large quantities of such fillers frequently produce undesirable properties in the final product. Probably the largest volume usage of S-B-S has been in footwear (93). Canvas footwear, such as tennis shoes, can be directly injection molded by using multistation machines. Unit soles can also be made by injection molding. These can then be cemented to leather or vinyl uppers. In both cases the products have good frictional properties, similar to those of conventionally vulcanized rubbers and much better than those of the flexible thermoplastics such as plasticized poly(vinyl chloride) (PVC). The products remain flexible under cold conditions because of the good low-temperature properties of the polybutadiene block.

In compounds based on S-EB-S polymers, polypropylene is a particularly valuable additive and serves both to improve processability and to extend the upper service temperature. Presumably, some of the improvement in service temperature can be attributed to the high melting temperature (ca. 165 °C) of the crystalline polypropylene.

S-EB-S polymers are very compatible with paraffinic or naphthenic oils because of their lower midblock solubility parameter. As much as 200 parts of some of these oils can be added without bleedout. The articles have excellent heat stability, and the resistance of pigmented stocks to degradation from outdoor exposure is very good. Several compounds based on S-EB-S and S-B-S block polymers have been developed for special purposes. These include pharmaceutical goods, wire and cable coatings, and automotive applications (94). In pharmaceutical applications, the ability of some molded products to withstand steam sterilization as

Table IV. Compounding S-B-S, S-I-S, $(S-B)_n-X$, and S-EB-S

Component	Hardness	Processability	Effect on Ozone Resistance	Cost	Other Properties
Oils	decreases	increases	—	decreases	sometimes decreases UV resistance
Polymers[a]	increases	generally increases	increases	generally decreases	often gives satin finish, frequently improves high-temperature properties
Fillers	some increase	variable	variable	decreases	often improve surface appearance

[a]Thermoplastic polymers such as S, E, P, and EVA.

well as their freedom from vulcanization residues makes them particularly attractive. The compounds intended for wire and cable coating can be extruded at high rates and have excellent electrical properties. Compounds produced for the automotive industry have fallen into two classes. Some intended for exterior use are rather hard. They have the advantage of being flexible and thus resistant to impact, and can be painted, by using a special primer, to resemble the metal body work. Softer compounds for use in other automotive applications such as flexible air ducts have also been developed.

Some A-B block polymers have also been developed for use in manufacturing mechanical goods by conventional vulcanization. These are usually materials in which one of the blocks is polystyrene and the other block is a tapered copolymer of styrene and butadiene. They are used in blends with other conventional rubbers, such as SBR and natural rubber, in which they improve the processability of the final product.

Polyurethane and polyether block copolymers are only available as fairly hard materials (70 Shore A to 75 Shore D). Articles molded from them are tough and have good resistance to abrasion and cut growth. Softer grades have good low-temperature flexibility. The polyethers are more highly crystalline and so have higher maximum service temperatures. The thermoplastic polyurethanes can easily be painted with flexible polyurethane paints to give them numerous applications in the automotive industry.

Thermoplastic polyolefin blends can be produced as quite soft materials (as low as 55 Shore A) or as harder (above 95 Shore A), flexible products. They are particularly useful when a combination of good weatherability and rubbery behavior is required. Applications include automotive parts, weather stripping, hose, and sporting goods. Because of their good dielectric and insulating properties, polyolefin blends may be particularly useful in wire and cable insulation.

Adhesives, Sealants, and Coatings. Other principal applications of thermoplastic rubbers are in adhesives, sealants, and coatings (95). A multiblock polyester (Du Pont Dyvac 5050) intended for hot melt applications has been introduced, and there are some hot melt applications of the multiblock polyurethanes. (Hot melt application eliminates the need for solvent. Block polymers are ideally suited for this application because they require no curing and develop their strength immediately upon cooling.) However, most of the published work has dealt with S-B-S, $(S-B)_n$-X, S-I-S, and S-EB-S. These block polymers are used to formulate both hot-melt and solution adhesives. Tackifying and reinforcing resins are added to achieve a desirable balance of properties, as are oils and fillers. Oils and resins that associate with the center rubber blocks give softer, stickier products, and resins that associate with the end polystyrene blocks increase hardness and strength. Generally, oils and low-melting resins that associate with the polystyrene blocks are to be avoided because they plasticize the domains and allow the polymer to flow under stress. An exception is in sealants for which a controlled amount of stress relaxation or plastic flow may be desirable. Because block polymers such as S-I-S, S-B-S, and $(S-B)_n$-X have unsaturated rubber blocks, they are susceptible to both oxidative and shear degradation. Thus, it is important that compositions   containing them be adequately stabilized, and in many cases it is desirable to nitrogen-blanket these hot-melt adhesives both during mixing and in the applicator.

S-EB-S block copolymers have a saturated midblock and so can be compounded with suitable resins and oils to make adhesives and sealants that have improved stability during long-term or extreme temperature exposure. In addition, hot-melt adhesives formulated from these polymers are more stable under melt processing conditions and can be applied at high temperatures. Nitrogen-blanketing is not required. Stable adhesives and sealants with large amounts of oils and resins can be formulated from these polymers.

Other Applications. S-B-S block polymers blended with high-impact polystyrene give a super high-impact product. Significant increases in Izod and falling weight impact resistances may be achieved with minimum loss of hardness. With crystal grade polystyrenes, these block copolymers act as flexibilizers or low to medium impact improvers. In blends with polyolefins such as polyethylene or polypropylene, S-B-S block copolymers give improved low-temperature impact resistance. In particular they improve the impact and tear resistance of high-density polyethylene film.

As previously noted, S-EP block copolymers are used as viscosity index improvers in multigrade motor oils.

Thermoplastic polyurethanes and polyester/polyethers are polar materials; thus, their use in polymer blending usually is limited to blends with other polar polymers such as PVC, acrylonitrile-butadiene-styrene copolymers (ABS), and polyesters. However, at this time polymer blending is a fairly small market for these polymers.

Thermoplastic elastomers can be one of the components in multilayer extrusion. They can serve either to bond two dissimilar materials, for example, ABS and high impact polystyrene, or as a layer with physical properties that contribute to the film properties, for example, a film with an S-EB-S center and a polypropylene outer surface. In this case they yield a flexible film with a nontacky surface. In other cases, multilayered films can be constructed to control gas permeability.

Some new types of S-B block copolymers are now being used as low-shrink/low-profile additives in sheet molding compounds made from unsaturated polyesters.

The Future

The unusual and attractive properties of the block polymers already identified, and the almost limitless combinations of possible block polymer structures, argue for an unbounded future. The rapidly growing applications for the commercial thermoplastic rubber block polymers of Table III have confirmed the trend. To lend some credibility to our look at the future, however, we have restricted it to the area of A-B-A block polymers in which we have the most experience. Some of the future trends we suggest are higher service temperature, oxidative stability, better processability, solvent resistance,   flame   retardance,   electrical   conductivity, multicomponent polymer blends, super impact-resistant blends, and bridging components for exceptional dispersion.

In the case of block polymer-homopolymer blends, we feel that applications of this type will continue to grow. We suggest that this growth will generate new block polymers with lower melt viscosities and other properties that will optimize both the blending processes and the properties of the blended product.

Solvent resistance is a very desirable property for automotive and various other large-scale applications. Investigations in this area are already under way (63). We would expect in due course to see new commercial products in this area.

Another area which is growing rapidly is multicomponent polymer systems. As noted in the section on structure and properties, block polymers should contribute special properties to multipolymer systems.

With the use of "plastics" in automobiles predicted to increase significantly in the 1980s (96) one would expect the use of thermoplastic rubbers of all types to increase also. New thermoplastic rubbers with higher service temperatures and/or solvent resistance, particularly at lower prices than the present specialty materials, should accelerate this trend. It has been estimated (97) that the U.S. use of thermoplastic rubbers in automobiles will be greater than 0.5 billion lb in 1986.

Acknowledgments

A great many people participated in the thermoplastic rubber block polymer project within our company, too many to list here. The authors particularly wish to cite the contributions of the following people involved in the discovery and elucidation of the potential of the S-B-S and S-I-S block polymers: J. T. Bailey, A. R. Bean, E. T. Bishop, D. W. Fraga, D. J. Meier, W. R. Hendricks, R. Milkovich, F. D. Moss, and L. M. Porter.

Literature Cited

1.   "Block Copolymers"; Allport, D. C.; Janes, W. H., Eds.; John Wiley & Sons: New York, 1972.
2.   "Block and Graft Polymerization"; Ceresa, R. J., Ed.; John Wiley & Sons: New York, 1972.
3.   "Block and Graft Copolymers"; Burke, J. J.; Weiss, V., Eds.; Syracuse Univ. Press: Syracuse, N.Y., 1973.
4.   "Colloidal and Morphological Behavior of Block and Graft Copolymers"; Molau, G. E., Ed.; Plenum Press: New York, 1971.
5.   "Multiphase Polymers"; Cooper, S. L.; Estes, G. M., Eds.; ADVANCES IN CHEMISTRY SERIES No. 176, American Chemical Society: Washington, D.C., 1979.
6.   Eisenberg, A.; King, M. "Ion Containing Polymers"; Academic: New York, 1977.
7.   Noshay, A.; McGrath, J. E. "Block Copolymers--Overview and Critical Survey"; Academic: New York, 1977.

8.   "Polymer Alloys--Blends, Blocks, Grafts, and Interpenetrating
     Networks"; Klempner, D.; Frisch, K. C., Eds.; Plenum: New
     York, 1977.
9.   "Polymer Alloys II"; Klempner, D.; Frisch, K. C., Eds.; Plenum:
     New York, 1980.
10. Olabisi, O; Robeson, L. M.; Shaw, M. T. "Polymer-Polymer
     Miscibility"; Academic: New York, 1979.
11. "Polymer Blends"; Paul, D. R.; Newman, S., Eds.; Academic: New
     York, 1978; Vol. 1 and 2.
12. Platzer, N. A. J. "Copolymer, Polyblends, and Composites";
     ADVANCES IN CHEMISTRY SERIES No. 142, American Chemical
     Society: Washington, D.C., 1975.
13. Sperling, L. H. "Interpenetrating Polymer Networks and Related
     Materials"; Plenum: New York, 1981.
14. Lipatou, Yu. S.; Sergeena, L. M. "Interpenetrating Polymeric
     Networks"; Naukova Dumka: Kiev, 1979.
15. "Anionic Polymerization. Kinetics Mechanics and Synthesis";
     McGrath, J. E., Ed., ACS SYMPOSIUM SERIES No. 166, American
     Chemical Society: Washington, D.C., 1981.
16. "Block   Polymers";   Aggarwal,   S.   L.,   Ed.;
     Proc. Am. Chem. Soc. Symposium, New York, 1979; Plenum Press:
     New York, 1970.
17. "Recent Advances in Polymer Blends, Grafts, and Blocks";
     Sperling, L. H., Ed.; Proc. Am. Chem. Soc. Symposium, Chicago,
     1973; Plenum Press: New York, 1974.
18. "Block Copolymers"; Moacanin, J.; Holden, G.; Tschoegl, N. W.,
     Eds.; Symposium sponsored by Calif. Inst. Techn. and Am. Chem.
     Soc., Pasadena, Calif., June 5, 1967; J. Polym. Sci. (C), No.
     26, 1969.
19. Pillar, P. S.; Fielding-Russell, G. S. Polym. Prepr. 1973,
     14(1), 346.
20. Bailey, J. T.; Bishop, E. T.; Hendricks, W. R., Holden, G.;
     Legge, N. R. "Thermoplastic Elastomers"; presented at the
     Division of Rubber Chemistry, American Chemical Society,
     Philadelphia, October 20–22, 1965; Rubber Age, Oct. 1966.
21. Holden, G.; Bishop, E. T.; Legge, N. R. Thermoplastic
     Elastomers, Proc. International Rubber Conference, MacLaren and
     Sons, London, 1968, p. 287.
22. Kuhn, T. S. "The Structure of Scientific Revolutions"; Univ. of
     Chicago Press: Chicago, 1970.
23. Bolland, J. H.; Melville, H. W. Proc. First Rubber Tech. Conf.,
     London; W. Heffer: London, 1938; p. 239.
24. Melville, H. W. J. Chem. Soc. 1941, 414.
25. Baxendale,   J.   H.;   Evans,   M.   G.;   Parks,   G.   S.
     Trans. Faraday Soc. 1946, 42, 155.
26. Lundsted, L. G. J. Am. Oil Chem. Soc. 1951, 28, 294.
27. Mark, H. F. Textile Res. J. 1953, 23, 294.
28. Horne, S. E.; et al. Ind. Eng. Chem. 1956, 48, 784.
29. Stavely, F. W.; et al. Ind. Eng. Chem. 1956, 48, 778.
30. Szwarc, M.; Levy, M.; Milkovich, R. J. Am. Chem. Soc. 1956, 78,
     2656.
31. Szwarc, M. Nature 1956, 178, 1168.

32. Foreman, L. E. "Polymer Chemistry of Synthetic Elastomers," Part II; Kennedy & Tornquist, Eds.; John Wiley & Sons: New York, 1969; p. 497.
33. "Block and Graft Copolymers"; Ceresa, R. J., Ed.; Butterworths: Washington, D.C., 1962.
34. Ceresa, R. J. "Synthesis and Characterization of Natural Rubber Block" and "Graft Copolymers in Block and Graft Copolymerization"; Ceresa, R. J., Ed.; John Wiley: New York, 1973; Chap. 3.
35. Bateman, L. C. Ind. Eng. Chem. 1957, 49, 704.
36. Merrett, F. M. J. Polym. Sci. 1957, 24, 467.
37. Schollenberger, C. S.; Scott, H.; Moore, G. R. Rubber World 1958, 137, 549.
38. Crouch, W. W.; Short, J. N. Rubber Plastics Age 1961, 42, 276.
39. Railsback, H. E.; Beard, C. C.; Haws, J. R. Rubber Age 1964, 94, 583.
40. Porter, L. M. (Shell); U.S. Patent 3 149 182, 1964, filed 1957.
41. Bailey, J. T.; Bishop, E. T.; Hendricks, W. R.; Holden, G.; Legge, N. R. "Thermoplastic Elastomers, Physical Properties and Applications"; presented at Division of Rubber Chemistry, American Chemical Society, Philadelphia, 1965; Rubber Age, October 1966.
42. Cooper, S. L.; Tobolsky, A. V. Textile Res. J. 1966, 36, 800.
43. Fetters, L. J.; Morton, M. Macromolecules 1969, 2, 453.
44. Quack, G.; Fetters, L. J. Polym. Preprints Am. Chem. Soc. Div. Polym. Chem. 1977, 18(2), 558.
45. Young, R. N.; Fetters, L. J. Macromolecules 1978, 11, 899.
46. Bi, Li-K; Milkovich, R. Rubber World 1979, 180(7), 129.
47. Martin, M. K.; McGrath, J. E. Polym. Preprints Am. Chem. Soc. Div. Polym. Chem. 1981, 22(1), 212.
48. Morton, M.; Fetters, L. J.; Inormata, J.; Rubio, D. C.; Young, R. N. Rubber Chem. Tech. 1976, 49, 303.
49. Foss, R. P.; Jacobson, H. W.; Sharkey, W. H. Macromolecules 1977, 10(2), 287.
50. Beinert, G.; Lutz, P.; Franta, E.; Rempp, P. Makromol. Chem. 1977, 179, 551.
51. West German Patent 2 230 227, 1972.
52. See Ref. 15, pp. 207-8.
53. See Ref. 15, pp. 343-52.
54. See Ref. 15, pp. 427-38.
55. Dreyfuss, M. O.; Dreyfuss, P. Adv. Polym. Sci. 1967, 4, 528.
56. Kennedy, J. P.; Melby, E. G. Polym. Prepr. 1974, 15(2), 180.
57. Berger, G.; Levy, M.; Vofsi, D. J. Polym. Sci. 1966, B4, 183.
58. Yamashita, Y.; Nobutoki, K.; Nakamura, Y.; Hirota, M. Macromolecules 1971, 4, 548.
59. See Ref. 1, p. 20.
60. Buck, W. H.; Cellae, R. J. Polym. Prepr. 1973, 14, 98.
61. Witsiepe, W. K. Polym. Prepr. 1972, 13(1), 588.
62. Noshay, A.; Matzner, M.; Williams, T. C. Ind. Eng. Chem., Prod. Res. Dev. 1973 12, 268.
63. Matzner, M.; Noshay, A.; McGrath, J. E. Polym. Prepr. 1973, 14, 68.

64. Bamford, C. A.; Paprotny, J. Polymer 1972, 13, 208; presentation at NATO Advanced Studies Institute, Rensselaer Polytech. Inst., July 15, 1973.

65. Tobolsky, A. V.; Rembaum, A. J. Appl. Polym. Sci. 1964, 8, 307.

66. Allport, D. C.; Janes, W. H. "Block Copolymers"; John Wiley and Sons: New York; 1972, pp. 137-207.

67. Agouri, E.; Parlant, C.; Mornet, P.; Radeau, J.; Teitgen, J. F. Makromol. Chem. 1970, 137, 229.

68. Kennedy, J. P.; Charles, J. J.; Davidson, D. L. Polym. Prepr. 1973, 14, 974; See Kennedy, J. P.; Planthottam, S. S.; Ivan, B. J. Macromol. Sci.-Chem. 1982, A17(4), 637, and previous publications cited therein.

69. Falk, J. C.; Schlott, R. J.; Hoeg, D. F. J. Macromol. Sci.-Chem. 1973, A7(8), 1947.

70. DeLand, D. L.; Purden, J. R.; Schoneman, D. P. Chem. Engr. Progress 1967, 63(7), 118.

71. Milkovich, R.; Chiang, M. T. U.S. Patent 3 786 116, 1974.

72. Falk, J. C.; Schlott, R. J.; Hoeg, D. F. J. Macromol. Sci.-Chem. 1974, A7(8), 1969.

73. Halasa, A. F. Rubber Chem. Tech. 1981, 54, 627.

74. "Reverse Osmosis Membrane Research"; Lonsdale, H. K.; Podall, H. E., Eds.; Plenum Press: New York, 1972; p. 317.

75. Bishop, E. T.; O'Neill, W. P.; Winkler, D. E. National Heart Inst. Annual Rept., S-14097, 1969.

76. Bishop, E. T.; et al. National Heart Inst. Annual Rept., S-14097, 1969; Winkler, D. E. U.S. Patent 3 607 979, 1971.

77. Winkler, D. E. U.S. Patent 3 577 357, 1971.

78. Murtland, W. O. Elastomerics 1981, 29.

79. Molau, G. E. "Block Polymers"; Aggarwal, S. L., Ed.; Plenum Press: New York, 1970; p. 79.

80. Kato, K. Polym. Engr. Sci. 1967, 7, 38.

81. Vanzo, E. J. Polym. Sci. A1 1966, 4, 1727.

82. Kuleznev, V. N.; Dogadkin, B. A.; Kylkova, V. D. Koll Zhurn. 1968, 30, 255; Helfand, E.; Tagami, Y. Polym. Lett. 1971, 9, 741; Broutman, L. J.; Agarwal, B. D. Polym. Eng. Sci. 1974, 14, 581.

83. Bishop, E. T.; Davison, S. J. Polym. Sci. (C) 1969, 26, 59.

84. Smith, T. L.; Dickie, R. A. J. Polym. Sci. (C) 1969, 26, 163; Morton; M.; McGrath, J. E.; Juliano, P. C. J. Polym. Sci. (C) 1969, 26, 94.

85. Puett, D.; Smith, K. J.; Ciferri, A. J. Phys. Chem. 1965, 69, 141; Fischer, E.; Henderson, J. F. Rubber Chem. Tech. 1967, 40, 1313.

86. Kraus, G.; Grauer, G. T. J. Appl. Sci. 1967, 11, 2121; Arnold, K. R.; Meier, D. J. J. Appl. Polym. Sci. 1970, 14, 427.

87. Molau, G. E. J. Polym. Sci. 1965, 3, 4235.

88. Coran, A. Y.; Patel, R. Rubber Chem. Tech. 1980, 53, 141.

89. Ibid., p. 781.

90. Ibid., p. 892.

91. Coran, A. Y.; Patel, R.; Williams, D. Rubber Chem. Tech. 1982, 55, 116.

92. Chem. Eng. News 1982, 60, 20, 14.

93. Lantz, W. L.; Sanford, J. A.; Young, J. F. Paper presented at ACS Rubber Division Meeting in San Fransisco, Oct. 1976.
94. Holden, G. J. Elastomers Plastics 1982, 14, 148.
95. St. Clair, D. J. Rubber Chem. Tech. 1982, 55, 208.
96. Burlant, W. J. Paper presented at the May 1974 Meeting, Chemical Marketing Research Assoc., published in the Proceedings of the May 1974 Meeting, New York.
97. Auchter, J. F. "Outlook for Thermoplastic Elastomers in the 1980s"; paper presented to the Rubber Division, American Chemical Society, Oct. 1981.

Multicomponent Polymer Systems

NORBERT PLATZER

Springborn Laboratories, Inc., Enfield, CT 06082

Copolymers
  Random Copolymers
  Alternating Copolymers
  Graft Polymers
  Block Copolymers
  Network Copolymers and Interpenetrating Polymer Networks
Polyblends
  Compatible Polyblends
  Incompatible Polyblends
Composites
  Reinforced Thermoplastics
  Fillers in Rubber
  Fillers in Thermosets
  Reinforced Thermosets

Multicomponent polymer systems, frequently defined as polymer hybrids or polymer alloys, represent physical or chemical polymeric blends. Chemical polymeric blends are linked together by covalent bonds formed of random, alternating, graft, block, or network copolymers. These blends may also be linked topologically with no covalent bonds as interpenetrating polymer networks, or they may not be linked at all except physically through polyblending. In addition, the copolymers may be thermoplastics or thermosets, and the polyblends may be compatible or incompatible. Furthermore, distinction is made between single-phase homogeneous systems and multiphase heterogeneous systems.

The term multicomponent polymer system is frequently extended to composites comprised of a polymeric matrix in which a filler or reinforcing material is dispersed. The large-volume commodity plastics known for decades are generally homopolymers. In the process of modifying the properties of these homopolymers, a large number of specialty and value-added plastics have been introduced.

0097–6156/85/0285–0219$06.00/0
© 1985 American Chemical Society

These polymers are not new, primarily, but are multicomponent systems of well-established polymeric materials.

Copolymers

Random Copolymers. Originally, only random copolymers (see Figure 1) could be made by charging two monomers simultaneously and using free-radical initiators. Typical examples from the 1930s are the first synthetic rubber materials, which were the random copolymers of butadiene with either styrene or acrylonitrile, and the first fully synthetic fiber, a random copolymer of vinyl chloride with vinyl acetate.

A random copolymer of styrene and acrylonitrile exhibited a 20 °C higher use temperature than that of polystyrene; a random copolymer of styrene with butyl acrylate was more flexible than the styrene homopolymer.

Scientists soon realized that the composition of these copolymers prepared from unsaturated monomers is governed by different rules than those controlling condensation reactions. The first attempt to treat copolymerization in a systematic way was made by F. T. Wall ($\underline{3}$) in 1941. He concluded that the relative rates of monomer addition to growing radicals were dependent on the nature of the monomers and on their relative proportion, as expressed in the top equation of Figure 2. There, M_1/M_2 is the ratio of monomers in the feed, m_1/m_2 is the ratio of added monomers in the resulting copolymer, and r_1/r_2 is the relative rate of the two monomers.

Soon afterwards scientists observed that Wall's equation did not hold for numerous random copolymer systems such as styrene/acrylonitrile (SAN) and styrene/methyl methacrylate (SMMA). In such cases the copolymer depletes the monomer with respect to either reactant and thus causes a drift in copolymer composition with conversion, as illustrated in the graph of Figure 2. The point at which the curve crosses the straight line was designated as the "azeotropic" composition. The original equation had to be modified, as shown at the bottom of Figure 2. In order to produce a copolymer of a composition that differed from that of the azeotropic composition, feeding of the faster reacting monomer in portions during the copolymerization reaction was necessary ($\underline{5}$). By this controlled feeding of the monomers, regulation of the sequence distribution in random copolymers and production of SAN of high acrylonitrile content were possible ($\underline{6}$).

Taking into account polarity, steric factors, and resonance stabilization, T. Alfrey and C. C. Price ($\underline{7}$) developed a Q-e scheme and predicted monomer reactivity. The effect of polarity on vinyl monomer copolymerization was recognized by F. R. Mayo and coworkers ($\underline{8}$), who distinguished between monomers of average activity and those acting as electron donors or acceptors. By combining these theories with experimental data, calculation of product probabilities of various monomer combinations and determination of monomer reactivity parameters were possible.

Random copolymers have grown to be the most versatile, economical, and easily synthesized types of copolymers. A wide variety of free-radical, ionic-addition, and ring-opening polymerization techniques, as well as many step-growth reactions, are employed.

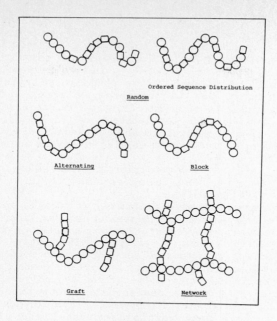

Figure 1.   Types of copolymers.

$$\frac{m_1}{m_2} = \frac{r_1}{r_2} \cdot \frac{[M_1]}{[M_2]}$$

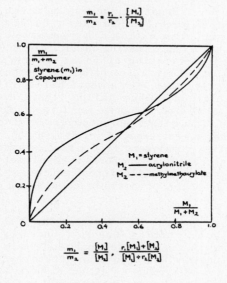

$$\frac{m_1}{m_2} = \frac{[M_1]}{[M_2]} \cdot \frac{r_1[M_1]+[M_2]}{[M_1]+r_2[M_2]}$$

Figure 2.   Copolymerization of SAN and SMMA.

During the past four years, linear low-density polyethylene (LLDPE) has probably become the most important of the thermoplastic copolymers. In contrast to the customary practice of producing branched ethylene homopolymer in a high-pressure reaction, a system of copolymerizing ethylene with $\alpha-C_{4-8}$ olefins at low pressure is used to make LLPDE copolymer. This random copolymerization is commercially carried out in gas-phase, slurry, and solution processes in the presence of a transition metal catalyst; 1-butene, 1-hexene, 4-methyl-1-pentene, or 1-octene are choices of comonomer. In the face of plant overcapacity and idle equipment existing at this time, LLDPE can also be made in high-pressure autoclaves and tubular reactors.

Other large-volume random copolymers are ethylene/vinyl acetate (EVA), ethylene/propylene/diene monomer (EPDM), and vinyl chloride/acrylonitrile copolymer.

Uniform random copolymers generally display a single-phase morphology. The ordered sequence distribution (see Figure 1) is too short to induce microphase separation in amorphous systems. For example, a transparent impact polystyrene is obtainable by the order sequence distribution of butadiene between styrene segments (9).

Furthermore, monomers from which crystalline homopolymer can be produced, such as high-density polyethylene and polypropylene, can be copolymerized to produce resins with controllably reduced crystallinity and thus greater transparency. The ethylene/propylene copolymers may range from partially crystalline plastics to amorphous elastomers.

Alternating Copolymers. Alternating copolymers, comprised of monomers in uniform alternating succession along the chain (see Figure 1), are rather rare because they require highly specific copolymerization reactivity ratios. Styrene/maleic anhydride is a copolymer that is highly alternating in nature.

The rate of polymerization of polar monomers, for example, maleic anhydride, acrylonitrile, or methyl methacrylate, can be enhanced by complexing them with a metal halide (zinc or vanadium chloride) or an organoaluminum halide (ethyl aluminum sesquichloride). These complexed monomers participate in a one-electron transfer reaction with either an uncomplexed monomer or another electron-donor monomer, for example, olefin, diene, or styrene, and thus form alternating copolymers (11) with free-radical initiators. An alternating styrene/acrylonitrile copolymer (12) has been prepared by free-radical initiation of equimolar mixtures of the monomers in the presence of nitrile-complexing agents such as aluminum alkyls.

Graft Polymers. At the Fall 1950 American Chemical Society meeting in Chicago, the term "grafting" was introduced by T. Alfrey, and shortly thereafter it was reported and adopted by H. F. Mark (13). In graft polymerization (see Figure 1) a preformed polymer having residual double or polar groups is either dispersed or dissolved in the monomer in the presence or absence of a solvent. Onto this backbone sequences of the monomer are grafted, generally in a free-radical reaction.

Grafting has become the most widely used technique for toughening brittle polymers with an elastomer. Graft polymers

include impact polystyrene, acrylonitrile/butadiene/styrene (ABS),
and methacrylate/butadiene/styrene (MBS) and have been treated in a
number of reviews and books (14).

Impact polystyrene is produced commercially in three steps:
solid polybutadiene rubber is cut up and dispersed as small particles in styrene monomer; mass prepolymerization; and completion of
the polymerization either in mass or aqueous suspension. During the
prepolymerization step, styrene starts to polymerize by itself by
forming droplets of polystyrene upon phase separation. When equal
phase volumes are attained, phase inversion occurs (15). The
droplets of polystyrene become the continuous phase in which the
rubber particles are dispersed. Impact strength increases with
rubber particle size and concentration, whereas gloss and rigidity
decrease.

The type and configuration of the dispersed rubber also have a
significant influence on properties. For example, four configurations of polybutadiene exist, as shown in Figure 3. Currently,
polybutadiene having a medium *cis*-1,4 configuration of about 36% and
made by using a butyllithium catalyst is the rubber most widely used
in impact polystyrene. This *cis*-1,4 configuration is characterized
by a lower glass transition temperature (-108 °C) than that of the
trans-1,4 configuration (-14 °C) and thus results in satisfactory
impact strength at low temperatures. This rubber also contains
about 12% 1,2-(vinyl) configuration which is more subject to cross-
linking and oxidative degradation.

It has also been shown (16) that grafting first takes place on
the vinyl sites and later on the less exposed 1,4 configurations. A
polybutadiene of high vinyl configuration (70%) has been evaluated
in Japan (17) for making photodegradable impact polystyrene for
disposable containers. This polybutadiene was made by anionic
polymerization with a special aluminum/chromium catalyst.

A polybutadiene with 60% *trans*-1,4, 20% *cis*-1,4, and 20% vinyl
configuration is made in an emulsion polymerization system for use
as a substrate in the manufacture of ABS. The rubber particles are
less than one-tenth the size of the rubber particles used in impact
polystyrene, as shown in Figure 4. With the large rubber particles,
grafting takes place first inside the voids and later as warts on
the outside (18). With the use of small solid particles, the graft
polymer forms a shell around the rubbery core. This shell grows
with increasing grafting level and provides a better separation of
the particles and more uniform distribution in the SAN matrix.

A higher cross-linking density of the larger particles leads to
an irregularly shaped graft shell and to a finer distribution of the
grafted polystyrene inside the rubber particles. Impact strength
plotted versus grafting level goes through a maximum, the position
of which depends on the size of the rubber particles. A steady rise
in melt viscosity is observed with increasing grafting. Differences
in yield strength and elongation are the result of the reduction of
the shrinkage stress inside the rubber particles arising from the
incorporation of rigid graft polymer. Constant stresses at failure
are the result of the viscoelastic behavior of the crazes inside the
matrix, a behavior pattern that is unaffected by particle size.

To obtain better outdoor weatherability, polybutadiene rubber is
frequently replaced as the substrate in impact polystyrene and ABS
by EPDM, EVA (19), or polybutyl acrylate. Grafting of styrene,

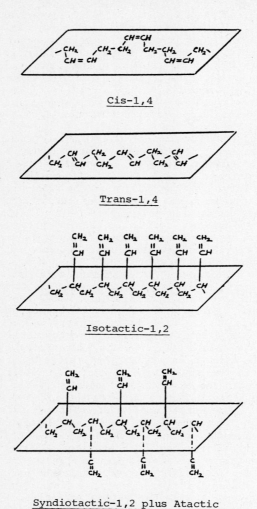

Figure 3.   Polybutadiene rubber configurations.

methyl methacrylate, or vinyl acetate in free-radical reaction onto cellulose results in thermoplastics in which the second-order transition temperature can be lowered to 77–137 °C, far below the decomposition temperature of unmodified cellulose (20).

Graft copolymers consisting of a functional and a structural segment were prepared by Y. Yamashita (21) by using the recently developed "macromonomer" technique in which styrene and methyl methacrylate form double-bond-terminated macromonomers. The two monomers were polymerized in the presence of a free-radical initiator and a chain transfer agent such as thioglycolic acid to form carboxylic-terminated prepolymers that were reacted with glycidyl methacrylate. Free-radical copolymerization of these macromonomers with various functional monomers proceeded easily and resulted in comblike graft polymers.

Better control of grafting and less homopolymer formation is achievable in ionic reactions than can be obtained in free-radical reactions. Anionic grafting via backbone initiation has been demonstrated (22) with caprolactam on macromolecular ester sites of styrene/methyl methacrylate copolymers. Cationic grafting of isobutylene onto poly(vinyl chloride) with the aid of aluminum alkyl has been carried out by J. P. Kennedy (23).

Block Copolymers. The manufacture of block copolymers became possible in 1956 by M. Szwarc's discovery (24) of "living" polymers prepared in homogeneous anionic polymerization. Diblock, triblock, and multiblock copolymers are produced ionically in the presence of sodium naphthalene, butyllithium, or Ziegler-type catalysts.

The simplest diblock polymer, styrene/butadiene block polymer, is formed when the two monomers are charged into a batch reaction along with the catalysts. The reactivity ratios are such that the butadiene polymerizes first and with almost total exclusion of any styrene present. Only after all of the butadiene monomer has been consumed does the bulk of the styrene enter the polymer chain.

Because no termination step exists, styrene/butadiene/styrene (SBS) triblock polymers can easily be made by charging styrene, butadiene, and again styrene, in succession, to the catalyst. Because the polystyrene blocks behave like styrene homopolymer, these triblock polymers are not suitable for use in automobile tires, but they lend themselves well in polyblends and as the backbone in the graft polymerization of impact polystyrene and ABS (25). Transparent impact polystyrene has been made with sequentially prepared triblocks containing 10–30% polybutadiene (26).

It is claimed that styrene/butadiene diblock polymers bring about an improvement in the hardness, strength, and processability of polybutadiene elastomers (27), as well as an improvement in the ozone resistance of neoprene rubber (28). Styrene diblock polymers have also been made with isoprene, α-methylstyrene, methyl methacrylate, vinylpyridine, and α-olefins. Block copolymers of ethylene, propylene, and other α-olefins with each other have been made as well. Heteroatom block copolymers based on styrene or other hydrocarbons and alkylene oxides, phenylene oxides, lactones, amides, imides, sulfides, or siloxanes have been prepared.

In SBS triblock polymers a microphase separation may occur, as illustrated in Figure 5 (29). The rigid polystyrene end segments

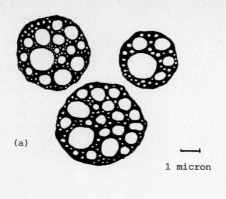

(a)

1 micron

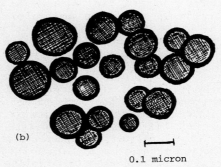

(b)

0.1 micron

Figure 4. Rubber particles (a) in impact polystyrene and made by
mass (solution) polymerization; (b) with grafted shell in
ABS and made by emulsion polymerization.

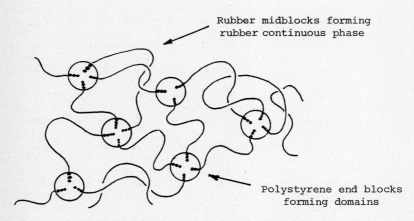

Rubber midblocks forming
rubber continuous phase

Polystyrene end blocks
forming domains

Figure 5. Block copolymer: styrene/butadiene/styrene.

are joined by elastomer polybutadiene center sections. The polystyrene segments associate with each other to produce large
aggregates or "domains." At room temperature these block polymers
behave almost like cross-linked rubber. When heated above the glass
transition temperature of polystyrene (100 °C), the domains soften
and may be disrupted by applied stress, and thus the block copolymer
flows. This process is reversible, and the thermoplastic elastomer
regains its original tensile strength, elongation, and resilience.
These block copolymers can be injection molded into articles such as
shoe soles without the need of a cure step.

Triblock polymers are generally made in sequential addition
processes with the aid of sodium naphthalene or alkyllithium. They
can also be produced in a "coupling" process wherein a diblock is
prepared first and is coupled with the aid of phosgene or alkyl
dihalides to form the ABA triblock. Tapered blocks can be made by
starting with a polystyrene block and then copolymerizing a mixture
of styrene and diene.

Radial or star-shaped SBS triblock polymers have been prepared
by the addition of divinylbenzene (30); in a coupling reaction with
silicon tetrachloride (31); or with epoxidized polybutadiene or
epoxidized soybean oil (32).

The presence of hard domains and intersegment chemical linkages
and their behavior during stretching have also been observed in
multiblock copolymers, as illustrated in Figure 6 (33). Spandex
fibers are segmented polyurethanes consisting of alternating soft
and hard blocks coupled together in condensation reactions (34).

A flexible linear polyglycol having a molecular weight of 500–
4000 can be prepared; it can be a polyester or a polyether, but it
must have reactive hydroxyl groups at both ends. These groups are
reacted with a diisocyanate (e.g., 2,6-toluene diisocyanate) to form
the soft-segment prepolymer. In the following step the hard
segments are formed by reacting the prepolymer with low molecular
weight glycols or diamines. This reaction results in a polyurethane
or polyurea having hydrogen bonding sites that provide the tie
points responsible for the long-range elasticity.

Other multiblock copolymers contain siloxanes as blocks, and
others are based on cross-linked epoxy resin systems (35).

Network Copolymers and Interpenetrating Polymer Networks. Network
copolymers (see Figure 1) can be made by cross-linking one linear
homopolymer with a second linear homopolymer. This cross-linking is
achievable by generating free-radical sites on a preformed polymer
in the presence of a monomer. Grafting takes place on these sites,
and combination termination results in cross-linking. For example,
cross-linked copolymers of polycarbonate and polychloroprene have
been prepared in the presence of molybdenum and manganese carbonyls
as initiators (36).

In contrast to network copolymers, interpenetrating polymer
networks (IPNs) (2, 37, 38) are generally composed of two or more
cross-linked polymers with no covalent bonds between them. They are
often incompatible polymers in a network form, and at least one is
polymerized and/or cross-linked in the immediate presence of the
other with the result of a complete entanglement of the chains.

Most IPNs are heterogeneous systems comprised of one rubber and
one glassy phase. This combination results in a synergistic effect

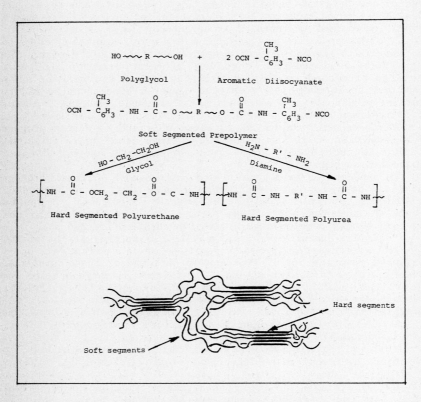

Figure 6.   Segregation of hard segments.

on mechanical properties and thus yields either high impact strength or reinforcement that are both dependent on phase continuity.

Thermoset IPNs are characterized by suppressed creep and flow and were pioneered by L. H. Sperling (38a), K. C. Frisch (38b), and coworkers, in 1969. They are subdivided into four types:

1. Sequential IPNs are started with one polymer that is already cross-linked. It is swollen, and a monomer along with an initiator and a cross-linking agent are added and polymerized in place. A typical example (37) is an elastomer such as butadiene or ethyl acrylate cross-linked with divinylbenzene as the first polymer. The monomer that is polymerized in the swollen polymer is styrene or methyl methacrylate cross-linked with divinylbenzene or tetraethylene glycol dimethacrylate. The initiator is potassium for butadiene, and benzoin for the other monomers.

2. Simultaneous IPNs are formed by homogeneously mixing together monomers, prepolymers, linear polymers, initiators, and cross-linkers. The monomers and prepolymers are simultaneously polymerized by independent reactions that differ enough to avoid interfering with each other. For example, a polyurethane/polymethacrylate and a polyurethane/polystyrene were made in a process in which both monomers were prepolymerized, dissolved together, and reacted to form an IPN. Another urethane system was made from castor oil reacted with toluene diisocyanate and sebacic acid polyesters. The resultant urethane prepolymer was then mixed with polystyrene to form an IPN.

3. Composites having one cross-linked polymer and one linear polymer are categorized as semi-IPNs. Among them are SBR/polystyrene, polyurethane/silicone, and silicone/nylon systems.

4. Homo-IPNs are interpenetrating networks made up of polymers that are identical but retain their specific characteristics. An example is an epoxy/epoxy IPN.

In contrast to the thermoset IPNs, thermoplastic IPNs can be processed in standard extrusion and molding equipment. They are made either by mechanical blending or chemically by sequential polymerization.

One of the first mechanical thermoplastic IPNs was that prepared by Exxon Chemical and made by melt blending ethylene-propylene (E-P) copolymer or EPDM with polypropylene in the presence of a small amount of peroxide as cross-linking agent. Recently, Shell Chemical (39) started to market Kraton IPN, a mechanical polyblend of a styrene-ethylene/butadiene-styrene (S-EB-S) block copolymer with poly(butylene terephthalate). It is specified that the block copolymer should have two polystyrene end blocks, each having a molecular weight of 25,000; the EB center block should have a molecular weight of 100,000; and the butadiene should have 42% 1,2-vinyl content. The block copolymer was selectively hydrogenated so that more than 95% of the aliphatic double bonds were reduced, and less than 5% of the aromatic bonds were reduced.

A chemically blended thermoplastic IPN was developed by Petrarch Systems (38). It consists of polyurethane, polyamide (40), or polyester elastomer as matrix and a reactive silicone component. When the melt stage is reached by using conventional injection molding or extrusion equipment, the silicones begin to react with one another in the presence of Pt catalyst and form the semi-IPN. These materials are said to exhibit tensile and tear strengths three to five times those of silicones.

Polyblends

Polyblends are made by intimately mixing two or more polymers on mill rolls, in extruders, or in Banbury or other mixing devices. The polyblends are admixtures of either two rigid polymers or two elastomeric polymers, or combinations of the two types. Their properties, and therefore their end uses, are strongly dependent on the degree of compatibility of the components.

Compatible Polyblends. When the polymeric materials are compatible in all ratios, and/or all are soluble in each other, they are generally termed polyalloys. Very few pairs of polymers are completely compatible. The best known example is the polyblend of poly(phenylene oxide) (poly-2,6-dimethyl-1,4-phenylene oxide) with high-impact polystyrene (41), which is sold under the trade name of Noryl. It is believed that the two polymers have essentially identical solubility parameters. Other examples include blends of amorphous polycaprolactone with poly(vinyl chloride) (PVC) and butadiene/acrylonitrile rubber with PVC; the compatibility is a result of the "acid-base" interaction between the polar substituents (1). These compatible blends exhibit physical properties that are intermediate to those of the components.

Incompatible Polyblends. Most polyblends are heterogeneous and consist of a polymeric matrix in which another polymer is embedded. In contrast to copolymers in which the components are linked by strong covalent bonds, the components in polyblends adhere together only through van der Waals forces, dipole interaction, or hydrogen bonding. We distinguish among four types of polyblends, as indicated in Table I. These are prepared to improve certain physical properties. Blends that display an intermediate degree of compatibility are usually based on random copolymers rather than matched compositions and have been termed mechanically compatible (42). Polyblends in which the two polymers are entangled in a partial or interlocking network are called interpenetrating polymer networks; these were discussed earlier.

Polyblends with Rigid Matrix and Soft Dispersed Phase. This combination is selected to toughen brittle polymers with elastomers. The first impact polystyrene was made by blending a polystyrene and a SB rubber latex, coagulating, and processing together. Miscibility (43) between the two components plays an important role. A more uniform polyblend of higher impact strength can be obtained by blending styrene homopolymer either with a graft polymer of styrene on polybutadiene or with an SBS terblock polymer.
    ABS went through an analogous development. Originally a nitrile rubber latex and a SAN random copolymer latex were produced separately, blended, coagulated, and compounded together. A more

miscible ABS can be manufactured by blending SAN with a graft polymer of styrene and acrylonitrile on polybutadiene.

A polyblend of superior impact strength has been claimed by Phillips Petroleum (44) and is obtained by blending SAN with a styrene/butadiene/caprolactone terblock polymer. Impact PVC is manufactured by blending the homopolymer with an elastomeric compound such as ABS, MBS, or acrylonitrile/methacrylate/butadiene/styrene (AMBS).

The impact resistance of polypropylene at low temperature has been improved by polyblending with EPDM or E-P rubber to make possible the application of this material in the automotive industry. The low-temperature properties of polyamides such as nylon 6 and nylon 66 have been improved by polyblending with ethylene copolymers or specially grafted polybutadiene (45).

Polycarbonate is being blended with ABS to lower the Vicat softening temperature and to make it more thermoformable (46). It is also blended with poly(butylene terephthalate) to increase stress-crack resistance.

Cellulose esters such as cellulose acetobutyrate and cellulose propionate, which originally required plasticizers, are today blended with EVA and EVA graft polymer, respectively, to convert them into thermoplastic products (47).

Polyblends with Rigid Matrix and Rigid Dispersed Phase. Polyblends in which both phases are rigid are made with the intention of improving melt flow and mechanical properties or for the purpose of reducing shrinkage. During hardening of unsaturated polyesters in the presence of styrene, a shrinkage of 6–9% takes place. By dispersing a thermoplastic resin in the styrene monomer, the shrinkage can be reduced to 0.01–0.1% (48).

Saturated polyesters such as poly(ethylene terephthalate) and poly(butylene terephthalate) are characterized by high rigidity, good impact resistance, and low creep at room temperature. By polyblending with special graft polymers (49), it became possible to maintain these fine properties at -20 °C and -40 °C without sacrificing chemical resistance.

A transparent rigid vinyl bottle compound is produced by blending PVC homopolymer with a methyl methacrylate/butadiene/styrene graft polymer of equal refractive index.

Polyblends of polyamides with polyethylene and of polyacetal with polytetrafluoroethylene have found application as bearing materials.

Polyblends with Soft Matrix and Soft or Rigid Dispersed Phase. Polyblends in which both components are soft are mixtures of various elastomers. For example, treads of automobile tires are made of a polyblend of SBR with either natural rubber or *cis*-polybutadiene.

The addition of a rigid polymer to a soft matrix results in an increase in the modulus of elasticity of the total polyblend and is known as rubber reinforcement. It also raises both tensile strength and tear strength, and reduces abrasion, cut growth, and flex cracking.

Composites

In contrast to polyblends, composites consist of a polymeric matrix in which a solid filler or reinforcing material is dispersed. An exact differentiation between filler and reinforcing material is sometimes difficult (50). The so-called "aspect ratio," that is, the ratio of length to thickness, of the added material determined whether it acts as a filler or as a reinforcing material.

Solid additives in the shape of spheres, cubes, or platelets generally act as a filler (extender) and, with the exception of raising stiffness, do not improve the mechanical properties of the composite. With very strong adhesive forces between filler surface and polymer chains, however, a filler may also provide reinforcement, for example, carbon black in rubber or uncoated calcium carbonate in polyamides.

Fibers having an aspect ratio of above 10, such as glass fibers, asbestos, wollastonite, cellulose fibers, carbon fibers, and whiskers, act as reinforcing agents. Composites may be classified into four groups, as shown in Table II. In modern composites, such as in the case of epoxy resins reinforced with carbon fibers (51), the polymer matrix has only a secondary function: to separate the individual fibers from one another and to transfer energy to the fiber surface; the fibers almost completely withstand the mechanical load.

Filled and reinforced polymeric materials may be classified into four groups, as shown in Table II.

Reinforced Thermoplastics. Thermoplastic composites are reinforced with either mineral particles such as calcium carbonate, kaolin, ground wollastonite, feldspar, and mica flakes or with short fine fibers of glass and asbestos, so that the mix can still be injection molded (52). To obtain good adhesion between the inorganic materials and the matrix, multifunctional silanes are added as coupling agents and provide the necessary binding force of the same magnitude as the secondary forces in polyblends. This coupling also leads to more uniform distribution of stress at the interface and shields against water penetration. The polymeric matrix itself must have a higher elongation than the reinforcing material in order to use the strengthening effect completely. By the addition of 20–40% short glass fibers, thermoplastics approach the strength of die-cast metals, as shown in Table III (53).

Fillers in Rubber. Carbon black and calcium silicate are able to reinforce rubber. For example, the tensile strength of an SBR vulcanizate can be raised from 350 to 3500 lb/in.^{2} by compounding with 50% of its weight of carbon black (54). The activity of the carbon black depends on particle size and shape, porosity, and number of active sites, which are less than 5% of the total surface. Elastomers of a polar nature, such as chloroprene or nitrile rubber, will interact more strongly with filler surfaces having dipoles, such as –OH and –COOH groups or chlorine atoms.

Table I. Polyblends

Matrix	Dispersed Phase	Improvement	Examples
Rigid	Soft	Tougher	Impact PS ABS Rigid PVC
Rigid	Rigid	Faster melt flow	PVC + ABS PPO + PS
Soft	Soft	Longer wear	Natural rubber + *cis*-polybutadiene
Soft	Rigid	Higher modulus	SBR + PS

Table II. Composites

Matrix	Reinforcing Materials, Filler
Thermoplastic	Short glass fibers Glass beads Minerals
Vulcanized rubber	Carbon black
Phenolic	Wood flour
Polyester, Polyepoxide	Long glass fibers Grass mat and cloth Graphite

<u>Fillers in Thermosets</u>. Sixty-five years ago, in a paper presented before the American Chemical Society, L. H. Baekeland (55) discussed the usefulness of phenol-formaldehyde resins that, when compounded with wood flour, could be molded. Wood flour, ground nut shells, α-cellulose, or paper add bulk to phenolics, melamine, or urea-formaldehyde resins and make them easier to fabricate and less expensive.

Asbestos, either powdered or in fiber form, imparts heat resistance; aluminum trihydrate and talcum act as flame retarders; mica gives excellent electrical properties; and aluminum powder improves heat transfer. Powdered minerals such as silica or china clay are often used to improve water resistance. One novel area is the use of hollow glass spheres in rigid polyurethane foam, known as "diafoam," to achieve 30–70% higher flexural and compression moduli. This foam is used as "structural foam" in furniture.

Reinforced Thermosets. In contrast to fillers that do not enhance the strength of thermosets, glass fiber, mat, and cloth are capable of increasing the strength of phenolics, unsaturated polyesters, and epoxies significantly, illustrated in Table IV. Novel composites are approaching the strength of steel. Reinforcement with graphite, boron, or the new Kevlar-49 fiber results in a product that surpasses steel (56). A hydrated aluminum carbonate fiber acts as a flame retarder by decomposing at 300 °C and releasing carbon dioxide and water.

The stiffness of the composite is even more influenced by the reinforcing material. Table V indicates that the modulus of elasticity of glass fiber reinforced thermoplastics reaches 1 million $lb/in.^2$, and that of reinforced polyesters goes to 5 million $lb/in.^2$. Polyimides and polyamide-imides having heat deflection temperatures between 250 and 350 °C are used with graphite fibers and quartz fabrics in heat-resistant composites. Reinforced polybenzimidazoles, polyquinoxalines, polybenzoxazoles, polybenzothiazoles, and other semiladder polymers show excellent mechanical properties in the range from –260 °C to +605 °C.

Table III. Tensile Strength of Reinforced Thermoplastics

| Matrix | $K(lb/in.^2)$ | |
	Unreinforced	Reinforced with Short Glass Fibers
Polyethylene, linear	1.0–3.5	7–11
Polypropylene	4.3–5.5	8–9
Polyoxymethylene	10	10.5–12.5
Polystyrene	6–8	11.5–15
SAN	9.5–11	14–19
ABS	5–7	15–19
Poly(butylene terephthalate)	8	17.3
Polycarbonate	8–9.5	18.5–20
Nylon 66	9–12	20–30
Polysulfone	10	18–19
Die-case metals		
Zinc	41	
Aluminum	33–47	
Magnesium	34	

Table IV. Tensile Strength of Reinforced Thermosets

Reinforcement, Filler	$K(lb/in.^2)$			
	Phenolic	Poly-ester	Poly-epoxide	Reinforcing Fiber
Unreinforced	7–8	2–13	5–15	—
Filled – Wood flour	6–9	—	—	—
Reinforced				
E–Glass – Strand	5–18	4–17	10–20	450–500
– Mat	5–12	20–25	14–30	—
– Cloth	40–60	30–70	20–60	—
Graphite (Thornel)	—	—	120	400
Boron	—	—	200	300
S–Glass	—	—	3–0	650–700

Table V. Rigidity of Structural Materials

Material	Modulus of Elasticity (10^6 $lb/in.^2$)
Rigid thermoplastics	0.1–0.6
Reinforced thermoplastics	0.8–2.0
Polyester reinforced with E–glass fiber	0.8–2.5
Magnesium	6.5
Epoxy reinforced with S–glass fiber	8
Poly-*p*-benzamide fiber (Kevlar 49)	9
Aluminum	10
E–glass fiber	10.5
S–glass fiber	12.5
Epoxy reinforced with graphite fiber	25
Stainless steel fiber	29
Epoxy reinforced with boron fiber	35
Graphite fiber	50
Boron fiber	60
Graphite whiskers	140

Literature Cited

1. Noshay, A.; McGrath, J. E. "Block Copolymers - Overview and Critical Survey"; Academic Press: New York, 1977.
2. Klempner, D.; Frisch, K. C. "Polymer Alloys I and II: Blends, Blocks, Grafts, and Interpenetrating Networks"; Plenum Press: New York, 1977, 1979.
3. Wall, F. T. J. Am. Chem. Soc. 1941, 63, 1862.
4. Ham, G. E. "Copolymerization"; J. Wiley & Sons: New York, 1964.
5. May, F. R.; Lewis, F. M. J. Am. Chem. Soc. 1944, 66, 1594. Alfrey, T.; Goldfinger, G. J. Chem. Phys. 1944, 12-205. Wall, F. T. J. Am. Chem. Soc. 1944, 66, 2050.
6. Hendy, B. N. Polymer Preprints, ACS Meeting Spring 1974, 84.
7. Alfrey, T.; Price, C. C. J. Polym. Sci. 1947, 2, 101.
8. Mayo, F. R.; Lewis, F. M.; Walling, C. J. Am. Chem. Soc. 1948, 70, 1529.
9. Firestone Tire & Rubber Company; FR Patent 2 190 834, 1973.
10. Harwood, H. J. American Chemical Society, Polymer Division, Symposium on Alternating Copolymers, New York, 1973 (Preprints).
11. Gaylord, N. G.; Takahashi, A. ADVANCED IN CHEMISTRY SERIES No. 91, American Chemical Society: Washington, D.C., 1969; p. 94.
12. Johnston, N. W. American Chemical Society, Polymer Division, Symposium on Alternating Copolymers, New York, 1973 (Preprints).
13. Mark, H. F. Rec. Chem. Prog. 1951, 12(3), 139.
14. Ceresa, R. J. "Block and Graft Copolymers"; Butterworth: London, 1962. Battaerd, H.; Tregear, G. W. "Graft Copolymers"; Wiley: New York, 1967. Stannett, V. J. Macromol. Sci. Chem. 1970, 4(5), 1177. Platzer, N. ADVANCES IN CHEMISTRY SERIES No. 99, American Chemical Society: Washington, D.C., 1971.
15. Molau, G. E.; Keskula, H. J. Polym. Sci. A-1, 1966, 1595.
16. Stein, D. J. American Chemical Society, Polymer Division, Spring Meeting, 1974 (Preprints).
17. Japan Synthetic Rubber Company; JA Patent 574 705 and 574 706, 1974.
18. Jenne, H. Kunststoffe 1972, 62, 616. Stabenow, J.; Haaf, F. Angew. Makromol. Chem. 1973, 20/30, 1-23.
19. Bartl, H., et al. ADVANCES IN CHEMISTRY SERIES No. 92, American Chemical Society: Washington, D.C., 1969; p. 477. J. Appl. Polym. Sci. 1981, 36, 165. Severini, F., et al. ADVANCES IN CHEMISTRY SERIES No. 142, 1975; p. 201.
20. Arthur, J. C. "Advances in Macromolecular Chemistry," Vol. 2; Academic Press: London, 1970; J. Macromol. Sci. Chem. 1972, A-4, 1052.
21. Yamashita, Y. J. Appl. Polym. Sci. 1981, 36, 193.
22. Matzner, M.; Schober, D. L.; McGrath, J. E. American Chemical Society, Polymer Division, Fall Meeting, 1971 (Preprints).
23. Kennedy, J. P., et al. "Recent Advances in Polymer Blends, Grafts, and Blocks"; Sperling, L. H., Ed.; Plenum Press: New York, 1974.
24. Swarc, M., et al. J. Am. Chem. Soc. 1956, 78, 2656. Nature 1956, 278, 1168. "Carbanions, Living Polymers, and Electron Transfer Processes"; Wiley: New York, 1968.

25. Phillips Petroleum Company; GB Patent 1 192 471 and 1 192 472, 1970.
26. Chemische Werke Huels AG; DT OLS 2 026 308, 1971.
27. Dunlop Company Ltd.; FR Patent 1 567 877, 1969.
28. Phillips Petroleum Company; U.S. Patent 3 417 044, 1968.
29. Hendricks, W. R.; Enders, R. J. "Rubber Technology"; Van Nostrand-Reinhold: Princeton, NJ, 1973; p. 517.
30. Eschwey, H., et al. Makromol. Chem. 1973, 173, 235.
31. Morton, M., et al. J. Polym. Sci. 1962, 57, 471.
32. Phillips Petroleum Company; U.S. Patent 3 859 250, 1975.
33. Cooper, S. L.; Tobolsky, A. F. Text. Res. J. 1966, 36, 800.
34. Rinke, H. Angew. Chem. Int. Ed. 1962, 1, 419.
35. Noshay, A.; Robeson, L. M. J. Polym. Sci., Polym. Chem. Ed. 1974, 12(3), 689.
36. Bamford, C. H.; Eastmond, G. C. American Chemical Society, Polymer Division, Spring 1974 Meeting (Preprints).
37. Sperling, L. H. "Interpenetrating Polymer Networks"; Plenum Press: New York, 1981.
38. School, R. Rubber Plastics News, October 24, 1983.
38a. Sperling, L. H.; Friedman, D. W. J. Polym. Sci (A-2) 1969, 7, 425.
38b. Frisch, H. L.; Klempner, D.; Frisch, K. C. J. Polym. Sci. (B) 1969, 7, 75; J. Polym. Sci. (A-2) 1970, 8, 921.
39. Shell Oil Company; U.S. Patent 4 101 605, 1978.
40. Keuerleber, R. Kunststoffe 1983, 73, 509.
41. General Electric Company; U.S. Patent 3 383 435, 1968.
42. Matzner, M., et al. Ind. Chim. Belge 1973, 38, 1104. American Chemical Society, Polymer Division, Fall Meeting, 1973 (Preprints).
43. Paul, D. R.; Newman, S. "Polymer Blends"; Academic Press: New York, 1978.
44. Phillips Petroleum Company; U.S. Patent 3 789 034, 1974.
45. Michael, D. Kunststoffe 1980, 70, 629.
46. Nouvertné, 2.; Peters, H.; Beicher, H. Plastverarbeiter 1982, 33, 1074.
47. Leuschke, C.; Wandel, M. Plastverarbeiter 1982, 33, 1095.
48. Busch, W.; Schulz, H. Kunststoffe 1971, 61, 602.
49. Fischer, W.; Gehrke, J.; Rempel, D. Kunststoffe 1980, 70, 650.
50. Schlumpf, H. Kunststoffe 1983, 73, 511.
51. Morley, J. M. (Shell Research BV); American Chemical Society, Organic Coatings & Plastics Chemistry Division, Spring 1978 Meeting (Preprints).
52. Deanin, R. D. American Chemical Society, Organic Coatings & Plastics Chemistry Division, Fall 1977 Meeting (Preprints).
53. Platzer, N. Ind. Eng. Chem. 1969, 61, 10.
54. Boonstra, B. B. "Rubber Technology"; Van Nostrand-Reinhold: Princeton, NJ, 1973.
55. Baekeland, L. H. Ind. Eng. Chem. 1909, 1(3), 149.
56. Gloor, W. H. ADVANCES IN CHEMISTRY SERIES No. 99, American Chemical Society: Washington, D.C., 1971; p. 482.

PHYSICAL PHENOMENA OF POLYMERS

Structure–Property Relationships in Polymers

TURNER ALFREY, JR.[1]

The Dow Chemical Company, Midland, MI 48640

T$_M$ and T$_g$: Dependence on Molecular Structure
Linear Viscoelasticity
Large–Strain Rubber Elasticity
Non–Newtonian Fluids
Behavior of Glassy Amorphous Polymers
Behavior of Crystalline Polymers
Summary

The common central structural feature of organic macromolecules is the chain of covalently bonded atoms. Bond lengths and bond angles are rather rigidly fixed, but restricted rotation about single bonds permits a polymer chain to assume a wide range of three–dimensional conformations. At elevated temperatures, bond rotation is frequent; the polymer chain wriggles rapidly from one conformation to another. This micro–Brownian motion confers flexibility upon a macroscopic specimen. At low temperatures, the chains are immobilized, and the specimen is hardened by either of two mechanisms: crystallization (packing into a crystal lattice), or vitrification (forming a glassy amorphous solid). The crystalline melting point, T$_M$, and the glass transition temperature, T$_g$, are important characteristics of a given polymer.

The molecular structure of a particular polymer has two aspects, chemical composition and molecular architecture. The term "chemical composition" refers to the local molecular structure––nature of the units that make up the chains (including the stereochemical structure of these units). "Molecular architecture" refers to molecular structure in–the–large (average molecular weight, molecular–weight distribution, branching, cross–linking, etc.). For network polymers it refers to average molecular weight between cross–links, number and lengths of dangling tails, and many elusive aspects of network topology.

[1] Deceased

0097–6156/85/0285–0241$06.00/0
© 1985 American Chemical Society

In approaching the formidable problem of structure-property relationship, the first and simplest step is to examine the relationships between molecular structure and the values of T_M and T_g. Next is the more difficult task of characterizing the quantitative mechanical behaviors (and relating them to structure) of polymers in each of various regimes: high-temperature viscoelastic fluids; glassy amorphous solids; semicrystalline solids containing flexible amorphous regions (between T_g and T_M); semicrystalline solids containing glassy amorphous regions; highly crystalline solids; metastable, supercooled amorphous polymers; rubbery elastic networks; etc. In some of these regimes, "structure" refers simply to molecular structure; in others (notably glassy and crystalline states), properties depend not only on molecular structure but also on supramolecular structure—molecular orientation, crystalline morphology, etc.

T_M and T_g: Dependence on Molecular Structure

The crystalline melting point and the glass transition temperature of a polymer, in themselves, provide a rough characterization of the polymer properties; they also provide reference points for the various regimes within which the quantitative evaluation of properties must be made. How do T_M and T_g depend on molecular structure?

In addressing this question, we will consider a single molecular architecture—high molecular weight, linear chains. We shall arbitrarily classify such chains into five broad structural classes:

    Class I   -- Perfectly repeating "matched pearl necklace
    Class II   -- Random copolymers
    Class III -- d-ℓ and *cis-trans* "copolymers"
    Class IV   -- Block copolymers
    Class V   -- Short-unit chains that assume helicalconformations

Class I chains, because of their structural regularity, usually pack efficiently into a crystalline lattice. They can crystallize and usually exhibit a well-defined T_M and T_g. (The prototype—linear polyethylene—crystallizes so rapidly that it cannot be trapped in the glassy amorphous state; consequently, its T_g has been a matter of disagreement.) The crystalline melting point of linear polyethylene is approximately 140 °C. If methylene groups of polyethylene are replaced, sparsely and regularly, by other moieties, higher or lower values of T_M are observed. Two factors govern T_M: chain flexibility and interchain forces. Flexible units (such as ether, ester, or sulfide) result in lowered melting points. Rigid units (such as p-phenylene) result in higher melting points. Strong intermolecular forces (such as hydrogen bonds) yield high melting points (1).

Within Class I, the same structural features that promote high crystalline melting points also yield high values of T_g. Consequently, there exists a rough correlation between T_M and T_g for this class of polymers (2). By controlling chain stiffness and intermolecular forces, polymers with high T_M and T_g or low T_M and T_g can readily be designed; but it is not possible to <u>independently</u> control these two characteristic temperatures.

Class II polymers--random copolymers--fit less neatly into crystal lattices. Melting points are depressed, and the degree of crystallization is reduced. (A few special exceptions exist, in which the two monomer units are sufficiently matched in geometry that they can interchangeably occupy sites in a common lattice.) Because vitrification does not involve fitting into a crystal lattice, the <u>glass</u> temperatures of copolymers are not depressed by the chain irregularity. Consequently, random copolymers do not follow the T_M-T_g correlation characteristic of Class I polymers (<u>3</u>).

Class III polymers are primarily made up of vinyl and diene addition polymers. When a vinyl monomer, CH_2CHX, is subjected to addition polymerization a stereochemical problem is encountered at every second carbon atom of the chain. The substituent X can extend above or below the plane of the extended zigzag chain, corresponding to a d- or 1-configuration of the chain carbon atom in question (<u>4</u>). When the addition polymerization is carried out with a stereospecific catalyst, the polymer may be a regular structure: "isotactic" (repeating *dddddd*) or "syndiotactic" (perfectly alternating *d1d1d1*) (<u>5</u>). On the other hand, free-radical addition polymerization tends to produce a rather random ("atactic") copolymer of the d- and 1-configuration: *dd1d11d1dd1d11d*, etc. Consequently, vinyl polymers produced by free-radical polymerization tend to be permanently amorphous or at most to exhibit only a small amount of crystallinity.

In the case of dienes, a single pure monomer can enter the polymer chain in several different manners. The simplest example is butadiene, CH_2=CH-CH=CH_2, which upon polymerization can convert to a 1,2-chain unit (with a pendent vinyl group), or to a *cis*-1,4- or a *trans*-1,4-chain unit:

1,2-unit   *cis*-1,4 unit   *trans*-1,4 unit

Polybutadiene formed by high-temperature, free-radical addition polymerization is a copolymer of these three kinds of structural units. With isoprene (2-methylbutadiene), the number of ways the unit can enter the polymer chain is still larger; for example, the 1,2-unit with a pendent vinyl group is structurally different from the 3,4-unit with a pendent isopropenyl group:

$$\begin{array}{c} CH_3 \\ | \\ -CH_2-C- \\ | \\ CH=CH_2 \end{array} \qquad\qquad \begin{array}{c} H \\ | \\ -CH_2-C- \\ | \\ C=CH_2 \\ | \\ CH_3 \end{array}$$

1,2–unit   3,4–unit

Synthetic polyisoprene, prepared by free-radical polymerization of isoprene monomer, is a copolymer of six structurally distinct kinds of isoprene chain units. Unlike natural rubber, which is a regularly repeating Class I structure (*cis*-1,4), such synthetic polyisoprene does not crystallize. On the other hand, by the use of the appropriate stereospecific catalyst, isoprene monomer can be converted to a regular Class I polymer with the same structure as natural rubber (6).

Block copolymers (Class IV) are made up of two (or more) different monomer units, arranged in long blocks of each type of unit. For example, a chain consisting of a block of 500 A units following by a block of 500 B units and another block of 500 A units is an ABA triblock copolymer. If an A block corresponds to a Class I chain structure, it can crystallize in the normal poly(A) crystal lattice and can exhibit a T_M that is only slightly depressed compared to the poly(A) homopolymer. Even if the individual blocks are noncrystallizing atactic addition polymers, they are ordinarily mutually immiscible (if long) and undergo microsegregation into separate microphases, or "domains." These domains may develop into regular geometrical arrays, the form of which depends upon the relative volume fractions of the individual blocks. If the volume fractions are approximately equal, a laminar domain morphology emerges, with laminar thickness depending upon block lengths. If the B blocks constitute the major part of the copolymer, the B phase tends to be continuous, with cylindrical or spherical A domains dispersed within it in a regular fashion. The properties of such a block copolymer depend upon the composition and length of each block and the domain morphology assumed by the chains. Because of the microsegregation, the individual components exhibit their own characteristic T_g and T_M values (slightly modified). Thus, a segregated block copolymer will normally exhibit two distinct glass transitions, in contrast to the single intermediate glass transition commonly seen in random copolymers (7).

Whereas polyethylene, polyamides, and polyesters assume an extended planar zigzag conformation in the crystal lattice, many short-unit polymers twist into some helical conformation (Class V). In isotactic polyolefins, the extended planar conformation is sterically forbidden; by twisting into a regular helix, the chain relieves the steric strain. If the angular twist of each unit (relative to its predecessor) is a rational fraction of one revolution, then the spatial orientation of the methyl groups will exhibit a definite repeat distance. If the individual twist angle is $2\pi/n$, successive methyl groups will be oriented at the angles $2\pi/n$, $4\pi/n$, $6\pi/n$, etc., and the orientation will repeat with a periodicity of n groups. If the individual twist angle is $4\pi/n$ with

n odd, the chain will go through two helical turns before repeating.
More generally, if the twist angle per group (measured in
revolutions) is given by the irreducible fraction m/n, then the
substituent group orientation would repeat after n units, and m
complete turns appearing in the repeat sequence. The helical
conformation is effectively a rod and packs parallel to neighboring
rods in the crystal lattice (8).

Although we have emphasized the effects of chemical composition
upon the crystallization and vitrification processes, molecular
architecture also influences these processes. Branching and cross-
linking introduce points of irregularity that cannot easily fit into
a crystal lattice. This can reduce the degree of crystallinity, the
value of T_M, and the rate of crystallization. Thus, branched
polyethylene is considerably less crystalline than linear
polyethylene, and consequently softer and less dense. Vulcanized
natural rubber crystallizes much more slowly than unvulcanized, and
a high degree of vulcanization can completely prevent
crystallization (9). Introduction of cross-links into a glassy
amorphous polymer increases the value of T_g (10).

Linear Viscoelasticity

When we progress from the foregoing qualitative discussion of
structure–property relationships to the quantitative specification
of mechanical properties, we enter a jungle that has been only
partially explored. The most convenient point of departure into
this large and complex subject is provided by the topic of "linear
viscoelasticity." Linear viscoelasticity represents a relatively
simple extension of classical (small-strain) theory of elasticity.
In situations where linear viscoelasticity applies, the mechanical
properties can be determined from a few experiments and can be
specified in any of several equivalent formulations (11).

The accurate applicability of linear viscoelasticity is limited
to certain restricted situations: amorphous polymers, temperatures
near or above the glass temperature, homogeneous, isotropic
materials, small strains, and absence of mechanical failure
phenomena. Thus, the theory of linear viscoelasticity is of limited
direct applicability to the problems encounted in the fabrication
and end use of polymeric materials (since most of these problems
involve either large strains, crystalline polymers, amorphous
polymers in a glass state failure phenomena, or some combination of
these disqualifying features). Even so, linear viscoelasticity is a
most important subject in polymer materials science--directly
applicable in a minority of practical problems, but indirectly
useful (as a point of reference) in a much wider range of problems.

In an un-cross-linked amorphous polymer, above its glass
temperature, the molecular chains are continuously wriggling from
one conformation to another. If a mechanical stress is imposed on
such a system of wriggling chains, it can respond in three distinct
ways: instantaneous elastic response; retarded (conformational)
elastic response: or viscous flow. Actually, in order to fit
experimental data adequately, the retarded elastic element must be
expanded into a whole series of such elements, some with shorter and
some with longer response times. The local "kinkiness" of the
chains can be straightened out (by stress) more rapidly than the

larger scale convolutions. The time scales of these various retarded elastic contributions range over many orders of magnitude from the fastest to the slowest.

In spite of these complications, the viscoelastic response of an amorphous polymer to small stresses turns out to be a relatively simple subject because of two helpful features: (1) the behavior is linear in the stress, which permits the application of the powerful superposition principle; and (2) the behavior often follows a time-temperature equivalence principle, which permits the rapid viscoelastic response at high temperatures and the slow response at low temperatures to be condensed in a single master curve.

The superposition principle makes it possible to calculate the mechanical response of an amorphous polymer to a wide range of loading sequences from a limited amount of experimental information. Thus, from a single complete creep curve in pure shear or pure tension at a single load, it is possible, in principle, to calculate the response to combined stresses and time-dependent stresses (e.g., sinusoidal). Going still further, problems involving nonhomogeneous time-dependent stresses in viscoelastic objects can be solved by means of the superposition principle. The two common types of boundary-value problems in elasticity theory (surface forces or surface displacements specified) generalize simply to the analogous viscoelastic problems (surface forces or displacements specified as functions of both position and time.) (12).

The time-temperature equivalence principle makes it possible to predict the viscoelastic properties of an amorphous polymer at one temperature from measurements made at other temperatures. The major effect of a temperature increase is to increase the rates of the various modes of retarded conformational elastic response, that is, to reduce the retarding viscosity values in the spring-dashpot model. This appears as a shift of the creep function along the log t scale to shorter times. A secondary effect of increasing temperature is to increase the elastic moduli slightly because an equilibrium conformational modulus tends to be proportional to the absolute temperature (13).

By use of the time-temperature equivalence principle, the viscoelastic response of a given polymeric material over a wide temperature range can be accommodated in a single master curve. By use the superposition principle, this master curve can be used to estimate the time-dependent response to time-dependent stresses in simple tensile or shear specimens or to nonhomogeneous time-dependent stresses arising in stressed objects and structures.

The relationship between molecular structure and viscoelastic properties involves both chemical composition and molecular architecture. The short-time (low-temperature) behavior is relatively insensitive to molecular architecture, but master creep curves for different architectures diverge strongly at long times (high temperatures). The curve for a network polymer approaches a limiting asymptote and corresponds to equilibrium rubber elasticity; that of a linear polymer increases to infinity in a limiting steady-state viscous flow. The equilibrium rubber modulus is related to the density of cross-links. To a first approximation, $\underline{G} = \underline{K} \cdot \underline{T} \cdot \nu$, where ν designates cross-link density. In the vicinity of the gel point, Flory showed that it was necessary to correct for the wasted dangling tails that are attached to the network because they cannot

carry load at equilibrium (14). Likewise, the melt viscosity of a
linear polymer is strongly dependent upon chain length. A log-log
plot of melt viscosity versus molecular weight commonly exhibits two
straight-line sections, with a slope of unity or somewhat higher in
the low molecular weight section and a slope of about 3.4 in the
high molecular weight section (15). The change in slope has been
attributed to the onset of molecular entanglement (16).

At a given temperature, two polymers of similar architecture but
different compositions exhibit creep curves of similar shape, but
different locations along the log t axis. When compared at
"corresponding" temperatures, relative to their respective glass
temperatures, their behaviors are very similar.

In a crude sense, the viscoelastic properties of a given polymer
can be correlated with two numbers—one that reflects its chemical
composition and one that characterizes its molecular architecture.
The value of T_g conveniently serves the first role. The molecular
architecture of a linear polymer can be roughly specified by the
average chain length; that of a network polymer by the network
density, or by the average molecular weight between cross-links.
Precise correlation of properties with structure must, of course, go
deeper than this: molecular-weight distribution must be considered;
also, in a polymer such as poly(octyl methacrylate), the alkyl side
group not only influences T_g but also occupies space and reduces the
number of chains per unit volume. Ferry (17) has considered such
matters in detail.

Overall, the regime of linear viscoelasticity is characterized
by reasonable success in establishing structure-property
relationships. The properties themselves are unambiguously and
simply specifiable. The relevant structural features are largely
recognizable aspects of molecular structure. Molecular theories
exist that provide a bridge between the molecular structure and the
macroscopic viscoelastic properties.

Large-Strain Rubber Elasticity

The equilibrium small-strain elastic behavior of an "incompressible"
rubbery network polymer can be specified by a single number—either
the shear modulus G or the Young's modulus E (which for an
incompressible elastomer is equal to 3G). This modulus being known,
the stress-strain behavior in uniaxial tension, biaxial tension,
shear, or compression can be calculated in a simple manner. (If
compressibility is taken into account, two moduli are required: G
and the bulk modulus B.) The relation between elastic properties
and molecular architecture becomes a simple relation between two
numbers: the shear modulus and the cross-link density (or the
cross-link density corrected for the dangling tails). There can be
some ambiguity as to how closely the "effective" cross-link density
(calculated from the elastic modulus) approaches the "chemical"
cross-link density (estimated from some chemical measure of cross-
linking); however, in many elastomers the "chemical" cross-link
density is not known with sufficient accuracy to make this a major
concern.

When we proceed to large elastic strains, the problem becomes
more complex. The stress-strain relation in uniaxial tension
becomes nonlinear. It could be linearized by a proper choice of the

measure of deformation and the measure of stress, but a satisfactory treatment must also be consistent with the multiaxial large-strain elastic behavior. One general approach to this problem has been through the use of a strain-energy function, W (18). This is a scalar function of the three extension ratios λ_1, λ_2 and λ_3. If $W(\lambda_1, \lambda_2, \lambda_3)$ is known, the deviatoric stresses s_1, s_2, and s_3 can be calculated (as functions of λ_1, λ_2, and λ_3). The problem then becomes that of finding the proper form of the scalar function $W(\lambda_1, \lambda_2, \lambda_3)$. Various choices have been suggested and tested. One of the most popular is the Mooney-Rivlin equation, which introduces a second elastic parameter. For uniaxial tension, the Mooney-Rivlin equation can be written:

$$\text{true stress} = 2C_1(\lambda^2 - \frac{1}{\lambda}) - 2C_2(\lambda - \frac{1}{\lambda^2})$$

corresponding to a strain-energy function of the form:

$$W = C_1[I_1 - 3] + C_2[I_2 - 3]$$

where I_1 and I_2 are invariants of the strain tensor.

A multiparameter property equation calls for a multi-parameter structure specification. As one attempts to go beyond the effective cross-link density (corrected for dangling tails), it becomes difficult to identify the precise structure features responsible for the observed elastic properties. Some of these structural features are probably related to network topology. When cross-links are introduced into a strained polymer, or in a solvent-swollen state, the resulting network has different properties from a network formed in an unstrained, unswollen condition—even if the average molecular weight between cross-links is the same (19). No structure specification couched only in terms of the connecting chains as network elements is likely to capture the significant differences among such networks. An adequate structure specification probably must involve the closed loops of the network and their topological patterns; such aspects of structure are very difficult to establish (20).

Non-Newtonian Fluids

At sufficiently high temperatures, a linear polymer behaves as an elastic fluid. At very low stress levels, the steady-state flow behavior is Newtonian; shear rate is directly proportional to shear stress. At higher stress levels, the elastic component of deformation contributes large elastic strains. The chain molecules are appreciably oriented by the flow process. Not only is there a transient elastic effect during the approach to steady flow and following cessation of flow but also the molecular orientation strongly affects the steady-state relation between shear stress and shear rate. In the low-shear region, this steady-state behavior can be expressed by a constant—the "zero-shear" viscosity (or its reciprocal, the fluidity). In the high-shear region, a nonlinear function is required to specify the relation between shear stress and shear rate. This can be formulated in various ways:

$$\dot{\varepsilon} = f(\tau)$$

$$\tau = F(\dot{\varepsilon})$$

or

$$\tau = \eta_{app}(\dot{\varepsilon})$$

with η_{app} a nonlinear function of $\dot{\varepsilon}$ or τ. By symmetry, $f(\tau)$ must be an odd function—that is, $f(-\tau) = -f(+\tau)$.

A power law expression provides a useful approximation to the flow curves for many molten polymers over a fairly wide range of shear rate (21).

$$\dot{\varepsilon} = k\tau^n$$

or

$$\tau = K\dot{\varepsilon}^m$$

As written, the power law does not satisfy the requirement of being an odd function. If negative values of $\dot{\varepsilon}$ and τ are to be accommodated, the expression should be written in terms of absolute values: $\left|\dot{\varepsilon}\right| = k\left|\tau\right|^n$.

The steady-state flow behavior is not only nonlinear but also characterized by the development of <u>normal</u> stresses, which are completely absent in a simple Newtonian fluid. Thus, a steady shear flow in the x-y plane, $\dot{\varepsilon}_{xy}$, not only leads to a (nonlinear) shearing stress, τ_{xy} but also leads to normal stresses, σ_{xx}, σ_{yy}, and σ_{zz}. Associated with these normal stresses are many distinctive phenomena exhibited by polymeric fluids, such as the Weissenberg effect, where a polymer being stirred by a turning shaft tends to climb up the shaft instead of being thrown outward by centrifugal action (22).

Thus, non-Newtonian polymeric fluids differ from simple Newtonian liquids in several ways: they exhibit transient effects in approaching steady-state flow, the steady-state flow is nonlinear, and it is accompanied by normal stress effects. Consequently, many parameters are needed to specify the fluid properties. The relation of these parameters with molecular structure is only partially understood but the form of the molecular-weight distribution and the degree of branching of the chains, as well as the average molecular weight, must be considered. The structure-property relationships in this melt-flow regime are most important with respect to the efficient melt-processing of thermoplastic polymers. This supplies a strong incentive to the development of more complete understanding of melt properties, molecular structures, and their interrelationships.

Behavior of Glassy Amorphous Polymers

At very low strain levels, a glassy amorphous polymer behaves as a simple linear elastic solid, with a high Young's modulus (e.g., 4 x 10^{10} dyn/cm^2). When forced beyond this linear regime, a variety of nonlinear, irreversible responses can occur: macroscopically brittle fracture, shear yielding (either uniform or localized), crazing, or some combination of these (23). The stress level at which onset of any of these modes of response occurs depends upon many variables: molecular structure (both composition and architecture), temperature, geometrical character of the stress, rate of loading, contact with deleterious environmental agents, etc.

Change in these variables can result in a switch from one mechanism of response to another.

Consider first the geometrical character of the stress. A multiaxial stress can be characterized by the three principal stresses S_1, S_2, and S_3, listed in descending value. Shear yielding depends primarily upon the difference between S_1 and S_3. The Tresca yield condition, $S_1 - S_3 = Y$ (applicable to metals), has been modified for polymers (for which the shear yield stress Y increases with hydrostatic pressure) (24).

The onset of crazing follows a completely different stress criterion, as reported by Sternstein (25). Crazing is favored by high tensile stress and a positive mean normal stress. Brittle fracture follows still another stress criterion. The mode of response to a particular type of stress depends upon which critical threshold is first crossed.

The shear yield stress depends upon temperature and strain rate. The stress levels for crazing and fracture also depend upon temperature and strain rate, but to different degrees. Thus a change in temperature or strain rate can shift the mode of response (26).

The molecular mechanisms of these various responses and their relationships with structure are only partially understood. One thing, however, is certain--we must go beyond molecular structure and consider supramolecular structure as well. THe critical stress levels for yielding, crazing, and fracture depend strongly (and differently) upon the molecular orientation of a specimen. In the case of polystyrene at room temperature, uniaxial orientation can provide a marked increased in tensile strength, toughness, and craze resistance in the direction of orientation and a marked loss in these properties in the transverse direction (27). Biaxial orientation can confer strength and toughness in all directions in the plane (28).

Behavior of Crystalline Polymers

Crystalline polymers, when forced beyond a limited linear regime, also can exhibit a variety of irreversible, nonlinear responses to stress. Again, the mechanical behavior depends upon molecular structure but also upon supramolecular structure--morphology and orientation. A given polymer can exhibit many different kinds of morphology, depending upon the history of temperature and stress encountered in processing. Among the recognized morphologies--each with its own distinctive pattern of properties--are the following.

1. Spherulitic morphology (commonly developed when a polymer crystallizes from an unstressed melt) (8)
2. Drawn fibrillar morphology (developed when a spherulitic polymer is stretched below its melting point and the original lamellar crystallites are fragmented and rearranged into an oriented fibrous structure) (29)
3. "Shish-kebab" morphology (a different highly oriented morphology which develops when an oriented melt is crystallized) (30)
4. Extended chain crystals (ECC) (formed when polymer crystallizes under high hydrostatic pressure, or a crystalline polymer is annealed under pressure) (31)

5. Oriented, extended chain crystals (formed by the combination of hydrostatic pressure and orientation) (32)
6. "Accordion" morphology or "hard-elastic" fibers (formed by appropriate sequences of tensile stress and temperature (33, 34)
7. Various intermediate morphologies (35)

For a given polymer, the mechanical properties--modulus, tensile strength, yield stress, etc.--can show order-of-magnitude differences in these various morphologies. Also, molecular structure influences properties, both directly and also indirectly, as it influences the development of a particular morphology (36).

In spite of the above diversity of oriented crystalline morphologies, Samuels has shown that the structural state can sometimes be adequately characterized by the crystalline and amorphous orientation factors (37). For polypropylene samples prepared with different draw ratios, draw temperatures, shrinkage temperatures, etc., simple property correlation with these two orientation factors was observed: ". . . these results suggest that different fabrication processes are simply different paths along which the sample is moved to equivalent structural states. Thus, general structure-property correlations are achieved by concentrating on the final structural state of the sample and not on the path by which the state was reached." Where applicable, this is a most useful approach; however, when radically different fabrication processes and radically different morphologies are compared, the definition of "structural state" must include more subtle features than the crystalline and amorphous orientation factors.

Summary

The structure-property relationships of polymers include the dependence of T_M and T_g on molecular structure, and the quantitative stress-strain-temperature-time behaviors exhibited in the various regimes relative to T_M and T_g. These quantitative behaviors, and their dependence on structure, are most completely developed in the regime of linear viscoelasticity (including the special cases of small-strain rubber elasticity and low shear rate viscous flow). Large-strain elasticity and high shear rate flow are somewhat more complicated, but are still correlated with molecular structure. In glassy amorphous polymers and crystalline polymers, supramolecular structure (for example, orientation) as well as molecular structure must be considered in developing structure-property relationships. Because molecular structure is primarily established during polymerization, and supramolecular structure is established during subsequent fabrication operations, the mechanical performance of such polymers depends upon the conditions of fabrication as well as of polymerization.

Literature Cited

1. Hill, R.; Walker, E. E. *J. Polym. Sci.* 1948, 3, 609.
2. Boyer, R. F. *Rubber Chem. Tech.* 1963, 36, 1303.

3. Alfrey, T.; Gurnee, E. F. "Organic Polymers"; Prentice-Hall: New York, 1967.
4. Billmeyer, F. W. "Textbook of Polymer Science"; Wiley: New York, 1966.
5. Natta, G.; Danusso, F. J. Polym. Sci. 1959, 34, 3.
6. Stavely, F. W.; Horne, S. E. Ind. Eng. Chem. 1956, 48, 778, 784.
7. Burke, J. J.; Weiss, V. "Block and Graft Copolymers"; Syracuse University Press: New York, 1973.
8. Geil, P. H. "Polymer Single Crystals"; Interscience: New York, 1963.
9. Holt, W. L.; McPherson, A. T. J. Res. Natl. Bur. Stand. 1936, 17, 657.
10. Boundy, R. H.; Boyer, R. F. "Styrene"; Reinhold: New York, 1952.
11. Gross, B. "Mathematical Structure of the Theories of Viscoelasticity"; Hermann et Cie; 1953.
12. Eirich, F. R. "Rheology", Vol. 1; Academic: New York, 1956; Chapter 11.
13. Williams, M. L.; Landel, R. F.; Ferry, J. D. J. Am. Chem. Soc. 1955, 77, 3701.
14. Flory, P. J. "Principles of Polymer Chemistry"; Cornell University Press: New York, 1953.
15. Fox, T. G.; Allen, V. R. J. Chem. Phys. 1964, 41, 344.
16. Bueche, F. J. Chem. Phys. 1956, 25, 599.
17. Ferry, J. D. "Viscoelastic Properties of Polymers"; Wiley: New York, 1961.
18. Treloar, L. R. G. "Physics of Rubber Elasticity"; Oxford: England, 1958.
19. Halpin, J. C.; Mandelkern, L. Polym. Lett. 1964, 2, 139.
20. Alfrey, T.; Lloyd, W. G. J. Polym. Sci. 1962, 62, 159.
21. DeWael, A. J. Oil Coll. Chem. Assn. 1924, 4, 33.
22. Weissenberg, K. Nature 1947, 159, 310.
23. Schultz, J. M. "Polymer Materials Science"; Prentice-Hall: New York, 1974.
24. Whitney, W.; Andrews, R. D. J. Polym. Sci. 1967, C-16, 2891.
25. Sternstein, S. S.; Ongchin, L. Polym. Pre. 1969, September.
26. Sultan, J. N.; McGarry, F. J. Polym. Eng. Sci. 1974, 14, 282.
27. Cleereman, K. J.; Karam, H. H.; Williams, J. L. Mod. Plast. 1953, 30 (5), 119.
28. Thomas, L. S.; Cleereman, K. J. Soc. Plast. Eng. J. 1972, 28 (4), 6.
29. Peterlin, A. J. Macromol. Sci. 1973, B8, 83.
30. Andrews, A. H. Angew. Chem. Intern. Ed. 1974, 13, 113.
31. Geil, P. H.; Anderson, R. F.; Wunderlich, B.; Aarakawa, T. J. Polym. Sci. A2, 1964, 3707.
32. Southern, J.H.; Porter, R. S. J. Appl. Polym. Sci. 1970, 14, 2305.
33. Clark, E. S.; Garber, C. A. Int. J. Polym. Mater. 1974, 1, 31.
34. Quynn, R. G.; Brody, H. J. Macromol. Sci.-Phys. B5, 1971, 721.
35. Clark, E. S.; Scott, L. S. Polym. Eng. Sci. 1944, 14, 682.
36. Capaccio, G.; Ward, I. M. Polymer 1974, 15, 233.
37. Samuels, R. J. J. Macromol. Sci. B8, 1973, 41.

Transport Properties of Polymers

D. R. PAUL

Department of Chemical Engineering, University of Texas, Austin, TX 78712

Thermodynamics of Solubility
Mathematics of Diffusion
Factors Affecting Solubility and Transport
   Crystallinity, Fillers, and Morphology
   Temperature and Transitions
   Penetrant Size
   Penetrant Concentration–Plasticization
   Polymer Molecular Structure
   Relaxation–Controlled Transport
Applications of Transport Concepts
   Barrier Materials
   Devolatilization
   Additive Migration
   Dyeing
   Membrane Separations
   Controlled–Release Technology

The transport of small molecules, such as gases, vapors, or liquids, in polymers has been of intense interest to engineers and scientists for many years. Early studies of gas transport in elastomers were motivated by concerns about gas loss or interchange from automobile inner tubes, rubber balloons, cable insulation, and foam rubber (1). The introduction of butyl rubber greatly reduced some of these problems because of its much lower permeability than natural rubber. Subsequent interest focused on the transport in plastics as they became more widely used for packaging applications that required certain barrier characteristics. This interest is illustrated by the more recent strong economic and safety incentives to replace glass with plastics for carbonated beverage bottles (2) and the subsequent search for "high barrier" polymers for these applications. Concerns about migration of additives, solvents, and

0097–6156/85/0285–0253$06.75/0
© 1985 American Chemical Society

residual monomers have introduced additional reasons for interest in
this subject.

The purpose of this chapter is to outline some of the basic
issues and principles that pertain to the transport of small
molecules in polymers and to introduce a few examples in which
transport properties of polymers may be important or lead to new
uses for polymers.

Thermodynamics of Solubility

In order to be transported through bulk polymers, small molecules
must first dissolve in them. Consequently, the practical indicators
of transport rates such as permeability coefficients include a
thermodynamic part to characterize solubility.

The extent to which small molecules will be sorbed into a
polymer at equilibrium depends on the entropy and enthalpy of mixing
and the activity of these molecules in the environment with which
the polymer is equilibrated. These points are best illustrated by
considering a polymer surrounded by a vapor of component 1 at a
partial pressure of p_1. If the saturation vapor pressure of pure 1
is p_1^*, then the activity in the vapor phase is $a_1 = p_1/p_1^*$. Flory
(3) has developed a thermodynamic model for mixing small molecules
of molar volume $\tilde{V}_1$ with large polymer molecules of molar volume $\tilde{V}_2$.
This model combines an estimate for the entropy of mixing with a
measure of the enthalpy of mixing expressed in terms of an
interaction parameter χ_1 and results in the following expression for
phase equilibrium:

$$\ln a_1 = \ln \phi_1 + (1 - \frac{\tilde{V}_1}{\tilde{V}_2})\phi_2 + \chi_1 \phi_2^2 \tag{1}$$

where ϕ_i are volume fractions. For binary systems, $\phi_1 + \phi_2 = 1$.
Equation 1 is one form of what is now called the Flory-Huggins
equation. When the polymer has a relatively large molecular weight,
the ratio of molar volumes is small compared to unity and can be
neglected. Figure 1 shows the extent of sorption of component 1
into the polymer as a function of its activity for various values of
the Flory-Huggins interaction parameter as calculated from Equation
1. Note that the extent of sorption increases as χ_1 becomes
smaller, and for $\chi_1 < 0.5$ the polymer would dissolve in liquid 1.
For small activities, the extent of sorption is proportional to p_1;
however, the concentration of sorbed vapor curves upward at higher
activities.

Ideally, χ_1 is a measure of the heat of mixing of the sorbed
molecules with the polymer; the less endothermic this process is,
the greater the extent of sorption will be. For nonpolar components, the interaction parameter can be estimated by using
solubility parameter theory (4)

$$X_1 = \frac{\tilde{V}_1}{RT} (\delta_1 - \delta_2)^2 \qquad (2)$$

where the solubility parameter for each component, δ_i, is the square root of the cohesive energy density of that component. (See (5) for a tabulation of values.)

For vapors at very low activities or for gases, the limiting relationship

$$C = Sp_1 \qquad (3)$$

known as Henry's law is useful for describing the sorption isotherm in liquids or rubbery polymers. Here, C is the concentration of gas or vapor in equilibrium with the gas phase at partial pressure p_1, and S is a solubility coefficient. S increases as the gas becomes more condensible, that is, has a higher boiling point or critical temperature. The following empirical relation has proved useful (6)

$$\log S = \log S^o + m(\varepsilon/k) \qquad (4)$$

where ε/k is the Lennard–Jones potential well depth describing the cohesive forces between gas molecules and is approximately proportional to the critical temperature (7). The parameter m is approximately 0.01 K^{-1} (ε/k has units of K) and S^o ranges from 0.005 to about 0.02 $cm^3(STP)/cm^3$ atm and depends on the polymer (6).

Gas sorption in polymers below their glass transition temperature (T_g) is more complex as shown in Figure 2 for CO_2 in polycarbonate (T_g = 145 °C). Here, the isotherm is not straight such as that suggested by Henry's law (Equation 3) and is curved in the opposite manner to that predicted by the Flory–Huggins theory (see Figure 1). Isotherms such as those in Figure 2 are described by a model envisioning dual sorption mechanisms (8, 9):

$$C = k_D p + \frac{C'_H b p}{1 + b p} \qquad (5)$$

The first term is simply Henry's law (k_D may be equated to S in Equation 3), and the second term is of the Langmuir form and characterized by capacity, C'_H, and affinity, b, parameters. The latter mechanism is believed to result from sorption into nonequilibrium regions of free volume existing in the glassy state, and the size of this term increases in proportion to the value of T_g relative to the observation temperature (6). Vapors sorbed into glassy polymers may exhibit isotherms that have the dual sorption shape at low activities and change to the Flory–Huggins form at high activities (10).

The solubility of solids, for example, an antioxidant, in a polymer are affected by the additional contribution of their free energy of fusion. This effect may be combined with Equation 1 to obtain the following prediction of solubility, ϕ_1, at unit activity of the solid (11):

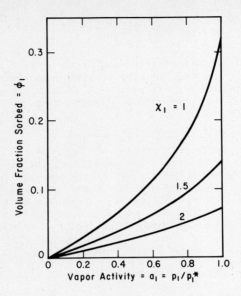

Figure 1.   Vapor sorption isotherms in a polymer as computed from
             Flory-Huggins theory by using interaction parameters
             shown.

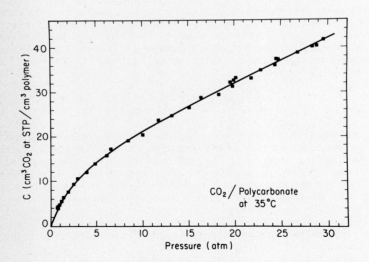

Figure 2.   Sorption isotherm for CO_2 in glassy polycarbonate at
             35 °C.

$$- \ln \phi_1 = \frac{\Delta H_f}{RT} (1 - \frac{T}{T_m}) + (1 - \frac{\tilde{V}_1}{\tilde{V}_2}) + \chi_1 \tag{6}$$

where ΔH_f and T_m are the heat of fusion and melting point of component 1, and it has been assumed that ϕ_2 is approximately 1.

Mathematics of Diffusion

The rate of transport is described by Fick's first law ($\underline{12}$) which for most purposes can be written as

$$N_1 = - D\frac{\partial C}{\partial x} \tag{7}$$

where N_1 is the flux of species 1, and D is the diffusion coefficient which depends on the nature of both components, temperature, and possibly concentration of species 1, C. For description of unsteady-state problems, it is necessary to combine Equation 7 with a differential mass balance to obtain Fick's second law:

$$\frac{\partial C}{\partial t} = - \frac{\partial N_1}{\partial x} = \frac{\partial}{\partial x} (D \frac{\partial C}{\partial x}) \tag{8}$$

Many solutions to Equations 7 and 8 with different boundary conditions have been compiled ($\underline{13}$), and information about the diffusion coefficients for polymer systems is extensive ($\underline{14}$). When D is a constant, Equation 8 can be simplified by factoring this coefficient outside the differential operator.

The kinetics of sorption of a penetrant into a polymer film of thickness ℓ serves to illustrate problems of nonsteady-state diffusion. At $t < 0$, $C = 0$ for all x; whereas, at $t > 0$, the surfaces at $x = 0$ and ℓ assume the value C_∞ which after enough time is the value achieved for all x, that is, equilibrium. If M_t denotes the total amount of permeant that has entered the film at time t, and M_∞ denotes the amount at equilibrium = $A \ell C_\infty$, then the solution to Equation 8 for these boundary conditions is

$$\frac{M_t}{M_\infty} = 1 - \sum_{n=0}^{\infty} \frac{8}{(2n+1)^2 \pi^2} e^{-D(2n+1)^2 \pi^2 t/\ell^2} \tag{9}$$

This equation, typical of solutions to Fick's second law, is difficult to manipulate, and certain approximate results valid near the beginning or end of this process such as the following

$$\frac{M_t}{M_\infty} = 4 \left(\frac{Dt}{\pi \ell^2} \right)^{1/2} \quad \text{for } 0 < \frac{M_t}{M_\infty} < 0.6 \tag{10}$$

$$\frac{M_t}{M_\infty} = 1 - \frac{8}{\pi^2} e^{-\pi^2 Dt/\ell^2} \quad \text{for } 0.4 < \frac{M_t}{M_\infty} < 1 \tag{11}$$

are helpful for quick calculations. Note that M_t initially is proportional to $\sqrt{t}$ but approaches M_∞ asymptotically in accordance with an exponential function.

Another common situation is transient permeation illustrated in Figure 3. Initially, $C = 0$ for all x but is increased to and held at C_o at $x = 0$ thereafter. The response observed is the amount of penetrant Q_t that has emerged from the downstream surface after time t. There is a time lag θ associated with the transient buildup of a steady concentration profile, after which a steady rate of permeation is established as given by Equation 7.

$$\left(\frac{dQ_t}{dt}\right)_{ss} = \frac{ADC_o}{\ell} \tag{12}$$

when D is constant. In this case, the time lag is

$$\theta = \frac{\ell^2}{6D} \tag{13}$$

A complete expression for Q_t is available but will not be given here (13). For gases, we may define the following permeability coefficient

$$P \equiv \frac{\ell}{p_o A}\left(\frac{dQ_t}{dt}\right)_{ss} = \frac{DC_o}{P_o} \tag{14}$$

When Henry's law applies, that is, Equation 3,

$$P = DS \tag{15}$$

Thus, from a transient permeation experiment we can get both D and S separately; whereas, a steady-state experiment only gives their product.

Factors Affecting Solubility and Transport

Crystallinity, Fillers, and Morphology. The solubility of low molecular weight compounds is extremely small in the crystallites of polymers in comparison to that in the amorphous regions of the same polymer (15). Thus, equilibrium sorption in semicrystalline polymers is less than that for corresponding completely amorphous ones. For the same reasons crystalline polymers are more chemical resistant than amorphous ones. As a good approximation for gases, the Henry's law solubility coefficient of a semicrystalline polymer is related to that for the same polymer in the amorphous state, S_a, by the following:

$$S = \phi_a S_a \tag{16}$$

where ϕ_a is the volume fraction of the polymer that exists in the amorphous state. The latter is related to the volume fraction that is crystalline, or the crystallinity, by $\phi_a = 1 - \phi_c$. Similar considerations apply to nonsorbing fillers frequently used in polymers, for example, glass fibers and minerals.

Crystals or filler particles that do not sorb the penetrant will obviously be impermeable to them and, thus, reduce transport rates in the composite. Figure 4 illustrates this effect for matrices in which such particles are arranged in an ordered and in a random manner. The rate of transport in such systems will be slower than when such particles are absent because of the reduced area for transport and the resulting more tortuous path for permeation. The effective diffusion coefficient for such cases will be a factor κ smaller than in the pure amorphous material D_a. The permeability coefficient in a crystalline polymer should then be given by

$$P_c = \phi_a \kappa S_a D_a \tag{17}$$

Various theories and empirical expressions are available (14, 16) for estimating κ for special situations. Obviously, this factor will depend on the volume fraction of impermeable particles, their shape, and their arrangement in space, that is, morphology. Crystalline polymers are much better barriers to permeation than are equivalent amorphous polymers by virtue of the obstruction to transport caused by their crystallites. Often their resistance to permeation can be further improved by stretching or drawing so that the crystals are converted from a random arrangement to a more ordered array such as that illustrated in Figure 4.

Temperature and Transitions. Diffusion in solids and liquids is extremely affected by temperature. Like reaction rates, diffusion may be thought of as an activated process following an expression of the Arrhenius form

$$D = D_o \, e^{-E/RT} \tag{18}$$

where the activation energy E depends on the polymer and size of the penetrant and is of the order of 10 kcal/mol. Because of these considerations it is convenient to plot experimental data in the form of log D versus $1/T$ as illustrated in Figure 5 except here the $1/T$ scale is reversed so that temperature increases along the abscissa. Far from any transitions, these plots are linear and thus show agreement with Equation 18.

The left part of Figure 5 is for diffusion of cyclopropane in high-density polyethylene (17), a highly crystalline polymer that melts at 135 °C. Upon cooling below T_m, D decreases in a dramatic manner because of the development of crystals that impede diffusion. Further below T_m, another Arrhenius form is established. Note that the lines above and below T_m are not parallel which shows that the effect of crystallinity is more than a simple geometric obstruction to diffusion in this case. The right part of Figure 5 is for diffusion of CO_2 in an acrylonitrile/methyl acrylate (AN/MA) copolymer (18) having a glass transition near 65 °C. The diffusion coefficient shows a change in slope on the Arrhenius plot in this vicinity as is often observed on traversing the T_g. Such breaks in

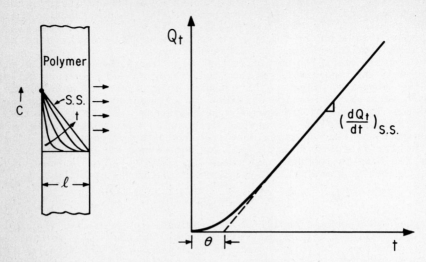

Figure 3.   Transient permeation experiment. Q_t is the amount of
             substance that has permeated through the polymer film of
             thickness ℓ in time t.   Illustration on left shows
             concentration profiles as a function of time.

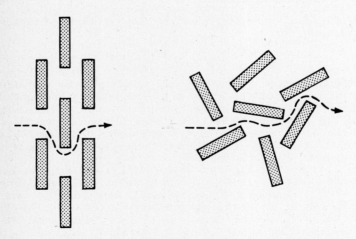

Figure 4.   Diffusion through media containing impermeable particles
             (ordered on the left and random on the right).

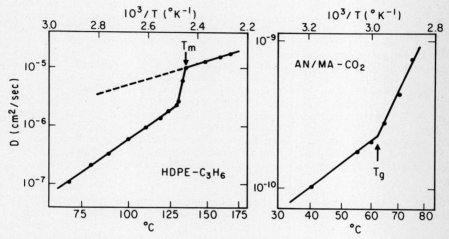

Figure 5. Effect of temperature and transitions on penetrant diffusion coefficients. Left: cyclopropane in high-density polyethylene to illustrate effect of melting point. Right: CO_2 in AN/MA copolymer to show effect of glass transition. Note: reversed $1/T$ scale is used.

the curve are not always observed especially for small gas molecules (19). D is actually higher below T_g than what one would estimate from extrapolating data from above T_g.

The AN/MA copolymer mentioned earlier is of interest technologically because of the very good barrier properties exhibited by polymers containing acrylonitrile--compare the D for this copolymer with those for polyethylene on the left. However, because pure polyacrylonitrile is essentially impossible to melt process, it is necessary to copolymerize with other monomers to obtain processable materials with the result being a sacrifice of barrier properties.

Penetrant Size. One view of diffusion in polymers envisions penetrant molecules hopping in random steps following the opening of a hole or tunnel of sufficient size for the penetrant molecule to pass, and the energy associated with such a passage is the activation energy in Equation 18. Somewhat different models have also been proposed (14); however, they would all agree that the size of the penetrant molecule should greatly affect the value of D. This fact is nicely demonstrated by the extensive data in Figure 6 for various small molecules in poly(vinyl chloride) (20). Here the size of the various penetrant molecules is expressed by the van der Waals volume, b, obtained from pressure-volume-temperature behavior in the gaseous state. Note that D varies in this plot by nearly 10 orders of magnitude. Special techniques are required to study systems with diffusion coefficients less than about 10^{-10} cm^2/s.

Penetrant Concentration-Plasticization. Earlier, in seeking solutions to Fick's laws, the diffusion coefficient was assumed to be a constant independent of penetrant concentration. This assumption is good for gases and other molecules of very limited solubility. However, for vapors or liquids that may sorb in significant amounts (see Figure 1) this assumption usually is not the case (14). Addition of small molecules in more than dilute amounts changes the environment of the polymer segments and causes their motions to be more rapid. This condition is called plasticization and may be thought of as resulting from an increase in free volume. As a result of this, diffusion of the penetrant molecule is more rapid or D is increased. Thus, in general, D depends on concentration, and sometimes in a very strong manner, for example, exponentially:

$$D = D_{C=0} \, e^{\gamma C} \qquad\qquad\qquad (19)$$

A discussion of various models for concentration dependent diffusion is available (21), and compilations of solutions to Fick's laws for some of these cases exist (13, 14).

Polymer Molecular Structure. We may expect polymer molecular structure to have an enormous effect on the rate of transport of small molecules through them. This fact may be most easily appreciated by examining gas transport because solubility of these molecules is quite low and, thus, not affected by plasticization effects as mentioned earlier. Table I gives the oxygen permeability in a wide range of polymers at 25 °C as collected from a variety of published (22) and unpublished sources. Among the olefins listed,

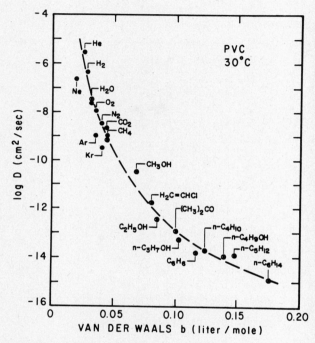

Figure 6.   Effect of penetrant size, expresssed as van der Waals
             volume, on diffusion coefficient in poly(vinyl chloride)
             at 30 °C.   (Reproduced with permission from Reference
             20.)

Table I. Oxygen Permeability in Polymers of Various
          Molecular Structures at 25 °C

Polymertype	Polymer	$P \times 10^{10}$ $(cm^3(STP) \, cm/cm^2 \, scm \, Hg)$
Olefins	polypropylene	1.4
	low-density polyethylene ($\rho=0.92g/cm^3$)	3.0
	high-density polyethylene ($\rho=0.96g/cm^3$)	0.58
	ethylene-propylene elastomer	14
Rubber	natural $\{CH_2CH=CH \, CH_2\}$	25
	butyl $\{CH_2-C(CH_3)_2\}$	1.3
	silicone $\{O-Si(CH_3)_2\}$	650
Halogenated	poly(vinyl fluoride)	0.18
	poly(vinylidene fluoride)	0.028
	poly(vinyl chloride)	0.055
	poly(vinylidene chloride)	0.0035
Highly polar	polyacrylonitrile	0.00025
	poly(vinyl alcohol), dry	$\sim 10^{-6}$

crystallinity ranges from none for the ethylene-propylene copolymer to very high for polypropylene and high-density polyethylene. This range is a major factor in variation of oxygen permeation shown by these materials. More than a 400-fold spread exists in permeability among the three rubbers shown in Table I. Butyl rubber, basically a polymer composed of isobutylene with small amounts of the comonomer isoprene to give unsaturation for vulcanization, is widely known for its low rate of gas transport believed to occur because of the sluggish segmental motions of its chain. These sluggish motions result from the steric hindrance caused by the two pendent methyl groups on every other chain carbon atom. Although silicone rubber has a similar substitution pattern, the ether oxygen connecting these units gives rise to facile segmental mobility. Silicone rubber is one of the most permeable polymers known.

Halogenation reduces permeability as Table I shows. Single halogen substitution, poly(vinyl fluoride) and poly(vinyl chloride) results in polymers less permeable than any polyolefin, even though polyolefins may be considerably more crystalline. Double substitution as in poly(vinylidene fluoride) and poly(vinylidene chloride) produces even more dramatic reductions in permeation rates partly because these polymers can be rather crystalline, but mainly because of the similar effect of double substitution on the same backbone carbon as seen for butyl rubber. The two highly polar polymers listed in Table I have extraordinarily low oxygen permeabilities because of the strong dipolar forces in polyacrylonitrile and the strong hydrogen bonding in poly(vinyl alcohol). The very low permeability for poly(vinyl alcohol) only exists when the polymer is dry. Poly(vinyl alcohol) is, of course, quite hygroscopic; sorbed moisture breaks up the strong bonding between chains, or plasticizes the polymer, and causes large increases in oxygen permeation.

Table II shows the properties of four structurally similar polymers (some are isomeric) and the sorption and transport properties for a simple gas, argon, in them (23). By comparison, the Henry's law solubility coefficient, S, varies relatively little among these polymers when contrasted to the nearly 200-fold variation in diffusion coefficients that exists. For these polymers, the transport parameters change in the direction one would expect on the basis of the cohesive energy densities (CED) of these materials, that is, transport slows down as the strength of the cohesive force field increases. However, this trend is only qualitative and is by no means general when polymers of diverse structures are considered. In fact, no general way is yet known to predict permeation rates through polymers from knowledge of their molecular structures. Some free volume correlations (24) may be used for very rough approximations; however, such prediction remains one of the unsolved problems in applied polymer science. Although molecular mobility and cohesive forces are considered to be contributing parameters, as they are in determining other properties, it is instructive to contemplate the following example (9). Poly(phenylene oxide) has rigid aromatic units in its backbone and exhibits a T_g of 220 °C, but it is greater than an order of magnitude more permeable than butyl rubber which has a T_g of −76 °C.

Table II. Solubility and Transport of Argon at 30 °C in Four Structurally Similar Polymers

Polymer	T_g (°C)	CED (cal/cm^3)	$P \times 10^{10}$ $\frac{cm^3(STP)\ cm}{cm^2\ sec\ cm\ Hg}$	$D \times 10^8$ (cm^2/sec)	$S \times 10^4$ $\frac{cm^3(STP)}{cm^3\ cm\ Hg}$
poly(vinyl methyl ketone)	20	127.5	0.024	0.32	7.38
poly(vinyl acetate)	28	109.2	0.20	1.6	12.6
poly(methyl acrylate)	3	102.5	0.50	5.7	8.80
poly(vinyl methyl ether)	−23	81.3	2.24	60.6	3.70

Relaxation-Controlled Transport. In the discussions thus far, it has been assumed that locally the mechanical state of the polymer-penetrant system adjusts rapidly to the changes imposed and that simple diffusion is the rate-controlling mechanism of penetrant transport. However, for glassy polymers these conditions may not be met when certain vapors or liquids are sorbed (25). For example, sorption of methanol into poly(methyl methacrylate) sheets sets up conditions of constrained swelling and generates internal stresses (26) that change the mode of transport to what has been called Case II behavior or "anomalous" diffusion. In this situation, time-dependent structural rearrangements or relaxations become the dominant process and give rise to deviations from Fickean behavior. Case II transport is characterized by a sharp front moving into the polymer at constant velocity. Behind the front exists swollen polymer essentially at equilibrium penetrant concentration, and in advance of the front is unpenetrated glassy polymer. As a result, mass increase from sorption varies linearly in time in contrast to the square-root dependence expected from Fick's laws (Equation 10). Sorption may be accompanied by crazing or even catastrophic cracking in some cases.

Applications of Transport Concepts

The sorption and transport characteristics of polymers are important issues in selecting the proper polymer for certain applications. This behavior may be tailored to the application by processing technique, chemical modification, or physical design. The following are selected illustrations chosen to suggest the wide range of applications for which such characteristics may be critical or form the basis for a useful product.

Barrier Materials. Polymers are often used as barriers to keep small molecules in or to keep them out. One common example is rubber tubes for tires or more recently the inner liner of tubeless tires. The purpose of such a material is to contain air under pressure to maintain tire inflation. From the data in Table I, it is clear that butyl rubber is a much better material for this purpose than natural rubber. Because of this, butyl rubber has entirely displaced natural rubber from this market. Aside from its prohibitive price, silicone rubber would be totally unsatisfactory for this use because of its high gas permeability.

Foodstuffs are packaged in a variety of polymeric film products. One purpose is to keep oxygen out in order to prevent decay. From Table I, it is clear that poly(vinylidene chloride)-based polymers would be a good choice of materials because of their low oxygen permeability. For this reason, poly(vinylidene chloride) forms the basis of one of the very commonly used foodwraps. Polyolefin films are also used for food packaging even though they do not have exceptionally good resistance to gas permeation; however, because of their nonpolar character they are good barriers to water vapor transport. This characteristic is an important one for avoiding dehydration of foods, especially vegetables and fruits.

Recently, there has been a trend to package carbonated beverages in plastic bottles because of their lighter weight and increased safety from breakage. In this application, the polymer chosen must be a good barrier in order to avoid loss of carbonation by CO_2 permeation and to prevent atmospheric oxygen from entering the bottle and spoiling the contents. Although dry poly(vinyl alcohol) is an excellent barrier to gas transport, it could not be contemplated for this use because it would swell or dissolve in the water that is to be contained. Acrylonitrile-based polymers were originally considered for this use because of their excellent barrier properties, but they were abandoned because of concerns about the potential health hazards associated with leaching of residual monomer from polymerization into the bottle contents. Today, bottles for carbonated beverages are based on poly(ethylene terephthalate) which appears to be the best choice in terms of all of the characteristics needed for this application. There is some loss of CO_2 from the bottle during prolonged storage, and Figure 7 shows a calculation for a typical 2-L bottle. The calculation shows the CO_2 pressure loss broken down into that which occurs by sorption of gas into the bottle wall and by eventual permeation from the exterior of the bottle surface. The former accounts for a significant fraction of the loss; hence, one must be concerned with solubility as well as transport behavior.

Devolatilization. Following polymerization of many polymers, removal of residual monomer or solvents is necessary. Ultimately this removal becomes controlled by diffusion from the polymer no matter what type of process is employed (27). Many processes use a steam-stripping operation for this purpose, and others employ devolatilization in a vented extruder. Removal of residual vinyl chloride monomer from poly(vinyl chloride) was an issue of much concern following the discovery that this monomer is a carcinogen.

Additive Migration. A variety of additives such as antioxidants and UV stabilizers are used in polymers to facilitate their processing or to prolong their useful life. These low molecular weight additives may be lost by migration during processing, and thus their benefit is lost, or they may migrate into foodstuffs when these polymers are used as packaging materials. For several years, the National Bureau of Standards has conducted an extensive program to develop rational models to evaluate these possibilities by using principles of solubility and diffusion for making regulatory decisions (28).

Dyeing. Most fibers are colored by a dyeing operation during which a dye molecule diffuses into the fiber structure (29). Naturally, the time required for this process to occur is an important issue and is one of the factors considered in developing fibers. To attract enough dye into the fiber to develop the depth of shades desired and to hold it in the fiber during laundering, many synthetic fibers have ionic dye sites to attract oppositely charged dye molecules. Other mechanisms such as reactive and disperse dyeing are also used. To accept a cationic or basic dye, the fiber must have strong acid groups in its structure, whereas anionic or acid dyes require basic groups. Condensation polymers such as nylon have unreacted end groups that can serve this purpose. Addition polymers such as the acrylics may have sulfonic acid groups at the chain ends resulting from redox initiation. Alternatively, ionic monomers may be copolymerized into the polymer for this purpose.

Membrane Separations. Separation processes using polymeric membranes (30) have become important techniques because of their simplicity and low consumption of energy in comparison to alternatives such as distillation. Membranes for ultrafiltration are porous, and no diffusive transport actually occurs through the polymer itself. However, for separation at the molecular level, diffusion through the polymer provides a possible mechanism for selective passage of the desired small molecule. Reverse osmosis for desalination of water can occur by this mechanism, and large commercial processes using this technique are now in operation.

Reverse osmosis is simply the application of pressure on a solution in excess of the osmotic pressure to create a driving force that reverses the direction of osmotic transfer of the solvent, usually water. The transport behavior can be analyzed elegantly by using general theories of irreversible thermodynamics; however, a simplified solution-diffusion model accounts quite well for the actual details and mechanism in most reverse osmosis systems. Most successful membranes for this purpose sorb approximately 5 to 15% water at equilibrium. A thermodynamic analysis shows that the application of a pressure difference, Δp, to the water on the two sides of the membrane induces a differential concentration of water within the membrane at its two faces in accordance with the following (31):

$$\Delta C_w = \frac{C_w^* \, V_w \, \Delta p}{RT} \tag{20}$$

where V_w is the molar volume of water, and C_w^* is the equilibrium concentration of water in the membrane. The gradient of water in the membrane leads to a diffusive flux of water, J_w, in accordance with Fick's law:

$$J_w = D_w \frac{\Delta C_w}{\ell} \tag{21}$$

Because for salt water, the osmotic pressure difference, $\Delta\pi$, acts in opposition to the applied hydraulic pressure, the water flux becomes

$$J_w = \frac{D_w V_w C_w^*}{RT\ell} (\Delta p - \Delta\pi) \tag{22}$$

where D_w is the diffusion coefficient for water in the polymer. The solute or salt flux is given by the following relation:

$$J_s = D_s K_s (C_{so}^L - C_{s\ell}^L)/\ell \tag{23}$$

provided the pressure differential does not contribute significantly to the driving force for its transport, as is the usual case. The terms are defined as follows: K_s is the partition coefficient for salt between polymer and liquid phases, D_s is the salt diffusion coefficient in polymer, C_s^L is the salt concentration in liquid phase with subscripts o and ℓ denoting upstream and downstream, respectively. The selectivity characteristics of a reverse osmosis membrane system in terms of a rejection coefficient can be defined as follows:

$$R \equiv 1 - \frac{C_{s\ell}^L}{C_{so}^L} \qquad (24)$$

The salt content on the downstream side of the membrane is determined by the relative salt and water fluxes through the membrane. A combination of results from the solution–diffusion model outlined earlier with the definition of the rejection coefficient gives

$$R = (1 + \frac{D_s K_s RT}{D_w V_w C_w *(\Delta p - \Delta \pi)})^{-1} \qquad (25)$$

Note that the rejection is independent of the membrane thickness, but the flux of water or productivity increases as the membrane becomes thinner.

Reverse osmosis was not commercially practical until techniques for increasing productivity were developed. The principal discovery (32) involved a casting procedure that results in asymmetric membranes having a thin dense layer of polymer, approximately 0.2μ thick, supported on a porous sublayer as illustrated in Figure 8. These membranes are called Loeb membranes (33). Current commercial membranes of this type are made of cellulose acetate, aromatic polyamides, and certain composites that achieve water fluxes of the order of 1.0 m^3/m^2 day with NaCl rejections of 99% or more (27). As seen in Equation 25, rejection increases with applied pressure.

A further advance in membrane technology was made by learning how to spin hollow fibers with walls of the Loeb-type structure. These fibers can be assembled into modules resembling shell–and–tube heat exchangers such as illustrated in Figure 9. This process results in a large surface area per unit of volume and greatly reduces the capital cost of such units in comparison to other module configurations (30). Similar devices are now available commercially for gas separations (34) and are economical in comparison to other processes because of their low energy consumption. The intrinsic separation factor for a polymer membrane for two gases is given by the ratio of the permeability coefficients of the two gases:

$$\alpha_{12} = \frac{P_1}{P_2} \qquad (26)$$

Polymers with fixed ionic charges may be used for ion exchange. When formed into membranes, these materials essentially prevent the passage of ions of the same charge as those fixed to the polymer

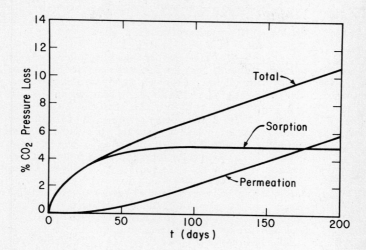

Figure 7.   Calculated CO_2 pressure loss from plastic carbonated beverage bottle. Total loss is comprised of CO_2 that has left the exterior surface of bottle wall plus that still contained in the wall.

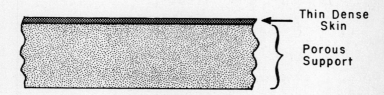

Figure 8.   Schematic diagram of Loeb–type membrane.

backbone and allow the passage of ions of opposite sign (30).
Alternation of anionic and cationic membranes as shown in Figure 10
permits the use of an electrical potential to drive the process
known as electrodialysis which can be used to desalinate water or to
produce a concentrated electrolyte mixture from a more dilute one.
Electrochemical membrane processes are presently being used in
processes for producing chlorine and caustic soda (30).

Controlled-Release Technology. Conventionally, chemicals such as
drugs, pesticides, fertilizers, and herbicides are administered in a
periodic fashion and thus cause temporal concentration variations.
The concentration extremes may range from dangerously high to
ineffectively low levels and achieve the optimum level for only
short periods of time. This cyclic application inefficiently uses
these chemicals and results in added expense and possibly certain
side effects. During the last two decades, the concept of
controlled delivery or release has emerged as a means of solving
many of these problems. Often this technology involves the use of
polymers (35).

Numerous concepts using polymers as the rate-controlling element
have been developed and commercialized (27, 30, 35). Figure 11
illustrates one approach that will deliver a solute such as a drug
at a constant rate for very long times depending on the design.
This instrument is a reservoir-type device in which the solute in
the form of a suspension is enclosed in a membrane that controls its
release. Because the reservoir contains solid drug particles at
unit activity, the driving force for transport remains fixed, and
the flux through the membrane is constant until all of the drug
particles have been dissolved. By selection of the polymer and the
geometry of the membrane, the designer can engineer a product that
meets the many different requirements of the application.

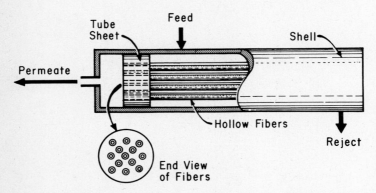

Figure 9.   Hollow-fiber membrane module for separation processes.

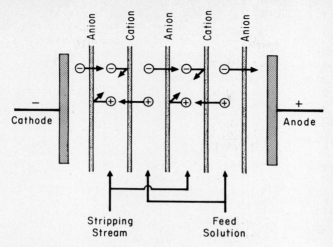

Figure 10.   Electrodialysis. Cation and anion designations refer to membranes that will exchange cations or anions, respectively.

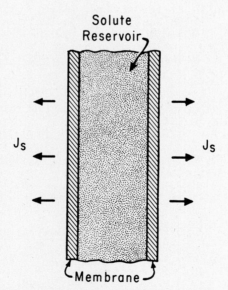

Figure 11.   Reservoir-type membrane device for controlled release.

Literature Cited

1. van Amerongen, G. J. Rubber Chem. Tech. 1964, 37, 1065.
2. Fenelon, P. J. Polym. Eng. Sci. 1973, 13, 440.
3. Flory, P. J. "Principles of Polymer Chemistry"; Cornell University Press: Ithaca, New York, 1962.
4. Hildebrand, J. H.; Scott, R. L. "The Solubility on Non-electrolytes", 3rd ed.; Reinhold: New York, 1950.
5. Brandrup, J.; Immergut, E. H., Eds.; "Polymer Handbook", 2nd ed.; John Wiley: New York, 1975.
6. Toi, K.; Morel, G.; Paul, D. R. J. Appl. Polym. Sci. 1982, 27, 2997.
7. Hirshfelder, J. O.; Curtiss, C. F.; Bird, R. B. "Molecular Theory of Gases and Liquids"; John Wiley: New York, 1954.
8. Vieth, W. R., Howell, J. M.; Hsieh, J. H. J. Membrane Sci. 1976, 1, 177.
9. Paul, D. R. Ber. Bunsenges. Phys. Chem. 1979, 83, 294.
10. Berens, A. R. Polym. Eng. Sci. 1980, 20, 95.
11. Billingham, N. C.; Calvert, P. D.; Manke, A. S. J. Appl. Polym. Sci. 1981, 26, 3543.
12. Bird, R. B.; Stewart, W. E.; Lightfoot, E. N. "Transport Phenomena"; John Wiley: New York, 1960.
13. Crank, J. "Mathematics of Diffusion"; Oxford Press (Clarendon): London, 1956.
14. Crank, J.; Park, G. S. "Diffusion in Polymers"; Academic Press: New York, 1968.
15. Michaels, A. S.; Bixler, H. J. J. Polym. Sci. 1961, 50, 393.
16. Hopfenberg, H. B.; Paul, D. R. In "Polymer Blends"; Paul, D. R.; Newman, S., Eds.; Academic Press: New York, 1978; Vol. I, Chap. 10.
17. Lowell, P. N.; McCrum, N. G. J. Polym. Sci.: Part A-2 1971, 9, 1935.
18. Yasuda, H.; Hirotsu, T. J. Appl. Polym. Sci. 1977, 21, 105.
19. Hopfenberg, H. B.; Stannett, V. In "The Physics of Glassy Polymers"; Haward, R. N., Ed.; Applied Science Publishers: London, 1973; p. 504.
20. Berens, A. R.; Hopfenberg, H. B. J. Membrane Sci. 1982, 10, 2830.
21. Rogers, C. E.; Machin, D. CRC Crit. Rev. Macromol. Sci. 1972, 1, 245.
22. Yasuda, H.; Stannett, V. Sec. III, p. 229, in Ref. 5.
23. Allen, S. M.; Stannett, V.; Hopfenberg, H. B. Polymer 1981, 22, 912.
24. Lee, W. M. Polym. Eng. Sci. 1980, 20, 65.
25. Hopfenberg, H. B.; Frisch, H. L. J. Polym. Sci., Part B 1969, 7, 905.
26. Thomas, N. C.; Windle, A. H. Polymer 1981, 22, 627.
27. Stannett, V. T.; Koros, W. J.; Paul, D. R.; Lonsdale, H. K.; Baker, R. W. Adv. Polym. Sci. 1979, 32, 69.
28. Smith, L. E.; Chang, S. S.; McCrakin, F. L.; Sanchez, I. C.; Senich, G. A. "Models for the Migration of Additives in Polyolefins"; NBSIR 80-199, 1980.
29. Rattee, I. D.; Breuer, M. M. "The Physical Chemistry of Dye Adsorption"; Academic Press: London, 1974.
30. Paul, D. R.; Morel, G. "Kirk-Othmer: Encyclopedia of Chemical Technology," 3rd ed.; Wiley: New York, 1981; Vol. 15, p. 92.

31. Paul, D. R. <u>Sep. Purif. Methods</u> 1976, <u>5</u>, 33.
32. Loeb, S.; Sourirajan, S. ADVANCES IN CHEMISTRY SERIES No. 38, American Chemical Society: Washington, D. C., 1962; p. 117.
33. Kesting, R. E. "Synthetic Polymeric Membranes"; McGraw-Hill: New York, 1971.
34. Henis, J. M. S.; Tripodi, M. K. <u>J. Membrane Sci</u>. 1981, <u>8</u>, 233.
35. Paul, D. R.; Harris, F. W., Eds.; "Controlled Release Polymeric Formulations"; ACS SYMPOSIUM SERIES No. 33, American Chemical Society: Washington, D.C., 1976.

Fracture Mechanics of Polymers

K. L. DeVRIES and R. J. NUISMER

Department of Mechanical and Industrial Engineering, University of Utah, Salt Lake City, UT 84112

Analysis of Failure
Failure of "Flawless" Materials
Fracture Mechanics
   Griffith Theory
   Stress Intensity Factors
   Fracture Energy
   Viscoelastic Effects
   Examples
Fatigue
Conclusion

Failure in materials is a complex phenomenon that is not well understood and is currently the subject of considerable research effort. Failure analysis in polymers is particularly complex because properties are not only functions of the chemical nature of the polymer but also highly dependent on the details of its physical structure. For example, polymers that are chemically very similar but have different structures can have moduli varying from less than 10^6 to more than 10^{11} N/m^2 (10^2 to 10^7 $1b/in.^2$). At the lower end of this scale (e.g., for elastomers) the resistance to deformation is due to entropic forces (1, 2), and at the other extreme (extended chain configurations) it is hypothesized that the very high modulus in the chain direction is due to the inherent stiffness of the polymer chain. Although less dramatic, morphology can also have a large effect on strength. Depending on the intricacies of structure, chemically similar polymers can have practical tensile strengths in the following ranges: 10^7 to 10^8 N/m^2 for rubbers; 3 x 10^7 to 2 x 10^8 N/m^2 for glassy and semicrystalline plastics; 2 x 10^8 to 10^9 N/m^2 for highly oriented fibers and films in the direction of orientation; and as much as 4 x 10^9 N/m^2 for extended chain fibers. In the latter two cases the strength and stiffness is highly anisotropic.

0097–6156/85/0285–0277$08.00/0
© 1985 American Chemical Society

Failure in polymers is further complicated by several other factors. For example, time plays a more significant role in the properties of polymers than in most other materials. Not only are polymers generally viscoelastic but they also often exhibit chemical and physical aging. Viscoelasticity, as the name implies, indicates the material exhibits a combination of viscous and elastic properties. Chemical and physical aging are the titles given to designate the changes in chemical and physical structure that occur as the polymer ages. These time effects all manifest themselves in changes in the resistance of the polymer to failure.

Polymers are often used in conjunction with other materials as composites. The most familiar types are the fiber-filled composites such as fiberglass and the carbon-, boron-, or Kevlar-filled advanced composites. Granular-filled composites such as those using clay as extenders in some plastics or ground quartz added to dental plastics to increase their wear resistance also represent an important class of materials. In these cases the presence of a second phase as well as the interface between polymer and filler increases the complexity of the failure analysis.

Polymers are also commonly used as adhesives. The important study of fracture in these systems is complicated by the interface(s) and the constraints put on the adhesive by the adherend(s).

To further discuss failure, one must first decide what constitutes failure. Although it is common to associate failure with a part breaking into two or more pieces, conditions other than fracture can make a part inoperative or render its operation unsatisfactory. Other forms of failure include excessive elastic deformation, plastic flow and deformation, wear, or even loss of attractive appearance. At times several of these modes may combine to produce the resultant failure. For example, plastic deformation often precedes fracture and may, in fact, nucleate the microscopic cracks that ultimately coalesce into a macroscopic fracture. This chapter will concentrate largely on fracture and will only touch briefly on other types of failure.

Analysis of Failure

The failure behavior of materials is often found to fall rather naturally into two ranges as represented schematically in Figure 1. This figure shows typical behavior observed in a great many materials (polymer and otherwise) as a function of the size of flaws present in the material. The flaws may be naturally occurring or introduced by machining, wear, scratches, etc. Flaws are commonly introduced when parts are joined, particularly by mechanical means. Failure, for example, often initiates at the sites of the holes associated with bolts, pins, screws, rivets, etc., or where welds meet virgin material. The important thing to note is that when the flaws are above a certain critical size, they cause stress concentrations that serve as nuclei from which cracks propagate and thus lower the strength of the material. For materials with flaws below this critical value it is typically observed that the stress to produce general yielding is reached before the stress required to propagate the cracks in a brittle (or quasi-brittle) manner. Thus, the material does not recognize flaws less than the critical size and behaves as if a flaw of at least the critical size is always

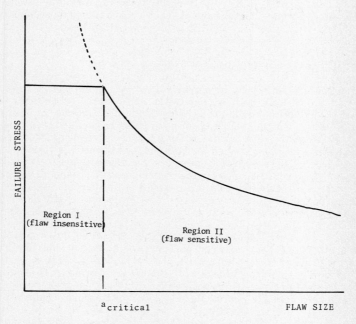

Figure 1.   Failure stress as a function of flaw size.

present. For this reason a_{cr} is often called the "inherent" flaw size of a material that is macroscopically "flawless." Near the critical flaw size plastic deformation may nucleate cracks of a critical size that can then propagate under the applied stress. Several models have been proposed for the development of such microcracks (3, 4). Analysis of the case to the right (Region II) in Figure 1 is generally classified as "fracture mechanics", and we will call the other case the failure of "flawless" materials. Although we plan to concentrate largely on fracture mechanics, no treatment of failure analysis would be complete without some discussion of the "flawless" approach.

Failure of "Flawless" Materials

Before recognition of the importance of inherent flaws in a material, the analyst relied upon one of several stress or strain criteria to predict conditions for failure. These criteria are still useful as failure predictors when flaws are less than the critical size. Typical criteria of this type include maximum principal stress, maximum shear stress, maximum octahedral shear stress, and others depending generally on experimental evidence and experience (5, 6). These criteria hypothesize, respectively, that failure occurs in the structural element of interest when the major principal stress reaches a critical value, when the maximum shear stress reaches a critical value, or when the octahedral shear stress reaches a critical value.

The critical values are generally obtained from a standard tensile test. Once the critical values are obtained the application of any (or all) of these criteria in conjunction with a dependable stress analysis is straightforward. Here we demonstrate the method by a simple example. Let us assume that it is desired to determine the torque required to cause failure of a 25 mm in diameter shaft constructed of an homogeneous isotropic plastic with a failure stress in tension, σ_o, of 7×10^7 N/m^2. Assume further that the modulus of elasticity, E, for this plastic is given by 3×10^8 N/m^2, and that is has a Poisson ratio of 0.3. We will explore the prediction of the three criteria just discussed.

Mechanics of materials texts (5) give the maximum shear stress, τ, induced in a round shaft by a torque, T, as follows:

$$\tau = \frac{TR}{(\pi R^4 / 2)} = \frac{2T}{\pi R^3} \tag{1}$$

where τ is the induced shear stress, and R is the shaft radius. Stresses are of course a second order symmetric tensor and therefore in general will have six components. The principal stresses and their directions can be determined in general from "stress resolution" by the eigenvalue and eigenvector operations (7), although in simple cases such as we are analyzing here elementary texts outline less sophisticated methods (5).

In a cylindrical frame of reference with one axis parallel to the shaft only one nonzero component is in the stress tensor, that is, the τ given in Equation 1. In this case stress resolution shows that the principal stresses are $\sigma_1 = |\tau|$, $\sigma_2 = -|\tau|$, and $\sigma_3 = 0$.

The shear stress on the octahedral planes (the eight planes making equal angles with the principal planes) is found to be related to the principal stresses by (6)

$$\tau_{oct} = \frac{1}{3}[(\sigma_1 - \sigma_2)^2 + (\sigma_2 - \sigma_3)^2 + (\sigma_1 - \sigma_3)^2]^{1/2} \qquad (2)$$

In the case of interest here for which $\sigma_1 = |\tau|$, $\sigma_2 = -|\tau|$, and $\sigma_3 = 0$ Equation 2 becomes

$$\tau_{oct} = \frac{1}{3}[(2\tau)^2 + \tau^2 + \tau^2]^{1/2} = \frac{\sqrt{6}}{3}\tau \qquad (3)$$

The maximum principal stress criterion would predict failure when

$$\sigma_1 = |\tau| = \frac{2T}{\pi R^3} = 7 \times 10^7 \text{ N/m}^2 \qquad (4)$$

or

$$T = 215 \text{ Nm} \qquad (5)$$

The maximum shear stress criterion, also called Tresca's criterion, would predict failure when the shear stress in the shaft equals the shear yield stress (determined by a tensile test). By stress resolution, the shear stress in a tensile test is equal to the normal stress divided by two. Hence the shear stress to produce yield for the material of interest here would be $\sigma_o/2$, so that failure of the shaft is predicted when

$$\tau = \frac{2T}{\pi R^3} = \left(\frac{7 \times 10^7}{2}\right) \qquad (6)$$

or

$$T = 107 \text{ Nm} \qquad (7)$$

The octahedral shear stress criterion, also called the maximum energy of distortion criterion or the Von Mises' theory, would predict the following torque at yield. For the shaft the octahedral shear stress would be (from Equation 3)

$$\tau_{oct} = \frac{\sqrt{6}}{3}\tau \qquad (8)$$

and for the standard tensile specimen (note that for pure tension $\sigma_1 = \sigma_o$, $\sigma_2 = \sigma_3 = 0$)

$$\tau_{oct} = \frac{1}{3} \sqrt{\sigma_o^2 + \sigma_o^2} = \frac{\sqrt{2}\sigma_o}{3} \tag{9}$$

Equating Equations 8 and 9 and using Equation 1 gives

$$\tau = \frac{2T}{\pi R^3} = \frac{\sigma_o}{\sqrt{3}} = 40.4 \text{ MN/m}^2 \tag{10}$$

or

$$T = 124 \text{ Nm} \tag{11}$$

Now that three separate values for the failure torque have been found for this shaft, the logical question is which (if any) of the answers is correct. The answer to this question depends very strongly on the nature of the material investigated. For very brittle materials (e.g., cast unplasticized polystyrene), experiments have shown that the maximum principal stress criterion gives quite reasonable results. For ductile materials such as molded nylon, experimental evidence indicates that either the maximum shear stress or octahedral shear stress criterion is more appropriate.

The octahedral shear stress criterion has some appeal for materials that deform by dislocation motion in which the slip planes are randomly oriented. Dislocation motion is dependent on the resolved shear stress in the plane of the dislocation and in its direction of motion (4). The stress required to initiate this motion is called the critical resolved shear stress. The octahedral shear stress might be viewed as the "root mean square" shear stress and hence an "average" of the shear stresses on these randomly oriented planes. It seems reasonable, therefore, to assume that slip would initiate when this stress reaches a critical value at least for polycrystalline metals. The role of dislocations on plastic deformation in polymers (even semicrystalline ones) has not been established. Nevertheless, slip is known to occur during polymer yielding and suggests the use of either the maximum shear stress or the octahedral shear stress criterion. The predictions of these two criteria are very close and never differ by more than 15%. The maximum shear stress criterion is always the more conservative of the two.

The failure criteria discussed to this point are for isotropic, homogeneous materials. Forming operations in polymers often result in chain orientations, phase segregation, etc. that can cause substantial anisotropy. For example, the strength in oriented fibers and films can differ by more than an order of magnitude in the directions parallel and perpendicular to the oriented chains (8). In such cases the criteria discussed here find little use. In the extreme case of fibers this result is of little consequence because the loading is almost exclusively in the oriented (strong) direction, and hence a simple tensile strength criterion is adequate. In other cases orientation is often present but the loading is still multiaxial. In principle it should be possible to determine the properties in various directions and establish a reasonable multiaxial stress criterion for such materials. However, no satisfactory methods are currently available in such cases, and the designer must often resort to prototype testing. This area is important and deserves careful study; many oriented films and fibers do fail by splitting, microfibral buckling, etc.

Fiber-reinforced composites are also highly anisotropic when fibers are aligned. In addition, they are nonhomogeneous and usually laminated with individual plies oriented in several different directions. Because of these difficulties, once again the isotropic failure criteria are of little use. However, reasonably successful failure criteria for individual plies have been developed (9-11) and applied to the failure of laminates by laminated plate theory (9, 12).

Fracture Mechanics

The type of analysis briefly outlined in the last section assumes that any flaws present are sufficiently small and/or uniformly distributed and that average stress criteria are reasonably valid. Such behavior might be viewed as corresponding to Region I in Figure 1.

However, there are conditions in which discrete flaws, substantially larger than the uniformly distributed flaws normally present, can exist in the material and dominate the failure behavior. These flaws may arise from localized corrosive attack, fabrication techniques, accidental surface nicks, cuts, scratches, or plastic deformation associated with fatigue and static loads, etc. Because they are discrete, often comparatively sharp, and larger than the surrounding voids, these flaws act as stress concentrators and give rise to very high local stresses and provide loci of fracture initiation.

If the flaw is not cracklike, ordinary stress concentration factors may be used in conjunction with the failure criteria of the preceding section to predict failure. If the flaws (inherent or introduced) are cracklike, however, stress concentration factors are useless because linear elasticity theory predicts an infinite concentration factor at the crack tip. For these situations, experiment shows a degradation in strength as schematically represented in Region II of Figure 1.

Griffith Theory. Griffith provided the first analysis of the degradation in strength as a function of flaw size by considering the problem of a small through-the-thickness line crack in a thin sheet of brittle material as illustrated in Figure 2 (13).

Theoretically, the stress at the crack tip is (mathematically) infinite for an elastic body (14) and thus gives rise to an infinite local stress at even small applied loadings, σ, a degree of concentration for which the relationships given in the last section are useless. Griffith circumvented this problem by considering changes in the strain energy (for a linear elastic material this energy is essentially the volume integral of the stresses squared divided by the appropriate moduli) of the plate which remains finite even in this case. In essence his proposal was that fracture would commence when the incremental loss of strain energy of deformation with increasing fracture area was equaled by the work required to create new fracture surface. Considering an infinite plate

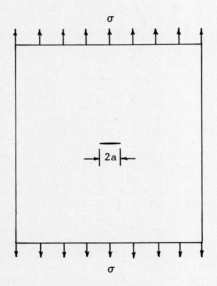

Figure 2.   Schematic of a through crack in an infinite plate, that is, a "Griffith Crack."

subjected to far field tension, and using the stress analysis of Inglis (15), he showed that the introduction of a crack of length 2a normal to the applied stress, σ, results in a change of strain energy, ΔU, in a plate of thickness, t, and Young's Modulus, E, given by

$$\Delta U = \frac{- \pi\sigma^2 a^2}{2E} t \tag{12}$$

where the change is from the energy stored in the uncracked plate.

This equation is equivalent to saying that the presence of a crack results in a decrease in the free energy of the system, that is, strain energy is released as a crack grows. With crack growth the strain energy release rate would be given by

$$\frac{\partial(\Delta U)}{\partial a} = \frac{- \pi\sigma^2 a}{E} t \tag{13}$$

The presence of a crack also increases the surface area. Griffith called the energy associated with a unit area of the crack the surface energy. Here we will call it the fracture energy for reasons that should become clear later. The fracture energy, Γ, associated with the crack may be expressed as

$$\Gamma = [4at] \gamma_c \tag{14}$$

where the bracketed quantity is the total surface area of the crack, and γ_c is the energy associated with the creation of a unit of fracture surface, that is, the specific fracture energy. (Note that in Equation 14 some authors prefer the use of the critical energy release rate, G_c, which is related to γ_c by $G_c = 2\gamma_c$.) Thermodynamically, one would anticipate fracture when

$$\frac{\partial}{\partial a}\left[\frac{\pi\sigma^2 a^2 t}{2E}\right] > \frac{\partial}{\partial a}[4at\gamma_c] \tag{15}$$

that is to say, the strain energy lost in a brittle material must be equal to or greater than the work needed to create the new fracture surface. From Equation 15 the critical applied stress for crack growth is determined as

$$\sigma_{cr} = \sqrt{\frac{4E\gamma_c}{\pi a}} \tag{16}$$

Strictly speaking, the Griffith expression (Equation 16) holds only for "brittle materials" for which the only energy required to create new surface is that required to rupture bonds (he experimented with glass). For most materials, including almost all polymers, this is not the case, as discussed later.

Stress Intensity Factors. A seemingly different approach to fracture on the basis of stress intensity factors, K_i, (4, 16, 17) is frequently encountered in the fracture mechanics literature. It is, therefore, appropriate to briefly discuss this approach and its relation to the Griffith theory.

We begin with a brief description of the mode of loading of a crack. Mode I loading is associated with displacements in which the crack surfaces in the vicinity of the crack tip move directly apart. Mode II is characterized by displacement in which the crack surfaces slide over one another perpendicular to the leading edge of the crack. In Mode III the crack faces slide over one another parallel to the leading edge of the crack. These are shown schematically in Figure 3.

By using the coordinates shown in Figure 4, the stresses in the vicinity of a crack in a linear elastic material under Mode I loading are found to be (14, 16)

$$\sigma_{xx} = \frac{K_I}{\sqrt{2\pi r}} \cos\frac{\theta}{2} (1 - \sin\frac{\theta}{2} \sin\frac{3\theta}{2}) \tag{17}$$

$$\sigma_{yy} = \frac{K_I}{\sqrt{2\pi r}} \cos\frac{\theta}{2} (1 + \sin\frac{\theta}{2} \sin\frac{3\theta}{2}) \tag{18}$$

$$\tau_{xy} = \frac{K_I}{\sqrt{2\pi r}} \sin\frac{\theta}{2} (\cos\frac{\theta}{2} \cos\frac{3\theta}{2}) \tag{19}$$

For plane stress, the displacements u and v near the crack tip in the x and y direction, respectively, are given by (14, 18)

$$u = \frac{K_I}{\mu} \frac{r}{2\pi} \cos\frac{\theta}{2} (\frac{1-\nu}{1+\nu} + \sin^2\frac{\theta}{2}) \tag{20}$$

$$v = \frac{K_I}{\mu} \frac{r}{2\pi} \sin\frac{\theta}{2} (\frac{2}{1+\nu} - \cos^2\frac{\theta}{2}) \tag{21}$$

where μ is the shear modulus, and ν is Poisson's ratio. K_I in these equations represents the stress intensity factor for the Mode I loading.

Similar expressions can be obtained for other loadings, crack geometries, etc. In fact, the value of K_i depends on the exact nature and dimensions of the crack, the type of loading, and other geometric factors, but the form of the crack tip stresses, that is, the r and θ dependency, is always the same. This fact that K_i "characterizes" the crack tip stresses has led to the proposal (16) that the value of K_i might be used as a failure criterion. It is hypothesized in this criterion that failure occurs, for example, in Mode I when K_I for any loading and crack geometry reaches a critical value, K_{Ic}. This critical stress intensity factor, K_{Ic} (now viewed as a material property), is sometimes called the fracture toughness

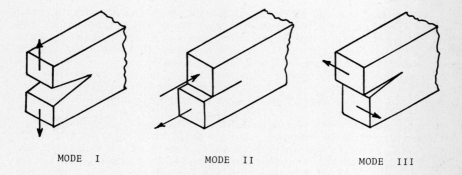

MODE I   MODE   II   MODE   III

Figure 3.   The three modes of loading near a crack tip.

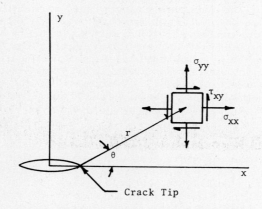

Figure 4.   Crack tip coordinates.

of a material. Its dimensions are stress x $\sqrt{length}$. It can be
shown that this criterion is in all respects equivalent to the
Griffith-derived theory (17), and in fact for plane stress

$$\gamma_c = \frac{K_{Ic}^2}{2E} \tag{22}$$

The energy method of Griffith and the stress intensity factor
approach of Irwin are, perhaps, the most basic and often used of the
fracture criteria. Therefore, they have been emphasized in this
article. However, the reader should be aware that other approaches
to the problem have gained considerable popularity in recent years,
foremost among them the J-integral (19) and the crack-opening
displacement methods (20). These two criteria have been extensively
applied to the fracture of metals for which crack tip plasticity is
significant.

In cases in which the materials under consideration are highly
anisotropic, such as oriented polymers or fiber-reinforced composite
materials, fracture mechanics principles remain valid, but their use
requires some modifications. For example, Equations 17-21 are no
longer valid but are replaced by similar equations involving the
material anisotropy (21). Complications are often encountered
because of the tendency of the cracks in such materials to be
subjected to mixed-mode loadings and to propagate in a nonself-
similar (not in the original crack plane) manner. Fracture in
fiber-reinforced composites is further complicated by the large
number of failure modes exhibited, such as fracture of the matrix
and fibers, debonding of the fiber-matrix interface, and
delamination of the individual plies. Reviews of the use of
fracture mechanics with composite materials have been given, for
example, in References 22 and 23. Discussions of the applicability
of fracture mechanics to fiber-reinforced composite material failure
and alternatives to its use in certain cases are given in References
24 and 25.

The combination of the criteria from the last section (Region I
in Figure 1) and this section (Region II) permits the designer to
determine a maximum allowable design stress for either region. Once
it is recognized that these criteria are complementary rather than
competing, it is possible to approach design against failure in a
more direct manner. In fact, for brittle materials use of either of
these criteria over the entire range of flaw sizes in Figure 1 is
possible (25). Thus, evaluating the failure criteria from the last
section at a small fixed distance from the crack tip fracture
mechanics is essentially recovered, and assuming an inherent flaw of
size a_{cr} to always be present even in an ostensibly unflawed
material fracture mechanics can be used as a failure criterion for
all cases.

The application of linear elastic fracture mechanics is in
principle straightforward, albeit at times very complicated in
application. Say, for example, a structural element is to be
constructed of a given material. The value of γ_c (or K_{Ic}) can be
determined from tests on standard specimens or perhaps obtained from
handbooks of materials properties. If the designer can now perform
a stress analysis for the part under the loads in question for its

inherent (or introduced) flaws, the loads at which the crack(s) will propagate can be determined. The literature contains many examples of such calculations as well as modifications in the use of fracture mechanics. For example, the concepts of fracture mechanics have been applied to adhesive bonds and adhesively bonded systems (26-29).

At one time the stress analysis required for the use of fracture mechanics would have posed a formidable problem for many practical applications. This situation is becoming progressively less of a problem as more sophisticated and better numerical methods for stress analysis become available (30). For example, the use of finite element techniques in the fracture mechanics analysis of adhesive bonds has been explored (31).

<u>Fracture Energy</u>. Considerable experimentation, analysis, theoretical development, and speculation on the exact nature of the fracture energy (or, equivalently, the critical stress intensity factor) has occurred. If fracture simply represented the rupture of a plane of atoms, the determination or calculation of γ_c would be a comparatively straightforward process, somewhat analogous to the determination from first principles of the surface tension of a fluid. Such is not the case, however. Not only is the fracture surface itself generally not a plane, but considerable damage occurs at distances somewhat removed from the actual fracture surface. This damage represents energy dissipation and hence may have an important effect on the energy that must be drawn from the free energy of the system. These damage mechanisms, therefore, may contribute significantly to the magnitude of the fracture energy.

Irwin (32) and Orowan (33) modified the original Griffith formulation by adding a term to account for the plastic energy dissipation as the crack grows. They postulated the conditions for crack growth as

$$\delta W + \delta U > \delta T + \delta \psi = \delta \Gamma \tag{23}$$

where δW is the change in external work, δU is the elastic strain energy, δT is the increase in surface energy (true), and $\delta \psi$ is the plastic flow energy dissipated during the crack growth, δa. Note that $\delta \Gamma$, which we will call the fracture energy, includes all the terms required to create the fracture surface. For plastics (as well as most metals) $\delta \psi$ is so much larger than δT that δT is generally considered negligible. If one assumes the energy dissipated in crack growth is proportional to the new crack area, a specific fracture energy, γ_c (or critical energy release rate, G_c), can be defined such that $\delta \Gamma = \gamma_c \delta a$. γ_c, of course, has the units of energy per unit area, that is, J/m^2, in.-lb/in.^{2}, etc. If we assume γ_c is a material property that can be obtained from standard tests, the problem of linear elastic fracture mechanics becomes one of performing a stress analysis and solving for the change in strain energy for an incremental crack growth, δa.

Continuing in this line of thought, Williams (34) has suggested a useful formulation of the Griffith approach. He suggested viewing γ_c as a compilation of effects. That is,

$$\gamma_c = \gamma_B + \gamma_{KE} + \gamma_p + \gamma_{VE} \tag{24}$$

where B, KE, p, and VE represent the contributions to γ_c due to brittle (surface free energy), kinetic energy, plastic, and viscoelastic effects, respectively. Other contributions might likewise be added.

A number of attempts have been made to calculate the contribution of these various mechanisms to γ_c. Most of these attempts have not tried to isolate the exact molecular processes involved. Rather, they have treated the material as a continuum with elastic, plastic, or viscoelastic material properties. Dugdale, for example, calculated the plastic zone size for a sharp tensile crack under plane stress in a nonstrain-hardening material (35). The reader is referred to the original reference and any of a number of texts on fracture (17, 36) for a discussion of this model and its more modern ramifications (some include strain hardening). These approaches have enjoyed considerable success in predicting behavior and performance in a variety of materials including plastics (37). For other materials it appears that the development of cleavage cracks, coalescing of microcracks or voids, crazing, etc. in front of the crack are more responsible for dissipation (or crack blunting) than is plastic deformation (38).

In some cases a material to effectively increase γ_c (or K_{Ic}) can be designed. For example, certain fibers introduced as fillers in a material can provide barriers to crack growth. Proper matching of properties, for example, relative strength of filler and matrix, adhesion between filler and matrix, etc., can drastically affect the amount of energy dissipated as a crack proceeds through a material. Likewise, orienting the molecular chains in a polymer can significantly increase the resistance to crack growth in one direction and typically decrease it in other directions.

Fillers are not always reinforcing, that is, they do not always increase the resistance to crack growth. Particularly large particles in glassy polymers may act as crack nuclei and result in a decrease in strength. This situation occurs also in unidirectional composites in which the fibers greatly increase the resistance to resin failure in longitudinal tension but may lead to a significant decrease in resin strength in the transverse direction.

Environmental factors can also affect the energy required to propagate a crack. For example, in metals, certain environments can synergistically interact with stress to produce the intergranular fracture often associated with stress corrosion. Although not as extensively studied, similar effects are present in polymers. The effect of various environments on the adhesive fracture energy of Solithane (a polyurethane manufactured by Thiokol) bonded to aluminum has been studied (26). Table I gives the results for the adhesive fracture energy when the interfacial crack is exposed to various environments. The materials are listed in order of ascending solvent power.

<u>Viscoelastic Effects</u>. Time plays a very important role in the properties of polymers. For metals, creep is generally significant only at relatively high temperatures, typically greater than half their absolute melting temperature. However, many polymers demonstrate time dependent behavior not only at room temperature but even at cryogenic temperatures (38).

For some materials, the principle of time-temperature superposition has been a convenient and useful concept. It implies that increasing the temperature is equivalent in its effects on properties to increasing the time (decreasing the loading rate).

The principle of time-temperature superposition derives from an equation of the form (39, 40)

$$\eta = Ae^{1/\phi} \tag{25}$$

This equation relates the material viscosity η to the free volume ϕ. It is commonly assumed that the free volume ϕ is related to the temperature by

$$\phi = \phi_0 + \alpha (T - T_0) \tag{26}$$

where ϕ_0 is the free volume at a reference temperature T_0. Taking the viscosity at T_0 as η_0 and dividing Equation 25 by this viscosity yields

$$\frac{\eta}{\eta_0} = e^{(1/\phi - 1/\phi_0)} \tag{27}$$

or

$$\ln \frac{\eta}{\eta_0} = \frac{1}{\phi} - \frac{1}{\phi_0} \tag{28}$$

η/η_0 is commonly called the time-temperature shift factor, a_T, and the reference temperature, T_0, is usually taken as the glass transition temperature, T_g. Making these substitutions and using Equation 26 yields

$$\ln a_T = \frac{1}{\phi_g + \alpha (T - T_g)} - \frac{1}{\phi_g} = - \frac{1/\phi_g (T - T_g)}{\phi_g/\alpha + T - T_g} \tag{29}$$

Changing to log base 10 and renaming some of the material constants one obtains

$$\log a_T = \frac{C_1 (T - T_g)}{C_2 + T - T_g} \tag{30}$$

This is the well-known Williams–Landel–Ferry (WLF) equation (39). The constants C_1 and C_2 were reported to be -17.44 and 51.6, respectively, for a large number of polymeric materials.

Treloar (2) points out that in addition to the "horizontal" shift represented by a_T, a "vertical" shift might also be required. He reports this shift to be the ratio of the reference temperature to the test temperature, and it is applied by multiplying the measured property, P, by this factor, that is, $P(T_S/T)$.

Application of these shift factors is illustrated by a schematic plot of the material property P shown in Figure 5. Note that generally all the data are presented for a limited time period

Table I. Adhesive Fracture Energy in a Blister Test in
Various Environments

Environment	Adhesive Fracture Energy	
	(in.-lb/in.^{2})	(J/m^2)
air	0.58 (±0.06)	102 (±11)
water	0.31 (±0.03)	54 (±5)
acetone	0.15 (±0.02)	26 (±4)
benzene	0.06 (±0.01)	11 (±2)

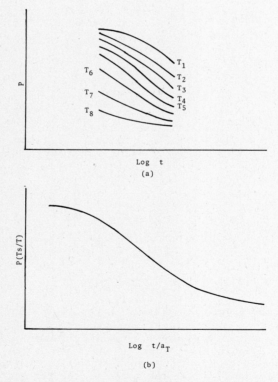

Figure 5.   Schematic of (a) experimental data obtained at various
temperatures and (b) master curve of material property.

because of practical considerations. For example, to determine property values, P, corresponding to those shown for T_8 but measured at T_1 might take years or longer. Alternatively, an attempt to determine the P corresponding to T_1 but measured at T_8 might requirement measurements in such a short time as to be experimentally impossible. Time-temperature superposition, when valid, facilitates "accelerated testing." That is, one might predict long-term service performance from short-term laboratory experiments at elevated temperatures. The key to this is the word "valid." Too often these procedures are used for inappropriate materials and for far too long of a time. Plots that predict "safe loads" for periods exceeding hundreds of years are not uncommon. The validity of these extrapolations based on purely physical effects for such long periods of time in which chemical aging effects become increasingly pronounced is questionable.

The elastic modulus is perhaps the property that has enjoyed the most popularity (and success) in the use of time-temperature shift factors. The concept has, however, also been used in the analysis of other viscosity-related phenomena including rupture and fracture, although rupture in Region I (the nonfracture mechanics region) of Figure 1 has been most extensively studied. One common approach has been to construct failure curves such as those shown schematically in Figure 6. Note the similarity in the general shape of this curve to that of the relaxation modulus. An excellent review of the time-temperature shift of rupture was given as early as 1964 (41). Reference 37 is also of interest.

More recently considerable interest in incorporating temperature and time effects into fracture mechanics has been shown. From the Griffith relation (Equation 16) one notes that in addition to the flaw size, a, two material properties, E and γ_c, enter the equation. In considering this fact somewhat superficially, it is tempting to simply replace these two constant properties for an elastic material by their time-dependent analogs for viscoelastic materials. That is, replace E by the relaxation modulus E(t) (or more commonly $1/C(t)$ where C(t) is the creep compliance), and because the energy to form a unit fracture surface might be expected to depend on time-dependent molecular processes and viscosity, let the specific fracture energy also be a function of time. On the basis of these assumptions, Equation 16 might be rewritten as

$$\sigma_{cr}(t) = \sqrt{\frac{4\gamma_c(t)}{\pi\, C(t)\, a}} \qquad (31)$$

It has been shown (42, 43) that this equation is not theoretically sound in that it is not derivable (as was initially though) from the principles of mechanics and viscoelasticity. Nevertheless, "empirically" it does appear to be a convenient way of predicting fracture in some systems. As cases in point, Figures 7 and 8 reproduce the results of Bennett et al. (44) and Gent (45), respectively. In both cases, the a_T for shifting the fracture energy (or what Gent calls the work of adhesion) is identical to the a_T for shifting the relaxation modulus.

More rigorous attempts to include viscoelastic effects into fracture mechanics predictions have recently been made, for example,

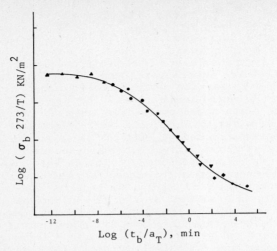

Figure 6.   Plot of rupture behavior of a class of polymer; t_b is the time it ruptures at stress σ_b; a_T is the time-temperature shift factor. Different data symbols are used to indicate that data is taken at different temperatures analogous to Figure 5. Data for such plots are generally taken from constant-strain rate tests at varying rates and temperatures. (See Chapter III B by Landel and Fedors in Reference 41.)

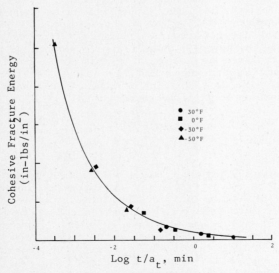

Figure 7.   Experimentally determined time-temperature shift of the cohesive fracture energy for model solid propellent (44).

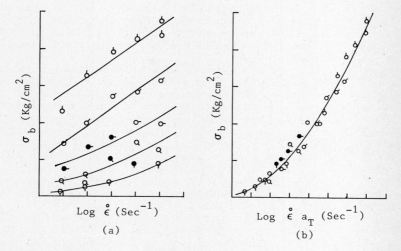

Figure 8.   (a) Tensile breaking stress (directly proportional to
work of adhesion) for a thin adhesive layer versus rate
of extension, $\dot{\epsilon}$ (b) Results shown in (a) replotted
against the effective rate of extension, $\dot{\epsilon}$, a_T, at
23 °C (45).

ρ = +50 °C α = +22 °C $0-$ = 0 °C δ = -20 °C

δ = -30 °C   0 = Adhesive Failure   ● – Internal Rupture
or Cohesive
Failure

References 46 and 47. These approaches have been based on models of cracks in viscoelastic materials in which finite crack tip stresses are obtained by considering a small failure zone or damage region to exist ahead of the crack, somewhat analogous to the Dugdale model (35) for crack tip plasticity. The extension of this approach to nonlinear viscoelastic material behavior and its relation to a generalized J-integral criterion for crack growth has recently been reported (48, 49).

Examples. To illustrate the use of fracture mechanics concepts, we will present several examples. Because these examples are intended to demonstrate the methods involved, they have been necessarily oversimplified; the reader is cautioned against using them as exact patterns for actual fracture design and is referred to the references cited at the end of this chapter for additional details on practical problems.

In considering the first example, assume we are investigating the addition of a glass filler to a glassy polymer to increase its abrasive wear resistance. This procedure is done, for example, in some dental restorative plastics. Assume the following experimental observations were made: the homopolymer tested in tension fails in a quasi-brittle manner at a stress of 60 MN/m^2; and adding 25% by volume of 40 μm glass filler reduced the tensile strength to 25 MN/m^2. The problem is to suggest means by which the strength might be increased to at least 80% of the strength of the homopolymer.

We note that the filler is certainly not reinforcing and in fact is acting as a crack nuclei. A reasonable assumption to make is that the "effective" Griffith crack size would be proportional to the filler particle size. To check this hypothesis one might measure tensile strength for other size filler particles. Assume, for example, that we could obtain filler particles of similar shape in 30 μm and 60 μm sizes and fabricate tensile specimens with these fillers. Assume also that these fillers are observed to have tensile strengths of 30 MN/m^2 and 20 MN/m^2, respectively. Figure 9 shows a plot of the breaking strength as a function of the reciprocal of the square root of the filler size. The Griffith equation (Equation 16) would indicate that if the flaw size was proportional to the filler size, this plot should yield a straight line passing through the origin. Our hypothesis appears, therefore, to be reasonable.

Extrapolation of the line in Figure 9 to 48 MN/m^2 (i.e., 80% of 60 MN/m^2) indicates that the square root of the reciprocal of the filler size should be 0.31, corresponding to a filler size of 10 μm. Then, on the basis of the hypothesis that the filler acts as a Griffith crack having an effective length proportional to the filler size, a strength of 48 MN/m^2 should result from the use of a filler of 10 μm.

There are some potential problems in the approach just discussed. First, the smaller particles might aggregate and thereby give rise to a larger effective flaw size. Secondly, it is not clear that these smaller particles effectively increase the wear resistance. However, these questions could be answered by a few fairly simple experiments. Fracture mechanics analysis has pointed the way for systematic design of this composite material.

The second example demonstrates how data from a standard fracture mechanics test might be used to predict failure for other loadings, flaw geometries, etc. A tensile test of a notched specimen will be used as our reference, and the behavior of a beam loaded in pure bending with different length central cracks will be analyzed.

Assume we have a beam of rectangular cross section 1/2 in. in thickness and 2 in. in depth. The beam is constructed of a high-strength, quasi-brittle material with properties $E = 0.5 \times 10^6$ lb/in.^{2} and $\nu = 0.3$, and is subjected to a pure bending moment. A crack that runs through the entire thickness and depth of the beam on its tensile edge is discovered. The beam will continue to be used at a reduced service load instead of using a replacement. The problem then is to find the maximum bending moment that can be applied to the member by using a factor of safety of three on crack growth.

To begin the fracture analysis of the beam one must first determine the specific fracture energy of the material, γ_c. For this determination a standard specimen must be constructed and tested. A large number of different configurations are available for this purpose. A convenient choice is a notched cylindrical tensile specimen proposed by Srawley and Brown (50) and shown in Figure 10. A natural crack is introduced at the root of the sharp machined notch by cyclic fatiguing of the sample. The value of the specific fracture energy is given by

$$\gamma_c = \frac{1 - \nu^2}{2E} \frac{1.63 \ P^2 D}{d^4} [0.172 - 0.8 (\frac{d}{D} - 0.65)^2]$$

$$\approx .126 \frac{P^2 D}{Ed^4} \tag{32}$$

where P is the applied load at fracture. (A slight correction of the crack depth is suggested, but for this case it would be small.) Samples of the same material as the beam with a 0.5-in. diameter are fabricated, and a series of these samples are tested with resulting failure loads of $P = 360 \pm 30$ lb. From Equation 32, γ_c is found to be 1.05 ± 0.17 in.-lb/in.^{2}.

For the beam under pure bending moment, M, the value of γ is given by Srawley and Brown as

$$\gamma \approx \frac{14.2}{W^3 E} (\frac{M}{B})^2 \left[34.7 \frac{a}{W} - 55.2 (\frac{a}{W})^2 + 196 (\frac{a}{W})^3 \right] \tag{33}$$

where B is the thickness of our bending member, W is the depth, and a is the crack depth. (A slight correction to the crack depth is suggested, but for this case it would be small.) For the case of interest here, the moment necessary to cause the beam to fail is

$$1.05 \pm 0.17 = \frac{14.2 \ M^2}{2^3 \ (0.5 \times 10^6) \ (.5)^2} [34.7(.2) - 55.2(.2)^2 + 196 \ (.2)^3] \tag{34}$$

or

$$M = 108 \pm 8 \ \text{in.-lb} \tag{35}$$

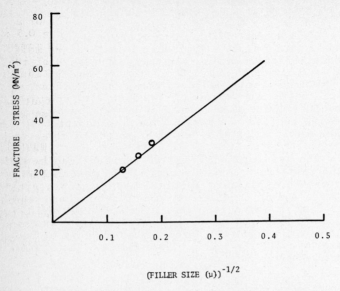

Figure 9.   Fracture stress versus the reciprocal of the square root
             of filler size.

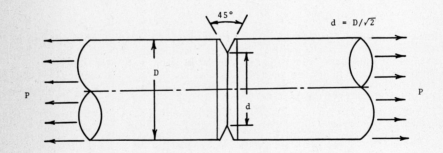

Figure 10.   Notched cylindrical tensile specimen analyzed by Srawley
             and Brown for determination of fracture energy (50).

Therefore, to prevent the crack from growing (with a safety factor of three) the maximum bending moment must not exceed ~ 35 in.-lb.

Because of the time dependence of polymer properties, it is important that the time scale used in our standard tests corresponds exactly with that of the moment loading. That is, if the moment is to be a sustained load, then the tensile test should also be at very long loading, for example, in the equilibrium modulus region.

Many stress analysis problems cannot be solved in closed form. This situation is particularly true for the complex geometries often associated with cracks. No treatise on fracture mechanics would be complete without some mention of how such problems might be treated. Therefore, a third illustrative example has been chosen to serve this purpose.

Consider a solid propellant rocket motor constructed of a rubber-based propellant bonded to a comparatively rigid fiber-wound motor case. The primary loading of the propellant during launching is the intertial force resulting from the acceleration of the rocket. Now assume that, although analysis shows the stresses are quite reasonable at the bond line, fabrication of the rocket motor without some small regions of debond at this interface is nearly impossible. The stress near the tip of these adhesive cracks is usually very large (mathematically singular), and hence if the debonds are large enough they may act as failure nuclei. We assume these regions can be detected and their size can be measured by some nondestructive method (e.g., ultrasonics). To reject all motors with such debonded regions would be prohibitively expensive. On the other hand, use of rockets that fail at the propellant-case interface during launch would be dangerous and potentially even more costly. The question is then what size interfacial flaw can be tolerated. In other words, our problem is to establish a quality assurance criterion for motor acceptance or rejection/repair.

The first step in the determination of the size of debond crack that can be tolerated in the rocket motor is to evaluate the "strength" of the adhesive bond. In this regard, we note that the Griffith failure criterion essentially states that the energy required to propagate a crack (increase its area) must come from the energy released from the stress field as the crack grows. The energy released from the stress field per unit area of crack growth is often called the energy release rate (dU/dA), which is essentially equivalent to the γ defined earlier. This thermodynamic balance might, therefore, be used to predict either adhesive or cohesive crack propagation. To distinguish between the two cases, one might designate the energy required to "create" a unit area of debond between two materials as γ_a, and that required to produce a unit area of new crack in a single material as γ_c. With these considerations, it is proposed that the strength of the rocket motor adhesive bond with its contained areas of debond is best quantified by measuring γ_a for the bond between the solid propellant and the rocket motor case.

In principle, any of a large number of test specimen geometries can be used to determine the value of γ_a. A number of these have been analyzed, and test methods, equations, etc. can be found in the literature (26). It is very important that the test specimen used in the determination of γ_a receives as near the same treatment (surface treatment, cure conditions, etc.) as possible as the actual motor. Also, test methods should duplicate as closely as possible the mode of loading of the actual debond crack being considered (e.g., tensile or shear mode at the crack tip) because the value of γ_a has been found to be very dependent on this mode. If uncertainty as to the exact stress state at the crack tip exists, then it is usually conservative to assume Mode I because for most materials γ_a is lowest in this mode.

Now assume that in the example of interest here, preliminary analysis of the motor case shows that at the most likely points of debonding, Mode I loading dominates adhesive crack behavior. Two test methods that could appropriately be used to determine γ_a in Mode I are the blister test and the double cantilever beam (26). If one assumes that the first is chosen, then a layer of the adhesive (the solid propellant) is cast on a plate of the material from which the motor case is made. This plate contains a hole that has been fitted with a pressure connection. This connection is used to pressurize (with air, inert gases, fluids, etc.) the region between the plate and propellant, and thereby produces a blister. The value of γ_a can be determined from measurement of the pressure at which this blister grows as a function of its diameter (26). In our case, let us assume a series of tests yields a "least value" for γ_a of 17 J/m^2.

We next direct our attention to the analysis of the motor case. Because of the complex stress state in the motor, closed-form solutions for the energy release rate are not available for this case as they were for the previous example. Thus, one must resort to either experimental or numerical methods to determine the stress-strain state and associated energy. If an appropriate finite element stress analysis code is available, then the desired energy release rate can be obtained as follows. A crack is introduced at the point of interest. The finite element code is used to solve the boundary value problem for the stress and strain state. If linear material behavior for the rocket motor is assumed, then the stresses would be proportional to the acceleration, and the associated strain energy would be proportional to the square of the acceleration. The appropriate acceleration, boundary conditions, and assumed crack size are used to calculate the stored energy in the system. The crack is allowed to increase (grow) in size in small steps, and the numerical calculations are repeated in succession. Successive values of the stored energy may be subtracted to obtain the energy released as the crack grows, that is, the energy release rate.

In the problem of interest here assume that such calculations have been made and that Figure 11 shows a plot of the energy released, ΔU, per unit increase in crack area, ΔA, versus the diameter of the debond crack. This figure shows that at a flaw size of 16 mm, the energy release rate corresponds to the experimentally determined value of 17 J/m^2. This result implies that cracks less than 16 mm should be stable, and larger cracks should grow under the assumed acceleration load. In response to the original problem, we may now select a permissible flaw size for quality control. If, for example, a factor of safety of two was used, all rocket motors in which interfacial flaws of 8 mm or larger are detected would be rejected or repaired.

To be precise, cohesive fracture refers to crack growth within an homogeneous material, whereas adhesive fracture refers to crack

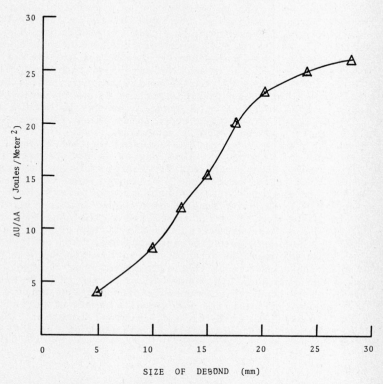

Figure 11.   Energy release rate as a function of adhesive crack size
for a hypothetical rocket motor.

growth along the interface between two materials. In practice, however, the distinction is not so clear. That is, the growth of a debond at an adhesive interface often takes place both within the adhesive and along the interface and alternates locations as it grows. Thus, a mixture of cohesive and adhesive fracture is obtained. Adhesive and cohesive fracture with their associated energies, γ_a and γ_c, can then be regarded as complementary failure mechanisms. Whichever mechanism takes place at a preexisting debond depends upon the interaction of the applied loads, the system constraints, the crack geometry, etc. That is, as the loading is increased, if the energy release rate for cohesive fracture away from the bond line and into the adhesive or adherend exceeds the γ_c of either material before the energy release rate for propagation along the interface reaches γ_a, then cohesive fracture should take place. In practice, however, this distinction is not always made, and γ_a is often used to denote the specific energy to create new "debond" surface, that is, surface essentially along the original interface as opposed to into the bulk adherends.

An additional comment concerns the referral in this example to the words "region of interest." This region is not necessarily easily located. The location of maximum energy release rate need not correspond to a region of maximum "average" stress. An example is provided by the analysis of a cone-shaped adhesive joint (26). In this case of pullout of a truncated cone bonded into a similarly shaped conical hole with a debond implanted at the cone end, the region of maximum stress occurs at the tip of the debond crack. Experimentally, however, failure initiated near the cone center rather than from propagation of the implanted debond. Further analysis showed the reason to be that even microscopic debonds near the center of the cone bond line have larger energy release rates than the implanted debond. In the example of the rocket motor case it might be necessary to analyze debonds at a number of potential failure sites in order to find the most critical location. Such investigation can add significantly to the cost of the failure analysis.

Fatigue

The problem of fatigue in polymers has also been attacked from the standpoint of fracture mechanics. Space will not allow a comprehensive review here. Rather, we refer the interested reader to the work of Manson and Hertzberg (51) for discussions of cohesive crack growth under cyclic loading and to the studies of Mostovoy and Ripling (29) for fatigue of cracks in adhesive bonds. These authors and others have shown that plots of the logarithm of crack growth per cycle (i.e., log da/dN) versus the stress intensity factor (or the fracture energy) are straight lines for a large number of polymers and adhesives. In general, a critical stress intensity factor exists below which the cracks do not grow. This factor is analogous to the fatigue limit in more conventional fatigue analysis.

Conclusion

The approaches outlined in this chapter should not be viewed as a fixed and rigid set of rules and formulas to be used in a particular situation. Rather, the approaches should be seen as a philosophy of approach to complicated problems in which the solution may often need to be tailor made for the specific problem of interest. Throughout this process, one should not lose sight of the fact that materials can fail in a number of different ways, and hence the analysis approach may need to explore several different regions and failure criteria.

Acknowledgments

Portions of this work were supported by the Polymer Program and the Applied Mechanics Program of the National Science Foundation under grants NSF–DMR–7925390 and NSF–CME–7915197, respectively.

Literature Cited

1. Flory, P. J. "Statistical Mechanics of Chain Molecules"; John Wiley–Interscience: New York, 1969.
2. Treloar, L. R. G. "The Physics of Rubber Elasticity"; Oxford Press: London, 1958.
3. Flinn, R. A.; Trogan, P. K. "Engineering Materials and Their Applications," 2nd ed.; Houghton Mifflin: Boston, 1981.
4. McClintock, F. A.; Argon, A. S. "Mechanical Behavior of Materials"; Addison–Wesley: Reading, Mass., 1966.
5. Popov, E. P. "Mechanics of Materials," 2nd ed.; Prentice–Hall: Englewood Cliffs, N.J., 1976.
6. Boresi, A. P.; Sidebottom, O. M.; Seely, F. B.; Smith, J. O "Advanced Mechanics of Materials," 3rd ed.; John Wiley: New York, 1978.
7. Fung, Y. C. "Foundations of Solid Mechanics"; Prentice–Hall: Englewood Cliffs, N.J., 1965.
8. "Plastic Deformation of Polymers"; Peterlin, A., Ed.; Dekker: New York, 1971.
9. Jones, R. M. "Mechanics of Composite Materials"; John Wiley–Interscience: New York, 1979.
10. Christensen, R. M. "Mechanics of Composite Materials"; John Wiley–Interscience: New York, 1979.
11. Wu, E. M. In "Composite Materials"; Sendeckyj, G. P., Ed.; Academic Press: New York, 1974; Vol. 2, p. 353.
12. Nuismer, R. J. Proc. Int. Conf. on Composite Materials–III 1980, Paris, France, p. 436.
13. Griffith, A. A. Phil. Trans. Royal Society 1920, A221, 163.
14. Williams, M. L. J. Appl. Mech. 1957, 24, 109.
15. Inglis, C. E. Trans. Naval Arch. 1913, 60, 219.
16. Irwin, G. R. Handbuch der Physik 1958, 6, 551.
17. Tetelman, A. S.; McEvily, A. J. "Fracture of Structural Materials"; John Wiley: New York, 1967.
18. Kobayashi, A. S. "Experimental Techniques in Fracture Mechanics"; Iowa State University Press: Ames, Iowa, 1973.
19. Rice, J. R. J. Appl. Mech. 1968, 35, 379.
20. Cottrell, A. H. Iron and Steel Inst. Spec. Rep. 1961, 69, 281.

21. Sih, G. C.; Paris, P. C.; Irwin, G. R. Int. J. Fracture
    Mechanics 1965, 1, 189.
22. Zweben, C. Eng. Fracture Mechs. 1974, 6, 1.
23. Kanninen, M. F.; Rybicki, E. F.; Brinson, H. F. Composites
    1977, 8, 17.
24. Whitney, J. M.; Nuismer, R. J.J. Composite Materials 1974, 8,
    253.
25. Nuismer, R. J.; Whitney, J. M. "Fracture Mechanics of
    Composites"; ASTM STP-593: Philadelphia, Pa., 1975; p. 117.
26. Anderson, G. P.; Bennett, S. J.; DeVries, K. L. "Analysis and
    Testing of Adhesive Bonds"; Academic Press: New York, 1977.
27. Gent, A. N. Rubber Chem. and Tech. 1974, 47, 202.
28. Bascom, W. D.; Cottington, R. L. J. Adhes. 1976, 1, 333.
29. Mostovoy, S.; Ripling, E. J. "Final Report - Fracturing
    Characteristics of Adhesive Joints"; Materials Res. Lab:
    Chicago, 1975.
30. Cook, R. D. "Concepts and Applications of Finite Element
    Analysis"; John Wiley: New York, 1981.
31. Anderson, G. P.; DeVries, K. L.; Williams, M. L. Int. J. Fract.
    1973, 9, 421.
32. Irwin, G. R. J. Appl. Mech.-Trans. ASME 1957, 24, 361.
33. Orowan, E. Rep. Prog. Phys. 1949, 12, 185.
34. Williams, M. L. Proc. U.S. Nat. Congr. Applied Mechanics 1966,
    p. 451.
35. Dugdale, D. S. J. Mech. Phys. Solid 1960, 8, 100.
36. Paris, P. C.; Sih, G. C. "Fracture Toughness Testing and Its
    Applications"; ASTM: Philadelphia, Pa., 1965.
37. Williams, J. G. "Stress Analysis of Polymers"; Longman:
    London, 1973.
38. Baer, E. "Engineering Design for Plastics"; Reinhold: New
    York, 1964.
39. Williams, M. L.; Landel, R. L.; Ferry, J. D. J. Am. Chem. Soc.
    1955, 77, 3701.
40. McKinney, J. E.; Belcher, H. V.; Marvin, R. S. Trans. Soc.
    Rheol. 1960, 4, 347.
41. "Fracture Processes in Polymeric Solids"; Rosen, B., Ed.; John
    Wiley: New York, 1964.
42. Nuismer, R. J. J. Appl. Mechs. 1974, 41, 631.
43. McCartney, L. N. Int. J. of Fracture 1977, 13, 641.
44. Bennett, S. J.; Anderson, G. P.; Williams, M. L. J. Appl.
    Poly. Sci. 1970, 14, 735.
45. Gent, A. H. J. Poly. Sci. 1971, A-29, 283.
46. Knauss, W. G. In "The Mechanics of Fracture"; Erdogan, F., Ed.;
    ASME AMD, 1976; Vol. 19, p. 69.
47. Schapery, R. A. Int. J. of Fracture 1975, 11, 141.
48. Schapery, R. A. "Texas A&M University Report MM 3724-81-1";
    College Station, Texas, 1981.
49. Schapery, R. A. Composite Materials, Proc. Japan-U.S. Conf.
    1981, Tokyo, Japan, p. 163.
50. Srawley, J. E.; Brown, W. F. "Fracture Testing and Its
    Applications"; ASTM: Philadelphia, Pa., 1965.
51. Manson, J. A.; Hertzberg, R. W. "Fatigue of Engineering
    Plastics"; Academic Press: New York, 1980.

Flammability of Polymers

Y. P. KHANNA[1] and E. M. PEARCE[2]

[1] Corporate Research & Development, Allied Corporation, Morristown, NJ 07960
[2] Department of Chemistry, Polytechnic Institute of New York, Brooklyn, NY 11201

Polymer Flammability
  Principles of Flammability
  Flame Retardation
Polymer Structure and Flammability
Flame Retardation of Polymers
  Synergism in Flame Retardation
  Selection of Fire Retardants
  Flame Retardation of Polymeric Materials

The use of polymeric materials is growing in a number of areas such as home furnishings, domestic and industrial buildings, appliances, fabrics, and transportation vehicles. An expanded growth of polymers concurrent with the proliferation of safety standards being set by the government and private agencies has stressed that reducing the flammability of polymeric materials is of primary importance. Selection or design of a flame retardant system for a particular application is often difficult. A flame retardant polymer should have high resistance to ignition, low rate of combustion and smoke generation, low toxicity of product gases, retention of low flammability during use, acceptability in appearance and properties, no environmental or health safety impact, and little or no economic penalty. To select or design a polymeric material with desirable flammability properties, a familiarity with the principles of combustion, the relationship of flammability to polymer structure, and the modes of flame inhibition is essential.

This chapter reviews the basic principles of polymer flammability, the relationship of polymer flammability to polymer structure, and the general approaches to flame retardation. In addition, examples illustrating the concepts of polymer flammability and flame retardation are presented.

0097–6156/85/0285–0305$06.00/0

Polymer Flammability

This section deals briefly with the basics of polymer flammability
and the approaches to flame retardation. A familiarity with both of
these subjects is essential for designing polymeric materials of
desired flammability properties.

Principles of Flammability. The presence of fuel, heat, and oxygen
is necessary to initiate and sustain a fire. A basic flammability
cycle is diagrammed in Figure 1. When sufficiently heated by the
external ignition source, the polymeric material reaches a
characteristic temperature at which it begins to degrade. The
extent to which the molecular oxygen plays a role in this surface
decomposition (i.e., thermal or thermo-oxidative) depends on the
specific polymer used. Gaseous combustible products may then be
formed at a rate dependent upon factors such as the intensity of
external heat, temperature, and rate of polymer decomposition.
    Flammable gases (fuel) thus produced diffuse to the flame front
where a series of heat-generating, complex, free-radical chain
reactions take place in the presence of surrounding oxygen. Because
the flaming of organic polymers simply represents the oxidation of
hydrocarbons, it would be instructive to consider the well-
understood methane-oxygen combustion system. This system, shown in
Reactions 1-11, can serve as a model for the more complex polymer
flames. The mechanism in Reactions 1-11 indicates that methane
combustion is a complex free-radical chain reaction consisting of
propagation, chain-branching, and termination steps. Oxygen is
found only in the major branching step. In the branching step, two
chain-carrying radicals are produced for each one consumed, and this
condition accounts for the explosive nature of burning. The main
chain-carrying radicals in the propagation steps are the hydrogen
atom and hydroxyl radicals. This mechanism indicates that hydrogen,
not carbon, is the fuel species most responsible for flaming
hydrocarbons.
    After ignition and removal of the ignition source, combustion
becomes self-propagating if sufficient heat is generated and
radiated back to the material to continue the decomposition process.
The combustion process is governed by such variables as the rate of
heat generation, rate of heat transfer to the surface, surface area,
and rate of decomposition.

Flame Retardation. Polymer combustion, a highly complex process, is
composed of a vapor phase, in which the reactions responsible for
the formation and propagation of the flame take place, and a
condensed phase, in which fuel for the gas reactions is produced.
Flame retardancy, therefore, can be improved by appropriately .pa
modifying either one or both of these phases (2). The approaches
aimed at reducing the flammability of polymer systems can be grouped
into the following three categories:

Vapor Phase. In the vapor-phase approach, a flame-retardant or
modified polymer unit releases upon heat exposure a chemical agent
that inhibits free-radical reactions involved in the flame formation
and propagation. For example, HX (where X is halogen), produced by

HEAT

(Heat of Combustion Produces Fuel Through More Material Decomposition)

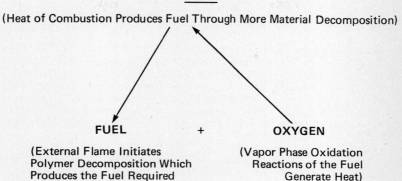

FUEL   +   **OXYGEN**

(External Flame Initiates   (Vapor Phase Oxidation
Polymer Decomposition Which   Reactions of the Fuel
Produces the Fuel Required   Generate Heat)
for Combustion)

Figure 1.   Schematic representation of the flammability cycle.

PROPAGATION

$CH_4 + HO \cdot$ ⟶ $CH_3 \cdot + H_2O$		1
$CH_4 + H \cdot$ ⟶ $CH_3 \cdot + H_2$		2
$CH_3 \cdot + O \cdot$ ⟶ $CH_2O + H \cdot$		3
$CH_2O + CH_3 \cdot$ ⟶ $CHO \cdot + CH_4$		4
$CH_2O + HO \cdot$ ⟶ $CHO \cdot + H_2O$		5
$CH_2O + H \cdot$ ⟶ $CHO \cdot + H_2$		6
$CH_2O + O \cdot$ ⟶ $CHO \cdot + HO \cdot$		7
$CHO \cdot$ ⟶ $CO + H \cdot$		8
$CO + HO \cdot$ ⟶ $CO_2 + H \cdot$		9

CHAIN BRANCHING

$$H \cdot + O_2 \longrightarrow HO \cdot + O \cdot \qquad 10$$

TERMINATION

$$H \cdot + R \cdot + M \longrightarrow RH + M^* \qquad 11$$

Reactions 1-11

the pyrolysis of a halogen-containing organic material in the polymer, acts as a radical scavenger and replaces less reactive halogen atoms with active chain carriers ($\underline{3}$, $\underline{4}$):

$$H^{\cdot} + HX \text{ ------> } H_2 + X^{\cdot}$$

$$HO^{\cdot} + HX \text{ ------> } H_2O + X^{\cdot}$$

Halogenated flame retardants such as chlorinated paraffins, chlorocycloaliphatics, and chloro- and bromoaromatic additives, which are commonly employed in flame-retarding plastics, are postulated to function primarily by a vapor-phase flame-inhibition mechanism. Flame retardation could be implemented by incorporating fire-retardant additives, impregnating the material with a flame-retardant substance, or using flame-retardant comonomers in the polymerization or grafting.

Condensed phase. In condensed-phase modification, the flame retardant alters the decomposition chemistry so that the transformation of the polymer to a char residue is favored. This result could be achieved with additives that catalyze char rather than flammable product formation or by designing polymer structures that favor char formation. Carbonization, which occurs at the cost of flammable product formation, also shields the residual substrate by interfering with the access of heat and oxygen. Phosphorus-based additives are typical examples of flame retardants that could act by a condensed phase mechanism.

Miscellaneous. These approaches include dilution of the polymer with nonflammable materials (for example, inorganic fillers), incorporation of materials that decompose to nonflammable gases such as carbon dioxide, and formulation of products that decompose endothermically. A typical example of such a flame retardant is aluminum oxide trihydrate ($Al_2O_3.3H_2O$). This type of material acts as a thermal sink to increase the heat capacity of the combusting system, lower the polymer surface temperature via endothermic events, and dilute the oxygen supply to the flame, thereby reducing the fuel concentration needed to sustain the flame.

 Of the several test methods for evaluating the burning behavior of different polymers, the limiting oxygen index (LOI) ($\underline{5}$) will be used here to illustrate the relative flammability of materials. This test measures the minimum concentration of oxygen in an oxygen-nitrogen atmosphere that is necessary to initiate and support a flame.

Polymer Structure and Flammability

The flammability of a particular polymer depends mainly upon its structure. The amounts of char and incombustible gases formed upon thermal decomposition determine to a great extent the flame resistance of a polymeric material. A material with LOI $\leq$ 26 is considered flammable ($\underline{6}$). Polymeric materials, on the basis of structure and LOI, can be broadly classified into three categories.

Table I presents the structures, thermal decomposition products, and LOI data for some representative polymers belonging to the three classes. (6, 7).

Class I incorporates aliphatic, cycloaliphatic, and partially aliphatic polymers containing some aromatic groups. The thermal decomposition of these materials gives rise to either small or no char residue. This absence of a pyrolysis residue is due to the high hydrogen content of these polymers, and as a result the entire decomposition product can be disproportionated into volatiles. High yields of monomers or other hydrocarbons from the thermal decomposition of these systems constitute the fuel needed for the flammability cycle shown in Figure 1. The cyano-containing polyacrylonitrile and the hydroxyl-containing polymers (cellulose and polyvinyl alcohol) must be considered as exceptions because they cyclize and, therefore, produce some char; but at the same time they give off much flammable product. The polymers based on structures similar to those included in Class I have LOI values lower than 26 and, therefore, are flammable.

Class II covers those materials that are known to be high-temperature polymers. Structurally they are characterized by the presence of either wholly aromatic or heterocyclic-aromatic functional units. An aromatic ring in the polymer backbone is not only relatively thermally unreactive but also increases the chain rigidity (8). Stiffer chains resist thermally induced vibrations, and, therefore, a higher temperature is required for their decomposition. These materials not only degrade at high temperatures but also undergo extensive cross-linking and cyclization reactions, and as a result their pyrolysis generates high char residues (20–70% at 700 °C). The generation of char residue is conveniently measured by thermogravimetric analysis (TGA) in inert atmosphere (9). Because the formation of char takes place at the expense of flammable products (that is, the elements are converted into carbonaceous involatile residue rather than volatile species), a reduction in flammability should be expected. In fact, van Krevelen (6) has demonstrated a significant correlation between the char residue and the oxygen index of polymers at 850 °C. As shown in Figure 2, char increases as oxygen index increases or flammability decreases. The high-temperature polymers of Class II such as those in Table I are considered flame retardant (LOI > 26). The lower flammability in this case is attributed to factors such as reduction in the amount of combustible volatiles, high energy requirements for continuous fuel generation, and insulating effects of the resultant char.

Class III consists of halogen-containing polymeric materials. Some of them form a small char residue, others form none at all. These polymers are inherently flame retardant because halogen radicals act as radical scavengers in the vapor phase and, therefore, inhibit combustion, as described earlier. Also the splitting-off of noncombustible gases such as HCl, Hf, and C_2F_4 seals the polymer surface from the combustion air and is thereby partly responsible for the flame retardancy. All these factors thus influence the interaction between pyrolysis and ignition.

Table I. Effect of Structure on Polymer Flammability Properties

Class	Name	Structure	Degradation Products	LOI
I	Polyethylene	$+CH_2-CH_2+_n$	Continuous Spectrum of Saturated and Unsaturated Hydrocarbons from C_2-C_{90}	17.4
	Polybutadiene	$+CH_2-CH=CH-CH_2+_n$	About 2% Monomer in Addition to Saturated and Unsaturated Hydrocarbons	18.3
	Polystyrene	$+CH_2-CH+_n$	About 41% Monomer, 2% Toluene, Remainder Dimer, Trimer and Tetramer	18.3
	Cellulose		H_2O and Small Amounts of CO_2, CO and a Tar Containing Principally Levoglucosan	19.9
	Polyethylene Terephthalate	$+OC-\bigcirc-COO-CH_2-CH_2-O+_n$	Acetaldehyde Major Product with CO_2, CO, C_2H_4, H_2O, CH_4, C_6H_6, Terephthalic Acid and More Complex Chain Fragments	20.6
	Nylon 66	$+H_2N-(CH_2)_6-HNCO-(CH_2)_4CO+_n$	H_2O, CO_2, Cyclopentanone, Traces of Saturated and Unsaturated Hydrocarbons	21.5

Polymer	Structure	Products	Value
Polycarbonate		Major Products CO_2, Bisphenol A; Minor Products CO, CH_4, 4-Alkyl Phenols; Much Char	29.4
Nomex®		High Char Yield ; Volatiles Contain CO_2, CO, H_2O, 1,3-Dicyanobenzene, 3-Cyanobenzoic Acid	29.8
Kynol®		High Char Yield; Volatiles Comprise Xylene (76%), Traces of Phenol. Cresol. Benzene	35.5
Polybenzimidazole		Very High Char Yield, Small Volatiles May Include NH_3, H_2 etc.	41.5
Polyvinyidene Fluoride		35% HF and High Yields of Products Involatile at 25°C, Some Carbonization	43.7
Polyvinyl Chloride		Quantitative Yields of HCl, Aliphatic and Aromatic Hydrocarbons	47.0
Polyvinyidene Chloride		High Yields of HCl	60.0
Polytetrafluoroethylene		>95% Monomer, 2–3% C_3F_6, No Larger Fragments	95.0

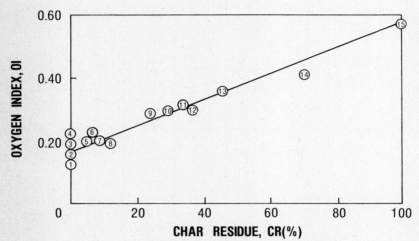

Figure 2.   Correlation between oxygen index and char residue.
1. polyformaldehyde; 2. polyethylene, polypropylene;
3. polystyrene, polyisoprene; 4. nylon; 5. cellulose;
6. poly(vinyl alcohol); 7. PETP; 8. polyacrylonitrile;
9. PPO; 10. polycarbonate; 11. Nomex; 12. polysulfone;
13. Kynol; 14. polyimide; 15. carbon.   (Reproduced
with permission from Ref. 6. Copyright 1975 IPC
Business Press.

Flame Retardation of Polymers

Much literature discusses the flame retardation of various polymeric materials (10–15). The techniques of reducing the flammability of polymers, in principle, are based on one or more of the three fundamental approaches described earlier. This section deals with the concept of synergism and its application in reducing flammability, selection of fire–retardant additives, and flame retarding some specific polymer systems.

Synergism in Flame Retardation. The effect of a mixture of two or more flame retardants may be additive, synergistic, or antagonistic. Synergism is the case in which the effect of two or more components taken together is greater than the sum of their individual effects. The concept of synergism is very important in fire retardation because it can lead to efficient flame retardants with less expensive polymer systems and minimal effects on other desirable properties. One of the classic illustrations of synergism observed in flame retardation is the addition of Sb_2O_3 to halogen–containing polymers. The reaction of a chlorine source with Sb_2O_3 produces $SbOCl$ as an intermediate with lower energy barrier, which then on thermal decomposition evolves $SbCl_3$, the actual flame retardant working by gas phase inhibition (16). The existence of phosphorus–halogen and nitrogen–phosphorus synergisms has also been suggested. Compounds based on the combinations of Sb–X (X is halogen), N–P, and P–X have been used as flame–retardant additives for thermoplastics.

Selection of Fire Retardants. The choice of flame retardants depends on the nature of the polymer, the method of processing, the proposed service conditions, and economic considerations. Although the processing, service, and economic factors are impor.pa tant, the flame–retardancy potential of an additive is of primary importance, and this factor can be readily evaluated by thermal analysis.

Einhorn (17) has described the use of TGA in selecting the appropriate fire retardants. The technique involves matching the degradation of candidate additives with that of the polymer under consideration. Figure 3 represents the TGA thermogram for a hypothetical polymer that exhibits a simple unimolecular degradation process. Point A represents the region of initial decomposition, and point B represents the temperature of maximum rate of degradation. Fire retardants are screened so that a material having thermal characteristics similar to those in Figure 3 is selected. The efficiency of matching the degradation curve of the polymer with the volatilization or degradation characteristics of the fire retardant has been cited as the key to effective flame retardancy (18). If the flame–retardant additive possesses low thermal stability compared to that of the polymer, it will be lost before its function is needed; if the additive has greater stability, it may be intact at the time its function is needed. The use of more than one fire–retardant additive depends on the thermal degradation characteristics of the flammable substrate.

Several manufacturers or suppliers of flame retardants have listed TGA weight–loss data to facilitate the selection of appropriate fire–retardant additives. These data are reported in Table II only for those additives with identifiable structures.

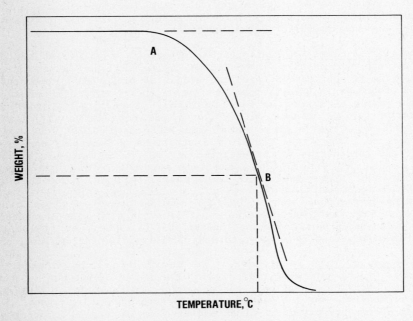

Figure 3.   TGA thermogram for a hypothetical polymer. (Reproduced
             with permission from Ref. 17. Copyright 1971 National
             Academy of Sciences.)

Table II. TGA Weight Loss Data for Various Flame Retardants

Flame retardant	TGA weight loss at (°C)		
	1%	5%	10%
Alumina, hydrated	—	—	290
Analine,2,4,6-tribromo-	121	156	174
Barium metaborate	200	350	1000
Benzene, hexabromo-	232	265	280
Benzene, pentabromoethyl-	180	217	232
Biphenyl, hexabromo-	—	—	299
Biphenyl, octabromo-	—	—	336
Bisphenol-A, tetrabromo-	245	284	298
Bisphenol-A, tetrabromo-, bis(methylether)	244	280	296
Bisphenol-A, tetrabromo-, bis(2,3-dibromopropylether)-	284	315	322
Bisphenol-A, tetrabromo-, bis(2-hydroxyethyl ether)-	284	322	337
Bisphenol-A, tetrabromo-, bis(allyl ether)-	220	245	261
Bisphenol-A, tetrabromo-, bis(2,3-dibromopropyl carbonate)-	—	—	328
Bisphenol-A, tetrabromo-, diacetate	247	278	283
Bisphenol-S, tetrabromo-, butenediol, dibromo-	240	—	—
Carbonate, bis(2,4,6-tribromophenyl)-	227	287	308
Cyclododecane, hexabromo-	—	230	255
Cyclohexane, pentabromo-, chloro-	175	200	235
Diphenylamine, decabromo-	260	277	351
Diphenyloxide, pentabromo-	—	247	—
Diphenyloxide, octabromo-	274	325	340
Diphenyloxide, decabromo-	317	357	373
Methylenedianiline, tetrabromo-	—	—	268
Neopentyl alcohol, tribromo-	—	—	160
Neopentyl glycol, dibromo-	115	135	150
Phosphate, tris(2-chlorothyl)-	—	170	190
Phosphate, tris(β-chloropropyl)-	—	155	175
Phosphate, tris(2,3-dibromopropyl)-	215	270	285
Phosphate, tris(dichloropropyl)-	—	210	225
Phosphonate, bis(2-chloroethyl)-vinyl-	—	135	155
Phosphonate, diethyl N,N-bis-(2-hydroxyethyl) amino methyl-	—	165	180
Phosphonate, dimethyl methyl-	—	60	75
Phosphonium bromide, ethylene bis,-tris(2-cyanoethyl)-	250	—	285
Phosphonium bromide, tetrakis(2-cyanoethyl)-	253	—	307

" Compiled from various product data literature.

Flame Retardation of Polymeric Materials. The suitability of a particular flame-retarding method is often determined by the nature of the polymer type (e.g., thermoplastic or thermoset). A brief and general discussion regarding the flame retardation of thermoplastics, thermosets, elastomers, and fibers follows.

Thermoplastics. Many flame-retardant chemicals have been developed for use in thermoplastics. Most of these flame retardants are of the additive type, usually halogen- and/or phosphorus-based compounds.

Some examples of the fire-retardant additives used in polyolefins (12) are hexabromocyclododecane, octabromodiphenyl, hexabromobiphenyl, chlorowax, chlorinated triphenyl, chlorendic acid, tris(tribromophenyl)phosphite, and tris(dibromopropyl)phosphate.

Touval (19) has described the use of antimony-halogen synergistic mixtures as flame retardants for polyethylene and polypropylene. Red phosphorus used as a flame retardant for high-density polyethylene (HDPE) has been described by Peters (20). He suggested that red phosphorus increased the thermo-oxidative stability of HDPE by scavenging oxygen at the polymer surface, thus retarding the oxidative degradation processes. This type of interaction between HDPE and phosphorus in the condensed phase led to a reduction in polymer flammability, as expected.

Incorporating halogens either as additives or as part of the polymer is the most common technique of flame retarding polystyrene. Preferably, the halogen should be part of the polymeric chain because the halogenated additives tend to be noncompatible and can cause several other side effects.

Prins et al. (21) described the lower flammability of polybromostyrene relative to that of polystyrene. On the basis of thermal analysis experiments, they suggested that bromine inhibited most of the oxidative chain reactions, and thus the combustion was not supported (vapor-phase mechanism). Khanna and Pearce (16) and Brauman (22) demonstrated that polystyrene could be flame retarded by appropriately modifying its structure with substituents that promote the char yield of the system (condensed-phased mechanism).

Poly(vinyl chloride) (PVC) has a high level of chlorine and as a result is considered inherently flame retardant. However, in many circumstances further improvement in flammability reduction of PVC is desired. An efficient way to enhance flame retardation of PVC is by the addition of Sb_2O_3 as a synergist. Additive retardants based on halogen and/or phosphorus are also employed.

Sobolev and Woycheshin (23) pointed out that the burning rate of PVC could be lowered by using $Al_2O_3 \cdot 3H_2O$ as a flame-retardant filler.

Thermosets. Fire retardancy in thermosetting polymers is achieved largely by the use of reactive fire retardants because the common fire-retarding additives lack permanence. The flammability of thermosetting materials can be reduced by the additions of inorganic fillers and/or reactive flame retardant components. Flame-retardant vinyl monomers or other cross-linking agents are also frequently employed.

A large number of halogen-containing reactive diols, polyols, anhydrides, and other functional groups containing intermediates have been used to produce flame-resistant unsaturated polyester resins. Flame-retardant polyester resins have been made by using bromostyrene as partial replacement of styrene for cross-linking (21).

As in polyester resins, reactive halogens containing fire-retardant chemicals are most often used in epoxy materials. Tetrabromobisphenol A is perhaps the most widely used component for flame-retarding epoxy resins. Nara and Matsuyama (24) and Nara et al. (25) described the thermal degradation and flame retardance of tetrabrominated bisphenol A diglycidyl ether compared to the nonbrominated structure. Their results indicate that bromine acts by vapor-phase as well as condensed-phase mechanisms of flame inhibition.

Fire retardancy in polyurethanes is primarily of commercial importance in foams. Foams are generally flame retarded by the incorporation of halogens, phosphorous, and/or nitrogenous compounds. Conley and Quinn (10) reviewed the fire retardation of various polyurethanes.

Elastomers. Many applications of rubbers such as tires, gaskets, and washers normally do not require flame resistance. When improvement in flammability is required, it can be achieved by the addition of halogen-containing materials, phosphorus compounds, oxides of antimony, and combinations of these materials. Rubbers containing chlorine and silicon atoms, for example, Neoprene and Silicone, have self-extinguishing properties. Rogers and Fruzzetti (10) described the flame retardance of elastomers.

Fibers. The flame retardation of fiber-forming polymers is generally achieved by the incorporation of additives based on halogen, phosphorus, and/or nitrogen. Use of antimony oxide as a synergist with halogen is also common.

The most common route for making cellulosic fibers flame retardant is the use of catalysts such as antimony trichloride and phosphoric acid that enhance the formation of char. Because of the major role of cellulosic fibers in the textile market, flame retardation of cellulose has been the subject of several reviews (10, 11, 15).

Wool has been regarded as a relatively safe fiber from the flammability point of view. However, it could be flame retarded to a higher degree if required. Hendrix et al. (26) suggested a large improvement in fire resistance of wool by treatment with 15% H_3PO_4. Beck et al. (27) showed that weak acidic materials, such as boric acid and dihydrogen phosphate, are effective additives for flame retarding wool by the condensed-phase mechanism (increased char residue).

Retardation of the combustion of nylon fibers has been reviewed by Pearce et al. (10). Several bromine and phosphorus compounds have been suggested to be effective for flame retarding nylon 6, but none of them is used in practice on a reasonable scale (12).

The flammability of poly(ethylene terephthalate) can be decreased by using bromo-compounds in conjunction with compounds such as Sb_2O_3 synergist, red phosphorus, and triphenylphosphine oxide. Flame retardation of poly(ethylene terephthalate) has been reviewed by Lawton and Setzer (10). Their review indicates that halogen and/or phosphorus compounds added to the melt prior to fiber spinning are useful flame retardants.

In the past decade several high-temperature fibers have been made commercially available. These materials based on aromatic and heterocyclic structures are inherently flame resistant; those based on heterocyclic structures originate from the heavy carbonaceous char that rapidly forms on the surface exposed to a flame. The flammability of even high-temperature fiber-forming polymers can be improved. Durette, made by chlorination or oxychlorination of Nomex, a polymer of m-phenyleneisophthalamide, exhibits much improvement in flammability properties over Nomex. Khanna and Pearce (28) described the reduction in flammability of aromatic polyamides by the use of halogen-substituted diamines. The mechanism of flame retardation in this case was reported to involve vapor phase as well as condensed phase. Burning resistance of these high temperature fibers can also be achieved by thermal treatments, which has been substantiated by Bingham and Hill (29) for Kynol (cross-linked phenolics) and Durette fabrics. Hirsch and Holsten (30) reported that the fabric made from poly(m-phenylenebis(m-bensamido)terephthalamide) could be stabilized thermally so that the product would not burn even if heated red hot in a gas flame. Such treatments are believed to involve cyclodehydration of the original polymer.

Polymer flammability continues to be an important field of research even though it is reasonably well understood at present. Current efforts will contribute to the design of safer polymeric materials of the future.

Literature Cited

1. Fristom, R. M.; Westenberg, A. A. "Flame Structure"; McGraw-Hill: New York, 1965.
2. Frisch, K. C. Int. J. Polym. Mater. 1979, 7, 113.
3. Factor, A. J. Chem. Educ. 1974, 51, 453.
4. Petrella, R. V. "Flame Retardant Polymeric Materials"; Lewin, M.; Atlas, S. M.; Pearce, E. M., Eds.; Plenum: New York, 1978; Vol. 2, p. 159.
5. Fenimore, C. P. Martin, F. J. Mod. Plast. 1966, 44(3), 141.
6. van Krevelen, D. W. Polymer 1975, 16, 615.
7. Grassie, N.; Scotney, A. "Polymer Handbook"; Brandrup, J.; Immergut, E. H., Eds.; Wiley: New York, 1975; Vol. II, p. 473.
8. Williams, D. J. "Polymer Science & Engineering"; Prentice Hall: New Jersey, 1971; p. 32.
9. Pearce, E. M.; Khanna, Y. P.; Raucher, D. "Thermal Analysis in Polymer Characterization"; Turi, E. A., Ed.; Academic: New York, 1981; Chap. 8, p. 793.
10. "Flame Retardant Polymeric Materials"; Lewin, M.; Atlas, S. M.; Pearce, E. M., Eds.; Plenum: New York, 1975; Vol. 1.
11. "Flame Retardant Polymeric Materials"; Lewin, M.; Atlas, S. M.; Pearce, E. M., Eds.; Plenum: New York, 1978; Vol. 2.
12. "Flame Retardancy of Polymeric Materials"; Kuryla, W. C.; Papa, A. J., Eds.; Dekker: New York, 1973; Vol. 1&2.

13. "Flame Retardancy of Polymeric Materials"; Kuryla, W. C.; Papa, A. J., Eds.; Dekker: New York, 1975; Vol. 3.
14. "Flame Retardancy of Polymeric Materials"; Kuryla, W. C.; Papa, A. J., Eds.; Dekker: New York, 1978; Vol. 4.
15. "Flame Retardancy of Polymeric Materials"; Kuryla, W. C.; Papa, A. J., Eds.; Dekker: New York, 1979; Vol. 5.
16. Khanna, Y. P.; Pearce, E. M. "Flame Retardant Polymeric Materials; Lewin, M.; Atlas, S. M.; Pearce, E. M., Eds.; Plenum: New York, 1978; Vol. 2, Chap. 2.
17. Einhorn, I. N. Reprint from "Fire Research Abstracts and Reviews"; National Academy of Sciences: Washington, D.C., 1971; Vol. 13, p. 3.
18. Pitts, J. J. J. Fire Flammability 1972, 3, 51.
19. Touval, I. "Flame Retardants for Plastics"; Presented at Flame Retardant Polymeric Materials course, Plastics Institute of America, Stevens Institute of Technology, Hoboken, New Jersey, 1975.
20. Peters, E. N. J. Appl. Polym. Sci. 1979, 24, 1457.
21. Prins, M.; Marom, G.; Levy, M. J. Appl. Polym. Sci. 1976, 20, 2971.
22. Brauman, S. K. J. Polym. Sci., Polym. Chem. Ed. 1979, 17, 1129.
23. Sobolev, I.; Woycheshin, E. A. in "Flammability of Solid Plastics"; Hilado, C. J., Ed.; Fire and Flammability Series; Technomic: Westport, Connecticut, 1974; Vol. 7, p. 295.
24. Nara, S.; Matsuyama, K. J. Macromol. Sci., Chem. 1971, 5(7), 1205.
25. Nara, S.; Kimura, T.; Matsuyama, K. Rev. Electr. Commun. Lab. 1972, 20, 159; Chem. Abst. 77, 75562h.
26. Hendrix, J. E.; Anderson, T. K.; Clayton, T. J.; Olson, E. S.; Barker, R. H. J. Fire Flammability 1970, 1, 107.
27. Beck, P. J.; Gordon, P. G.; Ingham, P. E. Text. Res. J. 1976, 46, 478.
28. Khanna, Y. P.; Pearce, E. M. J. Polym. Sci., Polym. Chem. Ed. 1981, 19, 2835.
29. Bingham, M. A.; Hill, B. J. J. Thermal Anal. 1975, 7, 347.
30. Hirsch, S. S.; Holsten, J. R. Polym. Prepr. 1968, 9, 1240.

POLYMER PRODUCTS AND THEIR USES

The Plastic Industry: Economics and End-Use Markets

R. A. MCCARTHY

Springborn Laboratories, Inc., Enfield, CT 06082

Sales History
Styrenics
Polyethylene
Polypropylene
Poly(vinyl chloride)
Engineering Plastics
Thermosets
Other Resins
Summary

The plastic industry of the 1980s is in a different situation than it was in 1973. At that time, annual plastic consumption had been growing at between 11 and 15% for the previous 50 years. Prices for polymers were rapidly escalating because of the recent removal of price controls, and, although some markets were being lost because of the higher prices, expansions in capacity amounting to 30% of the total were being planned. There was no doubt that the steady upward trend in the growth would continue because plastics were expected to maintain their relative competitiveness with other materials.

The Arab oil embargo of 1973 brought an end to this optimism. However, the short-term impact of the embargo, which further drove prices up because of increasing raw material feedstock costs, was softened because it came in the face of a strong worldwide economy that accepted increased prices and passed them on to the consumer. Plastic production continued to expand from new and modernized facilities using new technologies. The peak of polymer production of 40.6 billion lb was reached in 1979. The early 1980s saw increasing costs of capital worldwide, and inflation further pushed increasing feedstock and energy prices higher. The economy, worldwide, slowed and there began to appear declining forecasts for future plastic growth. Polymer consumption dropped as the major markets based on the housing and automotive industries declined.

0097–6156/85/0285–0323$06.00/0
© 1985 American Chemical Society

The recessionary economy, lowering price, and overcapacity caused the polymer producers to reassess their commitments and restructure their activities for a slower growth period. It was not until 1983 that the 1979 sales level was again reached.

Sales History

Table I reviews the industry performance in 1981. The value of shipments, after the all-time high in 1979, slumped badly in mid-1980. Early 1981 sales rose again to an all-time high for monthly sales; but a steady decline occurred after March and continued into early 1982, after which there was stabilization at reduced volumes and reduced selling prices. Recovery began in mid-1982 followed by a 15% increase in sales in 1983 and a 6.5% increase in 1984.

Table II provides information on the individual plastics that made up 1981 sales, insofar as data are available. Where there are fewer than three suppliers, no information is provided to the plastic trade association [Society of the Plastics Industry (SPI)] for compiling. The poor 1979-1983 sales record is evident. These plastics were used in industry as shown in Table III.

The 1982 resin sales total was 34,455 million lb, which is a drop of 6.5% from 1981, calculated by using equivalent compilation, and a 12.7% drop from 1979. The recovery brought strong growth and in 1984 resins sales were 43,400 million lbs. However, the 1982-1983 slump resulted in important changes in the industry. The box lists the 50 major plastic producers at the beginning of 1983 and significant changes that occurred in industry structure.

FIFTY LEADING PLASTIC PRODUCERS IN EARLY 1983

Air Products	Ciba Geigy	Georgia Pacific	Polysar
Allied	Cosden	Goodrich	Reichhold
American Cyanamid	Diamond Shamrock	Goodyear	Rohm & Haas
American Hoechst	Dow Chemical	Gulf	Shell
Amoco	DuPont/Conoco	Hammond	Shintech
Arco Chemical	Eastman Chemical	Hercules	Soltex
BASF-Wyandotte	El Paso	ICI	Tenneco
Borden	Ethyl	Mobay	Texstyrene
Borg-Warner	Exxon	Mobil Chemical	USI
Celanese	Formosa Plastics	Monsanto	USS Chemical
CertainTeed	General Electric	Northern Petro.	Union Carbide
Chemplex	General Tire	Occidental	Uniroyal
		Phillips	Upjohn

- Arco Chemical dropped out of both high and low density polyethylene (as Cities Service did in 1982). Their high density plant went to USI. Chemplex polyethylene went to Northern Petrochemical.

- Diamond Shamrock is essentially out of PVC following Firestone and Stauffer and Du Pont/Conoco.

- El Paso shut down an LDPE plant. Gulf is reported selling the propylene plant to Amoco.

Table I. The Industry in 1981 in Current Dollars

Parameter	Plastic Materials & Resins (SIC 3821)	Fabricated Prod. (SIC 3079)
Number of establishments (over 20 emp.)	487	4803
Number of employees	90.5	425
Value of shipments (industry data) billion	$16.4	$27.3
Value of shipments (product data) billion	$19.6	$32.8
Value of exports, billion	$ 2.6	$ 1.7

Sources: Survey of Industrial Purchasing Power, U.S. Dept. of Commerce
1983 Industrial Outlook, U.S. Dept. of Commerce

Table II. The Plastic Industry in 1983 – Sales by Resin in Million Pounds

Polymer Family	Major Producers	Estimated Capacity	1983 Sales	Major Process	Major Industry	1977-1983% Annual Growth
Epoxy	7	NA	334	coating	elec/electronics	0.0
Phenolic	32	NA	2484	adhesives	bldg. & construct.	NA
Unsaturated Polyester	21 [a]	NA	1084	FRP	bldg. & construct.	-1.8
Melamine & Urea	25	NA	1352	adhesives	bldg. & construct.	-3.4
Polyurethane	20	NA	1808	foams	furniture	-0.3
ABS & SAN	3	1415	831	inj. mold.	elec./electronics	-2.2
Engineering Thermoplastics [b]	16	NA	1485	inj. mold.	elec/electronics	2.4
HDPE	10	6690	5691	blow mold.	packaging	-2.4
LDPE & LLDPE	10	10400	8041	blown film	packaging	0.7
Nylon	11	430	321	inj. mold.	transporation	1.7
Polypropylene	12	5580	4434	inj. mold./fibers	pkg./home furnishings	3.6
Polystyrene	19	5700	3567	inj. mold.	packaging	-2.1
Polyvinyl Chloride	14	8400	6066	extrusion	bldg. & construct.	-0.9
Other [c]	21	NA	3221	coatings	home furnishings	NA

[a] Isocyanate & polyol producers

[b] Includes polyacetal, polycarbonate, polyphenylene sulfide, thermoplastic polyester, polysulfone, modified polyphenylene oxide, polyimide, polyamide-imide and fluoropolymers

[c] Includes acrylics (532 MM lbs.), cellulosics (114 MM lbs.), PVC copolymers (178 MM lbs.), Paint resins (699 MM lbs.), misc. thermosets including alkyds (350 MM lbs.), and 430 MM lbs of other resins

Source: Synthetic Organic Chemicals, U.S. International Trade Commission, SPI 1984 Edition Facts & Figures of U.S. Plastic Industry Modern Plastics

Table III. Major Market for Plastics 1977-1981

Major Market	1979	1980	1981	1982	1983	%/yr Change 1979-1983
Transportation	1934	1605	1573	1392	1896	-0.5
Packaging	10334	10003	10465	10497	11813	3.4
Building & construction	7573	6424	7259	7514	8552	3.1
Electric/electronic	3043	2453	2670	2275	2514	4.7
Furniture/home furnishings	1894	1646	1670	1556	2007	1.5
Consumer & institutional	3753	3553	3670	3269	3816	0.4
Industrial/machinery	517	391	393	241	318	-11.5
Adhesives, inks, & coatings	2794	2387	2572	2584	1800	-10.5
All others	3443	3054	3259	3232	3636	1.3
Exports	3432	3670	3425	3909	4150	4.9
Total	38717	35186	36956	35109	40502	1.1

Note: All values are in million pounds, on a dry weight basis, and include sales to resellers and compounders.
Source: SPI Committee on Resin Statistics

● Ethyl sold its PVC operations to Georgia Pacific who spun off all
  PVC to Georgia-Gulf. Great American, Rico and General
  Tire/Pantasote shut down PVC operations.

● General Electric dropped out of phenolics as did Monsanto.

● Hammond sold its polystyrene plant.

● Mobay dropped ABS as did Mobil Chemical and USS Chemical.

● Shell dropped polystyrene as did USS Chemical.

Downstream operations were sold by American Hoechst, Arco,
DuPont, Mobay, Northern Petrochemical, Phillips, Reichhold and
Tenneco.

On the positive side, Himont, Formosa Plastics, and Texstyrene
are new suppliers and some of the shutdown plants have been
purchased. Mobil Chemical and Eastman intend to expand in
polyethylene and new downstream plastic acquisitions have been made
by Ashland, El Paso, Goodyear and Hercules.

Since 1983, there has been cautious optimism that an economic
recovery is at hand (1). Real economic growth in the second quarter
was at 9.2% which followed a 2.5% growth in the first quarter. The
year could achieve a 6% growth overall with inflation under control,
the employment situation improving, housing starts up 50%, and
domestic auto production 30% ahead of 1982. Real gross national
product (GNP) growth is expected to average over 4%/year through
1987, and inflation rates should average about 5%/year. World oil
prices should be stable through 1984 and then rise at 0.5%/year
through 1987. Interest rates will remain high, and European
recovery will lag behind the United States, but housing starts and
domestic car production should remain strong.

These are good signs for the plastic industry in its
restructured mode, and they have influenced the future growth
forecasts for the individual plastics:

Styrenics (2)

Polystyrene and acrylonitrile-butadiene-styrene (ABS) are mature
products, and growth in the 1980s will not be as dramatic as in the
two previous decades. Solid polystyrene was down in sales 30% in
1984 from the 1979 high point. Expandable polystyrene (EPS) held
steady at 420--440 million lb over the early 1980s. Predictions for
future growth are for a 6% growth in 1985 and 4--5%/year thereafter.
EPS should have growth in the 8-9%/year range. A total for all
polystyrene of 4.1 billion lb is seen for 1988. New capacity from
Mobil in 1984 will ensure capacity for some time because utilization
in the recent past has been running from 76 to 80%. Some further
shakeout is still expected among the 11 largest suppliers.
Packaging and goods service items will continue to be the largest
markets.

ABS rebounded in 1983 and 1985, increasing in sales by 65% from
1982 levels. Growth will be in the 5%/year range over the next 5
years but will not reach 1979 levels until 1987-88. Utilization of
capacity is now back to 88%, since the suppliers have dwindled to

three. The automotive and pipe industries will never again be strong ABS markets, but ABS and alloys of ABS should compete with engineering plastics.

Polyethylene (3)

Polyethylene is also a mature market. In 1982 10,890 million lb of domestic consumption and 1821 million lb of export consumption were observed. Volume growth for the polyethylenes has been positive except in 1975 and 1980. A 4.3%/year growth is estimated from 1982–92 for domestic use, but a slumping export market should reduce overall growth to 3.1%/year. There are now 17 producers, many of whom have linear low-density polyethylene (LLDPE). This new type of polyethylene should be a factor in sustaining growth. Union Carbide and Dow plan expansions that will be needed to meet the 16.6 billion lb of consumption expected by 1992. Packaging will be the most important market, and it is expected that these resins will gain market share at the expense of paper, steel, and glass.

Polypropylene (4)

This product had overcapacity problems prior to the 1980s despite a 14.3%/year growth in sales volume from 1975 to 1980. There were strong markets in blow molding, fibers, film, sheet, and export, all growing in the 1970s at over 15%/year. Overall growth dropped to 3.9%/year in 1982 as most growth of polypropylene's markets flattened. However, predictions are for 6.2%/year growth for polypropylene in 1981–90, which will mean a 92% plant utilization by 1985. There are 12 suppliers, and they have been concentrating on technical innovations. Film, sheet, and blow molding should remain strong markets, and filled products should secure new applications. New products and new processes for polymerization should make polypropylene a better product in the 1980s.

Poly(vinyl chloride) (5)

In 1979, the industry shipped 6.2 billion lb of resin and utilization was 90%. This amount dropped to 5.4 billion lb in 1980 and 1982, but 1983 saw a return to 1979 levels. Construction is the moving force in poly(vinyl chloride) (PVC) sales and accounts for 54% of the total, and when it is strong, PVC sales are strong. The PVC industry overbuilt capacity. Since 1979, 2 billion lb of large reactor capacity were installed. Utilization dropped to 71% in 1982 but should be back up to 88% in 1985 when PVC demand is estimated to reach 6.85 billion lb. This is a 4–5%/year growth. However, PVC producers see problems ahead from imports from Mexico and Canada of finished products, and attacks by competititors and activists who seem to have targeted on PVC as a fire hazard. Research and development has suffered during the years of low profitability. There is much foreign ownership in PVC capacity (20%), and end uses such as pipe, rigid calendering, and wire and cable have become heavily foreign controlled. The 14 PVC suppliers will face changes as several more plants are expected to change hands.

Engineering Plastics

These materials (polyacetals, polycarbonates, thermoplastic polyesters, fluoroplastics, polyaryl esters, thermoplastic polyimides, polyphenylene sulfides, polysulfones but excluding ABS and nylon) were looked on, in the 1970s, as the market segment with high growth rates, high margins, exclusive proprietary positions, and good patent protection. Many commodity plastic producers became interested in developing such products to counteract the effects of slowing growth in their more conventional products. These resins did grow at a rate of about 11%/year, but this rate slowed to 4.7%/year in the last 5 years of the period. Volume was only 775 million lb in 1982, down 7% from 1981. In addition, the area became crowded with new suppliers, foreign and domestic, and alloys of existing engineering plastics are proliferating.

The engineering thermoplastics should reach 2,000 million lb in sales by 1988 and will provide an important part of the economic growth of the plastic industry because they are able to compete with metals better than the commodity plastics. Growth and profitability for engineering plastics will continue to be above that of the commodity plastics despite increasing entries.

Thermosets (6)

The major thermosetting resin with the exceptions of epoxies and ureas, already showed signs of maturity before the 1980s. The growth rate from 1972 to 1980 for these thermosetting resins was about 2%/year (7). Like PVC, these resin sales are closely related to the building and construction industry for bonding and adhesives. The improving housing market should bring this segment back to a 4%/year growth rate in 1982–88, which would mean 5.9 billion lb consumption in 1988.

Other Resins (6)

The polyurethanes, also like PVC, have had their share of problems with fire-related controversies and have suffered a 2%/year loss in sales over the past 5 years. The markets for flexible foam are close to saturation, but reaction injection molding (RIM) and rigid foam should have better growth rates. Flexible foam usage to 1988 should show a 5%/year growth, one-third of which will be due to RIM automotive usage. Growth in rigid foams should be in the 4.7%/year range. This would mean a total volume of 2,155 million lb in 1988.

There are many other resins, not specifically mentioned above, whose volume will remain roughly constant, or growing at 1%/year, over the 1982–86 period. Domestic growth will probably increase at a rate of 6-1/2%/year, but imports are expected to drop 12%/year which will result in little or no growth.

Summary

These figures are summarized in Table IV which gives an estimate for plastic resin usage from 1982 to 1988. This represents a 6.2%/year growth, but 1982 was a depressed sales year. Growth would be 4%/year if estimated from 1981. This rate is in line with expected

Table IV. 1982–1988 Estimated Growth in Consumption of Plastic Resins

Market Segment	1982	1988
Thermoplastic resins (ETP)	26,837	39,145
Thermosetting resins	4,670	5,900
Engineering thermoplastics (ETP)	1,185	2,000
Polyurethanes	1,483	2,155
Other resins	280	300
Total	34,455	49,500

Note: All values are in million pounds.
Source: Modern Plastics, SLI estimates

332   APPLIED POLYMER SCIENCE

growth in such market segments as transportation, building and construction, electrical/electronics, housewares, and toys. Only business machine housings are expected to see above average growth. Packaging and furniture growth are expected to be below average, and exports are expected to drop significantly.

Given the modest growth predicted to 1988, approximately 4.3%, the market value of plastic resins should increase from $18.1 billion (8) to $36.4 billion by 1988 in current 1982 dollars on the basis of a 10% increase in prices in 1983 (9) and an 8.5%/year increase per year thereafter (assuming 5%/year inflation). This is a real growth of 12.5%/year, and, compared to a real GNP growth of 3.5%/year, the plastic industry market value of plastic resins should grow at a 3.6 ratio to the GNP. This ratio compares to a negative ratio in the 1977–82 period and a 4.7 ratio for the 1972–77 period.

In summary, the growth of plastics will be slower than the 12.1%/year during 1960–70 and 7.1%/year for 1970–80 (8). Low-priced feedstocks and new fast production techniques will no longer be growth factors. Technical advances will be the major growth factor, and new technical developments in specialty plastics will be needed to maintain export markets. There will be more competitition from abroad in commodity plastics, and the demands of a more sophisticated fabrication industry will be great. The resin suppliers who successfully weather the 1980s will have to be innovative and efficient to a degree not previously required.

Literature Cited

1. Loos, K. D. "Economic Recovery: Oasis or Illusion," Ninth Plastic Planning Conference, New York, 1983.
2. Peppin, A. "The U.S. Styrenics Business," Ninth Plastic Planning Conference, New York, 1983.
3. Pastor, A. J. "Polyethylene," Ninth Plastic Planning Conference, New York, 1983.
4. Driscoll, J. J. "Polypropylene: A Light in the Sea of Darkness," Chemical Marketing Research Association: New York, 1982.
5. Disch, G. E. "PVC: What's New For Whom?," Ninth Plastic Planning Conference, New York, 1983.
6. Chem. Eng. News 1983, 33.
7. Predicasts, Inc., Cleveland, OH
8. U. S. Industrial Outlook 1983, U. S. Department of Commerce
9. Loos, K. D. Ibid.

Polyolefins

FRANK M. McMILLAN

99 Tara Road, Orinda, CA 94563

A good reason to look at history is to select those parts we should like to repeat. One part of science history that both polymer chemists and businessmen would surely like to see emulated is the unprecedented, explosive burst of creativity, invention, and successful development that occurred in the 1950s and gave the world a new class of polymers (stereoregular), a family of new plastics (linear and stereoregular polyolefins), a family of new synthetic rubbers, including the first duplication of a natural high polymer

0097-6156/85/0285-0333$08.25/0

(stereoregular polydienes), a new class of catalysts (Ziegler/Natta coordination catalysts), and a new nomenclature (isotactic, syndiotactic, atactic, etc.). This decade of discovery also spawned two Nobel prizes, thousands of patents, and thousands of scientific papers, not to mention dozens of lawsuits.

Because the stereoregular synthetic rubbers are the subject of a separate chapter, this discussion will be concerned only with the polyolefins. Although the newer and more exciting members of this family are stereoregular polyolefins, the first member (high-pressure polyethylene) is not a stereoregular polymer; moreover, the first synthetic stereoregular polymer (polyvinyl isobutyl ether) is not even a polyolefin.

Polymer science may be said to have begun in the 1920s with the Svedburg's pioneering work with the ultracentrifuge and Herman Staudinger's revolutionary theory that high polymers consist of giant, long-chain molecules rather than colloidal aggregates. After a decade of challenge and even ridicule, Staudinger's basic idea gradually was accepted, and further progress became possible.

An important advance was the recognition that cohesion in high polymers depends primarily on the entanglement of long chains and that additional strength and hardness are gained through any of three supplemental mechanisms (Figure 1): van der Waals attractions between polar groups, formation of crystalline domains among well-fitting chains, and chemical cross-linking.

Discovery of High-Pressure (Low-Density) Polyethylene

Had these and other basic relationships of composition and chain configuration to polymer properties been well understood fifty years ago, we could visualize a polymer chemist of that day--a day when Bakelite was well known but polystyrene and poly(vinyl chloride) (PVC) were novelties--setting out deliberately to design and synthesize a new high-utility, general-purpose, low-cost plastic with the aid of these principles.

He might have reasoned thusly: first, the polymer must have a high enough molecular weight for long-chain entanglement (but not so high as to make processing difficult). Secondly, strength and toughness should be added, not by introducing polar groups (for they would add to weight and to volume cost and would be apt to impair thermal stability), nor by introducing chemical cross-links (for that would require curing cycles, would probably introduce color, and would produce nonrecyclable scrap), but by the formation of crystalline domains. The advantages include no additives, no color problems, relatively sharp melting/molding range, fast and reversible molding by heating and cooling, and no extra cost. Properties could be varied considerably by varying the size and number of the crystalline domains, controlled in turn by the degree of regularity and hence the fit between adjacent chains.

Having forsworn polar substituents, the prescient polymer chemist would quickly see that by elimination, the logical composition would be a polymeric hydrocarbon. The chains would then have to be very smooth and well-fitting to achieve crystallization, because the forces of attraction between chains would be quite weak. The smoothest chain would be one of methylene groups. On consulting the literature, the chemist would be encouraged by the fact that the straight-chain paraffins become crystalline solids as soon as the chain length gets above 10 carbons, and he would note that polymethylene had, in fact, been made in the previous century by von Pechmann and others (1) by decomposing diazomethane (Figure 2). Thus encouraged, he would seek an economically feasible route to approximate such a polymer, and he would be led inevitably to ethylene as the most practical monomer. The remaining problem being how to force the relatively unreactive ethylene molecule to polymerize to a high polymer, he would logically turn to elevated temperatures and extreme pressure to force the reaction in the desired direction (Figure 2).

Is this a reasonable approximation of the manner in which high-pressure polyethylene was actually discovered? Not in the remotest degree, as we all know. J. C. Swallow of Imperial Chemical Industries (ICI), who was personally involved in that historic discovery, made the following observation:

> "As time goes on the story of inventions and discoveries tends to become idealized and overrationalized and presented in a way which suggests that everything follows logically from the first experiment. History shows that this rarely happens and that chance always plays a big part. What does, however, seem to be important is the significance of the recognition of a discovery which may often be more important than the discovery itself."(2)

We shall see how frequently the same comment could be made about subsequent discoveries.

Without repeating in detail the often told story of how this historic "first" in polyolefins came into being, let us note that it started with a bold decision on the part of ICI to explore the new frontiers of ultrapressure reactions opened up by the pioneering work of Bridgeman in the United States and Michels in Holland. Among 50 reactions tried by the ICI research group without success was the alkylation of benzaldehyde with ethylene, suggested by Sir Robert Robinson (3) and attempted by R. O. Gibson in March 1933. The intended reaction did not take place, but a trace of ethylene polymer was formed, thanks (as it was learned afterwards) to the fortuitous presence of peroxide in the benzaldehyde. Successful duplications of the polymerization had to await a second "happy accident"—a leak that resulted in the introduction of excess ethylene containing enough oxygen impurity to act as catalyst.

Although fortuitous circumstance played a key role more than once in these and subsequent events, full credit is due Fawcett, Gibson, Swallow, and others of ICI (4) for launching the research and for being alert enough to recognize the significance of the unexpected results and aggressive enough to pursue the lead to a successful conclusion. As a result of their efforts, polyethylene was available to insulate cables for radar and thus help win the Battle of Britain and World War II. The phenomenal postwar growth of polyethylene, resulting from the advantageous properties noted in the "predictions" made earlier, have made it a multibillion-pound plastic and a household word around the world.

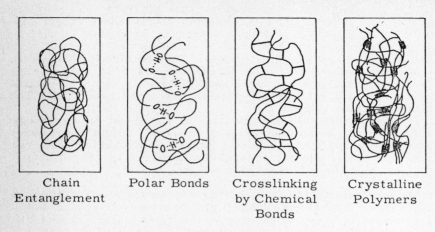

| Chain Entanglement | Polar Bonds | Crosslinking by Chemical Bonds | Crystalline Polymers |

Figure 1. Source of cohesion in long-chain polymers.

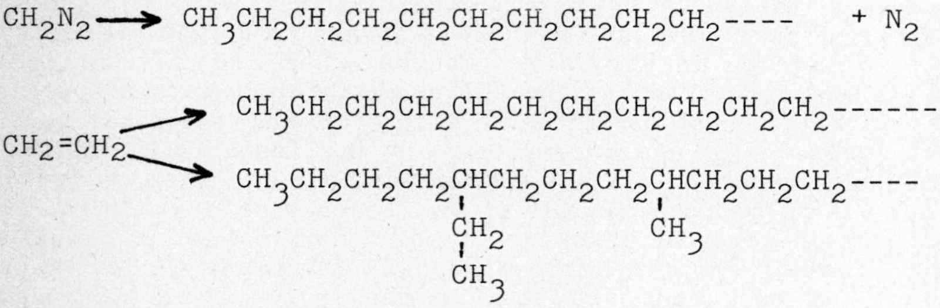

$$CH_2N_2 \longrightarrow CH_3CH_2CH_2CH_2CH_2CH_2CH_2CH_2CH_2CH_2 \text{----} \quad + N_2$$

$$CH_2=CH_2$$

$$CH_3CH_2CH_2CH_2CH_2CH_2CH_2CH_2CH_2CH_2CH_2 \text{------}$$

$$CH_3CH_2CH_2CH_2CHCH_2CH_2CH_2CHCH_2CH_2CH_2 \text{----}$$
$$\hspace{3.5cm} CH_2 \hspace{2.3cm} CH_3$$
$$\hspace{3.5cm} CH_3$$

Figure 2. Formation of polyethylene by decomposition of diazomethane and by polymerization of ethylene.

Polyisobutylene

Polyisobutylene is not usually even mentioned when polyolefins are discussed. Yet it is obviously a polyolefin, and it ranks at least a close second in the chronology of discovery. In the early 1930s, chemists at the I. G. Farben laboratories in Germany "stumbled on" the finding that BF_3 would explosively polymerize isobutylene to a "snowball" of high polymer (5). But the polymer is not stereoregular (no asymmetric carbon atom), crystallizes only slightly even when stretched, and has no unsaturation; consequently, it is neither a hard, crystalline plastic nor a vulcanizable elastomer. It has therefore found only limited commercial use. Copolymerization with a small amount of diene (butadiene or isoprene) introduces residual unsaturation and forms the well-known and highly useful vulcanizable synthetic rubber, "butyl rubber."

Discovery of Linear (High-Density) Polyethylene

Coming back to our hypothetical prescient polymer pioneer, we shall assume that, had he in fact invented high-pressure polyethylene, he would not have rested on his laurels, even after contributing to winning the war and creating one of the world's most useful plastics. Instead, he would doubtless perceive that a valuable variant of polyethylene would be one with more and stronger crystalline domains, because these would add hardness and strength and raise the melting point. He would by now have found out that high-pressure polyethylene forms imperfect crystals because the chains carry a number of short branches (Figure 2) so the right course would be obvious. He must find a way to polymerize ethylene without forming branches. Logically, this would require mild conditions and hence an exceptionally active and selective catalyst. This sort of reasoning would have formed a logical prelude to the discovery of linear polyethylene; but of course in actual fact it had nothing whatever to do with that discovery.

In 1953, Karl Ziegler, after many years of research on organometallic compounds, had discovered the "Aufbau" reaction, whereby ethylene was converted to low polymers (oligomers) by the action of aluminum alkyls (6). The polymers were strictly linear, but the potential significance of this fact was largely overlooked by the scientific and technical community. Then one of Ziegler's graduate students, Holzkamp, discovered through accidental contamination of the reaction system with nickel that this metal acted as a cocatalyst to stop the reaction at the dimer (1-butene) stage.

After that, according to Ziegler, it was simply a matter of systematically working through the periodic chart until a different cocatalyst was found that had the opposite effect: suppressing the displacement reaction and thus favoring continued chain growth and the production of high polymers. The best cocatalyst proved to be titanium, and thus was born the Ziegler catalyst system (titanium chloride and aluminum alkyl) and Ziegler polyethylene (7).

Fortuitous circumstance had once again rewarded alert observation and diligent follow-up. A critical factor, impressive if one considers the time and circumstances, was that Ziegler immediately recognized that his first crude polymer had properties distinctly superior to those of the well-known high-pressure polyethylene. On the strength of this fact and the allure of a low-pressure process, he undertook a vigorous program of patenting, publishing, and publicizing his discovery, with results that made him a millionaire Nobel laureate and Ziegler polyethylene an important commercial product in many countries. Unfortunately for Ziegler, who wrote his own patent application, he confined his coverage to ethylene—an expensive mistake that led to disagreements and litigation when stereoregular polyolefins were discovered.

Perhaps Ziegler's greatest good fortune was in timing. It seems no great feat today to recognize the merits of an all-linear polymer, but the 1950s were about the earliest years when the instant enthusiasm that greeted Ziegler could possibly have occurred. It is not surprising that von Pechmann's linear polymethylene roused no commercial interest and scarcely any scientific curiosity at the turn of the century; however, it is significant that Marvel made linear polyethylene with a metal alkyl catalyst in 1929, and that nothing came of it (8). Two major chemical companies looked at that process at that time and concluded that it had no commercial possibilities.

Ziegler's timing was also fortunate in bringing out a new product at a time when the chemical industry was in an aggressive, expansionist mood. Had he been a few years earlier, that wave would not yet have crested; had he been even a year or two later, he might have lost out to the other linear, low-pressure polyethylene processes that were developed independently and contemporaneously.

By a striking coincidence, during the same period (1953), Hogan and Banks of Phillips Petroleum were endeavoring to find a practical way of making gasoline-range hydrocarbons from ethylene. Like Ziegler, they started from known dimerization reactions and used catalyst systems such as cobalt on charcoal. In testing many combinations, they were nonplussed to find that one catalyst, chromium oxide supported on silica-alumina, caused the reaction tube to become plugged with a white solid.

Other workers had made similar observations earlier, but had discarded their plugged tubes and reported the experiment as a failure. Hogan and Banks did not discard their plugged tubes, but worked up the polymer to see what they had. What they had was linear polyethylene of almost perfect linearity (9).

In still another parallelism of timing and result, another research team at Standard Oil was investigating the same starting catalyst, cobalt on charcoal, at about the same time (actually, somewhat earlier [1951]), although for a different reaction. Their work led to still another accidental discovery of linear polyethylene (10).

Zletz was attempting to use the supported cobalt catalyst for alkylation with ethylene and found, to his surprise, that considerable solid polymer was formed. Like Hogan, he was not a trained polymer chemist, but had enough curiosity, initiative, and freedom to pursue the interesting bypath. Other metals and other supports were tested by Zletz and his coworkers and led eventually to improved catalysts such as molybdenum on alumina that formed the basis for development of a practical low-pressure polyethylene process.

These two competitive processes, discovered independently of
each other and of Ziegler, were developed aggressively by Phillips,
less so by Standard. The Phillips process is now the world's
leading producer of linear polyethylene. Once again, it may be said
that fruitful innovation was the result of an unexpected result
encountering a prepared mind.

Use of Comonomers

In the course of extensive market development, it was discovered
that perfect chain regularity, an ideal that could be approached
quite closely with the Phillips process, is not the ideal morphology
for commercial grades of polyethylene. Painful experience, to be
described more fully later (see section on Properties and
Applications) taught that a small and controlled amount of chain
irregularity is required to achieve the optimum degree of
crystallinity and the most useful balance of properties. This
condition was accomplished by the incorporation of small amounts of
a higher olefin, such as 1-butene, as comonomer, to introduce a
certain amount of short-chain branches that interfere with
crystallization to the desired degree.

Linear Low-Density Polyethylene

After reviewing the history of the development of polyolefin
processes, a thoughtful polymer chemist might decide that a
worthwhile goal would be to develop a product that would combine the
best features of low-density and linear, high-density polyethylenes,
or at least one that would be intermediate in properties between the
two extremes. A logical approach would be to attempt to employ a
Ziegler- or Phillips-type catalyst in the high-pressure process to
convert it from a free-radical to an ionic initiator. A change in
the type and amount of branching of the polymer chains might be
expected to result.

Whatever the inspiration, that exact result has, in fact, been
achieved. The Union Carbide Company was the first to announce in
the United States (in 1977) a "linear, low-density polyethylene"
(LLDPE) process and product, although somewhat similar products had
been made for some years by Du Pont in Canada. A competitive
process to the Carbide process was soon announced by Dow Chemical
Company. Several other companies in the United States and Europe
have subsequently either taken licenses from Carbide or developed
their own LLDPE processes and have either begun or planned
production. As of 1980, the status of various processes for
production of LLDPE throughout the world was reported to be as shown
in Table I.

In addition to Carbide and Dow, four other companies are
producing LLDPE in the United States: Exxon, Mobil, National
Petrochemicals, and El Paso (Table III). Others who have been
reported to be doing development work include ARCO, Cities Service
(now Occidental), Phillips, Soltex, and USI. Several additional
plants and plant expansions are expected in the near future.

A striking (and ironical) feature of the LLDPE development is
that it has given "Ziegler polyethylene" a new lease on life.
Whereas the Phillips process all but displaced the Ziegler process

Table I. Linear Low-Density Polyethylene Processes (<u>34</u>)

LLDPE Process Type	Status	Developing Companies	Pressure $(lb/in.^2)$	Catalyst[a]
Gas Phase Fluidized Bed	Commercial	Union Carbide	300	Cr
	Pilot Plant	Naphtachimie	350	Z
Gas Phase Stirred Bed	Pilot Plant	Amoco	300	Z
	Pilot Plant	Cities Service	300	Z
Liquid Phase Slurry	Commercial	Phillips	450	Cr
	Pilot Plant	Solvay	400	Z
Liquid Phase Solution	Commercial	Du Pont, Canada	400–600	Z
	Commercial	Dow	400–600	Z
	Commercial	Mitsui Petro-chemical	600	Z
	Pilot Plant	DSM	600	Z
High Pressure	Commercial	CdF	20,000	Z

[a]Z=Ziegler type, Cr=Chromium type

for high-density polyethylene in most countries, the majority of the
new (or converted old) LLDPE plants will use Ziegler-type catalysts.
   LLDPE is generally intermediate in properties between conventional low-density polyethylene (LDPE) made by high-pressure
processes and linear, high-density polyethylene (HDPE) of the
Ziegler or Phillips type. A qualitative comparison of the more
significant physical properties is given in Table II (some quantitative data are given in Table IV).
   The property differences cited in Table II are attributed
primarily to differences in the degree and kind of branching in the
polymer chains. LLDPE is reported to have fewer but longer branches
than LDPE. As with HDPE, the branching is further modified by the
introduction of higher α-olefins as comonomers (typically, about 8%
of 1-butene, 1-hexene, or 1-octene).
   LLDPE has been enthusiastically received in the marketplace
because of a number of favorable factors. One factor is the
advanced state of polyethylene technology, which made it easy to
recognize the advantages of the new product and to pursue its best
applications.   The originators were leading producers of
conventional polyethylene and were in good positions to mount
aggressive development, manufacturing, and marketing programs. Also
favorable was the possibility to convert some existing polyethylene
(and even polypropylene) plants to LLDPE production, with obvious
savings in time and costs. These factors, combined with an
attractive balance of physical properties, good processing
characteristics, and the availability of the process for licensing,
have resulted in rapid build up of production capacity, which in
1982 reached a total of 2.5 billion lb/yr in the United States.
Great optimism is evident in this figure because demand, although
growing at a record pace, is running behind capacity by a factor of
about two.
   Much of the growth in LLDPE demand will be at the expense of
other polyolefins, especially conventional high-pressure, LDPE. Its
proponents claim that LLDPE can replace LDPE in 70 to 80% of its
markets and that "the last conventional polyethylene plant has
been built." Another sign of the times is the fact that ICI,
originator of the high-pressure process, has now ceased production.
   Faced with shrinking demand for the conventional products,
several LDPE producers have undertaken production of special
copolymers (some using vinyl esters as comonomers) and modified
grades that more closely approach LLDPE in properties. Thus, with
the additional capabilities and stimulation provided by the LLDPE
processes, it has become possible to manufacture polyethylenes with
almost any desired degree and type of branching, molecular weight,
and molecular weight distribution. Control over these variables
enables products to be "tailor made" for optimum performance in a
host of different applications. As a consequence, the borders
between different types of polyethylene are becoming indistinct, and
the remaining gaps in the ranges of attainable properties are being
filled in. Much of the reported capacity for polyethylene
production must now be considered convertible from one type to
another.

Stereospecific Polymerization

If our imaginary chemist who could have achieved this sort of result by logical processes had actually done so, he would surely by now have gone as far as he could with ethylene. But if he read the literature thoughtfully, it would have impressed him that several of the pioneers of polymer science, starting with Staudinger in 1932 and later Huggins (11), Schildknecht (12), and Flory (13), had each recognized the possible existence of an additional degree of order, and hence of crystallizability, in high-polymer chains containing asymmetric carbon atoms. They thought that such substituted ethylene polymers as poly(vinyl bromide) (Figure 3) ought to be crystalline if each polymer chain were in an "all d" or "all l" steric configuration and that, because the known polymers were, in fact, noncrystalline, they must be in random order.

Schildknecht first made a new crystalline high polymer and ascribed its crystallinity to steric regularity. In 1947 he reported that poly(vinyl isobutyl ether) is either amorphous or highly crystalline, depending on conditions of polymerization, and he boldly proposed that the crystalline form had a regular steric structure (Figure 4): either dddddddddd or more likely (he thought), dldldldldldl (14). It was shown later by Natta (15) that the "all-d" structure (now known as isotactic) is the correct one.

With all these clues and portents in mind, it would have been but one more logical step for our imaginary chemist to conceive that a hydrocarbon polymer containing asymmetric carbon atoms could also exhibit a regular steric structure. Adjacent chains would then fit together exceptionally snugly to form stronger and higher melting crystals than those in linear polyethylene. Such a polymer would have correspondingly higher strength, hardness, softening temperature, and solvent resistance.

It would then have been logical for the chemist to settle on propylene as the simplest and most readily available monomer that could form such a polymer (Figure 5). It would have remained only to try to adapt a Ziegler-type catalyst or some other type that might overcome the tendency of propylene, in common with all allylic compounds, to form only low polymers.

Discovery of Polypropylene

Once again, in the real world, fortuity and acuity played far greater roles than cold reason. Natta had made an agreement that gave him access to Ziegler's new chemistry and catalysts, and he promptly set one of his chemists, Chini, to try Ziegler catalysts on propylene (Ziegler had already tried propylene, but under the conditions used the results were so poor that he was led to conclude, prematurely, that the higher olefins "would not work"). Natta was, of course, hoping for a high polymer, but what he expected was an amorphous, rubbery product, because the concept of stereospecificity was not in his mind at the beginning. In any event, he and his company sponsor, Montecatini, were at the time more interested in synthetic rubber than in plastics.

It was a great surprise, therefore, that the product of Chini's very first experiment, made on March 11, 1954, consisted mostly of a hard, highly crystalline thermoplastic. Natta made a simple,

Table II. Differences Between Linear Low-Density, High-Density, and Low-Density Polyethylenes (34, 35)

Property	LLDPE vs. LDPE	LLDPE vs. HDPE
Tensile strength	higher	lower
Elongation	higher	higher
Impact strength	higher	similar
Heat resistance	higher	lower
Stiffness	higher	lower
Warpage	lower	similar
Processability	harder	easier
Haze	worse	better
Gloss	worse	better
Clarity	worse	better
Melt strength	lower	lower
Melting range	narrower	narrower

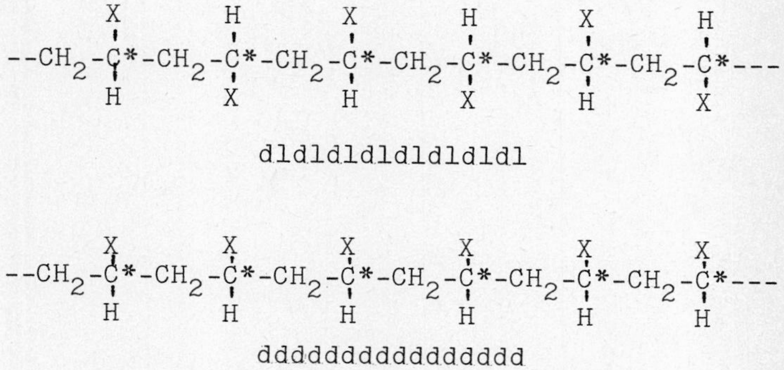

$$--CH_2-\overset{X}{\underset{H}{C}}*-CH_2-\overset{H}{\underset{X}{C}}*-CH_2-\overset{X}{\underset{H}{C}}*-CH_2-\overset{H}{\underset{X}{C}}*-CH_2-\overset{X}{\underset{H}{C}}*-CH_2-\overset{H}{\underset{X}{C}}*---$$

dldldldldldldldl

$$--CH_2-\overset{X}{\underset{H}{C}}*-CH_2-\overset{X}{\underset{H}{C}}*-CH_2-\overset{X}{\underset{H}{C}}*-CH_2-\overset{X}{\underset{H}{C}}*-CH_2-\overset{X}{\underset{H}{C}}*-CH_2-\overset{X}{\underset{H}{C}}*---$$

dddddddddddddddddd

Figure 3. Stereospecific structures possible with polymer chains containing asymmetric carbon atoms ($R = i-C_4H_9$).

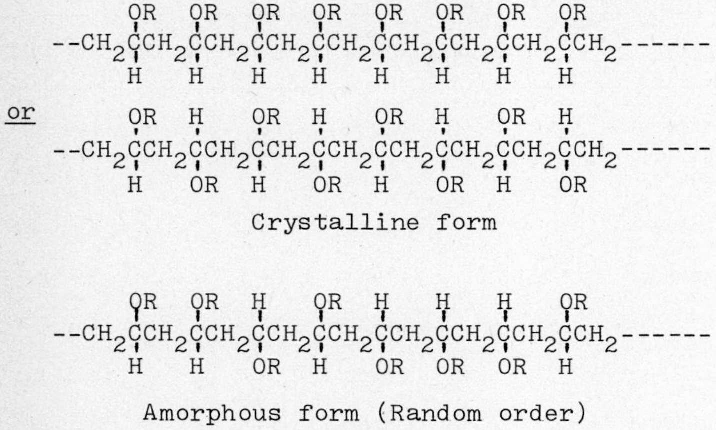

```
        OR   OR   OR   OR   OR   OR   OR   OR
      --CH2CCH2CCH2CCH2CCH2CCH2CCH2CCH2CCH2------
         H    H    H    H    H    H    H    H
or       OR   H    OR   H    OR   H    OR   H
      --CH2CCH2CCH2CCH2CCH2CCH2CCH2CCH2CCH2------
         H    OR   H    OR   H    OR   H    OR
```

Crystalline form

```
        OR   OR   H    OR   H    H    H    OR
      --CH2CCH2CCH2CCH2CCH2CCH2CCH2CCH2CCH2------
         H    H    OR   H    OR   OR   OR   H
```

Amorphous form (Random order)

Figure 4. Possible steric structures for polymers of vinyl isobutyl ether.

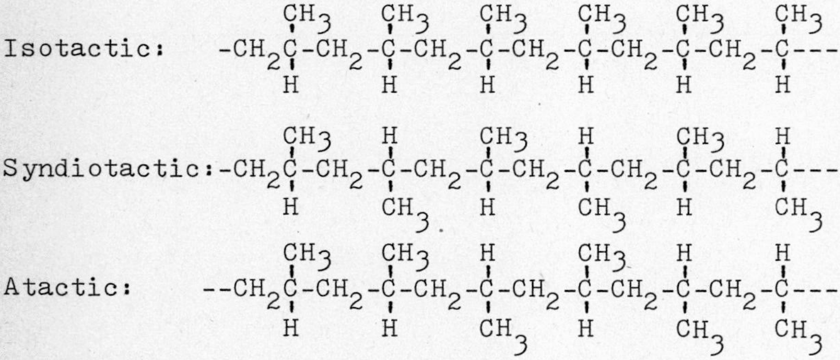

```
                  CH3   CH3   CH3   CH3   CH3   CH3
Isotactic:     -CH2-C-CH2-C-CH2-C-CH2-C-CH2-C-CH2-C---
                  H     H     H     H     H     H

                  CH3   H     CH3   H     CH3   H
Syndiotactic:-CH2-C-CH2-C-CH2-C-CH2-C-CH2-C-CH2-C---
                  H     CH3   H     CH3   H     CH3

                  CH3   CH3   H     CH3   H     H
Atactic:       --CH2-C-CH2-C-CH2-C-CH2-C-CH2-C-CH2-C---
                  H     H     CH3   H     CH3   CH3
```

Figure 5. Polypropylene.

laconic entry in his personal notebook: "Today we made
polypropylene." But his group lost no time in determining that the
crystallinity resulted from the remarkable fact that polypropylene
had a stereoregular, "all-d" structure, which Natta, at his wife's
suggestion, christened "isotactic" (16). Another regular form, with
the side groups appearing alternately on opposite sides of the main
chain (d1d1d1d1d1) was christened "syndiotactic," and the random
configuration was christened "atactic" (Figure 5).

Natta and his coworkers proceeded rapidly and brilliantly to
synthesize other stereoregular polyolefins (from 1-butene, styrene,
4-methyl-1-pentene, etc.) and to establish their chain
configurations and crystal habits. They found that the chain always
assumes a spiral configuration in the crystal, most commonly three
monomer units per complete revolution (the "threefold helix") (17).
This impressive body of work resulted in Natta's sharing the Nobel
Prize in chemistry with Ziegler in 1963 and in a race for commercial
production of polypropylene by numerous companies in many countries.

Completely isotactic polypropylene is too highly crystalline to
be useful for most applications. Depending on catalyst and
conditions, some amorphous atactic polymer is formed in all
practical commercial processes, and leaving a certain amount in the
product modifies the crystallinity to some extent. For applications
that demand very good impact strength and low-temperature
flexibility, however, something more is needed. The best approach
is again that taken with linear polyethylene: the introduction of
small amounts of higher α-olefins as comonomers. Copolymerization
with small amounts of ethylene is also used to improve impact
strength. As with all commercial polymers, nonpolymeric additives
are also employed to improve such properties as stability, clarity,
and processing characteristics.

Manufacturing Processes

A recent review of polyolefin production processes (18) provides
information on types of equipment used and operating conditions.
Summarized in the following sections are the major processes in
current use.

High-Pressure LDPE. The original ICI (and, later, Du Pont) process
uses steel autoclave reactors, operating at pressures ranging from
15,000 to 40,000 psig. and temperatures ranging from 150 to 250 °C,
with oxygen or a peroxide as a (free-radical) polymerization
initiator.

The Union Carbide Company in the United States and BASF
(Badische) in Germany developed processes using tubular reactors.
These processes give products that differ in molecular weight and
branching from the ICI process.

Low-Pressure, Linear HDPE. The Standard of Indiana process is a
solution process that uses a molybdena-on-alumina catalyst and
operates at pressures up to 1000 psig. and temperatures up to
300 °C.

The original Phillips process is a solution process that
operates in stirred reactors and employs a suspended solid catalyst.
The polymer solution is filtered for catalyst removal and then

stripped for solvent removal. The later Phillips "particle form" process is a slurry process in which the polymer precipitates as it forms. This process uses a circulating-loop reactor. Because of improved catalyst use efficiency, catalyst removal from the polymer is unnecessary.

The Ziegler process uses a finely divided catalyst (titanium halide and an aluminum alkyl) in a solvent and operates at temperatures ranging from 50 to 120 °C and pressures ranging from 150 to 300 psig. The product is treated to deactivate the catalyst and remove residues.

Gas-phase processes (monomer vapor polymerized in absence of liquid phase) have been developed by BASF and a number of other companies. These processes depend on high catalyst efficiency to obviate catalyst deactivation and removal.

LLDPE. Few process details have been revealed by the manufacturers, but from patent disclosures and other information it is apparent that LLDPE is made by a variety of processes: solution, slurry, and gas phase. For the solution process, catalysts of the Ziegler type, for example, a combination of titanium trichloride with aluminum trichloride and an aluminum alkyl, are used. Reaction temperature is approximately 160 °C, and pressure is approximately 300 psig. Catalyst productivity is quite high, and catalyst removal is unnecessary.

Slurry processes may use a combination of organo-aluminum and organo-magnesium compounds with titanium tetrachloride. Phillips-type catalysts (supported chromium compounds) are also used. Pressures and temperatures are moderate.

A hybrid process (ICI) uses two stirred autoclave stages and adds a free-radical initiator in the first (high-pressure) stage and a Ziegler-type catalyst in the second (lower-pressure) stage.

Polypropylene. Polypropylene is usually manufactured by a slurry process that uses a titanium trichloride-aluminum alkyl catalyst in a stirred reactor at temperatures of 60 to 90 °C and pressures of a few hundred pounds per square inch. The product slurry is flashed to remove propylene and treated to deactivate the catalyst and remove residues. If necessary, amorphous polymer is removed by solvent extraction.

Solution, bulk (no solvent) and gas-phase processes have also been developed and are in commercial use. Recent improvements in catalyst activity have reduced requirements to a point such that catalyst removal is no longer necessary in some processes.

A recently announced development promises to extend the technology used in making LLDPE to polypropylene manufacture. That objective is being pursued in a joint program of Shell Chemical, the second largest U.S. producer of polypropylene, and Union Carbide, one of the originators of LLDPE. The process operates in gas phase with a highly efficient catalyst and is expected to achieve substantial savings in capital and operating costs.

In 1984, this project had reached pilot plant stage, with commercial production targeted for the following year in a converted high-pressure polyethylene plant.

Other Polyolefins. Virtually every available α-olefin has been polymerized with Ziegler/Natta-type catalysts. The polymers are of considerable scientific interest (17), but few have achieved commercial status. The following olefins are of particular interest.

Poly(1-butene). The isotactic polymer is not particularly high melting (125-130 °C) (19, 24), but is exceptionally strong and tough. It is manufactured on a relatively small scale by Shell Chemical in the United States and by Hüls in Germany (18). Its applications include heavy-duty bags and piping. The polymer exhibits polymorphism (20) and undergoes a gradual phase transition that complicates its application technology. An examination of its unusual stress-strain curve suggests potential utility as a leather substitute (perhaps in either fiber mat or reticulated foam form), but this possibility apparently has not been actively pursued.

Poly(4-methyl-1-pentene). The isotactic polymer, as described by Natta (21), Sailors and Hogan (18), and Schildknecht (25), is a hard, high-melting, transparent, and chemically resistant plastic. It is manufactured by ICI in Japan (18) and is sold for specialty uses, notably laboratory and medical "glassware."

Commercial Production

Polyolefin manufacturing plants now exist in about 45 countries (18). As of early 1984, the United States had 11 each of high-pressure LDPE producers; linear HDPE producers; and polypropylene producers. In addition, there are eight producers of LLDPE with several more plants planned or under construction, and one producer of poly(1-butene). A (January 1984) list of producers and plant capacities is given in Table III. However, any such enumeration is rendered obsolete immediately by the numerous changes in corporate structure, strategies, and manufacturing capabilities that characterize the polyolefin industry today and for the near future. In addition to the expansions, conversions, and closings announced by current producers, several other companies are known to have plans for entering polyolefin production.

   Growth curves for LLDPE, polypropylene, and HDPE are shown in Figure 6. Starting a year or two later than HDPE, polypropylene production has grown at a comparable rate, and LLDPE production, which started two decades later, is setting a similar pace.

   In Figure 7, the rates of growth of the three major polyolefins are compared with those of the other two largest volume thermoplastics: PVC and polystyrene. For this comparison, the growth curves were shifted on the time axis to a common starting point, and year zero for each product is the year in which it first attained a volume of 50 million lb.

Several striking facts are made evident by these comparisons:

1. The polyolefins as a class have by far the highest growth rate and greatest total volume of all thermoplastics. Recession-induced fluctuations aside, there is little evidence of leveling off of total demand.

Table III.   U.S. Polyolefin Producers and Capacities (January 1984)

Producer	Polyolefin Capacity (thousands of metric tons per year)				
	LDPE	HDPE	LLDPE	Polypropylene	Total
Allied		320			320
American Hoechst		110			110
Amoco (Std. of Ind.)		160		230	390
ARCO (Atlantic Richfield)		155		180	335
Chemplex	190	130			320
Dow	440	170	315[a]		925
Du Pont	320	405			725
El Paso	170		15[b]	150	335
Exxon	265		270[a]	180	715
Gulf	410	260		180	850
Himont		15		590	605
Mobil	225		170[a]		395
Natl. Petrochemicals			250[a]		250
Northern Petro.	255		195[a]	100	550
Phillips		625		100	725
Shell				350	350
Soltex		340		90	430
Texas Eastman	180		100[b]	65	345
Union Carbide	180		610[a]		790
USI	330				330
USS Chemicals				235	235
Total	2965	2690	1925	2450	10030

[a] Plant can produce LDPE instead
[b] Plant can produce HDPE instead

(Reproduced with permission from Ref. 37.)

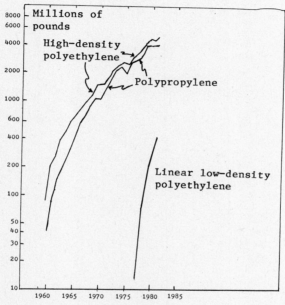

Figure 6.   U.S. production of high-density polyethylene, polypropylene, and linear low-density polyethylene.

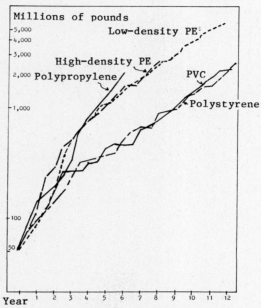

Figure 7.   Comparison of growth rates of major thermoplastics (Time axis adjusted to common origin).

2. The growth histories of the polyolefins have been remarkably similar, despite a spread of nearly half a century in the times of inception.
3. There is a near overlap of the growth curves for the polyolefins, in one class, and for PVC and polystyrene in another class. This coincidence is surprising in view of the disparities of properties, uses, and time scales, and suggests that there may be some controlling factors that are generally unrecognized.

Properties and Applications

Physical properties of the polyolefins are summarized in Table IV.
   Considerable variation is, of course, possible among different grades of the same polyolefin, and useful modifications are obtained by copolymerization, blending with other polymers, and compounding with various additives. As with other crystalline plastics, the polyolefins are not benefited by addition of liquid plasticizers. Such "foreign" molecules are generally not accepted by the crystal lattice, or, if incorporated, cause excessive weakening.
   Once the polymers have been made, the "chemistry" of polyolefins is essentially "nonchemistry," in the sense that everything is directed toward suppressing undesirable reactions. Thus, purification processes are applied to remove harmful catalyst residues, and a variety of stabilizers are added to prevent oxidation, metal-catalyzed decomposition, UV light degradation, thermal decomposition, etc. The effectiveness of these techniques can be gauged by the fact that unstabilized polyolefins cannot be fabricated in ordinary manufacturing equipment without significant degradation, and an unpigmented, unstabilized film will become brittle in a few days of exterior exposure; yet when suitably compounded formulations are used for applications such as indoor-outdoor carpeting or molded stadium seats, they withstand years of such exposure.
   The easy processability of the polyolefins permits ready fabrication by nearly all conventional processes. This versatility, plus their low cost, low density, and excellent balance of physical properties, gives them wide applicability in many different end-use fields. We now encounter polyolefins in our daily lives in the form of fibers (e.g., indoor-outdoor carpeting, underwear, and upholstery), film (packaging), blow-molded containers (squeeze bottles, detergent bottles), injection molded articles (housewares, appliance housings, auto parts), and extrusions (wire insulation, pipe). Statistics on major applications and accounts of new developments are published regularly (22).
   But what are not apparent in the published descriptions are the countless and grievous problems encountered and solved by applications research groups and process engineers to bring each new polyolefin into successful production and use. For example, polyethylene is subject to so-called "stress cracking" or "stress corrosion," which is worse the higher the degree of crystallinity. Linear polyethylene was also found to be prone to severe warping; moreover, moldings would even disintegrate if exposed to hot air for some hours. These problems reached crisis proportions in the early marketing phase and nearly caused abandonment of some polyethylene

Table IV. Typical Properties of Polyolefins

Property	High–Pressure LDPE	Linear HDPE	LLDPE	Polypropylene
Specific gravity (g/cm^3)	0.92	0.96	0.93	0.90
Modulus of elasticity (tens.) (10^5 lb/in.^{2})	0.2	1.0	NA	1.5
Tensile strength (lb/in.^{2})	1500	4000	3000	5000
Impact strength (ft–lb/in.)	very high	2	7	1
Softening temp. (°C)	100	120	110	150
Max. service temp. (°C)	80–90	110–120	90–105	130–140
Brittle temp. (°C)	very low	–90	–100	–10
Elec. resistance (ohm–cm)	10^{16}–10^{17}	10^{15}–10^{16}	NA	10^{16}–10^{17}
Stress crack time	21 days	3 min	24 h	high
Clarity	translucent	semi opaque	translucent to transparent[a]	translucent to transparent[a]

[a]oriented or nucleated

Note: Properties vary widely with different grades of each polyolefin.

ventures. The day was saved temporarily, though inelegantly, by the
onset of the "hula-hoop" craze, which provided for a time a
noncritical outlet for polymer. Eventually, permanent solutions
were found by the polymer chemists, who varied molecular weight
distribution and introduced a second monomer to modify crystallinity
and thus modify behavior.

Purists may have been disappointed to learn that the theoretical
ideal of a monodisperse, "chemically pure" stereoregular polymer is
not the practical ideal. But we are all beneficiaries of the
modifications and even "adulterations" that alleviated the
deficiencies that would otherwise have kept the pure polymers from
becoming useful plastics.

A fascinating bit of scientific detection was provided by
electron microscope studies of crystallites carried out at Bell
Laboratories by Keith and Padden (23). Their high-magnification
electron micrographs of very carefully prepared samples of highly
crystalline polyethylene revealed that the spaces between the
dendritic crystallites contain many thin filaments stretching from
one crystal to another. These appear to be bundles of oriented
molecules drawn out into fibers as material moved into the crystallites. Such filaments, or "tie molecules," are presumably what
holds the crystallized material together, and it is probably their
scission by oxidation that caused rapid loss of strength in highly
crystalline samples exposed to hot air.

Crises were also encountered when polypropylene was first
introduced: inherent high susceptibility to oxidation, relatively
high brittle temperature, void formation in moldings, etc. One by
one, these problems were surmounted by polymerization, processing,
or compounding variations worked out by numerous unsung heroes in
laboratories and plants. Their successes are recorded in the trade
and patent literature, but their enduring monument is the tremendous
growth history of polyolefin plastics.

Consumption of Polyolefins

Recent consumption patterns for polyethylene and polypropylene are
shown in Table V. Data for LLDPE and LDPE are combined in this
tabulation. Because of rapidly shifting market patterns and the
flexibility provided by dual-purpose plants, the ratio of LLDPE to
LDPE varies from time to time, but the long-term trend is strongly
upward.

Comparison of the total estimated consumption with production
capacities indicates that the polyolefin plants have been operating
at slightly over 80% of capacity. This represents some improvement
over previous years, and further improvement may result from the
conversion or closing of additional LDPE plants.

Certain fields of use are uniquely suited to polypropylene
because of its higher hardness, strength, and melting point, and
particularly because of its capability of being highly oriented by
drawing at temperatures just below the crystal melt point to produce
fibers, monofilaments, or oriented films. Drawing enhances and
realigns the crystalline domains so effectively that strength in the
draw direction is dramatically increased. In films, strength,
stiffness, and clarity are all improved by orientation, particularly
when it is two-dimensional.

Table V. U.S. Consumption of Polyolefins (1983) (36, 38)

Fabrication Method	Typical Products	Polyolefin Consumption (thousands of metric tons)		
		LLDPE,LDPE	HDPE	Polypropylene
Blow molding	Bottles, other containers, housewares, toys	20	817	39
Extrusion	Film (packaging and industrial), pipe, wire & cable, coated paper, fibers	2517	540	741
Injection molding	Containers, packages, housewares, toys, appliances, auto parts	289	548	539
Rotomolding	Large containers	43	25	
Other		298	248	284
Total U.S. Consumption		3167	2178	1603
Export		409	468	359
Total sales		3576	2646	1962
Total sales, all polyolefins		8184		

The melting point of polypropylene fiber is not quite high enough to permit ironing of fabrics; hence, its use in garments is limited, and the major fiber offtakes at present are in carpeting and upholstery. Another important "fiber" use is as a substitute for jute or other coarse fibers in woven bags. The "fibers" are made by drawing extruded, slit film until fibrillation occurs.

New markets are being created, and old ones expanded, at such rates that the exponential growth shown in Figure 7 seems likely to continue for some time. Stanford Research Institute has even predicted, on the basis of long-range extrapolations, that by the year 2000 U.S. production of polyolefins will reach the staggering total of almost 100 billion lb. It is predicted that polyolefins will then (as they do already) constitute well over half of all thermoplastics production.

New Monomers and Polymers

The original, simplest polyolefins, polyethylene and polypropylene, continue to dominate the scene, even after two decades, to such an extent that no other polyolefin even appears on the production charts. Nevertheless, a great many (we may assume all) available olefins have been tested, and many have been found capable of being converted to stereoregular polymers. As was mentioned above, poly(1-butene) and poly(4-methyl-1-pentene) are being offered commercially and may be expected to achieve significant volume in the future. Isotactic and syndiotactic polystyrene are of much theoretical interest (26) but are not yet commercial products.

As we have already seen, many of the deficiencies of the original stereopolyolefins have been corrected by judicious inclusion of small amounts of comonomers. The advancing techniques for making block and graft copolymers open extensive lines for additional improvements. One of the earliest and most noteworthy contributions in copolymerization is the ethylene-propylene copolymer rubbers (27). However, these have been largely superseded by terpolymer systems in which residual unsaturation has been introduced by inclusion of small amounts of a third monomer which is an unconjugated diolefin. These are no longer strictly polyolefins but fall clearly in the category of synthetic rubbers and are discussed in a separate chapter.

Growth of Understanding

On the scientific side, considerable progress has been made on two fronts: learning how the catalysts operate to make stereoregular polymers and learning how the properties of the polymers are controlled by chain shape, chain length, and, particularly, crystal habit.

Catalysts and Kinetics. Hundreds of variants and combinations of catalysts, cocatalysts, catalyst pretreatments, and reaction conditions have been discovered and described, mostly in the patent literature (28). It is now generally agreed that most coordination polymerizations are heterogeneous, but that some are clearly homogenous. The basic characteristic that distinguishes all Ziegler/Natta-type stereoregular polymerization catalysts is that

they involve coordination of the entering monomer with a transition
metal in an organometallic complex (typically, titanium trichloride
and aluminum triethyl), with monomer insertion occurring between the
metal atom and the growing chain. Thus, the polymer chain "grows
like a hair" (Figure 8) in marked contrast to free-radical
polymerization, in which monomer addition occurs at the opposite end
of the chain from the initiator.

The precise nature of the active sites and the exact mechanism
of the reaction are, after two decades of active research, still the
subjects of investigation and lively debate. The detailed
description of the polymerization process given 20 years ago by
Arlman and Cosee (29) still has current validity, although a great
deal more detail and some further insight have been added
subsequently. A review of catalyst structures and reaction
mechanisms is the subject of a separate chapter.

Structure vs. Properties. In the area of polymer behavior, knowledge of the effects of molecular weight distribution and of short-
and long-range branches has been developed for the polyolefins more
or less in common with other high polymers to which the same
principles apply. The factor of greatest interest and importance in
the morphology of polyolefins, however, is the crystalline domains,
because their size, number, structure, and transformations have a
dominant influence on rheology and physical properties. Here
progress has paralleled and benefited from the work with other
crystalline polymers, particularly polyethylene terephthalate
(PETP).

Since the pioneering (and initially controversial) work of
Keller (30), who made the startling proposal that the crystal axis
in the lamellae of oriented PETP fibers is perpendicular to the draw
axis, many workers have studied lamellar and spherulitic structures
in carefully prepared single crystals and in bulk samples. Much
light (if electron microscope illumination may be classed as light)
has been cast on how such crystals respond to stress and on how
their transformations are involved in the striking alteration of
physical properties that occur on uniaxial (fiber) or biaxial (film)
orientation. Studies such as those published by Geil (31) and the
work of Keith, et al. mentioned above (23), are the headwaters of a
stream of continuing contributions that has swelled to a volume
sufficient to support a specialized journal (32).

These explorers of the crystal domains have presented us with
fascinating pictures of a microworld full of macro surprises and
have brought our knowledge of this morphology to the point where it
should have predictive validity and should generate valuable new
technology. It is true of polypropylene as it is of metals that its
highly useful properties derive primarily from its microcrystalline
structure, and so do some of its deficiencies. Hence the ability to
modify that structure by stresses imposed under controlled
conditions, either external or internal--skill at manipulating
microcrystals--is a vital key to obtaining an optimum balance of
properties for a given application. A striking demonstration of
that fact is afforded by a comparison of isotactic with atactic
polypropylene, which is entirely amorphous and for which no large-
volume practical use has yet been found.

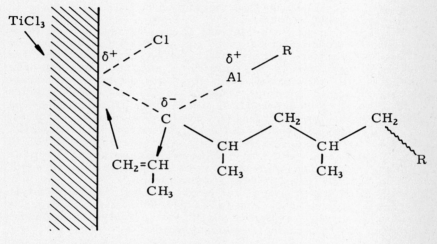

Figure 8.   Polymerization of propylene at catalyst surface (Al Et$_2$Cl + TiCl$_3$).

The Future

As polyolefins celebrate their 50th birthday, one might expect to observe the characteristics of stability, low growth rates, and absence of change or innovations that are commonly associated with maturity in either an industry or a scientific field. But the actual situation is far different, so much so that a recent publication by SRI International introduces the subject with the statement that recent advances have spotlighted polyolefins as one of the most exciting areas of development in the chemical industry, and that the outlook for the polyolefins industry through the next decade is one of change. The current wave of plant closings, expansions, sales, conversions, and modifications are evidence that the changes are well under way.

It is therefore pertinent to consider what directions future advances in scientific understanding and technological progress in the polyolefin field might take.

As has been pointed out repeatedly in this review, even 40 years ago enough was known about polymers for any number of chemists to have made the necessary mental connections between known principles and experimental facts to postulate and then demonstrate stereoregular polymerization. Today there are many more who, armed with the knowledge of what has happened in the past, beneficiaries of the legacy of understanding developed and bequeathed by Staudinger, Mark, Tobolsky, Huggins, Flory, Schildknecht, Ziegler, Natta, and a host of others, and equipped with all the powerful tools of modern science, are in a position to put mind and skill to work and at least make the hackneyed phrase, "tailor-made molecules," a reality by deliberately creating the next major advance in polyolefins or in some related field.

Enumeration of problems is far easier than solving them, but it is a necessary first step. The following are some questions, the answers to which might represent significant advances in knowledge and, eventually, in practical applications:

Polymerization Problems

1. Can effective stereospecific catalysts that work in aqueous systems be found?
2. Now that the difficult feat of controlling chain configuration has been mastered, can we achieve similarly precise control over chain length? Offhand, this would have seemed to be an easier task, but it has not yet been accomplished, except in respect to average chain lengths. Perhaps there has not been a sufficiently strong incentive, even though molecular weight average and distribution have long been recognized as of prime importance in many fields of technology. For example, recent work has shown that narrowing the molecular weight distribution of polypropylene improves its processing characteristics (33). What would be the effect of a much closer approach to monodisperse chain length?
3. What other monomers can be selected or synthesized that would give predictably unique and useful properties? Anyone can spot an asymmetric carbon atom, but the ability to visualize in

advance the specific configuration of the stereoregular form of a new hydrocarbon polymer (identity period, dimensions of the unit crystal cell, crystal melt point, etc.) is only poorly and partly evident. Can that capability be greatly expanded?

4.   Copolymerization is widely used but poorly controlled. Can precise control be established, for example, could we make a regularly alternating copolymer (ABABABABAB) in stereoregular form? Additional possibilities for advances in fully controlled copolymers, terpolymers, graft and block copolymers, etc. are too numerous to enumerate.

5.   Contributions to the basic understanding of coordination catalysis will bring us gradually closer to the day when we cannot only explain empirically discovered effects but predict results and deliberately design new catalyst systems.

6.   Is there some still unrecognized source of still higher order in hydrocarbon polymers that could be demonstrated and applied to yield a useful new class of polymers?

7.   Does the world have need, or even room, for another large-volume, general purpose plastic? It seems very unlikely that it does, but so it seemed also on the eve of the Ziegler–Natta discoveries.

8.   Years ago, it was predicted (by Herman Mark) that an important advance would come from the combination of thermoplastics technology with thermosetting technology, that is, by use of all four of the strengthening mechanisms named earlier (Figure 1) in a single polymer. Numerous laboratories have worked on various aspects of this approach, but the big breakthrough is apparently still to be made.

Postpolymerization Processing

1.   Manipulation of microcrystalline domains appears to provide such diverse phenomena and so many potentially useful effects that we may be sure they have not all been found; nor have those already found all been turned to maximum advantage. We now control spherulite growth to some extent by annealing, nucleation, etc. Solid-phase forming is a technique of perpetual promise, long known to be capable of producing usefully, even remarkably, enhanced properties in polyolefins, but is still awaiting large-scale commercial application. What further possibilities lie in this direction?

2.   Enormous industries are based on the empirically developed processes and moderately well-understood phenomena of one-dimensional orientation of polymers (fiber drawing); there is also some use and some understanding of two-dimensional orientation (biaxial film stretching, oriented blow molding). Is it out of the question to think that there might be some way of achieving three-dimensional orientation, with interesting scientific and valuable practical results? One thinks, for example, of controlled foaming at temperatures just below the crystal melt point.

3.   Learning to control or take advantage of the phase transitions that occur in certain stereospecific polymers (poly(1-butene), polystyrene) should lead to interesting and useful results.

Examples in these and other categories could be multiplied indefinitely, but these few will suffice. The fields of compounding, additives, blends, grafts, and composites all see continual small advances that eventually add up to major gains and will undoubtedly continue to do so. Similarly, new applications and new shaping techniques will continue to be developed (e.g., rotational molding is old hat but not exhausted, solid-phase forming has yet to come into its own, and polyolefin foams have enormous potentials that have scarcely been tapped).

So the imaginative chemist does not lack for opportunities to use his imagination or for new worlds to conquer. As we have seen, the unexpected result, the happy accident, has played a key role in nearly all of the major advances that have been made to date, and we may expect that fortune will continue to smile occasionally on the alert experimenter in the future.

Literature Cited

1. von Pechmann, H. Chem. Ber. 1898, 31, 2643.
   Bamberger, E.; Tschirner, F. ibid 1900, 33, 955.
   Hertzwig, J.; Schonbach, R. Monatsch. 1900, 33, 677.
   Meerwein, H.; Burneleit, W. Ber. 1928, 618, 1840.
2. Swallow, J. C. In "Polythene"; Renfrew and Morgan, Eds.; Interscience: New York; Chap. 1.
3. Robinson, R. personal communication.
4. Fawcett, E. W.; Gibson, R. O. J. Amer. Chem. Soc., 1934, 386.
5. Mark, H. "Giant Molecules"; Time, Inc.: New York, 1966; p. 129.
6. Ziegler, K. Brennstoff Chemie B-C Bd. 1954, 35(21-22), 321. Petr. Refiner 1955, 34, 111.
7. Ziegler, K.; Martin, H. Angew. Chem. 1955, 67(19/20), 541. McMillan, F. "The Chain Straighteners"; Macmillan: London, 1979; pp. 56-68.
8. Friedrich, M.; Marvel, C. S. J. Amer. Chem. Soc. 1930, 52, 376.
9. Clark, A.; Hogan, J. P.; Banks, R. L. Papers presented, American Chemical Society Petroleum Chem. Div., April, 1956, p. 211.
10. Peters, E. I.; Zletz, A.; Evering, B. L. Ind. Eng. Chem. 1957, 49, 1879.
11. Huggins, M. J. Amer. Chem. Soc. 1944, 66, 1991.
12. Schildknecht, C. "Vinyl and Related Polymers"; Wiley: New York, 1952.
13. Flory, P. "Principles of Polymer Chemistry"; Cornell Univ. Press, 1953.
14. Schildknecht, C. et al., Ind. Eng. Chem. 1948, 40, 2108. Ind. Eng. Chem. 1949, 41, 1998.
15. Natta, G. et al., J. Amer. Chem. Soc. 1955, 77, 708. J. Polym. Sci. 1955, 16, 143.
16. Natta, G.; Danusso, F. J. Polym. Sci. 1959, 34, 3.
17. "Stereoregular Polymers and Stereospecific Polymerization"; Natta, G.; Danusso, F., Eds.; Pergamon: New York, 1967.
18. Sailors, H. R.; Hogan, J. P. Macromolecular Sci.-Chem. 1981, A15(7), 1377.
19. Natta, G. Makromol. Chemie. 1955, 16, 213.
20. Natta, G. Ref. 17, p. 268.

21. Natta, G. <u>Makromol. Chemie.</u> 1960, <u>35</u>, 93.
22. See, for example, the January issues of <u>Modern Plastics.</u>
23. Keith, H. D. et al., <u>J. Polym. Sci.</u> 1966, <u>4</u>, 267.
24. Schildknecht, C. "Allyl Compounds and Their Polymers (Including Polyolefins)"; Wiley-Interscience: New York, 1973; p. 98.
25. <u>Ibid.</u>, p. 103.
26. Natta, G.; Corradini, P. <u>Rend. Accad. Naz. Lincei</u> 1955, (<u>8</u>), 18, 19. <u>Makromol. Chemie.</u> 1955,<u>16</u>, 77.
27. Garrett, <u>Rubber and Plastics Age</u> 1964, <u>45</u>, 1492. Crespi, G. et al., ibid., 1181.
28. Ref. 24, p. 68 et seq.
29. Arlman, E. J.; Cosee, P. <u>J. of Catalysis</u> 1964, <u>3</u>, 80-104.
30. Keller, A. <u>J. Polym. Sci.</u> 1955, <u>17</u>, 291, 351, 447.
31. Geil, P. H. "Polymer Single Crystals"; Interscience: New York, 1963. <u>Chem. and Eng. News</u> Aug. 16, 1955, p. 70.
32. <u>J. Macromolecular Sci.-Physics</u>; Geil, P. H., Ed.; Marcel Dekker: New York.
33. Schroeder, C. W. "Modern Polypropylene Resins"; Fiber Producer Conference, Greeneville, S.C., 1981.
34. <u>Chem. Age</u>, October 1980.   Chem Systems.
35. <u>Modern Plastics</u>, October 1980.
36. <u>Modern Plastics</u>, January 1983.
37. SRI International, World Photochemicals Program.
38. <u>Modern Plastics</u>, January 1984.

Polystyrene and Styrene Copolymers

WILLIAM DAVID WATSON and TED C. WALLACE

Dow Chemical Company, Freeport, TX 77541

History
Chemistry of Styrenic Polymers
Manufacturing Processes
Applications
Future
Conclusion

Styrenic-based polymers have evolved over the last 40 years into one of the major thermoplastics. The polystyrene family of polymers consists of polystyrene (PS), styrene-acrylonitrile copolymer (SAN), rubber modified PS (HIPS for high-impact polystyrene), and rubber modified SAN (ABS for acrylonitrile-butadiene-styrene). SAN and ABS contain approximately 25% by weight acrylonitrile. HIPS and ABS contain polybutadiene rubbers in amounts ranging from a few to 20% by weight. Compounding halogenated organics and inorganic oxides into rubber-modified styrenics produces a product with reduced potential to burn. A large number of copolymers of styrene with divinylbenzene, methyl methacrylate, α-methylstyrene, and maleic anhydride, for example, have been commercially produced. In addition, there are a large number of styrene butadiene block copolymers (SB blocks) that have found commercial use. This article will concentrate primarily on PS, SAN, HIPS, ABS, and SB block copolymers.

Polystyrene plastics have a wide range of properties as listed in Table I. PS has a glass transition temperature of 100 °C and is stable to thermal decomposition to 250 °C. Therefore, PS is easily fabricated and a versatile product as well as transparent. SAN has improved tensile yield, heat distortion, and solvent resistance because of the incorporation of acrylonitrile. However, both PS and SAN are brittle as shown by the low notched izod impact strength and the inability to elongate under stress as shown in Figure 1. Adding rubber produces products with dramatically improved impact strength and elongation. The properties of HIPS and ABS are very dependent

0097-6156/85/0285-0363$06.00/0
© 1985 American Chemical Society

Table I. Mechanical Properties for Main Classes of Styrene-Based Plastics (Compression Molded Specimens)

Property a,b	Polymer						
	PS	SAN	Medium IPS	High IPS	Medium ABS Type 1	Type 2	Std. ABS
Specific gravity	1.04	1.08	1.05	1.05	1.05	1.05	1.04
Butadiene rubber, weight %	--	--	4.85	7.0	6.5	14.5	19.0
Acrylonitrile, weight %	--	25	--	--	17	23.5	23.5
Vicat softening point, $^{\circ}$C	108	110	102	100	102	106	103
Tensile yield, lbf/in^2	6400	9500	3650	2600	3250	5500	4800
Ultimate elongation, %	1.5	2.0	30	40	40	12	8
Tensile modulus x 10^5, lbf/in^2	4.7	4.9	3.2	2.4	3.2	2.9	2.6
Izod impact, ft lbf/in of notch	0.25	0.30	1.3	1.5	1.5	4.5	8.0
Gardner impact	very low	very low	medium	high	high	very high	very high
Relative ease of fabrication	excellent	excellent	excellent	excellent	excellent	good	good

a ASTM methods.

b To convert lbf/in^2 to MPa, divide by 145.
To convert ft lbf/in to J/m, divide by 0.0187.

on the amount and type of rubber as well as many other variables.
ABS has improved impact strength, gloss, and solvent resistance
compared to HIPS. Because of the differences in refractive indices
between the rubber and the polystyrene phases, HIPS and ABS are
opaque. The properties of styrene–butadiene block copolymers depend
on their composition with high levels of butadiene giving an
elastomeric product and low levels of butadiene producing an impact
polystyrene. Because there is only one phase these products are
clear. The relative cost of manufacture is PS < HIPS < SB block <
SAN < ABS because of the cost of raw materials and manufacturing
differences. The phenomenal growth of styrene polymers in the
United States is shown in Table II. Examples of uses of these
products are PS for disposable tumblers, SAN for crisper trays, HIPS
for yogurt containers and toys, ABS for refrigerator liners and
telephones, and SB block copolymers for clear lids.

History

Styrene was first reported by Neuman in the late eighteenth century
from storax. Storax is a balsam derived from the trees of
Liquamber orientalis which are native to Asia Minor. His
experiments were confirmed years later. In 1839 Simon named the
product "styrol" and noted that after a few months it became
jellylike. In 1841 Gerhardt and Cahours arrived at the correct
formula, C_8H_8, and prepared the dibromide derivative. In 1845 Blyth
and Hofmann confirmed that a solid mass resulted when styrene was
heated. Additional work was done in which styrene was prepared from
cinnamic acid by decarboxylation. In 1866 Berthelot prepared
styrene from the reaction of benzene and ethylene in a hot tube.
Thus, at the turn of the century the following chemistry was known:

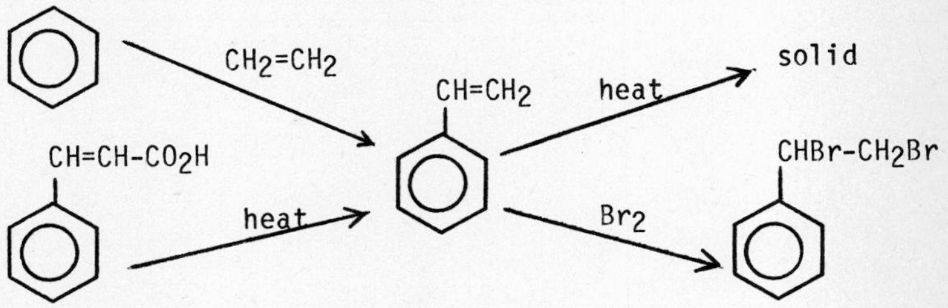

In 1925 Naugatuck Chemical Company built a commercial
styrene/polystyrene plant, but it only operated for a short time.
From this point I. G. Farben Industrie in Germany and Dow Chemical
in the United States pursued commercialization of styrene and
polystyrene. Both companies independently developed along similar
routes.

Dow became involved in making styrene and polystyrene in the
mid-1930s. Two excellent historical accounts are available (1, 2).
Styron 666 general purpose polystyrene was introduced in 1938, and
the first impact polystyrene, Styron 475, was introduced in 1948.
These early processes were batch, but a continuous process was
introduced in 1952. World War II increased the availability of
information concerning styrene and polystyrene because of the
cooperative U.S. effort to make styrene butadiene rubbers (SBR)

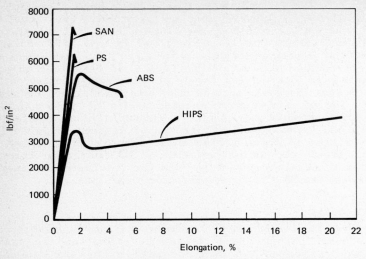

Figure 1. Relative stress—strain curve for polystyrene family.

Table II. U.S. Production of Styrene Polymers (millions of lb)

Year	General Purpose Polystyrene[a]	Rubber-Modified Polystyrene	SAN[b]	ABS[c]	Total
1945	23	--	--	--	23
1950	250	15	--	--	265
1955	320	115	3	8	446
1960	465	260	23	55	803
1965	830	642	38	240	1750
1970	1280	1215	58	510	3063
1975	1550	1630	102	840	4122
1980	1756	1765	111	920	4552

[a]Includes prime, off-grade, export, and polymer used for Styrofoam brand plastic foam as well as expandable polystyrene beads plus miscellaneous end uses.
[b]Styrene-Acrylonitrile copolymers mostly of ca. 25% acrylonitrile.
[c]Acrylonitrile-Butadiene-Styrene copolymers of all types.

rubbers. This situation was also true in Germany. Currently there are over 30 polystyrene manufacturers in the United States and Europe.

Chemistry of Styrenic Polymers

The purity of styrene is critical in producing high molecular weight polystyrene free of gels. Styrene manufacture begins with the alkylation of benzene with ethylene by using a Freidel Crafts catalyst to produce ethylbenzene. This alkylation is done at low conversions because of the activating effect of the ethyl group. The alkylation requires considerable recycle of benzene and disproportionation of the diethylbenzenes. The ethylbenzene is catalytically dehydrogenated in the presence of superheated steam at elevated temperatures and pressures:

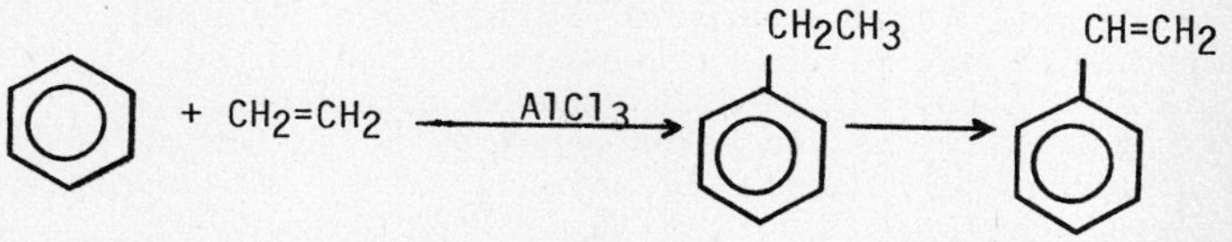

Impurities, chain transfer agents such as phenylacetylene, and cross-linking agents such as divinyl benzene need to be kept at extremely low levels.

Styrene can be polymerized radically either thermally or by using free-radical initiators, anionically or cationically. Thermally the reaction is initiated by a Diels Alder adduct in the following manner (3):

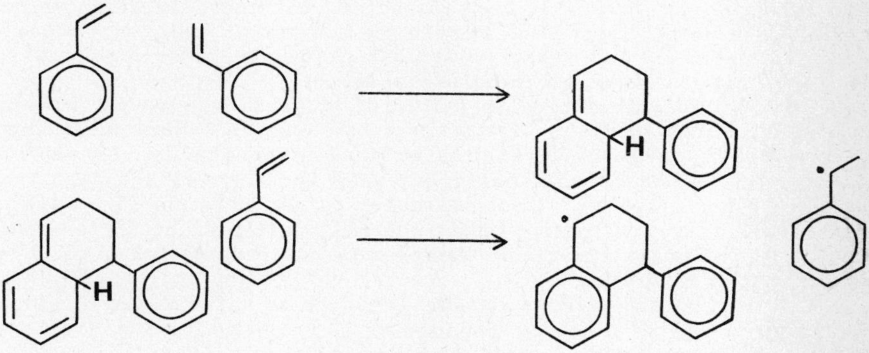

These two radicals initiate chain growth. Only the stereoisomer with the phenyl group in the axial position initiates the polymerization.

Free-radical initiated polymerization is normally done by using peroxy-type initiators such as benzoyl peroxide, tert-butyl perbenzoate, or difunctional initiators. The mechanism of these radical polymerizations proceeds via the following sequence:

I ----------> 2R·   Initiation

R· + S -------> R_n·   Propagation

R_n· + XH ----> PS_n + X·   Termination by Chain Transfer

R_m· + R_n· ---> PS_n + PS_m Termination by Disproportionation

R_m· + R_n· ---> PS_{n+m} Termination by Combination

The symbols used are I for initiator, R for the radical derived from the initiator, S for styrene, R_n· and R_m· for growing polystyrene radicals, XH for a source of hydrogen radical, and PS for polystyrene. Thus, polystyrene can be formed in the termination step by chain transfer, disproportionation, and combination. Temperature and chain transfer agents can be used to control molecular weight and molecular weight distribution. Polystyrene resulting from free-radical processes is amorphous.

Anionic polystyrene can be prepared by polymerizing styrene with butyl lithium, alkali metals, or soluble alkali metal complexes such as sodium naphthalene:

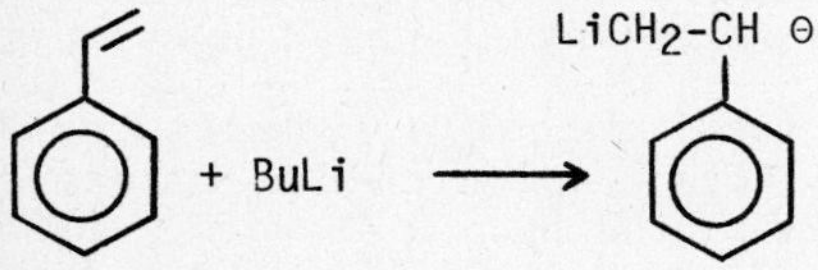

The propagation step follows as expected, but there is no inherent termination step. Therefore, very high molecular weight products with narrow molecular weight distribution can be made anionically. SB block copolymers are made anionically.

Cationic polymerization is initiated by acids such as perchloric acid, boron trifluoride, or aluminum trichloride. High molecular weight polystyrene is difficult to make cationically because of chain transfer reactions that occur with the monomer and with the commonly used solvents. Thus, molecular weight and molecular weight distributions can be controlled by the polymerization conditions and the method of polymerization. Examples of desired molecular weights are shown in Figure 2.

Isotactic polystyrene can be prepared by using a Ziegler-type catalyst. It is of interest because of its high melting point (240 °C). Syndiotactic polystyrene is unknown.

The reactivity ratios for the free-radical copolymerization of styrene (r_1 = 0.4) and acrylonitrile (r_2 = 0.04) result in uneven incorporation of each monomer into the copolymer as seen in Figure 3. Thus, most SAN and ABS polymers are made at the crossover point (A in Figure 3) to avoid composition drift.

The addition of rubber significantly complicates the picture. Rubbers commonly used for HIPS and ABS have various microstructures depending on the method of manufacture (4):

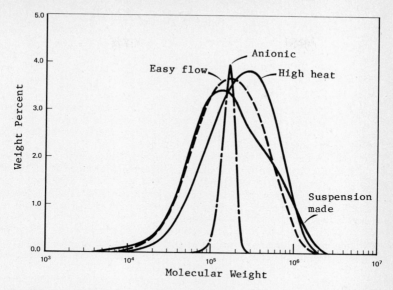

Figure 2. Molecular weights distributions for representative polystyrenes.

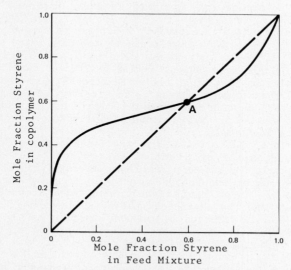

Figure 3. Copolymer composition curve for styrene-acrylonitrile

cis   trans   1,2-vinyl

The glass transition temperature of the rubbers used as impact modifiers must be below −50 °C to give good impact strength over a broad temperature range.   In addition, SB block copolymers can also be used as impact modifiers.

These rubbers are most effective as impact modifiers if they are grafted to the polystyrene rigid phase. Hydrogen abstraction at the allylic site by an alkoxy radical from the peroxide initiator and subsequent reaction with a growing polystyrene or SAN chain produces the graft (5–8). Grafting is important for particle sizing, but first, phase inversion must be discussed.

Polybutadiene rubbers are soluble in styrene. As the polymerization proceeds phase separation occurs, and the solution turns opaque because of the difference in refractive indices between the two phases.   Initially polybutadiene in styrene is the continuous phase, and polystyrene in styrene is the discontinuous phase. When the phase volumes are equal and sufficient shearing agitation exists, phase inversion occurs. After this point polystyrene in styrene is the continuous phase, and polybutadiene in styrene is the discontinuous phase. Phase inversion is represented in Figure 4. A change in viscosity is also observed at phase inversion (6).

Rubber particle size is extremely important to make an optimized impact product. Particles that are both too small and too large cause a loss of impact strength. The ability to form stable particles of optimum size depends on the graft that functions as an oil in oil emulsion. This might better be referred to as an emulsion of two incompatible organic phases.   To size the rubber particles, shearing agitation must be provided. If it is not provided, phase inversion does not occur, and a cross–linked continuous phase that produces gel is the result.

Measurements of particle size in the final product can be done with a Coulter Counter or electron microscopy by using osmium tetraoxide to stain the rubber particles. This staining also allows visual observation of the rubber morphology as shown for HIPS and ABS in Figure 5. The polybutadiene rubber is stained dark with osmium tetroxide and the continuous polystyrene phase is the light background.   What is observed is somewhat circular rubber particles with polystyrene occlusions. These occlusions extend the effective rubber phase volume for better use of the rubber. Particle size varies as well as the density of the rubber particle.

All of these variables are important to designing a product for a specific use. For example, a glossy HIPS product needs to balance the small particle size against impact strength (10). Resistance to environmental stress crack agents is improved by high rubber content, large particle size, high matrix molecular weight, and the choice of plasticizer (11).

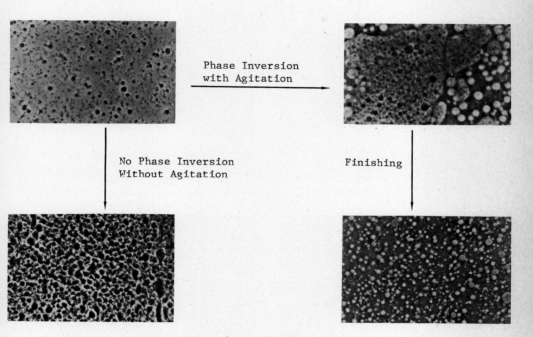

Phase Inversion
with Agitation

No Phase Inversion
Without Agitation

Finishing

Figure 4. Formation of rubber particles by phase inversion of polymeric oil-in-oil emulsion. Phase contrast micrograph (rubber phase is light) according to G. E. Molare and H. Keskkula.

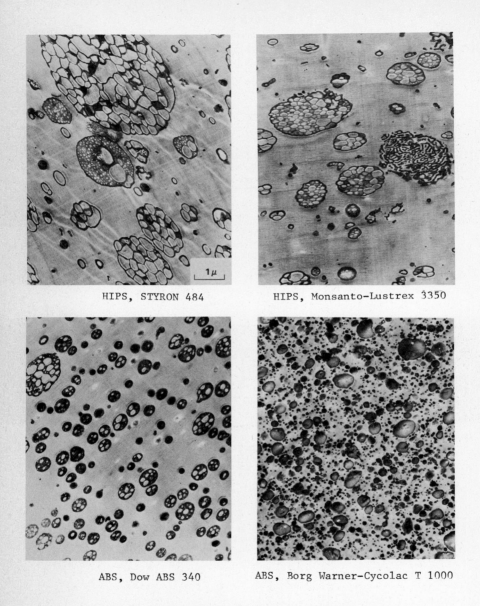

HIPS, STYRON 484   HIPS, Monsanto-Lustrex 3350

ABS, Dow ABS 340   ABS, Borg Warner-Cycolac T 1000

Figure 5. Transmission electron micrograph of osmium
tetroxide stand--HIPS and ABS.

SB block copolymers are made anionically. These copolymers can be diblocks, triblocks, and radial block copolymers with different degrees of tapering. Kraton was introduced in the mid-1960s by Shell. Another major manufacturer is Phillips with Solprene and K-Resins. These products can be used as thermoplastic elastomers or as impact modifiers. One of the most interesting aspects of these resins is the different types of morphologies that can be obtained as shown in Figure 6 (12–15).

Manufacturing Processes

Economics of making polystyrene dictates large, cost-efficient, continuous processes. Batch polymerization is important from a historical point of view and is still used in SB latex emulsion polymerization, anionic SB block copolymerizations, and suspension polymerization to make styrene-divinylbenzene copolymer or polystyrene expandable beads. In fact suspension polymerization of styrene made some of the best quality polystyrene although the process is no longer economically feasible. Examples of continuous processes are shown in Figures 7 and 8. These reactors are designed to remove the heat of reaction (-17.4 cal/mol) from a highly viscous medium. Reactor designs for polystyrene have recently been reviewed (16). Reactors such as the one shown in Figure 7 are long and cylindrical with multiple heat transfer tubes and an agitator. Styrene, initiator, and diluents such as toluene or ethylbenzene are fed into the reactors. The diluents are used to reduce the viscosity in the third stage. The temperatures in the reactors are gradually increased from approximately 100 °C to 180 °C during which time the solids increase. The polymer is fed to a devolatilizer where the diluent and any residual styrene is removed for recycle. The polystyrene is stranded, cooled, and cut into pellets for sale. Additives are usually added to improve product properties and processability. Such additives are plasticizers such as mineral oil to control flow properties, antioxidants such as hindered phenols to prevent yellowing, and mold release agents such as stearates to prevent mold sticking. Certain of these additives that do not interfere with the reactor such as mineral oil can be added in the feed or post added while additives that do interfere such as antioxidants are usually post added.

The use of rubber again complicates the picture. First the rubber must be dissolved in styrene. This process is slow and must be done as discussed earlier with enough shearing agitation to effect phase inversion and to adequately size the rubber particles. In HIPS manufacture phase inversion normally occurs in the first stage. Because of the higher amounts of rubber in ABS, phase inversion occurs in the second stage. The type and the amount of graft depend on the microstructure of the polybutadiene as well as the conditions of the thermal or initiated polymerization of styrene (17, 18). Careful control of the amount of conversion and devolatilization conditions produces a product with the desired amount of cross-linking (19) which occurs primarily between the 1,2-vinyl portion of the polybutadiene and the polystyrene phase. Because of differences in reactivity ratios, the cross-linking does not occur until almost complete styrene conversion. Figure 8 shows a continuous stirred tank reactor that uses evaporative cooling to achieve uniform temperature control.

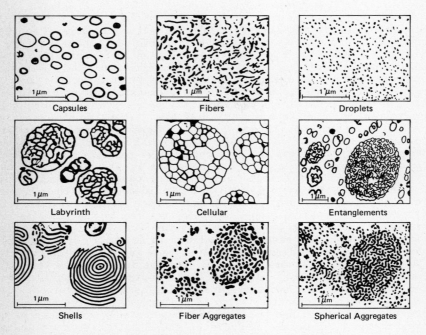

Figure 6. Various particle structures of rubber in impact polystyrene.

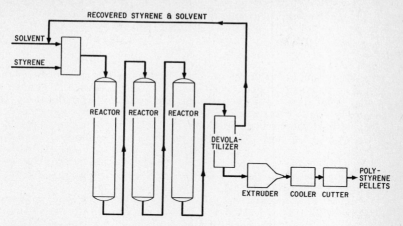

Figure 7. Diagram of a continuous process for styrene polymerization.

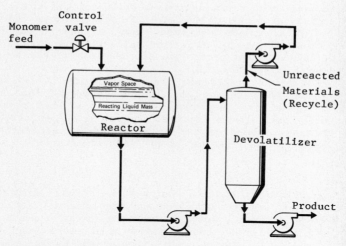

Figure 8. Diagram of a process to manufacture polystyrene that uses a continuous stirred tank (CSTR) reactor.

By changing variables numerous products can be manufactured. Control of molecular weight, composition, additives, grafting, particle sizing, morphology, and cross-linking produces products with a wide variety of physical properties.

Applications

One of the major attributes of the polystyrene family of resins is its ease of fabrication. Ease of plasticization, good melt strength, and low coefficient of thermal expansion make these polymers ideal for the relatively low-cost single screw extrusion/thermoforming process. This processing technology continues to grow, and 40% of all styrene polymers and copolymers are processed by this technique (20). The advent of low-cost computers has made closed loop control of the extrusion system commonplace and thus has yielded more uniform gauge control at higher output rates (21).

The other most common fabrication technique for the polystyrene family is injection molding, most frequently accomplished by reciprocating screw and screw preplasticiator machines (22).

Major applications for styrene plastics are summarized in Table III (23). The packaging and serviceware (disposables) markets predominate, and account for approximately 50% of the total. One of the most rapidly growing portions of these markets is in low-density (usually 1-10 lb/ft^3) polystyrene foams, either in the form of extruded foam sheet or expanded polystyrene beads (EPS). Projections indicate that production of these foams will be greater than 2000 metric tons (24).

The use of cellular styrene plastics for insulation has been widespread for many years (25). More recently, polystyrene "structural" foams have been used, especially in wood replacement applications. Such use is expected to grow in the future, particularly as wood becomes less available and greater demands are placed on more efficient use of "short" plastic materials.

Future

Although the styrene plastics industry is relatively mature, it is still an area for fruitful research and development. For example, an area of major current concern is the development of products that are difficult to ignite and yield little smoke when burned. Existing approaches to the flammability problem involve the use of halogen-containing additives plus antimony oxide synergists (26). In the long term, products may be made with decreased flammability potential via polymerization with halogen-containing comonomers, by alloying with fire-resistant engineering thermoplastics, or by copolymerizing with monomers that induce charring at the flame front by cross-linking.

Although impact modifiers have been investigated since the birth of the rubber-modified plastics industry, they still remain an active area of investigation. The earlier impact polymers were modified with polybutadiene and polystyrene-co-butadiene made via emulsion polymerization. As the industry matured, "solution" polybutadiene and to a minor extent styrene/butadiene block

Table III. Major Applications for Polystyrene (millions of lb)

Market	1960	1965	1970	1975	1980	1985[a]
Packaging	100	345	600	750	1,055	1,350
Appliances	93	121	150	140	110	115
Radio, TV, Electronics	32	55	138	200	235	275
Housewares, Furnishings, Furniture	95	184	338	325	320	365
Recreation	75	212	285	170	210	235
Serviceware	10	40	150	210	290	360
Miscellaneous	150	263	334	330	530	560
Total	555	1,220	1,995	2,125	2,750	3,260

[a]estimated

copolymers became widely used as reinforcing agents. The solution or "diene" type rubbers, prepared by alkyl lithium initiated polymerization of butadiene (27), tend to be "cleaner" than emulsion elastomers and have distinctly lower glass transition temperatures than their emulsion-prepared counterparts. These properties lead to better low-temperature impact properties in the rubber-modified styrene plastics. Likewise, stereospecific or high cis-1,4-polybutadiene-modified plastics exhibit similar desirable properties (28). Another example is the blending of special block copolymers and GP polystyrene (29-31). Considerable research is being done with specific thermoplastic elastomers used as compatibilizers for alloys of normally incompatible polymers such as polystyrene and polypropylene (32).

Additional opportunities in the styrene plastics industry exist for the development of products having such unique properties as high heat resistance and optical transparency. Arco produces the Dylark family of heat-resistant styrene plastics, which are copolymers of styrene-maleic anhydride. These products have a good balance of mechanical properties and have a heat distortion (under ASTM D68) of 234 °F (33) which is markedly higher than homo polystyrene.

Perhaps the best known member of the "heat-resistant" styrene plastics family is General Electric's Noryl (34). Noryl is an alloy of poly(phenylene oxide) and high-impact polystyrene. Heat deflection temperatures for Noryl range as high as 300 °F. The balance of mechanical properties is excellent, although processability is more difficult than for conventional styrene plastics.

Phillips Petroleum Company now manufactures the K-Resin family of optically transparent impact polystyrene (35). These resins, made via anionic polymerization techniques, owe their transparency to the extremely small size of the dispersed rubber phase. The principal use of K-Resin is in packaging applications.

The development of unique techniques for fabrication of styrene plastics into end-use items is also a fruitful area of research and development. The use of coextrusion, for example, is now becoming widespread. Coextrusion allows one to combine the properties of various polymers into a single layered structure (36). For example, the multilayered packaging structure in Figure 9 combines the structural properties of high-impact polystyrene and the oxygen barrier properties of Saran with the moisture barrier, food contact surface, and heat seal properties of polyethylene. Such film is made via coextrusion through a feedblock of the type illustrated in Figure 10. One can make film having hundreds of layers via this technique. It is also possible to use "scrap" in coextrusion processes by "hiding" it in a central layer. The 1980s should see widespread use of the all-plastic can based on complex structures that include polystyrene (37). An example of such a can is an aseptic packaging system that was developed jointly by Erca SA and Cobelplast of France and is typically a thermoformable seven-layer sheet: polyethylene/polypropylene/glue/poly(vinylidene chloride)/glue/PS/pigmented PS (38).

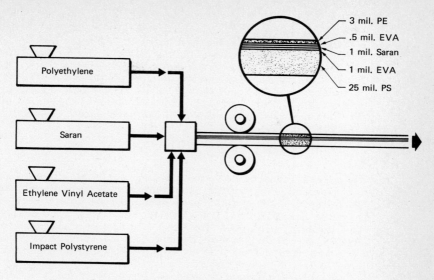

Figure 9. Multilayer structure made via coextrusion.

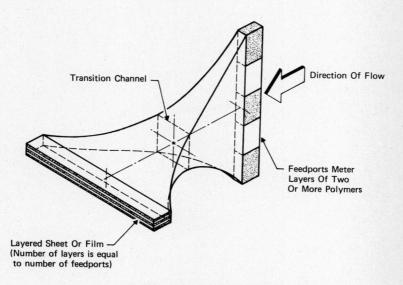

Figure 10. Coextrusion sheet and container.

An emerging and extremely significant fabrication technique for plastic materials, including styrene plastics, is "scrapless thermoforming" (39). As illustrated in Figure 11, it is possible to extrude (or coextrude) sheet, cut blanks from this sheet, and then forge the blank into a container without generating scrap. An additional benefit of this fabrication technique is the improved toughness and stress crack resistance that is derived from the orientation induced in the forging process. Continued use of fabrication techniques that minimize or eliminate scrap generation is expected in the future.

Conclusion

The styrene plastics industry has emerged over the past 30 years to become a major worldwide business. The industry has grown because the excellent balance of mechanical properties and processability of styrene plastics allow it to fill diverse market needs. The advent of workable industrial processes for both monomer and polymer and the fact that styrene plastics were made from once inexpensive raw materials have likewise contributed to the growth of the industry. In spite of the relative maturity of the science and the industry, styrene plastics remain a fruitful area for research. For example, the development of new materials having unique properties, such as fire and heat resistance, and the development of efficient energy and material-saving fabrication processes are expected to be the subject of extensive study in the future.

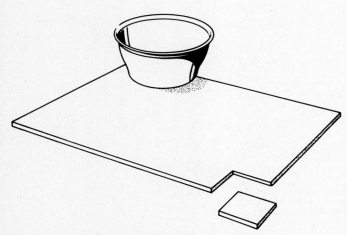

Figure 11.   Coextruded sheet and container formed via scrapless thermoforming process.

Acknowledgments

The authors would like to thank A. E. Platt of the Dow Chemical Company for preparing Figures 1-8 and for reading and commenting on the manuscript.

Literature Cited

1. Amos, J. L. Poly. Eng. and Sci. 1974, 14(1). 1.
2. Boyer, R. F. Macromol. Sci. - Chem. 1981, A15(7), 1411.
3. Olaj, O. F.; Kauffmann, H. F.; Breitenbach, J. W. Makromol. Chem. 1977, 178, 2702.
4. Hsieh, H. L.; Farrar, R. C.; Udipi, K. Chem. Tech. 1981, 11(10), 626.
5. Molau, G. E.; Keskkula, H. J. Polym. Sci. [A-1] 1966, 4, 1595.
6. Molau, G. E.; Keskkula, H. Appl. Polym. Symp. 1968, 7, 35.
7. Cameron, G. G.; Qureshi, M. Y. J. Polym. Sci., Polym. Chem. Ed. 1980, 18, 2143.
8. Kotaka, T. Makromol. Chem. 1976, 177, 159.
9. Amos, J. L.; McCurdy, J. L.; McIntire, O. R. U.S. Patent 2 694 692, 1954.
10. Lavengood, R. E. U.S. Patent 4 214 056, 1980.
11. Bubeck, R. A., Arends, C. B.; Hall, E. L.; Vander Sande, J. B. Polym. Eng. and Sci. 1981, 21(10), 624.
12. Schmitt, B. J. Angew. Chem. Int. Ed. Engl. 1979, 18, 273.
13. Riess, G.; Schlienger, M.; Marti, S. J. Macromol. Sci.-Phys. 1980, B17(2), 355.
14. Echte, A. Angew. Makromol. Chem. 1977, 58/59, 175.
15. Echte, A.; Gausepohl, H.; Lutje, J. Angew. Makromol. Chem. 1980, 90, 95.
16. Simon, R. H. M.; Chappelear, D. C. In "Polymerization Reactors and Processes"; ACS SYMPOSIUM SERIES 104, Henderson, J. N.; Bouton, T. C., Eds.; American Chemical Society: Washington, D.C., 1979; Chap. 4.
17. Fischer, J. P. Angew. Makromol. Chem. 1973, 33, 35.
18. Kamath, V. R. U.S. Patent 4 125 695, 1978.
19. Stein, D. J.; Fahrback, G.; Adler, H. ADVANCES IN CHEMISTRY SERIES No. 142; Platzer, N. A. J., Ed.; American Chemical Society: Washington, D.C., 1975; Chap. 14.
20. Modern Plastics 1982, 59(1), 82.
21. Plastics World 1979, 37(11), 40.
22. Lignon, J. Modern Plastics Encyclopedia 1980, 57(10A), 317.
23. Pearson, W. E. The Dow Chemical Company, March 1982.
24. Martino, R. Modern Plastics 1978, 55(8), 34.
25. Ingram, A. R.; Fogel, J. In "Plastic Foams"; Frisch, K. C.; Saunders, J. H., Eds.; Marcel Dekker: New York, 1973; Vol. II.26. "Flame Retardants for Plastics"; Multiclient Market Survey; Hull and Co.: Bronxville, N.Y., 1978.
26. "Flame Retardants for Plastics"; Multiclient Market Survey; Hull and Co.: Bronxville, N.Y., 1978.
27. Forman, L. E. "Polymer Chemistry of Synthetic Elastomers"; Kennedy, J. P.; Tornquist, E. G. M., Eds.; Interscience, 1969; Vol. II.
28. "Taktene 1202: An Impact Modifier for Polystyrene"; Handbook, Polysar Corp.: Sarnia, Ontario, Canada.

29. Durst, R. R. U.S. Patents: 3 906 057, 3 906 058, 3 907 929, 3 907 931, 1975.
30. Platzer, N. Chem Tech., October 1977, 634.
31. Aggarwal, S.; Livigni, R. Polym. Eng. and Sci. 1977, 17(8), 498.
32. Holden, G.; Govw, L. U.S. Patent 4 188 432, 1980.
33. Plastics Tech. 1981, Manuf. Handbook, 27(7), 557.
34. "Noryl Thermoplastic Resins"; Technical Bulletin, General Electric Company: Selkirk, N.Y.
35. "K-Resin Butadiene Styrene Polymers"; Technical Bulletin, Phillips Chemical Company, Bulletin 19430, 1980.
36. Schrenk, W.; Alfrey, T. "Polymer Blends"; Academic Press, 1978; Vol. 2, p. 129.
37. Terwilliger, F. Food and Drug Pkg. 1981, 46(1), 54.
38. Sneller, J. Modern Plastics 1981, 58(12), 54.
39. "Scrapless Forming Process"; Technical Bulletin, The Dow Chemical Co., Bulletin S-303-74-76, 1976.

General References

1. Boundy, R. H.; Boyer, R. F. "Styrene, Its Polymers, Copolymers, and Derivatives"; Reinhold: New York, 1952.
2. Bishop, R. B. "Practical Polymerization for Polystyrene"; Cahners: Boston, 1971.
3. Brighton, C. A.; Pritchard, G.; Skinner, G. A. "Styrene Polymers: Technology and Environmental Aspects"; Applied Science: London, 1979.
4. Boyer, R. F.; Keskkula, H.; Platt, A. E. In "Encyclopedia of Polymer Science and Technology"; John Wiley: New York, 1970; Vol. 13, pp. 128-447.
5. Basdekis, C. H. "ABS Plastics"; Reinhold: New York, 1964.
6. Bucknall, C. B. "Toughened Plastics"; Applied Science: London, 1977.

Poly(vinyl chloride)

ROY T. GOTTESMAN[1] and DONALD GOODMAN[2]

[1]The Vinyl Institute, A Division of the Society of the Plastics Industry, New York, NY 10017
[2]Tenneco Polymers, Inc., Flemington, NJ 08822

History of Production
Manufacture of Vinyl Chloride Monomer
Polymerization of Vinyl Chloride
Copolymerization of Vinyl Chloride
Characterization of Poly(vinyl chloride)
Physical Properties
Chemical Properties
Classification of Poly(vinyl chloride) Resins
Plasticizers
Heat Stabilizers
Processing Aids, Impact Modifiers, and Other Additives
Applications for Poly(vinyl chloride)
Vinyl Chloride Toxicity and Federal Regulations of the
   Poly(vinyl chloride) Industry

This chapter will cover the chemistry and technology of poly(vinyl chloride) (PVC) homopolymers having a repeating unit of $-CH_2-CHCl-$ as well as copolymers of vinyl chloride with smaller amounts of other unsaturated monomers such as vinyl acetate, ethylene, propylene, acrylates, and vinylidene chloride.

History of Production

Vinyl chloride was reported by Regnault in 1835 (1), and he reported its first polymerization three years later (2). Actually, later work showed that the Regnault polymer was poly(vinylidene chloride). It was not until 1872 that PVC was prepared by Bauman, who carried out polymerizations of various vinyl compounds in sealed tubes (3).
    Industrial development of PVC resins began about 50 years ago. Full commercial scale production started in Germany in 1931 and in the United States in the late 1930s. U.S. production was sparked by the observation by Waldo L. Semon at B. F. Goodrich in 1933 that PVC, when heated in the presence of a high boiling liquid (a plasticizer), became a flexible material that resembled rubber or

0097–6156/85/0285–0383$14.70/0
© 1985 American Chemical Society

leather. The annual production of PVC exceeded 1 million lb in
1935, and the pattern has been one of steady growth since then.
Today vinyl chloride polymers rank with polyethylene and styrene
resins among the most commercially significant polymers. Uses of
PVC are widespread--from automobile interiors, to the floors we walk
on, the packaging of the food we eat, the furniture we sit on, the
vinyl phonograph records we listen to, the garden hose we use, the
shower curtains in our bathroom, the dolls and toys our youngsters
play with, and last, but not least, the credit cards we use. In
1979, one hundred seven years after Bauman produced it in his
laboratory, U.S. production alone was over 6 billion lb. The
significant growth of this polymer is shown graphically in Figure 1
(4). Figure 1 also demonstrates the sensitivity of PVC production
to economic conditions. The significant growth of PVC peaked in
1979 and has since then shown signs of weakness due to a depressed
economy.
   The PVC producers in the United States at the start of 1982 are
listed in Table I. The name plate capacity of these plants is in
the order of 8 billion lb of PVC. The actual production rose from
5.5 billion lb in 1980 to approximately 6 billion lb in 1981. The
overcapacity that was felt in 1979 was accentuated during 1980 and
1981. Plants were run at a profitable 87% capacity in 1979 but were
less profitable in 1982 when operating at less than 75% capacity.
Capacity increases have peaked at a time of slack demand. PVC
consumption is very much tied into construction usages in the United
States, perhaps more so than other thermoplastics. With interest
rates high, construction activity is depressed, and PVC use
declines.
   Because of significant overcapacity and declining profit
margins, a predictable shakeout of PVC producers is occurring.
Considerable consolidation of companies has taken place. Smaller
companies with old plants are the most vulnerable. Expansion plans
have been postponed, and most construction activity is limited to
simple replacement of outdated equipment.
   All of this comes at a time when the versatility and energy
savings potential of PVC is most recognized. A recent article
demonstrated, for example, that PVC is still one of the least
expensive plastics, so that the cost per in.^{3} of fabricated matter
is among the lowest (Table II) (5). The functional versatility of
PVC is reviewed in Table III (6) in which the numerous use areas for
PVC are described. The energy required to produce PVC is also quite
low (7). Much less energy is needed relative to metals production,
particularly aluminum.

Manufacture of Vinyl Chloride Monomer

Two commercial feedstocks are used in manufacture of vinyl chloride:
ethylene or acetylene. Nearly 19% of the chlorine production in the
United States is used in vinyl chloride monomer (VCM) production.
Two other routes are not yet commercial.

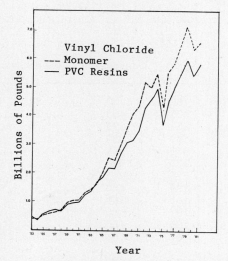

Figure 1.   PVC and VCM Production: 1953 to 1981.

Table I. PVC Producers and Capacities

| | In-place Nameplate Capacity as of 1/1/82 1000 Metric Tons | | | |
Supplier	Total	Suspension	Dispersion	Additions, 1982 or Later
B. F. Goodrich	780	700	80	
Tenneco	425	405	20	
Hooker	365	295	70	
Shintech	330	330		
Conoco	325	325		
Georgia–Pacific	320	320		
Borden	228	205	23	135
Diamond Shamrock	194	160	34	
Air Products	165	165		
Formosa	107	80	27	240
Certain Teed	100	100		
Ethyl	81	65	16	
General Tire & Rubber	70	70	--	11
Pantasote	65	65		
Union Carbide	65	65		
Stauffer	65	65	45	
Goodyear	45	--		
Talleyrand	45	45		
Great American	30	30		
Keysor	30	30		
Total	3835	3520	315	386

Table II. Cost Efficiency of Engineering Materials

Material	Dollars/lb	Cents/in.^{3}
Engineering Resins		
Acetal[a]	1.48	7.6
Nylon 6/6[a]	1.81	7.5
Nylon 11	2.96	11.1
Polyarylate	2.40	10.5
Poly(butylene terephthalate)	1.42–1.70	6.7–8.0
Polycarbonate	1.62	7.1
Poly(phenyl oxide) (modified)	1.33–2.37	5.1–9.1
Polysulfone	3.82	17.1
Reinforced PET, 30%[a]	1.41	8.0
Reinforced poly(phenylene sulfide), 40%	3.07–3.13	17.7–18.0
Other Plastic Resins		
ABS, high impact	0.97–1.14	3.7–4.3
Acrylic[a]	0.87	3.7
Polyethylene, high density[a]	0.55	1.9
Polypropylene	0.44–0.50	1.4–1.6
Polystyrene, high impact	0.50	1.9
Poly(vinyl chloride)	0.31	1.3–1.5
Metals		
Aluminum SAE-309 (380) ingot	0.88–0.89	8.4
Brass (#403) ingot	0.74	22.7
Magnesium AZ-63A ingot	1.43	9.3
Steel, CR carbon sheet	0.25	7.0
Zinc SAE-903 ingot	0.48–0.544	11.6–12.8

[a]Du Pont product basis. Source: Du Pont (based on list prices).

Table III. Major Markets for PVC (1981)

Market	Quantity[a]	Market	Quantity
Calendering		Extrusion	
Building & construction		Building & construction	
Flooring	68	Pipe & conduit	
Paneling	11	Pressure	354
Pool—pond liners	16	Water	354
Roof membranes	9	Gas	10
Other building	2	Irrigation	111
Transportation		Other	9
Auto upholstery/trim	32	Drain/waste/vent	104
Other upholstery/trim	8	Conduit	168
Auto tops	5	Sewer/drain	141
Packaging: sheet	36	Other	27
Electrical: tapes	4	Siding & accessories	82
Consumer & institutional		Window profiles	
Sporting recreation	10	All—vinyl windows	7
Toys	15	Composite windows	25
Baby pants	2	Mobile home skirt	7
Footwear	14	Gutters/downspouts	2
Handbags/cases	11	Foam moldings	14
Luggage	9	Weatherstripping	11
Bookbinding	2	Lighting	8
Tablecloths, mats	17	Transportation	
Hospital & health care	7	Vehicle floor mats	8
Credit cards	8	Bumper strips	4
Decorative film (adh.—back)	5	Packaging	
Stationery, novelties	2	Film	125
Tapes, labels, etc.	6	Sheet	15
Furniture/furnishings		Electrical:	
Upholstery	34	wire & cable	170
Shower curtains	5	Consumer & institutional	
Window shades/blinds/		Garden hose	18
awnings	5	Medical tubing	16
Waterbed sheet	4	Blood/solution bags	20
Wallcovering	14	Stationery, novelties	44
Other calendering	8	Appliances	13
Total calendering	369	Total extrusion	1473

[a]Thousand metric tons.

Market	Quantity	Market	Quantity
Blow molding: bottles	45	Dispersion coating	
Compression molding:		Building: flooring	61
sound records	449	Transportation	
Dispersion molding		Auto upholstery/trim	12
Transportation	16	Other upholstery/trim	2
Packaging: closures	14	Anticorrosion coatings	5
Consumer & institutional		Consumer & institutional	
Toys	3	Apparel/outerwear	6
Sporting/recreation	8	Luggage	4
Footwear	6	Tableclothes,mats	4
Handles, grips	6	Hospital & health care	3
Appliances	6	Furniture/furnishings	
Industrial: traffic cones	6	Upholstery	8
Adhesives, etc.		Window shades/blinds/	
Adhesives	5	awnings	6
Sealants	4	Wallcoverings	6
Miscellaneous	5	Carpet backing	7
Other dispersion	5	Other	15
Total dispersion		Total dispersion	
molding	84	coating	139
Injection molding			
Building & construction		Solution coating	
Pipe fittings	50	Packaging: cans	4
Other building	3	Adhesives/coatings	21
Transportation: bumper parts	5	Total	25
Electrical/electronics			
Plugs, connectors, etc.	30	Vinyl latexes;	
Appliances, bus. machines	17	adhesives/sealants	25
Consumer & institutional		Export	195
Footwear	20		
Hospital & health care	7	Grand total	2551
Other injection	15		
Total injection			
molding	147		

From Acetylene.

$$HC \equiv CH + HCl \xrightarrow[\substack{90-140\ °C}]{HgCl_2\ on\ charcoal} CH_2 = CHCl + Heat$$

The gas phase reaction of acetylene with hydrogen chloride uses mercuric chloride (8, 9) or other heavy metal halides as catalyst. It is important that the gas streams be dry and free from arsine, phosphine, or sulfur. Because ethylene is priced substantially lower than acetylene, most recent processes substitute ethylene for acetylene, and acetylene-based vinyl chloride plants have been disappearing.

From Ethylene. This process utilizes an oxychlorination reaction with ethylene and chlorine as feedstocks. In the process, three distinct reactions can be considered to be taking place:

$$CH_2=CH_2 + Cl_2 \longrightarrow \underset{\underset{Cl}{|}}{CH_2}-\underset{\underset{Cl}{|}}{CH_2} \tag{1}$$

EDC (1,2-dichloroethane)
(ethylene dichloride)

$$CH_2=CH_2 + 2HCl + 1/2\ O_2 \longrightarrow \underset{\underset{Cl}{|}}{CH_2}-\underset{\underset{Cl}{|}}{CH_2} + H_2O \tag{2}$$

The HCl consumed in this reaction is produced by decomposition of EDC:

$$\underset{\underset{Cl}{|}}{CH_2}-\underset{\underset{Cl}{|}}{CH_2} \xrightarrow[\substack{pumice\ or\\ hot\ tube}]{480-510\ °C} CH_2=CH\ Cl + HCl$$

The overall reaction can thus be seen to be

$$2CH_2=CH_2 + Cl_2 + 2HCl + 1/2O_2 \longrightarrow 2\underset{\underset{Cl}{|}}{CH_2}-\underset{\underset{Cl}{|}}{CH_2} + H_2O$$

$$2\underset{\underset{Cl}{|}}{CH_2}-\underset{\underset{Cl}{|}}{CH_2} \longrightarrow 2CH=CHCl + 2HCl$$

$$2CH_2=CH_2 + Cl_2 + 1/2O_2 \longrightarrow 2CH_2=CHCl + H_2O \tag{3}$$

The flow sheet for a balanced chlorination-oxychlorination of ethylene to vinyl chloride monomer is shown in Figure 2. Currently this process, with its variations involving fixed and fluid beds and different methods of heating and separation, dominates the commercial production of vinyl chloride with 93% of VCM being made by this route.

<u>Balanced Acetylene-Ethylene</u>. An advantage of the acetylene-based process is that HCl is not produced. Because HCl reacts with acetylene, a balanced acetylene-ethylene process has been described (10). This process represents a major step forward and reduces the amount of acetylene used with attendant raw material cost savings and use of by-product HCl. No new plants that use this process are being built in the United States. In 1970, Kureha Chemical Industry Company in Japan brought a plant onstream in which a self-developed process is used to produce ethylene and acetylene directly by cracking of crude oil. These are used captively in VCM production. Thermal cracking of the intermediate EDC is limited to 50% to avoid excessive formation of high boiling products. Reactions involved are the following:

$$CH_2=CH_2 + Cl_2 \longrightarrow \begin{matrix} CH_2=CH_2 \\ | \quad | \\ Cl \quad Cl \end{matrix}$$

$$\begin{matrix} CH_2-CH_2 \\ | \quad | \\ Cl \quad Cl \end{matrix} \xrightarrow[\substack{\text{pumice or} \\ \text{hot tube}}]{480-510 \ 'C} CH_2=CHCl + HCl$$

$$HC\equiv CH + HCl \longrightarrow CH_2=CHCl$$

$$CH_2=CH_2 + HC\equiv CH + Cl_2 \longrightarrow 2 \ CH_2=CHCl$$

<u>From Ethane</u>. Ethane is cheaper and more readily available than either ethylene or acetylene. The "Transcat" process involves cracking of a feedstock such as ethane to ethylene, which is chlorinated, oxychlorinated, and dehydrochlorinated simultaneously. Copper oxychloride acts as an oxygen carrier in this process and also functions in the recovery of HCl:

$$Cu_2Cl_2 + \tfrac{1}{2}O_2 \longrightarrow CuO \cdot CuCl_2$$
$$Cu_2Cl_2 + Cl_2 \longrightarrow 2CuCl_2$$
$$Cu_2Cl_2 + 2HCl + \tfrac{1}{2}O_2 \longrightarrow 2CuCl_2 + H_2O$$

High purity vinyl chloride is produced in an overall yield of 80 mol% based on ethane. The feed can contain ethane, ethylene, mixed ethylene-chlorination products, and HCl in various mixtures, and can thereby allow recovery of values from such materials. The flow sheet for the simultaneous chlorination, oxidation, and dehydrochlorination for producing vinyl chloride by the Transcat process is shown in Figure 3.

Vinyl chloride is a colorless, pleasant-sweet smelling gas at normal temperatures. It is soluble in aliphatic and aromatic hydrocarbons, esters, ketones, ethers, alcohols, and chlorinated solvents, but is essentially insoluble in water.

Vinyl chloride may be stored in ordinary steel cylinders, tank cars, and storage tanks. The monomer must be stored under pressure to maintain a liquid state. Vessels are loaded or unloaded by use of inert gas pressure or most commonly by using pumps.

In earlier production years, polymerization problems were encountered because of the presence of impurities such as acetylene,

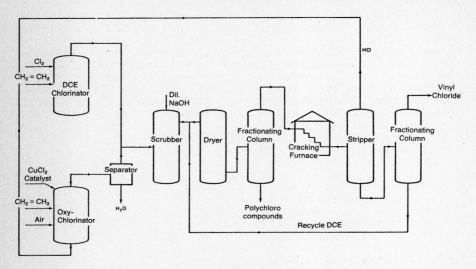

Figure 2.   Flowsheet: balanced chlorination-oxychlorination of
              ethylene to vinyl chloride.

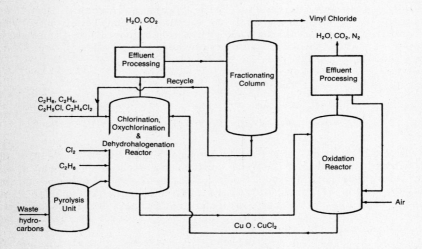

Figure 3.   Flowsheet: simultaneous chlorination, oxidation, and
              dehydrochlorination for vinyl chloride production
              (Transcat process).

iron, HCl, oxygen, water, and mercaptans, but these impurities have
largely been eliminated by improved manufacturing practices.
Properties of vinyl chloride monomer are given in Table IV.

Polymerization of Vinyl Chloride

Vinyl chloride can be polymerized by suspension, emulsion, bulk, or
solution techniques. The first two methods are the most important
in the United States. About 78% of the PVC produced by the United
States is made by the suspension process, and nearly all of the rest
is produced by the emulsion process. In Europe, suspension and
emulsion processes are used with approximately equal frequency.

Suspension Polymerization.

Suspension polymerization, also called
granular or pearl polymerization, involves the free-radical
catalyzed polymerization of discrete droplets of vinyl chloride
monomer suspended in water by agitation and a protective colloid
such as gelatin, methylcellulose, and poly(vinyl alcohol). Other
suspending agents such as maleic anhydride–vinyl acetate copolymers,
ethylcellulose, carboxymethylcellulose, alginates, or other
synthetic gums are also used. The quantity and type of suspension
system is very important in determining particle size, particle
shape, porosity, and size distribution. Level of suspending agent
is in the range of 0.02 to 0.5%, but details concerning these
systems are closely held commercial secrets. Modifying agents such
as buffers and pH adjusters must be carefully selected.

In a typical suspension process, the suspending agent is
dissolved in deionized and deaerated water and added to either a
glass–lined or stainless steel reaction vessel. The remainder of
the water, the catalyst or initiator, which is generally a monomer-
soluble organic peroxide, and other additives such as chain transfer
agent (used to control chain length or molecular weight), which are
generally chlorinated solvents such as trichloroethylene, are then
added. The reactor, which is jacketed and agitated, is sealed and
evacuated to remove oxygen, and liquefied monomer is pumped in. The
closed system is heated to ca. 120–130 $^{\circ}$F (50 $^{\circ}$C) with concomitant
increase in pressure to ca. 125 lb/in.^{2}. Temperature control to
±1 $^{\circ}$F is necessary in order to regulate the molecular weight.
Polymerization is continued at constant temperature until the
pressure drops about 10–30 lb/in.^{2} because of depletion of monomer
at which point the excess monomer is recovered, and the polymer
slurry is stripped to low residual VCM content, then cooled,
centrifuged, and dried. This pressure drop corresponds to a
conversion in the 80–92% range. Polymerization is generally
continued beyond the point at which 75% of the monomer has been
converted to polymer as indicated by pressure drop, and different
grades of resin are .pa produced by stripping off excess monomer at
different conversion levels. A simplified flow sheet for a
suspension polymerization process is shown in Figure 4.

Although the batch reactors originally used in suspension
polymerization were in the 1000- to 2200-gal capacity range, plants
were then built with larger sized reactors having 3750-gal and 5000-
gal capacity. Recently, Shit-Etsu in Japan pioneered exceedingly
large reactors (36,000 gal) of unique design capable of safe
operation, and this technology is being used in the United States by

Table IV.   Properties of Vinyl Chloride Monomer

Property	Value
Boiling point	−13.8 °C
Freezing point	−153.8 °C
Density at −20 °C	0.983 g/mL (8.21 lb/gal)
at 20 °C	0.910 g/mL (7.60 lb/gal)
Viscosity at −10 °C	0.25 cps
at −20 °C	0.274 cps
Index of refraction, n_D^{15}	1,398
Surface tension, −20 °C	22.27 dyn/cm
Molecular weight	62.50
Heat of polymerization	23 kcal/mol
Volume decrease on polymerization	35%
Flash point	−78 °C
Vapor pressure at 0 °C	25.1 lb/in.^{2}
at 20 °C	49.2 lb/in.^{2}
at 40 °C	87.6 lb/in.^{2}
Explosive limits in air	4–22% by volume

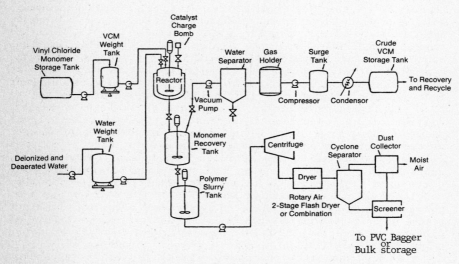

Figure 4.   Flowsheet: suspension polymerization of poly(vinyl chloride).

Tenneco at Pasadena, Texas; Shintech at Freeport, Texas; and Firestone (now Hooker Chemical Corp., a subsidiary of Occidental Petroleum) at Baton Rouge, Louisiana. Chemische Werke Huls in Germany is reported to be operating reactors having more than a 50,000-gal capacity.

The heat of polymerization of vinyl chloride to polymer is 22.9 kcal/mol corresponding to the evolution of 659.5 Btu/lb. Thus, a major factor in the successful operation of PVC manufacturing plants is the maximization of removal of exothermic heat balanced by kinetic control. This factor largely depends on the choice of catalyst systems to assist in obtaining uniform conversion rates.

A typical conversion curve for suspension polymerization is shown in Figure 5 as given by Barr (11). The conversion rate peaks after 12.25 h at approximately 20% conversion/h. The result is a peak heat load, a so-called "heat kick" that is difficult to dissipate through the walls of large reactors when cooling water is used. This heat load can be removed by use of refrigerated water or via a reflux condenser. Large reactor technology has advanced such that one half to one third of this time is required for polymerization.

As indicated earlier, monomer soluble initiators are used to catalyze this free-radical polymerization. Among these are lauroyl peroxide, di(2-ethylhexyl)peroxy dicarbonate, and 2,2'-azobisisobutyronitrile. By properly selecting the initiator it is possible to reduce or eliminate the induction period for the polymerization and obtain a uniform reaction rate. In addition to a quick start, the selection of initiators is governed by the reaction temperature. Usually the 10 h half-life temperature is a good indicator of this selection. A tabulation of current initiators used in PVC production and their half-life decomposition temperatures is given in Table V (12). Optimum reactor operation involves starting the reaction fast, maintaining an active controllable isothermal polymerization, and finishing without the characteristic exotherm ("heat kick"). To do this, it is usually necessary to use two or more initiators. Once a balance of initiators is found for a specific reaction temperature, the cooling water demand is predictable and controllable.

The molecular weight of the polymer formed is a function of the reaction temperature used during the polymerization. Lower temperatures favor increased molecular weights. The relationship between polymerization temperature and molecular weight of resin (as expressed by relative viscosity) is shown in Figure 6. In commercial practice, the molecular weight is controlled by reaction temperature, and the reaction rate is controlled by the selection of the initiator and its concentration.

Low molecular weight PVC homopolymer is made at higher reaction temperature. Generally less initiator, usually lauroyl peroxide or azobisisobutyronitrile (AIBN) is needed, and there is improved heat removal capacity because of the increased differential with the cooling water. Low molecular weight polymers are more difficult to strip of residual VCM and to dry down to low volatile levels.

Also, as polymerization temperature increases, the quality of the polymer begins to suffer, and 160-170 °F is generally considered the maximum safe operating temperature. If a still lower molecular

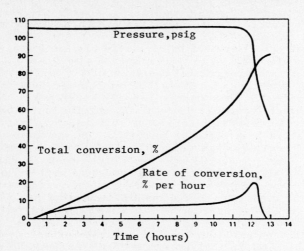

Figure 5.   Variation of conversion and pressure with time in a
             typical suspension polymerization.

Table V.   Decomposition Rates for PVC Initiators[a]

Product	Chemical Name	Half-life, °C 10 h	Half-life, °C 1 h	Activation Energy, kcal/mol
Lupersol 188M75	α–Cumyl peroxyneodecanoate	38	56	26.6
Lupersol 228Z	Acetyl cyclohexylsulfonyl peroxide	40	56	29.9
Lupersol TA–46	t-Amyl peroxyneodecanoate	46	64	28.7
Lupersol 47M75	α–Cumyl peroxypivalate	47	65	28.0
Lupersol 10M75	t-Butyl peroxyneodecanoate	49	66	28.5
Lupersol 223	Di-2-ethylhexyl peroxydicarbonate	49	66	30.5
Luperox IPP	Diisopropyl peroxydicarbonate	50	67	29.6
Lupersol 221	Di-n-propyl peroxydicarbonate	50	66	30.7
Lupersol 225	Di-sec-butyl peroxydicarbonate	50	67	27.6
Lupersol TA–54	t-Amyl peroxypivalate	54	74	25.7
Lupersol 11	t-Butyl peroxypivalate	57	75	29.7
Lupersol 219M75	Diisononanoyl peroxide	61	78	31.1
Decanox F	Didecanoyl peroxide	63	80	31.6
Alperox F	Dilauroyl peroxide	64	81	31.2

[a]Measured as 0.2 molar solutions in trichloroethylene.

weight is desired, use of chain transfer agents (regulators) at the higher temperatures is required.

Emulsion Polymerization. As noted with suspension polymerization, emulsion polymerization also involves the dispersion of VCM in an aqueous medium. As distinguished from suspension polymerization, however, the emulsion process involves the use of a surface active agent or soap as the emulsifier and a water-soluble catalyst or initiator instead of the monomer-soluble catalyst used in suspension processes. Although agitation is necessary, it is not as important as in the suspension processes because the emulsion is maintained by use of the soap and protective colloids to insure latex stability.

The emulsifier plays a significant role in emulsion polymerizations. At the concentrations used, most of the emulsifier exists in the form of micelles that consist of 50 to 100 molecules of emulsified soap. These micelles solubilize a portion of the vinyl chloride that is normally only slightly water soluble (0.09%) at 20 °C. The remainder of the monomer exists outside the micelles. The reaction starts when a free radical generated from the initiator enters the micelle, meets the monomer therein, and rapid polymerization occurs. As the reaction continues, the polymer particles grow larger than the micellar core and become engulfed by monomer molecules diffusing out from the monomer droplets. Harkins (13) has theorized that the monomer droplets are not the loci of polymerization. Rather, they supply monomer to the growing chain radical initiated in the soap micelle.

Smith and Ewart (13a, 13b) quantified the Harkins' theory by the equation $R_p = k_p M_p N/2$ where R_p is the rate of propagation, k_p is the rate constant for propagation, M_p is the monomer concentration in growing chain particles, and N is the number of polymer particles per unit volume. If M_p is the constant, this equation is reduced to $R_p = k_p N$. Thus, the rate of emulsion polymerization should solely be a function of the number of polymer particles. In actuality, the reaction rate increases up to 20–25% conversion because of the increase in the number of growing radical chains; then the rate steadies as does the number of polymer particles up to 70–80% conversion. Beyond this point, the rate drops off because of low monomer concentration. Thus, as Talamini (13c, 13d) has noted, available evidence indicates that emulsion polymerization of vinyl chloride does not resemble true emulsion polymerization as described by Smith and Ewart, but shows the general behavior of heterogeneous polymerization.

In the United States emulsion polymerizations are generally carried out in equipment quite similar to that used for batch suspension polymerization. After monomer recovery, the resin is generally isolated by spray drying. This process amounts to approximately 13% of the PVC produced in the United States. Outside of the United States, and particularly in Germany, most emulsion PVC is produced in continuous polymerizers by use of tall reaction towers. In a typical tower, which is the subject of a patent to BASF, the tower has a diameter of 5 ft 3 in. and is some 23 ft tall, and the reactor is agitated only at the top by a 4-ft by 18-in. paddle agitator at   50 r/min to avoid latex coagulation.

In a typical emulsion polymerization, emulsifier, initiator, and buffer are dissolved in deionized water and are fed continuously to the top of the reactor while monomer enters above the liquid level. These reactors are operated in parallel with 18–24 h residence time in each stage. The reaction temperature is maintained between 100 and 125 °F by use of brine circulation in the tower jacket. One of the reputed advantages of the tower reactor is the large L/D ratio in which monomer droplets carried down into more or less stagnant section of the tower will separate and coalesce rising to the top of the tower where there is agitation and the possibility for emulsification. This advantage is given as a major plus for the parallel tower type of operation as contrasted with a continuous series of cascade reactors.

At the conclusion of polymerization, unreacted monomer is recovered by vacuum stripping, then is compressed, condensed, and purified for recycle in the process. A stabilizer, usually sodium carbonate, is then added to the latex at a level of about 0.4%, and the stabilized latex is spray dried. Alternatively some processes involve drum drying following by grinding. In these procedures that involve total drying of the latex, any catalyze residues, emulsifier, buffer, or other additives during the process end up with the product. Particles from emulsion processes are about 1 μm in diameter, about 1/100 of those encountered in suspension polymerization.

Alternate processes that find little use in the United States include coagulation of the latex by use of added electrolytes, alcohols, ketones, heat, or shear. The coagulum is dewatered by any convenient filtration apparatus (thereby to remove the bulk of the emulsifier) and dried.

Emulsion polymerization leads to a very narrow particle size distribution, which is suitable for use in plastisol applications. Resins having both a monodisperse and bimodal particle size distribution are produced. Blending of latexes is sometimes practiced with the extensive use of seed–latex techniques. A simplified schematic flow sheet for a continuous emulsion polymerization process is given in Figure 7.

Bulk Polymerization. Bulk polymerization processes, unlike suspension and emulsion processes are conducted in the absence of diluents. Removal of heat was a definite problem until work by Produits Chimiques Pechiney–St. Gobain solved this problem by breaking up larger polymer blocks with heavy balls in a horizontal, rotating autoclave (14). The original Pechiney–St. Gobain process has since been improved and less than 10% of the total U.S. PVC capacity is produced by the licensees of the process: B. F. Goodrich Co. at Pedricktown, New Jersey and at Plaquemine, Louisiana; Hooker Chemical Corp., a subsidiary of Occidental Petroleum Corp. at Burlington, New Jersey; and Certain–Teed at Lake Charles, Louisiana.

The polymerization is a two–step batch operation initiated with 2,2'-azobisisobutyronitrile (AIBN) or lauroyl peroxide or with a mixture of AIBN and acetyl cyclohexyl sulfonyl peroxide (ACSP). The process has the obvious advantage of eliminating the need for protective colloids, emulsifier, buffers, and other additives. The reaction is started in a vertical reactor equipped with a high–speed agitator operating at 130 r/min at 140 °F and requires 3 h. The reaction is continued to 10% conversion, when the batch is dropped

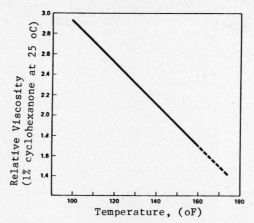

Figure 6.   Polymerization temperature–relative viscosity relationship on PVC homopolymers.

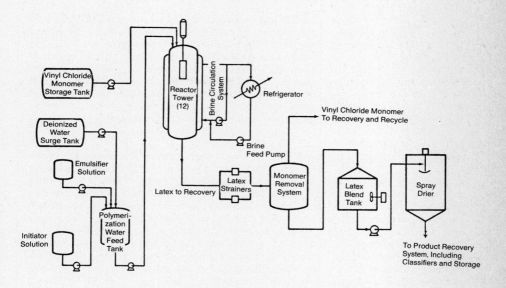

Figure 7.   Flowsheet: continuous emulsion polymerization of poly(vinyl chloride).

to a horizontal autoclave agitated with slowly rotating ribbon agitators as shown in Figure 8. The agitator system comprises two ribbons wound in spirals of different diameters and of opposite hands. The larger diameter spiral serves as a conveyor screw at the end of the reaction cycle and conveys the polymer to the product discharge port. Autoclaves ranging from 4200 to 21,000 gal are used in commercial operations, and it is understood that still larger autoclaves are being designed.

PVC is insoluble in VCM. The precipitated polymer tends to coagulate in the conversion range of 1–8%. As the conversion continues, the precipitated polymer absorbs more and more monomer, and at a 15–20% conversion, the reactor contains solid polymer swollen with monomer in a monomer atmosphere. The horizontal autoclave prevents the formation of large polymer blocks by breaking them up. Advantages claimed for this process are the higher bulk density, improved particle size distribution, and more rapid plasticizer absorption. Talamini and coworkers (15) have demonstrated that the bulk polymerization process is kinetically equivalent to suspension polymerization.

The autoclave is operated at an agitation speed of 30 r/min and initially is cooled to maintain the 140 °F polymerization temperature and a 130 lb/in.^{2} vinyl chloride pressure. As the reaction proceeds, hot water or cold water is circulated in the jacket of the autoclave to keep the autoclave pressure at 130 lb/in.^{2}.

Approximately 9.5 to 10 h are required for the second stage of the polymerization for a total reaction time of 12.5 to 13 h per batch. At this point, the autoclave is vented through a cyclone–bag filter combination (to remove entrained fine polymer) to the monomer recovery system. The residual monomer adsorbed in the polymer is removed by evacuating the autoclave twice and breaking the vacuum each time with nitrogen. The monomer stripped product is discharged through a port and transferred to product finishing where oversize material is reduced by grinding and milling to desired particle size specifications.

Solution Polymerization. Solution polymerization is over 45 years old, but only about 3% of the PVC produced in the United States is made this way. The solution process differs from the other processes already discussed in that a solvent is added to the polymerization system. The system may be heterogeneous, in which case the monomer is soluble but the polymer is insoluble. Examples are the use of hexane, butane, ethyl acetate, or cyclohexane as solvents. After addition of a peroxide initiator and heating to 40 °C, the polymerization starts and polymer precipitates out of solution as formed. In homogeneous reactions, both monomer and polymer are soluble therein. Examples are the use of dibutyl phthalate and tetrahydrofuran as solvents.

Solution polymerization is used almost exclusively for production of copolymers of vinyl chloride and vinyl acetate and terpolymers containing maleates. The vinyl acetate copolymers contain 10–25% acetate, are highly uniform with a narrow molecular weight range, and are valuable mainly because of their unique solubility and film-forming characteristics. They find use as solution coating resins where their high quality and uniformity command a premium price.

Copolymerization of Vinyl Chloride

In the copolymerization of two monomers, M_1 and M_2, the four propagation steps possible are

$$M_1^* + M_1 \xrightarrow{\quad k_{11} \quad} M_1^*$$

$$M_1^* + M_2 \xrightarrow{\quad k_{12} \quad} M_2^*$$

$$M_2^* + M_1 \xrightarrow{\quad k_{21} \quad} M_1^*$$

$$M_2^* + M_2 \xrightarrow{\quad k_{22} \quad} M_2^*$$

where M_1^* and M_2^* are growing polymer chains with M_1 and M_2, respectively, at the active growing ends.

The polymer composition is given by the reactivity ratios:

$$r_1 = \frac{k_{11}}{k_{12}}$$

$$r_2 = \frac{k_{22}}{k_{12}}$$

If r_1 and r_2 are less than 1, the polymers tend to alternate. If r_1 is greater than 1 and r_2 is less than 1, M_1 predominates in the polymer. To prevent forming a polymer with a wide distribution of composition, it is necessary that the more reactive monomer be added during the course of the polymerization. The reactivity ratios for vinyl chloride (M_1) and other monomers (M_2) are shown in Table VI.

Copolymers with all of the comonomers in Table VI have commercial value, but the copolymer containing vinyl acetate is the most important and is plasticized internally. Major uses include coatings, floor coverings, and phonograph records for which low melt viscosity, lower processing temperatures, freedom from external plasticizers, and solubility are required. Copolymerization with vinyl acetate reduces tensile strength, heat distortion temperature, abrasion resistance, chemical resistance, and heat stability. However, less heat stability is required because these copolymers can be processed at lower temperatures.

The composition of a vinyl chloride-vinyl acetate copolymer produced from a mixture of monomers is shown in Figure 9. Because r_1, the monomer reactivity ratio, is greater than one, and r_2 is less than one, the copolymer is richer in M_1 (vinyl chloride). Thus, if one were to copolymerize a mixture of monomers comprising 60% vinyl chloride, the resultant copolymer would contain approximately 75% vinyl chloride.

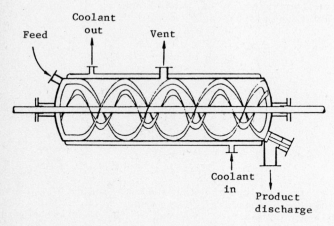

Figure 8.   Horizontal autoclave: Pechiney St. Gobain bulk
             polymerization process.

Table VI.   Monomer Reactivity Ratios, M_1 = Vinyl Chloride

M_2	r_1	r_2
Acrylonitrile	0.02	3.28
Butadiene	0.035	8.8
Dibutyl maleate	1.4	0.0
Diethyl fumarate	0.12	0.47
Diethyl maleate	0.8	0.0
Isobutylene	2.05	0.08
Maleic anhydride	0.296	0.008
Styrene	0.067	35.0
Vinyl acetate	1.68	0.23
Vinylidene chloride	0.2	1.8
N–Vinyl pyrrolidone	0.53	0.38

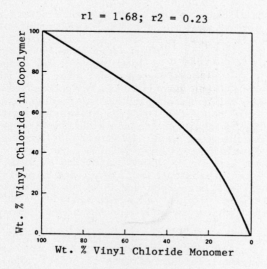

r1 = 1.68; r2 = 0.23

Figure 9.   Composition of copolymer produced from a vinyl chloride-vinyl acetate monomer mixture.

In attempts to reduce processing temperatures and increase solubility, copolymers of vinyl chloride and vinylidene chloride have been developed. These are soluble in ketone solvents, can be diluted with aromatic hydrocarbons, and are thus used in coatings applications. Such copolymers are useful as extender resins in plastisol applications for which rapid fusion at high temperatures is required.

Copolymers with 15% of maleate and fumarate esters afford good processing properties with only a mild impairment of physical properties. Heat distortion temperature is relatively high.

Copolymers with isobutyl vinyl ether were developed by BASF. These copolymers are soluble in toluene and xylene, compatible with nitrocellulose and other resins, and find use in coatings applications. They provide strong films with good adhesion to metal and possess excellent light stability and weathering properties.

Other copolymers that have improved processing characteristics at lower processing temperatures include a series of vinyl chloride-propylene copolymers that were introduced by Air Products and Chemicals and contain 3 to 10% propylene. The lower processing temperatures permit the use of lower quantities of stabilizers, the use of less-expensive nontoxic stabilizers, and higher processing rates. Such copolymers have Food and Drug Administration approval for food applications. Similar copolymers having like advantages and based on vinyl chloride-ethylene have been commercialized by Union Carbide.

The possibilities for copolymers with specialized performance characteristics depend on the ability to effect copolymerization with the other monomer. To date efforts to effect copolymerization with such monomers as styrene and butadiene have been unsuccessful (r_2=35 vs. r_1=0.067 for styrene, and r_2=8.8 vs. r_1=0.035 for butadiene). The possibilities nevertheless exist for modifying the basic polymer to effect such desirable properties as cross-linking, improvement in impact resistance, compatibility with processing aid modifiers, and UV stability.

Characterization of Poly(vinyl chloride)

The polymerization process results in polymer molecules with a variety of chain lengths or molecular weights. Thus, any measurement by experimental means leads to an average value. We should distinguish between two types of molecular weight averages:

1. Mn = number average molecular weight. This quantity is a measure of the number of molecules in a known mass and can be arrived at by osmometry and gel permeation chromatography.
2. Mw = weight average molecular weight. This quantity is measure of the effect that the mass of the molecule has on the average and is obtained by ultracentrifugation or gel permeation chromatography. Most PVC resins in commercial use have Mn ranging from 50,000 to somewhat above 100,000.

Mw is always greater than Mn except in the case of monodisperse polymer for which Mw = Mn.

The ratio of Mw/Mn is taken as a measure of the polydispersity or the breadth of molecular weight distribution of the polymer. The

molecular weight distribution for PVC produced commercially tends to
be narrower than for many other polymers produced via free-radical
polymerization because the broadening in distribution caused by
chain transfer to previously formed polymer molecules is largely
minimized by the precipitation of polymer from solution. Mw/Mn
usually lies between 2.0 and 2.3 and several investigators have
shown ratios in this range. Pezzin and coworkers (16) examined the
molecular weight distribution on six samples taken at degrees of
conversion ranging between 4 and 94% in a suspension polymerization
and concluded that the distribution does not change during the
course of the polymerization.

The molecular weight is generally determined by measurement of
viscosity of dilute solutions. The most widely used or preferred
solvents for PVC are tetrahydrofuran (THF), cyclohexanone, and
methyl ethyl ketone (MEK). PVC swells in aromatic hydrocarbons, but
is unaffected by alcohols or aliphatic hydrocarbons. Benzyl alcohol
at 155.4 °C and THF/water at 30 °C are the reported (17) theta
solvents for PVC, that is, one in which the dimensions of the
molecular coil are unperturbed by solvent effects. For such dilute
solutions, the viscosity and molecular weight are related by the
Mark-Houwink-Sakurada equation $[\eta] = KM^a$, where $[\eta]$ is the intrinsic
viscosity and M is the molecular weight. K and a are constants that
are determined by measurement of intrinsic viscosity and molecular
weight (obtained by an absolute technique such as light scattering,
ultracentrifugation, or osmometry).

The intrinsic viscosity $[\eta]$ is obtained by plotting the reduced
viscosity versus concentration and extrapolating to infinite
dilution. Reduced viscosity is given by the expression $\eta-\eta_0/\eta_0 C$
where η is the viscosity of solution, η_0 is the viscosity of
solvent, and C is the concentration in g/dL.

The following relationships have been determined by Bohdanecky
and coworkers (18):

$$[\eta] = 1.38 \times 10^{-4} M_w 0.78 \text{ (Cyclohexanone, 25 °C)}$$

$$[\eta] = 1.63 \times 10^{-4} M_w 0.77 \text{ (THF, 25 °C)}$$

and these provide reliable correlations for narrow fractions of
linear PVC where molecular weights are less than 10^5. Above 10^5,
discrepancies are noted, and these are attributed to aggregation.

McKinney (19) has developed the following expressions relating
number average and weight average molecular weights with intrinsic
viscosity:

$$M_n = 0.5 [\eta] \times 10^5$$

$$M_w = 0.9 [\eta] \times 10^5$$

If the molecular weight distribution differs from slightly greater
than 2, these equations lead to erroneous results.

Commercially, several types of viscosity measurements are used
to determine the molecular weight of PVC. These include, in
addition to intrinsic viscosity as previously defined, the
following:

1. Inherent viscosity (ASTM D1243-Method A)--viscosity is
    determined at 15 °C on a 0.2% solution of cyclohexanone.
2. Specific viscosity (ASTM D 1243-Method B)--viscosity is
    determined in nitrobenzene at 0.4% at 25 °C (used widely in
    Japan).
3. Relative viscosity--determined on a 1% solution in cyclohexanone
    at 25 °C.
4. Fikentscher K value--0.5% solution in cyclohexanone or ethylene
    dichloride (used widely in Europe).

The interrelationships between these viscosity methods are shown in
Figures 10, 11, 12, and 13.
    In addition to molecular weight and molecular weight distri-
bution, branching and crystallinity of PVC merit consideration.
    Branching in PVC, which occurs by a chain transfer mechanism, is
dependent on polymerization conditions. Thus, in a heterogeneous
system having high localized concentrations of polymer, branching
might be expected to be relatively high. At lower polymerization
temperatures, the degree of branching is low.
    The degree of branching is established by IR analysis on the
polyethylene produced by hydrogenation of PVC. The IR absorption
bands for a methyl group (which would occur at a branch point) at
1378 cm^{-1} and a methylene group at 1370 cm^{-1} are used for this
determination.
    In general, branching is not very prevalent in PVC polymerized
below room temperature (19a). The degree of branching in commercial
polymers has been found to range between 0.4 and 1.1 per 100 carbon
atoms. Changes in physical properties such as density, melting
point, and glass transition temperature have often been attributed
to branching, but in reality they are due to the syndiotacticity or
crystallinity of the polymer.
    It was Staudinger (19b) who first recognized the concept of
stereoregularity of vinyl polymers. PVC may be represented as a
chain in which the carbon atoms lie in the plane of the slide, and
atoms shown with a hatched line lie below the plane and those with
the bold line lie above the plane. Three configurations can thus
result:

1. Atactic-- No order to the configuration.

2. Isotactic--All chlorine atoms are in the same relative position.

3. Syndiotactic--Chlorine atoms are in an alternating sequence.

Although the isotactic and syndiotactic configurations are regular
and allow crystallization of the polymer chain to occur, most PVC is
atactic, and the regularity necessary for chains to pack and
therefore crystallize does not exist. Thus, PVC is not generally
classified as a crystalline polymer.
    As formed in conventional commercial processes, PVC is highly
amorphous with a very low degree of crystallinity associated with
its syndiotactic sequences. Thus, PVC produced in the normal
fashion contains about a 10% degree of crystallinity. However, if
the polymerization is carried out at very low temperatures, there is
a marked increase in crystallinity. Thus, polymer prepared at

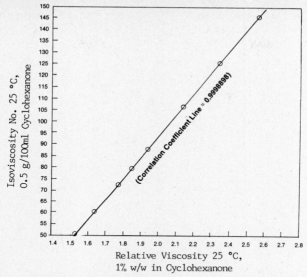

Figure 10. Relative viscosity vs. isoviscosity for PVC resins.

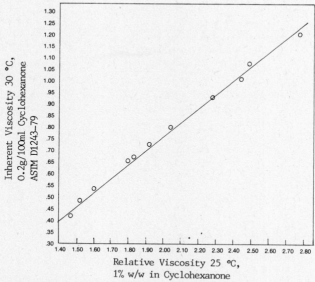

Figure 11. Relative viscosity vs. inherent viscosity for PVC resins.

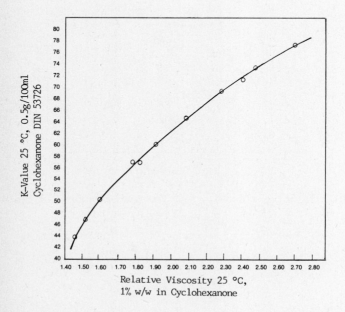

Figure 12.   Relative viscosity vs. K-value for PVC resins.

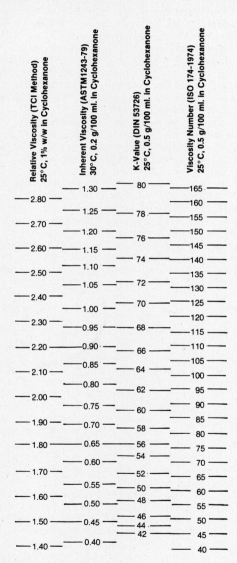

Figure 13.   Dilute solution viscosity correlation chart for PVC resins.

-15 °C has about 57% crystallinity and that produced at -75 °C has 85% crystallinity. The crystallinity is due to the presence of syndiotactic sequence.

NMR has also been used to study the degree of syndiotacticity in PVC as, for example, that prepared by the bulk polymerization process at temperatures ranging from -78 to 120 °C. By the analytical techniques used, an atactic polymer would have a fraction of syndiotactic diads, β equal to α, the fraction of isotactic diads, or equal to 0.5. Polymer prepared at or near commercial temperatures had a syndiotactic diad fraction β, equal to 0.51 to 0.53. However, the syndiotacticity increased as the temperature was decreased as shown in Figure 14.

Brief mention should be made as to ordering in the PVC polymer. If we consider the methylene group in vinyl chloride as the tail of the molecule and the carbon atom bearing the chlorine atom as the head of the molecule, three possibilities exist: head-to-tail, head-to-head, and tail-to-tail.

Addition polymerization leads preponderantly to the head-to-tail structure. Marvel (20) and coworkers carried out dechlorination of the polymer in boiling dioxane with zinc dust. In this procedure, chlorine is removed to yield cyclopropane rings and zinc chloride. The dechlorinated material contained 13-16% retained chlorine. Although this favors the head-to-tail arrangement, it does not preclude a certain amount of head-to-head or tail-to-tail product. However, the residual chlorine content corresponds to that expected from the random attack of zinc on a head-to-tail polymer structure.

Physical Properties

PVC is a hard, brittle polymer at room temperature, and, as such, would have little use were it not for the fact that it readily accepts a number of plasticizers. The resultant flexible plasticized products possess a broad spectrum of desirable properties. Typical properties of unmodified PVC resin are given in Table VII.

Glass Transition Temperature. The glass transition temperature, T_g, is that temperature at which an amorphous polymer changes from the hard or glassy state to a soft or rubbery material. Near this temperature certain properties change, for example, the modulus abruptly changes. Mechanical loss (ratio of energy dissipated to energy stored per cycle by a sample under oscillatory strain) exhibits a maximum at T_g. A plot of mechanical loss and the elastic component of shear modulus (obtained with a torsion pendulum at a frequency of ca. 1 c/s) versus temperature as determined on a commercial, suspension-polymerized product is shown in Figure 15 (21). The inflection point in the modulus versus temperature curve can be used to measure T_g as with an Instron tester. From Figure 15 it is seen that at room temperature and below, PVC is hard and brittle.

The effect on T_g of the polymerization temperature was studied by Reding, Walter, and Welch (22). Plotting their tabular data results in Figure 16, which shows that there is essentially a linear dependence of T_g on polymerization temperature. The variation of T_g with polymerization temperature is affected more by stereoregularity than by changes in head-to-head structure content.

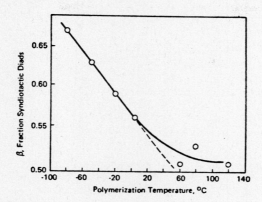

Figure 14.   Dependence of degree of syndiotacticity on polymerization temperature.

Table VII.   Typical Properties of Unmodified PVC Resin

Property	Value
Tensile strength, lb/in.^{2}	6500
Elongation at break, %	25
Tensile modulus, lb/in.^{2}	6×10^5 to 12×10^5
Shore hardness	D75 to 85
Heat capacity, 25 °C, cal/deg/g	0.226
Coefficient of thermal expansion, deg^{-1} (glass) (rubber)	5.2×10^{-4} 2.1×10^{-4}
Glass transition temperature, °C	70 to 80
Refractive index at 20 °C	1.54

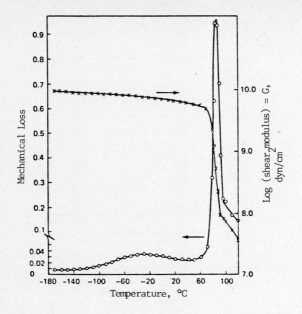

Figure 15. Mechanical loss (O) and shear modulus (X) vs. temperature for commercial suspension PVC.

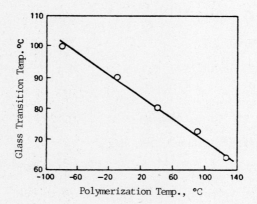

Figure 16. Glass transition temperature as a function of polymerization temperature.

Plasticizers have a marked effect on T_g. One would normally expect that the plasticized polymer would have a lower strength, hardness, and modulus than the unplasticized material. Actually incorporation of small amounts of a plasticizer such as di(2-ethylhexyl) phthalate (DOP) causes an increase in these values. At the 10–15% level of plasticizer, X-ray diffraction studies and IR analyses show an increase in the amount of crystallinity, probably due to an ordering of polymer chains. This result is the so-called "anti-plasticizer effect" and is dependent on the type of plasticizer. As the amount of plasticizer is increased, however, these properties (tensile strength and modulus) do decrease as shown in Figure 17. The effect on T_g as a result of increasing levels of DOP plasticizer is shown in Figure 18.

An increase in pressure results in an increase in T_g. When the polymer chains are subjected to pressure, more thermal energy is required to activate the energy loss mechanisms. Heydemann and Guicking (23) determined the specific volume of unplasticized and DOP plasticized PVC by dilatometry over the temperature range of −80 to 150 °C at hydrostatic pressures of 1–1000 atm. Figure 19 shows their results. Within the limits of accuracy, T_g appears to be a linear function of applied pressure.

Melting Point (Softening Point). As noted earlier, the tacticity or crystallinity of PVC is markedly affected by polymerization temperature, as is the melting point. Because polymer decomposition is too rapid at the elevated temperatures, the melting point of PVC cannot be determined by direct techniques. Reding and coworkers (22) used an extrapolation technique in which DOP was added to PVC at various concentrations, and stiffness modulus was measured as a function of temperature. The melting temperature on each plasticized sample was taken at the temperature at which the stiffness modulus was 10 lb/in.^{2}. By plotting the melting temperatures versus plasticizer content and extrapolating back to zero plasticizer, the melting point for the polymer was obtained. Their results shown in Figure 20 are in good agreement with those obtained by Nakajima and coworkers (24) who extrapolated the melting point of polymer-diluent systems as a function of thermodynamic interaction parameters.

The melting point of an all-syndiotactic or crystalline polymer would be very high, but is lowered by the amorphous chains of the polymer. When a plasticizer is added, it solvates the amorphous chains to produce a sheet that is flexible at room temperature. The sheet is strong and tough, however, because of the network of nonsolvated, high-melting crystalline portions.

Differential Thermal Analysis. Differential thermal analysis (DTA) is a technique in which differences in temperature between a sample under investigation and an inert reference material are measured as both materials are simultaneously heated at a uniform rate. Endotherms and exotherms due to melting, phase changes, or chemical changes are thus seen. Changes in heat capacity at T_g are often noted.

Matlack and Metzger (25) reported on a thermogram of PVC above room temperature, and their curve is shown in Figure 21. Point A shows a change in slope over the 65–80 °C range which is associated

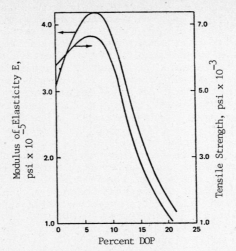

Figure 17.   Effect of plasticizer on modulus of elasticity and tensile strength.

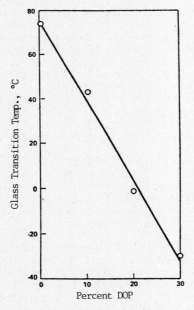

Figure 18.   Glass transition temperature of plasticized PVC.

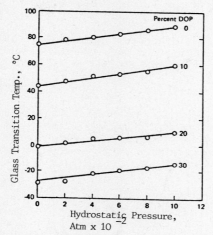

Figure 19. Effect of hydrostatic pressure on the glass transition temperature of unplasticized and DOP-plasticized PVC.

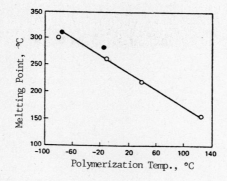

Figure 20. Melting point of PVC as a function of polymerization temperature.

with T_g. Point B, at which an endothermic response is noted between 165 and 210 °C, is the melting point. As noted earlier this will vary with temperature of polymerization. Point C, an exotherm at approximately 250 °C, is the point at which oxidative attack occurs and is dependent on the stabilizer system used. Point D is the temperature at which dehydrochlorination occurs (loss of HCl). Point E, an endotherm, is postulated as being due to decomposition.

In addition to DTA, other techniques used to detect thermal activity in PVC include differential scanning calorimetry, specific heat, thermal conductivity, and thermal diffusity.

Chemical Properties

PVC is generally insensitive to chemical attack in comparison to other commercial polymers. Contact with acids, for example, produces no deleterious effects. However, reaction with aqueous alkali at elevated temperatures or with organic amines adversely affects the stability of PVC.

Thermal Dehydrochlorination. PVC, per se, is unstable and, at moderate temperatures, undergoes decomposition with loss of HCl. The successful commercial development of PVC is largely due to the fairly complex technology involving development of stabilizers that neutralize the evolved HCl and result in improved color of the compounded polymer so that it can be processed like other thermoplastics.

The kinetics of the dehydrochlorination have been studied extensively, but there is no agreement on the mechanism. There is some support for a free-radical chain mechanism, although other investigators favor a unimolecular decomposition mechanism. A radical chain of any considerable kinetic length has been ruled out.

It was proposed very early in the history of PVC that HCl was lost primarily from positions where the chlorine atom was especially labile because of its location in an allylic position relative to the double bond. Such structures result from the loss of HCl which creates a double bond and thereby makes the adjacent chlorine atom labile. Subsequent loss of HCl appears to be catalyzed by HCl itself and is described as autocatalytic and produces alternate double bonds. Thus, a "zipper" reaction occurs as shown in Figure 22.

Bengough and Sharpe (26) measured the rate of dehydrohalogenation of laboratory-prepared samples in which catalyst residues were removed by purification. Nitrogen gas was swept over the degrading solution and thereby entrained the evolved HCl, which was measured analytically. An increased rate of HCl loss was noted with lower molecular weight polymers and was proportional to the number of moles of polymer present. This result suggests that the loss of HCl is connected in some way to the concentration of polymer chain ends.

The "zipper" mechanism is confirmed by the work of Baum and Wartman (27) who found that ozonolysis followed by hydrolysis of partially degraded PVC did not appreciably reduce its molecular weight. If the HCl loss had occurred in a random fashion within the polymer chain, a considerable drop in molecular weight would have been expected. In addition, when the resin was mildly chlorinated

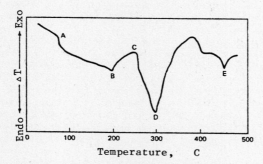

Figure 21. Differential thermal analysis curve of PVC.

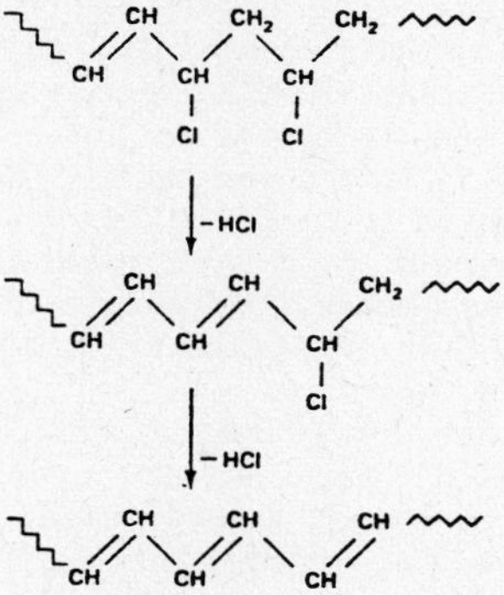

Figure 22. Dehydrohalogenation of PVC ("zipper" reaction).

to saturate double bonds at the chain ends but not to effect
substitution in the chain, the resultant polymer had a lower
dehydrohalogenation rate.

The presence of oxygen in the environment of degrading PVC
generally accelerates the loss of HCl. This result is believed to
be due to attack by oxygen at the conjugated polyene structures to
bleach them and simultaneously form oxygenated structures that
further labilize adjacent chlorine atoms.

Chemical Dehydrochlorination. The treatment of PVC with certain
chemical reactants results in complete loss of HCl and a product
with the empirical formula $(CH=CM)_n$, a polyene. This result occurs
when PVC is reacted with alcoholic or aqueous alkali, with calcium
hydroxide in 2-ethoxyethanol or calcium methylate in THF.

Because less than 0.5% of unreacted chlorine remains when a
dispersion of sodium amide in liquid ammonia is reacted with a 3%
solution of PVC in THF, it is hypothesized that the dehydrohalogenation does not occur in a random fashion. On the basis of
mathematical probability considerations, considerably higher
residual chlorine would be expected in a random attack. Thus,
chemical dehydrochlorination appears to proceed in a fashion similar
to thermal dehydrochlorination, that is, via a "zipper" sequence.

Radiation-Initiated Reactions. PVC is also degraded by actinic
irradiation resulting in discoloration, chain splitting, and cross-
linking. HCl is released during the process as measured directly or
as established by the fact that basic stabilizers are gradually
converted to their chloride form. It has been found that HCl is
only evolved when light of wavelengths shorter than 340 nm is used,
the amount of HCl formed increasing with decreasing light
wavelength. This photodegradation probably follows a radical
mechanism in which trace impurities and catalyst residues are
probably the source of initial radiation absorption and radical
formation. The unsaturated structures formed absorb more UV quanta
and decompose, and thereby continue the decomposition. Oxygen
affects this reaction, and molecular weight of the polymer decreases
probably because of chain-splitting, with the development of
carbonyl bands in the IR spectrum.

High energy irradiation may lead to HCl loss and formation of
color. Other effects of irradiation are cross-linking and/or chain
splitting as well as carbonyl development if oxygen is present.
Control of the cross-linking must be maintained so that chain
scission and color development are minimized. Electron beam cross-
linking and copolymerization of polymers including PVC have become
increasingly important tools in the plastics industry.
Manufacturers of communication wire have found that thicker and more
costly wrapped and lacquered wire can be selectively replaced with
PVC cross-linked insulation. The process involves irradiating
modified PVC insulated wire in a manner as depicted in Figure 23
(28).

Color Development. The absorption spectrum in the visible region of
PVC subjected to heat degradation shows several maxima that are
related to the degree of conjugation of the polyenes formed by
dehydrochlorination. Bengough and Varma (29) have studied the

visible absorption spectra of solutions of PVC subjected to
degradation at 198 °C. Their findings are shown in Figure 24. They
concluded that the absorption maxima noted corresponded to the
varying degrees of conjugated unsaturation.

Cross-linking. When it is heated in solution, as described above,
PVC not only loses HCl, but also undergoes cross-linking and thereby
results eventually in an insoluble gel. Initial stages of this
cross-linking reaction can be followed by measurement of intrinsic
viscosity. However, because of simultaneously occurring chain-
scission reactions and because the molecular weight of the branched
molecules formed cannot be reliably calculated,interpretation of
results is very difficult.
   According to the Flory-Stockmayer (29a) theory of gelation, the
gel point, corresponding to the formation of an infinite network, is
reached when one cross-link per initial polymer molecule present has
formed.

Hydrogenation. Approximately 97% of the chlorine atoms present in
PVC can be replaced by hydrogenations in the treatment of the
polymer with lithium aluminum hydride in THF. By use of this
reaction, which leads to a polyethylene, the amount of branching in
the polymer can be determined by IR analysis on the hydrogenated
material. The dehydrochlorinated resin that results from treating
PVC with ammonia under pressure can be hydrogenated.

Chlorination. Chlorination of PVC results in a resinous product
having about 73% chlorine, which corresponds to the introduction of
one chlorine per carbon atom. Such postchlorinated polymers have
achieved some industrial importance because they have lower
softening temperatures and increased solubility in a variety of
solvents. They have been used in synthetic fibers in Germany and in
solution coatings.
   By IR and X-ray analysis, it has been established that most of
the chlorine atoms substitute at methylene groups with only 25% of
the initial methylene groups remaining unmodified. Low-temperature
light-activated chlorination leads to more substitution of the
methylene group.
   If chlorination is carried out below 65 °C in the presence of a
diluent that swells but does not dissolve the polymer, the formation
of products with chlorines on all adjacent carbons is promoted.
Dannis and Ramp (30) have shown that products prepared by this
technique have higher T_g values and greater heat resistance than
products prepared at higher chlorination temperatures. The change
in T_g as a result of increased chlorine content in such low-
temperature chlorinations is shown in Figure 25. The high T_g shown
for poly(vinyl dichloride) is not that of poly(vinylidene chloride).

Classification of Poly(vinyl chloride) Resins

PVC resins are classified on the basis of their physical form and
the processing methods used.

General Purpose Resins. Homopolymer and copolymer resins produced
by suspension, emulsion, bulk, or solution techniques are included

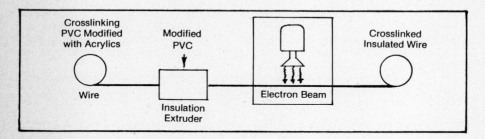

Figure 23.   Schematic outline of industrial PVC irradiation process.

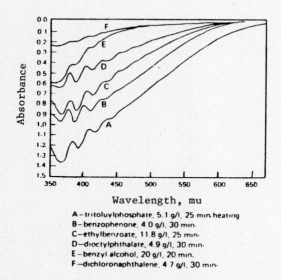

A — tritoluylphosphate, 5.1 g/l, 25 min heating
B — benzophenone, 4.0 g/l, 30 min.
C — ethylbenzoate, 11.8 g/l, 25 min.
D — dioctylphthalate, 4.9 g/l, 30 min.
E — benzyl alcohol, 20 g/l, 20 min.
F — dichloronaphthalene, 4 7 g/l, 30 min.

Figure 24.   Visible region absorption spectra: solutions of PVC
degraded at 198 °C in various solvents.

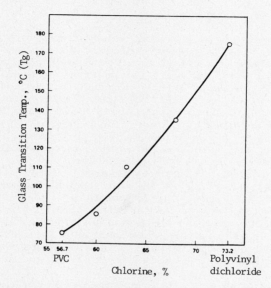

Figure 25. Dependence of glass transition temperature on chlorine
content in low-temperature chlorination of PVC.

in this category which comprises about 90% of all PVC used. These resins are converted in calendering, extrusion, injection molding, and blow molding processes.

In 1960, the American Society for Testing and Materials (ASTM) completed work on tests and standardizations and issued D1755, "Specification for Poly(Vinyl Chloride) Resins." In this specification, type GP refers to general purpose resins, and type D refers to dispersion resins (which will be discussed later). A similar specification for copolymer resins has been issued as ASTM D2427-78.

The following properties, determined by established ASTM test procedures, are of significance:

1. Molecular Weight--This property, which affects both processing and end use, was discussed earlier. Molecular weight is obtained from the ASTM dilute solution viscosity values.
2. Sieve Analysis--This analysis is important because particle size and distribution affect compounding, processing, and handling. Large particles having a relatively uniform size are generally preferred. Fines lead to dusting and uneven plasticizer sorption, and oversized particles do not absorb their full share of plasticizer, do not flux easily, and lead to gel particles or fish eyes. The average particle size for most commercial suspension resins is 50–150 μm.
3. Bulk Density--This property is generally desired to be high (spherical particles and uniform distribution) for maximum rates in compounding and processing equipment.
4. Plasticizer Absorption--The ability of a resin to absorb plasticizers depends on particle size and distribution, particle surface characteristics, and particle porosity. Highly absorptive resins, which generally have high particle porosity and irregular shapes, provide faster drying and minimum dry-blend time cycles.
5. Dry Flow--This property is a measure of ease with which the resin can be handled. Large particle size and spherical shape lead to good dry flow characteristics, which are important in avoiding hang-up and bridging in feed hoppers, screw conveyors, and feeders.
6. Electrical Conductivity--This factor, which is dependent on the amount and type of water extractables present in a resin, results from ionic and polar impurities that adversely affect use of the resin for electrical insulation. Resin with low electrical conductivity is desired for electrical applications.

Dispersion Resins. These materials used in plastisols or organosols have a very fine particle size, in the range of 0.1 to 2 μm and are made by emulsion polymerization followed by spray drying.

ASTM D1755-81 gives the classification for these resins. The following characteristics of dispersion resins are of significance:

1. Plastisol Viscosity--Brookfield viscosity (RVF) is used to determine viscosity at low shear rates by using a plastisol containing 60 parts per hundred parts resin (phr) DOP by ASTM D1823-66 (1979). The Castor-Severs viscometer is used to determine viscosity at high shear rates on the same plastisol

composition via ASTM D1824-66 (1980). In many cases, this instrument is being replaced by cone and plate instruments as well as high-speed rotational instruments.

2. Particle Size--Small particle size results in higher plastisol viscosity (due to greater surface area), but may also contribute to poorer viscosity stability. Larger particles fuse more slowly and may lead to poorer performance properties.

3. Plastisol Fusion--This property is a function of temperature, time, and solvating power of the plasticizer. This property may be measured by determining the clear point, that is, the temperature at which a clear solution is obtained when a low concentration of resin is suspended in a plasticizer and heated at a constant rate.

4. Air Release--For many applications, it is necessary in the processing of plastisols to apply a vacuum to remove air that may be entrapped as a result of mixing or that may have been contained in the resin. Rapid breaking of these air bubbles to insure a low level of foam height is desired. Air release rates and heights of bubble rise are used to determine this characteristic.

In addition to general purpose and dispersion resins, mention should be made of three other types:

1. Blending Resins--These are suspension polymerized homopolymers of vinylidene chloride copolymers that have a particle size in the 10-100 μm range and that are used as an extender in plastisol resins at levels up to 50% so as to reduce costs. The use of the larger particle size blending resin reduces viscosity and enables a higher resin loading in plastisols. Advantages of the use in plastisols, in addition to lower cost, are lower tackiness, drier hand, reduction of surface gloss, and a decrease in water absorption (due to lower amounts of emulsifier and protective colloids present).

2. Solution Resins--These are resins used in solution coating applications. Thus, particle properties and processing characteristics are relatively unimportant. Copolymers are generally used here because they are soluble in ketones, nitroparaffins, and chlorinated hydrocarbons and can be diluted with aromatic hydrocarbons such as toluene and xylene, depending on copolymer type and content. The best performance is obtained with highly uniform copolymers produced by careful polymerization control, and, if suspension or emulsion processes are used, they are washed to remove impurities. Generally, however, resins to be used for such coatings applications are made by solution polymerization, and thus the use of emulsifiers, protective colloids, buffers, and other impurities that are detrimental to clarity and metal adhesion is avoided.

3. Latexes--These are dispersions of homopolymers and copolymers in water that are produced by anionic emulsion polymerization and contain 50% solids. Particle size is low (0.2 μm) as is viscosity (20-30 centipoises). To produce film forming properties, homopolymer latexes must be plasticized. They are supplied either with or without plasticizer. Certain copolymer latexes such as those with acrylates are self-plasticized and do

not require the addition of plasticizer. Indeed at higher
comonomer levels, these latexes will fuse at or near room
temperature and thus sometimes result in surface tackiness.

Plasticizers

One of the most significant contributions to the development of PVC
as a commercially important plastic was the discovery by Waldo L.
Semon at B. F. Goodrich that the polymer could be flexibilized by
heating in the presence of certain esters--so called "plasticizers."
Plasticizers may be defined as high boiling liquids or low molecular
weight solids that are added to resins to alter their processing and
physical properties. Resin flexibility, softness, and elongation
are increased by a decrease in the intermolecular or van der Waals
forces, and, as noted earlier, the T_g is also decreased.
Plasticized PVC can be visualized as a polymeric mass in which
certain regions have been solvated and flexibilized by the
plasticizer, but which has other small, rigid, nonsolvated,
crystalline regions in a network. This characteristic imparts
toughness and strength to the plasticized polymer sheet.

The use of plasticizers results in improved low-temperature
stability, decreased tensile strength, and reduced processing
temperatures and melt viscosity. Plasticizers having solubility
parameters, polarity, and hydrogen bonding similar to those in PVC
are compatible with it and exhibit good solvating power. However,
when forces of attraction of either the plasticizers or resins are
greater for themselves than for each other, exudation, spewing, or
blooming results. Thus, glycol and glycerin are incompatible
because of their high polarity, and mineral oil and low molecular
weight polyethylene are incompatible because of their low polarity.

Plasticizers may be divided into two categories on the basis of
their solvating power and compatibility with resins. Primary
plasticizers can solvate resins and, within certain limits, retain
compatibility on aging. Secondary plasticizers have limited
solubility and compatibility and are, therefore, used only in
conjunction with a primary plasticizer. Plasticizer compatibility
is now largely related to cohesive energy density (which can be
calculated from the structure) or solubility parameter.

Plasticizers are also classified by molecular weight, ranging
from less than 500 for monomeric plasticizers and from 500-2000 for
polymeric plasticizers. The monomeric types are mono-, di-,and
triesters of acids or anhydrides such as phthalic, adipic, azelaic,
sebacic, citric, benzoic, trimellitic, and phosphoric acid or
anhydride with monohydric alcohols. Polymeric plasticizers are
based on adipic, azelaic, and sebacic diacids reacted with diols
such as propylene and butylene glycol. Although the efficiency of
polymeric plasticizers is not as good as monomeric plasticizers,
their low volatility and resistance to migration are outstanding.
Plasticizers such as epoxidized soybean oil or epoxy tallate are
intermediate in molecular weight and thus may be considered either
as polymeric or monomeric. A tabulation of commonly used
plasticizers is given in the box. The recent PVC consumption of
plasticizer by end use is shown in Table VIII (31), and the
consumption of plasticizers by type is in Table IX (31).

Table VIII. PVC Consumption of Plasticizers by End Use

Application	Consumption, 1000 Tons		
	1979	1980	1981
Film and sheet	127	132	136
Flooring	71	55	56
Molding and extrusion	118	109	111
Textile and paper coating	77	58	59
Wire and cable coating	105	93	95
Others[a]	109	92	95
Total	607	539	552

[a]Includes adhesives, protective coatings, paste processes, and miscellaneous calendering uses.

Table IX. Consumption of Plasticizers by Type

Plasticizer	Consumption, 1000 Tons		
	1979	1980	1981
Adipates	32	28	28
Azelates	5	5	5
Epoxy 59	55	57	
Phthalates	509	455	460
Polyesters	25	22	21
Trimellitates	14	13	13
Others[a]	108	97	100
Total	752	675	684

[a]Includes phosphates, glutarates, stearates, etc.

Commonly Used Plasticizers

Phthalates

      Butyl benzyl phthalate (BBP)
      Butyl octyl phthalate (BOP)
      Di(2-ethyl hexyl) phthalate (DOP)
      Di-isodecyl phthalate (DIDP)
      Ditridecyl phthalate (DTDP)
      Di(*n*-hexyl, *n*-octyl, *n*-decyl) phthalate (linear)

Phosphates

      Tricresyl phosphate (TCP)
      Tri(2-ethyl hexyl) phosphate (TOF)

Trimellitates

      Trioctyl trimellitate (TOTM)
      Tri-iso-octyl trimellitate (TIOTM)

Aliphatic Diesters

      Dioctyl adipate (DOA)
      n-Octyl decyl adipate
      Dibutyl sebacate

Benzoates

      Diethylene glycol dibenzoate
      Dipropylene glycol dibenzoate

Epoxides

      Epoxidized soybean oil
      Octyl epoxy tallate

Polymerics

      Low, medium, and high molecular weights

Heat Stabilizers

Initial work to process PVC resin into manufactured products led many investigators to believe that the polymer was not a useful one. Some of the performance shortcomings of PVC were overcome by the advent of plasticizers and development of copolymers, both of which were responsible for the subsequent commercial growth of PVC. Although resins with improved stability and processing characteristics have since been developed, the tremendous growth of vinyl plastics is in no small measure due to the development of effective heat stabilizers as well as efficient processing equipment.

As noted earlier, when PVC is exposed to certain conditions of heat and light, HCl is evolved and a chromophoric conjugated polyene structure results. Obvious degradation begins at about 200 °F and increases sharply with time and temperature. As the degradation proceeds, the color of the plastic changes from white (or colorless) to yellow through tan and brown to reddish-brown and finally to black. Color formation usually precedes serious deterioration of physical properties of the polymer. Thus, one requirement for a good stabilizer is its ability to prevent discoloration during the manufacture of a PVC compound.

A great deal of investigative work has been carried out to elucidate the mechanism of PVC stabilization, but no single theory is accepted by all authorities in this field. In general, however, it is the belief that the stabilizer reacts with the liberated HCl and thereby reduces the rate of further degradation of the polymer. The liberation of hydrogen chloride is a function of time, temperature, and the environment--the highest quantities being liberated in an oxygen atmosphere.

A wide variety of compositions are offered as commercial PVC stabilizers. Most of these are combinations of two or more active ingredients that perform synergistically. Such combinations are available in powder and liquid form. The basic components of PVC stabilizers may be classified as follows:

1. Underline: Lead Compounds--These were among the earliest stabilizers. Although quite effective as heat stabilizers, they are toxic, impart opacity to PVC compounds, and cause black coloration with time because of poor sulfide stain resistance. The primary use for lead compounds is in electric applications such as wire coatings. The principal lead compounds for this use are tribasic lead sulfate, basic lead sulfate silicate, basic lead carbonate, and basic lead phthalate. Use levels usually range from 3 to 8 parts-per-hundred parts of PVC (phr).
2. Metal Soaps--Barium and cadmium stabilizers are less toxic than lead and have lower cost on the basis of use levels, better sulfide stain resistance, and good clarity. They are a major class of heat stabilizers. Calcium and zinc soaps are also included in this classification because of their similar performance. The soaps used are based on a wide variety of anions such as octoates (2-ethylhexanoic acid derived), benzoates, laurates, stearates, and substituted phenolates.
    a. Barium compounds serve as the building blocks for a variety of stabilizer systems. They are seldom used alone because PVC stabilized solely with barium compounds rapidly becomes discolored.
    b. Cadmium compounds when used alone in PVC offer excellent initial color, but after a relatively short time at elevated temperatures cause sudden, severe degradation and color development. Advantage is taken of this early activity of cadmium compounds by using them in combination with barium compounds which offer long-term stability to provide the balance required for stabilizing PVC. The combined activity of cadmium and barium compounds has been shown to be synergistic.

    c. Zinc compounds have activity similar to cadmium compounds. However, although zinc compounds are responsible for early stability, once color develops these compounds cause rapid degradation of the polymer, which is difficult to control. Because of this result they are used only in minor amounts, in combination with barium and cadmium compounds, to further upgrade early color and improve the sulfide stain resistance of these systems. Zinc compounds also find use as nontoxic stabilizer (FDA sanctioned for use in food packaging) components in combination with calcium and magnesium materials.

    d. Calcium compounds function in a manner similar to barium-containing materials, but with greatly reduced efficiency. As a result, calcium-based materials are used only to a limited degree in general purpose stabilizers, but are a major component of nontoxic stabilizers.

Stabilizers containing barium, cadmium, calcium, and zinc compounds are generally used in PVC at levels of 1 to 4 phr.

3. <u>Auxiliary Organic Compounds</u>—These are of major significance in formulating barium-cadmium stabilizers and include the following:

    a. Organic phosphites may be used as integral parts of liquid stabilizers or may be added directly to the PVC composition during compounding, when solid stabilizer systems are used. Organic phosphites act as nonstaining antioxidants and may chelate the metallic salts formed during processing to result in greatly improved performance of the barium-cadmium and calcium-zinc classes of stabilizers.

    b. Polyhydric alcohols are used in the stabilization of rigid PVC, nontoxic systems, and vinyl asbestos floor tile compositions. They control the activity of the cadmium and zinc components and thereby extend the heat stability and thus provide improved color hold along with long-term stability. Examples of these materials are pentaerythritol, mannitol, and sorbitol.

    c. Nitrogeneous compounds such as melamine and dicyandiamide are used only in vinyl asbestos floor tile compositions, quite often in combination with polyhydric alcohols and metallic compounds. They provide heat stability and retard water growth (dimensional increase due to immersion in water) of vinyl asbestos materials.

    d. Epoxy plasticizers, derived from soybean oil and epoxidized iso-octyl tallate, are included as a standard part of most PVC formulations primarily as a supplement to barium-cadmium stabilizer systems. The normal level of usage is in the range of 3-8 phr. They extend the life of the metal stabilizer systems and offer some flexibility to the PVC compound.

4. <u>Organotin Compounds</u>—These offer good overall stabilization with excellent clarity and complete freedom from sulfide staining. They are more expensive than the common barium-cadmium types as a result of the higher tin metal prices. The commercially important tin stabilizers include various dibutyltin and dimethyltin salts of maleate half esters, alkyl mercaptans, and thioacids, as well as certain monoalkyltin derivatives. Certain

tin stabilizers (dioctyl derivatives) comply with FDA
requirements for specific food packing applications.

5. <u>Antimony Compounds</u>--These have gained much interest in recent
    years as potential low-cost replacements for tin stabilizers in
    rigid PVC pipe where they are used at extremely low levels. The
    use of antimony stabilizers is declining because these
    stabilizers show high degradation rates under demanding
    commercial processing conditions.

In selecting a heat stabilizer, the primary considerations include
the type of resin, exposure to the elements, high shear, and ultra-
high temperature conditions used in processing, electrical
properties, etc. Regardless of the final choice, the stabilizer
system must achieve the following:

a. Retard degradation by inhibiting onset of HCl formation.
b. Replace labile chlorines with groups having more stable bonding.
c. Absorb eliminated HCl in order to interfere with its
    autocatalytic action.
d. Interrupt double bond proliferation or add across double bonds.
e. Inhibit the oxidative process.
f. Scavenge free radicals or undesirable ionic species.
g. Counteract certain shear damage via lubrication.
h. Include components that enhance the effectiveness of other
    components in the system (e.g., phosphites used with zinc
    additives) as well as other auxiliary components such as blowing
    agents.

Stabilizers cannot be expected to completely prevent the
breakdown of PVC for an indefinite period. Rather, they serve to
retard degradation through the critical processing period and extend
the useful life of the finished product.

The market for heat stabilizers has kept pace with the growth of
the PVC industry. Current annual consumption is about 80 million lb
with the major part of the market made up of barium-cadmium
stabilizers. The overall make-up of the stabilizer market in 1981
is shown in Table X (<u>32</u>).

Processing Aids, Impact Modifiers, and Other Additives

<u>Processing Aids</u>. The processing of rigid PVC compounds based on
high molecular weight homopolymer resins is extremely difficult.
For example, surface imperfections are difficult to avoid in
calendering and extrusion. PVC resins have a tendency to soften
over a wide temperature range. However, often suitable melt
viscosities are only obtained at high temperatures at which thermal
stability is poor. Although lower molecular weight homopolymer or
vinyl acetate copolymer resins can be processed with less
difficulty, their heat stability and physical properties are
inferior to those of higher molecular weight resins. Frequently,
therefore, additives called processing aids are incorporated into
the PVC resin to achieve certain processing objectives.

These materials may be regarded as low melt viscosity,
compatible solid plasticizers. Examples are acrylic copolymers,
acrylonitrile butadiene resins, and chlorinated polyethylene. By

Table X. U.S. Consumption of Heat Stabilizers: 1979–1981

Stabilizer	Consumption, 1000 Tons		
	1979	1980	1981[a]
Barium/Cadmium (liquid & powder)	18,800	15,250	15,700
Tin	11,250	10,125	10,650
Lead	11,140	9,950	10,450
Calcium/Zinc	2,200	1,970	2,050
Antimony	520	430	450
Total	43,910	37,725	39,300

[a]Estimate.
Reproduced with permission from Ref. 32. Copyright 1981
Modern Plastics.

use of these, high molecular weight PVC homopolymer resins can be processed to take advantage of their better heat stability and higher heat distortion temperature. Processing aids increase processing rates, allow lower processing temperatures, promote fusion, impart smoothness and reduced surface gloss, improve roll release on calenders, reduce plate-out, increase hot tear resistance, and improve formability. On the negative side, they do reduce resistance to chemicals and solvents, and may adversely affect resistance to aging and UV light, reduce heat distortion temperature, and decrease impact strength somewhat if used a levels greater than 10 phr.

Impact Modifiers. These materials, which include acrylic copolymers, acrylonitrile-butadiene-styrene resins, ethylene-vinyl acetate copolymers, fumaric ester copolymers, chlorinated polyethylene, and certain graft copolymers (such as methyl methacrylate grafted onto a butadiene-styrene backbone) are used to protect rigid PVC from brittle fracture. They are used at a level of 5-15 phr. Impact strength is very important in numerous rigid applications, such as exterior building products, rigid panels, bottles, pipe, injection molded parts, and appliance housings.

The impact modifier can be considered to be a rubbery dispersed phase in the PVC. As such it can be considered to decelerate the propagation of cracks because it can be deformed and thus dissipate the energy of impact. If the refractive index of the impact modifier is close to that of the PVC resin, good clarity is obtained. As the particle size of the dispersed phase is decreased below 0.1 μm, clarity improves. Refractive index is temperature dependent, and the degree of clarity in the end product can vary with the particular modifier or service temperature. The efficiency of modifier dispersion, which is dependent on processing intensity and time has a marked effect on impact strength. Impact resistance is also temperature dependent. As the temperature approaches T_g, brittle fracture may occur. Impact modifiers have several disadvantages: high cost; reduced resistance to chemicals, solvents, and water; reduced tensile and flexural strength; impaired UV stability (with loss of impact strength on aging); and reduced heat distortion temperature. An extensive list of commercial impact modifiers for PVC and other plastics is presented in "Modern Plastics Encyclopedia" (33).

Fillers. Although fillers are primarily added to reduce costs, they also have several advantages, including opacity, improved electrical properties, resistance to UV light, resistance to blocking, improved dry blend characteristics, improved hardness, reduced tendency toward plate-out, and resistance to deformation under load at high temperature. On the negative side, fillers tend to reduce the physical and chemical properties of the resin, tensile and tear strength, degree of elongation, low-temperature performance, abrasion resistance, and resistance to attack by water and chemicals. They also reduce processing rates by increasing the viscosity of melts.

Calcined clay and precipitated or water ground calcium carbonates of less than 3 μm are the most common fillers used with PVC. Other fillers used are shown in Table XI.

Table XI. Fillers for PVC Compounds

Name	Chemical Type	Type of Particle	Specific Gravity at 25 °C	Oil Absorption, Percent[a]	Color	Refractive Index
Clay	Aluminum silicate	Platelike	2.6	44	White	1.56
Talc	Magnesium silicate	Flat	2.71	49	White	1.59
Mica	Potassium aluminum silicate	Thin plates	2.75	55	Off-white	1.60
Asbestos	Magnesium silicate	Short fiber	2.56	--	Gray	1.52
Whiting	Calcium carbonate	Crystalline	2.71	5-15	White	1.60
Calcium carbonate	Calcium carbonate	Crystalline	2.65	63	White	1.63
Diatomaceous earth	Silica	Diatom	2.0	81	Cream	1.46
Titanium dioxide	Anatase	Crystalline	3.8	24	White	2.55
Barytes	Barium sulfate	Granular crystals	4.46	7	White	1.64

[a]ASTM D218.

Fillers must be free of sulfur-containing impurities, iron, and zinc. In order to minimize reduction in clarity, fillers should have a fine particle size and an index of refraction close to that of the PVC resin. If opacity is desired, however, high refractive index fillers such as talc or calcium carbonate can be used to minimize the amounts of the more expensive titanium dioxide opacifying pigment required.

Light Stabilizers. As noted earlier, exposure of PVC to sunlight results in degradation due to the effect of UV light. Because numerous outdoor applications have been developed for PVC, such as extruded siding, rain gutters, building panels, auto seatcovers, automobile roof tops, tarpaulins, window shades, and lawn furniture, the use of light stabilizers is very significant. The first indication of degradation is visible discoloration followed by a loss in physical properties such as a loss in tensile strength and embrittlement.

UV stabilizers absorb UV light (below 355 nm where degradation is more severe), and the absorbed energy is dissipated as heat. Examples of commonly used UV stabilizers are benzophenones, benzotriazoles, salicylates, and substituted acrylonitriles. Substituted phenol esters, which are less expensive than conventional benzophenones, are transformed to hydroxybenzophenones when exposed to UV light. Improved light stability results when pigments such as titanium dioxide or carbon black are used as well as when epoxidized plasticizers and phosphites are incorporated. Use level in clear formulations is 0.2–0.5 phr, but the market is not large because of the high cost of these materials.

Fungicides. Although vinyl chloride polymers and copolymers are themselves resistant to microbiological deterioration, compounded polymers containing certain plasticizers and lubricants are attacked by fungi, bacteria, and other microorganisms. Adipates, azelates, sebacates, epoxy soybean oil, epoxy tallates, polymeric plasticizers, and lubricants are microbiologically susceptible. Although the microbiological attack may only manifest itself in a development of color--so called "pink staining," more serious degradation can occur and result in loss of tensile strength and elongation and a loss in electrical properties, exudation, and odor development.

Applications for which protection against such microbial attack is required include shower curatins, auto seatcovers, floor coverings, swimming pool liners, upholstery, wall coverings, ground cloths, tents, coated fabrics, clothing and apparel, and electrical insulation (particularly that used underground--in short any place where plastic is subjected to hot and humid conditions in contact with water or soil.

Materials used to protect against microbial attack include organotins, quaternary ammonium compounds, arsenicals, and copper compounds *N*-(Trichloromethylthio)phthalimide and quaternary ammonium compounds are particularly useful in preventing pink staining.

Flame Retardants. Because PVC has a high chlorine content, it is inherently nonburning or self-extinguishing. However, when more than 25 phr of a flammable plasticizer is compounded with the resin, the plasticized composition will burn as shown by a limiting oxygen index test. Certain phosphate esters such as tris(2-ethylhexyl)phosphate and diphenylisodecyl phosphate as well as certain halogenated hydrocarbons can be incorporated at the 10–20 phr level to provide a flame-retardant composition. Antimony oxide is also used at 1–3 phr levels at which the low-temperature properties of the phosphate esters are undesirable. It cannot be used in clear or translucent compositions because of opacity, and in these application boranes and barium metaborate are used.

Antioxidants. Certain components of heat stabilizers (polyols, phosphites) also serve as antioxidants. The protection against oxidative attack is not as great in PVC as it is in certain ethylenically unsaturated materials. However, PVC compounds may require protection in the high-temperature service as in electrical wire insulation; bisphenol A is often incorporated into the plasticizer for this purpose. Impact modifiers containing unsaturation such as acrylonitrile-butadiene-styrene polymers often require antioxidant protection. Hindered phenols such as butylated hydroxytoluene are often used for this purpose, especially when outdoor applications are involved.

Applications for Poly(vinyl chloride)

The development of the major markets for PVC was facilitated by the development of several significant processing methods for fabricating products from resin. A detailed discussion of these processing methods is beyond the scope of this paper, but an excellent summary is contained in "Poly(vinyl chloride)" (34).

The markets for PVC are manifold and are summarized in Table III along with information on the fabrication techniques used and the reasons why PVC was chosen for the indicated use (other than cost). In 1981, the U.S. consumption of PVC resins amounted to 2,551,000 metric tons or approximately 5.62 billion lb. Over one-half of this amount was made up of extruded products of which pipe and conduit are the major category. A summary of the principal processing methods and 1981 consumption is shown in Table XII and is further detailed in Table III.

Table III indicates some interesting statistics. After pipe and conduit wire and cable, packaging film and siding are the next in high-volume use. In addition, there are widely divergent uses ranging from baby pants to automobile upholstery to credit cards. PVC has remained a versatile, safe, functional plastic that is strong, lightweight, and moderate in energy requirements for its manufacture.

Vinyl Chloride Toxicity and Federal Regulation of the Poly(vinyl chloride) Industry

Early workers recognized the fire and explosive hazards associated with handling VCM. There was little work, if any, that indicated potential health hazards.

Over 20 years ago, Dow Chemical Company reported that long-term exposure of animals to 100 ppm vinyl chloride monomer for 4.5 to 6 mo at the equivalent of "normal working hours" resulted in slight

Table XII.   Principal Processing Methods and Consumption (1981) of PVC

Processing Method	Consumption (1981), 1000 Metric Tons	%
Extrusion	1,473	57.7
Calendering	369	14.5
Export	195	7.6
Injection Molding	147	5.8
Dispersion Coating	139	5.4
Dispersion Molding	84	3.3
Compression Molding	49	1.9
Blow Molding	45	1.8
Solution Coating	25	1.0
Vinyl Latexes	25	1.0
Total	2,551	100.0

liver damage. Although at that time governmental regulations allowed a maximum concentration of 500 ppm, Dow recommended that exposure levels be reduced to 50 ppm, a recommendation not accepted by the American Conference of Governmental Industrial Hygienists (ACGIH). In 1962, based on work at Yale University, ACGIH concluded that a level of 500 ppm offered adequate safety for human exposure and published this level as the recommended standard. In 1971, the Occupational Safety and Health Administration (OSHA) accepted an exposure level of 500 ppm as the national standard.

Because reports of finger abnormalities had been reported in the early 1960s, the Manufacturing Chemists Association (now Chemical Manufacturers Association) initiated a study in January 1967 at the University of Michigan on acroosteolysis, the softening of finger tip bones among workers, particularly those involved in cleaning polymerization vessels. The results of this study, in which the authors were unable to pinpoint the cause of this disease, were published in the "Archives of Environmental Health" in January 1971 and recommended animal studies be undertaken to develop an experimental model. Furthermore, a 50 ppm worker exposure limit was recommended at that time.

In May 1971, MCA (now CMA) having learned of work by P. L. Viola of the Regina Elena Institute of Cancer Research in Rome that exposure of rats to high levels (5000 ppm) of VCM resulted in cancers and tumors, recommended that U.S. VCM and PVC producers sponsor animal exposure and epidemiological research on VCM carcinogenicity. By March 1972, 17 U.S. companies agreed to financially support these studies. In the meantime, late in 1971, further studies were being initiated under the sponsorship of several European companies by Cesare Maltoni at the Institute of Oncology in Bologna, Italy. In mid-January 1973, a technical delegation visited Maltoni's laboratory to learn in more detail about his procedures and results. Their report to a meeting of technical representatives of companies participating in the MCA study in late January noted that Maltoni had found tumors in the ear canals, kidney, and liver of rats at concentrations of 500 and 250 ppm of VCM, but not at 50 ppm nor in controls. AT that time it was agreed that an epidemiological study of VCM/PVC workers would also be initiated. The animal exposure study was contracted for by MCA with Industrial Bio-Test Laboratories, Northbrook, Ill., on February 1, 1973, to include exposure of rats, mice, and hamsters to 2500, 200, and 50 ppm of VCM for 1 yr, with animals to be kept on observation for an additional year.

On January 22, 1974, OSHA was advised by the National Institute for Occupational Safety and Health (NIOSH) that B. F. Goodrich Chemical Company had reported that deaths of three employees from a rare liver cancer (angiosarcoma) may have been occupationally related. As a result, a fact-finding hearing on possible hazards involved with the manufacture of VCM was announced on January 30, 1974, and held on February 15, 1974. The hearing included information on the work of Maltoni on the exposure of rats, mice, and hamsters at 10,000; 6,000; 2,500; 250; and 50 ppm VCM concentrations for various periods of time. No tumors were observed in the animals exposed to 50 ppm. The hearing revealed that the dead B. F. Goodrich employees had an average exposure to vinyl chloride, at unknown concentrations, for approximately 19 yr. Some employees of Union Carbide, Firestone Tire and Rubber Company, and Goodyear Tire and Rubber Company were reported in post-hearing comments also to have had exposure to VCM and to have died as a result of angiosarcoma of the liver. Autopsies of the four deceased employees revealed their liver angiosarcoma tumors were histologically indistinguishable from those observed in Maltoni's experimental animals.

On April 5, 1974, OSHA, based on this information, promulgated an Emergency Temporary Standard that reduced the level from 500 ppm to a 50 ppm ceiling. Ten days later, on April 15, 1974, Industrial Bio-Test Laboratories, where the MCA animal study was being carried out, reported to OSHA, NIOSH, and EPA that 2 of 200 mice exposed to 50 ppm of VCM for 7 h/d, 5 d/wk, for approximately 7 mo had developed liver angiosarcoma. This result was not observed in rats or hamsters. OSHA therefore concluded that exposure even to 50 ppm might well constitute a serious health hazard.

OSHA Vinyl Chloride Standard. On May 10, 1974, OSHA announced a proposed permanent standard that included "no detectable" level for employee exposure to VCM in monomer, polymer, and fabricating operations by a sampling and analytical technique capable of detecting vinyl chloride concentrations of 1 ppm with an accuracy of 1 ppm ± 50%. Subsequently public hearings were held in Washington in June and July 1974 at which 35 hours of direct testimony by 44 panels of witnesses was presented, and 89 written submissions with comments on the proposed standard were received.

The final standard promulgated by OSHA effective October 5, 1974 [Section 1910.93 q (a) 39 FR 41848, December 3, 1974, and redesignated as Section 1910.1017 at 40 FR 23072, May 28, 1975] includes the following requirements for the control of employee exposure to VCM:

1. Permissible exposure limit--No employee may be exposed to concentrations greater than 1 ppm VCM averaged over any 8-h period and greater than 5 ppm VCM over any period not exceeding 15 min.
2. Monitoring--A program of initial monitoring and measurement to determine if any employee is exposed in excess of the action level of 0.5 ppm VCM over an 8-h workday.
3. Establishment of regulated areas where VCM concentrations are in excess of permissible exposure limit.
4. Methods of compliance to assure control at or below the permissible exposure by feasible engineering, work practice, or personal protective controls.
5. Respiratory protection including selection of respirators on the basis of exposure to VCM concentrations from less than 10 ppm to above 3600 ppm, and establishment and maintenance of a respiratory protection program.
6. Hazardous operations--Provision of respiratory equipment and protective garments for employees engaged in hazardous operations.
7. Emergency situations--Development of a written operational plan for each facility using VCM.
8. Training--Providing each employee with a training program on hazards and safe use of VCM, fire hazard, acute toxicity,

monitoring program, emergency procedures, medical surveillance program, and a review of the OSHA standard annually.

9. Medical surveillance--A general physical examination, with specific attention to detecting enlargement of the liver, spleen, or kidneys, must be provided to employees without cost every 6 mo when the employee has been employed in VCM of PVC manufacture for 10 yr or longer, and annually for all other employees.

10. Signs and labels--Regulated or hazardous operating areas must be posted with signs indicating the presence of a "cancer-suspect agent." Containers of PVC must be legibly labelled, "Poly(vinyl chloride) (or trade name) contains vinyl chloride. Vinyl chloride is a cancer-suspect agent."

11. Records and Reports--Record-keeping and reporting requirements are specified.

EPA Vinyl Chloride Standard. On October 21, 1976, the Environmental Protection Agency (EPA) promulgated the National Emission Standard for Vinyl Chloride (35) under the provisions of Section 112 of the Clean Air Act (42 U.S.C., Section 7412) which provides for national emission standards for hazardous air pollutants (NESHAP) that present a threat of increased mortality or serious irreversible illness. This standard provides emission standards applicable to ethylene dichloride, vinyl chloride, and PVC plants. In the case of PVC plants, this requires the following:

1. Exhaust gas discharged to the atmosphere from each reactor is not to exceed 10 ppm; reactor opening loss is not to exceed 0.00002 lb of vinyl chloride/lb of PVC; prohibition of manual vent discharges except for emergencies.

2. Concentration of VCM in all exhaust gases from the stripper is not to exceed 10 ppm.

3. The weighed average residual concentration of VCM in PVC resin after stripping may not exceed 2000 ppm for dispersion resins and 400 ppm for all other resins including latex resins, averaged separately for each type.

The full standard is found in 40 CFR 61.60 and requires semiannual reporting by producers to EPA.

Epidemiology Studies. An epidemiology study of 707 deaths in a population of 9677 men who had worked for 1 yr or more in jobs involving exposure to VCM through December 31, 1972, was reported on at a Conference to Reevaluate the Toxicity of Vinyl Chloride Monomer, Poly(vinyl chloride), and Structural Analogues held at the National Institutes of Health, March 20-21, 1980, by W. Clark Cooper (36). This study did not show a significant excess of deaths due to malignancies, although there did appear to be a significant excess of tumors of the brain and central nervous system. Except for this finding and the proven association with hepatic angiosarcoma, vinyl chloride does not appear to be associated with significant excess cancers of other sites.

This study was planned and analysis was under way before the cases of hepatic angiosarcoma (reported by Goodrich in 1974) had been diagnosed. An update is scheduled with inclusion of additional deaths in the cohort during 1973-79. With the refinements in manufacturing techniques and the significant reductions in VCM emissions and exposures that have resulted and as a consequence of required compliance with OSHA and EPA standards, future adverse human health effects due to VCM toxicity should have been largely eliminated.

Literature Cited

1. Regnault, V. Annalen 1835, 15, 63.
2. Regnault, V. Ann. Chem. Phys. 1838, 69, 151.
3. Baumann, E. Ann. Chem. Pharm. 1872, 163, 308.
4. Synthetic Organic Chemicals, U.S. Production and Sales, U.S. Tariff Commission (data for 1953-70); U.S. Production and Sales of Miscellaneous Chemicals, 1971 Prelim. Data, U.S. Tariff Commission; Annual Statistical Report, Plastics and Resin Materials, 1972, The Society of the Plastics Industry, Inc., April 1973 (data for 1972), CEH Manual of Current Indicators – Supplemental Data, p. 213, Feb., 1982, Chemical Economics Handbook – SRI International, Plastics and Resins, 580.1882M, 580.1882N, March 1981.
5. Chemical Week March 10, 1982, p. 5.
6. Modern Plastics 1982 59(1), 82-83.
7. Chemical Economics Handbook – SRI International, Plastics and Resins, 580.0070A, November 1980.
8. Barton, D. H. R.; Mugdan, M. J. Chem. Soc. Ind. 1950, 69, 75.
9. Wesselhoft, R. D.; Woods, J. M.; Smith Am. Inst. Chem. Eng. J. 1959, 5, 361.
10. Buckley, J. A. Chem. Eng. 1966, 73, 102.
11. Barr, J. T. "Poly(vinyl chloride)"; Advances in Petroleum Chemistry and Refining, Interscience Publishers: New York, 1963; Vol. 7, pp. 371-404.
12. Modern Plastics February 1981, 54-55.
13. Harkins, W. D. J. Am. Chem. Soc. 1947, 69, 1428.
13a. Smith, W. V.; Ewart, R. N. J. Chem. Phys. 1948, 16, 592.
13b. Smith, W. V. J. Amer. Chem. Soc. 1948, 70, 3695.
13c. Peggion, E.; Testa, F.; Talamini, G. Makromol. Chem. 1964, 71, 173.
13d. Vidotto, G.; Crosato, A.; Talamini, G. Makromol. Chem. 1970, 134(41), 3229.
14. French Patent 1 261 921 (April 17, 1961) on Continuous Bulk Polymerization to Compagnie de Saint-Gobain.
15. Crosato-Arnaldi, A.; Gasparini, P.; Talamini, G. Makromol. Chem. 1968, 117, 140.
16. Pezzin, G.; Talamini, G.; Vidotto, G. Makromol. Chem. 1961, 43, 12.
17. "Polymer Handbook," 2nd ed.; Brandrup, J.; Immergut, E. H., Eds.; Wiley Interscience, 1975; p. IV-168.
18. Bohdanecky, M.; Solc, K.; Kratochvil, P.; Kolinsky, M. J. Polym. Sci. A2, 1967, 5, 343.
19. McKinney, P. V. J. Appl. Polym. Sci. 1965, 9, 583.
19a. Boccato, G.; Rigo, A.; Talamini, G.; Zilio-Grandi, F. Makromol. Chem. 1967, 108, 218.

19b. Staudinger, H. in "Die Hochmolecularen Organischen Verbindung," Springer-Verlag, Berlin, 1932.
20. Marvel, C. S.; Sample, J. N.; Roy, M. F. J. Am. Chem. Soc. 1939, 61, 3241.
21. Koleske, J. V.; Wartman, L. H. "Poly(vinyl chloride)"; Gordon and Breach Science Publishers: New York, 1969; p. 51.
22. Reding, F. P.; Walter, E. R.; Welch, F. J. J. Polym. Sci. 1962, 56, 225.
23. Heydemann, P.; Guicking, H. D. Kolloid. Z. 1963, 193, 16.
24. Nakajima, A.; Hamada, H.; Hayashi, S. Makromol. Chem. 1966, 95, 40.
25. Matlack, J. D.; Metzger, A. P. Polym. Lett. 1966, 4, 875.
26. Bengough, W.; Sharpe, H. M. Makromol. Chem. 1963, 66, 31.
27. Baum, B.; Wartman, L. H. J. Polym. Sci 1958, 28, 537.
28. "Electron Beam Curing of Polymers"; SRI Report No. 116, L. A. Wasselle, July 1977.
29. Bengough, W.; Varma, I. K. Eur. Polym. J. 1966, 2, 61.
29a. Flory, P. J. in "Principles of Polymer Chemistry," Cornell University Press, Ithica, New York, 1953; Chapter 9.
30. Dannis, M. L.; Ramp, F. L. U.S. Patent 2 996 489 (August 15, 1961) assigned to The B. F. Goodrich Company.
31. Modern Plastics September 1981, 67.
32. Ibid., p. 70.
33. "Modern Plastics Encyclopedia"; 57(10A), 1980; pp. 576–86.
34. Sarvetnick, H. "Poly(vinyl chloride)"; Van Nostrand-Reinhold: New York, 1969, p. 145 ff.
35. 41 FR 46564 (Oct. 21, 1976).
36. Cooper, W. C. Environmental Health Perspectives 1981, 41, 101.

Fiber-Forming Polymers

MICHAEL J. DREWS, ROBERT H. BARKER, and JOHN D. HATCHER

School of Textiles, Clemson University, Clemson, SC 29631

Chemistry of Man—Made Fiber Formation
   Polyesters
   Nylons
   Spandex Polymers
   Rayon
   Acetates
   Acrylics and Modacrylics
   Polybenzimidazole and Carbon Fibers
Polymerization Process Technology
Fiber Formation
Postfiber Formation Chemistry and Technology
Future Trends

The chemistry and technology of man—made, fiber—forming polymers date back to 1885 when an artificial silk was patented by Chardonnet in France. Since then, these materials have progressed to become the focus of a major global industry with applications ranging from the everyday world of apparel to biomedical and advanced aerospace engineering concepts.

   For the purpose of discussion of the chemistry and technology of man—made, fiber—forming polymers, the term "synthetic fiber" will be used to denote all man—made fibers manufactured from noncellulosic raw materials. The term "cellulosics" will apply to those man—made fibers that are manufactured from cellulosic raw materials. The term "man—made fibers" will apply to all fibers except the naturally occurring cellulosic and protein fibers.

   The manufacture of all man—made fibers involves at least three distinct process steps. The first consists of the production of polymers or polymer derivatives suitable for spinning into fibers. In the second step, or spinning, a polymer melt or solution is extruded under pressure through the appropriate spinneret's orifice(s) to form the fiber or fibers. If only a single fiber is produced from a spinneret, it is referred to as monofilament. Multifilament spinnerets produce yarns. The third step is drawing,

0097–6156/85/0285–0441$07.75/0

which is the stretching, either hot or cold, of monofilament or yarn
to some multiple of its as-spun length. This step results in an
orientation of the polymer chains and crystallites with respect to
the fiber axis, and it is critical with respect to the ultimate
mechanical properties of the fiber.

Bundles of yarns from multiple spinnerets are referred to as
tow. If the tow is subsequently cut into short specified lengths,
it is referred to as staple tow, and the product is staple fiber.
Depending on the man-made fiber producer's production equipment,
one, two, or all three of the steps just discussed may be executed
in either a batchwise or continuous manner.

Table I shows the world-wide production of man-made fibers, the
U.S. production of man-made fibers, and the percentage of the
world's production of man-made fibers produced in the U.S. in 1983.
On the basis of these data, which are similar to the production
figures for the past five years, polyesters (11.20 billion lb),
polyamides (7.00 billion lb), cellulosics (6.62 billion lb), and the
acrylics (4.91 billion lb) account for over 95% of the world's man-
made fiber production.

In Table II the U.S. production of man-made fiber is further
divided into filament and staple fiber, and the U.S. consumption of
cotton and wool is included. On the basis of these data domestic
man-made fibers accounted for over 70% of all of the fiber consumed
in the U.S. in 1982. These numbers, as well as those in Table I,
clearly indicate the importance of the man-made fiber production
industry with respect to the U.S. and the world's textile industry.

The remainder of this chapter is concerned with briefly
summarizing the chemistry and technology of man-made fiber formation. An attempt has been made to place the emphasis on those
synthetic and cellulosic fibers that represent significance either
in terms of world production or, in the authors' view, in terms of
unusual or unique polymer chemistry. More detailed and
comprehensive general reviews on various aspects of these topics
have been published elsewhere (4-10).

Chemistry of Man-Made Fiber Formation

Of the fibers listed in Table II only the polyesters, polyamides,
spandexes, acetates, and rayon are discussed in this chapter. While
the acrylics and modacrylics are the third most important class of
commercial fibers; because their polymerization chemistry is also
discussed in other chapters concerned with vinyl addition emulsion
polymerizations, it will only be briefly summarized here. For the
same reason polypropylene polymerization chemistry is also not
covered in this section. However, two additional topics, carbon
fiber formation and polybenzimidazoles have been included on the
basis of the current interest in high-performance fibers for
composite materials.

Polyesters. Because of the hydrolytic instability of esters of
aliphatic acids, virtually all commercial polyesters are based on
aromatic acids (8). By far the most common is poly(ethylene
terephthalate) (PET):

Table I. U.S. and World-Wide Man-Made Fiber Production in 1983 (1)

Fiber	World-Wide, billion lb	U.S., billion lb	U.S., %
Polyester	11.20	3.54	31.6
Polyamides	7.00	2.42	34.6
Acrylics	4.91	0.67	13.6
Cellulosics	6.62	0.63	9.5
Other[a]	2.29	0.90	39.0

[a]Includes the olefin fibers as well as others.

Table II. Summary of U.S. Man-Made Fiber Production Data in 1982 (2)

Fiber Class	Yarns + Monofilaments, million lb	Staple + Tow
Acetate	195.2	4.0[a]
Rayon	46.6[b]	355.0
Nylon + Aramid	1,246.1	686.0
Acrylic + Modacrylic	--	624.1
Olefin	582.2	--
Olefin + Vinyon	--	138.2
Polyester	1,213.6	1,955.2
Textile glass fiber	899.2	--
Raw Wool	--	109.2
Raw cotton	--	2,491.1

[a]1981 Data and does not include cigarette tow production (3).
[b]1981 Data (3).

$$\{CH_2-CH_2-O-\overset{O}{\overset{\|}{C}}-\hexagon{O}-\overset{O}{\overset{\|}{C}}-O\}_x$$

The polarity of the ester linkage produces sufficient interchain forces to allow these polymers to develop good fiber-forming properties with number average molecular weights in the 15 to 25 thousand range.

PET is prepared commercially by either ester interchange (EI) or direct esterification processes (11, 12). For many years, the EI process was the only one practical because of the difficulties encountered in purification of terephthalic acid (TA). The dimethyl terephthalate (DMT) was much easier to purify, and therefore much more readily available as a raw material. In the normal EI batch process, ethylene glycol and DMT are mixed in a ratio ranging from 1.9 to 2.5 along with a transition metal catalyst at a level ranging from 50 to 150 ppm. Commonly, manganese acetate is used as the catalyst, although salts of calcium, magnesium, zinc, and titanium have also been employed. Temperatures are raised to 180 to 220 °C over a period of time, and the evolved methanol is removed by distillation. During this phase, bis(hydroxyethyl)terephthalate (BHET) (Structure I) is formed along with lesser amounts of dimer and trimer (Scheme I).

When the ester interchange step is completed, a phosphorus compound, usually an organic phosphate, phosphite, or polyphosphoric acid, is added to complex the EI catalyst and thus deactivate it. Antinomy oxide, or less commonly germanium or titanium oxides, is added at a level of 150 to 450 ppm to serve as a polycondensation (PC) catalyst. The mixture is heated to 280–295 °C, and a vacuum of 0.1 mmHg or less is applied to aid in the volatilization of the ethylene glycol formed during PC. The reaction is allowed to proceed until the increase in reaction viscosity, as determined by power consumption for stirring, is at a level that indicates that the desired molecular weight has been obtained. As the degree of polymerization increases, the inherent reactivity of the functional groups becomes less of a limiting factor than the ability to remove ethylene glycol from the highly viscous melt. Because of this situation, the PC reaction is usually carried out in equipment that maximizes the melt surface area by stirring and film generation on the reactor walls.

When the desired molecular weight has been achieved, the molten polymer is extruded as a rope, commonly referred to as spaghetti, which is quenched in a water bath, cut to chip, and dried. The dried chip is then melt spun.

As purified terephthalic acid became more available as a raw material, processes were developed based on direct esterification of the TA (Scheme II). These processes have the obvious advantages of an essentially single-stage process, although they are usually carried out in at least two phases. This method, however, allows the use of a single catalyst and greatly reduces the need for phosphorus stabilizers, although small amounts are almost always added as complexing agents for trace metal impurities. Being a single process, the direct esterification route is also much more amenable to continuous production, as well as integrated polymerization and spinning processes.

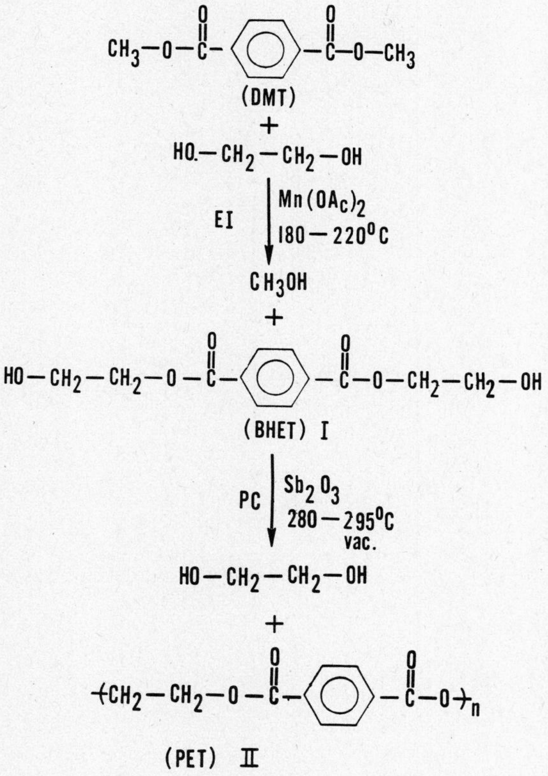

$$CH_3-O-\overset{\overset{\displaystyle O}{\|}}{C}-\langle\bigcirc\rangle-\overset{\overset{\displaystyle O}{\|}}{C}-O-CH_3$$

(DMT)

+

$$HO-CH_2-CH_2-OH$$

EI $\quad\Big\downarrow\quad\begin{array}{l}Mn(OAc)_2\\180-220^0C\end{array}$

$$CH_3OH$$

+

$$HO-CH_2-CH_2-O-\overset{\overset{\displaystyle O}{\|}}{C}-\langle\bigcirc\rangle-\overset{\overset{\displaystyle O}{\|}}{C}-O-CH_2-CH_2-OH$$

(BHET) I

PC $\quad\Big\downarrow\quad\begin{array}{l}Sb_2O_3\\280-295^0C\\vac.\end{array}$

$$HO-CH_2-CH_2-OH$$

+

$$\overset{}{+}CH_2-CH_2-O-\overset{\overset{\displaystyle O}{\|}}{C}-\langle\bigcirc\rangle-\overset{\overset{\displaystyle O}{\|}}{C}-O\overset{}{+}_n$$

(PET) II

Scheme I

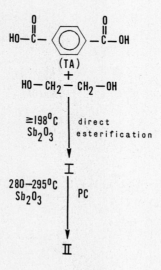

Scheme II

In the TA process, ethylene glycol and purified terephthalate acid are generally mixed in a ratio of less than 1.4 to 1 and heated in an initial phase with an antimony oxide catalyst to the boiling point of the ethylene glycol (198 °C). Because the reaction is relatively slow at this temperature, this step is often run under pressure to increase the achievable temperatures. As the reaction proceeds and the BHET is produced, the temperatures are increased and the pressure is lowered to allow distillation of the evolved ethylene glycol. This phase resembles the PC step of the EI process.

Once the desired molecular weights are achieved, the molten polymer is either extruded and chipped, as in the EI process, or extruded directly to yarn in a melt spinning operation.

In both the EI and TA processes, care must be taken to control the physical and chemical properties of the polymer. In particular, production of diethylene glycol (DEG) must be carefully monitored. High levels of DEG introduce weak links that make the polymer vulnerable to oxidative and thermal degradation and that disrupt crystallization and thus produce deleterious effects on physical properties. Low levels of DEG, however, are generally considered desirable because their disruption of the polymer order results in a more open structure with enhanced dyeing characteristics. In addition to DEG, concentrations of carboxyl, carbomethoxy, and free hydroxyl functions must be controlled. In many cases, small amounts of monobasic acid are added during polymerization to serve as a cap for free hydroxyl groups and thus produce greater thermal stability in the polymer melts. Exposure of the molten polymer to oxygen must also be carefully restricted because thermooxidative degradation results in poor physical properties, particularly color and strength.

A variety of chemically modified PET systems are produced for special purpose applications. Cationic dyeable PET is produced by including small amounts of 5-sulfoisophthalic acid. Flame-retardant polyesters may be produced by incorporating phosphorus- or bromine-containing additives or comonomers. The most common of the phosphorus compounds are the phosphonates, phosphates, and phosphazines. Among the bromine-containing systems, ethoxylated tetrabromobisphenol-A has been used as a coreactant. Inclusion of diacids larger than terephthalic introduces imperfections in polymer orientation and thus opens the structure to produce a deep-dyeing polyester. A number of systems of this type have been prepared commercially. Polymers with reduced physical strength have been prepared for use as a nonpilling polyester fiber.

In addition to PET, several other polyester systems have been evaluated for fiber use. Of these, only poly(cyclohexyl dimethylene terephthalate) (PCHDT) appears to have achieved commercialization. The PCHDT may be prepared either by direct esterification or EI processes with titanium alkoxide catalysts. The 1,4-bis(hydroxymethyl)cyclohexane, used as the diol in this polymer, is most commonly a mixture of <u>cis</u> and <u>trans</u> isomers in a ratio of approximately three to seven. Because of the greater steric requirements of the cyclohexyl ring, PCHDT has a higher viscosity and higher resilience than PET. It is also easier to dye. Its specific gravity is approximately 1.22 as compared to 1.38 for PET.

<u>Nylons</u>. Commercial nylons fall into two fundamental categories: those based on polyamides of amino acids and those resulting from the interaction of diamines with diacids (<u>8</u>, <u>13</u>–<u>15</u>). In the latter category, the most common is poly(hexamethylene diadipamide) (Structure III), commonly known as nylon 6,6. This material is usually prepared by the thermal decomposition of hexamethylene diammonium adipate, or nylon salt. This salt, which has the advantage of exact 1:1 stoichiometry, is readily prepared by adding dry adipic acid to an aqueous or methanolic solution of hexamethylene diamine. The nylon 6,6 salt is only slightly soluble in methanol and precipitates in a readily purifiable form. Polymerization of the salt may then be carried out either in batch or continuous processes. (Scheme III)

In the common commercial batch polymerization, the salt, as an aqueous solution or slurry, is charged into an autoclave that has previously been purged of oxygen. The batch is heated to the boiling point with a concurrent pressure increase to about 250 lb/in.^{2}. Steam is released from the autoclave as the temperature is increased over a period of time to approximately 270 °C. At this temperature, the polymer is above its melting point, and the autoclave pressure is slowly reduced to 1 atm, while holding the temperature constant. The melt is held under these conditions for about 1 h to complete the polymerization. The polymer is then extruded as a ribbon which is quenched in water and cut to flake or chip. Because the polymer melt viscosity is high during the latter stages of polymerization, the rate of water removal becomes a critical factor in determining polymerization rates. Because of this situation, measures are frequently introduced to facilitate removal of water. This removal may involve bubbling of inert gas through the melt, increased agitation, or application of a partial vacuum.

Nylon 6,6 may also be prepared by direct amidation of adipic acid and hexamethylene diamine. In this process, the molten monomers are added to a vessel in approximately stoichiometric proportions, but usually with a slight excess of acid. The exothermic salt formation produces sufficient heat to raise the batch temperature to approximately 200 °C, so polymerization proceeds rapidly. As the polymerization proceeds, the stoichiometry is adjusted to exactly equal molar portions of diacid and diamine, and the autoclave temperature and pressure are raised to 275 °C and 250 lb/in.^{2}, respectively. After a specified heating time, pressure is reduced, and the temperature is maintained to complete polymerization.

Although originally designed as a batch process, the direct amidation of the nylon 6,6 salt has been adapted to continuous polymerization by a wide variety of process modifications developed over the past 40 years. Many of these are quite different from an engineering standpoint, but all involve essentially the same chemistry as the batch system.

Of the wide variety of other polyamides based on aliphatic diamines and diacids, few have been commercialized for fiber application. A number, however, have been developed for commercial application in extruded and molded plastics.

A commercially important polyamide based on an alicyclic diamine is Qiana (Structure IV), which was introduced by Du Pont in 1968.

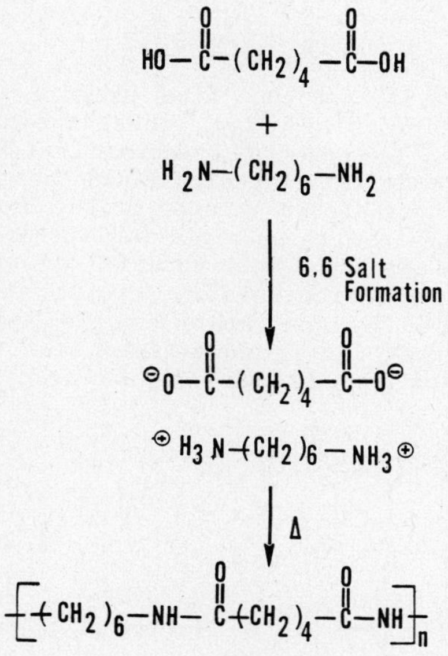

$$HO - \overset{\overset{\displaystyle O}{\|}}{C} - (CH_2)_4 - \overset{\overset{\displaystyle O}{\|}}{C} - OH$$

$$+$$

$$H_2N - (CH_2)_6 - NH_2$$

$$\downarrow \quad \text{6.6 Salt Formation}$$

$$^{\ominus}O - \overset{\overset{\displaystyle O}{\|}}{C} - (CH_2)_4 - \overset{\overset{\displaystyle O}{\|}}{C} - O^{\ominus}$$

$$^{\oplus}H_3N - (CH_2)_6 - NH_3^{\oplus}$$

$$\downarrow \quad \Delta$$

$$\left[(CH_2)_6 - NH - \overset{\overset{\displaystyle O}{\|}}{C} (CH_2)_4 - \overset{\overset{\displaystyle O}{\|}}{C} - NH \right]_n$$

III

Scheme III

It is produced by the melt condensation of dodecanoic acid and bis(4-aminocyclohexyl)methane. Many of Qiana's unusual properties, including its silklike feel and dyeability, result from the several isomeric forms of the diamine. The polymer is believed to be a mixture of the *trans-trans* and *cis-trans* diamine isomers in a ratio of approximately 4:1 (w:w) (13).

$$\left[\left(S\right)-CH_2-\left(S\right)-NH\overset{O}{\overset{\|}{C}}-(CH_2)_{10}-\overset{O}{\overset{\|}{C}}-NH\right]_n$$

IV

There are commercially important fiber-forming polyamides based on aromatic diacids and diamines (16). These materials, commonly referred to as aramids, are generally more difficult to make and process. However, the aramids offer significant advantages in terms of strength and thermal stability because of the high bond order in the chain backbone which results in rigid rodlike molecules capable of forming liquid crystals. The most notable members of this class are the <u>meta</u> and <u>para</u> isomeric aramids, structures V and VI. The <u>meta</u>-phenylenediamine isophthalamide polymer (V) is commonly encountered as the commercial product Nomex, and the para isomer (VI) is marketed under the trade name Kevlar. In both, the

V

VI

relatively low basicity of aromatic diamines, coupled with the reduced acidity of the aromatic acids, precludes the use of a nylon salt process similar to that used for nylon 6,6. To obtain requisite levels of activity, it is necessary to resort to the more reactive acid chlorides. Thus, both polymers are prepared from the diamine and the diacid chloride, usually in an interfacial polymerization to assure adequate stoichiometric control and thus achieve sufficient molecular weight.

In addition to the diacid-diamine polymers, a wide variety of aliphathic and aromatic polyamides from amino acid monomers are known and have been developed for commercialization. By far the most important member of this series is nylon 6, poly(ε-aminocaproamide) (Structure IX) (13). This material is prepared commercially by a ring-opening polymerization beginning with the readily available caprolactam (Structure VII). The process is adaptable to either batch systems by using autoclaves or continuous polymerization in appropriate column arrangements. Reaction is initiated by water or other nucleophiles to open a portion of the caprolactam to ε-aminocaproic acid. The rate of ring opening is

dependent upon the initial water concentration, acidity, and
temperature. The amino acid thus produced initiates the
polymerization by two competing pathways: polyaddition and PC, as
shown in Scheme IV.
   The addition occurs by attack of the free amino group on the
carbonyl of an unopened caprolactam ring to produce dimers and other
oligomers with structures similar to Structure VIII. Because these
materials possess a free amino group, they are capable of continuing
to interact with other caprolactam rings in a polyaddition reaction.
Competing with this reaction, however, is a direct amidation
reaction between the free amino group of one oligomer and the free
carboxyl of another. This reaction results in PC. The rate of
polyaddition depends upon end-group and cyclic monomer
concentration, and the rate of PC depends upon end-group
concentration and diffusion of water from the polymer melt (17).
The polyaddition reaction is reversible, thus the polymerization
cannot be driven to completion, and the polymer always is produced
in equilibrium with residual caprolactam. Under normal industrial
conditions, with temperatures in the range from 230 to 280 °C, the
equilibrium concentration of caprolactam is approximately 10%.
Because both residual caprolactam and low molecular weight cyclic
oligomers can cause problems in further processing, the crude
production polymer is normally extracted with hot water after
extruding and converting into chip. In order to prevent the
production of additional caprolactam in later stages of spinning and
processing, controlled amounts of monoacid are generally added
during the polymerization to cap free amino functions. These acids
also serve as catalysts in the initial ring-opening reaction and
stabilizers to control molecular weight during polymerization.

Spandex Polymers. Spandex polymers possess rubberlike elasticity
because of their segmented polyurethane structure. A variety of
commercial materials of this category are known, all of which
possess long, soft, flexible, noncrystalline, rubbery segments that
are generally hydroxyl-terminated polyethers or polyesters connected
by short, hard, polyurethane segments that are usually aromatic
(18). In one commercial process tetrahydrofuran is polymerized in a
ring-opening process to poly(tetramethylene glycol) (PTMG)
(Structure X) having a molecular weight of several thousand. The
hydroxyl functions of PTMG are then reacted with a diisocyanate,
such as diphenylmethane diisocyanate (MDI) (Structure XI), or, less
frequently, toluene diisocyanate (TDI) to produce a linear polymer
(Structure XII) having urethane groups joining the polyether
segments and terminated by isocyanate functions (Scheme V). Cross-
linking and polymer extension to produce rubber elasticity is
accomplished by either the reaction of excess diisocyanate with the
urethane groups to form allophanate linkages or by reactions of
hydrazine or other diamines with the dry spun fibers (Structure
XIII). This later reaction produces biuret cross-links for the
prevention of viscoelastic flow.
   In another commercial modification, glycols, usually a mixture
of ethylene and propylene glycols, are reacted with aliphatic
diacids such as adipic acid to produce hydroxyl-terminated
polyesters with molecular weights of approximately 2000. The
hydroxyl functions are then allowed to react with aromatic

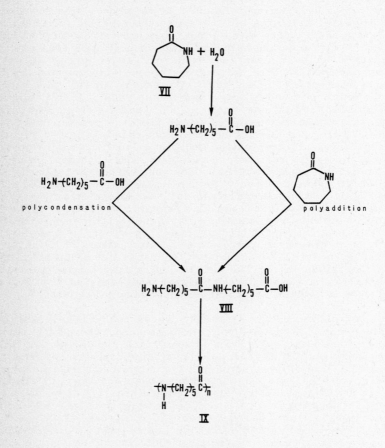

Scheme IV

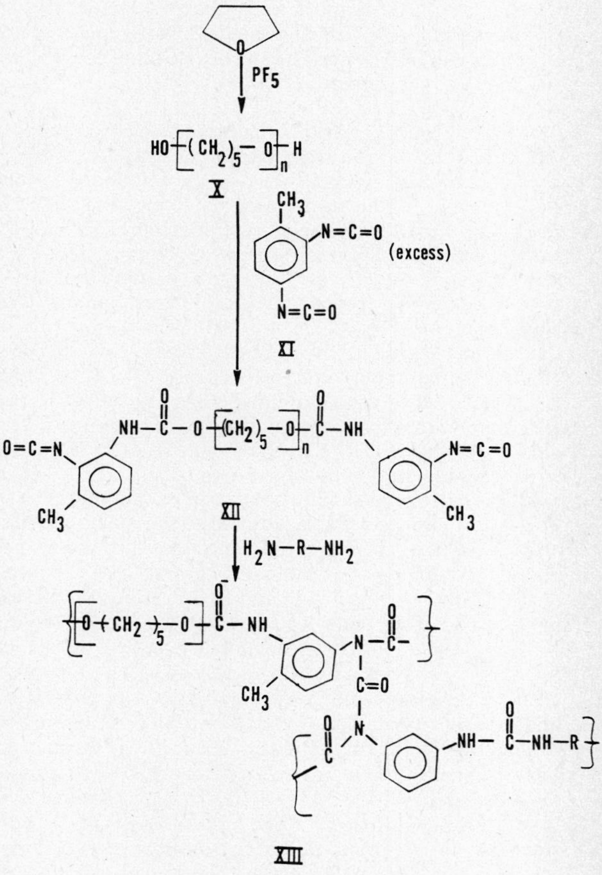
PF5
HO-[(CH2)5-O]n-H
X
CH3
N=C=O
(excess)
N=C=O
XI
O=C=N
NH-C-O-[(CH2)5-O]n-C-NH
N=C=O
CH3
CH3
XII
H2N-R-NH2
O-(CH2)5-O-C-NH
N-C-
C=O
CH3
NH-C-NH-R
XIII

Scheme V

diisocyanates, such as MDI, at elevated temperatures (120 °C). This
reaction produces a segmented polyester diisocyanate. Exposure of
this prepolymer to controlled amounts of water causes hydrolysis of
a portion of the diisocyanate links and concurrent production of
amino functionality. The resulting fiber, which contains free amino
and isocyanate links, is then cured for 1 h at 130 °C to extend the
polymer through urea linkages and cross-link through biuret and/or
allophanate structures. In fiber applications mixtures of ethylene
glycol and polyglycols can be used to balance "hard" and "soft"
segments in the polyurethane.

Rayon. Virtually all current commercial production of man-made
cellulosic fibers uses the viscose process, or some modification of
it (19,20) (Fig. 1). The basis for this process is the formation of a
metastable, water-soluble derivative from which cellulose can be
regenerated after the filament is formed. Similar technology is
used to manufacture cellophane film. Traditionally the process, as
outlined in the following, is a batch operation involving discrete
steps; more recently semicontinuous or continuous systems have been
developed.

Before derivatization, the dissolving cellulose pulp, in sheet
form, is soaked in aqueous sodium hydroxide of mercerizing strength,
squeezed to about 300% wet pickup, shredded to increase the surface
area (referred to at this stage as "white crumb"), and allowed to
age for several days in contact with air. This aging, in the
presence of oxygen and strong alkali, causes the cellulose to
undergo oxidative degradation to yield a product of more manageable
molecular weight. Because the degradation also reduces the strength
of the final product, this step is omitted in some processes for the
manufacture of high-performance fibers. After aging, the white
crumb is mixed with carbon disulfide, usually in a large churn.
This mixing allows the soda cellulose to react to form a xanthate,
CELL-O-C(S)-S$^-$Na$^+$. The resulting yellow crumb can then be
dispersed, and ultimately dissolved, in aqueous alkali. On
standing, the dissolved polymer undergoes a process referred to as
ripening which involves dexanthation and rexanthation to produce a
more random distribution of substituent groups along the chains.
The xanthate content is also reduced over time to alter the
coagulation properties of the solution. As with air aging, this
ripening step is frequently omitted when high product strength is
desired. After filtering and degassing, the ripened solution is
extruded into a bath of aqueous sulfuric acid and sodium sulfate,
usually containing a zinc salt and other inorganics. The contact
with the acid converts the xanthate functions into unstable xantheic
acid esters that spontaneously decompose to liberate and regenerate
CS_2 and cellulose. This process forms a skin around the filament
which acts as a semipermeable membrane. Diffusion of water out of
the filament is favored by the high ionic strength of the spin bath
and results in concentration and coagulation of the remaining
viscose. Concurrent diffusion of hydronium ion into the filament
neutralizes the viscose and regenerates the cellulose. In the area
between the coagulation point and the neutralization point the
polymer exists as a gel. It is only in this form that drawing can
occur. Because of this restriction, dopes for the production of
high-strength rayon may contain additives such as amines to slow the

neutralization and extend the opportunity for drawing and alignment of the chains. Increases in chain order may also be achieved by including Zn^{+2} salts in the spin baths. These ions coordinate with the xanthate sulfur on adjacent chains and form temporary cross-links.

Acetates. Two forms of cellulose acetate are of commercial importance for textile fibers (19, 20). Primary cellulose acetate, or triacetate, is almost completely esterified and has 10% or fewer free hydroxyls. Secondary acetate, or diacetate, is the more common fiber material. Because it has 25-30% free hydroxyls, it is more hydrophillic, soluble in more polar solvents, and less thermoplastic than triacetate. Both materials are made from cellulose pulp by preswelling with acetic acid and then esterification by using a sulfuric acid-acetic anhydride mixture. The reaction is carried to a point at which essentially all of the hydroxyls are converted to either acetate or sulfate esters. Further reaction allows transesterification of the sulfates to yield the triacetate. Hydrolysis of the mixed triester causes selective cleavage of the sulfates with partial removal of the acetates to yield the diacetate. The process is carried out in this manner because any attempt to go directly to the diacetate would result in a heterogeneous block copolymer system because cellulose in accessible regions would react to the triacetate stage before significant penetration of the crystalline regions could be accomplished.

After precipitation, neutralization, and drying of the product, it is dissolved in a volatile organic solvent and dry spun. The diacetate is soluble in acetone, but the less polar triacetate requires a solvent such a methylene chloride.

Acrylics and Modacrylics. The fiber-forming acrylic polymers are produced in aqueous media by using free-radical-initiated addition polymerizations. To provide dyestuff accessibility and sites for dye binding they contain small amounts (less than 15% by weight) of one or more comonomers such as 2-vinylpyridine, N-vinylpyrrolidone, acrylic acid, methallylsulfonic acid, and vinyl acetate, or acrylic esters. The modacrylic fibers contain significantly higher concentrations of comonomers than the acrylics (up to 65% by weight). The comonomers of choice are vinylidene chloride, vinyl chloride, and possibly acrylamide. The high halogen content of most modacrylics makes them relatively flame resistant but also lowers their melting points and reduces their heat stability relative to the acrylics (21).

Acrylic and modacrylic fibers are produced by either dry or wet spinning. As a result of the strong intermolecular attractions present in the acrylics, the only solvents that are suitable are those that are very polar and thus capable of disrupting these secondary valence bonds. These include N,N-dimethylformamide, dimethyl sulfone, dimethyl sulfoxide and dimethyl acetamide. Modacrylics, however, are soluble in more volatile, lower polarity solvents such as acetone. After spinning the residual solvent in acrylics must be removed by washing, and the fibers are drawn either dry (in a hot air oven or over-heated rolls at 80-110 °C) or wet (in steam or hot water at 70-100 °C). Finally the yarns must be dried and heat stabilized to control fiber shrinkage during subsequent processing.

Polybenzimidazole and Carbon Fibers. Polybenzimidazole (Structure XIV) fibers are specialty fibers with high melting points, very high decomposition temperatures (>500 °C), good strength, and extensibility. They are essentially nonflammable and have an unusually high moisture regain for a synthetic fiber (>10%). Because of the toxicity of the monomers, they must be manufactured in completely closed systems with special health and safety precautions. The most cited preparation involves the melt condensation polymerization of 3,3'-diaminobenzidine or its tetrahydrochloride and diphenyl isophthalate (21-23).

XIV

Carbon fibers have been the subject of extensive research and investigation (24). They are characterized by their high strength and moduli as well as high-temperature resistance. As a result, carbon fibers have been the fiber of choice for use as the reinforcing component in lightweight, high-performance composites. All carbon fibers are made by the carbonization of preformed fibers. These precursors may be acrylonitrile homopolymers, high-tenacity rayon filament yarns, or mesophase pitch fibers from coal or petroleum pitch. The three process steps common to the production of carbon fibers from the above precursors are preoxidation at temperatures ranging from 200 to 400 °C to prevent random chain-scission decompositions; carbonization at temperatures ranging from 1000 to 1400 °C; and graphitization at temperatures ranging from 1600 to 3000 °C to produce the final carbon fibers.

Polymerization Process Technology

In general terms a man-made fiber polymerization scheme can be classified as either a batch or a continuous process. In a pure batch process the polymerization step is carried out separately from fiber formation in reactors that receive discrete charges of monomer(s). In a continuous polymerization (CP), monomer is fed continually into the reactors, and polymer is continually removed downstream. For some polyamides and polyesters, fiber formation may or may not be an integral part of the CP line. Most modern polymerization schemes are continuous processes, and these are slowly replacing much of the older batch technology.

In the design of any polymerization reaction scheme, important consideration must be given to maximizing the production of high-quality chip. At the temperatures that favor high rates of production, reactor residence times must be minimized to avoid degradation of the polymer. Efficient recovery of unreacted monomer as well as low-energy process requirements are also important design criteria.

Of the fibers listed in Table II, polyester and polyamide polymers are made by hot melt polymerizations. Because the cellulosics are based on a natural polymer, they are not polymerized in production. The acrylics are emulsion polymerized, and the polyolefins may be polymerized either as a slurry in a hydrocarbon solvent, in bulk, or in gas-phase reactors. Although every manufacturer adapts his equipment to his physical plant and production needs, there are process steps characteristic of each type of man-made fiber polymerization technology. This fact is illustrated by two examples of hot melt polymerization processes from the recent patent literature; these are briefly discussed in the following and are presented schematically in Figures 2 and 3.

In Figure 2 a schematic representation for the continuous polymerization of PET from ethylene glycol and TA is shown (25). The primary esterifier is a multicompartment reaction vessel operated above the partial pressure of the glycol at the reaction temperature employed, which is typically around 180 °C.   The secondary esterifier is of similar design to the primary except that it is operated at a reduced pressure to remove water and glycol. The product of the secondary esterifier is fed to the low polymerizer where the polycondensation phase of the reaction begins. The low polymerizer is also a multizone reactor where the temperature and pressure are progressively increased and reduced, respectively. The product from the low polymerizer is fed into the high polymerizer where the temperature is brought to its final value of 280-295 °C, and the pressure further reduced. To maximize the molecular weight of the polymer while reducing degradation at these temperatures, the melt surface area is kept large in the high polymerizer by filling the reactor to only one-tenth to one fifth of its volume capacity. This condition also minimizes the dwell time of the melt in the reactor at these elevated temperatures.

A continuous process for polymerization of nylon 6,6 in which a fluidized bed solid state polymerization reactor is used as the high polymerizer is represented schematically in Figure 3 (26).   In this process the low molecular weight polymer is produced in a filled pipe reactor located just upstream of the spray drier. The liquid product of this step is then sprayed into a hot inert gas atmosphere where the water is flashed off and a fine powder is produced. This powder is fed into an opposed-flow, fluidized bed reactor at 200 °C where the high molecular weight polymer powder is generated at temperatures well below the 255 °C melting point of nylon 6,6.   The powder is then melted in the extruder and converted into fiber or chip.

Additional examples of polymerization processes can be found in a recently published review of fiber-forming polymerization patents by Robinson (27). A detailed comparison of batch and continuous polymerization for nylon 6,6 can be found in a review by Jacobs and Zimmerman (15).   In another review Short has summarized the current state of polypropylene polymerization technology and catalyst development (28).

Fiber Formation

The physical properties of the polymer often dictate the spinning method that must be used for fiber formation. For example, if the melt temperature is above the thermal degradation temperature, the polymers cannot be melt spun. Such polymers must be liquified with

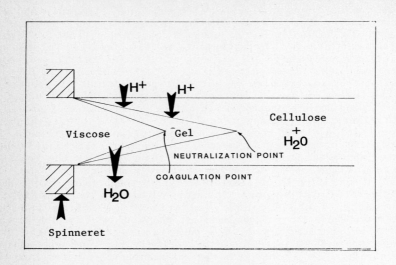

Figure 1.   Regeneration of viscose rayon.

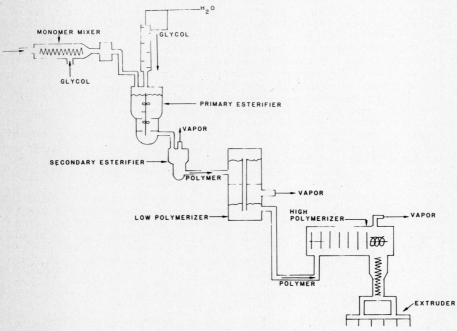

Figure 2.   Schematic of direct esterification, continuous polyester
            polymerizer (25).

solvents for spinning in either a wet or dry solvent removal system. These systems are similar to the melt spinning process in that flow of the polymer and solvent solutions must be established within a spinneret channel (extrusion). However, unlike melt spinning, wet and dry spinning require the precipitation of the polymer either in a solution bath for wet spinning or by evaporation of the solvent from the fiber filament in dry spinning. To be spinnable the polymer melt or solution must be capable of forming continuous, fluid threads when extruded. These fluid threads must be easily transformed into solid fibers that, after spin finish application, possess the physical properties required for subsequent processing (29). Polymers from Table II that are typically wet or dry spun are aramids, acrylics, modacrylics, and cellulosics. The polyesters, polyamides, and polyolefins are melt spun fibers.

During the past decade, significant advances have been made in the melt process spinning equipment, which is represented schematically in Figure 4. Improved extruder (melter) design has been achieved by a better understanding of the analytical treatment of non-Newtonian fluid-flow rheology (30, 31). Increases in production rates have been accomplished by shortening residence times and improving internal mixing. These changes have resulted in production rates that are more than double those typical for extruders during the 1970s. In this same period, there have been significant improvements in the filters that are necessary to prevent the spinneret's capillaries from becoming blocked. Originally these filters were sand and wire mesh screens, but developments in powdered metal and nonwoven stainless wire filters have improved filter life and efficiency. The development of large inline changeable filters upstream of the metering pumps in melt spinning has further extended by ten-fold the time required before shutdown for spinning filter changes.

Spinning speeds have also been increased dramatically during the past decade. In addition to increasing production, higher spinning speeds can be used to finish and orient the yarn while still on the extrusion/spinning line (spin drawing). In general, as the spinning speed increases, the air drag on the falling filaments begins to become significant. For example, at spinning speeds exceeding 3000 m/min, PET filaments are stretched by the air drag to induce a preorientation of the molecular chains in the direction of the fiber axis without significant crystallization. This partially oriented yarn (POY) is especially suitable for draw texturizing, and its commercialization has been called one of the more significant new developments in textile yarns (11). On a limited scale this process has been successfully extended to 10,000 m/min. To achieve higher speeds, which should result in spun yarns with even higher as-spun strength, it will probably be necessary for mechanical take-up to be replaced with more exotic winding devices.

Conventional, low-speed spun fibers must be further finished by drawing, although POY yarns are typically draw texturized. The drawing process adds strength by orientation of the molecular structures. Normal extensibility and proper tenacity are added by drawing the spun fibers to several times their as-spun length. The texturing process produces permanent crimp, loops, coils, or crinkles in the yarn that result in properties of stretch, bulk, absorbency, and improved hand (32). There are three basic types of

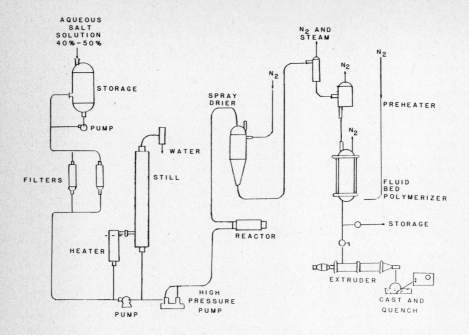

Figure 3.   Schematic of continuous solid state polymerizer for nylon 6,6 (26).

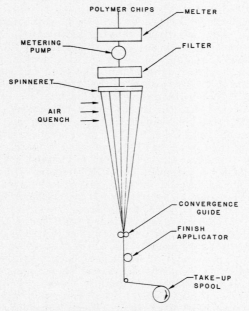

Figure 4.   Melt spinning process.

yarn: stretch, modified stretch, and bulk yarns. Stretch yarns are characterized by false twist or crimping. Modified stretch yarns are produced as stretch yarns but have had the amount of stretch reduced and the bulk increased by heat setting after the false twisting or crimping. Bulk yarns have little stretch with increased bulk.   Bulk yarns are produced by the stuffer box or edge-crimping methods of texturizing.

There have been several attempts to produce fibers without the use of the conventional extruding and spinneret devices.   In electrostatic spinning, for example, an electrostatic field is used to form the polymer into fiber strands (33).   This method holds promise, but will require significant development before commercial application can be achieved.

The production of bicomponent fibers increased considerably during the 1970s. Bicomponent fibers are filaments composed of two physically distinct phases consisting of different polymers. There are two significant types of bicomponent fibers: side-by-side and sheath-core (5).   The components are usually polymers of different chemical structure. The side-by-side fibers use polymers that have vastly different physical characteristics; this produces different shrinkage in the two side-by-side polymers and results in a wool-like bulk and crimp in the fibers. These fibers are made by feeding the different polymers into the spinneret at or near the capillary orifice.   The sheath-core fiber is usually produced to achieve overall strength with the core and wear resistance with the polymer of the sheath.   The sheath-core fibers are produced by spinning the two polymers through concentric capillary spinnerets (13).

An interesting variation of the bicomponent fiber idea is the use of air or micropores as the second phase. These fibers have found use in biomedical applications as filter media in artificial kidneys. Robinson has reviewed some recent patents in this area, as well as in the other segments of polymer extrusion, spinning, and processing (34).

Postfiber Formation Chemistry and Technology

The purpose of any post-fiber formation treatment is to modify an appearance or performance characteristic of a fiber or fabric for a specific application without limiting its general use. For example, most man-made fibers could be permanently colored (dope dyed) at the fiber extrusion stage.   However, in practice, except for polyolefins, almost all color is applied by using dyestuffs or pigments in a later processing step; therefore, the fiber manufacturer or consumer is not limited to a large quantity of permanently colored stock. Similar considerations apply to the wide variety of finishing treatments that may be used to impart specific properties to yarns and fabrics.

Although dyeing and resin finishing are by far the most important of the post-fiber formation manufacturing steps, the first finish to be applied to any man-made fiber is called a spin finish. It is applied as part of the fiber spinning process and usually consists of at least three components. These are a wax or heavy oil that acts as a lubricant to reduce fiber-to-fiber and fiber-to-machinery friction; an antistatic agent to reduce static charge build-up during the processing of hydrophobic polymers; and an emulsifier to prevent the separation of the hydrophobic lubricants and hydrophilic antistats in the spin finish. Spin finishes usually are temporary finishes that are removed by scouring prior to any subsequent dyeing or finishing steps.

Dyeing is one of the most universal of textile chemical processing steps. Dyes may be applied to staple fiber (stock dyeing), yarns (yarn dyeing), fabrics, or individual textile articles. Almost all dyestuffs are applied from aqueous solutions or dispersions, although there continues to be a low level of interest in solvent dyeing processing (35). The dyeing process may be either batch, in a variety of closed dyeing systems, or continuous on a continuous dye range. Dyeing conditions in batch processes may range from temperatures of 80 °C at atmospheric pressure for rayon to 130 °C under pressure for polyesters. In the continuous dyeing of polyester/cotton fabrics on a thermosol range, temperatures as high as 215 °C are not unusual. Table III lists the principal dyestuff application classes for the fibers in Table II, as well as some of the more important dyeing auxiliary textile chemicals.

The objective of postdyeing textile finishing is to impart improved performance characteristics in comparison to those inherent in a particular fiber or fabric construction. Some common examples are imparting flame resistance for protective clothing, water repellency for outerwear garments, and reduced static charge build-up for operating room gowns. In Table IV a brief summary of some finishing objectives and the broad chemical classes of textile chemicals used to achieve them are presented.

Except for certain dyeing procedures, there are three basic steps involved in most dyeing and finishing processes: application of the color or finish, predrying to remove the excess water, and curing to fix or chemically cross-link the dyestuff or finish, respectively. This procedure is commonly referred to as a pad–dry–cure operation.

These are illustrated by the schematic of a dye range for the continuous dyeing of cotton/polyester blend fabrics as shown in Figure 5. In this example two application classes of dyestuff (one from Table III for the cotton fiber and a disperse dye for the polyester) are applied simultaneously from the dyebath to the fabric. The fabric is predried with banks of IR heaters (either electric or refractory) and then the disperse dye, which is nonionic and can sublime, is thermally fixed or thermosoled into the polyester. The chemical pad and steamer parts of the range are necessary to provide the chemical components and the reactor for the fixation of the cellulosic dyes. For vat and sulfur dyes on the cotton, a typical sequence in the wash boxes might be a cold wash in box 1, warm washing in boxes 2 and 3, followed by an oxidation in the next two boxes, followed by hot washes and rinsing in the remaining boxes. Most commercial ranges have 8–10 boxes, rather than the four illustrated in Figure 5 (36). A typical range of the type shown in Figure 5 would dye 80–120 yd/min. On a finishing range, the drying cans would be replaced by a tenter frame, which is a horizontal forced air oven with the fabric passing through while tensioned at the edges.

Modifications that might be found on the range in Figure 5 or a similar range would be the use of a foam applicator head rather than

Table III. Dyestuffs and Chemical Auxiliaries Used
to Dye Man-Made Fibers

Dyestuff class	Class Fiber	Chemical Auxiliary (by class)[a]
Disperse	Polyesters Acetates Polyamides Acrylics	Dispersing agents (sulfonated lignins) Carriers/Accelerants (benzene and naphthalene derivatives) Surfactants[b]
Basic	Acrylics	Retarders (quarternary ammonium compounds, nonionic surfactants) Surfactants[b]
Acid	Polyamides	Leveling agents (nonionic surfactants) Surfactants[b]
Direct		Electrolytes ($NaCl, Na_2SO_4$)
Vat		--
Sulfur	Rayon	Surfactants[b] Electrolytes (NaCl)
Reactive		Electrolytes (NaCl)
Azoic		--
Pigments	Polyolefins (dope dyeing)	

[a]Only the most generally applied dyebath chemical auxiliaries have been listed for illustrative purposes.
[b]May be anionic, cationic, nonionic, or amphoteric depending on applications.

Table IV.   Summary of Some Wet Chemical Finishing Objectives
            and the Textile Chemicals Employed to Obtain Them

Fabric and/or Yarn Property Modified	Chemical Finishing Agent (by class)
Easy care/crease resistance/ wash-wear	N-methylol derivatives of ureas, triazones, carbamates, melamine
Handle	alkanol amides, polyethylene emulsions, polyacrylates, vinyl acetates, polyethers
Soil release	Fluorocarbons
Flame resistance	Phosphonium salts, organohalogens, w/wo Sb_2O_3, halogenated phosphates, vinyl phosphonates, sulfamates
Static dissipation	Poly(ethylene oxide) glycols, quaternary ammonium compounds
Water repellancy	Chlorinated paraffin waxes, fluorocarbons, silicone emulsions
Abrasion resistance	Reactive silicones, waxes
Pilling	Reactive silicones
Optical brightness	Stilbene derivatives, oxazoles, triazoles, triazines, phthalimides

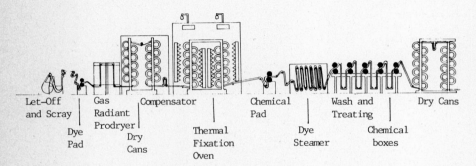

| Let-Off and Scray | Gas Radiant Prodryer | Compensator | | Chemical Pad | | Wash and Treating | | Dry Cans |
| Dye Pad | | Dry Cans | Thermal Fixation Oven | | Dye Steamer | | Chemical boxes | |

Figure 5.   Schematic of continuous dye range.

a dye pad, or vacuum slot extractors after the pads. These would be used to reduce the water that subsequently must be removed in the drying steps and thus would significantly reduce the energy requirements of the process. Radio frequency drying and the use of high-velocity air have recently been investigated as alternatives to conventional drying procedures (37, 38). These and other aspects of the chemistry and technology of dyeing and finishing have been extensively discussed and reviewed in the literature (39-47).

Future Trends

The chemistry and technology of man-made fibers has changed dramatically during the past 10 years as evidenced by the introduction of the TA process for polyester, the commercialization of the aramids, and the increasing use of solid state polymerization. As the man-made fiber industry matures, it becomes more difficult to predict whether this trend will continue. In addition, the development of new polymerization chemistry or the introduction of new fibers occurs only rarely. Nevertheless, some developments in new specialty fibers can almost certainly be expected. In particular, an elastomeric fiber based on polyester rather than urethane chemistry and an anisotropically spun polyester analogous to the aramids would seem to be natural targets for research and development. Some progress toward these goals has already been made.

Any significant advances in polymerization technology will probably result more from the increased sophistication of process control and monitoring devices rather than more major equipment design changes. By using tight process control to manipulate molecular weight distributions, reduce impurities and side reactions, and increase the accessible molecular weight range, it should be possible to more readily "tailor make" fibers for specific applications.

Fiber spinning is the one area of man-made fiber production in which a major change in technology is most likely to occur in the near future. Ultra high-speed spinning with nonmechanical devices could dramatically affect all phases of pre- and postfiber formation chemistry and technology.

Although environmental and energy considerations have had considerable impact on the textile dyeing and finishing industry, the emphasis in the recent past has been on equipment modification rather than new chemistry. This trend can be expected to continue as more and more dyeing and finishing processes are automated and process control is substantially improved.

In summary, it appears that the man-made fiber industry is likely to experience more changes in technology than in chemistry in the near future. However, the new technology may very well demand new chemistry and thus provide the chemist and engineer with new challenges.

Acknowledgment

The authors wish to thank the following colleagues for their encouragement, help, and suggestions in the preparation of this manuscript: Dr. R. Merrill, Dr. K. K. Likhyani, Dr. C. Chaisson,

Dr. K. Wagener, Dr. J. Dellinger, Mr. T. Montgomery, and Ms. J. Lucas-Drews.

Literature Cited

1. Textile Organon 1984, 55, 124.
2. O'Sullivan, D. A.; Storck, W. J. Chem. and Engin. News, Feb. 28, 1983, p. 24.
3. Textile Organon 1983, 53, 21.
4. Roberts, W. J. In "Encyclopedia of Chemical Technology"; Kirk-Othmer, Eds.; 3rd ed.; John Wiley & Sons: New York, 1982; Vol. 10, p. 148.
5. Moncrieff, R. W. "Man-Made Fibers"; John Wiley & Sons: New York, 1975.
6. Peters, R. H. "The Chemistry of Fibers"; Elsevier: New York, 1963; Vol. I.
7. Lenz, R. W. "Organic Chemistry of Synthetic High Polymers"; Interscience: New York, 1967.
8. Hicks, E. M., Jr., et al. Text. Prog. 1971, 3, 1.
9. Walczak, Z. K. "Formation of Synthetic Fibers"; Gordon and Breach: New York, 1977.
10. "Applied Fibre Science"; Happey, F., Ed.; Academic Press: London, 1977, 1978, 1979; Vol. I-III.
11. Davis, G. W.; Hill, E. S. "Polyester Fibers," in Ref. 4, 1982; Vol. 18, p. 531.
12. Ludewig, H. "Polyester Fibers, Chemistry, and Technology"; Wiley-Interscience: New York, 1979.
13. Saunders, J. H. "Nylon Fibers," in Ref. 4, 1983; Vol. 18, p. 373.
14. Hughes, A. J.; McIntyre, J. E. Text. Prog. 1976, 8, 18.
15. Jacobs, A. B.; Zimmerman, J. "High Polymers"; Schildknecht, C. W.; Skeist, I., Eds.; John Wiley & Sons: New York, 1977; p. 424.
16. Morgan, P. W. Macromolecules 1977, 10, 1381.
17. Reimschuessel, H. K. J. Polym. Sci., Macromol. Rev. 1977, 12, 65.
18. Peters, T. V. "Elastomeric Fibers," in Ref. 4, 1982; Vol. 10, p. 166.
19. "Rayon and Acetate, A Special Report"; ITT Rayonier: Stamford, Conn., 1979.
20. "High Polymers"; Bikales, N.; Segal, L., Eds.; Wiley-Interscience: New York, 1971; Vol. V.
21. Elias, H. G. "Macromolecules"; Plenum Press: New York, 1977; Vol. II.
22. Jackson, R. H. Text. Res. J. 1978, 48, 314.
23. Bus. Week. October 20, 1980.
24. Roberts, C. W. "Report to South Carolina State Development Board"; June 1978.
25. Schaefer, P. S.; Mason, P. A.; Yates, W. H. U.S. Patent 4 100 142, 1978.
26. Munroe, G. C. U.S. Patent 3 031 433, 1962.
27. Robinson, J. S. "Fiber-Forming Polymers, Recent Advances"; Noyes Data Corp.: Park Ridge, N.J., 1980.
28. Short, J. N. Indust. Res. and Devel. Sept. 1980, p. 109.

29.   Ziabicki, A. "Fundamentals of Fiber Formation"; John Wiley & Sons: New York, 1976.

30.   Middleman, S. "Fundamentals of Polymer Processing"; McGraw Hill: New York, 1977.

31.   Dean, A. F.; McCormick, W. In "Modern Plastics Encyclopedia"; McGraw-Hill: New York, 1977.

32.   "Draw-Textured Yarn Technology"; Wilkinson, G. D., Ed.; Monsanto Textiles Co., 1974.

33.   Larrondo, L.; Manley, R. J. Polym. Sci. 1981, 19, 909.

34.   "Spinning, Extruding, and Processing of Fibers, Recent Advances"; Robinson, J. S., Ed.; Noyes Data Corp.: Park Ridge, N.J., 1980.

35.   Cook, F. L. Book of Papers - 1982 AATCC Technical Conference, Atlantic City, N.J., 1982; p. 22.

36.   Smith, L. R.; Melton, O. E. Amer. Assoc. Textile Chem. & Colorists, J. 1983, 14, 113.

37.   Balmforth, D. Amer. Assoc. Textile Chem. & Colorists, J. 1983, 14, 126.

38.   Cook, F. L. Textile World, Aug. 1981, p. 77.

39.   Peters, R. H. "Textile Chemistry"; Elsevier Scientific: New York, 1975; Vol. III.

40.   Rys, P.; Zollinger, H. "Fundamentals of the Chemistry and Application of Dyes"; Wiley-Interscience: New York, 1972.

41.   Trotman, E. R. "Dyeing and Chemical Technology of Textile Fibers"; Whitstable Litho Ltd.: Kent, England, 1975.

42.   Vickerstaff, T. "The Physical Chemistry of Dyeing"; Interscience: New York, 1954.

43.   Valko, E. I. In the "Encyclopedia of Polymer Science and Technology"; John Wiley & Sons: New York, 1964; Vol. 5, p. 235.

44.   Valko, E. I. In "Man-Made Fibers, Science and Technology"; Mark, H. F.; Atlas, S. M., Cernia, E., Eds.; Interscience: New York, 1968; Vol. 3, p. 499.

45.   Marsh, J. T. "Self-Smoothing Fabrics"; Chapman and Hull: London, 1962.

46.   Tattersall, R. International Textile Machinery 1977, 79.

47.   Obenski, B. J. Amer. Dyestuff Reporter 1981, 70, 19.

Foamed Plastics

RUDOLPH D. DEANIN

Plastics Engineering Department, University of Lowell, Lowell, MA 01854

```
Processes
   General
   Blowing Agents
   Special Processes
Classes
Properties
   Unique Properties of Foamed Plastics
   Structural Features of Foamed Plastics
Leading Commercial Polymers
   Polyurethane
   Polystyrene
   Poly(vinyl chloride)
   Structural Foams
Applications and Markets
```

When bubbles of air or other gas are dispersed in a solid polymer matrix, the product is classified as a foamed plastic. Although such foams were often observed accidentally in the early development of commercial polymers, commercialization of foams began more slowly and has accelerated more recently. Paradoxically, phenolic resins were commercialized in 1908 only when Baekeland learned to prevent foaming by molding them under high pressure. Rubber latex was first converted to foam rubber in 1914, and foamed ebonite has been in service as insulation for over half a century. The 1930s saw the invention of foamed polystyrene in Sweden and the commercial production of foamed urea–formaldehyde insulation and azo blowing agents in Germany. During the 1940s, foamed polyethylene was invented in England, and foamed epoxies were invented in the United States; commercial production of foamed polystyrene and poly(vinyl chloride) began in the United States, and that of vinyls and polyurethanes began in Germany. Polyurethane foams came to the United States in the 1950s and grew rapidly to major commodity

0097–6156/85/0285–0469$07.50/0

status. Syntactic and structural foams appeared in the 1960s, followed by the rapid growth of reaction injection molding (RIM) of polyurethanes since the late 1970s. Current production of foamed plastics is thus about 3 billion lb/year and is growing much faster than total plastics production as knowledge of production techniques, properties, and optimum applications increases.

Much of this growth has been recorded periodically in Modern Plastics (particularly every January), the Journal of Cellular Plastics, and the annual "Modern Plastics Encyclopedia." Collection of information in textbook form began with Ferrigno in 1967, Benning in 1969, and Frisch and Saunders in 1972-73. More recently Meinecke and Clark, and most recently Hilyard, have made notable contributions to the understanding of foam properties. This chapter will summarize the processes by which foamed plastics are produced, the major classes, the relationships between foam structure and properties, the leading polymers used in foam production, and their major applications and markets.

Processes

General. The production of foamed plastics requires three successive steps: liquid state, bubbles of gas, and solidification. The way these three steps are carried out depends upon the type of polymer and the type of foam being produced.

Liquid State. The liquid state in most thermoplastics is produced by heating until molten; in vinyls it is generally obtained by dispersing resin particles in plasticizer to produce a plastisol; and in rubber it is obtained by use of latex. In thermosetting plastics, the monomers are reacted only up to low molecular weight reactive prepolymers, which are still liquid or at least readily fusible at low temperature.

Bubbles of Gas. Bubbles of gas may be produced physically or chemically. Physical blowing agents include permanent gases, such as air or nitrogen, which can be whipped into the liquid (frothing of urethanes, rubber latex, and vinyl plastisols) or compressed into the molten polymer (thermoplastic structural foams). Physical blowing agents also include volatile liquids, which are boiled by heating the thermoplastics (pentane in polystyrene) or by the exothermic reaction of thermosetting plastics (fluorocarbons in polyurethanes). Chemical blowing agents are primarily organic azo compounds, such as azobisformamide, which are decomposed by heating thermoplastics, and thereby produce nitrogen and other small gaseous molecules. Other chemical reactions used to produce foaming include the reaction of isocyanate with water to produce CO_2 in flexible urethane foams, and the reaction of sodium bicarbonate with citric acid in polystyrene to produce CO_2 and thus nucleate the primary foaming action of pentane. In general, during the "blowing step," surface tension and melt strength may become extremely critical.

Solidification. Solidification of the foamed liquid may be a physical or a chemical process. In thermoplastics solidification is generally accomplished by cooling the melt back to the solid state; in vinyl plastisols and rubber latex solidification is accomplished by more complex processes of physical gelation; it is sometimes aided by evaporation of solvents such as pentane from polystyrene. In thermosetting plastics, solidification is generally accomplished by polymerization and cross-linking up to infinite molecular weights.

Blowing Agents. Aside from air and nitrogen, the choice of blowing agents for each polymer and each process depends upon their specific characteristics, particularly the boiling points of volatile liquids (Table I) and the decomposition temperatures of organic azo compounds (Table II). Pentanes are most commonly used for polystyrene foamable beads, and fluorotrichloromethane is typical of the fluorocarbons used for rigid urethane foams. Azobisformamide formulations are the most widely used blowing agents for vinyl plastisols and most thermoplastic structural foams, but the newer high-temperature engineering thermoplastics sometimes demand more specialized blowing agents with much higher decomposition temperatures. Concentration ranges for chemical blowing agents may typically range from 0.05 to 15% depending on the application (Table III).

Special Processes. Several more recent developments have opened the possibility of special foam processes that may grow to tremendous importance and perhaps even change the entire concept of foamed plastics. These are structural foam, reaction injection molding (RIM), and syntactic foam.

Structural foam is produced by slight to moderate expansion of a normally solid plastic. Typically, enough blowing agent is used to reduce the density 10–50%, and the process is carried out to produce a foamed core surrounded by a solid skin. The polymers are usually thermoplastics, and the blowing agent is either compressed nitrogen or an organic azo compound. There are a number of patented processes and types of equipment (see Box 1). The major advantages are material saving, lower-pressure equipment, higher rigidity/weight ratio, and elimination of sink marks. This field is growing so fast that it is hard to say what percent of plastics is structural foam and what percent is completely solid.

Reaction injection molding (RIM) begins by metering two reactive liquid components (primarily polyurethane) to a mixing head and then injecting the mix into a large low-pressure mold. By using low-cost tooling and a high-speed process, large rigid or semirigid parts such as auto front ends are produced very rapidly and economically. Most of these are somewhat foamed. There is much current interest in extending the RIM process to other polymers, so it is hard to say at present in what direction this technique may continue to grow.

Syntactic foam is made by dispersing hollow microballoons into a liquid polymer and then solidifying it. Microballoons are typically hollow glass or hollow phenolic microspheres, and the most common liquid polymer is an epoxy prepolymer, which is then cured. Although some products are notably woodlike in their properties and machinability, primary applications are high-performance products such as deep-sea instrumentation.

These are typical of some of the novel foam processes that may be unknowingly changing the entire concept of foamed plastics.

Table I

Volatile Liquids for Blowing Agents

Name	Boiling Temperature °C
Pentane	30-38
Neopentane	10
Hexane	65-70
Isohexanes	55-62
Heptane	96-100
Isoheptanes	88-92
Benzene	80-82
Toluene	110-112
Methyl Chloride	-24
Methylene Chloride	40
Trichloroethylene	87
Dichloroethane (sym.)	84
Dichlorotetrafluoroethane	4
Trichlorofluoromethane	24
Trichlorotrifluoroethane	48
Dichlorodifluoromethane	-30

(Ref. 9, 1975-6, Pg. 127)

Table II

Organic Azo Compounds for Blowing Agents

Name	Decomposition Temperature °F	Gas Evolution cc/gm
Oxy-Bis-Benzene Sulfonyl Hydrazide	315-320	125
Azo-Bis-Formamide + Additives	329-419	140-230
Toluene Sulfonyl Semicarbazide	442-456	140
Trihydrazine Triazine	527	175
5-Phenyl Tetrazole	464-482	200
5-Phenyl Tetrazole Analogs	680-725	200

(Ref. 9, 1981-2, Pg. 194)

Table III

Concentration Ranges for Chemical Blowing Agents

Application	Concentration PHR
Compression Molding of Low-Density Foam	5-15
Casting of Soft Plastisol Foam	2-4
Extrusion of Structural Foam	0.2-1.0
Injection Molding of Structural Foam	0.3-0.5
Injection Molding: Elimination of Sink-Marks	0.05-0.1

(Ref. 9, 1976-7, Pg. 194)

Box 1

Structural Foam Processes

Impco Trac: Series of molds travel a closed path through stations for clamping, injection, solidification, and unloading.

Union Carbide: Extruder delivers melt + blowing gas into a pressurized accumulator, from which a short shot is injected into a low-pressure mold.

ICI Sandwich: 3-stage injection molding: short shot of solid polymer is followed by a second shot of foamable polymer and a third shot of solid polymer, forming a solid skin around a foamed interior.

Beloit Two-Component: Two extruders feed an injection mold simultaneously. One feeds solid polymer to form the skin, the other feeds foamable polymer to form the core of the finished molding.

USM: Full shot of foamable plastic is injected into a reduced-size mold cavity. Collapsible core or movable walls then expand the cavity to permit foam to expand.

TAF: Plasticating screw fills a high-pressure accumulator, which then fills a reduced-size mold cavity, which is then expanded by use of movable walls.

LIM: Reactive liquids are mixed batchwise in a small chamber and injected into a mold where they foam and cure rapidly.

RIM: Reactive liquids are mixed continuously and used to feed a series of molds in which the mixture foams and cures.

(Ref. 9, 1975-6, Pg. 322)

Classes

Although foam structures can cover a continuous spectrum, it is convenient to note several major classes of commercial importance.

The most important distinction is between closed- and open-cell foams. In closed-cell (unicellular) foams, each gas bubble is separated from the others by thin walls of polymer; these foams are optimal for flotation applications, structural rigidity, and thermal insulation. In open-cell foams, the cells are all interconnecting, and fluids and especially air can flow freely through the foam structure; these are optimal for sponge products and for soft flexible materials. In the extreme case, when the last few remaining cell walls (windows) have been chemically dissolved out of an open-cell foam, it is sometimes called "reticulated."

Another important distinction is based on density. For flotation, sandwich construction, thermal insulation, and economy, very low-density foams are preferred, often as low as 1 lb/ft^3 or less. Vinyl, polyolefin, and syntactic foams are generally of medium density. Structural foams are of medium to fairly high density and have a graded structure from solid skin to fairly low-density core. RIM foams are similar, often exhibiting a fairly high density.

In commercial practice, these distinctions separate most foams into discrete classifications. In theoretical analysis, it is difficult to use a single process to produce a complete spectrum of structures, and it is difficult to relate properties to the complete spectrum of structure in a single unambiguous homologous series. It is hoped that such continuous analysis will be more feasible in the future.

Properties

Unique Properties of Foamed Plastics. Foamed plastics have certain unique properties that distinguish them from solid polymers and are particularly useful in practical applications. Basically, these properties result from their composite structure—a continuous phase of polymer that is of relatively high modulus, and a gas phase of negligible modulus, which may be either dispersed as single cells in a closed-cell foam, or continuous and interpenetrating in an open-cell foam.

Buoyancy of closed-cell foams results from the low density of the gas cells. High ratio of rigidity/weight and strength/weight is particularly notable in closed-cell foams and in sandwich construction and structural foams; on the other hand, softness and flexibility in open-cell foams result from the low viscosity of the continuous gas phase. Impact cushioning is achieved paradoxically in both flexible and rigid foams. In soft flexible foams, the mechanism of cushioning is self-evident; however, in rigid foams, it may be explained by energy absorption in crushing the thin cell walls, or in flexural hysteresis of the stiffly flexible cellular structure.

Acoustic absorption in open-cell foams is very useful in quieting reflected noise. High thermal and high electrical insulation are due to the very low conductivity of the ubiquitous gas phase, and are definitely best at lowest density; even choice of the

blowing gas becomes important, with fluorocarbons generally giving
the highest thermal insulation.

Water absorption of sponges results from the ease with which
water can replace gas in flexible open-cell foams, because both
fluids can flow freely on flexing.

Economics are often cited as an important factor in the use of
foam, either through improved property/weight ratios as cited
earlier, or simply through the ability of the processor to expand
the solid polymer and make it go considerably further at little
extra cost.

For any or all of these reasons, foamed plastics are now growing
at a much faster rate than the plastics industry as a whole. At the
same time, current notions of the basic relationships between foam
structure and properties must be refined considerably to facilitate
the proper production and use of foamed plastics in an even greater
range of applications.

Structural Features of Foamed Plastics. The critical structural
features of foamed plastics may by itemized as follows: polymer,
gas, density, open/closed-cell ratio, cell size, and anisotropy. It
is important to consider how each of these structural features
affects important properties of the foamed plastic.

Polymer. There is a natural tendency for polymer scientists to
assume that the relationships between polymer structure and properties carry over from the solid polymer to its foams as well.   To
some extent this is true. For example, the effects of plasticizers
in vinyls and the effects of cross-linking in urethanes are well
understood in the solid polymers, and they have parallel effects in
the foamed plastics. In general, the properties of polymers in the
solid phase can be reflected in their foams in the following ways.

Mechanical properties of solid polymers generally become
considerably "softer" in the corresponding foams. This is particularly true in open-cell foams. In closed-cell foams, properties
per unit area or unit volume are also generally somewhat softer; but
properties per unit weight in the expanded polymer may be
considerably harder and stronger than in the solid polymer because
of the principles of sandwich structures.

Thermal mechanical behavior of foamed plastics parallels that of
their solid polymers, but may be shifted to lower temperatures
because of mechanical softening (described earlier). Thermal
stability also parallels the solid polymers, but is somewhat lower
because foams expose a high surface/volume ratio to aggressive
environments such as hot oxygen. Flammability similarly parallels
the solid polymers but is somewhat higher because of the high
surface/volume ratio of foamed plastics exposed to oxygen. On the
other hand, application of flame-retardant additives is sometimes
easier in the foam, and incorporation of flame-retardant blowing
agent gases such as fluorocarbons in closed-cell foams can actually
increase the flame retardance of the polymer.

Chemical resistance of foams is generally similar to their solid
polymers, but somewhat lower because of the high surface/volume
ratio exposed to chemical attack, and also because of orientation
strains frozen in during expansion. Aging is similar but somewhat
faster, as discussed earlier, for thermal stability. Permeability

is similar but higher because of the small amount of polymer that must be permeated per unit thickness; in open-cell foams it is possible to achieve high flow-through rates for filtration applications. Toxicity is the same for solid and foamed polymers, except that leaching of monomeric additives from foams will be faster because of their high surface-to-volume ratio.

Thus, in all these ways, the structure-property relationships in solid polymers do carry over qualitatively into their foams as well, but often with considerable quantitative modification. In terms of practical properties of maximum importance, other structural features of the foam, particularly gas and density, may be much more important than the particular polymer itself.

Gas. The role of the gas in foam properties is usually obvious but too often overlooked in the scientific and engineering design of materials and end products.

In low-to-medium-density foams, the gas phase occupies most of the volume of the foam and, therefore, plays a dominating role in controlling many properties. Most of the effects of the gas phase are due to its low density, low viscosity, and low conductivity. Some of the effects are evident in any type of foam, and other effects depend primarily on whether the foam is an open- or closed-cell structure.

The low density of the gas directly lowers the density of the foam, which is important in lightweight products. In closed-cell foams, the low density of the gas produces the buoyancy of the product, which is important in many marine products.

The effect of the gas on mechanical properties depends on the paradox of closed- versus open-cell construction. In a closed-cell foam, mechanical deformation compresses the gas, so the gas contributes to rigidity and strength of the plastic product. In an open-cell foam, on the other hand, the fluidity of air permits it to rush out when the foam is deformed and to rush back when the deformation is released; therefore, the foam contributes to the softness, flexibility, and resilience of the plastic product.

Thermal conduction occurs primarily through the solid polymer; it is the gas that provides thermal insulation. In closed-cell foams, restriction of convection insulates further; and choice of the gas, particularly fluorocarbons, produces the maximum insulating capacity. On the other hand, the gas in a closed-cell foam aggravates the coefficient of thermal expansion and contraction most seriously when it is used in refrigeration and condenses to a low-volume liquid. If the gas in a closed cell is a flame-retardant fluorocarbon rather than air, it may contribute to overall flame retardance of the polymeric material.

Electrical conduction also occurs primarily through the solid polymer. Relatively speaking, the gas is the primary electrical insulator, so foaming of plastic insulation can greatly increase its insulating capacity. This applies not only to resistance and conduction, but also to dielectric constant and loss as well. On the other hand, electrical breakdown may be aggravated by the gas cells in a foam, perhaps by acting as imperfections at which discharge can occur.

Diffusion of blowing gas out of a closed-cell foam during normal aging gradually replaces the original gas by normal air; thus, high-performance properties of fluorocarbon gases, such as thermal insulation and flame retardance, may be lost unless the total foam is completely sealed from the atmosphere. On the other hand, easy replacement of gas by water in an open-cell foam is responsible for the function of plastic sponges.

Density.   Because polymer and gas often have opposite effects on foam properties, density represents not only the relative amounts of the two substances in the foam, but often also their relative contributions to properties as well. Thus, density is usually the most important structural feature in foam structure-property relationships, much more significant than the particular polymer that happens to be used. This is an extremely important concept that foam scientists and engineers must learn to use.

In most practical studies, a foam is manufactured over a narrow range of densities (D), properties (P) are measured and plotted, and a fairly straight-line relationship appears. This is adequate for the particular product-line involved, but it is very misleading in terms of the entire density range from very low to completely solid, and in terms of the basic relationships and mechanisms involved. In more precise studies, particularly over a wide or complete density range, the form of the relationship is generally found to be $P = KD^n$, where the values of K and n depend on the specific polymer and the specific property being measured (Figure 1, Table IV). In more recent work, these experimental observations have gained further strength from theoretical mechanistic analysis as well.

A number of important properties correlate directly with increasing density. Those mentioned most often include

1.   Modulus resulting directly from the solid polymer.
2.   Strength, again resulting from the solid polymer.
3.   Ultimate elongation, apparently depending on the amount of material present to permit ductile flow, or perhaps being inverse to the number of gas-bubble flaws, which can initiate failure.
4.   Minimum load at which creep will become serious.
5.   Hysteresis loss, a function of the solid polymer.
6.   Thermal dimensional stability, being better for the solid polymer than for the gas.
7.   Thermal conduction, primarily through the solid polymer rather than through the gas.

Conversely, a number of important properties are inverse to density. Those mentioned most often include

1.   Buoyancy, which is the difference between the density of water and the density of the foam.
2.   Softness, flexibility, and "rubberiness," particularly in open-cell foams. These properties are due to the negligible modulus and high fluidity of the gas phase.
3.   Resilience and recovery from both slow and impact deformation, primarily in open-cell foams, again because of the fluidity of air as it rushes back into the foam after the deforming force is released.

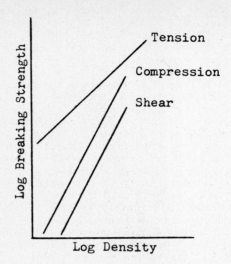

Figure 1. Strength vs. density in rigid urethane foams (5).

Table V

Experimental Values for K and n for Rigid Urethane Foams

Property in PSI vs. PCF	K	n
Compressive Strength	12.8	1.54
Flexural Strength	19.0	1.36
Tensile Strength	23.0	1.20
Shear Strength	14.9	1.16
Compressive Modulus	293.8	1.62
Flexural Modulus	186.3	1.75
Tensile Modulus	573.5	1.15
Shear Modulus	169.9	1.39

(Ref. 4, Vol. II, Pg. 487)

4. Shock absorbance, both in flexible open-cell foams because of the rush of escaping air, and in rigid foams because of the conversion of kinetic energy into potential energy during crushing of cell walls to create new surface.
5. Thermal dimensional instability of closed-cell foams due to expansion and contraction of gas and especially to condensation at very low temperatures in refrigeration. Also shrinkage during oven aging, due to permeation and loss of gas through the thin cell walls.
6. Thermal insulation due to the low conductivity of the gas phase, particularly in closed-cell foams, which restrict convection, and particularly when they are filled with an especially low-conductivity gas such as fluorocarbon.
7. Electrical insulation due to the low conductivity and low polarizability of the gas phase.
8. Permeability (Figure 2) due to the thinness of the remaining polymer walls acting as membranes, and of course particularly in open-cell foams when this property is desired.
9. Water absorption, particularly in flexible open-cell foams, for sponge-type applications.

All of these properties, and many more, depend more upon the density of the foam than upon any other structural feature; therefore, density should always be considered as the primary factor in the design of both materials and end products.

Open/Closed-Cell Ratio. The ratio of open to closed cells in a foam has important effects on many important properties. Although poor measurement techniques have reduced many studies from quantitative to simply qualitative, and although many foam processes produce only a semicontrollable mixture of open and closed cells, the basic relationships are of major theoretical and practical significance.

In general, the percent of open cells correlates directly with many important properties

1. Softness, flexibility, and cushioning result from the free flow of air out of the foam when it is deformed by mechanical forces (Figure 3). These properties are a major factor in the comfort of clothing and furniture.
2. Rebound and recovery depend on the ease with which air can rush back into the foam when the mechanical deformation is removed.
3. Mechanical properties in general correlate better with those of the solid polymer in an open-cell foam; however, they depend more on gas pressure in a closed-cell foam.
[Reticulation by chemically dissolving away the last traces of any "cell windows" at all removes weak flaws that could initiate premature failure, and thus ultimate strength (Figure 4) and elongation (Figure 5) are increased. However, this is the opposite of the normal effects of closed-cell walls, which also increase strength and elongation. Thus, the two concepts are paradoxical and must not be confused.]
4. Acoustic absorbance and insulation are accomplished by dispersion of sound waves in the open cells.
5. Thermal conductivity increases directly with increased convection through an open-cell foam.

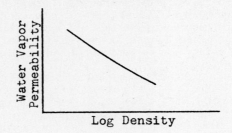

Figure 2. Permeability vs. density (2).

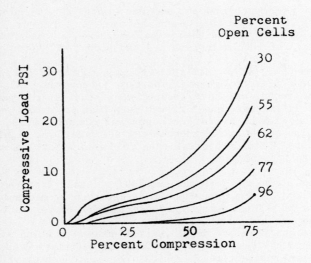

Figure 3. Compressibility vs. open cells in polyethylene foams (7).

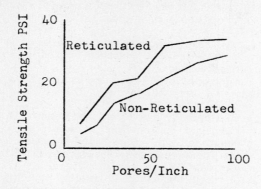

Figure 4. Tensile strength of flexible polyester urethane (2).

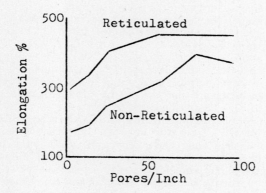

Figure 5. Elongation of flexible polyester urethane (2).

6. Permeability is easier and faster through an open-cell foam than
    through the thin walls of a closed-cell foam (Figure 6).
7. Water absorption in sponge-type applications and use as a filter
    medium depends on a high percent of open cells to permit easy
    passage of liquid water through the foam.

The gas in an open-cell foam is normally air, unless the entire foam
is enclosed in an impermeable wall and then filled with a particular
gas for a particular purpose.

Conversely, the percent of closed cells in a foam also correlates with many important properties

1. Buoyancy of the low-density gas is only effective if closed-cell
    walls protect it against replacement by water.
2. Rigidity and strength of the polymer are enhanced by mechanical
    deformation, which increases gas pressure and thus increases the
    total load-bearing capacity of the structure.
3. Compressive set, creep, mechanical damping, and hysteresis all
    occur because, under mechanical stress, gas permeates out of the
    closed-cell walls to reduce the stress. When the mechanical
    stress is released, the gas permeates back only slowly and often
    incompletely.
4. Thermal insulation improves (Figure 7) because closed cells
    reduce conduction by convection; thus, only radiation and solid
    conduction remain as conductive mechanisms.
5. Shrinkage during oven aging is much worse in closed-cell foams
    because the gas that permeates out of the cells cannot easily
    return when aging is over.

Thus, the mechanistic and practical significance of open versus
closed cells is very important, and deserves more precise
development of measurements and correlations with properties.

Cell Size. Microscopic examination shows that the cell size and
size distribution vary greatly between different foamed plastics.
It is natural to assume that the properties will vary accordingly.
Unfortunately, the exact relationships have proved elusive for two
reasons: any attempt to vary cell size experimentally invariably
changes other structural features simultaneously; thus, it is hard
to separate these independent variables; and there is a natural
predilection to prefer smaller cell size as evidence of better
experimental technique, and this sometimes blinds researchers to the
hard evidence before them. Despite these difficulties, certain
relationships between cell size and practical properties are
frequently reported.

Larger cell size is often reported to correlate with the
following properties

1. Modulus and stiffness, presumably due to the thicker cell walls.
2. Shock absorption, presumably by these thicker cell walls.
3. Buckling failure during compression.
4. Higher thermal conductivity by both convection and radiation.
5. Permeability, presumably because the permeating gas only has to
    go through a few cell walls and can travel freely through large
    open spaces within the cells.

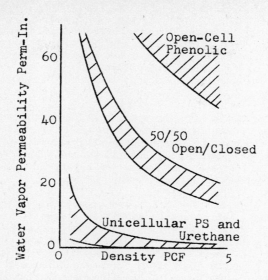

Figure 6. Effect of open/closed cell ratio on permeability (2).

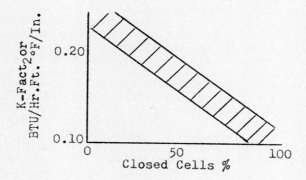

Figure 7. Thermal conductivity of rigid polyurethane foams (2).

Conversely, smaller cell size is often reported to correlate with the following properties

1. Higher strength, because of the larger number of cell walls that must be ruptured before failure will occur (Figure 8).
2. Higher ultimate elongation for the same reason.
3. Gradual failure during compression.
4. High thermal insulation because small cells reduce conduction by convection.

On the nihilistic side, however, some researchers have failed to find any significant correlation between cell size and practical properties. The specific questions must thus remain open until more precise studies, in which cell size is the only structural variable, are undertaken. Beyond this point, cell-size distribution may well prove to be another significant subvariable.

Anisotropy. When foam formation and expansion occur freely in all directions, the cell structure should be symmetrical and isotropic, and properties should similarly be isotropic in all directions. In many processes, however, the liquid is poured, and the gas bubbles that form are allowed to rise freely in the vertical direction. This process produces foam cells that are elongated vertically in the direction of foam rise. Naturally, this will have a severely anisotropic effect on many properties. Although this effect is sometimes predictable and is quite easily measurable, an unfortunate confusion in nomenclature has prevented scientists and engineers from making better use of it in product design. For clarity, any property test should indicate whether the test was taken in the direction of foam rise or across the direction of foam rise. A simple diagram could be used in reporting all results and should help clarify the meaning and facilitate use of the knowledge.

In general, compressive loading in the direction of foam rise must bear directly on the cell walls, which thus act as columns to support the load and give greater compressive modulus and strength in this direction. Loading perpendicular to the direction of foam rise will tend to fold the foam easily like an accordion (Figure 9).

Likewise, permeation through a foam in the direction of foam rise requires the permeating gas molecules to travel through a relatively small number of cell walls, in between which they can travel freely for relatively long distances down the length of each elongated cell. Therefore, permeability in this direction will be high. Permeation transverse to the direction of foam rise will force the gas molecules to pass through very many cell walls and only allow them short travel distances from wall to wall; therefore, permeability in this direction will be low (Figure 10).

Careful measurement and reporting should readily expand the current understanding of such anisotropic effects.

Leading Commercial Polymers

Standard texts have given good thorough descriptions of the individual commercial foamed plastics, so only brief review is needed here to put them into proper perspective.

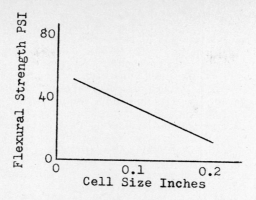

Figure 8. Flexural strength vs. cell size of polystyrene foam (<u>10</u>).

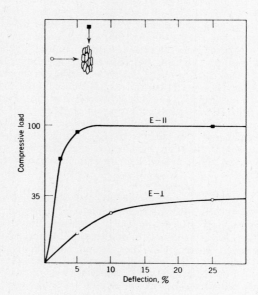

Figure 9. Effect of anisotropy on compressive behavior (<u>2</u>).

Polyurethane. Polyurethanes are formed by mixing liquid aliphatic polyols with liquid aromatic polyisocyanates, which react readily. These mixes gel in several minutes at room temperature, but often benefit from final heat cure. Most polyurethanes are foamed by introducing and expanding gas bubbles when the reacting liquids have reached the optimum viscosity, then the cross-linking reaction is completed to stabilize the foam structure.

For flexible foams, the polyols are primarily long-chain polyoxypropylene diols, with a small amount of triol to produce cross-linking and to ensure rubbery resilience; the polyisocyanate is primarily toluene diisocyanate (see Box 2). Gas bubbles are produced by the reaction of a measured amount of water in the polyol with a measured excess of isocyanate; in some formulations, low-boiling fluorocarbons may be used as auxiliary physical blowing agents. Open-cell structure is produced either by mechanical crushing or by chemical formulation to burst the cell walls before they cure. Such flexible urethane foams have largely replaced foam rubber and have taken a commanding position in automotive and furniture markets.

For rigid foams, the polyols have higher functionality, generally from three to six hydroxyls on a small compact backbone, and the polyisocyanate is generally two or three aromatic isocyanate groups joined by methylene bridges, to produce high-cross-link density (see Box 3). Blowing agents are primarily fluorotrichloromethane and mixed fluorocarbons, which are boiled by the exotherm of the cure reaction. The resulting closed-cell structure provides maximum thermal insulation, especially if the fluorocarbon gases can be permanently sealed into the total structure; and the closed-cell rigid foam also contributes to the overall mechanical rigidity and strength of the total structure. Such rigid urethane foams have thus become a major factor in thermal insulation for building construction and refrigeration.

More recently, structural foam and RIM have been produced by low to moderate foaming of semirigid to rigid urethane systems and have made a major contribution to furniture and automotive construction. Only the future will tell how widely such developments may continue to grow.

Polystyrene. Low-density polystyrene foam products are made by two different methods. Molded products such as cups and packaging are made by suspension polymerizing to tiny beads, swelling these with 5% pentane, pre-expanding with steam, and then molding the pre-expanded beads in a steam chest to expand them further and fuse them into the final shape. Extruded food-packaging sheet and insulating board is made by melting polystyrene pellets in a vented extruder, then injecting pentane or compressed nitrogen gas through the "vent," and then extruding through a die and sizing units to produce continuous sheet, board, or other profile. The combination of low density, rigidity, thermal insulation, nontoxicity, good color, and chemical resistance (Table V) has long been popular in mass markets such as packaging and building.

Poly(vinyl chloride). When plasticized poly(vinyl chloride) (PVC) is laminated onto cloth, for clothing and upholstery, the solid vinyl has a harsh feel often judged inferior by the consumer. When

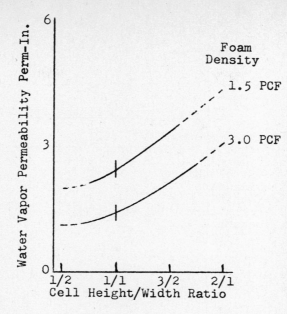

Figure 10.   Effect of Anisotropy on Permeability (2).

```
                              Box 2

                  Typical Flexible Urethane Foam

Polyol (Trifunctional) MW 3000                          100
Toluene Diisocyanate                                     46
Organotin Catalyst                                        0.4
Silicone Surfactant                                       1
Tertiary Amine Catalyst                                   0.2
Water                                                     3.6
Monofluorotrichloromethane                               0-15

Density PCF                                               1.4
Tensile Strength, PSI                                    14
Elongation %                                            220
Tear Strength Lb./In.                                     2.2
Compression Set (Method B) %
      50% Deflection                                      6
      90% Deflection                                      5
Indent Load Deflection Lb.
      25% Deflection                                     30
      65% Deflection                                     57
      Ratio 65% Deflection/25% Deflection                 1.9
Rebound Falling Ball %                                   38

                     (Ref. 4, Pg. 152)
```

Box 3

Typical Rigid Urethane Foam

Polyether Polyol (Hydroxyl No. 460)	100
N,N,N',N',-Tetrakis(2-Hydroxypropyl)Ethylenediamine	8
Triethylene Diamine	0.3
N,N-Dimethyl Ethanolamine	0.5
Dibutyl Tin Dilaurate	0.02
Silicone Surfactant	1.5
Trichlorofluoromethane	38
Toluene Diisocyanate	107
NCO/OH Ratio	1.03
Feed Temperature °F	80
Mold Temperature °F	125
Tack-Free Time Sec.	150
Density PCF	2.35
Closed Cells %	95
Compression Load Perpendicular to Rise PSI	
At Yield Point	23
At 10% Deflection	30
Compression Modulus PSI	600
Flexural Strength PSI	60
Flexural Modulus PSI	900
Shear Strength PSI	22
Shear Modulus PSI	210
Moisture Vapor Permeability Perm-In.	2
Water Absorption PSF Surface Area	0.06
K Factor, Cut Foam, BTU/Hr.Ft.^{2}°F/In.	
Initial	0.112
After 10 Days at 60°C	0.123
Dimensional Stability % Volume Change	
24 Hr. at -22°F	<1
24 Hr. at 158°F	<1
24 Hr. at 230°F	4
24 Hr. at 158°F, 95-100% RH	8
72 Hr. at 100°F, 95-100% RH	1

(Ref. 2, Pg. 161)

the plasticized vinyl layer is moderately foamed to medium-low density, the resulting soft rich feel is much more comfortable and pleasing to the consumer. This texture is most often accomplished by chemical blowing of plastisol (Table VI). The PVC is polymerized in emulsion, then spray dried to submicron particles of high molecular weight with a fused skin. These particles are stirred into a nonsolvent plasticizer, such as diisooctyl phthalate, to form a fluid paste or "plastisol." Azobisformamide is added as chemical blowing agent, and the plastisol is spread on a moving web of cloth reinforcement. This mixture then passes through an oven, and the heating process decomposes the blowing agent, foams the plastisol, and simultaneously fuses the foamed plastisol into a foamed homogeneous plastic mass. This mass is a low-medium density mixture of open and closed cells, which may not be optimum, but is the best that can be managed in present technology. It is widely used in automotive, clothing and accessories, and upholstery products.

Structural Foams. In recent years, there has been a growing tendency to introduce a low to moderate amount of foaming during molding and extrusion of solid thermoplastics, either by injection of compressed nitrogen or by addition of small amounts of chemical blowing agents such as azobisformamide. Very low degrees of foaming are useful simply to help mold-filling and to eliminate sink marks. Larger amounts of foaming may help economize on the weight of solid resin needed to make a product, or actually may produce high-performance lightweight parts on the basis of a modification of sandwich structure technology (Figure 11, Table VII). In addition, many structural foam processes operate at lower pressure than solid molding, and thus require less costly equipment and molds, with the result that capital and operating costs are lower. The extent of this growing technology is not quantitatively documented, but it constitutes a major and growing trend in the overall plastics field (Table VIII).

Applications and Markets

The uses of foamed plastics may be analyzed according to materials, markets, or types of applications (Table IX). As materials, polyurethanes are most often used, particularly flexible foams, with polystyrene second in use. Of the markets for which they are used furniture is the largest, followed by packaging, building insulation, the automotive market, and refrigerator insulation. In type of application, soft cushioning is largest, thermal insulation next, and rigid shock-absorbing packaging next, with smaller amounts in structural and other applications. However, only polyurethane and polystyrene foams are reported consistently, so most other more specialized foamed plastics would require more incisive market analysis to detail their usage. There is no doubt that the entire field is growing rapidly and playing an ever greater role in the total plastics picture.

Table V

Typical Polystyrene Foam Properties

Property	Value
Density PCF	2
Tensile Strength PSI	55
Compressive Stress at 10% Deflection PSI	40
Thermal Conductivity BTU/Sq.Ft./Hr./°F/In.	0.24
Coefficient of Thermal Expansion 10^{-5} In./In./°F	3.5
Dielectric Constant 1mHz	1.04
Dissipation Factor 1mHz	0.0002
Water Vapor Transmission Perm-In.	1
Water Absorption % in 96 Hr.	2

(Ref. 9, 1981-2, Pg. 546)

Table VI

Typical Vinyl Plastisol Foam Formulations

Ingredients	Rug Backing	Flooring	Protective Strip-Coating
Polyvinyl Chloride	100	100	100
Dioctyl Phthalate	20	20	20
Butyl Benzyl Phthalate	30	15	25
Epoxy Plasticizer	5	5	5
Secondary Plasticizer	20		
Ba-Cd Liquid Stabilizer	7	3	7
Calcium Carbonate		15	
Barium Petroleum Sulfonate	1	1	1
Sodium Bicarbonate		5	
Azodicarbonamide	3		2
Drying Agent	15	5	10
Density PCF	18	35	25

(Ref. 2, Pg. 405)

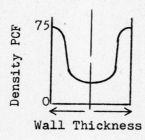

Figure 11. Density Gradient in
Rigid Urethane Structural Foam ([1]).

Table VII

Typical Rigid Urethane Structural Foams

	0.4	0.5	0.6
Density Gm/Cc			
PCF	25	31	37.5
Flexural Modulus 10^5 PSI	1	1.3	1.6
Flexural Strength PSI	3200	4600	6000
Compressive Strength PSI	800	1800	2800
Tensile Strength PSI	1600	2500	3400
Heat Deflection Temp. °F/66 PSI	180	192	205

(Ref. 1, Pg. 179)

Table VIII

Typical Thermoplastic Structural Foams

Polymer	Polycarbonate	Noryl "PPO"	Poly-ester	Poly-sulfone	ABS	Polyeth-ylene	Poly-styrene	Polypro-pylene
Specific Gravity	0.80	0.80	1.10	0.80	0.80	0.60	0.70	0.80
Flex. Str. 10^3 PSI	10	6	20	6.3	3.7	2.73	4.5	3.2
Flex. Mod. 10^5 PSI	3.0	2.4	10	2.13	1.25	1.20	2.10	1.20
Compr. Str. 10^3 PSI	7.5	5.5	12.5		1	1.3	3.3	1.6
Tens. Str. 10^3 PSI	5.5	3.3	11.8	4.55	2.7	1.3	1.3	2.0
HDT °F/264 PSI	270	205	380	341	180	94	160	132

(Ref. 1, Pg. 180)

Table IX

Use of Foamed Plastics

Materials	Thousands of Metric Tons
Flexible Polyurethane	531
Rigid Polyurethane	256
Polystyrene	368

Markets	
Furniture & Rug Underlay	334
Packaging	277
Building Insulation	231
Automotive	182
Refrigeration	73

Type of Application	
Soft Cushioning	531
Insulation	304
Rigid Packaging	277

(Ref. 8)

Literature Cited

1. Benjamin, B. S. "Structural Design with Plastics"; Van Nostrand Reinhold: New York, 1982.
2. Benning, C. J. "Plastic Foams"; Wiley–Interscience: New York, 1969.
3. Ferrigno, T. H. "Rigid Plastic Foams"; Reinhold: New York, 1967.
4. Frisch, K. C.; Saunders, J. H. "Plastic Foams"; Dekker: New York, 1972–73.
5. Hilyard, N. C. "Mechanics of Cellular Plastics"; MacMillan: New York, 1982.
6. Journal of Cellular Plastics.
7. Meinecke, E. A.; Clark, R. C. "Mechanical Properties of Polymeric Foams"; Technomic: Westport, CT, 1973.
8. Anom, I. Mod. Plast. 1982, 59(1), 82–85.
9. "Modern Plastics Encyclopedia"; McGraw–Hill: New York, Annual.
10. Griffin, J. D.; Skochdopole, R. E. "Engineering Design for Plastics"; Baer, E., Ed.; Reinhold: New York, 1964; Chap. 15.

Engineering Thermoplastics: Chemistry and Technology

DANIEL W. FOX and EDWARD N. PETERS

General Electric Company, Pittsfield, MA 01201

Nylon
Polyacetals
Polycarbonates
Other Polycarbonates and Related Polymers
    Polyestercarbonates
    Polyarylates
    Tetramethylbisphenol A Polycarbonate
Phenylene Ether-Based Resins
Polysulfones
Thermoplastic Polyesters
Poly(phenylene sulfide)
Polyetherimide
Blends and Alloys
Conclusion

Engineering polymers comprise a special, high-performance segment of synthetic plastic materials that offer premium properties. When properly formulated, they may be shaped into mechanically functional, semiprecision parts or structural components. The term "mechanically functional" implies that the parts will continue to function even if they are subjected to factors such as mechanical stress, impact. flexure, vibration, sliding friction, temperature extremes, and hostile environments.

As substitutes for metal in the construction of mechanical apparatus, engineering plastics offer advantages such as corrosion resistance, transparency, lightness, self-lubrication, and economy in fabricating and decorating. Replacement of metals by plastic is favored as the physical properties and operating temperature ranges of plastics improve and the cost of metals and their fabrication increases. Plastics applications in transportation, a major growth opportunity, have been greatly accelerated by the current awareness of the interplay of vehicle weight and fuel requirements.

0097-6156/85/0285-0495$06.00/0
© 1985 American Chemical Society

The ability to replace metal in many areas has resulted in tremendous growth in engineering thermoplastics. The consumption of engineering plastics increased from 10 million to more than 1 billion lbs from 1953 to 1982. Engineering polymers are the fastest growing segment of the plastics industry with an anticipated growth rate ranging from 12 to 15%. This chapter focuses on the development of engineering thermoplastics during the past 30 years.

Nylon

Nylon, the first commercial thermoplastic engineering polymer, is the prototype for the whole family of polyamides. Nylon 6,6 began at Du Pont with the polymer experiments of Wallace Carothers in 1928, and made its commercial debut as a fiber in 1938 and as a molding compound in 1941. By 1953, 10 million lbs of nylon 6,6 molding compound represented the entire annual engineering plastic sales.

Nylon was a new concept in plastics for several reasons (1). Because it was crystalline, nylon underwent a sharp transition from solid to melt, thus it had a relatively high service temperature. A combination of toughness, rigidity, and "lubrication-free" performance peculiarly suited it to mechanical bearing and gear applications. Nylon acquired the reputation of a quality material by showing that a thermoplastic could be tough, as well as stiff, and do jobs better than metals in some cases. This performance gave nylon the label "an engineering thermoplastic."

Nylon 6,6, derived from the condensation of a six-carbon diamine and a six-carbon dibasic acid and normally chain terminated with monofunctional reactants, presented some unusual characteristics (Figure 1) (2). The crystallinity and polarity of the molecule permitted dipole association that conveyed on relatively low molecular weight polymers the properties normally associated with much higher molecular weight amorphous polymers. At its crystalline melting point the polymer collapsed into a rather low-viscosity fluid in a manner resembling the melting of paraffin wax. It lacked the typical familiar broad thermal plastic range that is normally encountered in going from a glassy solid to a softer solid to a very viscous taffy stage. This factor led to some complications in molding because very close tolerances were required in mold construction, and very precise temperature and pressure monitoring was necessary to prevent flash or inadvertent leaking of the mobile melt. Early molders of nylon were highly skilled--they had to be because the industry was young.

Nylons based on ω-aminocarboxylic acids, although briefly investigated by Carothers, were commercialized first in Germany around 1939 (Figure 1). Of particular interest to the plastic industry is nylon 6 (based on caprolactam), which became available in 1946 in Europe. It was initially introduced to the United States in 1954 by Allied Chemical Company for fiber purposes. Polycaprolactam is crystalline, has a lower melting point than nylon 6,6, and has been successfully applied as a molding compound.

The key features of nylon are fast crystallization, which means fast mold cycling; high degree of solvent resistance; toughness;

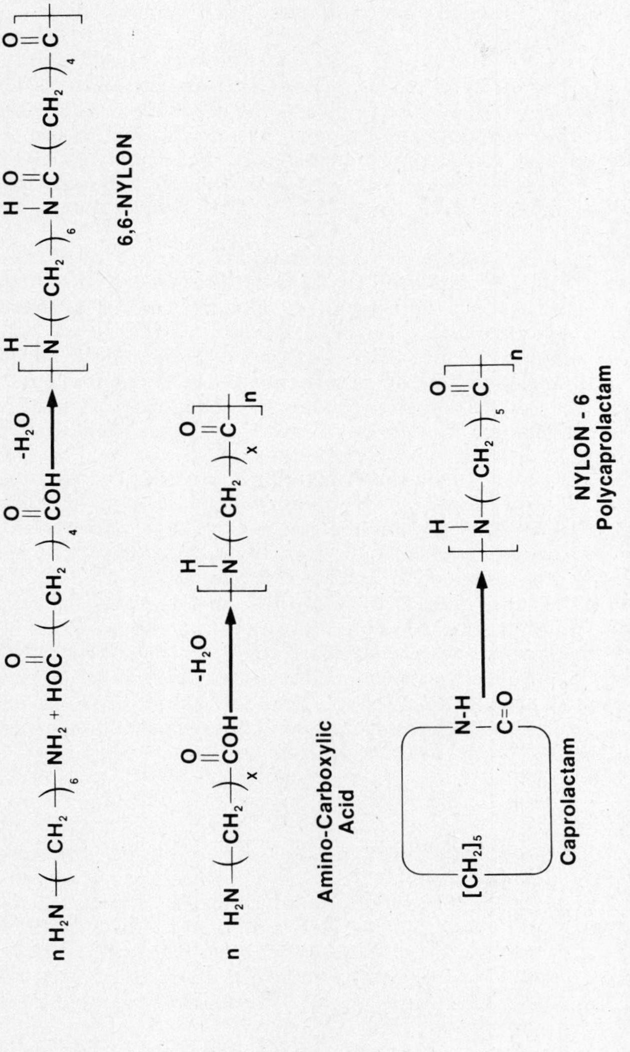

Figure 1. Nylons

lubricity; fatigue resistance; and excellent flexural-mechanical properties that vary with degree of water plasticization. Deficiencies include a tendency to creep under applied load.

The most important technological breakthroughs of the past decade are mineral-filling technology, which results in high modulus and high-temperature dimensional stability, fire retardancy, and impact modification as exemplified by Du Pont's ST technology (super tough nylon compounds derived from patented impact modification technology).

Many different varieties of polyamides have been produced by varying the monomer composition. Variations include nylon 6,9; nylon 6,10; and nylon 6,12 (made from the 9-, 10-, and 12-carbon dicarboxylic acids, respectively); and nylon 11 and nylon 12 (via the self-condensation of 11-aminoundecanoic acid and lauryllactam, respectively). These specialty nylons exhibit lower moisture absorption—only one-third or one-fourth that of nylon 6 or nylon 6,6.

When unsymmetrical monomers are used, the normal ability of the polymer to crystallize can be disrupted; amorphous (transparent) nylons can then be formed. For example, the polyamide prepared from the condensation of terephthalic acid with a mixture of 2,2,4- and 2,4,4-trimethylhexamethylenediamines (Figure 2) was developed at W. R. Grace and Company, and later produced under license by Dynamit Nobel AG. This polyamide is sold under the tradename Trogamid T.

Another amorphous nylon was developed at Emser Werke AG and is based on aliphatic, as well as cycloaliphatic, amines and terephthalic acid. It is marketed in Europe under the tradename Grilamid 55 and until recently was distributed in the United States by Union Carbide under the tradename Amidel. These amorphous nylons are not as tough as nylon 6 or 6,6, but they do offer transparency and good chemical resistance in some environments.

Nylons prepared from aromatic diamines and diacids (aramids) can lead to very high-heat amorphous nylons such as Du Pont's Nomex materials [poly(m-phenyleneisophthalamide)], or the highly oriented crystalline Kevlar fibers [poly(p-phenyleneterephthalamide)] derived from liquid-crystalline technology. Both of these aramids are sold as fibers. They have excellent inherent flame-retardant properties, and Kevlar exhibits a very high modulus.

Polyacetals

After nylon, the next engineering polymers to be commercially introduced were polyacetals (3). The basic polyformaldehyde structure was explored rather thoroughly by H. Staudinger in the late 1920s and early 1930s, but he failed to produce a sufficiently high molecular weight polymer with requisite thermal stability to permit processing (4). Pure formaldehyde could be readily polymerized, but the polymer equally, readily, spontaneously depolymerized—unzipped.

In 1947, Du Pont began a development program on the polymerization and stabilization of formaldehyde and its polymer. Twelve years later, Du Pont brought the unzipping tendency under control with proprietary stabilizers and commercially announced Delrin polyacetal polymer (Figure 3). The key to the stabilization of polyformaldehyde resins appears to be a blocking of the terminal

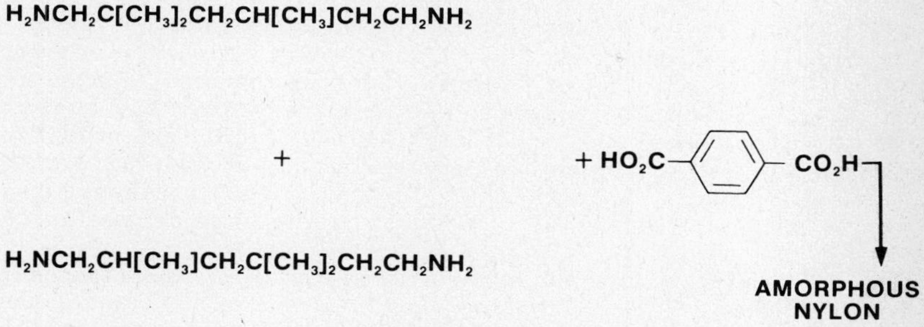

$$H_2NCH_2C[CH_3]_2CH_2CH[CH_3]CH_2CH_2NH_2$$

$$+$$

$$+ HO_2C \text{—} \bigcirc \text{—} CO_2H \text{—}$$

$$H_2NCH_2CH[CH_3]CH_2C[CH_3]_2CH_2CH_2NH_2$$

**AMORPHOUS
NYLON**

Figure 2. Amorphous nylon

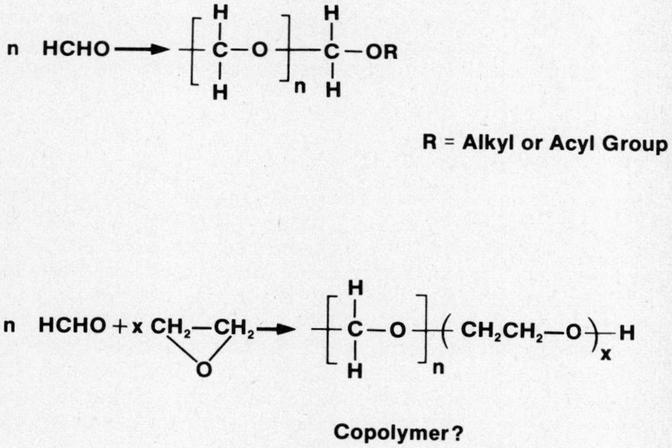

$$n \ HCHO \longrightarrow \left[\begin{array}{c} H \\ | \\ C-O \\ | \\ H \end{array} \right]_n \begin{array}{c} H \\ | \\ C-OR \\ | \\ H \end{array}$$

R = Alkyl or Acyl Group

$$n \ HCHO + x \ CH_2\text{—}CH_2 \longrightarrow \left[\begin{array}{c} H \\ | \\ C-O \\ | \\ H \end{array} \right]_n \left(CH_2CH_2-O \right)_x H$$

Copolymer?

Figure 3. Polyformaldehydes

hydroxyl (OH) groups that participate in, or trigger, an unzipping
action. The hydroxyl groups may apparently be blocked by post-
etherification or esterification.

Celanese joined Du Pont in the market with their proprietary
Celcon polyacetal polymer within a year (Figure 3). Celanese
managed to obtain basic patent coverage, despite Du Pont's prior
filing, on the basis of a copolymer variation that led to an
enhanced stabilization against thermal depolymerization

Both Celanese and Du Pont aimed their products directly at metal
replacement. Items such as plumbing hardware, pumps, gears, and
bearings were immediate targets. In many respects, the acetals
resemble nylons. They are highly crystalline, rigid, and cold-flow
resistant, solvent resistant, fatigue resistant, mechanically tough
and strong, and self-lubricating. They also tend to absorb less
water and are not plasticized by water to the same degree as the
polyamides.

Rapid crystallization of acetals from the melt contributes to
fast mold cycles. Crystallization also causes a significant amount
of mold shrinkage. Thus, it is necessary to compensate in mold
design for dimensional changes that occur during the transformation
from a hot, low-density, amorphous melt to a more dense, crystalline
solid.

Polyacetals are used mainly in liquid handling (plumbing)
equipment and miscellaneous hardware. With the exception of an
enhanced ability to metallize for hardware appearance, few technical
advancements have been announced. Key deficiencies of polyacetals
are a tendency to thermally unzip and an essentially unmodifiable
flammability.

Polycarbonates

The aromatic polycarbonates were the next engineering polymers to be
introduced (5, 6). Researchers at Bayer AG (Germany) and General
Electric Company independently discovered the same unique, super-
tough, heat-resistant, transparent, and amorphous polymer in 1953.
When the companies became aware of each others activities,
agreements were reached that enabled both parties to continue
independent commercialization activities without concern for
possible subsequent adverse patent findings. The General Electric
Company's polycarbonate called Lexan was introduced in the United
States in 1959 at about the same time as the polyacetals, and a
commercial plant was brought on stream in 1960.

Polycarbonates of numerous bisphenols have been extensively
studied. However, most commercial polycarbonates are derived from
bisphenol A. At first, both direct-reaction and melt-
transesterification processes were employed (Figure 4). In direct-
reaction processes, phosgene reacts directly with bisphenol A to
produce a polymer in a solution. In transesterification, phosgene
is first reacted with phenol to produce diphenyl carbonate, which in
turn reacts with bisphenol A to regenerate phenol for recycle and
molten, solvent-free polymer. Transesterification is reported to be
the least expensive route. It was phased out, however, because of
its unsuitability to produce a wide range of products.

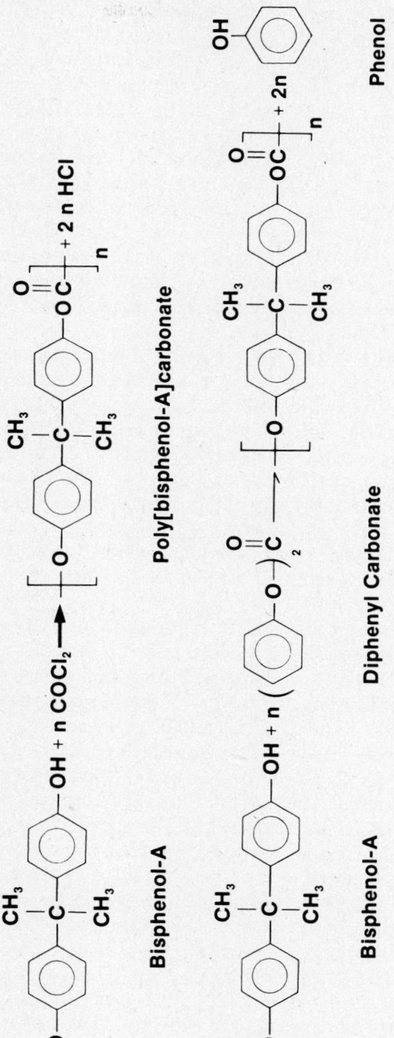

Figure 4. Polycarbonates

Today, most polycarbonate is produced by an interfacial adaptation of the reaction in equation 4 (7). The bisphenol plus 1-3 mol% monofunctional phenol, which controls molecular weight, is dissolved or slurried in aqueous sodium hydroxide; methylene chloride is added as a polymer solvent, a tertiary amine is added as a catalyst, and phosgene gas is dispersed in the rapidly stirred mixture. Additional caustic solution is added as needed to maintain basicity. The growing polymer dissolves in the methylene chloride, and the phenolic content of the aqueous phase diminishes.

The polycarbonates do not crystallize readily. They are essentially glassy or amorphous with unusually high softening points for glassy amorphous polymers. The glass transition temperature (T_g) of these polymers approaches 150 °C, and they have some mechanical strength or integrity up to this temperature. At temperatures below 100 °C they exhibit very little creep. Thermal-use properties of amorphous polymers are related to their T_g values, whereas thermal-use properties of crystalline polymers tend to be more related to crystal melting points.

The polycarbonates, like the nylons and acetals, were directed toward metal replacement applications. Transparency gave them another dimension—an early application was wing lights for supersonic military aircraft. Skin temperatures were becoming hot enough to curl the acrylics they were replacing. Their toughness over a very wide temperature range led to applications from nonbreakable windows and auto taillights to boat and pump impellers, food machinery parts, and astronaut helmets. Inasmuch as polycarbonates are polyesters, they do have some susceptibility to attack by bases and high-temperature moisture. They are soluble in a variety of organic solvents.

Glassfiber-filled versions of polycarbonates are available, and this combination is particularly well suited to compete with metal parts. As in the case of other semiamorphous polymers, glass acts as a stiffening and strengthening agent but does not raise operating temperatures significantly. In crystalline polymers, fillers tend to act as a pseudo cross-link or crutch to bridge the soft, amorphous regions that have T_g-dependent properties, thereby permitting the plastic to maintain structural integrity up to its crystalline melting point. Without filler, crystalline polymers tend to creep under static load at relatively low temperatures because their T_g values are generally comparatively low.

Recent advances in polycarbonates include further enhancement of normally good fire resistance; color elimination (an improvement approaching glass) and development of scratch-resistant coatings for glazing applications; branched material for blow-molding bottles; optimization of structural foam molding and compounds; selective copolymerization; and development of many different commercial polyblends.

Other Polycarbonates and Related Polymers

The early polycarbonate syntheses that were practiced enabled facile incorporation of aromatic carboxylic acids in the polymers to produce copolyestercarbonates. Subsequently, a wide range of polymers varying from 100% polycarbonate to 100% polyester was produced. The trouble with most of these early product variations, especially those comprised of all-aromatic polyesters, was the very high melt viscosity that necessitated very high processing temperatures. The plastic fabricators were not prepared to cope with polycarbonate processing when the polycarbonates were introduced, and they certainly did not need even tougher problems. Since then advances in technology and equipment have made the processing of such higher temperature polymers feasible. Recent activity in this area has ranged from the copolyestercarbonates to aromatic polyesters.

Polyestercarbonates. General Electric Company commercially introduced copolyestercarbonates in 1981. These copolymers provide the outstanding balance of tensile, flexural, and impact properties inherent in bisphenol A polycarbonate resins while increasing the heat-deflection temperature to greater than 160 °C. These polymers were designed to bridge the thermal performance gap between polycarbonate and other high-performance materials such as polysulfone and all-aromatic polyesters. Potential applications include structural and optical components in lighting, appliances, housewares, and specialty applications requiring higher heat resistance or a combination of clarity and heat resistance.

Polyarylates. The so-called polyarylates are a class of aromatic polyesters. They are prepared from bisphenols and dicarboxylic acids. The first commercially available polyarylate was the U-polymer of Unitika Ltd. (Japan). This material is marketed in the United States by Union Carbide as Ardel. Bayer AG likewise has introduced a bisphenol-A based polyarylate designated APE polymer. The basic U-polymer is derived from the reaction of bisphenol A and a mixture of terephthaloyl and isophthaloyl dichlorides in the molar ratio 2:1:1 (Figure 5). Hooker Chemical and Plastics introduced and later withdrew a similar polyarylate Durel that was prepared by the transesterification reaction of diphenyl isophthalates and terephthalates with bisphenol A. The polyarylates are transparent and, like polycarbonates, have high toughness and similar mechanical properties. Being amorphous polymers like polycarbonates, the polyarylates are susceptible to stress cracking, particularly from aromatic hydrocarbons, ketones, and esters. However, the thermal properties, particularly continuous-use temperature range and heat-deflection temperature, are higher. The UV resistance of polyarylates is reported to be excellent. Ardel molding conditions are similar to those employed for polysulfone, except that polyarylates require a 25-30 °F higher temperature. Suggested applications are in electronic and electrical hardware, tinted glazing, solar energy, mine-safety devices, and transportation components.

Tetramethylbisphenol A Polycarbonate. A new polycarbonate has been introduced in Europe by Bayer AG (Figure 6). It is based on tetramethylbisphenol A (TMBPA). The monomer is produced by condensing two molecules of 2,6-dimethylphenol, which is the monomer for General Electric's poly(2,6-dimethyl-1,4-phenylene ether) polymers, with acetone. The polycarbonate from tetramethylbisphenol A resembles the dimethylphenylene ether polymers in their unusually high T_g:207 °C for the polycarbonate and 215 °C for the polyether.

Both polymers have excellent hydrolytic stability—one because it is a polyether and the other because it sterically hinders hydrolytic action. The polycarbonate, like the polyether, is reported to be compatible with polystyrene. The high T_g would suggest dimensional stability and utility at very high temperatures, but thermal-oxidative limitations will probably favor blend applications at moderately high continuous-use temperatures.

Phenylene Ether-Based Resins

In 1956, A. Hay of General Electric discovered a convenient catalytic oxidative coupling route to high molecular weight aromatic ethers (8, 9). These polymers could be made by bubbling oxygen through a copper-amine catalyzed solution of phenolic monomer at room temperature (Figure 7). A wide variety of phenolic compounds were explored, but the cleanest reactions resulted from those that contained small, electron-donor substituents in the two ortho positions. Work quickly focused on 2,6-dimethylphenol because it was the most readily synthesized. The poly(phenylene oxide) (PPO) ether derived from 2,6-dimethylphenol had excellent hydrolytic resistance, an extremely high T_g (215 °C), a high melting range (260-275 °C), a very high melt viscosity, and a pronounced tendency to oxidize and gel at process temperatures. The polymer was called PPO resin in anticipation of developing the coupling process to produce unsubstituted poly(phenylene oxide) from phenol. When this objective and the program was redirected to 2,6-dimethylphenol, a parallel effort was initiated to develop a synthetic route to this monomer. Thus, both monomer and polymer commercialization proceeded simultaneously. They went on stream in 1964. The very high melt viscosity and required high processing temperature, coupled with a high degree of oxidative susceptibility, resulted in a polymer that essentially defied thermal processing. That is, if the temperature were raised sufficiently to induce a useful degree of flow, a trace of oxygen could lead to cross-linking and gelation. The phenylene ether-based polymers might never have achieved commercial success if their total compatibility with styrenic polymers had not been discovered at an early stage of development. This fortuitous and rather rare compatibility provided the basis for what is now known as the Noryl family of polymers.

Noryl resins became the world's most successful and best known polymer blends or alloys because combinations of PPO resins and styrene polymers tend to assume the best features of each:

1.   PPO resins with very high heat distortion temperatures (HDTs) can readily raise the HDT of styrenics to over 100 °C, which is a significant temperature because this qualifies the product for all boiling water applications.
2.   Styrene polymers, with ease of processing and well-established impact modification, balance the refractory nature of PPO resins.
3.   PPO resins bring fire retardance to the system.
4.   Both PPO resins and styrene polymers have excellent water resistance and outstanding electrical properties.

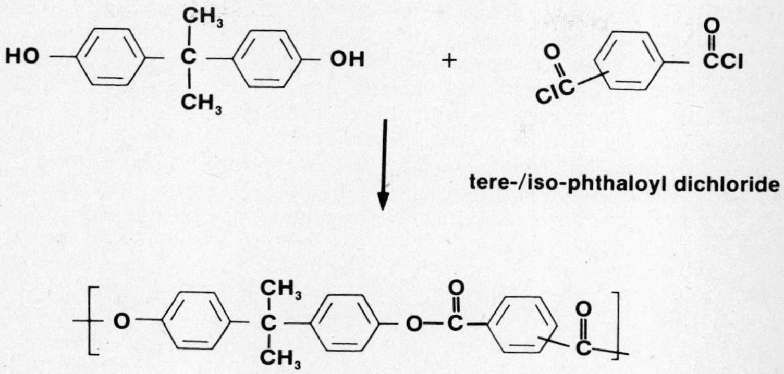

tere-/iso-phthaloyl dichloride

Figure 5. Polyarylate

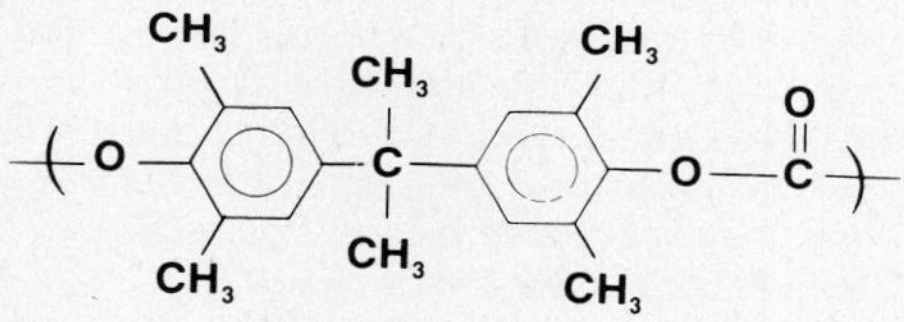

Figure 6. Tetramethyl bisphenol A polycarbonate

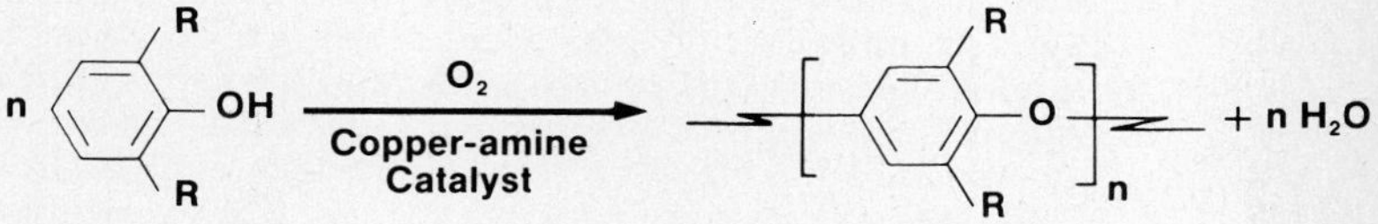

Figure 7. Phenylene ether based resin

The first applications were those requiring autoclaving (medical equipment) or outstanding electrical properties at elevated temperatures. As compounding, stabilization, and processing skills improved, markets for Noryl expanded to include office equipment, electronic components, automotive parts, water distribution systems, and general metal replacement.

Noryl phenylene ether-based resins are relatively resistant to burning, and their burn resistance can be increased by judicious compounding. They may be modified with glass and other mineral fillers. Because of low moisture absorption, dimensional stability, and ability to be used over a wide temperature range, Noryl phenylene ether-based resins are especially adaptable to metallizing. However, like most amorphous polymers, they show poor solvent resistance.

Polysulfones

Polyarylsulfones are a class of high-use temperature thermoplastics that characteristically exhibit excellent thermal-oxidative resistance, good solvent resistance, hydrolytic stability, and creep resistance (10). In 1965, Union Carbide announced a thermoplastic polysulfone based on dichlorodiphenylsulfone and bisphenol A (11). This polysulfone became commercially available in 1966 and was designated as Udel polysulfone. Since 1966, Imperial Chemical Industry (ICI), Minnesota Mining and Manufacturing (3-M), and Union Carbide have commercialized polyarylsulfones that contain only aromatic moieties in the polymer structure. These materials have been designated Victrex polyethersulfone (ICI), Astrel 360 (3-M), and Radel polyphenylsulfone (Union Carbide).

There are two primary methods for the synthesis of polyarylsulfones: nucleophilic displacement reactions and Friedel-Crafts processes. Udel polysulfone is made by the nucleophilic displacement of the chloride on bis(p-chlorophenyl) sulfone by the anhydrous disodium salt of bisphenol A (Figure 8). The reaction is conducted in a dipolar aprotic solvent reported to be dimethyl sulfoxide. The process is said to be a little demanding, and considerable attention is required in the various unit operations. Victrex and Radel are also made via nucleophilic substitution (Figure 9). The Friedel-Crafts reaction that uses diphenylether-4,4'-disulfonylchloride, biphenyl, and a Lewis-acid catalyst such as $FeCl_3$ is employed to produce Astrel 360 (Figure 10). The prices of Udel and Victrex are high in comparison to other engineering thermoplastics; the prices of Radel and Astrel are very high as a consequence of limited availability.

Polyarylsulfones are somewhat polar aromatic ethers with outstanding oxidation resistance, hydrolytic stability, and high heat-distortion temperature. Like polycarbonates, the amorphous character, high heat-distortion temperatures, high melting points, and very high melt viscosities of polyarylsulfones present severe processing problems. The processability of Udel polysulfone resin has been significantly improved since the earlier offering, but it is still not very easy to fabricate. Polyethersulfone is even higher melting than polysulfone, and it is more difficult to process.

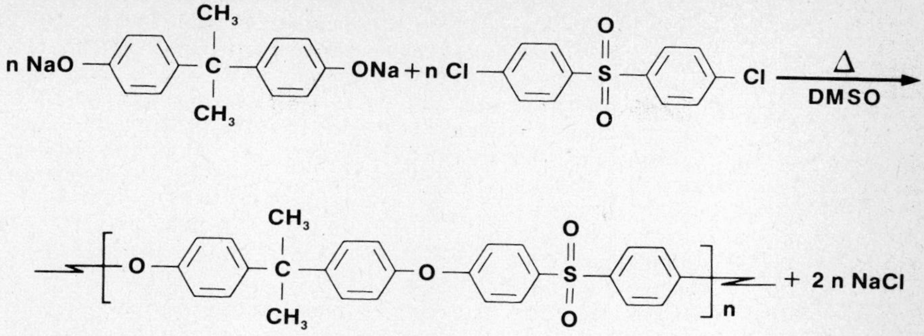

Figure 8. Polysulfone

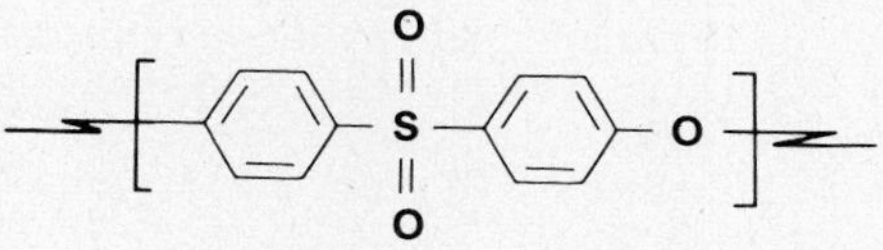

Polyethersulfone

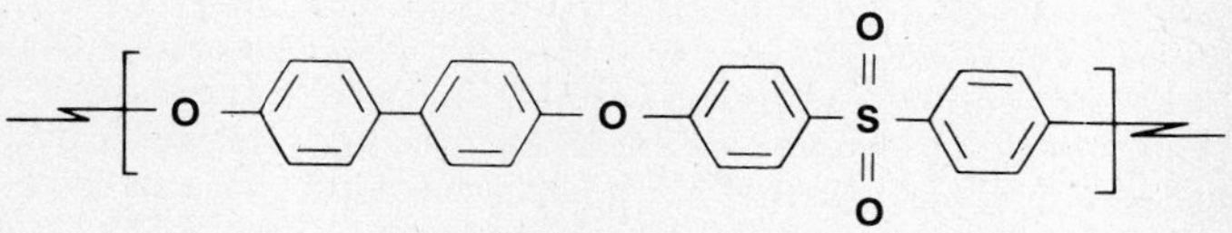

Polyphenylsulfone

Figure 9. Victrex and Radel

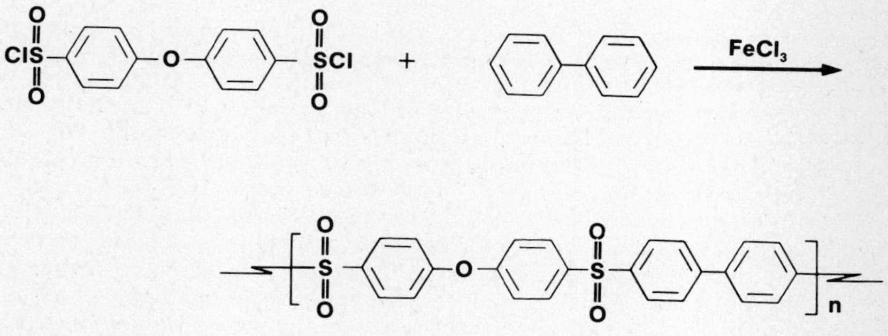

Figure 10. Astrel polysulfone

The hydrolytic stability and very high thermal endurance of this plastic in conjunction with a good balance of mechanical properties suit it for hot water and food handling equipment, range components, TV applications, alkaline battery cases, and film for hot transparencies. The unmodified product is transparent with a slightly yellow tint. Low flammability and low smoke suite it for aircraft and transportation applications. As with the other amorphous polymers, susceptibility to attack by organic solvents is a deficiency.

Thermoplastic Polyesters

Thermoplastic polyesters are currently the hottest topic in the engineering polymer field (12). They had their beginning in 1941 when J. R. Whinfield and J. T. Dickson discovered terephthalate-based polyesters. Earlier, J. W. Hill and W. R. Carothers had examined aliphatic polyesters and found them inadequate as fiber precursors because of their low melting points. The aliphatic polyesters were bypassed by polyamides with much higher crystalline melting points. Whinfield and Dickson, in a subsequent investigation of polyesters as fiber precursors, substituted terephthalic acid for the previously investigated aliphatic dibasic acids and discovered high melting crystalline polymers. These polymers were developed by ICI, Du Pont, and others into the familiar polyester fibers and films.

Whinfield and Dickson quickly realized that the polymer based on ethylene glycol and terephthalic acid was the best suited for fibers (Figure 11). They did, however, make and describe several other polyesters including poly(butanediol terephthalate) (PBT). Many years later a number of polyester fiber producers became interested in PBT. One producer explained that he was interested in PBT because it made a fiber that resembled nylon. Because nylon was becoming popular as a carpet yarn, and because he was not in the nylon business, he considered PBT a means of competing in carpet yarns.

While the fiber producers were busily expanding their fiber activities, a number of companies were simultaneously trying to adapt poly(ethylene terephthalate) (PET) to molding in much the same way as the nylons had been made to double in brass as molding compound and fiber. This objective has been particularly attractive because the manufacturing capacity for PET in the United States has surpassed 5 billion lbs per year, and all the economy of scale has been obtained to yield high-performance polymers at commodity prices.

A number of companies have tried to promote PET as a molding compound. In 1966, the first injection molding grades of PET were introduced; however, these early materials were not very successful. The primary problem was that PET does not crystallize very rapidly; a molded object composed of a crystallizable polymer caught in an amorphous or partially crystallized state would be rather useless. In service such a part could crystallize, shrink, distort, crack, or fail. The obvious solution was to use hot molds and hold the parts in the mold until the crystallization process was completed. Postannealing also permits continued crystallization. These approaches, especially with glass fiber incorporation, led to

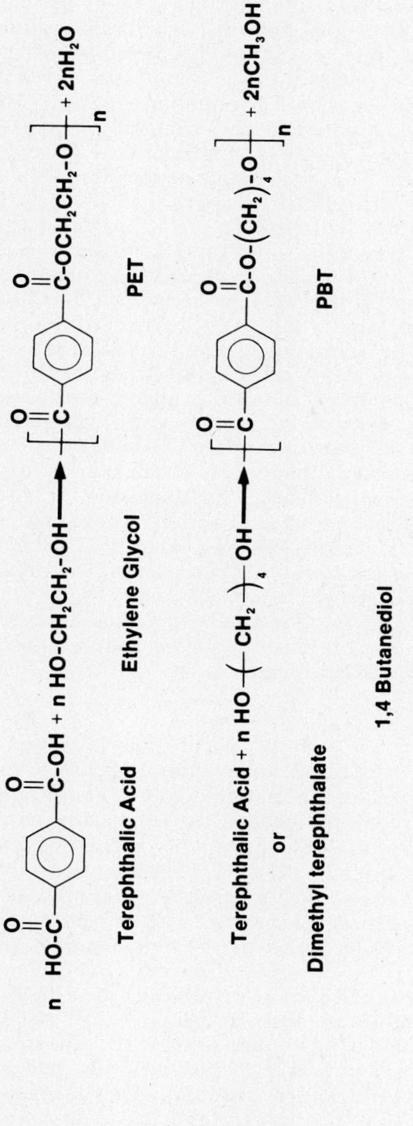

Figure 11. Polygycol terephthalates

acceptable parts at economically unacceptable molding cycles. Alternately, some developers tried to use very low molecular weight PET-glass products that crystallized more rapidly; however, because of their low molecular weights, these products lacked essential properties. A very broad search has been conducted for such things as nucleating agents and crystallization accelerators. An improved PET injection molding compound was introduced by Du Pont in 1978 under the tradename Rynite. A number of other companies have followed Du Pont into the market. The PET-based molding compounds are gaining acceptance at a substantial rate, but actual volume is relatively small because of the recent beginning. While other companies sought means of increasing the rate of crystallization of PET, Celanese chemists turned their attention to PBT and found that it met all the requirements for a molding compound.

The basic composition of matter patents had long since expired when Celanese sampled the market in 1970 with a glass fiber reinforced PBT product designated X-917. This product was subsequently called Celanex polyester molding compound. Eastman Kodak followed Celanese early in 1971, and General Electric followed Eastman Kodak later in the same year with Valox PBT polyester resin. Since that time a dozen or more additional companies around the world have entered (and some have subsequently exited) the business.

Basically, PBT seems to have a unique and favorable balance of properties between nylons and acetal resins. It has relatively low moisture absorption, extremely good self-lubrication, fatigue resistance, solvent resistance, and good maintenance of mechanical properties at elevated temperatures. Maintenance of properties up to its crystal melting point is excellent if it is reinforced with glass fiber. Very fast molding cycles with cold to moderately heated molds complete the picture. Key markets include "under the hood" automotive applications such as ignition systems and carburetion, which require thermal and solvent resistance; electrical and electronic applications; power tools; small and large appliance components; and athletic goods.

Poly(phenylene sulfide)

The first poly(phenylene sulfides) were made in 1897 by the Friedel-Crafts reaction of sulfur and benzene ($\underline{13}$). Researchers at Dow Chemical, in the early 1950s, succeeded in producing high molecular weight linear poly(phenylene sulfide) by means of the Ullmann condensation of alkali metal salts of p-bromothiophenol ($\underline{14}$). Various other early attempts have been reported, all of which resulted in amorphous resinous materials that decomposed between 200–300 °C. These materials were probably highly branched and partially cross-linked.

In 1973, Phillips Petroleum introduced linear and branched products under the tradename Ryton ($\underline{15}$). The materials were prepared by reacting 1,4-dichlorobenzene with sodium sulfide in a dipolar aprotic solvent (Figure 12). The polymer precipitates out of solution as a crystalline white powder. The polymer exhibits a T_g at 85 °C and melts at 285 °C. Continued heating in air leads to cross-linking.

Ryton is characterized by high heat resistance, excellent chemical resistance, low friction coefficient, good abrasion resistance, and good electrical properties. Physical characteristics include high flexural modulus, very low elongation, and generally poor impact strength. Glass, glass-mineral, and carbon fiber reinforced grades that have high strength and rigidity are available. The unreinforced resin is used only in coatings. The reinforced materials are finding applications in aerospace technology, pump systems, electrical and electronic equipment, appliances, and automotive vehicles and machines.

Poly(phenylene sulfides) are reported to be somewhat difficult to process because of their very high melting temperatures and relatively poor flow characteristics, and because some chemistry appears to continue during the fabrication step. Molded pieces have limited regrindability. Annealing of molded parts enhances mechanical properties but leads to almost total loss of thermal plastic character.

Polyetherimide

The newest engineering thermoplastic is a polyetherimide that was formally announced by General Electric Company in 1982 (16). This amorphous polymer is designated Ultem resin and resulted from the research work of a team headed by J. G. Wirth in the early 1970s (9). The early laboratory process involved a costly and difficult synthesis. Further development resulted in a number of breakthroughs that led to a simplified, cost-effective production process. The final step of the process involves the imidization of a diacid anhydride with *m*-phenylene diamine (Figure 13).

Polyetherimide offers an impressive collection of attributes such as high heat resistance, stiffness, impact strength, transparency, high mechanical strength, good electrical properties, high flame resistance, low smoke generation, and broad chemical resistance. In addition to its unique combination of properties matching those of high-priced specialty plastics, polyetherimide exhibits the processability of traditional engineering thermoplastics, although higher melt temperatures are required. The excellent thermal stability is demonstrated by the maintenance of stable melt viscosity after multiple regrinds and remolding. The processing window is nearly 100 °C, and polyetherimide can be processed on most existing equipment. Furthermore, this excellent flow resin can be used for the molding of complicated parts and thin sections (as thin as 5 mil). Polyetherimide is suitable for use in internal components of microwave ovens; electrical and electronic products; and automotive, appliance, and aerospace, and transportation applications.

Blends and Alloys

An interesting trend that appears to presage the wave of the future in engineering thermoplastics is the current focus on polymer blending and alloying. Metals and their alloys have been coeval with the spread of civilization. Early man used available metals in their naturally occurring state. The progress of civilization was literally determined by man's ability to modify natural metals

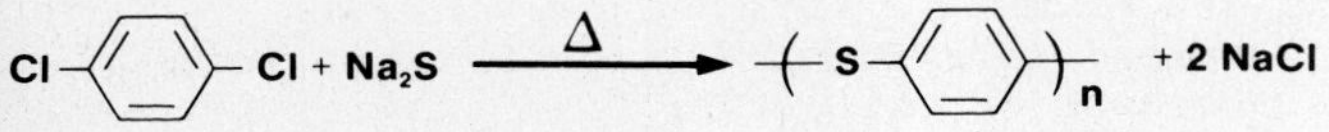

Figure 12. Poly(phenylene sulfide)

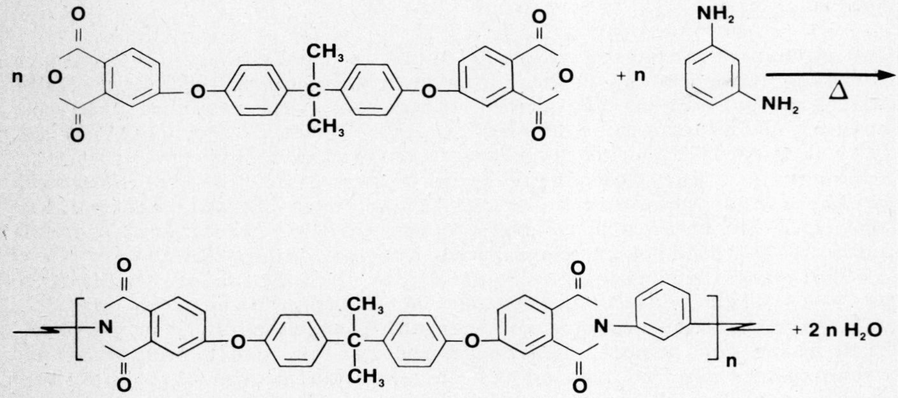

Figure 13. Polyetherimide

through alloying to induce the properties necessary for increasingly sophisticated tools. Indeed, societies that learned to exploit blends of metals developed distinct advantages over their monometallic neighbors; this was exemplified by the advent of bronze and later steel.

Metal alloying has probably made the transition from art to science in the past century. The basic ingredients are essentially known and fixed in number. The plastic age began in 1909 with the discovery by Leo Baekeland of synthetic phenol-formaldehyde resin. As pointed out in the beginning of this chapter, the engineering plastic age began 30 years ago. While metal alloy components are essentially fixed, polymer alloy components are unlimited from a technical standpoint, but somewhat fixed from an economic point of view. It is still possible to make totally new and useful polymers if their value will support the cost of synthesizing new monomers and polymers. However, it is much more economically attractive to try to combine available polymers to produce desirable and novel alloys. The available degrees of freedom make the opportunity challenging and provide almost infinite possibilities. Variables include base polymers (20 or more), impact modifiers, additives, and fillers.

It is hard to say where or when the concept of polymer alloys was born, but within General Electric Company, Robert Finhold was writing and talking about alloys of PPO resin with styrene polymers in the very early 1960s. Since then other blends have been introduced. Union Carbide has its polysulfone-polyester blends (Mindel) that exhibit improved processing and reduced cost vis-a-vis polysulfone by itself. Unitika has a broad line of polyarylate blends. Bayer offers polycarbonate-acrylonitrile-butadiene-styrene and tetramethylbisphenol A polycarbonate-polystyrene blends. Most recently, General Electric Company introduced a new polymer blend, Xenoy, into the European market. This blend is composed of two polymers, plus impact modification, to be used for a front-end, rear-end exterior bumper system for automobiles (17). It combines the key mechanical properties of impact strength, dimensional stability, and high modulus with chemical resistance. Clearly, the trend toward alloys and blends is gaining momentum.

Conclusion

The future of engineering plastics looks bright. Those industries served by these plastics, and many others who use traditional materials such as metals, glass, and ceramics, will look to the benefits of engineering polymers to provide them with cost-effective materials to help overcome the pressures of spiralling costs. The world economy will continue to influence technical trends. The commercialization of any new engineering polymer based on a new monomer, although not impossible, is unlikely. Rather, the major thrust will take place in molecular shuffling with existing monomers, blend activity, and processing improvements.

Literature Cited

1. Kohan, M. I. "Nylon Plastics"; SPE Monograph, Wiley: 1973;
    p. 2.
2. Welgos, R. J. "Kirk-Othmer Encyclopedia of Chemical Tech-
    nology," 3rd ed.; Interscience: New York, 1982; Vol. 18,
    pp. 406-25.
3. Persak, K. J.; Blair, L. M. "Kirk-Othmer Encyclopedia of
    Chemical Technology," 3rd ed.; Interscience: New York, 1978;
    Vol. 1, pp. 112-23.
4. Staudinger, H. "Die Hochmolekularen Organischen Verbindungen
    Kautschuk und Cellulose"; Springer-Verlag: Berlin, 1932.
5. Fox, D. W. "Kirk-Othmer Encyclopedia of Chemical Technology,"
    3rd ed.; Interscience: New York, 1982; Vol. 18, pp. 479-94.
6. Schnell, H. "Chemistry and Physics of Polycarbonates"; Wiley-
    Interscience: New York, 1964.
7. Vernaleken, H. in "Interfacial Synthesis"; Millich, F. and
    Carraher, C., Jr., Eds.; Dekker: New York, 1977; Vol. II,
    Chap. 13.
8. Hay, A. S. J. Polym. Sci. 1962, 58, 581.
9. White, D. M.; Cooper, G. D. "Kirk-Othmer Encyclopedia of
    Chemical Technology," 3rd ed.; Interscience: New York, 1982;
    Vol. 18, pp. 595-615.
10. Ballintyn, N. J. "Kirk-Othmer Encyclopedia of Chemical
    Technology," 3rd ed.; Interscience: New York, 1982; Vol. 18,
    pp. 832-48.
11. Johnson, R. N.; Farnham, A. G.; Clendinning, R. A.; Hale, W.
    F.; Merriam, C. N. J. Polym. Sci., Part A-1 1967, 5, 2375.
12. Jaquiss, D. B. G.; Borman, W. F. H.; Campbell, R. W. "Kirk-
    Othmer Encyclopedia of Chemical Technology," 3rd ed.;
    Interscience: New York, 1982; Vol. 18, pp. 549-74.
13. Genvresse, P. Bull. Soc. Chim. Fr. 1897, 17, 599.
14. Lenz, R. W.; Handlovits, C. E.; Smith, H. A. J. Polym. Sci.
    1962, 58, 351.
15. Short, J. M.; Hill, H. W. CHEMTECH 1972, 2, 481.
16. Floryan, D. E.; Serfaty, I. W. "Modern Plastics"; 1982; p. 146.
17. "Plastics World"; 1982; p. 8.

Polymers and the Technology of Electrical Insulation

J. H. LUPINSKI

General Electric Company, Corporate Research and Development, Schenectady, NY 12301

General Description
Historical Development
Direct Conductor Insulations
Other Insulating Resins
Structural Materials
Polymers for Electronic Applications

General Description

The function of electrical insulation is to isolate conductors of different electrical potential from one another. This potential difference is commonly expressed in volts; the greater the voltage difference, the greater the stress on the electrical insulation. Because conductors carry the currents that make electrical equipment work, the tendency is to pack as many of them as possible in the available space. Consequently, electrical insulation is only allotted a small fraction of the total volume of a device, and thus high demands are put on small amounts of insulating material that is frequently present in rather vulnerable shapes, for example, thin coatings and films.

The total spectrum of electrically insulating materials covers dielectric solids, dielectric liquids such as oils, silicone fluids, polychlorinated biphenyls (PCBs) and dielectric gases (SF_6, freons). This chapter will primarily be devoted to an overview of that part of the dielectric solids area in which polymers are the dominant ingredient, although inorganic materials may be present as fillers. After a few notes about historical developments, the material to be discussed in this chapter is organized as indicated in the table of contents. Emphasis will be placed on application aspects and polymer properties not likely mentioned in other chapters. (References to further general reading are given at the end of this chapter.) However, there will be occasional references to the basic chemistry of some polymers used in special circumstances. Applications in which the polymer system does not perform a major

0097-6156/85/0285-0515$06.00/0

dielectric function, for example, battery cases, in which polypropylene primarily provides a structurally sound container for the acid and the battery plates, fall outside the scope of this chapter.

Numerous polymer compositions are applied as electrical insulation. Some of them are only used in small volumes for special applications. The materials used in relatively large volume in 1980 are polyethylene (high density + low density), 211,000 tons; poly(vinyl chloride) (PVC), 177,000 tons; and polyesters, 77,000 tons.

Because the chemistry of several of the polymers used in electrical insulation has already been described in other chapters, it will not be repeated here. However, there is one chemical feature that is quite characteristic for polymeric electrical insulations, namely the cross-linked form.

Essentially all of the magnet wire and cable insulations, the potting, encapsulating, and impregnating resins, are—in their final forms and shapes—cross-linked polymer systems. A few exceptions are found among polymeric materials such as some polyesters, polypropylene, and polyimide films which are essentially thermoplastic materials and free of cross-links. The reason why cross-linked polymers are so common among insulating materials is that such polymers are insoluble and infusible and provide for excellent solvent resistance and good thermal capabilities, that is, no melting and flowing during overheating of electrical equipment.

In general, the total amount of polymeric electrical insulation is only a minute fraction of all the materials that constitute a transformer, a motor, or other rotating equipment. Yet it is often a failure in a few dimes worth of insulation that causes a complex piece of equipment to fail. It is therefore essential that the quality of the material is strictly controlled, and it is equally important that a qualified material be applied by reliable, well-proven application methods. Good insulation is a key to electrical equipment with long-term reliability in spite of the fact that it is often functioning under adverse conditions such as exposure to oil and solvents (hermetic motors for refrigerators) or to extreme conditions of temperature and humidity (transformers in outdoor installation, traction motors for locomotives, and off-highway vehicles used in mining operations).

In addition to the general testing for mechanical, thermal, and chemical properties which is quite common for plastics in nonelectrical applications, the insulating materials have to pass stringent electrical tests that are designed to show imperfections and nonuniformities that are not readily detected by other test methods. For example, in an insulated conductor that has variations in its 2-mil-thick insulating layer as shown schematically in Figure 1, a test probe (not shown) at an electrical potential of 10 V with respect to the conductor at ground potential will readily reveal the pinhole at A but it will not show the variations in insulation thickness at B, C, and D. However, a test probe at 100 V will also break down the thin layer at B and thus reveal one of the thin spots. Common test voltages of 1500 to 3000 V can readily find imperfections such as those shown at C and D. Specifications require that only a very small number of imperfections can be tolerated per unit length of conductor.

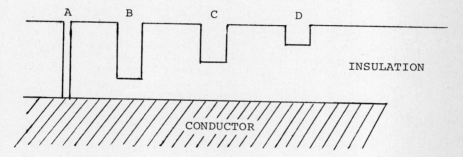

Figure 1. Schematic representation of imperfections and non-uniformities occurring in electrical insulation.

Today's role of polymers in electrical insulation is quite extensive and complex. This has not always been the case, and it is interesting to consider how the present state of development was reached.

Historical Development

Prior to the development of polymer insulated wire, the separation of conductors was achieved by wrapping the conductor with nonconducting fibrous material such as cotton and silk. Subsequently, such wrapped insulators were impregnated with resinous materials of natural origin. As our understanding of polymer chemistry advanced, synthetic resins were applied to the bare conductors without the need for fibrous wrappings to keep conductors separated. Presently, all magnet wire is made by film coating the bare wire with synthetic enamels.

However, by no means has the system of wrapped wire, impregnated by resins, disappeared. This so-called served wire is made with more modern materials such as mica, glass fibers, and specially treated paper, and it is used in high-voltage equipment where a relatively large separation between the conductors is required. Frequently the wrapping is applied over already insulated magnet wire, and the total construction is subsequently impregnated with liquid resin that is cured to form a rigid structure. These insulating materials must have excellent mechanical and thermal properties.

Whereas the advancement of resin materials science resulted in application of insulation to conductors by the enameling process (film coating), the advancement of polymer processing technology allowed extrusion of molten resinous materials directly to the bare conductors. This technology is primarily used to produce single or multistrand PVC coated ("hook-up") wire for application in radios, TV sets, and other electronic equipment. Polyolefin insulated multistrand cables are used for underground power distribution.

In addition, advanced polymer processing technology also provided for structural materials such as sheets and films, tubing, laminates, and other composite materials. Table I gives a general overview of the types of application for the various polymer systems.

In some applications such as in underground or underwater cable, electrical insulation has to perform under relatively constant temperatures; in other applications the equipment is subjected to wide ranging thermal cycles, for example, distribution transformers and direct current traction motors for locomotives. Both are exposed to cold winter and hot summer weather. Internal heating because of overload conditions enhances the effect of high ambient temperatures in the summer.

Electrical insulations are characterized by their capability to function at certain temperatures. A widely accepted thermal classification is described as follows:

1. Class 90--Unimpregnated materials based on natural products such as cellulose (e.g., cotton and paper) or animal protein (silk) for use below 90 °C.

TABLE I
Polymers and Their Electrical Applications

Application	Polyolefins	Epoxy Resins	Poly-esters unsaturated	Poly-esters saturated	Polyesterimides	Polyamides	Polyamideimides	Polyimides	Polyvinyl formal	Polyurethanes	Polyvinyl Chloride	Fluoro Polymers	Phenolic Resins	Melamine Resins	Silicone Resins	Cellulose Derivatives
Magnet Wire Enamels		•		•	•	•	•	•	•	•						
Cable Insulation	•										•	•				
VPI*		•	•													
Potting		•				•				•	•				•	•
Encapsulation		•				•				•	•				•	
Varnishes		•	•	•			•	•		•		•	•		•	
Tubing	•			•				•				•				
Sleeving		•		•				•		•	•	•			•	
Films & Tapes	•			•				•				•				
Sheets		•		•									•	•		
Heat Shrinkable Materials	•			•							•	•			•	
Papers						•										•
Composites		•		•		•		•					•	•		

*VPI = Vacuum Pressure Impregnation

2. <u>Class 105</u> (Class A)--Impregnated materials based on the natural products given in Class 90. Impregnation may be achieved with resins or dielectric fluids and is not to be used above 105 °C.
3. <u>Class 130</u> (Class B)--Combinations of inorganic materials such as mica or glass fiber with suitable binders for 130 °C temperature.
4. <u>Class 155</u> (Class F)--Combinations as for Class 130, except that the binders qualify for a 155 °C temperature.
5. <u>Class 180</u> (Class H)--Combinations of materials such as mica, glass fibers, and silica with suitable binders such as silicone resins to qualify for a 180 °C temperature index.
6. <u>Class 200</u>--Materials such as mica, porcelain, glass, and quartz with or without inorganic or organic binders for 200 °C.
7. <u>Class 220</u>--Materials and combination of materials that by experience or accepted tests can be shown to have a 220 °C temperature index.
8. <u>Over 220 Class</u>--Inorganic materials such as mica, glass, and quartz and combinations with binders that qualify for a temperature index higher than 220 °C.

The materials mentioned with each class are examples only; they are not intended to limit a class to these particular materials.

Most of the insulating materials development was tied to applications in which sizable amounts of insulation were required, for example, in traditional electrical equipment such as motors and transformers for which the insulation thickness is in the order of mils.

Recently, however, polymers are also being evaluated for microelectronic applications for which much smaller quantities and much thinner layers are required. Such applications introduce new demands on the materials. These polymers have to have very low impurity levels, and they must be readily processable to give adhering films having thicknesses of only several thousand angstroms and superior electrical integrity (no pinholes and uniform overall thickness).

Direct Conductor Insulations

<u>Magnet Wire Insulation</u>. Magnet wire insulation is coated on bare wire by repeatedly applying liquid enamel to the wire by means of dies, followed by evaporation of the solvent and exposing the formed thin film to high enough temperatures to establish a cross-linked polymer system. In this process, the wire travels at fairly high speeds through an enamel applicator and then through a furnace. Because of the high wire speeds needed for productivity, only a thin film can be made without blisters and blemishes in each enamel application. To obtain the required insulation thickness, the wire is repeatedly taken through the same process steps. In general, the number of coating passes varies from 6 to 16. An advantage of the multiple passes is that imperfections occurring in the earlier coating layers are likely to be covered by the material applied in subsequent passes to produce insulating films having excellent electrical integrity. It should also be recognized that each layer has been exposed to a different thermal treatment: the first layer is reheated again during each subsequent pass through the furnace,

and each following layer is subjected to fewer heating passes until the final layer is exposed to only one heating cycle. The result of this manufacturing method is that the insulation forms a set of concentric cylinders around the conductor wire. In some cases, cross sections of the magnet wire show the concentric rings in the insulation. Magnet wire also has to survive the rigors of high-speed coil winding, a key step in the manufacturing process for motors. To test for the flexibility needed for this application, a wire sample is first elongated by 25% and then wound around a mandrel one, two, and three times the wire diameter. To pass the test, the wire insulation must still adhere to the wire and be free of cracks.

Polyurethane Enamels. Small volumes of this enamel type are used in applications for which the temperature does not rise above approximately 105 °C. These enamels are made from hydroxyl-bearing polyesters and blocked isocyanates. During the heating cycle in the enameling process the isocyanates become unblocked and react with hydroxyl groups of the polyesters to form a cross-linked polyurethane film. Wires so insulated can be directly soldered, without prior removal of the polyurethane insulation.

Poly(vinyl formal) Enamels. Poly(vinyl formal) is made by hydrolysis of poly(vinyl acetate) followed by reaction of the hydrolyzed polymer with formaldehyde to give a polymer with units as follows:

$$
\begin{array}{c}
\text{H}_2 \\
\text{H}_2 \quad \text{H} \quad \diagup \text{C} \diagdown \quad \text{H} \\
-\text{C}-\text{C} \qquad\qquad \text{C}- \\
\quad\; | \qquad\qquad\quad | \\
\quad\; \text{O} \qquad\qquad\quad \text{O} \\
\qquad \diagdown \quad \diagup \\
\qquad\quad \text{C} \\
\qquad\quad \text{H}_2
\end{array}
$$

Usually the polymer contains residual unreacted acetate and hydroxyl groups, and enamel formulations always contain a second resin, frequently a phenol formaldehyde resin. Although it is a low-temperature material (105 °C class), it is widely used in liquid-filled transformers because of its excellent hydrolytic stability. Because cellulosic-type insulation and structural materials present in transformers degrade slowly during the life of the devices, water and other degradation products formed would hydrolyze other common magnet wire insulation such as polyesters and polyester imides at higher temperatures, but poly(vinyl formal) resin is not affected. Poly(vinyl formal) (Formvar) is manufactured by Monsanto, and it is available in several different molecular weights.

Polyester Enamels. The various commercially available enamel formulations contain a polymer made from ethylene glycol and phthalic acids as the main resin ingredients. Cross-linking features are built into the polymer by incorporation of trifunctional building blocks such as trimellitic acid, glycerine, or tris(2-hydroxyethyl)isocyanurate. Other cross-linking agents such as blocked isocyanates may also be present. In the liquid enamel formulations the polymers are still present in their linear form, but the compositions are designed so that cross-linking occurs on the wire at elevated temperatures during the enameling process.

Typically, polyester wire insulation is rated as Class 155. This insulation is frequently overcoated with nylon to provide a surface with good lubricity which is needed for application in high-speed coil winding. Magnet wire with this type of insulation is used in the manufacture of electric motors for nonhermetic applications.

Polyester Imide Enamels. The chemistry of the polyester imide enamels differs from that of the polyesters in that methylene dianiline is one of the basic building blocks. Reaction with anhydrides such as trimellitic anhydride provides for the imide content, and the remaining acid group of the trimellitic anhydride is available to form ester linkages.

Wire enameled with polyester imides can be used for applications in which it is occasionally exposed to temperatures in the 155–180 °C range. One of the most demanding environments is in hermetic motors used in refrigerators where the insulating enamel film is exposed at elevated temperatures to oil and refrigerants under pressure.

Polyamide Imide Enamels. Aromatic polyamide imide insulations are made from aromatic acids and aromatic diamines. The polyamide imide polymers are primarily present in the form of polyamic acids in the N-methylpyrrolidone solutions. During the enameling process the polyamic acids undergo imidization and become essentially fully imidized. Whereas the polyester and polyester imide enamels contain cresylic acids as the major solvent, the polyamide imide enamels are mainly formulated with N-methylpyrrolidone as solvent.

These polyamide imide films have excellent thermal, mechanical, and electrical properties, and they are used to upgrade other enamel films. Other attractive features of polyamide imide film insulations are a low coefficient of friction and good solvent resistance. These enamels are rated for use at 200 °C.

Polyimide Enamels. Polyimide enamels are really polyamic acids in N-methylpyrrolidone solution. The high-temperature capability (220 °C rating) is closely related to the chemical structure of the polymers which are made from pyromellitic anhydride and oxydianilin. The only oxidizable sites on these polymers are aromatic hydrogens (see structure of repeat unit on page 530). Strictly speaking, these polymers are polyether imides rather than straight polyimides, and they are not normally formulated to cross-link during the enameling conditions. After closure of the imide rings the polymer becomes insoluble and infusible and thus shows the properties of a cross-linked polymer although it is still in a linear, noncross-linked configuration.

Dual Constructions. In addition to the "basic" magnet wire insulations in which the conductor wire is coated with a single enamel, there are a number of cases where two different enamels are used: one overcoated over the other. Examples include polyurethane-base insulation overcoated with poly(vinyl butyral) or with nylon, or polyester-base insulation overcoated with nylon or with polyamide imide.

A polyester insulated wire which by itself would not qualify for application in hermetic motors can be made to qualify for such applications by a thin overcoat of polyamide imide film. In general, the polyamide imide protects the underlaying polyester from exposure to deleterious outside influences and elevated temperatures.

Trends in Magnet Wire Insulation. As long as energy and materials were inexpensive, the traditional magnet wire enameling method, which dates back to the 1920s, was an acceptable and widely used industrial process.

Because most enamels contain not more than 30% solids, 70% of the enamel is bulk that does not contribute to the final product. Instead, it adds to the cost of transportation and handling, and at the site of application the solvent is evaporated. Because of the low solvent concentration in the hot exhaust fumes formed in wire furnaces, it is not economical to recover the solvents by condensation. Long before our present concern with protection of the environment, it became common practice to incinerate the solvent vapors prior to release into the atmosphere.

As a first step toward conservation of energy, the solvent-bearing exhaust fumes are now frequently recirculated into the wire furnace where combustion of the solvents supplies part of the heat and thus reduces the fuel requirements of the operation. Other research and development efforts are directed toward new magnet wire enamel formulations such as high-solids systems with solids contents ranging from approximately 40–50% solids to enamels that are applied as melts (ca. 90% solids) or as powders in electrostatic powder coating operations. Water-based enamels have also been prepared. Most of these new enamel types are in the developmental stages, and large scale industrial application of some of these may occur during the 1980s.

Extrusion-Applied Insulations. The polymers used in extrusion applications can be divided into two classes: low-temperature applications and high-temperature applications. Polymers in the first category are poly(vinyl chloride), polyethylene, polypropylene, and their copolymers along with other elastomers. Polymers in the second category are mainly halocarbons such as Teflon polytetrafluoroethylene (which requires special extrusion or application conditions), fluoroethylene-propylene copolymer (FEP), perfluoroalkoxy-modified polytetrafluoroethylene (PFA), poly(ethylene-tetrafluoroethylene) (ETFE), poly(vinylidene fluoride) (PVF_2) (borderline temperature of 135 °C), and poly(ethylene-chlorotrifluoroethylene). Extrusion conditions for wire and cable insulations have to be tailored to resin composition, conductor size, and need for cross-linking of the insulating layer.

Poly(vinyl chloride). The application of poly(vinyl chloride) as wire and cable insulation occurred after the development of flexible formulations in the mid-1930s. Typical compositions for insulation purposes contain from 50 to 70% poly(vinyl chloride); the remainder is plasticizers, fillers, lubricants, and stabilizers. These formulations provide a combination of attractive features such as good mechanical properties, resistance to oil and moisture, and low flammability. Its main applications are in building wire for connection of outlets and switches with main panels, in equipment and apparatus cables, and in low-temperature electronic applications; it is also found in flexible and retractable cords, in telephone wires, and in power cables.

Polyethylene. Both low-density and high-density polyethylene are used in uncross-linked form as wire insulation. However, because of its lower thermal capabilities, its main use is restricted to applications close to room temperature such as in coaxial cables, telephone, radio, and television.

Cross-linked polyethylene and cross-linked ethylene-propylene copolymer, however, can be exposed to higher temperatures, that is, up to 130 °C. Cross-linking can be achieved by compounding with peroxide catalysts or by ionizing radiation. Heavy, multiconductor power cable and high-voltage cable are usually made with peroxide-catalyzed polyolefins. A comparison of the property profiles of both the polyethylene homopolymer and the ethylene-propylene copolymer compositions suggests a slightly better performance by the homopolymer.

In addition to abrupt dielectric failure due to electrical stresses exceeding the critical value, cable insulation of this type may also fail as a result of slowly progressing, localized deteriorations. This phenomenon is known as treeing because of the tree-shaped deterioration patterns that develop as a result of prolonged electrical stresses below the critical level. Because it was discovered that inorganic fillers tend to retard treeing, most cable insulation compositions contain such fillers, along with other additives which are frequently incorporated, for example, antioxidants, processing aids, and flame retardants.

Polypropylene. Like polyethylene, polypropylene and propylene-ethylene copolymers are used as insulation for telephone wires. However, polypropylene itself is also used in the fabrication of components for low-voltage automotive devices such as switches, connectors, and terminal blocks. As will be shown later, polypropylene plays another important role as an insulating material in an application where it is applied as a free film rather than as a formulation for coating or extrusion on wires and cables.

Fluoropolymers. Melt processable fluoropolymers such as Teflon FEP, Tefzel ETFE, poly(vinylidene fluoride) (Kynar), and ethylene-chlorotrifluoroethylene copolymer (Halar) are suitable for wire insulation in special applications because they combine good physical properties with low flammability. They are used for instrumentation cable in process-control rooms, as well as for computer and aircraft wiring and in military applications. The minimum wall thickness of these insulations is 0.009 in., and it increases with conductor size.

Miscellaneous Extrusion-Applied Polymers. As mentioned earlier, there is a tendency to develop solventless magnet wire enamel formulations, and extrudable polymer systems would fulfill that requirement. There have been reports about extrusion of thin coatings of polyesters over copper wire. At this point, the state of the art allows extrusion of thin insulating films only with thermoplastic materials. The reliable extrusion of uniform and concentric insulating films of approximately 0.001-0.002 in. wall thickness is already an improvement over the more traditional extrusions of polyethylene, poly(vinyl chloride), and several fluoropolymers in much greater wall thicknesses. Because cross-linked insulation is ultimately required for most magnet wire applications, further materials development needs to be done to provide polymer compositions that are both extrudable as thin films and can be cross-linked in an economical process suitable for large-scale industrial application.

Other Insulating Resins

Many polymeric systems have been adapted for applications in which they become an integral part of electrical equipment. Polymers may be used, for example, to protect components of electrical systems or to enhance the mechanical and electrical integrity of electrical apparatus. These applications are referred to as vacuum pressure impregnation, potting, encapsulation, and casting; some of these are illustrated by the following examples.

Resins for Vacuum Pressure Impregnation. When coils for motors are wound, there is always some free space between the wires in the windings. Such free space, filled with air, allows for vibration of the wires that leads to noise and to mechanical damage. In high-voltage machines the air space between the wires can also lead to corona discharges that damage the insulation of the magnet wire. None of these conditions are desirable, and the resulting problems are successfully addressed by impregnation with suitable resins to completely rigidize the coil. Impregnation is achieved by placing the parts in a vacuum tank and evacuating the assembly. The parts are then flooded with the liquid vacuum pressure impregnation (VPI) resin from a storage tank connected to the vacuum tank. After immersion in the liquid resin, atmospheric pressure is restored to the vacuum tank and thus forces the resin into the voids. The excess liquid resin is returned to the storage tank, and the impregnated motor is transferred to an oven where the resin is cured at elevated temperatures. To insure complete impregnation, this procedure may be repeated several times.

Polymer chemistry has advanced to the point where resin systems for these applications can be tailor made to provide for all the requirements of the resin properties before, during, and after application. For example, before application, the resin should be available as a low-viscosity liquid with a good pot life (3-6 months at room temperature) and show little increase in viscosity when exposed to normal ambient air and moisture. It should have a short but controllable gel time and a short cure time at moderate temperatures. After application, it should have all the properties needed to realize good quality electric motors. For example, it should adhere well but not diminish the properties of other insulation already present (magnet wire insulation). Its dissipation factor and breakdown strength, as well as its mechanical and thermal properties, should be tailored for the particular applications.

There are two basic VPI resin formulations. One is based on epoxy resins, and the other is based on unsaturated polyesters. Both are typically one-part solventless compositions that are completely polymerizable. Volatile, nonpolymerizable solvents cannot be tolerated because these will lead to bubbling in the vacuum impregnation procedure. To control viscosity of the resin formulations, the epoxies contain low-viscosity, reactive diluents of the monoglycidyl ether type; the unsaturated polyester formulations contain styrene or vinyltoluene as reactive diluents. Both resin types contain catalysts that become active only at elevated temperatures to insure long-term stability at room temperature. The epoxy resins are frequently catalyzed with metal organic compounds such as titanium complexes, and peroxides are usually the main catalyst in the unsaturated polyester formulations. In addition to the ingredients mentioned here, the compositions may contain additives such as cocatalysts, activators, and accelerators. However, there are no particular fillers used in VPI resins.

Potting and Encapsulation. The difference between potting and encapsulation is not sharply defined. In both cases the electrical assembly is placed in a container that is then filled at atmospheric pressure with a liquid resin that,after curing, solidifies the whole assembly and provides excellent protection of the device against damaging outside influences. Containers (or shells) for potting and encapsulation are made from epoxies, alkyds, and other reinforced plastics, and they become part of the total structure. Encapsulation is sometimes achieved simply by dipping the electrical device in liquid resin and curing the resin without the use of containers.

The resin formulations for these applications are solventless systems, but contrary to the VPI resins, the compositions for potting and encapsulation are frequently two-part systems, that is, the resin part and the catalyst part are stored separately and mixed prior to use. New, advanced handling equipment allows for accurate metering, mixing, and easy dispensing of the two parts without the problems of premature gelling frequently experienced with earlier equipment. A whole spectrum of resin compositions can be used for potting and encapsulation applications. The required properties can be obtained by selection of the proper composition from the following list of polymers: depolymerized rubber, epoxies, ethyl cellulose formulations, PVC plastisols, polyamides, polybutadiene-based resins, polyesters, polyimide, polysulfide, silicone gels, room temperature curing resins (RTVs), and polyurethanes (foams & liquids).

Protective Coatings. Varnishes are used to provide protection for electrical equipment against moisture, chemicals, particulate contaminants, and mechanical abuse. The formulations are generally solvent-based, cross-linkable polymers that can be air dried or thermally cured to form insoluble coatings.

Varnishes employ many different polymers spanning a wide range of properties. Well-known varnishes are based on acrylics, alkyds, epoxies, phenolics, polybutadienes, polyesters, silicones, and urethanes.

Linear polymers such as polyamide imides and polyimides are sometimes used as varnishes; however, they produce coatings that are not cross-linked. For some applications for which rework and repairs have to be made, cross-linked coatings have a disadvantage. Soluble polymer systems that can be repaired are to be preferred in such cases.

Conformal Coatings. To protect printed circuitry and electronic devices from contaminants such as moisture, particulates, and fumes, they are covered with thin films of polymers. These polymers may be applied as "all solids" formulations or as solvent-containing systems, or, as in a special case, by polymerization of a vapor deposited monomer.

Key properties required of the polymer system before application are acceptable pot life; compatibility with other components; and viscosity, cure time, and temperature suitable for particular applications. After application and curing, the coatings should have low moisture absorption, low moisture permeability, and good resistance to other outside influences that may tend to deteriorate the performance of the device.

The most common conformal coatings are derived from polyure-thanes, acrylics, and epoxies; the more special formulations for high-temperature performance are based on silicones, diallyl-phthalate esters, and polyimides. An example of a vapor deposited conformal coating is Parylene. It is obtained by vapor deposition of p-xylylene, which is formed as a transient by dehydrogenation of p-xylene at high temperature, and polymerization on the surface of the object to be coated. Because p-xylylene monomer is not stable, it is advantageous to work with the cyclic dimer, di-p-xylylene (paracyclophane), which, upon heating under reduced pressure, will produce the transient monomer which converts to the polymer at low temperatures.

$$H_3C \underset{}{\overset{}{\bigcirc}} CH_3 \xrightarrow[\text{pyrolysis}]{-H_2} H_2C = \bigcirc = CH_2 \longrightarrow \left[\overset{H_2}{C} \bigcirc \overset{H_2}{C} - \right]_n$$

Similar polymerizations can be carried out with paracyclophane derivatives; the substituents may be attached either to the aromatic ring or to the aliphatic part of the molecule. Polymer films formed from the unsubstituted monomer are highly crystalline and, like polymers made from substituted monomers, have high melting and high softening points. In addition, thin films with very high integrity can be made to adhere to a large variety of substrates and thus provide a protective skin with a uniform thickness. The poly-p-xylylenes are used only in small quantities in special applications to protect electronic circuitry and devices.

Structural Materials

Films. The most prominent polymeric materials used as films for
electrical insulation purposes are polypropylene, poly(ethylene
terephthalate), and polyimide. Contrary to the polymeric insulating
systems discussed earlier for which optimum property profiles were
realized with cross-linked polymers, most of the polymeric films
used for insulation purposes are uncross-linked thermoplastics.
Polymer films for capacitor production are rated electrical grade to
indicate a high degree of cleanliness and low levels of impurities
and particulate contaminants.   Such impurities could easily
jeopardize thin film (0.0005 in.) performance under high-voltage
conditions. Special precautions have to be taken to deal with the
tendency of dielectric films to attract particulate matter that then
clings tenaciously to the films because of electrostatic attraction.
Strict quality control measures are required to produce a
consistently high-quality material.
        The physical properties of polypropylene, poly(ethylene
terephthalate), and polyimide are greatly influenced by orientation
of the polymer. Polypropylene and poly(ethylene terephthalate)
become highly crystallized upon stretching. Oriented, crystalline
poly(ethylene terephthalate), for example, has significantly higher
modulus, tensile strength, and continuous use temperature than its
extruded, unoriented analog (see Table II).
        Electrical grade polymer films have now largely replaced the
Kraft paper that was previously used in capacitor applications.
Polypropylene and poly(ethylene terephthalate) are both used in
capacitors as free films or metallized with a thin layer of
aluminum. Other, newer materials for capacitor applications are
made by metallizing polysulfone and polycarbonate films.

Polypropylene. The combination of low density, very high electrical
breakdown strength, thermal stability up to 105 °C, and inertness to
most liquids makes polypropylene an attractive capacitor dielectric.
Polypropylene films for this application are made by blow molding
during which the films are subjected to considerable stretching so
that in their final form they are biaxially oriented and highly
crystalline. Some films are also provided with a micro surface
roughness to facilitate impregnation of the capacitor coil by the
dielectric fluid in which the coils are immersed. This feature
mimicks the wicking of the dielectric fluid into the Kraft paper in
older capacitor constructions.

Poly(ethylene terephthalate). Poly(ethylene terephthalate) film is
a widely used industrial product. However, only a small fraction of
the total film production is used in electrical applications (see
also section on polymers for electronic applications).
        Slot liners, for example, are used in motor and generator
construction to separate and insulate coils of magnet wire from the
slot walls in the steel stator, and films are also used to separate
different coils of wire from one another (phase separation) in
motors. For these applications, materials with good electrical,
thermal, and mechanical properties are required. Special grades are
applied in capacitors and heat-shrinkable tubing.

Polyimide. The best-known fully aromatic polyimide is made from pyromellitic dianhydride and oxydianilin. When completely imidized, the polymer is essentially insoluble and infusible. Therefore, bulk forms of the material are not readily available, but the polyamic acid precursor is soluble in polar solvents such as *N*-methylpyrrolidone. Such solutions do not only serve as wire enamels, but they can also be used to cast polyamic acid films that, upon heating, convert to the polyimide. These films have exceptional thermal capability: they have a continuous use temperature of 250 °C in air, and they will tolerate brief excursions to much higher temperatures (350 °C). Because several key dielectric properties are also outstanding, this material is primarily used in specialty, high-temperature applications. In cases where the polyimide has to adhere to other parts or to itself, the film is provided with a backing of fluorinated ethylene-propylene copolymer that functions as heat-sealable adhesive. In this form it can be used as tape for conductor wrapping, which, when followed by heat sealing, provides a tight, protective, and insulating skin. (See Scheme I.)

Noncellulosic Paper. Another high-temperature, polymeric insulating material is Nomex resin, a fully aromatic polyamide made from *m*-phenylenediamine and isophthaloyl chloride. It is available as short fibers and as continuous filament, but not as extruded film. However, thin sheets can be made from the short fibers and binder particles by conventional paper-making techniques, followed by compaction (calendering). It is used as insulating sheets, slot liners or wrappings (tapes), and, in its uncompacted form, it may be saturated with curable resins to make rigid structures, laminations, etc. Nomex is a recognized Class H material (180 °C), but it may be used at temperatures as high as 220 °C.

Tapes. A great variety of tapes find application in electrical equipment. Some tapes contain filler materials in macroscopic form such as glass fibers, mica flakes, and cloth; others have finely divided filler particles or no fillers at all. In the heavily filled materials the polymeric binders are present in small fractions, and the major emphasis may be on their adhesive capabilities rather than on their properties as dielectric materials. Most of the polymers used in tapes have already been mentioned in connection with other insulation applications, for example, polyesters, aromatic polyamides, polyimides, and polypropylene. Other polymers frequently used for electrical tapes are vinyls, including poly(vinyl fluoride); these are particularly well suited as conformable tapes. Polytetrafluoroethylene (Teflon TFE) has also been fabricated into tape constructions, frequently in combination with adhesives to provide a bondable material.

Tubing and Sleeving. Tubing and sleeving are used to protect and further insulate conductors, contacts, and connections. Three major forms of tubular shape can readily be identified: extruded polymeric materials, reinforced structures made from braided fibers and overcoated with resin, and heat-shrinkable tubing. The first type of tubing can be made by extruding any polymeric material that has an acceptable property profile for functioning as an insulating

Table II. Properties of Oriented and Unoriented Polymers

Material	Form	
	Extruded	Oriented
Poly(ethylene terephthalate)		
Tensile Modulus (10^3 lb/in.^{2})	300	550
Tensile Strength (10^3 lb/in.^{2})	6	25
Polypropylene		
Tensile Modulus (10^3 lb/in.^{2})	100	350
Tensile Strength (10^3 lb/in.^{2})	5	25

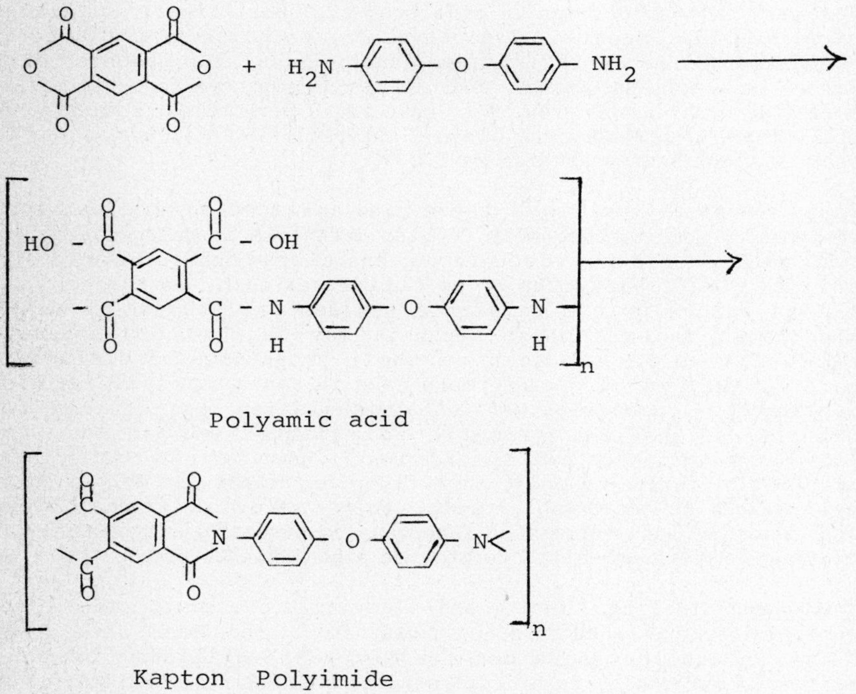

Polyamic acid

Kapton   Polyimide

Scheme I.

material under a particular set of thermal, mechanical, and electrical requirements.

Another method to produce tubing is to spirally wind ribbons of polymers and consolidate the spiral into tubing either by using adhesives or by heat sealing if the material permits. In this fashion tubing can be made from materials that cannot readily be extruded in tubular form but that are available as thin films such as aromatic polyimide.

In the second type of tubing, glass fibers are the most commonly used fibers for braiding, although aramid fibers are also used. Cotton and rayon fibers, which along with oleo resinous materials constituted earlier versions of tubing, have largely been replaced by materials with higher thermal capability. Common coatings are vinyls such as poly(vinyl chloride), acrylics, polyurethanes, polyesters, epoxies, silicones, polyimides, and halocarbons.

The third type of tubing is characterized by its ability to shrink to a significantly smaller diameter upon heating, and it is commonly known as heat-shrinkable tubing. Shrinking can be accomplished by two different methods. One uses linear, noncross-linked polymers that have been frozen with their molecules in an extended condition. Upon heating, the thermal motions of the polymer chains increase, and the matrix is allowed to relax and to return to its smaller volume, minimum energy configuration.

The second method uses cross-linked polymers that have been processed to produce an extended form that, upon heating, shrinks to smaller dimensions. Suitable polymers for this type of heat-shrinkable tubing are radiation cross-linked polyolefins and halocarbons including poly(vinyl chloride) and poly(vinylidene fluoride).

Selection of proper processing conditions leads to tubing with relatively greater shrinkage in the radial direction than in the axial direction.

Structures of two different materials have been made in which the heat-shrinkable tubing has a lining of a hot-melt adhesive. Upon heating, the combined melting of the adhesive and the shrinking of the tubing produce a tight and firmly bonded jacket around the object to be protected.

Composites and Laminates. In many of the applications for polymers in electrical equipment, the polymers are frequently used under circumstances in which a combination of dielectric and mechanical properties are required because of the structural function certain parts have to fulfill. Circuit breakers, meter housings, switches, connectors, plugs, circuit board, etc. are often made from filled or fiber-reinforced plastic materials. The polymers may be thermoplastics, for example, polypropylene, nylon, poly(butylene terephthalate), and acrylonitrile-butadiene-styrene terpolymer (ABS). Most of the thermoplastics are used in low-voltage applications (automotive) at temperatures close to ambient, although thermosets such as phenolics, polyesters, epoxies, epoxy novolacs, melamine resins, urea-formaldehyde resins and polyimides are found in applications in which higher voltages, higher temperatures, greater dimensional stability, or a combination of these properties are required. Most of these parts are produced by standard manufacturing methods such as molding and extrusion. Printed circuitry, however, requires more complicated processing before a complete product that meets a host of requirements is obtained.

Printed Circuitry. Printed circuitry, which represents a most intricate structure of polymers and conductors, is frequently made by laminating copper foil onto glass-fiber-reinforced epoxy composites. For certain specialty applications, more expensive fiber-reinforced polyimides are used.

The copper foil is subsequently covered with a photoresist layer that is patterned by lithography techniques to produce the desired circuitry. Areas of exposed, cross-linked resist protect the copper that is to remain on the board as circuitry. The unprotected copper that is not needed for the circuitry is etched away, usually with ferric chloride solution. At this point, the printed board must have the following characteristics: the substrate has to be structurally sound (no delamination), and the copper conductor patterns have to adhere well to the substrate; in addition, the assembly should have good dimensional stability, and it has to show good solderability (ability to withstand short exposures to liquid solder at temperatures around 250 °C or higher) to allow for the connection of components and devices to the circuitry.

Finally, in order to protect the circuitry against oxidation, the influence of high humidity, and contamination by particulate matter, it is covered with a protective coating.

Flexible printed circuitry is made in the same manner as the rigid boards, except that flexible, unfilled poly(ethylene terephthalate) or polyimide films are used as substrates.

Polymers for Electronic Applications

Many polymers are used in the fabrication of multilayer, high-density electronic circuitry consisting of silicon chips containing large numbers of integrated circuits. However, most of the polymers function in a temporary role in photoresist systems that are removed before the devices are completed.

Only a few polymers are presently under consideration for a permanent role as a dielectric layer in microelectronic circuitry. For example, there is a need for polymers that can replace the presently used SiO_2 and Si_3N_4 dielectric interlayers. The disadvantage of these inorganic coatings is that, although their dielectric properties are excellent over a wide temperature range, they reproduce the topography of the underlaying circuitry, although a flat layer is needed for deposition of the next level of circuitry. Polymeric coatings applied from solutions offer the desired planarization. These polymers will become an integral part of the circuitry. Because of subsequent high-temperature processing steps required to complete the device, only a few polymers can offer the necessary stability at elevated temperatures (350–400 °C in a nonoxidizing atmosphere).

The polymers considered for these applications are primarily polyimides. Although polyimides have been used as electrical insulation for the past 15 years, and a considerable know-how has been generated about these polymers and their applications as insulation, their use in microelectronic devices provides new challenges. Because the interlayers are approximately 2μm thick, coatings of extreme uniformity are required to provide the electrical integrity.

Another stringent demand is that the polymer have a very low concentration of ionic impurities because such impurities, when mobile, could interfere with the electronic function of the device. For example, the Na^+ ion concentration has to be less than 4 ppm. This requirement means that the Na^+ ion impurity content of the solvents for the polymer also has to be very low. In general, special grade materials are needed to fulfill the requirements for electronic applications.

Naturally, devices as delicate as microelectronic circuitry and not to be left exposed to the atmosphere because humidity and particulate contaminants could readily interfere with their performance. Several materials are being used as protective coatings or encapsulants. Depending on circumstances, so-called passivation layers may be applied as thin, conformal coatings before encapsulation. Passivation layers are frequently made from vapor-deposited inorganic oxides or glasses; however, polyimides are also being evaluated for this application. Passivation layers have to be patterned to allow for the connection of leads. Encapsulation of microelectronic circuitry can be carried out with materials similar to those discussed earlier in regard to the protection of printed circuitry.

Additional Reading

The amount of literature on the topics discussed in this chapter is enormous. Instead of citing a list half as long as the chapter itself, some selected sources are given for additional reading and consultation. These will provide quick access to many areas of interest.

1. "Digest of Literature on Dielectrics" – Annual review of the literature dealing with dielectrics. Published under auspices of the National Research Council of the National Academy of Sciences, Washington, D.C., National Academic Press.
2. "Electri•onics" – A monthly magazine devoted to materials components and methods used in electrical and electronic manufacturing. From time to time it features updated information on topics such as printed circuitry, wires and cables, and insulating materials, for example, films, papers, resins, and coatings. It is published by Lake Publishing Corporation, 17730 West Peterson Rd., Libertyville, IL 60048.
3. "Electrical Properties" – The electrical properties of polymeric dielectric materials are discussed by K. H. Mathes in a chapter of the above title in "Encyclopedia of Polymer Science and Technology," Vol. 5, p. 528, John Wiley & Sons, 1966.
4. "Data Banks" – Two volumes, "Plastics for Electronics" and "Film, Sheets, and Laminates," are desk-top data banks containing numerical values in table form of a multitude of polymeric materials. These volumes were published in 1979 by The International Plastics Selector, Inc., 2251 San Diego Ave., San Diego, CA 92110.

Medical Applications of Polymers

CHARLES G. GEBELEIN

Department of Chemistry, Youngstown State University, Youngstown, OH 44555

Polymer Biocompatibility
Tubular Prostheses, Plastic Surgery, and Related Applications
Orthopedic Applications
Cardiovascular Applications
Artificial Organs
Medication Applications

The application of polymeric materials in medicine is a fairly specialized area with a wide range of specific applications and requirements. Although the total volume of polymers used in this application may be small compared to the annual production of polyethylene, for example, the total amount of money spent annually on prosthetic and biomedical devices exceeds $16 billion in the United States alone. These applications include over a million dentures, nearly a half billion dental fillings, about six million contact lenses, over a million replacement joints (hip, knee, finger, etc.), about a half million plastic surgery operations (breast prosthesis, facial reconstruction, etc.), over 25,000 heart valves, and 60,000 pacemaker implantations. In addition, over 40,000 patients are on hemodialysis units (artificial kidney) on a regular basis, and over 90,000 coronary bypass operations (often using synthetic polymers) are performed each year (1).
    The types of polymers in current use are somewhat limited and are mainly poly(dimethylsiloxane), polyethylene, polytetrafluoroethylene, poly(methyl methacrylate) and derivatives, poly(ethylene terephthalate), poly(vinyl chloride) [plasticized] and some polyether polyurethane ureas. A large number of other polymers are being studied experimentally. Because of the wide diversity of uses, each of which has very specific requirements, it is not possible for any single polymer to be the only material used in medical applications. In the past, medical applications were usually attempted with a commercial grade of a polymer, but this

0097–6156/85/0285–0535$06.50/0
© 1985 American Chemical Society

approach has generally proven unsatisfactory for several reasons, including variability of the material and the presence of usually unknown amounts of various additives such as plasticizers, catalysts, and stabilizers. At the present time most biomedical applications of polymers utilize materials that are manufactured specifically for that purpose. This trend is certain to continue, and new polymers that have been developed specifically for biomedical use are already appearing on the market.

Interest in biomedical applications of polymers dates back over 50 years. This interest is due in part to the fact that most biomaterials present in the human body are macromolecules (proteins, nucleic acids, etc.). When tissues or organs containing such materials need complete or partial replacement, it is logical to replace them with synthetic polymeric materials whenever natural replacement materials are not readily available. Although ceramics and metals can be used in certain cases, most biomedical applications require the use of some synthetic polymer or a modified natural macromolecule. Several recent books describe the range of biomedical applications of polymers (2–12).

Polymer Biocompatibility

The properties of a polymer are of great importance in any application. For biomedical polymers the most important single property is probably biocompatibility. The term biocompatibility is actually not easy to define precisely because the term depends on the specific end use of the material. Biocompatibility refers to the interactions of living body tissues, compounds, and fluids (including the blood) with any implanted or contacting material (e.g., polymer, ceramic, or metal) and ultimately refers to the interactions of the human body with a biomedical device or prosthesis. In some cases an interaction is desirable, although in other cases inertness is sought. These interactions could be interactions of the body materials on the plastic (e.g., degradation, loss of function) or interactions of the polymeric material on the body (13–15). Although each system of polymer–body material interactions must be studied individually, these varied interactions can be summarized into three broad categories which will be discussed separately: polymer stability, general tissue–fluid interactions, and blood compatibility.

Polymer Stability. In most medical applications, a high level of polymeric stability is required during the lifetime of the application. The few exceptions to this generalization include such applications as erodable controlled-release drug systems, sutures, and resorbable bone plates. This restriction limits the range of polymers that can be used safely as implanted biomaterials or devices. Basically, polymers that rely on additives, such as plasticizers, for the achievement of their properties would usually prove unsatisfactory for internal applications, although external use might be possible. The stability problems in external biomedical polymer usage are similar to those encountered in many other applications, except that discoloration of an external prosthetic device (ear, finger, limb, etc.) would present greater aesthetic problems. Stabilizers, plasticizers, and the like can often be utilized in these external usages except for certain materials that might cause allergic or toxic reactions.

The internal environment in the human or animal body contains a variety of enzymes and other chemical agents that can promote polymer degradation. Polymers that contain ester or amide groups would be expected to show some hydrolysis under these conditions. Implanted polyamides have been observed to lose nearly half of their tensile strength after 17-24 months of implantation, whereas polyesters were usually more stable (16, 17). Polyester-type polyurethanes degrade more rapidly than the polyether polyurethanes in vivo (13). Some polyether polyurethane ureas have been developed that show very little degradation (16). Natural polymers, such as silk or cotton, degrade very rapidly (17, 19). Although the preceding systems can be considered to be examples of hydrolytic degradation promoted by physiological factors, the degradation processes are not limited to this pathway and can occur via oxidation-reduction, double bond scission, decarboxylation, dehydration, deamination, or even single bond cleavage. Polyethylene, for example, can lose as much as 30% of its tensile strength after 17 months of implantation (20). Cross-linking reactions apparently occur in some instances because the polymers sometimes become more brittle or hard after implantation; this effect has been observed with both polyethylene (16) and poly-tetrafluoroethylene (20).

Implanted polymeric materials can also adsorb and absorb from the body various chemicals that could also effect the properties of the polymer. Lipids (triglycerides, fatty acids, cholesterol, etc.) could act as plasticizers for some polymers and change their physical properties. Lipid absorption has been suggested to increase the degradation of silicone rubbers in heart valves (13), but this does not appear to be a factor in nonvascular implants. Poly(dimethylsiloxane) shows very little tensile strength loss after 17 months of implantation (16). Adsorbed proteins, or other materials, can modify the interactions of the body with the polymer; this effect has been observed with various plasma proteins and with heparin in connection with blood compatibility.

In addition to the chemical factors, implants can also be subjected to abrasion or stress under actual use conditions. Some examples would be replacement joints, tendons, and dental appliances. Essentially all polymeric materials show some chemical and/or mechanical degradation under actual physiological conditions, but often polymeric materials survive better than some ceramic or metallic materials.

General Tissue-Fluid Interactions. The exact response of the body to any implant depends not only on the chemical composition of the implant but also on the form of the polymer (sheet, fiber, foam, etc.), the shape of the implant, whether the implant can move, and the location of the implant within the body. (Infection can also occur if proper sterilization techniques are not used.) The reaction of the body can vary from a relatively benign acceptance of the implant to an outright rejection of the material with an attempt, by the body, to extrude the implant and/or to destroy the implant by chemical means. Chemical destruction is usually manifested by phagocytic or enzymatic activity. The formation of a fibrous membrane capsule around the implant occurs with essentially all polymers as well as most ceramic and metallic implants (1, 13, 21, 22).

Although these effects are highly specific, some generalizations can be made. Polymers that contain little or few polar groups or atoms in their repeat units tend to elicit minimal tissue response. For example, polyethylene and polytetrafluoroethylene do not evoke extensive tissue response compared with samples of a polyamide or polyacrylonitrile which have the same form and shape. Part of this chemical structure problem could arise from histological interaction of the hydrolysis products from the ester- or amide-type polymers, but part could also arise from immunological interactions. As the tissue response increases, inflammation increases, and large numbers of macrophages and giant cells arise in the vicinity of the implant. The more polar polymers also tend to promote adhesion of the tissues to the implant. This adhesion is normally considered disadvantageous.

With any given polymeric material, a rounded shape will tend to give less interactions with the body tissues than will a shape with rough or sharp edges. In a similar manner, surface smoothness tends to reduce adverse tissue interactions or adhesion. With these factors in mind, it is easy to understand why a sheet or bulk sample of a polymer elicits less tissue response than does a spongy or fibrous sample of the same material. The greatest response is normally due to a finely pulverized sample of the polymer and is primarily a surface area phenomenon for a given polymer. Movement of the implant causes increased tissue response due to irritation of the surrounding cells. The increased tissue response observed for bone cement fragments or joint material particles is probably due to a combination of these factors.

Adverse tissue response is usually higher for lower molecular weight polymers, and the presence of free monomer is usually cytotoxic. Many additives used in polymers give a strong, adverse tissue reaction due to a combination of their low molecular weight, higher mobility, moderate solubility in some body fluids, and their relatively reactive and toxic natures. For this reason, the use of plasticizers and/or stabilizers is ill-advised for biomedical polymers. In a similar manner, it is advantageous to remove the polymerization catalysts from the polymer before use as an implant material.

The screening of polymers for biocompatibility is difficult. Although animal testing is widely used, this practice is costly and not always indicative of the ultimate use in humans. Various pre-animal screening tests have been devised and show some promise. These include the IR assay of a pseudoextracellular fluid (PECF) extract of the polymer (23), tissue culture methods using animal or human cells (24, 25), and a cell suspension–culture method in which the ATP (adenosine triphosphate) concentration is measured by luminescence (25). A comprehensive array of tissue culture tests, rabbit muscle implantation, and rabbit blood hemolysis studies on the polymer and on various extracts has been proposed as a general screening procedure for materials to be used in artificial organs (26).

<u>Blood Compatibility</u>. The problem of blood contact occurs in all internal organs and in many extracorporeal devices, and imposes additional constraints on the polymers beyond the normal biocompatibility problems. Blood clot formation, or thrombus, is a basic defense mechanism designed into the body to prevent fatal bleeding and to isolate foreign matter from the blood. To an extent, blood clot formation can be prevented by the use of anticoagulants, such as heparin, but heparin could lead to bleeding problems in cases such as injury. Generally, this approach is used only in conjunction with extracorporeal devices and/or for limited time periods. In the other cases, it is desirable, and important, that the blood-contacting materials possess some degree of blood compatibility.

Although plastic materials do differ in their blood clotting tendencies, the reasons for this are not clearly understood. Several theories have been advanced to explain the observed blood compatibility differences, and these include wettability factors, surface zeta potentials, and protein adsorption phenomena. None of these theories works in every case. The wettability factor theory predicts an inverse relationship between the blood compatibility and the wettability of the polymer surface by the blood. This approach works for some polydimethylsiloxane and polyether polyurethane urea (PEUU) polymers, which are hydrophobic, but fails for the hydrophilic hydrogels. The surface zeta potential theory predicts greater blood compatibility with a highly negative value for the zeta potential and does seem to work for some ionic polymers and electrets, but fails with glass which is highly thrombogenic and also has a high negative zeta potential. The wettability and zeta potential theories utilize certain aspects of the behavior of natural blood vessels, which are wettable by the blood and have a highly negative surface zeta potential. These natural surfaces do not, however, appear to adsorb proteins from the blood, although synthetic surfaces do. The protein adsorption theory predicts thrombus formation when the adsorbed proteins are fibrinogen or globulin, although albumin adsorption does not promote clot formation. This theory has the virtue of working for such diverse materials as the hydrophilic hydrogels and the hydrophobic polyether polyurethane ureas (<u>1</u>, <u>8</u>, <u>27-30</u>).

Very few polymeric materials exhibit good blood compatibility. The best currently available materials include some hydrogels, some PEUUs, and some biolized materials. Biolization involves affixing some biologically inactivated natural tissue to the polymer surface (<u>31</u>).

Tubular Prostheses, Plastic Surgery, and Related Applications

<u>Tubular Applications</u>. Tubular prostheses and related devices include drains, catheters, and shunts, and these are always made from plastic materials such as natural rubber, silicone rubber, polyethylene, polytetrafluoroethylene, or plasticized poly(vinyl chloride). Many of these devices are designed for temporary use, but some permanent devices have been used for replacement of the trachea, ureter, bile ducts, and other body tubelike parts.

The hydrocephalus shunt is a relatively simple device that can be implanted for long time periods into the subarachnoid region of the head; it serves to drain away excess cerebrospinal fluid that might otherwise accumulate and cause hydrocephalus which could result in severe brain damage and/or death. This fluid is normally drained into a blood vessel in a lower part of the body. These devices usually consist of a silicone rubber tube with a one-way valve to prevent backflow of body fluids when the patient is in a horizontal position. This device is also used for postoperative drainage following brain surgery and for the relief of pressure due to inoperable tumors. Over 10,000 of these shunts are used annually (1).

Craniofacial Applications. Over 200,000 craniofacial reconstructions are performed each year. Although these are sometimes done with autogenous bone or tissue material, polymeric materials are usually the materials of choice. The cartilaginous material of the nose, ear, or chin can be replaced with silicone rubber of varying degrees of cross-linking to control the rigidity of the material. Polyethylene and polytetrafluoroethylene have also been used in this application. The alveolar ridge has been reconstructed by using poly(methyl methacrylate) or a polytetrafluoroethylene:carbon composite as well as various ceramic materials. Cranial bone reconstruction has been done with poly(methyl methacrylate) or nylon instead of metal plates. The polymeric materials offer the advantages of lightness and nonconductance of heat (1, 11). A recent craniofacial reconstructive technique makes use of a Dacron mesh that is impregnated with a polyester polyurethane. After this mesh is draped over a suitable solid mold, the prosthesis is cured thermally to produce a material with sufficient strength and rigidity for cranial or mandibular replacements (32).

Dental Polymers. Every year nearly a half billion dental fillings are done, and over a million dentures are constructed. Most of the materials used in each of these cases are polymeric. In addition, over 300,000 dental implants are made each year with either ceramics or polymers (1). The majority of the dental fillings and dentures are made from various copolymers of methyl methacrylate with other acrylics, although some other polymers, such as polyurethanes, vinyl chloride-vinyl acetate-methacrylate copolymers, vulcanized rubber, and epoxies, have been used to some extent. One major problem is aesthetics—the prosthesis must look natural and not discolor (by photoinduction or staining) to any great extent.

Most dental-filling and denture applications now involve composite materials to increase the hardness. A typical acrylic filling (seldom used now) has a hardness of about 15 KHN, although the modern thermosetting acrylic composites are about 80 KHN. (For comparison, tooth enamel is about 320 KHN and dentin is about 70 KHN.) The dental filling and denture materials consist of a mixture of monomer and polymer that is polymerized in the tooth or on a mold to make the finished system. The thermosetting acrylics contain a multifunctional monomer (such as diethylene glycol dimethacrylate) in addition to the methyl methacrylate, and are polymerized by using a redox catalyst system or UV light (11, 33, 34).

The materials used to make the impressions for dentures, crowns, or bridges are almost always polymeric. Natural polymers, such as agar-agar (a polygalactan) or alginates, have been used in this application for many years, but more recently various polysulfides, polysilicones, and polyethers have received wide acceptance. These synthetic systems consist of the primary rubbery polymer and a hardener that reacts with some terminal group on the polymer (such as an epoxide unit) to cross-link the system into a durable elastomer that can retain the impression of the mouth fairly faithfully (33, 34).

Soft Tissue Replacement. Much of the body is made of soft tissue, which includes the muscles, fatty deposits, and connective tissue. A satisfactory soft tissue replacement must match the physical characteristics of these tissues closely and must be able to maintain these properties for an indefinite period. Annually, over 200,000 breast prostheses and/or augmentations, 200,000 facial plastic surgeries, and 35,000 hernia repairs are performed. Poly(dimethylsiloxane), in various degrees of cross-linking, is by far the most commonly used synthetic polymer for this application. Some limited use has been found for some synthetic rubbers and polyurethanes. Polyethylene is frequently used for hernia repair. Although foams, sponges, and textiles might seem to be a good choice for this application, they fail in essentially all cases due to a high amount of fibrous tissue ingrowth that converts the prosthesis into a fairly rigid, hard lump.

The best-known example of this application is the mammary prosthesis, or replacement breast, which consists of a silicone rubber gel in a bag of silicone that has been preformed to match the contours of a natural breast after implantation. The replacement testicle uses the same basic materials in a different shape. Neither device performs any physiological function (1, 12).

Skin Replacements. Over 100,000 people are hospitalized each year with severe injury to large portions of the skin caused by fires or accidents. Immediate treatment is necessary to prevent the loss of body fluids and to prevent gross infection. When the amount of damage is fairly small, the treatment consists mainly of skin transplants, or grafts, from another part of the body. When the damage is about 50% or greater, grafts are impossible. Transplants from other people (homografts) or animals (xenografts) are rejected fairly rapidly (1-3 weeks). The most promising polymeric replacement material appears to be a composite system consisting of an inner collagen-glycosaminoglycan layer (1.5 mm) covered with a thin layer (0.5 mm) of a silicone rubber. The lower layer can be "seeded" with basal epithelial cells that eventually form a new skin layer as the lower layer is resorbed by the body. This new skin layer is nearly identical with the original, natural material, except for the lack of hair and glands (12, 35).

Temporary replacement skins have also been studied by using velour fabrics (usually a nylon or Dacron) backed by a polymeric film (silicone or protein) (36), synthetic polypeptides (37), collagen (38), and dextran hydrogels (39). These replacement materials are only for short term use and could be classed as wound dressings. Although they do prevent gross infection and body fluid loss, they do not duplicate any of the other functions of the skin. A true artificial skin does not yet exist.

Orthopedic Applications

Casts, Braces, and Bone Repair. Most broken bones are repaired by
resetting the break externally and then restricting motion of the
limb or surrounding area with a cast, brace, or splint.
Traditionally casts have been made with plaster, but plastic
materials have been used advantageously because of the great weight
reduction. These plastic casts are usually made from polyethylene
or polypropylene that is molded to match the exact contours of the
patient and are usually lined with a flexible polyester or
polyurethane foam for comfort. Plastic casts can also be made
sufficiently rigid to replace a brace, and thus will result in
greater comfort and more effective healing. Plastic strips and rods
have been used to replace metal or wooden splints.

In some cases it is necessary to set a fracture or repair a
disease degenerated bone by internal fixation. These procedures are
usually done with bone plates made from stainless steel or alloys of
cobalt or titanium. Several problems arise, however. The metallic
bone plates often corrode in the physiological environment and
sometimes exhibit fatigue fracture. A greater problem, however, is
that much bone resorption occurs by the body when the stress on the
bone is reduced by the stronger metal plates. Some experimental
studies have been made of various plastic bone plates with varying
degrees of strength and flexibility, and these seem to give good
bone healing with less resorption (40). Some experimental work is
also in progress with biodegradable polymers of poly(lactic acid)
and related materials as resorbable bone plates. The goal is to
have the polymer degrade and disappear as the bone heals; this
process would prevent bone resorption and enable complete recovery.

Joint Replacement. Frequently the joints in the human body must be
replaced because of disease or injury. Hundreds of designs have
been used in attempts to replace the wide variety of joints with
plastics, ceramics, and metals in many combinations. Most of these
attempts have had only limited success, but many joints can now be
replaced with a reasonably satisfactory prosthesis and thereby
restore much of the normal joint function. Essentially all of the
most successful replacement joints use a polymeric material.

Annually, over 400,000 finger joints are replaced with a
silicone rubber insert that consists of two triangular rods attached
to a concave hinge. The rods are inserted directly into the finger
bones and are usually held in place only by the tendons and
ligaments of the hand. This extremely simple prosthesis exists in
several designs and can survive over 10 million flexings. The
device does not promote bone resorption or bone damage, and the
failure rate is nearly zero. Recently poly(1,4-hexadiene) has been
used in experimental finger joint prostheses that also show
exceptional durability (12). Most other joint replacements are more
complex and have a much higher incidence of failure.

The natural hip joint, like the shoulder joint, consists of a
ball-and-socket connection. Many combinations of materials have
been tried for this and other joints. In general metal-metal joints
are unsatisfactory because of high corrosion and friction that
result in joint immobilization. All ceramic joints often fracture
too readily, and all plastic joints usually lack strength in at

least one portion of the joint (femoral section in the hip joint). The most successful joints have involved combinations of plastics, as sockets, pads or runners, and metals, as balls or rockers. The major type of hip prosthesis was devised by Charnley and consists of an acetabular socket made from high molecular weight, high density polyethylene (HDPE) and a metal femoral ball and shaft made from stainless steel or Vitallium alloy. Each portion of the prosthetic joint is placed into excavated regions in the acetabulum or femor and anchored in place with a bone cement that consists primarily of poly(methyl methacrylate) (41). Several plastics have been tried for the acetabular socket, including polytetrafluoroethylene, polyamides, polyesters, and acrylics, but none have worked as well as HDPE.

Although this hip prosthesis has shown good success, most implants have been for less active, older patients. As the hip wears, debris from the metal, plastic, and/or cement is scattered into the tissue surrounding the joint and causes irritation. In addition, the joint then has more "play" in it and deteriorates faster, thereby causing more irritation and a lack of steadyness in the joint action. The prosthesis is now being used in younger, more active people, but it is uncertain how long these will last under these more stressful conditions.

Knee joint prostheses are implanted into about 100,000 people each year. The knee is essentially a hinge joint, as is the elbow, but it is more difficult to design than the hip joint, and failure rates are much higher (often 20+%). The design problems arise because the tibia portion of the prosthesis must bear nearly all the weight of the body at an angle that arises from the geometry of the skeleton. This situation often causes bone fragmentation and subsequent loosening of the prosthesis, and wear disintegration is accelerated. Some of the earliest knee prostheses, such as the original Walldius prosthesis, were completely made of plastics (acrylics), but these wore poorly; the modern Walldius prosthesis is an articulated metal hinge made from a cobalt-chromium-molybdenum alloy.

Plastic-metal combinations are now used in most knee prostheses. The knee prosthesis and numerous variations of this style utilize metal runners attached to the bottom of the femur that ride on HDPE tracks that are cemented on the tibia. The two parts of this prosthesis are each cemented into the bones with poly(methyl methacrylate) bone cement and are maintained in the proper relative positions to each other largely through the cartilage, tendons, and ligaments of the knee region. The spherocentric knee prosthesis is a more complicated device that involves a ball-in-socket arrangement similar to the Charney hip prosthesis. This device also utilizes a metal ball in a HDPE socket (1, 42).

Knee-cap (patella) replacement has been done with a variety of materials, but most replacements since the early 1960s have used poly(dimethylsiloxane).

As noted earlier, many joint prostheses are held in place by a polymeric cement. This cement is usually a pasty mixture of a methyl methacrylate copolymer and monomeric methyl methacrylate with an added redox initiating system. The paste is pressed into the excavated bone region and polymerized in situ to produce a polymeric mass that then holds the prosthesis by mechanical entrapment. This polymerization is exothermic and thus causes necrosis (death) of much surrounding tissue. Although monomeric methyl methacrylate is relatively cytotoxic, the cement is considered, by orthopedic surgeons, to be the best available cementing system.

Artificial Limbs. Replacement of an entire limb is sometimes necessary. To a limited extent, these prostheses can be made with some level of function. The myoelectric signals from other intact body muscles, properly amplified, can signal and power the movement of the prosthetic devices. Generally, these devices contain a central metal shaft or tube and are covered with some plastic material to resemble the shape and color of the limb being replaced. Because the prosthesis has little direct contact with the body tissues, biocompatibility problems are minimal, but aesthetics and function do dictate the choice of materials. Low weight, light stability, and resistance to dirt and/or stain pickup are essential. Foamed plastics are often used because they are light and help simulate the feel of natural tissue, but an outer continuous layer is always used in these cases to prevent penetration of grease, moisture, etc.

Cardiovascular Applications

Heart Valves and Pacemakers. Pacemakers, which regulate the heart beat by electrical stimulation, have been used on humans since 1952, and implantable models have been used since 1958. The wires and electrodes are usually plastic coated for purposes of insulation, and the entire device is usually embedded in a plastic for protection from the body fluids. Over 60,000 of these pacemakers are placed in people each year.

Over 25,000 replacement heart valves are implanted each year. Of these, about 59% are aortic valves, 41% are mitral valves, and less than 0.5% are either pulmonary or tricuspid valves. Many different designs have been developed for replacement heart valves utilizing a ball in a cage, discs, or leaflets (12, 43, 44). Some of these prostheses are made from natural polymeric materials such as porcine heart valves or human dura mater that have been treated with glycerol or glutaraldehyde to reduce biological activity. Synthetic polymers include silicone rubber, polytetrafluoroethylene, and Delrin in the form of balls, discs, floats, or leaflets that are almost always confined in a cage composed of several struts. These struts are usually made of metal, but polypropylene and some other polymers have also been used. Much of the metal surfaces are covered with Dacron or Teflon fabric that serves as a suturing site and also reduces the wear of the disc, ball, etc. The polymers used are not truly blood compatible, and long-term anticoagulant use is necessary.

Blood Vessel Replacement. In 1980, about 110,000 coronary bypass operations were performed in the United States. Many other blood vessel replacements and/or repairs are done each year because of disease, accident, or other trauma. Although many of these replacements are made with autogenous materials (e.g., saphenous veins for arterial bypass operations), in cases of advanced coronary disease such use is not feasible, and some other material is necessary. Although metal and glass tubes were explored, flexible plastic tubes have been used for over 30 years. The most widely used vascular prostheses are made from knitted or woven Dacron and an expanded (foamed) polytetrafluoroethylene (PTFE). Neither material is actually blood compatible. In the case of the Dacron mesh, which has been used surgically since 1951, a clot is formed in the pores of the mesh and thereby blocks further blood leakage. This clot is gradually replaced by a neointima, which is a natural tissue essentially the same as the insides of the natural blood vessels, and this tissue is blood compatible. Similarly, neointima grows on the expanded foam surface of the PTFE polymers and makes the prosthesis blood compatible. The growth of neointima does, however, reduce the effective diameter of the prosthesis, and sizes below about 6 mm usually fail because of blockage. Some success (about 70%) has been obtained, however, with the expanded PTFE tubes as small as 4 mm in diameter (45, 46, 47).

Most blood vessels in the body are smaller than 4 mm, and no satisfactory replacement currently exists. The most promising materials seem to be certain polyether polyurethane ureas (PEUU) (48, 49) and some hydrogels (50). Both materials show good blood compatibility, and patency rates (in dogs) in excess of 75% have been reported for the PEUU system (49). Human studies have not been made, to date, with either material, but the PEUU material is about the same as that used in the artificial heart.

Heart Assist Devices. Heart attacks that require hospitalization occur in more than 500,000 people each year. In some cases, a mechanical device is needed to relieve the damaged heart of part of its pumping burden. Several such devices have been developed, and all involve polymers.

The simplest heart assist device is the intraaortic balloon pump (IABP) which consists of two small PEUU balloons mounted on a hollow catheter that is about 30 cm long. The distal occluding balloon is 18 mm in diameter, and the proximal pumping balloon is 14 mm in diameter. The device is inserted into the aorta via the femoral artery and connected to a pump that then expands and contracts the balloons in rhythm with the heart beat. This prosthesis is considered to be the best heart assist device because it is simple in design and easy to insert. Improved circulation does occur in about three-fourths of the patients, but the mortality rate still remains high (65-90%) in cases of refractory cardiogenic shock. The IABP is used over 15,000 times annually (51, 52).

The IABP cannot maintain an adequate blood flow in cases of severe heart damage or disease, and the use of a left ventricular assist device (LVAD) is required. Several models have been developed for LVADs for either implanted or extracorporeal use, but the devices are always externally powered, usually by an air pump. Although many materials, including silicone rubbers, poly(vinyl chloride), natural rubber, and Dacron, have been examined for this type of temporary device, the major one being used at this time is the PEUU material (53, 54, 55). Biolized poly(1,4-hexadiene) also shows some promise (31). Most designs consist of a rigid plastic chamber that contains a sac or bag made from PEUU that can be filled and emptied of blood readily when pumped by an external air supply. The blood is usually withdrawn into the LVAD from the left atrium of the heart and returned into the lower aorta. Although the devices can probably function for longer than 10 months, on the basis of animal tests, the right ventricule often collapses after prolonged use. In general, these LVADs are used for periods of 2 weeks or less. Right ventricular assist devices (RVADs) have also been developed but are used less frequently.

Total Artificial Heart. Although heart transplants have been performed since the pioneering work of Christian Barnard in 1967, this procedure always requires a donor heart, which may not always be available. In addition, the body does tend to reject any implanted organs as undesired foreign material, and close matching of the tissues is difficult at the best. Although various immunosuppressant drugs can minimize this problem, the patient becomes more susceptible to infectious disease. A total artificial heart (TAH) would offer at least a partial answer to both of these problems.

Essentially the TAH is a combination of a LVAD and a RVAD. Such devices have been used in animal experiments for more than 20 years and have been used to maintain the life of calves for at least 268 days. (Normally the calf outgrows the TAH before the device fails.) Temporary human use of a TAH was made in 1969 and in 1981 to sustain a patient's life until a suitable heart transplant donor could be located. In 1982, the first permanent TAH device was implanted in a human, Barney Clark, at the University of Utah, and he survived for 112 days with this TAH as the sole means of blood pumping. These TAH devices are being studied actively in many places worldwide, and more implants have been done recently (53, 56).

Essentially all the TAH devices have used the PEUU materials (53, 54), but some other materials such as biolized poly(1,4-hexadiene) show promise (31). In the past, polysilicones, natural rubber, plasticized poly(vinyl chloride), and other polymers were tried but were generally considered too thrombogenic or too weak. Better materials most likely will be developed in the future because much research work is in progress in the general area of cardiovascular materials. The lack of patient mobility probably poses the greatest difficulty for the present TAH devices. No doubt portable pumps will be developed, and perhaps implantable, and electrically powered devices (using rechargeable batteries) will be developed. Polymers will most likely be used in these developments as well.

Artificial Organs

In the broadest sense of the term, replacement limbs, cardiovascular replacements, skin substitutes, and other prostheses discussed earlier could be classed as an artificial organ because, by definition, an organ is a specialized collection of cells of tissues that are adapted for some special, specific function. In this section only the artificial kidney, lung, pancreas, and liver will be considered in any detail. As will be seen, these prostheses are not true artificial organs because they do not, as a rule, perform more than a single function of the organs they replace, even though the organ may have several important functions. In addition these "artificial organs" are often extracorporeal devices and are usually designed for short-term use. In all cases, polymeric materials play an important role (12).

Kidney. The device called an artificial kidney is actually an external hemodialysis system, first developed in the early 1940s, that washes the blood and removes waste products from the body. Over 40,000 patients are maintained by this device each year in the United States, and there are over 100,000 people worldwide undergoing routine dialysis. In addition, many others are placed on the hemodialysis unit for short-term treatment.

Perhaps the main reason for the tacit acceptance of the hemodializer stems from the fact that the human body contains two kidneys and can, usually, function satisfactorily with only one. This makes kidney donation and transplantation a more realistic operation than heart and other transplantations; thus, kidney transplantations have been performed on a fairly common basis since the initial transplant in 1954. With the aid of careful tissue matching, preferably with a relative, and some immunosuppressant drugs, this transplantation operation has a high (85+%) success rate. In practice, however, the donor organs are not nearly as available as the demand, and 100,000+ people use the hemodializer routinely until a transplant becomes possible. In many cases, the hemodializer is used for decades, and secondary disorders, such as anemia, hemolysis, hypertension, and psychiatric problems, sometimes develop.

Generally the hemodializer is connected to cannulae (often made from silicone rubber) that are implanted permanently into the blood vessels in the patient's nondominant arm. The blood from the patient is then passed through plastic tubing into the dialysis unit, which consists of a semipermeable membrane immersed in an aqueous salt solution. The tubing used is usually silicone rubber or poly(vinyl chloride). Either the plastic has been heparin coated or the patient is administered some anticoagulant to prevent blood clotting. Many designs have been made for the membrane unit such as coils, plates, or hollow fibers; the hollow fiber design is usually considered the best. In most designs a countercurrent circulation of the blood and the dialyzing fluid improves efficiency. Cellulosic derivatives are the most common materials used for the membrane, although polyacrylonitrile is sometimes used in the hollow-fiber devices. Neither polymer is actually blood compatible; many experimental polymers have been explored to some extent (57-60).

Originally hemodialysis had to be performed in a hospital, but in recent years home units have been developed, which reduces the cost to a great extent. Much research has centered on wearable artificial kidneys (WAK) which enable the patient to have fairly great mobility compared to the conventional units (59, 60). Only a limited amount of research is being done, however, to develop an implantable device that could truly be termed an artificial kidney.

Lung. No implantable, artificial lung exists, and transplantation of this organ is relatively rare. Much work has been done, however, on extracorporeal oxygenators, which are used in over 100,000 operations each year. These oxygenators add fresh oxygen to the blood and permit removal of carbon dioxide. Several designs have been made for these devices such as various bubblers, rotating discs, film plates, and membranes. Each design has, however, certain disadvantages. The bubblers (in which air is bubbled through the blood) and rotating discs (which rotate a thin film of blood into the air) give extensive blood damage, but they do have a good oxygenation capacity. Although these devices are still used, they probably will be replaced in the future by the more efficient membrane oxygenators. In principle, the membrane and film-plate oxygenators operate in the same basic way. The blood flows on one side of a plastic membrane film, and oxygenation occurs by diffusion, which can be regulated by the pressure gradient. In the membrane devices, the polymer is arranged in the form of a coiled hollow sheet and is relatively compact. For this reason, the membrane oxygenators require only a small priming volume of blood, whereas the film and disc devices require a large priming volume. Blood damage is lower with the membrane devices than with the others, and the oxygenation capacity is nearly as high as in the bubbler devices.

Very few polymers are actually used as the membrane material with poly(dimethylsiloxane) being the most common. This material has good gas permeability and can be obtained in the form of thin, pinhole-free, strong sheets. As noted earlier, this polymer is not blood compatible, and heparin, or some other anticoagulant, is either added to the blood or attached to the polymer surface. Because these oxygenators are for short-term use, this approach would not pose any great problem. Polytetrafluoroethylene has been used frequently in membrane and film oxygenators, in spite of relatively low gas permeabilities, mainly because large, strong, pinhole-free sheets of material were available. Some experimental polymers that show promise for membrane devices include the polyalkylsulfones and ethylcellulose perfluorobutyrate copolymers (1, 12, 61).

Pancreas. The major function of the pancreas is to produce digestive enzymes and some hormones, including insulin, which is secreted by the beta cells in the islets of Langerhans and controls the level of glucose in the blood. Lack of insulin is one of the causes of diabetes. The artificial pancreas is actually an infusion pump that can deliver insulin to the bloodstream at a controlled rate. These devices are usually attached to the outside of the body, but have been implanted in some cases. The polymer used is usually silicone rubber. These infusion pumps do permit better insulin concentration control, and thereby better glucose level control, than the insulin injection approach. This control can be vastly improved with a glucose sensor coupled with a microprocessor to match variations in the glucose level more accurately and rapidly. At present these glucose sensors are too large for convenient implantation, but implantation should be achieved during the 1980s, and the resulting pump-sensor-microprocessor device should duplicate this important function of the pancreas. The device will not, however, be able to synthesize insulin (1, 62, 63).

An alternate approach has been to encapsulate living beta cells into microcapsules. These microcapsules can then produce insulin on demand, but are protected from the body's immune system by the polymeric membrane. These experimental microcapsules are usually made from polyamide by the interfacial polymerization technique (64).

Liver. The liver performs a wide variety of chemical reactions in the body and is the main locus of detoxification. Successful liver transplantation is somewhat rare, and no true artificial liver seems likely in the near future. The process of hemoperfusion, which is sometimes termed an artificial liver, can be used to supplement or relieve the normal liver functions for short time periods. In this technique, the patient's blood is passed through a column or bed of some sorbent material that removes toxic chemicals from the blood. This technique is often used in cases of drug overdose, poisoning, and acute hepatitis. The sorbent material can be charcoal, ion-exchange resins, immobilized hepatic material, or liver material enclosed in artificial cells (microcapsules, usually made from a polyamide). The column is usually a plastic material, and plastic tubing is used to direct the blood flow to and from the device (1, 57, 64).

Miscellaneous. Many other organs sometimes become diseased or defective, and some artificial device has been used to replace them. For example, the gastrointestinal (GI) tract has often been replaced, totally or partially, by some type of plastic tubing. Such a prosthesis does not perform the normal GI tract functions but merely connects existent, nondiseased tubular parts in the body. Many materials have been used such as polyamides, polyesters, polysilicones, and polyethylene. In a similar manner, various ducts have been replaced by plastic tubing. Finally, the bladder, trachea, ureter, and similar organs have been replaced by nonfunctional plastic tubing (1).

Although not always considered to be an organ replacement, contact lenses are obtained by about six million people each year to improve vision. Currently, two basic types of contact lenses are used: the hard lens and the soft lens. The hard contact lenses are almost always made from a poly(methyl methacrylate) copolymer, and these have a permanent, fixed size and shape. The soft lenses, on the other hand, are made from copolymers of 2-hydroxyethyl methacrylate (HEMA) or N-vinylpyrrolidone (N-VP), which are cross-linked during the polymerization-fabrication process. These soft lenses are hydrogels that imbibe a high amount of water (30-70% depending on the material and the extent of cross-linking) when in use. This high water content permits better oxygen diffusion to the cornea and also permits more ready removal of metabolic waste products such as lactic acid than would occur with the hard lenses. This effect enables the soft lenses to be worn for longer time periods than are possible with the hard lenses. Some flexible lenses are also made from polysilicones; these have good oxygen permeability but are hydrophobic (65, 66).

The lens of the eye sometimes becomes opaque (cataract formation), apparently due to cross-linking of its proteinaceous material, and must be removed to restore vision. Intraocular lens implants are normally made from poly(methyl methacrylate), although other materials, including hydrogels, have been tried. In over 73% of the cases, vision of 20/40 or better can be achieved by these implants (67). The cornea is also subject to cataract formation and

then needs replacement to restore vision. This replacement is
usually done with a corneal transplant, although poly(methyl
methacrylate) and other plastics have been used in many cases (11,
66).

Medication Applications

Until recently, medication methodology was limited to injections or
oral administration of a therapeutic agent or drug. Although some
drugs do show some selectivity in their activity, most therapeutic
agents pervade the entire body and cause undesired side reactions
such as nausea, dizziness, vomiting, loss of hair, and skin
discoloration. In some cases these side effects can reach toxic
proportions. In the administration of a drug there is a desirable
concentration range for maximum therapeutic effect. Below this
level little, if any, useful drug action occurs, and above this
range the toxic conditions prevail. Because most side effects of a
drug arise from high drug levels and/or actions on organs or tissues
other than the diseased targets, improvements in medication could be
achieved by causing the drug activity to be exerted solely on the
targeted diseased organs or tissues. This effect would reduce both
sources of the toxic side effects. Recently, several new
developments in drug medication have involved polymers in one form
or another (3, 68, 69). These developments are discussed in the
next sections.

Biomedical Polypeptides. Proteins and polypeptides are polymeric in
nature, and their biological activities are extraordinarily
specific. These polypeptides are involved in a vast array of
chemical reactions that are necessary for good health. Thousands of
derivatives and analogs of these naturally occurring polypeptides
have been synthesized and examined for medical activity. These
derivatives and analogs are also often highly specific in their
activity, although they frequently exhibit activities much different
from the parent compounds. It would be beyond the scope of this
survey to attempt to cover this field in any detail; thus, only a
few illustrative examples will be cited.
    One of the best-known examples of this class is interferon (a
glycoprotein), which has shown much promise as an antiviral agent
and in the treatment of cancer (71). Numerous polymers (mostly
polyelectrolytes) are effective in inducing the production of
interferon in animals (72). Other well-known bioactive polypeptides
are insulin, adrenocorticotropic hormone (ACTH), human growth
hormone (HGH), prolactin, follicle stimulating hormone (FSH), and
luteinizing hormone; most of these have been synthesized, and some
derivatives have been prepared (73). Some small bioactive
polypeptides are the nonapeptides bradykinin, vasopressin, and
oxytocin; bioactive derivatives have been made for many of these
also (74).
    Recently, there has been much interest in a group of natural
polypeptides, called the enkephalins and endorphins, that exhibit
great analgesic activity. Some of the derivatives of these
polypeptides show even greater analgesic activity with very low
addicting properties and have much potential as new drugs (75, 76).
Several other polypeptides appear to have physiological and

psychological effects (77). Some of these highly specific polymers probably will become available as specialized drugs, with resultant improvements in medication.

Immobilized Enzymes and Analogs. The specific nature of enzymatic activity needs no documentation, nor does the fact that enzymes are regularly synthesized and metabolized in the body. Many diseases arise from the lack of one or more specific enzymes, and these are termed enzyme-deficient diseases. These diseases include phenylketonuria, tyrosinosis, alkaptonuria, and histidinemia. Most of these afflictions lead to mental retardation. In addition to treating these enzyme-deficient diseases, enzymes can serve other therapeutic functions such as the treatment of some cancers, Fabry's disease, heart attacks, and even poison ivy (78). Unfortunately, these added enzymes are rapidly metabolized by the body and must be replaced often. Much research has been conducted on immobilizing enzymes and other bioactive polymers, and on protecting these labile agents from the body's enzymolytic agents while still preserving the therapeutic effects (79, 80, 81).

An alternate approach is to prepare synthetic polymers with enzymelike activity (82, 83, 84). In a similar manner, analogs of nucleic acids (85, 86) and polysaccharides (87) have also been prepared, and some show potential therapeutic value.

Drug Release Systems. Polymers can be utilized to regulate the rate of drug release into the body in a variety of ways (3, 4, 69, 70, 88, 89). One method of regulation involves osmotic pressure-driven pumps that allow body fluids to diffuse through a membrane into a hollow chamber containing a drug, in solid or solution form, that is then expelled from the device through a small orifice. The rate of release can be controlled by means of the copolymer membrane composition (often a vinyl acetate-ethylene copolymer) and the concentration gradient across this membrane. These osmotic pumps are small enough to be implanted readily and can give fairly constant release rates for a variety of bioactive agents. Alternate designs include implantable pumps that can be controlled by an external power source (90, 91). The artificial pancreas, which was discussed earlier, is also an example of such a controlled-delivery device.

A second type of polymeric system for controlled drug release involves the use of biodegradable polymer from which a drug is released as the polymer erodes away. Several types of polymers have been used in this method such as poly(DL-lactide), aliphatic polyesters, maleic anhydride copolymers, and hydrogels with degradable cross-links (92, 93). The simplest examples of this system would involve a polymeric coating that would dissolve away and release the drug into the body; such systems have been on the market for several years as time-release tablets and capsules or enteric-coated drugs.

Most research on controlled release polymeric systems has, however, centered on compositions in which a drug is either encapsulated in the center of a polymeric membrane (reservoir type) or dispersed throughout the polymer (monolithic type). The drug diffuses through the polymeric material to the surface where it is released to the body fluids. Such systems have been used to give

controlled release of a wide variety of bioactive agents such as
gentamicin, tetracycline, inulin, insulin, serum albumin,
progestins, prostaglandins, scopolamine, steroidal hormones, and
antitumor agents. The polymeric materials used include silicones,
polyesters, polyamides, polyurethanes, vinyl copolymers, and
hydrogels (69, 89).

A related type of delivery is the release of a drug that is
attached to or incorporated into the polymer backbone chain. These
are discussed in the next section.

<u>Polymeric Drugs</u>. A polymeric drug can be defined as a material that
contains a therapeutic (drug) unit attached to the backbone chain as
a terminal or pendant group, incorporates a therapeutic unit into
the backbone chain, or is directly biologically active without the
presence of a specific attached therapeutic unit. Such systems are
under intense study in many laboratories, and hundreds of known drug
agents have been attached to or incorporated into polymer backbone
chains. These systems can operate either as a source for the
controlled release of the drug (essentially a monolithic-type
system), or they can be directly biologically active as a polymer;
both types of systems are known (3, 4, 68, 69, 70, 94).

Major examples of biologically active synthetic polymers are the
divinyl ether-maleic anhydride cyclic alternating copolymer (pyran
copolymer) (95, 96), carboxylic acid polymers and copolymers (97,
98), and the vinyl analogs of nucleic acids (85, 86, 99). These
polymers usually do not resemble any low molecular weight drug in
their repeat unit structures, but do exhibit powerful activities
against a variety of tumors, viruses, and bacteria. Some
organometallic polymers may also act as direct polymeric drugs, but
others do function as a release system for the attached or
incorporated drug (100).

Many examples have been synthesized in which a known drug agent
is attached to or incorporated into a polymeric chain, and these
have been reviewed (3, 68-70, 101-104). It is not always obvious
whether these systems operate directly or by the release of the drug
unit, but both types of behavior appear to occur. Direct activity
may involve endocytosis of the polymer molecule (102, 103), reaction
at cell membranes, or special transport mechanisms of the
macromolecule through the cell membranes. Although many homopolymer
drug systems are insoluble in the body fluids, it is possible to
synthesize copolymers that contain a solubilizing unit in addition
to the drug unit in the polymer. In addition, it may be possible to
attach some other unit to the polymer that might direct or guide the
entire molecule to a specific diseased tissue or organ in the body
(68, 102-106). These multifunctional polymeric drugs could be very
specific and effective in their activity and might eliminate most of
the toxic side effects that occur with many drugs.

Polymeric drugs are being studied for action against many
different types of disease such as cancer (97, 106-108).

Literature Cited

1. Gebelein, C. G. "Prosthetic and Biomedical Devices" in "Kirk-Othmer Encyclopedia of Chemical Technology," 3rd ed.; Wiley: New York, 1982; Vol. 19, pp. 275-313.
2. "Biomaterials: Interfacial Phenomena and Applications"; Cooper, S. L.; Peppas, N. A., Eds.; ACS SYMPOSIUM SERIES 199, American Chemical Society: Washington, D. C., 1982.
3. "Biomedical and Dental Applications of Polymers"; Gebelein, C. G.; Koblitz, F. F., Eds.; Plenum: New York, 1981.
4. "Biomedical Polymers: Polymeric Materials and Pharmaceuticals for Biomedical Use"; Goldberg, E. P.; Nakajima, A., Eds.; Academic: New York, 1980.
5. "Hydrogels for Medical and Related Applications"; Andrade, J. D., Ed.; ACS SYMPOSIUM SERIES 31, American Chemical Society: Washington, D. C., 1976.
6. Kronenthal, R. L.; Oser, Z.; Martin, E. "Polymers in Medicine and Surgery"; Plenum: New York, 1975.
7. "Biomedical Applications of Polymers"; Gregor, H. P., Ed.; Plenum: New York, 1975.
8. Bruck, S. D. "Blood Compatible Synthetic Polymers, An Introduction"; Charles C. Thomas: Springfield, Ill., 1974.
9. Boretos, J. W. "Concise Guide to Biomedical Polymers"; Charles C. Thomas: Springfield, Ill., 1973.
10. Bloch, B.; Hastings, G. W. "Plastic Materials in Surgery", 2nd ed.; Charles C. Thomas: Springfield, Ill., 1972.
11. Lee, H.; Neville, K. "Handbook of Biomedical Plastics"; Pasadena Technology Press: Pasadena, Cal., 1971.
12. "Polymeric Materials and Artificial Organs"; Gebelein, C. G., Ed.; ACS SYMPOSIUM SERIES No. 256; American Chemical Society: Washington, D. C., 1984.
13. Hench, L. L.; Ethridge, E. C. "Biomaterials, An Interfacial Approach"; Academic: New York, 1982.
14. Black, J. "Biological Performance of Materials. Fundamentals of Biocompatibility"; Dekker: New York, 1981.
15. Bruck, S. D. "Properties of Biomaterials in the Physiological Environment"; CRC Press: Boca Raton, Fla., 1980.
16. Leininger, R. I. "Plastics in Surgical Implants"; American Society for Testing Materials: Philadelphia, 1965.
17. Postlethwait, R. W. Ann. Surgery 1970, 171, 892.
18. Lyman, D. J. Ann. N. Y. Acad. Sci. 1968, 146, 30.
19. Halpern, B. D. Ann. N. Y. Acad. Sci. 1968, 146, 193.
20. Leininger, R. I.; Mirkovitch, V.; Beck, R. E.; Andrus, P. G.; Kolff, W. J. Trans. Am. Soc. Artif. Intern. Organs 1964, 10, 237.
21. Anderson, J. M. in Ref. 3 1981, 11.
22. Habel, M. B. Biomater. Med. Devices, Artif. Organs 1979, 7(2), 229.
23. Homsy, C. A.; Ansevin, K. D.; O'Bannon, W.; Thompson, S. A.; Hodge, R.; Estrella, M. E. J. Macromol. Sci Chem. 1970, A4 (3), 615.
24. Guess, W. L.; Rosenbluth, S. A.; Schmidt, B.; Autian, J. J. Pharm. Sci. 1965, 54, 1545.

25. Oser, Z.; Abodeely, R. A.; McGunnigle, R. G. Int. J. Polym. Mater. 1977, 5, 177.
26. Autian, J. Artificial Organs 1977, 1(1), 53.
27. Lyman, D. J.; Knutson, K. in Ref. 4 1980, 1.
28. Leininger, R. I. in Ref. 3 1981, 99.
29. Baier, R. E. Ann. N. Y. Acad. Sci. 1977, 283, 17.
30. Bagnell, R. Chem. Brit. 1978, 14, 598.
31. Murabayashi, S.; Nose, Y. in Ref. 3 1981, 111.
32. Habel, M. B.; Leake, D. L.; Maniscalo, J. E.; Kim, J. Plast. Reconstr. Surg. 1978, 61, 394.
33. Glenn, J. F. in Ref. 3 1981, 317.
34. Halpern, B. D.; Karo, W. in Ref. 3 1981, 337.
35. Dagalakis, N.; Flink, J.; Stasikelis, P.; Burke, J. F.; Yannas, I. V. J. Biomed. Mater. Res. 1980, 14, 511.
36. Spira, M.; Fissette, J.; Hall, C. W.; Hardy, S. B.; Gerow, F. J. J. Biomed. Mater. Res. 1969, 3, 213.
37. May, P. D. in "Biomaterials"; Bement, A. L., Jr., Ed.; University of Washington Press: Seattle, 1971; p. 257.
38. Chvapil, M. J. Biomed. Mater. Res. 1977, 11, 721.
39. Wang, P. W.; Samji, N. A. in Ref. 3 1981, 29.
40. Bradley, G. W.; McKenna, G. B.; Dunn, H. K.; Daniels, A. V.; Statton, W. O. J. Bone & Joint Surgery 1979, 61-A, 866.
41. Charnley, J. Plastics & Rubber 1976, 1(2), 59.
42. Sonstegard, D. A.; Matthews, L. S.; Kaufer, H. Sci. Am. 1978, 238(1), 44.
43. Myers, G. H.; Parsonnet, V. "Engineering in the Heart and Blood Vessels"; Wiley-Interscience: New York, 1969.
44. Lefrak, E. A.; Starr, A. "Cardiac Valve Prostheses"; Appleton-Century-Crofts: New York, 1979.
45. Bricker, D. L.; Beall, A. C., Jr.; DeBakey, M. E. Chest 1970, 58, 566.
46. Vaughan, G. D.; Mattos, K. L.; Feliciano, D. V.; Beall, A. C., Jr.; DeBakey, M. E. J. Trauma 1979, 19, 403.
47. Raithel, D.; Groitl, H. World J. Surgery 1980, 4, 223.
48. Lyman, D. J. Angew. Chem. Internat. Ed. 1974, 13, 108.
49. Lyman, D. J.; Seifert, K. B.; Knowlton, H.; Albo, D., Jr. in Ref. 3 1981, 163.
50. Ratner, B. D.; Hoffman, A. S.; Whiffen, J. D. J. Bioeng. 1978, 2, 313.
51. Kantrowitz, A.; Krakauer, J. S.; Zorzi, G.; Rubenfire, M.; Freed, P. S.; Phillips, S.; Lipsius, M.; Titone, C.; Cascade, P.; Jaron, D. Transplant. Proc. 1971, 3, 1459.
52. Bregman, D.; Nichols, A. B.; Weiss, M. B.; Powers, E. R.; Martin, E. C.; Casarella, W. J. Am. J. Cardiol. 1980, 46, 261.
53. Akutsu, T.; Yamamoto, N.; Serrato, M. A.; Denning, J.; Drummond, M. A. in Ref. 3 1981, 119.
54. Eskin, S. G.; Navarro, L. T.; Sybers, H. B.; O'Bannon, W.; DeBakey, M. E. in Ref. 3 1981, 143.
55. Pae, W. E., Jr.; Pierce, W. S. in "Surgery for the Complications of Myo-Cardial Infarction"; Moran, J. M.; Michaelis, L. L., Eds.; Grune & Stratton: New York, 1980; p. 411.
56. Jarvik, R. K. Sci. Am. 1981, 244(1), 74.
57. Chang, T. M. S. "Artificial Kidney, Artificial Liver and Artificial Cells"; Plenum: New York, 1978.

58. Gutcho, M. "Artificial Kidney Systems"; Noyes Data Corp.: Park Ridge, N. J., 1970.
59. Kolff, W. J. Artificial Organs 1977, 1(1), 8.
60. Kolff, W. J. Organic Coatings & Appl. Polym. Sci. Proc. 1983, 48, 134.
61. Gray, D. N. in Ref. 3 1981, 21.
62. Santiago, J. V.; Clemens, A. H.; Clarke, W. L.; Kipnis, D. M. Diabetes 1979, 28(1), 71.
63. Sanders, H. J. Chem. Eng. News 1981, 59, 30.
64. Chang, T. M. S. *CHEMTECH* 1975, 5, 80.
65. "Symposium on the Flexible Lens"; Bitonte, J. L.; Keates, R. H.; Bitonte, A. J., Eds.; C. V. Mosby: St. Louis, 1972.
66. "Symposium on Contact Lenses"; Black, C. J.; Buxton, J. N.; Gould, H. L.; Halberg, G. P.; Kaufman, H. E.; Mackie, I. A.; Rosenthal, P.; Sampson, W. G., Eds.; C. V. Mosby: St. Louis, 1973.
67. Langston, R. H. S. Artif. Organs 1978, 2(1), 55.
68. Gebelein, C. G. Polymer News 1978, 4, 163.
69. "Biological Activities of Polymers"; Carraher, C. E., Jr.; Gebelein, C. G., Eds.; ACS SYMPOSIUM SERIES 186, American Chemical Society: Washington, D.C., 1982.
70. "Polymeric Drugs"; Donaruma, L. G.; Vogl, O., Eds.; Academic: New York, 1978.
71. Marx, J. L. Science 1979, 204, 1183,1293.
72. Levy, H. B. in Ref. 70 1978, 305.
73. Ramachandran, J. in Ref. 69 1982, 119.
74. Walter, R.; Smith, C. W.; Roy, J.; Formento, A. J. Med. Chem. 1976, 19, 822.
75. Snyder, S. H. Chem. Eng. News 1977, 55, 26.
76. O'Sullivan, D. A. Chem. Eng. News 1976, 54, 26.
77. Gurin, J. Science 80 1980, (1), 28.
78. Cooney, D. A.; Stergis, G.; Jayaram, H. N. "Enzymes, Therapeutic" in "Kirk-Othmer Encyclopedia of Chemical Technology", 3rd ed.; Wiley: New York, 1980; Vol. 9, p. 225.
79. "Biomedical Applications of Immobilized Enzymes and Proteins"; Chang, T. M. S., Ed.; Plenum: New York, 1977.
80. "Immobilized Enzymes, Antigens, Antibodies, and Peptides"; Weetall, H. H., Ed.; Dekker: New York, 1975.
81. Manecke, G.; Schlunsen, J. in Ref. 70 1978, 39.
82. Lindsey, A. L. "Reviews in Macromolecular Chemistry"; Butler, G. B.; O'Driscoll, K. F., Eds.; Dekker: New York, 1970; p.1.
83. Imanishi, Y. J. Polym. Sci., Macromol. Revs. 1979, 14, 1.
84. Pavlisko, J. A.; Overberger, C. G. in Ref. 3 1981, 257.
85. Pitha, J. in Ref. 3 1981, 203.
86. Takemoto, K. in Ref. 70 1978, 103.
87. Schuerch, C. Adv. Polym. Sci. 1972, 10, 173.
88. Chien, Y. W. "Novel Drug Delivery Systems"; Dekker: New York, 1982.
89. "Controlled Release of Bioactive Materials"; Baker, R., Ed.; Academic: New York, 1980.
90. Zaffaroni, A. in Ref. 3 1981, 293.
91. Theeuwes, F.; Eckenhoff, B. in Ref. 89 1980, 61.
92. Mason, N. S.; Miles, C. S.; Sparks, R. E. in Ref. 3 1981, 279.
93. Heller, J.; Baker, R. W. in Ref. 89 1980, 1.

94. Donaruma, L. G.; Ottenbrite, R. M.; Vogl, O. "Anionic Polymeric
     Drugs"; Wiley: New York, 1980.
95. Butler, G. B. in Ref. 94 1980, 49.
96. Freeman, W. J.; Breslow, D. S. in Ref. 69 1982, 243.
97. Ottenbrite, R. M. in Ref. 69 1982, 205.
98. Ottenbrite, R. M. in Ref. 94 1980, 21.
99. Pitha, J. in Ref. 94 1980, 277.
100. Carraher, C. E., Jr. in Ref. 3 1981, 215.
101. Samour, C. M. *CHEMTECH* 1978, 8, 494.
102. Batz, H. G. Adv. Polym. Sci. 1977, 23, 25.
103. Ringsdorf, H. J. Polym. Sci. 1975, Symp. 51, 135.
104. Donaruma, L. G. Prog. Polym. Sci. 1974, 4, 1.
105. Gebelein, C. G.; Morgan, R. M.; Glowacky, R.; Baig, W. in Ref.
     3 1981, 191.
106. Goldberg, E. P. in Ref. 70 1978, 239.
107. Gebelein, C. G. in Ref. 69 1982, 193.
108. Hodnett, E. M. Polym. News 1983, 8, 323.

Resins for Aerospace

CLAYTON A. MAY

Arroyo Research and Consulting Corporation, 2661 Beach Road, H-67, Watsonville, CA 95076

Epoxy Resins
Phenolic Resins
Polyurethanes
High-Temperature Resins
Adhesives
Sealants
Thermoset Processing Science and Control

In 1979 at the American Chemical Society/Chemical Society of Japan Chemical Congress in Honolulu, the Division of Organic Coatings and Plastics Chemistry sponsored a symposium entitled "Resins for Aerospace." This was an acknowledgment and tribute to the fact that the aerospace industry is a heavy contributor to the use and understanding of thermoset resin systems. The papers from this symposium later appeared as a 35-chapter book (1). Topics ranged from the design and testing of urethane launch tube seals to a search for readily removable coatings for electronic applications.

In the aerospace industry, resinous polymers encompass a wide variety of hardware applications for aircraft, missiles, and space structures. In aircraft, resins are used as a matrix material for primary (flight-dependent) and secondary fiber-reinforced composite (FRC) structures, adhesives for the bonding of metal and composite hardware components, electronic circuit board materials, sealants, and radomes. Missile applications include equipment sections, motor cases, nose cones, carbon-carbon composites for engine nozzles, adhesive bonding, and electronics. As the exploration of outer space intensifies, applications will become even more exotic. FRC will be used to construct telescopes, antennas, satellites, and eventually housing and other platform structures where special properties such as weight, stiffness, and dimensional stability are important.

There is little question that FRC will be the largest user of resins in aerospace. As illustrated by Figure 1, all composite aircraft will soon be available on a commercial scale. Approxi-

Figure 1.   The Lear Fan "all composite" aircraft (photo courtesy of Lear Fan Corp.).

mately 95% of the airframe structure of the Lear Fan is made from
composite materials, primarily graphite/epoxy. This aircraft is
currently undergoing Federal Aviation Administration (FAA)
certification tests and should obtain approval in the near future.
Beech Aircraft has recently announced plans to build an "all
graphite" aircraft.

Although the consumption of these materials in aerospace is but
a small fraction of the resinous polymers used in industry, the
impact of the needs and applications is large. Virtually all modern
day structural adhesives have aerospace origins. The need for
elevated temperature performance resulted in polymers useful for
such diverse applications as electrical insulation and brake
linings. The search for coatings to resist rain errosion and UV
light contributed heavily to the technology of the polyurethanes.

Millions of dollars in hardware and human lives will soon be
dependent on aerospace structures fabricated from structural
adhesives and fiber-reinforced composites. This situation has
dictated a more scientific understanding of structural polymer
characterization and processing dynamics. Chemical variations in
starting raw material formulations can no longer be tolerated for
fear of unforeseen long-term degradation effects. Cure chemistry
must be understood and documented as proof of properly processed
assemblies. These needs have led to major advances in chromatography, thermal analysis, and spectroscopy of thermoset resin
systems and have opened the door toward a better understanding of
the scientific significance of cure kinetics and liquid and solid
state rheology.

It is the purpose of this chapter to discuss the types and uses
of resins for aerospace and also to document aerospace contributions
to the science and understanding of structural polymers.
Thermoplastics will not be a part of this discussion. They do have
aerospace applications, most notably, in the interior furnishings of
commercial aircraft. However, it is the thermoset resins that have
been the major contributor to aerospace hardware technology.

This chapter will deal with the chemistry and applications of
epoxies, phenolics, urethanes, and a variety of current vogue high-
temperature polymers. Applications in fiber-reinforced plastics
will be discussed in the individual sections on resin chemistry
where appropriate. Separate sections will deal with adhesives and
sealants. Adhesives are most important because, as early history
demonstrates, they led the way to the application of resins in
aerospace. A section is also included on silicone and polysulfide
sealants. Although these materials are elastomers rather than
resins, no discussion of aerospace polymers would be complete
without some mention. Some major thermosetting polymers have been
omitted from this review. Among these are the unsaturated
polyesters, melamines, ureas, and the vinyl esters. Although these
products do find their way into aerospace applications, the uses are
so small that a detailed discussion is not warranted.

Epoxy Resins

Of all the resins used by the aerospace industry, epoxy resins have,
by far, gained the widest acceptance. They offer a versatility that
is unattainable by any of the other materials that will be

discussed. As raw materials, they range in viscosity from water-thin liquids to high molecular weight solids. As such they are adaptable to a wide variety of manufacturing processes. For example, in the manufacture of fiber-reinforced plastic (FRP) structures, low viscosity would be required for a process such as filament winding or pultrusion, whereas a solid form of the resin would be prerequisite for the dry lay-up processing of laminates for electronic applications.

Cross-linking of these polymers is accomplished by combinations of resins and curing agents. The broad range of materials available for these purposes affords a wide variety of curing conditions and structural properties. Cures range from very short periods of time at room temperature to much longer periods at elevated temperatures for resin-curing agent combinations that are capable of elevated-temperature performance. When cured, polymer properties range from soft, flexible materials to hard, tough, chemical-resistant, and elevated temperature-resistant products. Characteristically, the epoxy resins shrink less during cure than most other thermoset resins, and no volatile by-products are generated during the cure. Epoxies also provide excellent electrical insulating characteristics, outstanding chemical and solvent resistance, and, above all, excellent adhesion. The most common aerospace applications of epoxy resins are in adhesives and FRP; however, they also find use in surface coatings and a variety of electrical and electronic applications.

The most commonly used epoxy resin is the diglycidyl ether of bisphenol A and its higher homologs. The pure diglycidyl ether (n = 0) is a low-melting crystalline solid. However, the commercial grades start as a liquid with a viscosity of approximately 40 P and range upward to values of n equal to 18 or more for some of the higher molecular weight coating grades.

There are three general chemical reactions basic to the curing of epoxy resins: the reaction of epoxides with amines, their reactions with carboxylic acids or anhydrides, and catalytic homopolymerizations. The most common curing agents are the amines wherein each of the amino hydrogens react with an oxirane ring. Depending on the number of amino hydrogens per molecule and the supporting chemical structure, a wide variety of properties and cures are available. Aliphatic amines afford rapid cures at low to modest temperatures, but only moderate elevated-temperature performance to 125 °C maximum. The aromatic amines are useful to around 175 °C and have excellent chemical and solvent resistance but require higher cure temperatures. Common practice is to cure the resin system at the desired use temperature. The cycloaliphatic amines offer an interesting compromise between the other two amines. They have useful properties in the 150 °C region but can be cured under relatively mild conditions, around 60 °C.

$$RNH_2 + CH_2\text{-}CH\text{-}R' \longrightarrow RNHCH_2\text{-}CH\text{-}R' \xrightarrow{CH_2\text{-}CH\text{-}R'} RN \begin{cases} CH_2\text{-}CH\text{-}R' \\ CH_2\text{-}CH\text{-}R' \end{cases}$$

Carboxylic acid and anhydride curing agents are used to a lesser extent in aerospace applications as compared to the amines which have mechanical properties and cure conditions that can be tailored to a wider variety of specific applications. Anhydrides tend to be somewhat brittle but offer useful service as high as 250 °C with novolac-type epoxy resins. In addition, the aliphatic dicarboxylic acid anhydrides give tough and sometimes flexible properties which are useful in encapsulation applications.

The curing reactions are highly complex, and the mechanisms have been the subject of numerous articles in the technical literature. The earliest mechanism, as proposed by Fisch and Hofmann (2), involved the reaction of the anhydride with an alcohol, either hydroxyl functionality of the resin itself or some hydroxyl-containing impurity such as moisture. The carboxyl group of the half-acid ester then reacts with an epoxide to form a second ester linkage and generates a new hydroxyl group as the reaction continues. When this mechanism prevails, optimum properties result at an anhydride:epoxide molar ratio of 0.85:1. This result indicates that a secondary reaction occurs during the cure. The most probable is an acid-catalyzed epoxide homopolymerization (see below).

$$R\begin{cases} C=O \\ C=O \end{cases}O + R'OH \longrightarrow R\begin{cases} C\text{-}OR' \\ C\text{-}OH \end{cases} + R''\text{-}CH\text{-}CH_2 \longrightarrow R\begin{cases} C\text{-}OR' \\ C\text{-}O\text{-}CH_2\text{-}CH\,R'' \end{cases} \xrightarrow{R\,(CO)_2\,O} etc.$$

It was later found that Lewis acids and bases were catalysts for the epoxide-anhydride reaction. The preferred stoichiometry with respect to anhydride and epoxy proved to be 1:1. This relationship led Fischer (3) to propose a mechanism wherein the first step was a reaction between the anhydride and the accelerator to form a carboxyl anion. This anion in turn reacts with an epoxide, and the reaction repeats itself.

$$R\begin{cases} C=O \\ C=O \end{cases}O + R_3N: \quad R\begin{cases} C\text{-}NR_3 \\ C\text{-}O^- \end{cases} + R'\,C\text{-}C \longrightarrow R\begin{cases} C\text{-}NR_3 \\ C\text{-}O\text{-}C\text{-}C\text{-}R' \\ \quad\quad O^- \end{cases} \xrightarrow{R\,(CO)_2O} etc.$$

The third curing reaction of importance to the aerospace industry is epoxide homopolymerization. The most prevalent is cationic polymerization induced by Lewis acids and may be illustrated as follows (4):

$$X^+Y^- + R\ \overset{O}{\overset{\diagup\ \diagdown}{CH-CH_2}} \quad Y^- \cdots X\ \overset{O^+}{\underset{\underset{R\ CH-CH_2}{\diagup\ \diagdown}}{}} \xrightarrow{+n\ R\overset{O}{\overset{\diagup\ \diagdown}{CH-CH_2}}} X\ (OCH_2CHR)_n\ \overset{O^+}{\underset{\underset{R\ CH-CH_2}{\diagup\ \diagdown}}{}} \cdots Y^-$$

Although this reaction does not lead to a high degree of polymerization with monoepoxides, the polyfunctionality of most epoxy resins leads to highly cross-linked structures and good thermal properties. The most important Lewis acid is boron trifluoride, most commonly used in the form of an amine salt which facilitates handling and controls the exothermic heat of reaction. Other compounds in this category include the halides of tin, aluminum, zinc, boron, silicon, iron, titanium, magnesium, and antimony. The fluoroborates of these metals have also been reported as catalysts. Lewis acids are also used to catalyze the amine-epoxide reaction as described in the following discussion.

The class of epoxy resins most commonly used in aerospace applications today are the glycidyl amines. Possessing good adhesive characteristics and a high degree of interlaminar shear, they find widespread use in FRP. The most popular is tetraglycidyl methylenedianiline (TGMDA) cured with diaminodiphenyl sulfone (DDS). The mechanisms of cure are highly complex. Because Lewis acid catalysts are many times included in the resin formulations, both amine-epoxide and homopolymerization reactions are important considerations when studying the cure chemistry. As many as nine different chemical reactions can be written to describe the cure. Morgan et al. ($\underline{5}$) have studied this system extensively and concluded that the presence of a BF_3 amine complex enhances the consumption of the epoxide group by homopolymerization during cure even when less than the stoichiometric amount of DDS is present in the resin system.

TETRAGLYCIDYLMETHYLENEDIANILINE

DIAMINODIPHENYLSULFONE

Epoxy resins are the most versatile product used in aerospace composite hardware. They are employed both alone and in combination with a variety of other resins to form a broad range of products. The only limitation to the number of available products rests in the ability of the formulator to meet the ever-increasing demands for improved structural properties. Greater detail on the use of epoxy resins for bonding can be found in the section on adhesives.

In addition to FRP and adhesives, a third important use of epoxy resins in aerospace is in surface coatings. Epoxy resin-polyamide combinations are used as primers under urethane top coats. The most general formulations involve the lower molecular weight solid grades of epoxy resins where n equals 2-4. The polyamide curing agents are the reaction product of an aliphatic amine such as diethylenetriamine with C_{18} dimerized or trimerized fatty acids.

Epoxy resins also find extensive use in the fabrication of printed circuit board laminates. These are of great importance to the aerospace industry. Most laminations are made by using dry lay-up techniques because flat surfaces are involved. Here again the lower weight, solid grades of resin are used. However, because long ambient temperature storing of the prepreg is desirable, latent curing agents such as dicyandiamide are most commonly employed. Epoxy novolacs are added to these formulations to improve elevated-temperature performance, and fire retardance is achieved by incorporating resins based on tetrabromobisphenol A.

Before leaving the subject of epoxy resins, toughening will be discussed briefly because it is an area of current research and development activity. Even though epoxy formulations are tougher than most high glass transition temperature (T_g) polymers, the aerospace community demands more. Normally, to toughen a material one adds a plasticizer. This addition, however, reduces the cross-link density of the system and lowers the T_g. Therefore, current attention has turned to the use of dispersions of small amounts of rubber or other elastomeric particles in the resin that arrest crack growth in a brittle matrix, thereby increasing the fracture toughness. In current technology this result is accomplished by incorporating relatively low molecular reactive liquid rubbers into the epoxy formulation (6, 7). Dynamic mechanical studies by Manzione and Gillham (8) have shown that by control of the rubber-epoxy compatibility and cure conditions a wide range of rubber particle morphologies can be obtained. These morphologies result in different stress-response mechanisms.

The elastomeric materials most commonly used for this purpose are carboxyl-terminated acrylonitrile-butadiene copolymers. The carboxyl groups react with the epoxy group to produce an epoxy-terminated rubber that promotes interfacial bonding in two-phase systems (9). By controlling the concentrations of M, X_1, X_2, Z, and U, a broad range of compatibilities can be achieved. Impact resistance and fracture toughness are increased with minimal sacrifice in molulus and elevated-temperature performance.

$$\underset{}{\overset{O}{\overset{\|}{HO-C-R}}}\left[(CH_2-CH=CH-CH_2)\overline{X_1}(CH_2-\underset{\underset{\overset{\|}{CH_2}}{CH}}{CH})\overline{X_2}(CH_2-\underset{C\equiv N}{CH})\overline{Z}(CH_2-\underset{\underset{OH}{C=O}}{CH})\overline{U}\right]R-\overset{O}{\overset{\|}{C}}-OH \quad M$$

Phenolic Resins

Phenolics are the oldest resins employed by the aerospace industry. Baekeland began his work on these materials around 1905, two years after the Wright brothers made their first airplane flight. By 1910 phenolic resins were being used commercially in laminates, moldings, and insulating varnishes. As discussed in the section on adhesive bonding, phenolic resins played an important early role as a resin for aerospace applications. They continue to be an important resin for adhesives, ablatives, and carbon-carbon composites. Their high char formation characteristics could well lead to a renewed importance in aerospace applications. Fire retardant resins, particularly in commercial aircraft, could well be a demanded prerequisite in the near future.

Both resole and novolac types of phenolic resin are used in aerospace applications. The resoles are generally made by using alkaline catalysts and an excess of formaldehyde. The resole structure is highly complex and involves both methylene and dimethyl ether bridges between the phenolic moieties. During cure both water and formaldehyde are evolved.

The novolac resins are prepared by using acidic catalysts and a deficiency of formaldehyde. Because this type of resin is less reactive, cross-linking is accomplished by the addition of a curing agent or catalyst. The most common is hexamethylenetetramine or "hexa." The curing agent serves as a latent source of formaldehyde. As in the case of the resoles, volatiles are emitted during the cure. The chemistry of the phenolic resin is old but complex and well documented in the literature (10).

As indicated earlier, in spite of a long history of association with the aerospace industry, the products find only limited use in specific applications. The evolution of volatiles during the cure complicates the formation of intricate shapes. When cured, the resins are highly cross-linked and expectedly brittle. They also oxidize readily in air at elevated temperatures. The main areas of aerospace application as indicated are adhesives, ablatives, and carbon–carbon composites.

In adhesives, because of their brittle nature, phenolics are generally formulated in add mixture with other polymers. They are used as the primary cross-linking agents for nitrile rubbers, poly(vinyl formal) and butyral resins, and epoxy resins.

The other main aerospace applications actually take advantage of the poor oxidation resistance. During the space race of the late 1950s and early 1960s, considerable research was conducted on the ablative characteristics of thermoset resins (11). Phenolics proved to be ideal materials providing one of the highest char yields and good short–term, elevated–temperature strength. High–density graphite fiber-reinforced laminates have proven to be an excellent ablative heat shield material (12). If the char formation is carried out at high pressures under carefully controlled conditions, excellent carbon–carbon composites result. These materials are used in reentry nose cones and rocket nozzles.

Polyurethanes

The polyurethanes, although not used in large volumes, have specific applications of value for aerospace. Accordingly, a brief discussion of the chemistry and the material applications is in order. There are three chemical reactions of importance: the reaction of hydroxyl groups with isocyanates to form urethanes (Reaction 1), the reaction of isocyanates with water to form amines (Reaction 2), and the reaction of amines with isocyanates to form ureas (Reaction 3).

$$R-N=C=O + R^1OH \longrightarrow R-\overset{H}{\underset{}{N}}-\overset{O}{\underset{}{C}}-OR^1 \tag{1}$$

Urethane

$$R-N=C=O + H_2O \longrightarrow R-NH_2+CO_2 \tag{2}$$

Amine

$$R-N=C=O + R^1NH_2 \longrightarrow R-\overset{H}{N}-\overset{O}{C}-\overset{H}{N}-R^1 \tag{3}$$

Urea

Although the polyurethanes form useful adhesive bonds, particularly between metals and elastomers, their use in the aerospace industry for bonding purposes is limited. Because polyurethanes tend to depolymerize above 120 °C and are subject to hydrolysis, and because aromatic urethanes autoxidize when exposed to thermal or UV light (13), epoxies are the preferred bonding material. Recently they were studied as launch seals for both land and sea missile launch tubes and were found to be superior to seals based on neoprene rubber (14). The chemical reaction for this application is proposed to be that between isocyanates and amines (Reaction 3).

The major aerospace use of isocyanate chemistry is in surface coatings. This use resulted from the discovery that aliphatic isocyanates were vastly superior to the conventional aromatic isocyanates in resistance to UV light. The preferred cross-linking agent is hexamethylene diisocyanate. Because this component is highly toxic, the formulations involve isocyanate-polyol precondensates applied over epoxy-polyamide primers. The precondensates generally employ an excess of the diisocyanate. Thus, because aircraft usually rely on ambient-temperature cures, Reactions 2 and 3 are of primary importance for coating applications.

High-Temperature Resins

No treatise on resins for aerospace would be complete without discussing high-temperature polymers. There has been a continuing major research effort in this area. Initial work was devoted to elevated-temperature performance and stability. Little or no attention was aimed at hardware fabrication. The result was a plethora of unique, almost impossible to handle, intractable polymers. Fortunately, it was quickly recognized that polymer cost, availability, and processibility were also important, and the necessary compromises began. Hergenorther and Johnston have recently published an excellent status report on high-temperature polymers (15).

The initial high-temperature resins of significance were the linear polyimides formed by the reaction of an aromatic dianhydride such as 3,3',4,4'-benzophenone tetracarboxylic dianhydride with bis(aminophenylene)diamines. These products processed poorly but did find a limited marketplace in the form of coatings, molding materials, and adhesives. Other thermoplastic high-temperature polymers of interest were the polybenzimidazoles made by the condensation of aromatic bis(o-diamines) with diphenyl aromatic dicarboxylates and the polyphenylquinoxalines from bis(o-diamines) and aromatic dibenzyls. The development of more processible thermoset high-temperature resins began when polyimides of lower molecular weight and improved flow were synthesized by end capping the polymer chain with monofunctional dicarboxylic anhydrides capable of providing cross-linking via vinyl polymerization. The most popular end-cap compounds were the anhydrides of maleic and Nadic acids.

Typical Polyimide Resin

The low molecular weight, improved processibility approach was developed further by the NASA-Lewis polymerization of monomeric reactants (PMR) concept (16). It is based on the fact that mixtures of p,p'-methylenedianiline with a selected concentration blend of the methyl or ethyl half esters of Nadic and benzophenone dianhydrides form a viscous mass yielding composite prepregs and adhesives with desirable tack and drape. Drape is a particularly important shop characteristic because it permits conformal movement of the prepregs on tool surfaces during the fabrication of the complex shapes required for aerospace applications. The main problem associated with this concept is that volatiles are evolved during the early processing stages because of the imidization reactions.

NASA-Langley provided a further improvement in processibility by replacing the methylenedianiline with a commercial aromatic diamine mixture that afforded even better handling characteristics (17). This type of product is formed by the reaction of formaldehyde with aniline as a precursor in isocyanate manufacture. Fluorine-containing thermoplastic polyimides and polyimides end capped with acetylene groups for cross-linking are also available as potential commercial high-temperature resins.

A current version of the low molecular weight concept is the use of bismaleimides made from aromatic diamines and/or mixtures thereof and maleic anhydride in admixture with other vinyl monomers. Several commercial formulations of this type are now available as the neat resins or as FRP prepregs. There is considerable current research in this area because of the favorable processibility. However, to date, FRP products of this type are outperformed by the PMR-type polyimides by about 50 °C at elevated temperatures.

The fabrication of high temperature performance composites is the area of most current interest. The favorite approaches are the PMR polyimides and bismaleimide/vinyl monomer combinations as matrix resins. One thermoplastic polyimide for adhesive bonding is commercially available, and research on the use of high-temperature resins for this purpose continues. Polyimide films and prepregs also are currently receiving considerable attention for the fabrication of printed circuit boards. Research in the area of high-temperature resin systems continues to be popular. The main deterrents to success remain: high-priced starting materials, limited availability, and difficult processibility.

Adhesives

If there is one area that can be singled out as the catalyst for the aerospace application of resins, it is adhesives. They have been used in the fabrication of aerospace hardware for over 60 years. Shortly after World War I, the Loughead brothers built an aircraft in Goleta, California, called the Sport. It was a bonded plywood veneer structure that employed a casein glue. The molds used to form the aerodynamic shape were made from concrete. This structure was followed shortly by a line of commercial aircraft called the Vega. This line was produced by the same process and served notice to the then fledgling aircraft industry that resins were to play an important role in the aerospace industry. Figure 2 shows this early "assembly line." During the 1920s the adhesive was changed to a phenolic resole that was cured by induction heating. This type of construction was still in use as late as 1938.

The next major advance in bonding technology was the vinyl-phenolic "redux" adhesive developed by de Bruyne (18) during World War II. It was used for the assembly of the De Havilland aircraft. Bonds were made by spreading a solution of a phenolic resole on the plywood parts to be bonded, sprinkling this coating with a powdered poly(vinyl formal) resin, and blowing off the excess powder. The solvent was allowed to evaporate, and the bonding process was completed by the application of heat and pressure. Although this technique sounds crude, consistent high-performance bonds resulted. The use of a high molecular weight thermoplastic to toughen a brittle thermoset was probably the first example of the commercial

Figure 2.   The Lockheed Vega production line, early 1920s. Note the
            similarity of these bonded wood structures to aircraft
            currently being made from aluminum (photo courtesy of
            Lockheed – California Company).

application of a polymer alloy and the start of structural adhesive
formulation technology.

During World War II, research in Germany led to the discovery of
the urethanes. Adhesives based on hydroxyl-containing polymers,
primarily synthetic hydroxyl-terminated polyesters, in combination
with isocyanates, were excellent for bonding rubber to metal
substrates. They were used to bond rubber tank treads to the metal
cogs of the drive chains. Simultaneously, epoxy resins were being
developed both in the United States and abroad (Switzerland). Thus,
during the late 1940s, a number of outstanding candidates for
structural bonding became available, and they are responsible for
many of the products in today's marketplace.

From these early efforts, epoxy resins emerged as the leading
candidate. They were tougher than the phenolics, and no volatiles
were evolved during the cure. Compared to the isocyanates they had
superior thermal stability and moisture resistance. The uses of
epoxy resins for adhesive bonding in aerospace applications are so
widespread that they would be impossible to describe in a
dissertation of this length. The resin-curing agent combinations
range into the hundreds. In addition, epoxies are used in
combination with other resins. Phenolic resins improve elevated-
temperature bond performance. The use of nylon adds peel strength
to the bond line and gives very high-strength bonds at ambient
temperatures. Vinyl resins are added to paste adhesives to improve
the green strength during bonding. Nitrile elastomers provide
toughness. The only limitation to the number of epoxy-based
adhesives available for the aerospace industry rests in the ability
of the formulator to meet the ever-increasing demands for improved
structural properties. Furthermore, the wide range of material
forms available to the formulator permits application of the
adhesives in the form of pastes, powders, and highly tacky to dry
tapes. As illustrated by the data in Table I, the choice of curing
agent alone can lead to uses covering a broad range of temperatures
(19).

Another noteworthy aerospace adhesive was developed during the
mid-1950s. It was a combination of a phenolic resole and an epoxy
resin that had a use temperature ranging up to 260 °C (500 °F) as
shown by the last entry in Table I. The formulation consisted of
the following (parts by weight (pbw)):

Solid diglycidyl ether of bisphenol A (n=2-3)	33
Phenolic resole (A-stage)	67
Aluminum powder	100
Dicyandiamide	6
Cu 8-hydroxyquinoline	1

The adhesive is manufactured in tape form by a hot-melt process. It
is a tacky solid at room temperature. The integrity is maintained
by using a finely woven glass fabric scrim as the carrier. This
process is an excellent example of the compromises required in the
technology of formulation. Some of the high-temperature performance
that is expected from the phenolic resole is sacrificed for the
improved bond strength and toughness afforded from the epoxy resin.
The filler is added to make the thermal coefficient of expansion of
the cured adhesive more metallic in nature. Dicyandiamide is the

Table I. Bond Strengths Expected From Major Curing Agent Types

Type of Curing Agent	Typical Cure Temp. °F (K)	Approximate tensile shear strength to expect on aluminum, lb/in.^{2} (kPa x 10^{-3})					
		-70 °F(216 K)	Room Temp.(297 K)	180 °F(355 K)	250 °F(394 K)	300 °F(422 K)	500 °F(533 K)
Polyamides, polysulfides	75 (297)	3000 (20.7)	4000 (27.6)	1000 (6.9)	—	—	—
Aliphatic amines	75 (297)	2000 (13.8)	3000 (20.7)	2500 (17.2)	—	—	—
	175–210 (353–372)	3000 (20.7)	3500 (24.1)	2500 (17.2)	500 (3.4)	—	—
Mixed aliphatic tertiary amines	175–210 (353–372)	3000 (20.7)	3500 (24.1)	3500 (24.1)	1000 (6.9)	—	—
Aromatic amines	275–325 (408–436)	3000 (20.7)	3500 (24.1)	4000 (27.6)	3500 (24.1)	2000 (13.8)	—
Insoluble latent amines	300–350 (422–450)	2500 (17.2)	3500 (24.1)	3500 (24.1)	3000 (20.7)	1500 (10.3)	—
Carboxylic acid anhydrides	200–300 (366–422)	2000 (13.8)	2500 (17.2)	3000 (20.7)	2500 (17.2)	1500 (10.3)	—
Phenolic resole resins	330–350 (439–450)	2500 (17.2)	2500 (17.2)	2000 (13.8)	2000 (13.8)	2000 (13.8)	1500 (10.3)

curing agent. It promotes cross-linking of the epoxy resin and coreaction with the resole. The copper compound was added to increase the elevated-temperature, bond-performance life of cured adhesive. The adhesive curing is accomplished by heating the bonded assembly for 1 h at 177 °C (350 °F).

If tougher adhesive bonds are required than are attainable from conventional epoxy resins, they can be modified. A typical example involves the carboxyl-terminated elastomer modifications discussed earlier. A modified base resin is first formed as follows (pbw):

Liquid diglycidyl ether of bisphenol A	100
Carboxyl-terminated nitrile rubber	10

The rubber and resin are coreacted by heating the mixture at around 100 °C in the presence of a Lewis base. Most commonly tertiary amines are used for this purpose. The morphology of the rubber dispersion is important to bond performance and is governed by the coreaction procedure. This material may be formulated into a useful aerospace paste adhesive by the following combination of materials (pbw):

Rubber-modified resin (above)	100
Melamine resin	2
Fumed silica	5
Dicyandiamide	8
3-(3,4-dichlorophenyl)-1,1-dimethylurea	1.5
Aluminum powder	50

Melamine acts as a flow-control agent for the resin. The fumed silica adds thixotrophy to the mixture which retards flow on vertical surfaces. Dicyandiamide in this case is a latent curing agent. Thus, the catalyzed adhesive exhibits long pot life on storage at ambient temperatures. The urea derivative is an accelerator for the cure. This formulation can be cured in the range of 110–120 °C whereas 177 °C is the normal curing temperature for dicyandiamide-based formulations.

Epoxies are not the only resins used in aerospace bonding. A common formulation is a nitrile/phenolic film adhesive. A typical formulation may consist of the following (pbw):

Nitrile rubber	100
Phenolic resin	100
Sulphur	2
Zinc oxide	5
Benzothiazyl disulfide	1.5
Carbon black	20

Here again is a formulation compromise between the brittle, high-temperature characteristics of the phenolic resin and the rubbery, low-temperature performance of the elastomer. Note also that the other ingredients are the type of material combination that is normally associated with nitrile elastomer vulcanization.

Representative of an ultra-high-temperature adhesive is the following polyimide-based formulation (pbw):

Linear condensation polyimide	100
Aluminum powder	80
Fumed silica	5
Arsenic thioarsenate ($AsAsS_4$)	2

Adhesives of this type are difficult to process because of their high melting and poor flow characteristics. They are normally applied from a solvent solution, in this case N-methylpyrrolidone. The solvent must be removed during the bonding operation to obtain maximum elevated-temperature performance. A technique commonly used is evaporation of the solvent by heating an open assembly prior to the final bonding operation. The final assembly is then accomplished through conventional heat and pressure applications. The arsenic thioarsenate is added as a stabilizer to retard elevated-temperature aging.

From this brief discussion it can be seen that the formulator of aerospace adhesives has an almost infinite arsenal of materials and combinations thereof by which useful products can be devised. A good summary of the many types of materials available is presented in Figure 3. As the data show, useful bond strengths can be obtained over a broad temperature range. Each different type of product displays its maximum strength in a different temperature region. This reflects the fact that, in addition to the inherent strength characteristics of a formulation, bond strength also depends on the glass transition temperature (T_g) of the basic resin system. Normally, an adhesive formulation gives optimum strength in the region of the T_g. It is in this region where thermal strains on a bond line are at a minimum (19).

Sealants

Polymeric sealants are important materials in the aerospace industry. Virtually every aircraft or space craft employs them in one form or another. The most important resins used for this purpose are the polysulfides and the polydimethyl siloxanes. Of the two materials, sealants based on the polysulfides are the most widely used.

The first step in the synthesis of the liquid polysulfides involves the reaction of bischloroethyl formal (Structure IV) and sodium polysulfide (Structure V) to form a di- and tripolysulfide mixed polymer (Structure VI). The resulting polymer is then

$$nClC_2H_4OCH_2OC_2H_4\,Cl + n\ Na_2S_{2.25} \longrightarrow (-C_2H_4OCH_2OC_2H_4S_{2.25})_n + 2\ n\ NaCl$$

$$\qquad\quad IV \qquad\qquad\qquad V \qquad\qquad\qquad VI$$

reacted with a splitting salt mixture of sodium sulfide and sodium acid sulfite to form a lower molecular weight polysulfide polymer (Structure VII). The concentration of the splitting salts is used to control the molecular weight of the liquid polysulfide polymer that is then formulated into the desired sealant.

$$HS \; \text{(} C_2H_4\text{-}O\text{-}CH_2\text{-}O\text{)}_n\text{-}C_2H_4\text{-}SS\text{)}_n\text{-}C_2H_4\text{-}O\text{-}CH_2\text{-}O\text{-}C_2H_4SH$$

VII

Sealants are used by the aircraft industry to form flexible adherent barriers that will withstand chemical attack. The major uses are seals around riveted joints, fileting on overlapping bonded surfaces, and sealing fuel tanks. They also find some application with electrical components for protection from moisture and vibration.

Polysulfide sealants are supplied as both one-package and two-package systems. A two-package system that uses lead dioxide as the curing agent is the most common overall. However, current aircraft systems are largely one-package systems where the preferred curing agents are zinc or calcium peroxide. Because of the complexity of sealant adhesion problems, a variety of primer formulations have been developed for various surfaces. Most are based on chlorinated rubbers and silanes or admixtures thereof (20). Although epoxy resins are also used as curing agents for the polysulfide elastomers, use in aerospace is limited. This particular combination is used primarily by the building trades.

The silicones (mainly polydimethyl siloxanes) are considerably more expensive than the polysulfides; therefore, silicone use is restricted to more specialized applications where specific attributes of the materials are required. For example, virtually every U.S. missile or satellite contains some form of silicone sealant because of its long use life, low volatile content, and heat resistance. They are also used around aircraft windows and doors. High-temperature fuel tank sealants based on fluorocarbon-containing polysiloxanes have also been the subject of considerable government-sponsored research (21).

Siloxane-based sealants cure either by condensation or addition reactions (22). One-package systems are available that cure in the presence of atmospheric moisture. The cross-linking reaction results in the elimination of either acetic acid or methyl ethyl ketone depending on the resin system. The more conventional two-package systems use metallic soaps such as dibutyl tin dilaurate to catalyze the cross-linking with the subsequent elimination of ethyl alcohol. Addition cures can also be accomplished with vinyl-containing polysiloxanes and a transition metal catalyst.

Thermoset Processing Science and Control

A recent and important aerospace contribution to the thermoset resin field has evolved from efforts to develop a scientifically sound basis for thermoset resin processing and control. It resulted from the exacting mechanical strength requirements placed on hardware by the industry. Over the past 10 years there has been considerable activity in the area of processing science. This situation is particularly true in the areas of adhesive bonding and FRP production. Because of the efforts of aerospace scientists, augmented by the universities, scientific techniques for solving material- and process-related problems are rapidly developing. Prior to this time bonding and lamination were almost entirely dependent on operator skills. Key changes made during processes were based on human decision and inadvertently led to fabrication errors and material misuse.

Fabrication of hardware based on the types of resins discussed herein is obviously a chemical process. To control a chemical process the starting raw materials must be precisely defined, and the course of the chemical process must be followed. In the aerospace industry this means the starting resin formulations must be consistent and precisely defined, and the chemical and physical changes that take place as a thermoset resin cures must be monitored and controlled.

Studies in the area of chemical definition of starting materials began more than 10 years ago (23). In addition to wet chemistry, the laboratory measurements currently used for this purpose are differential scanning calorimetry (DSC), high-performance liquid chromatography (HPLC), gel permeation chromatography (GPC), infrared spectroscopy (IR), and rheology.

Shown in Figure 4 is a thermogram of a catalyzed TGMDA/DDS prepreg formulation. It reveals considerable information about the material and how it cures. The T_g of the prepreg is a measure of the degree of B-staging. The presence of residual solvents can be detected as shown. The onset of the cure, the influence of accelerators, and the peak exotherm are other useful features of the thermogram. The area under the curve is the heat of reaction, a processing parameter. Although the method has been criticized for lack of precision, common errors in the prepreg composition and its processing can be observed by this technique. It is also a valuable aid in developing cure cycles for new resin systems.

One of the most powerful tools for qualitative and quantitative examination of thermoset resin formulations is chromatography. Shown in Figure 5 is an HPLC typical of the TGMDA/DDS prepreg resin system. The location of a peak on the elution plot indicates the presence of the desired components of the formulation. The area under a given peak is proportional to its concentration when standardized. Note also that a reaction peak appears on the chromatograph. This peak can be used for an accurate estimate of the extent of B-stage during prepreg fabrication and subsequent storage.

The area under the curing agent peak is not an accurate measure of its concentration in the starting formulation due to reaction with the resin during manufacture. The starting formulation concentration, however, can be calculated by using the reaction peak. With this particular formulation, the starting (total) concentration of the curing agent can also be measured precisely by IR analysis by using a peak that corresponds to the sulfone group (24). Subtractive analysis with Fourier transform IR (FTIR) has also been suggested as a future chemical characterization procedure (25). A combination of FTIR with HPLC appears to be another promising technique.

Rheology is a most recent addition to aerospace material and process technology. It can be used as an incoming material inspection tool to predetermine the processibility characteristics of a given thermoset resin system. Although liquid state rheology is currently the most widely used method, solid state measurements are useful in determining the extent of cure because they can be used to measure the T_g of a finished hardware item. Shown in Figure

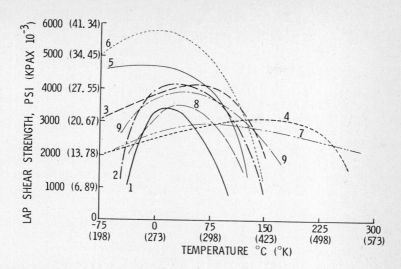

1. GENERAL PURPOSE EPOXY POLYAMIDE
2. ONE PART GENERAL PURPOSE EPOXY
3. AROMATIC AMINE CURED EPOXY
4. EPOXY PHENOLIC
5. RUBBER (NITRILE) MODIFIED EPOXY

6. NYLON EPOXY
7. POLYIMIDE
8. VINYL PHENOLIC
9. NITRILE PHENOLIC

Figure 3.   Lap shear strength of typical structural adhesives.

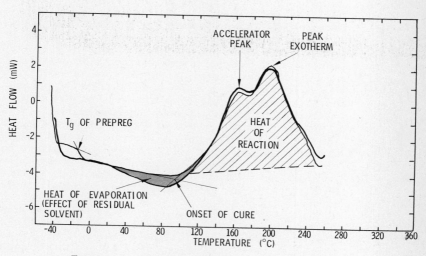

Figure 4.   DSC of an epoxy resin formulation.

6 are liquid state rheological properties of the TGMDA resin system
and its prepreg as a function of ambient temperature aging when a
typical hardware cure profile is used. On the basis of the neat
resin data it could be concluded that the proper viscosity for
consolidation pressure application can be controlled by adjusting
the hold cycle. Even the 60-day-old sample appears processible.
The 60-day-old prepreg, however, never attains a "viscosity" as low
as the unaged starting material. Thus, it may not be processible
with the particular cure cycle shown. The data demonstrate that
there is a strong resin-fiber interaction during lamination
processes.

In the area of liquid state rheology there is also considerable
research in progress on the development of mathematical models that
predict material behavior during composite fabrication cure
processes. For example, the viscosity of a thermosetting matrix can
be predicted for any cure cycle by using a mathematical model
developed from kinetic and rheological data (26).

In-process monitoring of chemical and physical change must be
accomplished by indirect means. It is virtually impossible to
measure these changes directly in the presses and autoclaves used
for bonding, lamination, and molding of thermoset polymers.
Although much work remains, alternating current dielectric
measurements offer much promise. In Table II, the mechanical-
electrical analogies developed by Cheng (27) show that the proper
electrical measurements should bear a mathematical relationship to
the rheological and chemical events in the thermoset-curing process.
A recent publication discusses the mathematical relationship between
dielectric relaxation and molecular viscosity (28).

As suggested by Kranbuehl (29), the dielectric signal chosen for
a given segment of a cure cycle should be that which gives the most
meaningful information on the chemical or physical state of
interest. Figure 7 is a case in point. In this figure, phase angle
(Ø), vector voltage (VV), and viscosity (η*) are plotted as a
function of time for the indicated cure cycle. Over the segment of
the cycle shown, the vector voltage and viscosity have very similar
shapes. The phase angle, although it probably contains meaningful
information, appears to be a poor means of process control.
Restated, the early stages of a cure when matrix or prepreg
viscosity is the most important control element, an electrical
analogy such as vector voltage or alternating current resistivity is
probably the most useful. Toward the end of a cure these electrical
parameters change little. At this point, interest focuses on the
completion of the cure. Thus, electrical changes that relate to
molecular segmental motions or dipole movement, such as phase angle
change or loss tangent, should be considered. As Kranbuehl further
points out, frequency should also be a consideration in developing a
dielectric cure-control system.

The aerospace industry has had a pronounced influence on
developments in thermoset resins that are also useful for other
industrial applications. The efforts of the aerospace industry in
the science of thermoset resin processing will afford techniques
that are applicable to a broad range of nonaerospace applications.
Fabrication of resins into aerospace products is but a step away
from being a complete scientifically controlled process from resin
formulation to finished hardware. As this work becomes a practical
reality, the entire thermoset resin industry will benefit.

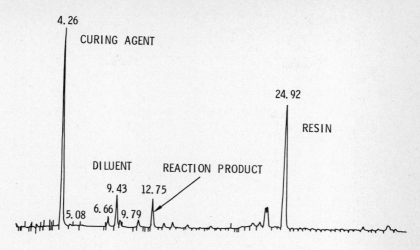

Figure 5.   HPLC of an epoxy prepreg formulation.

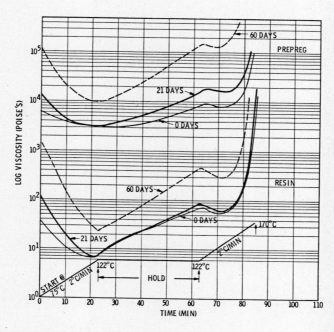

Figure 6.   Effect of aging on the rheology of a resin and prepreg.

Table II. Cheng's Mechanical–Electrical Analogy

Mechanical System	Electrical System
Force	Voltage
Velocity	Current
Displacement	Charge
Mass	Inductance
Damping coefficient	Resistance
Compliance	Capacitance
Mechanical loss	Loss tangent

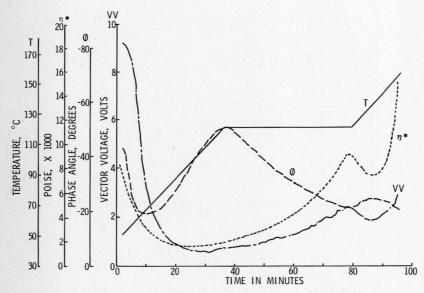

Figure 7. Dielectric properties and viscosity for a production cure cycle.

Literature Cited

1. "Resins for Aerospace"; May, C. A., Ed.; ACS SYMPOSIUM SERIES No. 132, American Chemical Society, Washington, D.C., 1980.
2. Fisch, W.; Hofmann, W. J. Polym. Sci. 1954, 12, 497.
3. Fischer, R. F. J. Polym. Sci. 1960, 44, 155.
4. Tanaka, Y.; Mika, T. F. In "Epoxy Resins, Chemistry and Technology"; May, C. A.; Tanaka, Y., Eds.; Marcel Dekker: New York, 1973; pp. 198-99.
5. Morgan, R. J.; Happe, J. A.; Mones, E. T. SAMPE Series 1983, 28, 596.
6. Rowe, E. H.; Seibert, A. R.; Drake, R. S. Mod. Plastics 1970, 47, 110.
7. Sultan, J. N.; McGarry, F. J. Polym. Eng. Sci. 1973, 13(1), 29.
8. Manzione, L. T.; Gillham, J. K. J. Appl. Polym. Sci. 1981, 26, 907.
9. Riew, C. K.; Rowe, E. H.; Siebert, A. R. "Toughness and Brittleness in Plastics"; Deanin, R.; Crugnola, A., Eds.; ADVANCES IN CHEMISTRY SERIES No. 154, American Chemical Society, Washington, D.C., 1976; p. 326.
10. Martin, R. W. "The Chemistry of Phenolic Resins"; Wiley: New York, 1956.
11. Schwartz, A. S. WADD Tech. Report 60-101, September 1960.
12. Neuse, E. W. Materials Science and Eng. 1973, 11, 121.
13. "Urethane Science and Technology"; Frisch, K. C.; Ragan, S. L., Eds.; Technomic: Stamford, Conn. 1972; Vol. 1.
14. Meier, J. F., et al., "Resins for Aerospace"; May, C. A., Ed.; ACS SYMPOSIUM SERIES No. 132, American Chemical Society, Washington, D.C., 1980; p. 151.
15. Hergenrother, P. M.; Johnston, M. J. "Resins for Aerospace"; May, C. A., Ed.; ACS SYMPOSIUM SERIES No. 132, American Chemical Society: Washington, D.C., 1980; p. 3.
16. Serafini, T. T.; Delvigs, P.; Lightsey, G. R. J. Appl. Polym. Sci. 1972, 16, 905.
17. St. Clair, T. L.; Jewell, R. A. National SAMPE Tech. Conf. Series 1976, 8, 82.
18. de Bruyne, N. A. U.S. Patent 2 499 134, 1950 (to Rohm and Haas Co.).
19. Dannenberg, H.; May, C. A. In "Treatise on Adhesion and Adhesives"; Patrick, R. L., Ed.; Marcel Dekker: New York, 1973; Vol. 3, Chap. 2.
20. Panek, J. R. "Handbook of Adhesives"; Skiest, J., Ed.; Van Nostrand Reinhold: New York, 1977; p. 372.
21. Rosser, R. W.; Parker, J. A. NASA Technical Memorandum, NASA TM X-62, December 1974; p. 401.
22. Gair, T. J.; Thimineur, R. J. "Modern Plastics Encyclopedia"; McGraw-Hill: New York, 1978-79; p. 102.
23. May, C. A.; Fritzen, J. S.; Whearty, D. K. AFML-TR-76-112, "Exploratory Development of Chemical Quality Assurance and Composition of Epoxy Formulations"; June 1976.
24. May, C. A.; Hadad, D. K.; Browning, C. E. Polym. Eng. and Sci. 1979, 19(8), 545.

25. May, C. A. _Pure and Appl. Chem._ 1983, _55_(_5_), 811.
26. Dusi, M. R.; May, C. A.; Seferis, J. C. "Chemorheology of Thermosetting Polymers"; May, C. A., Ed.; ACS SYMPOSIUM SERIES No. 227, American Chemical Society: Washington, D.C., 1983; Chap. 18.
27. Cheng, D. K. "Analysis of Linear Systems"; Addison-Wesly: Reading, Mass., 1961.
28. Baumgartner, W. E.; Ricker, T. _SAMPE J._ 1983, _19_(_4_), 18.
29. Kranbuehl, D. E.; Delos, S. E.; Jue, P. K. _National SAMPE Ser._ 1983, _28_, 608.

Polymer Processing

NICK R. SCHOTT and ROBERT MALLOY

Department of Plastics Engineering, University of Lowell, Lowell, MA 01854

Extrusion
  Screw Design
  Extruder Output Rate
  Energy Requirements
  Shear Rate Limitations
  Profiles
  Film Processing
  Thermoplastic Foam Extrusion
Injection Molding
Blow Molding
Thermoforming
Thermoset Processing
Urethane Processing
Mixing in Plastics Processing
State of Polymer Processing

Polymer processing is defined as the conversion of high polymers into useful products (1). It includes the following conversion operations: extrusion, injection molding, coating, blow molding, calendering, casting, blending and compounding, compression molding, thermoforming, and others. The processing operations are carried out either on a small or on a large scale by using complex technology that involves many unusual engineering problems.

The past 30 years have seen the rapid development of the polymer industry while at the same time a systematic engineering analysis approach was applied to the problems associated with the processing of the materials and the design of the machinery for doing it. Rheology, the science of deformation and flow, plays an important part in the understanding of the many phenomena that are encountered in polymer processing. The rheology of the polymer is changed via the addition or removal of heat. Hence, fluid flow and heat transfer form the basis for the analysis of these processes.

0097–6156/85/0285–0581$07.75/0
© 1985 American Chemical Society

Extrusion

On a volume basis, extrusion is the most important of all the
conversion operations. It applies to thermoplastics only since any
interruption of the process with thermosets can cause irreparable
machine damage. It is a continuous process in which a product is
formed where two dimensions are controlled. Many simple and also
complex geometric shapes are possible. In its most common form one
talks about plasticating extrusion. This process is represented
schematically in Figure 1. Plastication means that one starts with
a solid feed, which is brought to a melt state. The feed is most
commonly in pellet form, usually 1/8 of an inch in diameter, for
easy conveying of solids. The feed drops by gravity into the screw
channel, and from there it is conveyed by the screw rotation toward
the die end of the machine. The screw channel depth is decreasing
as one goes toward the die. This decrease in volume compacts the
solids as they are conveyed. Simultaneously, as the solids are
compacted they are also heated by the friction that is generated
between particles and also between metal surfaces and particles.
Soon a melt pool is formed in part of the channel, and further
frictional heat in the melt causes rapid heating and melting of the
whole mass. The frictional heating is called viscous dissipation.
Virtually all the heat required for the process is generated via
viscous dissipation. The electrical energy of the motor is
converted into mechanical energy, which in turn is converted into
heat. The electrical energy from the heater bands is used during
start up of the extruder to melt any material that is in the
machine. After start up the electrical heater bands essentially
make up the heat losses from the extruder through the barrel. The
feed throat is usually water cooled to prevent premature melting of
material, a phenomenon called "bridging."

Screw Design. Screw design plays an important part in extruder
performance. A screw is specified by two parameters, which are the
L:D ratio (length:diameter) and the compression ratio, CR, which is
the ratio of the volume of one flight in the feed section divided by
one flight in the metering section. The typical L:D ratio is 24:1.
Higher values give better mixing and a more homogeneous melt.
Values as high as 36:1 are commercially available, but the
mechanical strength of the screw limits one beyond this value.
Older extruders may have L:D ratios as low as 10:1. The trend in
the past 20 years has been toward longer, more specialized screw
design. Figure 2 shows a typical screw for a polyolefin-type resin.
The compression ratio varies from 2:1 to 5:1. Low compression
ratios are used with heat-sensitive materials such as vinyls where
thermal and shear degradation may be a problem. The higher
compression ratios are used with the polyolefins and the styrenics,
which are rather insensitive to degradation.

Extruder Output Rate. Flow and output rate of an extruder may be
predicted from Equation 1, which is derived from fluid dynamics (2-
4).

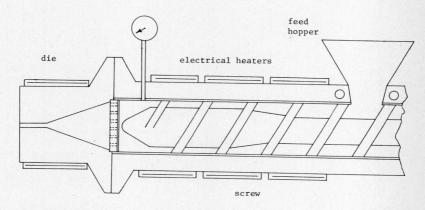

Figure 1.   Schematic of plasticating extrusion.

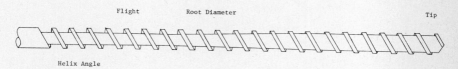

Figure 2.   Typical single-stage metering screw for polyolefins.

$$Q = AN - B\frac{\Delta P}{\eta} \tag{1}$$

where Q = volumetric extrusion rate, A and B = constants dependent on the screw geometry, N = screw speed, ΔP = pressure rise along the screw channel, and η = apparent viscosity of material in the channel. Equation 1 was developed for the metering section of the screw under the assumption of isothermal Newtonian flow. The conveying capability of the feed section is greater than the pumping capability of the metering section. Similarly, the melting and conveying rate in the transition section is greater than that of the metering section, and the output rate is controlled by the pumping capacity of the metering section. The conveying and melting mechanisms in the feed and transition sections are less completely understood. However, computer programs exist that model the plasticating extrusion process (5, 6). Equation 1 shows that the output rate is linearly dependent on the screw speed and decreases with the pressure rise in the melt. The barrel is designed to withstand pressures up to 10,000 psi. Most extrusion processes use much less melt pressure, which is determined by the die geometry, melt viscosity, and flow rate. Table I lists typical die pressures (maximum pressures) encountered in extrusion (7, 8).

Energy Requirements. As mentioned earlier, the energy requirements for extrusion processing are furnished by a drive motor whose energy input is converted into heat. An energy balance of the extruder is shown in Equation 2 (9):

$$Z + q = CQ\Delta T + Q\Delta P + H \tag{2}$$

where Z = mechanical power to the screw, q = electric power, C = mean volume specific heat of the plastic, Q = volumetric extrusion rate, ΔT = temperature rise in the plastic, ΔP = pressure rise in the extruder, and H = heat loss in the extruder from radiation, conduction, and convection.

The term $C\Delta T$ is the enthalpy content of the particular plastic at the appropriate extrusion temperature. Typical values of the heat capacity are shown in Table II. If one considers the heat loss term to be offset by the electrical heat, q, then Equation 2 reduces to

$$Z = CQ\Delta T + Q\Delta P \tag{3}$$

The term $Q\Delta P$ is the power required to pressurize the melt (pumping power) and represents less than 5% of the total power required for the process (9). Hence, as can be seen from Equation 3, the power requirement is essentially determined by the product $CQ\Delta T$. Typical processing temperatures are listed in Table III (10, 11, 12). Thus, it is found that high-density polyethylene requires the most power per pound of product while polystyrene requires the least. It should be noted that the average heat capacity values in Table II include the heat of fusion for the semicrystalline polymers such as high- and low-density polyethylene.

Table I. Typical Extruder Die Head Pressures

Extruder Form	Melt Pressure at Die	
	lb/in.^{2}	kg$_f$/cm^2
Monofilament	1000–3000	70–211
Sheet	500–1500	35–105
Wire coating	1500–8000	105–562
Pipe	500–1500	35–105
Cast film	500–1500	35–105
Layflat film	2000–6000	141–422

Table II. Mean Specific Heat Capacity of Polymers at Extrusion Temperatures (7).

Polymer	C_p (Btu/(lb °F) or kcal/(kg °C))
ABS	0.4
Acrylic	0.6
Nylon	0.65
Polyethylene	
Low density	0.75
High density	0.85
Polypropylene	0.7
Polystyrene	
General purpose	0.4
Impact	0.42
PVC	0.35
PC	

Tables II and III show that high-density polyethylene requires the most energy per pound for processing while a polystyrene or a PVC requires the least. In order to estimate the drive requirements, one can use the following approximations as indicated in Table IV.

Shear Rate Limitations. Extruders are run at relatively low screw speeds. Small extruders have variable drives to adjust screw speeds from about 20 to 150 rpm for extruders up to about 2 in. in diameter. As the screw diameter increases the screw speed must be reduced. The peripheral screw speed is given by Equation 4

$$V = \pi DN \tag{4}$$

where V = peripheral speed, D = screw diameter, and N = screw speed. The average shear rate in the screw channel is given by

$$\overline{\dot{\gamma}} = \frac{\pi DN}{h} \tag{5}$$

where h = screw channel depth.

The recommended peripheral speed is 35-65 ft/min for heat-sensitive materials and 100-150 ft/min for general materials. Commercial extruder sizes range from about 3/4 in. in diameter to about 10 in. in diameter. Larger sizes, up to 24 in., have been built, but these are usually for in plant use and of a specific design.

Profiles. The types of profiles that can be produced are determined by the die design and the secondary shaping that occurs in the haul off of the extrudate. Flow in the die occurs by pressure flow only as indicated by Equation 6.

$$Q = \frac{\Delta P}{K\eta} \tag{6}$$

where Q = volumetric flow rate through the die, ΔP = pressure drop across the die, K = die constant, and η = apparent melt viscosity at the shear rate in the die at the melt temperature. The die constant may be calculated for simple geometries such as slits, circular profiles, or annular shapes, which then allows one to calculate an output rate or a pressure drop via Equation 6. It is more difficult to predict the final shape and dimensions of the extrudate. This is due to the elastic properties of the melt as it emerges from the die. The elastic strain is recovered as the shear stress at the wall is removed. White and Huang (13) have made a theoretical and experimental study for predicting die swell in rectangular and trapezoidal dies. Theory allows one to predict the shape, but one usually underestimates the amount of swell. In practice a lot of the design is done by trial and error. One usually machines an inexpensive die via EDM (electrode discharge machining) and tries it out in the process. A post-forming tool forces the extrudate to cool to a desired shape that is somewhat different from the profile

Table III. Typical Melt Temperature Ranges for Processing of General-Purpose Grades of Thermoplastics (12)

Material	Processing Temperature Range	
	°C	°F
ABS	180–240	356–464
Acetal	185–225	365–437
Acrylic	180–250	356–482
Nylon	260–290	500–554
Polycarbonate	280–310	536–590
Polyethylene		
Low density	260–240	320–464
High density	200–280	392–536
Polypropylene	200–300	392–572
Polystyrene	180–260	356–500
PVC (rigid)	160–180	320–356

Table IV. Drive Power for Extruders (12)

Material	Drive Power	
	hp/(lb h)	kW/(kg h)
PVC	0.1–0.123	0.15–0.2
Polyolefins	0.123–0.2	0.2–0.3
Polysulfone	0.096–0.134	0.14–0.22
Polycarbonate	0.096–0.134	0.14–0.22

that would be obtained by die swell alone. Figure 3 illustrates the
effect of die swell for simple die geometries (14).

All flows in polymer melt processing are laminar. The typical
shear rates for extrusion are about 100–500 s^{-1}. Sufficiently long
dies are required to ensure a smooth extrudate. Short die lands
lead to melt fracture and excessive swell at high extrusion rates.
Glanvill (15) gives die design guidelines for pipe, sheet, film, and
blow molding dies.

In the extrusion process the haul off rate varies inversely with
the wall thickness of the extrudate. In wire coating the line
speeds range up to 5000 ft/min. Similarly, for extrusion coating
line speeds are several hundred to about 1500 ft/min. On the other
hand, pipe extrusion and heavy-gauge sheet extrusion are very slow
with speeds from 10 to 100 ft/min. Figure 4 shows two types of wire
coating dies. The design in Figure 4a is a pressure die where the
melt makes contact with the wire inside the die. The pressure
assures good adhesion between the substrate and the polymer. Figure
4b shows a die where the melt contacts the substrate outside the
die. This type of die is used for cable-jacketing applications
where a second coating is applied to a previous coating. A wire
coating operation is a good example of a crosshead extrusion line as
indicated by the diagram of Figure 5. In this operation the product
is hauled off at 90° to the direction of extrusion. Most resins
processed for wire coating are vinyls and polyethylenes. The
economics are, in many instances, determined by the production rate.
For vinyls, the rate in turn is highly influenced by the resin
formulation. This formulation has about 10 constituents, which
include various vinyl resins of different particle sizes,
plasticizers, fillers, processing aids, stabilizers, and internal or
external lubricants. Formulations can be compared and tested
against a reference formulation. In many instances the wire coating
extrudate can be foamed, with either a chemical or a physical
blowing agent. The foaming improves the dielectric properties and
gives better insulation with less plastic since the air in the cells
is a better dielectric. Polyethylene coatings are often cross-
linked. This can be done either by an electron beam use or by the
thermal decomposition of an organic peroxide, after the coated wire
emerges from the die. Cross-linked polyethylene affords additional
abrasion resistance and a higher service temperature.

Film Processing. Film can be extruded in two ways: blown (lay
flat) film and chill-roll cast film.   The tubular film process is
schematically shown in Figure 6.

This process is capable of producing films less than 0.5 mil in
thickness to about 20 mil thick. Lower thicknesses can be made with
this process as compared to cast film, but cast film usually gives
better optical properties for polyethylenes since quenching the film
suppresses the size of the crystallites and thus reduces haze. Many
die designs are available for the process. The spider leg die is
shown in Figure 7a. This die is used for heat-sensitive materials
such as PVC. The die has little hang-up and can readily be cleaned.
However, the spider leg support of the mandrel in the die body
causes weld lines, which can lead to film failure if the melt is too
cold and the weld line does not properly knit. The spiral die
design shown in Figure 7b does not create weld lines. The melt

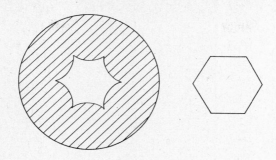

Figure 3.   Shapes of die and extrudate due to die swell (elastic recovery).

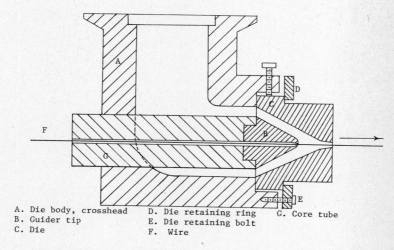

A. Die body, crosshead   D. Die retaining ring   G. Core tube
B. Guider tip   E. Die retaining bolt
C. Die   F. Wire

Figure 4a.   Dies for wire and cable coating (pressure coating).

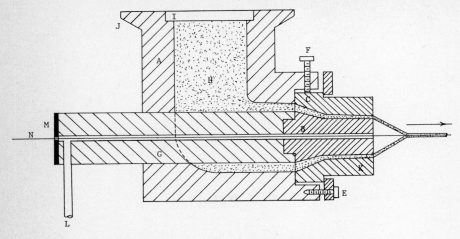

A. Die body, crosshead   H. Molten plastic
B. Guider, male die part   I. Seat for breaker plate
C. Die, die bushing, female die J. Ring for attachment to extruder
   part   K. Die land
D. Die retaining ring   L. Vacuum connection
E. Die retaining bolt   M. Wire guide and vacuum seal
F. Die centering bolt   N. Wire
G. Core tube

Figure 4b. Wire coating tubing die.

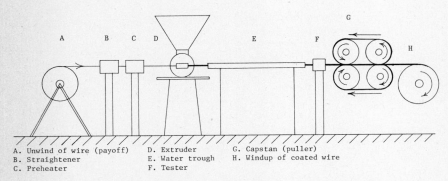

A. Unwind of wire (payoff)   D. Extruder   G. Capstan (puller)
B. Straightener   E. Water trough   H. Windup of coated wire
C. Preheater   F. Tester

Figure 5.   Crosshead extrusion for wire coating.

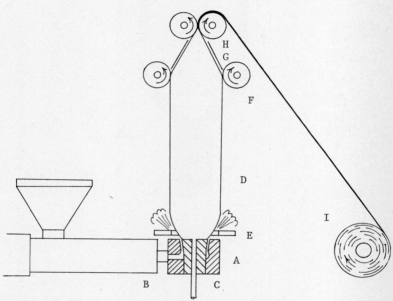

A. Blown film die   D. Plastic tube (bubble) G. Collapsing frame
B. Die inlet   E. Air ring for cooling   H. Pull rolls
C. Air hole and valve   F. Guide rolls   I. Windup roll

Figure 6.   Schematic of blown film process.

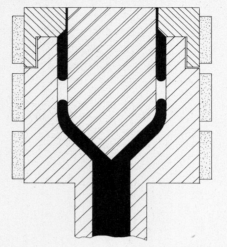

Figure 7a.   Spider leg blown film die.

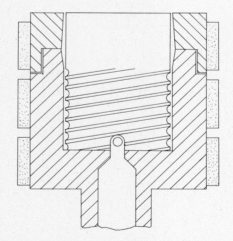

Figure 7b.   Spiral die design for blown film.

spirals up and forms a continuous bubble without knits. Also, this die may be rotated as the film is hauled up, to distribute all gauge variations uniformly across the width of the roll. The shrink and mechanical properties of the film are determined by the degree of orientation in each direction. The draw down (the ratio of die opening to final film thickness) determines the orientation in the machine direction. The blow up ratio (ratio of tubular film bubble diameter to die diameter) determines the orientation in the transverse direction. The orientation is adjusted for the type of end product in mind. For example, a balanced shrink film would have the same degree of orientation in both directions. Glanvill (16) gives guidelines for matching the extruder size to the die diameter for lay flat film or to the die width for cast film. Data are also given for setting the die gap versus film thickness and the effects of die geometry on haze and shrinkage. Recent developments in blown film extrusion have been coextrusion of multilayer film (17). In this process up to three bubbles are blown simultaneously one inside the other. Each film has a different orientation and could even be of a different color or composition. The layers are contacted at the frost line (height at which a melt turns opaque due to crystallization or solidification) to achieve lamination. The product shows improved mechanical and barrier properties for packaging applications. Production rates of the process can also be improved with higher bubble cooling rates (18). Dual air rings were introduced in 1975 and 1976 and have been improved since. They are used with the new linear low density polyethylenes that have an inherent poor melt strength between the die lip and the frost line. Internal bubble cooling also increases the production rate. In this process the air inside the bubble is continuously replaced with cool air. Thus, heat transfer takes place from two surfaces, the haul off speed can be increased, and the optics of the film are improved. The introduction of linear low polyethylene in the past 2 years (1979–1980) has also led to die design modifications. Due to their processing characteristics, the die gap has been increased. These resins can tolerate higher shear rates as reported in Plastics Technology (19).

The cast film extrusion process is schematically shown in Figure 8. Usually it is difficult to achieve a film thickness below 0.5 mils. The process is best suited for films up to 20 mils thick. Above 20 mils one cannot arbitrarily distinguish between the film and the sheet extrusion process, which is shown in Figure 9. Sheet extrusion finds many applications. Usually thickness up to 0.375 in. are standard, but heavier gauges can be produced with special die and haul off designs. Dies for making sheets up to 8 ft wide are available. Die design is important to assure uniformity across the width of the sheet. Due to the elastic nature of the melt, there is a tendency for the sheet to shrink at the edges as it emerges from the die. The thick bead that is formed must be trimmed as the sheet is hauled off. In continuous thermoforming the sheet may be fed directly to the thermoformer. This is particularly true for the drinking cup business and disposable dishes, lids, etc. Usually this is a high-speed integrated process that is computer controlled. Coextrusion of cast film and sheet is available. The applications will increase since special properties can be

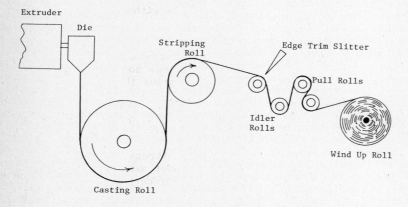

Figure 8.   Cast film extrusion process.

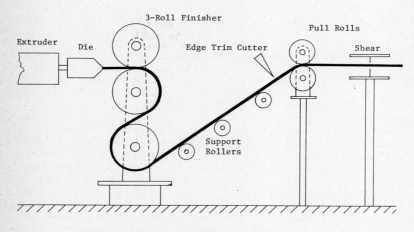

Figure 9.   Sheet extrusion process.

incorporated into various layers to save money on lower cost interior resins and improve the properties of the overall product.

The past 25 years have seen a tremendous improvement in the extrusion process (19). Productivity has increased 250–300% in this time. The process has evolved from an art to a science. The increased output is attributed to improvements in screw design and haul off equipment for increased cost effectiveness (70–80% lower than in the fifties). Processing costs have decreased due to energy efficiency. Plastic processing energy efficiency has increased 300–400% since the fifties. Typical values for today's extruders are (20) 8–9 lb h^{-1} hp^{-1} for LDPE and 10–11 lb h^{-1} hp^{-1} for PVC.

Thermoplastic Foam Extrusion. Foamed plastics find applications as rigid profiles, pipe, sheet, packaging material, and thermal insulation. Polystyrene (PS) finds the widest application in foamed products while poly(vinyl chloride) PVC, low–density polyethylene (LDPE), and acrylonitrile–butadiene–styrene (ABS) (21) are also used in large quantities. All common thermoplastics can be foamed by various techniques as described hereafter.

The most widely used technique employs a chemical blowing agent (CBA), such as azodicarbonamide, that decomposes at a specific temperature to release nitrogen gas for cell formation. CBA is used in minute amounts (approximately 1/4–1%) for wire coating or in larger amounts up to 20% by weight for much lower density products. CBAs can be tumbled as a powder onto the resin pellets, or more preferably, they are pumped as a liquid concentrate into the extruder feed throat. One can also use a foam agent concentrate where the blowing agent is melt compounded into a resin but at a temperature below the decomposition point of the CBA. The concentrate is dry tumbled with regular resin to achieve the proper density of the product. One can shift the decomposition point of the CBA by changing its particle size, which affects .pa heat transfer. Smaller particles will decompose at a lower temperature. Also, an activator can be used, which serves to reduce the decomposition temperature.

Conventional single–screw extruders may be used for the foaming process, but premature decomposition of the CBA, which causes a loss of nitrogen through the feed throat, is difficult to avoid. The object is to reach the decomposition temperature in the metering section of the screw and allow the nitrogen to go into solution (22). The high melt pressure keeps the gas in solution until the melt exits through the die and the pressure is released. The die should have a steep pressure gradient to prevent premature foaming. Foaming in the die leads to tear and rupture of the cell structure (23). The foam density and the surface quality are affected by the melt temperature. A low temperature leads to a high viscosity and incomplete foaming (high density) because the gas cells cannot expand sufficiently. If the temperature is too high, the viscosity is too low, gas escapes, and the cellular structure is destroyed and the density will increase again.

There are two patented techniques for sizing and cooling the extrudate. In the Gatto process, the part is extruded from an undersized die and allowed to foam outside the die as it enters the water–cooled vacuum calibrator, which determines the final shape (24). A skin of high–density material can be formed if air cooling

is applied to the extrudate surface. Cooling must be controlled
since overcooling will restrain foaming.

The French patented Ugine Kuhlman process uses a die with an
internal mandrel, as shown in Figure 10. The outer skin is quickly
frozen via a water-cooled sizing die while the mandrel forms an
internal void in the extrudate that is filled by the inward foaming
action. The process creates a thick-skinned part with a foamed
core. The process is very important for structural members such as
window frames using PVC as a wood replacement.

Low-density polystyrene foams down to 2.5 lb/ft^3 can be made for
packaging applications via the use of physical blowing agents.
These blowing agents change from a dissolved liquid to a gas as the
melt emerges from the die to foam the extrudate. Pentane and
fluorocarbons such as Freon R11 and R12 are commonly used. The
blowing agent can be dissolved in the polymer pellets prior to
processing, or it may be injected as a liquid into a tandem extruder
system. The first extruder is a plasticating extruder to melt the
pellets while the second extruder is a melt extruder that is one
size larger. It pumps and cools the melt as it goes into the die to
control foaming. The second extruder, since it is larger, adds very
little heat to the melt via shear heating, which is also known as
viscous dissipation. To aid the foaming process, about 1/2% talc is
incorporated, which acts as a nucleating agent for the growing gas
bubbles. Citric acid with sodium bicarbonate is also used as a
nucleating agent. The blowing agent is in solution in the melt.
The first extruder has a port to add the blowing agent. Intensive
mixing is required to disperse the blowing agent and make it go into
solution under the conditions of high pressure (3000 psi or higher)
and a melt temperature of 425–450 °F. The second extruder is run at
265–300 °F (24), and the melt cools further quite suddenly since the
blowing agent will flash vaporize at the die exit and use its heat
of vaporization to cool the melt. Control of the melt viscosity via
temperature is the key to good foam. If the extrudate is to be
thermoformed in a second operation, the foamed extrudate must first
be stored to allow diffusion of the pentane or Freon out of the
cells and the diffusion of air into the cells.

Injection Molding

Injection molding is a cyclic discontinuous process used to form
three-dimensional plastic parts (25, 26). Both thermoplastics and
thermosets are injection molded. The process is shown schematically
in Figure 11 for a reciprocating screw injection molder. This type
of a machine has a screw rotating in a barrel similar to an
extruder. However, here the melt flow is discontinuous and
controlled via a check valve at the tip of the screw. Material is
melted in the extruder, and the melt is accumulated in front of the
screw tip until sufficient melt is at hand to fill the mold cavity
(cavities).

Once the melt is accumulated, the screw stops rotating and a
hydraulic cylinder moves the screw forward to inject the melt into
the cavity. The mold remains closed while the part cools. At the
same time, the screw is rotating again to plasticate melt for the
next shot. The cooling time is determined by the type of resin and
the part thickness. The machine is specified by two main variables:

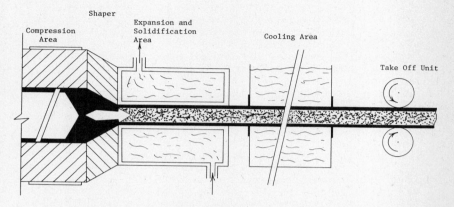

Figure 10.   The "Celuka process" for foamed extrusion.

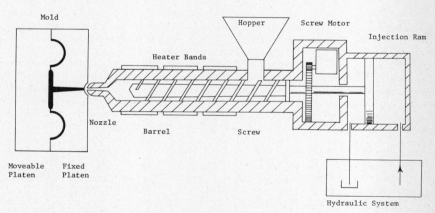

Figure 11.   Injection molding process for reciprocating screw process.

(1) the shot size and (2) the clamping force. The shot size is the maximum volume of melt that can be injected at one time. Typically it is reported in cubic inches or more commonly in ounces of polystyrene. Machine sizes vary from less than 1 to about 500 oz. The clamping force is the force required to keep the halves of the mold closed while the melt is being injected and the part is cooled. The force is reported in tons. Machines vary in size from 20 to 5000 tons. Most machines that are sold are in the 75–300-ton range. The clamping force depends on the projected area of the molded parts. A rule of thumb is that one uses 2–5 tons/in.^{2} of projected area. Higher values are used for filled materials, high–viscosity materials, and parts with small gates, long flow paths, and thin walls.

Two major developments have taken place in injection molding in the past 25 years (19). The first was the introduction of the reciprocating screw machine in the early sixties. The older plunger machines were limited in their plasticating rate by poor heat transfer from the barrel into the melt, mixing was virtually nonexistent, and pressure losses during injection were excessive. The screw machine gave an order of magnitude improvement in the molded part when the same material was used.

Process control has been the second revolutionary advance in injection molding. Prior to 1970 only machine variables such as hydraulic pressure, barrel temperatures, and ram position were controlled with simple electromechanical controls. The past 10 years have seen the use of microprocessor control of the melt variables using solid–state systems. The degree of sophistication varies from simple peak cavity pressure control to shot size, melt pressure, and cushion control (27, 28). These microprocessors can be programmed to control both the machine variables and the melt variables. Some microprocessor–based controls give card or tape programs that can be changed with push–bottom entries. The units also give diagnostics and status reports and help in the set up of new molds. The data from the microprocessor in turn can be used for management reporting. For example, one can report on resin consumption, job running time, machine down time, the number of parts produced per hour, and the estimated time required to fill an order.

Other developments have not been as dramatic, but they have nevertheless contributed significantly to the process efficiency. The electric screw drives were replaced by hydraulic motors with a gear box on the reciprocating screw unit. In the future the hydraulic motor may in turn be replaced by a direct hydraulic drive (19). The introduction of stack molds has also increased productivity. In a stack mold, two cavity plates are placed on top of each other so that the projected area is the same and twice as many parts can be formed with the same clamping tonnage. This technique works best with thin wall, flat parts with short cycle times.

Energy savings in production has been a major concern. The clamping system is selected much more carefully (29). Toggle (mechanical) clamps are used on machines of less than 400 tons. These mechanical systems are fast and low in energy consumption but give less speed control in closing than straight hydraulic units. The hybrid hydromechanical clamping systems are very common on larger machines.   These machines use the mechanical toggle for the high-speed low-pressure closing of the mold and the hydraulics for the final mold lock up.

Energy savings and scrap reduction were introduced by using heated sprue bushings, insulated runners, and hot runner systems. In a conventional system, the sprue and runners are separated from the part when the mold is opened either via a stripper plate or manually by the operator.   The sprues and runners are then reground and used with virgin resin in subsequent molding operations.   The hot sprue busting keeps the material molten all the time.   A spring-loaded shut off keeps melt from flowing into the cavity when the mold is open.   In an insulated runner system, the runner is very large in diameter (approximately 3/4 in.) and never completely solidifies from shot to shot.   The runner core remains molten since it is insulated by the solidified outer skin.   In a hot runner system a cartridge heater keeps the material in the runner melted. Usually the melt is in an annular conduit with the cartridge heater forming the core.   Material handling and robotics have also contributed to the processing efficiency.   For large molding operations the resin can be ordered by the truck load (40,000 lb) or the car load (120,000 lb) provided rail service is available (30). The bulk use saves about   2-3¢/lb.   The material is stored in silos and vacuum or air loaded into bins and storage hoppers for processing.   The automatic loading eliminates direct contamination and assures that the machine does not run out of material. Automatic product handling is also an advance of the past 10 years. Robots can remove the molded parts to secondary operations or to a packaging station.   Special handling lines have been set up for handling high-speed, high-volume moldings such as lids or caps. Also, the entire molding plant design has been rationalized.   New plants are built to allow for the progressive flow of raw materials at one end to emerge as finished product for shipping at the other end (31).

Many resins are hygroscopic and require stringent drying conditions prior to processing.   This is particularly true of thermoplastic polyesters, polycarbonates, and nylons (32, 33).   In the past decade efficient drying systems have been developed to predry these resins.   The resins are air-dried by using heated dehumidified air with a drew point of −20 °F or lower.   The drier is operated in conjunction with a hopper loader to keep the resin dry and eliminate manual handling.   Materials that have less stringent drying conditions may be processed via a vented barrel injection molding machine (34).   Vented barrel extrusion has been used for the past 20 years, but vented barrel injection molding has come about only in the seventies.   The moisture is vaporized and goes into solution in the melt as the resin is plasticated.   As the resin is passed past the vent port in the barrel, the moisture and other volatiles come out of solution and are removed via a vacuum line.

Progress has also been made in part cooling.   Portable and centralized chillers have been developed to cool the part in a minimum time.   These closed loop systems use water or water-ethylene glycol combinations as the heat transfer medium.   The application of heat transfer principles allows one to computer model the cooling process (35).   For the new engineering resins such as polyamide-imide and polyphenylene sulfide, one must use mold heating.   The use of heat transfer oils is required since the boiling point of water-ethylene glycol mixtures is exceeded. Another improvement in mold cooling was the introduction of a patented device called Logic Seal (36). Essentially the cooling channels are run at a pressure below atmospheric. This prevents leaks that may occur with broken water lines, and also one gains freedom in the lay out of the support pillars, knockout pins, and vents since they may cut through a cooling channel without causing a leak.

Melt homogeneity in injection molding was further enhanced with the introduction of motionless mixers in the early seventies (37). These devices are placed in the nozzle of the injection molding machine. The mixers enhance the radial uniformity of temperature and composition of the melt. The devices are good for getting better cavity filling and also for saving on color concentrate.

The middle seventies saw the advent of high-speed injection molding. For thin wall parts, the production rate was limited by the hydraulic response of the machine. Nitrogen assist accumulators help inject the melt to reduce the overall cycle time. The early seventies also saw the introduction of co-injection molding. In this process a first melt is injected and cooled. The cavity is rotated and the second melt is injected to form the finished part. Usually both injections use the same material but a different color. The process is extensively used for calculator keys where the different colors eliminate a printing step.

Structural foam injection molding is a major development of the seventies. The process was introduced in the late sixties, but major use has come about in the seventies. Structural foams have a solid skin and a cellular core with density reductions of 0–30% over the unfoamed resin. A major advantage of the foamed parts is the increased rigidity over a similar part of equal weight. The part thickness is usually 1/8–1/4 in. Thinner walls do not achieve foaming while thicker parts have long uneconomical cycle times. Foam molding may be carried out on a conventional machine. A chemical blowing agent is added as in foam extrusion. The melt is injected as a short shot, i.e., insufficient melt to fill the cavity. The melt expands as the gas comes out of solution in the cavity. The surface finish of the part is less than perfect. As gas bubbles on the melt front surface contact the cold wall during flow, they break and the surface freezes against the cold mold wall. A priming and painting operation is required to achieve a class A finish. The best mechanical properties are achieved with a uniform cell size and a density that varies in a parabolic profile from the center to the solid skin. Best results are obtained if the melt is injected in the shortest time (usually less than 1 s). Some very large parts such as warehouse pallets are molded with nitrogen as a physical blowing agent (Union Carbide process). A special machine incorporates the gas into the melt, which is forced into an accumulator and then injected via up to 24 nozzles. A manifold system distributes the melt to the nozzles. The process is considered low pressure (200–500 psi) for the melt and the clamping tonnage is less than 0.25 tons/in.^{2}. Aluminum tooling is most common, and because of the large part size, special molding machines are built for economic reasons.

Blow Molding

The forming of hollow containers is done via extrusion blow molding and injection blow molding. In extrusion, which is historically the older process, a parison or hollow melted tube is extruded from a mandrel die. When the parison is of sufficient length, it is cut off and clamped between two halves of a mold that closes around it. Once the mold is closed, air is introduced to force the expansion of the parison against the walls of the mold. A schematic of the process is shown in Figure 12. Prior to 1958 all extrusion blow molding was done on proprietary machinery. The introduction of German machines by Kautex (19) opened the market to all processors. The extruders must have an L:D ratio of at least 20:1 and a compression ratio of 3 or 4:1 for polyolefins and polystyrene, whereas PVC has a CR of 2:1. Many resins can be formed, and melt temperature guidelines are given by Glanvill (38). The temperature of the melt must be controlled to give sufficient melt strength for the parison. The process is very versatile since large containers such as 55-gal drums may be molded. For large parts a melt accumulator is required, which then is used to quickly form the parison. HDPE has been the most versatile resin, but PVC may be the resin for high growth potential in the eighties (39).

The melt parison is extruded from an annular die. The wall thickness of the parison depends both on the annular gap setting and on the shear rate of the melt in the die. The melt will swell after the parison exits from the die and the die swell increases with increasing shear rate. The die swell is also a function of the temperature, the type of polymer, and its elastic melt properties. Usually the shear rate in the die varies from 10 to as high as 700 s^{-1}. As a rule of thumb one uses a die land length 8 times the annular gap. The parison should normally not be blown up beyond 3:1 (i.e., bottle:parison diameter).

The blowing pressure is usually below 100 psi except for acetal, which can go as high as 150 psi. These pressures allow one to go to aluminum tooling. The clamping force is set at 1.25 times the blowing pressure times the projected area. Other mold materials used are zinc alloy, steel, and beryllium copper.

A problem associated with large parts is "parison droop." This is the deformation of the parison under its own weight, which leads to walls that are too thin in the blown article. The advent of solid-state microprocessor-controlled parison programmers that vary the parison wall thickness via a hydraulically actuated mandrel have reduced or eliminated this problem. The system can save up to 30% in part weight and also in cycle time due to faster heat transfer during cooling. The cycle time can be further reduced by cooling the blown part from the inside as well as the outside. For inside cooling refrigerated air is coinjected with water vapor. The water vapor turns into ice crystals, which reevaporate when they hit the part wall and cause rapid cooling via the phase change.

In addition to the accumulator head, a number of other machine advances have taken place in the past 20 years. These are in-mold prefinish, automated flash trimming, leak detection, and conveying to filling station. Also, dual accumulators allow up to four parisons to be extruded simultaneously and molded on the clamping unit.

Injection blow molding is a new process that has been developed in the past 10 years. The process is usually combined with an orientation step to give axial as well as radial orientation for better mechanical properties. The beverage bottle market is the biggest application using PET resin (40). The process is limited to smaller bottles, 64 oz and smaller. Also, one cannot mold a handle onto the part, and the bottle has to be either round or square. Aside from these limitations, the key advantages are a precision finish and an elimination of trimming, a molded high-tolerance thread, and superior mechanical properties over an extrusion molded part. A common version of the process is shown in Figure 13. In the first stage a preform is injection molded. The preform is moved to a second station and brought to a proper temperature for orientation. A blow pin first stretches the parison to orient it in the axial direction and then blows the container by blowing air inside the parison. The next station is a cooling station while the last station is a part removal station.

Future advances for the injection blow molding process are in tooling for bottles with special tabs, the use of process control to profile the melt injection pressure, and the use of microprocessor control when the volume of machines on the market makes it economical.

Thermoforming

Thermoforming is a process where a sheet material is shaped into a three-dimensional part under the influence of heat and pressure. The process is a relatively old one that has been around for the past 40 years. The major improvements have occurred in throughput rates for lids and caps that are used in the packaging industry. For example, present machines can produce 113,000 coffee cup lids/h on a roll-fed machine in which 110 lids are formed at a time when a 3.5-s cycle is used (19).

The basic thermoforming process has been modified so that there are dozens of variations. These include vacuum forming, pressure forming, plug-assist forming, twin sheet forming, and many others. The size of the machines has increased so that today parts up to 15 ft by 30 ft can be formed on a single station machine (41). The current rotary and continuous machines have decreased their cycle time by switching to mechanical and electromechanical drives instead of pneumatic ones. Also, the tooling has improved with increased heat transfer capabilities to cool the part after forming. Most continuous formers use roll-fed plastic sheets that are as thin as possible. The parts are formed on aluminum tooling that has internal cooling lines for fast heat transfer. Wood plaster and aluminum-filled epoxy molds are seldom used for production runs because of the limited heat transfer.

Trends for the future are the use of improved process controls. Temperature and time controls are now solid state and may be microprocessor based in the future. Temperature control of the heating process is further enhanced by the use of ceramic heaters and "shading" where part of the radiant heat is blocked over a specific area of the sheet. Further trends are for continued energy savings in the process (42). The use of blowers to convectively heat one side of the sheet is one concept. A second

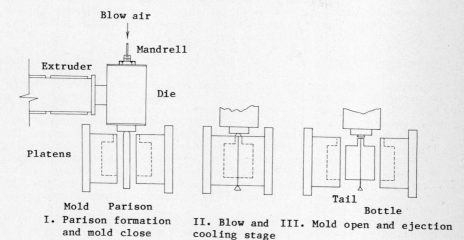

Figure 12. Extrusion blow molding process.

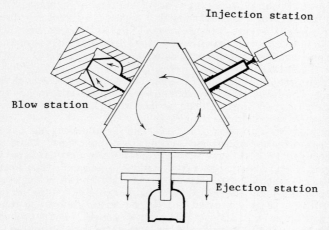

Figure 13. Injection blow molding process.

concept would use the hot sheet from the extruder so that a preheating step for thermoforming would be eliminated. Finally, savings of materials can save energy when thinner sheets of material with equivalent properties can be formed in less time with less energy. The important materials are foamed and unfoamed polystyrene, ABS, polyethylene, polypropylene, and the engineering resins. Coextruded sheets of the combination of materials may become important in the future.

Thermoset Processing

Thermoset processing has long lagged behind the growth of thermoplastics because of high labor costs and long curing cycles. The most common usage of reinforced thermosets is still fiberglass-polyester composites (43). In the past 20 years some processes have been automated and resins have been developed for faster cures. The late sixties saw the introduction of fiber resin forced bulk molding compounds for injection molding. The patented breach-lock injection molding machine by New Britain Plastics Machine Corporation led to the acceptance of that process. The late sixties saw the development of fast-curing phenolic compounds followed by polyester and epoxy compounds.

Sheet molding compounds for use in compression molding have found large-scale applications in the automotive industry. For military and aerospace applications, the hand lay up methods were supplanted by prepreg (B stage) tapes that are, in many cases, applied by automatic tape laying machines. Pulltrusion is a process for making continuous profiles of a thermoset-continuous fiber reinforced composite. High-strength S glass is commonly used as is the Kevlar aramid fiber. Experimental pulltrusions are under way with graphite fibers with epoxy and polyester resins. The pulltrusion process can use either a heated die, an RF dielectric cure, or a UV-initiated cure.

Filament winding of continuous roving for making pressure vessels and pipes has also shown many advances. These advances are concerned with fiber payout and the use of collapsible mandrels for continuous pipe production.

A major innovation over the hand lay up or spray up fiberglass-polyester process has been the introduction of the "resin transfer molding" process (RTM). This process uses low-cost tooling in which a fiberglass preform is produced by chopping strands of glass fibers and blowing the cut fibers against a wire form with a little binder. The form is placed in the mold, and after mold closure a polyester resin is pumped into the mold. The gel time is less than 10 min, and parts can be molded in one to several hours. The process may be competitive with SMC, thermoforming, and structural foam.

Urethane Processing

Advances in urethane processing involve changes in starting chemicals as well as improvements in processes. The chemicals have led to the use of low-cost polyether polyols, high-resilience (HR) "cold cure" foams, safer to handle than prepolymers, catalysts, and chain extenders. A major process improvement was the use of flat-top bun technology for low-density urethane foams. This technology was developed in Europe and appeared in the United States in 1972. Prior to that date, about 14% of the bun had to be trimmed and the scrap had to be chopped and incorporated into other end products.

The major development for urethane molding has been the RIM (reaction injection molding) process (44). The major application has been for automotive parts such as bumpers, fenders, and shrouds, but furniture is also becoming economical (45). The process involves the injection of a reactive liquid mixture into a closed mold where a chemical cure and expansion take place. The process was developed in Germany in the middle sixties but did not become commercially significant until the 1970s. It is anticipated that RIM will be the fastest growing part of the plastics industry for the rest of this century. The cycle time for parts is 1–10 min and the advantages of the process are as follows: large part size; complex part capability; corrosion-resistant paintable surface; lightweight parts; variable properties via variable chemistry; lower capital costs due to a low-pressure process.

The heart of the RIM process is a special high-pressure impingement mixing of the reactive components. The streams impinge at high pressure (3000 psi) to give turbulent mixing in the head without the need for a mechanical mixer. The mixing head design also eliminates flushing the reactants with a solvent. The process will show further evolvement as chopped fibers are being tried as a reinforcement in the process (43) known as RRIM.

Mixing in Plastics Processing

Although mixing is not a processing step to form a finished product, it nevertheless is so important that it should be covered as a separate category. Mixing is important because most polymers in their virgin polymerized form are modified via additives to enhance properties or reduce costs. Because of the high melt viscosity, all of the mixing takes place in laminar flow. A review of the equipment is given by Schoengood (46). The developments of the past 20 years have shown that the batch mixing devices such as Banburys, sigma blade mixers, and pony mixers have been replaced by continuous mixing devices such as twin-screw extruders, special single-screw compounding extruders, the Farrel continuous mixer (FCM), and devices such as the Transfermix. The advantage of continuous devices is the reduction of labor costs and the elimination of batch to batch variations. Twin-screw extruders come in many designs; an analysis and evaluation are given by Rauwendaal (47) and Janssen (48).

The FCM and the twin-screw extruder are probably the most important melt compounding devices. Both, however, are limited in the amount of melt pressure they can generate and are thus often combined with single-screw extruders. Batch mixing of solids is done on ribbon blenders, V shell blenders, drum tumblers, and Henschel mixers. Their principles of operation are reviewed in McCabe and Smith (49) and by Schoengood (46). Mixing of solids is distributive in nature in which the particle distribution of components is randomixed. Intensive melt mixing, on the other hand, leads to a dispersive mixing process where the particle distribution is randomized and at the same time the particle size is reduced.

Motionless mixers were a new class of mixing devices that were introduced in the late sixties. These devices are placed in line and give radial mixing to a stream to homogenize composition and temperature (50). They are used in extrusion, injection molding, and fiber spinning to achieve a radially homogeneous stream.

State of Polymer Processing

Polymer processing has evolved both as a technology and as a science. The first attempts at a scientific treatment of the process calculations were given by McKelvey (1) and by Bernhardt (3). The importance of the technology led to more coverage of the processing analysis. Up to date tests include the books by Tadmor and Gogos (4) and Middleman (2). A review of the processing field concerning its theoretical treatment is given by Suh and Sung (51), while a practical treatment of the technology is given by Frados (30). The analysis of polymer processing has been aided by the ability to measure the rheological properties of the melt. The viscoelastic nature of the melt and the ability to measure and characterize these properties have led to a better understanding of the processes. Technological developments have come primarily from Europe. There is a need in the United States for rigorous formal training in the machinery and process design of plastics. Some universities award a degree in polymer or plastic engineering while in others an option in a traditional discipline such as chemical engineering is possible.

Literature Cited

1. McKelvey, J. M. "Polymer Processing"; Wiley: New York, 1962; p. vii.
2. Middleman, S. "Fundamentals of Polymer Processing"; McGraw-Hill: New York, 1977; pp. 123-9.
3. Bernhardt, E. C., Ed. "Processing of Thermoplastic Materials"; Reinhold: New York, 1959.
4. Tadmor, Z.; Gogos, C. "Principles of Polymer Processing"; Wiley: New York, 1979; pp. 351-60.
5. Tadmor, Z.; Klein, I. "Engineering Principles of Plasticating Extrusion"; Van Nostrand-Reinhold: New York, 1970.
6. Klein, I.; Marshall, D. I. "Computer Programs for Plastics Engineers"; Reinhold: New York, 1968.
7. Glanvill, A. B. "Plastics Engineer's Handbook"; Industrial Press: New York, 1974; p. 76.
8. Willer, A. M. Plast. Des. Process. 1968, 8, 29.
9. Maddock, B. SPE ANTEC 1958, 14.
10. "Modern Plastics Encyclopedia"; McGraw-Hill: New York, 1980-1981.
11. Plast. Technol. Special Plastics Manufacturing Handbook and Buyers Guide Issue, 1979, 25.
12. Glanvill, op. cit., p. 16.
13. White, J. L.; Huang, D. Polym. Eng. Sci. 1981, 21, 1101.
14. Miles, D.C.; Briston, J. H. "Polymer Technology"; Chemical Publishing: New York, 1965; p. 375.
15. Glanvill, op. cit., pp. 74, 75, 117-9.
16. Glanvill, op. cit., p. 96.

17. Sneller, J. Mod. Plast. 1981, 58, 36.
18. Cole, R. J. Plast. Eng. 1982, 38, 35.
19. Staff Plast. Technol. 1980, 26, 53.
20. Spaulding, L. D. Plast. Eng. 1982, 38, 25.
21. Bush, F. R.; Rollefson, G. C. Mod. Plast. 1981, 58, 80.
22. Lasman, H. R. SPE J. 1962, 18, 1184.
23. Schott, N. R.; Weininger, C. SPE ANTEC 1977, 23, 549.
24. Sansone, L. F. SPE Extrusion Div. Newslett. 1982, 11, 3.
25. Rubin, I. "Injection Molding Theory and Practice"; Wiley: New York, 1972.
26. Tadmor, Z.; Gogos, C. "Principles of Polymer Processing"; Wiley: New York, 1979.
27. Smith, D. L. SPE ANTEC 1975, 21, 225.
28. Davis, M. A. Plast. Eng. 1977, 33, 26.
29. Schott, N. R.; Darby, H. In "Energy Conservation in Textile and Polymer Processing"; Vigo, T. L.; Nowacki, L. J., Eds.; ACS SYMPOSIUM SERIES No. 106, American Chemical Society: Washington, D.C., 1979; pp. 9-20.
30. Frados, J. "Society of the Plastics Industry Handbook," 4th ed.; Van Nostrand-Reinhold: New York, 1976.
31. "Injection Molding Operations," 2nd ed.; Husky Injection Molding Systems, Ltd.: Bolton, Ontario, Canada.
32. Williams, D. C. SPE ANTEC 1976, 22, 579.
33. Calland, W. N. SPE ANTEC 1971, 17, 306.
34. DeCapite, R.; Gudermuth, C. S. "The Vented Reciprocating Screw Plasticator"; HPM Corporation: Mt. Gilead, OH 43338.
35. Prasad, A. SPE ANTEC 1974, 20, 532.
36. Logic Devices, Inc., Bethel, CT 06801.
37. Chen, S. J.; Schott, N. R.; Kiang, J. K. SPE ANTEC 1974, 20, 168.
38. Glanvill, op. cit., p. 116.
39. Proctor, D. E. Plast. Eng. 1981, 37, 41.
40. Smoluk, G. Mod. Plast. 1981, 58, 45.
41. Sneller, J. Mod. Plast. 1981, 58, 38.
42. Smoluk, G. Mod. Plast. 1981, 58, 42.
43. Lubin, G., Ed. "Handbook of Fiberglass and Advanced Plastics Composites"; Van Nostrand-Reinhold: New York, 1969.
44. Becker, W. E., Ed. "Reaction Injection Molding"; Van Nostrand-Reinhold: New York, 1979.
45. Sneller, J. "Reaction Injection Molding"; Van Nostrand-Reinhold: New York, 1979.
46. Schoengood, A. SPE J. 1973, 29, 21.
47. Rauwendaal, C. J. Polym. Eng. Sci. 1981, 21, 1092.
48. Janssen, L. "Twin Screw Extrusion"; Elsevier Scientific Publishing Co.: New York, 1978.
49. McCabe, W. L.; Smith, J. C. "Unit Operations of Chemical Engineering"; 3rd ed.; McGraw-Hill: New York, 1976; pp. 895-910.
50. Schott, N. R.; Weinstein, B.; LaBombard, D. Chem. Eng. Prog. 1975, 71, 54.
51. Suh, N. P.; Sung, N., Eds. "Science and Technology of Polymer Processing"; MIT Press: Cambridge, Mass., 1979.

PLASTICIZERS, SOLUTIONS, AND SOLVENTS

Plasticizers

J. K. SEARS, N. W. TOUCHETTE, and J. R. DARBY[1]

Monsanto Polymer Products Company, St. Louis, MO 63167

Historical Perspective
General Theory of Plasticization
Plasticizer Compatibility
Compatibility Stability
Fusion Properties of Plasticizers
Plasticizer Concentration Effects
Heat Stability
Odor Development
Low-Temperature Flexibility, Volatility, Extraction, and Migration
Flammability
Plasticizers for Outdoor Durability
Toxicity and Environmental Concerns

The technology of plasticizers fostered an enormous growth in use of plastics. It provided dozens of industries with a progressively widened spectrum of useful new materials whose flexibility (and concomitant physical properties) could be tailored to a need. It preceded by many years the appearance of synthetic commercial polymers that were inherently highly flexible.

Plasticizers today are used in commercially important amounts in about 35 different polymers with poly(vinyl chloride) (PVC) taking the largest share. Beyond the starting point in cellulosics, plasticizers have become highly important in poly(vinyl butyral), acrylic polymers (especially for surface coatings) and poly(vinyl acetate) (coatings and adhesives) and even in thermosets such as epoxies. Plasticizers are also used in crystalline polymers such as nylon, polypropylene, and poly(vinylidene chloride). Traditionally, extenders for rubber and synthetic elastomers were considered outside the field of plasticizers; yet many such

[1]Current address: Darby Consulting, Inc., St. Louis, MO 63119

extenders are also secondary plasticizers for PVC. Secondary plasticizers are materials not compatible with a resin alone but which become compatible when used with a primary plasticizer. Their principal use is for cost reduction, although at times certain performance properties are improved. Elastomers like ethylene/propylene rubber require high-quality extender oils that approach plasticizers in nature. End-use demands of certain polymers require special plasticizers such as solid plasticizers in hot-melt adhesives. These plasticizers must melt at a temperature just below the temperature of application. They become very inefficient plasticizers or may even revert to solids after the bond is formed. Plasticizers of different kinds have found uses in widely divergent applications: in rocket propellants, in smokeless powder, and in products based on wood and paper. Many are used in nonplastic applications as chemicals or for their per se physical properties--such as optical coupling fluids, adipates and sebacates for lubricants, and phosphate esters for flame-retardant functional fluids.

Worldwide use of plasticizers by the early 1980s was about 7 billion pounds, with the United States' consumption accounting for about 30%. Extensive technical data on evaluation and performance of plasticizers as well as theory of mechanisms have been published (1).

Historical Perspective

Development of plasticizer technology started from very early roots and progressed in step with commercialization of plasticizable resins. Along the way, much of the physical chemistry of plasticization was unraveled and a great many commercial pitfalls were discovered along with the knowledge of how to avoid them.

The history of plasticizers for man-made resins goes back to 1846 when Schoenbein prepared cellulose nitrate, which provided the technology to make a resin that was amenable to plasticization (2). It was first plasticized by Alexander Parkes when he made "Parkesine," the forerunner of Celluloid (3). Parkes went on to produce various articles of plasticized cellulose nitrate. To modify their flexibility and hardness, he tried plasticizing the resin with oils, gums, paraffins, stearine, tar, glycerine, and other substances and varied proportions of those to his "pyroxyline." Cottonseed or castor oils were the preferred plasticizers cited in his master patent issued in 1865.

Parkes, however, became the first financial casualty of unsatisfactory plasticization. His process involved dissolving cellulose nitrate in solvent, mixing in castor oil, and driving off the solvent. But his steps to drive off the solvent were ineffective; the process left behind what in effect was a "volatile plasticizer" and results were disastrous. As this evaporated, the integrity of the plasticized resin deteriorated, resulting in combs and other articles that were shrunken, twisted, and distorted. Customer reaction was predictably unfavorable and forced liquidation of the enterprise. And with this financial collapse, Parkes established a principle: A plasticizer is a high-boiling "solvent" dimensional change.

In 1870, John Wesley Hyatt patented use of an excellent plasticizer for cellulose nitrate: camphor (4). This led to the successful launching of Celluloid. The combination of cellulose nitrate and camphor produced such an outstanding polymer composition that it stood as the standard of comparison for later plasticized compositions. Researchers began to study cellulose acetate as a less-flammable polymer base and to examine hundreds of substances to find a plasticizer that was analogous to camphor's performance in Celluloid. This search successively embraced other polymers as they became known.

In 1910, the Celluloid Company patented use of triphenyl phosphate in combination with cellulose acetate to circumvent the inherent flammability danger of cellulose nitrate (5). Since then, use of phosphate plasticizers to reduce flammability of polymer compounds has become widespread. Other plasticizer types have become standards for their effect on other specific properties.

This early history brings the art to the 1920s. By this time tricresyl phosphate and phthalate esters (dimethyl, diethyl, dibutyl) were in vogue for plasticizing naturally occurring shellac and waxes and the useful cellulosics. Many synthetic polymers had been prepared, and the most promising were poised for commercialization. In 1929, Kyrides applied for his basic patent on di-2-ethylhexyl phthalate (DOP) (6). This phthalate ester burgeoned to become the largest volume plasticizer to be sold up through today. During this period, synthetic thermoplastic polymers came into commercial production: poly(vinyl acetate) in 1925 and cellulose acetate in 1927. Each resin that successively became commercially available stimulated the search for new and improved plasticizers.

In 1931, poly(vinyl chloride) became available. Unplasticized, it appeared to be the most unpromising of all emerging synthetic resins. It was insoluble in common solvents; it could not be molded without thermal decomposition. When heated, it gave off irritating fumes of hydrochloric acid. It turned black after a few days of exposure to sunlight. But this new resin was unusually tough. Commercial success of PVC was assured when Waldo Semon found that upon heating this intractable resin would readily dissolve in such solvents as o-nitrodiphenyl ether, dibutyl phthalate, or tricresyl phosphate and upon cooling form a "rubber-like composition." His vision was keen and broad enough to foresee many of the current uses of plasticized PVC, and these were described in his key patent issued in 1933 (7).

In 1934, about 56 different plasticizers were available commercially, and today some 8 of these are still produced. By 1943, roughly 20,000 were cited in the technical literature and 150 were in commercial manufacture (1). Roughly 60 of these early products survived both the trials of economics and the rigors of change in technology and were still among the approximate 300 plasticizers available in the mid-1960s.

Compounding technology and performance evaluation became more sophisticated. Formulators learned to measure a plasticizer's basic performance more accurately, evaluate its permanence more fully, and assess the commercial value of optimizing particular property modifications. Customers increasingly demanded end products with enduring integrity, through their own standards as well as to meet

specifications issued by agencies such as Underwriters'
Laboratories. Currently, about 460 plasticizers are available with
perhaps 100 being of commercial importance (8).
     The initial design and the commercial survival of a plasticizer
are highly dependent upon imparting a desirable property to the
finished product. The history of this technology is the story of
progressive solutions of technical and commercial problems toward a
goal of marketing excellent-quality products at the most reasonable
price.
     The preponderant portion of plasticizers today is used in PVC.
Consequently, PVC will be the reference polymer used in the balance
of this text to illustrate theory, mechanics, and action of
plasticizers and in relationship to practical applications.

General Theory of Plasticization

By 1950 two theories had evolved to account for the major
"flexibilizing" effect of plasticization: the "lubricity" theory
and the "gel" theory (9). In addition, the "free volume" theory
devised to explain fluid flow was also being adapted to plasticization (10).
     The lubricity theory explains the resistance of a polymer to
deformation. Stiffness and rigidity are explained as the resistance
of intermolecular friction. The plasticizer acts as a lubricant to
facilitate movement of macromolecules over each other, thus giving
the resin an internal lubricity. The gel theory is applied to
predominantly amorphous polymers. It proposes that their rigidity
and resistance to flex are due to an internal three-dimensional
honeycomb structure or gel. The spatial dimensions of the cell in a
brittle resin are small because their centers of attraction are
closely spaced and deformation cannot be accommodated by internal
movement in the cell-locked mass. Thus, the elasticity limit is
low. Conversely, a thermoplastic or thermosetting polymer with
widely separated points of attachment between its macromolecules is
flexible without plasticization.
     PVC has many points of attachment along the chain. Introduction
of a plasticizer separates the macromolecules, breaks the
attachments, and masks the many centers of force for intermolecular
attraction, thus producing an effect similar to what exists in a
polymer with fewer points of attachment.
     Plasticizers of different types have various affinities for
different resins and cohere to macromolecules by forces of different
magnitudes. None are permanently bound. A dynamic equilibrium
between solvation and desolvation is assumed to be established
whereby there is a continuous exchange with one plasticizer molecule
attaching itself to a given force center only to be replaced by
another. The proportion of total centers of force so masked is thus
dependent upon such factors as plasticizer concentration and type,
temperature, and pressure. In partially crystalline resins,
plasticization affects primarily amorphous regions, or regions of
crystal imperfections. The free volume theory gives a mathematical
and thermodynamic perspective to plasticization and clarifies the
mechanism of earlier theories. An increase in free volume between
and around the macromolecules, achieved by heating or by addition of

smaller molecules, enables the macromolecules to move, so softens or plasticizes the polymer (11).

Plasticizer Compatibility

Compatibility signifies the ability of two or more substances to mix intimately to form a homogeneous composition with useful plastic properties. The anomalies of why one solvent will dissolve a given resin, another will only swell it, and a third will leave the resin unaffected have promoted a considerable amount of empirical work and extensive theoretical explanation. Applying thermodynamic theory to real solutions led to several: the Hildebrand solubility parameter δ based on cohesive energy (12), the Flory-Huggins parameter, χ (13), and several newer parameters with sharper precision but which are more difficult to apply.

Cohesive energy density (CED) is a measure of the intensity of a pure substance's intermolecular attraction and is equated with the energy required to vaporize 1 cm^3. The square root of CED provides the useful Hildebrand solubility parameter δ. By this approach, compatibility is predicted when the δ of the polymer and the δ of its intended "solvent" do not differ more than ± 1.5 $(cal/cm^3)^{1/2}$.

By the method of Small (14), a procedure was developed to predict compatibility without determining the energy of vaporization. Small developed "molar attraction constants" for the more common atom groupings in solvents and in macromolecules. By adding these constants for the resin and doing the same for the plasticizer and then dividing these totals by their respective molar volumes, one can obtain solubility parameters (11, 15). When these are used in Hildebrand's comparison, compatibility on the order of the 90% confidence limit can be predicted.

Crowley, Teague, and Lowe (16) recognized four types of interacting forces between molecules: London forces (dispersion), Keesom forces (dipole-dipole), Debye forces (dipole-induced dipole), and Lewis forces (acid-base type, the most important being hydrogen bonding). The solubility parameter is the gross resultant measure of the effect of all four. However, because solubility parameters with the same numerical value can represent different portions of these forces, a similarity between solubility parameters of resin and plasticizer does not invariably assure miscibility. A refinement is needed. Figure 1 illustrates a three-parameter system for predicting compatibility applied to cellulose acetate butyrate. This shows that plasticizers with given dipole moments, hydrogen-bonding tendencies, and solubility parameters that all fall within the solid area will be compatible with the resin.

Because the dielectric constant, DK (ε), is strongly influenced by both dipole moment and hydrogen bonding, this single, simple measurement also provides a predictor of compatibility (17). The classical plasticizers for PVC fall between DK = 4 and 10. A two-parameter system, then, using the Hildebrand solubility parameter (δ) and the dielectric constant gives an improved estimation of compatibility. When each falls within the prescribed limits endemic to the particular resin, that plasticizer will be compatible. Figure 2 illustrates various types of plasticizers and shows their relation to the compatibility area for PVC.

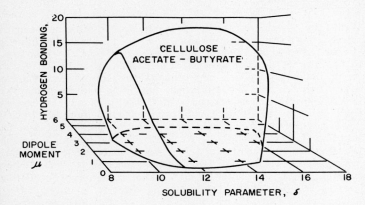

Figure 1. Three-parameter compatibility diagram for cellulose acetate butyrate (16).

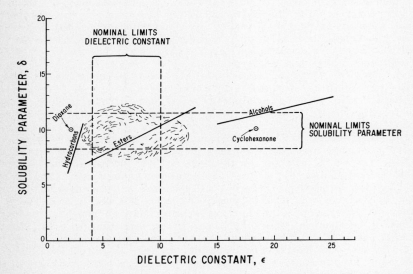

Figure 2. Compatibility of plasticizers with PVC. The common plasticizers lie approximately in the area between solubility parameters 8.4 and 11.4 and dielectric constants 4 and 10, as shown by the cloudy area. The area shrinks as the molecular weight of the plasticizer increases (17).

Some effective PVC solvents possess dielectric constants outside the DK = 4-10 range stipulated for compatibility, but these are characterized by low molecular weight. Figure 3 illustrates how low molecular weight organic compounds (solvents) can be compatible in the resin over a broader DK range than similarly structured high molecular weight compounds (plasticizers). Likewise, the solubility parameter range (as established by Hildebrand) narrows for high molecular weight compounds.

Anagnostopoulos and Coran developed a simplified method of quantitizing the interaction of a polymer and plasticizer by developing a method to establish a chi (χ) polymer-plasticizer interaction parameter (18, 19). It is done by observing a grain of resin in a drop of plasticizer at various temperatures on a microscope hot stage. The solvation temperature profile of a PVC homopolymer in DOP obtained by this technique is shown in Figure 4. Observation of the temperature at which a resin dissolves completely in the plasticizer is determined visually. The χ value is calculated from the depressed melting temperature and Flory's equation (13):

$$\frac{1}{T_M} = \frac{1}{T_M^{\circ}} + \frac{RV_u}{\Delta H_u V_1} (U_1 - \chi U_1^2)$$

T_M = depressed melting temperature, T_M° = melting temperature of the pure polymer, V_u = molar volume of the polymer unit, V_1 = molar volume of the diluent, ΔH_u = heat of fusion per mole of polymer units, U_1 = volume fraction of diluent, and χ = polymer-diluent interaction parameter.

The curve for the χ value as a function of alcohol chain in di-n-alkyl phthalate plasticizers is shown in Figure 5. According to theory, compatibility is assured when the χ value is below 0.5; above this the plasticizer becomes incompatible. In substantiation, dimethyl phthalate and di-n-dodecyl phthalate are both incompatible.

Compatibility Stability

In practice, theories and predictive measurements of compatibility have been frustrated by chemical changes and/or unforeseen physical environments that promoted phase separation, exudation, or serious surface tackiness. Illustrating with a vinyl composition, the di-2-ethylhexyl ester of the isoprene/maleic anhydride adduct (Figure 6) fulfills all the theoretical requirements for plasticizer compatibility. It possesses a structure much like DOP but contains unsaturated aliphatic groups. Aging exposure of a plasticized specimen behind a north window produced a tacky surface in a short time. The product exuded was not the initial compatible ester but a photooxidized, chemically changed product that collected on the surface and could not return to the vinyl matrix. Exudation was aggravated by the type of stabilizer (epoxy, the worst) and accelerated by mild light exposure.

Because exudation caused serious dissatisfaction with plasticized vinyl products during the 1950s, safeguards were developed to prevent such recurrences. Iodine numbers run on potential

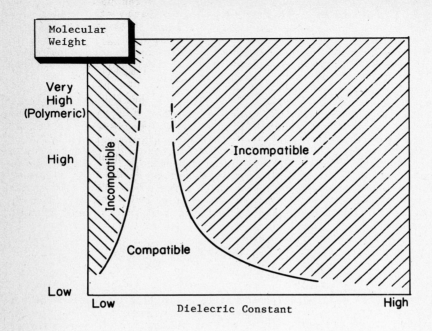

Figure 3.   Influence of plasticizer molecular weight on PVC
            compatibility related to dielectric constant (17).

plasticizers could predict this phenomenon and forewarn formulators.
In consequence, PVC plasticizer markets dwindled for esters of
oleic, ricinoleic, and tetrahydrophthalic acids. Although ingenious
stabilization systems were developed, none were completely
satisfactory, and chemically stable plasticizers such as phthalates
surged in popularity.

A compatibility stability idiosyncrasy of diisodecyl adipate
(DIDA) could pose a troublesome problem. A PVC plastisol formulated
with DIDA and fused in an oven for 10 min at 175 °C will exude; but
if the fusion period is lengthened to 0.5 h, it will not. One
explanation is the dielectric constant of DIDA is below 4; it should
be incompatible and therefore exude. However, upon extensive
heating in air this oxidation-susceptible plasticizer builds up
peroxides, undergoes chain scission and weight loss, and changes in
compatibility. As shown in Table I, if the oxidation phenomenon is
negated in per se DIDA by the presence of an antioxidant, bisphenol
A, the plasticizer remains unchanged in dielectric constant and
predictably would be incompatible (20).

A rash of exudation problems plagued the marketplace as a result
of pressure incompatibility. Vinyl-coated wire tightly wrapped

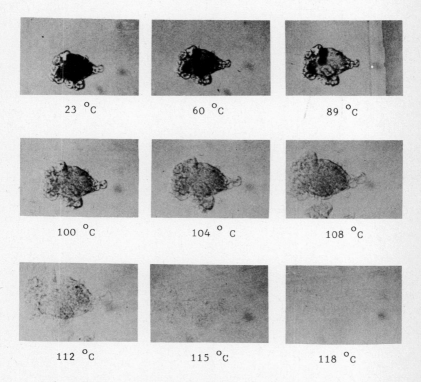

Figure 4.   PVC resin particles heated in DOP (18).

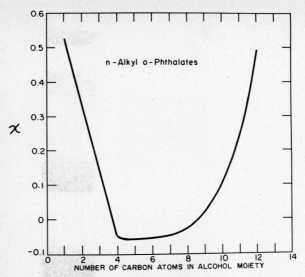

Figure 5.   Phthalate ester compatibility with PVC (19).

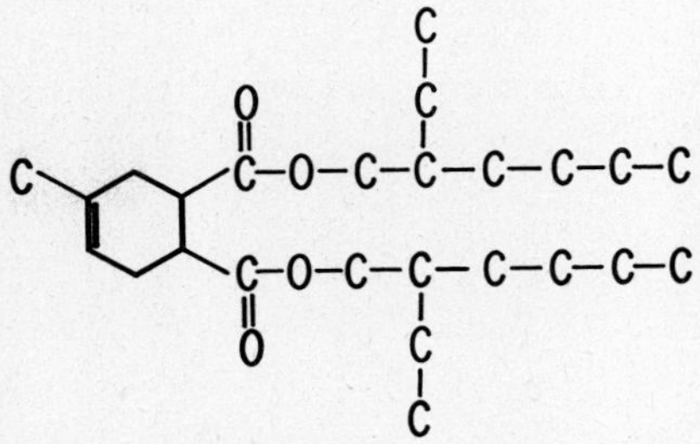

Figure 6.   Di-2-ethylhexyl ester of the isoprene/maleic anhydride
             adduct.

Table I.   Influence of Antioxidant upon Volatility and
            Dielectric Constant of DIDA upon Heating

	DIDA		DIDA + BPA
Days at 105 °C	0	10	10
Weight loss, %	0	20	2
Dielectric constant	3.7	4.3	3.7

around insulators was found to exude plasticizer at the loop. Vinyl floor tiles at the bottom of large stacks showed exudation whereas those tiles at the top of the stack did not. Some upholstery compositions, creased and folded tightly, exuded channels of plasticizer within the fold. To assess such pressure incompatibility, ASTM developed a compression test method for compatibility evaluation (21). By this test, the concentration of secondary plasticizer that will be tolerated by the primary plasticizer can be meaningfully assessed. Such pressure testing proved prudent in evaluating use of chlorinated branched-chain paraffins containing labile chlorine groups. Upon heat processing, these secondary plasticizers liberated HCl and lowered the compatibility of the entire plasticizer system. After thorough heat exposure, the ASTM pressure test can indicate the ultimate compatibility of a plasticizing system in which some compatibilizing groups are thermally destroyed.

Insight into another aspect of potential environmental incompatibility is illustrated when a DOP-plasticized sheet is suspended above a reservoir of hexane as shown in Figure 7. In this atmosphere, the plasticized composition soon begins to increase in weight as hexane is absorbed by the DOP. The weight increases to a given level, surface exudation begins, and finally, drops consisting of hexane and plasticizer drip from the film. When removed from the hexane atmosphere, the incompatibility disappears as hexane evaporates. When returned to the hexane environment, the exudation drip off resumes. Thus DOP, the general standard of compatibility for vinyl plasticizers, shows significant environmental incompatibility in this particular atmosphere. In contrast with DOP, polyester plasticizers do not exude in a hexane atmosphere, as shown in Figure 8, and show only a slight initial weight gain.

Conversely, a DOP composition subjected to water vapor (100% relative humidity at 60 °C) for a whole year shows no exudation or appreciable weight change. But with a polymeric plasticizer of the type shown in Figure 9, environmental incompatibility occurs. In a period of several days at 60 °C, a mixture of water and plasticizer exudes on the surface. When the composition is returned to an environment of 50% relative humidity at room temperature, the signs of incompatibility disappear. Some commercial polymeric plasticizers will exude even at room temperature under 100% relative humidity but will also return to compatibility with a relative humidity of 50%. The polymeric plasticizers absorb water to a degree such that the dielectric constant is no longer optimal for compatibility and thus they exude. Prolonged exposure of polymeric-plasticized compositions to high temperatures and water vapor may eventually cause hydrolysis, but this is not the same phenomenon that can be observed in the initial stages of incompatibility. An ASTM test was developed to determine the compatibility stability of a plasticizer in the presence of water vapor (22).

These environmental compatibility phenomena illustrate the mobile nature of much of the plasticizer within the resin's gel structure. They show the necessity of testing a proposed composition strenuously in all foreseeable environments of intended use. Among such performance tests are exposure to hexane, water vapor, soapy water extraction, and rub-off examinations. The mobility and nature of the plasticizer explain the easy absorption

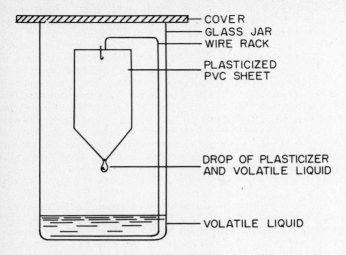

Figure 7.   Environmental compatibility test.

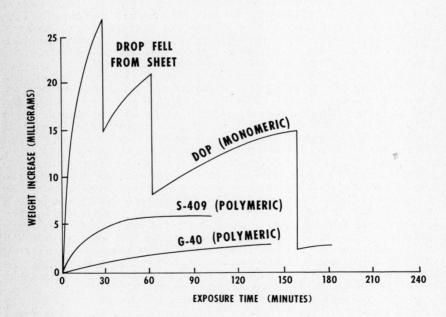

Figure 8.   Environmental effect of *n*-hexane vapor upon weight of a
4-mil film containing 50 phr of plasticizer.

of gases, odors, cigarette smoke, stains, soils, etc., by particular compositions. They explain why such a wide range of plasticizer and resin compositions has had to be "tailored" for the job it is expected to do. Because of this, PVC compositions have been developed to give this polymer the widest range of physical properties of any plastic and to provide the desired performance under the most diverse conditions.

Fusion Properties of Plasticizers

A valuable property imparted by the plasticizer is elongation. Optimization of this property as well as others depends upon adequate fusion time and temperature for thorough plasticization.

   Plasticizers differ significantly in their fusion rates and effects on physical properties. Figure 10 illustrates the differences in solvation between three phthalate plasticizers observed at various temperatures in contact with PVC on a hot-stage microscope. Elongation buildup at 150 °C as a function of fusion time, for three different plasticizers, is illustrated in Figure 11.

   These data underscore the principle that a vinyl composition must be completely fused before physical property measurements become meaningful (23). Relative fusion rates have been determined for a number of plasticizers, and the differences show that a preponderance of aromatic moieties in the plasticizer promotes fusion and favors processing. Thus, butyl benzyl phthalate fuses at a lower temperature than DOP; diisodecyl phthalate fuses at a higher temperature. The relative fusion rates published for a variety of plasticizers provided valuable information for vinyl processors (1). Each plasticizer system for PVC must reach a specific minimum temperature in order to achieve complete plasticization.

   In actual production PVC fusion can be accomplished in several stages in either the hot compounding or the plastisol techniques. Prior to the actual fusion step by hot compounding, a dry blend may be formed where plasticizer is totally absorbed into resin to yield a dry powder (24-26). Fusion is completed in an intensive mixer such as an extruder or Banbury mixer (27). Substitution for a portion of slow-fusing plasticizer, a small amount of fast-fusing plasticizer can produce shorter processing cycles, less work, higher output, or improved appearance than in unmodified formulations (28-31). In plastisols a relatively stable fluid dispersion of resin and plasticizer is prepared at room temperature. Plasticizer influence on fusion is the same in plastisols as in hot compounding, with the added complication of viscosity increase of plastisols (32, 33). Proper choice of plasticizer and resin types gives the balance of gelation-fusion-melt viscosity characteristics necessary for optimum cell formation in chemically blown foam (34, 35).

Plasticizer Concentration Effects

Plasticizing a PVC resin normally reduces modulus, tensile strength, hardness, and low-temperature flexibility while improving elongation. Figure 12 illustrates changes in mechanical properties with increasing amounts of plasticizer in the composition. Initial low concentrations (1-15%) increase modulus and tensile strength but

Stearic ————— Propylene ————⟨ Adipic ————— Propylene ————⟩ Stearic
Acid   Glycol   Acid   Glycol   n   Acid

$$C_{17}H_{35}-\overset{O}{\overset{\|}{C}}-O-CH_2-\overset{CH_3}{\underset{|}{CH}}-O\left(\overset{O}{\overset{\|}{C}}-(CH_2)_4-\overset{O}{\overset{\|}{C}}-O-CH_2-\overset{CH_3}{\underset{|}{CH}}-O\right)_n\overset{O}{\overset{\|}{C}}-C_{17}H_{35}$$

Figure 9.   A linear polyester plasticizer. General-purpose
polymeric plasticizers have molecular weights of about
2000.

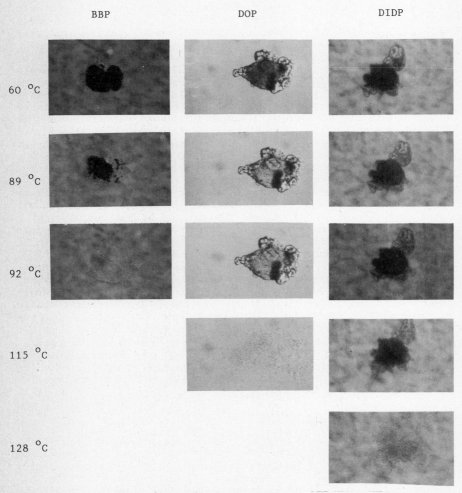

Figure 10.   Relative fusion temperatures of PVC in phthalate
plasticizers (18).

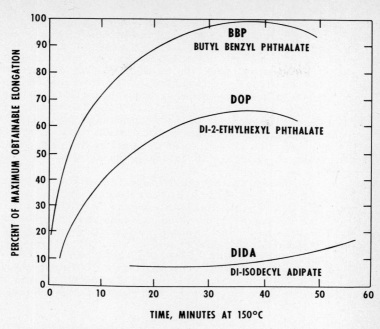

Figure 11. Elongation buildup of PVC plastisols with different plasticizers (23).

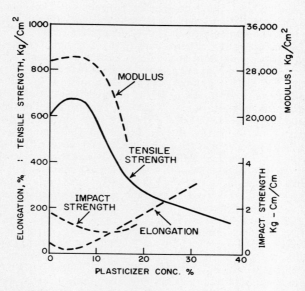

Figure 12. Physical property change of PVC with increasing plasticizer concentration (37).

diminish elongation (36, 37). This effect at low plasticizer concentration, now called antiplasticization, occurs in many resins. In strictly amorphous ones, this stiffening appears to result from a loss in free volume as the macromolecules or their segments initially have the freedom to adjust to more compact arrangements, sometimes guided by dipole attraction or by hydrogen bonding. Addition of more plasticizer eventually adds enough free volume to achieve softness and distensibility. In PVC the decrease in free volume during antiplasticization is accompanied by sufficient molecular order that a marked increase in crystallinity occurs (36). Polycarbonates show this effect to an unusually high degree, and tensile enhancement is valuable in building the strength of strands or fibers (38).

Figure 13 illustrates the elongation profile of the composition when three different plasticizers are used at increasing concentrations (37). Dioctyl adipate emerges from the antiplasticization area at relatively low concentrations and the elongation property builds steeply. The composition responds intermediately with increasing amounts of dioctyl phthalate, and dicyclohexyl phthalate is quite inefficient. When different plasticizers are substituted in a vinyl composition that must still maintain specific mechanical properties, the concentration of the substitutions must be adjusted to accommodate each particular plasticizer's "efficiency factor."

Heat Stability

The necessity to maintain narrow temperature ranges and to reduce processing time at elevated temperatures to avoid heat degradation was the bane of vinyl product manufacturing until midcentury. Lead stabilizers introduced in the 1940s provided good heat stability for processing opaque compositions for electrical insulation. However, substantial improvement in heat stability for both clear and opaque vinyl compositions was achieved only with the advent of epoxidized plasticizers and their synergistic effect with Ba/Cd stabilizers (39). Prior to this, esters of organic acids aided the fusion process and provided a measure of heat stability but only as contrasted to the degradation of the unplasticized resin.

Development of epoxidized soya oil, as an acid-absorbing and low-volatile flexibilizing agent, provided the quantum jump in heat stability. High concentrations of epoxidized soya oil could be used as a replacement for much of the primary plasticizers with attendant improvement in the composition's thermal tolerance.

The problem of heat stability was particularly acute with compositions formulated with phosphate ester plasticizers to achieve improved flame retardance. The thermal degradation products of phosphate esters are acidic and catalyze both ester decomposition and PVC degradation. For example, 2-ethylhexyl diphenyl phosphate begins to decompose thermally at 170 °C, a conventional vinyl-processing temperature, as shown in Figure 14. Diphenyl phosphoric acid decomposes PVC, so its development (along with other more acidic degradation products) during processing must be inhibited. Epoxidized esters stabilize phosphate plasticizers by delaying the appearance of acidic products thereby prolonging the time to initial decomposition.

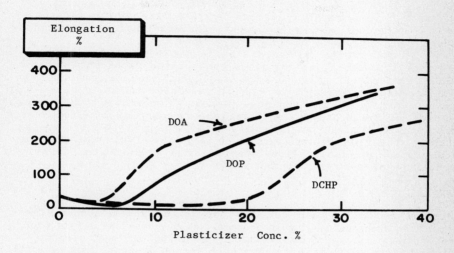

Figure 13. PVC elongation change with increasing concentrations of different plasticizers (37).

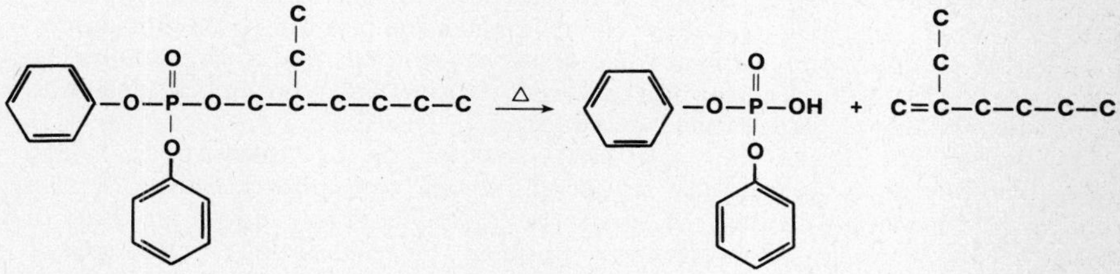

Figure 14. Thermal degradation of an alkyl aryl phosphate.

Recognition that heat stabilization of the plasticizer as well as the PVC resin was necessary greatly assisted researchers working on the problem. Epoxy compounds perform both functions. Today, they are part of many stabilization systems for vinyl compositions, so much so that worldwide use approaches 400 million pounds. In addition to vinyl stabilization, they are also used in phosphate ester functional fluids (i.e., fire-resistant hydraulic fluids) where acid development must also be avoided.

The manufacture of epoxy stabilizers begins with soya and linseed oils or oleate and other unsaturated fatty acid esters. When epoxidation is incomplete, residual unsaturation remains, leading to poor compatibility-stability. This results in exudation on products in the marketplace. Makers of epoxidized fatty acid esters strive to prevent such failure by ensuring more complete epoxidation and supplying products with the lowest possible iodine number. Vinyl formulators further protect themselves by diminishing the proportion of epoxidized stabilizer to an amount ca. 3% while maintaining plasticizer concentration at ca. 35%.

Odor Development

Plasticized vinyl can develop unexpected shortcomings after manufacture and upon aging in storage. Among these are strong odor, color change, weight loss, exudation, bad taste, and poor electrical properties. These effects have been termed "latent degradation" since the real damage was initiated during processing. Late-appearing effects can often be traced to the formation of peroxides. These are colorless and odorless like the original plasticizer but later decompose at room temperature to cause unforeseeable effects (40).

Figure 15 shows in stylized form the oxidation of a typical plasticizer. Initially, there is an induction period when uptake of oxygen is slight and formation of peroxides or hydroperoxides is slow. After this, an autocatalytic stage is reached where the peroxides catalyze additional oxidation.

Vinyl stabilizers can aggravate oxidation or inhibit it. The same is true for pigments, some of which are prooxidants while others are antioxidants. To assure satisfactory odor in use, a chemically stable plasticizer should be selected; the vinyl composition should be well stabilized during processing and antioxidants added to the formulation, if not already present in the "stabilizer package" used. To assess potential odor development, it is prudent to test every new vinyl formulation by heating in a closed bottle at 50 °C for several days.

Low-Temperature Flexibility, Volatility, Extraction, And Migration

A major property improvement achieved through plasticization is the enormous reduction in the glass transition temperature, T_g, of PVC so that pliability and impact resistance of the plasticized composition is retained over a wide temperature range. Lowering of T_g is so important in the free volume theory that it is considered by some synonymous with plasticization. Many essential commercial applications of flexible vinyl depend upon this property modification. Without low-temperature flexibility, some of the

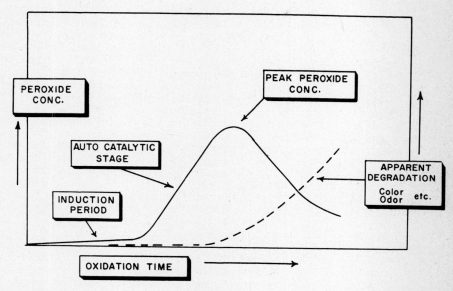

Figure 15. Stylized oxidation pattern of a typical plasticizer (40).

largest markets for plasticized vinyl would not exist. Among these are automotive upholstery, wire and cable insulation, refrigerated food wrap, winter apparel, and shoe soles.

Measurement of low-temperature flexibility was significantly advanced by the method of Clash and Berg. This assessment of torsional modulus was adopted as an ASTM procedure (41). By its use, it can be shown that plasticizing with linear molecular structures enhances the low-temperature flex most efficiently (i.e., lower plasticizer concentration has a more pronounced effect); chain branching and ringed aromatic moieties show lower efficiency. Increasing any compatible plasticizer's concentration in the composition lowers the temperature at which brittleness develops. However, this formulation approach may sacrifice the optimum in other properties, such as tensile strength, modulus, or hand.

Taking DOP as the reference standard of performance for low-temperature flex efficiency, other commercial plasticizers that are significantly better performers include linear dialkyl phthalates and alkyl esters of sebacic, azelaic, adipic, phosphoric, and epoxidized oleic acids. Significantly poorer performers are tricresyl phosphate, diphenyl phthalate, butyl benzyl phthalate, and branched dialkyl phthalates.

The viscosity index of a plasticizer per se is a predictor of its low-temperature flexibility efficiency. A small viscosity change with change in temperature indicates it will confer flexibility at low temperatures better than a plasticizer that shows a greater change in viscosity with temperature change. All the performances rated above conform to this principle. The low-temperature performance of plasticizer mixtures can be closely predicted simply by interpolations based upon their individual performance and the concentration in which they will be used. Formulation techniques have been computerized; by such guidance a wide range of property modifications can be achieved with mixtures and specified property values can be maintained with the indicated substitutions and concentrations (42).

Early researchers who worked on flexible vinyl formulations were convinced that the technique of external plasticizing was doomed because of inadequate "permanence" in the composition. Although good initial properties were attainable, it was felt these could not be maintained. Consequently, a great amount of study was devoted to flexibilizing by "internal plasticization" through copolymerization. Resins so developed, however, are usually found unable to meet the wide range of property demands and therefore in turn require additional external plasticization.

Notwithstanding, plasticizing technology has come a long way to overcome a composition's vulnerability to property change due to loss of plasticizer. Such loss can stem from volatility, leaching or chemical degradation during outdoor aging, water or oil extraction, fungal attack, chemical decomposition, or migration to a contacting surface such as an adhesive or rubber foam. Forestalling such loss through proper selection of plasticizer, adroit formulating and processing techniques, and adequate stabilizing challenged the art of plasticization. As a result, techniques have been developed to assure sufficient longevity of essential properties in the most aggressive environments, and markets have thereby expanded.

Dibutyl phthalate imparts excellent initial properties to PVC compositions, but its high volatility makes it relatively useless for applications requiring durability. A 4-mil film, for example, stiffens at room temperature in 3 months. Films plasticized with DOP or other long-chain alkyl phthalates do not suffer from this impermanence.

Only in exceptional short-lived applications are volatile plasticizers useful. Certain dihexyl phthalates are sufficiently low in volatility to be satisfactory for use in noncritical applications. Volatility loss from vinyl compositions is directly related to a plasticizer's vapor pressure and oxidation resistance. These are dependent upon molecular weight and structure. The effect of molecular structure can be illustrated with two isomeric dioctyl phthalates. Figure 16 shows that the straight chain ester with half the volatility confers much improved low-temperature flexibility compared with the branched structure.

PVC compositions that retain their flexibility perform exceptionally well for wire and cable insulation, and their use for communication wire burgeoned because of their superiority over rubber and other earlier materials. Retention of the elongation property under long-term, high-temperature exposure is a key factor for various applications. Polymeric plasticizers and trimellitate esters such as tris-2-ethylhexyl, isononyl, and *n*-octyl *n*-decyl derivatives show very low volatility '. High molecular weight phthalates approach and overlap their excellent permanence. Retention of elongation compared with volatility is shown in Figure 17. These results were obtained with electrical insulation type compounds, all formulated to equivalent initial modulus. These high-temperature conditions used to assess volatility effects are so severe that internally plasticized polymers can fail on property retention because of oxidative cleavage.

Antioxidants, typified by bisphenol A, can be added to compositions plasticized by "permanent-type" plasticizers of low volatility when marketed for electrical applications. This is another example of an application where both the PVC resin and plasticizer require heat stabilization (43).

Polymeric plasticizers offer a three-fold profile of permanence: low volatility, low oil extraction, and good migration resistance. Trimellitates are even lower in volatility and show low extraction by water but are not oil resistant.

Flammability

Because of its high chlorine content, unplasticized PVC burns with extreme difficulty. Combustibility of plasticized PVC is therefore dependent upon the type and the concentration of plasticizer present. Plasticizers such as phthalates, adipates, and polyesters add fuel value. Both triaryl and alkyl aryl phosphates help inhibit burning of plasticized PVC and can be blended with other types of plasticizers to achieve an acceptable balance of flame resistance and physical properties. In the mid-1950s flammability testing was restricted to rather simple vertical burn test methods. However, in the latter 1950s and early 1960s, flammability of many polymer systems began to receive attention from a variety of agencies and organizations. Test methods and equipment began a rapid development

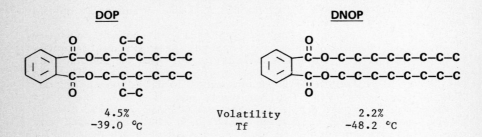

Figure 16.   Effect of branching of isomeric dioctyl phthalates in
PVC.

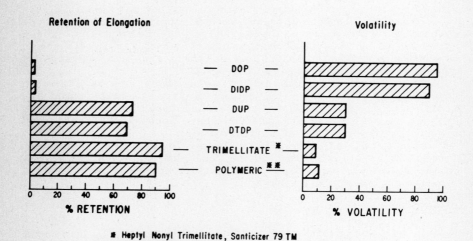

＊ Heptyl Nonyl Trimellitate, Santicizer 79 TM
＊＊ Polyester, Santicizer 429

Figure 17.   Retention of elongation of PVC electrical insulation
with different plasticizers; oven aging (7 days at 136 °C).

to measure flame or ignition resistance of polymers. Later, smoke characteristics assumed an important role in assessing the performance of a polymer system (44).

The mechanism of burning for polymers is believed to take place through thermal pyrolysis of the solid plastic to produce gases that act as fuel for the fire (45). Fire retardants work in both the condensed and the vapor phase to interrupt melting of the polymer and burning of the gases. Triaryl phosphates function well in the vapor phase. Alkyl aryl phosphates are believed to decompose in the flame front to form polyphosphoric acid, which stays in the condensed phase to form char, which reduces flammability and smoke evolution (46, 47).

Highly chlorinated paraffins can be used as secondary plasticizers to reduce flammability and smoke (48). Nonplasticizer additives to assist in flame-retarding plasticized PVC include antimony oxide, alumina trihydrate, zinc borate, and magnesium carbonate (49).

Plasticizers for Outdoor Durability

Plasticizers usually improve light stability and weather resistance of unpigmented PVC. The more common carboxylic acid ester type plasticizers vary from fair to good in this respect. Their primary effect is physical separation of the PVC molecules to retard the spread of resin degradation, but in some cases there is probably chemical inhibition. Other types such as phosphates and epoxies have special chemical effects that can be synergistically beneficial. A small amount of phosphate plasticizer used with a phthalate frequently adds a measure of light stability and fungus resistance (50-52). Epoxy compounds varying from inefficient epoxy resin plasticizers (e.g., bisphenol A diglycidyl ether) to the moderately flexibilizing epoxidized vegetable oils (e.g., ESO, epoxidized soya oil) and to the very efficient octyl epoxystearates and tallates are all used for their excellent heat-stabilizing ability. This carries over into improved outdoor durability.

Artificial weathering devices have helped clarify mechanisms and effects, and although there is no one-to-one correlation with outdoor weathering, they are still very useful for quick screening and for specification of PVC compounds.

Figure 18 shows some of the effects of formulation on the change in physical properties during artificial weathering. The unstabilized sheet began to degrade immediately. The material of Curve 2 lost gloss in the first 1500 h and then degraded rapidly. The UV absorber prolonged service life to 4500 h, equivalent to 3-4 years of continuous weathering in south Florida. This specimen had also retained 70-75% of its original tensile strength and elongation.

Outdoors in south Florida a clear film of rigid PVC, containing stabilizers but no plasticizer, will darken upon exposure to the sun in 2-4 months. By contrast, a clear plasticized PVC film, similarly stabilized but containing no UV absorber, will endure for 10-20 months depending on the plasticizer choice and concentration.

Figure 19 shows clear and pigmented weathered vinyl sheets all plasticized with the general-purpose plasticizer DOP. The clear control developed brown spots from UV degradation of the PVC resin

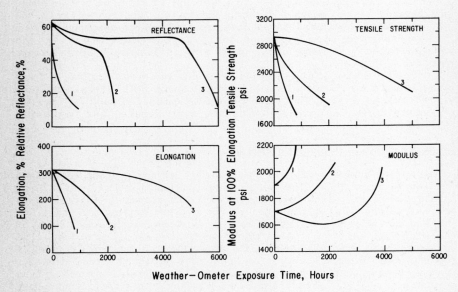

Figure 18. Artificial weathering of plasticized PVC 1 mm (40 mil)
thick. Plasticizer: 50 phr DOP. Stabilizer systems:
Curve 1, no stabilizer; Curve 2, ESO 3 phr,
Ba/Cd/Zn/organic phosphite 2.25 phr; Curve 3, same as
for Curve 2 plus 1 phr of 2-hydroxy-4-
methoxybenzophenone UV absorber. Weathering: Carbon
Arc Weather-Ometer, black panel temperature 145 °F
(63 °C), water spray 18 min every 2 h (52).

Exposure

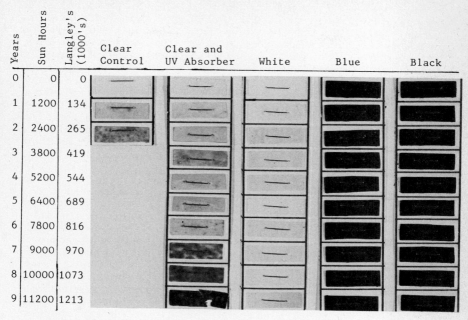

Years	Sun Hours	Langley's (1000's)	Clear Control	Clear and UV Absorber	White	Blue	Black
0	0	0					
1	1200	134					
2	2400	265					
3	3800	419					
4	5200	544					
5	6400	689					
6	7800	816					
7	9000	970					
8	10000	1073					
9	11200	1213					

Figure 19. Weathered plasticized PVC sheet 0.5 mm (20 mil) thick. Basic composition, clear control; PVC 100 parts; DOP 48.5 or 50 phr; ESO 3 phr; Ba/Cd/Zn/phosphite stabilizer 2.25 phr; stearic acid lubricant 0.5 phr. 2-hydroxy-4-methoxybenzophenone, 1.5 phr, was used in the second clear and the white composition. White contained 5 phr of rutile TiO_2, blue 0.9 phr of phthalocyanine blue, and black 1 phr of channel black and 3 phr of $CaCO_3$. Weathering: Miami, Fla., 45° from horizontal, facing due south, direct to sun, starting September 20, 1963 (51, 53). Reproduced with permission from Ref. 61. Copyright 1982 John Wiley and Sons.

in 1.5 years. The UV absorber extended the time to spotting to 5 years, but fungus or mildew had become a problem by 3 years. The decrease in fungus afterward is a result of changes at the surface. The white, blue, and black sheets looked good at 9 years but were getting stiff. Figure 20 shows failure of the two clear compositions in three thicknesses by four different criteria—time to development of five small spots in the sample area, time until the sheet will crack when folded 180°, change in brittleness temperature (modified ASTM D-746), and finally time when the sheet was too brittle to withstand the vigorous winds and rain of Florida and stay on the exposure rack. The shapes of the cold crack curves of Figure 20 and reflectance curves of Figure 18 show the sensitive relationship between cracking and degradation at the surface. A rate of degradation higher at the surface of clear film than throughout may at times be related to incipient incompatibility and ease of migration of the plasticizer type (53, 54). The thicker black sheets were stiffened but were still tough after 15 years. White and blue sheets were intermediate between clear and black.

The effect of structure change in general purpose plasticizers is evident in Figure 21. The number of easily oxidized tertiary hydrogen sites increases from the nearly linear di C_7-C_{11} alkyl phthalate with an average of less than one (methyl) branch per molecule to DIDP with an average of six (mostly methyl) branches per molecule. A decrease in weather resistance tends to follow the same pattern. DOP with its ethyl branching in the 2-position is more resistant to photodegradation than would be predicted. Stabilizers with the organic part of the system tailored for exceptionally good compatibility are required to develop the excellent weather resistance of long-chain linear phthalates. With about 2 phr of metal stabilizer, clear film and sheet can last 7–10 years or more in south Florida. Stabilizer concentrations are frequently adjusted to achieve even longer service life.

Fungus growth can become a problem in warm moist environments when plasticizers containing fatty acids or aliphatic dicarboxylic acid moieties are present to serve as food for fungus growth (55). Although this includes adipates, azelates, sebacates, and polymerics, the ubiquitous epoxidized vegetable oils are the most common food (56). Fungus growth can be controlled in large measure by choice of plasticizer, use of an epoxy resin stabilizer, selection of a fungus-resistant metal stabilizer system, and choice of a PVC synthesized with fungus-resistant surfactants and protective colloids. Phthalate and phosphate plasticizers are relatively inert to fungus. Sulfonamide plasticizers are fungicidal and may be used in small amounts for fungus control. Commercial arsenic and some totally organic fungicides have been devised to withstand the temperatures of PVC processing and are useful for the most aggressive environments. For underground uses (soil burial) as in pond and irrigation ditch liners, fungus and bacterial growth may occur. If this growth is restricted to the surface, physical performance is not impaired. Triaryl phosphates and nonmigrating polymeric plasticizers have shown unusual value in this area (57).

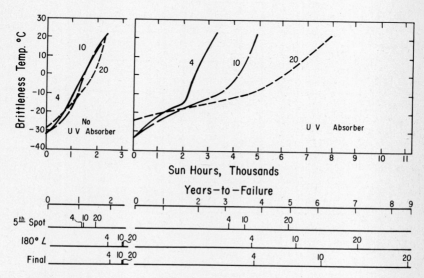

Figure 20.   Weathering of clear plasticized PVC of three
thicknesses.   Compositions and weathering conditions are
those of Figure 21 at 4-, 10-, and 20-mil (0.1, 0.25,
and 0.5 mm) thickness (51, 53).   Reproduced with
permission from Ref. 62.   Copyright 1982 John Wiley and
Sons.

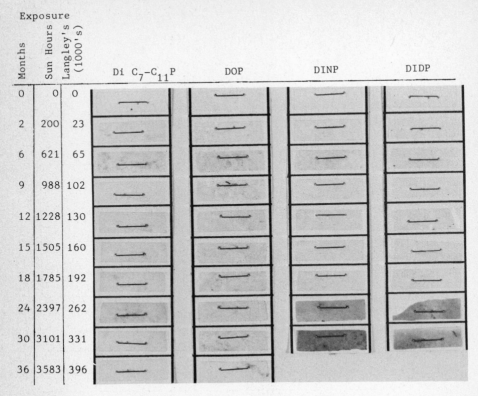

Exposure						
Months	Sun Hours	Langley's (1000's)	Di C$_7$-C$_{11}$P	DOP	DINP	DIDP
0	0	0				
2	200	23				
6	621	65				
9	988	102				
12	1228	130				
15	1505	160				
18	1785	192				
24	2397	262				
30	3101	331				
36	3583	396				

Figure 21.   Effect of alkyl chain branching on weathering of PVC
plasticized with general—purpose phthalate plasticizers.
Formulation: PVC 100 parts; plasticizer 50 phr; ESO 3
phr, Ba/Cd/Zn/phosphite stabilizer system 2.25 phr;
stearic acid lubricant 0.5 phr; 2-hydroxy-4-
octoxybenzophenone UV absorber 1 phr. Plasticizers:
diheptyl, nonyl, undecyl phthalate; di-2-ethylhexyl
phthalate, diisononyl phthalate, and diisodecyl
phthalate. Weathering as in Figure 21 but starting July
1975 (51, 53).   Reproduced with permission from Ref. 63.
Copyright 1982 John Wiley and Sons.

Toxicity and Environmental Concerns

Government regulatory activities beginning in the late 1970s have touched many chemicals including plasticizers and in particular phthalates and adipates. Even earlier, domestic manufacture of polychlorinated biphenyl (PCB) was discontinued because of adverse effects on wildlife due to their persistence.

Evidence on phthalate and adipate toxicity spans a spectrum from a long history of commercial use without health hazard to recent rodent bioassay screening tests that suggest DOP and DOA may be carcinogenic. In early 1982, the EPA and Chemical Manufacturers Association (CMA) agreed to a comprehensive testing program to gain a better understanding of the cause of liver cancer in rats exposed to DOP and DOA in the National Cancer Institute's bioassays and to determine if these products represent a hazard to humans.

The scientific literature contains hundreds of reports of health and ecological tests to confirm the safety of phthalates. In response to a "reporting rule" issued in 1978 by EPA, industry supplied 216 health and safety studies. Several industry-wide symposia from 1972 through 1981 have reported on toxicity and safety of phthalates (58-60). The 1981 meeting in Washington, D.C., was a 3-day "Conference on Phthalates" cosponsored by the National Toxicology Program and Interagency Regulatory Liaison Group. The first day's program presented by industry consisted of papers describing the function of plasticizers, common and unique properties of different plasticizers, and the effect of environmental exposure to phthalates upon workers and the general public. The second day of the program was devoted to a summary of recent toxicity test results and the third day to testing initiatives planned, under way, or needed. Information shared at this conference played a significant role in reaching a better understanding of the environmental and human safety properties of phthalates.

Literature Cited

1. Darby, J. R.; Sears, J. K. In "Encyclopedia of Polymer Science and Technology"; Bikales, N. M., Ed.; Interscience: New York, 1969; Vol. 10, pp. 228-306; In "Kirk-Othmer Encyclopedia of Chemical Technology," 2nd ed.; Standen, Anthony, Ed.; Interscience: New York, 1968, Vol. 15, pp. 720-89. Sears, J. K.; Touchette, N. W. In "Kirk-Othmer Encyclopedia of Chemical Technology," 3rd ed.; Wiley: New York, 1982; Vol. 18, pp. 111-83.

2. Zimmer, Fritz "Nitrocellulose Ester Lacquers: Their Composition, Application and Use"; Chapman and Hall: London, 1934; 246 pp., trans. by Cameron, H. K.

3. Kaufman, Morris "The Chemistry and Industrial Production of Polyvinyl Chloride: The History of PVC"; Gordon and Breach Science Publishers: New York, 1969; 191 pp.

4. Hyatt, John W.; Hyatt, Isaiah S. U.S. Patent 105 338, 1870.

5. Brown, Bruce K.; Crawford, Francis M. "A Survey of Nitrocellulose Lacquer"; The Chemical Catalog Co.: New York, 1928; 368 pp. This is an annotated worldwide bibliography of patents from about 1855.

6. Kyrides, Lucas P. U.S. Patent 1 923 938, 1933.
7. Semon, Waldo L. U.S. Patent 1 929 453, 1933.
8. "Modern Plastics Encyclopedia"; McGraw-Hill: New York, 1980–81; Vol. 57, No. 10-A, pp. 699–709.
9. Doolittle, Arthur K. "The Technology of Solvents and Plasticizers"; Wiley: New York, 1954; 1056 pp.
10. Kaeble, D. H. In "Rheology Theory and Application"; Eirich, F. R., Ed.; Academic Press: New York, 1969; Vol. 5, Chap. 5.
11. Sears, J. K.; Darby, J. R. "The Technology of Plasticizers"; Wiley: New York, 1982; 1166 pp.
12. Hildebrand, J. H.; Scott, R. L. "The Solubility of Non-Electrolytes," 3rd ed.; Reinhold: New York, 1950.
13. Flory, P. J. "Principles of Polymer Chemistry"; Cornell University Press: Ithaca, N.Y., 1953.
14. Small, P. A. _J. Appl. Chem._ 1953, _3_, 71–80.
15. Sears, J. K. "Plasticizers"; Madson, W. H., Ed.; Federation Series on Coatings Technology, Federation of Societies for Paint Technology: Philadelphia, Pa., 1974; Unit 22, 103 pp.
16. Crowley, J. D.; Teague, G. S. Lowe; J. W. _J. Paint Technol._ 1966, _38_, 269–80.
17. Darby, J. R.; Touchette, N. W.; Sears, J. K. _Polym. Eng. Sci._ 1967, _7_, 295–309.
18. Anagnostopoulos, C. E.; Coran, A. Y.; Gamrath, H. R. _Mod. Plast._ 1965, _43(2)_, 141.
19. Anagnostopoulos, C. E.; Coran, A. Y. _J. Polym. Sci._ 1962, _57_, 1–11.
20. Darby, J. R.; Touchette, N. W.; Sears, J. K. _J. Appl. Polym. Sci._ 1970, _14_, 53–64.
21. "Compatibility of Plasticizers in Poly(Vinyl Chloride) Plastics Under Compression," ASTM D 3291—74, 1980.
22. "Plasticizer Compatibility in Poly(Vinyl Chloride) (PVC) Compounds Under Humid Conditions," ASTM D 2383—69, 1981.
23. Graham, P. R.; Darby, J. R. _SPE J._ 1961, _17_, 91–95.
24. Carleton, L. T.; Mischuk, E. _J. Appl. Polym. Sci._ 1964, _8_, 1221–55.
25. Wingrave, J. A. _J. Vinyl Tech._ 1980, _2(3)_, 204–8.
26. Glass, J. E.; Fields, J. W. _J. Appl. Polym. Sci._ 1972, _16_, 2269–90.
27. Roesler, F.; Metz, K. M. In "Encyclopedia of PVC"; Nass, L. I., Ed.; Marcel Dekker: New York, 1977; Vol. 2, Chap. 20, pp. 1039–1120.
28. Cannon, J. A.; Emge, D. E.; Payne, M. T. _J. Vinyl Tech._ 1980, _2(4)_, 250–3.
29. Arendt, W. D. _Plast. Eng._ 1979, _35_, 46–49.
30. Bergen, H. S.; Darby, J. R. _Ind. Eng. Chem._ 1951, _43_, 2404–12.
31. Paul, D. H. _J. Vinyl Tech._ 1979, _1(2)_, 92–97.
32. vanVeersen, G. J.; Dijkers, J. L. C. _Ger. Plast._ 1974, _64_, 10–12; trans. from _Kunststoffe_ 1974, _64_, 292–6.
33. Renshaw, J. T.; Gabbard, J. D. _Polym.-Plast. Technol. Eng._ 1978, _10(2)_, 131–44.
34. Joyce, S. F.; Garlich, R. N. _SPE J._ 1972, _28_, 46–48.
35. Renshaw, J. T. "Parameters Affecting the Processability of Vinyl Foams," _Polym.-Plast. Technol. Eng._ 1976, _6(2)_, 137–56.

36. Horsley, R. A. In "Progress in Plastics 1957"; Morgan, Phillip, Ed.; Iliffe and Sons: London, and Philosophical Library: New York, 1957; pp. 77-88.
37. Ghersa, P. Mod. Plast. 1958, 36(2), 135.
38. Jackson, W. J.; Caldwell, J. R. In "Antiplasticizers for Bisphenol Polycarbonates", "Plasticizer and Plasticization Processes"; Gould, R. F., Ed., American Chemical Society: Washington, D. C., 1965; Chap. 14.
39. Cowell, Elmer; Darby, J. R. U.S. Patent 2 671 064, 1954.
40. Sears, K.; Darby, J. R. SPE J. 1962, 18, 671-7.
41. Clash, R. F., Jr.; Berg, R. M. Mod. Plast. 1944, 21(11), 119-24, 160.
42. Tang, Y. P.; Harris, E. B. "Computer-Aided Vinyl Formulation," SPE 25th Annual Technical Conference, Detroit, Mich., Technical Papers, 1967, Vol. 11, pp. 422-7.
43. Fischer, W.; Vanderbilt, B. M. Mod. Plast. 1956, 33(9), 164-72.
44. Keeney, C. N. J. Vinyl Tech. 1979, 1(3), 172-5.
45. Warren, P. C. SPE J. 1971, 27, 17-22.
46. Keeney, C. N. Adhesives Age 1978, 21(3), 25-30.
47. Hamilton, J. P. Mod. Plast. 1972, 49, 86-89.
48. Price, R. V. J. Vinyl Tech. 1979, 1(2), 98-106.
49. Mathis, T. C.; Morgan, A. W. U.S. Patent 3 869 420, 1975.
50. Darby, J. R.; Sears, J. K.; Touchette, N. W. SPE J. 1971, 27,(2) 32-36, (4) 74-79, and errata.
51. Orem, J. H.; Sears, J. K. J. Vinyl Tech. 1979, 1(2), 79-83.
52. Graham, P. R.; Darby, J. R.; Katlafsky, B. J. Chem. Eng. Data 1959, 4(4), 372-8.
53. Reference 11, pp. 739-60.
54. Wartman, L. H. Ind. Eng. Chem. 1955, 47, 1013-9.
55. Berk, S.; Ebert, H.; Teitell, L. Ind. Eng. Chem. 1957, 49, 1115.
56. Darby, J. R. SPE J. 1964, 20, 738-43.
57. DeCoste, J. B. Ind. Eng. Chem. Prod. Res. Dev. 1968, 7(4) 238-47.
58. "Environmental Health Perspectives," Experimental Issue No. 3, January 1973 (papers presented at the conference on PAE's sponsored by the National Institute of Environmental Health Sciences, Department of Health, Education and Welfare, at Pinehurst, N.C., September 1972).
59. "Flexible Vinyls and Human Safety: An Objective Analysis," SPE Regional Technical Conference, Monticello, N.Y., March 1973, 115 pp.
60. "Conference on Phthalates," sponsored by National Toxicology Program and Interagency Regulatory Liaison Group, Washington, D.C., June 9-11, 1981, 3 volumes, 700 pp.
61. Ibid. Ref. 11, p. 753.
62. Ibid., p. 755.
63. Ibid., p. 748.

Theories of Solvency and Solution

P. E. RIDER

Department of Chemistry, University of Northern Iowa, Cedar Falls, IA 50614

Solubility Characteristics of Solvents and Polymers
Assumptions -- Pure and Applied
  A Volume Change on Mixing
  The Geometric Mean Assumption
  Interaction Energies
The van der Waals Equation and Free Volume Effects
More Formal Treatments of Solutions
Practical Characterization of Polymers and Solvents
Additional Reading

Sixty years ago "The Solubility of Nonelectrolytes" by Hildebrand and Scott was published. Their concepts of regular solutions and of the solubility parameter have been extremely important to the treatment of practical solution problems in the coatings and polymer chemistry field. These concepts have been generalized and broadened over the years to meet the needs of the industrial chemists who must deal with matters associated with solution formation.

When the first edition of "Solubility of Nonelectrolytes" appeared in 1924, the state of the art was summarized by the ancient rule that "like dissolves like." By 1936 the second edition appeared, and the treatment of solutions was becoming more sophisticated. The work of London on the nature of the dispersion forces between nonpolar molecules and that of Debye on polarity and dipole moment now loomed large and "the special effects of hydrogen bonding upon solubility had begun to be appreciated" (1). The third edition, appearing in 1950, devoted a whole chapter to the then new field of high polymer solutions and united the ideas of regular solutions and solubility parameter with the statistical treatment of these solutions developed independently by Flory (2, 3) and Huggins (4, 5).

Concurrent with these developments of the solubility parameter concept, other approaches to the treatment of solution thermodynamic problems were being undertaken. A review of some of these approaches has been given by Prigogine (6). Increasing emphasis,

0097–6156/85/0285–0643$06.00/0

particularly after 1950, was placed on the need for theoretical rigor. In this search for a rigorous and fundamental treatment of solutions, problems of awesome complexity arose. Even for the simplest systems such as mixtures of cryogenic fluids, physical and mathematical approximations were introduced and often reduced the procedures involved in applying rigorous theory to exercises in parametric curve fitting.

For the coatings and polymer chemist, the problems involved in formulating and using a rigorous theory of polymer solutions would appear to be insurmountable. Not only are polymer molecules large and complex, but also strong specific interactions such as polar and hydrogen bonding effects invariably occur in all systems of practical interest. To even attempt these problems, polymer and coatings chemists must be optimists. This situation is not unique to the solution field. Dushman and Seitz (7), in an early paper on quantum theory of valence, quote van Vleck and Sherman as follows: "The pessimist....is eternally worried because the omitted terms in the approximations are usually rather large, so that any pretence of rigor is lacking. The optimist replies that the approximate calculations do nevertheless give one an excellent "steer" and a very good idea of 'how things go,' permitting the systematization and understanding of what would otherwise be a maze of experimental data codified by purely empirical rules....It is futile to argue whether the optimist or the pessimist is right....One thing is clear. In the absence of rigorous computations, it is obviously advantageous to use as many methods of approximation as possible." Thus, in spite of these inherent difficulties in treating solutions, great progress has been made by the optimists; progress, indeed, that throws considerable light on the nature of liquids and solutions. These successes have arisen from the modification and extension of Hildebrand's original solubility parameter concept to complex polymer-solvent systems. In this review, then, emphasis will be placed on solubility parameters and the development of the concept as it has grown over the years far beyond the limitations originally imposed by the theory of regular solutions. Some of the reasons for the applicability of the solubility parameter concept will be described.

Solubility Characteristics of Solvents and Polymers

In 1955 H. Burrell (8) summarized the problems and needs in this field in the following way: "There is no way in which solubilities can be predicted with reasonable assurance except by laborious trial and error....some simple way to predict solubility would be invaluable to paint formulators." He further noted that, ideally, solubility characterization might involve a single numerical constant and that it would be desirable "to have the solubility factors additive in such a way that the solvency of mixtures of solvents could be predicted, since virtually all coating compositions require a properly balanced mixture of solvents."

The only feasible guides at hand at the time for Burrell were the concepts propounded by Hildebrand and Scott in their book "The Solubility of Nonelectrolytes" (1) and specifically the recognition that under certain restrictions the compatibility of two components could be determined by the magnitude of the change of energy on mixing, ΔE^M, given by

$$\Delta E^M = V_m O_1 O_2 (\delta_1 - \delta_2)^2 \tag{1}$$

where V_m is the molar volume of the mixture, O_1 and O_2 are the volume fractions of the two components, and δ_1 and δ_2 are their "solubility parameters." These parameters are defined as

$$\delta^2 = \Delta E^V / V = C.E.D. \tag{2}$$

where ΔE^V is the energy of vaporization of a component, and V is the molar volume of that component. The solubility parameter is thus equal to the square root of the cohesive energy density, CED. Burrell was well aware of the theoretical restrictions on the use of Equation 1, and specifically that Equation 1 was not meant to apply to mixtures in which specific intermolecular interactions occur (e.g., polar and hydrogen bonding effects). Furthermore, there were assumptions in the development of Equation 1 that, as will be discussed later, are critical and that made prospects for the successful use of solubility parameters in complex coating systems dim indeed.

In spite of the many potential reasons why solubility parameters would fail, the fact remains that Burrell's original paper in 1955 was a great success. It showed that order could be brought out of the chaos of industrial experience. Not only could data be systematized post facto, but some progress toward prediction of polymer solubility was made, although many problems still remained. Specifically, the predictions of polymer compatibility with nonsolvent mixtures were successful in only 20% of the examples tried.

Burrell's work showed the path to be followed, and in the 15 years following his 1955 paper, a great deal of progress was made toward putting his work on a more fundamental and quantitative basis. As Burrell suggested, there was a need to quantify polar and hydrogen bonding effects. All the contributions made in this period cannot be covered, but some are particularly noteworthy. Teas, for instance, in his "Graphic Analysis of Resin Solubilities" (9) positioned organic liquids on a triangular chart in such a manner as to indicate the relative strength of the dispersion (London or van der Waals) forces, polar forces, and hydrogen bonding forces. This chart served to characterize quite accurately and with fair confidence the ability of liquids and liquid mixtures to dissolve various polymers. In all approaches of this type the problem was to find a meaningful measure of the relative importance of polar and hydrogen bonding contributions to the solubility parameter. Hildebrand and Scott had already proposed (1) that for polar systems the total solubility parameter, δ, could be given as the sum of a polar and nonpolar contribution, that is,

$$\delta^2 = \Delta E^V / V = \delta_d^2 + \delta_p^2 \tag{3}$$

but the equation was not extensively tested. Gardon (10) also discussed the influence of polarity on solubility parameter, and Lieberman (11) discussed the quantification of a hydrogen bonding parameter for resin solvents. Some of the anomalies reported by Lieberman were discussed by Crowley (12) with reference to his three-dimensional approach to polymer-solvent compatibility problems. Crowley, Teague, and Lowe (13, 14) proposed this three-dimensional approach using as the three axes of a polymer-solvent compatibility chart the solubility parameter as defined by Equation 2, a spectroscopically based hydrogen bonding parameter, and the dipole moment of the solvent. In a further refinement, Nelson, Figurelli, Walsham, and Edwards (15) used a fractional polarity parameter instead of a dipole moment, and a modified spectroscopically based hydrogen bonding parameter that offsets donors against acceptors by assigning negative values to donors and positive values to acceptors. Thus, the calculated average hydrogen bonding parameter for a solvent blend represents the net hydrogen bond accepting capacity of that blend. The great power of such a three-dimensional approach is illustrated by the wide acceptance of the procedures industrially and by the success in reformulating coating formulations subject to environmental protection rules (15).

Instead of parameters such as dipole moment and a spectroscopic hydrogen bonding index to characterize polar and hydrogen bonding effects, Hansen (16-19) extended Equation 3 by assuming the total energy of vaporization could be written as

$$\Delta E^V = \Delta E_d^V + \Delta E_p^V + \Delta E_h^V \tag{4}$$

That is, he assumed that the total energy of vaporization is the sum of the energies required to overcome dispersion force interactions (ΔE_d^V), polar interactions (ΔE_p^V), and the energy required to break the hydrogen bonds in the liquid (ΔE_h^V). This means that the total solubility parameter, δ^2, as defined in Equation 2 can be written as

$$\delta^2 = \delta_d^2 + \delta_p^2 + \delta_h^2 \tag{5}$$

The separation of δ^2 into the three components employs the homomorph concept to determine δ_d as described by Blanks and Prausnitz (20) followed by a subsequent semiempirical estimate of δ_p^2 and δ_h^2 separately.

Summaries of three-dimensional solubility parameters are given by Hansen and Beerbower (21) and by Hoy (22). A problem that now arises, however, and that is of real significance, is the fact that the three-dimensional parameters tabulated by Hansen and Beerbower and by Hoy do not always agree. This disagreement is illustrated in Table I for three common solvents, chosen at random from the source tables. An additional problem, noted by Hansen, is that the homomorph concept for estimating δ_d fails in cases of solvents containing chlorine or sulfur atoms, and, in addition, homomorphs for cyclic compounds are hard to find.

The solubility parameter concept has been proven, in the hard world of industrial application, to have a validity and generality far beyond expectation. It thus has become important from the more fundamental scientific standpoint to try to understand the reasons for these outstanding achievements.

An excellent review of the solubility parameter concept by Barton (23) is available that covers the theoretical and practical treatments of this important tool for the solution chemist.

Assumptions -- Pure and Implied

In examining the reasons for the fundamental use of the solubility parameter concept, it must be recognized that independent of any assumptions in the derivation of Equation 1 and the definition of δ^2 in Equation 2 there is a computational and experimental problem. This problem is often overlooked. Consider mixing two components, each having molar volumes of 100 cm^3/mol and energies of vaporization of 8000 and 9000 cal/mol. From Equations 1 and 2, ΔE^M would be computed as 13.6 cal/mol. If ΔE^V for the first component is in error by only 5%, so that ΔE^V is 7600 cal/mol instead of 8000 cal/mol, the computed value of ΔE_M will be 19.2 cal/mol, almost 50% higher than the earlier value of 13.6 cal/mol. Yet it is rare indeed that ΔE^V is known to within 5%. Table II, compiled by T. P. Nelson (24) shows the range of ΔE^V reported by different observers for five common solvents. Even for these materials deviations ranging from 5% to more than 10% are seen. There is, thus, relatively little experimental data on solvents good enough to test Equation 1 quantitatively. This problem is frequent in dealing with solutions for which the computed quantity of interest (e.g., ΔE^M) depends on the small difference between the large numbers characterizing the components of the mixture. (See also Bagley, Nelson, and Scigliano (25) for a further discussion of this point.)

With these comments in mind, the assumptions involved in the use of Equation 1 as the criterion for compatibility can be examined. Many of the points to be raised have been considered in more detail by Hildebrand and Scott (1, 26) and by Hildebrand, Prausnitz, and Scott (27), but more recent experimental evidence bearing on the subject justifies a reiteration of this earlier material with an emphasis on what now appear to be the critical problems.

A Volume Change on Mixing. The criterion for miscibility of two materials mixed at constant temperature (T) and pressure (p) is that the change in the Gibbs free energy on mixing, ΔG^M, must be less than or equal to zero. ΔG^M is related to the associated changes in entropy, $\Delta S^M_{T,p}$, energy, $\Delta E^M_{T,p}$, and volume, $\Delta V^M_{T,p}$, by the equation

$$\Delta G^M_{T,p} = \Delta E^M_{T,p} + P\Delta V^M - T\Delta S^M_{T,p} \tag{6}$$

$$= \Delta H^M_{T,P} - T\Delta S^M_{T,P} \tag{7}$$

where H = E + PV by definition.

If mixing is carried out at constant T and V, the analogous criterion for miscibility is that the change in Helmholtz free energy, $\Delta A^M_{T,V}$, must be less than or equal to zero. By definition of A,

$$\Delta A^M_{T,V} = \Delta E^M_{T,V} - T\Delta S^M_{T,V} \tag{8}$$

In using the magnitude of $\Delta E^M_{T,V}$ of Equation 1 as a criterion of compatibility, it has been assumed that the term $- T\Delta S^M$ is always negative and hence that the sign of $\Delta A_{T,V}$ is determined by the

Table I.   Comparison of the Total Solubility Parameter,δ, and
           the Nonpolar, Polar, and Hydrogen Bonding Solubility
           Parameters as Tabulated by Hansen and Beerbower (21)
           and by Hoy (22)

Solvent	Table	Solubility Parameters at 25 °C			
		δ	δ_d	δ_p	δ_h
Pyridine	Hansen & Beerbower	10.65	9.3	4.3	2.9
	Hoy	10.62	8.62	4.95	3.734
Styrene	Hansen & Beerbower	9.33	9.1	0.5	2.0
	Hoy	9.33	8.20	4.45	0.0
n-Butyl alcohol	Hansen & Beerbower	11.31	7.80	2.8	7.7
	Hoy	11.60	7.82	4.05	7.547

Table II.   Variation in the Energies of Vaporization of Some
            Common Alcohols as Reported in the Literature.
            (Compiled by T. P. Nelson 24)

Alcohol	ΔE^V (cal/mol)		
	High	Low	Difference
Methanol	8480	7645	835
Ethanol	9948	9405	543
1-Propanol	10,971	10,520	451
2-Propanol	11,260	9930	1330
1-Butanol	12,030	11,320	710

magnitude of Δ_T^M, $_V$ compared to $T\Delta S_T^M$, $_V$. Furthermore, because the question of compatibility is usually a constant (T,P) problem, it has been assumed that $\Delta V_{T,p}^M$ equals 0 so that $\Delta A_{T,V}^M$ and $\Delta G_{T,p}^M$ can be equated and so that we can take $\Delta E_{T,V}^M = \Delta E_{T,p}^M - \Delta H_{T,p}^M$. In actuality ΔV^M is not usually zero for mixtures of organic solvents and polymers. This volume change turns out to be absolutely critical and affects both the entropy change (as will be described later) and $\Delta H_{T,p}^M$, the quantity directly accessible by experiment. Specifically ($\underline{1}$)

$$\Delta H_{T,p}^M = \Delta E_{T,V}^M - T(\partial P/\partial T)_V \, \Delta V^M \tag{9}$$

Because $\Delta E_{T,V}^M$ is commonly of the order of 50 cal/mol for equimolar mixtures of organic molecules and $T(\partial P/\partial T)_V$ is of the order of 100 cal/cm^3, it is evident that because ΔV^M is of the order of 0.5 cm^3/mol, the correction term in Equation 9 is comparable in magnitude to $\Delta E_{T,V}^M$.

Thus, volume changes on mixing are critical and cannot be neglected in any quantitative check of theory. (A review of volume changes on mixing for binary liquid mixtures has been given by Battino ($\underline{28}$)).

The Geometric Mean Assumption. In setting up a model to predict $\Delta E_{T,V}^M$ in Equation 8, a liquid lattice is assumed, and the expression obtained by Scatchard ($\underline{29}$) was

$$\Delta E_{T,V}^M = V_m \phi_1 \phi_2 (C_{11} + C_{22} - 2C_{12}) \tag{10}$$

Here C_{11} and C_{22} are measures of the energies of interaction of component 1 with itself and component 2 with itself, respectively. C_{12} is a mixture property, a measure of the interaction of component 1 with component 2. The geometric mean assumption is that

$$C_{12} = \sqrt{C_{11}C_{22}} \tag{11}$$

which reduces Equation 10 to Equation 1 and eliminates the mixture parameter C_{12}.

This assumption is without theoretical or experimental justification and is made solely for convenience. The effects of deviations of C_{12} from the geometric mean on $\Delta E_{T,V}^M$ are discussed by Hildebrand, Prausnitz, and Scott ($\underline{27}$). Even deviations in C_{12} of only 1% to 5% from the geometric mean average can affect not only the magnitude but even the sign of the calculated $\Delta_{T,V}^M$.

Interaction Energies. Surprisingly, little work on how C_{11}, C_{22}, and C_{12} should be evaluated has been done. Intuitively, the choice of cohesive energy density as a measure of the c_{ii} terms seems reasonable. Nevertheless, Hildebrand and Scott recognized that the thermodynamic quantity $(\partial E/\partial V)_T$, termed the internal pressure of the liquid, P_i, could well be a more meaningful measure of the c_{ii} terms. The internal pressure is accessible experimentally not only on pure components but also on mixtures, and the measurement can be extended to solids. The internal pressure concept arises from the thermodynamic equation of state that is obtained as follows:

$$dE = Tds - PdV \tag{12}$$

$$P_i = \left(\frac{\partial E}{\partial V_T}\right) = T\left(\frac{\partial S}{\partial V_T}\right) - P \tag{13}$$

$$P_i = \left(\frac{\partial E}{\partial V_T}\right) = T\left(\frac{\partial P}{\partial T_V}\right) - P \tag{14}$$

Equation 14 follows from Equation 13 because of the thermodynamic identity $(\partial S/\partial V)_T = (\partial P/\partial T)_V$. Because $(\partial P/\partial T)_V$ can be determined directly, the internal pressure can be measured. Note that the term $T(\partial P/\partial T)_V$, and hence (P_i+P), also arises in Equation 9 in accounting for the effect of volume changes that occur on mixing.

In spite of the importance of P_i studies in understanding liquid state behavior, there has been suprisingly little done since Hildebrand and coworkers in 1928. It has been over 50 years since their initial work, and since that time direct measurements of P_i for only about 50 materials have been made. The experimental data available show that P_i and cohesive energy density are not equivalent, as shown in the review by Allen, Gee, and Wilson (30). They tabulated, for a wide range of liquids, values of n, the ratio of P_i to CED, following Hildebrand and Carter (31), and n was found to vary from 0.32 to 1.3.

Bagley et al. (32) proposed that instead of taking the ratio of P_i to CED, the difference between these quantities should be examined. Subsequent papers showed that this approach provided a very simple interpretation of the internal pressure of liquids (25, 33) even for strongly interacting systems in which polar and hydrogen bonding effects are predominant. These results showed unambiguously that, first of all, energies in the liquid state are additive. This relationship justifies the fundamental separation of dispersion, polar, and hydrogen bonding effects made by workers in the coatings industry and, more specifically, the quantitative expression of this separation as given in Equation 5 by Hansen.

The quantities C_{11} and C_{22}, by this interpretation, thus become directly measurable experimentally through the internal pressures of the pure components at total system pressures not too far removed from atmospheric. Furthermore, from this interpretation, a two-dimensional solubility parameter concept emerges. One of these, δ_v, is a solubility parameter evaluated from P_i and includes the volume dependent terms in the total liquid state energy expression; the second is termed a residual solubility parameter, δ_r, evaluated as the difference between P_iV for a component and ΔE^V. Both δ_v and δ_r are thus directly measured on the pure components (25) and are related to Hansen's three-dimensional solubility parameters by Equations 15 and 16.

$$\delta_v^2 = \delta_d^2 + \delta_p^2 \tag{15}$$

$$\delta_r^2 = \delta_h^2 \tag{16}$$

These equations have been confirmed experimentally. On this basis two-dimensional compatibility diagrams can be made as indicated in Figure 1 for poly(vinyl acetate) (34). In this figure, each point represents a solvent of known δ_v and δ_r. Closed circles represent solvents dissolving the polymer; open circles designate solvents swelling the polymer. Solvents not visibly affecting the polymer are designated by crosses. The "zone of compatibility" is enclosed by the circle, and the center of this circle designates δ_v and δ_r for the polymer. Experimentally, it is found that such two-dimensional compatibility diagrams show fewer "misplaced solvents" (i.e., fewer nonsolvents inside the compatibility zone and solvents "outside") than yielded by the three-dimensional procedures (34).

Note here that no comment has been made about the value of C_{12}. As will be shown, C_{12} becomes an unnecessary quantity by the interpretation in this section. Its place is taken by ΔV^M as will now be shown.

The van der Waals Equation and Free Volume Effects

The experimental results for P_i and the interpretation of these results discussed earlier lead to the conclusion that a modification of the van der Waals equation holds for organic liquids. Thus

$$(P + (\frac{\partial E}{\partial V})_T) \quad (V-b) = BRT \tag{17}$$

has been suggested, where B is a measure of the number of external kinetic degrees of freedom of the molecule as discussed by Haward (35). The use of this equation with B equal to 1 has been described by Bagley and Scigliano (36) and by Bagley, Nelson, and Scigliano (37). The quantity b is considered a "quasi-lattice" volume, and (V-b) is then the "free volume," v_f, for the system. Bagley, Nelson, and Scigliano (37) used their experimental results to show that on mixing two components the thermodynamics of the mixing is described to the accuracy of the data by

$$H^E = P_{im}V_m - (x_1P_{i1}V_1 + x_2P_{i2}V_2) \tag{18}$$

$$S^E = -R \ [x_1 \ln v_{f1} + x_2 \ln v_{f2} - \ln v_{fm}] \tag{19}$$

where the subscripts 1 and 2 refer to the pure components, and the subscript m refers to the mixture. In addition, v_f is defined as

$$v_f = RT/(P + P_i) \tag{20}$$

and P_{im} is related to volume change on mixing by

$$P_{im} = \frac{RT}{x_1 v_{f1} + x_2 v_{f2} + \Delta V^M} \tag{21}$$

Equations 18-21 show the interplay of entropic, energetic, and volume changes on mixing. P_{im} is related to ΔV^M. Internal pressure and ΔV^M each play a part in contributing both to the energetic and

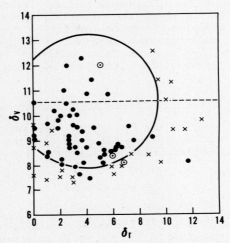

Figure 1. A two-dimensional compatibility diagram for poly(vinyl acetate). Solvents represented by closed circles dissolve the polymer; open circles designate solvents swelling the polymer; solvents not visibly affecting the polymer are designated by crosses (after Scigliano (34)).

to the entropic effects on mixing ($\underline{1}$). The solvent interaction term C_{12} disappears, and its place is taken either by P_{im} or by ΔV^M. Volume changes on mixing, as known to Hildebrand and Scott (Equation 9) are critical in these quantitative relationships.

It does not at present seem possible to determine ΔV^M from a knowledge of pure component parameters alone. The problem is one of packing of the solution components. At the present time ΔV^M is to be regarded as an empirical parameter which must be used in any quantitative treatment of solutions. Semi-quantitatively, the use of two-dimensional compatibility diagrams is a sound method for systematizing polymer/solvent behavior, and it seems reasonable at this stage (and is confirmed experimentally in a limited number of examples) to attribute the misplacement of solvents on these diagrams to abnormal volume changes on mixing.

Using the concepts of this section, Scigliano ($\underline{34}$) has succeeded in calculating the Flory-Huggins parameter, x, as a function of concentration. This is illustrated in Figure 2 for the system polyvinylacetate-benzene at 30 °C. Not only can x be predicted but the entropic and enthalpic contributions, x^H and x^S, can also be computed.

More Formal Treatments of Solutions

The approach used by Flory and Huggins more than 30 years ago to describe polymer solution behavior has proven over the years to be a fundamentally sound and useful approach, although the Flory-Huggins "constant" is found in general to vary with both concentration and temperature and must therefore be regarded as a fitting parameter ($\underline{38}$-$\underline{40}$). Numerous attempts have been made over the years to improve the theory, and considerable progress has been made. Attention can be drawn here to recent papers by Huggins ($\underline{41}$, $\underline{42}$), Maron and Min-Shiu-Lee ($\underline{43}$-$\underline{45}$), Patterson and Delmas ($\underline{46}$), and Patterson and Bardin ($\underline{47}$). Flory in a most interesting and instructive paper has summarized his views on polymer solution theory ($\underline{48}$). The situation with these more formal approaches, however, is that supposedly "constant" parameters are found in general to vary with both concentration and temperature. It is often just as convenient to return to the original Flory-Huggins "constant" as the most useful parameter for correlating experimental data.

Engineering calculations are well illustrated by the successful and informative studies by Heil and Prausnitz in dealing with polymer/solvent systems showing strong interactions including hydrogen bonding ($\underline{49}$) and the treatment of polar systems as described by Blanks and Prausnitz ($\underline{20}$).

Some of Flory's more recent contributions are described in his interesting and illuminating Spiers Memorial Lecture ($\underline{48}$), but recent experimental studies by Chahal et al. ($\underline{51}$) indicate that "....the new Flory theory gives poor predictions as do other models consistent with the Prigogine corresponding states approach." There appears to be, as noted by Gaeckle et al. ($\underline{52}$) "an inadequacy in the treatment of the equation of state effect in these theories." However, as Abe and Flory note ($\underline{53}$), although enthalpies and free energies are only $\underline{fairly}$ well reproduced by their theory, excellent agreement is achieved by using observed excess volumes ($V^E = \Delta V^M$) in place of those calculated according to theory.

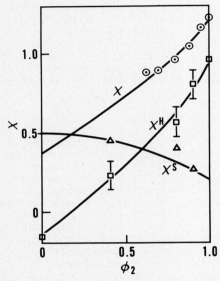

Figure 2. Calculated and experimental values of the Flory-Huggins
parameter x and its entropic and enthalpic components, x^H
and x^S, for the system poly(vinyl acetate)-benzene
(Scigliano (34)). The solid lines are computed and the
points are experimental by Makajima et al. (50).

Practical Characterization of Polymers and Solvents

A quantitative treatment of solution behavior thus requires pure
component internal pressure data plus a knowledge of volume changes
on mixing. Such data are rare.

Compatibility diagrams, though at best only semiquantitative, do
provide guidance in correlating and predicting the behavior of
polymer/solvent systems. Methods of determining the required
solubility parameters for solvents have been summarized by
Hildebrand and coworkers (1, 26, 27). Extensive tabulations of
three-dimensional parameters are provided by Hansen and Beerbower
(21) and Hoy (22). With such characterization information on
solvents, the parameters for polymers can be obtained by examining
compatibility diagrams as in procedures described by Crowley (12,
13), Hansen (16-19), Burrell (8), and Nelson et al. (15).

Other approaches to the solubility characterization of polymers
should be cited. Mangaraj (54) discusses the determination of the
CED of high polymers from swelling measurements. The method is
applied to polyacrylates and polymethacrylates by Mangaraj et al.
(55). The cohesive energy density can be estimated for polymers by
determining the intrinsic viscosity, [η], in a variety of solvents
of differing solubility parameter δ. Mangaraj et al. (56) show that
[η] varies with δ in a simple way described by

$$[\eta] = [\eta]_{max} \exp - V(\delta - \delta_0)^2 \eqno(22)$$

where δ_0, the solubility parameter of the solvent giving the maximum
value of [η], is taken as the solubility parameter characterizing
the polymer. It is rather surprising that this simple experimental
approach to polymer/solvent interaction has not been more
systematically exploited, although Manning and Rodriguez (57) and
Dondos and Benoit (58) have published recently on related effects.

There has been interest also in characterization of polymer/
solvent interactions by gas-liquid chromatographic measurements (59-
62). Microscopic observation has proven a useful tool in the work
of Anagnostopoulos et al. in studying a diluent interaction with
poly(vinyl chloride) (63, 64). There has been related work on
determining the CED of polymers from turbidimetric titration, for
example, that of Suh and Clark (65). Krause and Stroud published on
cloud point studies (66). Mechanical tests can also be employed to
this end (67).

Investigation of the solubility behavior of several organic
solvents, acting as solutes in solutions with high concentrations of
polymer have been carried out recently. These have involved the use
of gas-liquid chromatography and have included measurements of heats
of solution at infinite dilution to characterize such phenomena as
the polarizability of solutes, dipole interactions, hydrogen
bonding, and charge transfer complex formation (68, 69). Other
studies (70-72) have involved measured solute solubilities to
estimate Flory parameters that are used to estimate polymer
solubility parameters.

Rider (73, 74) has developed a method to treat the hydrogen
bonding interactions between solvents and polymers on the basis of a
two-parameter model that depicts the hydrogen bonding donating (b)
and accepting (C) tendencies of both solvent and polymer molecules.

These parameters are determined from experimental results that
include frequency shift data, solubility data, acid-base
characteristics, and other measured properties.

The b and C parameters can be used to estimate the enthalpy of
hydrogen bond formation between donor and an acceptor by using
Equation 22. This equation yields values in units of kcal/mol.

$$\Delta H = b \times C \tag{22}$$

Values calculated for a large number of systems agree quite well
with experimentally determined enthalpies.

By using parameters for solvents and polymers, a second equation
can be used to estimate the hydrogen bonding potential (HBP) that
exists between a solvent and a polymer. This HBP is shown in
Equation 23.

$$HBP = (b_{solv.} - b_{polm.})(C_{solv.} - C_{polm.}) \tag{23}$$

If the value of HBP is negative, formation should be favored. A
positive HBP value corresponds to the opposite case, in which
solution formation is not favored.

On the basis of a statistical analysis of several polymers and a
wide variety of solvents, this model tends to work fairly well,
although there are exceptions. A negative HBP value offers an
approximately 80% probability that either a solution will form or a
significant interaction approaching solubility will occur. A
positive HBP offers an approximately 70% probability that both
solvent and polymer will be insoluble.

The HBP approach can be used to construct "solubility maps" that
place solvents in regions, relative to the b and C parameters for a
polymer. Such placements predict whether or not solution formation
might be expected. The general character of this map is shown in
Figure 3. The regions of expected solubility behavior are
indicated. An example of using this approach for nitrocellulose
with 26 solvents is shown in Figure 4. The solvents appear as
circles, soluble cases are indicated by open circles, and insoluble
cases are indicated by filled circles. An open triangle corresponds
to some interaction between solvent and polymer.

The HBP approach works well for solvent blends also. The b and
C values for a blend are determined from the values for each
constituent weighted according to the mole fraction of the
constituent present.

Additional Reading
_____

In conclusion, the author wishes to call the reader's attention to a
recent publication of Barton (75). In it, he presents a thorough
and critical review of most of the approaches to solubility and
cohesion parameters described in this chapter and summarizes more
recent contributions contained in a large number of references
published since this chapter first appeared. In his treatment,
Barton provides extensive comparisons of the various methods and
techniques (including GLC, liquid, liquid-solid, and gel
chromatography) used to study solubility and cohesion parameters.
Numerous examples as well as detailed tabulations of pertinent data
are provided for a variety of systems. Barton also provides data

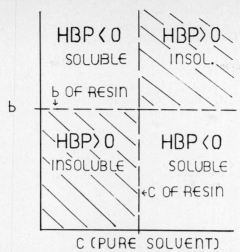

Figure 3.   Solubility map using $\underline{b}$ and $\underline{C}$ parameter approach. Shaded regions correspond to positive HBP values, and open regions correspond to negative HBP values between polymeric resin and solvents. (Reproduced with permission from Ref. 74. Copyright 1983 Society of Plastics Engineers.)

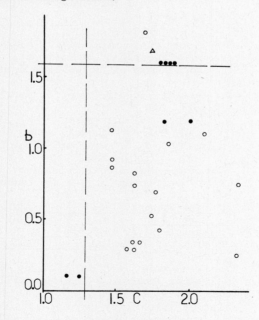

Figure 4.   Solubility map for nitrocellulose and 26 solvents. (Reproduced with permission from Ref. 74. Copyright 1983 Society of Plastics Engineers.)

from the older literature updated in terms of the use of SI units. This reference is highly recommended for those seeking a detailed and comprehensive treatment of solubility and cohesion parameters.

Acknowledgment

The author wishes to acknowledge the efforts and assistance of E. B. Bagley of the Agricultural Research Center, U.S. Department of Agriculture, in Peoria, Illinois, who was the original author of this chapter.

Literature Cited

1. Hildebrand, J. H.; Scott, R. L. "The Solubility of Nonelectrolytes," 3rd ed.; Reinhold: New York, 1950.
2. Flory, P. J. J. Chem. Phys. 1941, 9, 660.
3. Flory, P. J. J. Chem. Phys. 1942, 10, 51.
4. Huggins, M. L. J. Chem. Phys. 1941, 9, 440.
5. Huggins, M. L. J. Phys. Chem. 1942, 46, 151.
6. Prigogine, I. "The Molecular Theory of Solutions"; Interscience: New York, 1957.
7. Dushman, S.; Seitz, F. J. Phys. Chem. 1937, 41, 233.
8. Burrell, H. Official Digest, Fed. Soc. Paint Technol. 1955, 27(369), 726.
9. Teas, J. P. J. Paint Technol. 1968, 40(516), 19.
10. Gardon, J. L. Official Digest, J. Paint Technol. 1966, 38(492), 43.
11. Lieberman, E. P. Official Digest, Fed. Soc. Paint Technol. 1962, 34, 444.
12. Crowley, J. D. "Some Applications of Three-Dimensional Solubility Parameter Theories," presented at Kent State University, 1972, short course on "Solubility Parameters--Their Origin and Use."
13. Crowley, J. D.; Teague, G. S.; Lowe, J. W., Jr. J. Paint Technol. 1966, 38(492), 269.
14. Crowley, J. D.; Teague, G. S.; Lowe, J. W., Jr. J. Paint Technol. 1967, 39(504), 19.
15. Nelson, R. C.; Figurelli, V. F.; Walsham, J. G.; Edwards, G. D. J. Paint Technol. 1970, 42(550), 644.
16. Hansen, C. M. "The Three-Dimensional Solubility Parameter and Solvent Diffusion Coefficient"; Danish Technical Press: Copenhagen, 1967.
17. Hansen, C. W. J. Paint Technol. 1967, 39(505), 104.
18. Hansen, C. W. J. Paint Technol. 1967, 39(511), 505.
19. Hansen, C. W. J. Paint Technol. 1967, 39(511), 511.
20. Blanks, R. F.; Prausnitz, J. M. Ind. Eng. Chem. Fundam. 1964, 3, 1.
21. Hansen, C. W.; Beerbower, A. "Solubility Parameters," Encyclopedia of Chemical Technology Supplement Volume, 2nd ed.; John Wiley & Sons: New York, 1971.
22. Hoy, K. L. "Tables of Solubility Parameters"; Union Carbide Corp., Research & Development Department, South Charleston, W. Va., July 21, 1969.
23. Barton, A. M. F. Chem. Rev. 1975, 75, 731.
24. Nelson, T. P. private communication.

25. Bagley, E. B.; Nelson, T. P.; Scigliano, J. M. <u>J. Paint Technol</u>. 1971, <u>43</u>(<u>555</u>), 35.
26. Hildebrand, J. H.; Scott, R. L. "Regular Solutions"; Prentice-Hall: Englewood Cliffs, N.J., 1962.
27. Hildebrand, J. H.; Prausnitz, J. M.; Scott, R. L. "Regular and Related Solutions"; Van Nostrand Reinhold: New York, 1970.
28. Battino, R. <u>Chem. Rev</u>. 1971, <u>71</u>(<u>1</u>), 5.
29. Scatchard, G. <u>Chem. Rev</u>. 1931, <u>8</u>, 321.
30. Allen, G.; Gee, G.; Wilson, G. J. <u>Polymer</u> 1960, <u>1</u>(<u>4</u>), 456.
31. Hildebrand, J. H.; Carter, J. M. <u>J. Am. Chem. Soc</u>. 1932, <u>54</u>, 3592.
32. Bagley, E. B.; Nelson, T. P.; Barlow, J. W.; Chen, S. A. <u>Ind. Eng. Chem. Fundam</u>. 1970, <u>9</u>(<u>1</u>), 93.
33. Bagley, E. B.; Nelson, T. P.; Chen, S. A.; Barlow, J. W. <u>Ind. Eng. Chem. Fundam</u>. 1971, <u>10</u>(<u>1</u>), 27.
34. Scigliano, J. M. "Considerations of Complex Solutions Using Internal Pressure Measurements"; Doctoral Dissertation, Washington University, St. Louis, Mo., 1972.
35. Haward, R. M. <u>Trans. Faraday Soc</u>. 1966, <u>62</u>(<u>520</u>), 828.
36. Bagley, E. B.; Scigliano, J. M. <u>Polym. Eng. Sci</u>. 1971, <u>11</u>(<u>3</u>), 177.
37. Bagley, E. B.; Nelson, T. P.; Scigliano, J. M. <u>J. Phys. Chem</u>. 1973, <u>77</u>(<u>23</u>), 2794.
38. Booth, C.; Gee, G.; Jones, M. N.; Taylor, W. D. <u>Polymer</u> 1964, <u>5</u>, 353.
39. Brown, W. G.; Gee, G.; Taylor, W. D.; <u>Polymer</u> 1964, <u>5</u>, 362.
40. Allen, G.; Booth, C.; Gee, G.; Jones, M. N. <u>Polymer</u> 1964, <u>5</u>, 367.
41. Huggins, M. L. <u>J. Paint Technol</u>. 1972, <u>44</u>(<u>567</u>), 55.
42. Huggins, M. L. <u>Polym. J</u>. 1973, <u>4</u>(<u>5</u>), 502.
43. Maron, S. H.; Min-Shiu-Lee <u>J. Macromol. Sci. Phys</u>. 1973, <u>B7</u>(<u>1</u>), 29.
44. Maron, S. H.; Min-Shiu-Lee <u>J. Macromol. Sci. Phys</u>. 1973, <u>B7</u>(<u>1</u>), 47.
45. Maron, S. H.; Min-Shiu-Lee <u>J. Macromol. Sci. Phys</u>. 1973, <u>B7</u>(<u>1</u>), 61.
46. Patterson, D.; Delmas, G. <u>Trans. Faraday Soc</u>. 1969, <u>65</u>(<u>555</u>), 708.
47. Patterson, D.; Bardin, J. M. <u>Trans. Faraday Soc</u>. 1970, <u>66</u>(<u>566</u>), 321.
48. Flory, P. J. Discussions of the Faraday Soc., <u>Polymer Solutions</u> 1970, (<u>49</u>), 7.
49. Heil, J. F.; Prausnitz, J. M. <u>AIChE J</u>. 1966, <u>12</u>(<u>4</u>), 678.
50. Nakajima, A.; Yamakawa, H.; Sakurada, I. <u>J. Polym. Sci</u>. 1959, <u>35</u>, 489.
51. Chahal, R. S.; Kao, Wei-Pang; Patterson, D. <u>J. Chem. Soc., Faraday Trans</u>. 1973, <u>1</u>, <u>10</u>, 1834.
52. Gaeckle, D.; Kao, Wei-Pang, Patterson, D.; Rinfret, M.<u>J. Chem. Soc., Faraday Trans</u>. 1973, <u>1</u>, <u>10</u>, 1849.
53. Abe, A.; Flory, P. J. <u>J. Am. Chem. Soc</u>. 1966, <u>88</u>(<u>13</u>), 2887.
54. Mangaraj, D. <u>Makromol. Chem</u>. 1963, <u>65</u>, 29.
55. Mangaraj, D.; Patra, S.; Rashid, S. <u>Makromol. Chem</u>. 1963, <u>65</u>, 39.
56. Mangaraj, D.; Bhatnagar, S. K.; Rath, S. B. <u>Makromol. Chem</u>. 1963, <u>67</u>, 75.

57. Manning, A. J.; Rodriguez, F. J. Appl. Polym. Sci. 1973, 17, 1651.
58. Donos, A.; Benoit, H. Macromolecules 1973, 6, 242.
59. Patterson, D.; Tewari, Y. B.; Schreiber, H. P.; Guillet, J. E. Macromolecules 1971, 4, 356.
60. Tewari, Y. B.; Schreiber, H. P. Macromolecules 1972, 5, 329.
61. Summers, W. R.; Tewari, Y. B.; Schreiber, H. P. Macromolecules 1972, 5, 12.
62. Schreiber, H. P.; Tewari, Y. B.; Patterson, D. J. Polym. Sci., Part 2A 1973, 11, 15.
63. Anagnostopoulos, C. E.; Coran, A. Y.; Gamrath, H. R. J. Appl. Polym. Sci. 1960, 4, 181.
64. Anagnostopoulos, C. E.; Coran, A. Y. J. Polym. Sci. 1962, 57, 1.
65. Suh, K. W.; Clark, D. H. J. Polym. Sci., Part A 1967, 5, 1671.
66. Krause, S.; Stroud, D. E. J. Polym. Sci., Part A2 1973, 11, 2253.
67. Riser, G. R.; Palm, W. E. Polym. Eng. Sci. 1967, 7(2), 110.
68. Bonner, D. C. J. Macromol. Sci. Rev. 1975, C13, 263.
69. Karim, K. A.; Bonner, D. C. J. Appl. Polym. Sci. 1978, 22, 1277.
70. Braun, J. M.; Guillet, J. E. Adv. Polym. Sci. 1976, 21, 107.
71. Merk, W.; Lichtenthaler, R. N.; Prausnitz, J. M. J. Phys. Chem. 1980, 84, 1694.
72. Prausnitz, J. M.; Anderson, T. F.; Grens, E. A.; Eckert, C. A.; Hsieh, R.; O'Connell, J. P. "Computer Calculations for Multicomponent Vapor-Liquid and Liquid-Liquid Equilibrium"; Prentice-Hall: Englewood Cliffs, N.J., 1980.
73. Rider, P. E. J. Appl. Polym. Sci. 1980, 25, 2975.
74. Rider, P. E. Polym. Eng. Sci. 1983, 23, 810, 814-15.
75. Barton, A. F. M. "Handbook of Solubility Parameters and Other Cohesion Parameters"; CRC Press: Boca Raton, Fla., 1983.

Solvents

ROY W. TESS

P.O. Box 577, Fallbrook, CA 92028

Origin and Growth of the Solvents Industry
Methods of Manufacture of Solvents
Solvent Properties and Their Effect on Properties of Solutions
    and Coatings
  Effect of Choice of Solvent on Ultimate Mechanical Properties
    of Deposited Films
  Effect of Solvents on Properties of Latex and Electrodeposition
    Paints
  Surface Tension and Coating Properties
  Viscosity of Resin Solutions as Related to Solvent
    Properties
  Evaporation Rate of Solvents
  Evaporation of Solvent-Water Blends
  Flash Point of Solvents
Air Quality Regulations and Solvents
Formulation of Solvent Blends
Solvents for High-Solids Coatings

For purposes of this discussion solvents are considered to be liquid organic compounds that are used to dissolve and reduce the viscosity of various resins and other materials. Emphasis will be on solvents for use in surface coatings.

In noncoating applications solvents serve as extraction agents for grains, oilseeds, stumps, petroleum products, wood products, and minerals; ingredients and manufacturing aids for toiletries, cosmetics, and drugs; cleaning agents; aids to textile processing and dyeing; hydraulic and heat-transfer fluids; reaction solvents and components in various other end uses and processes.

The rise of the synthetic organic chemical industry has required an ever increasing share of solvents for use as chemical intermediates where solvency and other traditional solvent properties are immaterial. Examples include use of acetone for

0097-6156/85/0285-0661$11.00/0
© 1985 American Chemical Society

manufacture of methyl methacrylate, butanol for butyl acrylate and butylated urea and melamine resins, and xylenes for phthalic anhydride and terephthalic acid. Therefore, increased volumes of solvent production do not mainly reflect increased solvent usage in coatings.

In the 1920s the advent of the synthetic organic chemical industry led to the displacement of many of the older methods for chemical production that made use of fermentation and destructive distillation of natural materials.

Chemical technology advanced rapidly in the 1920s. To meet the need for a forum for discussion of coatings, a new section on paint was started in the American Chemical Society. As an illustration of a topic of the time, D. B. Keyes presented a paper entitled "Solvents and Automobile Lacquers" before the Section of Paint and Varnish Chemistry at Baltimore in 1925. The wealth of data and soundness of formulating principles in this paper may come as a surprise to modern lacquer formulators.

In this chapter, the wide scope of the topic of solvents precludes detailed discussion. Several books (5, 9, 13, 18–23, 30) listed in the Literature Cited section will provide much information for the scientist and technologist.

Origin and Growth of the Solvents Industry

The production of turpentine by distillation of gum from pine trees was the start of the solvents industry in the United States early in the eighteenth century. By 1900 the annual production was 30,000,000 gal and averaged this amount through many years. Turpentine was the solvent of choice for the paint industry, and 80% of the volume in 1922 was used in the paint and varnish industry (1).

As the second major source of solvents in the early 1900s, the fermentation of molasses and grain produced ethyl alcohol. This product had been made in this manner for thousands of years, but its use as a solvent in the United States was delayed until 1906 when denatured alcohol became tax-free if used as a solvent (2).

In the period of World War I, Weizman found that by use of a special strain of bacillus the fermentation of corn gave a mixture of acetone (3 parts), n-butyl alcohol (6 parts), and ethyl alcohol (1 part). Destructive distillation of wood constituted the third major source of solvents in this period; the products consisted of methyl alcohol, acetone, and acetic acid.

In 1925, hydrocarbon solvents derived from coal tar consisted of 5.1 million gal of toluene and 4.0 million gal of solvent naphtha including xylene (3). These volumes were much less than those of turpentine at the time.

The demand for solvents was a big factor in stimulating the production of synthetic chemicals in the 1920s. In 1925, 85% of all aliphatic chemicals were obtained by fermentation, 13% by the destructive distillation of wood, 2% from coal, and 0.1% from natural gas and petroleum (4). The total production of aliphatic chemicals in 1925 was 150 million pounds and reached 9 billion pounds in 1945. Volume of aromatics was about 300 million pounds in 1925. The only aliphatic chemicals made by entirely synthetic commercial routes by 1925 were isopropyl alcohol and three chlorinated hydrocarbons according to McClure and Bateman (4). Volumes of some major lacquer solvents in 1925 were ethyl acetate, 26.7 million pounds; butyl acetate, 16.5 million pounds; and amyl acetate, 1.7 million pounds (3). By 1945, 50% of aliphatic chemicals were derived from natural gas and petroleum, 21% from coal, and 28% from fermentation processes.

The tremendous increase in solvent production since 1925 can be illustrated by the following figures issued by the U.S. Tariff Commission/U.S. International Trade Commission where volumes are in million pounds:

	1925	1973	1981	1982
Acetone	13	1989	2144	1694
Methyl ethyl ketone	0	541	611	468
Isopropyl alcohol	30	1835	1669	1380
Methyl alcohol	5	7064	8577	7554

These figures are misleading as to solvent usage because an increasingly greater share of these products has been used to synthesize other chemicals. Moreover, even when used as solvents, many of the chemicals are used in noncoating applications. Incidentally, the sharp drop in production in 1982 probably was mainly due to the economic recession at that time.

It has been estimated by Doolittle and Holden (2) that, excluding petroleum hydrocarbons, the protective coatings industry in 1933 consumed about 40% of chemicals classified as solvents. By 1949 only 25% of these solvents were used in coatings (5), and by 1972 it was estimated by Stewart (6) that only 14% of oxygenated solvents (which include ketones, esters, alcohols, and glycol ethers) were consumed by the coatings industry. The total consumption of these solvents in coatings was estimated at 1.9 billion pounds for 1972.

Based on data of SRI International, the National Paint and Coatings Association (7) has reported the quantities of solvents used in manufacture of coatings, in thinning, in clean-up operations, and in dissolving resins supplied to the coatings industry (Table I). In 1981 the following quantities of solvents were used: 955 million pounds of aliphatic hydrocarbons, 1215 of aromatic hydrocarbons, 1782 of oxygenated solvents, 21 of chlorinated solvents, and 34 of miscellaneous solvents.

The decrease in solvent usage from 1973 to 1981 was mainly the result of limitations on solvents because of air pollution regulations.

The total U.S. production of solvents in 1973, 1981, and 1982 as reported by the U.S. Tariff Commission/U.S. International Trade Commission is shown in Table II. The sharp decrease in volume in 1982 probably was mainly due to the economic recession.

Methods of Manufacture of Solvents

The methods of manufacture of solvents are much too varied and extensive for discussion at any length in this paper. Early methods are discussed by F. C. Zeisberg in the silver anniversary volume of the American Institute of Chemical Engineers (8). Another

Table I. Estimated Consumption of Solvents in Paints and Coatings
(millions of lb)

Solvent	Consumption	
	1973	1981
Aliphatic hydrocarbons	1455	955
Toluene	690	580
Xylenes	575	465
Other aromatics	210	170
Butyl alcohols	110	122
Isopropyl alcohol	120	110
Ethyl alcohol	200	190
Other alcohols	60	58
Acetone	200	175
Methyl ethyl ketone	380	335
Methyl isobutyl ketone	122	100
Ethyl acetate	120	103
Butyl acetates	115	108
Propyl acetates	60	50
Other ketones and esters	109	100
Ethylene glycol	57	61
Propylene glycol	30	35
Glycol ethers and ether esters	230	235
Chlorinated solvents	15	21
Miscellaneous	40	34
	4898	4007

Note: Data from "NPCA Data Bank Program 1982"; SRI International,
September 1982.

Table II. U.S. Production of Major Oxygenated Solvents
(millions of lb)

Solvent	Production		
	1973	1981	1982
Alcohols			
Methanol	7064	8577	7554
Ethanol, synthetic	1961	1317	1023
Isopropyl alcohol	1835	1669	1380
n-Propyl alcohol	93	154	127
n-Butyl alcohol	519	809	730
Isobutyl alcohol	133	142	156
2-Ethylhexyl alcohol	402	389	325
Ketones			
Acetone	1989	2144	1694
Methyl ethyl ketone	541	611	468
Methyl isobutyl ketone	155	218	131
Diacetone alcohol[a]	51	24	20
Esters			
Ethyl acetate	221	277	235
Isopropyl acetate	45[a]	NR[b]	NR
n-Propyl acetate	35	55	57
n-Butyl acetate	81	124	121
Isobutyl acetate	37[a]	67	68
Glycol ethers			
Methoxyethanol	86	93	89
Methoxyethoxyethanol	14	29	28
Ethoxyethanol	190	206	178
Ethoxyethoxyethanol	28	32	28
Butoxyethanol	138	227	217
Butoxyethoxyethanol	25	50	48
All esters of polyhydric alcohols including (mainly) glycol ether esters			
As a group	220	NR	NR
Glycols			
Ethylene glycol	3278	4142	4309
Propylene glycol	502	473	400
Hexylene glycol	24	NR	NR

Note: Data from U.S. Tariff Commission/U.S. International Trade
Commission
[a]Volumes represent sales instead of production.
[b]NR means no report.

discussion of methods of synthesis is in the original and subsequent editions of the book "Solvents" by Durrans (9). The "Kirk-Othmer Encyclopedia" (10) covers much information on the manufacture of solvents of various types.

One of the most fascinating stories of the coatings industry involves the production of acetone, butanol, and ethanol by the Weizmann process (11, 12). Because the main objective was to produce acetone for explosives, the butanol piled up until it was found that butyl acetate was an excellent solvent for the new nitrocellulose lacquers. Commercial Solvents Corporation (of Maryland) was formed in 1919 to take over the fermentation plants operating at Terre Haute to make butanol and derivatives. The availability of butyl alcohol and the acetate was of major aid in the success of nitrocellulose lacquers in new automobile paints that permitted a reduction in the time required for painting automobiles from 23 days in 1920 to a matter of about 12 h in 1940 (13).

The first totally synthetic route to a solvent in the United States was the synthesis of isopropyl alcohol from propylene by Melco Chemical Corporation in 1917. In 1928 Union Carbide made acetone from isopropyl alcohol; the synthesis of acetone in the cumene-to-phenol process came much later and now is the source of about 85% of acetone production. In 1927 Du Pont began the synthesis of methanol. Synthetic ethyl alcohol was made from ethylene by Union Carbide in 1929. Specialized books on ethyl alcohol (14, 15) and isopropyl alcohol (16) give many details on the manufacture, properties, and uses of these major products.

The oxygenated solvent in greatest use in coatings today (about 335 million lb/yr) is methyl ethyl ketone. In 1932 Shell Chemical became the first commercial producer. It was made from n-butylene by hydration and subsequent dehydrogenation. Methyl ethyl ketone also is currently coproduced with acetic acid by the oxidation of n-butane by other manufacturers.

n-Butyl alcohol was first produced commercially by the fermentation process and later was made from acetaldehyde via the aldol route to crotonaldehyde and subsequent reduction. In World War II the Oxo process was developed in Germany and adapted to U.S. manufacture of n-butanol and isobutanol; propylene plus synthesis gas (CO and H_2) gives a mixture of n-butyraldehyde and isobutyraldehyde (3:1 ratio) which upon hydrogenation yields the corresponding alcohols. 2-Ethylhexanol is made via an aldol route from n-butyraldehyde. The Shell hydroformylation process for n-butanol is basically an oxo process run under conditions that lead to much higher ratios of n-butanol to isobutanol.

The hydrocarbon solvents are straight run distillates as well as products made by various refinery processes including distillation, extraction, alkylation, hydroforming, hydrogenation, hydrocracking, and hydrotreating. Hydrocarbons are classified and characterized by boiling range and composition, viz, various percentages of aliphatic, naphthenic, and aromatic constituents. Solvents that are predominantly aliphatic may be classified as shown in Table III.

Petroleum processes also yield aromatic hydrocarbons such as toluene, xylene, and aromatic concentrates or fractions with boiling ranges of 300-350 °F and 350-400 °F. Until World War II the major source of aromatic hydrocarbons was coal tar (17), but by 1975 only about 5% of aromatic hydrocarbons were derived from that source.

Solvent Properties and Their Effect on Properties of Solutions and Coatings

The power to dissolve resins is the foremost requirement of a solvent except in cases involving dispersions in nonaqueous solvents (NADs) or dispersions in water (latices, emulsions, and dispersions). Theories of solvency and solution are covered by Rider in the preceding chapter. The classic books by Hildebrand and Scott (18) and Hildebrand, Prausnitz, and Scott (19) discuss solubility and solutions in considerable depth. The monumental book by Doolittle (5) covers both theoretical and applied aspects of solvents. Several chapters in the Mattiello series published in 1941-46 deal with solvents; the chapter on lacquer solvents by Bogin (20) contains many early references on the development of solvent technology. Reynolds (21) covers the physical chemistry of petroleum solvents. Physical constants and properties of solvents are compiled by Mellan (31) as well as by Durrans (9). Patton (22) considers properties of solvents as related to paint flow. Some developments in applied and theoretical aspects of solvents are reported by Tess (23). Commercial brochures published by commercial suppliers of solvents contain much useful information; solvent charts listing major physical properties of commercially available solvents are quite useful (24).

Solvents for coatings are sometimes regarded as a source of unwanted expense and simply as a means to permit application of a paint material to a surface. The actual, but often unrealized, facts show that solvents go far beyond this simple function. Some desirable properties of solvent-based paints include the following:

1. Low surface tension causes excellent wetting of substrate which is required for good adhesion of the coating to the surface.
2. Good adhesion promotes good corrosion resistance, impact resistance, and chip resistance.
3. Low viscosity permits penetration of the microscopic rough areas of the substrate and aids adhesion.
4. Easy touch-up and repair of defects in dry film.
5. Low viscosity and controlled rheology and flow give smooth coatings with good gloss.
6. Evaporation rate of solvent and drying of coating can be regulated easily by choice of solvents in solvent blends. Proper evaporation rate helps to give smooth films and to minimize sagging, running, and popping of paint.
7. Evaporation of typical solvent blends is not greatly affected by changes in relative humidity.
8. Application of paints by air spray, electrostatic spray, roller coat, dip, and other means can be readily performed and also can be done in existing equipment.
9. Good acceptance of aluminum pigments for automotive metallic paints.
10. In architectural paints, continuous films are formed even in cold weather; water-stains are sealed effectively with no bleed-through; and fresh films are not removed by rain.
11. Liquid paints are not spoiled by bacteria and molds.
12. Solvent paints can produce wide range of thickness of coating.

13. Molecules of resin are in an extended state to yield maximum flexibility and maximum density of films along with good resistance to permeation by water and gases.
14. General high quality of films minimizes need for frequent repainting.

On the other hand, solvent-based paints have the disadvantage of flammability. Also, they require careful handling because of possible adverse health effects in some cases, and may require control of solvent vapors because of air pollution regulations.

Perhaps the two most important properties of solvents are evaporation rate and solvent power. Solvent power is related to various fundamental solution parameters as discussed in the chapter on theories of solvency and solution. However, solvent power also influences viscosity and the orientation of molecules which in turn affects many other properties.

Effect of Choice of Solvent on Ultimate Mechanical Properties of Deposited Films. Kauppi and Bass (25) cast ethylcellulose films from combinations of toluene and ethyl alcohol in various proportions. As shown by Figure 1, the best tensile strength and elongation occurred when about 70 to 90% toluene was used in the solvent blend. These mechanical properties degraded rapidly with increasing proportions of ethanol. Minimum viscosity of the solutions was attained when the solvent blend contained about 70% toluene. It seems reasonable to ascribe the superior film properties to the shape and orientation of the polymer molecules in the solution and in the final film. It is suggested that the superior films consist of molecules in relatively extended random configuration with a high degree of chain entanglement, whereas the weaker films consist of molecules in relatively more coiled states with fewer intermolecular enganglements.

By depositing ethylcellulose films from ethanol and toluene, Doolittle (26) confirmed the observations of Kauppi and Bass and also measured the density and permeability of the films to water vapor, oxygen, and nitrogen as shown in Table IV.

The film deposited from the toluene solution was lower in density and higher in gas transmission than the film from ethanol. As indicated by its lower density the film from toluene can be considered to have more free volume than the film from ethanol. The more rapid diffusion of gases through the less dense film is consistent with this viewpoint. These differences in properties can be attributed to differences in orientation and shape of the molecules as they are deposited from the solutions.

It is evident from these results, as well as from results cited later in this discussion, that the choice of solvent is of critical importance in achieving the best film properties from a resin system. The cheapest solvent combination may not and probably will not give best film properties. It is also evident that in the process of evaluating the merits of resins, conclusions could be erroneous unless the optimum solvent blend is chosen for each resin. Extrapolating these considerations further, it can be expected that the same resin could yield entirely different properties if it is applied to a surface in different forms, for example, as a solution, a dispersion in solvents, a dispersion in water, or as a powder.

Table III. General Classification of Aliphatic Hydrocarbon Solvents

Type	Boiling Range
Rubber and extraction solvents	140–200 °F
Lacquer diluents	200–250 °F
VM and P naphthas	240–300 °F
Mineral spirits	300–400 °F
Heavy solvents	400–600 °F

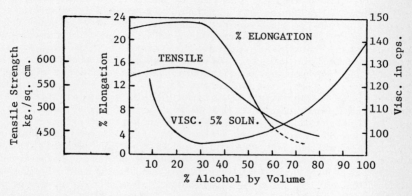

Figure 1. Effect of composition of toluene–ethanol solvent on ethylcellulose properties.

Table IV. Properties of Ethylcellulose Films

Property	Film from Ethanol	Film from Toluene
Density, g/cm^3	1.1454	1.1428
Permeability		
Water vapor, $mg/cm^2/h$	4.2	5.06
Oxygen, $mL/100 \ in.^2/day$	1101	1322
Nitrogen, $mL/100 \ in.^2/day$	402	436

Quite obviously, the solvent has a function far beyond that of aiding the delivery of a resin or paint to a surface.

Effect of Solvents on Properties of Latex and Electrodeposition Paints. Solvents are usually used in latex paints to enhance film integrity, strength, and flexibility; to obtain maximum smoothness of film; to aid coalescence of latex particles especially at low temperatures; to aid scrub resistance; and generally to enhance properties. Excessive quantities of solvents can cause pigment flocculation upon storage and ruin freeze-thaw stability of paint. Ester solvents have a tendency to hydrolyze; the generated acidity results in pH drift, change in viscosity of paint, and hydrolysis of poly(vinyl acetate) polymers whereupon acetic acid and gelatinous poly(vinyl alcohol) may be formed. Various types of solvents have been found to be useful. Glycol ethers such as the monobutyl ether of ethylene glycol, or preferably of diethylene glycol, are excellent coalescing agents for latex paints. Normally, about 2% weight of solvent based on total weight of paint is very effective. Other coalescents include hexylene glycol and stable esters. The coalescent is balanced with an equal weight of ethylene glycol (except when using hexylene glycol) to prevent or ameloriate degradation of freeze-thaw resistance often caused by coalescents; some coalescents degrade freeze-thaw stability to such a great extent that addition of glycol cannot overcome the problem.

Hoy (27) has examined some factors that influence the efficiency of coalescing aids. An important factor is the distribution of solvent between aqueous and polymer phases at the critical time during film formation. Water-soluble solvents may be lost from the coating through wicking action into the substrate.

The evaporation of water and solvent from latex films has been the subject of several studies. Sullivan (28) investigated the evaporation of water, ethylene glycol, and coalescent solvents (mostly butyl ethers of ethylene glycol and diethylene glycol) from cast films, at 25 °C and 50% relative humidity, of clear and pigmented acrylic latex polymers. Prior to film formation, evaporation of volatiles is controlled by surface resistance phenomena. After film formation, evaporation of volatiles may or may not be diffusion-controlled, depending upon the solvent involved and the interactions among the volatile materials. The presence of a continuous hydrophilic network throughout the latex film facilitates the evaporation of solvents, especially the more polar solvents that tend to evaporate more rapidly than the less polar solvents that partition more strongly to the polymer phase. Evaporation rates of solvent from pigmented paint films were minimal for paint at about the critical pigment volume concentration (cpvc); the pigment particles below the cpvc apparently acted as barriers to solvent passage through the film. The important conclusion was reached that the high-boiling solvents used in latex paints evaporated quite rapidly and that remarkably little of these solvents remained in the films for appreciable times.

An early investigation by Tess and Schmitz (29) on styrene-butadiene paints demonstrated that use of hexylene glycol improved tensile strength, elongation, leveling, and scrub resistance of latex paints. Electron micrographs of the films, obtained by a replica technique, showed that use of the coalescent resulted in much smoother films (Figures 2a, 2b, and 2c). These micrographs were made on alkyd-modified styrene-butadiene latex paints at 45% pigment volume concentration.

It has been shown by May (30) that solvents have several beneficial effects in electrodeposition coatings where they may comprise about 2-3% of the paint bath. One purpose in using solvents is to dissolve the resin to provide ease of handling during the preparation of the aqueous solutions. A second purpose is the attainment of smoother films. A third reason is that the solvent also can confer better water solubility characteristics on the resin and help maintain bath stability. The best solvent for any electrodeposition paint depends on the resin used. In general the monobutyl ethers of ethylene glycol and diethylene glycol were the most versatile of nine solvents tested, but in some cases good results were also obtained by use of 2-ethoxyethanol, hexylene glycol, diacetone alcohol, and possibly even *sec*-butyl alcohol or isopropyl alcohol.

Surface Tension and Coating Properties. Surface tension has a strong influence on several important coating properties such as gloss, surface texture, floating and flooding of pigments, and adhesion of films. The effects of surface tension are usually a consequence of the basic fact that a liquid of lower surface tension will wet and spread over another material of higher surface tension or surface free energy. The overall process can occur if the net change in free energy of the system is negative. Water has a high surface tension of 72.7 dyn/cm, but the addition of surfactants as may be present in latex paints will reduce this value considerably. Most oxygenated organic solvents have surface tensions of about 24–30 dyn/cm, aromatic hydrocarbons of about 28 dyn/cm, and aliphatic hydrocarbons of about 18-24 dyn/cm. Because resins have higher surface tensions than solvents, for example, 42.6 dyn/cm for polystyrene, 44.6 dyn/cm for an epoxy resin, and 36.5 dyn/cm for poly(vinyl acetate), it is evident that the dissolution of resins in solvent will reduce surface tension and aid the wetting of a substrate whether it be a metal, plastic, or other surface. Increase in temperature reduces the surface tension of a liquid, and advantage of this situation can be made in some instances. Zisman (32) has developed a method of measurement and the concept of the critical surface tension of a substrate as being equal to the (maximum value of) surface tension of a series of liquids that will spontaneously spread when applied to the surface.

When the solvent from paint films evaporates, there may develop a circulatory motion within cells (Benard cells) in the body of the liquid film. This vortex action was described by Bartell and Van Loo in 1925 and has been extensively reviewed and interpreted by Hansen and Pierce (33, 34). Each cell has an hexagonal shape, and the overall pattern of the dried film consists of raised ridges at the outer edges of closely packed hexagons. In each cell liquid moves upward at the center of the cell, spreads outward at the surface to the edges of the hexagon, and then moves downward to the depths of the film and completes the circular motion. As the liquid moves across the surface it cools and becomes more concentrated as solvent evaporates. As a result of both of these factors the surface tension increases from center to edge of each cell. Liquid

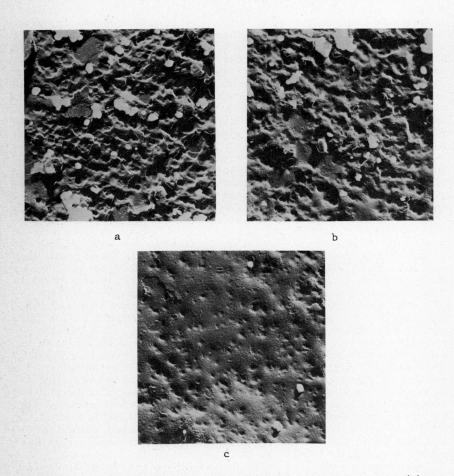

Figure 2. Electron micrographs of latex paint containing (a) no
hexylene glycol; (b) 20% hexylene glycol on polymer content,
equivalent to 0.31 lb HG per gal.; and (c) 40% hexylene glycol,
equivalent to 0.62 lb HG per gal.

builds up into a ridge at the perimeter of each cell and thereby forms a hexagon pattern in the dried film.

The vortex action in Benard cells can cause the defects known as flooding and floating because pigments of different sizes and weights will move at different velocities, will separate, and show nonuniformity of colors in films. The pigments also can be concentrated at certain local spots and thus leave pure binder at other spots that are vulnerable to film failure upon exposure.

Other surface irregularities of films including frosting, wrinkling, and cratering are related to surface tension effects in films. Dannenberg, Wagers, and Bradley (35) have shown that some defects are caused in many cases by a small particle of extraneous matter such as dust. Hahn (36) has discussed cratering and its relationship to surface tension in some detail.

Although the formation of surface patterns via the Benard cell may occasionally be useful to form special finishes, usually it is desirable to eliminate the situation. Use of higher boiling solvents will retard evaporation rate and the cooling effect that changes surface tension and propels the vortex action. Increase in paint viscosity will inhibit the action, as will decrease in film thickness. Addition of a surfactant will provide a more uniform value of surface tension and also retard evaporation and thereby help to inhibit the vortex motion in cells.

Surface tension forces are important in the process of film formation from latices. Brown (37) has developed equations that include surface tension as the prime force in the coalescence process. Use of solvent in the latex can change the modulus of the polymer, which is also a critical factor in regulation of coalescence temperature.

When clear films are applied and dried under humid conditions, frequently a whitish haze called blushing appears in the final film. An electron micrograph of a blushed film is shown in Figure 3 (38). The blushing was caused by the condensation of moisture droplets upon the film in humid air when the film became cold because of solvent evaporation. Liquid lacquer with a relatively low surface tension spontaneously flowed over the water droplets with a higher surface tension, the lacquer then formed a film, and the incompatible encapsulated water droplets eventually evaporated through a partially ruptured film.

Good adhesion depends on thorough wetting of the substrate by the coating and can be achieved quite readily by use of solutions of low surface tension. Because good adhesion depends on other factors such as the structure of the polymer molecule, good adhesion does not automatically result from good wetting, but it is a requisite for eventual good adhesion. If a given resin is applied in various forms such as a powder, dispersion, or solution, it should not be expected that equal adhesion results regardless of the form of the resin. As a rule, best adhesion of a paint may be expected when applied from solution.

Viscosity of Resin Solutions as Related to Solvent Properties. A principal function of solvents in coatings is to reduce the viscosity of the resinous binder so that the coating may be applied to the substrate. The proper application viscosity must be attained

at practical nonvolatile content, which may vary from 10 to 80% or
more by weight.

A solvent for a polymer is one that dissolves the polymer; a
good solvent is one in which the polymer exists in an uncoiled,
random configuration, and there is a high degree of attraction
between the polymer and solvent molecules. The best solvent for a
given resin may or may not give the lowest viscosity of the derived
polymer solution. Viscosity of solutions may be considered to have
a contribution from the neat (no dissolved resin) viscosity of the
solvent and a contribution from the dissolved polymer molecules.
The contribution of the polymer depends on its size and shape and
the degree of aggregation of the polymer molecules; in turn, the
configuration and degree of aggregation of the polymer molecules
depend on the nature of the solvent. Polymer chemists usually are
concerned with intrinsic viscosity, which is indicative of the size
and shape of an isolated polymer molecule in a particular solvent,
whereas coating chemists are more often concerned with neat
viscosity of solvents and measured viscosity of polymer solutions at
much higher concentrations of polymer.

η = measured viscosity as by capillary flow

η_0 = viscosity of solvent

η_r = η/η_0 = relative viscosity or viscosity ratio

η_{sp} = $(\eta - \eta_0)/\eta_0$ = $\eta_r - 1$ = specific viscosity

η_{red} = η_{sp}/c = reduced viscosity, where c = conc.

$[\eta]$ = $(\eta_{sp}/c)_{c=0}$ = $[(\ln \eta_r)/c]_{c=0}$ = intrinsic viscosity

In very dilute solutions in which no aggregation of polymer
molecules occurs, the polymer chain is coiled in a relatively poor
solvent but is extended in a random manner in a good solvent; hence,
the polymer molecule contributes greater viscosity in the latter
case, and the intrinsic viscosity of the polymer is higher than in a
poor solvent. As the concentration of solutions increases, however,
aggregation of polymer molecules occurs in relatively poor solvents,
with the result that at high concentrations (typical of lacquers and
varnishes) the best solvent usually gives the lowest viscosity.
Because the neat viscosity of solvents varies considerably, this
factor must be taken account of in predicting viscosities; in the
practical case where a great number of solvents (or solvent blends)
are good solvents for a given polymer, the measured viscosity of a
solution is roughly proportional to the neat viscosity of the
solvent.

The data of Reynolds and Gebhart (39) on the viscosity of an
easily soluble long oil alkyd illustrate some of the considerations
discussed (Figure 4). In this case solution viscosity is determined
by the neat viscosity of the solvent over a wide range of
concentrations. Above 20-30% concentration the strong solvents
(toluene, tetralin) still continue to dissolve the resin easily, but
the other solvents begin to permit aggregation of alkyd molecules
with accompanying greater increase in viscosity as concentration of

Figure 3.   Electron micrograph at 13,000 times magnification
showing moisture blush in nitrocellulose lacquer film.
Key: SBA, 7.0% w; BUOX: 13.0% w; water: 80.0% w;
temp, 25 °C; air flow, 20 l/min. Reproduced with perm-
permission from Ref. 38. Copyright 1967 Shell Chemical Co.

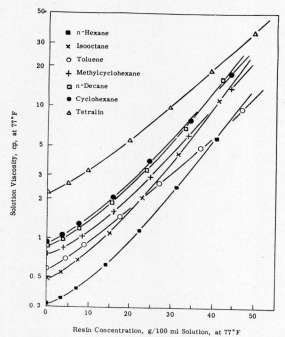

Figure 4.   Solution viscosity of long oil alkyd (Aeroplaz 1273,
73% soya) in various solvents as a function of concentration. Reproduced with permission from Ref. 39.
Copyright 1957 Federal Society of Paint Technology.

resin increases. McGuigan (40) has plotted nitrocellulose solution
viscosity versus neat solvent viscosity (Figure 5) for a large
number of strong oxygenated solvents and has found that these two
viscosity values can be correlated quite well for solutions at 8%
concentration. It is concluded that, in solutions of resins in good
solvents at modest concentrations, solution viscosity is determined
by neat solvent viscosity. Other workers have observed more or less
similar results as related in Reference 39.

Among many other correlations made by Reynolds and Gebhart,
Figure 6 shows a plot of reduced viscosity of three different alkyd
resins as a function of the solvent power (solubility parameter) of
solvents used in the viscosity measurements. Reduced viscosity was
calculated for alkyds at 40% concentration, which is a practical
concentration. Reduced viscosity is a measure of the contribution
of the resin to the solution viscosity. All three resins had
approximately the same molecular weight, and, therefore, the value
of the reduced viscosity is essentially an indication of the degree
of aggregation of molecules of each alkyd. As the solubility
parameter of the solvent decreases, it is evident that the degree of
aggregation increases modestly in the case of the long oil alkyds.
In the case of the short soya alkyd Beckosol 7 (41% oil), the
increase in aggregation becomes much greater as the solvent becomes
less powerful (lower solubility parameter). Beckosol 70 (Reichhold
product) is a 65% soya oil alkyd, and Aeroplaz 1273 is a 73% soya
oil alkyd.

Because neat solvent viscosity is so important in regulating
viscosity of resin solutions, the calculation of viscosity of blends
of solvents is a useful tool in formulating solvent systems for
coatings. As part of a computer program to formulate solvent
blends, Nelson et al. (41) calculated viscosity of blends by use of
the following equation:

$$\log \eta = \Sigma \, V_i \log \eta_i$$

where η is the viscosity of the blend, η_i is the viscosity of the
i^{th} component, and V_i is the volume fraction of the ith component.
In this equation the viscosity of donor–acceptor solvents (e.g.,
alcohols and glycol ethers) is not the viscosity of the pure
compounds, which have high viscosities because of hydrogen bonding;
the viscosity used is an effective viscosity, which is defined as
the viscosity that a solvent would exhibit if it did not involve
hydrogen bonding.

Rocklin and Edwards (42) elaborated on the work of Nelson et al.
and presented substantial data on blends containing interacting
solvents. They also extended the work to aqueous blends containing
up to 20% solvent. Thereafter, Rocklin and Barnes (43) further
modified the method of calculation to handle aqueous blends
containing up to 50% organic solvent. The calculations were based
on several assumptions: the viscosity of a blend is the sum of
contributions from all its components; the effect of each component
solvent on the blend is independent of the other solvents; and the
contribution of each solvent depends only on its interaction with
water. Correlation of experimental values with calculated values of
viscosity was excellent.

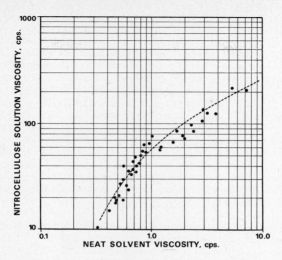

Figure 5.   Viscosity of nitrocellulose solutions vs. viscosity of
            neat solvent.

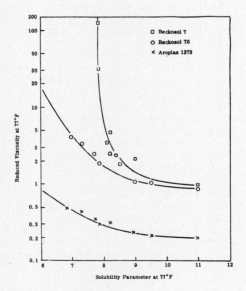

Figure 6.   Reduced viscosity of several alkyd resins as a function
            of solvent power (solubility parameter) of solvent at
            concentration of 40 g/mL. Copyright 1957 Federal Society
            of Paint Technology.

Coatings, especially high-solids coatings, often are heated to allow the coatings to be applied in ordinary equipment. The relationship of temperature and viscosity is of considerable practical importance. Hill, Kozlowski, and Sholes (44) used a simplified form of the Williams-Landel-Ferry (WLF) equation to analyze the temperature dependence of viscosities of solutions of polyester resins and etherified melamine-formaldehyde cross-linking resins. The dependence of viscosity on temperature was completely described by a parameter designated as the solution glass transition temperature (T_{gs}). Good predictions of viscosity at various resin concentrations and temperatures were made.

Sherwin, Koleske, and Taller (45), using several different polymers of low molecular weight, found straight-line relationships when they plotted the log of solution viscosity versus reciprocal temperature. In another phase of their investigation, they concluded that low molecular weight polymers in solution can be considered to assume conformations approaching rigid, impermeable spheres or ellipsoids.

Evaporation Rate of Solvents. Evaporation rate and solvency are the two most important properties of solvents and provide a good guide for development of useful formulas for various types of coatings.

In 1924, Gardner and Parks (46) described a simple method for determination of evaporation rates which consisted of placing a 1-2 g sample of solvent in a can lid and making periodic weighings of the lid in still air. Many investigators made good use of the method over many years. A comprehensive list of evaporation rates was published by Doolittle (47). Bent and Wik (48) arranged the cups holding solvent on a round table and placed a revolving fan at the center to speed up the process.

A major advance in technique was made by Curtis, Scheibli, and Bradley (49) in 1950 when they employed a jolly balance device (the Shell Thin Film Evaporometer) to follow weight loss of solvent when applied to a filter paper cone. The New York Paint and Varnish Production Club made a detailed study of the Evaporometer (50), made some modifications in the method, and concluded that the instrument gave results that were repeatable within one laboratory, reproducible at different locations, and in general provided better results than alternative methods investigated. The instrument has been improved over the years and now is produced in a highly automated model (51).

The practical measurement and meaning of evaporation rates of solvents are much more complicated than they appear to be at first glance, as Rocklin (52) has reported. It was shown that evaporation rates from filter paper, the most popular surface, sometimes differ considerably from evaporation rates from smooth surfaces such as an aluminum disk. In particular, evaporation of water and alcohols from filter paper substrate is relatively slow, probably because of hydrogen bonding forces between solvent and substrate.

Evaporation behavior often is reported as shown in Figure 7 where weight percent solvent evaporated is plotted against time. A convenient numerical value of evaporation rate is taken as the time in seconds required for 90% of the solvent to evaporate when tested under standard condition in the Evaporometer. It is also customary to report evaporation rate of a solvent relative to that of n-butyl acetate taken as 1. In this method, the 90% evaporation times for the solvents are used. Some values for evaporation rates are shown in Table V.

Evaporation of solvent blends can be quite complicated because of intermolecular interactions that cause them to deviate from ideal behavior as expressed by Raoult's Law. The dotted line in Figure 8 shows ideal behavior of an equal volume blend of two solvents, and the dotted line in Figure 9 shows the evaporation behavior of a solvent blend that has positive deviation from Raoult's Law.

The conditions under which azeotropes may form during evaporation of solvent blends have been discussed by Rudd and Tysall (53) and by Ellis and Goff (54), among others. Inspection of Figure 10 shows that highest vapor pressure of a blend of *n*-butanol and *n*-octane occurs at 30 mol% *n*-butanol, the azeotropic composition. Above this proportion of butanol in the liquid the vapor will be poorer in butanol; for example, a liquid containing 80 mol% butanol will be in equilibrium with a vapor containing about 38 mol% butanol, and evaporation of the mixture will leave a remaining liquid that becomes progressively richer in butanol. On the other hand a mixture containing less than 30 mol% butanol will leave a residual liquid progressively richer in octane as evaporation proceeds. A blend at the azeotropic composition will evaporate at constant ratio of the two solvents in both the vapor and residual liquid. Thus, some control over the residual solvent composition can be exercised by choice of solvents and their relative amounts in a blend (Figure 11).

Some factors that influence the rate of evaporation of a solvent include temperature, flow of air over sample, vapor pressure of the solvent, latent heat, specific heat, and molecular weight (53). Galstaun (55) in 1950 reported a thorough study of the evaporation of some hydrocarbon solvents and developed equations for evaporation rates as related to several factors including temperature drop of liquid as it evaporates. Sletmoe (56) developed equations for the evaporation of neat solvent blends. The total rate of evaporation was proposed to be equal to the sum of the rates for the individual solvent components:

$$\text{Total Rate} = C_1\,\gamma_1\,R_1{}^\circ + C_2\,\gamma_2\,R_2{}^\circ + \text{etc.}$$

where C is the concentration, γ is the escaping (activity) coefficient, and R is the rate of evaporation of the pure individual components. Sletmoe suggested means of calculating solvent balance (uniform evaporation rate of components during evaporation of blend) as an aid to formulating good solvent blends.

Hansen (57) pointed out that evaporation of a solvent from a polymer solution faced two barriers when cast on an impermeable substrate: resistance to solvent loss at the air-liquid interface and diffusion from within the film to the air interface. Evaporation of neat solvents as well as moderately dilute solutions is limited by resistance at the air interface, but as solvent concentration becomes low (5-10-15%), the rate-controlling step is diffusion through the film. Hansen pointed out that at the point when solvent loss changes to a diffusion-limited process, the concentration of solvent is sufficient to reduce the glass transition temperature, T_g, of the polymer to the film temperature.

Table V. Evaporation Rates and Boiling Points of Some Common Solvents

Solvent	Evap. Time, Seconds to 90% Evap.	Evap. Rate Relative to n-Butyl Acetate = 1	Boiling Point or Range
Acetone	82	5.59	56.1 °C
Tetrahydrofuran	97	4.72	66.0 °C
Lacquer diluent	101	4.6	95–104 °C; 204–219 °F
Ethyl Acetate, 99%	117	3.91	77.1 °C
Methyl Ethyl Ketone	121	3.79	79.6 °C
Isopropyl Acetate	134	3.42	88.7 °C
Methanol	221	2.07	64.5 °C
Toluene	225	2.00	110.6 °C
Ethyl Alcohol, 100%	278	1.60	78.3 °C
Methyl Isobutyl Ketone	282	1.62	116.2 °C
Isobutyl Acetate, 90%	305	1.50	117.2 °C
V M & P Naphtha	314	1.5	119–129 °C; 246–265 °F
Isopropyl Alcohol	319	1.44	82.3 °C
n-Butyl Acetate, 90–92%	458	1.00	118–128 °C
Methyl n-Butyl Ketone	482	1.05	127.5 °C
n-Propyl Alcohol	530	0.86	97.2 °C
s-Butyl Alcohol	563	0.81	99.5 °C
Isobutyl Alcohol	740	0.62	107.8 °C
Xylene	770	0.6	138–140 °C
EG[a] Monomethyl Ether	884	0.52	124.5 °C
Methyl n-Amyl Acetate	1,004	0.46	146.3 °C
Methyl Isoamyl Ketone	1,016	0.45	145.4 °C
n-Butyl Alcohol	1,076	0.43	117.7 °C
EG[a] Monoethyl Ether	1,213	0.38	135.1 °C
Cyclohexanone	1,566	0.29	156.7 °C
Methyl Isobutyl Carbinol	1,711	0.27	102.2 °C
Ethyl Amyl Ketone	1,770	0.26	160.5 °C
4-Methoxy-4-methylpentanone-2	1,880	0.24	159.1 °C
Diisobutyl Ketone	2,437	0.19	169.3 °C
EG[a] Monoethyl Ether Acetate	2,533	0.18	156.3 °C
Diacetone Alcohol	3,840	0.12	183.8 °C (extrap.)
Mineral Spirits	4,560	0.1	167–199 °C; 332–390 °F
EG[a] Monobutyl Ether	6,750	0.07	171.2 °C
Odorless Mineral Spirits	7,410	0.06	174–206 °C; 346–403 °F
Isophorone	20,000	0.02	215.2 °C
2-Ethylhexanol	25,700	0.02	184.8 °C
DEG[b] Monoethyl Ether	27,800	0.02	201.9 °C
DEG[b] Monobutyl Ether	150,000	<0.01	230.4 °C

[a]EG = Ethylene Glycol
[b]DEG = Diethylene Glycol

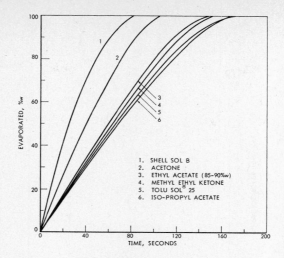

Figure 7.   Evaporation curves for solvents.

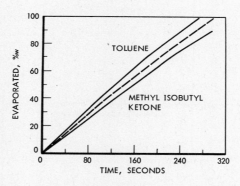

Figure 8.   Evaporation of an ideal solvent blend.

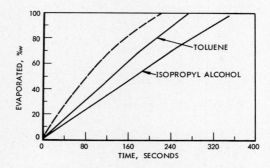

Figure 9.   Evaporation of blend with positive deviation.

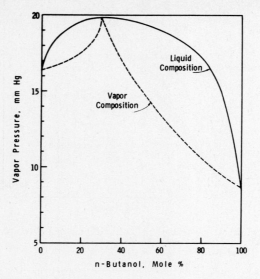

Figure 10.   Vapor pressure vs. composition of n-butanol-n-octane.
              Reproduced with permission from Ref. 54. Copyright 1972
              J. Paint Technol.

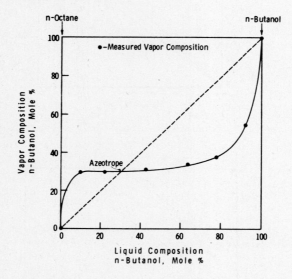

Figure 11.   Vapor-liquid equilibrium compositions of n-butanol-n-
              octane. Reproduced with permission from Ref. 54.
              Copyright 54 J. Paint Technol.

Therefore, a solution of a resin with a T_g below room temperature never undergoes evaporation by the diffusion process. Considerable amounts (5-10%) of solvent can remain in a film for years. Diffusion is extremely slow through cured films having high T_g.

During the diffusion stage of evaporation the amount of solvent remaining in the film can be plotted against time divided by the square of the film thickness; this plot eliminates the effect of varying thickness in experimental results and illustrates the fact that doubling the film thickness will quadruple the time to a given state of dry in the diffusion state of evaporation (57).

The rate of diffusion in the second stage of drying depends on the volatility and the shape and size of the solvent molecules. Branching of a chain retards diffusion to a very great extent as shown by Nunn and Newman (58). Isobutyl acetate with a relative evaporation rate of 1.7 evaporates much faster than n-butyl acetate (rate = 1.0); nevertheless, the film retains isobutyl acetate more tenaciously (second stage of evaporation) than n-butyl acetate as shown in Figure 12. Methyl isobutyl ketone (MIBK) (not shown in the figure) is held more tenaciously in the film than methyl n-butyl ketone (MNBK), in spite of the fact that the former has the lower boiling point. The evaporation rate of pure MNBK is 1.05, and that of MIBK is 1.62. Solvent retention of films can be measured by use of gas chromatography or by use of isotopically tagged solvents by the method of Murdock and Wirkus (59) or by hardness tests.

Ellis (60) pointed out that high-solids polyester coatings typically contain so little solvent that the T_g of the resin–solvent combination is above the temperature of the deposited film. Therefore, evaporation of solvent occurs very slowly solely by diffusion as the rate-limiting step; therefore, the film sags and runs before sufficient solvent evaporates to provide "set-up" of the film. A precise method was provided for determination of the "transition point" of a given resin–solvent combination. This point is the resin concentration at which solvent evaporation changes from surface-controlled to a diffusion-controlled process.

A mathematical model of solvent blend evaporation was developed by Walsham and Edwards (61). The model accounts for the nonideal behavior of solvent blends in terms of component activity coefficients. The model allows accurate prediction of blend evaporation time by computer calculations. The technique provides a means to follow residual solvent composition (solvent balance) as evaporation proceeds.

<u>Evaporation of Solvent–Water Blends</u>. The evaporation of blends of solvents and water presents several problems, the most prominent of which is the fact that the relative rates of evaporation of solvents and water in a given blend vary greatly under conditions of varying humidity. Under dry conditions the water evaporates relatively rapidly, but under high humidity the water evaporates relatively slowly in comparison to solvent. Usually it is desired that the water and solvent evaporate at rates such that the residual liquid remains constant in composition (perfect solvent balance). Often, enrichment of solvent in the residual liquid is desired, but depletion of solvent in the residual liquid might lead to resin kick-out or loss of ability of the resin to coalesce into a smooth, continuous film. One means of minimizing premature loss of

coalescing solvents under conditions of high humidity involves the
addition of more solvent to the formulation as humidity increases.

Rocklin (62) made a careful study of the effect of humidity and
other ambient conditions on the evaporation of ternary water-solvent
blends. The problem of evaporation of these systems is illustrated
by Figure 13 on the system water-butanol-butoxyethanol. This
particular figure shows trends only and is not intended for exact
predictions. In these studies, the author employed a "central
composite experimental design" that allowed a large number of
variables to be investigated simultaneously and economically by
using a modest number of experiments to achieve significant results.

A mathematical model of evaporation for blends of water and
solvents at various humidities was developed by Dillon (63). He
derived a concept called critical relative humidity (CRH) which is
defined as that humidity at which the composition of organic solvent
in an aqueous blend remains unchanged as evaporation proceeds. At
humidities higher than the CRH, the evaporation of water is retarded
so that the residual solvent blend becomes richer in water. At
humidities less than the CRH, the residual solvent becomes richer in
solvent. CRH values did not vary much with varying concentration
when high-boiling solvents were used; for example, CRH values for 2-
ethoxyethanol ranged from 87 to 84 at 5 to 30% solvent volume
concentration. Stratta, Dillon, and Semp (64) gave additional
background details of the computer simulation leading to the CRH
concept. They also presented experimental data on the determination
of activity coefficients and evaporation rates of some systems.

Rocklin and Bonner (65) developed a computer method that
predicts solvent balance and evaporation times of water-solvent
blends at any humidity with any number of water-soluble organic
solvents. The method also can be used for regular water-free
solvent blends but ignores humidity. Key considerations of the
method are the following: it uses the UNIFAC method for calculating
activity coefficients; it computes the actual evaporation
temperature on the filter paper substrate; it calculates evaporation
rates at the calculated temperature by using the activity
coefficients at that temperature; humidity is accommodated by
applying a correction factor to the water evaporation rate.
Experimental data on several systems verified the computer
calculations.

Flash Point of Solvents. The flash point of a solvent is an
important property because methods of shipping and handling solvents
and solvent-containing products such as paints are regulated
according to the flash point of the product. For example, it is
more expensive to package and ship products that fall in the
category of flash points below 100 °F than those that fall in the
category of flash points above 100 °F. Some Department of
Transportation (DOT) definitions relating to flash points are the
following (66):

1. A flammable liquid means any liquid, with some minor
   exceptions, having a flash point below 100 °F (37.8 °C).
2. A combustible liquid is one that has flash point at or above
   100 °F (37.8 °C) and below 200 °F (54.5 °C).

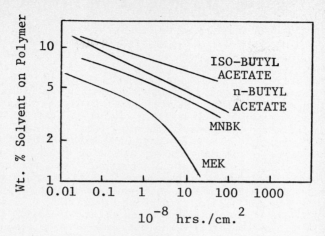

Figure 12.   Solvent retention of various solvents in a vinyl resin
film.

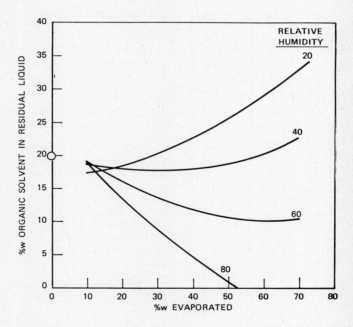

Figure 13.   Effect of relative humidity on solvent balance upon
evaporation of sec-butyl alcohol–2–butoxyethanol–water blend.
Reproduced with permission from Ref. 62.   Copyright 1978
Federation of Societies for Coatings Technology.

3.   A pyrophoric liquid is any liquid that ignites spontaneously in
     dry or moist air at or below 130 °F (54.5 °C).

The test methods stipulated for measuring flash points are the
Tag Closed Tester (ASTM D56-70), the Pensky-Martens Closed Tester
(ASTM D93-71), and the Setaflash Closed Tester (ASTM D3278-73).   Of
these methods, the Setaflash gives the greatest precision.   The
Golden Gate Society for Coatings Technology (66) found excellent
agreement between the Setaflash and Tag results for flash points.
In comparison with the Pensky-Martens method, the Setaflash method
gave flash points slightly lower in the case of some solvents and
paints, but the general agreement was very good.
     Brown, Newman, and Dobson (67) have investigated the fire hazard
of water-borne coatings.   Correlation of flash points with
combustibility of paints is not very good.   Some products that
cannot sustain combustion often have low flash points.   In a
detailed study of flash points of water-solvent blends, solutions of
resins in blends and of finished paints and inks, W. Hansen (68)
observed that some products with flash points of 21-55 °C varied
considerably in combustibility.   In the case of some products, the
initial flame went out spontaneously.
     Walsham (61) developed a computer-based method that gave
satisfactory predictions of flash points.   He defined an individual
solvent flash point index as an inverse function of its heat of
combustion and vapor pressure at the flash point.   Flash points of
mixtures were computed by trial and error as the temperature at
which the sum of weighted component indexes equals 1.0.   Solution
nonidealities were accounted for by component activity coefficients
calculated by a multicomponent extension of the Van Laar equations.

Air Quality Regulations and Solvents

The term smog refers to an atmospheric condition caused by
interaction of organic compounds, nitrogen oxides, and UV light, and
manifested by a combination of poor visibility, eye irritation, and
plant damage.   The existence of a serious smog problem led Los
Angeles County to conduct extensive investigations on its causes and
cure and issue regulations in attempts to rectify the situation (70,
71).   Rule 66 was the particular regulation designed to reduce
generation of smog from solvent emissions; it was based on smog
chamber tests and used eye irritation as the principal criterion,
but account was also taken of plant damage and the quantities of
various individual solvents actually used in the area.   Other
investigators (72, 73) also used smog chambers to study the chemical
reactions undergone by solvent vapors upon exposure to UV light
under controlled conditions.
     Rule 66 was adopted in 1966 and became effective according to a
designated schedule.   The rule was based on the premise that the
extent of smog produced depended upon the chemical structure of the
solvents.   Deemed to be conducive to smog production were aromatic
hydrocarbons except benzene, branched chain ketones, and especially
unsaturated compounds.   The regulation specified that solvent blends
were considered to be photochemically reactive or unreactive on the
basis of their composition as defined in the box on page 687.

Definition of Photochemically Reactive Solvent
by Rule 66 Section k

For the purpose of this rule, a photochemically reactive
solvent is any solvent with an aggregate of more than
20% of its total volume composed of the chemicals
classified below or which exceeds any of the following
individual percentage composition limitations, referred
to the total volume of solvent:

1. A combination of hydrocarbons, alcohols, aldehydes,
    esters, ethers, or ketones having an olefinic or
    cycloolefinic type of unsaturation: 5%;
2. A combination of aromatic compounds with eight or
    more carbon atoms to the molecule except
    ethylbenzene: 8%;
3. A combination of ethylbenzene, ketones having
    branched hydrocarbon structures, trichloroethylene,
    or toluene: 20%.

Whenever any organic solvent or any constituent of an
organic solvent may be classified from its chemical
structure into more than one of the above groups of
organic compounds, it shall be considered as a member of
the most reactive chemical group, that is, that group
having the least allowable percent of the total
volume of solvents.

Rule 66 (later called Rule 442 of the South Coast Air Quality Management District) also limited oven emissions from baked coatings. Incineration, absorption, or recovery of volatiles is needed to meet this restriction. In further regulations, use of high-solids, low-solvent, water-borne, and other types of coatings with low-volatile emissions was encouraged. The evolution of various air pollution regulations has been summarized (74, 75).

The main thrust of the Rule 66 type of regulation was to require the use of less reactive solvents in place of the more photochemically reactive solvents. However, more recent regulations are geared to limitation of the emissions of all solvents. Transport theory involves the concept that air pollution in a given area may result from emissions in other areas upwind. Accordingly, the more reactive solvents are deemed to yield pollutants in the immediate area where they are emitted, and the less-reactive solvents are considered to cause pollution later in downwind areas of the country. In this view all solvent emissions cause pollution, but differences in reactivity simply determine where the pollution occurs. Thus, the emphasis on control of pollution from coatings has shifted from restricting the composition of solvents to restricting the emissions of all solvents from coatings.

In November 1976, the Environmental Protection Agency (EPA) issued the important document entitled "Control Methods for Surface Coating Operations" (76). Thereafter, guidelines for solvent content for various types of paint for specific end uses have been issued periodically as shown in Table VI. It is important to note that the guidelines specify that a given volume of coating including solvent should contain no more than a given weight of solvent. Therefore, in meeting regulations, it is advantageous to use solvents of powerful solvency and low specific gravity. Tess (75) has summarized and interpreted these guidelines and has suggested options available to meet the guidelines.

Formulation of Solvent Blends

Much information on specific formulations of solvent blends has been published in commercial brochures, technical articles, and books such as those cited in this chapter (5, 20, 23, 31). Some techniques and aids used by formulators will be discussed here briefly.

Regions of solubility of resins in solvents can be expressed in terms of various solution parameters such as solubility parameter, hydrogen bonding index, fractional polarity, dipole moment, internal pressure, or the components of solubility parameter due to dispersion, polar, and hydrogen bonding forces. An example of the solubility of nitrocellulose in various solvents as a function of solubility parameter and fractional polarity is shown in Figure 14 (77). Solubility in mixtures of two solvents can be predicted by drawing a straight line between two different solvents and fixing a point according to the relative amounts of each solvent. Three and more component systems can be located in the solubility map by similar principles. In the case of some resins, hydrogen bonding forces are the predominant solubilizing forces, and it is advantageous to include some measure of these forces in a two-dimensional system. Of course, use of three solution parameters can

Table VI. EPA Guidelines for Maximum Volatile Organic Content
of Coatings

Guidelines, EPA Volume	Process	Limitation (g/L)	(lb/gal)
II	Can coating		
	Sheet basecoat & overvarnish;		
	two-piece can exterior	340	2.8
	Two-, three-piece can interior body		
	spray, two-piece can exterior end	510	4.2
	Side-seam spray	660	5.5
	End sealing compound	440	3.7
II	Coil coating	310	2.6
II	Fabric coating	350	2.9
II	Vinyl coating	450	3.8
II	Paper coating	350	2.9
II	Auto and light-duty truck coating		
	Prime	230	1.9
	Topcoat	340	2.8
	Repair	580	4.8
III	Metal furniture coating	360	3.0
IV	Magnet wire coating	200	1.7
V	Large appliance coating	340	2.8
VI	Miscellaneous metal parts	50-520	0.4-4.4
VII	Wood paneling		
	Printed interior	200	1.7
	Natural finish hardwood	380	3.2
	Class II hardboard	320	2.7

		Limitation of Volatiles
VIII	Inks	
	Rotogravure or flexographic	60% NVM
	Water-borne	75% vol. water and 25% volatiles
	Solvent type	70% overall reduction of volatiles or 60% NVM

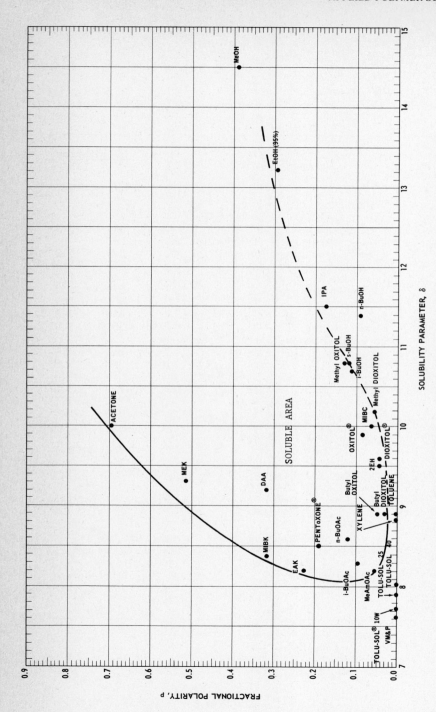

Figure 14. Estimated solubility map for 8 g RS 1/2 s nitrocellulose per 100 mL solvent at 25 °C.

easily be handled by use of a computer as will be discussed later, but is hard to project on paper.

Areas of solubility of resins in ternary solvent systems often are expressed in triangular diagrams as shown in Figure 15. An additional feature of this diagram is that the area containing up to 20% toluene is a complying ("exempt") solvent under Rule 66. It is often useful to express viscosity and volatility of resins in ternary solvent blends as shown in Figure 16. In this figure the time in seconds required for 90% of the solvent to evaporate is given as a measure of volatility, and the viscosity in centipoises is given as a measure of viscosity. Point A describes the composition of a solvent blend that has evaporation time of 160 s and viscosity of 85 cP.

A wealth of useful data on evaporation times can be expressed in a diagram as shown in Figure 17. The volatility of combinations of methyl ethyl ketone with various higher boilers has been investigated by Dante (78). Horizontal lines show volatility of methyl isobutyl ketone and methyl *n*-butyl ketone; from the volatility standpoint each of these can be replaced by compositions indicated by intersections of the horizontal lines with each of the curves.

The introduction of Rule 66 caused tremendous activity in the reformulation of solvent blends to conform with the regulation. An example of nitrocellulose lacquer solvents for both complying (exempt from control Rule 66) and noncomplying (nonexempt) formulations is shown in Table VII (79). Exempt solvent systems for alkyd resins have been discussed by Fink and Weigel (80), for nitrocellulose and vinyl resins by Crowley (81), for vinyl resins by Park (82) and by Burns and McKenna (83), and for epoxy resins by Somerville and Lopez (84).

In reformulating thousands of solvent formulations to comply with Rule 66, the coatings industry faced a tremendous task. To meet this problem, a combination of solution theory, solvent expertise, and computer technology proved to be of great help. Nelson, Figurelli, Walsham, and Edwards (41) described a two-step method of solvent selection by computer. The first program calculates average properties of an input solvent blend by using a data file that contains basic data on over 100 solvents. After a known solvent blend expressed in volume percent is fed to the computer, in a matter of seconds the following information is obtained:

1. Weight fraction of each component solvent.
2. Mole fraction of each solvent.
3. Volume fraction of each of the four categories of solvent segregated according to the classes in Rule 66.
4. Three solution parameters that are used to describe solvency of the solvent blend in fundamental terms: solubility parameter, fractional polarity, and hydrogen bonding index number (85).
5. Neat viscosity.
6. Specific gravity.
7. Evaporation time for each 10% increment evaporated.
8. Raw material cost.

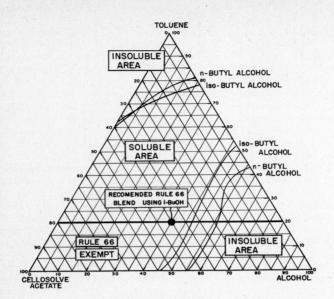

Figure 15.   Solubility of Epon 1007 (Shell Chemical) in blends of
              toluene, Cellosolve Acetate (Union Carbide), and
              various alcohols.

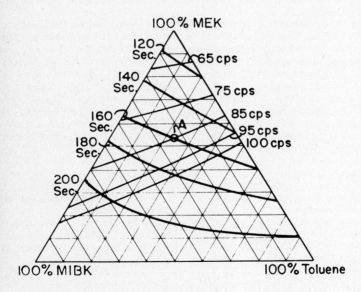

Figure 16.   Isoviscosity and isovolatility curves for a vinyl resin
              (Union Carbide Vinylite VYHH) at 20% solids.

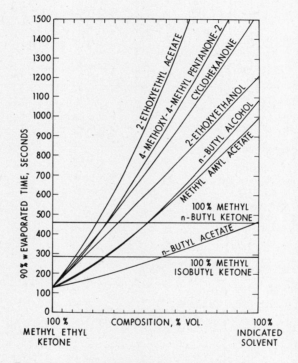

Figure 17.   Evaporation behavior of mixtures of methyl ethyl ketone with slower evaporating solvents.

Table VII. Solvent Systems for Nitrocellulose Furniture Lacquer

Solvent Composition, %v	Nonexempt Control	Rule 66 Exempt Replacements	
		1	2
Methyl ethyl ketone	7.0	8.8	8.8
Methyl isobutyl ketone	20.0	--	--
n-Butyl acetate	--	30.4	21.8
Ethyl amyl ketone	5.0	--	--
Methyl amyl acetate	--	--	4.8
2-Butoxyethanol	--	--	3.8
Isopropyl alcohol (inc. IPA ex. cotton)	16.0	17.8	17.8
Toluene	21.0	9.3	9.3
Xylene (inc. xylene ex. modifying resins)	15.0	8.0	8.0
Lacquer diluent (9% toluene)	--	25.7	25.7
Lacquer diluent (23% toluene)	16.0	--	--
Total	100.0	100.0	100.0

In the second program, constraints such as compliance with Rule 66, minimum values of the solution parameters, 90% evaporation time, and maximum value of neat viscosity are imposed on the selection of replacement formulas. This second program is a linear optimization program that selects a single solvent blend from a chosen list of 15-20 candidates, bounded by the imposed constraints and optimized in some function, usually cost. The output of the computer indicates a single solvent blend that meets the constraints at minimum cost. The suggested blend should be checked in the laboratory to ensure that the formulation meets the desired requirements.

Hansen (86) has modified the method of Nelson, et al. by incorporating solubility parameters for the three component forces in the solubility parameter, viz, dispersion, polar, and hydrogen bonding forces. The solvent selection procedure was designed for use by plant laboratories on a time sharing terminal. The enlistment of the computer in the selection of solvent blends has been a boon to the formulator, but the use of older methods is still very useful.

Solvents for High-Solids Coatings

Unless emissions of volatile organic compounds (VOC) from coatings are controlled by incineration, absorption, or other means, current EPA regulations require that only very limited quantities of solvent may be used in industrial coatings as shown in Table VII. The regulations sometimes can be met with suitable powder paints, water-borne paints, or 100% convertible paints, but solvent-based paints often are preferred because of their traditional good properties. Because only limited amounts of solvent are permitted, it is desirable that the solvents be as powerful as feasible. Also, because the regulations specify a maximum weight of solvent per unit volume of bulk paint, the solvent used should have as low a specific gravity as feasible.

Use of application methods that can use relatively high-viscosity paints is advantageous. Appropriate methods include roller coating, special electrostatic spray equipment, hot-spray techniques, and two-nozzle spray guns.

Because ketones have both high solvency power and low specific gravity, they are attractive solvents for high-solids coating systems. Some ketones and esters have been compared as solvents for an epoxy resin as shown by the data in Table VIII. The quantity of solvent in each case is maintained at 250 g/L of solution. Comparisons between pairs of ketone and ester solvents are made in the categories of low-boiling, medium-boiling, and high-boiling solvents. In each case the ketone solutions have substantially lower viscosity than the ester solution. This condition is due partly to the more powerful solvency of the ketones and partly to their lower specific gravity which permits more of the ketones by volume in the final solutions as shown by the data in Table VIII. In the cases of the low-boiling and the medium-boiling solvents, the ketones have lower viscosities in the neat state in the absence of resin.

Sprinkle (87) has pointed out the importance of controlling surface tension in high-solids coatings. As coatings increase in

Table VIII. Viscosity of Eponex DRH-151 Resin Solutions at 250 g/L of Solution

Solvent	Solvent Sp. G., 25 °C	Solution Visc., cp	g of Solvent/ L of Solution	% vol. Solvent	% vol. Solids	% wt. Solvent	% wt. Solids	Sp. G. Soln.
Methyl ethyl ketone	0.802	18	250	31.2	68.8	24.6	75.4	1.016
Ethyl acetate	0.897	26	250	27.9	72.1	23.7	76.3	1.053
Methyl isobutyl ketone	0.799	29	250	31.3	68.7	24.7	75.3	1.012
n–Butyl acetate	0.878	42	250	28.5	71.5	23.9	76.1	1.044
Ethyl isoamyl ketone	0.818	48	250	30.6	69.4	24.5	75.5	1.020
2-Ethoxyethyl acetate	0.970	68	250	25.8	74.2	23.3	76.7	1.073

Note: Eponex DRH-151 resin is a product of Shell Chemical Company.

solids content, the surface tension increases with the deleterious consequences of decreased sprayability, wettability, and possible increased presence of defects such as cratering. Selection of solvents with low surface tension can mitigate the increase in surface tension of high-solids coatings.

High-solids coatings of the solvent type have been discussed in other sections of this chapter. This field is active, but greatest emphasis appears to be focused on the resin component of high-solids solvent coatings. Nevertheless, proper solvent selection can materially help in the quest for high-solids coatings based on solvents.

Literature Cited

1. Powers, P. O. In "Protective and Decorative Coatings"; Mattiello, J. J., Ed.; Wiley and Sons: New York, 1941; Vol. I, Chap. 23.
2. Holden, H. C.; Doolittle, A. K. Ind. Eng. Chem. 1935, 27, 485-91.
3. U.S. Tariff Commission, Census of Dyes and Other Synthetic Organic Chemicals, 1925.
4. McClure, H. B.; Bateman, R. L. Chem. Eng. News 1947, 25(44), 3208-14; ibid. 1947, 25(45), 3286-89.
5. Doolittle, A. K. "The Technology of Solvents and Plasticizers"; John Wiley and Sons: New York, 1954; p. 2.
6. Stewart, A. C. address at Chemical Marketing Research Association, May 1972, New York; Chem. Marketing Reporter 1972, 201(19).
7. "NPCA Data Bank Program 1982"; SRI International, September 1982.
8. Zeisberg, F. C. In "Twenty-Five Years of Chemical Engineering Progress"; D. Van Nostrand Company: New York, 1933; Chap. IV.
9. Durrans, T. H. "Solvents"; D. Van Nostrand Company: New York, 1930; also later editions.
10. "Kirk-Othmer Encyclopedia of Chemical Technology"; Interscience: New York.
11. Clayton, E.; Clark, C. O. Proceedings of Journal of the Society of Dyers and Colorists 1931, 47(7), 183.
12. Kelly, F. C. "One Thing Leads to Another"; Houghton Mifflin Company: Boston, 1936.
13. Weed, F. G. In "Protective and Decorative Coatings"; Mattiello, J. J., Ed.; John Wiley and Sons: New York, 1943; Vol. III, Chap. 14.
14. "Ethyl Alcohol"; Enjay Chemical Co., 1962.
15. "Shell Ethyl Alcohol"; Shell Chemical Co., Bulletin IC:69-32, 1969.
16. "Isopropyl Alcohol"; Enjay Chemical Co., 1966.
17. Kenney, J. A., "Coke-Oven, Light-Oil Distillates in the Paint and Varnish Industry"; In "Protective and Decorative Coatings"; John Wiley and Sons: New York, 1941; Vol. I, Chap. 26.
18. Hildebrand, J. H.; Scott, R. L. "The Solubility of Nonelectrolytes," 3rd ed.; Dover Publications: New York, 1964; 1st, 2nd, 3rd eds., Reinhold Publishing Corporation: New York, 1924-1950.

19. Hildebrand, J. H.; Prausnitz, J. M.; Scott, R. L. "Regular and Related Solutions"; Van Nostrand Reinhold: New York, 1970.
20. Bogin, C. D. In "Protective and Decorative Coatings"; Matiello, J. J., Ed.; Wiley and Sons, 1941; Vol. I, Chap. 27.
21. Reynolds, W. W. "Physical Chemistry of Petroleum Solvents"; Reinhold: New York, 1963.
22. Patton, T. C. "Paint Flow and Pigment Dispersion"; Interscience: New York, 1966.
23. "Solvents Theory and Practice"; Tess, R. W., Ed.; ADVANCES IN CHEMISTRY SERIES No. 124, American Chemical Society, Washington, D.C., 1973.
24. "Solvent Properties Chart"; Shell Chemical Co., Bulletin SC:48-74, 1974.
25. Kauppi, T. A.; Bass, S. L. Ind. Eng. Chem. 1938, 30, 74.
26. Doolittle, A. K. Ref. 5, p. 855-57.
27. Hoy, K. L. J. Paint Technol. 1973, 45(579), 51.
28. Sullivan, D. A. J. Paint Technol. 1975, 47(610), 60.
29. Tess, R. W.; Schmitz, R. D. Off. Dig. Fed. Socs. Paint Technol. 1957, 29, 1346.
30. May, C. A. Chapter 10 in Ref. 23.
31. Mellan, I. "Industrial Solvents," 2nd ed.; Reinhold: New York, 1950.
32. Zisman, W. A. J. Paint Technol. 1972, 44(564), 41.
33. Hansen, C. M.; Pierce, P. E. "XII FATIPEC Congress Book"; 1974; pp. 91-99.
34. Hansen, C. M.; Pierce, P. E. Ind. Eng. Chem. Prod. Res. Develop. 1973, 12(1), 67.
35. Dannenberg, H.; Wagers, J.; Bradley, T. F. Ind. Eng. Chem. 1950, 42, 1594.
36. Hahn, F. J. J. Paint Technol. 1971 43(562), 58.
37. Brown, G. L. J. Polym. Sci. 1956, 22, 423.
38. "Test Methods and Techniques for the Surface Coatings Industry," 2nd ed.; Shell Chemical Co., Bulletin IC:67-53, 1967.
39. Reynolds, W. W.; Gebhart, H. J., Jr. Off. Dig. Fed. Soc. Paint Technol. 1957, 29(394), 1174.
40. McGuigan, J. P., private communication.
41. Nelson, R. C.; Figurelli, V. F.; Walsham, J. R.; Edwards, G. D. J. Paint Technol. 1970, 42(550), 644.
42. Rocklin, A. L.; Edwards, G. D. J. Coatings Technol. 1976, 48(620), 68.
43. Rocklin, A. L.; Barnes, J. A. J. Coatings Technol. 1980, 52(655), 23.
44. Hill, L. W.; Kozlowski, K.; Sholes, R. L. J. Coatings Technol. 1982, 54(692), 67.
45. Sherwin, M. A.; Koleske, J. V.; Taller, R. A. J. Coatings Technol. 1981, 53(683), 35.
46. Gardner, H. A.; Parks, H. C. Paint Manuf. Assoc. U.S. Tech. Circ. 218; 1924; p. 113.
47. Doolittle, A. K. Ind. Eng. Chem. 1935, 27. 1169.
48. Bent, F. A.; Wik, S. N. Ind. Eng. Chem. 1936, 28, 312.
49. Curtis, R. J.; Scheibli, J. R.; Bradley, T. F. Anal. Chem. 1950, 22, 538.
50. New York Paint and Varnish Production Club, Off. Dig. Fed. Soc. Paint Technol. 1956, 28, 1060; 1958, 30, 1230.

51. "Evaporation Rates of Solvents as Determined Using the Shell Automatic Thin Film Evaporometer"; Shell Chemical Co.; Bulletin IC:69-39, 1969.

52. Rocklin, A. L. J. Coatings Technol. 1976, 48(622), 45.

53. Rudd, H. W.; Tysall, L. A. J. Oil Colour Chem. Assn. 1949, 32, 546.

54. Ellis, H. W.; Goff, P. L. J. Paint Technol. 1972, 44(564), 79.

55. Galstaun, L. S. ASTM Bulletin 1950, 170, 60.

56. Sletmoe, G. M. J. Paint Technol. 1966, 38, 641; 1970, 42, 246.

57. Hansen, C. M. Ind. Eng. Chem. Prod. Res. Develop. 1970, 9(3), 282.

58. Nunn, C. J.; Newman, D. J. "XII FATIPEC Congress Book"; 1974; pp. 103-8.

59. Murdock, R. E.; Wirkus, W. J. Off. Dig. Fed. Soc. Paint Technol. 1963, 35, 1084.

60. Ellis, W. H. J. Coatings Technol. 1983, 55(696), 63.

61. Walsham, J. G.; Edwards, G. D. J. Paint Technol. 1971, 43(554), 64.

62. Rocklin, A. L. J. Coatings Technol. 1978, 50(646), 46.

63. Dillon, P. W. J. Coatings Technol. 1977, 49(634), 38.

64. Stratta, J. J.; Dillon, P. W.; Semp, R. H. J. Coatings Technol. 1978, 50(647), 39.

65. Rocklin, A. L.; Bonner, D. C. J. Coatings Technol. 1980, 52(670), 27.

66. Golden Gate Society for Coatings Technol., J. Coatings Technol. 1977, 49(627), 52.

67. Brown, R. A.; Newman, R. M.; Dobson, P. H., National Paint and Coatings Assn. Scientific Circ. 804, 1977.

68. Hansen, W. J. Coatings Technol. 1982, 54(694), 45.

69. Walsham, J. G., Chapter 5 in Ref. 23.

70. County of Los Angeles Air Pollution Control District, Rules and Regulations, 1973.

71. "Air Pollution Control Regulations - County of Los Angeles Air Pollution Control District"; Shell Chemical Co.; Technical Bulletin IC:73-15 SN, 1973.

72. Levy, A.; Miller, S. E. "The Role of Solvents in Photochemical Smog Formation"; Nat. Paint Varn. Lacquer Assn., Circ. 799, April 1970; also Chapter 6 in Ref. 23.

73. Laity, J. L.; Burstain, I. G.; Appel, B. R. "Photochemical Smog and the Atmospheric Reactions of Solvents"; Chapter 7 in Ref. 23.

74. Tess, R. W. In "Applied Polymer Science"; Craver, J. K.; Tess, R. W., Eds.; Organic Coatings and Plastics Chemistry Division, American Chemical Society: Washington, D.C., 1975; Chap. 44.

75. Tess, R. W. Organic Coatings and Plastics Chemistry 1979, 41, 536; Preprints of ACS Meeting, Sept. 1979.

76. "Control of Volatile Organic Emissions from Existing Stationary Sources"; U.S. Environmental Protection Agency, Office of Air Quality Planning and Standards, Pub. EPA 450/2-76-028 (OAQPS No. 1.2-067), Nov. 1976; Vol. I.

77. Shell Chemical Co., Technical Bulletins IC:67-64SN, 1967, and IC:68-1SN, 1968.

78. Dante, M. F. unpublished results.

79. "Rule 66 Exempt Solvents for Nitrocellulose Lacquers"; Shell Chemical Co., Bulletin IC:67-23SN, 1967.

80. Fink, C. K.; Weigel, J. E. Paint Varn. Prod. 1968, 58(2), 45.
81. Crowley, J. D. Paint Varn. Prod. 1971, 61(12), 35.
82. Park, R. A. Chapter 13 in Ref. 23.
83. Burns, R. J.; McKenna, L. A. Paint Varn. Prod. 1972, 62(2), 29.
84. Somerville, G. R.; Lopez, J. A. Chapter 12 in Ref. 23; Shell
     Chemical Co., Bulletin SC:69–71, 1969.
85. Nelson, R. C.; Hemwall, R. W.; Edwards, G. D. J. Paint Technol.
     1970, 42(550), 636.
86. Hansen, C. M. Chapter 4 in Ref. 23.
87. Sprinkle, G. J. Coatings Technol. 1981, 53(680), 67.

ANALYSIS AND PHYSICAL CHEMISTRY OF COATINGS AND RELATED PRODUCTS

Spectroscopic Methods in Research and Analysis of Coatings and Plastics

CLARA D. CRAVER

Chemir Laboratories, 761 West Kirkham, Glendale, MO 63122

Electronic Absorption and Emission Spectroscopy
   Visible Spectroscopy
   Ultraviolet Spectroscopy
   Luminescence Spectroscopy
   Electron Spectroscopy
   Atomic Identification and Analysis
Infrared and Raman Spectroscopy
   Infrared Spectroscopy
   Infrared Sample Preparation
   Internal Reflection Spectroscopy (IRS, ATR)
   Reflection Absorption (RAIR or IRRAS)
   Diffuse Reflectance (DRIFT)
   Pyrolysis
   Computer-Assisted Infrared
   Raman Spectroscopy
Nuclear Magnetic Resonance
Mass Spectrometry (MS)
Data Banks and Computer Retrieval

During the past five decades, spectroscopy has moved out from the laboratories of physicists and theoretical chemists into every area of analysis and chemical research. Applications vary from routine, single data point measurements for control of plant streams to structural analysis of complex molecules and conformational analysis of polymers. Enhancement of the sensitivity of all spectroscopic methods has been achieved through computer-assisted data handling. It is possible to find structural differences in polymers under different degrees of stress and to analyze for very low levels of chemical structures at surfaces and interfaces. For difficult structure determinations, data from several spectroscopic disciplines may be combined and require months or years of research.
    Each subcomponent or energy level that makes one kind of atom or molecule different from another gives rise to physical phenomena

0097–6156/85/0285–0703$10.00/0
© 1985 American Chemical Society

around which a measuring system can be devised to provide detection and characterization. From X-rays to microwaves the interaction of electromagnetic energy with matter provides chemists with powerful investigative tools.

Thus, excitation of orbital electrons can cause ultraviolet and visible absorption or fluorescence that characterizes the energy levels of atoms in specific bonding situations. Still more vigorous stimulation can cause atomic emission, which serves to identify metallic elements. Mass spectroscopy permits precise measurements of the mass, and therefore, the empirical formula of a molecule or its fragments. The magnetic moment from the spin of atomic nuclei permits the measurement of nuclear magnetic resonance and of the influence on it of closely associated atoms. The energy in the infrared (IR) region of the electromagnetic spectrum corresponds to the vibrational and rotational energy of atomic groups within molecules. IR is commonly observed as an absorption phenomenon, but analytical applications for IR emission are occasionally found. The Raman effect is an energy emission that corresponds to the same range of vibrational energies as observed by infrared spectroscopy.

Workers in the field of applied polymer science may require only a few or all of these techniques for quality control, pollution monitoring, trouble-shooting, or research on new products. Only the largest of research centers can justify maintaining a complete line of the most sensitive spectroscopic equipment. Even more important and costly is the maintenance of a staff of specialists in the many spectroscopic disciplines. These specialists need expertise in their own and related disciplines including instrumentation. They need to be familiar with the chemistry of the materials they work on, and they need a good analytical sense of representative sampling, standardization, repeatability, and possible interfering materials. They need judgment about how complete an analysis is required for a given problem. They need to maintain up-to-date spectral reference files relevant to the company's products and be familiar with data digitization and computer technology. Oftentimes they must have the engineering ability to help interface spectroscopic equipment with research experiments or plant control devices.

It is apparent that such ideally trained spectroscopists are rare. It is even rarer to find laboratory or company management policies that bring equipment, spectroscopist, and analytical problems together at the right time to gain maximum benefit from the potential that the science of spectroscopy offers. It is the purpose of this chapter to provide applied polymer specialists with an understanding of what the different spectroscopic methods can do for them.

The author hopes to provide enough perspective to help management in both large and small organizations decide which spectroscopic analytical tools it needs in-house and which ones it should obtain from specialized laboratories and research institutions. Abundant textbook references and an extensive applications bibliography are included to help the analytical chemist or polymer chemist who finds himself in the position of a do-it-yourself spectroscopist.

Myers and Long (1) devoted a chapter to each of the principal spectroscopic techniques applicable to coatings, and a "Plastics

Analysis Guide" (2) contains useful spectroscopic data and references as well as chemical data. Spectroscopic techniques are
included in books on the analysis of synthetic polymers (3) and
additives in plastics (4). This author's recent book on polymer
characterization (5) includes thermal and mechanical characterization methods and relates them to structural determinations by
spectroscopic methods. Symposium books published by the American
Chemical Society and commercial publishers often contain state-of-
the-art summaries and liberal literature citations. Current
literature coverage is provided in alternate years for fields of
application such as analysis of high polymers, coatings, rubber, and
surfactants and each spectroscopic discipline in the Reviews Issue
published each April in Analytical Chemistry by the American
Chemical Society.

Electronic Absorption and Emission Spectroscopy

The absorption of visible and ultraviolet radiation by a molecule
arises from energy transitions of orbital electrons. Basic
spectroscopy texts describe quantum theory considerations that
govern the types of chemical bonding in organic molecules which will
interact with the energy levels that correspond to visible and
ultraviolet wavelengths (6). Instrumentation and analytical methods
are comprehensively described (7-10).

Visible Spectroscopy. Visible spectroscopic methods are used to
quantify many of the sensitive color reactions so familiar to
chemists for qualitative analysis. In polymer systems these are
often tests for residual monomers or for quantitation of end groups
or hydrolysis or degradation products. Amino groups in polymers are
measured by a Schiff's base (11), carbonyl groups in hydrocarbon
polymers are estimated from 2,4-dinitrophenylhydrazones (12), a low
level of poly(vinyl alcohol) is determined in poly(vinyl chloride)
by an iodine-boric green complex (13), aldehyde and ketone groups
are analyzed in nylon 6 and 66 (14), and trace amounts of Si-H in
poly(organosiloxanes) (15).
    Colorimetric methods for metals depend on chelating or
complexing agents which yield soluble products with high absorpti-
vities. The selectivity of a given analysis may rely upon precise
pH adjustment of the solution for a single complexing agent to
distinguish among similar metals.
    A monthly index to coatings literature that includes analyses is
published in each issue of Paint Technology. The 1980 and 1982
Analytical Chemistry reviews of "Ultraviolet and Light Absorption
Spectrometry" (16, 17) follow a pattern similar to earlier reviews
including several pages of summaries of methods for metals,
nonmetals, and organic compounds. Data included are the matrix,
wavelength, reagent, molar absorptivity or concentration range, and
bibliographic reference. Additional references to
spectrophotometric methods will be found in Refs. 18-20. The
important field of color measurements is reported by F. W. Billmeyer
(21).

Ultraviolet Spectroscopy. UV excitation levels occur for less
highly conjugated molecules than the chromophores of the visible

region. The near-ultraviolet region extends from about 190 nm to the visible, and the shorter wavelength end of this region is marginally accessible depending upon instrumentation. At the shorter wavelength limit, monounsaturated compounds absorb, with a shift toward the visible that results from increased conjugation in related compounds. The systematic trend of these shifts makes it relatively straightforward to predict which systems can be analyzed spectrometrically. The data in Table I are typical of major compound classes. Certain groups called auxochromes, such as -OH and -NH$_2$, shift the absorption of a given chromophore to longer wavelengths.

Inspection of the data in Table I reveals that alkyl carbonyl compounds absorb weakly and that conjugated unsaturated compounds and aromatics absorb strongly. From this information many of the most useful applications of UV can be predicted. Aromatic plasticizers can be measured directly in alkyl ester polymers or in poly(vinyl chloride). Polystyrene is extracted from paper (23) or other products and analyzed quantitatively. Styrene monomer is analyzed in polystyrene (24), and monomer determinations, such as free acrylic acid in polyacrylic acid and ethyl acrylate in poly(ethyl acrylate), are reported (25). There is an important quality control method for nylon raw materials based on the fact that the unsaturated or aromatic impurities which interfere with the production of high-quality linear nylon polymer absorb more strongly than the predominantly alkyl functional groups such as hexamethylenediamine and adiponitrile. In this situation, highly sensitive analyses that are essential to production can be made routinely.

Ultraviolet or visible analytical methods generally can be developed for conjugated unsaturated compounds in the absence of any other materials that have strong absorption at or near the same wavelength. A computer-assisted method of spectral matching to determine polymers with up to six conjugated double bonds in degraded poly(vinyl chloride) has been developed (26). Reliable analytical methods are prescribed by committee E-13 of the American Society for Testing and Materials (27). Specific analyses approved for a variety of materials are reported annually in the "ASTM Book of Standards" for each classification of materials.

An important consideration for selecting UV or visible spectroscopy for a given analysis is the solubility of the sample in solvents transparent in the region to be used. Solvents useful through most of the UV and visible range are water, paraffinic hydrocarbons, cyclohexane, and methanol. Practical cutoffs of other common solvents are listed in Table II.

It is important that spectroscopically pure solvents be used for analytical procedures. The strong point of UV or visible methods is the high sensitivity they offer for trace components. Impure solvents may negate this advantage.

From the standpoint of compound identification, electronic absorption spectra are much less useful than IR or mass spectroscopy. For fingerprinting of individual compounds, large numbers of characteristic bands are generally necessary, and UV and visible spectra are more likely to have only one major and a few minor bands. Broad bands with considerable overlapping are a limiting

Table I. Position and Intensity of Typical Electronic Absorption Bands

Compound	λ/max[a]	ε/max[a]
Acyclic Structures		
Monoalkyl ethylenes	173–178	5,000
Dialkyl ethylenes	185–205	5,000
Ketones	195	1,000
Esters	205	50
Carboxylic acids	208	60
Butadiene	217	21,000
C=C–C=O	217	16,000
Hexatriene	258	35,000
Decapentaene	335	118,000
Aromatic Structures		
Benzene	198	8,000
Styrene	244	13,000
Diphenyl	246	20,000
<u>trans</u>–Stilbene	295	27,000
Azobenzene	319	20,000

[a] λ is expressed as nanometers (nm), and the band intensity, ε, is expressed as the molar absorption coefficient, which is the product of the absorptivity and molecular weight of a substance (<u>22</u>).

Table II. Useful Transparency Limits of Common Solvents

Solvent	Cutoff (nm)	Solvent	Cutoff (nm)
Pyridine	305	Chloroform	245
Tetrachloroethylene	290	Dichloromethane	235
Benzene	280	Ethyl ether	220
N,N–Dimethylformamide	270	Acetonitrile	215
Carbon tetrachloride	265	Alcohols, hydrocarbons	210

factor not only for qualitative analysis but also for quantitative analysis of mixtures.

The most useful feature of electronic spectroscopy is the high sensitivity it offers for the highly conjugated structures, which have strong bands in the ultraviolet and visible regions, in the presence of more saturated structures.

Luminescence Spectroscopy. Luminescence spectroscopy is based on the energy dissipating behavior of electronic systems of atoms, ions, and molecules. Two major subdivisions of luminescense are fluorescence and phosphorescence. The relationship between these phenomena is described by Smith (28) and by Fox and Price (29). A fluorescence spectrum can be produced as an apparent "mirror image" of the longest wavelength band system of the absorption spectrum. An additional energy transition may follow, i.e., at a lower rate, and produce phosphorescence. The diagram in Figure 1 shows the relationship between absorption, fluorescence, and phosphorescence. Instrumentally, time discrimination is provided by light choppers to separate fluorescence and phosphorescence signals.

An advantage of luminescence spectroscopy, as with all emission systems, is flexibility of geometry of the sample that permits front surface viewing. That is, opaque or geometrically irregular specimens may be sampled directly (28).

Applications of luminescence spectroscopy include very sensitive measurements for additives in synthetic rubber (30), brighteners on synthetic fibers (31, 32), and monitoring of photooxidation products in highly pure products (33). Investigation of molecular orientation and small molecular segments in polymer networks (34-36) is receiving increased attention. When excimer trap fluorescence is studied in aromatic systems, conformational changes can be followed as a function of environmental parameters such as temperature (37) or pressure (38) and the kinetics of phase separation in polymer blends can be evaluated in terms of solvents, mixing temperature and molecular weight (39). Polymerization reactions can be monitored with molecular probes that exhibit viscosity-dependent fluorescence (40, 41). A method for determining the number average molecular weight by fluorescence procedures has been reported (42). An extensive bibliography through 1980 is included in a book on fluorescent probes (43).

In contrast to the ultraviolet and visible absorption methods described earlier, details of the methodology of luminescence spectroscopy are not widely known and few standard methods have evolved. Books on theory and techniques are helpful (44) as are memoranda on applications from instrument manufacturers. ASTM Committee E-13.06 on Molecular Luminescence has had large task forces working for several years on practices for instrument testing, nomenclature, and analytical procedures. Recent symposia sponsored by that committee are the basis for two new books (45, 46).

Electron Spectroscopy. Electron spectroscopy is a rapidly developing field involving measurement of electrons ejected from a bombarded sample. It may be divided into categories according to the bombarding source, ultraviolet excitation (UPS), X-ray photoelectron spectroscopy (XPS), electron spectroscopy for chemical analysis (ESCA), and electron-gun excitation (Auger spectroscopy). Practical applications are being reported at a rapidly increasing rate for ESCA and Auger spectroscopy. They provide a powerful tool for studying surfaces up to about 20-Å depth and are therefore finding research applications in surface chemistry areas such as friction and wear, adhesion, and catalyst research.

There is a Journal of Electron Spectroscopy and Related Phenomena, and the Nobel Prize Award address by Siegbahn is widely available (47). Basic background in both theory and applications is available in a book edited by Brundle and Baker (48). An introductory monograph by Barker and Betteridge is focused on chemical and analytical aspects of electron spectroscopy (49) and broader aspects are discussed by Sevier (50). General articles relating to surface analysis (51-53), corrosion research (54), and catalysts (55-57) have been reported. Glass-fiber resin composites were studied (58) and adhesive transfer of polytetrafluoroethylene to metals was measured (59). A study of adhesion at the rubber-metal interface (60) used XPS as did studies of surface grafts of polypropylene, polyethylene and polyester (61, 62), polyimide films (63), and surface functional groups on polyethylene films (64).

While most of the applications important to polymer science have been on surfaces of solids, papers have been published on liquid-phase XPS (65) and sulfate groups on polystyrene latexes have been analyzed (66).

Atomic Identification and Analysis. Atomic emission and absorption spectroscopy and X-ray fluorescence and absorption are used for elemental analyses. These methods vary in their sensitivity and quantitative applicability. A summary of the usually accepted virtues and limitations of these methods is given in Table III.

A two-part special issue of Spectrochimica Acta (67, 68) is dedicated to atomic absorption spectroscopy. A book by Van Loon (69) and a chapter by Robinson (70) are good basic texts. An educational audiocassette/slide course by SAVANT (71) covers basic principles of atomic, emission, and fluorescence spectroscopy, and SAVANT has also produced a course on inductively coupled plasma (ICP). There are now abundant references on ICP including various aspects of inductively coupled plasmas to emission spectroscopy (72). Compilations of spectral tables (73) and spectral interferences (74) are valuable contributions to this technique. An easy to read and enlightening comparison of various methods of atomic spectroscopy has recently been published (75).

Analyses of pigments and fillers are the major coatings applications of atomic spectroscopy. Listings of the standards organizations, which recommend specific methods, are detailed in the 1983 review by Anderson and Vandeberg (76). Research applications are apt to be in areas of complexing or chelation characteristics (77) or adsorption, contamination or migration studies. An example is the use of X-ray analysis in establishing a method for predicting durability of paints under marine exposure. It was found that small changes in the level of toxin after short exposures could be extrapolated to predict lifetime persistence (78).

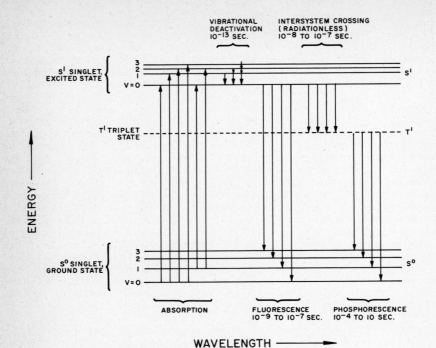

Figure 1.   Energy relationships of electric electronic
             transitions of a molecule.   Reproduced with
             permission from Ref. 28.   Copyright 1971 Plenum
             Press.

Table III.   Spectroscopic Methods of Elemental Analysis

Method	Atomic Phenomenon	Use
Atomic emission	Light emission from excited electronic states of atoms	General for qualitative identification of metals; simultaneous determinations
Atomic absorption	Absorption of atomic resonance line	Metals analyzed individually; for quantitative analysis
X-ray fluorescence	Reemission of X-rays from excited atoms	General for all elements above atomic no. 10
X-ray absorption	X-ray excitation of K and L shell electrons	Good quantitative method for heavier elements in presence of light elements

Infrared and Raman Spectroscopy

Infrared and Raman spectroscopy correspond to similar molecular energy phenomena--the vibrational energies of atoms or groups of atoms within molecules, and rotational energies.

Infrared spectroscopy may be observed as absorption or, less commonly, emission spectra for which the frequencies that interact with electromagnetic radiation are those that involve a change in the dipole moment of the molecule as the vibration takes place. This spectrum is observed in the region past the red end of visible radiation to the microwave region. The common "fingerprinting" region is from 5000 wavenumbers (cm^{-1}) to 200 cm^{-1}. This corresponds to 2-50 microns, or in SI units, 2-50 μm. Infrared spectroscopy is the most broadly applicable molecular spectroscopic tool and relates so intimately to the nature of the chemical bond that it is predicted that it will be a leading investigative tool for another 50 years (79).

Raman spectroscopy is the observation of an emitted pattern of frequency displacements from an exciting line caused by vibrations within a molecule that result in a change in polarizability. It can be observed by stimulation with electromagnetic radiation far removed from IR, e.g., UV or visible, although the observed frequency displacements correspond to the frequency of infrared radiation.

The Raman effect is weak, and its application was severely limited until an extremely strong exciting source, the laser, became available. The principal limitations of Raman spectroscopy now arise from the simultaneous phenomenon of fluorescence, which is a strong interfering signal in the spectrum of many commercial materials.

Infrared and Raman spectroscopy are complementary in structural determinations because some molecular vibrations that are inactive in the infrared (that is, do not result in a change in dipole moment and therefore do not cause an absorption band) do have a strong Raman line. The reverse is also true. Some bands that are weak or forbidden in the Raman spectrum are strong in the infrared spectrum. With the combined use of these techniques, the vibrational energies of a molecule can be fully described.

Infrared Spectroscopy. Infrared spectroscopy was first recognized to be a nearly universal tool for characterizing chemical structure after the monumental research of W. W. Coblentz, reported in a major publication in 1905 (80).

Research by many investigators over the next 30 years developed the well-known group frequencies which correlate with chemical structure for common groups such as OH, CH, and C=O. It was learned that the 7-15 μm region of the spectrum provided a fingerprint of individual molecules even for closely similar isomers.

Commercial spectrometers became available in the late 1930s. World War II spurred production of improved spectrometers and development of analytical methods to solve the isomer analysis needs of the petroleum, rubber, and chemical industries. Fortunately, the rock salt prism data obtained during those productive years in the application of infrared spectroscopy to commercial products and complex mixtures are as effective in the fingerprinting region for

most of these materials as data obtained on newer grating and
interferometric instruments, so that many of the methods developed
during that period are still in use today. The technical literature
of the 1940s and 1950s reflects the rapid development of methods and
their extension to polymers (81-83), crude oil identification (84),
analysis of inorganics (85), and characterization of coal and coal
hydrogenation products (86). In the polymer and organic coatings
field the wide range of applications is indicated by studies on
epoxy compounds (87), novolak and resole resins (88), polyurethanes
(89), polyethylene terephthalate (90), polyacrylonitrile (91),
polybutadiene and butadiene-styrene (92), curing studies on epoxy
resins (93) and elucidation of the effect of driers on the oxidative
drying of oils and varnishes (94).

Infrared spectroscopy continues to be one of the principal
techniques for structural analysis of polymers and for identifying
components of complex formulations. The distinctiveness of
important vinyl, alkyl, and aryl chemical structures in the infrared
such as ester, amide, nitrile, isocyanate, hydroxyls, amine, and
sulfone makes it ideal for the first gross characterization of
chemical types present and for following the reactions of these
functional groups in curing or degradation studies.

Up-to-date compendiums on applications of infrared spectroscopy
in applied polymer science are as follows. "An Infrared
Spectroscopy Atlas for the Coatings Industry" (95) describes
techniques, has liberal references to specific methods, and contains
high-quality grating reference spectra on paint components and
blended compositions. "Atlas of Polymer and Plastics Analysis," 2nd
ed., by Hummel and Scholl (96), has issued two volumes: Vol. 1,
Polymers; Vol. 3, on Additives and Processing Aids; Vol. 2, on
Plastics, Fibers, Rubbers, Resins, is in press. "Infrared Spectra
of Plasticizers and Other Additives," 2nd ed., published by The
Coblentz Society, Inc., is a high-quality IR reference spectrum
collection (97).

The identification of polymers is largely done by fingerprint
matching. Representative spectra of the most important commercial
polymers are included here for the convenience of the reader
(Figures 2-10). They are grouped by structure to emphasize the
features that chemically related polymers have in common and to
demonstrate the distinctive identifying characteristics of each
polymer class.

A recent book on theory and applications of vibrational
spectroscopy to polymers (98) by Painter, Coleman, and Koenig is a
major reference book, as is the new edition of the familiar Haslam
and Willis book (99) covering the combined chemical and
spectroscopic analysis of plastics and resins. An excellent chapter
in which G. Zerbi describes the basis for "Probing the Real
Structure of Chain Molecules by Vibrational Spectroscopy" is in an
ACS Advances in Chemistry book (100), along with polymer degradation
studies by FTIR (101, 102). Henniker has written on IR of
industrial polymers (103), Perkin-Elmer Corporation distributes a
technical applications sheet on infrared grating spectra of polymers
(104), and two excellent publications on infrared spectroscopy of
rubber and elastomers have been written by R. Hampton (105) and
W. C. Wake (106). Practical approaches to coatings analysis (1) and

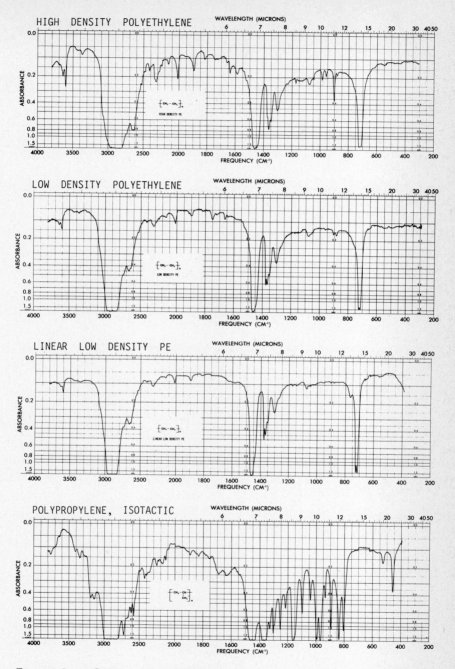

Figure 2.   Spectrum of high-density polyethylene, low-density
polyethylene, linear low-density polyethylene, and
isotactic polypropylene.

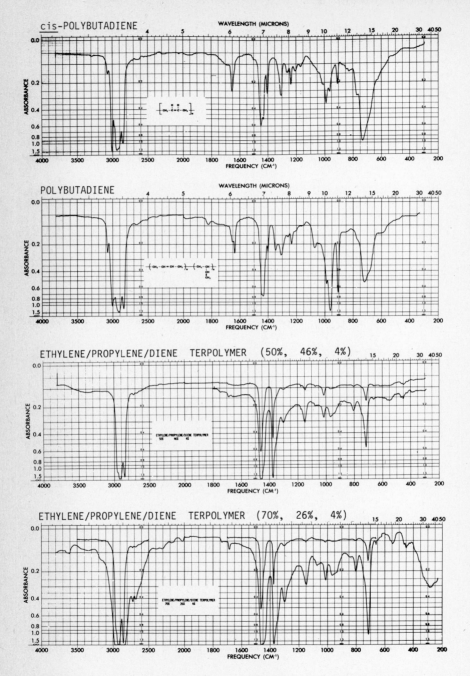

Figure 3.   Spectrum of *cis*-polybutadiene, polybutadiene, ethylene/propylene/diene terpolymer.

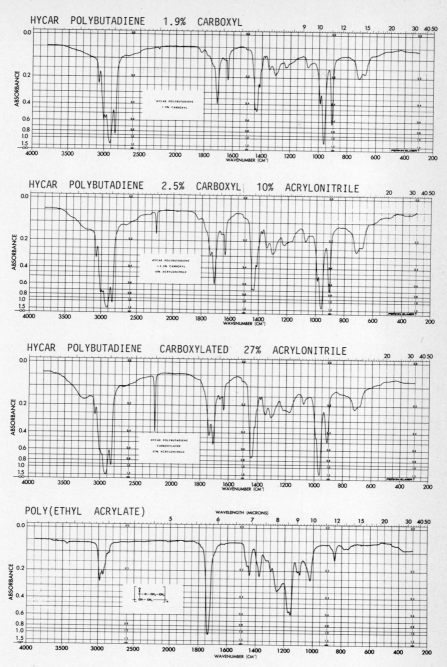

Figure 4. Spectra of Hycar polybutadiene/1.9% carboxyl, Hycar polybutadiene/2.5% carboxyl/10% acrylonitrile, carboxylated Hycar polybutadiene/27% acrylonitrile, and poly(ethyl acrylate).

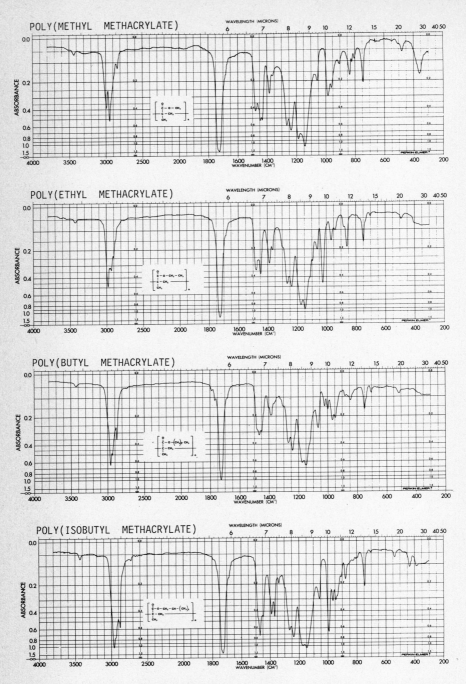

Figure 5.   Spectra of poly(methyl methacrylate), poly(ethyl
            methacrylate), poly(butyl methacrylate), and
            poly(isobutyl methacrylate).

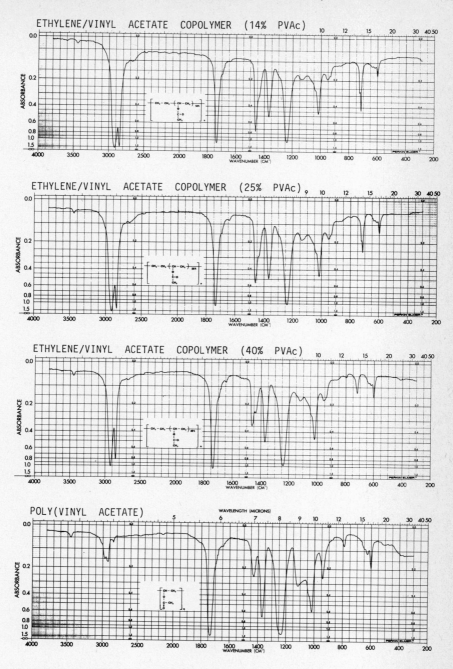

Figure 6. Spectra of ethylene/vinyl acetate copolymer (14%, 25%, and 40% PVAc) and poly(vinyl acetate).

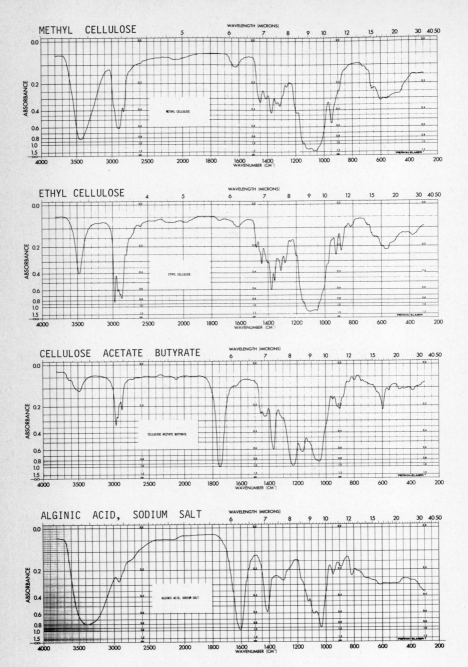

Figure 7.   Spectra of methylcellulose, ethylcellulose, cellulose acetate butyrate, and sodium alginate.

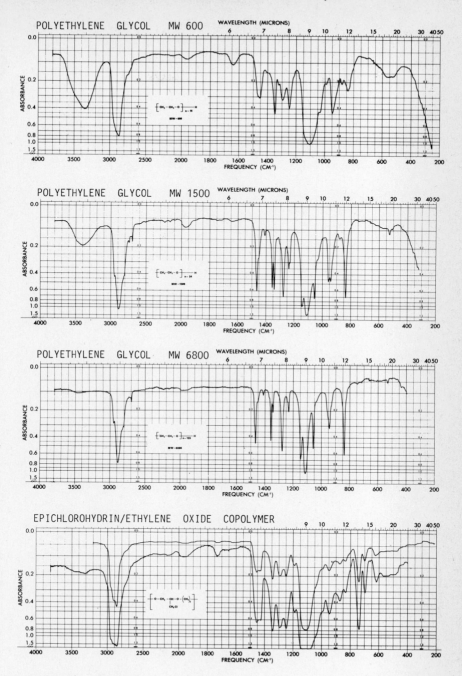

Figure 8. Spectra of polyethylene glycol (MW 600, 1500, and 6800) and epichlorohydrin/ethylene oxide copolymer.

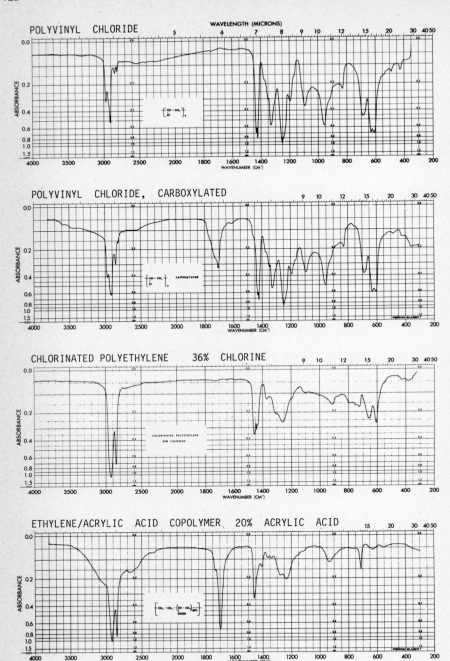

Figure 9.   Spectra of poly(vinyl chloride), carboxylated poly(vinyl chloride), chlorinated polyethylene (36% chlorine), and ethylene/acrylic acid copolymer (20% acrylic acid).

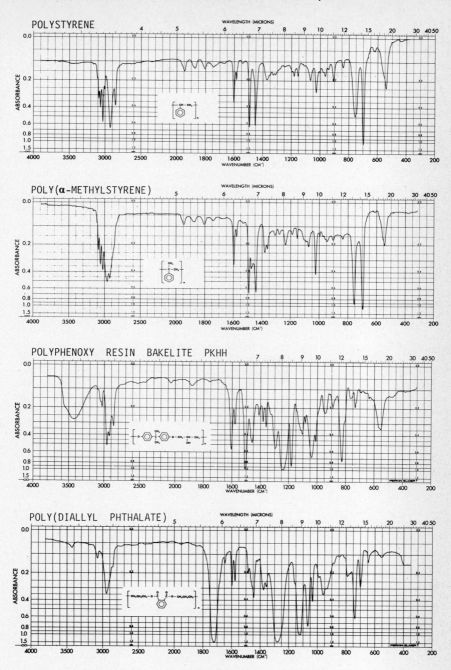

Figure 10. Spectra of polystyrene, poly(α-methylstyrene), polyphenoxy resin, Bakelite PKHH, and poly(diallyl phthalate).

applications of infrared and Raman spectroscopy to characterization and analysis of surfaces (107) have been reported.

The classical book on techniques by W. J. Potts (108) is still the book of choice for sample preparation and quantitative methods. Practical aspects of theory, sample handling, and application are well-described by A. L. Smith (109). ASTM E-13 is expected to release a new practice on qualitative analysis in 1985, and its "Manual on Practices in Molecular Spectroscopy" (27) covers nomenclature, instrument testing, microanalysis, internal reflection, and quantitative analysis.

Bellamy (110) and Colthup (111) are the best recognized and most extensive references on group frequency interpretation, and an easy-to-use introduction to group frequencies and a summary of major compound classes are in the Coblentz Society's "Desk Book" (112).

Infrared Sample Preparation. Details of sample preparation for obtaining infrared spectra are discussed for films, capillary layers, demountable cells, fixed thickness cells, pressed halide pellets, and oil mulls in the above references by Potts and Smith and in ASTM methods (27). Other more specialized techniques are described briefly below.

Internal Reflection Spectroscopy (IRS, ATR). The spectrum of the sample is determined at the interface of the specimen and a crystal in optical contact. It has found widespread use in the field of coatings and surfaces because of the ease of sampling and the opportunity it provides to investigate successive strata through a coating. Reports on internal reflection determination of contaminants on adhesive surfaces, exuded plasticizer, fiber surfaces, in situ tissue, and sensors in viscous plant streams are well referenced (1, 27, 107, 113, 114). Application literature is readily available from instrument companies: Foxboro Analabs, Norwalk, Conn.; Perkin-Elmer Corporation, Norwalk, Conn.; Beckman Instruments, Irvine, Calif.; Barnes Engineering Company, Stamford, Conn.; Harrick Scientific, Ossining, N.Y. An article focused on analysis of adhesives (113) points up some of the experimental difficulties in obtaining reproducible data.

Reflection Absorption (RAIR or IRRAS). A significant development in infrared spectroscopy during the 1960s has been reported in journals not commonly read by either polymer chemists or analytical chemists. It has developed in the fields of electrochemistry and catalysis and is called reflection-absorption spectroscopy. Papers by Greenler (115-118), Yates (119), and Hansen (120, 121) describe the theory and some experimental data, and Boerio and Gosselin report on its applications to polymers (122).

Essentially the technique involves transmission through ultrathin sample layers on metal plates and reflection at a glancing angle as diagrammed in Figure 11. Actual experimental layouts are highly varied as reference to the above-cited articles will delineate.

The observation that infrared spectra of ultrathin films on metal surfaces measured at a high angle of incidence give absorption intensities greater than that calculated from the sample thickness and number of reflections was observed and discussed in the late

1950s (123). Greenler has demonstrated theoretically that the spectral sensitivity at an interface can be enhanced a 1000-fold or more by choosing an angle of incidence in the range 80-88°. Some workers appear to prefer lower angles of incidence, but experimental work is in early stages. The advance of theory relating to ATR (114) and its importance in measuring electrochemical reactions at electrodes led to the present day understanding that promises that reflection-absorption may provide a powerful tool for studying molecules or a few bond lengths of a coating at metal interfaces.

Diffuse Reflectance (DRIFT). This technique has been so severely energy limited that rare use was made of it until computer-averaging capabilities prompted design work on improved sample-handling attachments. Simple attachments are now available for most new spectrometers, both Fourier transform and dispersive. Irregular sample shapes can be used, but better results are generally obtained on powders. Extensive coverage is given by Griffiths and Fuller (124), and its application with dispersive spectrophotomoters is described by Hannah and Anacreon (125).

Pyrolysis. This is best regarded as a highly valuable but "last resort" technique. It is valuable because it can supply spectra when other efforts to obtain them have failed, for example, on insoluble thermoset plastics and on carbon-filled cross-linked elastomers. It is used as a last resort because the collected pyrolyzate may not represent all of the components of the sample and the same sample may not give the same spectra in consecutive runs; thus, pyrolysis is useful, but some components of a blend or copolymer may be missed entirely and comonomer proportions on even a semiquantitative basis are difficult to achieve. As with other technique-sensitive analytical methods, it is possible to stan- dardize carefully in a highly repeatable system for consistent analysis if good calibration procedures have been used. A pyrolysis unit for maximum effectiveness should be evacuable, control temperature and time, and provide for collection of vapor-phase pyrolyzate and of condensed pyrolyzate. The "Pyro-Chem" unit, which was designed by Wilks Scientific and is now sold by Foxboro Analabs, fits these requirements. A pyrolysis unit by Chemical Data Systems, Inc., Oxford, Pa., offers these features as well as programmed temperature and controlled sweeping of the pyrolyzate for stepwise sampling. There are many useful references available on pyrolysis techniques (82, 126-128).

Computer-Assisted Infrared. The impact of computers on spectroscopy has been high. In vibrational spectroscopy, it was felt strongest in the 1970s when the revolution in computer technology brought prices down into the range where it was practical to have a dedicated computer as an integral part of a spectrometer.

Initially this development was most important in the field of Fourier transform spectroscopy. The Fourier transform (FT) spectrometer is essentially an interferometer, which measures the entire spectrum simultaneously instead of measuring sequential frequencies as is the case with dispersive spectrometers. The spectrum is computed from the interferogram and, in order to be practical, this computation requires a computer. Once the spectrum

is available in digital form, computerized data handling advantages
are so great that newer dispersive spectrometers are also computer
assisted. Repeated scans, computer averaging, and curve smoothing
routines yield a high signal/noise ratio that has permitted
extension of infrared spectroscopy into ultramicrosampling, diffuse
reflection, and photoacoustic spectroscopy (PAS).

Spectrum subtraction techniques are among the most popular
features of computerized IR and are proving especially applicable to
comparing closely similar polymers (129–132), and excellent accounts
of data processing of polymer spectra have been written by Koenig
(133, 134).

In the opposite direction from new theoretical developments and
elaborate instruments is the interesting engineering by Wilks
Scientific (now The Foxboro Co.) to produce portable spectrometers
containing variable long-path cells for sensitive monitoring of air
pollutants (Figure 12). The spectrometer is a small unit that may
be used as a single-beam point reader or connected to a recorder to
provide spectra similar to the commercial instruments of 1940. The
long-path cell is the bulkiest part of the apparatus as is shown in
Figure 12, and it is capable of measuring most gases at the current
OSHA limits (0.1–1000 ppm) or lower. It offers the big advantage of
giving instantaneous samplings around any given work area while
changes in processing variables or exhaust systems are being made.
Thus, it can be a valuable tool to the plant engineer in the process
of planning for or making clean-air installations.

Raman Spectroscopy. As mentioned earlier, the Raman effect is an
emission phenomenon, which means that front-surface sampling is
possible. An irregularly shaped solid may be used in the
spectrometer without processing it to make it flat, as required for
internal reflection, or without processing it to a film or powder
for transmission IR. Another sampling convenience is that Raman is
more easily applicable to water solutions than IR.

Specific circumstances in which Raman spectra may be more useful
than IR on an "as received" sample are as follows.

(1) Filled polymers or composites containing silica, clay, or
    similar materials may have less interference from the filler in
    the Raman spectrum than in the IR spectrum of the polymer
    because most fillers are poor Raman scatterers but give strong
    infrared bands that interfere with polymer identification. It
    may not be necessary to remove the filler in order to obtain a
    good Raman spectrum, but such a step is usually necessary for
    IR.
(2) Chunks or pieces of polymers can be examined directly by Raman,
    which is an advantage for thermosets and tough rubbery
    materials. Variations in crystallinity and amorphous regions
    or orientation produced by processing may be observed because
    sample preparation is not required.
(3) The intensity advantage of Raman for the C=C band that in IR is
    highly variable in intensity, in fact, being absent in
    molecules that are symmetrical around the C=C band, has been
    used by Grasselli et al. to study cross-linking in a
    polystyrene polymer (135). A similar curing study in styrene-
    based polyester resins using the vinyl C=C in the styrene

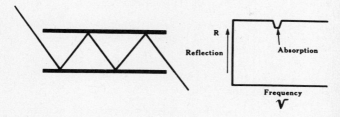

Figure 11.   Optical path and spectral effects in reflection-
             absorption spectroscopy.

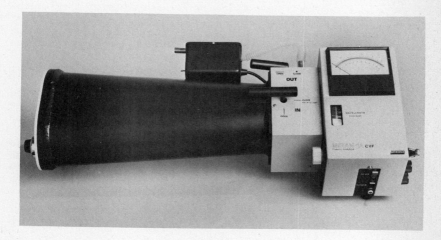

Figure 12.   Gas analyzer consisting of a variable path length gas
             cell and infrared spectrometer can be used for the
             analysis of any gas or vapor having infrared
             absorption in concentrations ranging from a few parts
             per million to several percent.

monomer relative to the aromatic ring band at 1590 cm^{-1} was reported by Koenig (136). Quantitative analysis of three-component systems is reported (137) and determination of thiol groups in sulfur polymers showed high sensitivity (138).

(4) Fifty-nine references on the application of Raman spectroscopy to polymers are cited in a recent popular book on applications of Raman (139). Polyacetylene and related polymers are being widely investigated at this time (140-143).

Sloane, Boerio and Koenig, and McGraw have described the sampling and other instrumental considerations for Raman spectra of polymers (144). Other reports on Raman investigations of polymers include molecular orientation in bulk polyethylene terephthalate (145), crystallinity of ethylene-propylene rubber (146), and the structure of unsaturated polyester resins cross-linked with styrene (147).

Orientation of polymers has been increasingly studied by Raman. PMMA orientation (148), oriented PP and PE (149), quantitative data on chain orientation of PP (150), and vibrational spectroscopic studies of polymer chain order have been reviewed (151).

The Raman spectrum of most compounds is not similar enough to the infrared spectrum for direct fingerprinting comparisons. However, a spectroscopist competent in infrared group frequencies can readily interpret Raman data. Sloane (144) gives examples of the most important group frequencies for which Raman spectra offer advantages over infrared. Nyquist and Kagel (107) compare infrared and Raman spectra of typical classes of organic compounds, and Dollish et al. (152) have compiled comprehensive data on Raman group frequencies. Colthup (111) treats infrared and Raman bands simultaneously. General spectral comparisons between infrared and Raman are made in Table IV. It should be recognized that the frequencies listed are in round numbers and that both band positions and intensities may vary significantly for individual compounds.

An excellent application of Raman to inorganics in the coatings field is the analysis of TiO$_2$ (153). An accurate quantitative Raman analysis for anatase in rutile was developed for quantities in the range of 0.03 and 10 wt%. The details of the analysis are shown in Figures 13-15. Figure 13 shows the Raman spectrum of anatase natural crystal. The Raman line at 143 cm^{-1} is so intense and sharp that it is proposed as an internal standard for Raman spectroscopy. In Figure 14 the spectral comparison is between a synthetic rutile crystal cube and a powder sample containing 99.34% rutile and 0.66% anatase. The linearity of the calibration curve for anatase below 1% is attested to by the data in Figure 15.

There is a tendency in discussing a new technique to point out its advantages and that is the approach taken here for laser Raman spectroscopy. It is worth emphasizing that fluorescence still presents a major sampling problem for most commercial materials in the Raman and that at this time infrared is much more widely applicable to applied polymer science. Infrared is generally the more effective tool for trace analyses and for quantitative data.

However, a phenomenon broadly described as the resonance Raman effect has been investigated with great interest in the past decade and may be expected to greatly increase sensitivities for some analyses. In this resonance Raman effect, the intensity of a Raman

Table IV. Comparison of Group Frequencies in Infrared and Raman Spectra

Structure	Frequency	Comments
OH	3300	Strong in IR; weak in Raman
NH, aliphatic	3300	Weak in IR; stronger in Raman
SH	2600	Weak in IR; strong in Raman
R–C≡C–R, R's equal	2200	Forbidden in IR; strong in Raman
–C≡N	2200	Variable in IR; stronger in Raman
–C=O	1700	Strong in IR; medium in Raman
–C=C–	1640	Medium to absent in IR; strong in Raman
–P=O	1270	Strong in IR; weak in Raman
C–S	800–570	Weak in IR; strong in Raman
S–S	550–500	Weak in IR; strong in Raman

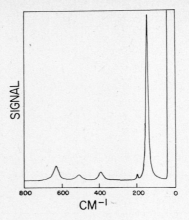

Figure 13.   Raman spectrum of anatase natural crystal. Laser
power was 450 mW at 4880 A.   Spectral slit width was
4 cm^{-1}. Reproduced with permission from Ref. 153.
Copyright 1972 Appl. Spectrosc.

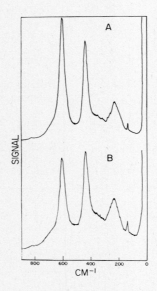

Figure 14.   Raman spectra of TiO$_2$. A, synthetic rutile single
crystal cube; B, powder sample, 99.34% rutile and
0.66% anatase. Reproduced with permission from Ref.
153.   Copyright 1972 Appl. Spectrosc.

signal may be increased orders of magnitude by exciting a sample with a laser having an emission wavelength within the electronic absorption band envelope of the compound being excited. A general discussion of the specialized technique was reported in European Spectroscopy News (154), and the many variations that have been developed on enhancement of Raman spectra are described in a book by Harvey (155).

Nuclear Magnetic Resonance

Some atomic nuclei behave like spinning magnets. If these nuclei are placed in a strong magnetic field, they may absorb impinging radio-frequency signal as a function of the field strength. At a particular radio frequency, the magnetic field strength at which a nucleus absorbs is a function not only of that particular isotopic nucleus but also of its immediate electronic environment, i.e., the nature of the chemical bonding of the atom.

Thus, for a hydrogen atom (proton resonance), the field strength at which the resonance occurs for a CH_2 group is different from the field strength at which a CH_3 group resonates. A nuclear magnetic resonance spectrum of ethanol shows different bands for the H of the OH group and for the CH_2 and the CH_3. Much more subtle chemical distinctions are, of course, possible with high-resolution NMR.

The nature of the nucleus is the most important consideration in whether a useful level of nuclear magnetic resonance occurs. The most sensitive nuclei are those that behave like strong magnets, have a high relative isotope abundance, and have nuclear charge distribution approaching spherical. In Table V, the nuclear resonance susceptibility of some common atoms is summarized.

As with infrared spectroscopy, Fourier transform techniques are used to provide instantaneous data or to accumulate scans for increased sensitivity on microsamples. Spectra have been obtained on as little as 10 μg with Fourier transform NMR. However, it is not sensitive to minor components in a mixture.

Recent fundamental books are Akitt's on Fourier transform and multinuclear NMR (156), Mehring's on high-resolution NMR in solids (157), and Levy's on methods and applications (158). Books directed more toward polymers are by Kaufman (159), Bovey (160, 161), and Iven (162), which contains useful reviews.

The greatest advantage NMR brings to polymer characterization is its specificity for identifying the chemical structures adjacent to a given group in a molecule. Thus, it can be used to differentiate between block and random copolymers and measure the tacticity of polymer chains. As examples of application, NMR characterizes polymers as to stereoregularity (163,164), monomer sequences (165-167), and microstructure (168-170) and is used to analyze monomers (171-174) and elucidate polymerization mechanisms (175-178). Of considerable interest is the use of pulsed NMR to measure cure in UV-irradiated coatings (179). The importance of NMR in polymer characterization is described in the proceeding of the 1983 Phillips Award Symposium honoring F. A. Bovey (180).

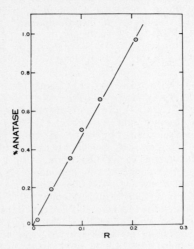

Figure 15. Calibration curve for Raman spectroscopic determination of low levels of anatase in TiO_2. $R = (I_{143}/I_{610}) - (I_{143}/I_{610})_{rutile}$. Reproduced with permission from Ref. 153. Copyright 1972 <u>Appl. Spectrosc.</u>

Table V. Usefulness of Common Nuclei in Natural Abundance for NMR Studies

Nucleus	Intrinsic Applicability of NMR
H^1 F^{19} P^{31}	Most useful
B^{11} N^{14} Si^{29}	Useful
C^{13}	Not abundant, but special instrumentation has made it available and very useful
H^2 O^{17} S^{33} N^{15}	Usable
C^{12} O^{16} S^{32} $Si^{28,30}$	Not usable

Mass Spectrometry (MS)

Mass spectrometry requires that the sample under investigation be volatile or produce volatile fragments under high vacuum. In the polymer field, it is used principally to analyze raw materials, residual monomers or solvents, degradation products, and low molecular weight oligomers. Instrument capabilities are increasing with the use of high-field magnets and special ionization techniques so that increasing applications to molecular weight determinations are being reported.

Mass spectral data were observed on polybutadiene samples with molecular weights in the range of 1000-3000 (181), on PEG and PPG oligomers (182), and on polystyrene oligomers (183) by field desorption MS.

Reviews on the analysis of polymers and polymer products have recently appeared (184-187). Typical applications are covered for analysis of raw materials for coatings (188-191) and determination of volatile components of commercial polymers (192). Mass spectrometry of thermally treated polymers (193) and of stressed polymers are treated (194) with instrumental and experimental details given in more detail than in corresponding journal references.

Pyrolysis, combined with gas or liquid chromatography and infrared and mass spectroscopy, is a powerful tool for studying polymer fragments. A 14-page bibliography covering selected reports from 1973 to 1980 is available at no cost (195), and the method is reported on by Liebman (5, 128).

Methods of extending MS to solid state are summarized by Gardella, Graham, and Hercules in a comprehensive report on laser desorption mass spectrometry (5).

This is another field in which the uninitiated can be overwhelmed by acronyms, so some of the most widely used ones are explained here: HRMS = high-resolution mass spectrometry; HRGC/HRMS = high-resolution gas chromatography/HRMS; LC/MS - liquid chromatography/mass spectrometry; CI = chemical ionization; FD = field desorption; FAB = fast atom bombardment; LDMS = laser desorption mass spectrometry.

Data Banks and Computer Retrieval

Both infrared and mass spectroscopy produce complex band patterns that require comparison to reference spectra for identification. Peak position data have been abstracted from published spectra in both disciplines, and computer programs to retrieve the data in the file are highly effective. As Raman spectroscopy becomes more prevalent the same approach may be expected for it.

The practical limit of these search systems is now only one of having adequate quality data bases available. Several technical groups address themselves to this problem. Among the leaders are the Joint Committee on Atomic and Molecular Physical Data, whose secretary is from the National Standard Reference Data Center of the National Bureau of Standards, Washington, D.C., and The Coblentz Society, Inc., whose secretary is at Perkin-Elmer Corporation, Norwalk, Conn.

Instrument companies have added spectral search systems as a spectrometer accessory for modest-sized files. The largest mass spectral files are accessible through the NIH-EPA Chemical Information System (CIS), and the largest IR on-line search system is available on Tymshare (196).

Literature Cited

1. Myers, R. R.; Long, J. S. "Treatise on Coatings, Vol. 2, Characterization of Coatings, Physical Techniques "; Marcel Dekker: New York, 1970.
2. Krause, A.; Lange, A.; Ezrin, M. "Plastics Analysis Guide"; Verlag: Munich, 1983.
3. Urbanski, J.; Czerwinski, W.; Janicka, K.; Majewska, F. "Handbook of Analysis of Synthetic Polymers and Plastics"; Wiley: New York, 1977.
4. Crompton, T. R. "Chemical Analysis of Additives in Plastics"; International Series in Analytical Chemistry; Pergamon: Oxford, England, 1978; Vol. 46.
5. Craver, C. D. "Polymer Characterization: Spectroscopic, Chromatographic and Physical Instrumental Methods"; American Chemical Society: Washington, D.C., 1983.
6. Bauman, R. P. "Absorption Spectroscopy"; Wiley: New York, 1962.
7. Meehan, E. J. "Treatise in Analytical Chemistry," 2nd ed.; Elving, P. J.; Meehan, E. J.; Kolthoff, I. M., Eds.; Wiley: New York, 1981.
8. Fleet, M. St. C. "Physical Aids to the Organic Chemist"; Elsevier: Amsterdam and New York, 1962.
9. Cahill, J. E. "Fundamentals of UV/Visible Spectrophotometry: An Outline"; Applications Data Bulletin, ADS 123, Perkin-Elmer Corp.: Norwalk, Conn., 1980.
10. Silverstein, R. M.; Bassler, G. C.; Morrill, T. C. "Spectrometric Identification of Organic Compounds," 4th ed.; Wiley: New York, 1981.
11. Esko, K.; Karllson, S.; Porath, J. Acta Chem. Scand. 1968, 22(10), 3342.
12. Belisle, J. Anal. Chim. Acta 1968, 43, 515.
13. Eliassaf, J. J. Polym. Sci., Part B 1972, 10, 697.
14. Schmitz, F. P.; Mueller, H.; Rossback, V. Kunststoffe 1979, 69(6), 321.
15. Rotzsche, H.; Prietz, U.; Diedrich, H.; Clauss, H.; Hahnewald, H. Plaste Kautsch 1978, 25(7), 390.
16. Hargis, L. G.; Howell, J. A. Anal. Chem. 1980, 52, 306R.
17. Howell, J. A.; Hargis, L. G. Anal. Chem. 1982, 54, 171R.
18. Snell, F. D.; Snell, C. T. "Colorimetric Methods of Analysis Including Some Turbidimetric and Nephelometric Methods," 3rd ed.; Van Nostrand: Princeton, N.J., 1971; Vol. 4AAA.
19. Mavrodinean, R.; Schultz, J. K.; Menis, O. "Accuracy in Spectrophotometry and Luminescence Measurements"; NBS Special Publication No. 378, U.S. Government Printing Office: Washington, D.C., 1973.
20. Burgess, C.; Knowles, A. "Techniques in Visible and Ultraviolet Spectrometry, Vol. 1: Standards in Absorption Spectrometry"; Chapman and Hall: London, 1981.

21.   Craver, J. K.; Tess, R. W. "Applied Polymer Science"; Division of Polymeric Materials: Science and Engineering, American Chemical Society: Washington, D.C., 1975.

22.   "Standard Definitions of Terms and Symbols Relating to Molecular Spectroscopy"; ASTM Designation E-131-71, American Society for Testing and Materials: Philadelphia, PA 19103.

23.   Spagnola, R. Appl. Spectrosc. 1974, 28(3), 259.

24.   Newell, J. E. Anal. Chem. 1951, 23, 445.

25.   Brunn, J.; Doerffel, K.; Much, H.; Zimmerman, G. Plaste Kautsch 1975, 22(6), 485.

26.   Daniels, V. D.; Rees, N. H. J. Polym. Sci., Polym. Chem. Ed. 1974, 12(9), 2115.

27.   "Manual on Practices in Molecular Spectroscopy," 4th ed.; ASTM Committee E-13, American Society for Testing and Materials: Philadelphia, PA, 1979.

28.   Smith, H. F. In "Polymer Characterization, Interdisciplinary Approaches"; Craver, C. D., Ed.; Plenum: New York, 1971.

29.   Fox, R. B.; Price, T. R. In "Polymer Characterization, Interdisciplinary Approaches"; Craver, C. D., Ed.; Plenum: New York, 1971.

30.   Drushel, H. V.; Sommers, A. L. Anal. Chem. 1964, 36, 836.

31.   Maruyama, T.; Kuroki, N.; Koniski, K. Kogyo Kagaku Zasshi 1965, 69, 2428.

32.   Maruyama, T.; Kuroki, N.; Kawaii, M.; Koniski, K. Kogyo Kagaku Zasshi 1966, 69, 86.

33.   Fox, R. B.; Price, T. R.; Cain, D. S. ADVANCES IN CHEMISTRY SERIES No. 87, American Chemical Society: Washington, D.C.; p. 72.

34.   Cozzens, R. F.; Fox, R. B. Polym. Preprints 1968, 9(1), 363.

35.   Niushiyima, Y.; Onogi, Y.; Asei, T. J. Polym. Sci., Part C 1966, 15, 237.

36.   Deshpandi, A. B.; Subramanian, R. V.; Kapur, S. L. Makromol. Chem. 1966, 98, 90.

37.   Frank, C. W. Macromolecules 1975, 8, 305.

38.   Fitzgibbon, P. D.; Frank, C. W. Macromolecules 1981, 14, 1650.

39.   Gelles, R.; Frank, C. W. Macromolecules 1983, 16, 1448.

40.   Loutfy, R. O. Macromolecules 1981, 14, 270.

41.   Loutfy, R. O. J. Polym. Sci., Polym. Phys. Ed. 1982, 20, 825.

42.   Baumbach, D. O. J. Polym. Sci., Polym. Lett. Ed. 1982, 20, 117.

43.   Beddard, G. S.; West, M. A. "Fluorescent Probes"; Academic: London, 1981.

44.   Hercules, D. M. "Fluorescence and Phosphorescence Analysis"; Interscience: New York, 1966.

45.   Eastwood, D. "New Directions in Luminescence"; STP822, ASTM: Philadelphia, 1983.

46.   Love, L. J. Cline; Eastwood, D. "Advances in Luminescence"; ASTM: Philadelphia, in press.

47.   Siegbahn, K. Prix Nobel 1982, 114; Science (Washington, D.C.) 1982, 217, 111; Rev. Mod. Phys. 1982, 54, 709.

48.   "Electron Spectroscopy: Theory and Technical Applications"; Brundle, C. R.; Baker, A. D., Eds.; Academic: New York, 1980; Vol. 4.

49.   Barker, A. D.; Betteridge, D. Int. Ser. Monogr. Anal. Chem. 1972, 53.

50.   Sevier, K. D. "Low Energy Electron Spectrometry"; Wiley-Interscience: New York, 1972.

51. Ignatiev, A.; Rhodin, T. N. Am. Lab. 1972, 4, 8.
52. Brundle, C. R. Surf. Sci. 1971, 27, 681.
53. Roberts, M. W. Adv. Catal. 1980, 29, 55.
54. Olefjord, I. Scand. Corros. Cong., Proc. 6th, Swed. Corros. Inst., Stockholm, Sweden, 1971.
55. Brinen, J. S.; Barr, T. L.; Davis, L. E. "Applied Surface Analysis ASTM STP 699"; American Society for Testing and Materials: Philadelphia, Penn., 1980; p. 24.
56. Briggs, D. Appl. Surf. Sci. 1980, 6, 188.
57. Wieserman, L. F.; Hercules, D. M. Appl. Spectrosc. 1982, 36, 361.
58. Wong, R. J. Adhesives 1972, 4, 171.
59. Pepper, S. V.; Buckley, D. H. NASA Technical Note, NASA TN D-6983, 1972.
60. VanOoij, W. J. Surf. Sci. 1977, 68, 1.
61. Bradley, A.; Cynba, M. Anal. Chem. 1975, 47, 1839.
62. Swartz, W. E. Anal. Chem. 1973, 45, 788A.
63. Leahy, H. J., Jr.; Campbell, D. S. Surf. Interface Anal. 1979, 1, 75.
64. Everhart, D. S.; Reilley, C. N. Anal. Chem. 1981, 53, 665.
65. Siegbahn, H.; Lundholm, M. J. Electron Spectrosc. Relat. Phenom. 1982, 28, 135.
66. Stone, W. E. E.; Stone-Masui, J. H. Sci. Technol. Polym. Colloid 1983, 2, 480, NATO ASI Ser. E.
67. Spectrochim. Acta, Part B. 1980 35B(11/12).
68. Spectrochim. Acta, Part B 1981, 36B(5).
69. Van Loon, J. C. "Analytical Atomic Absorption Spectroscopy: Selected Methods"; Academic: New York, 1980.
70. Robinson, J. W. In "Treatise on Analytical Chemistry," 2nd ed.; Elving, P. J.; Meehan. E. J., Kolthoff, Eds.; Wiley: New York, 1981; Vol. 1, p. 729.
71. Walsh, A. "Principles of Atomic Emission and Fluorescense, AA 101" (49 slides); SAVANT: Fullerton, CA, 1979.
72. Barnes, R. M. "Applications of Inductively Coupled Plasmas to Emission Spectroscopy"; Franklin Institute Press: Philadelphia, 1978,
73. Boumans, P. W. J. M. "Line Coincidence Tables for Inductively Coupled Plasma Atomic Emission Spectrometry"; Pergamon: New York, 1980; 2 Vol.
74. Parsons, M. L.; Forster, A. "An Atlas of Spectral Intereference in ICP Spectroscopy"; Plenum: New York, 1980.
75. Parsons, M. L.; Major, S.; Forster, A. R. Appl. Spectrosc. 1983. 37, 411.
76. Anderson, D. G.; Vandeberg, J. T. Anal. Chem. 1983, 55(5), 10.
77. Kim, W. Y.; Shin, H. C.; Maeng, K. S. Pollimo 1983, 7, 168.
78. Driscoll, C.; Freiman, A. Paint Technol. 1970, 42(549), 521.
79. Crawford, B. L. "The Future of Infrared Spectroscopy," an address on the occasion of the 25th Anniversary of the Fisk Infrared Institute, Nashville, Tenn., 1974.
80. Coblentz, W. W. "Investigations of Infrared Spectra"; Carnegie Institute of Washington: Washington, D.C., 1905. Reprinted by the Coblentz Society and the Perkin-Elmer Corp., Norwalk, Conn, 1962.
81. Hausdorff, H. "Analysis of Polymers by Infrared Spectroscopy" Pitts. Conf. Anal. Chem. Appl. Spectrosc. 1951.

82. Harms, D. L. Anal. Chem. 1953, 25(8), 1140.
83. Swann, M. H.; Exposito, G. E. Anal. Chem. 1954, 26 1054.
84. Smith-Craver, C. D.; Black, J. F. U.S. Patent 2 700 593, 1947.
85. Miller, F. A.; Wilkins, C. H. Anal. Chem. 1952, 24, 1253.
86. Friedel, R. A.; Queiser, J. A. Anal. Chem. 1956, 28, 22.
87. Schreve, O. D.; Heether, M. R.; Knight, H. B.; Swern, D. Anal. Chem. 1951, 23, 277.
88. Grisenthwaite, R. J.; Hunter, R. F. J. Appl. Chem. 1956, 6, 324.
89. Corish, P. J. Anal. Chem. 1959, 31, 1298.
90. Tobin, M. C. J. Phys. Chem. 1957, 61, 1392.
91. Adams, M. L.; Swann, M. H. Off. Dig. 1958, 30, 646.
92. Binder, J. L. Anal. Chem. 1954, 26, 1877.
93. Dannenberg, H.; Harp, W. E., Jr. Anal. Chem. 1956, 28, 86.
94. Smith-Craver, C. D.; Mueller, E. R. Ind. Eng. Chem. 1957, 49, 210.
95. "An Infrared Spectroscopic Atlas for the Coatings Industry"; Frederation of the Society for Paint Technology, Philadelphia, 1980.
96. Hummel, D. O.; Scholl, F. "Atlas of Polymer & Plastics Analysis," 2nd ed.; Carl Hanswer Verlag: Munich, Verlag Chemie International: New York, Vol. 1, "Polymers," 1978; Vol. 2, "Plastics, Fibers, Rubbers, Resins," 1984; Vol. 3, "Additives and Processing Aids," 1981.
97. Craver, C. D. "Infrared Spectra of Plasticizers and Other Additives," 2nd ed.; The Coblentz Society: Kirkwood, Mo., 1980.
98. Painter, P. C.; Coleman, M. M.; Koenig, J. L. "The Theory of Vibrational Spectroscopy and Its Application to Polymeric Materials"; Wiley: New York, 1982.
99. Haslam, J.; Willis, H. A.; Squirrel, D. C. M. "Identification and Analysis of Plastics," 2nd ed.; Iliffe: London, and Van Nostrand: Princeton, N.J., 1972.
100. Zerbi, G. "Polymer Characterization: Spectroscopic, Chromatographic, and Physical Instrumental Methods"; Craver, C. D., Ed.; ADVANCES IN CHEMISTRY SERIES No. 203, American Chemical Society: Washington, D.C., 1983; Chap. 29.
101. Pearce, E. M.; Bulkin, B. J.; Yeen Ng, M. "Polymer Characterization: Spectroscopic, Chromatographic, and Physical Instrumental Methods"; ADVANCES IN CHEMISTRY SERIES No. 203, American Chemical Society: Washington, D.C., 1983; Chap. 33.
102. Coleman, M. M.; Sivy, G. T. "Polymer Characterization: Spectroscopic, Chromatographic, and Physical Instrumental Methods"; ADVANCES IN CHEMISTRY SERIES No. 203, American Chemical Society: Washington, D.C., 1983; Chap. 32.
103. Henniker, J. C. "Infrared Spectrometry of Industrial Polymers"; Academic: New York, 1967.
104. Zeller, M. V.; Pattacini, S. C. "The Infrared Grating Spectra of Polymers"; Perkin-Elmer Instrument Corp.: Norwalk, Conn., 1973.
105. Hampton, R. R. Rubber Chem. Technol. 1972, 43(3), 546.
106. Wake, W. C. "Analysis of Rubber and Rubber-like Polymers"; Wiley-Interscience: New York, 1969.
107. Brame, E.; Grasselli, J. "Practical Spectroscopy"; Marcel Dekker: New York, 1974.

108. Potts, W. J., Jr. "Chemical Infrared Spectroscopy"; Wiley: New York, 1962; Vol. 1.
109. Smith, A. L. "Applied Infrared Spectroscopy"; Wiley: New York, 1979.
110. Bellamy, L. J. "The Infrared Spectra of Complex Molecules," 3rd ed.; Wiley: New York, 1975.
111. Colthup, N.; Daly, L.; Wiberly, S. "Introduction to Infrared and Raman Spectroscopy," 2nd ed.; Academic: New York, 1975.
112. Craver, C. D. "The Desk Book of Infrared Spectra," 2nd ed.; The Coblentz Society: Kirkwood, Mo., 1982.
113. Paralusz, C. M. J. Colloid Interface Sci. 1974, 47(3), 719.
114. Harrick, N. J. "Internal Reflection Spectroscopy"; Wiley Interscience: New York, 1967.
115. Greenler, R. J. Chem. Phys. 1966, 44, 310.
116. Ibid., 1969, 50, 1963.
117. Greenler, R. G.; Rahn, R. R.; Schwartz, J. P. J. Catalysis 1971, 23, 42.
118. Kottke, M. L.; Greenler, R. G.; Tompkins, H. G. Surf. Sci. 1972, 32, 231.
119. Yates, J. T., Jr. Chem. Eng. News 1974, 52(34), 19.
120. Hansen, W. N. Symp. Faraday Soc. 1970, No. 4, 27.
121. Hansen, W. N. J. Opt. Soc. Am. 1973, 63(7), 783.
122. Boerio, F. J.; Gosselin, C. A. "Polymer Characterization: Spectroscopic, Chromatographic, and Physical Instrumental Methods"; ADVANCES IN CHEMISTRY SERIES No. 203, American Chemical Society: Washington, D.C., 1983; p. 541.
123. Battelle Memorial Institute, personal communication, 1957.
124. Griffiths, P. R.; Fuller, M. P. Adv. Infrared Raman Spectrosc. 1982, 9, 63.
125. Hannah, R. W.; Anacreon, R. E. Appl. Spectrosc. 1983, 37(1), 75.
126. Csete, A.; Levi, D. W. "Literature Survey on Thermal Degradation, Thermal Oxidation and Thermal Analysis of High Polymers V"; Plastec, N.T.I.S.: Springfield, Va., 1976; Parts 1, 2.
127. Stevens, M. P. "Characterization and Analysis of Polymers by Gas Chromatography"; Marcel Dekker: New York, 1969.
128. Liebman, S. A.; Levy, E. F. "Polymer Characterization of Pyrolysis, GC-Mass Spectrometer-FTIR"; Marcel Dekker: New York, 1984; in press.
129. Koenig, J. L. Appl. Spectrosc. 1975, 29, 293.
130. D'Esposito, L.; Koenig, J. L. In "Fourier Transform Infrared Spectroscopy"; Ferraro, J. R.; Basile, L. J., Eds.; Academic: New York, 1978; Vol. 1, Chap. 2.
131. Painter, P. C.; Watzek, M.; Koenig, J. L. Polymer 1977, 18, 1169.
132. Gartan, A.; Carsson, D.; Wiles, D. M. Appl. Spectrosc. 1981, 35, 432.
133. Koenig, J. L. Acc. Chem. Res. 1981, 14, 171.
134. Koenig, J. L. Adv. Polym. Sci. 1984, 54, 87.
135. Grasselli, J. G.; Hazle, M. A. S.; Mooney, J. R.; Mehicic, M. Proc. 21st Coloq. Spectrosc. Int., 1979.
136. Koenig, J. L. Chem. Technol. 1972, 2, 411.
137. Sloane, H.; Bramston-Cook, R. Appl. Spectrosc. 1973, 27, 217.

138. Mukherjee, S. K.; Guenther, G. D.; Bhattacharya, A. K. Anal. Chem. 1978, 50, 1591.
139. Grasselli, J. G.; Snaveley, M. K.; Bulkin, B. J. "Chemical Applications of Raman Spectroscopy"; Wiley-Interscience: New York, 1981.
140. Lauchlan, L.; Chen, S. P.; Etemad, S.; Kletter, M.; Heeger, A. J.; MacDiarmid, A. G. Phys. Rev. B, Condens. Matter 1983, 27, 2301.
141. Galtier, M.; Benoit, C.; Montaner, A. Mol. Crystallogr. 1982, 83, 1141.
142. Zannoni, G.; Zerbi, G. Chem. Phys. Lett. 1982, 87, 50.
143. Ibid., p. 55.
144. Boerio, F. J.; Koenig, J. L.; Sloane, H.; McGraw, G. E. In "Polymer Characterization: Interdisciplinary Approaches"; Craver, C. D., Ed.; Plenum: New York, 1971; Chap. 1-3.
145. Derouault, J. L.; Hendra, P. J.; Cudby, M. E. A.; Willis, H. A. J. Chem. Soc., Chem. Commun. 1972, 1972, 1187.
146. Schreier, G.; Peitscher, G. Fresenius' Z. Anal. Chem. 1971, 258, 199.
147. Koenig, J. L.; Shih, P. T. K. J. Polym. Sci., Part A-2 1972, 10, 721.
148. Bower, D. I. "Structure and Properties of Oriented Polymers"; Applied Sciences: London, 1975.
149. Hendra, P. J.; Willis, H. A. Chem. Ind. (London) 1967, 2146, London; Chem. Commun., 1968, 225.
150. Satija, S. K.; Wang, C. H. J. Chem. Phys. 1978, 69(6), 2739.
151. Holland-Moritz, K. J. Appl. Polym. Sci., Appl. Polym. Symp. 1978, 34, 49.
152. Dollish, F. R.; Fateley, W. G.; Bentley, F. F. "Characteristic Raman Frequencies of Organic Compounds"; Wiley: New York, 1975.
153. Capwell, R. J.; Spagnola, F.; DeSesa, M. A. Appl. Spectrosc. 1972, 26(5), 537.
154. Van Haverbeke, L.; Herman, M. A. Practical Appl. Reson. Raman Spectrosc. ESN 1982, 40.
155. Harvey. A. B., Ed. "Chemical Applications of Non-Linear Spectroscopy"; Academic: New York, 1981.
156. Akitt, J. W. "NMR and Chemistry-An Introduction to the Fourier Transform-Multinuclear Era," 2nd ed.; Chapman and Hall: New York, 1983.
157. Mehring, M. "Principles of High Resolution NMR in Solids," 2nd ed.; Springer-Verlag: Berlin, 1983,
158. Levy, G. C., Ed. "NMR Spectroscopy: New Methods and Applications"; ACS SYMPOSIUM SERIES No. 191, American Chemical Society: Washington, D.C., 1982.
159. Kaufmann, E. N.; Shenoy, G. K., Eds. "Nuclear and Electron Resonance Spectroscopies Applied to Materials Science"; Elsevier: New York, 1981; Vol. 3.
160. Bovey, F. A. "Chain Structure and Conformation of Macromolecules"; Academic: New York, 1982.
161. Woodard, A. E.; Bovey, F. A., Eds. "Polymer Characterization by ESR and NMR"; American Chemical Society: Washington, D.C., 1980.
162. Ivin, K. J., Ed. "Structural Studies of Macromolecules by Spectroscopic Methods"; Wiley: New York, 1976.

163. Fujishige, S. Makromol. Chem. 1978, 179, 2251.
164. Spevacek, J. J. Polym. Sci., Polym. Phys. 1978, 16, 523.
165. Hirai, H.; Koinuma, H.; Tanabe, T.; Takeuchi, K.
     J. Polym. Sci., Polym. Chem. 1979, 17, 1339.
166. Logothetis, A. L.; McKenna, J. M. ACS Div. Polym. Chem., Pap.
     1978, 19, 528.
167. Keller, F.; Michajlov, M.; Stoeva, S. Acta. Polym. 1979,
     30(11), 694; Chem. Abstr. 1980, 92, 42579a.
168. Evans, D. L.; Weaver, J. L.; Mukherji, A. K.; Beatty, C. L.
     Anal. Chem. 1978, 50, 857.
169. Inoue, Y.; Konno, T. Makromol. Chem. 1978, 179, 1311.
170. Natansohn, A.; Maxim, S.; Feldman, D. Polymer 1979, 20, 629.
171. Zambrini, A. Pitture Vernici 1978, 54, 169.
172. Mozayeni, F. Appl. Spectrosc. 1979, 33, 520.
173. Barton, F. E., II; Himmelsbach, D. B.; Walters, D. B. J.
     Am. Oil Chem. Soc. 1978, 55, 574.
174. Harris, R. K.; Robins, M. L. Polymer 1978, 19, 1123.
175. Caze, C.; Loucheux, C. J. Macromol. Sci. 1979, A12, 1501.
176. Chiavarini, M.; Bigatto, R.; Conti, N. Agnew. Macromo. Chem.
     1978, 70, 49.
177. Fages, C.; Pham, Q. T. Makromol. Chem. 1978, 179, 1011.
178. Plochokca, K.; Harwood, H. J. ACS Div. Polym. Chem., Pap.
     1978, 19, 240.
179. Barrett, J. L. J. Radiat. Curing 1979, 6, 20.
180. "Organic Coatings and Applied Polymer Science Proceedings";
     American Chemical Society: Washington, D.C., 1983; Vol. 48,
     pp. 76–102, 192–216.
181. Craig, A. G.; Callis, P. G.; Derrick, P. J. Int. J. Mass
     Spectrom. Ion Phys. 1981, 38, 297.
182. Lattimer, R. P.; Hansen, G. E. Macromolecules 1980, 14, 776.
183. Lattimer, R. P.; Harmon, D. J.; Hansen, G. E. Anal. Chem.
     1980, 52, 1808.
184. Israel, S. C. "Flame–Retardant Polymer Materials"; Lewin, M.;
     Atlas, S. M.; Pearce, E. M., Eds.; Plenum: New York, 1982;
     Vol. 3, p. 201.
185. Foti, S.; Montaudo, G. "Analysis of Polymer Systems"; Bark, L.
     S.; Allen, N. S., Eds.; Applied Science Publishers: London,
     1982.
186. Luderwalt, I. Pure Appl. Chem. 1982, 54, 255.
187. Sedgwick, R. D. Dev. Polym. Charact. 1978, 1, 41.
188. Hase, A. T.; Anderegg, R. J. Am. Oil Chem. Soc. 1978, 55, 407.
189. Holmbom, B.; Era, V. J. Am. Oil Chem. Soc. 1978, 55, 342.
190. Pai, J. S.; Lomanno, S. S.; Nawar, W. W.; J. Am. Oil Chem.
     Soc. 1979, 56, 495.
191. Startin, J. R.; Gilbert, J.; McWeeny, D. J. J. Chromatogr.
     1978, 152, 495.
192. Ligon, V. W., Jr.; George, M. C. J. Polym. Sci., Polym. Chem.
     1978, 16.
193. Mol. G. J.; Gritter, R. J.; Adams, G. E. "Applications of
     Polymer Spectroscopy"; Brame, E. G., Jr., Ed.; Academic: New
     York, 1978; Chap. 16.
194. Grayson, M. A.; Wolf, C. J. In "Applications of Polymer
     Spectroscopy"; Brame, E. G., Jr., Ed.; Academic: New York,
     1978; Chap. 14.
195. Applications Laboratory, Chemical Data Systems, Inc., Oxford,
     Pa.
196. IRGO, C. D. Craver, Manager, 761 W. Kirkham, Glendale, Mo.

Electron Microscopy in Coatings and Plastics Research

L. H. PRINCEN

Northern Regional Research Center, Agricultural Research Service, U.S. Department of Agriculture, Peoria, IL 61604

Optical Microscopy
Transmission Electron Microscopy
Scanning Electron Microscopy
Peripheral Techniques
Trends and Future Projections

Almost all morphological and many phenomenological studies of coatings and plastics require some form of microscopy because their building blocks are so small in size. The dimensions of pigments, fibers, latex particles, film thicknesses, surface undulations, or emulsion droplets are often below the resolution of the naked eye or a hand lens, and need a good deal of magnification to become visible, either individually or collectively. During the early periods of modern and more scientific approach to coatings research and development, the only magnifying techniques available involved the various modes of optical microscopy. Although these modes included transmission, reflection, polarization, dark field, UV, and ultramicroscopy, there were many drawbacks to their use. Resolution and depth-of-focus especially limited their usefulness.

In 1932, the electron microscope was developed in Europe. It promised to become the microscopist's dream, with predicted magnifications in the order of 200,000X, and resolutions in the order of 10 Å. Still, another 12 years went by before the electron microscope was used occasionally in problems related to coatings and plastics. However, with the development of shadowing and replica techniques and with the introduction of commercial instruments, electron microscopy became commonplace in polymer research. Although specimen preparation remained tedious and image interpretation was often difficult, transmission electron microscopy increased our understanding in the field of polymers a great deal.

In 1965, the scanning electron microscope became commercially available, and almost overnight it became so popular that pictures of insects, moon dust, prehistoric pollen, and other materials could be admired almost daily in newspapers, magazines, and semi-scientific journals. Also the impact on scientific literature has been tremendous. The availability of research funds to purchase scientific equipment in the 1960s, the simplicity of sample preparation, and the spectacular, easily interpreted images were probably the main stimuli that increased the popularity of the scanning electron microscope. Despite this overall popularity and suitability, its use in coatings and polymer research was rather limited for some time.

In addition to these major forms of microscopy, there have been several developments of peripheral techniques that are important for the polymer scientist. They include X-ray spectroscopy, electron spectroscopy, and the electron microprobe. Their present and future impact has been evaluated. Also some predictions have been made on some expected or desired further developments in the field of microscopy.

Optical Microscopy

In the early 1900s, optical microscopy was well developed and was already being used regularly in the research laboratory and in industrial production. Various modes of imaging were available, such as transmission, reflection, polarizing, and ultramicroscopy, and all of them were successfully applied by the research scientist. Unfortunately, methods for recording the image on film and subsequent production of high-quality prints for publication were not well developed at that time. Often the researcher might have been tempted to apologize for the quality of the images appearing in print by saying, "I wish you could have been there to see for yourself what I mean."

In the middle 1920s were the years of improved artificial lighting (1-3), micromanipulation (4), microtoming and sectioning (5-7), ultramicroscopy (8-11), and pigment studies (12-14). The first books on industrial microscopy appeared at this time (15-17) and indicated a greater demand for information in this area than one might suspect from the small number of research articles published on the subject. In the early 1930s, optical microscopy was used regularly and successfully in both rubber and fiber research (18-21), but in the coatings field most work was restricted to pigment powders.

Of course, the optical microscope was not so well suited for paint research because of inherent resolution and depth-of-field limitations. For example, pigment particles are often well below 1μm in size; therefore, the most interesting features of their surfaces are below the resolving power attainable. Optical scientists have continually pushed for greater resolution through oil-immersion objectives (14), shorter wavelength UV radiation (22), or design of an instrument based on a new concept, namely, the scanning optical microscope (23). The limited depth-of-focus, however, is not a feature that can be improved easily, and until today, has remained a severe handicap in optical microscopy.

Another restriction to microscopy of coatings is related to their optical nature. Pigmented films are opaque to light, and clear films are often too transparent to reveal any details.

In the mid-1930s when both photographic and printing techniques for photomicrographs were greatly improved, U.S. research was not directed toward microscopy, but rather to development of heavy production equipment and manufacturing facilities. At this same time, a completely new instrument--the transmission electron microscope--was introduced in Europe and went almost unnoticed in the United States. It took many years before this development gap was overcome.

Transmission Electron Microscopy

Although the discovery in 1928 of the deflection of an electron beam in a magnetic field is seen by some as the birthdate of the electron microscope, the first working transmission electron microscopes were described by German investigators in 1932 (24, 25). This development was soon followed by other manuscripts in Russian (26), English (27), French (28), and Italian (29). However, rapid developments and improvements were coming almost entirely from Germany. For example, by the time that the first U.S. electron microscope, with a magnification of 40X, was described in the literature (30), German scientists were already thinking in terms of 200,000X (31), and the new technique was being applied to actual problems in colloid research (32), biology (33), and aerosol and pigment particle investigations (34).

Despite these rapid developments, transmission electron microscopy could not possibly contribute significantly to coatings and polymer research until commercial machines became available. The first Siemens microscope was marketed in 1938 (35), followed the next year by a more advanced unit. In this country, RCA marketed its first unit in 1941 (36). Even with commercial equipment available, many other problems had to be overcome. The main difficulty was that of sample preparation. Sections had to be extremely thin to be penetrated sufficiently by the electron beam. Often contrast was not sufficient to form a suitable image. In 1939, the shadowing technique was developed to enhance contrast (37). In 1941, the replica technique was introduced to simulate surfaces of materials normally opaque to electrons (38). Radiation damage to the specimen became recognized (39), and special techniques for viewing pigment powders were devised (40).

During the 1940s, transmission electron microscopes became generally available, and their handling and specimen preparation techniques were greatly improved. Scientific research in general increased rapidly, and synthetic polymers appeared at a record pace for fibers, plastics, and coatings. These changes expanded the use of both optical and electron microscopy as evidenced by the large number of publications. No major changes in microscopy technology took place during the 1950s and the first half of the 1960s. The major effort during these years was toward increased resolution. More powerful machines were built, with acceleration potentials as high as 1,000,000 kV. During these 25 years, many scientists were working on ion microscopy, by which magnifications of 1,000,000X can be achieved and individual atoms of heavy metals can be seen. Unfortunately, the limited usefulness of this form of microscopy excludes the study of polymers and coatings.

During the early years of electron microscopy, two scientists stand out because of their unorthodox approach to problem solving and instrument design. Their names are Max Knoll and Manfred von Ardenne. Only one year after he developed one of the first electron microscopes (25), Knoll was able to form images from the secondary electrons created by electron irradiation of metals and insulators (41). Knoll's main interest was in electron emission and cathode ray tubes during these years. In contrast, von Ardenne was mainly interested in electron microscopy at that time. In 1938, he calculated that the limit of resolution in the electron microscope is 10^{-6} mm at 50 kV, and that this limit may not be reached because of less than ideal conditions (42). That same year he constructed what he called an "electron scanning microscope" and obtained images from a crude television tube (43, 44). Knoll described a similar instrument soon afterwards (45). Von Ardenne recognized image imperfections in conventional transmission electron microscopy due to electron scattering in the object (43). He proposed a combination electron–X-ray microscope to eliminate this problem (46). He was extremely active in applying his microscopic techniques to polymer and colloid research (47, 48).

In the United States, the first transmission electron micrographs on coatings and related materials appeared in 1944 (40, 49). Although several researchers have contributed to coatings and polymer electron microscopy since then, E. G. Bobalek and his research staff should be mentioned especially. They were particularly active in this field during the mid–1950s (50–54).

Scanning Electron Microscopy

Although the concept of the scanning microscope was introduced early (23, 43), it was not until the early 1960s that scientists started looking seriously into this mode of imaging. Presumably, the main reason for this delay was caused by the strong emphasis placed on increased resolution that appeared to be the main goal in instrument development during the first 25 years. However, in the early 1960s a research group at Cambridge University, England, started building on the initial studies of Knoll and von Ardenne. Coupled with the improved technology of electron optics and electronics, this work culminated in 1965 in the first commercially available scanning electron microscope, the Cambridge Stereoscan.

Whereas basically image formation in the transmission electron microscope is similar to that in optical microscopy, complete with condensor, objective, field, and projector lenses, in scanning electron microscopy the image is formed in an entirely different fashion (55). Images can be produced from the reflected, secondary, conducted, or transmitted electrons, as well as from the X-rays and optical radiation that are produced by the target area upon radiation by the primary electron beam. Images are often spectacular, especially those made in the secondary mode. They often appear as if the object was inspected at close range under normal lighting conditions. Consequently, such images normally are easily interpreted. Sample preparation is simple: a small piece of the material to be studied is mounted on a platform, coated with a thin layer of metal, and placed in the microscope for examination. Because only the surface layer can be observed, no replicas have to be made. Internal structures can be observed by cutting or fracturing the sample. Magnifications from 10X to 200,000X can now be reached routinely, resolution is in the order of 30 Å, and the depth-of-focus can be more than 500 times that attained in optical microscopy.

The first nine years of active scanning electron microscopy have seen tremendous progress in instrument design. Solid state electronics has helped reduce the equipment from console to tabletop models; improved vacuum techniques and electron gun design have resulted in better resolution and additional modes of operation. Scanning electron microscopy became popular almost overnight, and many instruments were purchased and used worldwide to examine almost any kind of living or nonliving material. As many as 10 different brands of scanning electron microscopes became available in a few years.

Although the technique was applied to coatings and polymers early (56, 57), its use in these fields has not been so prevalent as would be expected. At present, the only drawback to scanning electron microscopy appears to be the cost of equipment and operation. The bibliography on scanning electron microscopy, compiled by O. C. Wells (58), includes less than 20 manuscripts that can be linked directly to the study of polymers, coatings, and plastics. To promote such use, a symposium on this subject was expressly organized and presented before the joint conference of the Chemical Institute of Canada and the American Chemical Society (ACS) at Toronto, Canada, in May 1970, with the ACS Division of Organic Coatings and Plastics Chemistry as prime sponsor (59). A follow-up symposium on the same topic was held at Dallas, Texas, in April 1973 (60). These symposium papers were representative of the work done in this field. Some additional recent techniques were described at the ACS meeting in 1981 at Atlanta (61).

A good example of the capability and usefulness of all three modes of microscopy is illustrated in Figure 1 with micrographs of rutile titanium dioxide pigment. Figure 1a is an optical micrograph at nearly maximum magnification. Often such an image is sufficient, for example, to determine degree of dispersion, and no time-consuming procedures and expensive equipment have to be used. If particle shape and size become important, such as in production processes, the scanning electron microscope is often the most suitable instrument, as exemplified in Figure 1b. If more surface detail is required, the transmission electron microscope with its increased resolution is the preferred instrument. Figure 1c shows such detail at an identical magnification used in Figure 1b. The example is just one of many that could be taken from the coatings and plastics research and production fields.

Peripheral Techniques

Although the production of X-rays upon bombardment of a target with an electron beam was known in the 1930s (46), not until the early 1950s was an instrument specifically designed to use this principle for analysis of materials at the microscopic level. The electron

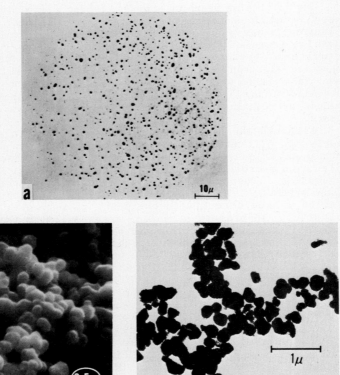

Figure 1.   Optical (a), scanning (b), and transmission (c) electron
micrographs of rutile titanium dioxide pigment show the
differences in information that can be obtained with the
three modes of microscopy.

microprobe has an electron beam that can be moved for mapping an area or kept stationary for spot analysis. Upon excitation with an electron beam, all chemical elements emit characteristic X-rays, which then can be analyzed according to their wavelength or their energy. When the electron beam is scanned over the target surface, the X-ray signal may be converted into an image similar to that made in the scanning electron microscope. If the image is produced from the X-rays of a given wavelength or energy, it will show the locations of a given chemical element in the target area examined. Images can be made at different wavelengths or energies to obtain a complete microanalysis of the target. Also, the beam may be kept stationary at any given spot in the target area, and an X-ray spectrum can be taken in a short time for a complete chemical analysis of that spot. A good review of the use of the microprobe in coatings research has been presented by Labana (62).

Upon development of the scanning electron microscope, it was soon combined with features of the microprobe through attachment of an X-ray detector to the new instrument. Unfortunately, the beam intensity and spot size on the target are both much less in the scanning electron microscope than in the microprobe; consequently, the generation of X-rays in the scanning electron microscope is less intense, and both elemental and spacial resolution are reduced. New developments in detector technology have improved the elemental resolution, and it has become possible to detect low elements, such as fluorine or oxygen; even lithium can be detected without problems in the microprobe. In coatings and polymer research, the most important elements are often carbon, nitrogen, and oxygen. Also, the physical features of coatings and plastics are often in the submicron size range, and again, resolution is too poor to produce good results. Despite these shortcomings of present-day X-ray attachments to scanning electron microscopes, this instrument combination has already produced excellent results in selected areas of coatings research (63).

Also the scanning transmission electron microscope (STEM) can be regarded as a peripheral instrument (64) that until now has had little impact on polymer research. This form of microscopy may become more important with greater availability. Resolution of better than 15 Å is already attainable in the STEM mode.

Trends and Future Projections

Although optical microscopes become more sophisticated every year, basically no improvements can be expected in magnifying power, resolution, or depth-of-focus in the conventional optical system. Perhaps, optical microscopy could be revolutionized by combining the scanning principle with laser optics. Even where electron microscopes are available, low-power binocular or high-power standard microscopes still are excellent supporting instruments in the coatings, fiber, or polymer laboratory. A quick look at low-power magnification or under special lighting may reveal features that might otherwise be overlooked.

Developments in transmission electron microscopy will not affect coatings and plastics research much in the near future. New instruments may have scanning and X-ray features included as options. New guns, lower vacuum, and solid state electronics have

improved stability and resolution without going to higher
accelerating voltages, but again, drastic changes cannot be
expected. However, these same developments will probably further
revolutionize scanning electron microscopy. Although this technique
became commercially available only 16 years ago, we have already
seen three generations of instruments, and many more will follow in
short order. Solid state electronics, computer-assisted imaging,
new high-vacuum techniques, new high-intensity electron guns,
improved detectors of emitted radiation, and new presentation modes
have already been introduced in a short time. These improvements
have already resulted in resolutions below 30 Å. Improved X-ray
detectors and data processors have also improved the elemental and
spacial resolution of this research mode to a range needed for most
polymer research. Also, electron spectroscopy will become more
prominent with these new developments. Mass production and
miniaturization have already reduced the price of basic equipment in
electron microscopy. In turn, independent laboratories are now
able to charge relatively low rates for contract work to those small
companies that cannot afford to maintain their own microscopy facility.

Literature Cited

1. Myers, W. M. Amer. Mineral. 1924, 9, 177.
2. Silverman, A. Ind. Eng. Chem. 1925, 17, 651.
3. Jonés, L. A. Paper Trade J. 1926, 83(16), 56.
4. Hauser, E. A. Z. Wiss. Mikrosk. 1924, 41, 465.
5. Bödecker, C. F. Dent. Cosmos 1926, 68, 860.
6. Spear, E. B.; Moore, R. L. Ind. Eng. Chem. 1925, 17, 894.
7. Garner, W. J. Soc. Dyers Colour. 1926, 42, 261.
8. Szegvari, A. Z. Phys. 1924, 21, 348.
9. Rocasolano, A. de G. Kolloid-Z., Special Issue, April 1, 1925, pp. 80-82
10. Tuarila, P. Kolloid-Z. 1928, 44, 11.
11. Jentsch, F. Z. Wiss. Mikrosk. 1930, 47, 145.
12. Green, H. Ind. Eng. Chem. 1924, 16, 667.
13. Endres, H. A. Ind. Eng. Chem. 1924, 16, 1149.
14. Spear, E. B.; Moore, R. L. Ind. Eng. Chem. 1925, 17, 936.
15. Hanausek, T. F. "The Microscopy of Technical Products"; J. Wiley and Sons: New York, 1926.
16. Ambronn, H. "Das Polarisationsmikroskip und seine Anwendung in der Kolloidchemie," Akad. Verlagsgesellschaft M.B.H.: Leipzig, 1926.
17. Poschl, V. "Technische Mikroskopie"; Enke: Stuttgart, 1928.
18. Schwalbe, C. G. Papierfabrikant 1930, 28, 809.
19. Schulze, B.; Göthel, E. Papierfabrikant 1934, 32, 110.
20. Roninger, F. H. Ind. Eng. Chem. Anal. Ed. 1933, 5, 251.
21. Hauser, E. A.; Le Beau, D. S. Kautschuk 1934, 10, 113.
22. Green, H. Ind. Eng. Chem. 1925, 17, 802.
23. Synge, E. H. Phil. Mag. 1928, [7]6, 356.
24. Bruche, E. Naturwissenschaften 1932, 20, 49.
25. Knoll, M.; Ruska, E. Z. Phys. 1932, 78, 318.
26. Malov, N. N. Usp. Fiz. Nauk 1933, 13, 367.
27. Calbick, C. J.; Davisson, C. J. Phys. Rev. 1933, 43, 764.
28. Boutaric, A. Nature 1935, 2953, 450.

29. Piontelli, I. R. Met. Ital. 1935, 27, 817.
30. McMillan, J. H.; Scott, G. H. Rev. Sci. Instrum. 1937, 8, 288.
31. Müller, E. W. Z. Phys. 1937, 106, 541.
32. Beischer, D.; Krause, F. Naturwissenschaften 1937, 25, 825
33. Krause, F. Naturwissenschaften 1937, 25, 817.
34. Friess, H.; Müller, H. O. Gasmaske 1939, 11, 1.
35. Anon., Engineering 1938, 146, 474.
36. Rhea, H. E. Wallerstein Lab. Commun. 1941, 4, 99.
37. Boersch, H. Z. Tech. Phys. 1939, 20, 347.
38. Zworykin, V. K.; Ramberg, E. G. J. Appl. Physiol. 1941, 12, 692.
39. Mahl, H. Kolloid-Z. 1942, 96, 7.
40. Fuller, M. L.; Brubaker, D. G.; Berger, R. W. J. Appl. Physiol. 1944, 15, 201.
41. Knoll, M.; Lubszynski, G. Phys. Z. 1933, 34, 671.
42. von Ardenne, M. Z. Phys. 1938, 108, 338.
43. von Ardenne, M. Z. Phys. 1938, 109, 553.
44. von Ardenne, M. Z. Tech. Phys. 1938, 19, 407.
45. Knoll, M.; Theile, R. Z. Phys. 1939, 113, 260.
46. von Ardenne, M. Naturwissenschaften 1939, 27, 485.
47. von Ardenne, M.; Beischer, D. Rubber Chem. Technol. 1941, 14, 15.
48. von Ardenne, M.; Beischer, D. Z. Phys. Chem. 1940, B45, 465.
49. Bennett, A. N. C.; Woodhall, W. Off. Dig. Fed. Soc. Paint Technol. 1944, 234, 130.
50. Bobalek, E. G.; LeBras, L. R. Off. Dig. 1954, 356, 860.
51. Bobalek, E. G.; LeBras, L. R.; von Fischer, W.; Powell, A. S. Off. Dig. 1955, 27, 984.
52. LeBras, L. R.; Bobalek, E. G.; von Fischer, W.; Powell, A. S. Off. Dig. 1955, 27, 607.
53. Cheever, G. D.; Bobalek, E. G. Ind. Eng. Chem. Fundam. 1964, 3(2), 89.
54. Bobalek, E. G. Off. Dig. 1962, 34, 1295.
55. Baker, F. L.; Princen, L. H. Encyl. Polym. Sci. Technol. 1971, 15, 498.
56. Brooks, L. E.; Sennett, P.; Harris, H. H. J. Paint Technol. 1967, 39, 473.
57. Heymann, K.; Riedel, W.; Blaschke, R.; Pfefferkorn, G. Kunststoffe 1968, 58, 309.
58. Wells, O. C. Proc. 5th Annu. Scanning Electron Microsc. Symp.; IITRI, 1972; p. 375.
59. "Scanning Electron Microscopy of Polymers and Coatings"; Princen, L. H., Ed.; J. Wiley-Interscience: New York, 1971.
60. "Scanning Electron Microscopy of Polymers and Coatings, II"; Princen, L. H., Ed.; J. Wiley-Interscience: New York, 1974.
61. "Spectroscopic and Instrumental Methods of Polymer Characterization"; Craver, C. D., Ed.; ADVANCES IN CHEMISTRY SERIES, American Chemical Society: Washington, D.C., 1982.
62. Labana, S. S. In "Scanning Electron Microscopy of Polymers and Coatings, II"; Princen, L. H., Ed.; J. Wiley-Interscience: New York. 1974.
63. Quach, A. In "Scanning Electron Microscopy of Polymers and Coatings, II"; Princen, L. H., Ed.; J. Wiley-Interscience: New York, 1974.
64. Crewe, A. V.; Isaacson, M.; Johnson, D. Rev. Sci. Instrum. 1971, 42, 411.

Rheology of Film-Forming Liquids

RAYMOND R. MYERS and CARL J. KNAUSS

Chemistry Department, Kent State University, Kent, OH 44242

Tests and Evaluation
  Drying Tests
  Evaluation of Mechanical Properties
Viscosity of Shear-Sensitive Materials
  Dispersed Systems
  Polymer Solutions
Viscoelasticity
  Viscous and Elastic Systems
  Time Dependence
  Rigidity Modulus Trends
  Morphological Changes during Drying
Physical Chemistry of Film Formation
  Drying from Solution
  High Solids Coatings
  Latex Drying
  Population Dynamics
Future Possibilities
  Latexes and Other Polymer Dispersions
  High Solids
Conclusions

Tests and Evaluation

Numerous methods are available for the conversion of liquids to solids, ranging from freezing to chemical reactions leading to cross-linking. Over 15,000 years ago paintings were dried in caves on both sides of the Pyrenees mountains by evaporating an aqueous matrix made from egg white or blood. Millenia later, oxidative polymerization of naturally occurring unsaturated oils allowed the Chinese to coat objects that were designed to be handled, and this technique is the one that carried us through the Industrial Revolution.

   Synthetic resin systems added a dimension to these ancient formulations: they were solutions that in many cases dried by

0097–6156/85/0285–0749$07.00/0

evaporation and then cured. Whether or not they needed to be baked, they were called thermoset systems. Then, largely because of rheological reasons, a high molecular weight resin dispersed in water came on the market for trade sales coatings. Under the name of latex, these coatings actually paced the ecology movement for about half of the coatings market. The other half, industrial coatings, had to wait a few decades for the development of aqueous systems.

Modern coatings continue to be developed in response to both economical and ecological pressures. Two promising routes are via aqueous solution/dispersion systems using polymers with carboxyl functionality temporarily neutralized by amines and via high-solids systems where small reactive molecules are applied in high concentrations.

Because the success of any of these systems depends on their ability to form hard, nearly impermeable, adherent films, the methods available to monitor and characterize this transition are important to the manufacture or application of coatings. One is dealing with high technology in the sense that the approaches used in colloid, solution, and polymer chemistry converge in the development of "compliance coatings."

Drying Tests. Measurements of the rheological changes that take place on drying and subsequent curing are quite numerous and have undergone a similar shift to high technology. Until recent years more reliance was placed on the "educated finger" of the paint technician than on any instrumental device. As long as the deposited film possessed the qualities of a liquid, it remained tacky. Thereafter, it was considered to be a solid.

The first instruments were based on rheology but did not make full use of it. The Gardner drying time recorder (1, 2) measured the time when a stylus riding on the film detected viscosity increases as the coating passed through the tacky stage in drying and finally encountered a surface dry to the touch. No measurement of the hardness of the film was made in this attempt to record little more than the time required to achieve a given state. Only after rheologists entered the field with determinations of the mechanical properties of the films was progress made on revealing the mechanisms of the various drying processes.

Early drying tests only monitored the gel-forming tendencies of liquid coatings. If the liquid increases in viscosity without reaching the gel state, its surface will be tacky. A piece of felt weighted by a small block will deposit strands on the film. No indication of the degree of rigidity of the film is given, and so this primitive test must be considered nonrheological in the sense that the only reading that makes sense is the time to achieve the felt-free state.

Drying tests that have been developed by rheologists are designed to characterize coatings at progressive stages of drying and curing rather than simply to monitor these changes. Two techniques that have been used for over a decade to gather drying data descriptive of the state of the film are described later in this chapter. They opened a new subdiscipline that lately has been referred to as chemorheology.

Evaluation of Mechanical Properties. An applied coating must form a tough and fairly impregnable film in order to be protective. Film formation always results in shrinkage, so that adhesive and/or cohesive loss often accompanies drying. Measurements must be capable of isolating these catastrophic effects from the rheological changes that one seeks to measure.

Evaluation of paints and paint ingredients can be done with conventional viscometers, in which the material is sheared between two mating surfaces. Constant shear rates or known frequencies of oscillation may be used. Raw materials are evaluated and formulation changes are monitored by conventional viscometric methods.

Two problems complicate the selection and use of rheometers in the study of supported films: (1) the film starts off the drying cycle with less pronounced mechanical properties than the substrate and (2) one free surface must be maintained in order for the coating to dry in the manner intended for surface coatings. The first of these problems renders it difficult to detect the onset of rigidity in a film residing on a rigid substrate, and the second problem rules out the selection of viscometers of standard design wherein a stress is applied at one surface and the strain rate response is read at the other. The coatings scientist cannot use the conventional gap-loading type of viscometer, for he must work with systems that have one free surface.

The first step in the solution of both of these problems requires the use of a cyclic shearing stress in measuring the two major components of a material in transition: viscosity and elasticity. These parameters measure the resistance of the material to flow and deformation, respectively. Most articles of commerce display both components simultaneously; they are said to be viscoelastic.

Because this chapter covers liquids, the first property discussed is the universal characteristic of all liquids, namely, viscosity.

Viscosity of Shear-Sensitive Materials

Dispersed Systems. Newtonian Behavior. Most film-forming liquids are dispersions. The rheology of these systems is simpler than the rheology of polymers and is treated first.

Primers on rheology all portray viscosity as resistance to flow. Practically every text (3, 4) shows that the stress σ needed to effect a given shear rate $\dot{\gamma}$ obeys the relation

$$\sigma = \eta\dot{\gamma} \tag{1}$$

where η is the viscosity. Dimensions of σ are pascals (Pa), where 1 Pa equals 1 (N/m^2). The cgs equivalent is dynes per square centimeter. Shear rate, $\dot{\gamma}$, is in s^{-1}; therefore, η is in pascal seconds or newton seconds per square meter.

Substances obeying this relation over wide ranges of shear rates are of little interest, at least to rheologists. Dilute dispersions tend to obey this law of Newton, in which case the viscosity is found from the viscosity of the medium η_0 by (5)

$$\eta = \eta_0(1 + 2.5\phi) \tag{2a}$$

where ϕ is the dimensionless volume fraction of the pigment or dispersed resin.

Equation 2a can be written in the form

$$\eta_{sp} = (\eta/\eta_0) - 1 = 2.5\phi \tag{2b}$$

where η_{sp}, the specific viscosity, is seen to bear a linear relation to ϕ. This equation is valid only at ϕ below 0.1 or even less.

Newtonian flow may be observed in suspensions of higher concentrations than those obeying Equation 2a. Higher-order powers of ϕ are needed, for which there are over 100 proposed variations of equations and coefficients. One is advised to start with the early work of Vand ([6], [7]) and Robinson ([8]-[10]) and then refer to the studies by Krieger ([11], [12]).

Shear-Sensitive Systems. In addition to hydrodynamic effects and simple viscous behavior, the act of pigmentation creates a certain amount of complex behavior ([13]). If the particles are fine, Brownian movement ([14]-[17]) and rotational diffusion ([14], [18], [19]) are among the phenomena that cause dispersed systems to display complex rheology. The role of van der Waals' forces in inducing flocculation ([20]) and the countervailing role of two electroviscous effects ([17], [21], [22]) in imparting stability, particularly in aqueous systems, have been noted. Steric repulsions appear to be the responsible factor in nonaqueous systems ([23], [24]). The adsorbed layer can be quite large ([25]-[28]), as detected by diffusion and density measurements of filled systems or by viscometry and normal stress differences ([29]).

One empirical advance that holds promise is the finding by Casson ([30]) that

$$\eta = \eta_0 \frac{1 - b\phi}{(1 - 1.75b\phi)^2} \tag{3}$$

where b is unity for spheres and independent of ϕ. It is a material parameter that converts ϕ to a larger value reflective of the amount of solvent immobilized (i.e., $b\phi$ is the sediment volume).

For particles in oil, b decreases with σ and also with ϕ. In fact, the controlling parameter is $\sigma\phi$. Generally, a plot of $1/b$ versus $\sigma\phi$ will produce a single curve in which b may drop from 5 at low stress to 2 at high stress where the immobilized layer is largely sloughed off.

As the pigment concentration is increased toward practical levels, practically no dispersion obeys Newton's law. Viscosity becomes a variable quantity in that it becomes dependent on the shear rate. Generally, this dependency is in the direction of a decrease in η with $\dot\gamma$. Flow curves of σ versus $\dot\gamma$ become concave toward the $\dot\gamma$ axis ([31]). When the pigment is acicular, with a length-to-diameter ratio greater than unity, shear sensitivity can be imparted at exceedingly low concentration due to the effect of orientation of the particles on the resistance to flow ([32], [33]).

Transient effects lead to viscosity increases that depend cyclically on the rate of shear.

The degree of concavity is a measure of the shortness, or butterlike quality, of the dispersion. The system is known as shear thinning (34). Naturally, a system displaying opposite concavity would be called shear thickening. Such terms as thixotropic and pseudoplastic are to be avoided, even though they appear frequently in the literature.

Viscometers designed to measure shear-sensitive systems have parallel shearing surfaces, as in rotational viscometers, or else they have surfaces that differ in angular velocity by an amount equal to the difference in clearance, as in cone-plate viscometers. These different gap-loading types are designed to assure a constant shear rate. Flow through a capillary results in a shear rate that varies from the center of the capillary to the wall (35) and is to be avoided in measuring shear-sensitive materials. Both σ and $\dot{\gamma}$ can be determined at the wall, but the material being measured will be subjected to $\dot{\gamma}$ values falling off to 0.

Polymer Solutions. All film-forming liquids are polymers or polymer solutions. The viscosity of these systems is constant only at extremely high dilution where the measurements leading to knowledge of molecular dimensions are made. At practical concentrations polymer solutions invariably become shear sensitive.

In the Newtonian range the viscosity of a polymer solution can be measured by passing the solution through a capillary and measuring the efflux time at a given temperature. At higher concentrations the parallel plate or cone/plate designs are used.

In contrast with Equation 2a, showing that dispersion viscosity is independent of molecular weight M, the viscosity of a polymer solution depends strongly on molecular weight. As M increases, the swept-out volume of the distended polymer molecule increases, so that ϕ of Equation 2b mounts rapidly. Instead of dealing with an unknown volume fraction, one uses concentration, c, in g/dL and evaluates the quantity, η_{sp}/c. For good measure, the value of this fraction is determined at various concentrations and extrapolated to zero c, giving the intrinsic viscosity $[\eta]$ in dL/g. The molecular weight dependence of the polymer's intrinsic viscosity therefore is related to its swept-out volume by (36).

$$[\eta] = KM^a \qquad (4)$$

where K (the Staudinger constant) runs around 10^{-4} and a is a fractional exponent expressing the extent of polymer-solvent interaction. Because of the variation and fractional value of a, K cannot be assigned units. As the solvent attraction decreases, the polymer molecule balls up; a approaches zero, and the viscosity approaches that of a dispersion.

A limited amount of structural information can be obtained via viscometry of polymer solutions. Molecular dimensions have been mentioned as the province of dilute solution viscosity. Flow properties of more concentrated solutions provide information on whether solvent drains freely from the polymer coils and on how solvent attractions compete with the interactions among macromolecules that are bound to occur (37, 38).

Proper treatment of polymer solutions requires one to start with equations of change (continuity, motion, and energy) and expressions for the heat flux vector and the stress tensor. Such treatments are beyond the scope of this chapter, but an excellent review article has appeared (39) that points out the importance of evaluating the stress tensor in place of simply finding a shear viscosity and observing its dependence on shear rate. Rheological equations of state, or "constitutive equations," can be derived from rheometric measurements in simple modes of flow, from continuum mechanics, and from molecular theory. Models are selected to describe a polymer molecule's orientability, extensibility, and internal modes of motion (40). Using a simple model consisting of beads connected by Hookean springs, one can write a constitutive equation with a time constant that characterizes the viscoelastic response of the polymer solution (41). Unfortunately, the ability of this equation to account for shear thinning requires further assumptions.

Pigmented polymer solutions that are the basis of conventional coatings display truly complex viscosity phenomena. If the pigment possesses a surface of high energy, it will immobilize solvent (giving ϕ coefficients greater than 2.5 in Equation 2a) or polymer (drastically extending the volume fraction). Interactions now become subject to steric and entropic factors (42); these factors, being of thermodynamic nature (43, 44), cause various temperature dependencies that are inextricably coupled with the shear rate dependencies described above.

Viscometric determinations of these phenomena all involve the destruction of the very structures that one sets out to measure and therefore are not emphasized in this chapter. It is more important to recognize the partial solidlike character of film formers at the outset and to treat them in a manner that elucidates both the solid and the liquid characteristics in the same nondestructive measurement. That is, one should measure viscosity and elasticity simultaneously.

Viscoelasticity

A limpid fluid such as water or paint thinner has been characterized adequately by its shear viscosity, η, in Equation 1.

On the other extreme, the dried coating is a solid of elasticity modulus, G. If the coating is not deformed extensively by the shear stress, it will obey a relation involving the deformation, γ (4):

$$\sigma = G\gamma \qquad (5a)$$

Deformation is expressed as a fraction such as displacement divided by the thickness of the specimen; it is dimensionless. Both σ and G have the dimensions of dynes per square centimeter or newtons per square meter.

Viscous and Elastic Systems. When both viscosity and elasticity are present in the material, an additive combination of the two responses is needed. In the case of a drying film, the stresses are added, giving the result for a viscoelastic body:

$$\sigma = \eta\dot{\gamma} + G\gamma \qquad (5b)$$

If the mass of the film is taken into account, part of the stress will be used to accelerate it (F = ma), so that an inertial term is added involving the area, A, over which the stress is applied:

$$\sigma = (m/A)\ddot{\gamma} + \eta\dot{\gamma} + G\gamma \tag{6}$$

Here $\ddot{\gamma}$ is the second time derivative of displacement, in reciprocal squared seconds.

Equation 6 describes a damped harmonic oscillator, suggesting that the coatings be studied in oscillation. In fact, this decision is forced on the experimenter by the inadequacy of gap-loading instruments. When applied to the situation represented by a free oscillation, Equation 6 is solved with $\sigma = 0$. If forced oscillation provides the stress, the left-hand side is represented by $\sigma_0 \sin \omega t$, where ω is the circular frequency. Examples are given later of both types.

Because η is time dependent and G is not, it behooves one to solve the second-order differential equation either in terms of a coefficient (η) or a modulus (G). Either route may be chosen, but one usually selects the modulus. In this case the response to the applied stress is measured as the strain rather than the strain rate.

Devices for making the oscillatory measurement are called rheometers rather than viscometers. The result is the complex modulus of rigidity G*, which contains the in-phase response, G', and the out-of-phase response, G". That is

$$G^* = G' + iG'' \tag{7}$$

Two unknowns require two measurements. For a free oscillator these measurements are the resonant frequency and the damping. For a forced oscillator the favored combination is the amplitude ratio and phase angle over a range of applied frequencies. This combination is not available for the evaluation of coatings because of the requirement that one surface be free. The two measurements described in the example below are damping and phase angle.

Time Dependence. When drying kinetics are studied and rheology is used as a monitor of the process, it is difficult to measure phase angles; as a result, one monitors drying by a single measurement of G' and makes approximations of the out-of-phase component, G".

Time dependence enters the data-gathering process in another way. The essential nature of kinetic studies is that they use time as the main variable, and the data are plotted against time. In viscoelastic studies time enters the equation by virtue of the dependence of the mechanical properties on the time of measurement. This problem is resolved by expressing all measurements in terms of the frequency, ω. A material on the borderline between liquid and solid behaves more liquidlike when the time of observation $1/\omega$ is increased, and more solidlike when $1/\omega$ is decreased. (See Figure 1.)

Rheologists make use of a dimensionless quantity known as the Deborah number (45, 46), D:

$$D = \tau/t \tag{8}$$

where τ is the relaxation time of the specimen. Expressed in terms of ω, D becomes the familiar $D = \omega\tau$.

Elasticity prevails when D exceeds unity, for the time required for relaxation of the applied stress is not exceeded. Of course, G is a function of time when measured statically. Under dynamic conditions $G'(\omega)$ takes its place, representing energy stored per cycle. There is also a $G''(\omega)$ corresponding to energy lost per cycle, and in many quarters one evaluates the dimensionless ratio G''/G', or tan δ, as a measure of the loss characteristics of a film.

When D falls below unity, viscous responses prevail (47). Both G' and G'' fall to low values under conditions of small τ or large t, but under these circumstances the contribution from G'' can be amplified by converting it into a coefficient rather than a modulus by the relation

$$\eta' = G''/\omega \tag{9}$$

Long times of observation correspond to small ω and, therefore, an increase in η' for a given level of G''.

Rigidity Modulus Trends. If one now superimposes the curing time dependency on the measuring time dependency, the result can be expressed by Figure 2 (48). The abscissa represents measuring time, and the various curves represent different times of drying.

Mechanical measurements do not range over many decades of frequency, so that the well-established principle of time-temperature equivalency (49) is used to fill in the gaps. The continuous plot of log modulus versus log frequency obeys a distinctive profile for a liquid (Systems 1 and 2), for lightly gelled solid (Systems 3 and 4), and for the hardened film (System 5).

The vertical line on the left of Figure 2 at a frequency of 1 Hz is best viewed as an indication of drastically increasing modulus levels at various degrees of cure; the levels range from flowing liquid G' that disappears from the scale (and would drop to zero if a log plot would permit) to hard solids whose dependence of modulus on frequency is virtually nonexistent. In between, the profiles have slopes ranging from zero at the plateaus to unity in the transition region.

By contrast, the vertical line on the right of Figure 2 at 10^7 Hz indicates hardly any change from polymer solution to dried film. What this region lacks in spread of G' values it gains in two ways: the measuring time is short enough to monitor even the most rapid of polymerizations and the wavelength of the cyclic stress is smaller than the thickness of the film. Both of these virtues are of sufficient importance to warrant the use of ultrasonic frequencies in characterizing films.

We have devised instruments to conduct measurements at both ends of the spectrum. They are described in various places (50–56) and therefore are referred to in only the briefest of terms in the following section. One is known as TBA, for torsional braid

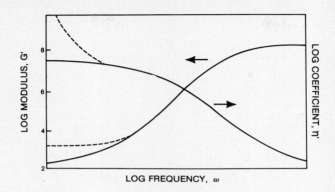

Figure 1.   Change from liquid to solid behavior as measurement
             frequency is increased. The dashed line portrays a
             cross-linked system.

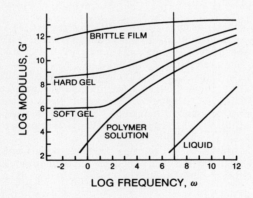

Figure 2.   Rigidity modulus trends for five types of coatings. The
             modulus depends on frequency of the measurement or
             inversely on the time of application of stress.
             Differences between essentially liquid films and those
             possessing varying amounts of structure can be observed
             at frequencies at which the relaxation time of the
             coating approximates the time required for the
             measurement. At a frequency of 1 Hz, shown by the
             slender vertical line at log ω = 0, variations of 10^{10}
             are displayed by G'.

analyzer (54), and the other is referred to as ARP, for attenuated reflection of pulses (55).

Morphological Changes during Drying. A recent development in understanding how the ambient conditions during cure affect the morphology of coatings was provided by Gillham (57–59). Borrowing from the teachings of metallurgy, he constructed a time–temperature–transition (TTT) diagram for each film–forming material, the general features of which appear in Figure 3.

Construction of the diagram was aided by exhaustive measurements using the low–frequency TBA device, where the locations and shifts of relaxation peaks were used as evidence of vitrification and gelation, complemented by the detection of postcure. In this diagram there are three characteristic temperatures: (1) resin T_g, the glass transition temperature of the starting material; (2) gel T_g, the temperature at which gelation and vitrification occur simultaneously; (3) $T_{g\infty}$, the glass transition temperature of the fully cured polymer of essentially infinite molecular weight.

Curing below $T_{g\infty}$ is unable to produce a thermally stable coating. Baking will result in further reaction and a merging of T_g and $T_{g\infty}$. Gelation before vitrification should be minimized in order to reduce the residual stresses present in the film.

Curing below the gel T_g will avoid the negative aspects of premature extensive gelation but will produce a "green" film whose properties deviate widely from those that are attainable by the polymer.

Curing below the resin T_g will be no cure at all. The glassy condition of the reactants will prevent reaction. This condition, naturally, does not prevail when one deals with liquid film formers, which are the subject of this chapter.

Bauer and Dickie (60) devised the concept of a cure window to explain in practical terms how to control film morphology. The cure window is the range of temperatures and cure times over which acceptable properties are obtained. Too low a temperature will produce green properties, and too high a temperature leads to decomposition.

The ultimate state of a coating is that of a hard solid. This state of matter falls in the province of this chapter and is covered here in a manner that completes the life history of a film–forming liquid.

Physical Chemistry of Film Formation

During the latter stages of drying, depending on whether the polymer is thermosetting or thermoplastic, cross–linking must be established so that the coating is not only rigid but also is converted into a composition that resists water. The static modulus of rigidity, G, depends on the molecular weight between cross–links more or less as follows (61):

$$G = K'/\overline{M}_c \qquad (10)$$

where $\overline{M}_c$ is the molecular weight between cross–links. The constant K' has a magnitude of about 25,000 g atm so that division by $\overline{M}_c$ in grams per mole produces G values of the order of hundreds of atmospheres (i.e., 10^8 dyn/cm^2 or 10^7 N/m^2).

When the degree of cross-linking is low ($\overline{M}_c$ is large), the glass transition temperature increases slightly as cross-linking proceeds. As $\overline{M}_c$ approaches values in the hundreds, T_g becomes a sensitive measure of cure. Equation 10 can be derived in its essential features from the kinetic theory of rubber elasticity (62), giving

$$G = \frac{\overline{r}^2}{\overline{r}_0^2} \frac{\rho RT}{\overline{M}_c} \left(1 - \frac{2\overline{M}_c}{\overline{M}_n} \right) \tag{11}$$

where $\overline{M}_n$ is the number-average molecular weight of the starting polymer, $\overline{r}^2/\overline{r}_0^2$ is the ratio of the mean square distance between network junctures to the corresponding distance of network chains in free space (generally assumed to be 1.0), ρ is polymer density, and RT is the thermal kinetic energy (62). G calculates out to be of the order of magnitude of R [8.314 x 10^7 ergs/(deg mol)].

The stress that leads to coating failure depends on the product of G and the extent of shrinkage. But coatings usually are applied to substrates that do not shrink, so that considerable ingenuity is needed to evaluate the effect played by dimensional changes that occur when a film dries. A cantilever device described by Gusman (63) and elaborated on by Corcoran (64), Figure 4, utilizes the deflection of a thin metal shim by the shrinkage of a coating applied to one side. Corcoran used a more general theory based on the deflection of a plate rather than a beam and arrived at the relation

$$\sigma = dEh^3/3cl^2(h + c)(1 - \nu) \tag{12}$$

where the lateral stress σ depends directly on the deflection, d, per unit length, l. Young's modulus E (where E $\simeq$ 3G) is used in place of G. The stress also depends on the thickness h of the plate and inversely on the coating thickness c. Calculation of σ requires a knowledge of Poisson's ratio ν of the plate, which can be assumed to be less than 0.5 but which must be gotten from handbooks.

Internal stresses develop early in the drying of cross-linkable coatings, and at a time that is critical to the subsequent development in film morphology and orientation (65). A cantilever device cannot detect early and subtle evidences of stress because the weight of the shim and its coating produces a deflection from the start. Croll solved this problem by devising a version with two suspension points located at distance l from each other and an overhang of distance l_0 at either end. (See Figure 5).

Calculations of the proportions of l_0 to l that would obliterate the effect of film weight on the horizontal bar revealed that $l_0 = 0.4564\,l$ for proper balance with the cantilever portion of the bar rather than $l_0 = 0.5\,l$ as less rigorous analyses led one to believe. Croll's design gave strain values (which convert to stress) that were independent of the film thickness. This independence is important not only in rating the effects of dry film thickness but also in allowing the entire drying process to be monitored without interference from the weight changes taking place in the coating.

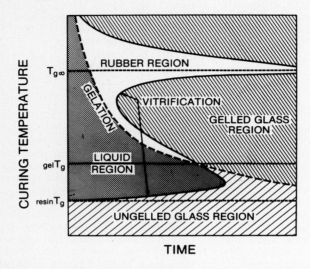

Figure 3.   Time–temperature transformation diagram for the curing of
             coatings, from Gillham (59).

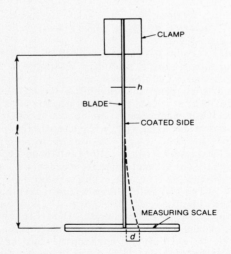

Figure 4.   Cantilever device for measuring internal stresses during
             cure (top view). The coating thickness is not indicated.

In effect, the physical chemistry of drying from solution is a problem in mechanics. In practice, instead of warping the substrate, a typical lacquer develops stresses (in excess of σ as calculated by Equation 12, because some stress relief is brought about by bending) that may exceed a threshold value, whereupon the system ruptures. Inasmuch as the system for all practical purposes is a two-dimensional laminated structure, rupture will occur at or near the interface. Laminar failure does not necessarily signify poor adhesion, for the failure may be cohesive as indicated by ESCA and SEM detection of residual polymer on separated surfaces (66). In some cases, failure is, indeed, adhesive as revealed by the absence of radioactive-tagged molecules (67).

In order to measure residual strain independently of the restraint of the substrate, one must separate the dried coating from the substrate and measure its dimensional changes. Croll (65) showed that epoxy coatings dried from solution shrank appreciably when they were released from tin-plated substrates by amalgamation. The shrinkage ranged from 1.4% to 1.8% when the epoxy content in methylcellulose was decreased from 70% to 30% in the applied film, indicating that the lengthened time for drying allowed gelation to occur while considerable residual solvent remained in the film. This conclusion was verified by the discovery that strains as high as 9% resulted from the use of a slowly evaporating solvent such as one of the higher glycol ethers.

Of further interest was the finding that residual strain was independent of applied coating thickness in the case of three low-boiling solvents but increased with thickness with high-boiling solvents (68). Pigmented systems also show a dependency on loading (69, 70).

Apart from oleoresinous systems that form films by cross-linking the neat polymer, a major portion of today's trade sales market is served by aqueous dispersions. Aqueous solution resin systems are being developed for industrial coatings. Aqueous resin dispersions made their entry via alkyd emulsions, followed a decade later by styrene-butadiene resins as an outgrowth of the synthetic rubber program. Later came poly(vinyl acetate) and acrylic latexes, long before social pressures were applied to reduce the volatile organic constituents from coatings as an ecology measure.

<u>Drying from Solution.</u> The simplest coatings are those that contain a polymer in a solvent. Organic solvent systems are hardy perennials destined to last for several more decades, until their insult to the environment requires the development of one or more of the new technologies described above. The physical chemistry of drying from solution therefore becomes one's first concern.

Wetting, of course, is a prime requirement; but since a coating formulator who has not achieved initial wetting of the substrate will not remain in business, drying poses a more fundamental rheological problem than an interfacial one, even though the interface is involved.

During the early stage the mechanical properties of polymer solutions are governed by their viscosity. For linear polymers the contribution of the film-forming ingredient to viscosity is expressed by the Staudinger relation, Equation 4, where *a* generally

lies between 0.6 and 0.8. From the intrinsic viscosity [η] (which
in reality expresses the size of the swollen polymer molecule), one
could work backward and compute the viscosity of the coating at any
given concentration, but this exercise would be both futile and
inaccurate. No coating dries at infinite dilution. It is of
greater significance to quantify the stresses that occur via the
development of rigidity in the film as the solvent evaporates, the
swollen molecules take over the entire volume, and shrinkage
persists after that point.

Two examples of drying of polymer from aqueous solution were
published by Myers and coworkers. Selecting a polymer that had
well-characterized molecular structure and fairly well-known
morphology, Tsutsui and Myers (71) allowed 7% solutions of
poly(acrylic acid) to dry on the ARP device with varying degrees of
neutralization α and learned that retention of water by the dried
film depended on α. As long as less than half of the carboxyl
groups were neutralized, the residual water content was determined
by the primary hydration number of carboxyl groups and carboxylate
ions. The gently sloping portion of the water retention curves (71)
of Figure 6 shows this phenomenon.

Increasing the degree of neutralization beyond α = 0.5 with
alkali introduced secondary hydration, as evidenced by the steeply
rising portions of the curves. A considerably reduced tendency
toward secondary hydration was observed with amines. Because water
retention and water sensitivity are possible sources of poor wet
adhesion, these findings have practical significance.

Water retention determines the mechanical properties of films in
general and of water-dispersible coatings in particular. Coincident
with the increased water retention of poly(acrylic acid), the
rigidity modulus of partially amine-neutralized solutions (0 < α <
1.0) of the polymer was higher than that of the zero or fully
neutralized polymer, as shown by the ARP drying curves (72) of
Figure 7. Rigidity is measured in terms of the attenuation Δ of the
shear pulse. Initially, water is tightly bound in the polymer
coils, which are spread somewhat by charges on the carboxylate ion;
later the spreading is more likely to produce water clusters that
plasticize the film. The result is a peak in modulus at
intermediate α.

Neutralization with the branched amine homologues 2-amino-1-
propanol and 2-amino-2-methyl-1-propanol required higher α to
achieve peak rigidity. The effect was not steric (72) and so
decreased hydrophilicity was suspected.

A copolymer of butyl acrylate and acrylic acid was synthesized
so as to approximate formulations used in waterborne formulation
practice without departing drastically from the acrylic acid
homopolymer. When 2-methyl-2-propanol solutions of these polymers
were diluted with water and then dried, the rigidity trends followed
the pattern (72) shown in Figure 8 and no evidence of secondary
hydration was present. Reference to the original articles will
reveal that the number of carboxylate triads should be minimized in
the copolymerization if one wishes to ensure that the marketed
product will be water insensitive.

Considerable activity has been reported in the development of
water-reducible coatings (73-76).

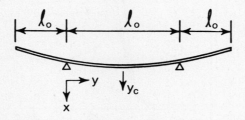

Figure 5. Overhanging beam device for measuring internal stresses during cure (side view).

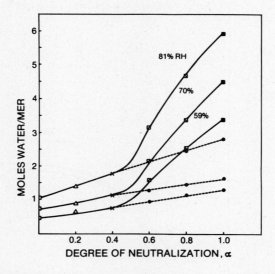

Figure 6. Water retention by poly(acrylic acid) films at various degrees of neutralization and three relative humidities. Temperature: 40 °C. NaOH used as the neutralizer.

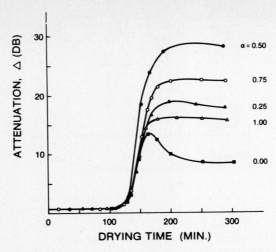

Figure 7.   Drying behavior of a poly(acrylic acid) 7% solution.
Attenuation is a direct measure of rigidity only in the
case of perfectly adherent films. Temperature: 35 °C.
Relative humidity: 50%.

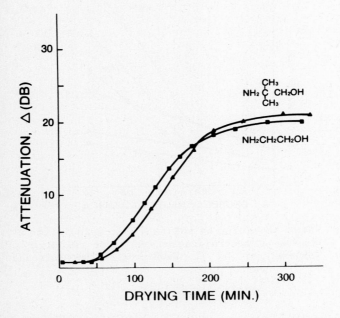

Figure 8.   Drying of a 1:12 acrylic acid-butyl methacrylate
copolymer in the presence of ethanolamine and its
dimethyl homologue. Water retention differences are of
no effect when the carboxyls are widely spaced (72).

<u>High-Solids Coatings</u>. Rheology is particularly important in the application and baking of high-solids coatings (<u>77</u>). The rules formulated from dilute solution rheology begin to fail when polymer concentrations above 70% are put to the test. In fact, the application of polymer chain statistics also fails because the compositions used at high concentrations are oligomeric, rather than polymeric, and the low molecular weights focus more emphasis on the chain ends and functional groups than is the case with high polymers. If one is dealing with a random polymer, a high probability exists of finding oligomeric species with functionalities of zero or one. Neither of these species can participate in cross-linking, and therefore neither one can impart rigidity to the dried film. These maverick species would serve as plasticizers.

Because of this barrier to fundamental studies involving oligomers, one cannot provide examples of research on the drying of high-solids coatings. This situation must be corrected, for high solids offer the most promise in the resolution of ecological and economic problems.

The Paint Research Institute of the Federation of Societies for Coatings Technology has started a program in high-solids coatings. It seeks answers in fundamental terms to such problems as these. (1) Sagging occurs and is accentuated by heating. In contrast to dealing with conventional coatings, one cannot rely on solvent release to counteract the viscosity drop that occurs on baking. (2) Stability is a serious problem. Viscosity increases on aging, cure response changes, and pigments flocculate. These difficulties arise because a high concentration of functional groups exists. (3) One cannot coat effectively due to high viscosity, high surface tension, and unacceptable electrical characteristics. (4) Cratering and intolerance of contaminants mar film integrity. Slow evaporation of solvent accentuates problems of air entrapment.

<u>Latex Drying</u>. Film formation from latexes poses a special problem in that one is not dealing with a continuum until a stage is reached at which most of the water has evaporated, the solid particles have undergone considerable distortion, and polymer begins to displace water at the interface. The residual stresses in this case are directed outward from the center of each particle in tiny domains. In the past, pigmentation and other means of achieving opacity exacerbated the problem (<u>78</u>).

The question of when the liquid coating has been converted to a solid wherein G takes over as the fundamental parameter is answered more precisely in the case of latex drying than in the case considered above of drying from solution. The particles of a latex have a finite G from the start, and as soon as they begin to deform from their spherical shape and begin to establish area contact with the substrate, as opposed to point contact, one is dealing with solid film coverage; a later stage in drying may involve cross-linking and is not basically a phase change.

The unique way in which the initial conversion is carried out is a matter of considerable scientific interest (<u>79</u>, <u>80</u>) and leads to some practical consequences. Spheres are made to coalesce into distorted polyhedra until only microscopic traces of their original outlines are visible; at the same time they may assimilate or collect in the interstices the various surfactants, defoamers, and other debris that existed in the original formulation.

A freshly applied latex displays a viscosity in the general range of the aqueous medium from which it is made, but soon after application it develops sufficient rigidity to resist sagging and the leveling of brush marks. Within minutes this development reaches an irreversible stage after which the latex particles can no longer be separated. At this stage, the coating still has to lose an additional 30% of its volume as water, and as a consequence of this loss, it continues to shrink. Latex particles that formerly have made only point contact with the substrate begin to flatten against the surface and establish areas of contact that start out as circles and end up as polygons (53). This deformation is carried out against the elastic resistance of the latex particles (80). In these cases, it is necessary to treat the modulus as the complex quantity G*, referred to, above, with its components G' and G''.

The rigidity modulus G' and damping modulus G'' go hand in hand for any given series of compositions.

In the course of drying, a film will undergo an increase in modulus of 10 or more decades. Figure 2 shows that the magnitude of this increase depends on where G' is measured on the frequency scale. A measuring frequency of 10^7 Hz (10 MHz) was used in conducting the studies on drying from aqueous dispersions. High-frequency devices have been reported (50, 51) as quite effective in studying subsequent curing and incipient adhesive failure. Low-frequency studies suffered for years from an operational problem until Kutz (81), in our laboratory, reduced to practice the concept of applying a coating to a braided substrate and measuring it in torsion (54). This device was perfected by Lewis and Gillham (82) and is known as the torsional braid analyzer (TBA).

Both extremes of frequency pose problems: low-frequency measurements cannot be made on coatings in rational terms and high-frequency measurements tend to obscure the differences between liquid and solid states of matter.

Population Dynamics. General. The replacement of liquid media by solid polymer and the eventual takeover by polymers as the interfacial species can be described in the same terms as one uses in describing the growth of a colony of bacteria. The science of population dynamics is remarkably well suited to describe the phenomenon. It is convenient to dissect a population growth curve into three segments: (1) An initial growth, after an induction period, in which growth appears to be exponential. (2) A steady growth that, in the case of populations approaching a condition roughly characterized as half saturation, appears to be linear. (3) An approach to saturation, characterized by a downward concavity in contrast to the upward concavity of the initial exponential growth curve.

The induction period in latex drying ends when the polymer spheres begin to flatten against the substrate as a consequence of their filling the interstices where water originally resided. About one-third of the latex volume is interstitial, so that by the time ϕ reaches 0.67 the population growth has started. At this point the percentage of substrate surface that contacts solid polymer is zero.

Circular "colonies" of polymer are inoculated at places where only point contact had existed, and these colonies grow outward in steadily increasing circular areas until they become tangent to each other; thereafter they grow as polyhedra until the entire surface is covered with hexagons.

The population growth law obeys the relation

$$\log \ [N/(1 - N)] = k(t - \theta) \tag{13}$$

where N is the population expressed as a fractional coverage, $0 < N < 1$, θ is the time required for the population to reach half saturation (N = 0.5), and k is the rate of growth.

Equation 13 is a sigmoid when N is plotted on a linear ordinate, for $N = (1 - N)e^{k(t-\theta)}$ or $N(1 + e^{k(t-\theta)}) = e^{k(t-\theta)}$. Thus

$$N = e^{k(t-\theta)}/(1 + e^{k(t-\theta)}) \tag{14}$$

For t = 0, the population approaches zero but remains finite; for t = ∞, N approaches unity asymptotically; for t = θ, N is 0.5, and the curve has a slope, dN/dt, of kN(1 - N) = 0.25k.

The parameters k and θ are characteristic of the kinetics of the drying system; the level of N, which is normalized in Equation 14, depends on the cross-link density or other means of achieving ordering of the polymer, with crystallinity or transition to the glassy state as the ultimate. The sigmoid departs at an early stage from the true exponential, as shown in Figure 9, passes through an inflection, and curves downward so that it asymptotically reaches saturation on drying. With N normalized to unity (as it is only in this summary) and with the parameters k and θ restricted to magnitudes consonant with the sharpness and duration of the drying curve, the liquid-to-solid conversion can be expressed in two practical terms: midpoint and spread. Deductions of molecular significance require application of Equation 11.

The use of population dynamics in evaluating the drying behavior of coatings was the outgrowth of a generalization by Price (83) regarding trends in natural phenomena and has resulted in several semiquantitative applications (50, 51, 55) to the drying of latexes. It has been applied to drying from solutions (70, 71) for which the mathematics have been based on the rate of diffusion of solvent (84) under two sets of conditions: surface boundary resistance at the beginning and internal diffusion resistance at the end of drying. The diffusion coefficient in the latter case depends exponentially on solvent concentration.

The Catastrophe. Population increases are not effected without hazard. If the initial rate is too steep or if N is too high, the sigmoid gives way to a peaked curve whose residual level is some fraction of the maximum population. This behavior, observed in latexes, has been interpreted (51) in terms of the percentage of substrate that is occupied by polymer. Most films break away from the substrate in the form of discrete islands, reflecting the domain character of the residual stresses (in contrast with the stretched membrane analogy). Both cohesive and adhesive failures are observed.

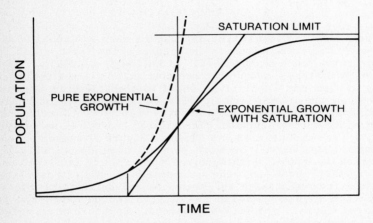

Figure 9.   Population growth curve. No physical phenomena, except
             perhaps the growth of the universe, follow the
             exponential law shown by the dashed line. The vertical
             line denotes the half-life τ of the growth. Coatings
             failure is manifested as a curve that peaks below the
             saturation limit and levels off at a lower value.

The role of the rigidity modulus in controlling the onset of the catastrophe is clear once the mechanism of drying has been specified. An array of high-modulus spheres (such as glass, for example) will not distort sufficiently to form the polygonal array needed to produce a continuum. Yet many useful coatings are designed to have glassy properties, and therefore, a means has been sought since the advent of latex coatings to improve the mar resistance of dried films by increasing their G. Ways in which this dilemma can be avoided are listed in the last section of this survey.

Because moduli are temperature dependent, the ambient conditions for drying determine whether a catastrophe will be encountered, signaling the onset of film failure. One of the practical consequences of this dependence is that high-temperature cures can be effected with materials whose rigidity at service temperatures is high enough to be serviceable.

Future Possibilities

Latexes and Other Polymer Dispersions. The superiority of dispersed systems in achieving low viscosity with polymers of high molecular weight will continue to attract formulators to aqueous dispersions (latexes), especially for trade sales paints, or to nonaqueous dispersions (NAD's) for systems that can be heated. Problems other than rheological will have to be overcome in using latexes, such as porosity of the film, difficulties in wetting surfaces, and pigmentation. Basic research will be needed to uncover the principles of solvent selection to allow film formation to proceed smoothly over a range of relative humidities. Other principles that must be understood are the effects of polyvalent ionic impurities in the water used, foaming tendencies, flash rusting, gloss, and defect formation.

Water-soluble/dispersible polymers, by contrast, present significant rheological problems. Unless one can surmount the tendency of polymers with nonpolar backbones and polar side groups to increase greatly in viscosity on dilution with water, these aqueous types are destined to be restricted to primers from which the polymer can be attached to the substrate out of dilute solution, as in electrodeposition, or else to systems that can be rid of solubilizers by baking or other means. They are not believed to be candidates for top coats, largely because of the problem of bonding to low-energy surfaces. This picture may change if specific interactions can be exploited, such as acid-base adduct formation.

High Solids. Great expectations are held for high-solids formulations, for industry's first coating was, in reality, a 100% solids system. Natural drying oils flowed easily, they pigmented easily, and they did not require baking in order to cross-link. The problem was chemical: the oxidatively dried unsaturated species were not resistant to water, oxidative embrittlement, yellowing, and the like.

Modern attempts to start with oligomers of low molecular weight and exploit their functionality in the creation of high polymers after deposition have not yet succeeded commercially. In the first place, every molecule should be at least difunctional, with a sizable portion carrying three functional groups to participate in cross-linking. Perhaps a version of alkyd chemistry, which incorporates polyols in a system containing dibasic acids, would lead to the development of reactive pairs that have matching functional groups whose reaction product would be more resistant than esters. Cross-linking must be more stringently controlled with high solids than with conventional systems (85).

High-solids research needs to be focused on two fronts. One is synthetic, where the distribution of species from random copolymerization cannot be tolerated in oligomers. The other is rheological. Concentrated solution rheology needs to be elucidated as well as it has for dispersions and for conventional solution coatings. For papers in this area, the works of Ericksen (86), Schoff (87), and Lamb (88) are recommended.

Rheology affords one last word in considering the future of latexes. It is well-known that impact resistance accrues from component incompatibility; latexes provide a built-in means of achieving composition gradients and interphases on a grand scale, simply by blending. A more interesting way of imparting heterogeneities is to coat particles of one composition by a sheath of another composition. Success has been achieved in sheathing rigid particles of polystyrene with a soft shell made from an acrylate. The result is a latex that derives its ability to coalesce from the sheath; the hard core remains the discontinuous phase.

If the situation could be reversed and a film former achieved with a hard sheath, the impact resistance would be phenomenal, for the soft cores would absorb energy while remaining intact.

Conclusions

The drying and subsequent curing of coatings involve a first-order conversion of a liquid to a solid, regardless of the chemical or physical transformations taking place. Increased emphasis on high-solids and aqueous coatings has focused attention on the delicate balance of mechanical and ultimate properties that must be achieved in a formulation in order to minimize the lateral stresses induced by the conversion.

Literature Cited

1. Gardner, P. N. U.S. Patent 2 280 483, 1942.
2. Sward, G. G. In "Paint Testing Manual," 13th ed.; Sward, G. G., Ed.; American Society of Testing Materials: Philadelphia, 1972; Chap. 4.3, pp. 268-80.
3. Billmeyer, F. W., Jr. "Textbook of Polymer Science," 2nd ed.; Wiley-Interscience: New York, 1971.
4. Reiner, M. "Deformation and Flow"; H. K. Lewis and Co.: London, 1949.
5. Einstein, A. Ann. Phys. 1906, 19, 289.
6. Vand, V. J. Phys. Colloid Chem. 1948, 52, 277-99.
7. Vand, V. J. Phys. Colloid Chem. 1948, 52, 300-13.
8. Robinson, J. V. J. Phys. Colloid Chem. 1949, 53, 1042-56.
9. Robinson, J. V. J. Phys. Colloid Chem. 1951, 55, 455-64.
10. Robinson, J. V. Trans. Soc. Rheol. 1957, 1, 15-24.
11. Krieger, I. M.; Maron, S. H. J. Colloid Sci. 1951, 6, 528-38.

12. Maron, S. H.; Madow, B. P.; Krieger, I. M. J. Colloid Sci. 1951, 6, 584-91.
13. Mewis, J. Proc. Int. Congr. Rheol., (8th Congr. ,Naples) 1980, p. 149.
14. Brenner, H. In "Progress in Heat and Mass Transfer"; Schowalter, W. R., et al., Eds.; Pergamon: Oxford, 1972; Vol. 5, p. 89.
15. Batchelor, G. K. J. Fluid Mech. 1977, 83, 97.
16. Giesekus, H. Rheol Acta 1962, 2, 50.
17. Russel, W. B. J. Rheol. 1980, 24(3), 287.
18. Scheraga, H. A. J. Chem. Phys. 1955, 23, 1526.
19. Hinch, E. J.; Leal, L. G. J. Fluid Mech. 1972, 52, 683.
20. Sonntag, H.; Strenge, K. In "Koagulation und Stabilitat Disperser Systeme"; VEB Deutscher Verlag der Wissenschaft: Berlin, 1970.
21. Russel, W. G. J. Fluid Mech. 1978, 85, 209.
22. Stone-Masui, J.; Watillon, J. J. Colloid Interface Sci. 1968, 28, 187.
23. Doroszkowski, A.; Lambourne, R. J. Colloid Interface Sci. 1968, 27, 214.
24. Graziano, F. R.; Cohen, R. E.; Medalia, A. I. Rheol. Acta. 1979, 18, 640.
25. Snook, I.; Van Megen, W. J. Colloid Interface Sci. 1976, 57, 40.
26. Van Vliet, T.; Lyklema, J.; Van den Tempel, M. J. Colloid Interface Sci. 1978, 65, 505.
27. Kumins, C. A.; Roteman, J. J. Polym. Sci. 1961, 55, 623, 699.
28. Kumins, C. A.; Roteman, J. J. Polym. Sci., Part A 1963, 1, 532.
29. Schoukens, G.; Mewis, J. J. Rheol. 1978, 22, 381.
30. Casson, N. J. Oil Colour Chem. Assoc. 1981, 64, 480.
31. Myers, R. R.; Miller, J. C.; Zettlemoyer, A. C. J. Appl. Phys. 1956, 27, 468-71.
32. Ivanov, Y.; Van de Ven, T. G. M.; Mason, S. G. J. Rheol. 1982, 26, 213.
33. Ivanov, Y.; Van de Ven, T. G. M. J. Rheol. 1982, 26, 231.
34. Myers, R. R.; Brookfield, D. R.; Eirich, F. R.; Ferry, J. D.; Traxler, R. N. Trans. Soc. Rheol. 1959, 3, 120-22.
35. Reiner, M. In "Rheology, Theory, and Applications"; Eirich, F. R., Ed.; Academic: New York, 1956; Vol. 1.
36. Staudinger, H.; Heuer, W. Chem. Ber. 1930, 63B, 222.
37. Rouse, P. E., Jr. J. Chem. Phys. 1953, 21, 1272.
38. Zimm, B. J. Chem. Phys. 1956, 24, 269.
39. Bird, R. B. J. Rheol. 1982, 26, 277.
40. Williams, M. C. A.I.Ch.E. J. 1975, 21, 1.
41. Giesekus, H. Rheol. Acta. 1966, 5, 29.
42. Vold, M. J. J. Colloid Sci. 1961, 16, 1.
43. Osmond, D. W., et al. J. Colloid Interface Sci. 1973, 42, 262.
44. Vincent, B. J. Colloid Interface Sci. 1973, 42, 270.
45. Judg. 5:5
46. Reiner, M. Phys. Today 1964, 17, 62.
47. Myers, R. R. J. Polym. Sci., Part C 1971, 35, 3.
48. Myers, R. R. J. Coat. Technol. 1976, 48(613), 27.
49. Williams, M. L.; Landel, R. F.; Ferry, J. D. J. Am. Chem. Soc. 1955, 77, 3701.

50. Myers, R. R.; Schultz, R. K. Off. Dig., Fed. Soc. Coat. Technol. 1962, 34, 801.
51. Myers, R. R.; Schultz, R. K. J. Appl. Polym. Sci. 1964, 8, 755.
52. Myers, R. R.; Klimek, J.; Knauss, C. J. J. Coat. Technol. 1966, 38(500), 479.
53. Myers, R. R.; Knauss, C. J. J. Coat. Technol. 1968, 40(523), 315.
54. Myers, R. R.; Knauss, C. J.; Schroff, R. N. Proc. Int. Congr. Rheol. (Fifth Congr., Tokyo) 1969, 1, 473.
55. Myers, R. R.; Knauss, C. J. In "Polymer Colloids"; Fitch, R. M., Ed.; Plenum: New York, 1971; p. 173.
56. Myers, R. R. In "Applied Polymer Science"; Craver, J. K.; Tess, R. W., Eds.; American Chemical Society: Washington, D.C., 1975; Chap. 22.
57. Gillham, J. K. Polym. Eng. Sci. 1979, 19, 319.
58. Doyle, M. J., et al. Polym. Eng. Sci. 1979, 19, 687.
59. Gillham, J. K. ACS Prepr. Div., Org. Coat. Plast. Chem. 1980, 43, 677.
60. Bauer, D. R.; Dickie, R. A. J. Coat. Technol. 1982, 54(685), 57.
61. Smith, T. L. In "American Institute of Physics Handbook"; Gray, D. E., Ed.; McGraw-Hill: New York, 1963; Sec. 2f.
62. Nielsen, L. E. J. Macromol. Sci.--Rev. Macromol. Chem. 1969, C3(1), 69.
63. Gusman, S. Off. Dig., Fed. Soc. Paint Technol. 1962, 34, 451, 884.
64. Corcoran, E. M. J. Paint Technol. 1969, 41(538), 635.
65. Croll, S. G. J. Oil Colour Chem. Assoc. 1980, 63(7), 271.
66. Wyatt, D. M., et al. J. Appl. Spectrosc. 1974, 28(5), 439.
67. Peterson, C. M. Kolloid Z. 1968, 222(2), 148.
68. Croll, S. G. J. Coat. Technol. 1981, 53(672), 85.
69. Perera, D. Y.; Vanden Eynde, D. J. Coat. Technol. 1981, 53(677), 39.
70. Perera, D. Y.; Vanden Eynde, D. J. Coat. Technol. 1981, 53(678), 41.
71. Tsutsui, K.; Myers, R. R. Ind. Eng. Chem. Prod. Res. Dev. 1980, 19, 310.
72. Raju, K. S.; Myers, R. R. J. Coat. Technol. 1981, 53(676), 31.
73. New England Society for Coatings Technology. J. Coat. Technol. 1982, 54(684), 63.
74. Kordomenos, P. I.; Nordstrom, J. D. J. Coat. Technol. 1982, 54(686), 33.
75. Wicks, Z. W., et al. J. Coat. Technol. 1982, 54(688), 57.
76. Woo, J. T. K., et al. J. Coat. Technol. 1982, 54(689), 41.
77. Myers, R. R. J. Coat. Technol. 1981, 53(675), 29.
78. Aronson, P. D. J. Oil Colour Chem. Assoc. 1974, 57(2), 59.
79. Bradford, E. B.; Vanderhoff, J. W. J. Macromol. Chem. 1966, 1, 335.
80. Brown, G. L. J. Polym. Sci. 1956, 22, 423.
81. Kutz, D. Senior Thesis, Lehigh University, Bethlehem, Pa., June 1959.
82. Lewis, A. F.; Gillham, J. K. J. Appl. Polym. Sci. 1962, 6, 422.
83. Price, D. J. "Science Since Babylon"; Yale University Press: New Haven, Conn., 1961.
84. Hansen, C. H. J. Oil Colour Chem. Assoc. 1968, 51, 27.
85. Bauer, D. R.; Dickie, R. A. J. Coat. Technol. 1982, 54(685), 57.
86. Ericksen, J. R. J. Coat. Technol. 1976, 48(629), 58.
87. Schoff, C. K. Prog. Org. Coat. 1976, 4, 189.
88. Cochrane, J.; Harrison, G.; Lamb, J.; Phillipa, D. W. Polymer 1980, 21, 837.

An Introduction to Corrosion Control by Organic Coatings

R. A. DICKIE

Ford Motor Company, Dearborn, MI 48121

Fundamentals of Corrosion Chemistry
   Principal Corrosion Reactions
   Theoretical Bases of Corrosion Control
   Types of Corrosion Phenomena
Methods of Studying Corrosion Control
   Goals of Testing and Classification of Test Methods
   Nonelectrochemical Methods
   Electrochemical and Electrical Methods
Barrier Characteristics of Coatings
   Adhesion of Organic Coatings
   Transport Properties of Coatings
   Other Film Properties
Corrosion Resistance of Painted Metals
   Metal Surface Preparation and Conversion Coatings
   Corrosion Control by Pigments
   Effects of Resin Structure on Corrosion Control

The term "corrosion" has been applied to a wide variety of processes that have in common a degradation of the chemical or physical state of a material upon exposure to an aggressive environment. More commonly, the term is restricted to chemical degradation of a metal by its environment. The technological and economic importance of corrosion, and hence of corrosion protection, are enormous. A National Bureau of Standards study ($\underline{1}$) estimates that the total cost of corrosion in the United States in 1975 was $70 billion, about 4.2% of the gross national product; of this amount, the study concludes that about 15% might be avoidable through the economic use of currently available technology. Organic coatings have long been used to provide corrosion protection. Smith ($\underline{2}$) mentions the use of melted pitch by the Romans and goes on to describe reports from the early nineteenth century of the application of varnish to smoked steel and of a composition of resin, olive oil, and ground bricks used to protect a water carrier from rust. This chapter presents an introduction to the fundamentals of corrosion chemistry and their

0097-6156/85/0285-0773$07.75/0
© 1985 American Chemical Society

application to the understanding of corrosion control by organic coatings. The scope of the subject precludes a comprehensive review; additional sources of information include well-known compendia on corrosion by Evans (3), Uhlig (4), and Shreir (5), the preprint and proceedings volumes of recent symposia (6), and, of course, the periodical literature (7). Many introductory treatments of corrosion and corrosion protection are also available (e.g., 8-13). Leidheiser (14) has recently surveyed current research on the corrosion of painted metals.

Fundamentals of Corrosion

Principal Corrosion Reactions. Organic coatings play many roles in corrosion protection. Although the primary function may be to prevent the onset of corrosion by isolating the substrate metal from the environment, control of the spread of corrosion from an initial corrosion site can also be of critical importance. An understanding of the nature of corrosion reactions is a necessary foundation to understanding their control. Metal/environment corrosion reactions are most often redox reactions occurring at a metal/nonmetal interface. The metal is oxidized, and the nonmetal, usually oxygen, is reduced. If the product of the metal oxidation forms an adherent and inert film, the corrosion process may be self-limiting, and the metal may appear to be relatively unreactive. If the oxide does not form an adherent film, or if the metal corrosion products are soluble in the corrosive medium, metallic corrosion proceeds.

In general terms, the oxidation of a metal M can be written as

$$M \longrightarrow M^{z+} + z \ e^-$$

and the reduction of a nonmetal Q can be written as

$$Q + z \ e^- \longrightarrow Q^{z-}$$

for the overall corrosion process

$$M + Q \longrightarrow MQ$$

The oxidation and reduction reactions occur at the same rate; there is no net accumulation of charge.

Corrosion can be represented in terms of a simple electrochemical cell as illustrated in Figure 1 for aqueous corrosion. In the electrochemical cell, the electrical circuit is completed by the transport of charge through the aqueous electrolyte. The source of potential may be, for example, a difference in electrode composition. For metals exposed to a corrosive environment, the anodic and cathodic reactions can occur at adjacent or widely separated sites on a single metal, or on different metals that are electrically connected. As in the electrochemical cell, the electrical circuit is typically completed by an aqueous electrolyte; sources of potential include differences in electrolyte concentration and in electrode composition.

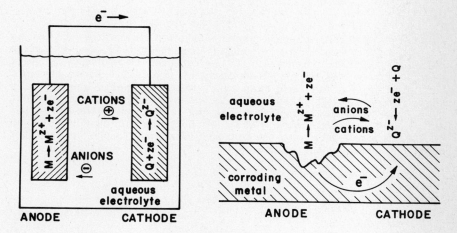

Figure 1.   Schematic comparison of a corrosion site (right) and a
simple electrochemical cell (left).

Theoretical Bases of Corrosion Control. This section provides a brief review of thermodynamic and kinetic considerations in the control of corrosion. Shreir (5) or one of the other general references (3–12) can be consulted for detailed discussion.

In thermodynamic terms, for the metal dissolution reaction to proceed, the free energy change associated with the oxidation reaction must be negative. In terms of the usual convention, the free energy change associated with reduction of the metal ion formed by metal dissolution must be positive. The free energy change ΔG is given by

$$\Delta G = -zeF$$

where z is the number of electrons involved in the reaction, F is the Faraday constant, and E, the potential, is given by

$$E = E^0 - (RT/zF)\ln(a_{M^0}a_{M^{z+}})$$

where a is the thermodynamic activity of the metal or its ion as indicated by the subscripts M^0 and M^{z+}, respectively, and E^0 is the standard electrode potential. E^0 is referred to hydrogen as zero; its sign is an indication of basicity or nobility. Under standard conditions, if E^0 is less than zero, the free energy change ΔG is positive, and metal dissolution is spontaneous. If E^0 is greater than zero, then ΔG is negative, and the reduction of the metal ion to metal is spontaneous. This situation corresponds to an electrochemical cell with the hydrogen electrode to the left and the metal electrode to the right:

$$Pt|H_2(1 \text{ atm})| H^+(a=1)| | X^+ | X$$

$$E = E_R - E_{X/X^+}$$

Standard potentials for a number of metals are given in Table I; the values given in many handbooks are oxidation potentials employing the opposite sign convention (the order of the electrochemical cell just shown is reversed).

Electrochemical potentials can arise from differences in electrolyte or electrode concentration as well as from differences in chemical composition. Thus, for example, there will be a difference in potential between two amalgam or alloy electrodes of the same basic type in which there is a difference in the activity of one of the alloy or amalgam constituents. The practical implication is that galvanic corrosion can occur between similar alloys of different composition. Crevice corrosion phenomena are often explainable in terms of differences in oxygen concentration.

The corrosion of iron is one of the most widespread and technologically important corrosion phenomena. Overall the formation of ferric oxide from metallic iron in the presence of water and oxygen can be written as:

$$4 \text{ Fe} + 2 \text{ H}_2O + 3 \text{ O}_2 \longrightarrow 2 \text{ Fe}_2O_3.H_2O$$

Table I.   Standard Electrode Potentials
           in Acidic Solutions

Reaction	E^0 (volts)
$Mg^{+2} + 2e = Mg$	-2.37
$Al^{+3} + 3e = Al$	-1.66
$Zn^{+2} + 2e = Zn$	-0.736
$Fe^{+2} + 2e = Fe$	-0.440
$Sn^{+2} + 2e = Sn$	-0.136
$Pb^{+2} + 2e = Pb$	-0.126
$Fe^{+3} + 3e = Fe$	-0.036
$2 H^+ + 2e = H_2$	0
$Cu^{+2} + e = Cu^+$	0.153
$Cu^{+2} + 2e = Cu$	0.337
$Fe^{+3} + e = Fe^{+2}$	0.771
$Ag^+ + e = Ag$	0.7991

In most cases, complex mixtures of hydrated oxides are formed. The anodic dissolution of iron results first in the formation of ferrous ion:

$$Fe \longrightarrow Fe^{+2} + 2e^-$$

Ferric species are formed in subsequent hydrolysis and oxidation reactions including, for example, the following:

$$Fe^{+2} + H_2O \longrightarrow FeOH^+ + H^+$$

$$Fe^{+2} \longrightarrow Fe^{+3} + e^-$$

In general, the pH at sites of anodic dissolution decreases. The process of iron corrosion is in fact extremely complicated; Pourbaix (15) gives some 29 equilibria for the corrosion of iron in aqueous media.

The cathodic reduction reactions commonly observed are the hydrogen evolution reactions:

$$H_3O^+ + e^- \longrightarrow 1/2\ H_2 + H_2O \text{ (acid solutions)}$$

$$H_2O + e^- \longrightarrow 1/2\ H_2 + OH^- \text{ (neutral/basic solutions)}$$

and the oxygen reduction reactions:

$$1/2\ O_2 + 2\ H_3O^+ + 2\ e^- \longrightarrow 3\ H_2O \text{ (acid solutions)}$$

$$1/2\ O_2 + H_2O + 2\ e^- \longrightarrow 2\ OH^- \text{ (neutral/basic solutions)}$$

In general, the pH increases at cathodic reaction sites. As the pH increases, the hydrogen evolution reaction becomes kinetically more difficult, and the oxygen reduction reaction more important. Oxygen transport is often rate limiting, especially at high rates of corrosion in aqueous solutions.

The many corrosion equilibria involving iron can be conveniently summarized in an equilibrium or Pourbaix diagram (15). These diagrams map regions of stability for various ionic and solid species as functions of pH and E^0 on the basis of equilibrium potentials. A simplified equilibrium diagram is shown in Figure 2. At sufficiently negative potential, iron is thermodynamically stable, and corrosion cannot occur. The domain of stability of water is defined by the dashed lines a and b; below a,

$$H_2O + e^- \longrightarrow 1/2\ H_2 + OH^-$$

and above b,

$$H_2O \longrightarrow 1/2\ O_2 + 2\ H^+ + 2\ e^-$$

Because there is no overlap of the domains of stability for water and iron, iron is normally susceptible to corrosion in the presence of water and oxygen. At sufficiently positive potential and high pH, passivation can be achieved by formation of a protective oxide layer (anodic protection). Immunity from corrosion can be achieved

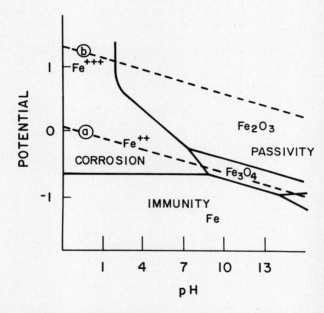

Figure 2.   Simplified equilibrium diagram for iron in water. The
            dashed lines a and b denote the limits of stability of
            water. (Reproduced with permission from Ref. 15.
            Copyright 1966 Pergamon)

by lowering the potential (cathodic protection)--for example, by coupling to a more reactive metal. Although the equilibrium diagrams are useful for qualitative considerations of the corrosion of metals, they give no indication of the rates of corrosion. They are valid only for the environment for which they have been constructed; the diagram reproduced in Figure 2 does not, for example, take into account the effects of complexing agents or electrolytes on the corrosion of iron.

The description of corrosion kinetics in electrochemical terms is based on the use of potential-current diagrams and a consideration of polarization effects. The equilibrium or reversible potentials involved in the construction of equilibrium diagrams assume that there is no net transfer of charge (the anodic and cathodic currents are approximately zero). When the current flow is not zero, the anodic and cathodic potentials of the corrosion cell differ from their equilibrium values; the anodic potential becomes more positive, and the cathodic potential becomes more negative. The voltage difference, or polarization, can be due to cell resistance (resistance polarization); to the depletion of a reactant or the build-up of a product at an electrode surface (concentration polarization); or to a slow step in an electrode reaction (activation polarization).

Potential current diagrams for the anodic and cathodic reactions involved in a typical metal dissolution process are schematically illustrated in Figure 3A. As the imposed current is increased, the difference between the observed and equilibrium potentials (the electrochemical polarization) becomes progressively larger. In the absence of resistance and concentration effects, the observed polarization, termed activation polarization, is due to a slow step in an electrode process. Activation polarization is an intrinsic feature of electrode processes at nonzero current flow. Because there is no net accumulation of charge during an electrochemical reaction, the total anodic and cathodic currents must be equal. In Figure 3A, this situation occurs at the point at which the anodic and cathodic curves cross; the magnitude of the current at this point provides a measure of the rate of corrosion. If the medium has electrical resistance, R, the anodic and cathodic potentials will differ by IR, and the current (i.e., the rate of corrosion) will be reduced, as illustrated in Figure 3B. If the availability of, for example, the cathodic reactant is limited by diffusion (concentration polarization), as often occurs in the aqueous corrosion of metals, a reduction in current is also observed. In solution, the concentration polarization can be reduced, and the corrosion current increased, by stirring. Thus, in Figure 3C, under diffusion-controlled conditions, the corrosion current is I; with stirring the current increases, for example, to I'. The maximum current achievable by suppression of concentration polarization is I". Concentration polarization can be contrasted with the effect of a change in activity: if the equilibrium potential is increased (e.g., by employing oxygen instead of air), the corrosion rate will also increase (Figure 3D). This effect is, however, thermodynamic rather than kinetic. Finally, the question of galvanic corrosion also needs to be addressed. For two metals N and M, as illustrated in Figure 4 (adapted from Ref. 16), the individual corrosion rates are given by I(M) and I(N); M is the more

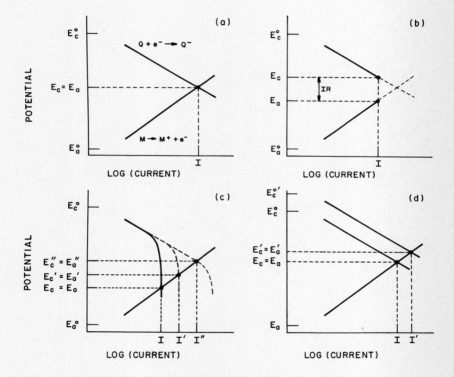

Figure 3.   Idealized potential–log(current) diagrams illustrating cell polarization effects. (Reproduced with permission from Ref. 10. Copyright 1975 Pergamon)

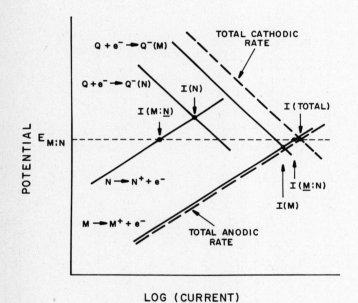

Figure 4. Idealized potential–log(current) diagram for a galvanic couple. (Reproduced with permission from Ref. 16. Copyright 1978 Society of Automotive Engineers)

active of the two metals. When coupled, the system assumes potential E(M:N), and the individual metal corrosion rates shift such that N corrodes at a rate I(M:N̲) < I(N), and M corrodes at a rate I(M̲:N) > I(M). Generally, the more active metal becomes the anode, and the less active one becomes the cathode, with corresponding changes in their corrosion rates.

Some of these concepts of the kinetics of metal corrosion are further illustrated in Figure 5 for iron in an aqueous saline medium. At pH 6.5 in the presence of oxygen, the dominant cathodic reaction is the reduction of oxygen to form hydroxide. The essentially vertical cathodic curve indicates that the reaction is diffusion controlled. At higher oxygen concentration and with good stirring, the rate of corrosion would be substantially higher than shown. If, however, the steel electrode were to be held at potentials below about –800 mV (vs. SCE), no iron dissolution would occur. At high pH, the situation is very different, even though the dominant cathodic reaction remains the formation of hydroxide. Below about –1000 mV (vs. SCE), the exact value depending on pH, there is no iron dissolution; this situation corresponds to the immune region of the Pourbaix diagram. Between approximately 600 and –700 mV (vs. SCE), the rate of corrosion is very low because of the formation of a passive oxide film; this region corresponds to the passive region of the Pourbaix diagram. During natural corrosion, anodic and cathodic reactions are polarized to the same potential, save for resistance effects. When anodic and cathodic sites become separated on the surface, the pH of cathodic sites increases, and the cathodic areas become passivated (no visible corrosion occurs in these areas). Iron dissolution continues at anodic sites which tend, as noted before, to become acidic.

Organic coatings control corrosion by resistance inhibition, by suppression of the anodic reaction, and by suppression of the cathodic reaction. Isolation of the reactive substrate from its environment by a high-resistance barrier reduces the rate of the corrosion reactions and is probably the most general mechanism of corrosion protection afforded by paint films. Resistance inhibition is affected by the presence of electrolytes in the paint film, or beneath it, by penetration by water, and by the adhesion and mechanical integrity of the film. Organic coatings also act as reservoirs for active metal pigments and for corrosion inhibitors. Suppression of the anodic reaction is achieved by pigmentation in some coatings, for example, in paints that supply electrons, such as zinc-rich paints, and in paints that help maintain a passive oxide film, for example chromate-pigment-containing paints. Suppression of the cathodic reaction is achieved by isolating the substrate from the cathodic reactants. Inorganic conversion coatings act to control the rate of corrosion in part by suppressing the cathodic reduction of oxygen.

Types of Corrosion Phenomena. The major categories of phenomena (5) include uniform, localized, and pitting corrosion; selective dissolution; and corrosion acting together with a mechanical phenomenon. In uniform corrosion, all areas corrode at the same rate. Examples of uniform corrosion include tarnishing and active dissolution of metals in acids. In localized corrosion some areas corrode more readily than others; crevice corrosion and filiform

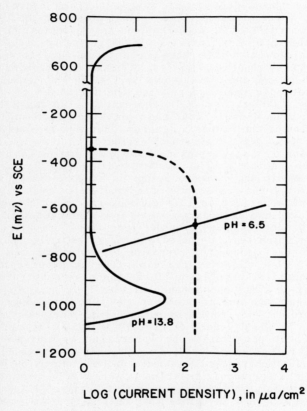

Figure 5.   Comparison of iron corrosion kinetics at neutral and
             alkaline pH.   (Reproduced with permission from Ref. 19.
             Copyright 1968 Journal of Coatings Technology)

corrosion are examples. Pitting is severe localized corrosion, and may be exemplified by the pitting of stainless steel in the presence of chloride ion. Selective dissolution is typically a problem encountered with alloys, for example, the dezincification of brass. Finally, the combined action of corrosion and a mechanical factor, as in stress corrosion cracking, may involve accelerated localized corrosive attack or mechanical fracture due to the synergistic action of corrosion and applied stress. Although all of these categories are of technological importance, and all may be alleviated in some measure by properly selected and applied coatings, localized corrosion phenomena are of particular importance. Furthermore, an understanding of localized corrosion phenomena and the formation of separated anodic and cathodic sites on metal surfaces is central to understanding paint performance in corrosion protection.

Important factors influencing localized corrosion phenomena include the formation and/or presence of solid corrosion products, and the effect of such products on oxygen availability; changes in pH at anodic and cathodic reaction sites; and the ratio of cathodic to anodic reaction site area. At the onset of corrosion, the anodic and cathodic reaction sites are likely to be close together, and may occur at the same sites sequentially; as a deposit of corrosion products forms, the availability of oxygen for the cathodic reaction may be locally restricted, and thus the anodic reaction will be favored. With subsequent changes in pH and an increase in the concentration of soluble corrosion products in anodic areas, oxygen availability can become more restricted and cause further separation and localization of reaction sites. Figure 6 (adapted from Ref. 17) illustrates a classic experiment that demonstrates the separation and localization of anodic and cathodic corrosion sites. A drop of salt water containing small amounts of phenolphthalein and potassium ferricyanide as indicators is placed on an iron surface. At first, there are random spots of blue (anodic sites: ferricyanide in the presence of ferrous ion) and pink (cathodic sites: phenolphthalein in the presence of hydroxide). If allowed to remain undisturbed, rings of color eventually form, blue at the center, then rust, and finally pink at the perimeter. The drop is an oxygen concentration cell, the availability of oxygen for the cathodic reaction being slightly greater at the edges than in the center.

The corrosion phenomena commonly observed on painted metals include cathodic delamination, anodic undercutting, and filiform corrosion. Cathodic delamination results when the alkali produced by the cathodic corrosion reaction disrupts the paint-metal interface. This phenomenon has long been observed on cathodically protected painted steel (18) and has also been demonstrated to be responsible for the loss of paint adhesion that often occurs adjacent to corrosion sites on painted steel (19). The localization and separation of anodic and cathodic sites associated with corrosion at a break in a paint film on steel are schematically illustrated in Figure 7.

Anodic undercutting does not significantly contribute to the loss of paint adhesion from steel under normal aqueous corrosion conditions; even when painted steel has been subjected to substantial anodic currents, little or no adhesion loss has been observed (19–21). Anodic undercutting has been reported to be of

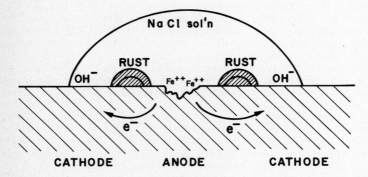

Figure 6.   Schematic illustration of corrosion in a drop of salt
            water.   (Reproduced with permission from Ref. 17.
            Copyright 1979 Elsevier–North Holland)

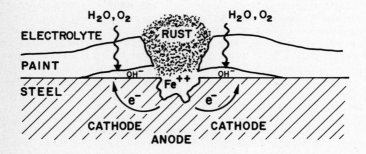

Figure 7.   Illustration of delamination mechanism at a corrosion
            site on painted steel.

importance for aluminum (22). Leidheiser (14) cites as an outstanding example the dissolution of the tin coating between the organic coating and the steel in a food container; as the tin dissolves, the coating separates from the substrate and loses its protective character.

Filiform corrosion is characterized by formation of interconnecting filaments of corrosion under a paint film upon exposure to a humid environment. Filiform corrosion typically occurs only when the relative humidity exceeds about 65%. The mechanism is complicated and has been the subject of considerable discussion in the literature (5, 14, 21, 23, 24). Basically, a localized corrosion mechanism is responsible (Figure 8). The head of the growing filament is anodic, and as a result the filiform corrosion process has been termed a specialized form of anodic undermining (14). Nevertheless, the initial disruption of paint adhesion is most likely a cathodic delamination (21). Perhaps the most interesting question--the origin of the filamentary form--has not been answered with certainty.

Methods of Studying Corrosion Control

Goals of Testing and Classification of Test Methods. The goals of corrosion testing are twofold: first, to evaluate the performance (and predict service history) of existing materials, and second, to direct the development of improved paint materials and processes. The evaluation of performance generally relies on tests that compare the performance of materials in simulated and accelerated service environments. The ideal test environment is a realistic one, with all relevant reactions and processes equally accelerated. Laboratory tests often do not live up to this ideal, however, and considerable caution must be exercised in the interpretation of results. Performance test results are generally evaluated by type (qualitative) and extent (quantitative) of failure. To direct development of improved paint materials and processes, tests based on an understanding of basic mechanisms of corrosion protection and failure are arguably more useful. Test environments in this case may be chosen to isolate individual factors, and the test results are evaluated by comparison with standards and/or theoretical models. Test methods have been discussed extensively in the literature; general reviews have been given by several authors (13, 25-27); electrochemical test methods have received special attention (28-31).

Nonelectrochemical Methods. Nonelectrochemical methods of studying corrosion include exposure tests of performance and primary film property measurements. Standard exposure tests include salt water immersion (3-5% aq. NaCl, usually at room temperature, sometimes oxygen saturated); cyclic immersion (e.g., salt water immersion alternated with drying periods); salt fog or spray (5% aq. NaCl fog, e.g., according to ASTM B117); the CASS test (5% aq. NaCl fog with $CuCl_2$ and acetic acid, according to ASTM B368); the Kesternich test (100% relative humidity to which SO_2 is added; c.f. DIN 50018); and SO_2 modified salt fog (ASTM B117 with SO_2 added). Cyclic exposure tests that incorporate freeze-thaw and dirt exposure have been proposed; these tests are more complicated than conventional laboratory corrosion tests, but have been reported to give better agreement with corrosion performance in service (32). Useful preliminary tests are water or humidity resistance (e.g., by water immersion, static humidity, or condensing humidity exposure) and, for many applications, physical resistance tests such as abrasion, stone peck, mandrel bend, and impact tests. Standard methods for most of these tests are described in the Annual Book of ASTM Standards (33). The results of exposure tests are typically the product of many physical and chemical interactions with the test environment and reflect combinations of film properties, so that interpretation of test results in molecular or mechanistic terms is at best difficult.

Primary film property measurements include measures of adhesion, of permeability, of mechanical properties, and of chemical resistance. Coating adhesion is a prerequisite for corrosion protection, and is assessed by a variety of methods including tape adhesion (ASTM D3359), scrape and parallel groove adhesion (ASTM D2197), and direct pull-off. Testing adhesion with adhesive tape provides a qualitative measure of paint adhesion; the method is often employed at the end of an exposure test or after controlled mechanical damage to the film (e.g., scribing a grid, or subjecting the painted specimen to an impact or stone impingement test). The direct pull-off method involves adhesive bonding of a test button to the sample; a special tool or a conventional tensile testing machine with a special fixture may then be used to measure the force required to remove the test button. Adhesion of paint films when dry is not, however, generally a reliable indicator of adhesion upon exposure to a humid environment (34). Both the tape adhesion and direct pull-off methods have drawbacks in application to measurement of the adhesion of films during or immediately following exposure to water; a comparative absorption method has been proposed by Funke and coworkers (25-26, 35). Many other techniques--scratch or stylus, ultrasonic, ultracentrifugal, blister, and abrasion--are described in the literature (36, 37). Acoustic emission has been used to study the effects of water immersion on coatings and to distinguish between failure mechanisms (38-40).

Permeability of paint films to water, oxygen, and ions has been studied by many workers with a variety of methods including volumetric, gravimetric, and capacitance methods for water; volumetric, barometric, and oxygen electrode based methods for oxygen; and radioactive tracer, conductivity, and ion sensitive electrodes for ions (25, 41, 42; see also 14, 13, 43, and the references cited therein).

Electrochemical and Electrical Methods. Electrochemical and electrical methods for studying film properties and corrosion phenomena have been extensively reviewed (29-31). Comparisons of corrosion test results with direct current measurements of conductivity suggest that visible corrosion is associated with film resistance less than about 1 Mohm/cm^2, but this condition may well correspond with the occurrence of virtual pores in the film allowing development of local conductive pathways. In studies of the equivalent alternating current resistance as a function of frequency, Kendig and Leidheiser (44) found that the development of a region of slope -1 on a log permittivity versus log frequency plot (indicative of local penetration by conducting electrolyte) corresponded with the onset of visible corrosion. In studies of the capacitance of coating films, Touhsaent and Leidheiser (45) observed increases in capacitance due to the uptake of water; interpretation of capacitance data relies on somewhat arbitrary models for the distribution of water, and is therefore necessarily somewhat uncertain. Leidheiser has discussed the topic at length (14, 29). Impedance measurements have been discussed (31); the progressive deterioration of paint films can be monitored by studying the changes in the frequency dependence of impedance. Corrosion potential measurements give an indication of the changes in the ratio of anodic to cathodic surface area; movement in the active direction suggests an increase in the anode/cathode ratio, with an increase in the rate of metal dissolution (46). Polarization curves have been determined for coated metals, but the flow of current is likely to affect the properties of the coating–metal interface (29).

Barrier Characteristics of Coatings

Adhesion of Organic Coatings. Although it is in principle possible for a nonadherent film to provide good barrier properties, practical paint systems must normally remain adherent to the materials to which they are applied for effective corrosion protection to be maintained. The initial strength of the adhesive bond between coating and substrate need not be particularly high. What is important is that after environmental exposure the coating should still be able to withstand the forces exerted on it in its intended service application. The adhesion of virtually all coatings is adversely affected by exposure to water or humid environments. Walker (34) found typical organic coatings to have initial dry adhesion of 20 to 40 MPa as measured in a direct pull–off test; after more or less prolonged exposure to humid environments, the adhesion was observed to drop to 5 to 15 MPa. Exposure at 65% relative humidity or less was found to have little or no effect on adhesion.

Funke and coworkers (35) have compared the moisture absorption characteristics of free and supported paint films, and have found that films that lose adhesion upon exposure to humid environments tend to absorb more water as supported films than as unsupported ones. These results are interpreted in terms of accumulation of moisture within an interfacial region. It should be noted that the moisture absorption results are subject to pronounced scatter, and relatively little weight can be attached to isolated measurements. The scatter in the results appears to be a property of variations in film properties, rather than an artifact of the experimental method (47).

Transport Properties of Coatings. There has been substantial disagreement over the years as to the transport of water, oxygen, and ionic species through paint films. It has been stated that under normal conditions, paint films are saturated with water for about half their useful life, and for the remainder the water content corresponds with an atmosphere of high humidity. The water permeation rate of coatings is sufficiently high that corrosion could proceed on steel surfaces as fast as if the steel were not coated were there no other limiting factors. Mayne (41) estimated that freely corroding unpainted steel consumes water at a rate of 0.4 to 6 x 10^{-5} g/cm^{2}d. The water transmission rates of organic films 0.1 mm thick typically range from 0.5 to 5 x 10^{-3} g/cm^{2}d. Oxygen transmission rates are approximately two orders of magnitude lower. Freely corroding unpainted steel consumes oxygen at a rate about 1 to 15 x 10^{-5} g/cm^{2}d. Thus, oxygen transmission and consumption rates are nearly comparable, and it has been argued that the rate of oxygen transport is rate limiting for the corrosion of painted steel (35). Other calculations (48) suggest, however, that—at least for thin (15 μm) paint films—there is ample oxygen available for the observed rates of corrosion, and that other limiting factors (e.g., the diffusion of cations to cathodic sites) must be operable.

One other factor is likely the high ionic resistance of organic coatings, which tends to prevent the formation of a complete electrolytic cell (c.f. the previous discussion of resistance inhibition). Conduction in polymer films is ionic and can be increased by the presence of underfilm electrolytes; by ionogenic groups in the coating; and by water and electrolytes external to the paint film (41). Underfilm electrolytes tend to promote osmotic blistering and thus result in underfilm corrosion and loss of paint adhesion. Painting of rusty surfaces is bad practice in part because of the electrolytes embedded in rust deposits. Ionogenic groups include soluble pigment components and ionizable resin functional groups. These groups can result in ion exchange processes and eventually reduce film resistance and promote transport of electrolytes through the film (42, 49).

Although both water and electrolyte can penetrate films, electrolyte tends to penetrate films only locally, as evidenced by studies of the electrolytic resistance of free films (41, 50, 51). The regions of direct electrolyte penetration have been shown to be small, localized, and randomly distributed. Furthermore, these regions have been demonstrated to correspond to initial sites of corrosion. The molecular origin of the direct electrolyte penetration regions is not entirely clear, but in some cases they have been associated with inhomogeneous cross-linking.

Pigmentation can also have a profound effect on the transport properties of organic coatings. Schematically illustrated in Figure 9 is the effect of pigment volume concentration (PVC) on blistering, rusting, and permeability. At low PVC, pigment is fully wet by resin, and at high PVC, there remain unfilled interstices between pigment particles. The transition between these regimes is centered on the critical pigment volume concentration (CPVC). The CPVC is the formula pigment volume concentration above which a significant void fraction is present in the film because of a deficiency of binder resin. Many film properties in addition to those illustrated in Figure 9 change substantially at the CPVC, including tensile strength, stain penetration, exterior durability, enamel hold-out, hiding power, and gloss (52-54). Despite these manifold changes in properties, determination of the CPVC by observation of the variation in film performance tends to be uncertain. Methods of determining CPVC based on density determinations have been described (53, 54). Barrier and inhibitive primers are best formulated at PVC < CPVC, in the range PVC/CPVC = 0.8 to 0.9 (55). Moisture

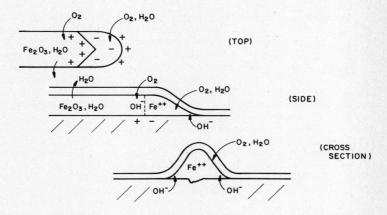

Figure 8.   Schematic illustration of filiform corrosion.   (Reproduced with permission from Ref. 4.   Copyright 1971 Wiley)

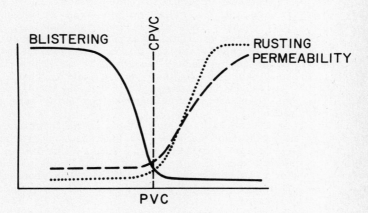

Figure 9.   Dependence of paint film properties on pigment volume concentration.   (Reproduced from Ref. 52.   Copyright 1949 American Chemical Society)

adsorption or absorption by pigments, pigment particle size distribution and particle aggregation, pigment particle shape, and reactions of pigments to form products of different solubility or volume are complicating factors.

Moisture exclusion, adsorption, and permeation are strongly affected by pigment-binder and pigment-water interaction. Strong pigment-binder interaction tends to result in restricted molecular mobility, reduced swelling, and reduced water transmission through the pigmented film. Strong water-pigment interaction tends to result in increased water update by the film and increased film permeability, especially at high PVC. Below the CPVC, strong pigment-binder interaction effectively increases the volume fraction inaccessible to water. Strong water-pigment interaction eases the passage of water (the diffusion coefficient is often strongly concentration dependent). Above the CPVC, unfilled interstices allow passage through the film (56).

Size distribution and particle shape will affect the way that pigment particles can pack, and hence will affect the CPVC. Large aggregates that are not broken up and wet out during pigment grinding can result in the presence of pores in the pigmented film to mimic the effect of pigmentation above the CPVC. Flake or micaceous pigments properly employed can significantly reduce the permeability of the paint film, as can the presence of chemically reactive pigments (such as zinc dust) that form voluminous corrosion products (41, 56).

Other Film Properties. The mechanical properties of polymer films are dominated by the glass transition phenomenon. The glass transition temperature, T_g, is that temperature at which the behavior of a polymer changes from glasslike (high modulus, often brittle) to rubberlike (low modulus, flexible). At temperatures below T_g, the molecular chains are immobile, and at temperatures above T_g, the molecular chains become increasingly mobile until (in the absence of cross-links) flow begins. An increase in the T_g of a coating film tends to reduce its permeability, but may result in the film having a higher level of internal stress and being more prone to cracking and fracture.

Coatings can, of course, suffer many forms of degradation other than corrosion and mechanical damage. Photooxidative processes, typically most severe upon exposure to UV light (290-400 nm) can result in bond rupture and thereby lead to a loss of gloss and eventually a loss of film integrity. Control of these processes is achieved by proper choice of resin pigment, and photostabilizers. Many resins are prone to hydrolysis, especially at high temperatures. The result can be an increase in resin polarity and permeability to electrolyte. Control is principally by resin design for greater chemical resistance. Biological processes (e.g., mildew) are also of concern; control is achieved principally by the use of biologically active additives (biocides and biostats). Miscellaneous environmental factors including industrial pollution, high temperature, and unvented fumes may play a significant role in coating degradation in specific applications.

Corrosion Resistance of Painted Metals

Metal Surface Preparation and Conversion Coatings. It is axiomatic that only with good preparation of the substrate surface can optimal paint performance and corrosion protection be obtained. For metallic substrates, adequate preparation may range from grit blasting or abrasive cleaning with subsequent application of wash primers or chemical conversion coatings to much milder chemical or physical treatments to remove gross surface contamination (57).

In industrial coating applications, inorganic phosphate conversion coatings are a commonly used surface pretreatment for metals. The objective of such treatments is the deposition of an insoluble metal phosphate on the surface to be painted. Bender et al. (58) and Gabe (9) have reviewed the literature and give references to earlier reviews and compendia on the subject. Secondary phosphates of divalent iron, manganese, and zinc are only sparingly soluble. In the phosphating process, the metal substrate is attacked, the pH rises at the metal surface, and amorphous metal phosphate is precipitated on the metal surface. Crystallization and crystal growth of the phosphate then occur. Finally, a process of crystal dissolution and reorganization sets in which modifies the porosity of the phosphate layer and decreases the exposed metal surface area. Overall, the reaction of primary zinc phosphate with steel may be written as:

$$Fe + 3 Zn(H_2PO_4)_2 \rightarrow Zn_3(PO_4)_2 + FeHPO_4 + 3 H_3PO_4 + H_2$$

In practice, depolarizers are added to the bath so that the cathodic reaction becomes the oxygen (or oxidizing agent) reduction reaction rather than the hydrogen evolution reaction. Bath compositions are proprietary, the reactions involved are very complex, and the performance of the coatings obtained are highly dependent on bath composition and deposition conditions as well as on the initial condition of the substrate (32, 59–62). Typically the coatings have a mixed crystal/amorphous structure; "zinc phosphate" is primarily tertiary zinc phosphate tetrahydrate and dihydrate, with small mounts of zinc-iron phosphate (phosphophyllite). Coating weights are typically 0.1 to 1 mg/cm^2; coating thickness is in the range 1 to 10 μm.

Oxide and chromate conversion coatings (used primarily for zinc, aluminum, tin, and magnesium) improve paint adhesion relative to that which is observed on untreated metal. These coatings are applied by a variety of proprietary processes (9, 57, 58).

Corrosion Control by Pigments. Pigmentation for protective coatings has been reviewed briefly by various authors (13, 41); Patton's Pigment Handbook (63) provides more comprehensive information. Chromate pigments have been reviewed by Wormald (64), and lead based pigments have been reviewed by Dunn (65). Recent developments--many aimed at elimination of chromates and lead from coatings--have been discussed by Wienand and Osterland (66). The influence of inert pigments on corrosion protective properties has been reviewed by Kresse (56). The requirements for an ideal corrosion inhibitor, as summarized by Leidheiser (67), are that the inhibitor be effective at pH values in the range of 4 to 10, and ideally in the range of 2 to 12; react with the metal surface such that a product is formed with much lower solubility than the unreacted inhibitor; have a low but sufficient solubility; form a film at the coating/substrate interface that does not reduce adhesion; be effective both as an anodic and cathodic inhibitor; and inhibit the two important cathodic reactions, $H_2O + 1/2 \ O_2 + 2 \ e^- = 2 \ OH^-$ and $2 \ H^+ + 2 \ e^- = H_2$.

Corrosion control by pigments relies on well-known principles of corrosion inhibition. Iron and steel exposed to air are quickly covered by an oxide film; aqueous electrolytes tend to break down this film, and further oxidation of the metal surface ensues. The role of anodic corrosion inhibitors is to supplement or to aid in the repair of the surface oxide film. Basic pigments may form soaps, for example, with linseed oil; autoxidation of these soaps may yield soluble inhibitors in the film. Some other pigments of limited solubility act directly as inhibitors. Active metal pigments supply electrons to the iron substrate and thus lower its potential and prevent metal dissolution.

Typical basic pigments include basic lead carbonate, basic lead sulfate, red lead, and zinc oxide. The soaps formed when these materials interact, for example, with linseed oil, are oxidized in the presence of water and oxygen to form mono- and dibasic straight chain C_7 to C_9 acids. Materials of this type (e.g., sodium and calcium azelate and pelargonate) are known to inhibit corrosion. Inhibition is associated with formation of complex ferric salts that reinforce the oxide film. Lead salts act at lower concentration than the sodium or calcium salts (3, 41).

Anodic passivation of steel surfaces can be efficiently achieved by metal chromates. Chromates of intermediate solubility (e.g., zinc chromate and strontium chromate) allow a compromise between mobility in the film and leaching from the film to be achieved. Chromates inhibit corrosion in aqueous systems by formation of a passivating oxide film. The effectiveness of chromate inhibitors in aqueous systems depends on the concentration of other ionic species in solution, for example, chloride. Synthetic resin composition can also significantly influence the effectiveness of chromate pigments. The effect appears to be related to the polarity of the resin (20); chromate pigments appear to be less effective in resins of low polarity.

Corrosion control by the incorporation of active metal pigments is a form of cathodic corrosion protection. The pigment must be less noble than the substrate; furthermore, the pigment particles must be in metallic (electronic) contact with each other and with the substrate. These requirements dictate the use of high levels of zinc (90 to 95% by weight in zinc-rich primers, 75 to 80% by weight in the older conventional zinc-dust paints) and thorough cleaning of the substrate. For protection of steel substrates, Zn, Al, and Mg appear to be attractive candidates as protective metal pigments. Of these, however, only Zn meets the requirement of electronic contact, as both Al and Mg are covered with a low conductivity oxide. Part of the effectiveness of the zinc pigment may be due to the formation of zinc hydroxide and other zinc corrosion products that tend to clog the pores of the coating film (13, 41, and references cited therein); there is, however, good evidence for true cathodic protection with these materials (68). When zinc-rich primers are

coated with paints having high electrical resistance, the effective
sacrificial area may be restricted to a small zone around the
damaged area. The primer may aid in healing the damaged area and in
preventing the spread of corrosion (68). A variety of binders, both
organic and inorganic, have been used; inorganic binders, for
example, ethyl silicate, are reported to give the best performance
(69, 70).

Effects of Resin Structure on Corrosion Control. Resin composition
plays a role in many different aspects of corrosion protection. The
effects of resin permeability and of adhesion in the presence of
water have been discussed in previous sections. Resin composition
also influences the chemistry of the interface between coating and
substrate during corrosion. From performance tests based on simple
electrochemical principles, it is known that the delamination of
otherwise strongly adherent organic coatings that often accompanies
corrosion is attributable to the action of cathodic alkali (19, 20).
The details of the chemical changes that occur at the interface have
been extensively studied (71-81); it is evident that degradation of
the organic resin can play a major role in the loss of adhesion of
coatings that are not readily displaced by water (72-77). The
degradation mechanism may involve hydrolysis of base labile bonds
such as esters, urethanes, or ureas by hydroxide. Use of paint
resins that do not have alkali sensitive bonds has been shown to
result in significant improvements in paint performance provided
that other important paint performance criteria are also met (e.g.,
resistance to humidity) (73, 75). The degradation of paint resin to
form ionic products probably greatly increases the ionic conductivity of the interfacial region, and thus expands the cathodic zone
of reaction and accelerates the overall corrosion process. It has
also been found that the oxide layer tends to increase in thickness
under the intact film (79-81), and it has been argued (24, 81) that
the initial step of delamination involves dissolution of the
interfacial iron oxide. This argument does not, however, take into
account the large decrease in rate of delamination observed with
changes in polymer structure designed to improve resistance of the
organic coating to alkali degradation (73, 745, 82). The mechanism
of corrosion-induced adhesion failure can range from simple
displacement by water to a process involving substantial chemical
changes in the coating, the substrate or both (77); the mechanism
that occurs in any given case is evidently a complicated function of
test environment, substrate and coating composition, and details of
interfacial structure and composition.

Summary

    The factors that influence the corrosion protective performance
of organic coatings can be broken into three groups: environmental
conditions, barrier properties of the coating system, and inhibitive
properties of the coating system. Typically, the corrosive
environment comprises water, oxygen, and an electrolyte. The
severity of the environment depends especially on the activity of
oxygen and electrolyte, and on pH. The initiation of corrosion is
influenced by the choice and preparation of the substrate, the
transport and permeability characteristics of the organic coating,

pigmentation (especially as it affects film transport characteristics), mechanical integrity of the coating, and coating adhesion (especially under conditions of the anticipated service environment). The propagation of corrosion from an initial site is affected by substrate preparation, pigmentation (especially the presence of active pigments), and the adhesion of the coating adjacent to corroding sites (in this respect, the composition of the coating in the interfacial region can play a significant role). An understanding of the electrochemistry of corrosion processes, of the transport processes in coating films, and of the chemistry and physics of the coating–substrate interface can be of substantial assistance in resolving corrosion related coating problems.

Literature Cited

1. NBS Special Publication 511-1, "Economic Effects of Metallic Corrosion in the United States," SD Stock No. SN-003-01926-7, 1978; NBS Special Publication 511-2, "Economic EFfect of Metallic Corrosion in the United States," Appendix B, SD Stock No. SN-003-01927-5, 1978,
2. Smith, C. A. Mod. Paint Coat. 1978, 68(3), 59.
3. Evans, U. R. "The Corrosion and Oxidation of Metals"; St. Martins Press: New York, 1960; Ibid., 1st Supplementary Volume, St. Martins Press: New York, 1968; Ibid., 2nd Supplementary Volume, Edward Arnold: London, 1976.
4. Uhlig, H. H. "Corrosion and Corrosion Control," 2nd ed.; Wiley: New York, 1971.
5. "Corrosion," 2nd ed.; Shreir, L. L., Ed.; Newnes–Butterworths: London, 1976.
6. Symposium on Interfacial Phenomena in Corrosion Protection Preprints, Am. Chem. Soc. Div. Org. Coat. Plast. Chem., 37(1), 1977; Symposium on Advances in Coating Metals for Corrosion Protection, Preprints, Am. Chem. Soc. Div. Org. Coat. Plast. Chem., 43, 1980; "Corrosion Control by Coatings"; Leidheiser, H., Jr., Ed.; Science Press: Princeton, 1979; "Corrosion Control by Organic Coatings"; Leidheiser, J., Jr., Ed.; National Association of Corrosion Engineers: Houston, 1981; "Automotive Corrosion by De-icing Salts"; Baboian, R., Ed.; National Association of Corrosion Engineers: Houston, 1981; "Adhesion Aspects of Polymeric Coatings"; Mittal, K. L., Eds.; Plenum: New York, 1983.
7. Of particular interest are The Journal of Coatings Technology, Progress in Organic Coatings, The Journal of the Oil and Colour Chemists Association, Farbe und Lack, Corrosion-NACE, Corrosion Science, and I&EC Product R&D.
8. Evans, U. R. "An Introduction to Metallic Corrosion," 2nd ed.; Edward Arnold: London, 1972.
9. Gabe, D. R. "Principles of Metal Surface Treatment and Protection," 2nd ed.; Pergamon: Oxford, 1978.
10. Scully, J. C. "The Fundamentals of Corrosion," 2nd ed.; Pergamon: Oxford, 1975.
11. West, J. M. "Electrodeposition and Corrosion Processes," 2nd ed.; Van Nostrand: London, 1972.

12. Bennynk, P. Janssen; Piens, M. Prog. Org. Coat. 1979, 7, 113–39.
13. Von Fraunhofer, J. A.; Boxall, J. "Protective Paint Coatings for Metals"; Portcullis Press: Redhill, Surrey, 1976.
14. Leidheiser, H. Corrosion NACE, 1982, 38(7), 374.
15. Pourbaix, M. "Atlas of Electrochemical Equilibria in Aqueous Solutions"; Pergamon Press: New York, 1966; and "Lectures on Electrochemical Corrosion"; Plenum: New York, 1971.
16. Hospadaruk, V. Paper No. 780909, SAE Proceedings P78 1978, 123.
17. Banfield, T. A. "The Protective Aspects of Marine Paints," Prog. Org. Coat. 1979, 7, 253.
18. Anderton, W. A. Off. Dig. 1964, 36, 1210.

19. Wiggle, R. R.; Smith, A. G.; Petrocelli, J. V. J. Paint Technol. 1968, 40(519), 174.
20. Smith, A. G.; Dickie, R. A. Ind. Eng. Chem. Prod. Res. Dev. 1978, 17, 42.
21. Funke, W. Prog. Org. Coat. 1981, 9, 29.
22. Koehler, E. L. In "Localized Corrosion"; Staehle, R. W.; Brown, B. F.; Kruger, J.; Agarwal, A., Eds.; National Association of Corrosion Engineers: Houston, 1974; p. 117.
23. Hoch, G. M. In "Localized Corrosion"; Staehle, R. W.; Brown, B. F.; Kruger, J.; Agarwal, A., Eds.; National Association of Corrosion Engineers: Houston, 1974; p. 134.
24. van der Berg, W.; van Laar, J. A. W.; Suurmond, J. In "Advances in Organic Coatings Science and Technology"; Parfitt, G. D.; Patsis, A. V., Eds.; Technomic: Westport, Conn., 1979; Vol. 1, p. 188.
25. Funke, W. J. Oil Colour Chem. Assoc. 1979, 62 63–67.
26. Funke, W. Farbe und Lack 1978, 84, 380; Funke, W.; Machunsky, E.; Handloser, G. Ibid., 1979, 84, 498; Funke, W.; Zatloukal, H. Ibid., 1979, 84, 584.
27. Rowe, L. C.; Chance, R. L. In "Automotive Corrosion by De-icing Salts"; Baboian, R., Ed.; National Association of Corrosion Engineers: Houston, 1981; p. 133.
28. Baboian, R. "Electrochemical Techniques for Corrosion"; National Association of Corrosion Engineers: Katy, Tex., 1977.
29. Leidheiser, J., Jr. Prog. Org. Coat. 1979, 7, 79.
30. Sato, Y. Prog. Org. Coat. 1981, 9, 85.
31. Szauer, T. Prog. Org. Coat. 1982, 10, 171.
32. Hospadaruk, V.; Huff, J.; Zurilla, R. W.; Greenwood, H. T. Paper No. 780186, Trans. SAE 1978, 87, 755.
33. "Annual Book of ASTM Standards"; Part 27, "Paint--Tests for Formulated Products and Applied Coatings"; American Society for Testing and Materials: Philadelphia, 1982.
34. Walker, P. Paint Tech. 1967, 31(8), 22; Ibid. 1967, 31(9), 15.
35. Funke, W.; Haagen, H. Ind. Eng. Chem. Prod. Res. Dev. 1978, 17, 50.
36. Mittal, K. L. In "Adhesion Aspects of Polymeric Coatings"; Mittal, K. L., Ed.; Plenum: New York, 1983.

37. "Adhesion Measurement of Thin Films, Thick Films, and Bulk Coatings"; Mittal, K. L., Ed.; ASTM STP No. 40, American Society for Testing and Materials: Philadelphia, 1978.
38. Rawlings, R. D.; Strivens, T. A. J. Oil Colour Chem. Assoc. 1980, 63, 412.
39. Rooum, J. A.; Rawlings, R. D. J. Mater. Sci. 1982, 17, 1745.
40. Rooum, J. A.; Rawlings, R. D. J. Coat. Technol. 1982, 54(695), 43.
41. Mayne, J. E. O. In "Corrosion," 2nd ed.; Shreir, L. L., Ed.; Newnes–Butterworths: London, 1976; Vol. 2, 15:24.
42. Svoboda, M.; Mleziva, Prog. Org. Coat. 1973–74, 2, 207.
43. Dickie, R. A.; Smith, A. G. CHEMTECH 1980, 10(1), 31.
44. Kendig, M. W.; Leidheiser, H., Jr. J. Electrochem. Soc. 1976, 123, 982.
45. Touhsaent, R. E.; Leidheiser, H., Jr. Corrosion 1972, 28, 435.
46. Wolstenholme, J. Corrosion Sci. 1973, 13, 521.
47. Corti, H.; Fernandez–Prini, R. Prog. Org. Coat. 1982, 10, 5.
48. Leidheiser, H., Jr. "Corrosion Control Through a Better Understanding of the Metallic Substrate/Organic Coating Interface"; AD/A 095420, National Technical Information Service: Springfield, Va., 1980.
49. Kumins, C. A. J. Polym. Sci. C 1965, 10, 1.
50. Kinsella, E. M.; Mayne, J. E. O. Brit. Polym. J. 1969, 1, 173; Mayne, J. E. O.; Scantlebury, J. D. Brit. Polym. J. 1970, 2, 240.
51. Mayne, J. E. O.; Mills, D. J. J. Oil Colour Chem. Assoc. 1975, 58, 155.
52. Asbeck, W. K.; Van Loo, M. Ind. Eng. Chem. 1949, 41, 1470.
53. Pierce, P. E.; Holsworth, R. M. Off. Dig. 1965, 37(482), 272.
54. Stieg, F. B. J. Coat. Technol. 1983, 55(696), 111.
55. Hare, C. H. "Anti–corrosive Barrier and Inhibitive Primers"; Unit 27, Federation Series on Coatings Technology, Federation of Societies for Coatings Technology: Philadelphia, 1978.
56. Kresse, P. Farbe Lack 1977, 83, 85.
57. Hare, C. H. "Corrosion and the Preparation of Metallic Surfaces for Painting"; Unit 26, Federation Series on Coatings Technology, Federation of Societies for Coatings Technology: Philadelphia, 1978.
58. Bender, H. S.; Cheever, G. D.; Wojtkowiak, J. J. Prog. Org. Coat. 1980, 8, 241.
59. Zurilla, R. W.; Hospadaruk, V. Paper No. 780187, Trans. SAE 1978, 87, 762.
60. Kargol, J. A.; Jordan, D. L.; Palermo, A. R. Paper No. 271, "Corrosion/82"; National Association of Corrosion Engineers: Houston, 1982.
61. Wenz, R. P.; Klaus, J. J. Mater. Perform. 1981, 20(12), 9.
62. Wojtkowiak, J. J.; Bender, H. S. J. Coat. Technol. 1978, 50(642), 86.
63. Patton, T. C. "Pigment Handbook"; J. Wiley: New York, 1973; Vols. I, II, and III.
64. Wormald, G. In "Treatise on Coatings"; Myers, R. R.; Long, J. S., Eds.; Marcel Dekker: New York, 1975; Vol. 3, Part I, p. 306.

65. Dunn, E. J. In "Treatise on Coatings"; Myers, R. R.; Long, J. S., Eds.; Marcel Dekker: New York, 1975; Vol. 3, Part I, p. 355.
66. Wienand, H.; Ostertag, W. Farbe Lack 1982, 88(3), 183.
67. Liedheiser, J., Jr. J. Coat. Technol. 1981, 53(678), 29.
68. Fernandez-Prini, R.; Kaputa, S. J. Oil Colour Chem. Assoc. 1979, 62, 93.
69. Schmid, E. V. Farbe Lack 1982, 88, 435.
70. Kapse, G. W.; Rani, Km. Bela J. Oil. Colour Chem. Assoc. 1980, 63, 70.
71. Castle, J. E.; Watts, J. F. In "Corrosion Control by Organic Coatings"; Leidheiser, H., Jr., Ed.; National Association of Corrosion Engineers: Houston, 1981; p. 78.
72. Hammond, J. S.; Holubka, J. W.; Dickie, R. A. J. Coat. Technol. 1979, 51(655), 45.
73. Holubka, J. W.; Hammond, J. S.; deVries, J. E.; Dickie, R. A. J. Coat. Technol. 1980, 52(670), 63.
74. Dickie, R. A.; Hammond, J. S.; Holubka, J. W. Ind. Eng. Chem. Prod. Res. Dev. 1981, 20, 339.
75. Hammond, J. W.; Holubka, J. W.; deVries, J. E.; Dickie, R. A. Corrosion Sci. 1981, 21, 239.
76. deVries, J. E.; Holubka, J. W.; Dickie, R. A. Ind. Eng. Chem. Prod. Res. Dev. 1983, 22, 256.
77. Holubka, J. W.; deVries, J. E.; Dickie, R. A. Ind. Eng. Chem. Prod. Res. Dev. 1984, 23, 63.
78. Ritter, J. J.; Kruger, J. Surf. Sci. 1980, 96, 364.
79. Ritter, J. J.; Rodriguez, M. J. Corrosion-NACE 1982, 38(4), 223.
80. Ritter, J. J.; Kruger, J. In "Corrosion Control by Organic Coatings"; Leidheiser, H., Jr., Ed.; National Association of Corrosion Engineers, Houston, 1981, p. 28.
81. Ritter, J. J. J. Coat. Technol. 1982, 54(695), 51.
82. Dickie, R. A. In "Adhesion Aspects of Polymeric Coatings"; Mittal, K. L., Ed.; Plenum: New York, 1983.

APPLICATION AND CURE OF COATINGS

Methods of Application of Coatings

EMERY P. MILLER

641 East 80th Street, Indianapolis, IN 46240

Coating Application Methods
  General
  Brushing
  Dipping
  Flow Coating
  Curtain Coating
  Roll Coating
  Air Spraying
  Hydrostatic Spraying (Airless Spraying)
  Electrostatic Spraying
  Powder Coating
  Electrodeposition Coating Process
Conclusion

Any coating operation is concerned with the deposition of a layer of material upon a surface for the purpose of protecting or decorating that surface or modifying its functional characteristics. The nature of such an operation is bound to be closely related to many other manufacturing steps that in combination make up the complete system of which coating is but one step. Such a system will involve a step in which the surface to be coated is adequately prepared to receive the coating (cleaning, wiping, pretreating, priming). It will also involve the selection of the coating material, the application of the material, and the curing of the applied layer (drying, baking, or irradiating) so as to fully develop the characteristics of the coating. Here we are concerned primarily with the application step and are assuming that all other steps in the process have been appropriately handled.

The ultimate choice of the application method itself will further be determined by an analysis of many factors. Whether the material to be applied is solid, liquid, or powdered; thick or thin; clear, metallic, or wrinkled; or waterborne or oil based are all

0097–6156/85/0285–0803$06.75/0

factors that will bear upon the nature of the selected application
method. The nature of the surface is also important. Flat surfaces
may well require a method that is quite different from that which
would be used on contoured ones, and certainly the coating of
different materials (wood, aluminum, steel, etc.) will place
different demands upon the application method. Other factors, such
as the nature of the film required, the speed of application needed,
the economy demanded, the space allotted, and the curing cycle
required will have to be considered in the selection of the method.
Whether the product demands a single or several colors may be the
deciding factor in method selection. In addition, factors may
become important that at first glance may appear to be completely
unrelated. Ecological pressures as well as job safety
considerations have recently come to play a major role in
determining the final nature of a coating application operation.
Generally speaking, as new demands have come into existence, new
methods have been developed aimed at coping with these new demands.
The old methods remain in use in those cases where they are
adequately applicable.

Coating Application Methods

General. There are numerous methods available for applying a
coating layer to a surface. Coatings can be applied by plating,
metal vacuum vaporizing, pressure layering, and anodizing, but in
this discussion we will be limiting our considerations to those
methods that are used to apply resin or plastic films. Some of the
more established methods in this category will simply be mentioned
while greater detail will be devoted to the newer, less well-known
techniques.

Brushing. Brushing was perhaps the earliest, specifically
recognized application technique (1). It remains with us today as
the simplest, most direct, most economical means. There is more to
brushing than is at first apparent. Great care in the selection and
preparation of materials must be exercised, and the brushing
equipment must be chosen carefully if a good application is to be
made by even the most adept artisan. The process is relatively slow
and can only be used on today's industrial high-production lines
where the areas to be coated are small and uniquely located.
Certain coating applications can only be effectively done by
brushing, and one should not hesitate to use brushing where such a
method is indicated even though it may appear to be out of place in
today's fast-moving technology.

Dipping. Dipping is another method of coating that continues to
serve industry even though it is somewhat archaic. In this method
the coating is placed in a suitable container and the parts are
simply immersed into the liquid to wet all exposed surfaces. The
part is then removed from the bath and allowed to drain. Being
direct and uncluttered with equipment requirements, this process
appears to be simple but such is not the case if a good film is to
be obtained. Since the film is formed as the coating material flows
from the part, careful controls must be exercised over the condition
of the material in the tank, the drain time, and the drain position

of the part. The part may have to be repositioned after dipping in order to free material from pockets and trapping locations. The material may tend to run and sag particularly near surface discontinuities such as seams and holes. By controlling all elements of the operation carefully so that consistency exists, it is possible to obtain satisfactory films by this method. Because it can be easily automated and ensures complete coverage of the part, dipping is used today for the application of simple, not too critical coatings (2). Single color primers are applied by dipping on high-production lines where a variety of parts are to be primed and where some subsequent sanding can be done before the application of a finish coat by some other method. The flow lines and runs that are inherent with this method are thus not apparent on the finished surface. Although dipping is very efficient in the application step, its overall efficiency can be lowered if a nonuniform film, i.e., light at the top and heavy on the bottom, results from the flow of the applied layer. Dipping requires a large quantity of the coating material to be in process and so does not readily adapt to operations requiring different colors. To maintain such volumes at constant viscosity and evaporation rate presents many problems. Because dipping operations involve large quantities of exposed potentially flammable materials and long drain areas where drips accumulate, they have usually been considered highly hazardous areas that industry would just as soon avoid. The introduction of waterborne coatings that are usable with this method has greatly reduced this hazard and has widened the scope of application of the dipping process.

Flow Coating. The flow coating method is closely related to dipping in that the surfaces of the part are wetted with an excess of material and allowed to hang in fixed position until drainage is complete. In flow coating, however, the material is applied to the surface as the part is carried through an enclosure in which the material is showered over and about the part from either fixed or moving nozzles that are the outlets of a pumped recirculating system. With this technique only external exposed surfaces are coated so a potential material saving is presented to the user who requires only an external coating. Because the part can be carried straight through the enclosure without having to be lowered as in the dipping process, greater simplicity in the path of the conveyor travel is possible. The recirculation of the coating material allows a smaller volume of the material to be in use at any moment than is required in the dipping process. Because the material is showered over the part as a coarse spray, there is a considerable amount of solvent lost to the surroundings. Great care must be exercised to maintain those characteristics of the material that influence its flow over the surface during draining and film formation. After the flow coater, the path of the conveyor is sometimes enclosed so that the part can be held in a constant atmosphere of solvent to aid in controlling flow. Automatic methods to control the viscosity of the material are also used to help maintain the pumped material in a consistent condition. When such techniques are employed, flow coating can be made to be a consistently reliable and reproducible method. It is being used on many high-production lines. Since the same difficulties exist as with dipping so far as film uniformity and runs and sags are concerned, the use of the flow coating process is often limited to rough one-coat applications or to applications that are followed by some sanding or sag removal process. Articles such as bed springs, automobile frames, and springs are representative of parts being coated by this method. Figure 1 in the chapter "Appliance Coatings" illustrates a commercial flow coating operation.

Curtain Coating. As opposed to dipping and flow coating, which are highly flexible with regard to the type of item being coated, curtain coating is generally considered to be limited to the coating of flat surfaces. In the practice of this method a crossbar type of feed conveyor is arranged to carry the sheets to be coated in a horizontal position to the coater location. Here the feed conveyor ends, and after a suitable gap, a removal conveyor begins. This conveyor is of the same crossbar type as the feed conveyor. Located above the conveyor level at the gap between the two is a V-shaped trough that has a slit along its lower edge. This trough is as wide as the conveyors and is a closed container. The coating material to be applied to the surface is pumped from a supply reservoir to the inside of the trough under reasonable pressure. It flows from the slit in a downward direction in the form of a liquid sheet or curtain. Below the conveyor level a catch trough is located. The flowing curtain flows into this, and the material is then recirculated to the reservoir. Under this circulation the curtain is constantly maintained with uniformity and without breaks. A sheet of the material to be coated is placed on the delivery conveyor and carried at a controlled rate across the gap and thus through the curtain. During passage the curtain falls upon the horizontal surface and is carried away by the sheet at the same rate at which it would normally fall into the lower trough. The surface in this manner has a wet film of coating applied to it. The coated sheet is held in a horizontal position on the removal conveyor or on storage racks until the coating dries or is cured into a solid film. This method has found its greatest use in the wood industry where it is used to apply clear coatings to plywood sheet and such items as cabinet doors. It is obviously very efficient and fast. In such installations the coating operation sometimes is followed by a radiation curing process so that the liquid film applied can be converted to a solid film in just a few seconds. Considerable control of all parts of a curtain coating operation is a must if the integrity of the liquid curtain is to be maintained. If the curtain is allowed to break, a bare strip will be produced on the surface of the product. Continuous integral curtains are more easily obtained with oil or solvent based materials than with waterborne materials so curtain coating is usually done with these types. Although this technique is mainly used to coat the upper exposed surface of the object, special methods have been developed that allow the edge surfaces to also be coated. These latter techniques are not always simple or effective.

Curtain coaters have recently appeared on the market that offer the user more control over the operation. These so-called Omega curtain coaters promise a more effective use of the process and a wider application (3).

Roll Coating. Roll coating is the most widely used method for coating flat surfaces; Figure 1 shows a commercial installation in operation. The basic elements of this technique involve spreading the material to be applied upon a roll and then transferring this material to the surface to be coated by bringing the roll into contact with the surface. The exact nature of the equipment by which this transfer is accomplished varies from application to application. Most individuals have become acquainted with the nature of roll coating as it is done domestically, but much more complicated arrangements are used industrially. In all applications, however, a wetted roll is used as a transfer device to carry the material from supply to surface. In some uses the coating roll rotates in the material bath and the surface to be coated moves over the roll at another position on its circumference to pick up the material. Such an arrangement is often used to coat continuous sheet. After being applied, the coating material can be leveled and the excess removed by using a doctor blade in contact with the surface after the roll operation (4).

In other arrangements the material is passed from the surface of the primary roll to a transfer roll and then to the surface as in a printing process. Such arrangement is used to get material onto both sides simultaneously when continuous sheet is to be coated. In some applications the transfer rolls are power rotated so that a slipping occurs between the roll surface and the surface being coated. The material is thus wiped into surface irregularities. Roll coating in one or the other of its many modifications is used to coat metal strips that are eventually converted into metal house siding. It is also used to coat discrete wood or metal panels. In many steel mills where precoated stock is being produced for today's metal fabricating market, roll coating is the application method employed. The process is extremely efficient and, although appearing to be simple, produces acceptable coatings only when operated by technicians skilled and experienced in its use. Careful control of the many involved variables is a must. The nature of the roll surfaces, the properties of the coating materials, the speed of application, roll pressures, and the temperature and humidity of the application area all are influential in determining consistency of coating quality. Roll coating is perhaps the only method that can be used for the high-speed application of reasonable films to fast-moving continuous wide sheets of material.

Air Spraying. Air spraying provides to industry a method of creating on a surface, regardless of its contour, a plastic coating that has uniformity and a high-quality appearance. In addition, spray coating is very fast and, therefore, is a method readily adapted to today's high-speed production. In this method a gun is used to develop a directed spray of the coating material. In the gun, air openings are arranged relative to a fluid opening so that as the material issues from its orifice it is impinged upon by the air streams. This impact of the air on the liquid converts the liquid into the small particles that form the spray. The forward force of the atomizing air streams directs the particles and carries them onto the surface to be coated. They land there, accumulate to form a wet layer, flow together, and level to form the desired smooth film (5).

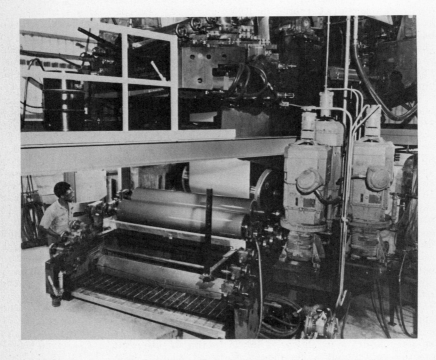

Figure 1. Coil coating operation. Roll Coater, Inc., Kingsbury, Ind.; 72—in. Wean United coil coating line; finish coater.

To get complete coverage of the surface, the gun is moved about and directed to various areas by the sprayman in case the operation is a hand spray one. When arranged as an automatic application, the gun or guns may be mounted on a mechanical manipulation device to produce the desired motion over the surface. The part is often carried on a conveyor to be moved past the spray gun position to further create the desired relative movement of the surface with respect to the spray.

Air spray has become well-known in the finishing industry and is widely used. Most all types of material can be applied by this method, and when the material is properly adjusted in viscosity and solvent evaporation rate, almost any desired film weight can be formed on the surface. Air spray guns are designed so that the degree of atomization, the fan pattern shape, and the rate of material delivery are all easily adjusted by the operator. The spray can also be turned on and off promptly, consistent with the demands of the operation. This flexibility is highly desirable if the trained sprayman is to produce a high-quality coating.

Air spray coating, although very fast, can be quite inefficient. Under ideal conditions where a large flat surface is being coated, no more than 60% of the coating material that leaves the supply becomes deposited on the surface. On open-type wire products spraying may result in the loss of all but 15% of the material. This spray loss, known in the trade as overspray, becomes dispersed in the surrounding atmosphere and is the main reason why spraying is not used more on outdoor applications. Its presence also explains why any industrial spray operation is always conducted within a suitable enclosure or spray booth. The oversprayed solids are usually trapped in the filters or washers of the spray booth ventilating system but are seldom reprocessed as usable material. The solvent component of the oversprayed material is exhausted with the booth air. The hydrocarbon vapor emissions have led the air pollution people to place restrictions on paint spraying operations conducted in concentrated industrial areas. Despite these obvious difficulties, the air spray technique is the one that is most widely used in industry at the present time.

The great flexibility of the air spray gun is its chief advantage. The nature of the atomization produced with a given material is easily controlled by the operator. He can also control the distribution of this spray in the pattern being produced by the gun. He simply adjusts a valve on the gun that modifies the ratio of material to air and the atomization can be made finer or coarser at his will. The control of another valve allows him to readily change the pattern shape from round to a flat fan.

Hydrostatic Spraying (Airless Spraying). The advantage of applying a coating to a surface by depositing on that surface the material from a spray of finely divided droplets has led coaters to try a variety of methods of atomizing the material into such a spray. Hydrostatic spraying is one such alternate approach to the problem. In this method the coating in a suitable system is subjected to a relatively high hydrostatic pressure--800-1500 $lb/in.^2$. The material is conducted to a suitable gun that has at its forward end a very small valved opening (equivalent diameter in the neighborhood of 0.005-0.01 in.). When the valve is opened, the liquid coating under pressure is released through the small opening and is disrupted into a fine spray. The popular aerosol spray cans operate on this same principle. If the opening is correctly shaped, the resultant spray leaves the gun in a fan-shaped spray distribution of material. Each small particle is mechanically projected away from the gun. When the gun is pointed at an oppositely disposed surface, that surface can be spray coated just as with an air spray device. The major difference between the two devices is that the absence of the atomizing air in the hydrostatic gun allows the spray to be "softer" and hence not so apt to be oversprayed from the surface. The application efficiency resulting from the use of such a device has the possibility of being higher.

The hydrostatic sprayer has certain inherent characteristics that make it less flexible and therefore less readily accepted as a production tool. The orifice size, being preselected with each gun, combines with the fluid pressure used to determine both the atomization and the rate of delivery of the coating material. The orifice shape also automatically determines the shape of the spray fan and its forward velocity. The operator has no way of easily adjusting these variables that are so fundamental to his readily obtaining a good coating. The pressure must be changed or a new orifice placed in the gun if such a change is to be made. This basic inflexibility is the reason hydrostatic spraying has had only limited application. It has found use in those cases where the coating requirements have been somewhat secondary to such needs as speed of application and increased efficiency. The application of primers to metal castings and outside maintenance painting are typical examples of its use. Both hand usage and automatic usage have been made with hydrostatic atomizers.

The high pressure used on the liquids being atomized by these devices causes the liquid to issue from the orifice at high velocity. This velocity is high enough to produce skin injection if the orifice is held close enough to skin surface. Numerous individuals have experienced this injection when the gun is operated while a part of the body is close to the orifice. Any operator of such guns should be mindful of this possibility and guard against it.

In an effort to gain the advantages of such pressure atomizers and have them safe, users often heat the coating material before introducing it into the gun. Heating allows one to obtain comparable atomization at lower pressures and thus avoids the possibility of an injection. Still another pressure atomizing gun has been introduced that has air outlet orifices near the fluid outlet nozzle. This air assists the atomization and thus also allows the liquid pressure to be reduced to a level where it is less apt to cause injection. Unfortunately the efficiency of application of these devices, although generally higher than that of air sprayers, still is not sufficiently high that overspray recovery booths can be omitted from the spraying systems. If this were possible, these sprayers would be much more attractive. However, this type of atomizer has proven to be an acceptable one that has features to recommend it for certain applications, and those charged with the responsibility of finishing any product should be mindful of its existence and its favorable characteristics.

Electrostatic Spraying. General. In the electrostatic spraying process the electrical force of attraction that exists between two bodies that are carrying opposite electrical charges is put to work. In this method the small atomized bodies of coating material that constitute the spray are charged to one electrical sign--positive or negative--while the surface to be coated is given the other sign. Under these conditions the particle is attracted to the surface and, if it is free to respond to this attraction, should be deposited on the surface. The electrostatic spraying process potentially should thus have no overspray, and its efficiency should approach 100%. The degree to which this ideal can be realized depends upon the arrangement of the equipment that is used and how effective this is in charging the particles so they will respond to the electrical force.

Process 1. In one type of electrostatic spraying arrangement, the so-called Process 1 (Figures 2 and 3), air spray guns are used to create the small coating material particles. The articles to be coated are carried on an electrically grounded conveyor. They are carried by this conveyor into a spray booth. In the booth a charging electrode is positioned alongside of the path of travel of the parts. It extends for several feet along the conveyor path and is separated from the article at a distance of about one foot. This electrode has points or fine wire elements distributed over its surface. It is also insulated from ground, and because it is spaced from the grounded article, it can be charged to a high voltage by being connected to a suitable supply. When it is so connected, a strong electric force field is established between it and the article on the grounded conveyor. The electrode is usually negative, so the article is positive with respect to it. Under the action of this field, the air adjacent the electrode becomes charged to both signs. The positively charged air quickly moves to the negative electrode while that which is negative moves away from the electrode toward the article and quickly fills all the space extending to the article.

When this condition exists, the spray guns that produce the particles of coating material are set up so as to spray the material into the electrode-article space toward and about the objects. These paint particles will collide with the charged air particles and will become charged. Once they are charged they will be electrically attracted to the object. In the process of being formed, these particles are given some mechanical direction that could or could not carry them directly toward the object. If a given particle is directed toward the object, the electrical force will simply aid in depositing it upon the object. If it is not mechanically directed toward the object, the electrical force will try to alter its flight path so that it will also be deposited. The extent to which the electrical force is able to redirect the potentially oversprayed particles to deposition on the article and thus reduce the amount of oversprayed material determines the efficiency improvement, which can be brought about by the use of this electrostatic spraying process. Whereas normal air or hydrostatic pressure spray methods allow articles to be coated with an efficiency of about 50%, the use of Electrostatic Process 1 will permit these same items to be coated with an efficiency of 75%.

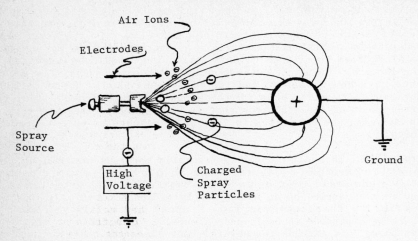

Figure 2.   Principles of electrostatic spraying process 1.

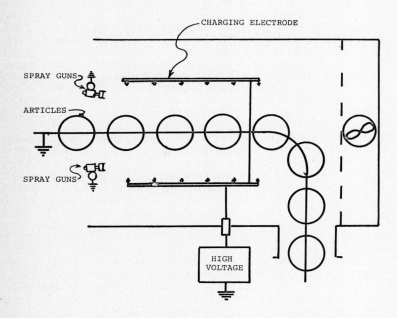

Figure 3.   Schematic of electrostatic spraying operation 1.

Half of the normally oversprayed material is thus captured as useful coating. To the average spray finishing operation a reduction of 25% in the amount of material required to do the job is already a sizable saving. Because of this possible saving, this electrostatic spray process has been used by industry to coat many types of items from automobiles to bicycles.

Process 1 was originally introduced as a completely automatic one. Subsequent development in the nature of the electrical charging circuits has made it possible to employ these same concepts in hand spray guns. Air atomizing-electrostatic depositing hand spray guns and hydraulic pressure atomizing-electrostatic depositing hand spray guns are now available for safe industrial use. In these guns the atomizing and pattern producing components are essentially the same as the corresponding elements in the nonelectrostatic guns. The electrostatic handgun has the particle charging electrode of the System 1 incorporated into it as an integral part of the front of the gun. This is usually a single point electrode that is capable of being charged safely to a sufficiently high electrical potential that it will charge the paint particles as they are formed and projected toward the article. When such guns are used by an operator, he moves the produced spray over and about the surface in the usual manner, directing the spray to deposition on the surface. That portion of the spray that is misdirected will, because it is charged, be redirected and deposited upon the surface. Overspray is reduced. Such spray redirecting action also allows the painting operation to be done with less care since surfaces that are not normally facing the operator will be painted by this electrostatically attracted material.

Process 2. It was recognized that the material that escaped deposition in Electrostatic Process 1 did so because the electrical forces were not able to adequately control the flight of the particles against the impulse imparted to them by the agency that was responsible for their formation. A second method, the so-called Electrostatic Process 2, was developed to overcome this deficiency. Figures 4 and 5 show schematic drawings of the principles of this method. Two photographs of installations for the process appear in the chapter "Applications Coatings." In this process the particles are formed, charged, and deposited all by the action of the electrostatic field. Its method of operation readily can be understood by realizing that certain basic electrical facts exist relative to the behavior of electrically charged bodies. These are that any two bodies that are at opposite potential--one positive and the other negative--are attracted toward each other. If they are free to move, they will respond to this attraction by moving toward each other. Likewise, any part of either body that is free to move will move toward the other body. The actual charge on a body will be concentrated on extended or sharp-edged portions of that body because that is where the charges on the body go in order to be as far as possible from other charges on the body that are of the same electrical sign. The force of attraction will be greatest on that part of the body where the charge is greatest. It follows, therefore, that the attraction will be greatest on those portions of the body that are at sharp extensions toward the other body.

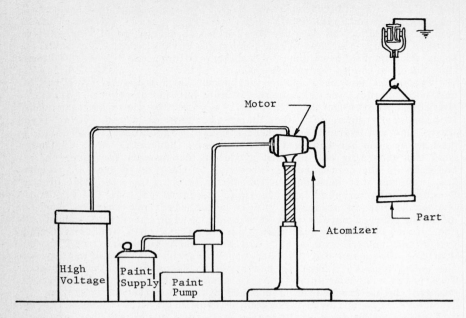

Figure 4.   Schematic of electrostatic bell spraying process 2.

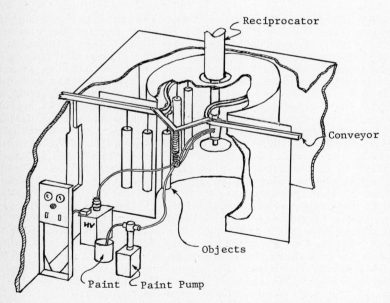

Figure 5.   Schematic of electrostatic disk spraying operation 2.

Understanding these concepts, we can now readily understand the operation of Electrostatic Spraying Process 2.   Two objects are positioned opposite to each other at a distance of about 1 ft.   One of these objects is the surface to be spray coated and the other is an "atomizing head" that is insulated from ground and that has a sharpened edge directed toward the surface.   An electrical voltage difference (approximately 100,000 V) is created between these two objects by connecting the atomizing head to the negative terminal of a high-voltage source.   The other terminal of the source is grounded as is the surface to be coated (6).

If under these conditions a liquid coating material is placed on the head so that it can flow to the sharpened edge, it will, when it arrives at the edge, become highly charged and will be attracted to the surface with a sizable force.   Under this attractive force, a portion of the liquid at the edge will be pulled out of the body of liquid as a cusp or liquid filament extending toward the surface. The small element of liquid at the end of such a cusp is very highly charged and, therefore, is experiencing a high attraction.   When the end portion of the filament is attracted to the surface with a force that exceeds the surface tension forces holding it in place, the portion will leave as a highly charged liquid particle.   This particle will fly directly to and be deposited on the surface.   Such a series of events occurs from each of the many cusps that are formed along the liquid edge on the atomizing head.   A composite spray of liquid particles leaving the head and depositing on the surface is thus formed.   Because particles do not leave the head until attracted and because they leave directed to a predetermined point, the deposition efficiency of this process is potentially 100%.

In the actual practice of this system of electrostatic spraying, the atomizing heads take a form and shape consistent with their intended use.   One such form is the rotating bell and another is the rotating disk.

Bell Process 2.   In the electrostatic bell modification, a member shaped like a bell or funnel is mounted on the hollow shaft of a rotation motor so that its wide mouth is directed toward the surface to be coated.   The bell edge at the widemouthed opening is sharpened.   The motor, shaft, and bell are all electrically conducting and are mounted on an insulating support.   When in use, they are connected to a high-voltage supply that raises them to a high voltage with respect to the surface.   The motor is energized and the bell set into rotation.   Liquid coating material is pumped at a controlled rate to the inside surface of the hollow shaft.   It flows to the inside surface of the bell at its apex.   Rotation causes the material to be formed into a film on the inside surface of the bell, and this flows forward to the outer sharpened edge.   It appears there as a thin extended exposed body of liquid.   At this point, it first comes under the action of the field.   The film is formed into a series of liquid extensions about the bell edge as above described.   From the ends of these cusps, a spray of highly charged particles is discharged.   This spray cloud is attracted to the surface where it forms a liquid layer.   By moving the surface on its support relative to the spray or by moving the atomizing head relative to the object by moving it on an automatic manipulator, the desired coating can be formed over the entire surface. In addition, film uniformity can be obtained much more readily if the objects are rotated on their point of support as they are carried past the spray station. The thickness of the deposited film is regulated by controlling the speed of the conveyor and carefully correlating this with the rate at which the coating material is delivered to the atomizer. Multiple bells may be positioned along the path of the conveyor so multiple applications can be made if heavier films are desired. The electrostatic bell system is being used by industry to finish all types of items. Single bell installations are used to finish small items such as toys while larger objects like refrigerators are finished by units employing as many as 24 or 30 bells. Practically speaking, there is no loss of coating material when this process is correctly installed and used. When solvent-reduced material is sprayed, this solvent leaves the liquid during spraying and curing so suitable ventilation must be provided about the unit to avoid solvent contamination of the surrounding atmosphere.

As originally introduced, Electrostatic Bell Process 2 was used primarily with solvent-based coating materials. The bells were rotated at the lowest possible rate consistent with obtaining good atomization of these materials because it was desired that the mechanical forces acting on the sprayed particles be held to a minimum. Rotation rates were on the order of 1800–3600 rpm. With the recent advent of high solids and waterborne materials that are much harder to atomize, it has become necessary to increase the rotation speed of the bells. Most bells used with this system today are rotated at 20,000–30,000 rpm. This increase does increase the size of the spray pattern emitted from the bell but does not materially decrease the overall transfer efficiency on those applications where the arrangement of the equipment has been made with this fact in mind.

As with all electrostatic systems, the object being coated with Bell Process 2 must be established as one electrode in the collection field. It must therefore have sufficient electrical conductivity to carry to ground the charge brought to it by the deposited coating material particles. As a result, this process finds its broadest use in the coating of metallic items. In .pa those cases where the process is used to coat items made of wood, plastic, or other nonconductors, the material must be somehow treated to render it conducting.

The concept of Bell System 2 has been incorporated into a handgun device. By suitable electrical design, it has been possible to place a handle on a bell-type atomizer so an operator can safely manipulate it over the surface to be coated consistent with producing a high-quality film on that surface. Such Electrostatic Handguns 2, because so little material escapes deposition when they are in use, are being used in commercial buildings for the in-place finishing of all types of metal office equipment. For example, a group of metal office furnishings can be cleaned and refinished at their point of use without the need for special precautions to avoid the overspraying of floors, walls, etc. Such units are also being used industrially to finish wrought iron furniture because they are at their best on such open tubular items where other spray methods are most inefficient.

Disk Process 2. The bell system produces a spray pattern that is doughnut shaped, reflecting the annular nature of the bell edge that is the origin of the spray. To blend several of these patterns together to produce a uniform film on an object surface often becomes difficult. To cope with this difficulty, the Electrostatic Disk Process 2 was developed. In this all the basic operational concepts of the bell system are utilized. The shape and construction of the atomizer are different. Here, as the name implies, the atomizer is a disk shaped element having a sharpened edge. The disk is mounted on the vertical shaft of a motor and positioned so it can be rotated with the face of the disk in a horizontal attitude. The disk and motor are positioned at the center of an almost complete loop in the parts-carrying conveyor. The disk and its motor are insulated from ground. The parts on the conveyor are grounded through their hooks. A high-voltage supply is connected to the disk to establish it at a high voltage (approximately 100,000 V) with respect to the grounded parts.

With the voltage applied and the disk rotating, coating material is pumped to the surface of the disk at its center. This material flows as a thin film to the outer sharpened edge of the disk. As before, this film edge is formed into a series of liquid filaments spaced about the circumference of the disk. These filaments become the source of a spray of charged particles. These travel radially outward from the disk and are attracted to any article on the conveyor within the loop. The conveyor motion moves the articles about the loop and so about the disk. They receive their coating in a sequential manner from all points on the disk circumference. If the parts are long as they hang on the conveyor, it is possible to mount the disk on a vertical stroke reciprocator and move it up and down along the loop axis as the parts are carried about the loop. The entire length of the parts is thus coated overall from a single disk. Each element of surface is subjected to the same spray pattern. Where such motion will simplify obtaining a uniform coating over the various surfaces of the parts, the parts can be rotated on their points of support as they are carried about the loop. The entire conveyor loop and the coating operation are surrounded by a cylindrical enclosure that acts as a spray booth for the operation. This enclosure is usually ventilated to contain the solvents escaping from the coating material.

As with the electrostatic bells, the originally installed disks were rotated at 1800-3600 rpm. Because of the introduction of new types of coating materials, presently installed disks are rotated at higher speeds (20,000-30,000 rpm). They are also smaller in diameter. Such equipment can handle both high-solid and waterborne materials.

The disk electrostatic system is perhaps the ultimate in electrostatic coating systems. It has a high application efficiency and great flexibility. It is being used to apply topcoat finishes to all types of household appliances, bicycle parts, toilet seats, and even golf balls. It represents the best means of getting the film-forming components of a liquid finish onto a surface so that they can be converted to a solid plastic film.

Powder Coating. General. The most ideal coating system would be one that would create a layer of plastic upon a surface with a minimum of material loss at a minimum of expense and with the simplest of application equipment. Until recently, most processes developed for application of coatings have contemplated the application of materials in liquid state because from such materials smooth, uniform continuous films of solids could most readily be obtained. Such liquid materials are always composed of film forming solids and diluting solvents. These solvents are used only as a means of preparing the solids for effective application to the surface. It would be most desirable if the materials could be applied as solids without dilution because then the expense of solvents could be saved and environmental pollution problems resulting from exhausted solvents would be avoided. The powder coating method of applying plastic films to surfaces allows this possibility (7).

Fluidized Bed. As originally practiced, the powder coating method was referred to as the fluidized bed method. In this technique, the solid resin to be applied is ground to the consistency of a fine powder that has reasonable uniformity of particle size. This powder is placed in the upper section of a two-section tank. The upper and lower sections of the tank are separated from each other by a porous plate. Air introduced into the lower section flows through the porous plate and aerates or fluidizes the powder in the upper section. The parts to be coated are heated to a temperature above the melting point of the resin. The heated part is then dipped into the fluidized powder. That powder that contacts the part fuses to it and hangs onto the surface. The part is then removed from the bed, and the retained sintered powder is further fused and flowed to a smooth film by heat retained in the part. If fusion is incomplete, the part can be put into a postheat oven and reheated to complete its flow. Heavy plastic coatings up to 50 mils thick can be produced on parts by this method. Wire baskets, wire fence, and similar parts are coated with polyvinyl chloride materials in large quantities by this technique.

Electrostatic Fluidized Bed. The powder in a fluidized bed can be charged by placing appropriate high-voltage electrodes within the bed adjacent to the porous plate that separates the two sections of the bed (Figure 6). Under the action of these charged elements, the powder in the bed becomes charged to an electrical sign that is the same as that of the electrode. A grounded article, which is at the opposite electrical sign of the electrode, when held over the bed will attract the powder to its surface. The powder actually jumps out of the bed onto the article without the article having to be immersed in the fluidized material. The article need not be heated with this technique because the charge retained by the particle after deposition will hold the particle in place for a sufficiently long time to allow the article to be carried into an oven where the retained powder is fused onto the surface as the coating.

Electrostatic Powder Spraying. The success of the electrostatic liquid application methods in the finishing field and the proven viability of the fluidized bed powder technique combined to suggest

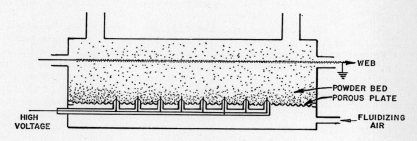

Figure 6.   Schematic of an electrostatic fluidized bed coating
device.

the possibility of using electrostatic methods for the application of powders. Electrostatic powder spray coating became a reality (Figure 7).

In this method the fine powder is placed in a fluidized bed. A venturi pump positioned below the surface of the expanded bed draws powder into the air stream flowing through a delivery hose. This powder-air mixture moves through the hose from the bed to an electrostatic powder gun. The gun is equipped at its forward powder outlet with a charging electrode similar to those used in the air atomizing-electrostatic charging liquid guns. As the powder leaves the gun it passes near the charging electrode and becomes charged. The carrying air stream transports the charged powder toward and about the article to be coated as a powder cloud. The particles in this cloud are attracted to the object surface. If the article has been preheated before being brought to the coating position, the deposited particles will melt upon contacting the surface and adhere to the surface. Subsequently, arriving particles will continue to adhere so long as the surface heat remains above the melt point of the resin. Heavy plastic films can thus be built upon such preheated parts. If flow is not complete when the spraying operation is stopped, the part can be postheated to complete fusion and flow.

If the part is introduced to the charged powder cloud while at ambient temperature, the powder particles will still be attracted to and deposited on the part surface. Peculiarly enough, under these conditions the powder will be retained on the part surface by the electrical attraction of the charged particles for the surface. No fusion at this point is required to accomplish adhesion. As the charged powder layer accumulates on the cold surface some of the charge is retained on the deposited powder. A charged surface is thus presented to oncoming particles. A condition will be reached at which further powder accumulation will stop. At this point the part is carried to an oven, its temperature raised to a point where the powder is melted, and the desired film produced. If the applied powder is a thermoplastic resin, heat need only be used to form the coating. If, however, the powder is a thermoset resin, the part may have to be raised to powder fusion temperature to produce melt and flow and then retained at an elevated temperature to allow the resin to complete its cure.

The electrostatic powder coating method has had great appeal to those industrial finishers who are having their liquid systems threatened by the air pollution regulatory agencies. Powder systems have also proved effective in certain applications on their own merits.

No better way exists to produce a thick, run- and sag-free plastic coating on an object. When the coating or powder applications are used, steps must always be carried out in a ventilated enclosure. The system is not 100% efficient so the oversprayed powder must be retained by a flowing air stream. This oversprayed powder--usually about 20-30% of that directed toward the object--is picked up and carried by the exhaust air to a filter system where it is removed from the air. Such a filter, or separation system, usually consists of a cyclone filter followed by a bag type filter. In some of the newer systems, the powder recovery system operates in conjunction with belts that move over

Figure 7. Electrostatic powder guns being used to apply powder coatings to aluminum chairs. Courtesy Ransburg–Gema.

the surface of the enclosing booth. The oversprayed powder collects on these belts and is removed from them by a vacuum system that carries the powder to a suitable separator. The recovered powder, if kept clean and uncontaminated, can be reintroduced into the coating system. The overall operation efficiency of such a powder system can be as high as 95%.

As appealing as it is, the electrostatic powder spray system of applying coatings has some inherent problems. It is difficult to change colors with such systems. The entire gun, booth, and collection system has to be purged of one color before another can be used if cross contamination of colors on the sprayed parts and in the collected powder is to be avoided. Fusion and flow of the applied powder to form thin (approximately 1 mil) smooth, continuous films are difficult to obtain particularly with pigmented powders. Heavier films are more readily obtained. A further difficulty is presented because it is difficult for the sprayman to know from examination of the applied powder film just what the fused film appearance will be. The handling of powdered materials also introduces some difficulties since they cannot be transported over large distances, from storage to use position, as can liquids. The powders being used must be at hand adjacent the application area. This causes considerable difficulty for the large-volume, multicolor user.

The principles employed in the electrostatic disk liquid application process have been applied to the application of powders (Figure 8). A disk located at the center of a loop in the parts carrying conveyor is adapted to distribute and charge the powder particles pumped to it from a fluidized bed supply. The charged powder leaves the disk edge and is carried toward the parts on the conveyor by an air stream and the attraction forces resulting from its charge. It accumulates on the part and is held there by the charge for a sufficiently long time to allow the part to be carried to an oven where fusion of the powder takes place.

As mentioned, electrostatic powder spray application methods are being used in industry. Many units of this type are being used to apply an exterior plastic coat to gas and oil pipe that will be placed under ground. Sewing machine body castings are given a heavy plastic coat of resin material to cover casting imperfections and to produce a smooth glossy external appearance. Automobile bodies have been coated and are being tested as part of a study to evaluate this system for the applications of top coats. As of 1982, one automobile manufacturer is using an epoxy powder as a primer surfacer on one of its truck lines, but there is no commercial application of powder topcoats to automobiles in the United States. Powder systems are being used to coat many other automotive components where the finish requirements are not so high. Washer and dryer components are being coated by this method as are the exposed parts of air conditioners. The process is satisfactory for many applications, and the range of its use will expand and become more definite as users become better acquainted with its merits and shortcomings.

Electrodeposition Coating Methods. A detailed description of electrodeposition is given by G. E. F. Brewer in a separate chapter of this book. In addition, a photograph of a commercial

Figure 8.   Electrostatic disk being used to apply powder coatings
            to sheet metal cabinets.   Courtesy G&R Electro–Powder
            Corp.

installation is given in the chapter on "Appliance Coatings" by T. J. Miranda. Only the essential elements of this process will be given here. In this technique waterborne material having extremely low vehicle solid content is placed in a tank. The object to be coated is immersed in the liquid so that all surfaces to be coated are wetted by the liquid. The object and the tank or other suitable immersed electrodes are connected to opposite terminals of a high-current low-voltage (up to 200 V) DC supply. Current passes through the solution and deposits on the object surface a semisolid plastic layer. One feature of this application is that the deposited layer due to its inherent resistive nature is self-limiting in thickness so most exposed surfaces will receive essentially the same thickness of deposited coating. The coated part is withdrawn from the bath and subjected to a shower bath of clear water to remove loose material. The part is then carried to an oven where the semisolid-deposited plastic layer is fused to a solid coating film. Although the process sounds very simple and straightforward, it really is quite complicated to apply with reproducibility and complete success. The coating material in the tank must be maintained in its best condition even though solids are being removed from it at all times during operation. Likewise, the deposition process brings about certain chemical changes in the bath that must be counteracted to maintain consistent operation. These difficulties have been overcome through the use of appropriate microfiltration processes. The heat generated by the passage of high currents through the bath is dissipated by the use of cooling systems in the bath circulation system. This is a very powerful application method having many admirable characteristics. It is finding wide use in the application of prime coats in the auto industry as well as in other areas where only a single coat is required. The system cannot be readily used for the application of topcoats since the prime coat layer is already sufficiently resistive to inhibit the operation of the process.

Conclusion

The various available coating methods have been discussed and the elements of their operation requirements presented. No effort has been made to be complete in any of these discussions since any one of them deserves a full chapter in their own right. No effort has been made to compare these methods for a particular application. Each new coating application that is encountered will have its own peculiar requirements that will have to be balanced against the capabilities of each method to arrive at a final choice. A great effort is being made to develop new coating materials to meet the ecological pressures being exerted against those material types that have been used for years. Waterborne materials have entered the market as substitutes for the solvent-reduced materials that have been widely used. Solvent coating material systems having as high as 80% film-forming solids have been developed and are being application tested. Two component materials that normally are liquid but which polymerize to solid films upon being mixed and applied are being used. The chlorinated "exempt" solvents are being presented to users as a means of meeting the demands of EPA while still using present coating systems. All these new materials and approaches will have to be applied industrially in order that they find effective use. Just which of the application methods described here will be able to be used with the new approaches and materials is still a matter to be considered. Each is a study in itself, which should bring together all interested parties and all information that is available. The reader is referred to Reference 9, which contains a full listing of suppliers.

It may be that new methods will have to be developed or these old ones modified to allow their use. It is likely that if any new material has advantages over presently used ones, application methods will be found. No ideal, universal application method exists. Each application must be evaluated in the light of requirements, restraints, and available methods before the best overall finishing system can be determined. Man's ingenuity knows no bounds in the finishing field as in all other branches of his endeavor. It conceivably could be that one of his thrusts, such as protecting surfaces by lamination procedures, precoated stock, or through the use of inert metal alloys, could effectively make it unnecessary to concern ourself with the application of coatings.

Literature Cited

1. "Finishing Materials and Methods"; Soderberg, G. A., Ed.; McKnight & McKnight: Bloomington, Illinois, 1952; Chap. 25.
2. Ind. Finishing 1944, 26, 46.
3. Manufact. Eng. 1981, Apr.
4. Ind. Finishing 1947, 23.
5. "ABC of Spray Painting Equipment", 3rd ed.; The DeVilbiss Co.: Toledo, Ohio, 1954.
6. "Electrostatics and Its Applications"; Moore, A. D., Ed.; Wiley: New York; Chap. 11.
7. "Fundamentals of Powder Coating"; Miller; Taft, Eds.; SME Publications: Dearborn, Michigan, 1974.
8. J. Paint Technol. 1973, 45 (587).
9. "1982 Directory"; Products Finishing; Gardner Publications: Cincinnati, Ohio, 1982.

Electrodeposition of Paint

GEORGE E. F. BREWER

Coating Consultants, Birmingham, MI 48010

```
Background
The Practice of Electrocoating
Equipment and the Industrial Painting Process
Advantages
```

In the late 1950s a group of chemical engineers of Ford Motor Co. began to experiment with a new paint application process that they called "electrocoating." The process resembles metal plating in as much as an electric direct current causes the paint to deposit on conductive surfaces (Figure 1). Quickly, the many advantages of the process were recognized and summarized as "Electrocoating... The greatest breakthrough since the invention of the spray gun...."

Background

Corrosion of metals results in enormous damage to our economy, and a great need exists for improved corrosion protection. A study of the deplorable junk piles reveals that discarded merchandise exhibits shiny, almost new looking surfaces in certain of its areas, while other sections of the same piece are completely corroded. Typical examples for the selective action of corrosion are automobile bodies. The roof of a car, the hood, and the trunk lid show usually very little corrosion, while other areas of a car body are completely destroyed.
  A study of cut open inner surfaces of new automotive bodies revealed that an incomplete paint coat existed inside those areas that subsequently corroded. The insufficient paint coat resulted from the inability of the spray painting process to reach highly recessed areas, like the insides of doors or the capillary recesses of butted joints or flanges. The dip-coating process does form a wet paint coat in all recessed areas, but during the subsequent

0097–6156/85/0285–0827$06.00/0
© 1985 American Chemical Society

paint bake the paint is partially removed through a phenomenon
called "reflux damage" or "solvent wash," due to temperature
differences between outer and inner layers of metal that cause vapor
condensation in certain areas of the merchandise.

What was needed was a painting process that allows the
application of corrosion-protective films even in most recessed
areas in virtual absence of solvents. The awareness of the
shortcomings of the then existing paint application processes led to
the concept and development of the electrocoating process, which is
a dip coating process using waterborne paint compositions. An
electric direct current deposits the paint solids (essentially
resins and pigments) on the electrically conductive surfaces of
merchandise. The migration of colloidal materials toward an
electrode of the opposite polarity is called electrophoresis, which
can be either cataphoresis (cathodic deposition) or anaphoresis
(anodic deposition).

The migration of colloidally dispersed particles in a direct
current field was reported as early as 1809. From then on the
phenomenon received attention only every 30 or 50 years as an
analytical method, culminating in 1948 when Arne Tiselius received
the Nobel Prize for his experimentations, particularly
electrophoretic separation of proteins.

Electrophoretic deposition seems not to have been observed
before 1905, and the earliest patented use of electrodeposition was
made in 1919 by Wheeler P. Davey, who deposited bituminous material
on electric wires as insulation. Subsequently, patents were granted
for electrodeposition of rubber latex (1923) and for deposition of
bees wax and other materials as protective coatings in food cans
(1937 and 1943). All of these processes for electrophoretic
deposition were using naturally occurring materials, and none of
these processes seem to have been in operation by 1950.

The technology that is now known as electrodeposition of
coatings, electropainting, electrocoating, etc., uses synthetic,
waterborne oligomers. (See References 1-10.)

The Practice of Electrocoating

Figure 1 shows the essential steps of the process: the merchandise
receives first the usual metal treatment and then enters the
electrocoating tank where it is electrically connected to one
polarity of a power source, while the tank or electrodes in the tank
are used as counterelectrodes. Within 1 or 2 min the desired film
thickness of usually 1 mil (25 μm) is formed. The current
consumption ranges from 2 to 3 A/ft^2. The process depends on the
existence in the coatings bath of about 5 to 20 wt % of film-forming
electrodepositable macroions of the general formula $RCOO^-$ for anodic
deposition and R_3NH^+ or R_3S^+ for cathodic deposition. The freshly
deposited coat consists of water-indispersible material, since the
action of the electric current has converted the macroions into
molecules, somewhat resembling the conversion of soluble metal ions
into insoluble metal atoms during the electroplating process. The
freshly painted merchandise is then rinsed with water for removal of
adhering bath droplets. More recently, some of the fluid is
separated from the paint by ultrafiltration, since the use of
ultrafiltrate as rinse fluid and subsequent return to the coating

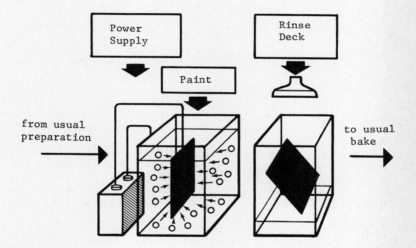

Figure 1.   The electrocoating process.

tank results in higher efficiency. The rinsed merchandise is then moved to the usual paint bake or cure, as indicated in Figure 1.

Thus, an existing metal finishing department can be converted from spray painting or dip coating by replacing the paint transfer equipment with an electrodeposition tank plus a rinse deck.

Electrocoating materials are acid oligomers RCOOH or more frequently $R(COOH)_n$ for anodic deposition or R_3N for cathodic deposition. These oligomers are water insoluble but can be dispersed in water through the action of acids or bases, used as external solubilizers. The solubilization and deposition processes can be symbolized as follows:

$$\begin{array}{ccccccc}
& & & \text{dispersion} & & & \\
\text{water} & & \text{aqueous} & \overrightarrow{} & \text{dispersed} & & \text{counter} \\
\text{insoluble} & + & \text{external} & \overleftarrow{} & \text{macroions} & + & \text{ions} \\
\text{oligomer} & & \text{solubilizer} & \text{deposition} & & & \\
\end{array}$$

$$RCOOH \;+\; BOH \;\rightleftharpoons\; RCOO^- \;+\; B^+ + aq$$

$$R_3N \;+\; HX \;\rightleftharpoons\; R_3NH^+ \;+\; X^- + aq$$

It may be noted that the resinous moiety "R" may contain the chemical groupings characteristic for practically any known film former, such as acrylics, alkyds, epoxies, phenolics, polyesters, etc. The electrodeposited cured films exhibit essentially the same properties as their basic resins. Thus, there is a large variety of formulations possible.

Electrocoating paints are usually sold as approximately 40 wt % dispersions of paint nonvolatiles in the presence of 10 wt % volatile organic cosolvents and 50 wt % water. The paint solid concentration in the coating tank is selected to give the best performance and varies for individual installations from 5 to 20 wt %.

As a guideline, most electrocoating paints require approximately 50 C for the deposition of 1 g of paint, or 1 Faraday for 2000 g. The typical molecular weight of these oligomers is approximately 10,000, and not all of the ionizable groups are neutralized, since a high degree of neutralization results in an increased "wash off" of the freshly deposited resin, while a low degree of neutralization reduces the stability of the aqueous dispersion.

The oligomers deposit on the electrode of opposite polarity, while the counterion is discarded or reused for solubilization of replenishment resin (see Figures 2 and 3).

The electrodeposition process is currently used for the coating of merchandise ranging from structural steel, automotive bodies, furniture, coil stock, appliances, and toys to nuts and bolts.

The use of cationic resins of the R_3NH^+ type is widely advocated, since their cured films result in high corrosion protection. A new type of cationic resin is based on "onium bases," particularly sulfonium bases, $R_3S^+ OH^-$. These do not require the control of an external solubilizer and lose their ionizable group during deposition on the cathode.

The properties of the final, polymerized film will largely depend upon the chemical nature of R. Thus, if the R carries many

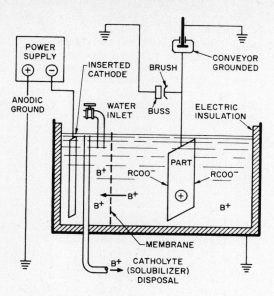

Figure 2.   Solubilizer removal (completely solubilized feed).

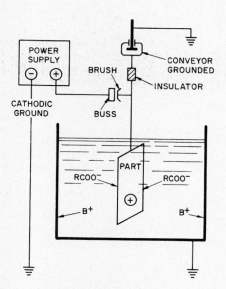

Figure 3.   Solubilizer reuse (solubilizer-deficient feed).

epoxy groups, the final film will have essentially the properties of epoxy spray or dip primers; if the R carries many acrylic groups, the electrodeposited film will have conventional acrylic coat properties.

The paint formulator has additional latitude in the choice of resins because dispersed macroions can be used as emulsifying agents for a variety of resins. These, as well as many pigments, will electrodeposit together with the macroions.

In addition to water dispersibility, the process requires other special properties: low redispersibility of the freshly deposited film, since the work piece, on removal from the bath, will be covered with bath droplets that tend to redissolve the fresh deposit; lower resin viscosity (Gardner Scale Z3, for instance), when compared with conventional paint resins, since a desirable electrodeposit is virtually free from viscosity reducing solvents; comparatively high electrical resistance (10^7–10^9 ohm cm) of the freshly deposited film to obtain "throwing power"; designing of the contact of counterion and the counter electrode to produce soluble products that do not interfere with the continuing operation. If, for instance, in the cathodic deposition of a film-former R_3NH^+, acetate ion CH_3COO^- is used as the counterion, water-soluble acetic acid will be generated near the anode, while oxalate ions will similarly give rise to the liberation of CO_2. In the case of anodic deposition of $RCOO^-$ as film-forming macroions, the counterions carry positive electric charges (B^+). Alkali metal ions, ammonia, and particularly organic amines are most widely used as counterions.

The most unconventional yet most important feature to be built into electrocoating materials is "throwing power", or the ability to form electrodeposited films of such uniform thickness that even the most recessed areas of a workpiece are covered.

Electrocoats are known to deposit first on areas that are closest to the counterelectrode. However, the electrodeposit is of high electrical resistance, and the current has to seek the nearest still available path, namely, bare metal, until all the metal surface is covered.

The original electrical resistance between two electrodes immersed in the coating bath is comparatively low, but the resistance rises during the deposition process to 100 and more times its original value, with a related drop in current flow. Thus, paint is deposited quickly until all bare metal areas are covered. The film build is self-limiting: areas of thinner film offer less resistance against current flow, so deposition continues there until the film thickness has evened out.

The application of higher voltage produces higher film thickness and more throwing power. There exists, however, for each individual paint a maximum applicable voltage: beyond this point the freshly deposited film ruptures, causing unsightly blemishes, lowered corrosion protection, etc.

Other factors that increase throwing power are higher bath conductivity, which makes it easier to get electricity into recesses, and higher electric equivalent weight of the paint, which gives more deposit per unit of electricity (coulomb).

Equipment and the Industrial Painting Process

Metallic objects are first cleaned to remove impurities and materials associated with metal forming, welding, temporary rust protection, etc. In many instances phosphate and/or chromate treatments are applied. Overall, any electrically conductive surface preparation will impart benefits to electropainted articles similar to the benefits imparted to spray painted or dip-coated articles. The workpieces go from pretreatment to the electrocoating tank either water wet or dried.

Most baths are operated at 5-20% nonvolatile concentration. A low solid level is desirable, since the paint loss on drag out is smaller and the paint solids dwell a shorter time in the bath (called faster "turnover"). The bath pH ranges from 3 to 6^+ for cathodic deposition and from 7 to 9 for anodic deposition.

The majority of electrocoating installations take the merchandise through two to nine stages of cleaning and phosphating under <u>continuous movement</u> through the electrodip, rinse, and bake, followed by spray coating if so desired.

Energy savings are accomplished by so called batch-type operation. Here one or more parts are located over a dip tank and enter through <u>vertical downward</u> motion. The parts exit vertically upward and are then moved horizontally until located over the next tank.

Very small parts, like fasteners, are coated in bulk, sometimes through the use of three short endless conveyors. The first conveyor receives the phosphated pieces and drops them into the electrocoating tank; a second conveyor located in the tank and completely submerged is electrically energized to accomplish the electrodeposition. The pieces drop then onto a third conveyor, which moves the pieces upward and out of the bath.

Symmetrically designed pieces, like cans, are sometimes held and rotated by a mandrel. An electrically energized and perforated tube supplies the paint to the vicinity of the surfaces to be coated.

Electrocoats are also applied by fast-running coil coaters (600-1000 ft/min).

The majority of installations uses conveyorized tanks. Recently, however, batch-type tanks are sometimes selected for small- and medium-sized production.

Tanks may be classed into two groups. (a) The tank wall is lined with an electrically insulating coat (Figure 2), while the counterelectrodes are inserted in the tank and then positioned according to size or shape of workpiece. (b) The tank wall is used as counterelectrode (Figure 3). When lined tanks are used (Figure 2), the workpieces can be grounded through the conveyor. In many cases, however, the hanging device (paint hook) carries an electric contactor (brush) sliding along a grounded rail (bus bar) to ensure electrical ground.

In any event, electrical insulation has to be provided between the positive and the negative sides of the system. In the case of the lined tank, it is the "liner" that insulates the bath fluid from the ground. If the entire tank wall is used as an electrode, it is grounded and an "insulating link" is provided that separates the upper, grounded section of the paint hook from its lower part. The lower insulated section of the hook is in electrical contact with the power source (Figure 3) and energized only inside the tank enclosure.

Pumps, draft tubes, line shafts, and ejector-nozzle systems capable of moving or turning over the entire bath volume in 6–30 min are used to prevent the paint from settling in the tank.

As a rule, 50–75 μm pore size filters are used to pass the entire paint volume through the filter in 30–120 min.

The feed materials are manufactured and shipped at paint solid concentrations ranging from 40% to 99+%. In some installations, the feed is metered into the tank in the form of two or more components, one component being the resin, the other component being a pigment slurry, etc.

To keep a bath in operating condition, <u>removal</u> of leftover solubilizer is accomplished through electrodialysis (Figure 2), ion exchange, or dialysis methods.

If the dispersed, electrodepositable resin is symbolized as $RCOO^- + Y^+$, it is implied that the $RCOO^-$ moiety will be removed from the bath in the form of a paint coat that is anodically deposited on the workpiece. If the solubilizer moiety Y^+ would remain in the bath, it would accumulate and eventually interfere with the coating operation.

The electrodialysis method uses tanks lined with an insulating layer. From 5 to 30 properly spaced counterelectrodes are surrounded by membranes (Figure 2), which separate the coating bath from the counterelectrode by means of a membrane permeable to the solubilizer Y^+. A plumbing system flushes the solubilizer out of the counterelectrode compartment (Figure 2).

A widely practiced method is the <u>reuse</u> of leftover solubilizer for the dispersion of the feed material. In this case, the feed consists of partially solubilized resin that is reacted with a solubilizer-rich bath through the use of an ultrasonic or a high-shear homogenizer.

An interesting variation of the electrocoating process, the so-called exhaustion method, has been developed for bulk coating of fasteners, electronic components, fine mechanical components, etc. These are used for very small tanks which use very little paint; thus, the cost of monitoring the tank for replenishment is comparatively expensive, while the cost of discarding the tank fill is low. Modified electrocoating paints are, therefore, placed into the tank at approximately 12% paint solids and 35% neutralization. The solids are gradually used up through painting until only 4% solids at 70% neutralization remain. The tank is then refilled with fresh paint at 12% solids.

The fill of an electrocoating dip tank is actually an inventory of diluted paint. Concentrated paint is added to the tank to replace the paint that is removed from the tank on the surfaces of painted merchandise. The time interval in which the paint additions to the tank equal the original fill is called the turnover period. Electrocoating tanks are reported to have turnover periods ranging from several days to several months. Since both freshly added paint and older paint are coated onto the surfaces of merchandise, sizable quantities of paint reside in the tank for long periods, as predictable from the law of averages.

Electrocoating tanks are heavily agitated to prevent pigment settling. Thus, high-shear stress resistance is required.

Furthermore, as in all dip processes, the paint is subject to a long residence period in the tanks requiring high resistance against saponification, oxidation, bacteria, etc. Experience tells that electrocoating baths should have a pumping stability exceeding four turnover periods.

In general, a higher pigment/binder ratio is observed in the anodically deposited films than exists in the alkaline bath. A typical bath may, for instance, contain 25 g of inorganic pigment combined with 75 g of resin, while the film deposited from this bath may contain 28 g of pigment with 72 g of resin. It is easy to maintain the bath in proper working condition by feeding a composition containing 28 g of pigment plus 72 g of resin. Similar differences between bath and film may exist regarding other bath components.

In other words, the bath is formulated to give the desired film properties, while the feed is formulated to replace the materials that were coated out or otherwise removed from the tank.

Rectifiers that deliver direct current of less than 5% ripple factor are usually specified. Various output voltage controls are in use, such as tap switches, induction regulators, saturable core reactors, etc. Voltages in the 50–500 V range are usually provided. The current requirement is calculated from the weight of coating to be applied in the available time. For example, if 1 lb of coating is to be deposited in 2 min, by using a paint that requires 75 C/g, then an average of 453 g x 75 C/g / 120 s = 283 A is required.

Entering merchandise demands a large amperage that diminishes as the insulating paint coat forms. The current fluctuations in the tank can be mitigated through the incorporation of current limiting devices. Suppose that an entering workpiece produces a peak current draw of 705 A; if the current draw is limited to 500 A, a smaller power source but a somewhat longer coating time will be required. Practically all of the applied electrical energy is converted into heat. Cooling equipment must be adequate to maintain the desired bath temperature, usually between 70 and 90 °F as specified by the paint suppliers.

The cleaned or pretreated workpieces may enter the tank carrying an electric charge (sometimes called "live entry," "energized entry," or "power in"). Figure 4 shows the time/amp relation observed in the case of "live entry" of workpieces.

Suppose that the conveyor line enters one workpiece into the tank with a charge of 250 V. Suppose further that it takes 30 s for complete submersion. The current draw will rise during that time (Area A). If the workpiece remains submerged in the bath for 1 min, then the current will diminish during that period (Area B) and then fall rapidly when the workpiece leaves the tank.

Workpieces may enter the tank without applied voltage and, therefore, without current draw (sometimes called "dead entry"). When the submerged pieces are then energized, a large current surge will be observed. This may be mitigated through gradually raising the applied voltage or through successively engaging different power sources with built-in maximum amperage settings in addition to maximum voltage settings (see right half of Figure 4).

Although "energized entry" at full-coating voltage is preferred, some electrocoating baths, on energized entry, produce merchandise showing peculiar patterns or streaks, called "hash marks" or

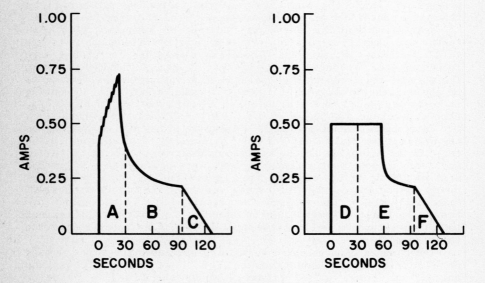

Figure 4.   Time–amperage study.

"wisps." In this case, pieces are entered into the bath while unenergized or low energized to eliminate these defects.

Users of the electrocoat process provide for various methods of energizing through the installation of two or more smaller power sources rather than one source of the full required size. Multiple power sources also protect against complete shutdown in case of breakdown of one of the power sources.

Amperage limitation, current cycling, or intermittent current application lengthens the required coating time, since it is the applied ampere seconds (Coulombs) that produce the electrodeposit.

Current consumption ranges from about 15 C/g of finished coat up to 150 C/g. After an initial amperage surge, the high electrical resistance of the freshly deposited film diminishes the current flow, resulting in an overall requirement of 2–4 A/ft^2 for 1–3 min, or between 1 and 3 kWh/100 ft^2.

The coating time ranges usually from 1 to 3 min. For some special work, such as wires, steel bands, etc., coating times as low as 6 s are reported.

The voltage requirement is largely dictated by the nature of the dispersed resin in the bath. Installations are usually operated at between 200 and 400 V, although some are reportedly operated as low as 50 V and others as high as 1000 V.

Freshly coated pieces, when lifted from the bath, carry bath droplets and even puddles of paint. A high concentration of paint solids exists in the vicinity of a workpiece that is being coated. It is estimated that an automotive body may carry about 1 gal of bath. At 10 wt % nonvolatiles this is approximately 1 lb of solids. Considering the migration of solids toward surfaces that are being coated, solid concentrations of up to 35% are expected in their vicinity. Thus, it is evident that the recovery of the lifted paint bath is necessary, and a lucrative way has been found in the form of an "ultrafiltrate rinse" or ultrafiltrate dip.

Ultrafiltration uses membranes that allow the passage of water and truly dissolved substances, such as solvents, solubilizers, salts (impurities), etc. Dispersed paint resins, pigments, etc., are retained by the membrane. One hundred or more gallons of bath pass on one side of the membrane under pressure, while 1 gal of clear aqueous fluid passes through the membrane. The fluid, called permeate or ultrafiltrate, is collected and used as rinse fluid. A three-stage rinse system recovers approximately 85% of the paint solids that were lifted from the bath.

Some of the ultrafiltrate (permeate) is occasionally discarded to remove low molecular weight ionized materials. The high biological oxygen demand of the permeate makes its disposal expensive. Subjecting the ultrafiltrate to reverse osmosis (RO) lowers the disposal cost for the RO concentrate, while the RO permeate is returned to the rinse tank.

The conventional type of bake oven is used. Electrodeposited coats are usually baked from 5 to 30 min at temperatures varying from 225 to 400 °F. The air velocity through the oven can be comparatively low, due to the very small quantities of organic volatiles in the paint coat. The time/temperature requirements are dictated by the resin system and are similar to those required for conventional dip or spray paints—usually 5–25 min at 225–400 °F air temperature. Ambient temperature drying electrocoats are on the market.

Advantages

The speedy acceptance of electrodeposition is due to the fact that it combines many advantages of other painting methods with new and desirable features: Formation of protective films in highly recessed areas such as cavities, box sections, creases, and flanges results in very much improved corrosion protection. High transfer efficiency results in up to 50% lower paint consumption. Use of water as practically the only carrier virtually eliminates the fire hazard. Low paint bath viscosity--approximately equal to that of water--results in ease of bath agitation and pumping and allows fast entry and drainage of merchandise. Freshly deposited coats allow rinsing and some handling immediately and have the lowest known tendency to sag or wash during cure. A second coat, or color coat, of waterborne or solventborne spray paint can be applied directly over the uncured or semicured electrocoat. Overall savings, including materials, labor, facilities investment, electrical power, etc., are reported to reach 20-50% when compared with spray, electrostatic spray, or dip-coat painting.

Literature Cited

1. Raney, M. W. "Electrodeposition and Radiation Curing of Coatings"; Noyes Data Corp.: Park Ridge, N.J., 1970.
2. Yeates, R. L. "Electropainting"; Robert Draper Ltd.: Teddington, England, 1970.
3. Brewer, G. E. F.; Hamilton, R. D. "Paint for Electrocoating"; ASTM Gardner-Sward Paint Testing Manual, 13th ed.; ASTM: Philadelphia, 1972; pp. 486–89.
4. "Electrodeposition of Coatings"; Brewer, G. E. F., Ed.; ADVANCES IN CHEMISTRY SERIES No. 119, American Chemical Society: Washington, D. C., 1973.
5. Machu, W. "Handbook of Electropainting Technology"; Electrochemical Publication, Ltd.: Ayr, Scotland KA7 1XB, 1978.
6. Brewer, G. E. F., Chairman. "Organic Coatings and Plastics Chemistry"; American Chemical Society: Washington, D.C., 1981; Vol. 45, pp. 1–22, 92–113.
7. "Advances in Electrophoretic Painting"; Chandler, R. H., Ed.; Bi- or Triannual Abstracts since 1966; Braintree: Essex, England.
8. "Electrodeposition Processes and Equipment"; Duffy, J. I., Ed.; Noyes Data Corp.: Park Ridge, N.J., 1982.
9. Kardomenos, P. I.; Nordstrom, J. D. J. Coat. Technol. 1982, 54(686), 33–41.
10. Robison, T., Organizer of "Electrocoat" Conferences (even years, since 1982), Products Finishing, Cincinnati, Ohio.

Curing Methods for Coatings

VINCENT D. McGINNISS[1] and GERALD W. GRUBER[2]

[1]Columbus Laboratories, Battelle, Columbus, OH 43201
[2]PPG Industries, Allison Park, PA 15101

IR Drying
Radio Frequency and Microwave Radiation
Light Energy Radiation
  Light Source
  Photoinitiators
  Free Radical Initiated Radiation Curable Coating
    Compositions
  Acid Intermediate Initiated Radiation Curable Coating
    Compositions
Electron Curing
Glow Discharge Polymerization

In conventional organic solvent based coating systems, a preformed functional polymer component is dissolved or dispersed [nonaqueous dispersion (NAD) technologies] in an organic solvent (30-80% solids), and a cross-linking oligomer and various flow agents, catalysts, pigments, etc., are added to make up a complete coating formulation. The coating formulation is applied to a substrate by conventional methods, for example, spray, roll coating, flow coating, etc., and subsequently cured in gas or infrared thermal oven equipment. Curing of conventional solvent-based coatings involves both solvent removal and thermal initiation of chemical reactions between the preformed functional polymers and cross-linking oligomers that are involved in developing final film properties through three-dimensional network formation of cross-link sites. Similar analysis can be established for waterborne polymer solutions or dispersions (emulsions) in that the solvent (water) must be removed followed by polymer-polymer or polymer-cross-linking oligomer interaction for formation of network structures (1-4) (Figure 1).

The development of high performance coating systems requires the use of special energy-related chemical reaction mechanisms in order to effect the formation of complex network structures. These network formation reactions take place between high molecular weight

0097–6156/85/0285–0839$06.00/0

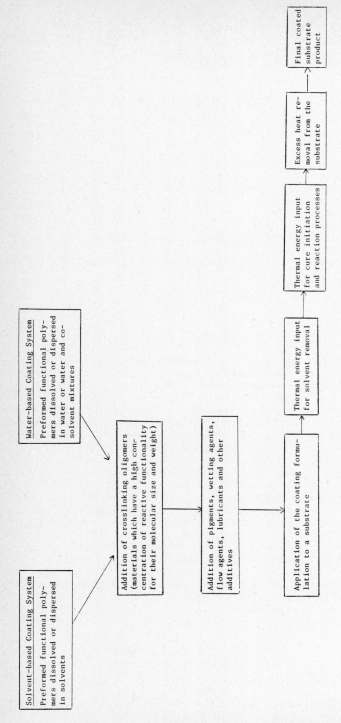

Figure 1. Conventional thermally cured coating systems and processes.

polymer chains containing reactive functional groups or between functional polymer chains and cross-linking oligomers that have a very high concentration of functional groups per molecular weight and molecular size dimensions (figure 2). Examples of typical functional polymer systems, cross-linking reactions, and mechanisms commonly used in the coating industry are listed in Table I.

The cross-linking or curing reactions associated with many of the polymer systems listed in Table I usually require the use of elevated thermal energies to effect formation of three-dimensional network structures and high-performance coating properties in a time frame useful enough to be commercially acceptable for a wide variety of industrial applications. A major step in reducing paint or coating drying times was the installation of industrial drying ovens. Present-day convection ovens are suitable for continuous or batch operations and can accommodate a wide variety of object sizes and shapes. However, diminishing natural gas supplies and environmental considerations are lending impetus to the development of alternative methods and alternative energy sources for curing coatings. Some of these alternatives to present-day convection oven drying of paints are listed in Table II (9).

In understanding how the various energy sources and curing methods work, it is helpful to recall that energies, on the order of a few electron volts, are required to break a chemical bond or raise a molecule to an electronically excited state. Thus, it is clear that infrared, microwave, and radio frequency radiation produces chemistry similar to what one would expect from heating a coating in a convection oven. Ultraviolet and accelerated electron chemistry can be characterized by bond breakage and subsequent polymerization resulting from electronic excitation or ionization processes.

IR Drying

IR is produced by any warm object. Its energy distribution depends upon the nature and temperature of the heated object. Dominant IR wavelength (emitted by a tungsten filament) as a function of temperature were 1.45, 1.15, 0.965, and 0.905 μm at 2000, 2500, 3000, and 3200 °C, respectively (10).

The intensity of the radiation is a function of the fourth power of the temperature of the heated object. This means that a radiant heat source operating at several hundred degrees will have a high rate of heat transfer toward a substrate at temperatures up to a few hundred degrees, providing the substrate is not highly reflective or transparent toward the emitted infrared. In practice, a filament is heated by gas or electricity and the radiant energy is directed toward the substrate by means of reflectors.

Since utilization of IR depends upon absorption, it becomes essentially a line of sight process. Of course, a highly conductive substrate (e.g., metal) will distribute the thermal energy out of the line of sight. Conversely, an insulating substrate (e.g., wood) will do little to distribute the heat, and hot spots can develop if exposure is nonuniform.

While most organic coatings absorb somewhere in the IR, there are advantages to its selectivity. Leather is largely transparent in the region from 1 to 1.5 μm. This makes it possible to selectively heat glue on the inside of a shoe and cause it to dry

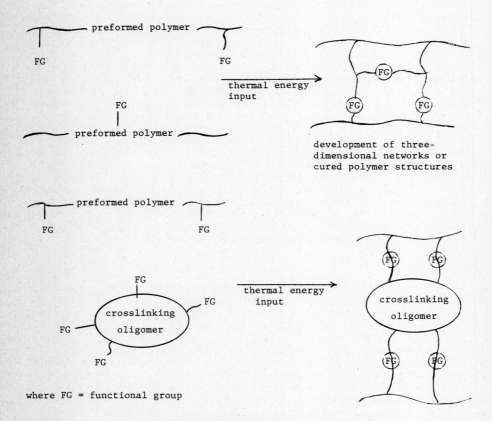

where FG = functional group

Figure 2. Idealized polymer curing (cross-linking) reactions.

Table I. Typical Polymer Cross-Linking Oligomer Curing Reactions Used in the Coatings Industry

Polymer Systems	Cross-Linking Oligomers	Functional Groups	Cross-Linking Reaction (Mechanisms)	Reference
Air-drying oils or polyester resins (single component)		Conjugated double bonds and other unsaturation sites	Air oxidation, peroxide formation, and coupling between polymer chains (free radical)	5
Polyesters, acrylics, epoxy (single—component systems)	Urea formaldehyde and melamine resins	Carboxyl and hydroxyl groups—methylol and ethers of methylol groups	Transesterification or transetherification reactions between polymers and cross-linking oligomers (condensation)	6
Epoxy (two—component mixtures)	Polyfunctional amines or polyfunctional carboxyl materials	Oxirane, primary, and secondary amines, carboxy groups	Addition followed by ring opening	7
Polyurethanes (two—component mixtures)	Polyfunctional iso—cyanates and hydroxyl—containing polymers	Isocyanates and hydroxyl groups	Addition of hydroxyl groups to the isocyanate to form urethane linkages	8
Polyurethanes (single—component system)	Block polyisocyanates and hydroxyl polymers	Phenol-blocked isocyanates and hydroxyl groups	Unblocking or removal of phenol to form a free isocynate followed by addition of hydroxyl groups to form urethane linkages	8

Table II. Alternative Energy Sources to Gas Fuel Convection
           Ovens for Cure of Coatings

Energy Source	Energy (eV)	Mechanism
Infrared	10^{-1}	Thermal
Microwave	10^{-3}	Thermal
Radio frequency	10^{-6}	Thermal
Visible	1	Electronic excitation
Ultraviolet	5	Electronic excitation
Electron beam	10^{5}	Ionization excitation
Gamma-ray	10^{8}	Ionization excitation

fast enough to be an inline process (10). This is perhaps the most unexploited attribute of IR. The emergence of infrared lasers has only recently provided highly monochromatic sources of IR energy with high efficiency, particularly in energy regions where thermal emitters are inefficient. For example, CO_2 lasers emit at 10.6 μm and have powers of several kilowatts (11).

Actually, this suggestion serves to introduce the subjects of microwave and radio frequency radiation. RF radiation is being used to cure adhesives on laminates for furniture and for drying coatings, while microwave has been used to vulcanize rubber, convert plastics, and dry inks.

Radio Frequency and Microwave Radiation

Radio frequency radiation and microwave radiation cause coatings to dry or cure by thermal activation. The most important mechanism of activation involves rotation of polar molecules so as to align their dipoles in an electric field. The rate at which electrical energy can be dissipated in a dielectric material is proportional to the frequency of the energy and to the square of the electric field strength. The relationship is expressed in the equation (12)

$$P = Kf(E')(\tan \delta)(E^2)$$

where P = rate of energy dissipation (watts), f = field frequency (hertz), E' = substrate dielectric constant, $\tan \delta$ = loss tangent or power factor, and E = field strength (volts per centimeter).

The dielectric constant and loss tangent are the substrate variables. The loss tangent is apparently related to variables such as the molecular weight, viscosity, and conductivity of the material to be dried. Some examples will serve to illustrate the difficulty in predicting the rate at which a material will be heated.

Water absorbs microwave energy about 10^3 times better than ice. This is apparently due to the crystal structure of ice inhibiting rotation of the water molecules. Distilled water is a polar insulator, and while it absorbs microwave rather well, a 0.5 molal sodium chloride-distilled water solution absorbs 50% more efficiently. Carbon is nonpolar, but its conductivity is such that it is a respectable microwave absorber. In practice, researchers have elucidated many of the principles involved in microwave and RF energy use and one can formulate an ink or coating with these principles in mind, but the ultimate test is the rate at which the formulated product interacts with microwave energies.

Microwave generation is somewhat expensive, but at least one economic analysis has suggested that RF radiation compares most favorably with IR in that increased energy efficiency leads to substantially lower energy costs (13).

Light Energy Radiation

The interaction of light energies (visible and ultraviolet) on photosensitive liquid materials leading to their hardening or curing has been known since ancient times but only until recently has this process found commercial success in major industries such as photoresist manufacture, printing, wood finishing, dental composites, metal decorating, vinyl floor coatings, and adhesives.

The concept of light energy radiation curing or photocuring of coatings can be divided into five basic segments: (1) A stable light source is needed--one capable of producing ultraviolet wavelengths of light or visible light (near- and far-UV, 200–400 nm; visible, 400–750 nm) with sufficient power or intensity to be commercially feasible. (2) A photoinitiator or photoactive catalyst capable of absorbing ultraviolet radiation, at appropriate wavelengths of energy emitted from the light source, is needed. (3) Active free radical intermediates or strong acid species must be produced through the action of light absorption by the photochemically active photoinitiator. The free radicals initiate polymerization of unsaturated monomers, oligomers, and polymers while the photochemically released strong acid compounds initiate ring opening polymerization reactions of epoxy resins and other monomers, oligomers, and polymers sensitive to cation-induced polymerization reaction mechanisms. (4) Unsaturated (high boiling acrylic, methacrylic) monomers, oligomers, cross-linkers, and low molecular weight polymers make up the fluid, low viscosity, free radical light-induced curable coating system while low molecular weight epoxy monomers and polymers make up the cationic light-induced curable coating systems. (5) After free radical or cationic initiation of the reactive liquid coating, the monomers, oligomers, and polymers propagate into a fully cured, cross-linked solid coating or film (14).

Light Source. Early UV processing of coatings involved the use of low-pressure mercury lamps for curing styrenated polyesters. Because these lamps emit only 1 W or less of ultraviolet per inch of arc length, they soon gave way to the so-called medium-pressure mercury arcs. In general, these lamps operate at 200 W/in. input power with UV output being in the region of 30–50 W (15). This type of lamp is offered by several manufacturers, and it has become the industry standard. These arcs have a lifetime of a few thousand hours, and about 50% of the input power emerges as infrared. But excessive substrate heating can be a problem. The arc must operate at greater than 600 °C, but at the same time the arc electrodes must be cooled to below 300 °C. Clearly temperature control around the arc is a serious problem. It has been stated that about 1000 CFM of air is required to cool three 40-in. lamps (16).

Temperature control around a mercury arc is necessary, and an obvious way to do this is to cool with air. Because oxygen inhibits free radical polymerizations, air cure does not afford the superior properties required for many end uses. Much progress has been made in this area, and indeed most of the present-day UV curing lines operate in air atmosphere; however, new developments in UV hardware are now providing alternatives. One alternative system utilizes water cooling of electrodes and lamp reflector housings so that minimal amounts of nitrogen can be used as an inert blanket in order to exclude oxygen and produce a very high quality coating finish. Other systems have made use of special infrared absorbing filters or combinations of inerted low-pressure lamps and medium-pressure lamps that operate in air. The low-pressure lamps require only nominal cooling, which allows for modest nitrogen use. In such a system, the low-pressure source dries the surface of a coating. This provides an oxygen barrier so that the final stage can achieve the bulk cure in air.

A high-intensity, electrodeless ultraviolet source has recently been introduced. This UV source is distinguished from other commercially available UV lamps by the fact that no electrodes are inside the lamp. Electrical energy is supplied in the form of radio frequency power that is coupled into an evacuated quartz tube containing mercury and other additives. As compared to conventional medium-pressure mercury arcs with electrodes, the electrodeless system has several attributes that make it unique (17). These are summarized as follows: (1) instant on-off; (2) modular lamp design; (3) increased lamp lifetime; (4) variable spectrum; (5) excellent energy conversion.

One of the disadvantages of electrode lamps is that they take about 3 min to warm up from a cold start and about 5 min to restart after being turned off. Electrodeless lamps need no more than 15 s to come to full power from a cold start or 60 s from a hot start. This hot start time can be reduced to instant starting (less than 1 s) if required. Presumably this would alleviate the need for shutters on commercial curing lines.

A unique characteristic of the electrodeless system is the modular lamp construction. The basic lamp is 10 in. in length, and because there are no electrodes, it emits from the entire length. Thus, a 40-in. web can be cured by butting four 10-in. modules next to each other.

The elimination of glass-metal seals results in a simpler, less stressed bulb. As a result, the electrodeless lamp potential lifetimes should be superior to the electrode type. In fact, a mechanism for failure has not yet been identified.

Because there are no electrodes in the radio frequency discharge lamp, less care need be taken to avoid fill materials that interact unfavorably with them. Additives can be used in small or large quantities to vary the spectral output distribution. The same basic lamp system and power supply can be used with different fills to provide a wide range of spectral characteristics, including sources with enriched high- or low-energy photon outputs. There is also a wider range of mercury pressures easily available with electrodeless lamps, so that various ratios of line to continuum radiation can be obtained. It would seem that these variables would be of considerable value in promoting the cure of titanium dioxide pigmented coatings.

Because the electrodeless lamp can be cooled uniformly, it can be made to run cooler than the electrode type, but at the same time it can be driven at higher power. This allows for lamp input powers up to 600 W/in. as compared to the conventional 200 W/in. of electrode lamps, without any serious degradation of lamp lifetime. Because the plasma inside has a higher electron temperature, the spectrum is shifted further toward ultraviolet. With the same fill, the same lines and continuum appear as with an electrode lamp but the output in the UV is much greater. The actual UV output from a 320 W/in. electrodeless lamp is 117 W/in. This represents a 36% energy efficiency as compared to about 20% for 200 W/in. electrode arcs. Of course, it must be added that considerable energy losses occur in converting ac to radio frequency.

Many other types of light sources can also be used for photopolymerization reactions, for example, low-pressure mercury arcs, flash lamps, fluorescent lamps, tungsten halide sources, and even lasers. A complete review of light sources used in photopolymerization reactions can be found in Reference 18.

Photoinitiators. Many theories of photoninitiated polymerization reactions with different light-sensitive catalysts have been reviewed in References 19-21. There are, however, two general classes of photoinitiators: (1) those that undergo direct photo-fragmentation upon exposure to UV or visible light irradiation and produce active free radical or cationic intermediates; (2) those that undergo hydrogen abstraction electron transfer reactions followed by rearrangement into a free radical or cationic intermediate.

It is also important to select photosensitizers and photoinitiators with absorption bands that overlap the emission spectra of the various commercial UV and visible light sources (only the light absorbed by a molecule evokes a photochemical reaction). Some of the most common photoinitiator systems in use today are as follows:

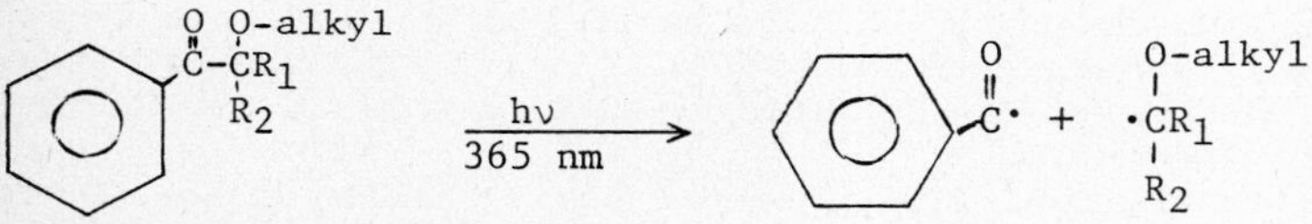

$$\text{derivative of acetophenone} \xrightarrow{\; h\nu \;\; 365 \text{ nm}\;} \text{free radical intermediates}$$

derivative of acetophenone
where R_1=H, alkyl, aryl, alkoxy
and R_2=H, alkyl, aryl

$$\text{Ar-}\overset{\overset{\displaystyle Ar}{|}}{\underset{\underset{\displaystyle Ar}{|}}{S}}{}^{+}X^{-} \xrightarrow{\; h\nu \;\; 365 \text{ nm}\;} Ar_2 S^{\cdot +} + Ar\cdot + X^{-}$$

radical cations, radicals and lewis acid intermddiates

where X^{-} = BF_4, PF_6, AsF_6, SbF_6,

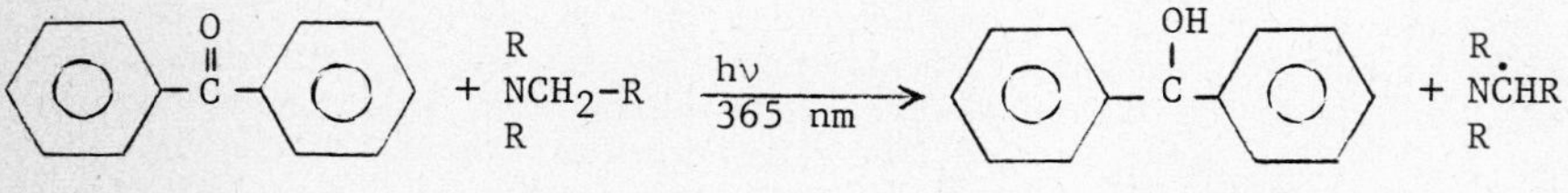

$$+ \; \underset{\underset{\displaystyle R}{|}}{\overset{\overset{\displaystyle R}{|}}{N}}CH_2\text{-}R \xrightarrow{\; h\nu \;\; 365 \text{ nm}\;} + \; \underset{\underset{\displaystyle R}{|}}{\overset{\overset{\displaystyle R\cdot}{|}}{N}}CHR$$

free radical intermediates

$$+ \; \underset{\underset{\displaystyle R}{|}}{\overset{\overset{\displaystyle R}{|}}{N}}CH_2\text{-}R \xrightarrow{\; h\nu \;\; 480 \text{ nm}\;} +$$

R
N-CHR
R

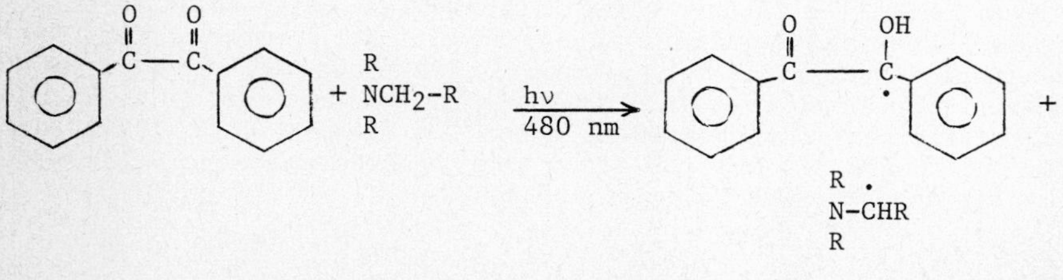

Other photoinitiator structures (free radical and acid intermediate generators) have been reviewed in References 22 and 23.

<u>Free Radical Initiated Radiation Curable Coating Compositions</u>. Conventional thermally cured coating systems are generally based on the following polymer backbone chemical structures:

(1)   Epoxy

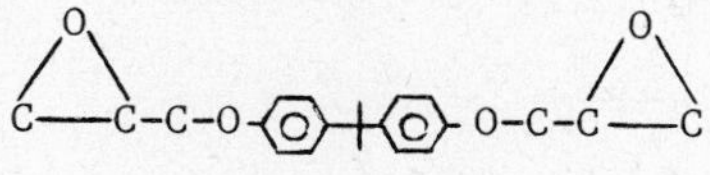

(2)   Urethane

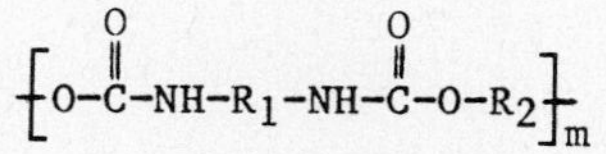

(3)   Polyester

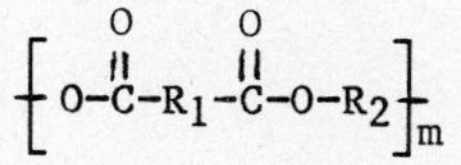

(4)   Acrylic

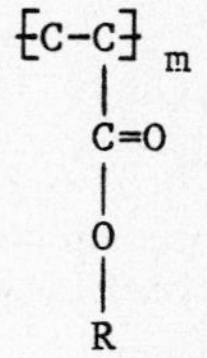

Present-day thermal curing coatings systems utilize these types of polymer structures as well as fillers and pigments dissolved or dispersed in an organic solvent for coatable application viscosities. These solvents are then thermally removed and the coating is cross-linked into a three-dimensional network by an energy-rich chemistry requiring a high degree of thermal energy to convert the polymers into useful commercial acceptable properties.

Radiation curable polymer systems are based on the same chemical structural design as the conventional polymer systems, but certain modifications are made in order to accommodate reactive unsaturation sites necessary for a radiation-induced free radical curing mechanism. Examples of these modifications of conventional polymer structures to form radiation curable polymers are as follows:

(1) Unsaturated (Acrylic/Methacrylic) Epoxy

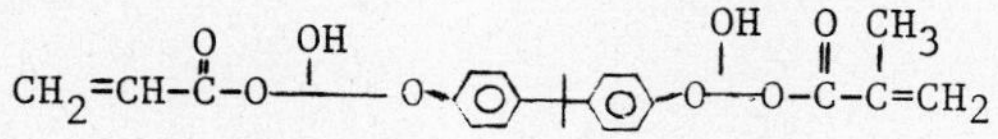

(2)   Unsaturated Urethane

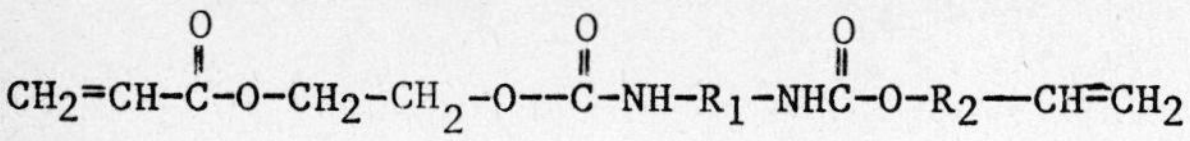

$$CH_2=CH-\overset{\overset{\textstyle O}{\|}}{C}-O-CH_2-CH_2-O--\overset{\overset{\textstyle O}{\|}}{C}-NH-R_1-NH\overset{\overset{\textstyle O}{\|}}{C}-O-R_2-CH=CH_2$$

(3)   Unsaturated Polyester

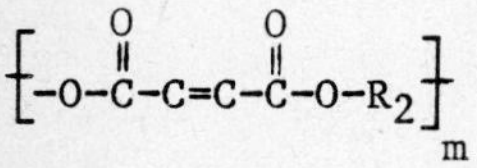

$$\left[-O-\overset{\overset{\textstyle O}{\|}}{C}-C=C-\overset{\overset{\textstyle O}{\|}}{C}-O-R_2 \right]_m$$

(4)   Unsaturated Acrylic

$$\left[C-C \right]_m$$
$$\begin{array}{ccc} C-O & OH & O \\ | & | & \| \\ O-CH_2-CH-CH_2-O-C-CH=CH_2 \end{array}$$

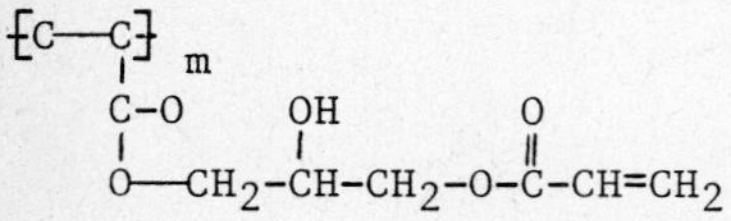

The monomer in radiation-curable coatings is the analog of the solvent in a conventional paint. Although it performs like a solvent by being a medium for all of the other ingredients and by providing the necessary liquid physical properties and rheology, it differs in that it enters into the copolymerization and is not lost on cure.

Most radiation-curable monomers contain single unsaturation sites and are high-boiling acrylic esters, although in the wood area some coatings use styrene as the monomer. Usually, where styrene is used, most or all of the polymer-polyester unsaturation is fumarate rather than acrylic.

Cross-linking oligomers in conventional thermosetting coatings formulations are usually melamine resins (acid, hydroxyl-transetherification cross-linking reactions), amine/amide hardeners (oxirane ring opening reactions), and blocked isocyanate prepolymers. Oligomers and cross-linking materials in radiation curing systems are similar to single vinyl functional monomers except they contain di-, tri-, or multifunctional unsaturation sites. These multifunctional components cause polymer propagation reactions to proceed into three-dimensional network structures of a cured film.

$$PI \xrightarrow{\quad h\nu \quad} PI\cdot$$

photoinitiator   free radical intermediate

PI· +   multifunctional unsaturated

monomers and polymers

three-dimensional network formation

Formulation of radiation curable coatings requires a balance in composition among three variables: single unsaturated functional monomers, cross-linking agents (multiunsaturated oligomers), and unsaturated polymers. For example, as the composition of a coating changes from monomer- to polymer-rich mixtures at a constant cross-linking oligomer concentration, the coating viscosity increases, the rate of cure for the coating may decrease, and the final film properties would be expected to have good adhesion and better extensibility. If the coating composition changes from a monomer-rich to cross-linking oligomer rich mixture at constant polymer concentration, then the rate of cure for the coating will be increased but the final cured film may be brittle and have very little adhesion to certain substrates (24).

The physical properties of the cured film depend upon the initial liquid formulation ingredients, their chemical structure, unsaturated reactivity, and individual component concentrations.

Acid Intermediate Initiated Radiation Curable Coating Compositions. In this process, photogenerated acid intermediates can cause ring opening reactions of various oxirane monomers and prepolymers that can then further polymerize into three-dimensional network structures.

$$Ar_3S^{+}X^{-} \xrightarrow[RH]{h\nu} Ar_2S + Ar\cdot + R\cdot + HX$$

Coating compositions for this process can contain cyclic ethers, vinyl ether monomers, organosilicone monomers, and a wide variety of mono-, di-, and polyepoxy functional materials. Further modifications to this process include the addition of free radical polymerizable monomers in combination with the cationic polymerizable monomers so that both curing processes may take place at the same time. The photoinitiator systems that generate acid catalyst intermediates also generate free radical intermediates or can be combined with other photosensitizer materials such that both ring opening and radical vinyl addition reactions can occur. A major advantage to this hybrid system (free radical vinyl addition and acid-catalyzed ring opening photochemical reactions) is that very fast tack-free cured surfaces can be achieved through the vinyl polymerization reaction while the slower curing ring opening reaction causes the development of excellent adhesive bonding capabilities of the cured film to the substrate.

Electron Curing

Accelerated electrons were first used to cure coatings in the 1930s. It was not until the late 1960s, however, that any serious commercial interest in such a process was apparent. In the past several years electron processing has been the perennial bridesmaid but seldom the bride. Many reasons have been cited for the failure of electrocuring to be accepted in the marketplace. Probably the dominant reasons, though, are the relatively large investment needed for an electron processing line and perhaps the thought that the ultimate equipment was not yet available. The past several years have seen the development of electron cure equipment that promises to reduce the investment cost, increase the efficiency, and simplify the system. Early devices were aptly described as producing electron beams, which were then scanned magnetically. For various reasons they operated at 300–500 keV. The large accelerating potential was undesirable for two reasons: (1) When an accelerated electron undergoes a collision, the energy is released as a highly penetrating X-ray that must be shielded, and the amount of shielding needed is related to the X-ray energy (25). A 300-keV beam is around 200 keV at the workpiece due to losses at the accelerator window and in the air gap between the window and the substrate. For efficient energy use, the coating thickness should be on the order of 8–12 mils.

The new generation of electron accelerators employs linear cathodes that need not be scanned and are specifically designed for coating processing. As a result, the energy at the workpiece is on the order of 100–125 keV and the unit density coating can be 4–8 mils and have optimum energy coupling. The linear cathode allows for reduced dose rates at the same total dose. This has clear mechanical advantages, while in principal the lower dose rate should have chemical advantages. The compact linear cathode devices reduce the shielding requirements such that the total electron cure package is compatible with many existing curing lines.

The time has come for electron curing. The economics are acceptable for high-volume uses; the equipment has been designed for coatings, and the chemistry for curing pigmented materials is superior to ultraviolet approaches. One other advantage of electron curing of coatings over light-induced curing reactions is the elimination of photoactive catalyst systems from the coating formulation. The energetic electrons from the processor are adsorbed directly in the coating itself where they create the initiating free radicals uniformly in depth.

Since electron energies of only 100eV or less are required to break chemical bonds and to ionize or excite components of the coating system, the shower of scattered electrons produced in the coating leads to a uniform population of free radicals (excited atoms or ions) throughout the coating, which then initiate the polymerization reaction. In the liquid-phase systems of interest here for coatings work, the polymerization process will propagate until the activity of the growing chain is terminated. These energetic electrons are capable of penetrating many different types of pigmented coatings and are capable of producing through cure down to the substrate–polymer coating interface. The coating materials used in electron curing processes are the same as those previously described for free radical initiated radiation curable coating compositions.

Glow Discharge Polymerization

Glow discharge polymerization (GDP) refers to the formation of polymers by means of an electric discharge. The breakdown of molecules in gas discharges has been known for 100 years (26, 27). Glow discharge polymers can be produced quite readily. One needs only a reaction vessel containing monomer and an electrical source of sufficient power to cause the monomer to glow. The details of how GDP works are a matter for speculation (22), but the energy input is large enough to break any chemical bond and as a result even materials like benzene can be polymerized (29). Because glow discharge polymerization is accomplished in an evacuated system, it is not ideal for continuous processing, although methods for doing so have been commercialized (30).

Literature Cited

1.   Blank, W. J. J. Coat. Technol. 1982, 54(687), 26.
2.   Barrett, K. E. J. "Dispersion Polymerization in Organic Media"; Wiley: New York, 1975.
3.   "Treatise on Coatings--Formulations"; Myers, R. R.; Long, J. S., Eds.; Marcel Dekker: New York, 1975; Vol. 4, Part 1.
4.   Schultz, A. R. "Encyclopedia of Polymer Science and Technology"; Interscience: New York, 1965; Vol. 4, p. 331.
5.   Patton, T. C. "Alkyd Resin Technology"; Interscience: New York, 1962.
6.   Widmer, G. "Encyclopedia of Polymer Science and Technology"; Interscience: New York, 1965; Vol. 2, p. 1.
7.   Lee, H.; Neville, K. "Handbook of Epoxy Resins"; McGraw-Hill: New York, 1967.
8.   "Treatise on Coatings--Film-Forming Compositions"; Myers, R. R.; Long, J. S., Eds.; Marcel Dekker: New York, 1967; Vol. 1, Part 1.
9.   Moore, N. L. "Radiation Drying of Paints and Inks"; Watford College of Technology: Hertis, 1973.
10. Summer, W. "Ultraviolet and Infrared Engineering"; Sir Isaac Pitman and Sons: London, 1962.
11. Ladstadter, E.; Hanus, H. D. preprints of OCCA Conference, June, 1973.
12. Readdy, A. F. "Plastics Fabrication by Ultraviolet, Infrared Induction, Dielectric and Microwave Radiation Methods"; Plastics Technical Evaluation Center, Picatinny Arsenal: Dover, N.J.

13. Prince, F.; Young, S. E. Prod. Finish. 1973, Sept., 29.
14. McGinniss, V. D. "National Symposium on Polymer in the Service of Man"; American Chemical Society: Washington, D.C., 1980; p. 175.
15. Calvert, J. G.; Pitts, N. J. "Photochemistry"; Wiley: New York, 1966.
16. Pray, R. W. Society of Manufacturing Engineers Technical Paper FC74–513, 1974.
17. Fusion Systems, Inc., Rockville, MD.
18. McGinniss, V. D. In "UV Curing Science and Technology"; Pappas, S. P., Ed.; Technology Marketing Corp.: Stamford, Conn., 1978, p. 229.
19. McGinniss, V. D. Photogr. Sci. Eng. 1979, 23(3), 124.
20. McGinniss, V. D. J. Radiat. Curing 1975, 2, 3.
21. McGinniss, V. D. "Photoinitiated Polymerization" In "Developments in Polymer Photochemistry–3"; Allen, N. S., Ed.; Applied Science Publishers: Essex, England, 1982; p. 1.
22. Perkins, W. C. J. Radiat. Curing 1981, 8(1), 16.
23. Crivello, J. V., et al. J. Radiat. Curing 1978, 5(1), 2.
24. McGinniss, V. D.; Kah, A. Polym. Eng. Sci. 1977, 17(7), 478.
25. Nablo, S. V., et al. J. Paint Technol. 1974, 46(593), 51.
26. DeWilde, P. Ber. Dtsch. Chem. Ges. 1874, 7, 4658.
27. Thenard, A. C. R. Hebd. Seances Acad. Sci. 1874, 78, 219.
28. Simonesca, C., et al. Eur. Polym. J. 1969, 5, 427.
29. For an excellent review of glow discharge polymers, see: Bloor, J. E. J. Radiat. Curing 1974, April, 21.
30. Williams, T.; Hayes, M. W. Nature (London) 1967, 216, 614.

POLYMER PRODUCTS AND THEIR USES IN COATINGS

The Coatings Industry: Economics and End-Use Markets

JOSEPH W. PRANE

213 Church Road, Elkins Park, PA 19117

Products of the Coatings Industry and Shipments
Trade Sales (TS)
Industrial Finishes (IF)
  Automotive Topcoats
  Coil Coatings
  Coatings Materials/Systems
Special-Purpose Coatings (SPC)
Pressures on the Coatings Industry
  Raw Materials and Finished Products
  Energy Costs
  Environmental Control
U.S. Coatings Suppliers

The subject of this chapter is the economics of the coatings industry in the United States. This industry comprises the manufacture, sales, and use of pigmented coatings (paints) and clear finishes. The focus is on this industry and its largely unsung role in the growth and health of the overall U.S. economy. Although relatively mature, the coatings industry is vital, vibrant, and progressive and continues to recognize and react positively to the needs of both its customers and the evolving regulatory structures that have been placed on it and other industries.

Products of the coatings industry--paints, varnishes, lacquers, enamels, chemical coatings, maintenance finishes, and other products both basic and sophisticated--have been used to protect, decorate, and provide functional properties to a host of surfaces and objects. The leveraged value of coatings is exemplified by the contribution of corrosion-resistant primers and of protective and decorative topcoats to the long life of structures, such as buildings and bridges, and to the appearance and durability of automobiles and appliances.

0097–6156/85/0285–0857$07.25/0
© 1985 American Chemical Society

Products of the Coatings Industry and Shipments

The types of products produced by the U.S. coatings industry are
shown in Table I. Sales of the industry, distributed by shipments
in the three general categories shown, are reported by the Bureau of
the Census, U.S. Department of Commerce, in their monthly "Current
Industrial Reports" M28F. Table II is a summary of these reports
for the period 1977–82, with projections to 1987. Figure 1 displays
these data in graphical form. Shipments for 1987 are estimated in
terms of constant 1982 dollars, since exact inflation factors cannot
be realistically predicted.
 The U.S. inflation rate started to abate at the end of 1982;
however, resumption of the inflation spiral cannot be entirely
discounted. The short-range effect of inflation can be seen in
Table III, where shipments in the first 3 months of 1982 are
compared to shipments in the same period in 1981. (In Tables II and
III, unit values, dollars per gallon, have been calculated, for
comparison purposes.)
 Examination of these tables leads to a number of general
conclusions concerning the current and future status of the U.S.
coatings industry.

Trade Sales (TS)

These architectural coatings include stock type or shelf goods
formulated for normal environmental conditions and for general
applications on new and existing residential, commercial,
institutional, and industrial structures. These products are
usually distributed through wholesale/retail channels and are
purchased by the general public, paint and building contractors,
government agencies, and others.
 The effect of inflation in the trade sales area has been felt in
the period 1979–82. Note in Table III the sharp increase in the
calculated unit value of architectural finishes--compared with the
static real unit value predicted for 1987 (as compared to 1982).
The decrease in housing starts in 1979–82, resulting in lower house
paint shipments, has been partially balanced by increased home
maintenance. The trade sales market is relatively mature; long
range, few major formulation changes are expected with little
increase in the real value of finished goods.
 Despite its maturity, the trade sales market has reasonably good
growth potential-because of the pent-up demand for new housing and
the proclivity of the buying public to engage in home maintenance
projects during times of recession.
 The trade sales market has led the way to the use of waterborne
(WB) finishes, mostly based on emulsion polymers. As shown in Table
IV, WB coatings represented over 79% of TS finishes (on a gallon
basis) in 1982; this is projected to increase to about 86% in 1987.

Industrial Finishes (IF)

This category of chemical coatings includes finishes that are
formulated specifically for original equipment manufacture (OEM) to
meet specified conditions of application and product requirements.
They are applied to products as part of the manufacturing process.

Table I. U.S. Coatings Industry: Product Types

Product Type	Use
Trade sales (architectural coatings)(TS)	"Do-it-yourself," over-the-counter sales Exterior solventborne Exterior waterborne Interior solventborne Interior waterborne Architectural lacquers
Industrial finishes (product coatings, OEM; chemical coatings, factory applied)(IF)	Automotive finishes (primers, sealers, topcoats) Truck and bus finishes Other transportation finishes, e.g., aircraft, railroad, etc. Marine coatings, including off-shore structures Appliance finishes Wood furniture and fixture finishes Wood and composition board flat stock finishes Sheet, strip, and coil coatings on metals Metal decorating, e.g., can, container, and closure coatings Machinery and equipment finishes Metal furniture and fixture coatings Paper and paperboard coatings Coatings for plastic shapes and films, e.g., packaging Insulating varnishes Magnet wire coatings Magnetic tape coatings
Special-purpose coatings (SPC)	Industrial maintenance paints --interior, exterior Metallic paints, e.g., aluminum, zinc, bronze, etc. Traffic paints Automobile and truck refinish coatings Machinery refinish coatings Marine refinish coatings Aerosol paints and clears Roof coatings Fire-retardant paints Multicolor paints

Table II. U.S. Shipments of Coatings

Product Type	Quantity	1977	1978	1979	1980	1981	1982	1987	AGR^2 (%) [a] 77/82	82/87
Total	mm gal.	990	1033	1069	1019	991	903	1060	-1.8	3.2
	$ mm[b]	5371	6008	7025	7636	8396	8297	9830	9.1	3.4
	$/gal.	5.43	5.82	6.57	7.49	8.47	9.19	9.27	11.1	0.1
TS	mm gal.	460	523	571	530	505	454	600	0	5.7
	$ mm[b]	2466	2900	3419	3641	3969	4052	5340	10.4	5.7
	$/gal.	5.36	5.54	5.99	6.87	7.86	8.95	8.90	10.8	-0.1
IF	mm gal.	370	350	332	298	302	269	240	-6.2	-2.3
	$ mm[b]	1961	2092	2284	2419	2737	2546	2400	5.4	-1.2
	$/gal.	5.30	5.98	6.88	8.12	9.06	9.46	10.00	12.3	1.1
SPC	mm gal.	160	160	166	191	184	180	220	2.4	4.1
	$ mm[b]	944	1016	1322	1576	1690	1699	2090	12.5	4.2
	$/gal.	5.90	6.35	7.96	8.25	9.18	9.44	9.50	9.9	0.1

[a]AGR represents annual growth rate.

[b]This value is at the manufacturer's levels; mm = million. Estimates for 1987 are based on constant 1982 dollars.

Source: Bureau of the Census (except for 1987 figures).

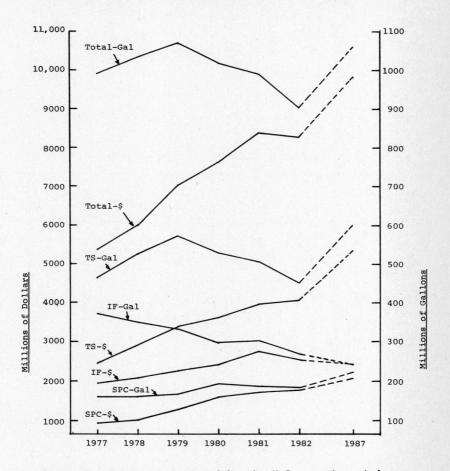

Figure 1. Products produced by the U.S. coatings industry. Abbreviations are defined in Table I.

Table III. U.S. Shipments of Coatings: Three—Month Data

Product Type	Quantity	1981 Jan—Mar	1982 Jan—Mar	% Change
Total	mm gal.	236.5	202.7	−14.3
	$ mm	1876.4	1823.2	−2.8
	$/gal.	7.93	8.99	13.4
TS	mm gal.	113.9	98.9	−13.2
	$ mm	833.8	853.6	2.4
	$/gal.	7.32	8.63	17.9
IF	mm gal.	78.2	66.2	−15.3
	$ mm	672.2	616.4	−8.3
	$/gal.	8.60	9.31	8.3
SPC	mm gal.	44.4	37.6	−15.3
	$ mm	370.4	353.2	−4.6
	$/gal.	8.34	9.39	12.6

Source: Bureau of the Census.

Table IV. Trade Sales Coatings by End Use—U.S. Shipments

Coating Type	1980 mm gal.	1980 %	1981 mm gal.	1981 %	1982 mm gal.	1982 %	1987 mm gal.	1987 %	AGR (%) 1982-87
Interior									
Waterborne	260	81.3	255	82.3	236	84.0	350	89.7	8.2
Solventborne	60	18.7	55	17.7	45	16.0	40	10.3	-2.3
Subtotal	320	100	310	100	281	100	390	100	6.8
% Interior		60.4		61.4		62.0		65.0	
Exterior									
Waterborne	140	66.7	135	69.2	125	72.3	165	78.6	5.7
Solventborne	70	33.3	60	30.8	48	27.7	45	21.4	-1.3
Subtotal	210	100	195	100	173	100	210	100	4.0
% Exterior		39.6		38.6		38.0		35.0	
Total	530		505		454		600		5.7
Waterborne	400	75.5	390	77.2	361	79.5	515	85.8	7.4
Solventborne	130	24.5	115	22.8	93	20.5	85	14.2	-1.8

Markets using IF, for example, automotive, have been in a severe recession, which continued through 1982. U.S. passenger car production in the calendar year 1982 was down to 5.4 million units; this was the lowest in over 12 years. However, in 1983 and beyond, automotive (with smaller cars), appliance and other IF markets should start to move forward. New formulations, for example, high solids, will be used to a greater extent; therefore, dollar growth will somewhat outpace gallonage growth.

The industrial finishes market has always been the forerunner of new technology in the coatings industry. Closer examination of the business, economics, and technology in this market segment will be illustrative of the continually growing importance of advanced applied scientific techniques and principles in the coatings industry.

Table V distributes U.S. shipments of industrial finishes into eight end-use areas for 1980–82, with projections to 1987. Declines in gallonage are indicated in markets for metal, wood, and transportation finishes. However, some of this is due to the projected greater use of high-solids (HS) finishes; overall, consumption of IF by dry gallon is expected to be about the same from 1982 to 1987, since the average nonvolatile content by volume is expected to increase from 40% to about 50%. Can coatings will be affected by state returnable-container laws, the upsurge in the use of glass bottles, and the growth of plastic containers for soft drinks.

Automotive Topcoats. The types of automotive topcoats that have been (or will be) used are shown in Table VI. Examination of these topcoats offers a good illustration of the projected progress of HS coatings.

TPL-SB are low-solids (12–18% by volume) solution coatings that have given brilliant metallic finishes for many years. Their properties were maintained when the major user, General Motors, went over to TPL-NAD. TSE-SB, the solution enamel used originally by Ford, Chrysler, and American Motors, is now virtually extinct as a topcoat for U.S. passenger cars. It has been replaced by TSE-NAD, which has higher solids and lower emission characteristics.

TSE-WB is used only by General Motors at three of its U.S. plants. This operation was decreed by the EPA as the best available technology (BAT) at that time. GM does not plan to introduce TSE-WB at any other of its assembly plants—primarily because of the high cost of equipment and the environmental changes required. GM continues to use acrylic lacquers at their other plants—but will be moving to HS-E by 1985. This will be a major move for GM—going away from their traditional TPL.

However, HS-E, as they are being developed, offer the best opportunity for automotive companies to meet the stringent VOC (volatile organic compound) requirements and transfer efficiency minima being mandated by the EPA. These HS-E are predominantly of the acrylic type—with a goal of 60% solids by volume as delivered to the spray gun. At this level, and with electrostatic spray application, EPA compliance should be ensured to at least 1985.

At one time, powder coatings (PC) appeared to be the answer to Detroit's prayer for nonpolluting topcoats. Both General Motors and Ford mounted extensive development programs with PC and coated many

Table V.   Industrial Finishes by End Use--U.S. Shipments

Finish	mm^a gal.				AGR^b,% 82/87
	1980	1981	1982	1987	
Metal finishes	80	78	69	60	-2.8
Can, container	35	35	30	26	-2.8
Prefinished (coil)	25	26	24	22	-1.7
Furniture and fixtures	10	9	8	6	-5.6
General metal	10	8	7	6	-3.0
Wood finishes	73	72	64	50	-4.8
Furniture and fixtures	53	52	46	30	-8.2
Prefinished	20	20	18	20	2.1
Transportation	60	55	46	36	-4.8
Automotive	45	43	37	28	-5.4
Truck, bus	6	5	4	3	-5.6
Aircraft, railroad	5	4	3	3	0
Marine	4	3	2	2	0
Machinery and equipment	30	30	27	23	-3.2
Appliance	10	9	8	6	-5.6
Packaging (paper, foil, plastic film)	20	25	24	28	3.1
Plastic parts	10	15	16	18	2.4
Electrical, electronic	8	10	9	10	2.1
Miscellaneous	7	8	6	9	8.4
Total	298	302	269	240	-2.3

^{a}mm = million.

^{b}AGR represents annual growth rate.

cars. They were able to demonstrate the technical feasibility of
the process. However, the need for many color application stations,
difficulty of color change, limited use of overspray PC, and
difficulty of achieving reliable aluminum flake orientation in
metallic colors served to doom the short-range potential of PC as a
viable and economical alternative to liquid automotive coatings.

Urethane enamels (PU) were also promoted as the ideal auto
topcoat--with their well-known properties of mar and abrasion
resistance, flexibility and toughness, and high gloss and gloss
retention (in aliphatic systems). Both General Motors and Ford have
demonstrated that excellent nonmetallic topcoats can be provided as
two-component, aliphatic PU systems. These can be applied in two-
headed mixing spray guns and cured at low temperatures (e.g., 150
°F) at rates that are consistent with assembly-line schedules. But,
as with PC, metallic colors are difficult to achieve reliably
because of the reduced flow of the system and the poor orientation
of the aluminum flakes. However, the overwhelming blockage in PU
usage remains the toxicity hazard associated with the residual
isocyanate content.

Both General Motors and Ford have shown that the use of robots
in applying PU finishes can obviate direct toxicity hazards to
workers. However, with the depressed auto business in the United
States in the last 4 years, their recent efforts have been directed
to coatings systems that were more compatible with their existing
assembly lines, for example, TSE and HS-E, and that provided reduced
emissions of VOC.

Subcontractors to the auto industry are using considerable
quantities of one-package PU enamels to finish plastic parts, for
example, soft EPDM and thermoplastic elastomer front fascia and RIM
and RRIM parts. These subcontractor operations are not yet subject
to the stringent emission controls mandated by the EPA on auto
assembly lines.

However, the use of PU finishes on automotive metal on assembly
lines will not come about unless or until the toxicity problem is
overcome. Urethane coating producers are active in the development
of new PU systems--some involving proprietary blocking agents--to
tackle this problem. Their efforts should show some success by
1985. Positive influences here include the growing use of base
coat/clear coat systems; clear PU topcoats are under active
consideration. In addition, vapor permeation curing (VPC) is under
investigation. Here, acrylic/urethane topcoats are cured rapidly
under ambient conditions in an amine catalyst atmosphere.

Table VII distributes automotive topcoat usage by type, with
predictions for 1987. The progression from low-solids, high-
emission coatings to high-solids systems is shown. Research and
development of HS-E are active at each of the auto companies and at
every major coatings supplier and raw materials producer.
Certainly, problems such as specific polymer design and cross-linker
choice still exist--as well as application, flow, and aluminum flake
orientation. However, current and future research and development
should reduce these problems rapidly to manageable levels.

Table VIII projects trends in overall auto topcoat usage as a
function of exterior surface area coated and nature of the surface.
The dramatic drop in topcoat consumption can be explained as
follows: The average dry film thickness of auto topcoats will

Table VI.   Nomenclature for Automotive Topcoats

Abbreviation	Types of Automotive Topcoats
TPL–SB	Thermoplastic acrylic lacquer, solventborne
TPL–NAD	Thermoplastic acrylic lacquer, nonaqueous dispersion
TSE–SB	Thermosetting acrylic enamel, solventborne
TSE–NAD	Thermosetting acrylic enamel, nonaqueous dispersion
TSE–WB	Thermosetting acrylic enamel, waterborne
HS–E	High–solid thermosetting enamel, solventborne
PC	Powder coating, thermosetting
PU	Urethane enamel, aliphatic, one or two parts

Table VII.   U.S. Automotive Topcoat Usage (Percent, Gallonage Basis)

Topcoat	1976	1979	1980	1982	1987[a]
TPL–SB	25				
TPL–NAD	25	56	57	50	10
TSE–SB	3				
TSE–NAD	42	35	31	29	17
TSE–WB	5	7	7	7	6
HS–E				7	50
PC	<1	<1	<1	<1	2
PU		2	5	7	15
Total %	100	100	100	100	100
Total (mm[b] gal.) (estimate)	21	19	14	12	12

[a]Projected.

[b]mm = million.

Table VIII. U.S. Automotive Topcoats: Trends

Factor	1976	1977	1978	1979	1980	1981	1982	1987[a]
Number of cars, in thousands	8506	9211	9170	8423	6373	6097	5376	7000
ft^2 of exterior surface per car	700	690	685	675	670	670	665	625
Percentage of surface (ft^2) that is plastic	5	6	8	10	12	13	14	15–20
gallons of top-coat per car	2.50	2.48	2.40	2.30	2.22	2.20	2.18	1.70

[a]Projected.

remain at about 1.8–2 mils. The trend is to higher solids coatings, applying the same dry film thickness, with lower gallonage consumption. There will be increased production of smaller cars in the United States, with reduced surface area to be coated. Greater use of electrostatic spray procedures for application of topcoats will take place. This results in improved <u>transfer efficiency</u>, with a higher percentage of paint leaving the spray gun being deposited onto the workpiece than with conventional air spray. There will be an increased percent of the exterior surface of the car that will be plastic. While most coatings for plastics will be PU, the auto companies will be attempting to use HS–E on these surfaces for improved assembly line efficiency and economics.

<u>Coil Coating</u>. Coil coating is one of the most efficient methods of applying primers, topcoats, graphics, and laminates to coils or rolls of metals. The prefinished metal, which can be shaped and formed, enables the customer to eliminate costly in-house paint lines, energy-consuming curing ovens, and expensive pollution-control equipment.

Figure 2 displays North American shipments of coil-coated stock from 1962 to 1982. Coil coating enjoyed an annual growth rate of about 15% to 1978; however, it declined slightly in 1979 as its major markets suffered from the leading edge of the recession. The major growth market for coil-coated aluminum was cans, ends, and tabs--with residential siding following. Dominant markets for coil-coated steel were industrial and rural buildings and passenger cars. The transportation market for coil-coated stock, currently depressed, should move forward sharply in the mid-1980s. Appliance markets should also prosper with such growing applications as refrigerator wrappers.

<u>Coatings Materials/Systems</u>. The IF segment of the coatings industry has made considerable progress from the early days of almost complete reliance on solventborne (SB) coatings. Table IX estimates the distribution of IF by coating material and system for 1982 and 1987. Figure 3 illustrates these data in a bar chart.

Conventional SB systems, with unlimited solvent use, are estimated to have accounted for 31% of the dollar volume of IF shipments in 1982; they are expected to decrease to 12% in 1987 as the full effects of environmental regulations are established. Conforming SB systems, that is, those conforming to the revised California Rule 66, CARB (California Air Resources Board) codes, and the evolving EPA regulations, will decrease at a lower rate, from 34% in 1982 to 22% in 1987. These include the use of certain chlorinated solvents (e.g., 1,1,1-trichloroethane and methylene chloride) that have been exempted by several regulatory authorities. However, problems with reaction with coating application equipment components made of aluminum may impede their full utilization.

Waterborne (WB) systems, already well established as conforming coatings, will increase from 20% in 1982 to 33% in 1987. Electrodeposition (ED), the preferred method for automotive priming, has been widely accepted.

Powder coating (PC), which can be totally nonpolluting, but which is somewhat energy intensive, will enjoy significant growth in

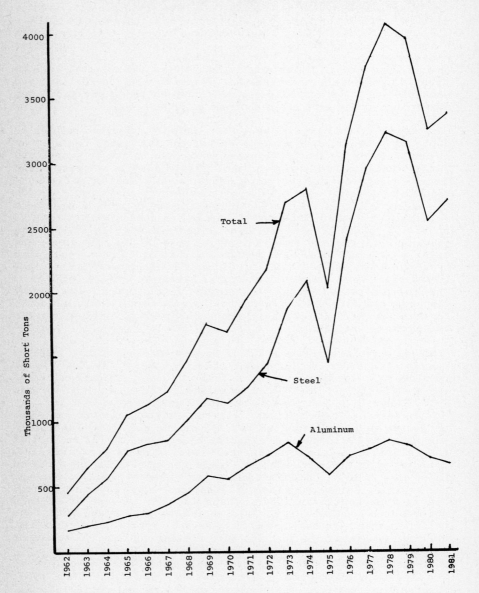

Figure 2.   Coil-coated stock shipments in the U.S., Canada, and Mexico from 1962 to 1981.

Table IX. U.S. Shipments of Industrial Finishes by Coatings Materials and Systems

Coatings Material/System	1982 $ mm	1982 %	1987a $ mm	1987a %	AGR (%) 1982–87
Solventborne					
Conventionalb	1670	65.6	830	34.6	-13.0
	800	31.4	300	12.5	-17.8
Conformingc	870	34.1	530	22.1	-9.4
Waterborne	500	19.6	800	33.3	9.9
Electrodeposition (ED)	100	3.9	150	6.3	8.4
Non-ED	400	15.7	650	27.0	10.2
Powder coatings	150	5.9	200	8.3	5.9
Radiation curable	70	2.8	120	5.0	11.4
Ultraviolet (UV)	55	2.2	70	2.9	4.9
Electron beam (EB)	15	0.6	50	2.1	27.2
High solids–liquidd	156	6.1	450	18.8	23.6
Total	2546	100	2400	100	-1.2

aThese values are in constant 1982 dollars.

bCoatings are up to 60% NV (nonvolatile content by weight) with unlimited solvent use.

cCoatings are up to 60% NV, conforming to Rule 66, EPA, and other regulations. These include NAD (nonaqueous dispersions) and coatings containing approved chlorinated solvents.

dCoatings are 60-100% NV, conventionally cured.

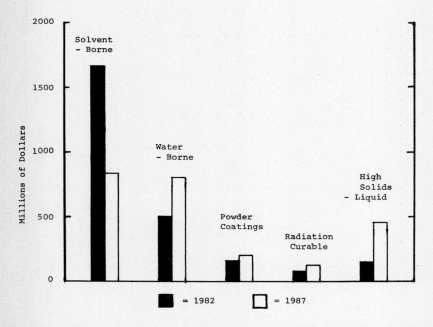

Figure 3. U.S. shipments of industrial finishes by coating material and system.

specialized areas, for example, electrical insulation, wire goods, pipe coating, and concrete reinforcing bars. Truck bodies are being powder coated on a limited basis.

Radiation-curable coatings (RC) should enjoy a dramatic growth as the full extent of their solvent-free, rapid, and ambient-curing properties is realized and materials and application systems are perfected. UV-curable coatings are currently the most important in this category; however, electron beam (EB) curable coatings should show excellent growth to 1987 and beyond.

High-solids (HS) coatings are increasing rapidly in importance (as previously discussed) because of their compatibility with existing application equipment and their ability to meet many emission regulations. Automotive and appliance applications for these systems should continue to grow well in the 1980s.

Special Purpose Coatings (SPC)

These coatings may be stock type or shelf goods. They differ from general architectural or product finishes in that they are formulated specifically for field application, usually refinishing, and for special applications and/or special conditions in the environment, for example, extreme temperatures or corrosive-chemical atmospheres. Many channels of distribution are used.

Markets for SPC stayed strong in the 1979-82 period since industrial maintenance activities continued despite recession. Other parts of this market, for example, marine finishes for off-shore rigging and increased use of auto finishes, will maintain good growth of SPC over the long range. The growth of these markets is estimated in Table X.

Pressures on the Coatings Industry

The coatings industry, like many others, has been ravaged since 1973 by sharp rises in the costs of materials used in the manufacture of products. Both producers and coatings customers have had to make major adjustments to accommodate the escalating costs (and frequent lack of availability) of the fuels and energy they require for processing. In addition, major expenditures have been made for equipment and systems to control environmental emissions and to minimize worker exposure to toxic materials.

However, in many ways, these pressures have strengthened the industry. Efficiency and productivity have been enhanced by the adoption of modern computer techniques for formulation, inventory control, scheduling, production, etc. New technologies have been developed and implemented (e.g., electrodeposition, autodeposition, high solids, powder coatings, radiation curables, metal prefinishing, and robots) that will serve the industry well into the future.

Raw Materials and Finished Products. Table XI lists prices for a selected number of key raw materials used in the coatings industry; they cover bulk prices in September for the period 1972-82. In most cases, prices (particularly for petrochemically derived materials) have more than doubled in this period, although they had stabilized in 1982. However, because of competitive pressures, prices for

Table X. Special-Purpose Coatings by End Use--U.S. Shipments

| Coating Type | mm gal. | | | | AGR (%) |
	1980	1981	1982	1987	1982–87
Industrial maintenance	60	60	60	70	3.1
Automotive and machinery refinish	40	40	40	50	4.6
Traffic paints	40	40	38	45	3.4
Aerosol paints	25	20	18	25	6.8
Marine refinish	10	10	10	15	8.4
Miscellaneous	16	14	14	16	2.7
Total	191	184	180	220	4.1

Table XI.   Selected Coatings Raw Material Prices: 1972–82

Raw Material	1972	1973	1979	1980	1981	1982
Pigments, extenders						
Titanium dioxide, rutile	27	28	53	63	75	75
Zinc oxide, American process	18.3	21	45	42.5	53	50
Red lead, 95%	20.5	21.2	70	51	54	36
Yellow iron oxide, synthetic medium	16.8	19	46	47.5	58.8	58.5
Chrome Yellow	46	50	105	122	135	135
Carbon Black, channel	12	60	70	70	70	64
Calcined clay, 2 μm	3.8	3.8	8.8	9.8	10.9	12
Oils						
Linseed oil, raw	9.5	23.5	32	30	32	26
Soybean oil, crude	10.1	25.1	28	25	21	17.5
Tung oil	10.5	30.5	55	45	62	63
Menhaden oil, light pressed	10.3	13.5	26	24	26	22
Dehydrated castor oil, G–H	29	51.5	57	67	62	61
Emulsion polymers						
Acrylic emulsion, 46% NV	14.8	14.8	29.3	36	43	46.8
Vinyl acetate copolymer, 55% NV	15	15	24	28	34	34
Resin						
Epoxy resin, liquid	38	41	93	93	116	116
Resin raw materials						
Tall oil fatty acids, 2% rosin	11.5	13.5	28	25	28	22
Phthalic anhydride	8.8	12	40	41	42	36
Glycerin, H. G.	22.5	22.5	54	62	81	81
Vinyl acetate monomer	8.8	9.5	37	33.5	37.5	32
Solvents, plasticizers						
Dibutyl phthalate	22	23	42	50.5	56	59
Mineral spirits, regular	3.1	3.2	10.9	13.8	20	20
Toluol	3.8	4.9	18	18	21	21
Xylol	4.9	4.9	18.8	18	21	21
Methyl ethyl ketone	8	10	23	35	42	42

Note: All values, in cents per pound, in September of the year
shown. All prices are list prices, bulk: liquids in tank cars;
solids by the truck load, bags.

Source: American Paint Journal.

finished coatings have not kept pace. This is illustrated in Figure 4, which shows wholesale price indices for paint materials and prepared paint. As shown, the annual growth rates, 1976 to 1982, were as follows: for paint materials, 8.2%; for prepared paint, 7.1%.

Prices are expected to vary during the 1980s with changes in petroleum prices. Materials from "renewable sources" may become available but, for all practical purposes, not before about 1990.

Energy Costs. Figure 5 shows 1972–1980 price indices for U.S. purchases of hydrocarbon feedstocks and of fuels and electric power. Note that hydrocarbon feedstock prices have increased 12-fold and that fuel and electric power prices have gone up by a factor of 6 since 1972 with devastating effects on U.S. industry, including the coatings industry. As stated earlier, fuel price rises may be abating.

This has spurred the use of HS coatings with low solvent content, waterborne coatings, radiation-curable coatings, and other low-energy curing systems, for example, polyesters, epoxies, and urethanes. In addition, curing ovens have been designed to maximize efficiency; pollution-control equipment, for example, incinerators, now frequently use regenerative heat-recovery systems to avoid the loss of the heat value of solvents and of effluent products.

Environmental Control. Three types of pollution can be attributable to operations associated with the coatings industry (as well as other, similar industries): air (from coatings production and application); water (process and waste water; discharged to sewers, streams, and lakes); solid (solid waste generation; to landfills).

In addition, a number of materials used in the industry can produce toxic reactions and other effects on people who handle these materials (and/or their effluent products) at every stage of the manufacture and application of coatings.

A large number of regulatory agencies now exist that serve to monitor, codify, and regulate such potential hazards. Among these are EPA, Environmental Protection Agency (air pollution); TSCA, Toxic Substances Control Agency (toxic substances and materials in production); OSHA, Occupational Safety and Health Administration (factory work conditions); CPSC, Consumer Products Safety Commission (consumer products); FDA, Food and Drug Administration (food and drug materials and additives; potential contaminants from coatings); DOT, Department of Transportation (control of hazards in shipping); RCRA, Resource Conservation and Recovery Act (management of solid and hazardous wastes); and Superfund, Comprehensive Environmental Response, Compensation and Liability Act of 1980.

The story of pollution control regulations in the United States has been extensively reported; however, it is briefly summarized here. The first and most famous regulation concerning solvent and other emissions in the coatings industry was "Rule 66" of the Los Angeles Air Pollution Control District, adopted July 26, 1966 (with a number of subsequent amendments). Rule 66 was based on tests of the photochemical activity of solvent emissions. The Clean Air Act of 1970 established the EPA; the act was further amended in 1977. The EPA, in their proposed regulations, proceeded to go beyond Rule 66 guidelines in their control strategies.

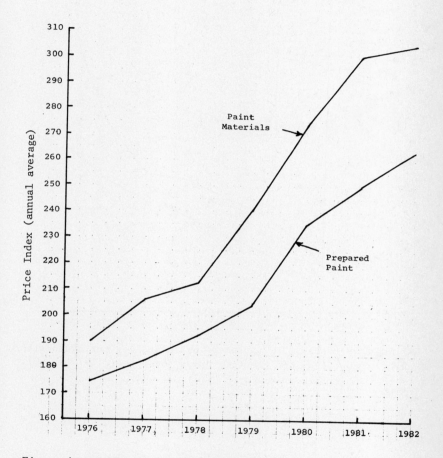

Figure 4. Wholesale price indices for paint materials and prepared paint (base year, 1967 = 100). Source: Bureau of Labor Statistics.

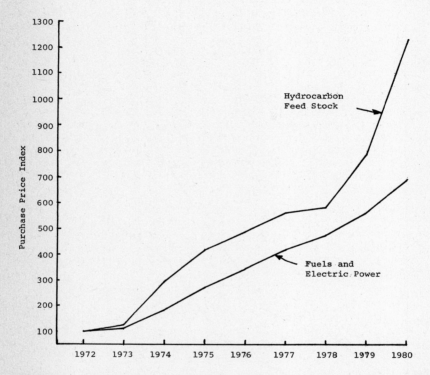

Figure 5.   Energy costs: U.S. purchase price indices for hydrocarbon
feedstocks and for fuels and electric power (base year, 1972 = 100).
Source: U.S. Department of Energy.

The California Air Resources Board (CARB) established the VOC concept for the delineation of acceptable coatings and application systems (VOC = volatile organic compounds, e.g., a maximum of 3.5 lb of controlled solvent/gal. of wet paint).

In 1978, the EPA established Control Techniques Guidelines (CTG), leading to requests for statements of Reasonably Attainable Control Technology (RACT). The goal of the EPA was to reduce VOC emissions to set limits or to achieve an 85% reduction of existing VOC. VOC levels were to be reduced by one or more of four methods: adjustment of LEL (lower explosive limit) of curing ovens; solvent scrubbing; solvent incineration; and carbon adsorption of solvents.

The EPA has actually been putting pressure on the coatings industry, using "Technology Forcing" techniques in their proposed regulations. These favor the substitution and/or replacement of SB coatings by WB systems, PC, and HS. As noted earlier, the coatings industry has generally responded to this challenge with the development and marketing of numerous conforming coating systems.

With their State Implementation Plans (SIP), EPA has mandated that each state submit pollution control plans for their districts by 1982 (with probable extension to 1987). EPA has also established an equivalence or "bubble" concept, which provides for umbrella control or grouping of individual emission sites in one area for the purpose of overall emission control.

U.S. Coatings Suppliers

Of the over 1300 companies in the United States that produce coatings, the top 10 accounted for almost 40% of total dollar sales in 1980 as shown in Table XII. These companies also controlled 42% of trade sales and 61% of industrial finish sales (as defined in Table XII).

Most of the companies listed are extremely active in development and marketing of the evolving IF coating species listed in Table IX, for example, WB, PC, HS, and RC. The top four companies are highlighted.

Sherwin-Williams, the largest U.S. coatings supplier overall, is the top U.S. producer of TS coatings. They have been leaders in industrial urethane finishes and are active in PC, WB, RC, and HS coatings. Recent research and development in the cathodic ED field has led to the introduction of several new products for the automotive and appliance industries.

PPG, the leader in ED coatings, was the innovator in cathodic ED and is currently the major factor in this field. They were also early suppliers of HS polyester systems, initially to the appliance industry. PPG has also been a leader in WB and RC coatings--both UV and EB. They have chosen not to be a factor in PC.

Du Pont has been active in all new coating systems and offers a variety of TS and IF through their Fabrics and Finishes Department. They have marketed a number of effective WB coatings (including ED), HS coatings, urethanes, and PC products.

Glidden has had a long history in the TS and IF business. They had an early involvement in anodic ED (together with Ford Motor Co.) for automotive applications. They are also active in WB, PC, HS, and RC coatings.

Table XII. U.S. Coating Producers: Sales, Fiscal 1980

Company	Coatings Sales		Estimated[a]	
	$ mm	% of corp.	% TS	% TF
Sherwin–Williams	600	55	70	30
PPG	545	19	30	70
Du Pont	385	1	60	40
Glidden (SCM)	310	21	60	40
Mobil Chemical (Mobil)	265	1	10	90
DeSoto (Sears, 31%)	226	67	50	50
Inmont (UTC)	225	41		100
Benjamin Moore	167	98	90	10
Grow Group	140	61	70	30
Insilco	135	21	100	
Total ($ mm)	2998		1524	1474
Total, industry ($ mm)	7636			

[a]For this table, the following definitions are used: Trade sales, TS = architectural coatings plus industrial maintenance; industrial finishes, IF = chemical coatings plus automotive and machinery refinish.

Source: "Kline Guide to the Paint Industry," 6th edition (except the last two columns).

Each of these four companies, through other divisions within their corporate structure, also offers products as raw materials or intermediates for the manufacture of coatings.

In summary, despite current depressed markets, the coating industry remains alive and well. The prospects for growth are evident; the markets are viable and attainable. Cooperative efforts among raw material suppliers, formulators, equipment suppliers, coatings users, and other components of coatings' "ecological cycle" should continue. All are dedicated to producing cost-effective coatings and coated substrates with minimal environmental and health impact. However, it is also essential to stress that coatings must be competitive with other types of surface treatments and that they must be salable to the end user at a reasonable profit for the coatings producer.

Appliance Coatings

THOMAS J. MIRANDA

Whirlpool Corporation, Benton Harbor, MI 49022

History
Appliance Markets
Coating Processes
   Cleaning
   Phosphating
   Flow Coating
   Electrocoating
   Topcoats
   Application of Topcoats
Appliance Coating Specifications
Recent Developments
   Flow Coat Conversion
   Electrocoating
   High Solids
   Powder Coatings
   Low Energy Cure Coatings
   Coatings to Replace Porcelain
Future

Appliances may be classified into three major types: white goods, brown goods, and traffic appliances. White goods are primarily those products associated with laundry and refrigeration such as washers, dryers, dishwashers, ranges, refrigerators, and freezers. The name was due to the primary use of white for these products. Brown goods include radios, television sets, furnaces, air conditioners, and home entertainment products, while traffic appliances include toasters, mixers, electric knives, fans, blenders, and other more portable items.

The larger appliances are generally prepared from metals, while the traffic appliances employ large amounts of plastics because of light weight, durability, design flexibility, productivity, and cost.

0097–6156/85/0285–0883$06.75/0
© 1985 American Chemical Society

Major appliances consume large quantities of steel and aluminum that must be coated to prevent corrosion. Cold roll steel and enameling iron (for porcelain) are the predominant metals used in large appliances.

Because of the environment to which appliances are exposed, coatings vary from the inert glass of porcelain to hard flexible organic finishes. Appliance coatings are exposed to detergents, water, cleaning agents, abrasion, impact, and food stains. For severe applications such as the automatic washer tub, basket, top, and lid, porcelain has been the preferred coating. In some refrigerator and freezer liners, porcelain had been used extensively but is being replaced by powder coatings and precoated steel or aluminum. The high temperatures used in ranges require porcelain enamels, and it is doubtful that organic coatings can replace porcelain for these applications.

In addition to providing protection for the substrate, the coating must above all present an attractive finish for showroom appeal and compliment the decorative scheme of the homeowner, meet rigid test specifications, and represent a favorable cost/benefit relationship for the manufacturer.

History

The history of major appliances goes back to the beginning of the twentieth century involving hand-operated washing machines, ice boxes, and wood-, coal-, gas-, and kerosene-fueled ranges. Less than a million homes were wired for electricity. The turning point came with the development of the universal electric motor, which led to the introduction of the vacuum cleaner. By 1909 vacuum cleaners and washers were being powered by a fractional horsepower motor; by 1920 the first million washers and vacuum cleaners had been sold (1).

With the advent of rural electrification, the major appliance industry took hold. From 10,000 refrigerators in 1920, the appliance industry now sells about 6 million refrigerators per year in the United States.

Early coatings for appliances were slow-drying varnishes based on shellac and later on natural oils and gums. These were replaced with faster drying phenolics modified with tung oil with excellent acid-alkali and corrosion resistance (2). With the advent of the nitrocellulose lacquers following World War I, rapid-drying topcoats became available that revolutionized the industry.

The development of the alkyd resin by Kienle (3) provided a system that lasted from the thirties until they were largely supplanted by the thermosetting acrylic. Although alkyds developed more build, they required long baking times and high temperatures, 200 °C or higher, compared to the nitrocellulose lacquers. Other resin types entered the picture such a vinyls, melamines, epoxies, and silicones. The alkyd resin became the major appliance topcoat as improvements in alkyd chemistry advanced. Melamine-modified alkyds improved the properties as well as lowered the bake schedule. Since the alkyds are based on natural oils, long-term yellowing and food staining were an ongoing problem.

A major revolution in appliance coatings came about with the discovery by Strain (4) of the thermosetting acrylic. This was made practical by Christenson (5), Sekmakas (6, 7), and others (8).

Thermosetting acrylics became the hallmark of appliance finishes for over a decade but became a victim of the environmental movement, which required lower emissions. To achieve lower emissions, the volume solids content had to be increased. However, increasing the solids content increased the viscosity beyond the capability of existing application equipment. This led to the development of alternative coatings designed to meet environmental regulations (9). The new approach to compliance included high solids based on polyester, electrocoating, and powder coatings (10). Concurrent with the development of these technologies was the emergence of high-speed turbines to deposit the higher viscosity liquids, cationic electrocoating for single-coat applications, and powder coatings capable of being deposited in thin films.

Appliance Markets

The appliance market consumes a large volume of coatings, adhesives, and plastics. In 1980 over 383 million units were sold (11). Of these, major appliances represented some 35 million units. Table I lists some major appliance shipments for 1980. The quantity of paint per unit for several appliances is shown in Table II. While the total gallonage varies, the per unit value provides a means for estimating the yearly consumption of coatings in this industry. Although several appliances such as the refrigerator and washer are mature products, the ongoing replacement market opportunity prevents these products from becoming obsolete.

New major appliances are difficult to develop. The last new major appliance to be marketed was the trash compactor, which followed the dryer introduced 35 years earlier. The appliance market exceeded the 10 billion dollar mark in 1980. The number of major producers has decreased in recent years through acquisitions and mergers, so that today there are five major producers who account for the biggest market share. For example, in refrigerators, four suppliers share 93% of the market.

Coating Processes

Coating of appliances involves several key steps on conveyorized lines. These are cleaning, metal preparation, priming, baking, topcoating, and baking.

Appliance coating lines are highly automated, consisting of overhead chain conveyors that carry parts throughout the various stages of processing; even part storage is accomplished on conveyors.

Cleaning. Cleaning is generally carried out in a multiple-stage operation, which includes treatment with zinc or iron phosphate. The objective is to remove oils used in forming parts as well as mill oils added during the manufacture of steel.

Alkaline cleaners and detergents are used in spray washers followed by hot water rinses. Water temperatures of 74 °C are required to melt fats and oils and to facilitate removal of smut or

Table I. Major Appliance Shipments 1980

Appliance	Unit (Millions)
Automatic washer	4.5
Dryers (gas and electric)	3.5
Dishwashers	3.5
Disposers	3.3
Trash compactors	0.29
Ranges (gas, electric, and microwave ovens)	7.8
Refrigerators	5.1
Freezers (chest and upright)	1.7
Air conditioners	6.0
Furnaces	2.3
Water heaters	5.3
Television sets	16.4

Source: Appliance, March 1981.

Table II. Quantity of Paint Used per Appliance

Appliance	gal./unit	Total (millions of gal.)
Room air conditioner	0.6	1.8
Gas dryer	0.5	0.385
Electric dryer	0.5	1.4
Automatic washer	0.4	1.96
Electric range	0.3	0.81
Refrigerator	0.6	3.48
Central air conditioner	0.2	0.42
Color TV	0.02	0.164
Total		10.419

Source: Appliance Manufacturer, 1976.

other dirt on steel. Since the energy crisis of 1974, there has been a significant effort to reduce the temperature of the cleaning process. Some systems operate at temperatures as low as 40–45 °C (13, 14). One of the difficulties of lower wash temperatures is that certain higher melting oils cannot be effectively removed in the washing cycle.

The cleaning and phosphating process is very energy intensive and consumes up to 50% of the process energy in an appliance plant. On the other hand, the cleaning process is the most critical step in the coating process and until recently has not received the proper attention by process engineers.

Methods for determining surface cleanliness are too involved for online evaluation, so that simple tests such as water break are employed. Recently, Buser reported a rapid method based on surface tension (15). Evaporative rate analysis (16) has also been used to determine surface cleanliness as well as more sophisticated methods employing scanning auger spectroscopy (17, 18).

Phosphating. Two types of phosphate treatment are used for treating steel, depending upon the degree of corrosion resistance required. Phosphate treatment imparts corrosion stability, prevents rust creep, and imparts the necessary tooth to promote adhesion of primers. For lower corrosion requirements such as for a furnace, dryer, or freezer, iron phosphate is used. Iron phosphate provides a smooth dense film and is deposited at a film weight of about 70–100 mg/ft^2. Being smooth, the amorphous iron phosphate coating is preferred for single–coat applications.

When greater corrosion protection is required, especially for laundry products or air conditioners, zinc phosphate treatment is preferred. Zinc phosphate is more coarse than iron and is used with a primer. With microcrystalline zinc phosphates, direct topcoating can be achieved. Film weights of 180–210 mg/ft^2 are used to ensure adequate protection.

Recently, some suppliers have converted zinc phosphate back to iron phosphate because of the reduced amount of sludge generated by iron phosphate. However, detergent resistance and corrosion resistance are lower for iron.

Following the phosphate treatment, the substrate is given a chrome rinse for passivating the surface. Because of environmental problems there has been a trend away from chrome to chrome–free rinses. In some cases where chrome rinses are used, Cr^{6+} is reduced to Cr^{3+} before discharging to waste treatment.

A typical system for cleaning and phosphating of refrigerators may involve the following steps: tank 1, cleaner at 160–170 °F (70–75 °C); tank 2, rinse at 145 °F (62–68 °C); tank 3, cleaner at 145–155 °F (62–68 °C); tank 4, rinse at 145 °F (62 °C); tank 5, zinc phosphate at 150–160 °C (65–70 °C); tank 6, cold water rinse; tank 7, chromic acid rinse; tank 8, deionized water rinse (ambient).

Flow Coating. The principal priming methods are dip or flow coating. For small parts and nonappearance parts, such as heat exchangers on freezers, dipping is the preferred method. For faster line speed, flow coating is the preferred method (Figure 1). Before the introduction of electrocoating (a dip process), flow coating was the

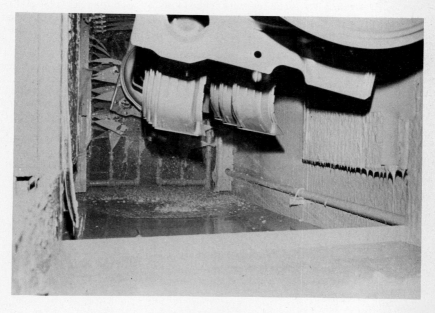

Figure 1.   Flow coating dryer parts.

preferred priming method.   In many cases the prime coat is actually
the only coat that the metal will receive. Examples include toe
plates, gussets, the supporting member, dryer drums, and bulkheads.
     Flow coaters permit fast line speed, use less coating than dip
coats, but require a solvent flash off.   The process is basically a
pumping system that floods the parts with paint from spray nozzles.
The moving ware then moves to a vapor chamber where bubbles can
break and the film can flow out.   Flow coating has several
disadvantages.   The film tends to wedge with fat edges on the bottom
of the part and heavy accumulations that can solvent pop in the
oven.   The presence of a vapor chamber for organic solvent presents
an explosion or fire hazard.   Finally, recessed edges are difficult
to cover, and puddling can occur in crevices.
     The flammability problem was resolved by converting to water-
borne flow coat primers.   The first waterborne appliance flow coat
primer was installed on a dryer drum line in 1972 (9).   The primer
was a water-soluble epoxy-modified acrylic that had excellent
hardness, abrasion resistance, and chemical resistance.
     Epoxy resins have provided excellent corrosion resistant primers
for appliances.   These can be made from epoxy resins esterified with
dehydrated castor oil fatty acids and cured with melamine.
Excellent detergent resistance and corrosion resistance are obtained
from this type of resin system.
     Another type of primer is based on a styrene/allyl alcohol
copolymer (Monsanto RJ-100).   This polymer can be modified with
fatty acids and cured with melamine resin (19).
     Reaction schemes for primers are shown as follows:

EPOXY RESIN

$$CH_2\text{-}CH\text{-}CH_2\text{-}O\text{-}\langle\bigcirc\rangle\text{-}\overset{CH_3}{\underset{CH_3}{C}}\text{-}\langle\bigcirc\rangle\text{-}O\text{-}CH_2\text{-}CH\text{-}CH_2 \quad + \; 2RCOOH \longrightarrow$$

$$R\text{-}\overset{O}{C}\text{-}O\text{-}CH_2\text{-}\overset{OH}{CH}\text{-}CH_2\text{-}O\text{-}\langle\bigcirc\rangle\text{-}\overset{CH_3}{\underset{CH_3}{C}}\text{-}\langle\bigcirc\rangle\text{-}O\text{-}CH_2\text{-}\overset{OH}{CH}\text{-}CH_2\text{-}O\text{-}\overset{O}{C}\text{-}R$$

RJ100 RESIN

$$-CH_2\text{-}CH\text{-}CH_2\text{-}CH\text{-} \quad + \; RCOOH \longrightarrow \quad -CH_2\text{-}CH\text{-}CH_2CH\text{-}$$

Electrocoating.   Major disadvantages of the flow coat system is the
inability to coat recesses and edges and the presence of thin spots
due to the solvent wash of deposited primer.   This leads to lower

detergent and salt spray resistance. A major improvement in appliance priming occurred with the introduction of electrocoating in 1962 on automatic washer cabinets. Electrocoating was introduced to the automotive industry by Brewer (20) to reduce corrosion. This highly automated system provides excellent coverage of edges and recesses and is not affected by solvent wash since the deposited coating is not soluble in rinse water.

Electrocoating vehicles were initially prepared from natural oils and maleic anhydride (21). Those adducts can then be solubilized by ammonia or other bases. Electrocoating resin systems were initially anodic because of ease of preparation and availability of acidic monomers. About 10 years after the introduction of anodic electrocoating, cathodic electrocoating was developed by Bosso and first successfully applied to air conditioner compressors (22, 23).

The results obtained from electrocoating were so good that flow coating's domination as a priming process was replaced for laundry cabinets and small parts. The remaining flow coat systems were either phased out or converted to waterborne types.

In many instances the flow coat or electrocoat primer serves as the finish coat as in dryer drums and bulkheads, air conditioner cabinets and trash compactor boxes, back plates, and other small parts.

Electrocoating requires a number of controls. First, conductivity must be controlled since the solubilizing base (or acid with cathodic types) accumulates in the tank, thereby increasing conductivity and affecting deposition. This can be controlled by adding makeup primer deficient in amine (or acid) or by using a process called ultrafiltration (24). As solids are depleted, a makeup charge consisting of resin, pigment, and additives is fed to the electrocoat tank.

Rinsing of ware provides an economy as electrocoating solids can be returned to the tank, providing high utilization of paint. The pH must also be carefully controlled to avoid kick out of the resin. Cooling must also be provided to control bath temperature.

Ultrafiltration not only provides high utility of coating material but also contributes to lower pollution.

Baking temperatures for electrocoating vehicles vary depending upon the polymer used and the degree of corrosion resistance required. Anodic primers are baked from 375 to 450 °F, while cathodic primers are baked as high as 450°F. Current research by manufacturers of coatings is directed toward lower bake temperatures without compromising physical properties.

Typical properties of a cationic appliance primer are shown in Table III, and an electrocoat tank is shown in Figure 2.

Topcoats. For some applications such as chest freezers or vacuum cleaners, alkyds are still employed. These evolved from oil-modified types to the oil-free polyesters that have good stain resistance, high gloss, and excellent flexibility. But until recently, the thermosetting acrylic was the hallmark of appliance coatings. Acrylics exhibit excellent hardness, adhesion, chemical resistance, light fastness, and resistance to stain, detergents, and salt. In addition, acrylics provide high gloss for refrigerators but are formulated at lower gloss for laundry products.

Prior to environmental regulations, the acrylic topcoats were applied at 34% volume solids and provided high build and good flow and leveling. With regulations, volume solids were pushed up to 62% volume solids to comply with EPA guidelines (25,26).

Thermosetting acrylics can be produced from a variety of monomers in varying percentage compositions; the systems utilize a variety of cross-linking mechanisms (Table IV). A typical thermosetting acrylic can be prepared from 15–80% styrene, 15–18% alkyl acrylate, and 5–10% acrylic acid (27). Acrylic acid provides the functionality for cross-linking with epoxy resins.

Acrylic acid copolymers can be modified with 1,2–butylene oxide to provide hydroxyl sites for cross-linking (28). The copolymer can then be cross-linked with amino resins.

Another approach is to prepolymerize an epoxy methacrylate or acrylate to provide an epoxy function that can be cross-linked by acids (29, 30). More elaborate cross-linking mechanisms include oxazolines (31), which can be formed in situ.

The methylolated acrylamide type has been widely used in appliance finishes. These are cross-linked with melamine or epoxy modifiers and form hard, chemically resistant finishes. Detailed discussion of thermosetting acrylics can be found in References 2 and 8. Recently, topcoats based upon high-solids liquid polyesters have replaced thermosetting acrylics in response to environmental

Table III. Properties of an Electrocoat Primer

Properties	Values
Coating	epoxy
Type	cationic/electrocoat
pH	6.2–6.5
Solids (w/w)	10–14%
Solvent	water, coupling solvents
Film	
Bake	20 min at 410 °F
Color	gray
Adhesion	pass
Detergent, 1.5% at 165 °F	1000 h
Impact, direct	40 in. lb
Thickness	0.5 mils
Chlorox, neat 72 h at 140 °F	pass
Hardness, pencil	4H

Figure 2. Electrocoating of washer cabinets.

Table IV. Types of Thermosetting Acrylics

Type	Functional Monomer	Cross-linker
Acid	$CH_2=CH–COOH$ Acrylic acid	Epoxy resin
Hydroxyl	$CH_2=CH–CO$ $O–CH_2–CH_2OH$ 2–Hydroxyethyl acrylate	Melamine
Epoxy	$CH_2=CH$ $COOCH_2–CH–CH_2$ (epoxide O) Glycidyl acrylate	Acid
Methylol	$CH_2=CH$ $CONHCH_2OH$ Methylolated acrylamide	Epoxy, melamine
Oxazoline	$–CH_2–CH–$ ring: C, O, N, CH_2, $C(CH_2OH)_2$ 	Melamine

constraints. Powder coatings are also making inroads into the acrylic market. These polyesters are based on neopentyl glycol, isophthalic acid, and trimethylolpropane and are cross-linked with epoxy resins (32). They possess excellent gloss, stain resistance, flexibility, and adhesion and have a pencil hardness of 2H or higher.

Application of Topcoats.   Appliance topcoats are applied by both hand and automatic electrostatic spray. For high-volume production on washer and dryer cabinets, a Ransburg No. 2 electrostatic disk typically is used. The paint is supplied to a rotating disk that is at a 90,000–110,000–V potential to the part being coated. As the paint is spun off the rotating disk (900–1800 rpm), it is broken up into droplets and charged. This provides excellent coverage, good wraparound, and an efficiency of application of over 90% utilization (Figure 3).

Another method for applying topcoats is the Ransburg No. 2 bell. This unit consists of a small rotating bell that breaks up and charges the paint as it leaves the periphery of the bell (Figure 4). Older bells had lower speeds, but newer high-speed Mini Bells operate at higher speeds and permit the use of newer high-solid coatings. Bells arranged on stands are an effective means for coating long parts such as refrigerator cabinets or doors. Unlike the disk, which is mounted vertically, the bell can operate from a number of positions and is more versatile.

Hand guns are used to touch up areas that are not adequately covered by the disk or bells and are also used to reinforce critical areas where more coating is desired.

Film thicknesses vary with the product. For primers, 0.4–0.5 mil is typical while topcoats are 0.7–1.5 mils applied over a primer. A total system of 1.2–2.5 mils is suitable for laundry products. On refrigeration products either a wet-on-wet topcoat or a primer and a thinner topcoat system is used, depending upon the condition of the metal. If there are scuff marks on the metal, the latter system is employed. Porcelain coatings are much thicker, for example, up to 10 mils.

Appliance Coating Specifications

Table V lists some specifications for a laundry appliance finish, which vary according to model as well as type of laundry appliance. For example, an automatic washer has a more severe detergent resistance requirement than a dryer. On the other hand, a refrigerator has no detergent requirement but must be stain resistant and chemical resistant to a variety of cleaners or foods that this product encounters.

For washers and dishwasher tubs, porcelain is required that has an entirely different specification. Some manufacturers use vinyl plastisols for dishwasher liners with good success.

A list of specifications for refrigerators is shown in Table VI. Note the emphasis on stain resistance due to the exposure requirements of the kitchen.

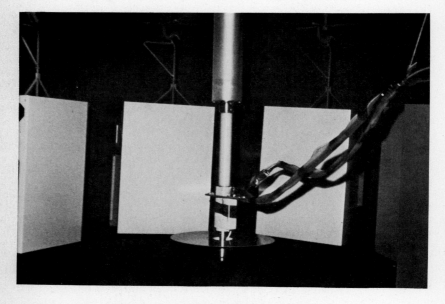

Figure 3.   Electrostatic disk.

igure 4.   Electrostatic bell coating refrigerator doors.

Table V. Laundry Specifications: Washer

Property	Value
Pencil hardness	2H
Adhesion, 1/16-in. crosshatch	No loss
Stain resistance, 24 h	None
Detergent immersion	250 h
Abrasion, taber, CS10, 1000g	15 mg/100 cycles
Humidity, 100%	1000 h
Salt spray, 5%	500 h
Flexibility	1 in. from small diameter of mandrel
Heat, 200 h at 200 °F	1.5 RD units
Impact, direct	40 in. lb
Grease, 50/50 oleic cottonseed oil	168 h
Gloss, 20°	50-75

Table VI.   Refrigerator Finish Specification

Property	Value
Pencil hardness minimum	2H
Abrasion, taber, CS10, 1000 g	60–100 mg
Intercoat resistance (Hoffman scratch)	2500 g
Flexibility, 180° bend, 1/4-in. mandrel	No flaking
Impact, direct/reverse	20/10
Adhesion crosshatch, 1/32 in.	Pass
Salt spray, 5%	500 h
Warm water soak, 120 °F	750 h
Humidity, 100%	1000 h
Gloss, 60° minimum (%)	85
Mold resistance	No growth
Weatherometer	300 h
Grease, 50/50 oleic acid/cottonseed oil grease fume, 60 min	5 weeks no change in gloss or hardness
Stain resistance (24-h exposure)	No color change
Anthraquinone Violet R	None
Butter	None
Cigarette burn	Slight
Cigarette smoke	None
Citric acid, 1%	None
Ethyl alcohol	None
Grease pencil	None
Lipstick	None
Mustard	None
Sodium hydroxide, 0.5%	None
Tape	None
Vinyl gasket	None

Recent Developments

Beginning in 1970 the appliance finish was subjected to a complete revolution brought about by government regulations of the Environmental Protection Agency (EPA) and by the energy crisis of 1974. The first attempt by EPA was to convert appliance coatings to nonexempt solvents based upon Rule 66 as practiced in California. However, this was not a reasonable approach since hydrocarbon emissions were still present. Nevertheless, coatings suppliers reformulated to exempt solvents at a great cost.

The advent of EPA did bring about a number of significant changes in appliance coatings. These are conversion of flow coating from solventborne to waterborne systems, replacement of flow coating with electrocoating, introduction of cationic electrocoating, introduction of high-solids coatings, introduction of powder coating, development of low-temperature baking systems, and development of porcelain replacement coatings.

Flow Coat Conversion. Solventborne flow coaters are a major source of hydrocarbon emissions. This source became an early target for a replacement system. An alternative to reduce emission was the gas incinerator, but high capital costs and gas consumption precluded its use.

The first successful waterborne flow coat for appliances was developed by the Glidden Division of SCM and applied to dryer drums and bulkheads. Emissions were reduced by 80%, and fire hazards associated with the solvent system were reduced significantly.

Electrocoating. Electrocoating is a significant improvement over flow coating. Recessed edges, corners, and holes are uniformly coated, and the system is highly automated. Anodic electrocoating is used as a primer for washer cabinets, small parts of dryer cabinets, and trash compactor components. A major innovation in electrocoating was the successful introduction of cationic electrocoating. The first application was for air conditioner compressors (23). This was followed by air conditioner cabinets replacing a two-coat system consisting of a flow coat primer and an acrylic topcoat. Corrosion resistance of 1000 h was achieved with a cationic epoxy electrocoating system.

Cationic electrocoating is now used on dryer drums, washer cabinets, and areas where superior corrosion resistance is required. Although cationic electrocoating is somewhat more expensive (10% higher than anodic), its advantages for certain applications justifies the higher cost.

After the successful introduction to appliances, the automotive industry began a major conversion to cationic electrocoating. Advantages of the system are good chemical resistance, excellent salt spray resistance, hardness of 4H+, flexibility, and low dissolution of phosphate at the anode (33).

Some problems were encountered with ultrafiltration, but they have since been resolved.

Cationic epoxy resin based systems can be prepared by reacting quaternary amine salts with epoxy resins to introduce the cationic functions. The polymers can then be cross-linked with blocked isocyanates at reasonable baking temperatures (34, 35).

$$Z-\overset{\displaystyle \underset{O}{\diagdown\!\!\diagup}}{CH}-CH_2 + R_1-\overset{\displaystyle R_3}{\underset{\displaystyle \underset{X^-}{H}}{N^+}}-R_2 \longrightarrow R_1-\overset{\displaystyle R_3}{\underset{\displaystyle \underset{\displaystyle \underset{\displaystyle \underset{Z}{CH-OH}}{CH_2}}{}}{N^+}}-R_2 \quad X^-$$

A survey of polymer compositions for cationic electrocoating was prepared by Kordomenos and Nordstrom (36).

Recently, cationic acrylic white enamels have been developed that have high gloss, hardness and abrasion resistance and provide a means for developing single coat applications. A recent application for this system is on freezer baskets.

For primers not requiring high corrosion resistance, anodic electrocoating will still be a viable choice.

High Solids. In an effort to reduce emissions, higher solids coatings were developed for the appliance industry. EPA regulations require no more than 2.8 lb of solvent/gal. of paint (0.34 kg of solvent/L of paint) by 1982. For the conventional thermosetting acrylic, raising volume solids increases the viscosity such that conventional application equipment is not capable of applying the coating.

Two major developments occurred that overcame the problem. Coating research laboratories developed high-solids low-viscosity coatings based on polyester intermediates or oligomers. Those lower molecular weight polymers have higher functionality. In addition, new melamines were developed to cross-link the more reactive systems (37).

Manufacturers have also recognized the need to develop application equipment to handle higher solids coatings. As a result, high-speed turbines have been developed by a number of firms. Depending upon the system, coatings having up to 80% solids can be applied with or without heaters (38). Turbines with speeds up to 60,000 rpm were developed in contrast to conventional disks operating at 1500 rpm. At 20,000 rpm these turbines can effectively atomize coatings having Ford No. 4 viscosities of 100 s.

Recently, oligomeric acrylics cross-linked with isocyanates or epoxy resins have become available for appliance finishes (39). High-solids coatings should be a major factor in the future of appliance finishes in that they permit the use of existing paint lines with little modification. At 62% volume solids, minibells can be used. Because of the bubble concept, an appliance coating employing a waterborne flow coat or electrocoat system can be used to offset a topcoat with slightly lower volume solids that can be applied on lower speed disks.

Powder Coatings. Powder coatings have made inroads into the appliance market (40). Their principal value is little or no solvent emission and high utilization of coating. Fluidized bed vinyl systems are used on wire goods such as dishwasher racks. Electrostatic powder spray using guns or disks are used on freezer and refrigerator liners, dryer drums, doors, and range cabinets.

An interesting application of powder coatings is in replacement of plating. Refrigerator shelves are coated with an epoxy powder replacing zinc plating, thereby eliminating wastewater disposal and providing a quality finish. On chest freezer liners, powder coatings have replaced porcelain, thus greatly reducing the high energy consumption of porcelain furnaces.

Acrylic, epoxy, and polyester powders are all used in appliance finishes. Properties of a typical acrylic as shown in Table VII (41, 42). For exterior exposure, the acrylic or polyester is preferred.

Low Energy Cure Coatings. With strong emphasis on energy conser-vation, appliance manufacturers sought to lower the energy required in the baking or curing process. One approach to energy-independent coatings is the urethane coating. Urethanes can be formulated to a variety of curing schedules, especially the blocked and multicomponent types.

Appliance finishes have been formulated and tested by using acrylic or polyester modified urethanes that can cure to a suitable hardness in 10 min at 150 °F and cure to an ultimate hardness of 2H in 168 h (43).

The major barriers to the application of urethanes in appliances is cost and perceived toxicity. However, the outstanding performance of urethanes may yet overcome the problems cited above.

Another approach to low energy cure coatings is the aliphatic epoxy systems (44) and the low fusing powders that cure at 250–275 °F. Self-cross-linking emulsions suggest another way of achieving low energy cure coatings.

Coating To Replace Porcelain. Porcelain is still widely used in appliances and is accepted as a quality finish. Porcelain frits are generally ground on site in large ball mills and let down to spray viscosity. Porcelain also requires zero carbon steel and a pickling process for proper adhesion.

Recent developments in porcelain include powdered frit and the use of cold rolled steel instead of enameling iron. Porcelain steel requires a high fusion temperature, 1500 °F. This has led appliance manufacturers to seek alternatives to porcelain.

The first successful application was on a range door, replacing porcelain with an acrylic system. Because of federal standards, the outer door temperature must be below 160 °F, which permits the use of organic coatings.

Chest freezer and refrigerator liners have been successfully coated with powders or precoated metals to eliminate porcelain with good success. The improved impact resistance results in significant savings from shipping damage.

Table VII (<u>41</u>). Properties of an Acrylic Powder

Properties	Values
Powder properties	
Specific gravity	1.2–1.8
Storage stability	1 year at 80 °F
Particle size	20–40 µm average
Application data	
Voltage	50–100 kV
Typical cure	15 min at 375 °F metal temperature
Minimum cure	330 °F metal temperature
Overbake color stability	100%--time at temperature
Film thickness range	0.7–3.0 mils
Film properties	
Gardner impact	20–100 in. 1b
Pencil hardness	2H–3H
Flexibility	Pass 1/8-in. mandrel
Salt spray	1/16-in. creepage at 1000 h
Adhesion	No failure with 1/16-in. squares
Weathering	Minimum change 500 h--Atlas Weatherometer

Note: All tests were on the B–1000 test panel at 1.5 mils.

For laundry applications, organic coatings are replacing porcelain on tops and lids and can conceivably be used on tubs and baskets since the technical feasibility of organic coatings in these applications has been demonstrated (10).

Future

Future coating trends for appliance coatings will be high-solids, waterborne including electrodeposition, and powder coating in that order. High-solids coatings permit the use of existing equipment whereas powder requires new capital equipment (45). Electrocoating and waterborne flow coats will be used for priming and in some cases as the topcoat. Cationic electrocoating will find favor in high-corrosion environments and for porcelain replacement. Porcelain usage will continue to be eroded in favor of high-performance coatings. Coil coatings will become an increasing factor in appliances. The use of coil reduces pollution, cleaning, and phosphating, and bake ovens. Higher cost of scrap and raw edges are major concerns. However, coil coatings are being used successfully on refrigerator liners, range shells, and dehumidifier cabinets. For large cabinets such as the refrigerator, special joining techniques are required to reduce damage to the finish since weld burns damage the finish.

An interesting approach to coil coating is glass transition forming in which the coated coil is heated above its glass temperature for forming. This prevents cracking and permits the use of a high-hardness coating with good scuff resistance. Dehumidifier covers are prepared in this manner (46).

Robots should play an increasing role in the coating of appliances such as microwave oven cavities and roundware where bounceback is a problem.

Greater emphasis will be placed upon the quality of metal surfaces. Methods for determining the quality of cleaning and phosphating on line will be implemented. A good example is the development of an infrared analysis technique for zinc phosphate by Cheever (47), which permits rapid determination of coating weight in a nondestructive manner. An instrument for this purpose was developed by Foxboro (48).

Literature Cited

1. Franklin, J. F. "Full-line Development, Related Mergers and Competition in the Major Appliance Industry During the 1980s"; University Microfilms: Ann Arbor, Mich., 1963; 64-6679.
2. Shur, E. G. "Treatise On Coatings"; Myers, R. R.; Long, J. S., Eds.; Marcel Dekker: New York, 1975; Vol. 4, Chap. 2.
3. Kienle, R. H. Ind. Eng. Chem. 1949, 41, 726.
4. Strain, D. E. U.S. Patent 2 173 005, Sept. 12, 1939, to E. I. du Pont de Nemours & Co.
5. Christenson, R. M. U.S. Patent 3 037 963, June 5, 1962, to PPG.
6. Sekmakas, K. U.S. Patent 3 163 615, 1964, to DeSoto.
7. Sekmakas, K.; Ansel, R.E.; Drunga, K. U.S. Patent 3 163 623, 1964, to DeSoto.
8. Brown, W. H.; Miranda, T. J. Off. Dig. 1964, 36(475), 92.

9.   Miranda, T. J. "Water Soluble Polymers"; Bikales, N. M., Ed.;
     Plenum: New York, 1973; p. 187.
10. Miranda, T. J. J. Coat. Technol. 1977, 49, 66.
11. Appliance 1981, March, 22.
12. Appliance Manufact. 1976, Jan. 92.
13. Varga, C. Ind. Finish. 1974, 50, 62.
14. Obrzut, J. J. Iron Age 1981, June 1, 1981, 43.
15. Buser, K. R. Org. Coat. Plast. Chem. Preprints 1980, 43, 498.
16. Anderson, J.; Root, D.; Greene, G. J. Paint Technol. 1968,
     40(523), 320.
17. Wojtkowiak, J.; Bender, H. S. Org. Coat. Plast. Chem. Preprints
     1980, 43, 513.
18. Zurilla, R. W.; Hospadaruk, V. SAE Technical Paper 780187,
     March 1978.
19. Monsanto Technical Bulletin 5035.
20. Brewer, G. E. F. J. Paint Technol. 1973, 45(587), 36.
21. Miranda, T. J. Off. Dig. 1965, 37(489), 62.
22. Wismer, M.; Bosso, J. F. Chem. Eng. 1971, 78(13), 114.
23. Miranda, T. J. Org. Coat. Plast. Chem. 1981, 45, 109.
24. LeBras, L. R.; Christenson, R. M. J. Paint Technol. 1972,
     44(566), 63.
25. Obrzut, J. J. Iron Age, 1981, Sept. 16, 51.
26. Schrantz, J. Ind. Finish. 1978, June, 24.
27. Segall, C. H.; Cameron, J. L. U.S. Patent 2 798 861, 1956, to
     Canadian Industries, Ltd.
28. Vasta, J. Belgian Patent 634 310, 1963, to Du Pont.
29. Simms, J. A. J. Appl. Polym. Sci. 1961, 5, 58.
30. Ravve, A.; Khamis, J. T. U.S. Patent 3 306 883, 1967, to
     Continental Can.
31. Miranda, T. J. J. Paint Technol. 1967, 39, 40.
32. Eastman Chemical Products Bulletin N-277, August 1981.
33. Anderson, D. G.; Murphy, E. J.; Tucci, J. J. Coat. Technol.
     1978, 50(646), 38.
34. Jerabek, R. D. U.S. Patent 3 799 854, 1974, to PPG Industries,
     Inc.
35. Bosso, J. F.; Sturni, L. C. U.S. Patent 4 101 486, 1978, to
     PPG Industries, Inc.
36. Kordomenos, P.; Nordstrom, J. D. Org. Coat. Plast. Chem. 1980,
     43, 154.
37. Blank, W. J. Coat. Technol. 1982, 54(687), 26.
38. Tholome, R.; Sorcinelli, G. Ind. Finish. 1977, 53(11), 30.
39. Brushwell, W. Am. Paint Coat. J. 1981, Nov. 23, 47.
40. Gribble, P. R. N.P.C.A. Chemical Coating Conference,
     Cincinnati, Ohio, May 10 and 11, 1978, Powder Coating Session,
     p. 47-61.
41. Glidden Chemical Coatings and Resins. Pulvalure 154.
42. Harris, S. T. "The Technology of Powder Coating"; Portcullis:
     London, 1976.
43. Consdorf, A. P. Appliance Manufact. 1977, May, 64.
44. Bauer, R. S. CHEMTECH 1980, 10(11), 692.
45. Petrovich, P. SME Technical Paper FC79-701, 1979.
46. "Glass Transition Forming Saves Painting Costs"; Precis. Met.
     1978, Nov., 69.
47. Cheever, G. D. J. Coat. Technol. 1978, 50(640), 78.
48. Infracoat 450 Analyzer, Foxboro Analytical, South Norwalk,
     Conn.

Polymer Coatings for Optical Fibers

L. L. BLYLER, JR.[1], and C. J. ALOISIO[2]

[1] Bell Laboratories, Murray Hill, NJ 07974
[2] Bell Laboratories, Norcross, GA 30071

Requirements of a Coating Material
Coating Application
Coating Properties and Fiber Performance
  Fiber Strength
  Microbending Loss
  Material Considerations
  Microbending Loss: Relation to Coating Properties
  Microbending Loss: Relation to Fiber Cabling

Optical fiber technology, developed during the 1970s, has ushered in the lightwave telecommunications era of the 1980s wherein the implementation of commercial systems has begun. The optical fiber, shown schematically in Figure 1, represents the medium over which light signals, consisting of streams of digital pulses that comprise voice, video, or data information, are transmitted. The fiber consists of a central core, usually composed of a highly transparent glass, and surrounded by cladding of lower refractive index that confines the light energy to propagate within the core by total internal reflection. The cladding material is commonly another glass, but polymer-clad fibers have also been developed for short to medium distance applications.

The outstanding requirement for the glass core is high transparency so that light propagation will occur with very little energy loss by absorption or scattering processes. Losses below 0.5dB/km at selected transmission wavelengths have now been routinely achieved (1, 2). The cladding material, whether glass or polymeric, does not need to be quite so transparent, but because the light energy propagating within the core penetrates the cladding to some extent, cladding material losses below approximately 1000 dB/km are a practical necessity. This requirement rules out all but the most transparent polymers as cladding materials. Most fibers of commercial importance have cores composed of silica that is doped

0097–6156/85/0285–0907$07.00/0

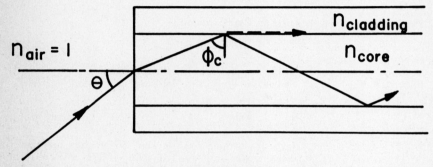

Figure 1.   Schematic of an optical fiber. All light rays that
             enter the fiber end face and strike the core-cladding
             interface at an angle $\phi > \phi_c$ (the critical angle), where
             $\sin \phi_c = n_{cladding}/n_{core}$, are totally internally
             reflected and propagate along the core. The fiber's
             acceptance angle, Φ, is given by $\sin \Phi = (n^2_{core} -
             n^2_{cladding})^{1/2}$, which is called the numerical aperture
             (NA). (Reproduced with permission from Ref. 28.
             Copyright 1981 Gordon and Breach Science Publishers.)

with oxides of germanium and phosphorus to raise the refractive index. The cladding may be either pure silica or a doped silica such as a fluorosilicate. Fibers based on multicomponent glasses such as sodium borosilicate and soda-lime silicate systems have also been developed (3). Polymer-clad fibers usually consist of a silica core clad with either a poly(dimethyl siloxane) resin or a fluorinated acrylic polymer. The refractive index, n, of the polymer must be lower than that of the silica core (n = 1.458), a requirement that restricts the choice of available cladding polymers quite severely. Owing to their higher transmission losses and inherently lower information-carrying capacity, polymer-clad fibers have limited utility for high-performance telecommunications applications, which are beginning to dominate the field. Consequently, we shall not deal with this fiber type further, and additional information is published elsewhere (4). We shall concentrate instead on the all-glass fiber type which, as we shall see, requires a polymer coating. In this case the polymer coating has no special optical wave guiding requirement or function.

Requirements of a Coating Material

Polymer coatings play a crucial role in protecting the glass fiber from physical damage. Figure 2 illustrates the reduction in the tensile strength of a silica fiber caused by stress concentration at a semielliptical surface crack of radius a_c (5). The highest tensile strength that can be practically realized in a carefully produced silica fiber at room temperature is about 5.5 GN/m^2 (800,000 psi). (This result implies that intrinsic structural defects in the glass surface are about 10^{-2} μm large.) However, a 1-μm flaw in the glass surface lowers the tensile strength by nearly an order of magnitude to 0.62 GN/m^2 (90,000 psi). Such tiny flaws are readily produced by abrasive contact of the fiber with another solid surface. Therefore, the optical fiber must be coated as it is drawn, before any contact with another surface can occur.

Figure 3 schematically depicts a typical fiber-drawing machine (6). A glass preform rod is advanced into the top of a tubular high-temperature (~2200 °C) furnace at a controlled rate. The rod softens in the furnace, and it is drawn into a fiber by a capstan at the base of the machine. The fiber passes through a diameter measurement and control-system unit (7) located directly below the furnace. This system utilizes a laser forward-scattering technique to monitor fiber diameter with high resolution (e.g., 0.2μm) at a rapid update rate (e.g. 500 Hz) and provides a signal for feedback control that adjusts the speed of the capstan to drive the fiber diameter toward the desired value. The coating apparatus is located between the fiber-diameter monitor and the capstan at the base of the machine.

The logistics of the fiber-coating operation have far-reaching implications for the choice of coating materials used. For example, the coating must be applied to the fiber without damaging the glass surface. This requirement is best met when the coating is applied as a moderately low-viscosity (5-50 P) liquid. The low viscosity not only allows the coating to be applied to the glass in a gentle manner but also affords the opportunity of filtering the liquid to remove micron-sized particles that might damage the glass surface.

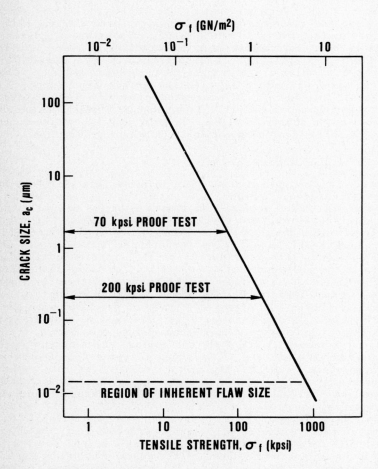

Figure 2.   Tensile strength of a silica fiber possessing a semielliptical surface crack of radius, a_c.

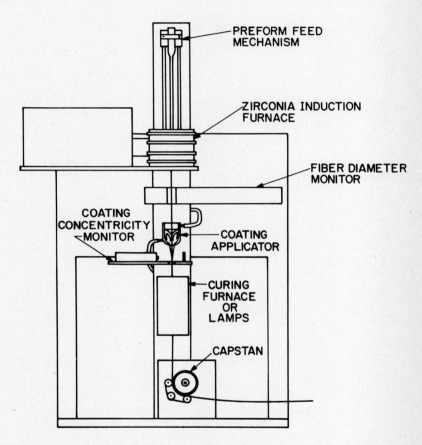

PREFORM FEED
MECHANISM

ZIRCONIA INDUCTION
FURNACE

FIBER DIAMETER
MONITOR

COATING
CONCENTRICITY
MONITOR

COATING
APPLICATOR

CURING
FURNACE
OR
LAMPS

CAPSTAN

Figure 3.   Schematic of an optical fiber drawing machine. (Reproduced with permission from Ref. 6. Copyright 1980
IEEE.)

The coating must also be solidified very rapidly once it is applied to the fiber so that it will offer protection when the capstan is reached. This requirement rules out solvent-containing formulations for all but the thinnest coatings, because solvent removal is a slow process. Solvent-free coating formulations that are rapidly cross-linked by thermal activation or by UV radiation can be used very successfully. Thermoplastic hot-melt systems that solidify quickly upon cooling are also viable. The material systems of choice for this application depend upon both coating-application and coating-performance considerations.

Coating Application

Figure 4 illustrates a simple applicator used for fiber coating. The fiber enters the coating resin contained in the applicator reservoir where it is wet by the liquid. The moving fiber produces a depressed meniscus in the liquid surface at the point of entry and drags the liquid downward into the conical coating die located at the base of the applicator. The drag flow into this converging channel produces a hydrodynamic pressure that drives a return flow or circulation upward along the walls of the die and reservoir. The hydrodynamic pressure also tends to stabilize the fiber position along the centerline of the die because excursions from this position produce a net restoring force. In practice this restoring force cannot be relied upon to center the fiber if its natural path is far from the centerline.

The alignment of the coating die with the fiber is one of the central problems of fiber coating. It is not possible to guide the fiber through the center of the die, because contact with a guide, even when it is immersed in the coating liquid, can damage the glass surface. Thus alignment of the hair-thin fiber (e.g., 125 μm od) within the die, which may be only slightly larger, requires special techniques, because the fiber is fixed in space at two points (the preform and the capstan) that are very remote from the coating applicator.

An early solution to this alignment problem was the use of flexible silicone rubber dies (8) that respond to the hydrodynamic forces generated in the converging flow field and tend to align themselves about the fiber. However, the flexible-die approach is inadequate for meeting close tolerances on coating eccentricity; therefore, a laser scattering technique (9) that has been developed enables coating concentricity to be continuously monitored by optical means. This practice allows the applicator to be mounted on a stage capable of micrometer adjustments so that the die may be accurately positioned about the fiber, and the coating eccentricity may be held within a few microns.

Another coating-application problem involves the meniscus formed at the point of entry of the fiber into the coating liquid at the free surface. As shown schematically in Figure 5, the air streamlines along the free surface of the moving liquid, and the moving fiber will result in a pressure increase at the tip of the meniscus. If this pressure exceeds the pressure-containment capability of the liquid, an air column forms, collapses, and reforms, and air bubbles are entrained in the liquid. These bubbles concentrate in the liquid in time, and large numbers may pass

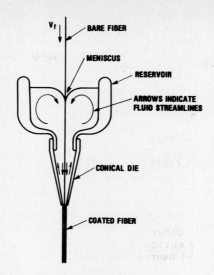

Figure 4.   Schematic of an applicator for coating optical fibers. (Reproduced with permission from Ref. 27.   Copyright 1982 Society of Plastics Engineers.)

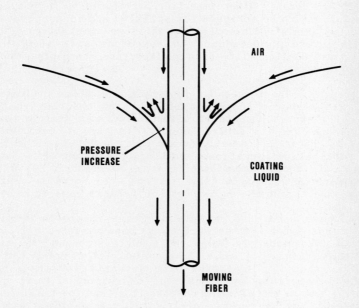

Figure 5.   Streamlines in the air and liquid at the coating liquid meniscus.

through the die where they become trapped in the coating. Large bubbles within the coating are considered to be coating defects that can affect fiber properties such as transmission loss.

As fiber draw speed is increased, severe instabilities of the meniscus occur. Ultimately, the air column formed around the fiber may propagate completely through the applicator to the exit of the die, resulting in uncoated or poorly coated fiber sections. Figures 6a and 6b show the air column propagating toward and through the exit of a conical die, and Figures 7a and 8a show defective coatings that result from this event. Frequently, sectioning of such coating defects reveal voids produced by entrained air during coating (Figures 7b and 8b).

A two-chamber coating applicator (Figure 9) has been developed to deal with meniscus stability problems (10). The device decouples the free surface at the top of the reservoir from the final coating operation. The first chamber (with the free surface) strips away large air bubbles. The second chamber is pressurized which produces a net flow into the first chamber and strips small bubbles from the fiber. The resultant coating is bubble free, smooth, and regular.

Coating Properties and Fiber Performance

Coating materials for optical fibers must be chosen with close attention paid to the manner in which their properties affect fiber performance. The ability to protect fiber strength and to provide resistance to excess transmission losses caused by microbending are the most important functions of the coating. These items will be dealt with in detail. Other important attributes of a good coating are the ability to be stripped for splicing and connectorization; thermal, oxidative, and hydrolytic stability; the ability to bond in cable structures; resistance to compounds such as water and gasoline; and handleability (low surface tack, toughness, abrasion resistance, etc.).

Fiber Strength. The successful implementation of an optical-fiber communication system depends critically on ensuring that each individual fiber is capable of withstanding some minimum tensile-stress level over its entire length. The stress requirement is governed by the maximum tensile strain that the fiber will encounter during fabrication operations, cabling, field installation, and service. The crucial importance of fiber strength is obvious when one considers that a single fiber failure will normally result in the loss of at least the equivalent of several hundred telephone voice circuits.

As we have seen, the failure of glass fibers in tension is commonly associated with surface flaws that cause stress concentrations and consequently lower the tensile strength from that of the pristine, unflawed glass. The occurrence of flaws on the glass surface is usually associated with defective or inadequately controlled materials or processing operations. Most of the processing and material requirements important for realizing high fiber strength with an absence of critically sized flaws have been studied, and progress has been made to bring the appropriate parameters under effective control for production. The principal requirements include the following items:

a b

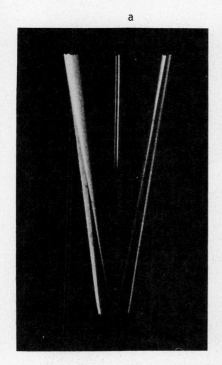 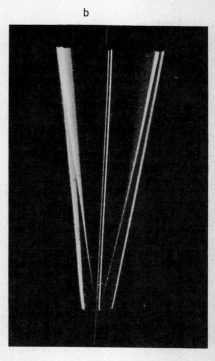

Figure 6.   a: Photograph of the air column surrounding an optical fiber propagating toward the end of a conical die. b: Photograph of the air column penetrating through the end of the die.   At this point no liquid is applied to the fiber.

a

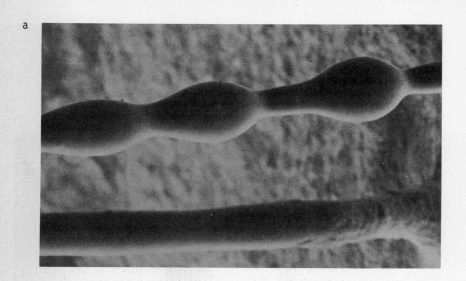

b

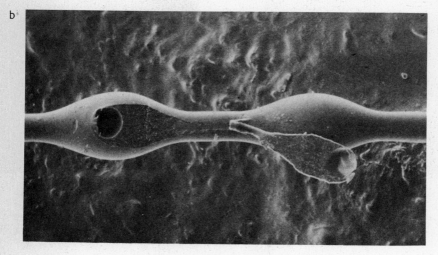

Figure 7.   a: Scanning electron micrograph of severe coating
             diameter fluctuations due to flow instabilities. b:
             Void   internal to   the diameter fluctuation.
             (Reproduced with permission from Ref. 27. Copyright
             1982 Society of Plastics Engineers.)

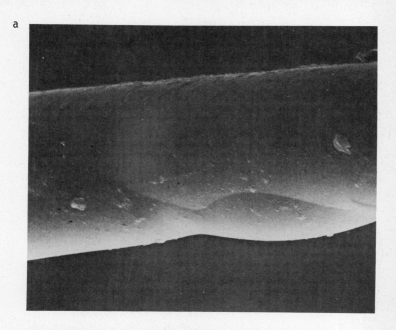

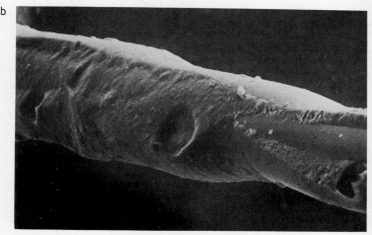

Figure 8.   a: Scanning electron micrograph of a poorly coated
fiber showing an uncoated region. b: Void within the
coated region. (Reproduced with permission from Ref.
27. Copyright 1982 Society of Plastics Engineers.)

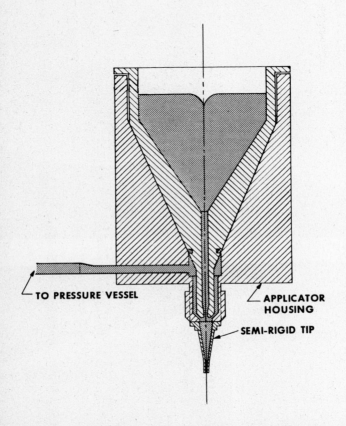

Figure 9.   Schematic of a two-chamber applicator that produces
smooth, bubble-free coatings. (Reproduced with permission from Ref. 10. Copyright 1982 Optical Society of
America.)

1. Use of a flaw-free glass preform. Surface flaws, such as scratches, on a glass preform may not be healed completely when the molten glass is drawn into a fiber.
2. Maintenance of drawing-furnace cleanliness. Particles that spall from refractory materials used in the furnace may become embedded in the molten glass.
3. Use of sufficiently high drawing-furnace temperature. Low drawing temperatures produce high glass viscosities that may promote the formation of surface defects or prevent the healing of existing preform flaws.
4. Maintenance of a dust-free drawing environment. Dust in the drawing environment may be convected into the furnace and become embedded in the molten glass.
5. Use of a suitable coating material and technique.

With regard to item 5, the two basic considerations governing fiber strength are the avoidance of damage to the glass surface during coating application and the quality of the applied coating. Quality factors relating to the coating influence fiber strength as follows:

1. Coating concentricity. A highly eccentric coating that is thin at a point on its circumference is susceptible to abrasive failure during normal handling operations.
2. Incomplete coatings. Coating flow instabilities may result in incompletely coated fiber sections, as described earlier (Figure 8a).
3. Coating mechanical integrity. Coatings with poor abrasion resistance, such as silicones, may fail during normal handling unless protected with a tougher jacket.
4. Particle contaminants in the coating. Particles in the coating resin that may abrade the fiber during or after fabrication must be removed by filtration.

The importance of carrying out all of these operations and precautions may be seen in Figure 10. This figure shows the tensile-strength distributions (Weibull plots) for several 1-km or longer optical fibers cut and tensile tested in 20-m gauge lengths. In each case represented by the open symbols, optimum processing conditions were employed except that one item was deliberately altered from optimum, for example, high draw tension, dust in the drawing environment, or the use of a flawed preform. In each case a significant weak tail was observed in the strength distribution. When all items were held in rigid control (filled symbols), a 3-km long optical fiber exhibiting only a high strength mode of failure, characteristic of an unflawed glass surface, was produced.

Microbending Loss. As a result of their small diameters, optical fibers bend very readily. This feature is advantageous in that the transmission medium is very flexible and easily routed. However, when the spatial period of the bending becomes small (approx. 1 mm or less), some of the light rays normally guided by the fiber are lost through radiation. Such small period distortions may occur when a fiber is wound on a spool under tension or when it is placed in cable structure. The phenomenon is termed microbending (11), and

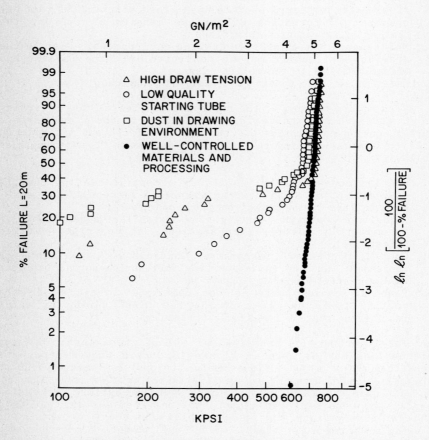

Figure 10. Weibull plots of tensile-strength distribution for kilometer lengths of fibers tested in 20-m gauge lengths. In each case one variable was intentionally altered and all others were optimized. (Reproduced with permission from Ref. 6. Copyright 1980 IEEE.)

in extreme cases it can result in very large transmission losses amounting to several decibels per kilometer, thereby seriously degrading the performance of an optical-fiber transmission link.

Material Considerations. Transmission losses due to microbending may be controlled or minimized through the proper choice of fiber coating materials and geometry (11, 12). Compliant coatings have been very successfully used (11-15) to cushion or buffer the fiber from stresses that induce microbending.   Such coatings are best utilized in a dual coating structure in which the compliant primary coating is surrounded by a high-modulus secondary coating.   The secondary coating forms a shell around the cushioned fiber and thereby further isolates the fiber from external stresses (11, 16). Thus primary polymer coatings with moduli in the rubbery range (10^6-10^7 Pa) are needed to achieve low microbending sensitivity.   Low glass transition temperatures (T_g) are also required to ensure adequate performance at low field temperatures, for example, −40 °C. Outstanding examples of such materials include silicone rubber systems based on poly(dimethyl siloxane) (T_g = −119 °C) and poly(methylphenyl siloxane) (T_g = −59 °C).   These materials are supplied as two-part liquid systems consisting of a vinyl-terminated polysiloxane resin, a tri- or tetrafunctional silane cross-linker, and a platinum catalyst (17).   After mixing, the pot life of these systems is usually a few hours at room temperature, but very rapid curing is attained at high temperatures (200-400 °C).   Fiber-coating operations at speeds of 1 m/s have been routinely achieved, but heat transfer difficulties are encountered at higher speeds.

As mentioned earlier, UV-curable resin formulations are very attractive for fiber coating because of the rapid cross-linking rates that are achievable.   Most commonly, epoxy- or urethane-acrylate resins are employed (18-22), and viscosity and cross-link density are controlled through the addition of reactive diluents. With these systems work has focused on producing low modulus, low T_g properties (20-22) through the incorporation of appropriate chemical constituents to enhance higher chain flexibility, for example, ether linkages.

Hot-melt thermoplastic elastomer systems (23, 24) are also effective coating materials.   These materials are generally based on copolymers that are comprised of hard (crystalline or glassy) and rubbery (amorphous) segments contained in separate phases.   The hard-phase regions form physical cross-links below their crystallization or vitrification temperature, and the system therefore has elastomeric properties.   The moduli and low-temperature characteristics of these materials can be tailored to compare reasonably well with silicone rubbers at −40 °C.   However, they are limited in high-temperature applicability because of enhanced creep or flow due to softening.

Figure 11 compares the low-temperature modulus characteristics of representative fiber-coating (buffer) materials of the three types described earlier.   The figure depicts the dynamic tensile modulus as a function of frequency at −40 °C.   The data were obtained on films of the materials by using a Rheometrics rotational dynamic spectrometer operated over the frequency range from 10^{-1} to 10^2 rad/s.   Curves obtained at different temperatures were shifted both horizontally and vertically in accord with established linear

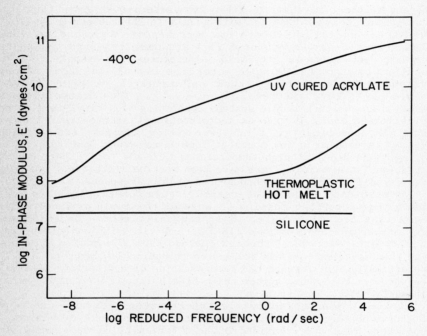

Figure 11.   Master curves of the dynamic tensile moduli at −40 °C
for three representative types of primary buffer
coatings for optical fibers.

viscoelastic superposition principles to produce a master curve at −40 °C. The results illustrate that rubbery moduli ($<10^7$ Pa) persist in the silicone and thermoplastic hot-melt materials, but the UV-cured acrylate material exhibits a broad glass-to-rubber transition at this temperature.

<u>Microbending Loss: Relation to Coating Properties</u>. The comparative microbending performance offered by a series of coatings that differ in modulus characteristics may be assessed by subjecting coated fibers to controlled microbending deformations. In one test (<u>6</u>, <u>12</u>) the coated fibers were wound on a microscopically rough drum under a set of controlled tensions, and the fiber losses were measured as a function of tension. In this test the fibers bend over the random microscopic perturbations on the drum surface, which causes excess losses. The loss versus tension curves are displayed in Figure 12.

The buffering action of a coating in this situation is determined by the relaxation modulus of the coating material. The relaxation modulus may be measured on a film cast from the material by carrying out tensile-stress relaxation measurements with a suitable apparatus such as a Rheovibron dynamic viscoelastometer operated in a static mode. Figure 13 (inset) displays such measurements for the four coating materials used on the fibers measured in Figure 12. The measurements were carried out at 23 °C at small tensile strains, where the materials exhibit linear viscoelastic behavior.

Figure 13 exhibits the results of the experiment in Figure 12 wherein the excess microbending losses of fibers coated with the four materials are correlated with the 30-min relaxation moduli of the coatings. The 30-min value was chosen because the loss measurements were carried out 30 min after tensioning the fibers. (When making such correlations, the coating properties and fiber performance must be characterized on the same time scale.) Coatings with time-dependent relaxation moduli below approximately 10^7 Pa are effective in providing protection from microbending losses (Figure 13). Fibers with higher modulus coatings exhibit very high microbending losses amounting to tens of decibels per kilometer. A modulus value of 10^7 Pa is considered to be at the upper limit of rubbery response. Thus for good microbending protection, a fiber coating should have a rubbery modulus $<10^7$ Pa in the appropriate time scale and temperature range of interest for a fiber-cable application.

<u>Microbending Loss: Relation to Fiber Cabling</u>. We shall consider here a particular type of fiber-cable structure known as an adhesive sandwich ribbon (<u>25</u>). In this structure 12 fibers are packaged in a linear array by sandwiching them between two adhesive-coated polyester tapes, as depicted in Figure 14. Within such a structure the fibers are susceptible to microbending when the ribbon is exposed to low temperatures. Microbending occurs because the large coefficients of thermal expansion of the various polymeric constituents of the ribbon (polyester tape, adhesive, fiber coating) cause the glass fibers, which have very low expansion coefficients, to experience axial compressive loading. When the compressive loads become high enough, the fibers respond by buckling, and microbending losses may result.

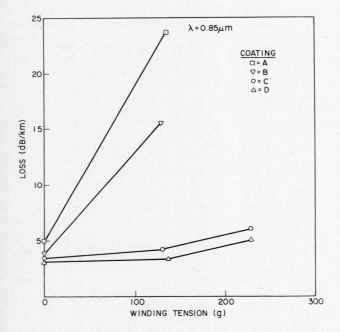

Figure 12.   Transmission losses of four optical fibers coated with
             different materials as a function of winding tension on
             a 25-cm diameter acrylic drum. The fibers are 110-μm
             cladding diameter, 8-μm core diameter single mode
             fibers. These fibers have a very small core-cladding
             refractive index difference and transmit only the
             fundamental, most axially directed light ray (mode).
             Coating thickness is 50 μm. (Reproduced with permission
             from Ref. 28. Copyright 1981 Gordon and Breach Science
             Publishers.)

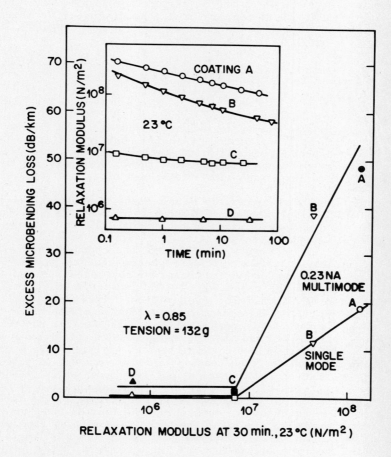

Figure 13.   Excess microbending losses measured in the experiment of
Figure 12 for various single mode and multimode fibers
as a function of the 30-min, 23 °C tensile relaxation
modulus of the coating material.   The relaxation
modulus–time relationships of the coating materials are
depicted in the inset.   (Reproduced with permission from
Ref. 6.   Copyright 1980 IEEE.)

A thermoviscoelastic analysis of the microbending losses produced in a fiber ribbon in response to a thermal history has been carried out (26). The important mechanisms contributing to compressive axial loading of the glass fibers are recovery of orientation (shrinkback) of the polyester tape coupled with stress relaxation at elevated temperatures, and differential thermal contraction of the ribbon constituents at low temperatures. The latter effect may be expressed as

$$\epsilon_f = \left[\frac{\sum_i \alpha_i E_i(t,T) A_i}{\sum_i E_i(t,T) A_i} - \alpha_f \right] \Delta T \tag{1}$$

where ϵ_f is the axial compressive strain induced on the fibers; α_i, $E_i(t,T)$, and A_i are the thermal–expansion coefficient, time– and temperature–dependent tensile relaxation modulus, and cross–sectional area of the i-th ribbon constituent, respectively; α_f is the fiber thermal–expansion coefficient; and ΔT is the temperature change.

We have used this expression to correlate average microbending losses for 12-fiber ribbons. Figure 15 shows data for three ribbons, each one containing fibers having a different coating (A, B, or C). The ribbons were stabilized for 48 hours at each of three temperatures (-9, -26, and -40 °C), and the changes in average transmission loss from their initial room–temperature values were measured. The loss changes represent microbending losses. The average microbending losses are plotted in Figure 15 against ϵ_f computed from Equation 1 for each ribbon. The time-dependent relaxation moduli for the ribbon constituents were determined from dynamic tensile modulus data obtained with a Rheometrics rotational dynamic spectrometer, and the thermal–expansion coefficients were measured with a Perkin–Elmer thermomechanical analyzer. The average microbending losses for each coating increase with ϵ_f, as expected (Figure 15). However, the curves do not superpose, which indicates that the properties of the individual coatings play a role in determining the level of compressive strain applied to the fibers, as indicated by Equation 1. Apparently the coating modulus also determines the spatial period of the buckling resulting from axial compression. Thus low–modulus coatings minimize microbending losses by allowing buckling periods that are large compared to the critical millimeter regime. Such coatings also cushion the fiber from distortion caused by nonuniformly applied lateral forces.

Coating C of Figure 15 represents the special case of a dual coating consisting of a low–modulus, rubbery, primary layer surrounded by a high–modulus jacket. As discussed earlier, its exceptional microbending–loss performance results from the cushioning effect of the rubber coupled with the flexural rigidity of the jacket (11, 16). A photomicrograph of this coated fiber is displayed in Figure 16.

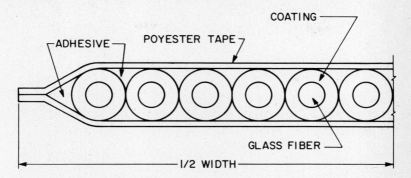

Figure 14.   Schematic of a 12-fiber ribbon.

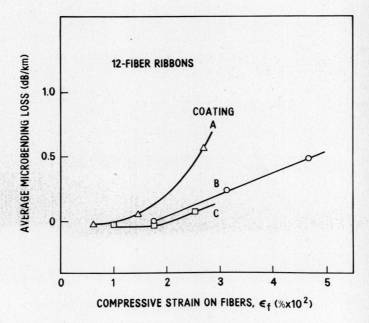

Figure 15.   Average microbending losses as a function of axial
             compressive strain on fibers in a ribbon for three
             different fiber coatings. (Reproduced with permission
             from Ref. 27.   Copyright 1982 Society of Plastics
             Engineers.)

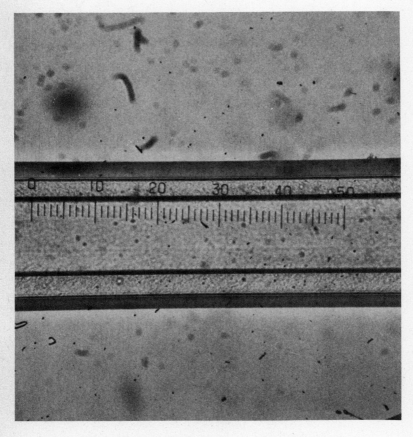

Figure 16.   Photomicrograph of a dual-coated fiber.

Literature Cited

1. Nagel, S. R.; MacChesney, J. B.; Walker, K. L. J. Quantum Electron., 1982, QE-18, 459.
2. Murata, H.; Inagaki, N. J. Quantum Electron., 1981, QE-17, 835.
3. Beales, K. J.; Day, C. R.; Dunn, A. G.; Partington, S. Proc. IEEE, 1980, 68, 1191.
4. Blyler, L. L., Jr.; Eichenbaum, B. R.; Schonhorn, H. in "Optical Fiber Telecommunications", Chynoweth, A. G.; Miller, S. E., Eds.; Academic: New York, 1979; Chap. 10.
5. Gulati, S. J.; King, C. B.; Quan, F. Proc. 27th Int. Wire and Cable Symp.; CORADCOM; Cherry Hill, NJ, Nov. 14-16, 1978; pp. 342-345.
6. Blyler, L. L., Jr.; Dimarcello, F. V. Proc. IEEE, 1980, 68, 1194.
7. Smithgall, D. H.; Watkins, L. S.; Frazee, R. E., Jr.; Appl. Opt., 1977, 16, 2395.
8. Hart, A. C., Jr.; Albarino, R. V. in Tech. Digest, Topical Meeting on Optical Fiber Transmission II, Williamsburg, VA, Optical Soc. of America, Feb. 22-24, 1977, TuB2.
9. Presby, H. M. Bell System Tech. J. 1976, 55, 1525. Eichenbaum, B. R. in Tech. Digest CLEOS/ICF-80, San Diego, CA, Optical Soc. America and IEEE, Feb. 26-28, 1980, paper TuBR4.
10. Lenahan, T. A.; Taylor, C. R.; Smith, J. V. Tech. Digest, Fifth Topical Meeting on Optical Fiber Communication, Phoenix, AZ, Optical Soc. of America/IEEE, April 13-15, 1982, paper WCC6.
11. Gloge, D. Bell System Tech. J. 1975, 54, 245.
12. Gardner, W. B. Bell System Tech. J., 1975, 54, 457.
13. Naruse, T.; Sugawara, Y.; Masuno, K.; Electron Lett., 1977, 13, 153.
14. Yoshida, K.; Nishimura, M.; Hondo, H.; Kuroha, T.; Murata, H.; Ishihara, K. in Preprints, Div. Organic Coatings and Plastics Chem., 1979, 40, 93.
15. Suzuki, H.; Osanai, H. in Preprints, Div. Organic Coatings and Plastics Chem. 1979, 40, 211.
16. Yoshizawa, N.; Yabuta, T.; Kojima, N.; Negishi, Y. Appl. Opt., 1981, 20, 3146.
17. Valles, E. M.; Macosko, C. W. Macromolecules, 1979, 12, 521.
18. Schonhorn, H.; Kurkjian, C. R.; Jaeger, R. E.; Vazirani, H. N.; Albarino, R. V.; DiMarcello, F. V. Appl. Phys. Lett. 1976, 29, 712.
19. Paek, U. C.; Schroeder, C. M. Appl. Opt., 1981, 20, 1230.
20. Ansel, R. E.; Stanton, J. J. in Advances in Ceramics, Vol. 2: Physics of Fiber Optics, Bendow, B.; Mitra, S. S., Eds.; American Ceramics Society: Columbus, Ohio, 1981; pp. 27-39.
21. Levy, N.; Massey, P. E. Polym. Eng. Sci., 1981, 21, 406.
22. Levy, N. Polym. Eng. Sci., 1981, 21, 978.
23. Miller, T. J.; Hart, A. C., Jr.; Vroom, W. I.; Bowden, M. J. Electron. Lett., 1978, 14, 603.
24. Blyler, L. L., Jr.; Hart, A. C., Jr.; Levy, A. C.; Santana, M. R.; Swift, L. L. Proc. 10th European Conf. on Optical Communications, Cannes, France, Sept. 21-24, 1982, pp. 245-249.
25. Saunders, M. J.; Parham, W. L. Bell System Tech. J., 1977, 56, 1013.

26. Brockway, G. S.; Mahr, P. F.; Santana, M. R. _Proc. 16th European Conf. on Optical Communications_, York, U.K., Sept. 16–19, 1980, pp. 338–341. Brockway, G. S.; Santana, M. R. _Bell Syst. Tech. J._, 1983, _62_, 993.

27. Aloisio, C. J.; Blyler, L. L., Jr. ANTEC '82, Society of Plastics Engineers, 1982, pp. 117–120.

28. Blyler, L. L., Jr. _Polym. News_, 1981, _86_.

Epoxy Resins

R. S. BAUER

Shell Development Company, Westhollow Research Center, Houston, TX 77001

Resin Types and Structure
Typical Bisphenol Epoxy Resins
Reactions of Epoxides and Curing Mechanisms
Addition Reactions of Amines
Addition Reactions of Polybasic Acids and Acid Anhydrides
Addition Reactions of Phenols and Mercaptans
Reactions of Alcohols
Catalytic Reactions
Curing Agents
Curing of Epoxy Resin Compositions
Applications of Epoxy Resins
Structural Applications
  Bonding and Adhesives
  Casting, Molding Compounds, and Tooling
  Flooring, Paving, and Aggregates
  Reinforced Plastics or Composites

The name epoxy resins has over the years become synonymous with performance; epoxy resins have established themselves as unique building blocks for high-performance coatings, adhesives, and reinforced plastics. Epoxy resins are a family of oligomeric materials that can be further reacted to form thermoset polymers possessing a high degree of chemical and solvent resistance, outstanding adhesion to a broad range of substrates, a low order of shrinkage on cure, impact resistance, flexibility, and good electrical properties.

Although epoxy resin sales have not achieved the volumes that commodity-type thermoplastic materials have achieved, the growth of epoxy resins has been impressive. Introduced commercially in the United States in the late 1940s, epoxy resins had reached annual sales of about 13 million lb by 1954. In 1981 total demand was about 319 million lb (1), and that figure is expected to grow to about 580 million lb in 1990 (Figure 1).

The U. S. markets in 1981 for epoxy resins were almost equally divided between protective coatings and structural end uses. Projections indicate, however, that by 1990 almost twice as much epoxy resin will be going into structural end uses than into coatings. This change is expected because of the anticipated increased usage of epoxy resins in electrical applications and fiber-reinforced structures. The breakdown of epoxy resin markets in the United States for 1980 is given in Figure 2 (1).

Resin Types and Structure

Epoxy resins are compounds or mixtures of compounds that are characterized by the presence of one or more epoxide or oxirane groups:

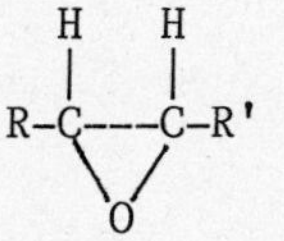

$$R-\underset{\underset{\displaystyle O}{}}{\overset{\displaystyle H}{C}}---\overset{\displaystyle H}{C}-R'$$

There are three major types of epoxy resins: cycloaliphatic epoxy resins (R and R' are part of a six-membered ring), epoxidized oils (R and R' are fragments of an unsaturated fatty acid, such as oleic acid in soybean oil), and glycidated resins (R is hydrogen and R' can be a polyhydroxyphenol or a polybasic acid). The first two types of epoxy resins are obtained by the direct oxidation of the corresponding olefin with a peracid as illustrated by the following:

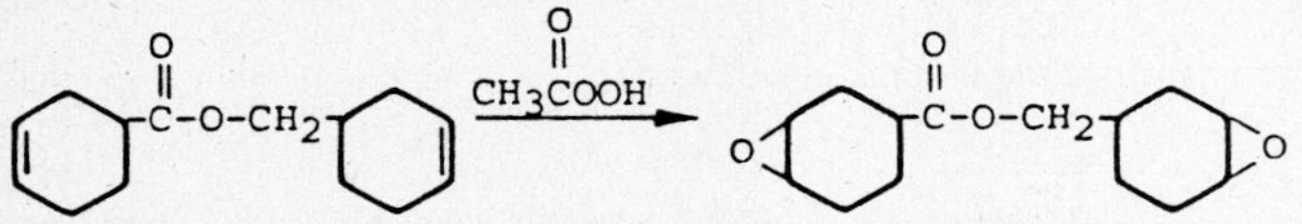

By far the most commercially significant of these resins, however, are the ones obtained by glycidation of bisphenol A with epichlorohydrin:

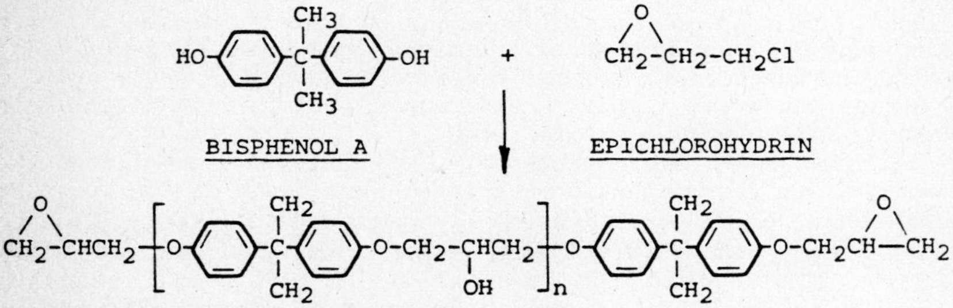

BISPHENOL A   EPICHLOROHYDRIN

Typical Bisphenol Epoxy Resins

As can be seen from the structure given for a typical bisphenol A based epoxy resin, a spectrum of products is available. Commercial resins are generally mixtures of oligomers with the average value of n varying from essentially 0 to approximately 25. Available products range from low viscosity, low molecular weight resins all the way up to hard, tough, higher molecular weight materials. A

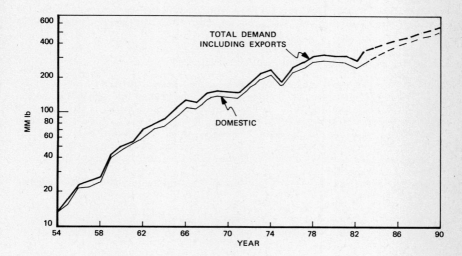

Figure 1.   Total demand forecast for epoxy resins produced in
the United States.

Coatings Applications, MM lb	
Transportation	24
Packaging	26
Maintenance + Marine	34
Appliance	5
Pipe	9
General Industrial	26
Others	7
	131

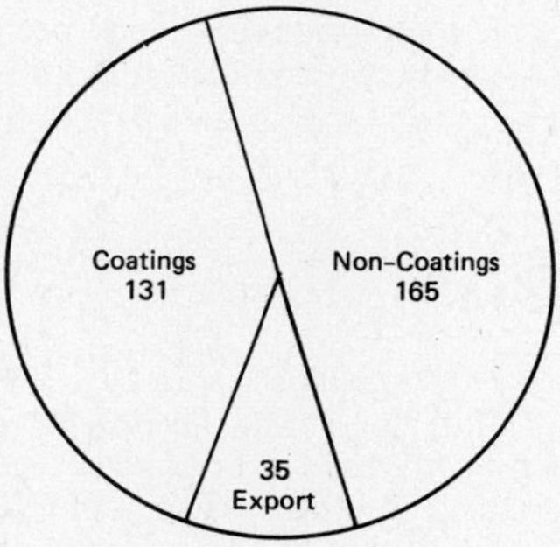

Coatings
131

Non-Coatings
165

35
Export

Non-Coating Applications, MM lb	
Electrical Laminates	53
Casting, Molding	24
Fiber Reinforced Composites	21
Floor/Paving/Aggregates	19
Adhesives	15
Others	33
	165

Figure 2.   Epoxy resin market.

list of the more common epoxy resins arranged in order of increasing molecular weight is shown in Table I. Although the structure drawn for epoxy resins depicts them as diepoxides, commercial epoxy resins are not 100% diepoxides. Other end groups can result from the manufacturing process, such as glycols derived from hydration of epoxide groups, unconverted chlorohydrin groups, and phenolic end groups from unreacted terminal bisphenol A molecules. The lower molecular weight resins may have a functionality of greater than 1.9 epoxies per molecule, but the solids usually have lower epoxy contents.

Reactions of Epoxides and Curing Mechanisms

Epoxy resins are reactive intermediates that, before they can be useful products, must be "cured" or cross-linked by polymerization into a three-dimensional infusible network through the use of coreactants (curing agents). Cross-linking of the resin can occur through the epoxide or hydroxyl groups, and proceeds basically by only two types of curing mechanisms: direct coupling of the resin molecules by a catalytic homopolymerization, or coupling through a reactive intermediate.

Most reactions used to cure epoxy resins occur with the epoxide ring:

$$\underset{O}{\overset{\diagdown}{C}}-\overset{\diagup}{\underset{}{C}}$$

The capability of this ring to react by a number of paths and with a variety of reactants gives epoxy resins their great versatility. The chemistry of most curing agents currently used with epoxy resins is based on polyaddition reactions that result in coupling as well as cross-linking. The more widely used curing agents are compounds containing active hydrogen (polyamines, polyacids, polymercaptans, polyphenols, etc.) that react as shown in Reaction 1 to form the corresponding β-hydroxyamine, β-ester, β-hydroxymercaptan, or β-hydroxyphenyl ether.

$$R-XH \;+\; \underset{O}{\overset{\diagdown}{C}}-\overset{\diagup}{\underset{}{C}} \;\longrightarrow\; R-X-\underset{\underset{OH}{|}}{\overset{|}{C}}-\overset{|}{\underset{|}{C}}- \qquad (1)$$

Epoxy resins and curing agents usually contain more than one reaction site per molecule, and the process of curing to form a three-dimensional network results from multiple reactions between epoxide molecules and curing agent. The specific reactions of the various reactants with epoxide groups have, in many cases, been studied in considerable detail and have been extensively reviewed elsewhere (2).

Addition Reactions of Amines

All amine functional curing agents, for example, aliphatic polyamines and their derivatives, modified aliphatic amines and aromatic amines, react with the epoxide ring by an addition reaction without formation of by-products. Shechter et al. (3) suggested that the amine addition reactions proceed in the following manner:

$$R-NH_2 \quad + \quad \underset{\diagdown O \diagup}{CH_2-CH-} \quad \longrightarrow \quad R-\underset{\underset{OH}{|}}{\overset{\overset{H}{|}}{N}}-CH_2-CH- \tag{2}$$

$$R-\underset{\underset{OH}{|}}{\overset{\overset{H}{|}}{N}}-CH_2-CH- \quad + \quad \underset{\diagdown O \diagup}{CH_2-CH-} \quad \longrightarrow \quad R-N \overset{CH_2-CH-\overset{|}{OH}}{\underset{\underset{OH}{|}}{-CH_2-CH-}} \tag{3}$$

$$\underset{\underset{OH}{|}}{-CH-} \quad + \quad \underset{\diagdown O \diagup}{CH_2-CH-} \quad \longrightarrow \quad \underset{\underset{OH}{|}}{\overset{\overset{|}{O-CH_2-CH-}}{-CH-}} \tag{4}$$

Subsequently, other workers including O'Neill and Cole (4) and Dannenberg (5) showed that Reactions 2 and 3 proceed to the exclusion of Reaction 4. The reactivity of a particular epoxide-amine system depends on the influence of the steric and electronic factors associated with each of the reactants. It has been known for some time that hydroxyls play an important role in the epoxide-amine reaction. For example, Shechter et al. (3) studied the reaction of diethylamine with phenylglycidyl ether in concentrated solutions. They showed that acetone and benzene decreased the rate of reaction in a manner consistent with the dilution of the reactants, but that solvents such as 2-propanol, water, and nitromethane accelerated the reaction (Figure 3). They also found that addition of 1 mol of phenol to this reaction accelerated it to an even greater extent that addition of 2-propanol or water.

The "modest" acceleration of the amine-epoxide reaction by nitromethane was ascribed to the influence of the high dielectric constant of the solvent. The greater influence of hydroxyl-containing compounds in accelerating this reaction has been suggested to result from the formation of a ternary intermediate complex of the reactants with hydroxyl-containing material, such as that proposed by Smith (6) or Mika and Tanaka (7):

$$R_2N \cdots \underset{\underset{\underset{HOX}{\vdots}}{\diagdown O \diagup}}{\overset{\overset{H}{|}}{\bullet}CH_2-CH-} \tag{5}$$

$$\text{SMITH}$$

Table I. Typical Properties of Bisphenol A–Based Epoxy Resins

Average Mol. Wt.	Average WPE[a]	Viscosity, P 25 °C	Softening Point, °C[b]
350	182	80	--
380	188	140	--
600	310	semisolid	40
900	475	solid	70
1400	900	solid	100
2900	1750	solid	130
3750	3200	solid	150

[a]WPE = weight per epoxide, i.e., grams of resins needed to provide 1 molar equivalent of epoxide. Also referred to as EEW (epoxide equivalent weight) and EMM (epoxy molar mass). All three items are interchangeable.

[b]Softening point by Durran's mercury method.

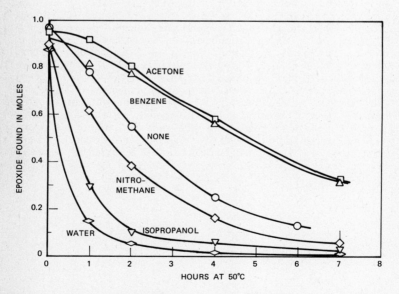

Figure 3.   Influence of reaction medium of amine–glycidyl ether reaction.

$$R_2\overset{\overset{\textstyle H}{|}}{\underset{\vdots}{N}}\cdots\cdots\cdots\cdots\bullet CH_2-CH-$$

$$\vdots\\ XOH\cdots\cdots\cdots\cdots\bullet O \qquad \text{MIKA AND TANAKA} \qquad (6)$$

Addition Reactions of Polybasic Acids and Acid Anhydrides

Acid anhydrides are probably second in importance to amine-type curing agents; however, polybasic acids have little application as curing agents. However, esterification of epoxides with fatty acids to produce resins for surface coatings has great commercial significance.

Shechter and Wynstra (8) proposed that the following reactions could occur with a carboxylic acid and an epoxide:

$$R-\underset{\underset{O}{\|}}{C}-OH \;+\; \underset{\underset{O}{\diagdown\diagup}}{CH_2}-CH- \;\longrightarrow\; R-\underset{\underset{O}{\|}}{C}-O-CH_2-\underset{\underset{OH}{|}}{CH}- \qquad (7)$$

$$R-\underset{\underset{O}{\|}}{C}-OH \;+\; \underset{\underset{OH}{|}}{-CH}- \;\longrightarrow\; R-\underset{\underset{O}{\|}}{C}-O-CH- \;+\; H_2O \qquad (8)$$

$$\underset{\underset{OH}{|}}{-CH}- \;+\; \underset{\underset{O}{\diagdown\diagup}}{CH_2}-CH- \;\longrightarrow\; \underset{\underset{\underset{\underset{OH}{|}}{O-CH_2-CH-}}{|}}{-CH}- \qquad (9)$$

$$H_2O \;+\; \underset{\underset{O}{\diagdown\diagup}}{CH_2}-CH- \;\longrightarrow\; HO-CH_2-\underset{\underset{OH}{|}}{CH}- \qquad (10)$$

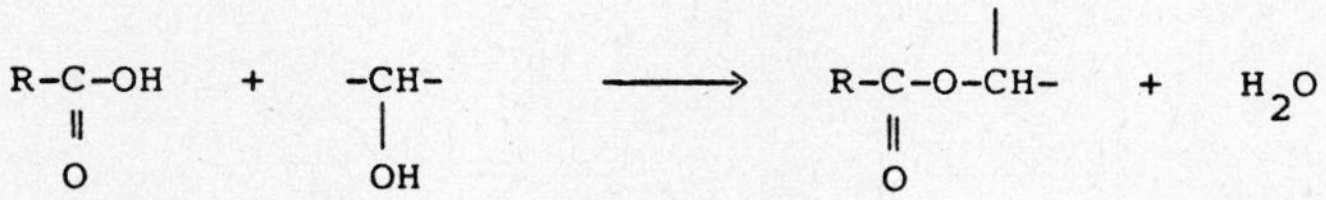

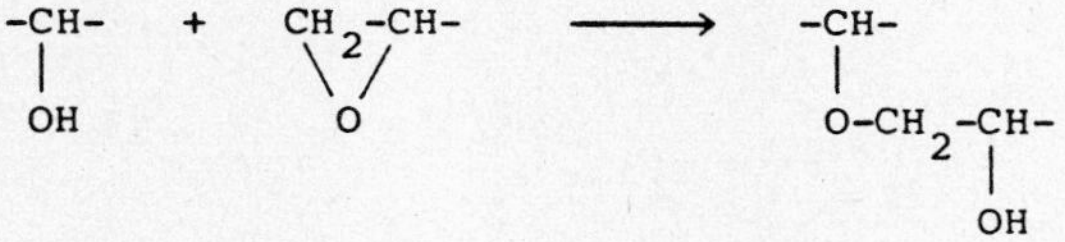

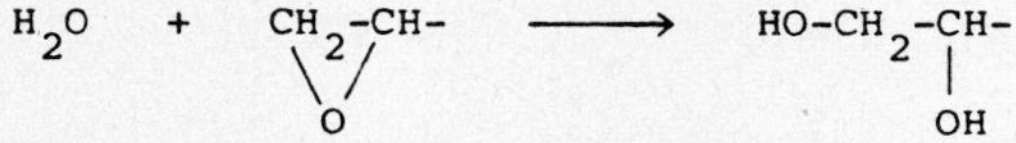

These workers showed that if water is removed during the reaction, Reactions 7–9 occurred in approximately the ratio 2:1:1 in an uncatalyzed system. A higher degree of selectivity for the hydroxy ester (Reaction 7) was observed to occur in the base-catalyzed reaction that was proposed to proceed as follows:

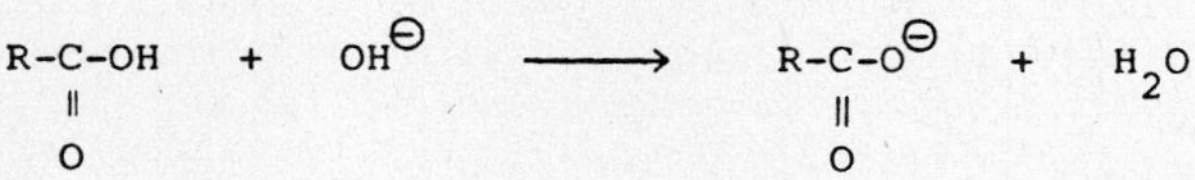

$$R-\underset{\underset{O}{\|}}{C}-OH \;+\; OH^{\ominus} \;\longrightarrow\; R-\underset{\underset{O}{\|}}{C}-O^{\ominus} \;+\; H_2O \qquad (11)$$

$$R-\underset{\underset{O}{\|}}{C}-O^{\ominus} \;+\; \underset{\underset{O}{\diagdown\diagup}}{CH_2}-CH- \;\longrightarrow\; R\underset{\underset{O}{\|}}{C}-O-CH_2-\underset{\underset{O^{\ominus}}{|}}{CH}- \qquad (12)$$

$$R-\underset{\underset{O}{\|}}{C}-O-CH_2-\underset{\underset{O^{\ominus}}{|}}{CH}- \;+\; R-\underset{\underset{O}{\|}}{C}-OH \;\longrightarrow\; R-\underset{\underset{O}{\|}}{C}-O-CH_2-\underset{\underset{OH}{|}}{CH}- \;+\; R-\underset{\underset{O}{\|}}{C}-O^{\ominus} \qquad (13)$$

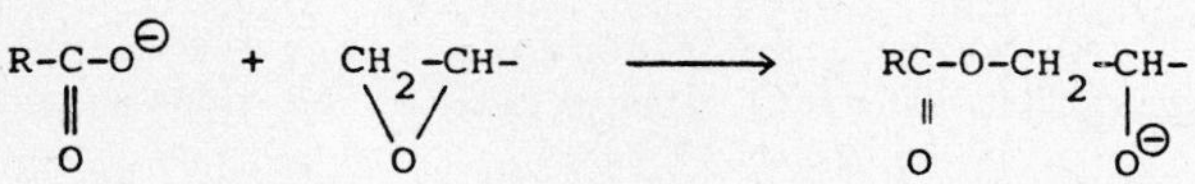

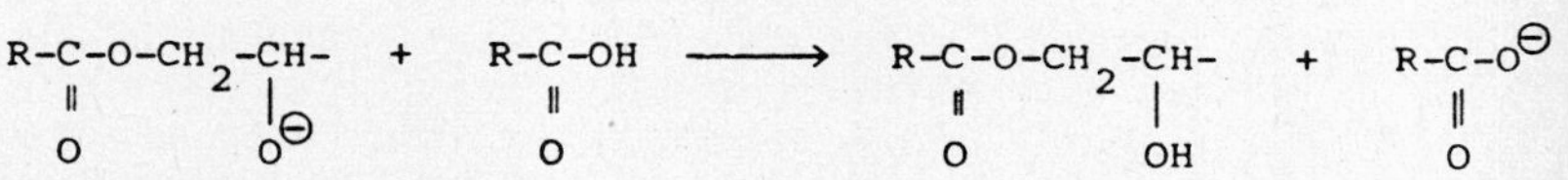

Although much more selective than the uncatalyzed reaction, the base–catalyzed reaction has some dependence on stoichiometry. At a ratio of epoxide to acid of 1:1, essentially all the product is the hydroxy ester. However, when an excess of epoxide groups is present, Reaction 7 proceeds until all the acid is consumed, after which the epoxide–hydroxyl reaction (Reaction 9) starts. This is illustrated in Figure 4.

The uncatalyzed reaction of acid anhydrides with epoxides is slow even at 200 °C; however, with either acidic or basic catalysts the reaction proceeds readily with the formation of ester linkages. The reaction of acid anhydrides with conventional commercial epoxy resins is probably initiated by water or hydroxyl and carboxyl compounds present in the mixture. The following sequence is illustrative of initiation by a hydroxyl-containing material:

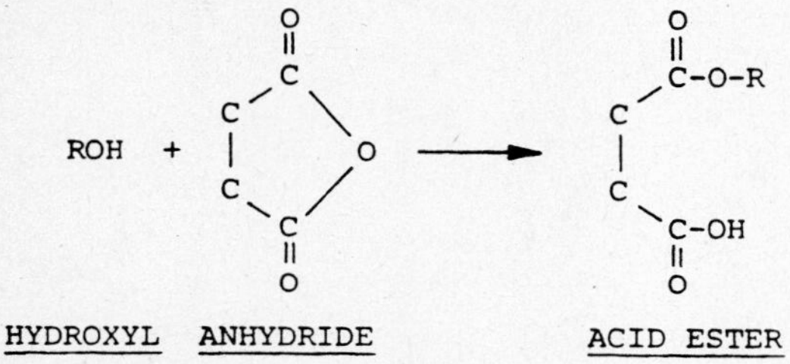

$$
\text{ROH} \ + \ \underset{\text{ANHYDRIDE}}{\left(\substack{\text{O}\\\|\\ \text{C}\\ \diagup \ \diagdown \\ \text{C} \qquad \text{O} \\ \diagdown \ \diagup \\ \text{C} \\ \| \\ \text{O}}\right)} \ \longrightarrow \ \underset{\text{ACID ESTER}}{\left(\substack{\text{O}\\\|\\ \text{C-O-R} \\ \diagup \\ \text{C} \\ | \\ \text{C} \\ \diagdown \\ \text{C-OH} \\ \| \\ \text{O}}\right)} \tag{14}
$$

HYDROXYL   ANHYDRIDE   ACID ESTER

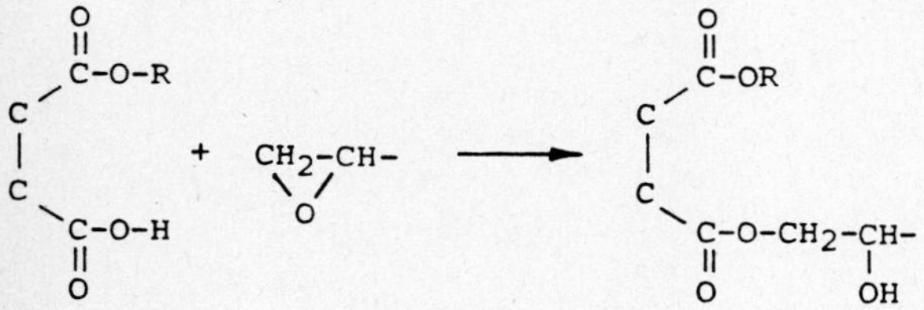

$$
\left(\substack{\text{O}\\\|\\ \text{C-O-R} \\ \diagup \\ \text{C} \\ | \\ \text{C} \\ \diagdown \\ \text{C-O-H} \\ \| \\ \text{O}}\right) \ + \ \underset{\substack{\diagdown \ / \\ \text{O}}}{\text{CH}_2\text{-CH-}} \ \longrightarrow \ \left(\substack{\text{O}\\\|\\ \text{C-OR} \\ \diagup \\ \text{C} \\ | \\ \text{C} \\ \diagdown \\ \text{C-O-CH}_2\text{-CH-} \\ \| \qquad\quad | \\ \text{O} \qquad\quad \text{OH}}\right) \tag{15}
$$

$$
\diagdown\!\!\diagup\!\!\text{CH-OH} \ + \ \underset{\substack{\diagdown \ / \\ \text{O}}}{\text{CH}_2\text{-CH-}} \ \longrightarrow \ \diagdown\!\!\diagup\!\!\text{CH-O-CH}_2\text{-CH-} \atop \qquad\qquad\qquad\qquad\quad | \atop \qquad\qquad\qquad\qquad\quad \text{OH} \tag{16}
$$

Thus the reaction is essentially a two-step process involving the opening of the anhydride by reaction with the hydroxyl-containing material to give the half acid ester (Reaction 14) and the resulting carboxyl group reacting with an epoxide to form a hydroxy diester (Reaction 15). The hydroxyl compound formed from Reaction 15 can then react with another anhydride, and so on. Evidence has been presented (9), however, that indicates that in this type of catalysis consumption of epoxy groups is faster than appearance of diester groups because of Reaction 16.

Base-catalyzed anhydride-epoxide reactions have been found to have greater selectivity toward diester formation. Shechter and Wynstra (8) showed that equal molar amounts of acetic anhydride and phenyl glycidyl ether with either potassium acetate or tertiary amines as a catalyst reacted very selectively. They proposed the following reaction mechanism for the acetate ion catalyzed reaction:

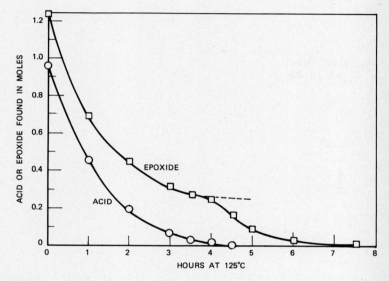

Figure 4.   Reaction of 1.25 mol of phenyl glycidyl ether with
1.00 mol of caprylic acid plus 0.20% (by weight) of
KOH catalyst.

$$CH_3-\underset{\underset{O}{\|}}{C}-O^{\ominus} \;+\; \underset{\diagdown O \diagup}{CH_2-CH-} \;\longrightarrow\; CH_3-\underset{\underset{O}{\|}}{C}-O-CH_2-\underset{\underset{O^{\ominus}}{|}}{CH-} \tag{17}$$

$$CH_3-\underset{\underset{O}{\|}}{C}-O-CH_2-\underset{\underset{O^{\ominus}}{|}}{CH-} \;+\; (CH_3-\underset{\underset{O}{\|}}{C})_2-O \;\longrightarrow\; CH_3-\underset{\underset{O}{\|}}{C}-O-CH_2-\underset{\underset{O-C-CH_3}{|}}{\underset{\underset{O}{\|}}{CH-}} \;+\; CH_3-\underset{\underset{O}{\|}}{C}-O^{\ominus} \tag{18}$$

Catalysis by tertiary amines has been proposed by Fischer (10) to proceed by the base opening the anhydride ring to form an internal salt that then reacts with an epoxide group to yield an alkoxide ester as shown in Reactions 19–21.

$$\begin{array}{c} \text{anhydride ring} \end{array} \;+\; R_3N \;\rightleftharpoons\; \begin{array}{c} C-NR_3^{\oplus} \text{ / } C-O^{\ominus} \end{array} \tag{19}$$

$$\begin{array}{c} C-NR_3^{\oplus} \text{ / } C-O^{\ominus} \end{array} \;+\; \underset{\diagdown O \diagup}{CH_2-CH-} \;\longrightarrow\; \begin{array}{c} C-NR_3^{\oplus} \text{ / } C-C-O-CH_2-CH-O^{\ominus} \end{array} \tag{20}$$

$$\begin{array}{c} C-NR_3^{\oplus} \text{ / } C-C-O-CH_2-CH-O^{\ominus} \end{array} \;+\; \begin{array}{c} \text{anhydride ring} \end{array} \;\longrightarrow\; \begin{array}{c} C-NR_3^{\oplus} \text{ / } C-C-O-CH_2-CH-O-C=O \text{ ... } C-C-O^{\ominus} \end{array} \tag{21}$$

Thus, the direct reaction between epoxide and anhydride has been found to be very slow, and the anhydride ring must first be opened before reaction can occur. Ring opening can result from reaction with hydroxyl groups present in commercial epoxy resins; addition of basic catalysts, such as tertiary amines and carboxylate ions; or addition of Lewis acids (not discussed here). Ring opening by hydroxyl functionality ultimately results in both esterification and etherification, whereas basic catalysts result predominantly in esterification.

The anhydride-epoxide reaction is complex because of the possibility of several reactions occurring simultaneously. Thus, appreciable etherification can result in undesirable amounts of unreacted anhydride and half-acid ester in the cured resin. On the basis of experimental results, Arnold (11) has suggested that for optimum properties anhydride to epoxide ratios of 1:1 are needed for tertiary amine catalysts, and 0.85:1 for no catalyst.

Addition Reactions of Phenols and Mercaptans

As with polybasic carboxylic acids, phenols have not achieved significant importance as curing agents; however, the reaction of phenols with epoxides is technologically important. For example, the reaction of bisphenols with the diglycidyl ethers of the bisphenol is used commercially to prepare higher molecular weight epoxy resins (12).

Shechter and Wynstra (8) proposed two possible types of reactions between phenol and a glycidyl ether. One involves direct reaction of the phenol with the epoxide; the other involves direct reaction of the aliphatic hydroxyl, generated from the epoxide-phenol reaction, with another epoxide as shown in Reactions 22 and 23:

$$-OH \ + \ \underset{\underset{O}{\diagdown\diagup}}{CH_2-CH-} \ \longrightarrow \ \underset{\underset{OH}{|}}{-O-CH_2-CH-} \tag{22}$$

$$\underset{\underset{OH}{|}}{-O-CH_2-CH-} \ + \ \underset{\underset{O}{\diagdown\diagup}}{CH_2-CH-} \ \longrightarrow \ \underset{\underset{\underset{OH}{|}}{O-CH_2-CH-}}{-O-CH_2-CH} \tag{23}$$

Using model systems, they found that without a catalyst no reaction occurred at 100 °C. At 200 °C epoxide disappeared at a much faster rate than phenol did (Figure 5); about 60% of the reaction was epoxide-phenol and the other 40% was epoxide-formed alcohol. Because alcohol was absent at the beginning of the reaction and only appeared when phenol reacted with epoxide, it was concluded that the phenol preferred to catalyze the epoxide-alcohol reaction rather than react itself.

The base-catalyzed reaction, however, proceeded readily at 100 °C with 0.2% mol of potassium hydroxide and exhibited a high degree of selectivity. As can be seen from Figure 6, disappearance of phenol and epoxide proceeded at the same rate throughout the course of the reaction. This phenomenon indicates that epoxide reacted with phenol to the essential exclusion of any epoxide-alcohol reaction. Shechter and Wynstra (8) proposed a mechanism in which the phenol first is ionized to phenoxide ion as shown in Reaction 24:

$$\langle\!\!\bigcirc\!\!\rangle\text{-OH} \ + \ KOH \ \longrightarrow \ \langle\!\!\bigcirc\!\!\rangle\text{-O}^{\ominus} \ + \ H_2O \ + \ K^{\oplus} \tag{24}$$

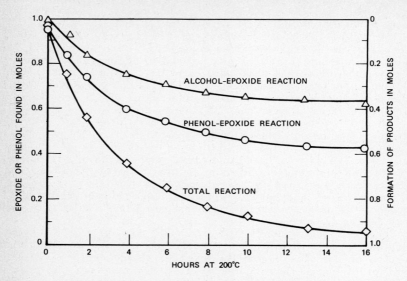

Figure 5.   Noncatalyzed reaction of equimolar amounts of phenol
             and phenyl glycidyl ether.

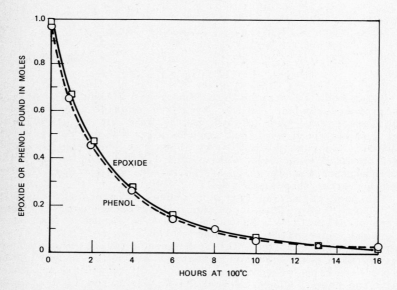

Figure 6.   KOH–catalyzed reaction of equimolar amounts of phenol
             and phenyl glycidyl ether plus 0.20% (by weight) of
             catalyst.

The phenoxide ion then attacks an epoxide as shown in Reaction 25:

$$\text{Ph-O}^{\ominus} + \text{CH}_2\text{-CH-} \longrightarrow \text{Ph-O-CH}_2\text{-CH-} \quad (25)$$

The highly basic alkoxide ion then immediately reacts with phenol to regenerate phenoxide to repeat the cycle (Reaction 26), and at the same time to exclude the possibility of side reactions taking place:

$$\text{Ph-O-CH}_2\text{-CH-} + \text{Ph-OH} \longrightarrow \text{Ph-O-CH}_2\text{-CH-} + \text{Ph-O}^{\ominus} \quad (26)$$

Shechter and Wynstra (8) also demonstrated that benzyldimethylamine was a somewhat more effective catalyst than potassium hydroxide, and the quaternary compound benzyltrimethylammonium hydroxide was even more powerful. Each reaction was highly selective.

Although the base-catalyzed reaction appears to be highly selective, more recent work (13, 14) has shown that the degree of selectivity is dependent on the catalyst, the addition of active hydrogen compound to the epoxide group, the temperature, and other variables. In reactions of diglycidyl ethers with difunctional phenols to produce higher molecular weight homologs, branched products or gelation may result if the reaction is not highly selective. Alvey (13) has developed a method of determining the relative amounts of alcohol side reactions, and he has also rated the selectivity of a number of catalysts.

Analogous to the phenols, mercaptans react with epoxide groups to form hydroxyl sulfides as shown in Reaction 27:

$$\text{RSH} + \text{CH}_2\text{-CH-} \longrightarrow \text{RS-CH}_2\text{-CH-} \quad (27)$$

The epoxide–mercaptan reaction is highly selective, and there appears to be no concomitant epoxide–alcohol reaction. Thus, to obtain a cross-linked network with a diepoxide, the functionality of the mercaptan must be greater than two.

The epoxide–mercaptan reaction can be accelerated by amines, which either react with the mercaptan to give a mercaptide ion (Reaction 28) that rapidly adds to the epoxide (Reaction 29), or by the amine first reacting with the epoxide (Reaction 31) to produce a reactive intermediate that then reacts with the mercaptan in a nucleophilic displacement reaction (Reaction 32).

$$\text{R'-SH} + \text{R}_3\text{N} \rightleftharpoons \text{R'-S}^{\ominus} + \text{R}_3\text{N}^{\oplus}\text{H} \quad (28)$$

$$\text{R'-S}^{\ominus} + \text{CH}_2\text{-CH-} \longrightarrow \text{R'-S-CH}_2\text{-CH-} \quad (29)$$

$$\text{R'-S-CH}_2\text{-CH-} + \text{R}_3\text{N}^{\oplus}\text{H} \longrightarrow \text{R'-S-CH}_2\text{-CH-} + \text{R}_3\text{N} \quad (30)$$

$$R_3N \ + \ \underset{\underset{O}{\diagup}}{CH_2{-}CH{-}} \ \longrightarrow \ R_3\overset{\oplus}{N}{-}CH_2{-}CH{-} \qquad (31)$$
$$\underset{\underset{\ominus}{|}}{O}$$

$$R'SH \ + \ R_3\overset{+}{N}{-}CH_2{-}CH{-} \ \longrightarrow \ R'S{-}CH_2{-}CH{-} \ + \ R_3N \qquad (32)$$
$$\underset{\underset{\ominus}{|}}{O} \qquad\qquad \underset{\underset{}{|}}{OH}$$

Reactions of Alcohols

The uncatalyzed epoxide–alcohol reaction was shown by Shechter and Wynstra (8) to be "rather sluggish"; a temperature of 200 °C is necessary to realize a conveniently rapid rate. The reaction can be catalyzed by either acid or base to yield primary and secondary alcohols that will further react with the free epoxide to form polyethers (Reaction 33).

$$ROH \ + \ \underset{\underset{O}{\diagup}}{CH_2{-}CH{-}} \ \longrightarrow \ \begin{cases} ROCH_2{-}\underset{\underset{OH}{|}}{CH}{-} \xrightarrow{\ CH_2{-}CH{-}\ } 2 \text{ Isomers} \\[2em] HOCH_2{-}\underset{\underset{OR}{|}}{CH}{-} \xrightarrow{\ CH_2{-}CH{-}\ } 2 \text{ Isomers} \end{cases} \qquad (33)$$

For example, in experiments with phenyl glycidyl ether and isopropyl alcohol (1:1 molar ratio) in which potassium hydroxide and benzyldimethylamine were used as catalysts, Shechter and Wynstra (8) found that after nearly complete consumption of the epoxide group approximately 80% of the charged alcohol was unreacted. This result indicates that the reaction was largely self-polymerization.

Acid catalysis of the epoxide–alcohol reaction is no more selective than base catalysis. The ratio of alcohols formed and the amount of polyether obtained vary with the type and amount of catalyst, epoxide-to-alcohol ratio, solvent, and reaction temperature.

Although the epoxide–alcohol reaction is used industrially to prepare certain glycidyl ethers, it is not an important curing reaction. Coatings based on the use of hydroxyl-functional phenol- and amino-formaldehyde resins to cure the higher homologs of the diglycidyl ethers of bisphenol A were among the first to attain commercial significance. The curing chemistry of both the phenolic and amino-epoxy systems is based on the reactions of the cross-linking resin with the hydroxyl groups of the epoxy resin (Reaction 34), the cross-linking resin with the epoxide group (Reaction 33), and the cross-linking resin with itself (19).

$$R\text{-(OH)}_n \quad + \quad -CH_2\text{-CH-CH}_2- \quad \longrightarrow \quad -CH_2\text{-CH-CH}_2- \quad + \quad H_2O \qquad (34)$$
$$\quad\quad\quad\quad\quad\quad\quad\quad |\quad\quad\quad\quad\quad\quad\quad\quad\quad\quad\quad\quad\quad |$$
$$\quad\quad\quad\quad\quad\quad\quad\quad OH\quad\quad\quad\quad\quad\quad\quad\quad\quad\quad O\text{-R-(OH)}_{n-1}$$

Catalytic Reactions

The reactions of the epoxide group involve addition reactions of epoxides with compounds having a labile hydrogen atom. Catalytic reactions are characterized by the reaction of the epoxide group with itself (homopolymerization). Although both Lewis-type bases and acids can catalyze homopolymerization by causing anionic and cationic propagation, respectively, the resultant structure is the same: a polyether. Catalytic polymerization of monoepoxides results in linear polymers, whereas diepoxides give a cross-linked network.

Anionic polymerization of epoxides can be induced by Lewis bases (usually tertiary amines) or by metal hydroxides. The amine-type catalysts are by far the most important type of catalyst for epoxide homopolymerization. The initiation of the polymerization of epoxides has been proposed by Narracott ($\underline{15}$) and Newey ($\underline{16}$) to result from the attack by the tertiary amine on the epoxide (Reaction 35), with the resulting alkoxide amine being the propagating species (Reaction 36).

$$R_3N \quad + \quad CH_2\text{-CH-} \quad \longrightarrow \quad R_3\overset{\oplus}{N}\text{-CH}_2\text{-CH-} \qquad (35)$$
$$\quad\quad\quad\quad\quad\quad\diagdown\!\!/\quad\quad\quad\quad\quad\quad\quad\quad\quad\quad\quad |$$
$$\quad\quad\quad\quad\quad\quad O\quad\quad\quad\quad\quad\quad\quad\quad\quad\quad\quad O^{\ominus}$$

$$R_3\overset{\oplus}{N}\text{-CH}_2\text{-CH-} \quad + \quad CH_2\text{-CH-} \quad \longrightarrow \quad R_3\overset{\oplus}{N}\text{-CH}_2\text{-CH} \qquad (36)$$
$$\quad\quad\quad |\quad\quad\quad\quad\quad\quad\diagdown\!\!/\quad\quad\quad\quad\quad\quad\quad\quad\quad\quad |$$
$$\quad\quad\quad O^{\ominus}\quad\quad\quad\quad\quad O\quad\quad\quad\quad\quad\quad\quad\quad O\text{-CH}_2\text{-CH-}$$
$$\quad\quad\quad\quad\quad\quad\quad\quad\quad\quad\quad\quad\quad\quad\quad\quad\quad\quad\quad\quad\quad\quad O^{\ominus}$$

Cationic polymerization of epoxides is initiated by Lewis acids, which are substances composed of atoms containing empty electron orbitals in their outer shells. Such atoms can form covalent bonds with atoms capable of donating pairs of electrons. Many inorganic halides are Lewis acids such as $AlCl_3$, $SbCl_5$, BF_3, $SnCl_4$, $TiCl_4$, and PF_5. In commercial practice the most important type of initiator for curing epoxies is BF_3, usually in the form of BF_3 complexes.

Lewis acids initiate polymerization through the formation of carbonium ions, and Plesch ($\underline{17}$) has proposed that a suitable coinitiator is necessary to produce these ions. The mechanism for the initiation of the homopolymerization of epoxy resins by BF_3 complexes has been proposed by Arnold ($\underline{11}$) to proceed as follows:

$$BF_3 \quad + \quad H_2O \quad \longrightarrow \quad BF_3OH^{\ominus} \quad + \quad H^{\oplus} \qquad (37)$$

$$BF_3 \quad + \quad R_2NH \quad \longrightarrow \quad BF_3NR_2^{\ominus} \quad + \quad H^{\oplus} \qquad (38)$$

$$BF_3OH^{\ominus}$$
$$\text{or} \quad + \quad H^{\oplus} \quad + \quad CH_2\text{-}CH\text{-} \quad \longrightarrow \quad CH_2\text{-}CH\text{-} \quad + \quad \text{or}$$
$$BF_3NR_2^{\ominus} \qquad\qquad\qquad \overset{\diagdown}{O}\diagup \qquad\qquad\qquad \overset{\diagdown}{O}\diagup \qquad\qquad\qquad BF_3OH^{\ominus}$$
$$BF_3NR_2^{\ominus}$$
$$\overset{\cdot\cdot}{\underset{H}{\cdot}}{}^{\oplus}$$

$$(39)$$

$$\longrightarrow \quad {}^{\oplus}CH_2\text{-}CH\text{-} \quad + \quad \begin{array}{c} BF_3OH^{\ominus} \\ \text{or} \\ BF_3NR_2^{\ominus} \end{array}$$
$$\underset{OH}{|}$$

The polymerization chain is then propagated by the resulting carbonium ion that is stabilized by interaction with the anion produced in Reaction 40:

$$-CH\text{-}CH_2^{\oplus}\cdots O \underset{CH_2}{\overset{CH_2}{\diagup}} \quad + \quad \begin{array}{c} BF_3OH^{\ominus} \\ \text{or} \\ BF_3NR_2^{\ominus} \end{array} \quad \longrightarrow$$
$$\underset{OH}{|}$$

$$(\text{-}CH\text{-}CH_2)_n\text{-}O\text{-}CH\text{-}CH_2^{\oplus} \quad + \quad \begin{array}{c} BF_3OH^{\ominus} \\ \text{or} \\ BF_3NR_2^{\ominus} \end{array} \quad \longrightarrow \quad \text{etc.}$$
$$\underset{OH}{|}$$

$$(40)$$

In recent years photoinitiated catalytic cures have been emerging as an important technology for applying solventless, low-temperature curing coatings. Initially photoinitiated or UV-initiated cure of epoxy resin systems employed free-radical polymerization of vinyl derivatives of epoxy resins because epoxy resins are not curable by typical free-radical chemistry. Therefore, the use of radical-generating photoinitiators is not applicable in directly effecting an epoxide cure. These free-radical cured systems are, generally, based on vinyl esters of diglycidyl ethers obtained as indicated in Reaction 41:

$$H_2C\text{-}CH \sim CH\text{-}CH_2 + CH_2\text{=}CH\text{-}C\text{-}OH \rightarrow H_2C\text{=}CH\text{-}C\text{-}O\text{-}CH_2\text{-}CH \sim CH\text{-}CH_2\text{-}O\text{-}C\text{-}CH\text{=}CH_2$$

$$(41)$$

These systems have the disadvantage that they are sensitive to oxygen and require blanketing with nitrogen. Also, they have high viscosities and must be used with vinyl monomers as diluents, many of which present health hazards.

As early as 1965 Licari and Crepean (20) reported the photo-induced polymerization of epoxide resins by diazonium tetra-fluoroborates for use in the encapsulation of electronic components and the preparation of circuit boards. The use of these materials in coatings was pioneered by Schlesinger and Watt (21–23). When irradiated with UV light, these materials produce BF_3, fluoroaromatic compounds, and nitrogen (Reaction 42):

$$AR-N \equiv N^{\oplus} BF_4^{\ominus} \xrightarrow[\text{(UV)}]{h\upsilon} AR-F + N_2 + BF_3 \qquad (42)$$

When BF_3 is produced in this fashion in an epoxy resin, it catalyzes the cationic polymerization of the resin as discussed earlier. Typically, a small amount of the diazonium compound is dissolved in the epoxy coating formulation and irradiated with UV light to form thin films (0.5 to 1 mil) deposited on metal, wood, or paper substrates. The high reactivity of the BF_3-type cure makes it possible to prepare hard solvent-resistant coatings in a few seconds exposure time under a standard 200-W/in. mercury vapor lamp.

The evolution of nitrogen on photolysis of the aryldiazonium salts appears to have limited the use of these systems to thin film applications such as container coatings and photoresists (23). Other efficient photoinitiators that do not produce highly volatile products have been disclosed (24-27). These systems are based on the photolysis of diaryliodonium and triarylsulfonium salts, Structures I and II, respectively. These salts are highly thermally stable salts that upon irradiation liberate strong Bronsted acids of the HX type (Reactions 43 and 44) that subsequently initiate cationic polymerization of the oxirane rings:

$$\begin{array}{cc} AR & AR' \\ | & | \\ I^{\oplus} \ X^{\ominus} & AR-S^{\oplus} \ X^{\ominus} \\ | & | \\ AR' & AR'' \\ (I) & (II) \end{array}$$

$$X^{\ominus} = BF_4^{\ominus}, \ ASF_6^{\ominus}, \ PF_6^{\ominus}, \ SbCl_6^{\ominus}, \ etc.$$

$$AR_2 I^{\oplus} PF_6^{\ominus} \xrightarrow[\text{YH (solvent)}]{h\upsilon} ARI + AR + Y + HPF_6 \qquad (43)$$

$$AR_3 S^{\oplus} ASF_6^{\ominus} \xrightarrow[\text{YH}]{h\upsilon} AR_2 S + AR + Y + HASF_6 \qquad (44)$$

Unlike free-radical propagation, photoinitiated cationic polymerizations of epoxides are unaffected by oxygen and thus require no blanketing by an inert atmosphere. However, water and basic materials present in UV-curable epoxy formulations can inhibit cationic cures and should be excluded.

Curing Agents

The reactions just discussed describe the chemistry by which epoxy resins are converted into cross-linked polymeric structures by a variety of reactants. This cross-linking or "curing" is accomplished through the use of coreactants or "curing agents." Curing agents are categorized into two broad classes: active hydrogen compounds, which cure by polyaddition reactions, and ionic catalysts.

Most of the curing agents currently used with epoxy resins cure by polyaddition reactions that result in both the coupling as well as cross-linking of the epoxy resin molecules. Although these reactions are generally based on one active hydrogen in the curing agent per epoxide group, practical systems are not always based on this stoichiometry because of homopolymerization of the epoxide and other side reactions that cannot be avoided and that, in fact, are sometimes desired. In contrast to the 1:1 stoichiometry generally required for active hydrogen–epoxide reactions, only catalytic amounts of Lewis acids or bases are required to cure an epoxy resin.

The number of coreactants developed over the years for epoxy resins is overwhelming. Selection of the coreactant is almost as important as that of the base resin and is usually dependent on the performance requirements of the final product and the constraints dictated by their method of fabrication. Although the following is far from a complete list, the most common curing agents can be classified as follows:

1. <u>Aliphatic polyamines and derivatives</u>. These include materials such as ethylenediamine, diethylenetriamine, triethylenetetramine, tetraethylenepentamine, and several cycloaliphatic amines. These curing agents as a class offer low viscosities and ambient temperature cures. However, the unmodified amines present certain handling hazards because of their high basicity and relatively high vapor pressure. Less hazardous derivatives of aliphatic amines are obtained from the reaction products of higher molecular weight fatty acids with aliphatic amines. Besides having lower vapor pressure, these "reactive polyamide" and amidoamine systems will cure epoxy resins at room temperature to give tougher, more flexible products.

2. <u>Modified aliphatic polyamines</u>. These are room temperature curing agents formed by reacting excess aliphatic amines with epoxy-containing materials to increase the molecular weight of the amine to reduce its vapor pressure. The performance properties of amine adduct cured systems are not significantly different from those of aliphatic polyamines.

3. <u>Aromatic amines</u>. These include materials such as 4,4'-methylenedianiline, m-phenylenediamine, and 4,4'-diaminodiphenylsulfone. Aromatic amines are less reactive than aliphatic amines and usually require curing temperatures as high as 300 °F. Thus, they offer systems that have a long pot life at room temperature and cure to products with excellent physical and chemical property retention up to 300 °F.

4. <u>Acid anhydrides</u>. These are the second most commonly used curing agents after polyamines. In general, acid anhydrides require curing at elevated temperatures, but offer the advantages of longer pot lives and better electrical properties than aromatic amines. Illustrative of some of the more commonly used acid anhydrides are phthalic, trimellitic, hexahydrophthalic, and methylnadic anhydride.

5. <u>Lewis acid and base type curing agents</u>. Cures with Lewis acid and base type curing agents proceed by homopolymerization of the epoxy group that is initiated by both Lewis acids and bases. These curing agents can provide long room-temperature pot life with rapid cures at elevated temperatures, and thus produce products with good electrical and physical properties at relatively high temperatures (150–170 °C).

6.   <u>Aminoplast and phenoplast resins.</u> This class represents a
broad range of melamine-, phenol-, and urea-formaldehyde resins
that cross-link by a combination of reactions through the
hydroxyl group of the epoxy resin, self-condensation, and
reaction through epoxide groups. These systems are cured at
relatively high temperatures (325 to 400 °F) and yield final
products with excellent chemical resistance. Also, a few
miscellaneous types of curing agents, such as dimercaptans,
dicyandiamide, dihydrazides, and guanamines, have found some
limited industrial applications. Except for dicyandiamide,
which finds extensive applications in electrical laminates and
in powder coatings, these materials account for only a small
volume of the total curing agent market.

Curing of Epoxy Resin Compositions

The properties that can ultimately be obtained from an epoxy resin
system depend on the nature of epoxy resin and curing agent as well
as the degree of cross-linking that is obtained during cure. The
degree of cross-linking is a function of stoichiometry of the epoxy
resin and curing agent, and the extent of the reaction achieved
during cure. To actually obtain a "cure," the reaction of the
curing agent with epoxy resin must result in a three-dimensional
network. A three-dimensional network is formed when one component
has a functionality greater than two, and the other component has a
functionality of not less than two, as follows:

$$
\begin{array}{lll}
4 \text{ H-N} \sim \text{X} \sim \text{N-H} & + & 6 \overset{O}{\overset{/\backslash}{\text{C-C}}} \sim \overset{O}{\overset{/\backslash}{\text{C-C}}} \longrightarrow \\
\quad | \qquad\quad | & & \\
\quad \text{R} \qquad\quad \text{H} & &
\end{array}
$$

Polyamine   Epoxy Resin

(45)

Primary and secondary aliphatic polyamines, their derivatives, and
modified aliphatic polyamines and aromatic amines react with and
cure epoxy resins as indicated earlier. The aliphatic systems
usually give adequate cures at room temperature (7 days above
60 °F); however, under most conditions aromatic amines are less
reactive and require curing temperatures of about 300 °F to give
optimum cured polymer properties.

The importance of stoichiometry in network formation can be illustrated with a difunctional epoxide and tetrafunctional curing agent, which are represented schematically in Figure 7. As can be seen, a spectrum of products is obtained in progressing from an excess of epoxide to an excess of a curing agent, such as a tetrafunctional amine. At an excess of epoxide groups over reactive sites on the curing agent, an epoxy–amine adduct is the predominate product. As the stoichiometry approaches one equivalent of epoxide per equivalent of reactive amine sites, the molecular weight approaches infinity, and a three–dimensional network polymer is obtained. As the ratio of curing agent to epoxide is increased, the product approaches a linear polymer (thermoplastic), and finally with an excess of curing agent, an amine–epoxide adduct is obtained.

On the molecular level the cure of an epoxy resin system involves the reaction of the epoxy groups (or hydroxyl groups in some cases) of the resin molecules with a curing agent to form molecules of ever increasing size until an infinite network of cross–linked resin and curing agent molecules is formed. As the chemical reactions proceed, the physical properties of the curing resin change with time from a fluid to a solid. More specifically, as the cure proceeds, the viscosity of the reacting system increases until gelation occurs, at which time the mass becomes an insoluble rubber. Further chemical reaction eventually converts the rubbery gel into a glassy solid (vitrification). Gelation corresponds to incipient formation of an infinite network of cross–linked polymer molecules, and vitrification involves a transformation from a liquid or a rubbery state to a glassy state as a result of an increase in molecular weight. Vitrification can quench further reaction.

In an attempt to understand the cure phenomena Gillham (28) has developed the concept of a state diagram. Such a time–temperature–transformation (TTT) diagram is given in Figure 8. It is a plot of the times required to reach gelation and vitrification during isothermal cures as a function of cure temperature, and it delineates the four distinct material states (liquid, gelled rubber, ungelled glass, and gelled glass) that are encountered during cure. Also displayed in the diagram are the three critical temperatures:

1. $T_{g\infty}$, the maximum glass transition temperature of the fully cured system;
2. $_{gel}T_g$, the isothermal temperature at which gelation and vitrification occur simultaneously; and
3. T_{go}, the glass transition temperature of the freshly mixed reactants.

Thus, during an isothermal cure at a temperature between $_{gel}T_g$ and $T_{g\infty}$, the resin will first cure and then vitrify. Once vitrification occurs, the curing reactions are usually quenched, which means the glass transition temperature (T_g) of the resin will equal the temperature of cure (such a material will not be fully cured). The resin, however, will not vitrify on isothermal cure if the cure temperature is above $T_{g\infty}$, the T_g of the fully cured resin. Above $T_{g\infty}$, the cure can proceed to completion, and the maximum T_g of the system is obtained. A more detailed discussion of the TTT cure diagram as it relates to epoxy resin curing can be found in the publications of Gillham (28, 29).

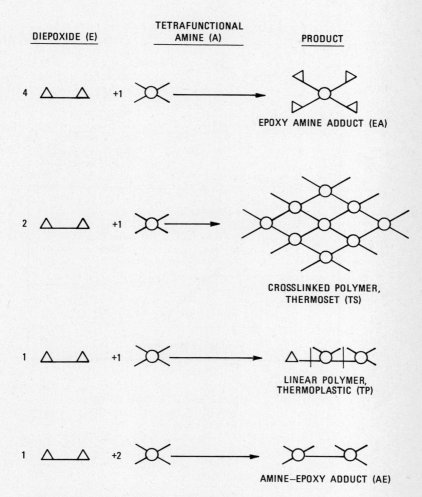

Figure 7. Schematic representation of network formation.

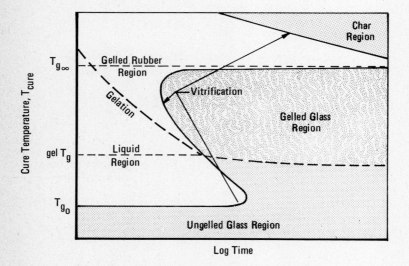

Figure 8.   Generalized TTT cure diagram.

Applications of Epoxy Resins

Epoxy resins have found a broad range of applications, mainly because the proper selection of resin, modifiers, and cross-linking agent allows the properties of the cured epoxy resin to be tailored to achieve specific performance characteristics. This versatility has been a major factor in the steady growth rate of epoxy resins over the years.

Besides this versatility feature, properly cured epoxy resins have other attributes:

1. Excellent chemical resistance, particularly to alkaline environments.
2. Outstanding adhesion to a variety of substrates.
3. Very high tensile, compressive, and flexural strengths.
4. Low shrinkage on cure.
5. Excellent electrical insulation properties and retention thereof on aging or exposure to difficult environments.
6. Remarkable resistance to corrosion.
7. A high degree of resistance to physical abuse.
8. Ability to cure over a wide range of temperatures.
9. Superior fatigue strength.

Even though epoxy resins are somewhat more expensive than many resinous materials used in coatings, epoxy resins have found commercial acceptance in a wide variety of high-performance protective and decorative surface coatings. Epoxy resin based coatings have been established as premium coatings because of their excellent chemical and corrosion resistance and their outstanding adhesion compared to other coating materials.

Epoxy resin coatings can be divided into two distinct types: those that are cured at ambient temperature and those that are heat cured. The first type is cross-linked through the oxirane ring by using polyamines, amine adducts, polyamides, polymercaptans, and catalytic cures. Heat-cured epoxy resin coatings are cross-linked through reaction of the hydroxyl groups, or in some cases a combination of the epoxy and hydroxyl functionality, by using anhydrides and polycarboxylic acids as well as melamine-, urea-, and phenol-formaldehyde resins.

Most people are familiar with the ambient cure or "two-package" type coating that is sold through retail stores. This coating was the first room-temperature curing coating to offer resistance properties previously obtainable only in baked industrial finishes. Coatings of this type are used extensively today as heavy duty industrial and marine maintenance coatings, tank linings, and floor toppings; for farm and construction equipment, and aircraft primers; and in do-it-yourself finishes for the homeowner. As the name implies, the "two-package" epoxy coatings are mixed just prior to use, and are characterized by a limited working life (pot life) after the curing agent is added. Commercial systems will have pot lives ranging from a few hours to a couple of days with typical working times ranging from approximately 8 to 12 h. By far the most common two-package coatings are based on epoxy resins having WPE values between 180 and 475 (see Table I) and cured with polyamines, amine adducts, and polyamides. The polyamine and amine adduct type cures provide better overall chemical resistance, but polyamides offer better film flexibility, water resistance, and are somewhat more forgiving of improper surface preparations.

Two other epoxy resin type coatings are classified as air dry, instead of ambient cure, because they do not involve a curing agent. The earliest of this type are the epoxy esters, which are simply higher molecular weight epoxy resins that have been esterified with unsaturated or drying type fatty acids, such as dehydrated castor acid. These esters are usually prepared from solid epoxy resins having WPE values in the range of 900 and can be considered as specialized polyols. As in conventional alkyd resin technology, these coatings are manufactured by esterifying the resin with a fatty acid at temperatures of 400–450 °F. Initially the fatty acid reacts with the oxirane ring at lower temperatures, and thereby forms a hydroxyl ester, and subsequently these hydroxyl groups and those already present in the resin backbone are esterified at the higher temperatures with the aid of esterification catalysts and azeotropic removal of the water formed. These epoxy ester vehicles are used in floor and gymnasium finishes, maintenance coatings, and metal decorating finishes, and air dry by oxidation of the unsaturated fatty acid, similar to the so-called "oil-based" paints.

A newer class of epoxy-derived finishes is based on very high molecular weight linear polyhydroxy ethers (these ethers are essentially bisphenol A epoxy resins where n is approximately 30 or greater) having molecular weights greater than 50,000. Unlike the systems described earlier, these coatings are usually not chemically cross-linked, but form tough, thermoplastic films upon evaporation of the solvent. However, on occasion, these coatings may be cross-linked with small amounts of aminoplast resins for baking applications, as described later. Coatings of this type offer extreme flexibility with good adhesion and chemical resistance and are used for air dry preconstruction primers as well as baked container coatings and automotive primers.

The higher molecular weight epoxy resins, that is, those with WPE values about 1750 or greater, generally are used in baking finishes. The concentration of oxirane groups is low, and cross-linking occurs principally through the hydroxyl functionality. Solid, high molecular weight epoxy resins coreacted with phenol-formaldehyde and aminoplast resins (melamine and urea-formaldehyde resins) through loss of water or alcohol to form ether linkages have been used for years for highly solvent- and chemical-resistant coatings. The phenolic converted systems are used as beverage and food can coatings, drum and tank linings, internal coatings for pipe, wire coatings, and impregnating varnishes. Although not as chemical resistant as the epoxy phenolic coatings, aminoplast resins are used in certain applications because of their better color and lower cure temperatures. Epoxy-aminoplast resin finishes are used widely as can linings; appliance primers; clear coatings for brass hardware, jewelry, and vacuum metalized plastics; foil coatings; and coatings for hospital and laboratory furniture.

Epoxy coatings obtain their excellent properties through reaction with curing agents. The curing agents reacting with epoxide and hydroxyl functionality of the epoxy resins result in highly chemical- and solvent-resistant films because all the bonds are relatively stable carbon-carbon, carbon-oxygen (ether), and carbon-nitrogen (amine) linkages. Many of the more common epoxy resin coating systems and their end uses are summarized in Figure 9.

Environmental pressures and energy concerns are now resulting in some rapid changes in epoxy coating technology. New curing agents and epoxy resins are currently being developed for more highly solid, solventless, and even water-borne coatings. Epoxy coatings are being applied as powders to completely eliminate solvent emissions, or solventless epoxy-acrylic esters are applied as 100% solids and then cured by initiation with UV light. Techniques are also being developed to reduce baking temperatures of the heat-cured type epoxy finishes, and the use of "two-package" systems in low bake applications is apparently increasing. As a whole, however, all these coatings are still based on the chemistry already developed and described earlier. The resulting coatings still retain the properties for which epoxy resins have become known.

Structural Applications

Structural or noncoating type applications of epoxy resins are much more fragmented and more difficult to simplify than the coatings area. There are, basically, four major structural-type end uses for epoxy resin systems: bonding and adhesives; castings, molding powders, and tooling; flooring, paving, and aggregate; and reinforced plastics or composites (electrical laminates and filament winding). Again epoxy resins have found such broad application in the structural area because of their versatility (see Figure 10). They can be modified into low viscosity liquids for easy casting and impregnating, or converted to solid compositions for ease of laminating or molding. Depending upon choice of curing agent, they can be made to cure either slowly (hours) or very quickly (minutes) at room temperature to give a variety of properties ranging from soft, flexible materials to hard, tough, chemical-resistant products.

Bonding and Adhesives. Epoxy resin adhesives are most commonly used as two-component liquids or pastes and cure at room or elevated temperature. This type of adhesive is cured with a polyamide or polyamine, or in the quick setting type with an amine-catalyzed polymercaptan, and is available to the householder at hardware stores. Although the bulk of adhesives are of this type, a great deal of the more sophisticated epoxy adhesives are supplied as supported tape, that is, a glass fabric tape impregnated with adhesive, or as nonsupported tape. These are "one-package" systems that contain a latent curing agent and epoxy resin. Latent curing agents such as dicyandiamide and boron trifluoride salts are used because they provide single-package stability and rapid cure at elevated temperatures. Adhesive systems of this type are used in the aircraft industry to replace many of the mechanical fasteners once used, and for automotive adhesives.

Casting, Molding Compounds, and Tooling. Epoxy resins maintain a dominant position in the electrical and electronic industry for casting resins, molding compounds, and potting resins. Their

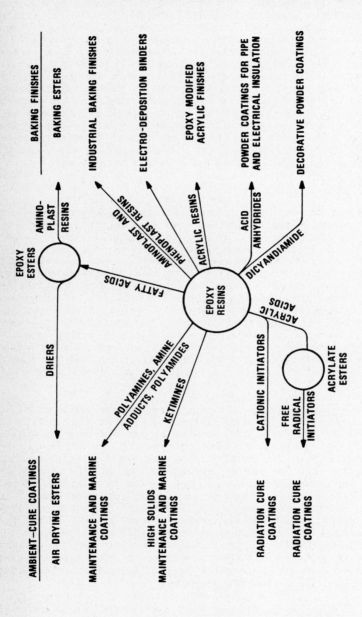

Figure 9. Epoxy resins in surface coating applications.

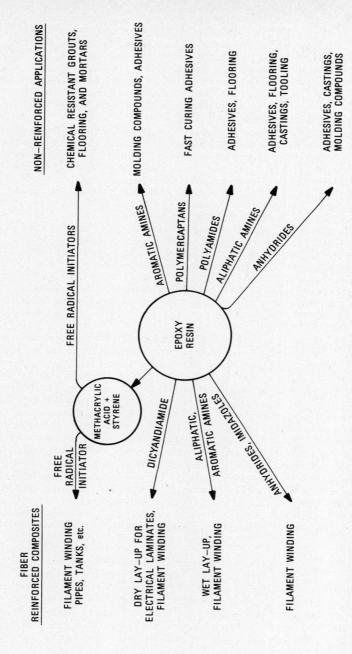

Figure 10. Epoxy resins in structural applications.

superior dielectric properties together with their low shrinkage upon cure and good adhesion make epoxy resins a natural for this end use.

Epoxy castings are made by pouring a resin and curing agent mixture into molds and curing to finished parts; subsequently, the mold is removed. In potting, the mold is retained as an integral part of the finished product. The epoxy systems are cast around parts that are in containers or housings. These applications use a liquid epoxy resin, usually cured at elevated temperatures, with aromatic amines or anhydrides that provide lower shrinkage, longer pot life, and lower exotherms on cure.

Molding compounds are based on solid epoxy resins cured with aromatic amines or anhydrides. They contain high filler loadings (as much as 50% by volume) and are manufactured by dispersing the components on a two-roll mill or in a heated kettle. The resulting mixture is cooled and then ground to the desired particle size. Currently, the major use of epoxy molding compounds is in the encapsulation of electrical components, the manufacture of pipe fittings, and certain aerospace applications.

Along with potting and encapsulating, one of the first two structural applications for epoxy resins was in tooling molds. For short production runs epoxy molds, drop hammers, stretch dies, jigs, and fixtures have replaced metal tools, especially in the aircraft industry and in automotive work. An epoxy tool can replace its metal counterpart on the basis of ease and speed of fabrication; however, a resin-based tool will not have the durability of metal in long-term production runs.

Flooring, Paving, and Aggregates. Two-package epoxy systems serve as binders for pothole patching; anti-skid surfaces; exposed aggregate; and industrial, seamless, and thinset terrazzo flooring. Typical formulations would be based on a liquid epoxy resin containing a reactive or nonreactive diluent, fillers, and special thickening agents; because a room-temperature cure is required, curing agents usually employed are amines or polyamides.

Epoxy terrazzo is being used in place of concrete flooring because it cures overnight and can be ground and polished the next day; its 1/4 to 3/8 in. thickness provides significant weight savings; it has superior chemical resistance; and it can be applied over many different surfaces. Industrial flooring provides anti-skid and wear surfaces and resistance to spills, such as in dairies and other food processing plants (18). A related application, which appears to be growing, is the use of epoxy resin systems as binders for exposed aggregate patios, swimming pool decks, walks, and wall panels.

Reinforced Plastics or Composites. After protective coatings, reinforced plastics represent the single largest market for epoxy resins (see Figure 2). The major offtake for this market is in the laminating industry, although filament wound applications account for the next largest segment of this market. Epoxy resin reinforced plastics exhibit low weight, good heat resistance, excellent corrosion resistance, and good mechanical and electrical properties (18).

Epoxy resin systems with fiber reinforcements are called "reinforced systems" or "composites." Composites are made by impregnating reinforcing fibers such as glass, synthetic polymer, or graphite fibers, by one of several processes with the desired epoxy resin system, and then curing in a heated mold or die. Epoxy composite systems are formulated with either liquid or solid resins with selection of the type of system dependent on the fabrication process, the cure temperature, and the final part application.

The several methods to convert resins and reinforcing fibers into composites are classified as either "wet lay-up" or "dry lay-up" methods. These processes are as follows:

1. **"Wet lay-up"** refers to a process in which liquid resin systems of low viscosity are used to impregnate the reinforcements, either before or after the reinforcements have been laid in place. The liquid resin penetrates the fibers and displaces the air. The distinctive feature of this method is that the composite object is shaped into its final configuration while the resin component of the lay-up is still liquid (uncured). Cure is effected after the lay-up is completely in place and conforms exactly to the mold.

2. **"Dry lay-up"** refers to a process in which the reinforcing material is preimpregnated with a resin solution. The solvent is then removed by heated air currents that also may partially cure the resin system. This removal produces a dry, resin-impregnated sheet (the "prepreg"). The prepreg stock is then cut and positioned in the desired configuration for cure. By application of heat under pressure, the resin softens and the curing agent is activated to complete the polymerization of the resin. Although either solid or liquid epoxy resins may be used for dry lay-up, use of low molecular weight solid resins is the most common. The pressure used in dry lay-ups gives a higher glass content and stronger composites than wet lay-up composites.

Filament winding processes use both wet and dry techniques. In this method, a continuous supply of roving or tape (formed from roving) is wound onto a suitable mandrel or mold. The glass fiber is either impregnated with the epoxy resin-curing agent system at the time of winding (wet winding), or has been preimpregnated and dried (as for dry lay-up laminating). This dry stock is wound onto the mandrel (dry winding). In either case, the mandrel may be stationary, but often it will revolve around one axis or more. The shapes of structures made by filament winding are not limited to surfaces of revolution, cylindrical shapes, and pipe; they may be round, oval, or even square.

Typically, a low molecular weight solid epoxy resin and latent curing agent, such as dicyandiamide dissolved in an appropriate solvent, are widely used in dry lay-up formulations for electrical laminates for computers, aerospace applications, and communications equipment. Wet lay-up systems are almost exclusively based on low-viscosity resins cured with aliphatic amines. These ambient cure resin systems are used primarily for manufacture of large chemical-resistant tanks, ducting, and scrubbers. Filament winding is used to produce lightweight chemical-resistant pipe that is generally produced from low molecular weight resin systems cured with anhydrides, aromatic amines, or imidazoles.

Although these applications will continue to use epoxy resins, another more recent approach to reinforced thermosetting epoxy resin systems is the use of epoxy-based vinyl ester resins. These vinyl esters are derived from the reaction of a bisphenol A epoxy resin with methacrylic acid. This reaction results in a vinyl-terminated resin having an epoxy backbone. Usually the epoxy vinyl esters are used in solution in a vinyl monomer such as styrene as a coreactant and viscosity reducer. Cross-linking is effected by free-radical generating initiators or high energy sources including UV- and electron-beam radiation. These resins provide the handling characteristics of polyesters resins, but the improvements in water, acid, base, solvent, and heat resistance over general purpose and isophthalic polyester resins. Their inherent chemical resistance and physical properties make the vinyl esters suitable for manufacture of equipment such as tanks, scrubbers, pipe, and fittings, or for tank and flue linings for application in a wide variety of chemical environments.

The scope of epoxy resin application is so broad that it has been possible to touch only upon a few major areas and types of epoxy resin systems. Epoxy resin as indicated in this article connotes not a single formulation but a wide range of compositions having properties specific to their chemical structure. Most of the important applications and typical curing agents used in the resin systems specific to an application are summarized in Figures 2, 9, and 10. The chemical versatility of the epoxy resins has been responsible for the diversified epoxy resin usage. It is this same versatility coupled with the ingenuity of chemists and engineers that ensures the continued growth of epoxy resins.

Literature Cited

1. Modern Plastics 1982, 59(1).
2. Tanaka, Y.; Mika, T. F. In "Epoxy Resins, Chemistry and Technology"; May, C. A.; Tanaka, Y., Eds.; Dekker: New York, 1973.
3. Shechter, L.; Wynstra, J.; Kurkjy, R. P. Ind. Eng. Chem. 1956, 48(1), 94.
4. O'Neill, L. A.; Cole, C. P. J. Appl. Chem. 1956, 6, 356.
5. Dannenberg, H. SPE Tran S. 1963, 3, 78.
6   Smith, I. T. Polymer 1961, 2, 95.
7. Tanaka, Y.; Mika, T. F. In "Epoxy Resins, Chemistry and Technology"; May, C. A.; Tanaka, Y., Eds.; Dekker: New York, 1973.
8. Shechter, L.; Wynstra, J. Ind. Eng. Chem. 1956, 48(1), 86.
9. Fisch, W.; Hofmann, W. J. Polym. Sci. 1954, 23, 497.
10. Fischer, R. F. J. Polym. Sci. 1960, 44, 155.
11. Arnold, R. J. Mod. Plastics 1964, 41, 149.
12. Somerville, G. F.; Parry, H. L. J. Paint Tech. 1970, 42(540).
13. Alvey, F. B. J. Appl. Polym. Sci. 1969, 13, 1473.
14. Sow, P. N.; Weber, C. D. J. Appl. Polym. Sci. 1973, 17, 2415.

15. Narracott, E. S. <u>Brit. Plastics</u> 1953, <u>26</u>, 120.
16. Newey, H. A. unpublished data.
17. Plesch, P. H. In "The Chemistry of Cationic Polymerization";
    Plesch, P. H., Ed.; Pergamon: Oxford, 1963.
18. Somerville, G. R.; Jones, P. D. In "Applied Polymer
    Science"; Craver, J. K., Ed.; American Chemical Society,
    Advanced in Chemistry Series: Washington, D. C., 1975.
19. Nylen, P.; Sunderland, E. In "Modern Surface Coatings";
    Interscience: London, 1965.
20. Licari, J. J.; Crepean, P. C. U. S. Patent 3 205 157 (1965).
21. Schlesinger, S. I. U. S. Patent 3 708 296, 1973.
22. Schlesinger, S. I. <u>Photo. Sci. Engr</u>. 1974, <u>18(4)</u>, 387.
23. Watt, W. R. In "Epoxy Resin Chemistry"; Bauer, R. S., Ed.;
    ACS SYMPOSIUM SERIES 114, American Chemical Society:
    Washington, D. C., 1979.
24. Smith, G. H. Belg. Patent 828 841, 1975.
25. Crivello, J. V.; Lam, J. H. W. In "Epoxy Resin Chemistry";
    Bauer, R. S., Ed.: ACS SYMPOSIUM SERIES 114, American Chemi-
    cal Society: Washington, D. C., 1979.
26. Crivello, J. V.; Lam, J. H. W. <u>J. Polym. Sci</u>. 1976, <u>56</u>, 383.
27. Crivello, J. V.; Lam, J. H. W. <u>Macromolecules</u> 1977, <u>10(6)</u>,
    1307.
28. Gillham, J. K. In "Developments in Polymer Characterization-
    3"; Dawkins, J. V., Ed.; Applied Science: England, 1982.
29. Enns, J. B.; Gillham, J. K. <u>J. Appl. Polym. Sci</u>. 1983, <u>28</u>,
    2567.

Chemistry and Technology of Polyamide Resins from Dimerized Fatty Acids

MICHAEL A. LAUDISE

Henkel Corporation, Minneapolis, MN 55413

```
Chemistry of Dimer-Based Polyamide Resins
Solid Polyamide Resins
    Properties
    Viscosity Characteristics
    Solubility
    Surface Activity
    Applications
Fluid Polyamide Resin Chemistry
Thermoset Coatings
Thermoset Adhesives
Thermosetting Systems for Casting and Laminating
```

Prior to 1900 the methodology of bodying oils for paint manufacture was well-known. However, the fatty acid chemistry involved was obscure until the early war years of 1940. Then an understanding started to emerge relative to the complex intermolecular condensation reactions occurring during oil bodying. These coupling reactions led to interesting high molecular weight polybasic fatty acids capable of entering into polymer formation. Continued study showed that these acids formed when unsaturated fatty acids containing one or more double bonds are polymerized. These studies also showed that the major products were complex dibasic acids containing 36 carbon atoms and smaller amounts of monocarboxylic, tricarboxylic, and higher carboxylic acids. It is these acid mixtures, either as such or refined to increase dibasic content, that are used to prepare polyamide resins via their reaction with polyamines.

The first polyamides from these polybasic acids or their corresponding esters were described by Bradley and Johnston (1) and subsequently by Falkenburg et al. (2). These chemists recognized that polyamide resins have unusual solubility in lower alcohols and that films, deposited from alcohol solution, have good water resistance, strong adhesion to various surfaces, and other

0097–6156/85/0285–0963$06.50/0

mechanical properties like good tensile strength and flexibility that contribute to toughness. This chapter will describe the chemistry of these resins and their properties and applications.

Polybasic acids are now available commercially in various purity grades. The chemistry of polybasic acids has been explained by Wheeler and coworkers (3-7). A common product of commerce contains 70–80% dibasic acid, 15–20% of higher functionality acids, and small amounts of monomeric acids. Much of the literature associated with commercial materials refer to it as "dimer" or mixed dimerized fatty acids. This terminology, accordingly, will be used here. The acid may be represented graphically as HOOCDCOOH where D is a 34–carbon radical.

Chemistry of Dimer–Based Polyamide Resins

The simplest polyamide resin results from the condensation of dimer acid with ethylenediamine. The structure of the resulting polymer is represented as $HO(-OC-D-CONH-RNH-)_nH$ where R is C_2H_4- and n may be 5–15.

Number–average molecular weights may be determined by use of a Mechrolab vapor pressure osmometer, utilizing a solution of polyamide resin in chloroform at 37 °C. The procedure is described in detail by Pasternak et al. (8). Typical molecular weight values for ethylenediamine–based polyamides with varying viscosities are listed in Table I. Viscosities of the melts were determined by utilizing a Brookfield viscometer, Model RVF, at 160 °C with a No. 3 spindle at 20 rpm.

The polyamide resins melting above 100 °C, which result from the condensation of ethylenediamine or related higher molecular weight diamines with dimer acid, find application in industry because of their film–forming and adhesive properties. These will be discussed in detail later.

If dimer acid is condensed with diethylenetriamine or similar polyalkylene polyamines, a liquid resin usually results that melts below 100 °C. The structure of such resins can be indicated diagrammatically as follows, although there is evidence that both primary and secondary amine groups are involved in resin formation.

$$HOOC-D-COOH$$

$$+$$

$$NH_2-C_2H_4-NH-C_2H_4-NH_2$$

$$\downarrow -H_2O$$

$$NH_2-C_2H_4-NH-C_2H_4--N--\left[\begin{array}{c} C--D--C-N-C_2H_4-NH-C_2H_4 \\ \| \quad\quad \| \ | \\ O \quad\quad O \ H \end{array}\right]-NH_2$$
$$\quad\quad\quad\quad\quad\quad\quad\quad\quad\quad\; H \quad\quad\quad\quad\quad\quad\quad\quad\quad\quad\quad\quad\quad\quad\quad\quad\quad n$$

Some molecular weights of this type of polyamide resin are listed in Table II. Viscosity was determined at 150 °C by utilizing a Brookfield viscometer, Model RVF, No. 3 spindle at 20 rpms. If

Table I. Typical Molecular Weight Values for
          Ethylenediamine–Based Polyamides

160 °C Melt Viscosity of Polyamide (P)	$\overline{M}_n$	n (Average)
15.1	3070	5–6
17.2	4000	6–7
22.0	4230	7–8

Table II. Molecular Weights of Polyamide Resins

150 °C Melt Viscosity of Polyamide (P)	$\overline{M}_n$	n (Average)
9.4	2780	4–5
11.7	2960	4–5

the monomeric fraction of dimer acid is condensed with a polyalkylene polyamine like diethylenetriamine, a new class of polyamides, sometimes referred to as fatty amido amines, results. These low-melting materials usually have viscosities below 10 P at 25 °C. The structure of such resins can be indicated diagrammatically as follows.

$$CH_3 (CH_2)n COOH$$

$$+$$

$$\overset{\displaystyle H}{\underset{\displaystyle |}{H_2N-C_2H_4-N---C_2H_4---NH_2}}$$

$$\downarrow -H_2O$$

$$CH_3(CH_2)_n\overset{\displaystyle O}{\overset{\displaystyle \|}{C}}--\underset{\displaystyle \underset{\displaystyle H}{|}}{N}--C_2H_4----N----C_2H_4----NH_2$$

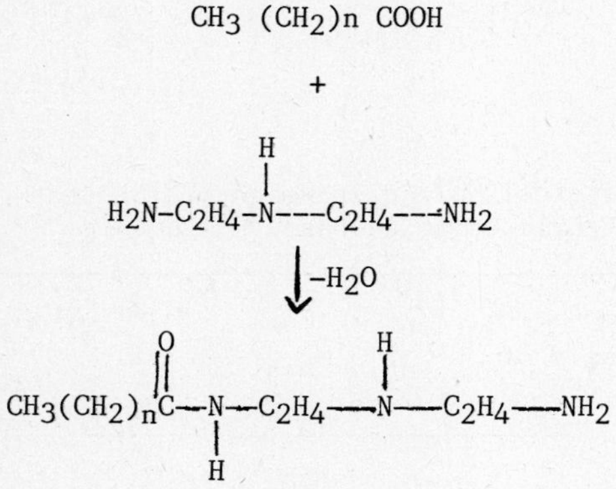

$$n = C_{16}-C_{19}$$

As may be seen from the above structures, the dimer-based liquid polyamides and monomeric fatty amido amines contain free amine groups that are available for further reaction. One such reaction is imidazoline formation, which occurs intermolecularly (9). The structure of imidazoline can be indicated diagrammatically as follows.

$$CH_3(CH_2)_n\overset{\displaystyle O\ H}{\overset{\displaystyle \|\ \|}{C-N}}-C_2H_4-NH_2$$

$$\downarrow -H_2O$$

$$CH_3(CH_2)_nC=N$$

$$HN \qquad CH_2$$

$$CH_2$$

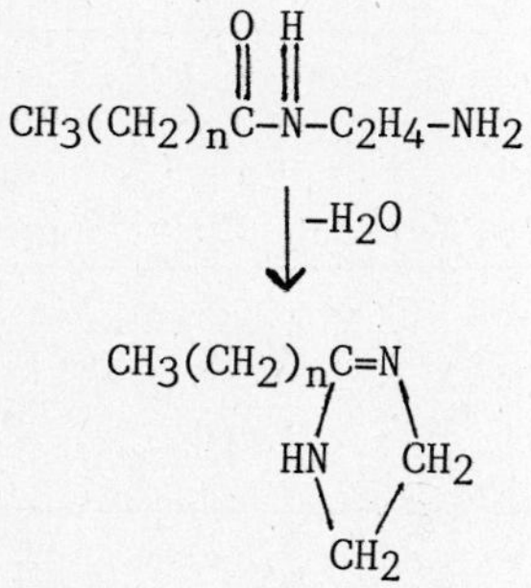

$$n = C_{16}-C_{19}$$

Still another reaction that accounts for the major application of liquid polyamides and fatty amido amines is a curing agent for epoxy resins. These reactions and related applications will be discussed in detail later.

The molecular weights of fatty polyamides generally encompass the range of 3000–15,000. Anderson and Wheeler (10) evolved a relationship for number-average molecular weight determined by end-group analysis and viscosity so that molecular weight can be estimated from solution viscosity measurements. From a practical point of view, solution and melt viscosities are more useful than molecular weight since viscosity values have important relationships to major application areas of coatings, inks, and adhesives.

The solid fatty dimer polyamides are related to nylons and, indeed, are considered nylons when based on highly purified dimer acid (11). When based on a less pure version of dimer acid, they are, of course, of considerably lower molecular weight than nylons and have a much lower order of crystallinity.

Solid Polyamide Resins

Properties. The solid polyamide resins are alcohol soluble and if properly formulated can supply some degree of hydrocarbon compatibility. Thus, they can be applied from solvent solution. However, they also can be applied as hot melts or from water dispersions. Water-based forms known as suspensoids have been described (12). Solid polyamides also may be finely divided for use as powders. The films are characterized by resistance to moisture, moisture vapor transmission, grease, and oils. They are resistant to many solvents and chemicals including aliphatic hydrocarbons and mineral and vegetable oils. They do not resist lacquer-type solvents and alcohols. They have a high degree of flexibility and maintain their flexibility upon aging. Also, they are heat sealable at relatively low temperatures and adhere to an unusually wide variety of substrates.

When fatty polyamides of different melting points are mixed, the resulting melting point is nonlinear with respect to composition as demonstrated in Figure 1. Thus, a formulator has the ability to make many resin alloys.

Viscosity Characteristics. As is typical of certain polyamides, viscosity decreases rapidly above the melting point. Figure 2 shows the relationship of temperature to viscosity in poises for three polyamide resins melting in the 105 °C range. These resins differ in molecular weight only. The important point to observe is that all three resins have a very low viscosity range of 10–15 P by the time temperature reaches 180 °C.

Because of this steep viscosity drop, solid polyamide resins are useful as hot melt adhesives. The resins may tend to "skin" or oxidize when exposed to air and high temperatures for long periods. However, this negative property can be circumvented by the use of antioxidants and by proper application equipment design.

Solubility. Solid polyamides are soluble in some alcohols and if properly formulated tolerate significant amounts of hydrocarbon-type solvents. Actually these formulated polyamides are more soluble in a combination of alcohol and hydrocarbon solvent than alcohol alone (13). Thus, ethyl alcohol, isopropyl alcohol, propyl alcohol, and butanol can be used in combination with hexane, mineral spirits, xylene, and toluene. Isopropyl acetate may also be used when nitrocellulose compatibility is required.

Surface Activity. As already indicated, the solid dimer polyamides adhere to numerous surfaces, have a high degree of flexibility, and

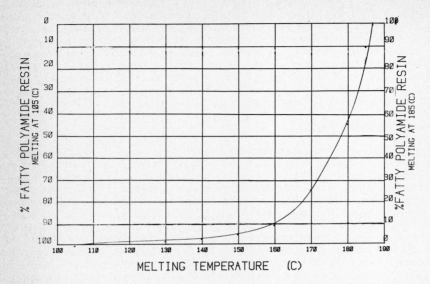

Figure 1. Ball and ring softening points of mixtures of resins
          melting at 150 and 185 °C.

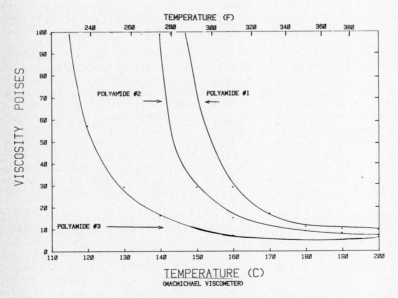

Figure 2. Relationship of temperature to viscosity of
          polyamide resins melting at about 105 °C.

have an excellent combination of mechanical properties. Because polyamide resins have a high fatty content coupled with highly polar amide, amine, and imidazoline groups, they have the structural characteristics of surface active agents. This surface activity is a particularly important property of liquid resins to be discussed later. The fatty characteristics also affect physical and chemical resistance properties such as compatibility with other resin types and melting point and water resistance. Indeed, fatty polyamides have considerably lower water absorption than classical nylons, presumably because the fatty molecular portions shield the polar amide and amine groups.

The fatty polyamides also demonstrate a high level of grease and oil resistance combined with a low order of irritation and toxicity (14).

Applications. Adhesives. In order for a solid polyamide to function effectively as a hot melt adhesive, it must possess high tensile strength in the glassy or solid state, melt in a range suitable for application by automated equipment, and have low enough viscosity when molten to wet surfaces to which it must adhere. High tensile strength requires moderate to high molecular weight polyamides while low melt viscosity calls for low molecular weights. Thus, an obvious compromise is necessary in molecular weight to achieve an appropriate balance of properties (15).

With the dimer polyamides, certain structural considerations interrelate with desirable adhesive properties. For example, even at moderate molecular weights around 10,000, the linear polyamides demonstrate a high degree of interchain attraction, largely attributed to hydrogen-bond formation (14). Such polyamides tend to offer high tensile strengths even though molecular weights are moderate. Yet, when molten, the polyamides flow readily because hydrogen bonding breakage permits the molecules to slide or flow past each other. Some examples of this type of polyamide are listed in Table III along with tensile data. Also, Figure 3 illustrates viscosity profiles over a relatively wide temperature range for several polyamides. These polymers flow readily when molten, set quickly when cooled, and demonstrate high tensile strengths when glassy or solid. Thus, in general, dimer polyamides useful as the main component of hot melt adhesives are linear with the linearity contributing to tensile strength properties.

Another distinguishing characteristic of dimer polyamide hot melt adhesives, which almost puts them in a class by themselves as engineering hot melts, is their unusually strong adhesion to many surfaces including most metals, leather, paper, many fabrics, plastic films including polyolefins, wood, and glass (15). It is this characteristic that dictates polyamide hot melt use in such demanding automated applications as side seam sealants for metal cans (16-19); long-lasting construction adhesives for leather, ABS, and vinyl (20); and lap seam adhesives for beer cans and sealants and insulators for electronic parts (21).

Coatings. The solid dimer polyamide resins are useful as protective coatings and inks on many substrates, especially flexibile ones like polyolefin film, paper, fiberboard, and metal foils. Thus, paper converters use these resins by applying them from solvent by roller-

Table III. Some Properties of Macromelt Polyamide Resin Hot Melts

Product	Softening Point	Hot Melt[a] Application Temperature	Polymer Tensile Strength (psi)	Polymer (% Elongation)
Macromelt 6200	100 °C 212 °F	205 °C 401 °F	1700	550
Macromelt 6212	110 °C 230 °C	195 °C 383 °C	1800	450
Macromelt 6240	140 °C 284 °F	240 °C 464 °F	800	900
Macromelt 6264	160 °C 320 °F	210 °C 410 °F	1300	450
Macromelt 6300	200 °C 392 °F	250 °C 482 °F	3000	500

Note: Weight per Gallon: Macromelt resins weigh very close to
       8 lb/gal.

[a]Melt viscosity of 35 poises was chosen as typical. Temperatures
shown achieve this viscosity for each particular resin.

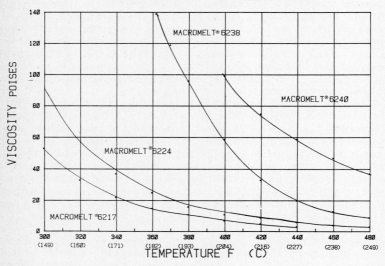

Figure 3. Macromelt polyamide resins: typical viscosity/tem-
           perature curves.

coating, rotogravure, or flexographic printing techniques. They demonstrate excellent weatherability and adhesion properties when properly pigmented.

Beside solvent application techniques, these resins can be applied as hot melts and from water dispersions. Water dispersions (12, 14) may be readily prepared by converting amine end groups to salts such as substituted ammonium acetates. These intrinsic emulsifiers readily cause water dispersions called suspensoids. Films from suspensoids must be heat fused in order to achieve barrier properties.

Thixotropic Coatings. The hydrocarbon solubility characteristics of solid dimer polyamide resins impart thixotropic properties. This property was extended into oil-based alkyd coatings by Winkler (22) and has been discussed in a number of publications (25, 26). Best results are obtained when the polyamide is incorporated into the alkyd vehicle by heating, either during or after formation, so a homogeneous system results. The nature of polyamide interaction with an oil vehicle like an alkyd is not fully understood. However, 2-5% of solid polyamide is usually required to achieve desired thixotropy. Above 5% there is a danger of coating gloss drop.

As one would expect, the thixotropy induced by solid polyamides markedly reduces coating pigment settling and makes possible the application of thicker coats. Additional benefits include less absorption on porous substrates, better hiding from improvements in pigment dispersion quality, improved flow, improved leveling, and longer wet edge times without effecting sag resistance.

Inks. Solid dimer polyamides occupy a dominant position as vehicles for flexographic inks. They are additionally used as clear overprint varnishes and in rotogravure inks. Polyamide-epoxy combinations, to be discussed later, are also useful as silk screening inks, particularly on difficult-to-wet surfaces such as molded polyethylene.

Solid dimer polyamide overprint varnishes are prepared by dissolving the polyamide in propanol and/or lower boiling related solvents. Solvents are selected on the basis of requirements for viscosity, solvent release, speed of dry, gelation resistance, and other factors. Solubility can often be enhanced by adding small amounts of water. Resin solids are usually in the 25-40 vol % range. However, newer polyamide technology has allowed for 50-60 volume solids (23). Most formulations include minor amounts of plasticizers, waxes, slip agents, and antiblocking agents.

Solid dimer polyamides used as vehicles for flexographic inks (14, 24) possess the required alcohol solubility to make it possible to use them on presses with rubber rolls. Their adhesion to films such as polyethylene, Saran, and Mylar is excellent, and their flexibility ensures that they will function properly on these flexible substrates. In addition, they have excellent gloss and good blocking resistance. Thus, they are commonly used on polyolefin, Saran, cellophane, Mylar, cellulose acetate, and PVC films.

Rotogravure polyamide ink formulations are particularly useful for the printing of gold colors on paper and cellophane. These formulations can contain nitrocellulose resins. Solvent combinations include alcohols, hydrocarbon, ketones, and esters.

Fluid Polyamide Resin Chemistry

The reactive dimer polyamides and fatty amido amines are liquid resins with many of the properties described above for solid polyamides. In addition, they are amine-like in that they are basic and form salts. They are more soluble than solid resins, although alcohols are still primary solvents.

The liquid dimer polyamides were first described by Bradley (25), but their reactivity with epoxy resins was first recognized by Renfrew and Wittcoff (26). In reaction with epoxy resins the amine-containing dimer polyamides contribute to water resistance, increased flexibility and adhesion, and most importantly corrosion inhibition properties.

Their pendent amine groups react readily with terminal epoxy groups to form cross-linking β-hydroxy amino and polyether reaction products. These reactions are illustrated in the following equations:

$$-CH-CH_2 + RNH_2 \rightarrow -CH-CH_2-NHR$$
$$\overset{\backslash}{\underset{O}{/}} \qquad 1° \text{ Amine} \qquad \overset{OH}{|}$$

$$-CH-CH_2 + R^1RNH \rightarrow -CH-CH_2-NRR'$$
$$\overset{\backslash}{\underset{O}{/}} \qquad 2° \text{ Amine} \qquad \overset{OH}{|}$$

$$-CH-CH_2 + R''R'RN \rightarrow -[O-CH_2-CH_2]_n-$$
$$\overset{\backslash}{\underset{O}{/}} \qquad 3° \text{ Amine} \qquad \text{Polyether}$$

The reaction mechanism of liquid dimer polyamides and fatty amido amines with epoxy resins has been studied by Peerman et al. (27), who employed infrared spectroscopic analysis to determine reaction rates. They showed that the terminal epoxy content of a blend of amino-containing polyamide and epoxy resin disappeared more rapidly at 150 °C than does the epoxy content of blends of epoxy resin with triethylenetetramine or tris[(dimethylamino)-methyl]phenol. Both of these compounds are well-known for their fast cure at ambient temperatures. Correspondingly, the liquid polyamide or fatty amido amines-epoxy combinations cure slower than the other two systems at ambient conditions.

The reason for these surprising results is postulated by utilizing a concept of immobilization in the cross-linking mechanism of epoxy resins. The aliphatic di- and polyamines react rapidly at ambient conditions. At the higher temperature, these systems are postulated to have reacted so rapidly that the cross-linked network becomes rigid and immobile long before complete epoxy reaction occurs (Figure 4). Thus, although increased curing temperature reduces viscosity, it also accelerates a state of immobility, which can have a pronounced effect on certain physical properties such as hardness, heat distortion, and flexural strength. Hardness effects caused by changing epoxy to polyamide mixtures are indicated by the data in Table IV.

The fluid dimer polyamides and fatty amido amines also react with phenolic resins (23). These reactions are significantly different from those of epoxy resins. With the heat-reactive phenolic resins, the aminopolyamide portions react with methylol groups. A carbon-nitrogen bond or cross-link is formed and a volatile byproduct, water, is produced. This reaction requires external heat to remove water. At temperatures near 150 °C the reaction proceeds smoothly. Since curing at elevated temperatures is required, the pot life or shelf life at room temperature is relatively long. The liquid dimer polyamide and fatty amido amines also react with alpha, beta unsaturated acids and esters (29) and with polyesters (30). The unsaturated esters reduce viscosity, lengthen useful pot life, and reduce heat of reaction. Thus, they are useful diluents when low viscosity is desired.

Thermoset Coatings

By far the most important coatings that make use of liquid amino-containing dimer polyamides and fatty amido amines are what is referred to as epoxy-polyamide coatings. These coatings are used widely as maintenance paints.

Coating formulations and properties have been detailed in technical bulletins (31). The epoxy-polyamide system is popular because it provides an unusual degree of inherent corrosion resistance. This will be discussed in detail later. The system is unusually tolerant as compared to related systems, such as amine cured, since the ratio of components is not particularly critical. Tolerance is also demonstrated because it may be applied to wet surfaces and to surfaces with tightly bound rust. Indeed, formulations are available that may be applied under water to structures such as submarines and off-shore oil well riggings. Both the corrosion resistance and the tolerance relative to application on poorly prepared wet surfaces are believed to be functions of the surface activity of the polyamide resin. Related also to the surface activity are the unusually strong adhesive properties that the system demonstrates with a broad range of substrates.

In addition, the film has a long life and is resistant to the water, solvents, chemicals, acids, and alkali normally found in chemical and industrial environments. The epoxy-polyamide coatings have less resistance to strong solvents than do amine-cured systems, and their heat resistance tends to be somewhat lower. The properties of these coatings have been described in detail in several publications (32, 33).

The dramatic corrosion resistance of epoxy-polyamide coatings is demonstrated by the data in a Government military specification, Mil-P-24441 (ships). This specification coating, at a relatively high solids of 61% by volume, is formulated entirely without

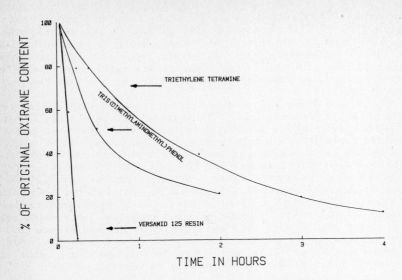

Figure 4.   Comparative rates of cure of Versamid 125 resin.

Table IV.   Effect of Resin Ratio on Hardness of Cured Blends

Ratio of Versamid Resin to Epoxy Resins	Versamid 140: Shell 815 Epoxy Resins	Versamid 140: Shell 828 Epoxy Resins	Versamid 125: Shell 815 Epoxy Resins
30:70	60.63[a]	67.69[a]	
40:60	54.56[a]	61.63[a]	60.65[b]
50:50	47.50[a]	55.56[a]	20.25[b]
60:40	20.23[a]	30.34[a]	90[b]
65:35	90.95[b]	90.95[b]	85[b]
70:30	65.70[b]	75[b]	50[b]
75:25	35.40[b]	50.55[b]	30[b]

Note: Versamid is a registered trademark of Henkel Corp.
[a]Barcol Impressor Model GYZJ–935.

[b]Shore Durometer A.

corrosion-inhibiting pigments. During the development of the specification, it was found that corrosion-inhibiting pigments were not necessary because the polyamide vehicle itself imparted adequate corrosion resistance. In addition, the paint described in this specification, presumably because of its surface activity, is said to wet the surfaces found in marine craft and similar structures, particularly where cracks, corners, welds, and edges are involved. The coating adheres well to poorly prepared surfaces, which makes it useful in bilges that are cleaned only by degreasing and wire brushing but not by sandblasting. The excellent wetting of the epoxy-polyamide coating is demonstrated by the fact that it may be applied under water.

The corrosion-inhibiting character of the epoxy-polyamide systems has been studied by Anderson et al. (34) and Wittcoff and Baldwin (32, 33), who showed that polyamide resins are corrosion inhibitors, protecting steel from attack by mineral acids and other corrosive materials. Similarly, the polyamide resin, as well as epoxy polyamide resin adducts in minute quantities, decreases the flow of electrical current in a corrosion cell containing a steel anode. Presumably, then, the polyamide polarizes the anode and prevents its corrosion, which reflects itself in decreased electrical output.

Practical evidence of the corrosion-inhibiting properties of epoxy-polyamide paints was provided by Glaser and Floyd (35), who showed by Florida tidewater exposure tests on steel substrates that these paints offer protection equal to and in most cases better than paints containing corrosion-inhibiting pigments.

Baldwin et al. (36) have shown that liquid dimer polyamides and their epoxy adducts adsorb onto alumina surfaces by chemisorption coincident with the formation of ionic salt linkages. The chemisorption was proved by calorimetric studies that measured heat of adsorption. The values were sufficiently high to suggest that chemical linkages were formed. That these linkages were in fact salt linkages was confirmed by infrared studies. Thus, it is postulated that the liquid polyamide adsorbs onto the alumina surfaces to form a strong ionically bonded initial layer that is then capable of adsorbing additional layers via secondary valence forces. Amines used in amine-cured epoxy systems, such as diethylenetriamine, adsorb initially with a heat of adsorption well below that required for chemisorption. They are subsequently incapable of attracting additional layers of molecules once the initial adsorption has taken place. A typical adsorption curve is shown in Figure 5. See Table V for typical calorimetric data.

The studies also showed that when there is an excess of vehicle--as occurs in normal painting--the polyamide resin will adsorb preferentially over the epoxy resin even when a certain amount of epoxy resin has reacted with the polyamide during induction or the mixing period required for good painting practices. It is this initially adsorbed and tightly bonded layer that provides corrosion resistance by forming a tightly adhering hydrophobic skin that interferes with typical corroding mechanisms.

Although the painting of aluminum surfaces that are actually believed to be aluminum oxide is common practice with epoxy-polyamide paints, a considerably larger amount is consumed for painting iron. Accordingly, the above adsorption and colorimetric

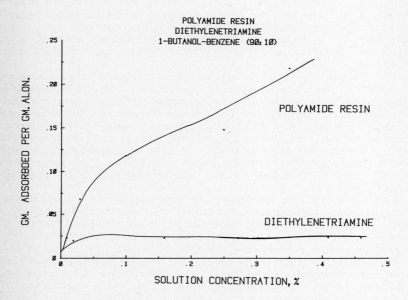

Figure 5.   Adsorption isotherms.

Table V.   Calorimetric Data

Compound of Alumina	kcal/equiv of Compound Adsorbed
Polyamide resin	32.0
Diethylenetriamine	4.5
Epoxy resin	0.0

experiments were repeated with iron oxide with essentially the same results. Finally, these experiments were repeated on wet surfaces to show that water does not interfere markedly with chemisorption.

In addition to maintenance coatings, the liquid dimer polyamides and fatty amido amine–epoxy systems can be formulated into mastic-like materials. When these are sand or aggregate filled, they are useful for flooring, road patching, and decorative building panels (37). Another mastic-type coating known as "splash zone" coatings (37) is useful for the protection of so-called splash zone on off-shore oil well riggings and other partially submerged structures. These coatings are applied under water on sandblasted steel. Sandblasting also can be done under water.

The epoxy–polyamide system may also provide a base for paper and board coatings applied by rotogravure or roller-coating techniques. It is useful for coating plastics and particularly valuable as a resin system for zinc-rich paints. It may be formulated with exempt solvents specified under Rule 66 (37). Higher solid formulations, particularly those formulated with fatty amido amines, are available. These formulations reduce solvent pollution and allow for higher film builds (37, 38).

Thermoset Adhesives

The chemistry of the thermosetting adhesives is similar to that for epoxy–polyamide coatings. The most suitable polyamides are highly branched liquids with reactive amino groups. Furthermore, their molecular weights should be controlled to fall within a low to moderate range. As a result, these dimer polyamides and fatty amido amines range from low-viscosity liquids of 2–10 P at room temperature to gum-like semisolids. All may be referred to as fluid materials, where ease of handling and ease of mixing are strong considerations. Most often these epoxy–polyamide adhesives are applied without solvent addition. However, solvents can be used where extremely low viscosity and deep substrate penetration are requirements.

As one would expect, both the epoxy and polyamide resins contribute heavily to final adhesive properties. In particular, the polyamide component (39) contributes excellent substrate wetting, strong specific adhesion, water resistance, and mechanical toughness. Epoxy–polyamide adhesives are particularly noted for their unusual strength, durability, and aging properties (40). Thus, they exhibit outstanding tensile and peel strengths both initially and after long-term aging.

These adhesives have been found to adhere strongly to metals, glass, wood, ceramics, masonry, asphalt, leather, and plastics like polystyrene, phenolics, polycarbonates, ABS, cellulose acetate, polyesters, rubbers, and some polyolefins. In general, the most favorable results are noted in the bonding of steel and aluminum, perhaps because the bond strengths are more easily observed before substrate failure.

Pressure is not required to form strong adhesive bonds. Only enough pressure to assure adhesive contact or displace excess adhesive is needed. These adhesives can be used at virtually any

temperature above 50 °F. If the application permits, a cure
temperature of about 150 °C is often preferred for quick use and
optimum bond strengths.

Fillers are often useful in epoxy-polyamide adhesives (41).
Fillers aid flow control—particularly at elevated temperatures.
Other properties contributed by fillers are exhibited in the cured
system. For example, they can increase the modulus of elasticity,
modify the coefficient of thermal expansion, increase heat
resistance, and even affect bond strength. Of course, the excellent
wetting properties of polyamides increase the dispersion efficiency
of these fillers. The type of filler selected and the amount used
also have a pronounced effect on adhesive quality. A few of these
effects are illustrated in Tables VI and VII (37). (See also Table
VIII.)

While most applications are found in the industrial world where
high strengths, low shrinkage, strong adhesion, excellent
electrical, and surface wetting characteristics are optimum
engineering qualities, a growing interest has developed in consumer
adhesive kits. They have been found useful for repairing toys,
skis, furniture, automobiles, garden tools, driveways, and a host of
other household items.

Thermosetting Systems for Casting and Laminating

The liquid dimer polyamides and fatty amido amines, here again, are
used. The chemistry of reaction and general use considerations are
similar to those for adhesives. Castings and laminates based on
epoxy-polyamides are noted for low shrinkage during cure, excellent
mechanical strengths, and good electrical properties. As mentioned
before, considerable property variations can be achieved by
adjusting the epoxy to polyamide mixture ratio (37).

In Table IX are illustrated some of the properties of cured
castings. Some of the electrical properties are listed in Table X.
While heat distortion values are not high, they are often adequate
and are usually compensated for by other excellent mechanical
properties. Furthermore, in glass and olefinic fiber laminates, the
wetting properties attributed to polyamides are outstanding, thus
minimizing voids and promoting mechanical properties.

Table VI. Effect of the Amount and Type of Filler on Tensile
Shear Strength of Versamid Epoxy Adhesives
(Versamid resin 115:Modified Epoxy Resin, 1:1)

Type	Quantity	Tensile Shear Strength (psi)[a]
Tabular alumina	20	3200
	60	3080
	100	2960
Calcium carbonate	20	2730
	60	3200
	100	3400
Kaolin clay	60	2790
	100	3280
Atomized aluminum	20	4080
	100	3800
Montmorillonite clay	60	3380
	100	2850

Note: Cured 20 min at 150 °C on aircraft aluminum with epoxy
resin = 190 epoxide equivalent weight (eew).

[a] ASTM D1002-72.

Table VII.   Effect of Filler on Peel Strength

Filler	Peel Strength (1b/in.) ASTM D903–49
None	28.5
20 parts of nylon powder	41.8
50 parts of aluminum flakes	40.8
20–50 parts of kaolin	25.4
Glass cloth	55.8

Note: Material bonded, aluminum alloy sheet, 0.5 mm; pretreatment, pickling method; cure, 300 °F, 20 min; base resins, Versamid resin 115:epoxy 60:40[a]. These same tables, including Table VIII, illustrate the general property ranges to expect from epoxy–polyamide adhesives under defined cure and testing conditions (42).

[a]Epoxy resin = 190 eew.

Table VIII.   Adhesive Tests (Versamid Resin 115:Epoxy Resin: Tabular Alumina, 40:60:20)

Conditions	Ultimate Tensile Shear before Exposure (av of 5) (psi)	Ultimate Tensile Shear after Exposure (av of 5) (psi)
250-h of salt spray	3130	2050
7 days in 2-propanol	3200	3000
7 days in hydrocarbon	3270	3300
60 h in weatherometer	2440	2680
7 days in JP4 fuel	2440	2680
30 days in tapwater	3130	2600

Table IX. Physical Properties of Polyamide-Epoxy Resin Castings

Versamid Resin	Versamid Resin: Epoxy Resin Ratio	ASTM D-638 Tensile Strength (psi)	ASTM D-790 Flexural Ultimate (psi)	ASTM D-790 Flexural Modulus (psi)	ASTM D-695 Compressive Strength (psi)	MIL-I-169230 Mechanical Shock (ft/lb)	ASTM D-470 Water Absorption (Wt %)
115	45:55	5700	9,300	2.3×10^5	12,000	11.9	0.055
115	50:50	4700	8,300	3.0×10^5	7,300	28.3	
125	35:65	7900	15,000	3.1×10^5	10,000	18.9	0.060
125	40:60	6600	12,300	2.5×10^5	9,100	23.4	
140	30:70	8900	14,500	3.1×10^5	13,700	28.3	0.093
140	30:70	8800	14,500	3.1×10^5	11,500	15.2	

Table X. Electrical Properties of Castings

Properties	ASTM Method	Versamid Resin 115: Epoxy Resin, 50:50	Versamid Resin 125: Epoxy Resin, 40:60
Dielectric Strength (step by step v/mil) (1/8 in. thick specimens)	D149	470	430
Volume resistivity (ohm cm)	D257–52T	1.5×10^{14}	1.1×10^{14}
Arc resistance (s)	D495	76	82
Power factor	D150		
60 cycles		0.0090	0.0085
10^3 cycles		0.0108	0.0108
10^6 cycles		0.0170	0.0213
Dielectric constant	D150		
60 cycles		3.20	3.37
10^3 cycles		3.14	3.34
10^6 cycles		3.01	3.08
Loss factor	D150		
60 cycles		0.0357	0.0285
10^3 cycles		0.0339	0.0359
10^6 cycles		0.0572	0.0656

Note: Cure schedule, gelled at 25 °C plus 3 h at 120 °C; Test conditions, 23.6 °C, 50% relative humidity, conditioned 96 h; Epoxy resin = 190 eew.

Literature Cited

1. Bradley, T.; Johnston, W. Ind. Eng. Chem. 1944, 16, 90.
2. Folkenburg, L.; Teeter, H.; Skell, P.; Cowan, J. Oil Soap 1945, 22, 143.
3. Paschke, R. F.; Wheeler, D. H. J. Am. Oil Chem. Soc. 1949, 26, 278.
4. Wheeler, D. H. Off. Dig. 1951, No. 322, 661.
5. Paschke, R. F.; Peterson, L. E.; Wheeler, D. H. J. Am. Oil Chem. Soc. 1964, 41, 723.
6. Paschke, R. F.; Peterson, L. E.; Harrison, S. A.; Wheeler, D. H. J. Am. Oil Chem. Soc. 1964, 41, 56.
7. Harrison, S. A.; Wheeler, D. H. Minn. Chem. 1952, 4(5) 17.
8. Pasternak, R.; Brody, P.; Ehrmantraut, H. S.C.H.E.M.A., 13th Chemistry and Enzyme Congress, Frankfurt Main, June 1961.
9. Peerman, D.; Tolberg, W.; Wittcoff, H. J. Am. Chem. Soc. 1954, 76, 6085.
10. Anderson, R.; Wheeler, D. H. J. Am. Chem. Soc. 1948, 70, 760.
11. Golding, B. "Polymers and Resins"; Van Nostrand: New York, 1959; pp. 293-94, 314-25.
12. Wittcoff, H.; Renfrew, M. M.; Speyer, F. B. TAPPI 1952, 35, 21. Wittcoff, H. U.S. Patent 2 728 737, Dec 27, 1955; U.S. Patent 2 824 842, Feb 25, 1958.
13. Technical Bulletin, "Versamid 930," Henkel Corp., Minneapolis, MN. Hinden, E. R.; Whyzmuzis, P. D. "Solvent Theory and Practice"; ADVANCES IN CHEMISTRY SERIES No. 124, American Chemical Society: Washington, D.C., 1973; p. 168.
14. Floyd, D. E. "Polyamide Resins," 2nd ed.; Reinhold: New York, 1966.
15. Vertnik, L. R. Adhesive Age 1967, 10(7), 27.
16. Graves, J. H.; Wilson, G. G. U.S. Patent 2 839 219.
17. Graves, J. H.; Wilson, G. G. U.S. Patent 3 840 264.
18. Wilson, G. G. U.S. Patent 2 839 549.
19. Technical Bulletin, "Bonding and Laminating Aluminum"; Henkel Corp., Minneapolis, MN.
20. Technical Bulletin, "Bonding and Laminating Vinyl"; Henkel Corp., Minneapolis, MN.
21. Technical Bulletin, "Electrical Applications Bonding"; Henkel Corp., Minneapolis, MN.
22. Winkler, W. U.S. Patent 2 663 649 to T. F. Washburn Co., Dec 22, 1953.
23. Technical Bulletin, "Polyamide Resin GAX 11-514"; Henkel Corp.; Minneapolis, MN.
24. Technical Bulletin, "Versamid 750"; Henkel Corp., Minneapolis, MN.
25. Bradley, T. U.S. Patent 2 379 413 to American Cyanamid Co., June 28, 1940.
26. Renfrew, M. M.; Wittcoff, H. U.S. Patent 2 705 223 to General Mills, Inc., March 29, 1955.
27. Peerman, D.; Tolberg, W.; Floyd, D. Ind. Eng. Chem. 1957, 49, 1091.
28. Peerman, D.; Floyd, D. U.S. Patent 2 891 023 to General Mills, Inc., June 16, 1959.
29. Floyd, D. E.; Peerman, D. E. U.S. Patent 2 999 825 to General Mills, Inc., Sept 12, 1961.

30. Peerman, D. E.   U.S. Patent 2 889 292 to General Mills, Inc., June 2, 1959.
31. Technical Bulletin, "Versamid and Genamid, Polyamide and Amido Amine Resins"; Henkel Corp., Minneapolis, MN.
32. Wittcoff, H. J. Oil Colour Chem. Assoc. 1964, 47, 4273.
33. Wittcoff, H.; Baldwin, W. S. Proc., VIII Fatipec Cong. (Verlag Chemie, GMBH, Weinheim/Bergstr.), 1966, p. 69.
34. Anderson, D. L.; Floyd, D. E.; Glaser, D. W.; Wittcoff, H. Corrosion 1961, 17(12), 9.
35. Glaser, D. W.; Floyd, D. E. Off. Dig. 1960, 32(420), 108.
36. Baldwin, W. S.; Milun, A. J.; Wheeler, D. H.; Wittcoff, H.; Geonkoplis, C. J. J. Paint Technol. 1970, 42(550), 592.
37. Technical Bulletin, "Versamid and Genamid, Polyamide and Amido Amine Resins"; Henkel Corp., Minneapolis, MN.
38. Masciak, M. J. Materials Prot. Performance 1972, 11, 38.
39. Peerman, D. E.; Floyd, D. E.; Mitchell, W. S. Plast. Technol. 1956, 2, 25.
40. DeLolles, N. J. Adhesive Age 1977, Sept.
41. Peerman, D. E.; Floyd, D. E. SPE J. 1960, 16(7), 717.
42. "ASTM Annual Standards"; American Society for Testing and Materials: Philadelphia, 1973; Part 16.

Urethane Coatings

KURT C. FRISCH[1] and PANOS KORDOMENOS[2]

[1] Polymer Institute, University of Detroit, Detroit, MI 48221
[2] Ford Motor Company, Paint Research Center, Mt. Clemens, MI 48043

Raw Materials for Urethane Coatings
  Isocyanate Components
  Di- and Polyhydroxy Components
  Chain Extenders and Cross-linkers
  Catalysts
  Other Urethane Coating Components
Types of Urethane Coatings
  One-Package Urethane Alkyd (Oil-Modified Urethanes) Coatings
    (ASTM Type 1)
  One-Package Moisture-Cure Urethane Coatings (ASTM Type 2)
  Single-Package Blocked Adduct Urethane Coatings
    (ASTM Type 3)
  Two-Package Catalyst Urethane Coatings (ASTM Type 4)
  Two-Package Polyol Urethane Coatings (ASTM Type 5)
  Urethane Lacquers
  Two-Package Coatings: Isocyanate-Terminated Prepolymers
    Cured with Polyamines
  Two Package Coatings: Isocyanate-Terminated Prepolymers
    Cured with Ketimines
  Two-Package Coatings with Aminoformaldehyde Cure
  Aqueous Urethane Coating Systems
  One Hundred Percent Solids Coating
  Radiation-Cured Urethane Coatings
  High Temperature Resistant Urethane and Other Isocyanate-Based
    Coatings
  Miscellaneous Urethane Coating Systems
  Coatings from Urethane Interpenetrating Polymer Networks
    (IPNs)

Urethane coatings were first developed by Otto Bayer and his coworkers in the laboratories of the I. G. Farbenindustry, today's Farbenfabriken Bayer, in Leverkusen, West Germany. Although urethanes are characterized by the linkage $-NH-C(=O)-O-$, they may contain other functional groups such as ester, ether, urea, amide, and other groups.

There are a number of syntheses leading to the formation of urethane polymers. However, the most important commercial route is the isocyanate addition polymerization, the reaction between di- and polyfunctional hydroxyl compounds such as hydroxy-terminated polyethers or polyesters and di- or polyisocyanates. When difunctional reactants are being used, linear polyurethanes are produced and the reaction can be schematically represented as follows:

$$HO-R-OH + O=C-N=R'-N=C=O \ ----->$$

Polyether or   Diisocyanate
polyester

$$\{O-R-O-\overset{\overset{O}{\|}}{C}-NH-R'-NH-\overset{\overset{O}{\|}}{C}\}_n$$

Polyurethane

If the functionality of the hydroxyl or isocyanate component is increased to three or more, branched or cross-linked polymers are formed.

The earliest commercial urethane coatings were based on polyester-polyisocyanate systems that exhibited excellent abrasion resistance, toughness, and a wide range of mechanical strength properties. Most urethane coating systems in this country were first based on tolylene diisocyanate (TDI), while in Europe many systems based on 4,4'-methylene bis(phenyl isocyanate) (MDI) were developed. In order to avoid the use of free TDI, adducts of polyols such as trimethyolpropane or 1,2,6-hexanetriol with TDI were introduced, particularly for two-component coatings (1, 2).

Other types of isocyanate derivatives that find increasing application in the urethane coating industry are isocyanate-containing isocyanurates, which include also aliphatic, cycloaliphatic, and aromatic derivatives.

One-component urethane coatings with "blocked" isocyanate groups were developed by Bayer (1, 3) and Petersen (4). Application of heat to these "splitters" with regeneration of free isocyanate groups and fast curing has led to the acceptance of these coatings as wire enamels in the electrical industry, as well as for coatings in other industries.

"Urethane oils" or "uralkyds," one-component urethane coatings based on reaction products of drying oils with isocyanates, were early developments in West Germany, as well as in England and the United States (3-7). Because of low costs and ease of pigmentation, this type of coating has been extensively used in the United States. Closely related to uralkyds are urethane coatings, based on castor oil and castor oil derivatives, which have found use for maintenance and protective coatings (8-11).

In more recent years one- and two-component urethane coatings based on polyethers have made a significant penetration of the urethane coatings market. Due to low cost and wide latitude in physical properties ranging from very flexible to hard, tough and solvent-resistant coatings, these materials have found acceptance for flexible substrates such as leather, textile, paper, and rubber and for hard substrates, such as wood (e.g., bowling alleys and gymnasium floors), concrete, metals, and "seamless flooring" (12).

An important application of urethane coatings is in the fabric coatings field. The term "wet look" was very fashionable for some time and was applied to highly glossy urethane coatings and was later succeeded by the "suede look." These types of coatings were applied to apparel but were also employed in furniture fabrics.

Due to environmental regulations, a considerable amount of effort is being devoted to the reduction or elimination of the solvent content through the development of high-solids or 100% solids coatings, as well as water-based coatings, for example, urethane latices. Another solventless approach to this problem is urethane powder coating systems.

Raw Materials for Urethane Coatings

The principal components of commercially available urethane coatings are di- or polyisocyanates and di- or polyhydroxy compounds. Active hydrogen-containing compounds, especially diols and diamines, as well as alkanolamines, are also employed as chain extenders. In addition, various cross-linking agents such as neutral or tertiary amine based triols or tetrols are also used.

Isocyanate Components. Aromatic Di- and Polyisocyanates. The most important monomeric aromatic diisocyanates used for coatings are tolylene diisocyanate (TDI) and 4,4'-methylene bis(phenyl isocyanate) (MDI). Tolylene diisocyanate, a colorless liquid (bp 120 °C at 10 mmHg), is generally used in 80/20 blends of the 2,4- and 2,6-isomers. Pure 2,4-TDI isomer has also been employed for coatings.

The NCO group in the 4-position is about 8 times as reactive as the NCO group in the 2-position at room temperature (25 °C). With an increase of temperature, the activity of the ortho NCO group increases at a greater rate than that of the para NCO group until at 100 °C the ortho and para NCO groups are similar in reactivity. This disparity in activity at low temperatures can be conveniently used in the preparation of isocyanate-terminated prepolymers.

Low molecular weight isocyanates such as TDI have a high vapor pressure (vapor pressure of TDI = 2.3×10^{-2} mmHg at 25 °C) and are respiratory irritants and lachrymators. Prolonged inhalation of TDI vapors may lead to symptoms resembling asthma. The maximum allowable vapor concentration has been reduced to 0.005 ppm.

Good ventilation should be provided in working areas handling TDI and other isocyanates, and suitable gas masks should be used where the concentration of TDI exceeds the safe limit, especially for spray applications (13).

4,4'-Methylene bis(phenyl isocyanate) contains two equally reactive isocyanate groups: $OCN-C_6H_4-CH_2-C_6H_4NCO$. MDI, a solid, melting at 37 °C, has a tendency to dimerize at room temperature. Storage at 0 °C or lower minimizes dimerization as does storage at 40-50 °C in the liquid state. MDI has a lower vapor pressure (10^{-5} mmHg at 25 °C) than TDI and hence is less of an irritant. Nevertheless, contact with skin or eyes or breathing of vapor or dust should be avoided. In addition to the "pure" MDI, a liquid form of MDI (isomer mixture) is commercially available as well as a special grade of MDI containing carbodiimide groups (e.g., Isonate 143-L, Upjohn Co.).

A number of "crude" isocyanates (polymeric isocyanates), undistilled grades of MDI and TDI, are available in the market. Some of these products, such as various forms of crude MDI, have a functionality varying between 2 and 3. They have a lower reactivity and a lower vapor pressure than the corresponding pure isocyanate. They have found extensive use in one-shot rigid foams. However, they are also employed in coating, sealant, and adhesive applications. The Upjohn Chemical Division has published an excellent bulletin on the use and precautions in handling isocyanates, polyurethanes, and related materials (13A).

Both MDI and TDI, being aromatic diisocyanates, yield urethane polymers that tend to yellow on prolonged exposure to sunlight, presumably due to oxidation of some terminal aromatic amine (derived from these isocyanates). MDI also possesses a methylene group that is susceptible to oxidation via a proton abstraction mechanism involving autoxidation of the aromatic urethane groups to a quinoneimide structure as proposed by Schollenberger et al. (24, 25).

Beachell and Ngoc Son (16) have shown that the urethane group can be stabilized against thermal degradation and yellowing by substitution of the labile hydrogen atom by groups such as the methyl and benzyl groups. Similarly, Wilson (17) reported that urethanes, treated with monoisocyanates, ketene, ethylene oxide, and acetic anhydride, can be stabilized to a large extent against yellowing.

Indications are that both oxidation and ultraviolet radiant energy may be involved in the processes responsible for yellowing of urethane coatings based on aromatic isocyanates.

Although p-phenylene diisocyanate (PPDI) has been synthesized as early as 1913 (18), the Akzo Corp. has developed more recently a modified Hofmann process for the preparation of PPDI (19). A comparison of the reactivity of PPDI in comparison with MDI and NDI (1,5-naphthalene diisocyanate) as well as with aliphatic diisocyanates (H_{12}MDI, IPDI, and CHDI--see Aliphatic Isocyanates) was reported by Wong and Frisch (20).

Aliphatic Isocyanates. In recent years major emphasis has been placed upon efforts to develop aliphatic isocyanates to impart light stability as well as improved hydrolytic and thermal stability.

The first commercial aliphatic diisocyanate used was 1,6-hexamethylene diisocyanate (HDI), OCN $(CH_2)_6$ NCO, a colorless liquid, boiling at 127 °C at 10 mmHg; it is less reactive than either TDI or MDI in the absence of catalysts. However, certain metal catalysts, such as tin, lead, bismuth, zinc, iron, cobalt, etc., activate the aliphatic isocyanate groups (21).

HDI produces urethane coatings with better resistance to discoloration, hydrolysis, and heat degradation than TDI (22, 23). However, due to the high vapor pressure of HDI and its inherent danger to human exposure, a higher molecular weight modification of HDI was developed by Bayer A. G. by reacting 3 mol of HDI with 1 mol of water, forming an isocyanate-terminated biuret:

$$
3 \begin{bmatrix} \text{NCO} \\ | \\ (\text{CH}_2)_6 \\ | \\ \text{NCO} \end{bmatrix} + \text{H}_2\text{O} \longrightarrow \begin{array}{c} \text{NH--(CH}_2)_6\text{--NCO} \\ | \\ \text{C=O} \\ | \\ \text{N--(CH}_2)_6\text{--NCO} \\ | \\ \text{C=O} \\ | \\ \text{NH--(Ch}_2)_6\text{--NCO} \end{array}
$$

This triisocyanate has been reported to impart good light stability and weather resistance when used in urethane coatings and is among the most widely used aliphatic isocyanates (23, 24). More recently Bayer A. G. introduced another product based on HDI. By partial trimerization of HDI in the presence of a catalyst, an isocyanurate ring containing isocyanate is formed that exhibits superior thermooxidative stability and weatherability:

$$
3 \begin{bmatrix} \text{NCO} \\ | \\ (\text{CH}_2)_6 \\ | \\ \text{NCO} \end{bmatrix} \xrightarrow[\Delta]{\text{trimerization cat.}}
$$

A number of other aliphatic diisocyanates that impart excellent color stability to urethane coatings have been introduced into the U.S. market. Foremost among these are 4,4'-methylene bis(cyclohexyl isocyanate) (H_{12}MDI) and isophorone diisocyanate (25, 26). An isocyanurate ring containing isocyanate produced by trimerization of IPDI is also commercially available (T-1890).

H_{12}MDI exists in three isomeric forms, the trans-trans, trans-cis, and cis-cis forms. Due to the fact the cis isomer is more sterically hindered, the trans isomer reacts considerably faster with a hydroxyl group than the cis isomer (27). As with other aliphatic isocyanates, catalysis (generally organotin catalysts) is usually used for both the prepolymer preparation and the cure reaction (cross-linking).

Other aliphatic diisocyanates that are being used commercially in urethane coatings are 3-(isocyanatomethyl)-3,5,5-trimethylcyclohexyl isocyanate (Veba-Chemie A.G.) (28), "dimeryl" diisocyanate derived from dimerized linoleic acid (Henkel Corp.) (29), and xylylene diisocyanate (XDI) (Takeda Chemical Co.) (30). It is interesting to note that no catalysts are required for the reaction of XDI with hydroxyl compounds and that its reactivity is similar to that of TDI.

An investigation by Frisch et al. (31) has shown that combinations of aliphatic and aromatic diisocyanates (e.g., XDI and TDI) impart good light stability to urethane coatings.

More recently, Takeda Chemical Co. has introduced 1,3-bis(isocyanatomethyl)cyclohexane (H₆XDI) for coating applications. Both one- and two-component coatings made from H₆XDI as well as from adducts of H₆XDI and trimethylolpropane were reported by Tanaka and Nasu (32).

Similarly, excellent two-component urethane coatings were obtained from the adduct of trimethylolpropane and 4-bis(isocyanatomethyl)cyclohexane (BDI), which was developed by Suntech (Sun Oil Co.) (33).

An interesting process for the preparation of polymeric isocyanates was reported by Wright and Harwell (34). It consists of the Hofmann degradation from the copolymers of methacrylamide with either butyl acrylate or styrene:

$$\left[\begin{array}{c} CH \\ | \\ -CH_2-C-- \\ | \\ C=O \\ | \\ NH_2 \end{array}\right]_x \left[\begin{array}{c} \\ -CH_2-CH-- \\ | \\ R \end{array}\right]_y \xrightarrow{\text{NaOCl}} \left[\begin{array}{c} CH_3 \\ | \\ -CH_2-C-- \\ | \\ NCO \end{array}\right]_x \left[\begin{array}{c} \\ -CH_2-CH-- \\ | \\ R \end{array}\right]_y$$

$$\text{where } R = C_6H_5- \text{ or } -\overset{\displaystyle O}{\overset{\displaystyle \|}{C}}-OC_4H_9$$

These isocyanates were used in coating applications although the reactivity of the isocyanate groups is very low, even in the presence of tin catalysts.

Similarly to PPDI, trans-1,4-cyclohexyl diisocyanate (CHDI), was prepared by the process described by Zengel (19). Both PPDI and CHDI are currently available from Akzo in developmental quantities.

The Dow Chemical Co. has recently been offering isocyanatoethyl methacrylate (IEM):

$$\begin{array}{cc} CH_3 & O \\ | & \| \\ CH_2=C- & C-O-CH_2-CH_2-NCO \end{array}$$

IEM can be used to prepare urethane prepolymers containing terminal or pendent methacrylic groups that can be cured either by means of free radical initiators or by radiation using UV or electron beam radiation. Alternately, IEM can be copolymerized with other acrylic or vinyl monomers and then cured via the isocyanate group (e.g., by moisture cure or by reaction with active hydrogen compounds). IEM can be used for the preparation of one- and two-component urethane coatings and adhesives as well as for cross-linking of other types of coatings containing active hydrogen-containing resin intermediates (35, 36).

New types of aralkyl diisocyanates have been reported by Arendt et al. (37). These are m- and p-tetramethylxylene diisocyanates:

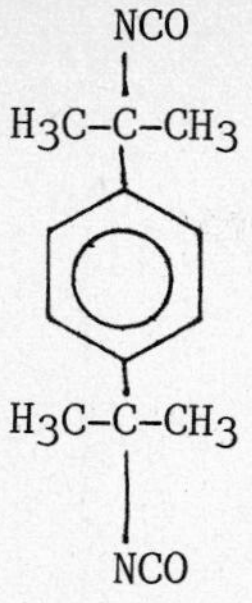

p-TMXDI

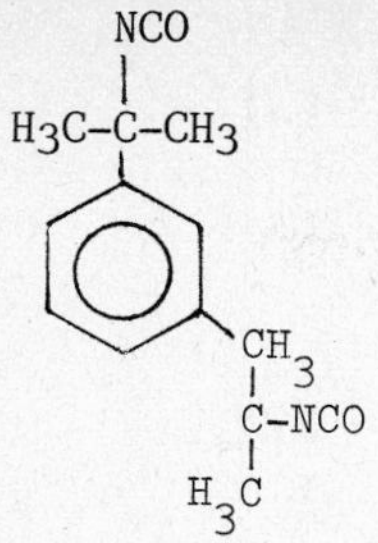

m-TMXDI

Similar to other aliphatic isocyanate groups, tin catalysts such as dimethyltin dilaurate and tetrabutyldiacetyldistannoxane and lead octoate were found to be effective for reaction of these isocyanates with hydroxyl compounds.

Isocyanate Dimers and Trimers.   Aromatic isocyanates have a tendency to dimerize readily

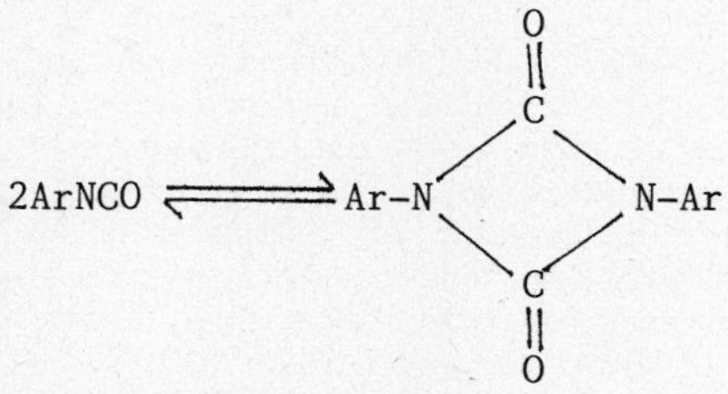

The rate of self-polymerization of isocyanates to dimers (uretidine diones) depends upon the electronic or steric influences of ring substituents. Ortho substitution greatly retards dimerization of the NCO groups, with the ortho NCO slower to dimerize. This dimerization is catalyzed strongly by trialkyl phosphines (38, 39) and more mildly by tertiary amines, such as pyridine (40, 41). MDI dimerizes slowly on standing at room temperature even without catalysts but is stable at low or at slightly elevated temperatures (40–50 °C).

Isocyanate dimerization is an equilibrium reaction. Dissociation of the dimer occurs only at elevated temperatures (42). In the absence of catalysts, temperatures as high as 175 °C are required to completely dissociate the dimer of 2,4-TDI, although initial dissociation was observed to occur at 150 °C (43). The preparation of aliphatic isocyanate dimers has also been reported (44).

Both aliphatic and aromatic isocyanates can trimerize to yield isocyanates:

$$
3R-NCO \ \text{-----}> \quad
\begin{array}{c}
O \\
\parallel \\
C \\
R-N \diagdown\diagup N-R \\
\mid \qquad\qquad \mid \\
O=C \qquad\quad C=O \\
\diagdown\diagup \\
N \\
\mid \\
R
\end{array}
$$

In contrast to dimer formation, trimerization is not an equilibrium
reaction since trimers are stable in the range of 150–200 °C. Ortho
substitution of an aromatic isocyanate greatly reduces the ease of
trimerization. Many catalysts have been reported for the
trimerization of aliphatic and aromatic isocyanates (39, 45–57).

   Nicholas and Gmitter (58) studied the relative effectiveness of
a number of trimerization catalysts (see Table I). Isocyanurate
coatings based on TDI and HDI have been reported by Kubitza (59).
In addition, Sandler (60) prepared polyisocyanurate adhesives from
isocyanate–terminated prepolymers employing calcium naphthenate as a
trimerization catalyst.

Isocyanate–Terminated Adducts and Polymers. Monomeric diisocyanates
such as TDI, MDI, or H_{12}MDI, because of their irritant
characteristics, are seldom used in an unreacted form for the
compounding of urethane coatings. They are normally converted into
isocyanate–terminated polymers or adducts of polyols such as
hydroxyl–terminated polyesters and polyethers, castor oil, etc.

   Adducts of TDI to monomeric polyols such as trimethylolpropane
are frequently used. When reacted at a NCO:OH ratio of 2:1, the
following product is formed: $C_2H_5C[CH_2OCONHC_6H_3(CH_3)NCO]_3$. Because
of its low vapor pressure, this triisocyanate adduct does not have
the irritant properties of TDI and can be handled relatively safely
in coating formulations.

   Other types of adducts are formed by reacting hydroxyl–
terminated polyethers or polyesters with TDI. When these polyols
are reacted at an NCO:OH ratio of 2:1, these adducts, also referred
to as prepolymers, have the schematic structure, depending upon the
functionality of the polyol, shown in Figure 1.

   The diisocyanate and polyols can be reacted in various
combinations by using diols, triols, and tetrols, or combinations of
these in which the NCO:OH ratio is lower than 2. Typical idealized
structures are shown in Figure 2. The adducts and polymers can be
produced with a relatively low content of free unreacted
disocyanate. Because of the volatility and irritant characteristics
of diisocyanates such as TDI, it is important to keep the unreacted
TDI to a minimum, particularly for spray applications. A vacuum
stripping step is often recommended to eliminate this hazard.

Blocked Isocyanates. For certain applications, such as wire
enamels, coil coatings, fabric coatings, powder coatings, cationic
electrodeposition coatings, etc., "blocked" isocyanates are being

Table I.   Activity of Catalysts in Trimerization of Phenyl
Isocyanate

Catalyst	Relative Reactivity
2,4,6-Tris[(dimethylamino)methyl]phenol[a]	1.0
2,4,6-Tris[(dimethylamino)methyl]phenol: diglycidyl ether of bisphenol A (1:1)	1.2
Mixture of o- and p-(dimethylamino)methyl-phenyl[b]	1.5
N,N',N"-Tris[(dimethylamino)propyl]-sym-hexahydrotriazine	2.4
N,N',N"-Tris[(dimethylamino)propyl]-sym-hexahydrotriazine:diglycidyl ether of bisphenol A (1:1)	7.7
Benzyltrimethylammonium hydroxide in dimethyl sulfoxide (as 25% solution)	8.5
Benzyltrimethylammonium methoxide	4.0
Sodium methoxide in dimethyl formamide (as saturated solution)	25.0

[a]DMP-30 (Rohm & Haas).
[b]DMP-10 (Rohm & Haas).

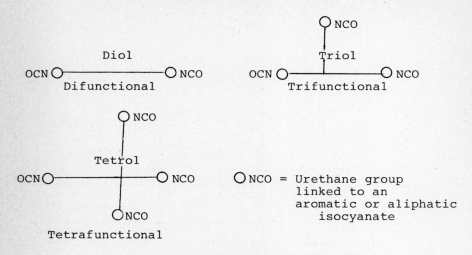

Figure 1.   Schematic structure of prepolymers.

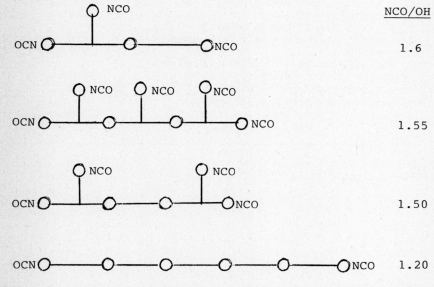

Figure 2.   Typical idealized structures of diisocyante-polyol combinations in which the NCO:OH ratio is lower than 2.

used in order to provide one-component coatings with hydroxyl-
containing or other active hydrogen-containing components. The
"blocking" reaction is reversible according to the classic reaction
of the adduct from trimethylolpropane and TDI with phenol:

$$C_2H_5C(CH_2OCONH-R-NCO)_3 + 3C_6H_5OH$$

$$C_2H_5C(CH_2OCONH-R-NHCOOC_6H_5)_3$$

Application of heat (~150 °C) regenerates the free isocyanate, which
is then capable of reacting with the hydroxyl component.

Both aliphatic and aromatic isocyanates can be blocked by a
variety of blocking agents. These include alcohols, phenols,
oximes, lactams, β-dicarbonyl compounds, hydroxamic acid esters,
bisulfite addition compounds, hydroxylamines, and esters of *p*-
hydroxybenzoic acid and salicylic acid. Excellent reviews of
blocked isocyanates have been written by Wicks (61). Perhaps the
most widely used blocking agents at present are phenol, branched
alcohols, 2-butanone oxime (methylethyl ketoxime), and ε-
caprolactam. The kinetics of the unblocking reaction of aliphatic
and aromatic isocyanurate ring containing blocked isocyanates were
studied by Kordomenos et al. (61B) using isothermal thermogravimetric analysis.

It should be pointed out that a number of blocking agents,
notably β-dicarbonyl compounds such as malonic esters and
acetoacetic esters, give abnormal products (62, 63), for example,
mixtures of esters and amides, presumably via ketene intermediates.

More recently, newer types of blocking agents have been reported
such as *N*-hydroxyphthalimide and *N*-hydroxysuccinimide (64), aromatic
triazoles (65), and substituted imidazolines and
tetrahydropyrimidines (66). Ulrich et al. (67-69) have demonstrated
that macrocyclic ureas will dissociate thermally to yield amido
isocyanates:

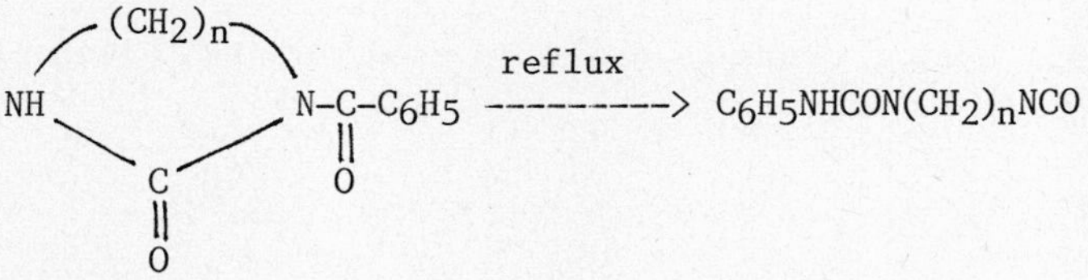

It was found that the reaction proceeds when n = 4 or 5 but not when
n = 2 or 3. Difunctional macrocyclic urea amides were made by
reacting the macrocyclic urea with diacid chlorides such as
isophthaloyl chloride. The utility of these cross-linkers was
proven in solvent coatings, acrylic powder coatings, and electrodeposition coatings.

Catalysis plays an important role in the deblocking or thermal
dissociation of the blocked isocyanates. Notably organometallic
compounds and tertiary amines are capable of lowering both the
deblocking temperature and time as compared to the uncatalyzed
systems. Wicks (61) has pointed out that since most of the
deblocking reactions are carried out in the presence of hydroxyl

compounds, the role of the catalysts in the deblocking reactions may involve catalysis of the alcoholysis reaction between the aryl urethane and the hydroxyl group of the second component.

In addition to tin compounds such as dibutyltin dilaurate and dibutyltin diacetate, various other metal compounds have been claimed to be effective deblocking catalysts such as zinc naphthenate (70), lead naphthenate, bismuth salts, and titanates (71), and Ca, Mg, Sr, or Ba salts of hexanoic, octanoic, naphthenic, or linolenic acid (72) and metal acetyl acetonates (73). Also combinations of organotin compounds and quaternary ammonium salts have been claimed to have a synergistic effect in lowering the deblocking temperature (74).

The use of blocked isocyanates for aqueous systems is of special interest. The sodium bisulfite blocking of hexamethylene diisocyanate was already described in Petersen's classic paper on blocked isocyanates (4). Its application in cross-linking paper (75) and acrylamide copolymers (76) has been described. The stability of bisulfite-blocked aromatic isocyanates in water can present a difficulty, and various stabilizer systems have been proposed. Peters and Reddie (77) have made an extensive study of the various factors affecting stability and have found that optimum stability was obtained at pH of 2–3 in aqueous ethanol with excess hydrogen peroxide.

Sodium bisulfite-blocked isocyanate-terminated prepolymers are extensively used as shrink-resistant finishes for wool. Polycarbamoylsulfonates (PCS) are produced from NCO-terminated prepolymers with bisulfite salts in aqueous alcohol as the solvent, shown schematically as follows (78–80):

$$R(NCO)_n + NaHSO_3 \xrightarrow[\text{EtOH}]{H_2O} R(NHCOSO_3^- \ Na^+)_n$$

NCO–terminal   PCS
prepolymer

The principal curing reaction of PCS is the hydrolysis to form urea cross-links:

$$2RNHCOSO_3^-Na^+ \xrightarrow{H_2O} RNHCONHR + CO_2 + 2NaHSO_3$$

Other water-soluble, blocked isocyanate cross-linkers have been prepared by reaction of diisocyanate with α,α-dimethylolpropionic acid followed by blocking with caprolactam and subsequent solubilization in water by means of a tertiary amine (81) or 2-ethylimidazole (82).

Blocked or partially blocked isocyanates have been extensively employed in cationic electrodeposition. Polymeric compositions employed for cationic electrodepositable coatings have been reviewed by Kordomenos and Nordstrom (82A). Matsunaga et al. (83) were the first to synthesize cationic thermosettable urethane resins. The system described was prepared from an isocyanato-terminated polyurethane prepolymer by reacting the free isocyanato groups with a tertiary amine that had at least three hydroxyl groups. The polyurethane prepolymer was prepared by reacting a diisocyanate with a diol at a ratio of NCO:OH = 1.5-2.0. One of the examples given was synthesized according to the following scheme:

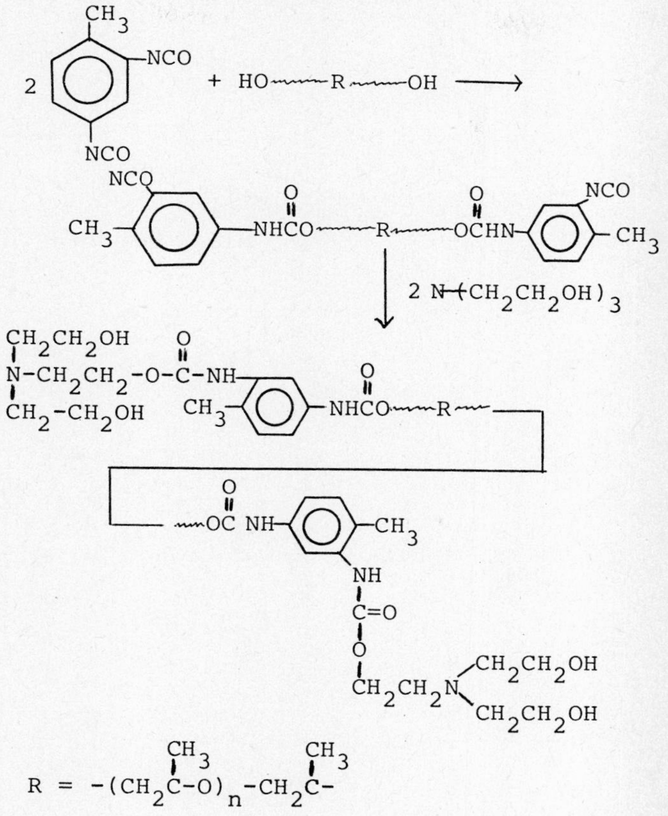

$$R = -(CH_2\overset{\overset{\displaystyle CH_3}{|}}{C}-O)_n-CH_2\overset{\overset{\displaystyle CH_3}{|}}{C}-$$

Pampouchidis et al. (83A,B) also synthesized urethane-based cationic electrodeposition resins.

In one of the examples, 1 mol of isophorone-diisocyanate was reacted with 1 mol of dimethylpropanolamine and the remaining isocyanato groups were reacted with 0.5 mol of tripropylene glycol monomethacrylate and 0.5 mol of pentaerythritol trimethacrylate according to the scheme on the following page.
This reaction took place in the presence of polymerization inhibitors.

The majority of the currently used coating systems is based on copolymers prepared from epoxy resins, polyfunctional amines, and partially blocked isocyanates (84-87). The resins are generally solubilized by the formation of cationic derivatives of amines by aminolysis of the carbamate in the partially blocked isocyanate. The principal applications of these cationic electrocoating systems are as automotive primers, in appliances, as primers and one-coat systems, and in various other industrial applications. The main

R = CH$_2$CH$-$($-$OCH$_2$CH$-$)$_2$$-OC-$C=CH$_2$
           | || |
           CH$_3$ O   CH$_3$

or

HOCH$_2$$-C-$($-CH_2$OCC=CH$_2$)$_3$
                        || |
                        O   CH$_3$

advantages of these coatings are in their outstanding corrosion
resistance and high throw power that is attainable with these
systems.

Acrylic cationic electrocoating systems employing blocked
aliphatic isocyanates are finding application where a combination of
excellent durability or light and color stability with corrosion
resistance is desired, e.g., in agricultural implements (86).

Di- and Polyhydroxy Components. By far the most important active
hydrogen-containing components of urethane coatings are di- and
polyhydroxy compounds, often referred to briefly as polyols.

Hydroxyl-Terminated Polyesters. The first hydroxyl-containing materials employed in urethane coatings were hydroxyl-terminated polyesters. The principal components of these polyesters are adipic acid, phthalic anhydride, "Dimer" acid (dimerized linoleic acid, Emery Industries), and monomeric glycols and triols. The most widely used glycols are ethylene, propylene, 1,2-butylene, 1,4-butylene, neopentyl, and 1,6-hexylene glycol. The most extensively used triols are trimethylolpropane, glycerol, and 1,2,6-hexanetriol. For elastic coatings exhibiting high elongation, elasticity, good low-temperature flexibility, and also relatively low chemical resistance, linear or slightly branched polyesters (i.e., polyesters containing some triol in addition to diols) are used. More highly branched polyesters, yielding more cross-linked films, are suitable for rigid coatings possessing excellent chemical resistance, good heat resistance, high hardness, and low elongation. Low acid number (usually 0.6 or less) and very low water content are requirements for urethane-grade polyesters. The introduction of aromatic components such as phthalic anhydride or isophthalic acid increases temperature resistance and imparts greater rigidity to the coating.

Polyesters derived from ε-caprolactone and ethylene glycol and triols, such as trimethylolpropane, have been employed for the preparation of different types of urethane coatings, including nonyellowing systems (88, 89). These types of polyesters exhibit improved hydrolytic stability and low-temperature properties as compared to adipate polyester based urethanes.

It also has been shown that azelate esters can also be used in place of adipate polyesters in thermoplastic polyurethanes, which also indicated improved hydrolytic stability as well as somewhat improved low-temperature properties as compared to those of the corresponding adipate polyester based urethanes (90). The tendency of polyester-based urethanes to hydrolyze, particularly at high humidities and elevated temperatures, can partly be reduced by incorporation of small amounts of polymeric carbodiimides (91). Alternately, the use of diisocyanate-containing carbodiimide groups helps to confer greater hydrolytic stability to polyester urethanes (92).

Hydroxyl-terminated, aliphatic and cycloaliphatic polycarbonates have also been introduced by PPG Industries and Beatrice Foods. Muller (93) had previously reported the use of poly(1,6-hexanediol-carbonates) in urethane elastomers.

A variety of urethane coating systems has been introduced based on hydroxyl-containing acrylic and methacrylic ester copolymers that contain hydroxyethyl acrylates or methacrylates and that are generally cured with aliphatic di- or polyisocyanates. Depending upon the type and relative amounts, acrylate or other saturated esters, for example, adipate-isophthalic esters, are also used in combination with aliphatic polyisocyanates.

Castor Oil and Transesterification Derivatives of Castor Oil and Other Oils. Castor oil, consisting mainly of a triglyceride of ricinoleic (12-hydroxyoleic) acid (90% ricinoleic acid, 10% nonhydroxy acids, largely oleic and linoleic acids), is a frequent hydroxyl component of urethane coatings. On the basis of the hydroxyl number, castor oil is 70% trifunctional and 30% difunctional. The presence of castor oil in urethane polymers imparts excellent water resistance and good weathering resistance. By transesterification (alcoholysis) of castor oil with polyols such as glycerol, mono- and diglycerides are obtained as follows:

$$
\begin{array}{ccccc}
\overset{\overset{\textstyle O}{\|}}{CH_2-O-C-R} & CH_2OH & & \overset{\overset{\textstyle O}{\|}}{CH_2-O-C-R} & \overset{\overset{\textstyle O}{\|}}{CH_2-O-C-R} \\[2em]
\overset{\overset{\textstyle O}{\|}}{CH-O-C-R} \; + & CHOH & \xrightarrow[\text{catalyst}]{\text{heat}} & CH-OH \quad + & \overset{\overset{\textstyle O}{\|}}{CH-O-C-R} \\[2em]
\overset{\overset{\textstyle O}{\|}}{CH_2-O-C-R} & CH_2OH & & CH_2-OH & CH_2OH \\[1em]
\text{Castor Oil} & \text{Glycerol} & & \text{Monoglyceride} & \text{Diglyceride}
\end{array}
$$

$$
R = -(CH_2)_7-CH{=}CH-CH_2-\overset{\overset{\textstyle OH}{|}}{CH}-(CH_2)_5-CH_3
$$

Transesterification of polyols results in castor oil polyols with lower or higher functionality. Oxidation of castor oil, by blowing air or oxygen through the oil, results in polymerization that yields products with increased viscosity, specific gravity, and saponification value of the oil. The bodied oils have been reported to be more useful in urethane coatings than the untreated castor oil (10).

Closely related to the castor oil polyols are polyols derived from the transesterification of certain drying oils, such as linseed, oiticica, and soya oils. The mono- and diglycerides resulting from this reaction are then utilized in the manufacture of uralkyds.

Polyether Diols and Polyols. Among the most widely used hydroxy compounds for use in urethane coatings and by far the most versatile are hydroxyl-terminated polyethers. They have been widely accepted in industry not only because of generally lower price than polyesters but also because of the possibility in designing urethane coatings possessing a wide spectrum of physical properties. Polyethers are generally prepared either by the addition of alkylene oxides to glycols, polyols, or di- or polyamines or by polymerization (ring opening) of tetrahydrofuran.

Polyether polyols are prepared commercially by the base-catalyzed addition of alkylene oxides such as propylene, ethylene, and butylene oxide to di- or polyfunctional alcohols. Since, for most applications, it is desirable to have hydrophobic urethane compositions, propylene oxide is usually used alone or in combination with small amounts (generally less than 10%) of ethylene oxide. The alcohols used in the manufacture of polyethers include glycols (e.g., propylene glycol) for diols, glycerol, trimethylolpropane, and 1,2,6-hexanetriol for triols, pentaerythritol and α-methyl glucoside for tetrols, sorbitol for hexitols, and sucrose for octols. The base-catalyzed addition of propylene oxide to diol or polyol initiators leads primarily to the formation of terminal secondary alcohol groups. However, a certain amount of primary alcohol groups (about 5-10% of all hydroxyls) is formed in the oxirane ring opening. In addition, formation of unsaturated monohydroxy polyethers occurs to some extent, with the degree of unsaturation tending to increase as the reaction temperature is raised and as the chain length increases (94-96).

Besides poly(oxypropylene) glycols, also referred to as polypropylene glycols, random or block copolymers of propylene oxide and ethylene oxide can also be used in coatings (97).

Poly(1,4-oxybutylene) glycols, also referred to as poly(oxytetramethylene) glycols, are obtained by polymerization of tetrahydrofuran with Lewis acid catalysts such as BF_3, $SnCl_4$, $SbCl_5$, $SbCl_3$, PF_3, and SO_2Cl_2. Their chemical structure $HO(CH_2CH_2CH_2CH_2O)_nH$ indicates not only the presence of terminal primary hydroxyl groups but also the tendency of the longer aliphatic chain to align and to crystallize. This results usually in higher strength properties than materials with amorphous chains of the same equivalent weight (98).

Random copolymers of various alkylene oxides and copolymers of tetrahydrofuran and alkylene oxides can also be used (99). Hydroxyl-terminated polythioethers can be employed where oil resistance and low-temperature flexibility are required.

In addition to neutral polyethers, there are certain nitrogen-containing polyethers that may be used in the preparation of coatings. The initiators are usually compounds containing amine nitrogen. 2-Methylpiperazine or alkyldialkanolamines such as methyldiethanolamine are typical initiators for basic polyether diols. Trialkanolamines such as triisopropanolamine are primary materials for basic polyether triols. Ethylenediamine is widely used as an initiator for basic polyether tetrols. N,N,N',N'-Tetrakis(2-hydroxypropyl)ethylenediamine (Quadrol Polyol, BASF-Wyandotte) is the adduct of 4 mol of propylene oxide to ethylenediamine and serves both as a cross-linking agent or as base material for higher molecular weight basic polyethers by further addition of propylene and ethylene oxide (100). Due to the presence of the tertiary amino nitrogen in these polyethers, the reactivity of the hydroxyl groups toward isocyanate groups is greatly enhanced and this fact is made use of for fast curing two-component urethane coatings.

Hydroxyl-Terminated Hydrocarbon Polymers. Although low molecular weight hydroxyl-terminated polybutadienes have previously been reported (101), homopolymers and copolymers of butadiene having terminal hydroxyl groups have been commercially available (102). The homopolymers have a molecular weight range of 2500-3500 and the hydroxyl functionality varies from 2.1 to 2.6.

$$\text{HO---(CH}_2) \quad \overset{\text{CH=CH}}{\diagup \diagdown} \quad \{\text{CH}_2)\text{--}(\text{CH}_2\text{--CH})\text{--}(\text{CH}_2 \quad \overset{\text{CH=CH}}{\diagup} \quad \overset{\text{CH}_2)\text{---OH}}{}$$

$$\underset{\text{CH=CH}_2}{|}$$

Butadiene homopolymer

Hydroxyl-terminated copolymers of butadiene with styrene or acrylonitrile may be represented schematically as

$$\text{HO--}\left[\ (\text{CH}_2\text{---CH=CH---CH}_2)_a \text{------}(\underset{\underset{X}{|}}{\text{CH--CH}_2})_b\text{---}\right]_n\text{OH}$$

where X is phenyl for the styrene-butadiene copolymer (a = 0.75, b = 0.25, n = 40–50) and is CN for the acrylonitrile-butadiene copolymer (a = 0.85, b = 0.15, n = 55–65).

The hydroxyl-terminated homo- and copolymers of butadiene have been utilized in the preparation or modification of urethane elastomers and sealants (102). Owing to the presence of unsaturation in these polymers, reinforcement with certain fillers such as carbon black can be achieved. Hydroxyl-terminated polyisobutylenes have also been used for the preparation of NCO-terminated prepolymers (103). Very recently (104), α,ω-di-(hydroxy)polyisobutylenes have been prepared by a different route. Polyurethane films prepared from these polyisobutylene glycols have exhibited excellent hydrolytic stability and very low water absorption (105). However, these materials are not yet commercially available.

Hydroxyl-Containing Graft Copolymers and Polyurea Dispersions. Graft copolymers of polyethers containing polyacrylonitrile chains have become available commercially, primarily for the manufacture of flexible foams, although they have also been evaluated in other applications (106–108).

These "polymer-polyols" are made by the in situ polymerization of vinyl monomers such as acrylonitrile (although grafting with other monomers has also been reported in a liquid polyol solution, e.g., polyether triol of molecular weight 3000) to give stable dispersions of the polymeric portion in the liquid polyol. Grafting is carried out with azobis(isobutyronitrile) or dibenzoyl peroxide as initiators at 80–90 °C. A polymer-polyol containing about 20% acrylonitrile appeared to be the best compromise between polyol viscosity and urethane foam properties (108).

Critchfield et al. (109) have also reported the use of acrylonitrile-styrene graft polyols in urethane polymers. The presence of styrene prevents the tendency toward yellowing of the acrylonitrile graft copolymers, which is due to cyclization of the polyacrylonitrile chains at elevated temperatures. Similar types of graft polyols have been reported elsewhere (110).

PHD (polyharnstoff dispersion) or PUD (polyurea dispersion) polyols have been introduced by Bayer-Mobay as an alternative to polymer-polyols. They are generally being prepared by dissolving first either hydrazine or primary amines in polyols (usually ethylene oxide capped poly(oxypropylene-oxyethylene) triols) and then reacting them with stoichiometric amounts of TDI although other diisocyanates can be used. Thus, polyurea dispersions of about 20% are formed in the respective polyols (111).

Both polymer-polyols as well as PUD polyols can be considered "reinforced" polyols since a number of physical properties as well as processing characteristics are enhanced. Both of the above types of polyols have been widely used in RIM and HR-foams but are finding applications also in other urethane fields.

Hydroxyl-Terminated Urethane Polyethers. Hydroxyl-terminated urethane polyethers are polyether polyols extended with diisocyanates to medium high molecular weight polymers employing a low NCO:OH ratio.

The procedure of preparation requires a closed reactor equipped with stirrer and N_2 blanket. The polyols and isocyanates are charged at such a ratio that NCO:OH is lower than 1. The reaction proceeds at 80–90 °C, often in the presence of 0.02% of a metal catalyst such as stannous octoate or dibutyltin dilaurate. The catalyst is normally added after the exothermic reaction ceases.

There are two specific groups in this series of products: (a) neutral urethane polyethers (112) that have to be activated with catalysts when used as second components in urethane coatings, which can be easily transformed into isocyanate-terminated polymers when reacted with additional isocyanate; (b) basic urethane polyethers (113, 114) having hydroxyls activated by tertiary amine nitrogen present in the chemical chain of the product and that are very reactive second components of urethane coatings. They can also be converted into isocyanate-terminated polymers, but their stability is very limited (12–24 h). They are stable when immediately blocked by means of phenol, alcohols, or other suitable blocking agents.

The molecular weight of these hydroxyl-terminated urethane polyethers can be varied greatly, although the majority of these products, used in coatings, range in molecular weight approximately from 1000 to 3000 (115, 116).

Some typical structures of hydroxyl-terminated urethane polyethers are as follows:

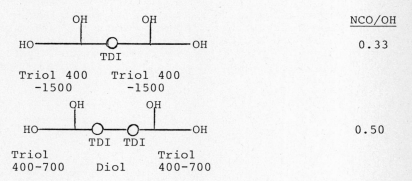

Basic:

Quadrol   Diol   Quadrol   0.40

<u>Chain Extenders and Cross-linkers</u>. In addition to the two principal components of most urethane coatings, isocyanate and polyol components, a number of di- or polyfunctional, active hydrogen components may be used as chain extenders or cross-linkers. The most important classes of compounds for this use are diols or polyols (monomers or oligomers), diamines, and alkanolamines. Typical examples of diols are ethylene, diethylene, dipropylene glycol, 1,4-butanediol, 1,6-hexanediol, neopentyl glycol, bis(hydroxyethyl) derivative of hydroquinone (HQEE), bis(hydroxyethyl)- and bis(2-hydroxypropyl)bisphenol A, and some low molecular weight propylene oxide adducts of trimethylolpropane, glycerol, and 1,2,6-hexanetriol. Examples of simple triols are glycerol and trimethylolpropane.

In addition, bis(2-hydroxypropyl) isophthalate (Arco Chemical Co.) and bis(2-hydroxyethyl) carbamate (Jefferson Chemical Co.) have been introduced into the market.

Typical examples of aliphatic diamines include ethylene- and hexamethylenediamines. Cycloaliphatic diamines include isophoronediamine and 1,4-cyclohexyldiamine.

The most important aromatic diamines used primarily in elastomeric coatings include "hindered" diamines, i.e., aromatic diamines containing substituents ortho to the amine groups such as chlorine atoms; 4,4'-methylenebis(o-chloroaniline) (formerly MOCA, E. I. Du Pont de Nemours & Co.) has been widely used as a chain extender and curing agent. However, due to severe restrictions placed upon MOCA by OSHA, considerable work was undertaken to find suitable replacements. Among the latter are the dimethyl ester of 4,4'methylenedianthranilic acid (MBMA), "Cyanacure" (American Cyanamid Co.), "Caytur 21" (E. I. Du Pont de Nemours & Co.), and "Polacure" (Polaroid Corp.) (<u>117</u>). Various aromatic diamines including phenylenediamines, tolylenediamines, and

methylenebis(aniline), as well as blends of these along with other
diamines, have been employed, although unsubstituted aromatic amines
are considerably more reactive than hindered aromatic amines.
    Of considerable importance are the alkanolamines. Some of these
such as Quadrol (see Di- and Polyhydroxy Components) have been
mentioned previously. Other di- and trialkanolamines used include
triethanolamine, bis(2-hydroxyethyl)- and bis(2-hydroxy-
propyl)aniline, and bisalkanol derivatives of piperazines. Similar
derivatives that are commercially available include N,N',N''-tris(2-
hydroxyethyl)isocyanurate (Allied Chemical Corp.), N,N'-[bis(2-
hydroxyethyl)]dimethylhydantoin (Glyco Chemicals Co.), and
dimethylolpropionic acid (International Minerals & Chemicals Co.).
    This listing of chain extenders or cross-linkers is obviously
incomplete since many other types of chain extenders may be used
where special end use requirements may have to be met, for example,
for flame retardance. In this case, halogenated diols, for example,
2,3-dibromo-2-butane-1,4-diol (GAF Corp.), or phosphate or
phosphonate group-containing diols, for example, Fyrol 6 (Stauffer
Chemical Co.), may be used.

Catalysts. Catalysis plays an important role in urethane coating
formulations. Catalysts may be employed in the preparation of
prepolymers (primarily those based on aliphatic isocyanates) but are
most widely used in the curing reaction, depending upon whether a
moisture cure ($NCO-H_2O$ reaction) or a cure involving $-NCO$ and $OH-$
groups. The most effective catalysts for the isocyanate-water
reaction are tertiary amines, while metal catalysts are by far the
most efficient catalysts for the isocyanate-hydroxyl reaction. In
particular, tin, mercury, lead, and zinc catalysts are used for the
latter reaction, although tertiary amines are also effective though
to a lesser degree. The catalytic activity of tertiary amines
generally increases with lack of shielding of the tertiary amine
nitrogen and with increasing basicity of the amine. Combinations of
both tertiary amines and metal catalysts may be employed and exhibit
generally a synergistic action. Even combinations of two amines may
have a synergistic effect, particularly if the two amines are so
chosen as to differ in their mode of activity, as indicated above.
For uralkyds, common paint driers such as lead, manganese, and
cobalt salts of organic acids are used.
    A great number of both tertiary amines and metal catalysts are
available, and a detailed discussion of these as well as their
mechanisms is beyond the scope of this paper. The reader is
referred to several reviews on the subject of catalysis in
isocyanate reactions (118-120).

Other Urethane Coating Components. Due to the limited scope of this
paper, no efforts are made to describe in detail the following
components, which may be present in some but not all urethane
coating formulations: (a) solvents; (b) pigments; (c) ultraviolet
absorbers, light stabilizers, and antioxidants; (d) flow agents;
(e) bodying agents; (f) flatting agents. Below are listed some
brief comments regarding the type and use of these materials.

Solvents. Of vital importance is the fact that no solvents may be
used that contain any free active hydrogen-containing group

including water (moisture). Aromatic and oxygenated solvent blends
are most frequently used in urethane coating formulations. The
effect of solvents on the reaction rate of the isocyanate–alcohol
reaction is quite pronounced (121). The reaction rates are
dependent upon the hydrogen–bonding character and the dielectric
constant of the solvent.

Pigments. An obvious prerequisite for the selection of pigments in
urethane coating formulations is their non–reactivity toward
isocyanates. Many pigments exhibit an affinity for moisture, and
the presence of some reactive chemical groups limits the use of some
pigments in urethane formulations.
     Uralkyds that contain no free isocyanate groups can be pigmented
by conventional means. However, in most cases, special techniques
have to be employed to remove moisture from the pigments. Since
oven–dried pigments readily absorb moisture, it is best to regrind
or disperse the pigments in the polyol component and remove the
moisture in vacuo or by azeotropic distillation. Other methods have
been used to dry pigments such as slurrying the pigment in a solvent
containing a diisocyanate or adding molecular sieves to the
pigmented bases (122, 123). Occasionally wetting agents may be
added to aid in the dispersion of the pigments and may prevent
pigmented urethane coatings from chalking.

Ultraviolet Absorbers, Light Stabilizers, and Antioxidants.
Ultraviolet absorbers and light stabilizers can be added to urethane
coatings in order to increase their resistance to discoloration and
to outdoor weathering. UV absorbers are employed to eliminate the
radiation in the region of 334 and 411 μm that sensitize the
polyurethanes and promote an autoxidation process. Light
stabilizers on the other hand do not absorb UV radiations but
inhibit or control the degradation by other means.
     UV absorbers perform by absorbing UV light energy that they
transform into molecular vibrations, rotations, and eventually heat.
In the case of hydroxybenzophenones, this takes place by forming a
tautomeric structure (123A):

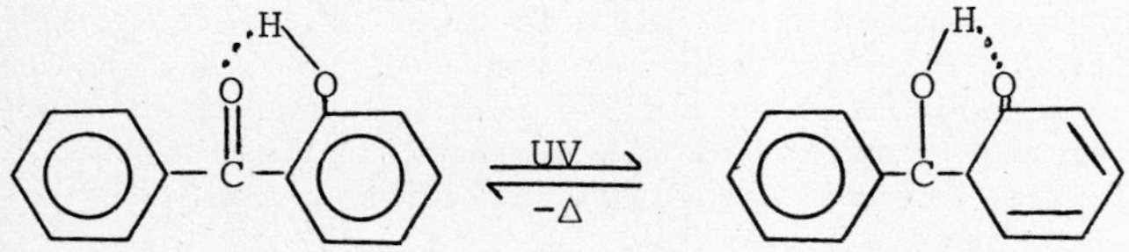

According to Berner and Kreibich (123B), there are four major
classes of commercial UV absorbers: 2–hydroxybenzophenones, 2–(2'-
hydroxyphenyl)benzotriazoles, oxalic anilides, and cinnamates.
     The light stabilizer category, hindered amine light stabilizers
(HALS) are very effective stabilizers for urethane coatings. These
compounds, developed primarily by Ciba–Geigy, do not act as UV
absorbers but act by inhibiting the degradation of the binder that
is already in an excited state.

$$\left[X_1{-}N \underset{CH_3}{\overset{CH_3}{\diagdown}} \underset{CH_3}{\overset{CH_3}{\diagup}} \underset{CH_3}{\diagup} \overset{CH_3}{\diagdown} \underset{O}{\overset{H}{\diagdown}} \right]_n {-} X_2$$

Several nickel compounds also are efficient light stabilizers. A typical representative is Argastab 2002 (Ciba–Geigy).

$$\left[\underset{(CH_3)_3C}{\overset{(CH_3)_3C}{}} \text{HO} {-} \hexagon {-} CH_2{-}\overset{O}{\underset{C_2H_5O}{P}}{-}O \right]_2 {-} Ni$$

This compound acts primarily as a "quencher" by absorbing energy from the excited binder molecules and dissipating it as heat.

Antioxidants inhibit oxidation and are usually employed in combination with UV absorbers. Most antioxidants used for urethane coatings fall into the following classes: substituted phenols, aromatic amines, condensation products of aminophenols with aldehydes, thio compounds, and phosphites. Often combinations of antioxidants from the above categories exhibit a synergistic effect. More recently Mathur et al. (125) carried out an extensive study involving the use of one to six stabilizers, selected from the above antioxidants and UV stabilizers, employing several independent test methods, to study the thermooxidation of urethane and urethane–urea films.

Flow Agents. Different types of flow agents are used to enhance leveling and to avoid cratering and bubbling in urethane coatings. Examples of typical flow agents for most urethane coatings are cellulose acetate butyrate, certain silicones and silicone–alkyd resins, zinc naphthenate, and others. Flow agents are of particular importance in urethane powder coatings in order to avoid formation of "orange peel." Typical flow agents include silicones, cellulose acetate butyrate, and certain acrylic copolymers.

Bodying Agents. Bodying agents in low concentrations are being employed to increase the viscosity of urethane coatings, to help eliminate bubbling at a high–solids content, and to control sagging when the coatings are applied on vertical surfaces. Certain types of cellulose acetate butyrate or polyvinylbutyral may be added in a 10% solution of oxygenated solvents.

In the case of urethane emulsions, many of the commonly employed thickeners for resin emulsions may be used, such as polyacrylic acids, polyacrylamides, etc. This is particularly useful in textile coating or adhesive applications where "hold–out" properties, for example, to permit the coating to stay on the surface rather than to penetrate the substrate, are required.

Flatting Agents. Generally urethane coatings exhibit a very high gloss. Flatting agents are frequently employed to reduce the gloss. The most widely used flatting agents are finely divided silicas, diatomaceous earth, and certain metal stearates, such as aluminum and zinc stearates.

Types of Urethane Coatings

Historically, the most commonly used urethane coatings have been grouped into five ASTM categories. Because these coatings have been reviewed in detail in many articles and chapters (126–130), they are listed below together with some brief comments.

One-Package Urethane Alkyd (Oil–Modified Urethanes) Coatings (ASTM Type 1). One-package urethane alkyd coatings, also referred to as uralkyds or urethane oils, are made by transesterification of drying oils with polyols, followed by the reaction of the resulting di- or monoglycerides with a diisocyanate. Curing of the uralkyds occurs by oxidation of the double bonds present in the fatty acid portion of the coating system.

One-Package Moisture-Cure Urethane Coatings (ASTM Type 2). One-package moisture-cure urethane coatings consist of isocyanate-terminated polymers cured by atmospheric moisture. The rate of cure depends upon the humidity in the air and the presence of certain tertiary amine catalysts that accelerate the NCO–water reaction. A major difficulty with this type of coating system is pigmentation.

Single-Package Blocked Adduct Urethane Coatings (ASTM Type 3). Single-package blocked adduct urethane coatings are based on blocked isocyanate components in a blend with hydroxyl-bearing materials, which are stable in a single package and which cure after deblocking at elevated temperatures, about 150-160 °C, and at lower temperatures and shorter curing times, when catalyzed. See also Blocked Isocyanates.

Two-Package Catalyst Urethane Coatings (ASTM Type 4). Two-package catalyst urethane coatings are similar to Type 2, consisting of an isocyanate-terminated polymer to which a catalyst is added as a second component prior to application. Two groups of catalysts are usually employed: (1) reactive catalysts, containing hydroxyl groups such as alkanolamines or (2) nonreactive catalysts such as tertiary amines and metal salts of carboxylic acids.

Two-Package Polyol Urethane Coatings (ASTM Type 5). Two-package polyol urethane coatings consist of isocyanate-terminated adducts of polymers that are cured by reaction with di- or polyfunctional hydroxyl-containing materials. The latter may consist of low- to medium-weight polyols with a polyester, polyether, polyether urethane, or castor oil backbone. When the two components (OH- and NCO-) are mixed together, they have only a limited pot life. Therefore, the components are mixed prior to application. Catalysts may be used to speed up the cure either for room temperature or oven cure.

In addition to the five groups of urethane coatings described here, other coating systems have been developed that are described as follows.

Urethane Lacquers. These include fully polymerized, thermoplastic coatings of relatively high molecular weight, dissolved in suitable solvents. The polymers are prepared by reaction of the polyisocyanates with a polyol until no free isocyanate remains. The resulting polymer is then dissolved in solvents such as *N*-methylpyrrolidone, dimethylformamide, tetrahydrofuran, or solvent blends such as methyl cellosolve/cellosolve acetate/xylene. The films are formed by mere solvent evaporation. One serious disadvantage is the generally low solids content of these coating systems. These lacquers find extensive use as coatings over flexible substrates such as textiles, flexible foams, microcellular foams, rubber, and other elastomeric products.

Two-Package Coatings: Isocyanate-Terminated Prepolymers Cured with Polyamines. Due to the fact that isocyanates react very rapidly with amines to form urea groups, the success of preparing suitable coating systems depends upon the following alternatives: (1) the use of hindered aromatic amines to slow down the urea formation; (2) the use of aliphatic isocyanates, for example, $H_{12}MDI$, in the prepolymer formation, followed by reaction with an aromatic diamine (aliphatic isocyanate groups react considerably slower than aromatic ones); (3) the use of aromatic diamines in spray systems where short pot lives can be used.

In all of the above described cases, urethane-urea block copolymers are formed that generally are characterized by higher strength properties and higher hardness when compared to coatings made from the same prepolymer but cured with polyols. A disadvantage of these types of coatings is the tendency toward yellowing upon weathering (129). The use of MOCA and other hindered diamines (as a replacement for MOCA) leads to coatings with a pot life of 0.5 to 1 h. 4,4'-Methylene bis(aniline), due to its greater reactivity, is used either in spray systems with aromatic diisocyanate-based prepolymers or with prepolymers made from aliphatic isocyanates. The latter are reported to yield coating systems with good properties and an acceptable pot life (131).

In addition to diamines, hydrazine (132, 133) has also been reported to give coatings with good abrasion resistance. Likewise, guanidine (134) has been used to cure isocyanate-terminated prepolymers, yielding highly cross-linked coatings due to the functionality of guanidine.

Two-Package Coatings: Isocyanate-Terminated Prepolymers Cured with Ketimines. Instead of diamines, ketimines can be employed to cure aromatic or aliphatic polyisocyanates (135, 136). Ketimines are generally produced from primary amines and a ketone. Since ketimines are not very reactive with aliphatic isocyanates, the presence of moisture is necessary in order to hydrolyze the ketimine to form amines, which further react with the isocyanate groups to yield urea linkages. The addition of drying agents that act as scavengers for moisture to prevent premature unblocking of the ketimines has been reported (137, 138). The curing mechanism between ketimines and polyisocyanates depends upon the amount of moisture to bring about hydrolysis of the ketimines, and several reactions are possible as shown by Furukawa et al. (139). Hence, side reactions may cause the formation of coatings having inferior properties and a tendency to yellow.

Two-Package Coatings with Aminoformaldehyde Cure. One component consists of a hydroxyl-terminated urethane prepolymer while the other component is an alkyl ether (usually methyl or butyl ether) of a methylolmelamine or methylolurea derivative (melamine- or urea-formaldehyde condensation products or derivatives thereof) (140, 141). Various catalysts may be employed to accelerate the heat cure ($\sim$125 °C) and to lower the curing temperature. Curing results by splitting off of the respective alcohol, as shown:

$$\rangle N-CH_2-O \; \overline{|\; R \; + \; HO\;|}-R_1-OCO-NH-R_2-\!\!\sqrt{}\!\!\sqrt{}\!\!\sqrt{} \; -----\rangle$$

$$\rangle N-CH_2-O-R_1-OCONH-R_2-\!\!\sqrt{}\!\!\sqrt{}\!\!\sqrt{} \; + \; R-OH$$

The storage stability of these coating systems is excellent, and a wide range of properties can be obtained through variation in the raw materials.

Aqueous Urethane Coating Systems. Published literature on the composition and applications of water-based urethane systems, for example, emulsions or latices, has been rather limited until a few years ago when the use of these systems began to increase substantially. Both nonionic and ionic systems have been prepared. Altscher (142) described the use of emulsions of blocked urethane systems containing from 5 to 20% of solvent for use in various textile applications.

Urethane-urea latices, consisting of small particles of high molecular weight polymers dispersed in water, are prepared by chain extension of isocyanate-terminated prepolymers in aqueous diamine solutions, employing either nonionic or ionic surfactants (143-145).

Ionic polyurethanes include both urethane polyelectrolytes and urethane ionomers. The principal difference between ionomers and polyelectrolytes is that the latter consist essentially of ionic monomer units while ionomers contain only a relatively small number of ionic monomer units (e.g., acid groups) that can be cross-linked by means of metal salts. It is interesting to note that when polyurethanes assume positive or negative charges, significant changes occur in their mechanical properties and solubility characteristics. In general, the strength properties are increased and the usually hydrophobic urethane polymers may become highly hydrophilic at a high concentration of charges. Most properties of ionic polyurethanes may be accounted for by (a) the primary structure, for example, the main chain, the nature and sequence of the segments, and the concentration of ionic groups; (b) hydrogen bonding; (c) ionic association; and (d) association of hydrophobic parts of polyurethane segments due to van der Waals' forces (146).

Dieterich et al. (147, 148) reviewed in depth aqueous ionic and nonionic polyurethane systems and described different methods for the preparation of these polymers.

Suskind (149) described the formation of film-forming cationic urethane latices. The isocyanate-terminated prepolymer derived from either a polyester or polyether diol and tolylene diisocyanate was first chain-extended with an alkyldiethanolamine to yield a relatively low molecular weight urethane capable of further chain-extending reactions:

$$\overset{\displaystyle R \qquad R}{\underset{\displaystyle}{3OCN \sim\!\!\sim NCO + 2RN(CH_2CH_2OH)_2 \longrightarrow OCN \sim\!\!\sim N \sim\!\!\sim N \sim\!\!\sim NCO}}$$

Emulsification occurs when the partially extended urethane is added with high-speed mixing to 3% aqueous acetic acid. Curing of the latex takes place either by reaction of water (150, 151) with the terminal isocyanate groups or by reaction with water-soluble diamines:

$$OCN\sim\!\!\sim\overset{R}{\underset{H}{N^+}}\sim\!\!\sim\overset{R}{\underset{H}{N^+}}\sim\!\!\sim NCO \xrightarrow{H_2O} \sim\!\!\sim\overset{R}{\underset{H}{N^+}}\sim\!\!\sim NH-\overset{O}{\overset{\|}{C}}-NH\sim\!\!\sim\overset{R}{\underset{H}{N^+}}\sim\!\!\sim + CO_2$$

$$OCN\sim\!\!\sim\overset{R}{\underset{H}{N^+}}\sim\!\!\sim\overset{R}{\underset{H}{N^+}}\sim\!\!\sim NCO \xrightarrow{H_2N-R'-NH_2} \sim\!\!\sim\overset{R}{\underset{H}{N^+}}\sim\!\!\sim NH-\overset{O}{\overset{\|}{C}}-NH-R'NH-\overset{O}{\overset{\|}{C}}-NH\sim\!\!\sim\overset{R}{N^+}$$

When triethanolamine is used as a third component in the preliminary chain extension step, at an NCO:OH ratio of 1.5, the resulting product consists of both linear and branched units containing terminal isocyanate groups. Further chain extension, as shown, results in cross-linked structures:

$$\sim\!\!\sim\overset{R}{\underset{H}{N^+}}\sim\!\!\sim NH-\overset{O}{\overset{\|}{C}}-NH\sim\!\!\sim\overset{R}{N^+}\sim\!\!\sim\overset{R}{\underset{H}{N^+}}$$

It has been assumed (149) that volatilization of the acetic acid occurs during the drying cycle of the urethane latices, which then no longer contain hydrophilic amine acetate groups. Rembaum and coworkers (152-156) also reported the preparation of cationic polyurethanes and various applications of these polymers, which were termed "ionene polymers" because they contained ionic amines.

Dieterich et al. (146) and Taft and Mohar (157) have prepared excellent reviews of urethane ionomers. One of the most important characteristics of urethane ionomers is the ease with which they form stable water dispersions without the use of emulsifiers (158-164). These dispersions consist of colloidal two-phase systems that can be readily prepared by adding a solution of the urethane ionomer in polar solvents, such as methyl ethyl ketone or tetrahydrofuran, to water with good agitation, followed by removal of the solvent by distillation. When chemical cross-linking is also carried out during or after the emulsion process, precipitation of the dispersion occurs. However, in many instances, the solid particles may be redispersed by simple agitation. Dispersible polyurethane powders have thus been obtained (165).

Because of its low boiling point and low toxicity, acetone is particularly suitable for preparing polyurethane dispersions. This procedure is therefore called the "acetone process." Although this process is rather uneconomical, it has certain advantages such as a wide scope for variation in structure and particle size, high quality of end products, and good reproducibility.

More recently Dieterich and Reiff (166) have described the formation of aqueous urethane dispersions by the dispersion of ionomer melts with subsequent polycondensation in two-phase systems. The principle of this procedure consists of reacting molten ionic modified polyester or polyether prepolymers containing NCO groups with urea to yield bis(biuret), followed by methylolation by means of aqueous formaldehyde in a homogeneous phase, and the resulting plasticized melt of methylolated ionic urethane bis(biurets) dispersed in water at 50–130 °C. These steps can be represented schematically as follows:

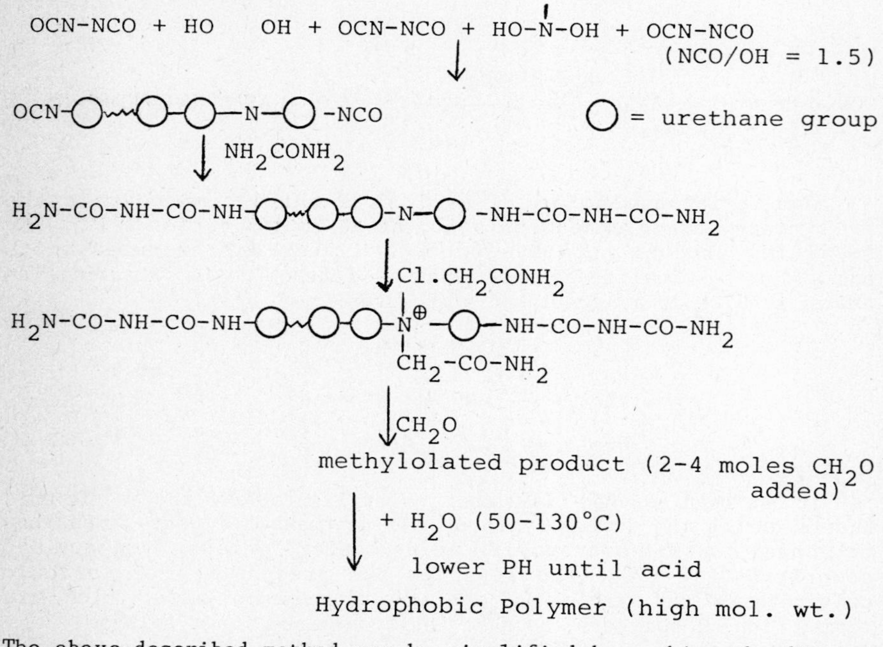

OCN–NCO + HO   OH + OCN–NCO + HO–N–OH + OCN–NCO
                                              (NCO/OH = 1.5)

OCN–◯⌇◯–◯–N–◯–NCO   ◯ = urethane group

         ↓ NH_2CONH_2

H_2N–CO–NH–CO–NH–◯⌇◯–◯–N–◯–NH–CO–NH–CO–NH_2

         ↓ $Cl.CH_2CONH_2$

H_2N–CO–NH–CO–NH–◯⌇◯–◯–N$^{\oplus}$–◯–NH–CO–NH–CO–NH_2
                                    |
                           CH_2–CO–NH_2

         ↓ CH_2O

        methylolated product (2–4 moles CH_2O added)

         ↓ + H_2O (50–130°C)

            lower PH until acid

        Hydrophobic Polymer (high mol. wt.)

The above-described method can be simplified by masking the diamine with a ketone to form a diketimine. Ketimines can be mixed with NCO-prepolymers without a reaction occurring. This mixture can be mixed with water to form a dispersion; subsequently the ketimine hydrolyzes to form a diamine that reacts with the NCO-prepolymer to form a chain-extended, high molecular weight polyurethane dispersion. Similarly, NCO-prepolymers can be mixed with a ketazine, aldazine, or hydrazone before dispersing in water. Hydrolysis of these compounds produces hydrazine, whose chain extends the NCO-prepolymer.

Due to their elasticity, abrasion resistance, good cold flexibility, and tenacity, aqueous urethane dispersions are eminently suitable for coating textiles, paper, and leather, bonding of nonwoven fabrics, and surface coating of wood and other substrates. Textiles coatings with these products can be used in the manufacture of garments such as jackets, overcoats, hats, and leatherlike products for use in apparel, handbags, and luggage (167).

Perhaps the most widely used aqueous urethane coating systems at present are the anionic polyurethane dispersions. They are usually prepared by reaction between a difunctional polyol, an acidic solubilizing group, and a diisocyanate to form a low molecular weight prepolymer. The prepolymer is neutralized with a suitable base and chain extended with a diamine. The resulting polyurethanes consist generally of colloidal dispersions of small particle size (less than 0.1 μm). Taub (168) has reviewed in detail the various components and formulations for anionic polyurethane dispersions. Many different acidic solubilizing compounds such as sulfonic acids and hydroxy-containing carboxylic acids have been evaluated in anionic urethane dispersions. Dimethylolpropionic acid finds wide application, and the level of the acid content has considerable influence on the colloidal nature of the system. Because it is desirable for the dispersions to have a relatively low viscosity, a solvent is usually employed that may consist of a highly polar water-soluble solvent such as N-methylpyrrolidone, tetrahydrofuran, methyl ethyl ketone, etc. The solvent may either be left in the final system or vacuum-stripped to yield a solvent-free product. Neutralization of the carboxyl-containing prepolymer is carried out with bases such as sodium hydroxide or tertiary amines. Triethylamine is usually employed for air-drying systems, while tertiary amines such as N,N-diethylethanolamine are preferred for systems that are baked after application. The amount of water used in the dispersion is very important both from a processing standpoint and for final consistency of the product and particle size. The final step in the preparation of the anionic dispersions is the chain extension of the neutralized isocyanate prepolymer in water with a diamine such as ethylenediamine, hydrazine, isophoronediamine, and methylenebis(4-cyclohexylamine) (168). Commercial applications of these anionic urethane dispersions include vinyl fabric, topcoats, floor polishes, printing inks, coatings for plastics, and resilient floor coverings (168).

One Hundred Percent Solids Coatings. As indicated in the introduction, environmental restrictions have led to increasing emphasis on either high-solids or even 100% solids urethane coatings. The latter may consist of either solventless liquid coatings or solid powder coatings. Relatively little information is currently available on these types of coatings.

Solventless, elastomeric coatings were reported by Bonk et al. (169). The sprayable compositions were based mainly on diethylene glycol adipates, Isonate 143-L, and various diamine chain extenders and curatives. Airless Binks spray guns were employed in the application of these coatings. Bieneman (170) also described a liquid 100% solids coating based on crude MDI and a polyol. A major problem with most liquid 100% solids coating is the relatively high viscosity and related difficulties in application.

More recently (171) a two-component 100% solids prepolymer system, based on TDI and poly(oxypropylene) glycols, has been disclosed by Alva-Tech, Inc. This system can be run on a conventional urethane or vinyl plastisol casting line and has been suggested as a replacement for vinyl coatings for fabrics.

Relatively few urethane powder coating systems have become commercially available. Urethane powder coatings based on polyester polyols and ε-caprolactam-blocked diisocyanate have yielded good films when baked at 200 °C for 30 min (172). Both ε-caprolactam-blocked aliphatic and aromatic isocyanates together with appropriate solid polyester systems are now available (Mobay Chemical Co.). Instead of ε-caprolactam-blocked aliphatic isocyanates, adiponitrile carbonate (ADNC) (see Miscellaneous Urethane Coating Systems) can be used with polyesters in urethane powder coating systems (173). Other powder coating systems employ solid polyisocyanates together with solid polyols (174).

Radiation-Cured Urethane Coatings. Urethane polymers containing allyl or vinyl groups can be cured by ultraviolet or electron beam irradiation. Johnson and Labana (175) employed electron beam irradiation to cure the reaction product of 2 mol of hydroxyethyl acrylate with TDI:

$$2CH_2=CH-CO-O-CH_2-CH_2OH + TDI \longrightarrow$$

$$[CH_2=CH-CO-O-CH_2-CH_2-O-OC-NH]_2C_6H_3(CH_3)$$

Other types of curing systems for allyl-terminated urethane polymers were described by Kehr and Wszolek (176). Polyether and polyester urethane-ene systems were prepared by reaction of polyether or polyester diols with allyl isocyanate or by reaction of isocyanate-terminated prepolymers with allyl alcohol. These urethane-ene systems can be cured by branched monomeric polythiols such as pentaerythritol tetrakis(thioglycolate): $C(CH_2OCOCH_2SH)_4$.

Curing can be carried out by mixing 100 parts of a polymeric allylic diene with 10 parts of tetrathiol, adding 0.05 parts of a subjecting hindered phenol antioxidant, and subjecting the mixture to radiation beam curing (2 MeV electrons, Van de Graaf electron accelerator).

Another curing method employing photocuring rate accelerators used the same mixtures of polymeric allylic diene and tetrathiol in the presence of 1 part of a photosensitizer such as benzophenone or dibenzosuberone and 0.05 part of a phenolic antioxidant. Ultraviolet radiation of about 3600 Å is optimum for curing the thiol-ene system. Thus films or coatings can be cured in a few seconds depending upon the photosensitizer and the intensity of the radiation source (176). Photopolymers consisting of urethane-cinnamates have been described by Thomas (77).

Chang (178) reported the preparation of thermoset coatings by curing a mixture consisting of a hydroxyl-containing polyester having a terminal acrylic or α-substituted acrylic group, an organic diisocyanate, and a polyol by means of high energy ionizing radiation. One or more copolymerizable ethylenic monomers may be included in these systems.

Smith et al. (179) described the preparation of a hard and adherent coating by irradiation of an unsaturated polyurethane-polycaprolactone resin.

An excellent review on radiation curing including the use of urethane systems was presented recently by Tu (180).

High Temperature Resistant Urethane and Other Isocyanate-Based Coatings. There has been an increasing tendency in recent years to raise the temperature resistance of urethane coatings for use in a variety of industries, especially for electric and electronic industries. The chemical approaches to this need follow closely those in other urethane applications, in particular rigid foams and elastomers (181-183). These consist primarily in the introduction of heterocyclic rings by ring formation of isocyanate with other reactive groups employing usually specific catalyst systems. They include oxazolidones, imides, hydantoins, parabanic acids, imidazolones, isocyanurates, and other heterocycles. Very recently, Kordomenos et al. (184) reported the synthesis and structure-property relationships of oxazolidone-containing coatings including poly(oxazolidones), poly(oxazolidone-isocyanurates), poly(oxazolidone-epoxides), and poly(oxazolidone-urethane-isocyanurates). The oxazolidone ring formation was carried out by reaction of aromatic or aliphatic diisocyanates with epoxides using zinc catalysts (185). When the NCO:epoxide ratio was kept at 1:1, thermoplastic poly(oxazolidones) were formed:

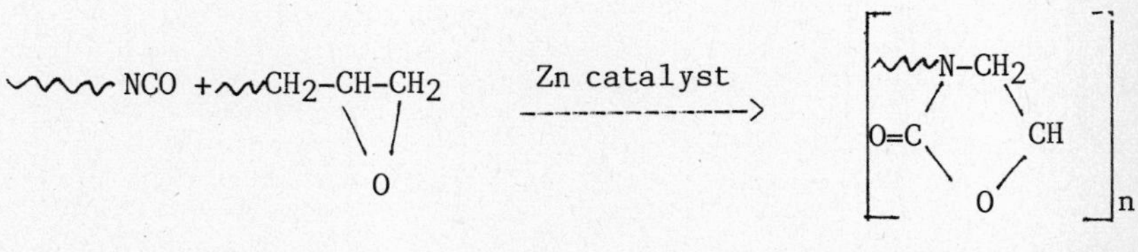

Poly(oxazolidone)

When an NCO:epoxy ratio of 2:1 was used, an NCO-terminated, oxazolidone-containing prepolymer was formed that was then trimerized by employing a trimerization catalyst to yield poly(oxazolidone-isocyanurates) as shown below:

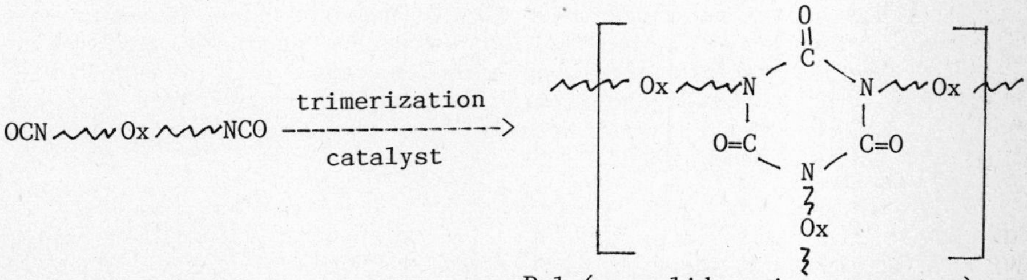

Poly(oxazolidone-isocyanurates)

Epoxy-terminated oxazolidone prepolymers were prepared by using a ratio of NCO:epoxide of 1:2. These prepolymers could then be cured with conventional epoxy hardeners or catalysts:

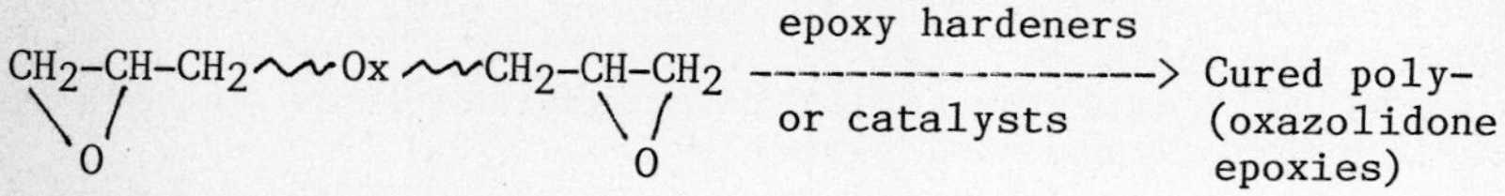

$$
\begin{array}{c}
\text{CH}_2\text{-CH-CH}_2 \sim\!\!\sim \text{Ox} \sim\!\!\sim \text{CH}_2\text{-CH-CH}_2 \\
\diagdown \diagup \qquad\qquad\qquad \diagdown \diagup \\
\text{O} \qquad\qquad\qquad\qquad \text{O}
\end{array}
\quad
\begin{array}{c}
\text{epoxy hardeners} \\
\text{-----------------} \longrightarrow \\
\text{or catalysts}
\end{array}
\quad
\begin{array}{l}
\text{Cured poly-} \\
\text{(oxazolidone} \\
\text{epoxies)}
\end{array}
$$

Epoxy-terminated oxazolidone prepolymer

Poly(oxazolidone-urethane-isocyanurates) were prepared from an excess of isocyanate-terminated oxazolidone prepolymers and poly(oxypropylene) glycols of different chain lengths via the formation of NCO-terminated poly(oxazolidone-urethanes) followed by trimerization of the terminal isocyanate groups as shown in the following reaction scheme:

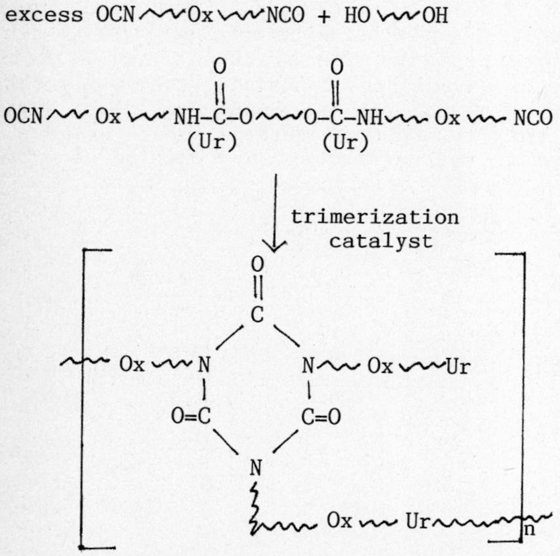

excess OCN$\sim\!\!\sim$Ox$\sim\!\!\sim$NCO + HO$\sim\!\!\sim$OH

$$
\begin{array}{ccc}
& \text{O} & \text{O} \\
& \parallel & \parallel \\
\text{OCN}\sim\!\!\sim \text{Ox} \sim\!\!\sim \text{NH-C-O}\sim\!\!\sim \text{O-C-NH}\sim\!\!\sim \text{Ox} \sim\!\!\sim \text{NCO} \\
& \text{(Ur)} & \text{(Ur)}
\end{array}
$$

trimerization catalyst

Poly(oxazolidone-urethane-isocyanurates)

The thermal stability of the oxazolidone-containing polymers, as determined by TGA, increased with increased concentration of the oxazolidone rings and increasing cross-link density (184). The hardness, stress-strain properties, and solvent resistance of the oxazolidone-containing polymers without the urethane linkages were very good while the Gardner impact resistance was only moderate to low. The introduction of urethane linkages in these polymers significantly improved the impact resistance but decreased somewhat the thermal stability depending upon the chain length of the polyether diol used (184).

Miscellaneous Urethane Coating Systems. Light-stable urethane coating systems have been developed from 3,3-tetramethylenedi(1,4,2-dioxazol-5-one) (ADNC, Arco Chemical Co.). The structure of ADNC is as follows:

$$N=C-(CH_2)_4-C=N$$

The preparation of ADNC and technology associated with nitrile carbonates, in general, are disclosed in a number of U.S. patents (186, 187). ADNC reacts with diols or polyols in the presence of catalysts such as certain metal catalysts, for example, tin catalysts, either alone or in combination with tertiary amine catalysts, to form the corresponding urethanes with evolution of carbon dioxide:

$$ADNC + HO-R-OH \xrightarrow[\text{cat.}]{\Delta} \{O-\overset{O}{\overset{\|}{C}}-NH-(CH_2)_4-NH-\overset{O}{\overset{\|}{C}}-O-R\}_n + CO_2$$

Light-stable urethane coatings with excellent physical properties prepared by the reaction of ADNC with either polyols or with a combination of polyols and epoxy resins in the presence of suitable catalysts have been described by Frisch et al. (188).

Urethane group containing-coatings can also be obtained from other types of compounds capable of generating isocyanates. These are aminimides, especially those based on methacrylic acid (189–191):

$$CH_2=\overset{\overset{\displaystyle CH_3}{|}}{C}----\overset{\overset{\displaystyle O}{\|}}{C}-\overset{-}{N}-\overset{+}{N}(CH_3)_2 \qquad R=CH_3,\ -CH_2CH_2OH,\ -CH_2-\overset{\overset{\displaystyle OH}{|}}{CH}-CH_2-OH$$

Aminimides decompose on heating to yield isocyanate groups as follows:

$$\sim\!\!\!\sim\!\!\!\sim\overset{O}{\overset{\|}{C}}-\overset{-}{N}-\overset{+}{N}(CH_3)_2 \xrightarrow{\Delta} \sim\!\!\!\sim\!\!\!\sim NCO + R-N(CH_3)_2$$

Aminimides can also be copolymerized with other vinyl monomers, yielding acrylic resins. Hence, either urethane homopolymers or urethane copolymers can thus be obtained.

A relative new type of moisture-cure urethane coating consists in the formation of terminal oxazolidine group containing systems that are formed by reaction of carbonyl compounds, for example, formaldehyde or isobutyraldehyde with ethanolamines (Rohm & Haas). The reaction of these aldehydes with diethanolamine is shown:

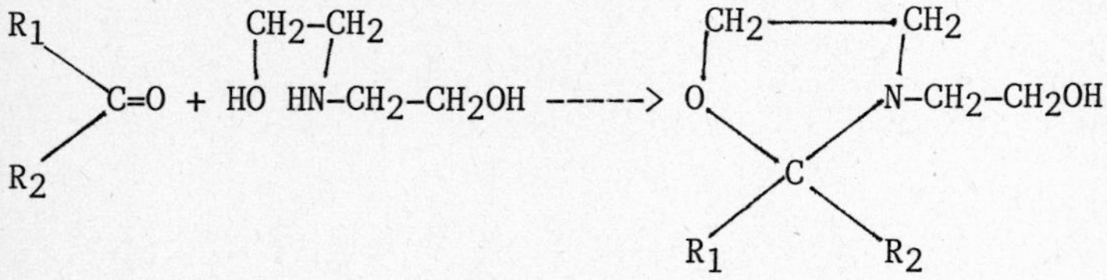

Oxazolidine

R_1 = H

R_2 = H or $(CH_3)_2$

This oxazolidine derivative can be introduced into urethane, polyester, acrylic, or other polymeric systems to contain terminal oxazolidine groups. The oxazolidine ring is moisture sensitive and opens up to re-form the alkanolamine group. One-package, storage-stable oxazolidine coating systems can be prepared from these oxazolidine-containing polymers by the addition of aliphatic di- or polyisocyanates and a tin catalyst, for example, dibutyltin dilaurate. Curing proceeds via ring opening of the oxazolidine ring with subsequent reaction of the liberated hydroxyl groups with the isocyanate groups.

The introduction of organosilicone groups into urethane polymers has been reported by a number of investigators (192–201). Polyurethanes based on silanediols with hydroxyl groups directly linked to the silicon atoms are quite moisture sensitive and are readily hydrolyzed by water. On the other hand, organosilicon glycols with hydroxyl groups located on carbon atoms were found to be stable, and no rupture of either the siloxane or the carbon silicon bonds was observed during the synthesis of these polymers (193–201).

Polyurethane coatings have been prepared from silicon-containing hydroxy compounds on the basis of symmetrical bis(hydroxymethyl)decamethylpentasiloxane (200), diphenylsilanediol (200), high molecular weight organosilicon polyols (198) or the type

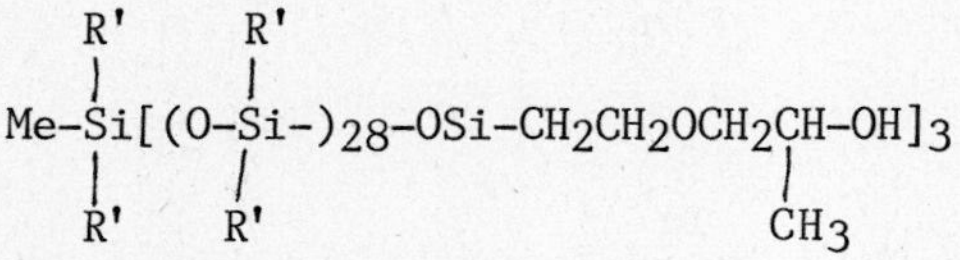

$$Me-Si[(O-Si-)_{28}-OSi-CH_2CH_2OCH_2CH-OH]_3$$

with R' substituents on the silicon atoms and CH_3 on the final carbon.

and some glycoxysilanes with di- and polyisocyanates (201).

Smetankina and Karbovskaya (196) have studied the effect of introducing carbofunctional organosilicon glycols into urethane coatings. The hydroxy-containing components had the general formula

$$\text{HO-(CH}_2)_n \overset{\overset{\displaystyle R'}{\displaystyle |}}{\underset{\underset{\displaystyle R'}{\displaystyle |}}{\text{Si}}}\text{-O-}\overset{\overset{\displaystyle R'}{\displaystyle |}}{\underset{\underset{\displaystyle R'}{\displaystyle |}}{\text{Si}}}\text{-(CH}_2)_n\text{-OH}$$

The polyisocyanate adduct based on trimethylolpropane and 65/35 TDI was used as the cross-linking component. Moisture-cure coatings were obtained by reacting the glycols with the polyisocyanate adduct in a ratio of NCO:OH = 2. The films were cured at 80 °C for 2–6 h in contact with moist air. The resulting coatings are characterized by high hardness, good adhesion, and high thermal resistance. The weight losses at 150 °C over a period of 100 h were only 1.6–6.5%, and at 200 °C over a period of 40 h, the weight loss was 15%.

In order to make these coatings more elastic, up to 62.5% of polyethers (based on the equivalent hydroxyl number) were introduced into varnish compositions based on the following organosilicon glycol

$$\text{HOCH}_2\text{-Si-O-Si-CH}_2\text{OH}$$

with CH_3 groups on each Si (CH$_3$ above and CH$_3$ below each Si).

and an isocyanate adduct (196). The polyethers used were a poly(oxypropylene) adduct of glycerol of MW 2875 (DTA–30), poly(oxypropylene) glycol of MW 2000 (D–35), and a poly(oxytetramethylene) glycol, containing 25% hydroxypropylene units, of MW 1400. The dependence of the weight loss in the coatings at 150 °C based on organosilicon glycol, a mixture of the Si–containing glycol with DTA–30 polyether, and the polyether as the sole hydroxy component was determined (196). It was found that the tensile strength of the films increased with the introduction of these organosilicon glycols into polyurethane coatings as compared to films based solely on polyethers as the hydroxy component. The tensile strength values changed with the chemical nature of the organosilicon glycols, for example, with the length of the aliphatic chain between the silicon atom and the hydroxy groups. A significant increase in the tensile strength of the films was observed when an organosilicon glycol with n = 4 was introduced.

Chip–Resistant Urethane Coatings. Polyurethane coating compositions have recently been used by the automotive industry to protect the lower body side of vehicles from stone chipping and corrosion. These coatings are usually applied at relatively high thickness (usually in excess of 4 mils) on top of electrocoated primer and below conventional spray primer and enamel. In the past, vinyl coatings have been used primarily for this type of application. These coatings were used at 20–mil thickness with an airless gun and their application required "masking" and "taping" of the upper portion of the vehicle. Polyurethane chip–resistant coatings have a distinct advantage over vinyl coatings with regard to smoothness and overall appearance characteristics and due to the fact that they require no masking and taping. The composition of these chip-resistant polyurethanes is based on a flexible polyether backbone terminated with blocked isocyanate groups that is cured with a diamine. Similar compositions are described in the work of Berndt et al. (202).

Vapor Permeation Curing. This technique is based on the reaction of a reactive isocyanate group in NCO-terminated polymers or adducts with a hydroxyl group containing compound or resin cured with either ammonia or an amine present in the vapor phase.

McInnes and Bolton (203) disclosed the use of ammonia and primary monofunctional or difunctional aliphatic amines in vapor form for the rapid curing of isocyanate-terminated prepolymers on various substrates, primarily for use in printing inks but made mention also for use in coatings, adhesives, and caulking compounds.

Taft et al. (204) found that the reaction of an aromatic hydroxyl (phenolic) functional polymer with an isocyanate can be significantly promoted by means of a tertiary amine catalyst. While a phenolic resin admixed with a polyisocyanate without a catalyst is relatively stable for a period of 2–48 h depending upon the reactivity of the resulting isocyanate prepolymer, when exposed as a film (40–100 μm) to a gaseous tertiary amine atmosphere, it cures almost instantaneously. The presence of o-methylol groups (as in saligenin or benzylic ether functionality in phenolic resins) increases the hydrogen-bonding polarization of the phenolic OH bond by a tertiary amine such as triethylamine, thus forming a stronger nucleophile available for attack on the electrophilic isocyanate group. Taft et al. (204) also found that certain benzylic ether phenolic resins prepared from phenol and formaldehyde by means of metallic ion catalysts (205) were ideally suited for vapor-permeating curing (VPC). A typical VPC coating process consists in passing the coated film panels from solvent solutions through an enclosed apparatus with a conveyor that contains a saturated atmosphere of a tertiary amine, for example, triethylamine in air or nitrogen. The tertiary amine vapors can be circulated in an an enclosed air space. Cure cycles depend upon the rate of permeation of the amine vapor through the film that varies also with the thickness of the film as well as the type of isocyanate used (aromatic vs. aliphatic). The major advantages of VPC curing is the relatively low cost capital equipment and room temperature curing at fairly high speeds (at least for aromatic isocyanate based systems) for both clear and pigmented films.

Coatings from Urethane Interpenetrating Polymer Networks (IPNs).
Interpenetrating polymer networks (206–211) are relatively novel types of polymer alloys consisting of two or more cross-linked polymers. They are more or less intimate mixtures of two or more distinct, cross-linked polymer networks held together by permanent entanglements with only accidental covalent bonds between the polymers, for example, they are polymeric "catenanes" (212, 213). They are produced either by swelling a cross-linked polymer with monomer and cross-linking agent of another polymer and curing the swollen polymer in situ (209–211, 214) or by blending the linear polymers, prepolymers, or monomers in some liquid forms (latex (207,

<u>208</u>, <u>215</u>, <u>216</u>), solution (<u>217</u>, <u>218</u>), or bulk (<u>219</u>, <u>220</u>)), together with cross-linking agents, evaporating the vehicle (if any), and curing simultaneously the component polymers.

IPNs synthesized to date exhibit varying degrees of phase separation dependent principally on the compatibility of the polymers. Complete compatibility is not necessary to achieve phase mixing, since the permanent entanglements (catenation) can effectively prevent phase separation. The combination of varied chemical types of polymeric networks in different compositions, often resulting in controlled different morphologies, has produced IPNs with synergistic behavior. For example, if one polymer is glassy and the other is elastomeric at room temperature, one obtains either a reinforced rubber or a high-impact plastic depending on which phase is continuous (<u>14</u>). In the case of more complete phase mixing, enhancement in numerous mechanical properties is due to the increased physical cross-link density due to this interpenetration. IPNs have been synthesized with intermediate maximum vs. network composition occurring in bulk for many physical and thermal properties.

Frisch et al. (<u>207</u>, <u>208</u>, <u>217-219</u>, <u>223</u>, <u>224</u>) have prepared many different types of urethane-based IPN coatings and adhesives employing the simultaneous polymerization technique (SIN-IPNs). The urethane polymers usually constituted the rubbery networks while the glassy networks consisted of polyacrylates or methacrylates, epoxides, unsaturated polyesters, polystyrene, etc. In all of the full IPN films and coatings examined so far, a maximum in tensile strength was observed as well as a significant enhancement in impact strength. Some of the IPNs examined also exhibited greatly improved thermal resistance as determined by TGA analysis, although this phenomenon was not observed for each IPN studied. Barrier and surface properties of IPN films consisting of aliphatic isocyanate based urethanes and epoxies were also investigated (<u>225</u>, <u>226</u>).

Sperling et al. (<u>227</u>) have developed a two-layer coating system termed "Silent Paint" that is capable of attenuating noise and vibration over a broad temperature range. It consists of an undercoat damping material and a stiff overcoat that behaves as a constraining layer. The damping undercoat layer is composed of a methacrylic/acrylic, semicompatible IPN in latex form that contains thickeners and additives to impart antioxidant and mildew resistant properties. The constraining layer employed consists of an epoxy resin with 5% by weight of Fibex fibers (E. I. Du Pont de Nemours & Co.) to impart a higher modulus. The damping system consists of a composite of these two layers. This system exhibits significant damping behavior from −30 to over +70 °C and can be used in the form of coatings over various substrate configurations.

Literature Cited

1. Bayer, O. <u>Farbe and Lacke</u> 1958, <u>64</u>, 235.
2. Pratt, B. C.; Rothrock, H. S. U. S. Patent 2 358 475, 1944.
3. Bayer, O. <u>Angew. Chem.</u> 1947, <u>A59</u>, 257.
4. Petersen, S. <u>Ann. Chem.</u> 1949, <u>562</u>, 205.
5. Pansing, H. E. <u>Off. Dig</u>. 1958, <u>30</u>(396), 37.
6. Robinson, E. B.; Waters, R. B. <u>J. Oil Colour Chem. Assoc.</u>
    1951, <u>34</u>, 361.

7. Sanderson, F. T. Am. Paint J. 1965, 50(24), 80.
8. Bailey, M. E. SPE-J. 1958, 14(2), 41.
9. Metz, H. M.; Ehrlich, A.; Smith, M. K.; Patton, T. C. Paint, Oil Chem. Rev. 1958, 121(8), 6.
10. Patton, T. C.; Metz, H. M. Off. Dig. 1960, 32(421), 202.
11. Wells, E. R. Paint Varnish Prod. 1961, 51(3), 41.
12. Bieneman, R. A. In "Polyurethane Technology"; Bruins, P. F., Ed.; Wiley-Interscience: New York, 1969.
13. Manufacturing Chemist's Association. Chemical Safety Data Sheet 5D-73, "Properties and Essential Information for Safe Handling and Use of Tolylene Diisocyanate," Washington, D.C., 1959.
13A. The Upjohn Co., Chemicals Division. "Precautions for the Proper Usage of Polyurethanes, Polyisocyanurates and Related Materials," Technical Bulletin 107, 2nd revised ed., Kalamazoo, MI, Jan 1981.
14. Schollenberger, C. S.; Dinbergs, K. SPE Trans. 1961, 1(1), 31.
15. Schollenberger, C. S.; Scott, H.; Moore, G. R. Rubber World 1958, 137, 549.
16. Beachell, H. C.; Ngoc Sun, C. P. J. Appl. Polym. Sci. 1964, 8, 1089.
17. Wilson, C. L. U.S. Patent 2 921 866, 1960.
18. Pyman, F. L. J. Chem. Soc. 1913, 103, 852.
19. Zengel, H. G. "Cellular and Non-Cellular Polyurethanes," International Conference SPI and FSK, Strasbourg, France, June 1980, Preprint Book, Carl Hanser Verlag, pp. 315-25.
20. Wong, S. W.; Frisch, K. C. In "Advances in Urethane Science and Technology"; Frisch, K. C.; Klempner, D., Eds.; Technomic: Westport, Conn., 1981; Vol. 8, p. 75.
21. Britain, J. W.; Gemeinhardt, P. G. J. Appl. Polym. Sci. 1960, 4, 207.
22. Wagner, K.; Mennicken, G. FATIPEC 1962, 6, 289-92.
23. Mennicken, G. J. J. Oil Colour Chem. Assoc. 1966, 49, 639.
24. Gruber, H. J. J. Oil Colour Chem. Assoc. 1965, 48, 1069.
25. Bunge, W.; Mielke, K. H.; Moller, F. U.S. Patent 2 886 555, 1959.
26. Cross, J. M.; Metzger, S. H.; Campbell, C. D. Br. Patent 1 037 340, 1966.
27. Kaplan, M.; Wooster, G. J. Paint Technol. 1959, 41, 551.
28. Anonymous. Chem. Week, 1967, 10, 13.
29. Henkel Corp., Bulletin CDS-3-65, July 23, 1965.
30. Kanzawa. T.; Naito, K. Jpn. Chem. Q. 1967, 3, 38.
31. Frisch, K. C.; Rumao, L. P.; Mickiewicz, A. J. J. Paint Technol. 1970, 42, 461.
32. Tanaka, M.; Nasu, K. In "International Progress in Urethanes"; Ashida, K.; Frisch, K. C., Eds.; Technomic: Westport, Conn, 1980; Vol. 2, pp. 27-39.
33. Frisch, K. C.; Lin, T. B.; Driscoll, G. paper presented at the Society of Coating Technology Meeting, St. Louis, Mo., Oct 1979.
34. Wright, H. J.; Harwell, K. E. J. Appl. Polym. Sci. 1976, 20, 3305.
35. Frisch, K. C., Jr.; Bozzelli, J. W.; Thomas, M. R. paper presented at the Polyurethane Symposium, ACS Division of Organic Coatings and Plastics, Las Vegas, Nev., Aug 1981.

36. Regulski, T.; Thomas, M. R. ACS Preprints 1983, 48, 998. Ibid. 1983, 48, 1003.
37. Arendt, V. D.; Logan, R. E.; Saxon, R. J. Cell. Plast. 1982, 18, 376.
38. Blair, J. S.; Smith, G. E. P. J. Am. Chem. Soc. 1934, 56, 907.
39. Frentzel, W. Ber. 1888, 21, 411.
40. Lyons, J. M.; Thompson, R. H. J. Chem. Soc 1950, 1971.
41. Raiford, L. C.; Freyermuth, H. B. J. Org. Chem. 1943, 8, 230.
42. Arnold, R. G.; Nelson, J. A.; Verbanc, J. J. Chem. Rev. 1947, 57, 47.
43. Bayer, O. Bios Final Report No. 719, "Interview with Professor Otto Bayer," July 1946.
44. Farbenfabriken Bayer. Br. Patent 1 153 815, 1969.
45. Hofmann, A. W. Ber. 1885, 18, 764.
46. Michael, A. Ber. 1905, 38, 22.
47. Balon, W. J. U.S. Patent 2 801 244, 1957.
48. Havekoss, H. "Polymerization of Diisocyanates", Q.P.B. Report 73894, Frames 4709–18, Apr 10, 1942.
49. Hofmann, A. W. Ber. 1970, 3, 761.
50. Laakso, T. M.; Reynolds, D. D. J. Am. Chem. Soc. 1957, 79, 5717.
51. Kleiner, H.; Havekoss, H.; Spulak, F. V. German Patent 872 618, 1941.
52. Jones, I. I.; Savill, N. G. J. Chem. Soc. 1957, 4392.
53. Kresta, J. E.; Shen, C. S.; Frisch, K. C. Makromol. Chem. 1977, 178, 2495.
54. Kresta, J. E.; Lin, I. S.; Hsieh, K. H.; Frisch, K. C. ACS Org. Coat. Plast. Preprints 1979, 40, 911.
55. Kresta, J. E.; Chang, R. J.; Kathiriya, S.; Frisch, K. C. Makromol. Chem. 1979, 180, 1081.
56. Bechara, I.; et al. U.S. Patent 3 892 687, 1975; U.S. Patent 3 993 652, 1976; U.S. Patent 4 040 992, 1977; U.S. Patent 4 116 879, 1978 (all to Air Products & Chemicals, Inc.).
57. Bechara, I.; Mascoli, R. J. Cell. Plast. 1979, 15(6), 321.
58. Nicholas, L.; Gmitter, G. T. J. Cell. Plast. 1965, 1, 85.
59. Kubitza, W. Paint Varnish Prod. 1969, 59(12), 36.
60. Sandler, S. R. J. Appl. Polym. Sci. 1967, 11, 811.
61A. Wicks, Z. W., Jr. Prog. Org. Coat. 1975, 3, 73–99.
61B. Kordomenos, P. I.; Dervan, A. H.; Kresta, J. E. J. Coat. Technol. 1982, 54, 43.
62. Wicks, Z. W., Jr.; Kostyk, B. W. J. Coat. Technol. 1977, 49, 77.
63. Wicks, Z. W., Jr.; Wu, K. H. J. Org. Chem. 1980, 45, 2446.
64. Kurita, K.; Imajo, H.; Iwakura, Y. J. Org. Chem. 1978, 43, 2918.
65. Wong, S. W.; Damusis, A.; Frisch, K. C.; Jacobs, R. L.; Long, J. W. J. Elast. Plast. 1979, 11, 15.
66. Gras, R.; Oberdorf, J.; Wolf, E. German Offen. 2 729 704, 1979; German Offen. 2 738 270, 1979; German Offen. 2 777 421, 1979. Gras, R. German Offen. 2 830 590, 1980.
67. Ulrich, H.; Tucker, B. W.; Richter, R. H. J. Org. Chem. 1978, 43, 1544.
68. Richter, R. H.; Tucker, B. W.; Ulrich, H. U.S. Patent 4 154 931, 1979.

69. Ulrich, H. ACS Div. Org. Coat. Plast. Chem., Preprints 1980,
     43, 950.
70. Mijs, W. J.; Reesink, J. B.; Groenenboom, J. C.; Vollmer, J. P.
     J. Coat. Technol. 1978, 50, 58.
71. Wyandotte Chemicals Corp., British Patent 994 348; Chem. Abstr.
     1965, 63, 5904.
72. Levy, J. F.; Kusean, J. U.S. Patent 3 705 119; Chem. Abstr.
     1973, 78, 59887.
73. Altynbaev, I. A.; Borisov, S. F.; Rogov, N. G.; Romanova,
     E. G.; Fedoseev, M. S. USSR Patent 324 250, 1971; Chem. Abstr.
     1972, 77, 35556.
74. Duncan, J. S.; Elmer, O. C. U.S. Patent 3 668 186; Chem.
     Abstr. 1972, 77, 76117.
75. Morak, A. J.; Ward, K., Jr.; Johnson, D. C. TAPPI 1970, 53,
     2278.
76. Norgami, S. Japanese Patent 6 805 221; Chem. Abstr. 1968, 69,
     20029.
77. Peters, D. E.; Reddie, R. N. Angew. Makromol. Chem. 1975, 75,
     99.
78. Guise, G. B.; Jackson, M. B.; Maclaren, J. A. Aust. J. Chem.
     1972, 25, 2583.
79. Guise, G. B. J. Appl. Polym. Sci. 1977, 21, 3427; Polym. News
     1981, 7, 149.
80. De Boos, A. G.; White, M. A. Melliand Textilber. 1980, 61, 267.
81. Murayama, S.; Nishizawa, H.; Ebisawa, K.; Tanuma, T.; Taraka,
     S. Japanese Kokai 1976, 75, 139 829; Chem. Abstr. 1976, 84,
     61331.
82. Nishizawa, H.; Tanaka, S.; Hashiya, K. Japanese Kokai, 1978,
     78, 37 663; Chem. Abstr. 1978, 89, 112584.
82A. Kordomenos, P. I.; Nordstrom, J. D. J. Coat. Technol. 1982, 54,
     33.
83. Matsumaga, Y.; Hoshino, Y.; Kobayashi, Y. U.S. Patent
     3 823 118, 1974.
83A. Pampouchidis, A. G.; Honig, H. U.S. Patent 4 179 425, 1979;
     U.S. Patent 4 176 099, 1979.
83B. Honig, H.; Pampouchidis, G. U.S. Patent 4 174 332, 1979.
84. Jerabek, R. D.; Marchetti, J. R. U.S. Patent 3 947 338, 1976;
     U.S. Patent 3 935 087, 1976; Jerabek, R. D. U.S. Patent
     4 031 051, 1977; U.S. Patent 3 799 854, 1974.
85. Bosso, J. F.; Sturni, L. C. U.S. Patent 4 101 486, 1976.
86. Marchetti, J. R.; Jerabek, R. D. U.S. Patent 3 975 250, 1976.
     Jerabek, R. D. U.S. Patent 3 925 180, 1975.
87. Wismer, M.; Pierce, P. E.; Bosso, J. F.; Christenson, R. M.;
     Jerabek, R. D.; Zwack, R. R. ACS Div. Org. Coat. Plast.
     Chem., Preprints 1981, 45, 1.
88. Comstock, L. R.; Milligan, C. L.; Monter, R. P. J. Paint
     Technol. 1972, 44(573), 63.
89. Comstock, L. R.; Gerkin, R. M.; Milligan, C. L.; Monter, R. P.
     J. Paint Technol. 1972, 44(574), 75.
90. Dieter, J. A.; Frisch, K. C.; Shanafelt, G. K.; Devanney, M. T.
     Rubber Age 1974, 106(7), 49.
91. Neumann, W.; Peter, J.; Holtschmidt, H.; Kallert, W. Fourth
     International Rubber Technology Conference, London, May 1962.
92. Rausch, K. W., Jr.; McClellan, T. R.; d'Ancicco, V. V.; Sayigh,
     A. A. R. Rubber Age 1967, 99(3), 78.

93. Müller, E. Angew. Makromol. Chem. 1970, 14, 75.
94. Saunders, J. H.; Frisch, K. C. "Polyurethanes"; Interscience: New York, 1962; Part I.
95. Simons, D. M.; Verbanc, J. J. J. Polym. Sci. 1960, 44, 303.
96. St. Pierre, L. E.; Price, C. C. J. Am. Chem. Soc. 1956, 78, 3432.
97. Lundsted, L. G. U.S. Patent 2 674 619, 1954.
98. Meerwein, H.; Delfs, D.; Morschel, H. Angew. Chem. 1960, 72, 927.
99. Murbach, W. J.; Adicoff, A. Ind. Eng. Chem. 1960, 52, 772.
100. Lundsted, L. G.; Schulz, W. F. U.S. Patent 2 697 118, 1954.
101. Hayashi, K.; Marvel, C. S. J. Polym. Sci. 1964, A2, 2571.
102. Verdol, J. A.; Carrow, D. J.; Ryan, P. W.; Kuncl, K. L. Rubber Age 1966, 98(7), 570.
103. Lapp, R. L.; Serniuk, G. E.; Minckles, L. S. Rubber Chem. Technol. 1970, 43(5), 1154.
104. Ivan, B.; Kennedy, J. P.; Chang, V. S. C. J. Polym. Sci., Polym. Chem. Ed. 1980, 18, 3177.
105. Chang, V. S. C.; Kennedy, J. P. Polym. Bull. 1982, 8, 69.
106. Stamberger, P. U.S. Patent 3 304 273, 1967.
107. Von Bonin, W.; Piechota, H. British Patent 1 040 452, 1966.
108. Kuryla, W. C.; Critchfield, F. E.; Platt, L. W.; Stamberger, P. J. Cell. Plast. 1966, 2, 84.
109. Critchfield, F. E.; Koleske, J. V.; Priest, D. C. Rubber Chem. Technol. 1972, 45, 1467.
110. BASF Wyandotte Corp. Urethane Applications Research Bulletin, "Pluracol Polyol 637. A Graft Polyol for Slab Stock Foam," Jan 8, 1974.
111. Phillips, B. A.; Taylor, R. P. paper presented at the Rubber Division, ACS Meeting, Atlanta, GA, March 1979.
112. Damusis, A. U.S. Patents 3 049 515 and 3 049 516, 1962.
113. Damusis, A. U.S. Patent 3 049 514, 1962.
114. Damusis, A. U.S. Patent 3 105 063, 1963.
115. Damusis, A.; McClellan, J. M.; Frisch, K. C. Ind. Eng. Chem. 1959, 51, 1386.
116. Damusis, A.; McClellan, J. M.; Frisch, K. C. Off. Dig. 1960, 32, 251.
117. Frisch, K. C. Rubber Chem. Technol. 1980, 53(1), 126.
118. Farkas, A.; Mills, G. A. In "Advances in Catalysis"; Academic: New York, 1962; Vol. 13.
119. Saunders, J. H.; Frisch, K. C. In "Polyurethanes"; Wiley-Interscience: New York, 1962; Part I, Chap. IV.
120. Frisch, K. C.; Rumao, L. P. J. Macromol. Sci. Rev., Macromol. Chem. 1971, C5, 105.
121. Ephraim, S.; Woodward, A. E.; Mesrobian, R. B. J. Am. Chem. Soc. 1958, 80, 1326.
122. Bieneman, R. A.; Baldin, E. J.; Markoff, M. K. Off. Dig. 1960, 32, 273.
123. Remington, W. J. E. I. Du Pont de Nemours & Co., Elastomer Chemicals Department, "Pigmenting Prepolymer-Type Urethane Coatings," Aug 1961.
123A. Buchachenko, A. L.; Golubev, V. A.; Medzhidar, A. A.; Rozantsev, E. G. Teor. Eksp. Khim. 1965, 1, 249.

123B. Berner, G.; Kreibich, V. A. "6th International Conference in Organic Coatings Science and Technology"; Parfitt, G. D., and Patsis, A. V., eds.; Technomic Publishing: Westport, Conn., 1982; p. 334.

124. Hill, H. E. Off. Dig. 1964, 36(468), 64.

125. Mathur, G. N.; Kresta, J. E.; Frisch, K. C. In "Advances in Urethane Science and Technology"; Frisch, K. C.; Reegen, S. L., Eds.; Technomic: Westport, Conn., 1978; Vol. 6, 103.

126. Damusis, A.; Frisch, K. C. In "Film Forming Compositions"; Myers, R. R.; Long, J. S., Eds.; Marcel Dekker: New York, 1967; Part I.

127. Saunders, J. H.; Frisch, K. C. "Polyurethanes"; Wiley-Interscience: New York, 1964; Part II, Chap. X.

128. Johnson, K. "Polyurethane Coatings"; Noyes Data Corp.: Parkridge, N.J., 1972.

129. Buist, J. M.; Gudgeon, H. "Advances in Polyurethane Technology"; Wiley-Interscience: New York, 1968,

130. Chang, W.; Scriven, R.; Peffer, J.; Porter, S. Ind. Eng. Chem. Prod. Res. Dev. 1973, 12(4), 278.

131. O'Shaughnessy, F.; Hoeschle, G. K. Rubber Chem. Technol. 1971, 44, 52.

132. Brennan, J. P. British Patent 1 215 922, 1970.

133. Farbenfabriken Bayer. Netherlands Patent 7 104 911, 1971.

134. Trapasso, L. U.S. Patent 3 627 735, 1971.

135. Nazy, J. R.; Stoker, K. B. U.S. Patent 3 493 543, 1970.

136. Haggin, G. A.; Kengon, R. W.; Kerrigen, V. U.S. Patent 3 567 692, 1971.

137. Scheibelhoffer, A. S. U.S. Patent 3 547 127, 1971.

138. Jackson, W. E. U.S. Patent 3 645 907, 1972.

139. Harada, K.; Mizol, Y.; Furukawa, J.; Yamashita, S. Makromol. Chem. 1970, 132.

140. Allied Chemical Corp. British Patent 1 090 449, 1967.

141. Kaplan, M. In "Advances in Urethane Science and Technology"; Frisch, K. C.; Reegen, S. L., Eds.; Technomic: Stamford, Conn, 1971; Vol. 1, Chap. 8.

142. Altscher, S. Am. Dyestuff Reporter 1965, 54(24), 32.

143. Axelrod, S. L. U.S. Patent 3 148 173, 1964; U.S. Patent 3 281 397, 1966; U.S. Patent 3 294 724, 1966.

144. McClellan, J. M.; MacGugan, I. C. Rubber Age. 1967, 100(3), 66.

145. McClellan, J. M.; Axelrood, S. L.; Grace, O. M. U.S. Patent 3 410 817, 1968.

146. Dieterich, D.; Keberle, W.; Witt, H. Angew. Chem. 1970, 82(2), 53.

147. Dieterich, D.; Keberle, W.; Wuest, R. J. J. Oil Colour Chem. Assoc. 1970, 53, 363.

148. Dieterich, D. Angew. Makromol. Chem. 1981, 98, 133.

149. Suskind, S. P. J. Appl. Polym. Sci. 1965, 9, 2451.

150. Hill, F. B., Jr. U.S. Patent 2 726 219, 1955.

151. Mallonee, J. E. U.S. Patent 2 968 575, 1961.

152. Rembaum, A. J. Macromol. Sci. 1969, A3(1), 87.

153. Rembaum, A.; Baumgarten, W.; Eisenberg, A. J. Polym. Sci. 1968, B6, 159.

154. Somoano, R.; Yen, S. P. S.; Rembaum, A. J. Polym. Sci, 1970, B8, 467.

155. Rembaum, A.; Rile, H.; Somoano, R. J. Polym. Sci. 1970, B8, 457.
156. Rembaum, A.; Yen, S. P. S.; Landel, R. F.; Shen, M. J. Macromol. Sci. 1970, A4(3), 715.
157. Taft, D. D.; Mohar, A. F. J. Paint Technol. 1970, 42, 674.
158. Dieterich, D.; Bayer, O.; Peter, J. German Patent 1 184 946, 1962.
159. Dieterich, D.; Bayer, O. British Patent 1 078 202, 1965.
160. Keberle, W.; Dieterich, D. British Patent 1 076 688, 1966.
161. Keberle, W.; Dieterich, D.; Bayer, O. German Patent 1 237 306, 1964.
162. Keberle, W.; Dieterich, D. British Patent 1 076 909, 1966.
163. Dieterich, D.; Müller, E.; Bayer, O. German Patent 1 178 586, 1962; German Patent 1 179 363, 1963.
164. Keberle, W.; Müller, E. British Patent 1 146 890, 1969.
165. Witt, H.; Dieterich, D. German Patent 1 282 962, 1966.
166. Dieterich, D.; Reiff, H. Angew. Makromol. Chem. 1972, 26, 85; Adv. Urethane Sci. Technol. 1976, 4, 112.
167. Neumaier, H. H. J. Coat. Fabrics 1974, 3, 181,
168. Taub, B. Recent Advances in Coatings Conference, State University of New York, New Paltz, and Polymer Institute, University of Detroit, Lake Mohonk, N.Y., April 1982.
169. Bonk, H. W.; Ulrich, H.; Sayigh, A. A. R. J. Elastoplast. 1974, 4, 259.
170. Bieneman, R. J. Paint Technol. 1971, 43, 92.
171. Chem. Week. 1974, Nov 27.
172. Dhein, R.; Rudolph, H.; Kreuder, H.; Gebauer, H. German Patent 1 957 483, 1971.
173. Damusis, A.; Yoon, H. K.; Frisch, K. C. unpublished results.
174. Beck & Co. French Patent 2 033 348, 1970.
175. Johnson, O. B.; Labana, S. S. U.S. Patent 3 660 143, 1972.
176. Kehr, C. L.; Wszolek, W. R. ACS Org. Coat. Plast. Preprints 1973, 33(1), 295.
177. Thomas, D. C. U.S. Patent 3 655 625, 1972.
178. Chang, W. H. Canadian Patent 3 655 625, 1972.
179. Smith, O. W.; Weizel, J. E.; Trecker, D. J. German Patent 2 103 870, 1971.
180. Tu, S. T. "Recent Advances in Radiation Curing," The 1978 Modern Engineering Technology Seminar, Taiwan, China, July 1978.
181. Frisch, K. C.; Kresta, J. E. In "International Progress in Urethanes"; Technomic: Westport, Conn., 1977; Vol. 1, p. 191.
182. Ulrich, H. J. Cell. Plast. 1981, 17(1), 31.
183. Frisch, K. C.; Ashida, K. U.S. Patents 3 792 236 and 3 817 938, 1974 (to Mitsubishi Chemical Industry, Ltd.).
184. Kordomenos, P. I.; Frisch, K. C.; Kresta, J. E. J. Coat. Technol. 1983, 55(700), 49. Ibid., p. 59.
185. Kordomenos, P. I.; Frisch, K. C.; Ashida, K. U.S. Patent 4 066 624, 1978 (to Mitsubishi Chemical Industry, Ltd.).
186. Burk, E. H., Jr.; Wolgemuth, L. G.; Kutts, H. W. U.S. Patent 3 531 425, 1970.
187. Burk, E. H., Jr.; Carlos, D. D. U.S. Patent 3 423 449, 1969.
188. Frisch, K. C.; Reegen, S. L.; Dieter, J. A. U.S. Patent 3 813 365, 1974.

189. Slagel, R. C.; Bloomquist, A. E. Can. J. Chem. 1967, 45, 2625.
190. Culbertson, B. M.; Slagel, R. C. J. Polym. Sci, 1968, A1(6), 363.
191. Culbertson, B. M.; Sedor, E. A.; Slagel, R. C. Macromol. 1968, 1, 254.
192. Andrianov, K. A.; Astakhin, V. V. Zhur. Obshch. Khim. 1959, 29, 2698; Chem. Abstr. 1960, 54, 10835.
193. Andrianov. K. A.; Makarova, L. I. Vysokamol. Soedin. 1961, 3, 996; Chem. Abstr. 1962, 56, 4914.
194. Andrianov, K. A.; Makarova, L. I.; Zharkova, N. M. Vysokomol. Soedin, 1960, 2, 1378; Chem. Abstr. 1962, 55, 21643.
195. Andrianov, K. A.; Pakhomov, V. I.; Lapteva, N. E. Plast. Massy 1961, No. 11, 17; Chem. Abstr. 1962, 56, 10177.
196. Smetankina, N. P.; Karbovskaya, L. E. Zhur. Obshch. Khim. 1968, 38, 911; Chem. Abstr. 1968, 69, 77331.
197. Greber, G.; Jager, S. Makromol. Chem. 1962, 57, 150.
198. French Patent 1 352 325, 1964 (to Dow Corning Corp.).
199. Speier, J. L. U.S. Patent 2 527 590, 1950 (to Dow Corning Corp.).
200. Upson, R. W. U.S. Patent 2 511 310, 1950 (to E. I. Du Pont de Nemours & Co.).
201. Paint Manuf. 1963, 33(7), 261.
202. Berndt, G.; et al. U.S. Patent 4 299 868, 1981.
203. McInnes, A. D.; Bolton, R. J. U.S. Patent 3 874 898, 1975 (to A. C. Hatrick Chemicals Pty., Ltd.).
204. Taft, D.; Robins, J.; Bayne, S. C.; Poluzzi, A. FATIPEC Congr. 1972, 11, 335.
205. Robins, J. U.S. Patent 3 485 797, 1969 (to Ashland Oil Co.).
206. Millar, J. R. J. Chem. Soc. 1960, 1311.
207. Frisch, H. L.; Klempner, D.; Frisch, K. C. J. Polym. Sci., Polym. Lett. 1969, 7, 775.
208. Frisch, H. L.; Klempner, D.; Frisch, K. C. J. Polym. Sci. (A2), 1970, 8, 921.
209. Sperling, L. H.; Friedman, D. W. J. Polym. Sci. (A2), 1969, 7, 425.
210. Sperling, L. H.; Taylor, D. W.; Kirkpatrick, M. L.; George, H. F.; Bardman, D. R. J. Appl. Polym. Sci. 1970, 14, 73.
211. Sperling, L. H.; George, H. F.; Huelck, V.; Thomas, D. A. J. Appl. Polym. Sci. 1970, 14, 2815.
212. Frisch, H. L.; Klempner, D. Adv. Macromol. Chem. 1970, 2, 149.
213. Frisch, H. L.; Wasserman, E. J. Am. Chem. Soc. 1961, 83, 3789.
214. Sperling, L. H.; Thomas, D. A.; Covitch, M. J.; Curtius, A. J. Polym. Eng. Sci. 1972, 12, 101. Sperling, L. H.; Thomas, D. A.; Huelck, V. Macromol. 1972, 5, 340.
215. Klempner, D.; Frisch, H. L.; Frisch, K. C. J. Elastoplast. 1971, 3, 2.
216. Sperling, L. H.; Thomas, D. A.; Lorenz, J. E.; Nagel, E. J. J. Appl. Polym. Sci, 1975, 19, 2225.
217. Frisch, K. C.; Frisch, H. L.; Klempner, D.; Mukherjee, S. K. J. Appl. Polym. Sci. 1974, 18, 689. Frisch, K. C.; Klempner, D.; Antczak, T.; Frisch, H. L. J. Appl. Polym. Sci. 1974, 18, 683.

218. Frisch, K. C.; Klempner, D.; Migdal, S.; Frisch, H. L.; Ghiradella, H. Polym. Eng. Sci. 1974, 14, 76. Frisch, H. L.; Frisch, K. C.; Klempner, D. Polym. Eng. Sci. 1974, 14, 562.

219. Kim, S. C.; Klempner, D.; Frisch, K. C.; Frisch, H. L.; Ghiradella, H. Polym. Eng. Sci. 1975, 15, 339.

220. Sperling, L. H.; Arnts, R. R. J. Appl. Polym. Sci 1971, 15, 2731. Touhsaent, R. E.; Thomas, D. A.; Sperling, L. H. J. Polym. Sci. 1974, 46C, 175.

221. Kim, S. C.; Klempner, D.; Frisch, K. C., Radigan, W.; Frisch, H. L.; Macromolecules 1976, 9, 258. Kim, S. C.; Klempner, D.; Frisch, K. C.; Frisch, H. L. Macromolecules, 9, 263. Ibid. 1977, 10, 1187. Ibid. 1977, 10, 1191.

222. Kim, S. C.; Klempner, D.; Frisch, K. C.; Frisch, H. L. J. Appl. Polym. Sci. 1977, 21, 1289.

223. Frisch, K. C.; Klempner, D.; Migdal, S.; Frisch, H. L.; Dunlop, A. P. J. Polym. Sci, 1975, 19, 1893.

224. Frisch, H. L.; Frisch, K. C.; Klempner, D. Pure Appl. Chem. 1981, 53, 1557.

225. Frisch, H. L.; Klempner, D.; Yoon, H. K.; Frisch, K. C. In "Polymer Alloys II"; Klempner, D.; Frisch, K. C., Eds.; Plenum: New York, 1979; p. 203.

226. Frisch, H. L.; Frisch, K. C. Prog. Org. Coat. 1979, 7, 105.

227. Sperling, L. H.; Chiu, T. W.; Gramlich, R. G.; Thomas, D. A. J. Paint Technol. 1974, 46(588), 47.

Chemistry and Technology of Acrylic Resins for Coatings

W. H. BRENDLEY and R. D. BAKULE

Rohm and Haas Company, Spring House, PA 19477

History of the Synthesis of Acrylic Monomers and Polymers
Relation between the Properties and Composition of Acrylic
    Coating Polymers
  General
  Resistance and Durability Properties
Development of Thermoplastic Acrylic Coatings
Development of Thermosetting Acrylic Coatings
Recent Trends in Acrylic Coatings
  Approaches to Low-Emission Acrylic Coatings
  Aqueous Thermoplastic Emulsions
  Aqueous Thermosetting Emulsions
  Water-Reducible Polymers
  Solubilizable Dispersions
  High-Solids Coatings
  Oligomers/Poligomers
  Powder Coatings
  Radiation-Cured Coatings

For the past 50 years, protective and decorative coatings based on acrylic and methacrylic polymers have found increasing use in a variety of industrial and trade sale applications. This expanded usage has occurred because of the versatile and unique properties of these polymers and because of the development of large-scale commercial processes for a variety of monomers. This paper reviews the history of the development of thermoplastic and thermosetting acrylic polymers, relates their chemical and physical properties to the unique virtues they possess as coatings, and looks into the future of acrylics in coatings. Several examples of the application of acrylic technology to specific market segments--automotive and appliance finishes--illustrate how acrylic polymers can be tailored to meet specific end use requirements.

0097–6156/85/0285–1031$06.50/0
© 1985 American Chemical Society

History of the Synthesis of Acrylic Monomers and Polymers (1-4)

Although acrylic acid was first synthesized in 1843 and polymerized in 1847, the properties of derived monomers and polymers were not extensively investigated until the classic work of Otto Rohm, beginning in 1901. The first general synthetic process for manufacturing acrylic esters was accomplished in 1927 in Germany and was based on the reactions

A. $CH_2=CH_2 + HClO \quad \text{--------->} \quad ClH_2CCH_2OH$

    $ClH_2CCH_2OH + NaCN \quad \text{--------->} \quad NCCH_2CH_2OH + NaCl$

    $NC-CH_2CH_2-OH + ROH + 0.5H_2SO_4 \quad \text{--------->} \quad CH_2=CH-CO_2R + 0.5(NH_4)_2SO_4$

developed by Rohm and his associate Otto Haas. Before the development was completed, however, Haas had immigrated to the United States, where he developed the first commercial production of methyl and ethyl acrylate in the United States in the early 1930s using the ethylene-cyanide process.

    Processes for a variety of acrylic monomers were developed in the ensuing decades by Rohm and Haas and other manufacturers.

Process	Manufacturer

B. $CH_3\overset{\text{O}}{\overset{\|}{C}}CH_3 + HCN \text{ -----> } CH_3\underset{\underset{CN}{|}}{\overset{\overset{OH}{|}}{C}}-CH_3$ Rohm and Haas Co. (1933)

$(CH_3)_2\underset{\underset{CN}{|}}{\overset{\overset{OH}{|}}{C}} + H_2O + ROH + 0.5H_2SO_4 \text{ -----> }$

$(CH_3)_2-\underset{}{\overset{\overset{OH}{|}}{C}}-CO_2R + 0.5(NH_4)_2SO_4$

$(CH_3)_2-\overset{\overset{OH}{|}}{C}-CO_2R \overset{P_2O_5}{\text{----->}} CH_2=\overset{\overset{CH_3}{|}}{C}-CO_2R + H_2O$

C. $HC\equiv CH + ROH + CO \overset{Ni(CO)_4}{\underset{HCl}{\text{------->}}} CH_2=CHCO_2R$ Rohm and Haas Co. (1948)

D. 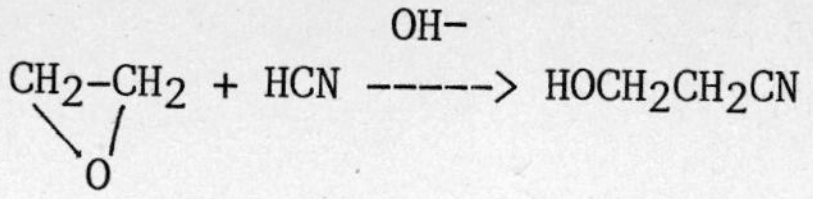 CH_2-CH_2 (epoxide, O bridge) $+ HCN \xrightarrow{OH-} HOCH_2CH_2CN$ Union Carbide (1949)

$HOCH_2CH_2CN + ROH + H^+ \longrightarrow$
$CH_2=CHCO_2R + NH_4^+$

E. $HC\equiv CH + CO + H_2O \xrightarrow{NiX} CH_2=CHCO_2H$ BASF (1956)

F. $CH_2=CHCN \xrightarrow{H_2SO_4} CH_2=CH-CONH_2 \xrightarrow[H_2SO_4]{ROH}$ Ugilor

$CH_2=CH-CO_2R$

G. 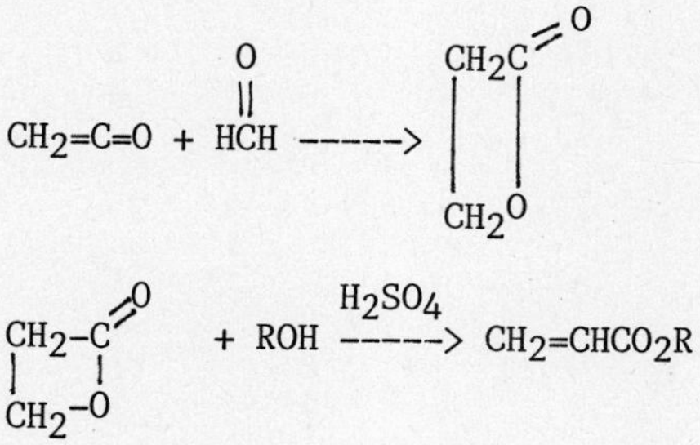 $CH_2=C=O + HCH \text{ (}O\text{)} \longrightarrow$ (β-propiolactone ring: $CH_2C=O$ / CH_2O) Celanese (1958)

(β-propiolactone: CH_2-C (=O), CH_2-O) $+ ROH \xrightarrow{H_2SO_4} CH_2=CHCO_2R$

H. $CH_3C(CH_3)(OH)-CN + H_2SO_4 \longrightarrow CH_2=C(CH_3)-CNH_2 \cdot H_2SO_4$ (with $C=O$) Rohm and Haas Co.

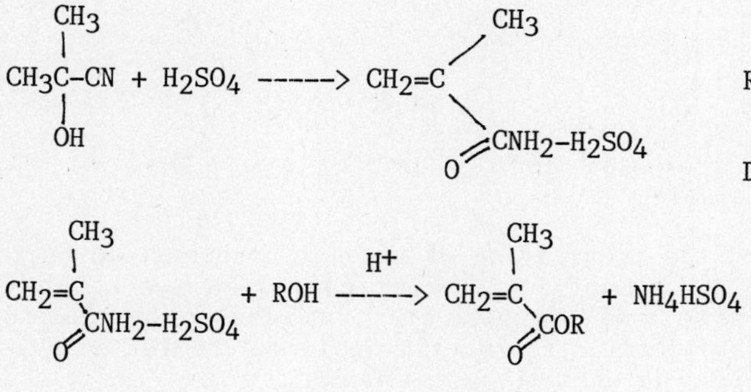

Du Pont

$CH_2=C(CH_3)-CNH_2 \cdot H_2SO_4$ (with $C=O$) $+ ROH \xrightarrow{H^+} CH_2=C(CH_3)-COR$ (with $C=O$) $+ NH_4HSO_4$ Imperial Chemical Industries

I. $CH_2=CHCH_3 + \dfrac{3O_2}{2} \longrightarrow CH_2=CHCO_2H + H_2O$ Union Carbide (1969)

J. $CH_2=C-CH_3$ $\xrightarrow[\text{[O]}]{\text{Mo-Bi-Fe-Li}}$ $CH_2-C\begin{smallmatrix}CH_3\\CHO\end{smallmatrix}$ Nippon Geon (1975)

with CH_3 substituent on the left $CH_2=C-CH_3$

$CH_2-C-CHO$ (with CH_3) $\xrightarrow[\text{[O]}]{MO_{12}P_2VCs_2Sr_{0.5}}$ $CH_2=C-CO_2H$ (with CH_3)

In addition to these primary methods for synthesizing acrylic monomers, many other esters and functional monomers are prepared by further reactions of these monomers. Two common methods for obtaining new monomers are the following.

K. Direct esterification of the acid:

$$CH_2=C\begin{smallmatrix}R_1\\CO_2H\end{smallmatrix} \quad + R_2OH \quad\longrightarrow\quad CH_2=C\begin{smallmatrix}R_1\\CO_2R_2\end{smallmatrix} \quad + H_2O$$

(where R_1 = H or CH_3 and R_2 is alkyl)

L. Transesterification of the lower esters:

$$CH_2=C\begin{smallmatrix}R_1\\CO_2R_2\end{smallmatrix} \quad + R_3OH \quad\longrightarrow\quad CH_2=C\begin{smallmatrix}R_1\\CO_2R_3\end{smallmatrix} \quad + R_2OH$$

(where R_1 = H or CH_3 and R_2 and R_3 are alkyl or other functional groups)

Further details on the preparation of acrylic monomers can be found in References 1, 2, 5, and 6. The essential point is that an extremely wide variety of monomers are available on a commercial scale. This variety permits the acrylic polymer chemist an unusual freedom in designing a polymer to meet a set of end use requirements.

Relation Between the Properties and Composition of Acrylic Coating Polymers

General. The solution and film properties of an acrylic coatings polymer are determined by (1) the molecular weight, (2) the nature of the polymer solution, and (3) the composition of the polymer backbone.

The effect of molecular weight is easy to visualize. Film formation in any solution coating depends either upon the formation of primary chemical bonds (thermosetting coatings) or upon entanglement of the polymer chains by secondary chemical interactions (thermoplastic coatings). In the case of thermoplastic coatings, it is evident that the longer the chains (i.e., the higher the molecular weight) the more thoroughly the chains will be entangled and the more coherent the film will be. As a result, the coating film will be tougher and more resistant to degradation. There is, however, a practical limit. After the molecular weight of the polymer reaches a value of about 90,000, little improvement in properties accrues with increasing molecular weight.

On the other hand, the viscosity of the coatings solution increases expotentially with the molecular weight (8).

$$\ln\,(\eta_p/\eta_s) \simeq KM^a C$$

where η_p and η_s are the viscosities of the polymer solution and solvent, respectively, M equals the molecular weight, C equals the polymer concentration, K equals a constant, and a equals a constant in the range of 0.5–1.5. Because of this sensitive dependence, a balance must be struck between the desirability of high molecular weight and the need to maintain a tractable viscosity at a reasonable applications solids. These considerations do not apply to emulsion coatings, which are dispersions of discrete polymer particles suspended in an aqueous continuum and whose viscosity is determined essentially by the continuum phase.

Acrylic polymers used in coatings are composed primarily of polymethacrylates and polyacrylates:

$$\begin{array}{cc}
CH_3 & H \\
| & | \\
(CH_2-C)_n & (CH_2-C)_n \\
| & | \\
C=O & C=O \\
| & | \\
O & O \\
| & | \\
R & R
\end{array}$$

        Polymethacrylate   Polyacrylate

and their properties are strongly influenced by three factors: (A) the presence of CH_3 or H on the α–carbon, (B) the length of the ester side chain, R, and (C) the presence of functionality in the ester side chain.

(A) The presence of a methyl group in the α–position results in a hinderance to the segmental rotation of the polymer backbone. As a result, polymethacrylates are invariably harder, less extensive polymers than the corresponding acrylates (7) (Table I).

(B) Similarly, as the length of the ester side chain increases, segmental rotations in the side chain, as well as an increase in the specific volume of the polymers, result in freer segmental motion within the polymer chain, with a concomitant decrease in tensile strength and an increase in the extensibility of the polymers (Table I). When one considers that commercial acrylic polymers are almost always copolymers of several monomers, it is obvious that a wide range of strength and flexibility can be achieved.

(C) It is well-known that the presence of certain types of functionality in the ester side chain imparts specific characteristic film properties to the acrylic coating (7). Some of these characteristics are summarized in Table II. By knowing the end use requirements of flexibility, adhesion, hardness, etc., the acrylic coatings polymer chemist can often tailor-make a polymer to fit the defined needs of the application.

The understanding of the relation between the macroscopic film properties of an acrylic coating and the polymer composition has been advanced significantly by the successful application and interpretation of two physical models: (1) the glass transition temperature, T_g; (2) the solubility parameter, δ.

The glass transition temperature, T_g, or more precisely the temperature range, is the temperature at which significant segmental rotation in the backbone and/or side chain is thermally excited. As the temperature of the coating increases and passes through this region, the properties of the polymer film change dramatically. Below the T_g the film is relatively glassy, rigid, and hard. Above the T_g the film becomes more rubbery, flexible, and softer, because the polymer segments can respond to stresses impressed on them.

Because the same molecular process, thermal excitation of segmental motion of the polymer chains, determines both the glass transition temperature and the tensile strength and extensibility of the polymer, it is not surprising to find a close correlation between the T_g and these mechanical properties of polymers (Table I).

Burrell (9) has given an excellent exposition of the application of glass transition temperatures to coatings. One example will illustrate how this concept unifies the interpretation of the performance of various types of coatings films.

In Figure 1, the temperature dependence of the hardness of several thermoplastic and thermosetting films is represented schematically.

The polymer having $T_g < 25$ °C is relatively soft and flexible at ambient conditions, whereas the polymer having $T_g > 25$ °C tends to be hard and brittle. The introduction of cross-linking will have a minimal effect on hardness at temperatures less than T_g since segmental motion does not occur to any great extent. On the other hand, at temperatures above T_g, cross-linking significantly increases the film hardness. By understanding the factors influencing the T_g of an acrylic polymer, one can correlate and predict a variety of properties of actual coating systems (10-12).

The solubility parameter concept invented by Hildebrand (13) and applied to polymers by Burrell (14) has also proved invaluable in understanding, correlating, and predicting the solubility and compatibility of acrylic polymers. The solubility parameter, δ, is a consequence of regular solution theory, which assumes that the entropy of mixing the components of a solution is the same as that for an ideal solution, for example, mixing is random, and that the enthalpy (heat) of mixing can be calculated from the model by making simplifying assumptions about the nature of the molecular interactions involved. The theory predicts that a polymer will be miscible with another polymer, or solvent, if the enthalpy of mixing

Table I. Properties of Polymethacrylates and Polyacrylates

Compound	Tensile Strength (psi)	Elongation (%)	T_g (°C)
Polymethacrylate			
Methyl	9000	4	105
Ethyl	5000	7	65
Isobutyl	3500	5	48
n–Butyl	1000	250	20
Polyacrylate			
Methyl	1000	750	9
Ethyl	33	1800	−22
n–Butyl	3	2000	−54

Table II. Effect of Various Monomers on Film Properties

Film Property	Contributing Monomers
Exterior durability	Methyacrylates and acrylates
Hardness	Methyl methacrylate; styrene; methacrylic and acrylic acid
Flexibility	Ethyl acrylate; butyl acrylate; 2-ethylhexyl acrylate
Stain resistance	Short–chain methacrylates
Water resistance	Methyl methacrylate; styrene; long–chain methacrylates and acrylates
Mar resistance	Methacrylamide; acrylonitrile
Solvent and grease resistance	Acrylonitrile; methacrylamide; methacrylic acid
Adhesion to metals	Methacrylic/acrylic acid

of the two materials is essentially zero. This condition is guaranteed if the solubility parameters of the components are equal. Thus, polymers and solvents having similar solubility parameters are miscible. The practical consequence of this can be illustrated by using the data in Tables III and IV.

Polymers containing long alkyl side chains are likely to have good resistance to water and alcohol since the solubility parameters of the polymer and solvents are quite different. Conversely, polar polymers, such as polyacrylonitrile, are predicted to show good resistance to attack by aliphatic hydrocarbons. By the same token, the longer alkyl chain acrylics are expected to be more soluble in aliphatic solvents since solubility parameter of the former polymers is more nearly equal to that of aliphatic hydrocarbons. These and related predictions of this theory have been experimentally verified innumerable times by coating chemists and formulators.

Resistance and Durability Properties. Acrylics generally have excellent durability properties. They resist discoloration when exposed to elevated temperatures and are not easily attacked by acids or bases. The reasons for these properties are to be found in the chemical nature of the polymer backbone. In the first place, the main polymer backbone is comprised entirely of C–C single bonds that are relatively inert and not susceptible to hydrolysis like ester, ether, or amide linkages. Even though the ester side chains can be hydrolyzed, with difficulty, such attack does not result in scission of the polymer backbone, and so, the film maintains its integrity. There are significant differences to be found in the durability of various acrylics (7, 15). These differences are summarized in Table V. The general superiority of methacrylates over acrylates results because the free radical intermediate contributing to chain scission $-(CH_2-C-CO_2R)-$ is more readily formed in acrylates than in methacrylates. The increasing durability of the acrylates with increasing ester chain length results from the greater flexibility and hydrophobicity of the softer polymers. They more readily withstand dimensional changes in the substrate without cracking and are more water repellant. Similar considerations also apply to copolymers. Experimentally, it is found that the exterior durability of certain copolymers plasticized to have the same T_g (15) decreases in the order

$$BMA \simeq MMA > MMA/BA > MMA/EA$$

in agreement with the trends shown in Table V. In emulsion coatings where film formation occurs by coalescence of the polymer particles, adequate film formation must take place if durability is to be obtained. A water-based coating applied in a hot, dry environment may show much poorer durability than the same coating applied in a warm, wet environment. By the same token, if the environment is too cold (less than the T_g of the polymer), poor film formation also results (15).

In addition to the chemical inertness cited above, acrylics show superior durability because the polymers are transparent in the spectral region between 3500 and 3000 Å, which is the most photochemically active region of the solar spectrum. Modification of acrylics with polymers or pigments that absorb in this region,

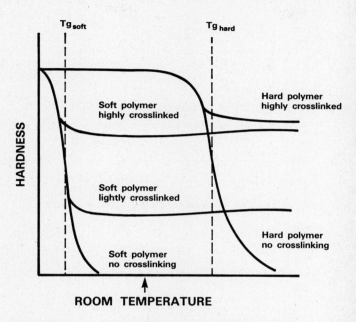

Figure 1. Hardness-temperature relationships for polymers of varying T_g and degree of cross-linking.

Table III. Solubility Parameters of Solvents

Solvent	δ	Solvent	δ
Mineral spirits	7.0	Xylene	9.0
Ether	7.4	Methyl ethyl ketone	9.3
Carbon tetrachloride	8.6	Ethanol	12.7
Toluene	8.9	Water	23.4

Table IV. Values for Acrylic Homopolymers (Small's Method)

Polymer	δ	Polymer	δ
Poly(methyl methacrylate)	9.4	Poly(methyl acrylate)	9.7
Poly(ethyl methacrylate)	9.0	Poly(ethyl acrylate)	9.2
Poly(n–butyl methacrylate)	8.7	Poly(n–butyl acrylate)	8.7
		Poly(acrylonitrile)	12.5

Table V. General Durability Characteristics of Acrylic Homopolymers

| Substituent | Durability Characteristics | |
	Methacrylate	Acrylate
Methyl	Very good	Poor
Ethyl	Excellent	Fair
Isobutyl	Excellent	Good
n–Butyl	Excellent	Excellent

for example, alkyds and TiO$_2$, invariably reduces the exterior durability of the coating.

Development of Thermoplastic Acrylic Coatings

Prior to the mid-1950s, all thermoplastic acrylics were classic solution polymers prepared in suitable organic solvents. They were employed in a variety of applications, including general industrial finishes and appliance enamels and coatings for a variety of wooden, metallic, and plastic substrates.

The history of the introduction of thermoplastic acrylics into automotive finishes serves to illustrate the general growth of acrylic technology during the 1950s (16) and the ability of these polymers to be modified to fill sophisticated requirements.

Prior to 1955, nitrocellulose lacquer was the dominant finish on automobiles. These finishes dried quickly, had a good appearance, were easy to repair, but lacked durability. Rapid loss of gloss required frequent polishing, which eventually eroded the coating. In 1956 General Motors introduced a thermoplastic acrylic lacquer developed by Du Pont. Because the coating, based on poly(methyl methacrylate), showed excellent durability, high pigmentation was not required as was the case in nictrocellulose lacquers. As a result, new metallic and "glamour" finishes could be used that were not practical in the old finishes. The acrylics were not perfect, however. They were deficient in adhesion and flexibility. Initially, these deficiencies were obviated by the addition of plasticizers and the use of special primers. Eventually, the adhesion deficiency was eliminated by the incorporation of special adhesion-promoting monomers into the backbone itself.

Like the older nitrocellulose lacquers, these acrylic finishes had to be polished to achieve maximum gloss. This costly operation was eliminated by the development of new technology commonly referred to as "bake-sand-bake." In this technique the lacquer is applied and baked long enough to allow repair of imperfections by sanding and recoating and then the entire body is baked at a higher temperature. The combined effect of retained solvent and plasticizers causes the finish to reflow, filling in all sanding scratches and leaving a smooth, glossy finish that requires no further polishing.

Because the new acrylic finishes required minimal polishing by the auto owner, the finishes were subject to water spotting. When water, or bird droppings, dried on the hood and trunk in the hot sun, permanent etches were left on the finish. Incorporation of longer ester chain methacrylates into the polymer backbone resolved this problem by rendering the coating more hydrophobic and also imparted improved flexibility without loss of hardness, when used in conjunction with special plasticizers.

Development of Thermosetting Acrylic Coatings

The oldest thermosetting vehicle used by man is porcelain enamel, which is an aqueous dispersion of silica, sand, and other components fused to produce a hard, ceramic finish. These high-quality finishes are still used extensively in bathtubs and household appliances such as range tops and hot water heaters where extreme heat and chemical resistance is required. However, the expense, lack of flexibility, poor adhesion except to a few substrates, and intense heat required to fuse the finish limit the use of porcelain to a few special applications.

In the decade following World War II, increased consumer demand and the resultant introduction of high-speed automated assembly line production of consumer goods demanded new coatings that dried rapidly or could be applied prior to fabrication.

During this period four major types of acrylic thermosetting systems were developed (17). These systems had the common feature that certain functionality was incorporated into the pendant ester group on the acrylic backbone that could be reached with other functionality forming primary chemical bonds. The first types of systems are compared in Table VI. Within a few years numerous chemical schemes for cross-linking acrylic polymers were developed (18) commercially. A few of these are illustrated in Table VII. In Table VIII, the relative performance of several acrylic thermosetting systems is compared with that of the older alkyd melamine. The superior durability of acrylics (T/P and T/S) and good metallic pigment control led to the rapid replacement of alkyd/melamine in U.S. automotive finishes. In appliance coatings where toughness and chemical resistance are crucial, acid cross-linked epoxies replaced alkyd melamines and provided the added benefit of resistance to yellowing by the finish.

There are strengths and weaknesses among the various acrylic thermosetting systems. For example, acid epoxies and urethane cross-linked systems produce no volatile byproducts. Cure temperatures differ widely (see Table VIII). These and other factors determine the acceptability of a particular system for a given application and allow the user considerable latitude in choosing an acrylic that best meets his requirements.

In summary, there are several distinct advantages of acrylics in thermosetting vehicles: (1) the variety of functional monomers that can be incorporated into the polymer to enable reaction with several cross-linking agents and provide specific properties such as adhesion to diverse substrates, efficient pigment wetting, and controlled cross-link density; (2) the variety of nonfunctional monomers available for imparting a desirable balance of hardness, toughness, and resistance properties; (3) the excellent inherent durability of acrylics in exterior and chemically harsh environments.

From World War II until the mid-sixties, solvent-based thermosetting acrylics cross-linked with nitrogen resins, epoxies, urethanes, and other modifiers exhibited superior properties and in many cases became the standard "best state of the art" coating in a variety of applications including automotive finishes, appliance finishes, coil coatings for exterior siding, can coatings, and metal decorating.

Recent Trends in Acrylic Coatings

Since the mid-1960s, extraneous forces have brought about the need for new technology in coatings. These extraneous forces resulted from a growing concern about the emission of organic solvents into the atmosphere and the levels of volatile emissions to which workers

Table VI. First Thermosetting Acrylic Resins

Acrylic Functionality	Coreactant Functionality	Reaction Products	Manufacturer

Row 1:
$\sim\sim$COH (with O double bond) ; $\sim\sim$HC—CH$_2$ (epoxide) ; $\sim\sim$C-O-CH$_2$-CH$\sim\sim$ (with O double bond and OH) ; Canadian Industries, Ltd.

Row 2:
2$\sim\sim$C-NH-CH$_2$OH (with O double bond) (OR) ; ------------> ; $\sim\sim$CNHCH$_2$NHC$\sim\sim$ (with two O double bonds) ; Pittsburgh Plate Glass (Duracron)

Row 3:
$\sim\sim$COC$_2$H$_4$-OH (with O double bond) ; $\sim\sim$HC—CH$_2$ (epoxide) ; $\sim\sim$CH-CH$_2$-OC$_2$H$_4$OC$\sim\sim$ (with OH and O double bond) ;

Row 4:
$\sim\sim$C-OR'CH-CH$_2$ (with O double bond and epoxide O) ; R-H = polyacid, polyamine, polyhydroxyl ; $\sim\sim$C-OR-CH-CH$_2$R$\sim\sim$ (with O double bond and OH) ; Du Pont

Table VII. Other Typical Structures for Thermosetting Acrylics

Acrylic Functionality	Coreactant Functionality	Reaction Product(s)

Table VIII. Comparison of the Properties of Several Thermoset Acrylic Polymers

Acrylic Polymers	Temperature(°F)	Toughness	Exterior Durability	Chemical Resistance
Alkyd melamine	250	2	4	3
Acid epoxy	375	1	3	1
Acrylic melamine	250	3	1	2
Methylolamide	325	2	2	2
Urethane	80	1	2	2

Note: 1 = best; 4 = worst. All values relative to those of acrylic polymers.

were exposed. These concerns were first codified in what have become known as "Rule 66" and "OSHA." These regulatory pressures demanded coatings that contained less than 20% by volume of a select group of photochemically reactive solvents and demanded new guarantees that workers were not exposed to hazardous levels of toxic materials.

In the 1970s, the EPA assumed a leading role in pollution control. With the realization that practically all solvents exhibited some level of photochemical activity, new rules--state and national--were focused on the total solvent content of paints as applied. The index used--VOC (volatile organic compound)--did not differentiate between solvents or coalescents but merely set limits in pounds of emissions per gallon of paint (excluding water) for varying industrial applications.

Further, the dislocations in the world oil supply and the resultant discontinuities in petrochemical prices increased the cost of traditional low-solids solventborne systems and the cost of curing higher temperature thermosetting systems as well. The coatings industry responded to these pressures with new technology: low-emission coatings.

Approaches to Low-Emission Acrylic Coatings. There are two general approaches to low-emission acrylics: (1) water-based coatings; (2) high-solid coatings. Each of these general approaches can be further divided into more specific categories. Aqueous thermosetting and thermoplastic acrylic coatings (19, 20): (1) aqueous thermoplastic emulsions; (2) aqueous thermosetting emulsions; (3) colloidal or solubilizable dispersions; (4) water-reducible thermosetting polymers. High-solid coatings: (1) thermosetting oligomers/poligomers; (2) thermosetting and thermoplastic powder; (3) radiation curable thermosetting coatings. Each of these new technologies has its specific advantages and its peculiar problems and disadvantages that are summarized in Table IX. Each of these new technologies for low-emission coatings will be treated briefly.

Aqueous Thermoplastic Emulsions. The technology of thermoplastic water-based acrylic emulsion coatings was first developed for trade sale application where consumer pressure demanded water-based vehicles. The mechanism of film formation in thermoplastic emulsions limits the range of hardness that can be designed into the polymer backbone without the use of excessive coalescent levels. High molecular weight vehicles with a T_g of 40-80 °C are suitable for industrial coatings and can be formulated to yield fast-drying coatings at low VOC levels. The higher the polymer T_g, however, the more serious a problem proper coalescence becomes and the more serious the negative impact of high relative humidity on film formation (21). Since the mid-70s, however, thermoplastic acrylic latices over a broad T_g range have been used successfully in trade sale/maintenance/industrial coating applications.

Aqueous Thermosetting Emulsions. The basic concept of a thermosetting emulsion is a logical extension of solvent-based thermosetting chemistry (22, 23). Reaction functionality is copolymerized into the polymer backbone, and after application, the

Table IX.   Low-Emission Coatings: Advantages and Disadvantages

Coatings	Major Advantages	Major Disadvantages
Aqueous thermoplastic emulsions	Fast air-dry/flow-temperature forced dry	Film formation guidelines must be followed
Aqueous thermosetting emulsions	Good resistance/durability properties	Film thickness limitations
Colloidal dispersion	Excellent rheology, film appearance	Require solubilizing amines
Water reducible	Properties closest to those of traditional solvent-based systems	Polymerized in water-miscible solvents; hence, organic solvents still present
Oligomers/poligomers	Low MW moieties cross-linked after application; as a result, high-solids application is possible	Application problems different than those of polymer coatings
Powder coatings	No solvent or water	Electrostatic spray efficiency < 100%; dust, color change limitations
UV curing	100% solids systems	In present practice, some volatiles are present, monomers and activators may be toxic

cross-linking reaction is thermally activated between comonomers or blended reactants. A wide range of properties for a variety of industrial applications has been formulated into commercial systems that have been available for several years (24–26). Major problems of emulsion coatings are the rheology and drying difficulties associated with the use of water as the primary solvent (19). The poor rheology is a result of the flat viscosity-solid profile at polymer concentrations less than the critical coalescence concentration compared to the relatively steep increase in the viscosity exhibited by a solution polymer at low solids (Figure 2). This can be controlled by vehicle and formulation design. The undesirable drying characteristics of emulsion coatings--skinning and blistering--have the same origin. Once the critical coalescence concentration is reached, during the drying process, the viscosity increases rapidly, resulting in encapsulation of water under the film. In thermosetting emulsions this trapped water erupts during baking unless sufficient air-drying is permitted. This limits practical film thickness.

Water-Reducible Polymers. Water-reducible polymers have been developed in recent years in an effort to overcome the deficiencies of aqueous emulsions and to make the handling properties of water-based systems more like those of traditional solvent systems. In principle, any solution acrylic polymer can be rendered water reducible by polymerizing the monomers in a water-miscible organic solvent and by increasing the acrylic or methacrylic acid content so that the addition of water and some solubilizing amine will maintain the polymer in solution.

The major limitation of this approach to a low-emission coating is the fact that in order to meet Rule 66 requirements the polymer solids must be high and, as a consequence, the molecular weight must be kept low. As a result, water-reducible polymers are restricted to thermosetting vehicles if adequate performance properties are to be maintained. In applications where thermosetting vehicles are required, this is no particular disadvantage since carboxyl and hydroxyl functionality is already present in the backbone to render the polymer water soluble.

Solubilizable Dispersions (19). The chemistry of solubilizable acrylic dispersions is a hybrid of emulsion and water-reducible technology. These polymers are synthesized by emulsion techniques but contain acidic or basic functionality that renders them water soluble upon neutralization with an appropriate titrant. For example, if the solubilizing functionality is acidic, the polymer will behave like an emulsion below a certain critical pH range, like a highly swollen emulsion within the critical pH range, and like a true water-soluble polymer at sufficiently high pH values. Such polymers offer a favorable balance of properties for many coating applications.

A comparison of the properties of the three types of water-based polymers is presented in Table X.

It is generally accepted that water-based coatings represent the best short-term solution to the new problems facing the coating industry.

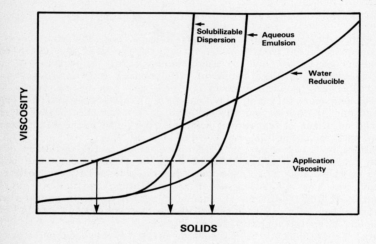

Figure 2.   Comparison of viscosity versus solid relationship between
the water-based acrylic resins.

Table X. Application Characteristics of Aqueous Acrylic Polymers

Property	Aqueous Dispersion	Solubilizable Dispersion	Water Reducible
Solids at application viscosity	Highest	Intermediate	Lowest
Durability	Excellent	Excellent (-)	Very good
Resistance properties	Excellent	Good-excellent	Fair-good
Viscosity control	Some require external thickeners	Thickened by addition of cosolvent	Governed by polymer MW
MFT restraint	Hard films require coalescent	Forms hard films with minimum coalescent	None
Formulation complexity	Complex	Intermediate	Simple
Pigment dispersibility	Requires external dispersant	Good-excellent	Excellent
Application difficulties	Many	Some	Few
Specular gloss	Lowest	More like water reducible	Highest

High-Solids Coatings (8). The second general approach to low-emission coatings--high-solids coatings--can be defined as any solvent-based coating that conforms to existing statutory requirements.

This definition varies with industry grouping but generally refers to 55-75+% volume solids coatings. High-solids coatings can be subdivided into three general categories--oligomers/poligomers, powder coatings, and radiation-cure coatings. The first of these can be considered an extension of existing traditional solvent-based coatings, although not without substantial technological input. The second two appear destined to fill specialty niches in the marketplace.

Oligomers/Poligomers. Oligomers are low molecular weight entities (MW $\sim 10^2 - 10^4$). By virtue of their low molecular weight, they can be applied at high application solids and subsequently cross-linked to form a coherent film. By virtue of their low molecular weight, oligomers are also efficient pigment dispersants, but their rheological properties are somewhat different than those of polymeric vehicles.

The development of this approach to low-emission coatings has been advanced by the recent perfection of production methods for producing narrow molecular weight distributions. It has been found that unless the molecular weight distribution is quite narrow, the cured film does not develop adequate properties.

In principle, any cross-linking reaction can be employed in oligomeric coatings, but the desire to reduce energy requirements has spurred interest in the perfection of novel curing mechanisms such as moisture curing. One such system, based on isocyanate-oxazolidine chemistry (27), has been successfully developed. The simplified chemistry involves the reactions

$$\begin{array}{c} CH_2-CH_2 \\ \diagup \qquad \diagdown \\ \sim\sim N \qquad\qquad O \; + \; H_2O \; \xrightarrow{\hspace{3cm}} \; \sim\sim N \qquad OH \; + \; R_1R_2C=O \\ \diagdown \qquad \diagup \\ C \\ \diagup \quad \diagdown \\ R_1 \qquad R_2 \end{array}$$

$$\}-NH-CH_2-CH_2-OH \; + \; 2RNCO \; \xrightarrow{\hspace{2cm}} \; \} \begin{array}{l} \overset{O}{\overset{\|}{-N-CH_2-CH_2-OC-NHR}} \\ \; | \\ C-NHR \\ \| \\ O \end{array}$$

By proper selection of the radicals R_1 and R_2, the rate of the isocyanate–oxazolidine reaction can be made much faster than that of the hydrolysis reaction of the isocyanate alone.

Poligomers fill the gap between oligomers and conventional solvent-based vehicles. With MW $\sim 5 \times 10^3 - 5 \times 10^4$, they can be applied at high enough application solids to meet many current emission standards. As a result of their fairly low molecular weight, they must ordinarily be cross-linked. In many applications they can be handled by conventional techniques.

Powder Coatings. Powder coatings, as the name implies, are solid particles of thermoplastic or thermosetting polymers, with pigment already incorporated, that are deposited on the substrate by electrostatic spray and subsequently fused into a continuous film by heat. When perfected, such coatings offer significant advantages in that no solvents or liquid vehicle are required. On the other hand, at present, deposition efficiency is less than 100% and the usual requirement that a single production line be capable of handling more than one color makes collection and recycling of the effluent powder a difficult problem.

This approach no doubt will find a place in future industrial coating applications. It is ironic that these experimental powder coatings mimic the oldest thermosetting finish—porcelain enamels.

Radiation–Cured Coatings. Ultraviolet-cured and electron beam cured coatings are presently being used in certain applications such as board coatings and printing inks. The basic chemistry of such systems involves the initiation of the cross-linking reaction by ultraviolet radiation or electron bombardment. The polymerization then proceeds rapidly to completion. The transparency of acrylics in the ultraviolet region makes them an ideal vehicle for this type of application. While these are potentially zero emission coatings, present systems contain monomers, oligomers, and photoinitiators, some of which present volatility and toxicity concerns.

Other problems such as the volume change that occurs during the cross-linking reaction and the difficulty of obtaining penetration of the ultraviolet radiation through pigmented films are not entirely resolved. This latter problem is not experienced in electron beam curing, and consequently this approach will probably find wider application in industrial coatings, although capital equipment costs are higher.

In summary, acrylic industrial coatings have been employed for almost half a century in applications where durability and resistance to the environment are paramount. Over the years, numerous chemical technologies have been developed to achieve specific properties. Because of their versatility and their inherent resistance properties, acrylic polymers have played a central role in the development of coating technology. The growth of new technologies for low-emission coatings has increased this role even more.

Literature Cited

1. Luskin, L. S.; Myers, R. J. "Encyclopedia of Polymer Science and Technology"; Wiley: New York, 1964; Vol. 1, pp. 177–444.

2. Riddle, E. H. "Monomeric Acrylic Esters"; Reinhold: New York, 1954.

3. Salkind, M.; Riddle, E. H.; Keefer, R. W. Ind. Eng. Chem. 1959, 51, 1232. Ibid., 1328.

4. Allyn, Gerould "Acrylic Resins"; Federation of Society for Paint Technology: Philadelphia, 1971; Federation Series on Coatings Technology, Vol. 17.

5. Sittig, M. "Acrylic Acid and Esters"; Noyes Development Corp.: Park Ridge, N.J., 1965.

6. Yokum, R. H. "Functional Monomers, Their Preparation, Polymerization, and Applications"; Dekker: New York, 1973; Vol. 1, 2.

7. Brendley, W. H. Paint Varnish Prod. 1973, 63, 19.

8. Mercurio, A.; Lewis, S. N. J. Paint Technol. 1975, 47, 37.

9. Burrell, H. Off. Dig. 1962, 34(445), 131.

10. Akay, M.; Bryan, S. J.; White, E. F. T. J. Oil Colour Chem. Assoc. 1973, 56, 86.

11. Kelley, F. N.; Bueche, F. J. Polym. Sci. 1961, 50, 549.

12. Mercurio, A. Off. Dig. 1961, 987.

13. Hildebrand, J. L.; Prausnitz, John M.; Scott, R. L. "Regular and Related Solutions"; Van Nostrand–Reinhold: New York, 1970.

14. Burrell, H. Interchem. Rev. 1955, 14(3), 31. Burrell, H. J. Paint Technol. 1968, 40, 197. Burrell, H. Off. Dig. 1955, 27, 728.

15. Harren, R. E.; Mercurio, A. "Acrylic Coatings--Design for Maximum Weatherability"; Chicago Coatings Symposium, April 1974.

16. Beckwith, N. P. Paint Varnish Prod. 1973, 63, 15.

17. Gerhart, H. L. Off. Dig. 1961, 680.

18. Piggot, K. E. J. Oil Colour Chem. Assoc. 1963, 46, 1009.

19. Brendley, W. H.; Haag, T. H. "Nonpolluting Coatings and Coating Processes"; Plenum: New York, 1973.

20. McEwan, I. H. J. Paint Technol. 1973, 45, 33.

21. Leasure, E. L.; Finegan, P. M.; Calder, G. V. J. Water Borne Coat 1977, 1977, 4.

22. Klein, D. H.; Elms, W. J. J. Paint Technol. 1973, 45, 68.

23. Blank, W. J.; Hensley, W. L. J. Paint Technol. 1974, 46, 46.

24. Harren, R. E.; Flynn, R. W.; Levantin, A. M. Off. Dig. 1965, 37, 511.

25. Yunaska, M. R.; Gallagher, J. E. Resin Rev. 1969, 19, 3.

26. Bufkin, G.; Grawe, J. R. J. Coat. Technol. 1978, 50, 41, 67, 83. Ibid., 50, 70. Ibid., 50, 65. Ibid. 1979, 51, 34.

27. Emmons, W. D.; Mercurio, A.; Lewis, S. N. J. Coat. Technol. 1977, 49, 65.

Cellulose Acetate and Related Esters

LARRY G. CURTIS and JAMES D. CROWLEY[1]

Eastman Chemical Products, Inc., Kingsport, TN 37662

Reactivity
  Esterification and Hydrolysis
  Importance of Hydroxyl Functionality
  Viscosity Blending
Cellulose Acetate (CA)
Cellulose Acetate Propionate (CAP)
Cellulose Acetate Butyrate (CAB)
Future of Organic Acid Esters of Cellulose

Cellulose ranks high among nature's more abundant products. Because it is the primary structural material of plant life, billions of tons of cellulose are created annually through photosynthesis. Over one million tons of cellulose are used annually by chemical industries in the manufacture of textile fibers, photographic products, inks, adhesives, explosives, plastics, and coatings.

The chemical structure of cellulose is relatively simple (Figure 1). The simplicity lies in repetitive utilization of the anhydroglucose unit ($C_6H_{10}O_5$) as the building block for the chain structure. The cellulose structure contains 31.48% by weight of hydroxyl groups: one primary and two secondary hydroxyl groups per anhydroglucose unit. These functional groups play the important role in chemical modification of cellulose in reactions with organic acids and acid anhydrides to produce organic esters of cellulose.

The first organic ester of cellulose was cellulose acetate, prepared by Schutzenberger in 1865 by heating cotton and acetic anhydride to about 180 °C in a sealed tube until the cotton dissolved (1). In 1879, Franchimont acetylated cotton at lower temperatures with the aid of a sulfuric acid catalyst (2). Miles, in 1903, described the preparation of partially hydrolyzed cellulose acetate, which was easily distinguished from the fully acetylated

[1]Current address: 3800 Hemlock Park Dr., Kingsport, TN 37663.

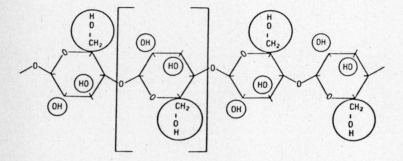

Figure 1.   Molecular representation of cellulose.

product by its solubility in common inexpensive solvents (3). This development led almost immediately to commercial applications, particularly in areas where flame resistance was important. Eastman Kodak Co. cast photographic film base from cellulose acetate as early as 1908; however, commercial prominence was not achieved until World War I when it replaced flammable cellulose nitrate in the manufacture of airplane dopes.

In the era following the war, cellulose acetate lacquers found many uses, but in most applications its performance was something less than satisfactory. Although cellulose acetate lacquers were excellent in some properties such as flame resistance, high melting point, toughness, and clarity, they did not have the level of water resistance or the desired range of solubility and compatibility with modifying resins for more widespread coatings applications. The general idea of using mixed esters such as acetonitrate, acetopropionate, and acetobutyrate to overcome the deficiencies of cellulose acetate is quite old (4), and as a result of a systematic study of the aliphatic series of cellulose esters and mixed esters by Malm and co-workers, it was learned that the cellulose acetate propionates and the cellulose acetate butyrates provided the desirable properties of the cellulose acetates plus improved solubility and compatibility, water resistance, flexibility, and resistance to weathering (5, 6). The butyrates were commercialized in 1933 for plastic molding, and the propionates were marketed in 1939 for the same purpose.

Lacquer applications for the butyrates and propionates were slow to develop primarily because the supplier did not market low-viscosity products attractive to the coatings industry. Both esters had poor compatibility with alkyd resins and would not tolerate resin modification without sacrificing many of their desirable characteristics. At the same time, cellulose nitrate was available in several viscosity grades attractive to the coatings industry and could be modified to a large degree with alkyd resins or natural resins without appreciably altering the performance of the cellulose nitrate polymer. Applications of the mixed organic esters of cellulose were those requiring the high performance of the unmodified ester such as toughness, resistance to ultraviolet light, chemical resistance, etc. A typical use for cellulose acetate butyrate (CAB) requiring all these attributes was airplane dopes. Kline, Malberg, and Reinhart compared several cellulose derivatives in many solvent compositions in tests on the tautness and durability of coated aircraft fabrics (7, 8). It was not until 1954 that a general-purpose, low-viscosity coating grade cellulose acetate butyrate was introduced. A coating grade of cellulose acetate propionate (CAP) was made commercially available in 1969.

In the United States, a wide range of organic acid esters of cellulose are marketed for the coatings industry (Table I). While some are used without modification, such as CA-398-10 for flashbulb coatings, most cellulose esters are employed as modifying resins in multicomponent coating formulations.

Other than the esters described, the principal organic cellulose ester manufactured is cellulose acetate acid phthalate. This product is insoluble in aqueous acidic media, but because of the free acid groups present in the pendant groups, it is soluble in aqueous basic media. It is used for enteric pill coatings.

Table I. Organic Acid Esters of Cellulose for Coatings Applications

Cellulose Ester Type	Viscosity ASTM (A), (s)	Average Acyl Content (%)			Hydroxyl Content (%)	Melting Range (°C)
		Acetyl	Propionyl	Butyryl		
Cellulose acetate butyrate						
CAB-551-0.01	0.01	2.0		53	1.5	110–125
CAB-551-0.2	2.0	2.0		52	1.8	130–140
CAB-531-1	1.0	2.8		50	1.7	135–150
CAB-500-1	1.0	5.0		49	0.5	165–175
CAB-500-5	5.0	5.0		49	0.8	165–175
CAB-553-0.4	0.4	2.0		47	4.3	150–160
CAB-381-0.1	0.1	13.0		37	1.5	155–165
CAB-381-0.5	0.5	13.0		37	1.5	155–165
CAB-381-2	2.0	13.0		37	1.5	170–185
CAB-381-20	20.0	13.0		36	2.0	195–205
CAB-171-15S	15.0	29.5		17	1.5	230–240
Cellulose acetate propionate						
CAP-482-0.5	0.5	2.5	45		2.8	185–210
CAP-482-20	20.0	2.5	46		2.0	185–210
CAP-504-0.2	0.2	2.5	40		5.0	185–210
Cellulose acetate						
CA-398-3	3.0	39.8			3.5	230–250
CA-398-6	6.0	39.8			3.5	230–250
CA-398-10	10.0	39.8			3.5	230–250
CA-398-30	30.0	39.8			3.5	230–250
CA-394-60s	60.0	39.4			4.0	240–260

Reactivity

Esters or mixed esters of monobasic acids containing up to four carbon atoms are prepared with little difficulty by the cellulose-anhydride reaction; however, synthesis of higher esters by this method is usually unsatisfactory since the reactivity of cellulose with organic acids and anhydrides diminishes rapidly as the number of carbon atoms in the acid or anhydride group increases. The use of strong acid catalysts can improve reactivity; however, the catalyst not only promotes esterification but causes depolymerization and degradation of the cellulose structure. Extended reaction time under these conditions leads to products with very low molecular weight and high color. Esters higher than butyrates are prepared by procedures other than direct esterification, for example, by the reaction of cellulose with an acid chloride in pyridine. Products of such reactions are necessarily expensive with use restricted to specialty applications.

For a general review of methods that can be used for the preparation of organic cellulose esters, the report by Malm and Hyatt is recommended (5). Cyrot (9) has collected information on over 100 cellulose esters, in some cases giving properties as well as methods of preparation.

Esterification and Hydrolysis. In the course of the production of organic esters of cellulose, esterification is performed by mixing cellulose with the appropriate organic acids, anhydrides, and catalysts. Normally, the reaction of this mixture proceeds rapidly and is permitted to continue until the three hydroxyl groups on each anhydroglucose unit have been replaced with the acyl group of the organic acid or mixture of acids.

Polymer chemists have found the fully acetylated product, called a triester, to be of limited value in the plastics and coatings industries. Some free hydroxyl groups along the polymer chain are necessary to effect solubility, flexibility, compatibility, toughness, etc. Acylation to a predetermined degree less than the triester is not feasible if a uniform, soluble product is desired. Therefore, the cellulose is fully acylated and then hydrolyzed back to the desired hydroxyl level. The hydrolysis reaction is relatively slow and may be controlled and terminated as required. Commercial cellulose esters may contain up to 5% by weight of hydroxyl obtained in this manner.

Figure 2 represents a general flow diagram of the cellulose ester manufacturing process. Prior to esterification, chemical-grade cellulose, derived from wood pulp or cotton linters, is activated in order to increase reactivity. Activation is a process wherein the cellulose is water treated to swell and separate the normally closely packed fibers of the product. Once activation is complete, the water is flushed out and replaced by either acetic, butyric, or propionic acid, depending on the type of cellulose ester being manufactured. The sole purpose of this procedure is to improve the accessability of the hydroxyl groups of the cellulose to the acid anhydrides used in the esterification process.

Esterification of the activated cellulose occurs rapidly when combined with the appropriate anhydride or mixtures of anhydrides and catalyzed with sulfuric acid. Control of the exothermic reaction is maintained by the amount of catalyst used and by external reactor cooling. If catalyst levels and temperatures exceed the limits prescribed for each cellulose ester type, polymer chain degradation will occur, resulting in low molecular weight and possible product discoloration.

The length of the esterification reaction depends on the type ester being manufactured. Products containing only butyryl modification require longer reaction time than those with acetyl or propionyl modification since reactivity decreases as length of the acyl carbon chain increases. Similarly, when higher molecular weight products are desired, lower catalyst levels must be used, resulting in increased reaction time.

Upon completion of the esterification reaction, the triester, or totally esterified cellulose, is transferred to a hydrolysis reactor and combined with water to destroy unreacted anhydrides and to effect hydrolysis. The reaction is carried out at elevated temperatures under the influence of sulfuric acid catalysis and continued until control measurements indicate the desired hydroxyl content of the ester has been attained. The hydrolysis reaction is terminated by the addition of magnesium carbonate, which neutralizes the sulfuric acid, releases carbon dioxide, and forms insoluble magnesium sulfate crystals that are subsequently removed during the filtration process.

Precipitation of the filtered water-acid solution of the hydrolyzed cellulose ester is accomplished by controlled addition to agitated demineralized water. Additional water is added as precipitation proceeds to prevent acid buildup to the point of redissolving the product. Normally, the cellulose ester is precipitated as a fibrous flake, hammer-milled to the desired particle size, and washed neutral with demineralized water.

Water is removed from the product by a two-step operation. The first step, centrifugation, lowers the water content of the cellulose ester to about 65 wt %, whereupon the product is transferred to vacuum dryers for continued drying to a moisture level not exceeding 3%.

Upon completion of drying, the free flowing powder is passed through screens to remove oversized particles that are retained for reprocessing. The final product is packaged in various type containers as required.

During the cellulose ester manufacturing process, acetic, propionic, and butyric acid mixtures are continuously formed due to reaction of respective anhydrides with water that is either generated or added at various manufacturing stages. Except for evaporative loss, all acids are recovered by extraction from aqueous solution with a low-boiling organic solvent and separated by fractional distillation. The separated acids are subsequently converted to the anhydride for reuse in the cellulose ester manufacturing process.

Importance of Hydroxyl Functionality. The hydroxyl content of each specific cellulose ester type is a very important chemical variable. Relatively small changes may dramatically affect such properties as solubility or compatibility of the ester with other resins. Thermoplastic coatings formulated with cellulose esters with varying hydroxyl content exhibit varying degrees of compatibility with corresponding durability.   Figure 3 illustrates the effect of hydroxyl content on the gloss retention of an acrylic lacquer exposed in Florida for 12 months, black box, 5° from horizontal, facing south.

Moisture resistance of the cellulose ester is altered by the degree of hydrolysis.   The greater the hydroxyl content, the more hydrophylic are films prepared from it.   Thus, if any degree of moisture resistance is required of a high-hydroxyl ester, it is necessary to chemically react the hydroxyl with some reactive constituent.   The hydroxyl group is the primary reactive point of a cellulose ester, and there is ample evidence that this group is capable of behaving in the same manner as any alcohol hydroxyl moiety and will react with commonly used coatings resins such as isocyanates and amino resins.   As the hydroxyl content of the cellulose ester increases, the degree of reactivity with other resins and the resultant cross-linked density increase.   Besides increasing moisture resistance, cross-linking usually improves heat resistance, film hardness, and resistance to solvents.   It is not unusual for one to observe a 50–100% increase in hardness when the film is thoroughly cross-linked.   A study was conducted in which cellulose acetate butyrate esters with varying hydroxyl contents were reacted with a butylated urea-formaldehyde resin and the resulting films tested for solubility (10).   Figure 4 shows the results of this experiment.

High-hydroxyl cellulosics such as CAB-553-0.4, containing 4.3% OH, appeared to cure to excellent solvent resistance under the conditions cited even when the solutions were uncatalyzed.   Most of the cellulosics, however, require an acid catalyst in the curing compositions to reach maximum solvent resistance and hardness.   The degree of hydrolysis selected for each cellulose ester type is that which provides the best combination of solubility, compatibility, moisture resistance, weathering, and other film properties.   Outlined in Table II are the average equivalent weights of a number of organic cellulose esters.   The equivalent weight expresses the amount of cellulose ester required to supply 1 equiv of hydroxyl moiety.   For example, approximately 1000 g of CAB-531-1 (1.7% hydroxyl) will contain 1 equiv or 17 g of hydroxyl groups.   This information is necessary when calculating the stoichiometry of coatings when cross-linking resins such as polyisocyanates are used.

In some instances such as in isocyanate cross-linked coatings, the amount of moisture present must be taken into account in formulating calculations.   Cellulose esters are packaged with very low moisture content; however, if exposed to the atmosphere, cellulose ester powder or flake will regain moisture.   The amount gained is dependent upon the relative humidity and the cellulose ester type, as illustrated in Table III.   To ensure a consistent composition, all cellulose esters should be stored in low relative humidity surroundings or kept in a closed container.

Viscosity Blending.   As shown in Table I, a wide range of cellulose esters are commercially available.   Each type may be identified by acetyl content alone, as in the case of cellulose acetate, or, for mixed esters, by the weight percent of the ester contributed by butyryl or propionyl groups following acylation.   Further

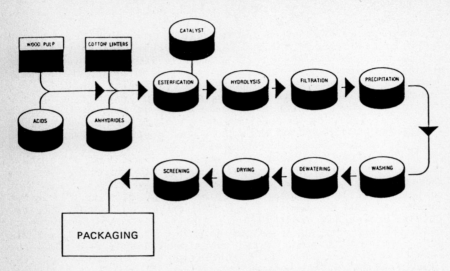

Figure 2.   Manufacturing steps in the production of cellulose esters.

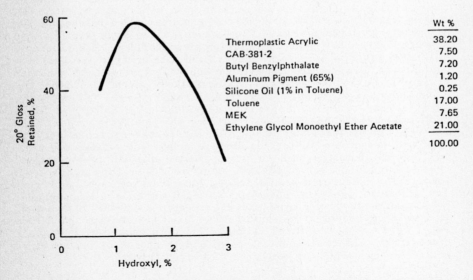

	Wt %
Thermoplastic Acrylic	38.20
CAB-381-2	7.50
Butyl Benzylphthalate	7.20
Aluminum Pigment (65%)	1.20
Silicone Oil (1% in Toluene)	0.25
Toluene	17.00
MEK	7.65
Ethylene Glycol Monoethyl Ether Acetate	21.00
	100.00

Figure 3.   20 Gloss retention of CAB-acrylic coating on 12-month
            Florida exposure.

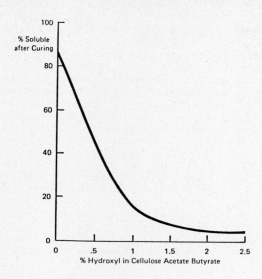

Figure 4. Acetone solubility versus hydroxyl content of cured CAB films.

Table II.   Average Equivalent Weights of a Number of Cellulose Esters

Cellulose Ester Type	% Hydroxyl	equiv wt[a]
CA–398–3	3.5	486
CAP–482–0.5	2.8	607
CAP–504–0.2	5.0	340
CAB–381–0.5	1.5	1130
CAB–500–1	0.5	3400
CAB–531–1	1.7	1000
CAB–551–0.2	1.8	944
CAB–553–0.4	4.3	395

[a]Equivalent weight = 1700/% hydroxyl.

identification may also be made within certain ester types by viscosity ranges. The type chosen for a specific application will depend upon such factors as solubility, compatibility, or viscosity.

The viscosities of the commercially available cellulose esters vary widely. For the mixed esters that include the acetate butyrates and acetate propionates, the viscosity ranges from a low of approximately 0.01 to 20 ASTM s. The viscosity range for the cellulose acetate esters extends from 3.0 to 60.0 s. Within each type ester, for example, CAB-381, the viscosity is primarily related to the molecular weight. As molecular weight increases, such properties as film hardness and flexibility are improved; however, some sacrifice in solubility and resin compatibility may be observed in the higher viscosity range. Cellulose ester viscosities reported in Table I are determined in accordance with ASTM procedures D817-65 and D871-63, respectively, for the mixed esters and cellulose acetate. Within each cellulose ester type, viscosities intermediate to those reported for individual esters may be obtained by blending. It is suggested that esters to be blended have viscosities adjacent to the desired viscosity instead of using the extreme viscosity values.

Blending is a logarithmic function and, as an example shown in Figure 5, CAB-381-20 (Ester "A"), which has an ASTM viscosity of 20 s, may be blended with CAB-381-2 (Ester "B") having a 2-s viscosity, and a 40/60 weight ratio to obtain an effective viscosity range of 5 s.

Cellulose Acetate (CA)

The cellulose acetate most suitable for coatings have acetyl contents ranging from about 38 to 40% and a hydroxyl range between 3.0 and 4.0%. The lacquer-type cellulose acetates possess certain fundamental characteristics that have made them difficult to formulate as coatings, yet it is exactly these characteristics that make them valuable in certain areas of the protective coating field. The cellulose acetates are soluble only in very strong solvents (Table IV). They respond well to latent solvents, but they have very low tolerance for hydrocarbons. Compatibility with resins is poor, only a dozen or so commercially available resins are compatible. Only very active plasticizers like dimethyl phthalate, triacetin, or dimethoxyethyl phthalate will remain in the film without exudation.

The hardness and melting point of cellulose acetate are high. These characteristics would seem to limit the use of cellulose acetate; however, these esters exhibit desirable properties for a number of applications such as solvent- and grease-resistant coatings for paper products; wire and cloth coatings; dopes and cement for model airplanes; lacquers for electrical insulation and the manufacture of capacitors, etc.; coatings for flashbulbs where toughness and high melting point are required; barrier and release coatings for pressure-sensitive tapes; and protective coatings for plastic products.

Cellulose acetate weathers well and is nonyellowing on long-term exposure to the sun; thus, it finds use in coatings for signs and decals. The transparency of cellulose acetate coatings and their ability to transmit sunlight, particularly the beneficial

Table III. Moisture Regain of Cellulose Esters upon Storage

Cellulose Ester Type	% Hydroxyl	% Relative Humidity			
		25	50	75	95
CA–389–6	3.5	1.5	3.7	7.0	13.1
CAP–482–0.5	2.8	0.7	1.0	2.8	4.5
CAB–381–0.5	1.5	0.4	0.9	1.8	2.6
CAB–531–1	2.0	0.4	0.8	1.5	2.2
CAB–553–0.4	4.3	0.2	1.2	2.4	4.4

Note: All values are in grams of water per 100 g of dry ester.

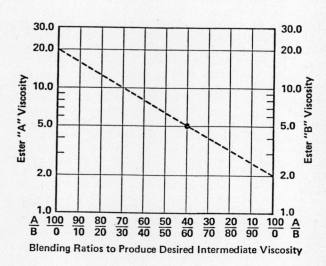

Figure 5. Cellulose ester viscosity blending chart.

ultraviolet rays, have led to extensive use of these film formers in the coating of wire screening for the production of windows for greenhouses and similar structures. The good physical strength of a cellulose acetate coating yields a tough, tear-resistant window. Solution grades of cellulose acetate are used to make membranes for reverse osmosis applications such as artificial kidneys, desalination of water, or purification of industrial waste. A typical formulation for cellulose acetate lacquer is given in Table V.

Cellulose acetate is compatible with cellulose nitrate, but little benefit has resulted from using the two together in a coating composition. Limited use has been made of cellulose acetate in reactive compositions even though it is compatible with urea-formaldehyde resins.

Cellulose Acetate Propionate (CAP)

During the manufacture of CAP, the acetyl/propionyl ratio can be varied widely, but if the acetyl content is more than a few percent, the product loses solubility and compatibility and performs more like cellulose acetate. Thus, there are only two commercial grades of CAP: a general-purpose coating and ink grade, CAP-482-0.5, and an alcohol-soluble product, CAP-504-0.2, which is useful in flexographic inks and overprints.

The cellulose acetate propionates are intermediate in properties between the cellulose acetates and the cellulose acetate butyrates, resembling the cellulose acetate butyrates in solubility and compatibility. Like the acetates, the propionates have practically no odor and thus can be used in applications where low odor is a requirement. These properties make the propionates especially useful in inks, overprints, plastic, and paper coatings, and various reprographic processes. CAP-482-0.5 requires moderately strong solvents to effect solution (Table VI).

Both the propionates may be cross-linked with reactive resins to form thermoset systems that offer the benefits of shelf stability, low-temperature cure, and resistance to high heat, chemicals, and solvents. CAP-504-0.2 is especially useful in this type of system because of its high hydroxyl content--5.0% by weight. This permits it to be dissolved in solvent systems composed primarily of alcohols, thereby increasing the stability or shelf life of the solutions. CAP-504-0.2 is a low-viscosity ester soluble in a wide range of solvents but is particularly interesting because it is soluble in blends of ethyl alcohol, isopropyl alcohol, or n-propyl alcohol with water (Figure 6). CAP-504-0.2 forms clear, hard, glossy films from these solvent mixtures. Methyl alcohol is the only alcohol that will dissolve this ester without the addition of a small amount of water. Paper coated with thermoplastic coatings incorporating CAP-504-0.2 may be repulped under mild conditions.

Both CAP-482-0.5 and CAP-504-0.2 form films that have excellent resistance to penetration by oils and grease. Overprints of these unmodified film formers were applied to paper at 0.1-mil dry film thickness and allowed to dry overnight. Penetration test materials containing a red dye (corn oil, castor oil, hot lard, mineral oil, peanut oil, and water) were placed on the overprints and allowed to remain in contact for 24 h. None of the test materials stained the

Table IV.   Solubility as Indicated by the Solution Viscosity
            of CA-398-3

Solvent	Solution Viscosity at 15% concn [cP (mPa s)]
Acetone	189
Methyl acetate	335
Methyl ethyl ketone	750
Ethylene glycol monomethyl ether acetate	2830
Isophorone	4880
Cyclohexanone	6620
Diacetone alcohol	9650
Ethyl acetate	Borderline
n-Butyl acetate	Insoluble

Table V.   Cellulose Acetate Paper Lacquer

Components	Quantity (wt %)
CA-398-3	18.0
Acetyl triethyl citrate	4.5
Acetone	52.0
Ethyl alcohol	9.5
Diacetone alcohol	7.0
Toluene	9.0
	100.0

Table VI. Solubility as Indicated by the Solution Viscosity
           of CAP–482–0.5

Solvent	Solution Viscosity at 15% concn [cP (mPa s)]
Acetone	46
Ethyl acetate	132
Nitromethane	282
Isopropyl acetate (95%)	338
Ethylene glycol monobutyl ether acetate	442
2–Nitropropane	720
Ethylene glycol monoethyl ether acetate	1350
Isobutyl isobutyrate	Insoluble

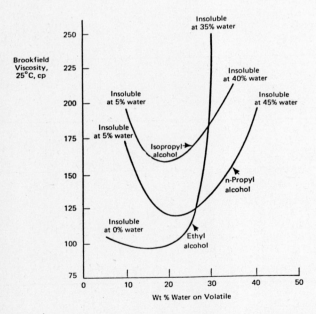

Figure 6. The effect of water content on viscosity and solubility of
           CAP–504–0.2 solutions (15 wt % solids).

paper through the propionate films, whereas, when control films of 0.5-s cellulose nitrate, vinyl chloride-acetate copolymer, and polyamide resins were tested in the same manner for only 2 h, the test material penetrated the films and stained the paper.

When the thermosetting overprint lacquer No. 1 (Table VII) was applied to aluminum foil and cured at 275 °F (135 °C) for 10 s, heat resistance of the coating was in excess of 500 °F (260 °C). The thermoplastic overprint lacquer No. 2, when dissolved in an alcohol-water blend, may be applied directly to the wet lithographic ink in order to improve gloss and permit the use of less expensive inks, eliminate starch sprays ordinarily used to prevent setoff, improve scuff resistance, and eliminate hydrocarbon effluents by replacing heat-set inks with oxidizing inks.

A very important property of polymers used in printing inks and overprints, especially for plastic substrates, is their ability to release solvents. Some polymers retain a small percent of solvent, thereby causing poor film characteristics or an objectionable residual odor. The propionates have been found to have excellent solvent-release characteristics and are used in laminating inks for plastic because of this property. Figure 7 compares solvent release of CAP-504-0.2 with that of other polymers used in plastic coatings.

Cellulose Acetate Butyrate (CAB)

Because there is a wide range of butyryl, acetyl, and hydroxyl levels available in the butyrate, there is also a wide range of properties. Low-butyryl ester, CAB-171, is very similar to cellulose acetate in solubility, compatibility, and performance and is used where toughness, durability, and grease resistance are required. An ester of medium butyryl level such as CAB-381 is widely soluble and compatible with plasticizers and resins. It serves in many coating applications including wood finishes, automotive topcoats, rubber and plastic coatings, cloth coatings, and glass coatings and is also used in inks, hot melts, and adhesives. It is frequently employed as a medium in which to disperse pigments on differential speed two-roll mills. This product has found use in so many different types of coatings that it is commonly referred to as the "general-purpose" butyrate.

An ester with the highest practical butyryl level, CAB-551, was first prepared in 1967 for use as an additive in thermosetting acrylics and polyester enamels to improve dry-to-touch time, reduce cratering, provide a better pigment dispersion medium, improve intercoat adhesion, and provide reflow capabilities to thermosetting automotive enamels. It is also useful in thermosetting powder coatings to provide a pigment dispersion medium, flow control, and anticaking properties to the powder. Although the commercial product, CAB-551-0.2, is not fully soluble in styrene, lower viscosity grades on the order of 0.01-s viscosity are soluble and provide flow control and pigment control to UV-cured coatings and inks. This very low viscosity ester performs a useful function in powder coatings where it reduces cratering of the finish and improves the caking resistance of the powder at elevated temperatures. Because of its low T_g, 80–100 °C, and low melting point, 110–130 °C, it contributes to good flow out during fusion of the powder.

Table VII. Overprints Based on CAP-504-0.2

Components	Quantity (wt %)	
	1	2
CAP-504-0.2	12.5	20.0
Urea-formaldehyde resin (60%)	20.8	
p-Toluenesulfonic acid	0.6	
Ethyl alcohol (anhydrous)	59.5	72.0
Ethyl acetate (99%)	6.6	8.0
	100.0	100.0
Viscosity [cP (mPa s)]	82	163
% nonvolatile	25.58	20

Note: Formula 2 may be prepared in blends of 80 parts of alcohol
       to 20 parts of water.

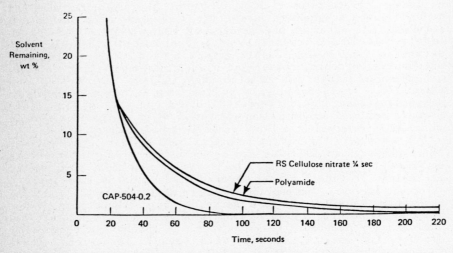

Figure 7. Solvent release of resins dissolved in an ethyl alcohol
(anhyd)/ethyl acetate blend (70/30) (equal film thickness-films cast
on aluminum foil).

CAB-500-1, with only 0.5% hydroxyl, was designed for hot melt strippable coatings where moisture resistance was one of the primary goals. CAB-553, with a 4.3% hydroxyl, is alcohol soluble and especially useful in converting finishes requiring hydroxyl functionality.

When used alone or modified with low levels of plasticizer and/or resin, CAB may perform well in certain applications such as wire coatings or airplane dopes, but film shrinkage and poor adhesion are problems that prevent its use in many general-purpose applications. Modification with higher levels of resin will not solve the shrinkage or adhesion problems and may compromise the desirable properties of the butyrate; for example, the films reflect the properties of the modifiers more than the butyrate. Thus, practically all the butyrates are used in combination with other film formers to impart some desirable properties to the primary film former. Depending on the polymer, the butyrates may be added for one or more of the following reasons: improved flow control; reduced cratering; quicker-dust free time; better viscosity control; increased intercoat adhesion; improved sprayability; better pigment control; increased cold crack resistance; reduced blocking; pigment dispersing medium; reduced solvent crazing; improved holdout; improved color stability; increased stain resistance; improved slip characteristics; and improved toughness.

The level of CAB used with the polymer to obtain improved performance will vary from less than 1% to 50%. In some cases, one-tenth of 1% CAB will eliminate craters and improve flow out, whereas 15-25% is required to give good solvent release from a thermoplastic acrylic coating and up to 50% butyrate on the resin may be required to obtain good performance with an oil-free polyester resin.

As an illustration, coating formulators blend up to 10% cellulose acetate butyrate ester with urethane elastomers to improve the slip characteristics of the coating, raise the blocking or printing temperature, give the coating a more pleasant feel, reduce the dirt pickup, and reduce the clinging tendency of the film. While aiding in these properties, the use of CAB with urethane elastomers generally has some detrimental effect on other properties such as low-temperature flexibility and/or drape and softness of the coating. The importance of these properties will determine the amount of CAB to be used with the urethane elastomer. An example of use of CAB in a thermoplastic urethane cloth coating is shown in Table VIII.

In a thermoplastic acrylic automotive coating, a CAB is used as the pigment dispersion medium and at 15% concentration based on polymer weight provides better sprayability, increased cold crack resistance, better pigment control, better solvent release, and improved exterior durability. (See Table IX.)

CAB is added to thermosetting acrylic enamels for improved flow control, quicker dust-free time, increased intercoat adhesion, and better pigment control (Table X).

Future of Organic Acid Esters of Cellulose

With the coating industry coming under federal and state regulations limiting the type and amount of solvent that can be released into the air, with regulations being written that require new technology,

Table VIII. Thermoplastic Urethane Cloth Coating

Components	Quantity (wt %)
Urethane elastomer (25%)[a]	87.4
CAB–381–0.5	2.4
Slip additive	2.9
Pigment[b]	
Toluene	5.8
Ethyl alcohol	1.5
	100.0
% nonvolatile	27.1
Viscosity [cP (mPa s)]	10,400
Urethane elastomer/CAB ratio	9/1

[a]Such as QI–10 from K. J. Quinn Co.

[b]Concentration will vary with color.

Table IX. Acrylic/CAB Automotive Lacquer

Components	Quantity (wt %)
Acrylic resin (40% in toluene)[a]	52.5
CAB–381–2[b]	4.5
Butyl benzyl phthalate	4.5
Ethylene glycol monoethyl ether acetate	8.0
Acetone	6.5
Isopropyl alcohol	8.0
Toluene	7.0
Xylene	3.0
Methyl isobutyl ketone	6.0
	100.0
% nonvolatile	30.0
% CAB on binder	15.0
Acrylic/CAB/plasticizer	70/15/15

[a]Such as Acryloid B–66 from Rohm and Haas.

[b]Other viscosity grades of this ester type may be used.

Table X.   Thermosetting Acrylic/CAB Automotive Enamel

Components	Quantity (wt %)
Thermosetting acrylic resin (50%)[a]	46.4
Melamine resin (55%)[b]	18.2
CAB–551–0.2[c]	7.2
Aluminum pigment (65%)	1.6
Ethylene glycol monoethyl ether acetate	9.4
Toluene	10.7
n–Butyl alcohol	6.5
	100.0
Spray solids	36

[a]Such as Acryloid AT–56 from Rohm and Haas.

[b]Such as Cargill 3382.

[c]In this application, the CAB is dissolved at 25% solids in an 80/20 toluene/ethyl alcohol blend before addition to the enamel.

with raw material shortages, and with an energy crisis that may limit the amount of energy that can be used to dry or to cure coatings, it is difficult to foresee the future of any product designed primarily for use in the coating industry. Yet, it may be these very regulations and material and energy shortages that make the future of the organic acid esters of cellulosic materials promising.

A few years ago, these materials were used only where their toughness, grease resistance, and UV resistance were important. Today they are being used extensively as additives to other polymer systems to solve problems such as cratering, solvent retention, picture framing, pigment agglomerations, blocking, solvent crazing, poor intercoat adhesion, etc.--problems that are common within the coatings industry.

The organic acid esters of cellulose are being used where exempt solvents are required. They can be readily formulated to meet environmental requirements. When formulated as lacquers, they can be rapidly air-dried. As converting finishes, they can be cured at rather low temperatures or even at ambient temperatures. Thus, little or no energy is required to form the film.

When used as an additive in a converting system, the cellulose ester does not have to be a high-performance film former but can be very low viscosity and still perform the function expected of it. Upon application, the presence of CAB in the coating gives it certain lacquerlike properties including rapid dry-to-touch time. During conversion, the cellulose ester performs as a polyol and converts with the balance of the reactive film former.

We expect the important new development in the organic acid esters of cellulose to be a process improvement and further modification of derivatives now known rather than an introduction of radically new types of cellulose esters. We expect further legislation and emerging technology in the coatings field to make the organic acid esters of cellulose more attractive to the coating chemist.

Literature Cited

1. Schutzenberger, P. C. R. Hebd. Seances Acad. Sci. 1865, 61, 485.
2 Franchimont, A. C. R. Hebd. Seances Acad. Sci. 1879, 89, 711.
3. Miles, G. W. U.S. Patent 838 350, Dec. 11, 1906; Chem. Abstr. 1907, 1, 653.
4. Cross, C. F.; Bevans, E. J. "Research on Cellulose"; Longmans, Green and Co.: New York, 1906; Ser. II, p. 90.
5. Malm, C. J.; Hiatt, G. H. "Cellulose and Cellulose Derivatives," 2nd ed.; Interscience: New York, 1955; Vol. 5, Part 2, p. 798.
6. "Cellulose and Cellulose Derivatives," 2nd ed.; Ott, E.; Spurlin, H. M.; Grafflin, Mildred W., Eds.; Interscience: New York, 1955; Vol. 5, Parts 1-3.
7. Kline, G. M.; Malberg, C. G. Ind. Eng. Chem. 1938, 30, 542.
8. Reinhart, F. W.; Kline, G. M. Film Forming Plast. 1939, 31, 1522.
9. Cyrot, J. Bull. Inst. Textile Fr. 1959, 85, 29-56.
10. Salo, M. Off. Dig. 1959, 31 (Sept).

Nitrocellulose, Ethylcellulose, and Water-Soluble Cellulose Ethers

RUFUS F. WINT[1,3] and KATHRYN G. SHAW[2]

[1]Hercules, Inc., 910 Market Street, Wilmington, DE 19899
[2]Hercules, Inc., Research Center, Wilmington, DE 19899

Chronology of Developments in Cellulose Chemistry

From a brief study of the important developments in the history of cellulose chemistry, it becomes apparent that very little progress was made in elucidating the structure of cellulose from its initial isolation by Payen in 1840 until 1920 (Table I), largely because of the lack of analytical instrumentation capable of defining molecular structure. However, beginning with the application of X-ray analysis to this problem in 1920 by Scherrer (Table II), great strides were made in determining the basic structure of cellulose. Scherrer's studies "showed that cellulose fibers were a composite of ordered and disordered regions" and initiated the extensive studies of crystallinity. These studies were successful in delineating the polymeric structure, in explaining the mechanical properties of fibers, and in revealing what physical attributes of polymeric systems are responsible for fiber-forming properties.

"The development of the primary-valence chain theory of cellulose polymers was the next and possibly the most important guidepost to complete understanding of cellulose chemistry and to the development of polymer theory in general. This development

[3]Current address: 800 Princeton Road, Greenville, DE 19807

Table I.   Scientific Developments in Cellulose Chemistry 1940–1920

Year	Development
1840	Payen--cellulose isolated and described
1850	
	F. Schultz
	Pelouze
1860	
	J. Erdman--constitution of cellulose
1870	
	Franchimont
	Cross and Bevan
1880	
1890	
	E. Schultz--cellulose nomenclature; defined hemicellulose
1900	Franchimont--acetylosis of cellulose to cellobiose
	Nishikawa--first applied to cellulose

Table II.   Scientific Developments in Cellulose Chemistry 1920–40

Year	Development
	Ambronn--beginning of micellar theory
1920	
	Scherrer--X-ray analysis
	Herzog--dimensions of crystallites
	Freudenberg--cellobiose yield as cellulose structure proof
	Staudinger--relationship of viscosity and D.P.
	Haworth--proof of structure of cellobiose
1930	
	Freudenberg and Kuhn--kinetic proof of cellulose structure
	Herzog, Hess, Freudenberg, and Staudinger--D.P. measurements
	Frey-Wyssling and Kratky--fibrillar theory
	Davidson--cellulose oxidation and concept of accessible regions

was not the result of the individual effort of any one research school, but rather the summation of the knowledge of many. Freudenberg's degradation of cellulose to cellobiose in the early 1920s, Haworth's elucidation of the chemical structure of cellobiose, and Staudinger's studies [on the relationship between viscosity and degree of polymerization (D.P.)] led to the postulation that cellulose was composed of polymer. The concurrent determination of the unit cell dimension by Meyer and Mark when amalgamated with earlier X-ray information led to the enunciation of the modern fibrillar theory of fiber structure by Frey-Wyssling and Krakty in the early 1930s. It was the concept of crystallinity in linear polymers and its relationship to strength and elongation that guided Carothers in his initial polymer investigations, which culminated in the development of the largely polymer investigations, which culminated in the development of the so-called "miracle fibers" of today" (1). Research in the period 1930-40 was largely directed toward testing, proving, and modifying the initial hypothesis growing out of the two major advances made during the previous decade. However, as Table III indicates, the development and use of the electron microscope in 1940 permitted investigators to determine the dimensions and average number of chains that are associated in the clearly defined crystalline regions of the fiber. Since the structure and reactivity of cellulose were reasonably well defined by the early 1950s, most of the research over the next two and a half decades was directed toward its further derivation for specific commercial applications. Today we know that the cellulose fiber is composed of 35-45% amorphous and 55-65% crystalline regions, the crystalline regions consisting of very ordered chains associated through strong hydrogen bonding and linked or hinged together by the amorphous sections (2). A representation of this structure is depicted in Figure 1. Modern-day methods of molecular weight (MW) determination such as measurement of intrinsic viscosity, osmotic pressure, sedimentation by ultracentrifuge, and light scattering have produced the source-dependent MW and corresponding D.P. values shown in Table IV (3).

Organosoluble Esters and Ethers

Chemistry and Chronology of Nitrocellulose. The charted history of currently known cellulose derivatives begins in 1832 with the nitration of cellulose via concentrated nitric acid by Braconnot (Table V). "However, in 1945 Schonbein nitrated cellulose with a mixture of nitric and sulfuric acids, and he is generally credited with the discovery of nitrocellulose" (4). The early history of this particular polymer is associated largely with the militaries of the European nations. Its eventual use in other applications was advanced when, in 1866 Abel demonstrated that its stability could be enormously improved by removing the acid retained from manufacture. While military uses and raw material sources were developing, nitrocellulose was finding a place in the plastics industry. The key to the development of the first plastic, a nitrocellulose plastic, was the trial-and-error discovery of camphor as a plasticizer in 1872. In 1884 Wilson

Table III. Scientific Developments in Cellulose Chemistry 1940–80

Year	Development
1940	Electron microscope investigations initiated Hess--mechanical degradation studies Husemann and Schulz--weak-link theory Nickerson and Mark--measurements of crystallinity Pacsu--weak-link theory Frey-Wyssling--showed existence of individual fibrils Degradation of cellulose and derivatives
1950	Reaction uniformity in heterogeneous and homogeneous reactions
1950–80	Further derivatization for commercial applications

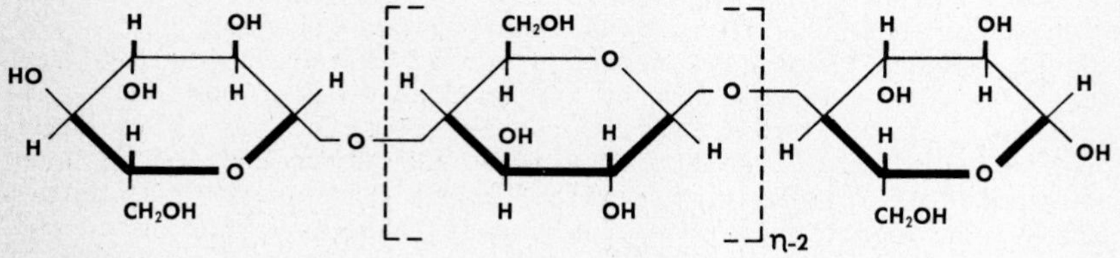

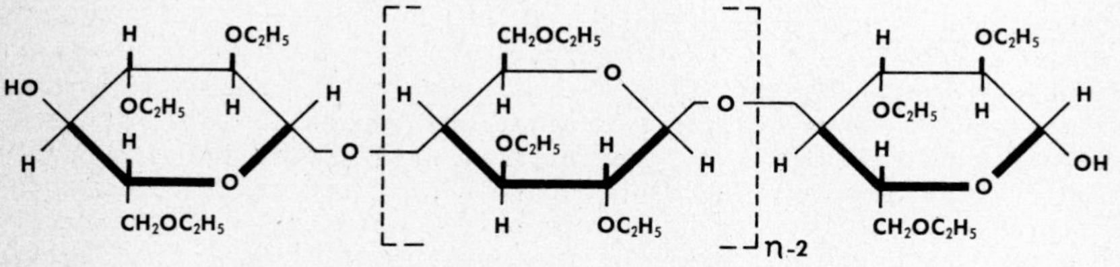

Figure 1. (a) Structural representation of cellulose. (b) Structural formula of ethylcellulose with complete (54.88%) ethoxyl substitution.

and Green obtained their patent on pyroxylin solutions, the precursors of today's lacquers.

During World War I, nitrocellulose was made from cotton linters in the United States. The unavailability of cotton linters in Germany during this period and a cotton crop shortage in the United States in 1918 instigated research on the use of cellulose from wood pulp, which was plentiful. The researchers were partly successful. However, completely satisfactory and economically feasible processes were not developed until the 1930–40 period. By the end of World War I, a combination of economic and technological situations led to the application of nitrocellulose in lacquers. The war economy left the United States with an overabundant supply of chemical cellulose, a large supply of nitrocellulose no longer needed in munitions, and a great supply of butyl alcohol and acetone. The rapidly expanding automobile industry needed an easily applied, quick-drying, protective coating. The need was satisfied by 1923, and an automobile with a nitrocellulose lacquer finish first appeared (Table VI). This result received a great assist from two process developments; in 1922 the first actual low-viscosity nitrocellulose was produced by batch digestion (5), and in 1928 the Milliken digester made this operation continuous (6).

Contrary to many people's beliefs, nitrocellulose lacquer technology has been dynamic during the past forty years (7). A series of innovative formulation design changes have resulted in nitrocellulose lacquers with widely different properties. For the most part, these changes have derived from problems that developed in the use of conventional lacquers.

Most conventional lacquers are completely thermoplastic in nature, although some of the new lacquer types do cure chemically. Others impart markedly different protective or decorative properties to the applied film. A list of the key innovations that have occurred in the lacquer industry over the past 40 years is shown in Table VII.

During the 1930s a surge of interest developed in minimizing or eliminating the cost of the organic solvents required to apply a lacquer. At this time it required 4 or 5 lb of volatile solvents/lb of applied lacquer film. This resulted in a substantial effort to produce lacquer emulsions (8). Emulsion lacquers have a number of advantages. They can be formulated with a higher nonvolatile content, do not penetrate porous substrates, and are dilutable with water, application equipment can be cleaned with water, and finally, it is easier to make high-flash-point lacquer emulsions than high-flash-point lacquers. On the negative side, they dry somewhat more slowly and have somewhat lower gloss than conventional lacquers. Since their development, the leather finishing industry has used both clear and pigmented lacquer emulsions extensively.

Because protective coatings for automobiles was a tremendous market, other competitive finishes challenged nitrocellulose. The advent of alkyd resins in the 1930s was prophesied to sound the death knell of nitrocellulose finishes because of the advantages alkyd-amine finishes showed in higher nonvolatile content improved exterior durability, and lower solvent costs. The type of thinking that produced this prophecy led to two

Table IV. Cellulose: MW and D.P. Values Versus Source

Source	MW Range	D.P. Range
Native cellulose	600,000–1,500,000	3,500–10,000
Chemical cottons	80,000–730,000	500–4,500
Wood pulps	80,000–340,000	500–2,100

Table V. Chronology of Nitrocellulose and Ethylcellulose 1830–1900

Year	Researcher	Development
1832	Braconnot	Nitration of cellulose
1846	Maynard	Nitrocellulose solution
1858	Parkes	Nitrocellulose lacquers
1866	Abel	Stabilization of nitrocellulose
1844	Wilson and Green	Pyroxylin solution

Table VI. Chronology of Nitrocellulose and Ethylcellulose 1900–70

Year	Development
1905	Suida--etherification of cellulose
1915	Nitrocellulose from wood pulp
1923	Automotive industry adopts nitrocellulose lacquer
1928	Continuous digestion of nitrocellulose
1935	Ethylcellulose--first commercial production
1956	Continuous nitration of cellulose

Table VII. Innovations in Nitrocellulose Lacquers

Time	Problem or Need	Solution
1935–40	Solvents undesirable	Emulsions
1940–45	Low nonvolatile content	High-solids lacquers (from 18–26°)
1945–50	Cost of volatile portion	Hot-spray lacquers
1950–55	Sensitivity to alcohol and cosmetics	Catalyzed lacquers
1955–60	Single application, multicolor finish	Multicolor lacquer enamel
1960–65	Lack of elongation	Nitrocellulose–ethylene–vinyl acetate Copolymer lacquers
1965–70	Solvent and mar resistance	Nitrocellulose–urethane super lacquers
1970–75	Desire for aqueous	Waterborne nitrocellulose coatings
1975–80	Low-volatile organic compound (VOC) lacquers	Compliance solvent 1,1,1-trichloroethane

countermoves in the lacquer industry in the 1940s that resulted in the development of "high-solid lacquers." In a comprehensive study on this subject, Hercules, Inc., reported that the following five factors outstandingly influenced the permissible nonvolatile content of nitrocellulose lacquers at selected viscosity levels: (1) molecular weight (viscosity of nitrocellulose); (2) solvent blend; (3) ratio of resin to nitrocellulose; (4) choice of resin; (5) temperature of application. Consequently, high-performance formulations appeared containing 21-30% nonvolatile content for clear compositions and 28-36% for pigmented compositions (9). While their nonvolatile contents remained below those of alkyd-amine finishes, the reduced tendency for nitrocellulose lacquers to sag when spray-applied to vertical surfaces made possible the application of films of equivalent thickness.

Still higher solids lacquers were produced in the late 1940s by hot spraying. The principle behind the spraying of lacquers at elevated temperatures is a simple one--that of reducing the viscosity of the lacquer by heating it instead of by thinning it with solvents, so that the nonvolatile content is not reduced. By this means, up to 50% higher solids were attained, or alternatively, the economic use of higher viscosity grades of nitrocellulose was made possible. Other advantages also became evident, such as the elimination of blushing, an improvement in flowout or leveling, and a reduction in overspray.

All of this progress at increasing their solids content still left nitrocellulose lacquers inferior to alkyd-amine coatings in alcohol and cosmetic resistance, two properties very important in furniture finishes. In the early 1950s, after several years of effort, these problems were largely resolved by the development of a new type of lacquer called "catalyzed lacquer" (10). Catalyzed lacquers were so named because of their reliance upon catalysts to effect a selective polymerization of the nonvolatile components int he system that typically contained nitrocellulose, plasticizer, and an alkyd-amine resin. Nitrocellulose polymers characteristically played the role of skeletal reinforcing members in the three-dimensional film matrix, thereby giving it ample early toughness and rigidity.

During the early 1950s, a new nitrocellulose-based finish having a unique decorative effect appeared on both factory-finished products and the interior and exterior walls of buildings. It became known as "multicolor lacquer enamel" (11). Simply described, a multicolor lacquer is two or more pigmented nitrocellulose lacquers suspended in water and stabilized to prevent blending of the different colored particles (12). These particles of varying shape remain separate during application and drying, and they are large enough to form a distinct color pattern (13). The advantages of using multicolor lacquer enamels include the following (14): (1) the ability in a single, safe application to apply a thick, durable finish of two, three, or more different colors, in a variety of textures; (2) the ability to finish a porous or rough surface with a coating that appears smoother than the original surface; (3) the ability to finish adjacent surfaces of different absorbent properties; applied heavily, the lacquer tends to obscure the difference between the original surfaces.

By 1963 the innovations in lacquer design described above had increased lacquer's acceptance so that it ranked second among all industrial finishes, trailing only alkyd finishes and leading phenolic, amine, acrylic, vinyl, and epoxy finishes. During this same period, new coatings based on various vinyl acetate copolymers began replacing nitrocellulose lacquers on certain flexible substrates. The inherent high rate of elongation of vinyl polymers made this possible even though films produced from these polymers had much lower tensile strength values. The incompatibility between binary blends of nitrocellulose and commercially available vinyl polymers made mixed coatings with a better balance of properties impossible. Then, in 1964, a patent was filed that disclosed that lacquers based on blends of nitrocellulose and ethylene-vinyl acetate copolymers having a vinyl acetate content of 37.0% or higher produced lacquers with vinyl-type flexibility and elongation for use on paper, foil, and other flexible substrates (15). When modified with ester resins such as sucrose isobutyrate and sucrose benzoate, high-performance coatings for wood and metal substrates also resulted.

Late in the 1960s, it became apparent that catalyzed lacquers, the hardest, most durable nitrocellulose lacquers commercially available for coating wood, had less resistance to alcohol, cosmetics, and chemicals than did the best quality alkyd-amine compositions or high-pressure melamine overlays. This situation was soon remedied by the development of nitrocellulose-urethane "super lacquers" based on nitrocellulose-isocyanate prepolymers (16). Nitrocellulose as manufactured contains free hydroxyl groups that can provide sites for true cross-linking with an isocyanate-bonded homogeneous coating. Laboratory studies confirmed the theoretical reaction:

$$2\ \underset{\text{nitrocellulose}}{\underline{\text{HOR}}}\ +\ \underset{\text{polyisocyanate}}{\underline{\text{OCN-R'NCO}}}\ \xrightarrow{\ \text{catalyst}\ }\ \underset{\text{nitrocellulose urethane}}{\underline{RO\overset{\displaystyle O}{\overset{\|}{-C}}-NH-R'-NH-\overset{\displaystyle O}{\overset{\|}{C}}-OR}}$$

Compatibility of nitrocellulose with a wide variety of commercial urethane resins was readily established. Laboratory work was needed, however, to develop useful formulations with excellent properties. The aliphatic-type isocyanate resins now available produce industrial coatings with good initial color and good color stability when combined with nitrocellulose.

Nitrocellulose was commercially available only in forms dampened or wet with alcohols or water. With the type of chemical reaction involved, it was obvious that such hydroxyl-containing solvents would be unsuitable for the practical application of this new technology. As refinements of the basic chemically reacted coatings were found, a new technique was developed whereby nitrocellulose could be produced dampened with isocyanate-grade toluene.

Nitrocellulose–urethane super lacquers for furniture and wood substrates combine the fast speed of dry contributed by nitrocellulose with the toughness and durability of urethane. They appeal for school, hospital, kitchen, casual, and contract furniture. The nitrocellulose–urethane lacquers applied over the proper wood sealers dry rapidly and cross-link when air- or force-dried. They develop hardness, abrasion, and impact resistance equivalent to or better than most commercial finishing systems. They have excellent resistance to penetration by solvents and staining by nail polish and nail polish removers, dry-cleaning solvents, writing inks, waxes, crayons, lipstick, mustard, and Mercurochrome. They are unaffected by soaps, alkali detergents, and household cleaners (17). Because of the time required by the cross-linking reaction, recoatability can be achieved commensurate with typical factory conditions.

The durability and solvent resistance of nitrocellulose–urethane finishes have prompted studies of these coatings for metal substrates. Excellent adhesion to steel and aluminum has been demonstrated. Pigmented systems dry rapidly and have high initial gloss and scratch resistance.

Thus, a new vista had been created for nitrocellulose, a workhorse in the coatings industry. Tough, resistant nitrocellulose–urethane coatings probably will not be used to replace conventional lacquers. Instead, new opportunities have been created to formulate plastic-like materials that can be spray-applied and air- or force-dried. This new chemical technique has been created to help solve the needs that are constantly recurring in our ever-changing world.

By the early 1970s, ecological pressures on the protective coatings industry were threatening major coating formulation changes. Los Angeles Air Pollution Regulation 442 came and then spread across the country in the same or modified forms (18). Lacquers readily adapted. However, the possible restriction of effluent solvents from a coating operation posed a more serious threat. After powder coatings had been thoroughly tested in research laboratories and had been relegated to the role of specialty coatings, enthusiasm for waterborne coatings soared. Again lacquers adapted. The lacquer emulsion technology of the 1935–40 era that resulted in both clear and pigmented oil in water emulsions used to protect and decorate leather was revived and refined (19). Recent data indicate that nitrocellulose coatings or inks wherein the internal phase consists of discrete particles, 0.2–0.3 m^3 in size: (1) can be produced in conventional high-speed equipment, (2) have good application properties, (3) can be formulated for fast air-dry or for low-temperature ovens, (4) can produce films with properties comparable to those of solvent-based lacquers, and (5) can reduce solvent needs by 50% or more. In spite of these technological advantages, lacquer emulsions proved less foolproof than their solvent applied counterparts. While their performance under laboratory conditions was successful, they have not yet been adopted on a plant scale. The principal disadvantage was slower dry time, which meant lower production per unit of time. This, in turn, caused a cost differential that the competitive market environment could not tolerate.

The failure of waterborne lacquers to match both the cost and performance of conventional lacquers caused users to seek alternative methods of lowering volatile organic compound (VOC) emissions. When the U.S. Environmental Protection Agency (EPA) included methylene chloride and 1,1,1-trichloroethane in a limited list of nontoxic compounds, immediate interest developed in evaluating these products as lacquer solvents. In a short time, lacquer formulas containing 1,1,1-trichloroethane appeared in the trade, and soon thereafter reports of lower VOC content lacquers attaining commercial use were heard. Some literature shows 35-46% reductions in VOC (20).

Continuing efforts to lower the VOC of lacquers and improvements in application equipment caused both coating manufacturers and users of lacquers to evaluate electrostatic spray application. On the basis of experience with other coated products, the EPA requested a 50% transfer efficiency from users of coatings. Their studies showed that conventional spray application results in 30% transfer efficiency. Thus, they feel that 50% is a reasonable, significant, and attainable improvement. Some studies of electrostatic application of lacquers suggest that this unsupported judgment by EPA may be achievable, and some literature (21) tends to support this judgment.

Technology of Ethylcellulose and Ethyl(hydroxyethyl)cellulose. The two major commercial organosoluble cellulose ethers, ethylcellulose and ethyl(hydroxyethyl)cellulose (EHEC), are film-forming polymers distinguished by unusual properties and versatility (22). They contribute to the basic film properties of special types of inks, coatings, and adhesives. Their special utility results from the following performance properties (23): (1) high impact resistance, flexibility, toughness, and the retention of these properties at extreme ranges of temperature; (2) solubility in a wide range of solvents allows for economy through solvent choice; EHEC, in particular, has good solubility in aliphatic-rich solvent mixtures; (3) broad compatibility with resins, plasticizers, oils, waxes, and tars; (4) excellent electrical properties; (5) thermoplastic characteristics essential for injection, extrusion, lamination, and calendering operations, for compounding and application of hot melts and for the application and heat sealing of paper coatings; (6) FDA acceptability as food additives; (7) good initial color and resistance to UV discoloration; (8) low densities that allow for more economic application coverage; (9) good resistance to strong alkalis, salt solutions, and ozone; good resistance to oxidation at temperatures below the softening point; and (10) no taste or odor. Of the properties previously listed, solubility in lean or low (Kauri-butanol) KB solvents such as heptane provides ink and coating manufacturers with best cost/performance balance. This is illustrated by the data that follow.

The effect of lean solvent (heptane) solubility of EHEC as well as the cosolvent efficiency of small additions of isopropyl alcohol is illustrated in Table VIII. About 4% isopropyl alcohol is sufficient to make heptane dissolve EHEC, yielding uniform clear solutions at a nonvolatile concentration as high as 20%.

The viscosity data in Figure 2 illustrate graphically the

Table VIII.   Effect of 2-Propanol Used as a Cosolvent for EHEC

2-Propanol, %	None	1	2	3	4	5	10
Heptane, %	100	99	98	97	96	95	90
Appearance	swollen particles	partially soluble, dispersed	hazy	slightly hazy	clear	clear	clear
Viscosity at 25 °C 5% concn (cps)	30	150	400	340	300	265	220

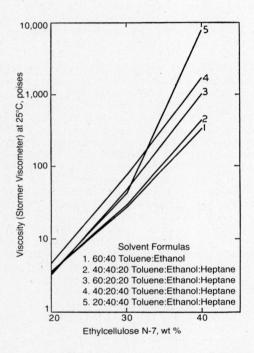

Figure 2.   In   solvent formulas for   ethylcellulose   coating
compositions,   part of the toluene (up to $33^{1}/3\%$,   but
not as much as $66^{2}/3\%$) can be replaced by heptane
without causing any appreciable increase in viscosity;
similar replacement of the alcohol by heptane causes a
much greater viscosity increase (Curves 1, 4, and 5).

tolerance   of   an   ethylcellulose   solution   in   a   60:40
toluene:ethanol blend for an aliphatic hydrocarbon, for example,
heptane.   A more complete listing of the physical properties of
N-type ethylcellulose appears in Table IX.

Water-Soluble Cellulose Ethers

Basic Functions in Water-Based Coatings.   In contrast to the
organosoluble derivatives just described, the water-soluble
derivatives of cellulose are not used in coatings primarily as
film formers or as additives to augment either the drying rate or
the durability of the applied coating. Most of the advantages of
the water-soluble derivatives are realized during the manufacture,   storage,   and   application of the aqueous coating.   They
function as thickeners, rheological control agents, and protective colloids (24).   That is, as thickeners they impart to the
coating the body and consistency necessary for solid suspension
during storage and transfer efficiency prior to application.   By
increasing the coating's viscosity, they inherently contribute to
the development and control of other desirable applied properties
such as ease of brushing, rolling, or spraying. As rheological
or flow control agents, they govern postapplication properties
such as adequate flow and leveling. As protective colloids, they
increase storage stability through partial adsorption on the
surfaces of the various solids in the system such as the pigments
and binder (25).   This helps to minimize particle-particle
interactions that might otherwise result in problems such as
syneresis, hard settling of the pigment, gelation, color drift,
or premature coalescence of the binder due to excessive heat or
cold during storage.   Their superior combination of performance
properties, along with their good economy of use relative to that
of other natural or wholly synthetic polymers, has enabled the
cellulose ethers to become the largest group of water-soluble
polymers used today for thickening trade-sale, water-based
coatings.   The bulk of the discussion concerning their role in
aqueous coatings will be directed toward latex trade-sales
paints.   From this discussion, it is hoped that the reader will
gain a clearer understanding of cellulose ethers and their
potential uses in other water-based coatings.

Chronology of Water Dispersible Cellulose Ethers.   Table   V
earlier revealed that most of the work before 1900 on derivati-
zing cellulose was concerned with its nitrate esters.   A flurry
of activity on structure determination was begun in earnest in
1920.   During this same time period, the reactivity of cellulose
toward various etherifying agents was initiated (Table X) (26).
As a result of Suida's preliminary though inconclusive work in
1905 of methylating cellulose, Denham and Woodhouse and
Lilienfield separately prepared and isolated methylcellulose in
1912. The first hydroxyalkyl derivative, (hydroxyethyl)cellulose
or HEC, was proposed by Hubert in 1920.   The next year, Jansen
realized the first carboxyalkyl derivative, (carboxymethyl)cellu-
lose, more commonly known as CMC (27).   Commercialization of
these three derivatives in the United States did not begin,
however, for another 17 years, when, in 1937, HEC was first

Table IX. Physical Properties of Ethylcellulose
           (48.0–49.5% Ethoxyl)

Physical Properties	Value
Bulking value, lb/gal., in granular form	2.602.8
Bulking value, gal./lb, in solution	0.099–0.104
Color, Hazen, in solution, ASTM D 365	2–5
Discoloration by sunlight	very slight
Electrical properties	
Dielectric constant at 25 °C, 1 Mc	2.8–3.9
Dielectric constant at 25 °C, 1 kc	3.0–4.1
Dielectric constant at 25 °C, 60 c	2.5–4.0
Dielectric strength, V/mil, 10-mil film,	
ASTM D 149–64, step by step	1500
Power factor at 25 °C, 1 kc	0.002–0.02
Power factor at 25 °C, 60 c	0.005–0.02
Volume resistivity, ohm/cm	10^{12}–10^{14}
Elongation at rupture, %, 3-mil film, conditioned	
at 77 °F, and 50% relative humidity	7–30
Flexibility, folding endurance, MIT double folds,	
3-mil film	160–2000
Hardness index, Sward, 3-mil film	52–61
Light transmission, practically complete, A	3100–4000 Å
Light transmission, better than 50% complete, A	2800–3100 Å
Moisture absorption, by film in 24 h at 80%	
relative humidity, %	2
Odor, flake	slight
Refractive index, cast film	1.47
Softening point, °C (°F)	152–162 (305–323)
Specific gravity	1.14
Specific volume, in.^{3}/lb (M^3/kg)	
in solution	23.9 (0.0008604)
Taste	none
Tensile strength, lb/in.^{2} (M/Pa), 3-mil film,	
dry	6800–10,500
	(46.9–72.4)
Tensile strength, wet (% of dry strength)	80–85
Water vapor transmission, g/(m^2 24 h),	
3-mil film, ASTM E 96–66, Procedure E	890

marketed by the Union Carbide Corp.   Methylcellulose followed in 1939 when the Dow Chemical Co. began its manufacture.   During the next 35 years, Dow commercially introduced several hydroxypropyl and hydroxyethyl mixed ethers of methylcellulose.   Hercules entered the area of water-soluble cellulose ethers in 1943 with the first commercial production of CMC (28).   Next to follow from Hercules were HEC in 1962 and (hydroxypropyl)cellulose or HPC in 1969.   Today one of these derivatives, HEC, dominates the water-based coating industry.   CMC, HPMC, and HPC are also used in these coatings, but less frequently due to limitations. CMC has poor tolerance for organic solvents.   Both HPMC and HPC have limited solubility in water at elevated temperatures.   Each derivative has a combination of advantages and disadvantages that make it best suited for particular applications.

Chemistry of the Major Ethers.   The   basic commercial method   of preparation for each of these ethers is shown:

$$cellulose + alkali + water \longrightarrow alkali\ cellulose$$

$$1. \quad R\text{-}X \longrightarrow alkyl\ cellulose\ ether$$

alkali
cellulose +

$$2. \quad R'\text{-}\overset{\overset{\displaystyle O}{\|}}{C}H\text{-}CH_2 \longrightarrow (hydroxyalkyl)cellulose\ ether$$

$$3. \quad X\text{-}R\text{-}COOH \longrightarrow (carboxyalkyl)cellulose\ ether$$

R = alkyl, R' = alkyl or hydrogen, and λ = halogen.   It consists of swelling the cellulose, obtained from bleached wood pulp or cotton linters, with a concentrated aqueous solution, usually sodium hydroxide.   The appropriate reagent then reacts with the alkali cellulose under varying conditions of time, temperature, and pressure, depending upon the reactivity and physical form of the reagent.   Thus, methyl chloride is used for the preparation of methylcellulose; propylene oxide for (hydroxypropyl)cellulose; and ethylene oxide for (hydroxyethyl)cellulose.   (Carboxy-methyl)cellulose is formed by the reaction of monochloroacetic acid or its sodium salt with alkali cellulose.   The effective utilization of these various reaction conditions allows products of varying degrees of substitution (DS) to be produced.   The DS is defined as the average number of hydroxyl groups on the anhydroglucose unit that have been substituted.   The range of DS for each derivative yielding water solubility is shown in Table λI.   For substituents that can chain out such as ethylene oxide, there is also another measurement, moles of substitution (MS). This is defined for (hydroxyethyl)cellulose as the average number of moles of ethylene oxide per anhydroglucose unit.   Those HECs having MS levels of at least 1.8-3.0 are commercially most significant in water-based coatings, especially trade-sale latex paints.   The dual ranges of commercial substitution for the HPMCs in the same application are greater, the methoxyl DS varying from 0.4 to 2.0, and the hydroxypropyl MS from 0.07 to 1.4 depending on the particular product selected.   Although Hercules manufac-tures CMC in four DS types (typically 0.4, 0.7, 0.9, and 1.2),

Table X. Chronology of Water-Soluble Cellulose Ethers

Cellulose Ether	Synthesis and Isolation	U.S. Commercialization
(Hydroxyethyl)cellulose	1920	1937
Methylcellulose	1912	1939
Sodium (carboxymethyl)cellulose	1921	1943
(Hydroxypropyl)methylcellulose	1927	1948
(Hydroxypropyl)cellulose	1960	1969

Table XI. Cellulose Ethers: DS Range for Cold-Water Solubility

Cellulose Ether	DS Ranges	
	Cold-Water Solubility	Commercial Production
Methylcellulose	1.3–2.6	1.6–2.0
(Hydroxypropyl)methylcellulose	0.2–2.1 M	0.4–2.0 M
	0.05–1.6 HP (MS)	0.07–1.4 HP (MS)
(Hydroxyethyl)cellulose	1.3 to >5 (MS)	1.5–3.0 (MS)
(Hydroxypropyl)cellulose	2.0 to >5 (MS)	3.2–4.5 (MS)
Sodium (carboxymethyl)cellulose	≥ 0.4	0.4–1.4

the highest DS type seems to contribute most to improved flow and leveling in latex semigloss paints. Hercules' HPC, also of interest for flow improvement in latex semigloss systems, is commercially produced at an MS level greater than 3.0, the most typical range being between 3.2 and 4.5 (29).

Each of these ethers is available in a number of viscosity types of varying thickening efficiency as Figure 3 illustrates. The ultra-high viscosity types, usually using chemical cotton as a furnish, may have a degree of polymerization (D.P.) as high as 4500 and achieve viscosities of 5000 cps at 1% solids in water. The moderate- and low-viscosity types may be produced by beginning with a lower D.P. furnish such as wood pulp and/or oxidatively cleaving the glucosidic linkages in the polymer backbone with a reagent like hydrogen peroxide. These types may have D.P.s as low as 150 and yield aqueous solution viscosities of only 100 cps at 10% solids.

Both the HECs and the HPMCs are available in rapid-dispersing grades that may be added to water at neutral pH without lumping. Thickening then begins typically in 6-20 min depending upon the amount of glyoxal or other temporary cross-linking agent that was applied to the surface (Figure 4). More rapid thickening may be obtained by raising the pH to 9 or higher by the addition of a small amount of alkali, preferably a fugitive base like ammonium hydroxide. Methylcellulose, HPC, and CMC are not readily surface treatable. These, however, may be added to water without lumping either by preslurrying them in a small amount of an organic, water-miscible, nonsolvent like ethylene glycol or by sifting them slowly into the vortex of vigorously stirred water. Alternatively, methylcellulose and HPC may also be dissolved by preslurrying the ether in hot water (55-60 °C) and then allowing the liquid to cool to room temperature.

An order of their relative hydrophilic-hydrophobic properties in aqueous solution may be seen in Table XII where a typical DS or MS type of each derivative is compared (30). The extremely hydrophilic nature of anionic CMC relative to the four nonionic gums is readily apparent from the high percentage of moisture absorbed by its dried films after exposure at 50% relative humidity and 25 °C for 1 week. All five derivatives display cold-water solubility but vary in their behavior at elevated temperatures. The more hydrophilic CMC and HEC remain soluble even above boiling, while the particular HPMC and methylcellulose shown gel 65 and 56 °C, respectively. HPC differs from all of these by not merely gelling but precipitating from solution as a highly swollen floc at 40 °C. This behavior is a relative measure of how much more weakly water is bound to these more hydrophobic derivatives. Their tolerances to various salts also differ according to their ability to bind water. A CMC solution is the most tolerant of salts of mono- and divalent nonheavy metals but precipitates in the presence of trivalent cations such as aluminum. Of the nonionic ethers, only the more hydrophilic HEC is soluble in a 26% salt solution. Its compatibility in a 10% $Al_2(SO_4)_3$ solution illustrates the advantage of achieving solubility via nonionic hydrogen-bonding groups as opposed to anionic groups in the case of CMC. HPC, because of its high DS and relatively high hydrophobic functionality, has the lowest

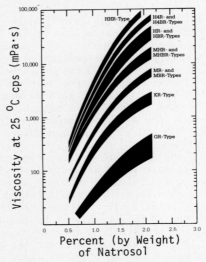

Figure 3. Effect of concentration on viscosity of aqueous solutions of Natrosol.

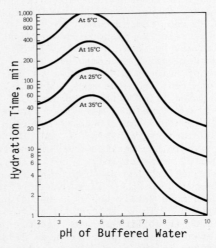

Figure 4. Effect of pH and temperature on hydration time of Natrosol 250 HR.

Table XII.   Solution Properties of Cellulose Ethers

Cellulose Ethers	NaCMC	HEC	HPMC	MC	HPC
Degree of Substitution	0.82	1.3	1.22 M	1.68	2
M.S.	--	2.80	0.16 HP	--	4.0
Equilibrium Moisture (%)	16.1	5.9	6.0	5.2	2.5
Solubility					
Cold Water	S	S	S	S	S
Hot Water	$100^{\circ}C$	$100^{\circ}C$	Gels, $65^{\circ}C$	Gels, $65^{\circ}C$	ppts, $40^{\circ}C$
5% NaCL	S	S	S	S	S
10% NaCL	S	S	S	S	I
26% NaCL	S	S	I	I	I
10% Na_2CO_3	S	I	I	I	I
10% $Al_2(SO_4)_3$	I	S	I	I	I
70% Ethanol	I	S	S	S	S
100% Ethanol	I	I	I	I	S
Polar Organics	none	V.few	few	few	Many
Surface tension					
(dynes/cm)	71	64	44–50	47–53	45

Note: (1% solutions). Key: I = insoluble, S = soluble.

salt tolerance and lowest hot-water solubility. However, these molecular characteristics work to its advantage in giving it the best solubility in polar organic solvents like ethanol and methanol, propylene glycol, ethylene glycol monoethyl ether, morpholine, and cyclohexanone. Like its methylated counterparts, HPC possesses appreciable surface activity in aqueous solution, reducing the surface tension of water to 45 dyn/cm. In contrast, the more hydrophilic HECs and CMCs exhibit only slight and no surface activity, respectively.

<u>Chronology of the Major Cellulosic Ethers in Latex Paints</u>. Today's aqueous emulsion or latex paints find their origin in the water-based casein paste paints first commercially produced around the turn of the century (<u>31</u>). Casein is a phosphate-containing, alkali-soluble protein. It was widely used as a binder but also functioned as an emulsifying agent and as a water-soluble thickener. In this sense, casein can be considered the forerunner of today's water-soluble cellulose ethers. Gradually, the amount of drying oil in the casein paint was increased, and the transition from casein paint to the emulsion paint took place. From their introduction in 1934 until the present, oil-based alkyd emulsion paints have played a major role in the protective coating industry. Originally developed for application on interior walls and exterior masonry, these coatings were comparatively low in cost and gave excellent durability, flexibility, gloss retention, solvent resistance, heat resistance, and color retention (<u>32</u>). A number of disadvantages associated with them, however, greatly influenced coating chemists to try to achieve the above advantages with aqueous emulsions. Proper credit for the first large-scale use of aqueous synthetic-resin emulsions must go to the Germans, who prepared poly(vinyl acetate) latexes just prior to World War II to replace solvent-based emulsion paints made scarce by a shortage of drying oils (<u>33</u>). Latex paints based on styrene and butadiene were introduced domestically in 1948 for use primarily on interior surfaces such as walls, wallboard, plaster, and cinderblock (<u>34</u>). Casein was the preferred thickener for styrene-butadiene latex paints; it thickened efficiently and gave alkyd-like flow properties. Casein began losing the market to cellulosics with the introduction of poly(vinyl acetate) paints in the 1950s (<u>35</u>). One hundred percent acrylic latexes also appeared on the market in the early 1950s. From this point on, the use of both poly(vinyl acetate) and acrylic emulsion paints for exterior masonry and interior surfaces climbed steadily. Concurrently, due to their high thickening efficiency, cellulose ethers became the predominant thickener.

Today, the latex paint industry produces primarily acrylic and vinyl-acrylic (vinyl acetate-acrylic copolymer) paints. The vinyl-acrylic latexes are used most commonly in interior flat paints, while the acrylic latexes are used most frequently with exterior paints and interior gloss or semigloss paints. In 1981, approximately $30-40 million worth of cellulose ethers was sold to the paint industry to produce over 1 billion gallons of paint in the United States (<u>36</u>).

<u>Technology of the Cellulose Ethers in Latex Paints</u>.   Earlier in
the   discussion,   the basic functions of these polymers in water-
based   coatings   were   briefly   described.   This   section   will
describe some of the structural parameters of these ethers, and
their influence on several important latex paint properties will
be illustrated with Natrosol (hydroxyethyl)cellulose.

Methods   of   Incorporation.   The cellulose ether can be incorporated into a latex paint at several stages of the manufacturing
process.   The method varies among manufacturers, depending on the
type of equipment and process used, as well as the paint formulation.   Many add all or part of the thickener to the premix. This
technique   increases the viscosity of the grind and makes it a
more efficient dispersing medium.   Normally for flat paints, one-
third to half of the thickener is added to the premix with the
remainder being postadded to the letdown to assist in final
viscosity adjustments.   The addition ratios depend primarily on
the   amount of viscosity and shear resistance contributed to the
grind by the shape, quantity, and water demand of the pigments in
the grind as well as the MW of the thickener.   The grind should
be viscous enough so that the shear stresses developed during the
grinding produce a sufficiently fine dispersion to achieve
desirable properties such as good hiding power.   Another option
is to add the entire thickener during the letdown phase, as is
typically done with semigloss and gloss paints.   During the paint
manufacture and subsequent paint storage, the thickener acts as a
protective colloid by adsorbing on or "coating" the particles of
pigment and latex.   This reduces their mobility and imparts
essential   storage   properties to the paint such as   good
mechanical, heat, freeze-thaw, and chemical stability.
      Cellulose ethers can be added during the paint manufacture as
a dry powder, stock solution, or slurry.   As with most water-
soluble polymers, correct incorporation of the cellulosics is
important in determining the production time and product quality.
The rate of dissolution of the cellulosic is directly related to
the degree of agglomeration that occurs during mixing.   The key
is to be able to disperse the particles before they begin
hydrating.   The   surface-treated grades were developed   to
temporarily delay hydration until the particles have been wetted
and well dispersed.   Several other precautions can be taken to
assure trouble-free usage of the cellulose ethers. The particles
may be sifted into a well-agitated grind. The pH at the point of
addition should be 7 or lower.   Alkaline ingredients such as
pigment dispersants should be added after the cellulosic is
dispersed.   Sufficient time should be allowed for hydration
before the latex is added because the latex can coat the
particles and prevent complete hydration.
      Increasing usage of pigment slurries has made it difficult
for some paint manufacturers to add sufficient amounts of
thickener when using a typical 2-3% solution.   When pigment
slurries are used, large amounts of water, 50% or greater by
weight, are added to the paint as part of the slurry. There may
not be enough water remaining in the paint formula to make a
dilute solution of the cellulosic.   An alternative is to add the
cellulosic as a slurry, which allows incorporation at a higher

solids concentration. Surface-treated grades may be slurried in water but must be added to the paint immediately so that the cellulose ether does not have time to hydrate and thicken the slurry to an unusable mixture. Glycols and coalescing agents are commonly used to slurry cellulosics for the manufacture of latex paints. The cellulosic may swell, particularly in propylene glycol or ethylene glycol, and hence should not be stored for prolonged periods.

Rheological Properties. The prime advantage of using a binder in emulsion form, namely, the attainment of a high MW resin at moderately high solids having a low viscosity in water, is a disadvantage from an applicator's viewpoint. Since the resin is present in a dispersed form, the aqueous or continuous phase controls the viscosity of the emulsion (37). The primary purpose of a thickener in a latex paint system is to control the rheology, or flow properties. Cellulosic thickeners influence the rheology primarily by thickening the aqueous phase. Interactions do occur, however, between the thickener, the pigment, the latex, and possibly the surfactants, which have considerable bearing on the performance of the paint. This section describes the effect of the molecular weight or viscosity type of Natrosol on the paint properties such as leveling, sag resistance, film build, and spatter resistance (38).

It is necessary to review some basic solution and suspension rheology in order to understand the reasons for the effect of the molecular weight of Natrosol on paint rheology. Solutions of cellulosic ethers exhibit pseudoplastic behavior, i.e., the apparent viscosity decreases with increasing shear. The viscosity returns to its original value when the shear is removed. The amount of pseudoplasticity is very molecular weight dependent; the lower the cellulosic molecular weight, the less the viscosity varies under increasing shear, as depicted in Figure 5.

Suspension rheology can be used to study systems such as latex paints. In paints, bridging networks can be formed when the shear rate is low enough. These interactions dominate the rheology and cause the viscosity to be higher than it would otherwise be. When shear is applied to this network system, however, these bonds are broken. The entanglements of the thickener polymer in the free liquid then dominate the rheology and contribute to its viscosity and elasticity. The solids that are present increase the viscosity only by a volume-filling effect.

Figures 5 and 6 compare the effect of the cellulosic molecular weight on solution viscosity curves and paint viscosity curves (prepared to a constant Stormer viscosity). These figures indicate that the solution viscosity curves correspond to the paint viscosity curves in the high-shear-rate region. This occurs because under high shear, the interaction between paint components is reduced so that solution rheology dominates. Film build is directly related to viscosity during the high-shear processes of brush and roll application. Clearly, just as in water alone, the thickener having the lowest MW produces the highest applied viscosity or greatest degree of brush drag.

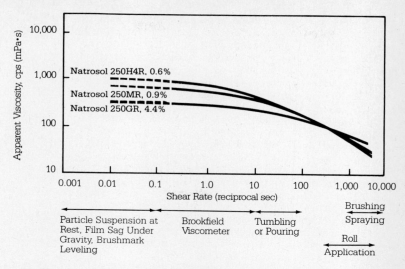

Figure 5.   Influence of molecular weight and shear rate on viscosity of solutions of Natrosol.

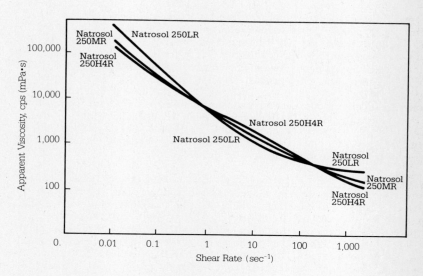

Figure 6.   Influence of molecular weight of Natrosol and shear rate on paint viscosities.

Greater film build promotes better hiding. Spatter resistance, also a high-shear application property, does not depend on thickener viscosity but does depend on the elasticity of the aqueous phase. Higher molecular weight thickeners are more elastic; therefore, lower molecular weights are more resistant to spatter during roll application.

At low shear rates, the paint viscosity curves diverge from the behavior of the solution viscosity curves. Interactions, such as those between the latex and thickener, dominate the paint rheology under low-shear conditions. The interactions increase as the thickener concentration increases. Therefore, lower molecular weight thickeners, which must be used at high concentrations, will have more interactions with the latex and cause increased viscosity. The leveling process of a paint film occurs at very low shear rates. The preferred viscosity type for good leveling are those having high molecular weights. Improvements in leveling will generally decrease sag resistance. Therefore, low molecular weight polymers offer the best sag resistance properties. Other factors, such as the leveling properties of the latex and the rate of water absorption into the substrate, are equally important in determining the leveling/sag resistance properties of a paint. The molecular weight effect on rheological properties is summarized in Table XIII.

Cellulose ethers can also have a profound effect on the dispersion state of both the hiding and tint pigments in a latex paint. Their relative degrees of adsorption onto the surface of a semigloss-grade titanium dioxide in an aqueous paste (50:20, titanium dioxide:water) can be quantified by determining the amount of anionic dispersant needed to displace each of them. Table XIV shows these values for an HPMC (Me DS = 1.5, HP DS = 0.3) and Natrosol at two different MS levels. Approximately 50% more dispersant was required to displace the methylated derivative, while the lower MS Natrosol took 100% more, relative to the same D.P. Natrosol of MS = 2.5 at the same solution concentration. High D.P. CMCs of DS = 0.7 are known to require even greater amounts of dispersant, whereas CMCs of both high D.P. and high DS (1.2) have been found to be dispersants. Thus, the choice of both thickener and pigments necessitates close examination of the formula's dispersant demand in order to achieve optimum hiding and color development.

Future Outlook for Cellulose Ethers. Cellulose ethers fulfill many of the requirements demanded of rheological modifiers in latex paints. Cellulosics, particularly (hydroxyethyl)cellulose, have the advantage of being "all-purpose" thickeners. HEC thickens all types of latexes efficiently and will work well in an entire paint line. With the wide versatility and good overall balance of properties, (hydroxyethyl)cellulose will continue to be used by paint manufacturers to ensure quality paint performance. The sales of architectural coatings are expected to grow at a rate of 2–4% per year (39). The volume of cellulose ethers sold to the latex paint industry is expected to decrease slightly or remain constant. In recent years, synthetic thickeners that offer good rheological properties have been introduced to the paint industry. However, these have many formulation constraints

Table XIII.   Effect of Natrosol Molecular Weight on Paint
              Properties

Property	Best MW Thickener	Reason
Spatter resistance	low	Low elasticity
Film build	low	High high-shear viscosity due to solution rheology
Leveling	high	Less polymer creates few interactions
Thickening efficiency	high	Less polymer is needed to achieve desired KU

Table XIV.   Effect of Dispersant and Thickener on Pigment
             Dispersion Using the Fluidity Titration Technique

wt % of Thickener in Test Solution[a]	wt % of Tamol° 731 in Test Solution[b]	Titrant vol to Fluidity, mL[c]
None (control)	0.6	13.0
0.6% Natrosol 250HR	0.6	13.0
0.6% Natrosol 250KR	0.6	13.0
0.6% Natrosol 250LR	0.6	13.0
0.6% (hydroxypropyl)methylcellulose (medium-high viscosity type)	0.6	19.0
0.6% (hydroxypropyl)methylcellulose (medium-high viscosity type)	0.8	13.0
0.6% Natrosol 180HR	0.6	24.0
0.6% Natrosol 180HR	1.1	13.0

[a]Added to a standard paste consisting of 50 g of Ti-Pure R900
(E. I. du Pont de Nemours & Co.) titanium dioxide and 20 g of
distilled water.

[b]Registered trademark of Rohm & Haas Co.

[c]Titrant volumes greater than 13.0 cc indicate pigment flocculation due to the thickener. The greater the volume, the greater the
quantity of dispersant necessary to displace the thickener from the
pigment surface and fluidize the paste.

and production difficulties. As of the end of 1982, the synthetic thickeners are primarily used as speciality items. Significant breakthroughs in technology will be required to expand their
usage. Cellulosics should continue to play a major role as
rheological modifiers in latex paints.

Literature Cited

1. Skolnik, Herman; McBurney, Lane F. Tappi 1954, 37(9), 190A-
    193A.
2. Savage, A. B. "Encyclopedia of Polymer Science and
    Technology"; Interscience: New York, 1965; Vol. 3, p. 459.
3. Hamilton, J. Kelvin; Mitchell, R. L. "Encyclopedia of
    Chemical Technology," 2nd ed.; Interscience: New York,
    1964; Vol. 4, p. 595.
4. Barsha, J. "Cellulose and Cellulose Derivatives";
    Interscience: New York, 1954; Vol. 5, Part II, Chap. IλB,
    p. 713.
5. Yeager, J. R. paper presented to the Technical Committee of
    the National Paint and Coatings Association, Hershey, Pa.,
    Sept 15, 1960.
6. Yeager, J. R. Hercules Chem. 1961, No. 41, 2.
7. Arne, Frances, Ed. Chem. Eng. 1963, 70(4), 90.
8. Creasy, J. Prod. Finish. 1956, 9(7), 81-86.
9. "High-Solids Lacquers," Form 500-497, Hercules, Wilmington,
    Del., p. 5.
10. "Catalyzed Lacquers for Wood Finishing," CSL-148, Hercules,
    Wilmington, Del., p. 1.
11. "Lacquer, Multicolored Dispersion Type (For Spray
    Application)," Federal Specification TT-L-45, Jan. 17, 1958.
12. Zola, John C. U.S. Patent 2 591 904, 1952.
13. "Multicolor Lacquer--Less Flammable than Conventional
    Types," CSL-101A, Hercules, Wilmington, Del., p. 1.
14. Campbell, John R. Materials Methods 1954, 40, 86-89.
15. Unger, James G. U.S. Patent 3 321 420, 1967.
16. Hercules Chem. 1969, No. 59, 31.
17. "Nitrocellulose-Urethane Super Lacquers," CSL-202, Hercules,
    Wilmington, Del., p. 1.
18. "Nitrocellulose Lacquers Acceptable Under Rule 66," CSL-
    196, Hercules, Wilmington, Del., p. 1.
19. "Preparation Procedures for Nitrocellulose Waterborne
    Coatings or Inks," CSL-225, Hercules, Wilmington, Del., p.
    1.
20. "Formulating Fast-Dry Lacquers for Electrostatic Coatings,"
    CSL-188, Hercules, Wilmington, Del., p. 1.
21. "Low-VOC (Volatile Organic Compound) Lacquers Formulated
    with Chlorinated Solvents," CSL-198, Hercules, Wilmington,
    Del., p. 1.
22. Hamilton, Eugene C.; Early, Lawrence W. Fed. Ser. Coat.
    Technol. 1972, Unit 21.
23. Klug, E. D. "Encyclopedia of Chemical Technology," 2nd ed.;
    Interscience: New York, 1964; Vol. 4, pp. 616-52.
24. "Natrosol Controls the Flow Properties of Paints," Bulletin
    VC800-10, Hercules, Wilmington, Del., 1980; p. 1.

25. Lindenfors, D. <u>Farbe Lack</u> 1973, <u>No. 3</u>, 44–56.
26. Skolnik, Herman; McBurney, Lane F. <u>Tappi</u> 1954, <u>37(9)</u>, 192A.
27. Savage, A. B. "Encyclopedia of Polymer Science and Technology"; Interscience: New York, 1965; Vol. 3, p. 459.
28. Batdorf, J. B.; Rossman, J. M. In "Industrial Gums"; Whistler, R. L., Ed.; Academic: New York, 1973; Chap. 31, p. 696.
29. Ibid., p. 650.
30. Klug, E. D. <u>J. Polym. Sci. C, Polym. Symp</u>. 1971, <u>No. 36</u>, 493.
31. Nylen, Paul; Sunderland, Edward, "Modern Surface Coatings"; Interscience: New York, 1965; p. 663.
32. Martens, Charles R. "Technology of Paints, Varnishes, and Lacquers"; Reinhold: New York, 1968; Chap. 4, p. 33.
33. Nylen, Paul; Sunderland, Edward "Modern Surface Coatings"; Interscience: New York, 1965; Chap. 18, p. 664.
34. Allyn, Gerould "Emulsions and Emulsion Technology"; Lissant, Kenneth J., Ed.; Dekker: New York, 1974; Part I, Chap. 7, p. 353.
35. Salzberg, H. K. <u>Am. Paint J.</u> 1964, <u>48(45)</u>, 101.
36. Levine, Ralph M. <u>Am. Paint Coat. J.</u> 1981, <u>66(14)</u>, 122–35.
37. Martens, Charles R. "Technology of Paints, Varnishes;" Reinhold: New York, 1968; Chap. 28, p. 513.
38. "Natrosol Controls the Flow Properties of Paints," Bulletin VC800–10, Hercules, Wilmington, Del., 1980, pp. 14–18.
39. <u>Am. Paint Coat. J.</u> 1982, 12–16.

Amino Resins

L. L. WILLIAMS[1], I. H. UPDEGRAFF[2], and J. C. PETROPOULOS[1,3]

[1]Stamford Research Laboratories, Polymer Products Division, American Cyanamid Company, Stamford, CT 06904-0060
[2]407 Den Road, Stamford, CT 06903

Chemistry
Manufacturing
Applications
Trends in Amino Resins
Future Trends
Summary

The term "amino resins" is normally applied to condensation products of formaldehyde and polyfunctional amides and amidines such as urea and melamine. The bulk of amino resins in use today are based on these two compounds. These highly functional, reactive materials find a broad range of applications. In the plastics industry, these include molding compounds, coating resins, and adhesives; other traditional markets are as treating agents for textiles and paper. The objective of this paper is to provide a broad outline of the uses and chemistry of amino resins with major emphasis on the coatings area and a lesser amount on molding compounds. There are a number of leading references on the chemistry and applications of amino resins (1-4).

Although the chemistry of the reaction of formaldehyde with urea and other amino compounds was investigated much earlier, the first useful product did not come on the market until the 1920s. The first commercial application for amino resins was in molding compounds and utilized a resin made with an equimolar combination of urea and thiourea, the invention of Edmond C. Rossiter (5). The Beetle brand name was applied to the new molding compound and has remained prominent in amino resins ever since. It is of special interest that the very first commercial product based on amino resins should be a complex formulation such as a molding compound.

[3]Deceased

0097-6156/85/0285-1101$06.00/0

Adhesive resins based on aminos are much simpler and now are far larger volume products than molding compounds. The reason that the more complicated product came first is that the amino molding compound was unique for its time. In contrast to the phenolics, the amino resin could be supplied in pastel and transluscent colors. It had no objectionable phenolic odor yet had a hard, lustrous surface that was not easily stained.

Melamine resins did not appear until a few years before World War II, first in Europe and later in America (6, 7). At that time melamine was prepared from dicyandiamide (dicy), but in recent years a process utilizing urea has been shown to be more economical. The various urea-based processes (high and low pressure) have completely supplanted the process using dicy. Once introduced, melamines moved rapidly into applications already established by urea, and corresponding formulations for molding, laminating, coating, gluing, and textile finishing were soon available from a number of manufacturers. The remarkable stability of the symmetrical triazine ring made these products resistant to chemical change once the resin had been cured to the cross-linked, insoluble state. Melamine has naturally remained more expensive than its precursor urea and so finds use in those applications demanding superior performance. Molded plastic dinnerware takes advantage of the exceptional hardness and water-resistance offered by the triazine resin. Melamine cross-linkers for coatings provide hardness, water-resistance, and freedom from yellowing.

The only other triazine resins of commercial importance are based on benzoguanamine, 2-phenyl-4,6-diamino-1,3,5-s-triazine. Its use is small relative to that of urea and melamine, representing only a few million pounds per year. Benzoguanamines are used in surface coatings because they provide better adhesion than melamine; however, their poor resistance to ultraviolet light compared to that of melamine has always limited their use. In recent years the sharply increased price of benzoguanamine relative to that of melamine has further reduced its importance as an item of commerce.

Prior to the rapid rise of thermoplastics following World War II, aminoplastics served a broad, diversified market. They continue to show a moderate growth today, but this is a result of good growth in some areas and contraction in others. Growth of overall amino resin volume was at a rate of 7% from 1969 to 1979 (8), reaching a total volume of 1.6 billion pounds in the latter year. However, the major component of this growth has been the urea–formaldehyde resins used as adhesives in various wood products, such as chip and flakeboard. More than 60% of the total volume of aminos goes to this end use. In the higher technology areas, molding compounds and protective coatings, growth has been much less. Amino molding compounds particularly show little or no growth, while use of aminos in protective coatings is still slowly expanding. The major reason for the growth in aminos in coatings has been that baking finishes are incorporating higher percentages of cross-linkers, rather than any significant growth in coatings volumes. While coatings employ the largest number of different aminos, the volume in the United States still does not exceed 100 million pounds on a real basis. In spite of their modest volume, amino coating resins are the segment of this class of materials showing the greatest technological change (see Table I).

Table I.   Pattern of Consumption for Urea and Melamine Resins

| | Thousands of Metric Tons | |
Market	1982	1983
Adhesives for fibrous and granulated wood	345	415
Adhesives for laminating	12	17
Adhesives for plywood	27	20
Molding compounds	28	34
Paper treatment and coating resins	30	33
Protective coatings	29	35
Textile treatment and coating resins	27	31
Export	8	10
Other	6	7
	512	602

Note: As this table shows, the largest outlet for amino resins by far is their use as adhesives or binders for reconstituted wood products made from sawdust and wood chips. Urea-formaldehyde resin is most commonly used. Melamine-formaldehyde resin can provide improved water resistance and may be combined with the urea resin to provide an improved product. Molding compounds are about the next most important outlet for amino resins. It is approximately evenly divided between urea and melamine. The primary use for urea moldings is in the electrical field, while the most important area for molded melamine plastic is dinnerware.

A factor that may affect the growth of amino resins is the question of formaldehyde toxicity. Like many other chemicals, aminoformaldehyde resins have encountered questions of product safety in recent years. This has been largely due to reports that formaldehyde causes nasal tumors in rats, although epidemiological studies of workers exposed to formaldehyde have shown no confirmation of this.

The controversy surrounding formaldehyde is complicated by the fact that its odor is obnoxious to humans in very small concentrations. Thus, most humans will not tolerate formaldehyde at a concentration of 1 ppm, far below the 15 ppm that was shown to cause tumors in rats. Thus far, the evidence has not led to any reduction in the already fairly low 3-ppm OSHA limit on formaldehyde exposure. As noted, a lower exposure level is self-enforcing because of the highly unpleasant odor of formaldehyde.

The one area of amino resin usage that has been severely affected by formaldehyde toxicity is the use of foamed urea-formaldehyde as insulation. The material has many good qualities for this application: it can be injected readily into the walls of existing housing and cured in place, providing a good "R" factor. However, in practice there were too many cases where enough formaldehyde was given off by the cured foam to make houses unlivable. Had more technical sophistication been used in the industry this might well have been avoided, but at this point use of urea-formaldehyde foam has been outlawed in some states and is widely rejected by consumers. No early revival of this use of aminos appears possible.

The impact of formaldehyde on the use of amino resins in other areas has been much less. There is pressure to reduce the level of free formaldehyde in the urea-formaldehyde-bonded construction materials. However, it appears likely that this will be resolved technically with either better resins or a higher degree of cure. This use is sensitive to the free formaldehyde issue because it can affect the end user, the people living in the homes built with the material. Other uses of aminoformaldehyde resins are less likely to be affected. Molded articles, laminates, and surface coatings are generally highly cured materials where any release of formaldehyde occurs in the factory where the resins are cured. In such environments control of formaldehyde levels is a straightforward problem of providing adequate ventilation in the work areas. Although formaldehyde will probably have a further dampening effect on growth of amino resins, it does not seem likely to greatly reduce the use of this type of chemical.

Given that aminoformaldehyde resins will probably continue as a major item of commerce, a brief review of their chemistry is appropriate.

Chemistry

Amino resin chemistry goes back to the 1880s. The first mention of a urea-formaldehyde reaction product was by Tollens (9) in 1884 in a report of work done by an associate named Holzer. Very few references relating to amino resins can be found in the chemical literature prior to 1900. Ludy (10) in 1889 and Carl Goldschmidt (11) in 1896 mentioned products that were similar to the material prepared by Holzer. The next significant contribution was the isolation of mono- and dimethylolurea by Einhorn and Hamburger (12) in 1908. Ten years later, Hans John (13), an Austrian citizen, was granted a patent for an adhesive made from urea and formaldehyde. During the early 1920s Hans Goldschmidt and Oskar Neuss (14) were experimenting with urea resins in Germany, while Kurt Ripper and Fritz Pollak (15, 16) were making amino resins in Austria. As mentioned above, Rossiter (5) introduced the first commercial product, his Beetle molding compound, in England about this time (1926).

Two main reactions are involved in the formation and cure of amino resins. The first is a simple addition reaction of formaldehyde to the amino compound, as illustrated in Equation 1, that shows the formation of a typical methylol intermediate.

$$R-NH_2 + HCHO \rightleftharpoons R-NH-CH_2-OH \tag{1}$$

The second important reaction is a condensation involving the liberation of a molecule of water between a methylol group and an active hydrogen. This is the polymerization reaction that links the amino molecules together into a chain or network structure. It is illustrated in Equation 2.

$$R-NH-CH_2-OH + H_2N-R \rightleftharpoons R-NH-CH_2-HN-R + H_2O \tag{2}$$

Since both of the hydrogen atoms of the $-NH_2$ group are replaceable, each amino group has a functionality of two. Therefore, an amino compound having only one amino group can form only linear polymers with formaldehyde. Amino compounds such as urea and melamine that have more than one amino group per molecule can form cross-linked three-dimensional polymer networks with formaldehyde.

The methylolation reaction may be catalyzed by either acids or bases. Under acid conditions, the formaldehyde probably picks up a proton forming a carbonium ion that then attacks the unshared pair of electrons on the amino nitrogen. In alkaline solution, the amino compound may react with OH^- ion to form an amino anion that can then react with formaldehyde. These reaction mechanisms are illustrated in Equations 3–8.

$$CH_2O + H^+ \rightleftharpoons {}^+CH_2OH \tag{3}$$

$$RNH_2 + {}^+CH_2OH \rightleftharpoons RN^+H_2CH_2OH \tag{4}$$

$$RN^+H_2CH_2OH \rightleftharpoons RNHCH_2OH + H^+ \tag{5}$$

$$RNH_2 + OH^- \rightleftharpoons RN^-H + H_2O \tag{6}$$

$$RN^-H + CH_2O \rightleftharpoons RNHCH_2O^- \tag{7}$$

$$RNHCH_2O^- + H_2O \rightleftharpoons RNHCH_2OH + OH^- \tag{8}$$

The methylol compounds formed are fairly stable under neutral or alkaline conditions, and many amino resins (adhesives, for example) are simply mixtures of these methylolated monomers and low molecular weight polymers. When an acidic catalyst is added, the resin cures to an insoluble, cross-linked solid.

Unlike the methylolations reaction, the condensation (curing) reaction is catalyzed by acids only. Apparently the methylol group is protonated and a molecule of water is lost to give the intermediate carbonium-immonium ion. This then reacts with an amino group to form a methylene link. These reactions are illustrated in Equations 9-12.

$$RNHCH_2OH + H^+ \rightleftharpoons RNHCH_2O^+H_2 \tag{9}$$

$$RNHCH_2O^+H_2 \rightleftharpoons RNHCH_2^+ + H_2O \tag{10}$$

$$RNHCH_2^+ + H_2NR \rightleftharpoons RNHCH_2H_2N^+R \tag{11}$$

$$RNHCH_2H_2N^+R \rightleftharpoons RNHCH_2HNR + H^+ \tag{12}$$

A closely related variation on this mechanism has recently been suggested (17). It is depicted in Equations 13-16.

$$RNHCH_2OH + H^+ \rightleftharpoons RNHCH_2O\overset{+}{H}_2 \tag{13}$$

$$RNHCH_2O\overset{+}{H}_2 \rightleftharpoons RNCH_2 + H^+ + H_2O \tag{14}$$

$$RN-CH_2 + RNH_2 \rightleftharpoons RN^-CH_2N\overset{+}{H}_2R \tag{15}$$

$$RN^-CH_2N\overset{+}{H}_2R \rightleftharpoons RNHCH_2NHR \tag{16}$$

Distinguishing between these two mechanisms is difficult; however, recent evidence in the coating area suggests that the second route is important, especially for the cure of certain types of amino resins.

In addition to the two main reactions, methylolation and condensation, there are a number of other reactions that are important for the manufacture and use of amino resins. For example, two methylol groups may combine to produce a dimethylene ether linkage and liberate a molecule of water.

$$2RNHCH_2OH + H^+ \rightleftharpoons RNHCH_2OCH_2NHR + H_2O + H^+ \tag{17}$$

The dimethylene ether so formed is less stable than the diamino-methylene bridge and may rearrange to form a methylene line and liberate a molecule of formaldehyde.

$$RNHCH_2OCH_2NHR \longrightarrow RNHCH_2NHR + CH_2O \tag{18}$$

The simple methylol compounds and the low molecular weight polymers obtained from urea and melamine are soluble in water. They are quite suitable for the manufacture of adhesives, molding compounds, and some kinds of textile treating resins. However, amino resins for coating applications require compatibility with the film-forming polymer resins with which they must react. Furthermore, even where compatible, the free methylol compounds are often too reactive and too unstable for use in a coating formulation that may have to be stored for some time before use. Reacting the free methylol groups with an alcohol to convert them to alkoxymethyl groups overcomes both problems. This is shown in Equation 19.

$$RNHCH_2OH + HOR' \rightleftharpoons RNHCH_2OR' + H_2O \tag{19}$$

With the replacement of the hydrogen of the methylol compound with an alkyl group, the material becomes much more soluble in organic solvents and is much more stable. This condensation reaction is also catalyzed by acids and is usually carried out in the presence of a considerable excess of the alcohol to suppress the competing self-condensation reaction. After neutralization of the acid catalyst, the excess alcohol may be stripped out or left as a solvent for the amino resin.

The mechanism of the alkylation reaction is similar to the curing reaction shown in Equations 9–12. The methylol group becomes protonated and dissociates to form a carbonium ion intermediate that may react with alcohol to produce an alkoxymethyl group or water to revert to the starting material. As one might expect, the amount of water in the reaction mixture should be kept to a minimum, since the relative amounts of alcohol and water will determine the position of the final equilibrium. Probable steps in the alkylation reaction are as follows:

$$RNHCH_2OH + H^+ \rightleftharpoons RNHCH_2O^+H_2 \tag{20}$$

$$RNHCH_2O^+H_2 \rightleftharpoons RNHCH_2^+ + H_2O \tag{21}$$

$$RNHCH_2^+ \rightleftharpoons RNH^+ CH_2 \tag{22}$$

$$RNHCH_2^+ + HOR' \rightleftharpoons RNHCH_2OR' + H^+ \tag{23}$$

In theory, another way of achieving the desired organic solvent compatibility is to employ an amino compound having an organic solubilizing group in the molecule, such as benzoguanamine. With one of the $-NH_2$ groups of melamine replaced by a phenyl group, benzoguanamine-formaldehyde has some oil solubility without modification. Nevertheless, benzoguanamine resins are generally modified with alcohols to provide a still greater range of compatibility with solventborne surface coatings. These benzoguanamine resins are used because they provide a high degree of detergent resistance together with good ductility and excellent adhesion to metal. However, benzoguanamines have never been a major factor because they are more expensive and also because they lack the outdoor durability of melamines.

Manufacturing

Having considered the chemical reactions involved in the preparation of amino resins, a few illustrations will show how these transformations are utilized in the manufacture of molding compounds and coating resins.

Amino resins are generally made by a batch process. Formaldehyde is usually used as an aqueous solution (formalin) because this is the lowest cost form and much easier to handle than the solid paraformaldehyde. If the water content of the formalin is undesirable, paraformaldehyde or an alcohol solution of formaldehyde may be used. In general, the reactants are charged to a kettle, the pH of the mixture adjusted, and the batch heated to bring about chemical reaction. When the desired degree of reaction has been reached, the conditions may be altered to stop the reaction and recover the product or proceed to the next step in the manufacturing process.

A simple amino resin for use in a molding compound might be made by combining melamine with 37% formalin in the ratio of 2 mols of formaldehyde for each mole of melamine. The reaction conditions should be neutral or slightly alkaline pH at a temperature of 60 °C. The reaction should be continued until some polymeric product has been formed so that the dimethylolmelamine does not crystallize when the solution is cooled. Polymer formation can be detected by adding a small amount of the reaction mixture (1 or 2 mL) to a test tube half-full of water. Polymer is indicated by the degree of turbidity in the solution. When the proper degree of reaction has been reached, the resin syrup is pumped to a dough mixer where it is combined with α-cellulose pulp, approximately 1 part of cellulose to each 3 parts of resin solids. The wet spongy mass formed in the dough mixer is then spread on trays and dried in a controlled-humidity drying oven to produce a hard, brittle popcorn-like intermediate. This material may be coarsely ground and sent to storage. To make the molding material, it is only necessary to combine the cellulose-melamine resin intermediate with suitable catalyst, stabilizer, colorants, and mold lubricants. This is best done in a ball mill. The materials must be ground for several hours to achieve the uniform fine dispersion needed to get the desired decorative appearance in the molded article. The molding compound may be used as a powder or it may be densified under heat and pressure to produce a granular product that is easier to handle in the molding operation.

In contrast to this complex procedure, a butylated urea-formaldehyde resin for use in the formulation of fast curing baking enamels may be made as follows (18): Urea, paraformaldehyde, and butanol are charged in the mole ratio 1.0 urea/2.12 formaldehyde/1.5 butanol. Triethanolamine is added to make the solution alkaline (about 1% of the weight of the urea), and the mixture is heated to reflux and held until all the paraformaldehyde has dissolved. Phthalic anhydride is then added to give a pH of 4.0, and the water is azeotroped off until the batch temperature reaches 117 °C. The mixture is cooled and diluted to the desired solid content with solvent.

A highly methylated melamine-formaldehyde textile treating resin may be made as follows (19): A methanol solution of formaldehyde (55% formaldehyde/35% methanol/10% water) is charged to a reaction kettle and adjusted to a pH of 9.0-9.5 with sodium hydroxide. Melamine is then added to give a ratio of 1 mole of melamine for each 6.5 mol of formaldehyde. The mixture is heated to reflux and held for 0.5 h. The reaction mixture is then cooled to 35 °C and more methanol is added to bring the ratio of methanol to melamine up to 11 mol to 1 mol. With the batch at 35 °C, enough sulfuric acid is added to reduce the pH to 1.0. After the reaction mixture is held at this temperature and pH for 1 h, the batch is neutralized with 50% sodium hydroxide and the excess methanol stripped off to give a product containing 60% solids. The product is then clarified by filtration. A highly methylated resin, such as this, may be used for imparting crease resistance to cotton. It might also be used in certain water-soluble coating formulations.

Applications

Having discussed the chemistry and manufacture of amino resins, we return now to their uses. Applications for amino resins fall logically into three broad classes. First, those in which the amino is the major component. Adhesives, laminating resins, and molding compounds are the important examples. Second, there are those in which the amino resin is used as a treatment and represents only a very minor part of the finished article. Paper and textile treatments are the best known examples of this type of application. Third, there are coatings applications where the amino resin acts as a cross-linking agent for another polymer system, such as alkyds, vinyl or acrylic copolymers, or polyesters having functional groups that may react with the amino.

In the first type of applications, the reaction is the acid-catalyzed self-condensation, giving a three-dimensional polymer network bonded together with methylene bridges. In the second group, treatment of cellulose, either as paper or as woven cotton cloth, there is evidence that both polycondensation and the cross-linking of cellulose are important. This is especially true for wash-and-wear treatment of cotton cloth. In contrast, the treatment of paper to improve wet strength is a surface phenomenon. Best results are obtained with a colloidal solution of a melamine-formaldehyde condensate. The surface of the cellulose fiber adsorbs the resin before the sheet of paper is made; when the paper is dried, the resin cures to an insoluble network on the surface. Some grafting on the cellulose surface probably takes place due to the high concentration of hydroxyl groups in the fiber.

The third type of application is the one that is involved in surface coatings and the one in which we are primarily interested. The utilization of amino resins in surface coatings is based on the same simple reactions mentioned in the discussion of resin preparation. However, this chemistry can be varied to produce a wide range of properties. Amino-containing coating vehicles generally consist of two main ingredients, a polymer portion, which may vary from 50% to 90%, and the amino cross-linking agent, making up the balance. The polymer generally contains reactive sites, typically hydroxyl, carboxyl, or amide groups. The cross-linking agent contains condensation sites, either $-NCH_2OH$ or $-NCH_2OR$ groupings. A typical structural element of a conventional butylated melamine resin is I. Such resins, along with the analagous butylated ureas, were the predominant amino resins used in coatings for many years.

$$\text{ROCH}_2-\underset{\underset{\underset{\text{ROCH}_2}{N}}{N}}{\overset{\overset{\overset{N}{N}}{N}}{N}}-\text{CH}_2\text{OH}$$

I

$$R = C_4H_9$$

As a coating composition containing a polymer and an amino resin is heated under acidic conditions, reaction occurs between the active hydrogen groups and the condensable groups of the aminoplast as follows:

$$BH + -NCH_2OR \rightleftharpoons N-CH_2B + HOR \tag{24}$$

Since both the amino resin and the polymer have multiple reactive sites, application of sufficient heat will yield a cross-linked structure with the degree of cure depending on the functionality of the system and on how far to the right the reaction (24) is driven.

This, in its very simplest form, is the chemistry of amino coatings. There are, of course, an enormous variety of types and reactivities of both amino compounds and polymers. While it would be useless to try to enumerate the variations possible, some systematization of the reactivity of amino resins is necessary to understand the trends in the coating industry.

There are a number of possible configurations for the substituents on melamine coating resins:

$$-N\!\!\begin{smallmatrix}H\\H\end{smallmatrix} \qquad\qquad -N\!\!\begin{smallmatrix}CH_2OH\\H\end{smallmatrix} \qquad\qquad -N\!\!\begin{smallmatrix}CH_2OH\\CH_2OH\end{smallmatrix}$$

II   III   IV

$$-N\!\!\begin{smallmatrix}H\\CH_2OR\end{smallmatrix} \qquad\qquad -N\!\!\begin{smallmatrix}CH_2OR\\CH_2OH\end{smallmatrix} \qquad\qquad -N\!\!\begin{smallmatrix}CH_2OR\\CH_2OR\end{smallmatrix}$$

V   VI   VII

In general, in melamine cross-linking agents for coatings, the first three groupings are of negligible importance. Most of the useful materials have at least 4 mols of combined formaldehyde, more than half of which are further reacted with an alcohol. Thus, statistically II-IV are present in only minor amounts. The curing reactions of melamine cross-linking agents mainly involve the chemistry of structures V-VII (17).

Structure VII is basically the only significant grouping in the hexa(methoxymethyl)melamine (VIII) type of cross-linking agent that is becoming increasingly important commercially. With this resin the chemistry is relatively simple.

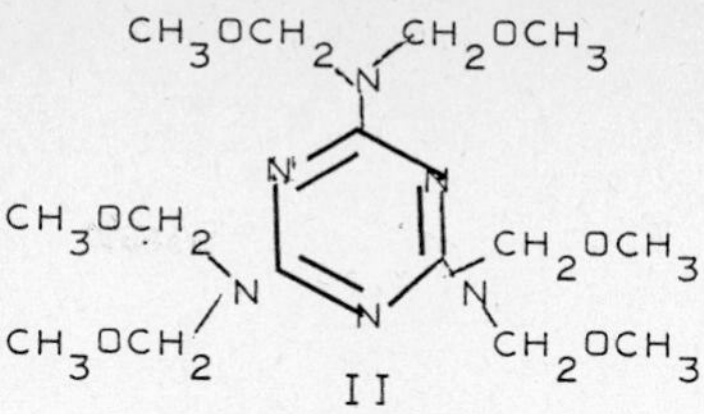

II

VIII

The important reaction is

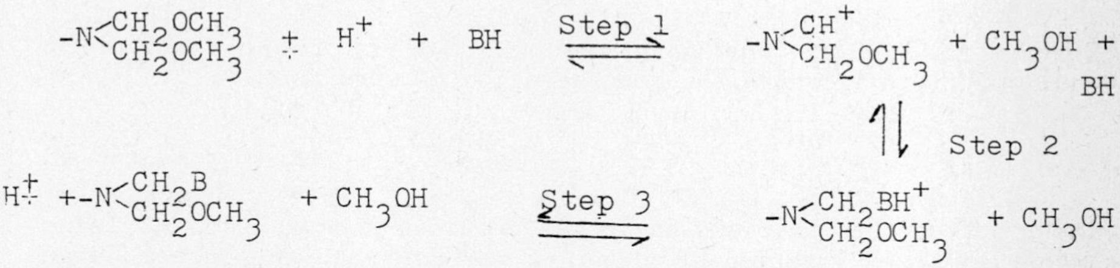

$$-N\begin{smallmatrix}CH_2OCH_3\\CH_2OCH_3\end{smallmatrix} + H^+ + BH \xrightarrow{\text{Step 1}} -N\begin{smallmatrix}CH^+\\CH_2OCH_3\end{smallmatrix} + CH_3OH + BH$$

Step 2

$$H^+ + -N\begin{smallmatrix}CH_2B\\CH_2OCH_3\end{smallmatrix} + CH_3OH \xrightarrow{\text{Step 3}} -N\begin{smallmatrix}CH_2BH^+\\CH_2OCH_3\end{smallmatrix} + CH_3OH$$

Strong acid catalyst is required for this reaction to go to any extent. Similar chemistry can occur when R is butyl or methyl, but the cure speed is always slower, presumably because the loss of butanol is not as fast so that the reverse reaction in Step 1 is more likely. Another characteristic of melamines with the (dialkoxymethyl)amino grouping as the main functionality is their selectivity. Except under conditions of high acidity or high temperature, this type of material reacts mainly with the hydroxyl, carboxyl, or amide functionality of the backbone polymer in contrast to reacting with other melamine molecules.

Resins that contain groupings V and VI are more complex in their reactivity. Their cross-linking reactions are also acid-catalyzed, but in this case a weak acid is sufficient. Polymers with moderate acid numbers will catalyze the curing reactions without the need for additional catalyst.

Recent work in the development of resins with high concentrations of grouping V have led to some revisions in the thinking about the mechanism of reaction of the partially methylolated and alkylated type of melamine (17). For many years it was felt that the key to reactivity was the presence of the methylol $-NCH_2OH$ grouping. It was generally believed that the reaction followed the same carbonium ion route as grouping VII but that the carbonium ion formed more readily from the methylol than from the alkoxymethyl group.

A key finding that refutes this mechanism is that resins in which grouping V predominates are more reactive than those high in grouping VI. Another important finding was that resins high in methylol groups lose formaldehyde during cross-linking in an amount similar to the original methylol content (17). This has led to the thesis that the actual reaction course is as follows:

$$-N\begin{smallmatrix}CH_2OR\\H\end{smallmatrix} + H^+ + BH \xrightarrow{\text{Step 1}} -N\begin{smallmatrix}H+\\CH_2OR\\H\end{smallmatrix} + BH$$

$$\Big\updownarrow \text{Step 2}$$

$$-N\begin{smallmatrix}CH_2BH^+\\=\end{smallmatrix} + ROH \xrightarrow{\text{Step 3}} \qquad -N=CH_2 + ROH + H^+$$
$$+ H^+ \qquad\qquad\qquad\qquad\qquad + BH$$

$$\text{Step 4} \;\Big\updownarrow$$

$$-N\begin{smallmatrix}CH_2B\\H\end{smallmatrix} + ROH + H^+$$

In the case where the main grouping is VI, there is a prior equilibrium to increase the concentration of structure V:

$$-N\begin{smallmatrix}CH_2OR\\CH_2OH\end{smallmatrix} \rightleftharpoons -N\begin{smallmatrix}CH_2OR\\H\end{smallmatrix} + CH_2O$$

The reaction then proceeds as in the previous scheme. This explains both the slower reaction of group VI versus V and why such a large proportion of the free methylol groups is lost as formaldehyde during the cross-linking. Because this type of resin has substantial −OH and −NH functionality, self-condensation takes place much more readily than in resins with only grouping VII. The melamines of this class can act as the "BH" component in the scheme.

Trends in Amino Resins

Developments in coatings technology have been reflected in the types of amino resins used. Melamines have grown in general at the expense of ureas, and the higher solids, more monomeric melamines have grown at the expense of conventional polymeric melamines.

For many years the polymeric, butylated melamines were the standard cross-linking agents used for coatings. They were relatively easy to manufacture because the condensation reaction could be driven by azeotropic distillation of water/butanol. Since water and butanol are only partially soluble in each other, the distillate could be separated simply by decantation and the butanol-rich fraction recycled in subsequent batches. As long as butanol was inexpensive, discarding the aqueous fraction, which contained about 10% butanol, was not a great economic penalty.

In the last 15 years the growth products in coating resins have been the high-solids melamines, which are predominantly the methylated rather than butylated resins. These cross-linking agents are generally more difficult to manufacture than butylated resins were. Large excesses of methanol have to be used, and recovery and recycle of raw materials are a necessity for economical manufacturing. Nonetheless, these products are more economical than the older type even though they are more expensive on an "as is" basis. Lower weight loss and elimination of expensive solvents are the key reasons for this (21).

Future Trends

As stated above, over the course of the last 20 years there has been growth of the high-solids cross-linking agents at the expense of the low-solid butylated aminos.   This trend began slowly but has accelerated in the last 5 years, fueled by changing coating technology, and the better overall economics of the high-solids resins.

Industrial coating technology is changing, and five areas are growing: (1) high solids; (2) waterborne; (3) electrodeposition; (4) powder; (5) radiation cure.   Of these five areas, aminos are very important in the first two.

High-solids technology is emerging as a very important technology for many types of coatings—automotive, container, coil, and general industrial.   The higher solids aminos fit well into this type of technology (20).   Because they are of low molecular weight, are of moderate viscosity, and carry little or no solvent, they offer many advantages in high-solids formulations.   Because this type of coating tends to use higher levels of amino resin per pound of coating, the growth of this technology may increase the total volume of amino resins even though coating volume shows very low growth.   Of all the markets where higher solids will impact amino resin volume, automotive promises to be the most important.   It appears highly likely that in the intermediate term higher solid acrylic/melamine systems will capture most of the automotive topcoat and primer-surfacer coatings.   This would mean a major increase in high-solids cross-linkers.

In general, high solids seems likely to be the most important of the newer technologies.   In part this is because it is the least different of the five from the conventional technology.   In its simplest terms high solids is achieved by lowering the molecular weight of the conventional types of backbones and then increasing the amount of cross-linker.   However, it is also important that high solids offers generally more economical components that the other contending technologies.

Waterborne coatings are also major users of the higher solids melamines.   One important factor here is the greater stability of the higher solids, fully alkylated resins, which are exemplified by hexa(methoxymethyl)melamine (VIII).   The lower organic volatiles given off by methylated resins (versus butylated) also favor their use in waterborne coatings, which is frequently adopted to reduce organic emissions from coating installations.   Although waterborne coatings seem to be losing out as automotive finishes (except for primers), they are growing factors in container and wood coatings.

Electrodepositable coatings are, of course, a class of waterborne coatings, so in a sense their growth is a special part of the trend to water.   Initially electrodepositable coatings were mainly of the anodic type.   That is, the vehicles were negatively charged and deposited on the anode in the coating process.   In this type of system melamine resins of the high-solid type were quite important.   Most of the resins used were of the fully alkylated type and had both methyl and higher alkyl groups to increase the hydrophobicity of the resin.   More recently, cathodic electrocoating has become the favored system.   The speed with which cathodic electrocoating captured the automotive primer business was dramatic and highlighted the shift to cationic systems. Aminos are much less important in the existing cationic systems, and their applicability in general is doubtful. This is because many cathodic systems utilize acidic baths, in which aminos tend to be unstable, and cure under basic conditions, where aminos tend to be unreactive. Thus, electrodeposition of systems cured with aminos will almost certainly grow much more slowly than electrodeposition as a whole.

In the areas of powder and radiation-cured coatings, aminos have no particular applicability. While in principle they could be used in powder, they have proven only marginally useful. Sintering caused by aminos has proven to be a major problem in powder formulations. Higher melting aminos have not solved the problem. The organic volatiles given off in curing are also more of a problem in powder, which seems to work best when addition rather than condensation reactions are involved in cross-linking.

Radiation cure seems an unlikely application for aminos. In most cases vinyl polymerization activated by ultraviolet or other radiation is used as the cure reaction. Thus, aminos are not naturally useful and need modification to introduce polymerizable double bonds. This is possible (21), but so far such technology has not found much practical application.

Summary

In the future aminos are likely to grow slowly for some time. Since the overall volume is so strongly tied to the huge quantities of urea-formaldehyde used in wood products, the course of this business is crucial. As noted, toxicity of formaldehyde is a concern, and the ultimate resolution of this issue is uncertain. However, the economy of urea-formaldehyde for this application is proven, and it appears probable that improved technology will sustain the use of these resins in this market. We may see modified resins but their elimination is unlikely.

Amino molding compounds do not appear to be a likely growth area. The technology is mature, and no new markets have appeared for many years. The only significant use of melamine molding compounds is for dinnerware, a slow growth market. Urea-formaldehyde molding compounds have certain markets in electrical devices and closures. These too are not expected to show any great growth.

Use of aminos in laminating applications is also a modest growth area, tied largely to building cycles. The type of resins used in laminating has changed little in the last 20 years.

Coatings are likely to be a growth area for aminos. As noted previously, the technology of the resins used in this application is the most dynamic. Also, the newer amino resins are well suited to certain growth areas of new coating technology, high solids and waterborne. As a result a moderate growth of amino usage in coatings is likely in the next 5-10 years.

Literature Cited

1. Vale, C. P.; Taylor, W. G. K. "Aminoplastics"; Iliffe: London, 1964.
2. Widmer, G. "Encyclopedia of Polymer Science and Technology"; Interscience: New York, 1965; Vol. 2, pp. 1-94.

3. Wohnsiedler, H. P. "Kirk-Othmer Encyclopedia of Chemical Technology," 2nd ed.; Interscience: New York, 1963; Vol. 2, pp. 225-58.
4. Blaise, J. F. "Amino Resins"; Reinhold: New York, 1959; pp. 190-97.
5. Rossiter, E. C. British Patents 248 477, 1924; 258 950, 1925; and 266 028, 1925.
6. Henkel and Co. German Patent 647 303, 1935; British Patent 455 008, 1936.
7. Gams, A.; Widmer, G.; Fisch, W. Helv. Chim. Acta 1941, 24, 302E-319E.
8. "Kline Guide to Chemical Industry," 4th ed.; Charles Kline and Co.: Fairfield, N.J., 1980.
9. Tollens, B. Ber. Dtsch. Chem. Ges. 1884, 17, 659.
10. Lüdy, E. J. Chem. Soc. 1889, 56, 1059; Monatsch 1889, 10, 259.
11. Goldschmidt, C. Ber. Dtsch. Chem. Ges. 1896, 29, 2438.
12. Einhorn, A.; Hamburger, A. Ber. Dtsch. Chem. Ges. 1908, 41, 24.
13. John, H. British Patent 151 016, 1918; U.S. Patent 1 355 834, 1920.
14. Goldschmidt, H.; Neuss, O. British Patents 187 605, 1921; 202 651, 1922; and 208 761, 1922.
15. Pollak, F. U.S. Patent 1 458 543, 1923; British Patents 171 094, 1920; 181, 014, 1921; 193 420, 1922; 201 906, 1922; 206 512, 1922; 213 567, 1923; 238 904, 1924; and 248 729, 1925.
16. Ripper, K. U.S. Patent 1 460 606, 1923.
17. Blank, W. J. J. Coat. Technol. 1979, 51(656), 61-70.
18. Lindlaw, W. Celanese Chemical Co., New York, Technical Bulletin "The Preparation of Butylated Urea-Formaldehyde and Butylated Melamine Formaldehyde Resins Using Celanese Formcel and Celanese Paraformaldehyde," Table XIIIA.
19. Celanese Chemical Co., New York, Technical Bulletin S-23-8, 1967, and supplement to Technical Bulletin S-23-8, 1968, Example VIII.
20. Albrecht, N. J.; Blank, W. J. Proc. 6th Int. Conf. Org. Coat. Sci. Technol. 1980, 4.
21. Blank, W. J. J. Coat. Technol. 1982, 54(687), 26-41.

Resins and Additives Containing Silicon

SHELBY F. THAMES

Department of Polymer Science, University of Southern Mississippi, Hattiesburg, MS 39401

Inductive Effects and (p-d)π Bonding
Bond Strengths
Silicon-Halogen Bond
Silicon-Carbon Bond
Double Bond to Silicon
Silicon-Hydrogen Bond
Silicon-Oxygen Bond
Silicon-Nitrogen Bond
Characteristics of Silicone Polymers
Silicon-Containing Polymers
Silicones as Water Repellents
Pigmentation
Silicones as Additives

The synthesis of tetraethylsilane, the first organosilicon compound in which silicon was bonded to carbon, was accomplished in 1863. This reality, when combined with the fact that the carbon and silicon atoms are members of the same periodic group, provided impetus for the idea that there might exist a branch of chemistry in which silicon would take the place of carbon.

The pursuit of this idea was the cornerstone of the most careful and extensive series of investigations in organosilicon chemistry. These investigations were initiated by F. S. Kipping (1). His series of investigations led him ultimately to conclude that the field of organosilicon chemistry did not match that of carbon and that the differences between the two fields were greater than the similarities as the following will indicate (2):

1. Both carbon and silicon have a normal covalency of four.
2. Both carbon and silicon have their normal bonding tetrahedrally arranged.
3. However, silicon is larger (20%) and heavier than carbon.
4. Silicon is less electronegative (more electro positive).
5. Under favorable conditions silicon may have a coordination number greater than 4; for example, silicon can utilize its 3d orbitals for bonding.

In the first two relationships, carbon and silicon exhibit striking similarities, while the latter comparisons confirm marked

0097–6156/85/0285–1117$07.00/0
© 1985 American Chemical Society

differences between the two elements. For the most part these differences can be explained by considering two fundamental properties of the silicon atom: (1) its low electronegativity and (2) its vacant 3d orbital. The electronegativity of silicon on the Pauling scale is 1.8 while that for carbon is 2.5 (Table I). Thus, silicon in relation to carbon is an electron donor, and therefore silicon bonds have a greater degree of ionic character than do their carbon analogues (Table I).

It must be remembered, therefore, that the silicon atom is the most electropositive partner of not only the Si–C bond but also the Si–H bond. Thus, ionic cleavages of such bonds as Si–O, Si–C, and Si–H proceed by electrophilic attack (E$^+$) on silicon's bonding mate and nucleophilic attack (Nu$^-$) on silicon itself (Scheme I).

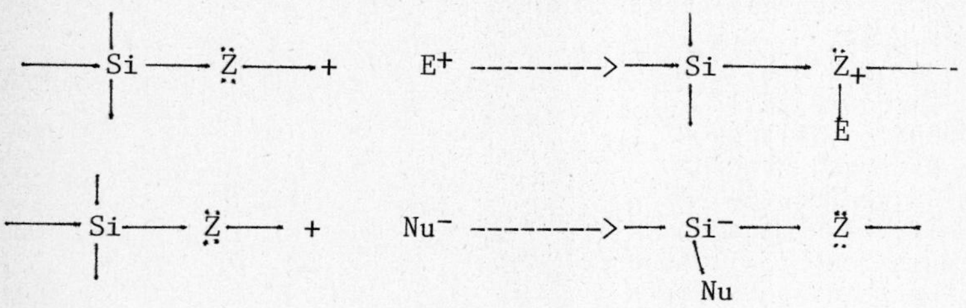

$$\begin{array}{c} | \\ ---Si--- \ddot{Z}--- \\ | \end{array} + \quad E^+ \quad --------> \begin{array}{c} | \\ ---Si--- \ddot{Z}_+--- \\ | \quad\quad | \\ \quad\quad E \end{array}$$

$$\begin{array}{c} | \\ ---Si--- \ddot{Z}--- \\ | \end{array} + \quad Nu^- \quad --------> \begin{array}{c} | \\ --- Si^--- \ddot{Z} --- \\ \backslash \\ Nu \end{array}$$

Scheme I

Inductive Effects and (p–d)π Bonding

The electropositive effect of silicon gives rise to markedly higher basicities of trimethylsilyl–substituted aliphatic amines (3) as compared to their carbon analogues. This is, of course, the result of a positive inductive effect where competing (p–d)π bonding is not possible. This was confirmed by the comparative studies of Sommer and Rockett (4) with silicon- and non-silicon-containing amines. In each case the silicon containing amine was the stronger base (Scheme II). This inductive effect diminishes as the number of carbon atoms insulating silicon from nitrogen <u>increases</u>, as shown by the decreases in the ratio of K_{b_1}/K_{b_2}.

Amine 1	Amine 2	K_{b_1}/K_{b_2}	
CH$_3$)$_3$–Si–CH$_2$–NH$_2$	CH$_3$–NH$_2$	1.82	Note decreasing influence on
CH$_3$)$_3$Si–CH$_2$CH$_2$NH$_2$	CH$_3$CH$_2$NH$_2$	1.73	nitrogen basicity as the number of
CH$_3$)$_3$Si–CH$_2$CH$_2$CH$_2$NH$_2$	CH$_3$CH$_2$CH$_2$NH$_2$	1.14	methylene groups between silicon and nitrogen increases

Scheme II

Consider, for example, Compound III as compared to Compound IV using II as a reference with no inductive or electronic effects (Scheme III, with chemical shifts in parentheses).

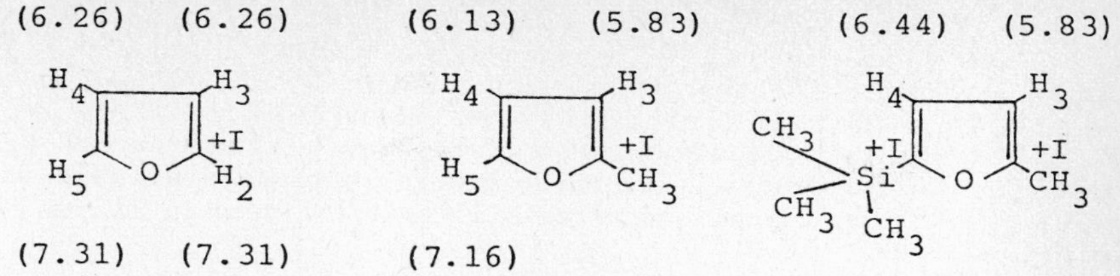

(6.26)   (6.26)   (6.13)   (5.83)   (6.44)   (5.83)

(7.31)   (7.31)   (7.16)

II   III   IV

Scheme III

Bond Strengths

The strength of the carbon–carbon bond and the silicon–carbon bond is 82.6 and 76.0 kcal/mol, respectively, and thus the latter is slightly more reactive toward homolytic cleavage. This is shown by the greater ease of thermal decomposition of the tetraalkylsilanes (5) as compared to their carbon analogues. Heterolytic cleavage, on the other hand, is a different matter, and the fact that the silicon atom is larger than carbon by approximately 20%, less electronegative, and capable of a greater maximum coordination number than the carbon atom makes the silicon–carbon bond considerably more reactive than the carbon–carbon bond toward a number of reagents (6, 7).

Table I.   Electronegativities of Selected Elements and Ionic Character and Ionic Bond Energies of Si–X Bonds

Element	Electro-negativity	Ionic Character (%)	Ionic Bond Energy (kcal/mol)
Si	1.8		
C	2.5	12	222.9
H	2.1	2	249.8
N	3.0	30	
O	3.5	50	242.4
S	2.5	12	192.7
F	4.0	70	237.4
Cl	3.0	30	190.3
Br	2.8	22	179.0
J	2.4	8	167.4

Silicon–Halogen Bond

The silicon–halogen bonds, on the other hand, occupy a position of great importance in organosilicon chemistry as most all organosilicon compounds must be synthesized ultimately through halosilane intermediates. The three general methods available for the preparation of halosilanes from silicon are listed in Scheme IV.

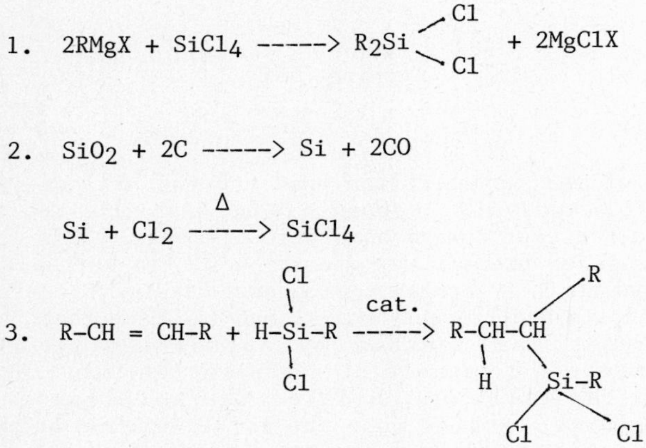

1. $2RMgX + SiCl_4 \longrightarrow R_2Si\begin{smallmatrix} Cl \\ \\ Cl \end{smallmatrix} + 2MgClX$

2. $SiO_2 + 2C \longrightarrow Si + 2CO$

    $Si + Cl_2 \overset{\Delta}{\longrightarrow} SiCl_4$

3. $R-CH = CH-R + \underset{\underset{Cl}{|}}{\overset{\overset{Cl}{|}}{H-Si-R}} \overset{cat.}{\longrightarrow} R-CH-\underset{H}{\overset{|}{CH}}\begin{smallmatrix} R \\ \diagup \\ \diagdown Si-R \\ \diagup \diagdown \\ Cl \quad Cl \end{smallmatrix}$

Scheme IV

They include the Grignard alkylation process originally discovered by Friedel and Crafts (1), the direct method developed by Rochow (8) and hydrosilylation (9).

Once prepared, the silicon halides are much more reactive than their carbon analogues toward polar reagents. Thus, while carbon tetrachloride (CCl_4) and chloroform ($CHCl_3$) are stable toward aqueous solvents, silicon tetrachloride ($SiCl_4$), and trichlorosilane ($SiHCl_3$) are hydrolyzed rapidly, even in moist air.

Silicon–Carbon Bond

The Si–C bond is most frequently formed via the Grignard synthesis route and was the first industrial method of preparing organosilanes, Equation 1.

$$SiCl_4 + CH_3---MgBr \longrightarrow (CH_3)_4Si + MgBrCl \qquad (1)$$

Steric factors do play a role in product identity and yields, as exemplified by the following competitive reaction, Equation 2.

$(C_2H_5)_3SiCl + CH_3MgBr + C_2H_5MgBr \longrightarrow (C_2H_5)_3Si-CH_3 \text{ (only)}$

$(CH_3)_3SiCl + (C_2H_5)_3SiCl + C_2H_5MgBr \longrightarrow (CH_3)_3SiC_2H_5 +$

$(C_2H_5)_4Si \qquad (2)$

Furthermore, cyclic products are prepared by the Grignard synthesis route, Equation 3.

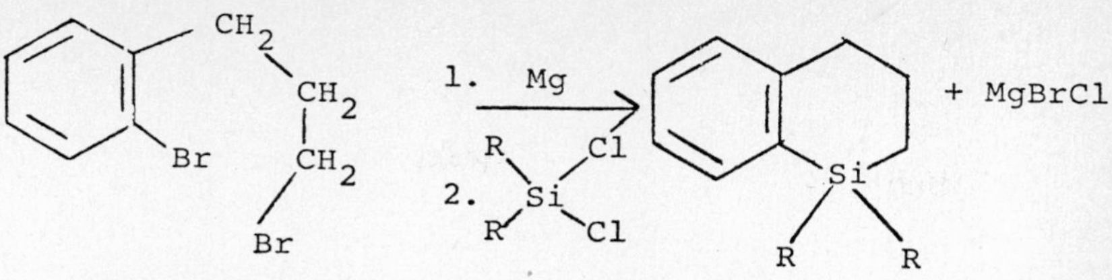

(3)

In addition to Mg, several other organometallics such as organo-sodiums, -lithiums, -aluminums, -mercury, and -zincs can be employed for the formation of organosilicon compounds.

The Si-C bond is, in general, more polar than the C-C bond in that a bond of approximately 12% ionic character is formed. Thus, aryltrimethylsilanes are cleaved by acids (6) under rather mild conditions, while tertiary butyl aryls are resistant and are not cleaved by even strong acids. Similarly, the silicon–carbon bond is also more susceptible to basic cleavage (7) than the carbon–carbon bond, and any substituent that tends to increase the polarity of the bond has a greater effect on the reactivity of a silicon–carbon than on its carbon–carbon analogue.

Furthermore, the molecular weights of silicone polymers are higher than their viscosities indicate. This is a function of the low intermolecular forces characteristic of organosilicones.

In addition, the cleavage of aryl-silicon bonds by <u>neutral</u> reagents delineates even more the vast differences between the fields of organosilicon and carbon chemistry.

The carbon-silicon bond is also more reactive toward halogenation than is the carbon-carbon bond. This has been demonstrated by attempts to brominate or iodinate trimethyl-phenylsilane (Scheme V).

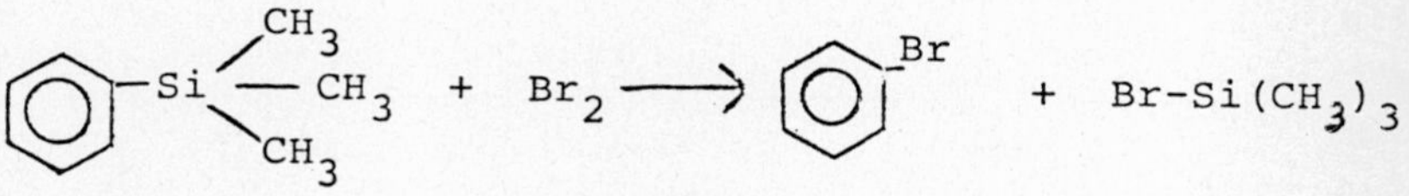

Scheme V

Anderson and Webster (10) showed the pyridyl–silicon bond of 2-(trimethylsilyl)pyridine to be susceptible to cleavage by alcohol and water to form unsubstituted pyridine. Thames and coworkers (11) showed that neutral reagents, including aldehydes, acid halides, chloroformates, and anhydrides, have the ability to cleave several silicon–carbon bonds on select heterocycles (Schemes VI-VIII). However, the silicon–carbon bond of nonheterocycles is not susceptible to neutral reagent cleavage, and too, the full extent of heterocyclic cleavage reactions is not known.

Similar contrast is seen in the behavior of the two bonds toward anhydrous aluminum chloride. In carbon chemistry, the Friedel-Crafts reaction is one of the most commonly used methods of forming carbon–carbon bonds. In organosilicon chemistry, however, aluminum chloride is a convenient means of cleaving the silicon–carbon bond (12). Even anhydrous ferric chloride cleaves silicon–carbon bonds under mild conditions.

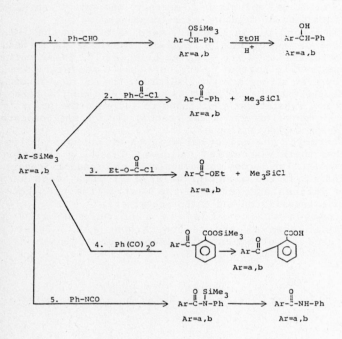

Scheme VI

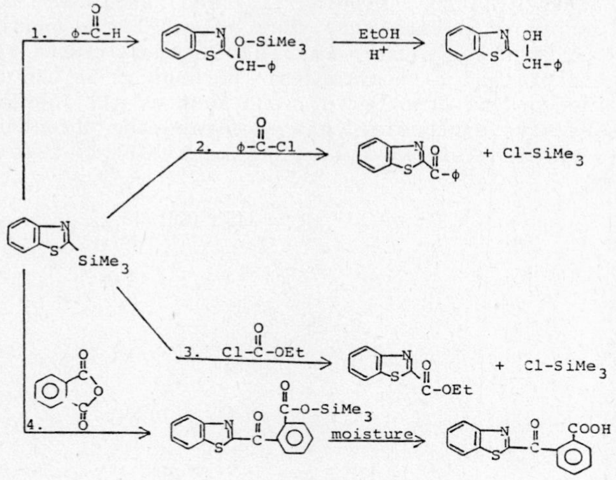

Scheme VII

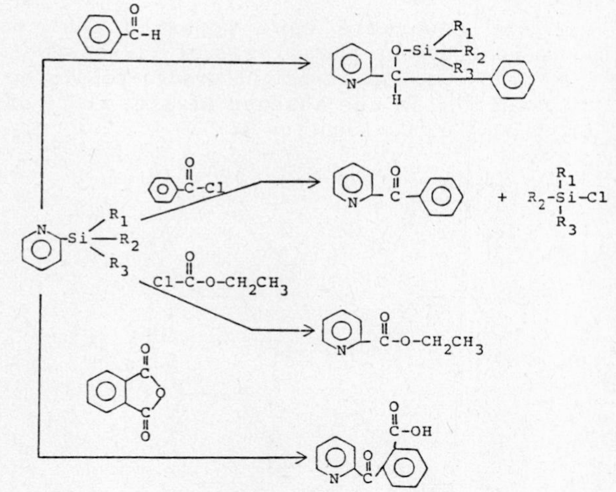

Scheme VIII

Double Bond to Silicon

It is also most significant that there was, until publication of the work by A. G. Brook (13) and Robert West (14), a complete absence of evidence for the existence of any organosilicon compound containing a double bond to silicon. Kipping (15) had concluded in the early years of organosilicon chemistry that an ethylenic bonding between carbon and silicon was either impossible or it could be produced only under exceptional circumstances. Perhaps it is the latter case that makes the work of Brook et al. and West et al. possible. Brook and co-workers have synthesized and confirmed the structure of 2-(1-adamantyl)-2-trimethylsiloxy-1,1-bis(trimethylsilyl)-1-silaethylene, Equation 4.

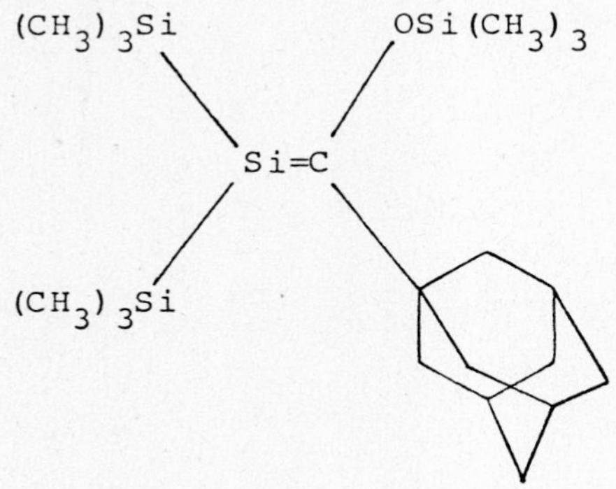

$$(CH_3)_3Si \qquad\qquad OSi(CH_3)_3$$

$$Si=C$$

$$(CH_3)_3Si$$

(4)

Likewise, West and coworkers have reported the synthesis of tetramesityldisilene by the photolysis of 2,2-bis(mesityl)hexamethyltrisilane. The compound is bright orange-yellow and exists as a crystallizing solid. In the absence of air, it is stable up to its melting point of 176 °C, Equation 5.

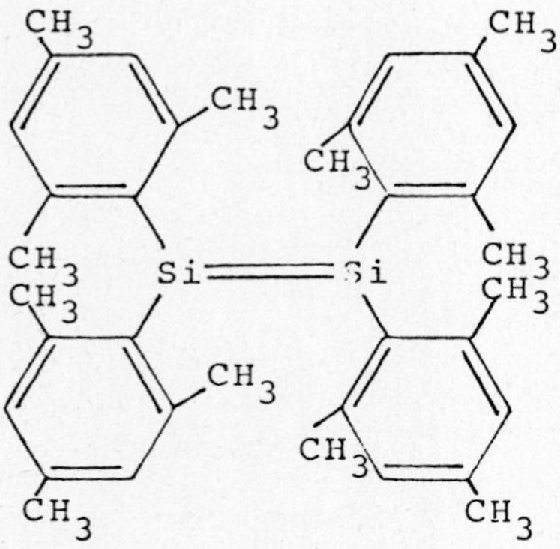

$$Si === Si$$

(5)

Furthermore, vinylsilanes are a reality and undergo an anti-Markownikoff addition of HBr as a result of the $(p-d)\pi$ bond influence, Equation 6.

$$R_3Si-CH=CH_2 + HBr ---> R_3Si-CH_2-CH_2-Br \qquad\qquad (6)$$

Silicon–Hydrogen Bond

In organic chemistry the element most commonly associated with carbon is hydrogen, yet in organosilicon chemistry the number of compounds containing the silicon–hydrogen bond is almost a negligible portion of the total. This fact gives rise to one of the most important differences between the two branches of chemistry in which there is much greater reactivity toward polar reagents of the silicon–hydrogen bond as compared with the carbon–hydrogen bond. Thus, in contrast to the inertness of methane (CH_4), silane (SiH_4) is spontaneously flammable and is hydrolyzed readily by water and even more readily by aqueous acids and bases while halogen acids react with it to give mixtures of halosilanes and hydrogen, Equation 7.

$$HSiCl_3 \xrightarrow{\quad Cl_2 \quad} SiCl_4$$

$$R_3SiH \xrightarrow{\quad X_2 \quad} R_3Si\text{-}X + HX$$

$$RSiH_3 \xrightarrow{\quad X_2 \quad} RSiH_2X \xrightarrow{\quad X_2 \quad} R\text{-}SiHX_2 \tag{7}$$

Furthermore, under certain conditions the Si–H bond is known to add to ethylenic double bonds, and such a reaction is termed hydrosilylation, Equation 8.

$$Si\text{-}H + \underset{\substack{| \\ R \; R}}{\overset{\substack{R \; R \\ |}}{C=C}} \xrightarrow{\quad H_2PtCl_6.6H_2O \quad} \underset{\substack{| \quad | \\ R \quad R}}{\overset{\substack{R \quad R \\ | \quad |}}{Si\text{----}C - C - H}} \tag{8}$$

In the Si–H bonding the silicon atom is more electropositive than hydrogen and possesses 2% ionic character. Thus, attack of the Si bond takes place via electrophilic attack on hydrogen or by nucleophilic attack on silicon.

The most widely applicable method for the synthesis of silicon hydrides on a laboratory scale is the reduction of silicon–halogen, –nitrogen, or –oxygen bonds.

Silicon–Oxygen Bond

Compounds containing silicon–oxygen bonds occupy a position of special importance in organosilicon chemistry. Many industrially important materials contain chains of silicon–oxygen bonds, which provide for molecules of very high molecular weight having high chemical and electrical resistance, low temperature coefficient of viscosity, and strong water-repellent properties. Aldehydes give high polymers that are analagous to the polysiloxanes with the exception that they contain an unsubstituted hydrogen atom in the chain (Scheme IX). However, these carbon compounds depolymerize so readily that they have no industrial importance.

$$\text{Scheme IX}$$

A common species containing a silicon–oxygen bond is that of the silanols and silanediols, which are analogous to alcohols and gem-diols in the carbon series. Both the silanols, Equation 9, and silanediols, Equation 10, undergo intermolecular condensation with the elimination of water. This reaction is both acid and base catalyzed and occurs so readily with alkyl- and mixed alkylarylsilanols that the uncondensed silanols are frequently quite difficult to prepare.

$$2R_3Si-OH \longrightarrow R_3Si-O-SiR_3 + HOH \tag{9}$$

$$\tag{10}$$

Furthermore, the stability of two hydroxyl groups on a silicon atom is much greater than two hydroxyl groups on a carbon gem-diol, as these carbon compounds dehydrate to give ketones, Equation 11, or aldehydes, Equation 12, while the corresponding silanediols give rise to the formation of polysiloxanes via intermolecular condensation.

$$\tag{11}$$

$$\underset{H}{\overset{R}{\diagdown}} \underset{OH}{\overset{OH}{\diagup}} C \longrightarrow \underset{H}{\overset{R}{\diagdown}} C=O \;+\; HOH \qquad (12)$$

$$\text{aldehyde}$$

Other Si-O bonds are produced in the following manner, Equation 13.

$$R_3Si-X \;+\; R'OH \;\longrightarrow\; R_3Si-O-R' \;+\; HX$$

$$R_2SiX_2 \;+\; 2R'OH \;\longrightarrow\; R_2Si-(OR')_2 \;+\; 2HX$$

$$SiCl_4 \;+\; 4R'OH \;\longrightarrow\; Si(OR')_4 \;+\; 4HCl \qquad (13)$$

R' may be a variety of substituents of both the aliphatic and aromatic types.

The Si-O bond can itself be cleaved by appropriate reagents, as in Equation 14.

$$\equiv Si-OR + R'OH \underset{\longleftarrow}{\overset{\longrightarrow}{}} \equiv SiO-R' + ROH \qquad (14)$$

This reaction can be shifted to the right by either removing the alcohol formed in the condensation (ROH) or by using a large excess of the alcohol (R'OH). Such condensation reaction rates are greatest with the larger number of hydroxyl groups per silicon atom; for example

$$R_3SI-OH \;<\; R_2Si(OH)_2 \;<\; RSi(OH)_3$$

Silicon-Nitrogen Bond

The silicon to nitrogen bond, due to its greater ease of hydrolysis to silanols and amines, does not play the important role in organosilicon chemistry that the carbon-nitrogen bond plays. For instance, the silicon-nitrogen compounds or silazanes hydrolyze about as readily as do the halosilanes, but the products are the relatively mild amines instead of the strongly corrosive halogen acids. Similarly, Si-S, Si-O-C-, and Si-metal bonds also hydrolyze with relative ease (Scheme X).

$$\equiv Si-N\diagdown \;+\; HOH \longrightarrow \equiv Si-OH \;+\; H-N\diagdown$$

$$\equiv Si-Cl \;+\; HOH \longrightarrow \equiv Si-OH \;+\; HCl$$

$$\text{Scheme X}$$

Likewise, the Si-N bond can be split by H_2S although with greater difficulty than with oxygen-containing reagents, Equation 15

$$R_3Si-N-R' + H_2S \longrightarrow R_3Si-S-SiR_3 + R_3---SiSH \qquad (15)$$

hydrogen halides, Equation 16, or

$$R_3Si\text{-}NR_2 \quad + \quad HX \text{--------}\!> R_3Si\text{-}X + HNR'_2$$
$$R_2Si(NR_2)_2 \quad + \quad HX \text{--------}\!> R_2SiX_2 + 2H\text{-}NR_2$$
$$RSi(NR_2)_3 \quad + HX \text{--------}\!> R\text{-}SiX_3 + 3HNR_2 \qquad (16)$$

acid halides to give halosilane, Equation 17.

$$(R_3Si)_2\text{-}N\text{-}H + SO_2Cl_2 \text{--------}\!> R_3Si\text{-}Cl + (R_3SiN)_2\ SO_2 \qquad (17)$$

Silicon-nitrogen bonds are formed in the following manner, Equation 18.

$$\geq SiX + 2HNR_2 \text{----}\!> \geq Si\text{-}NR_2 + HNR_2\text{-}HX$$
$$RH_2SiCl + NH_3 \text{--------}\!> (RH_2Si)_3N + 3HCl \qquad (18)$$

Such a reaction is controlled by the bulk of the chlorosilane. For instance, at a higher substitution of the silicon atom, only silazanes are formed, Equation 19.

$$2R_3SiCl + NH_3 \text{--------}\!> R_3Si\text{-}\underset{\displaystyle H}{\underset{|}{N}}\text{-}SiR_3 + 2HCl \qquad (19)$$

As with siloxanes, silylamines can be prepared by the "reamination reaction," Equation 20.

$$R_3Si\text{-}\underset{\displaystyle R'}{\underset{|}{N}}\text{-}R' + H\text{-}N\begin{smallmatrix} R'' \\ R'' \end{smallmatrix} \rightleftharpoons R_3\text{-}Si\text{-}\underset{\displaystyle R''}{\underset{\backslash}{N}}\text{-}R'' + H\text{-}N\begin{smallmatrix} R' \\ R' \end{smallmatrix} \qquad (20)$$

As this reaction is reversible, it can be used successfully only if the amine formed can be removed by distillation, thereby shifting the reaction to the right.

An important silicon-nitrogen compound is hexamethyldisilazane or HMDS. This reagent is an effective silylating agent and has found wide applicability assisting in the volatilization of temperature-sensitive alcohol. The silylation process is characterized by the following generalized equation (Equation 21):

$$(CH_3)_3Si\text{-}N\text{-}Si(CH_3)_3 + \underset{\text{nonvolatile alcohol}}{2HO\text{-}R} \text{--}\!> \underset{\text{volatile silyl ether}}{2(CH_3)_3\text{-}Si\text{-}O\text{-}R} + NH_3 \uparrow \qquad (21)$$

The reaction can be run neat, at low temperatures, and provides for high conversion to the silyl ethers.

Characteristics of Silicone Polymers

Silicone resins traditionally possess (1) enhanced stability to heat and greater resistance to oxidation than typical organic polymers, (2) retention of physical properties over a wider temperature range than hydrocarbon polymers and their properties change more slowly with temperature, (3) higher water repellency, and (4) incompatibility with organic materials and thus do not dissolve in or adhere to such materials.

These properties allow them to be employed in a myriad of applications in addition to surface coatings, including plastic moldings (16), electrical insulators (17), cleaning substances, shoe-, furniture-, automotive, and floor waxes (18), in virological practice (19), cloth coatings, and mold release agents, biomedical applications, and hydraulic fluids and lubricants (20).

Silicon–Containing Polymers

A vast majority of the commercially available silicon–containing polymers today are polysiloxanes or silicones. In the same way that aldehydes and alcohols are characterized by the presence of the $-CHO-H$ and $\equiv C-OH$ moieties, respectively, a siloxane is a compound containing the bond $\equiv Si-O-Si\equiv$, and <u>silicones</u> are materials consisting of alternating silicon and oxygen atoms in which most silicon atoms are bound to at least one monofunctional organic radical. Polymers and copolymers of R_1, R_2, and R_3 monomers are termed poly(siloxanes). Linear poly(siloxanes)

$$
\text{containing more than 1000} \quad \begin{array}{c} CH_3 \\ | \\ (Si-O) \\ \backslash \\ CH_3 \end{array}
$$

units make up the class of silicone rubbers. End capping linear poly(siloxanes) of less than 1000 $(Si(CH_3)_2-O)$ units make up the category of silicone fluids, $Me_3Si-[OSi(CH_3)_2]_n-OSiMe_3$. Silicone resins, on the other hand, are branched poly(siloxanes), Equation 22.

$$
\left(\begin{array}{ccc}
Ph & Me & \overset{\displaystyle H}{\underset{\displaystyle |}{OH}} \\
| & | & | \\
Si-O-Si-O-Si-O \\
| & | & | \\
Me & Me & \underset{\displaystyle X}{\overset{\displaystyle |}{O}}
\end{array}\right)
\tag{22}
$$

Such polymers are produced via the hydrolysis and subsequent condensation of chlorosilanes. The monomers typically employed are mono-, di- and/or trichlorosilanes containing alkyl and/or aryl groups, such as the methyl and phenyl moieties.

It was Ladenburg who, in the late 1800s, prepared poly(diethylsiloxane), the first silicone polymer, via the hydrolysis of diethyldiethoxysilane (21). It was not until 1927, however, that Kipping (22) reported that the materials were macromolecular in structure. Such polymers were not made commercially available on a large scale until the 1940s when Dow Corning Corp., General Electric Co., and the Plaskon Division of Libby-Owens-Ford Glass Co. began producing polysiloxanes.

Generally, the properties obtained with a given silicone polymer depend upon its molecular weight, the organic substituents attached, and the chlorine content of the organochlorosilane precursor(s). The silicone oils are typically linear poly(dimethylsiloxane) obtained by hydrolysis of dimethyldichlorosilane containing some trimethylchlorosilane, such that the silanol ends are capped with nonreactive trimethylsilyl groups, Equation 23.

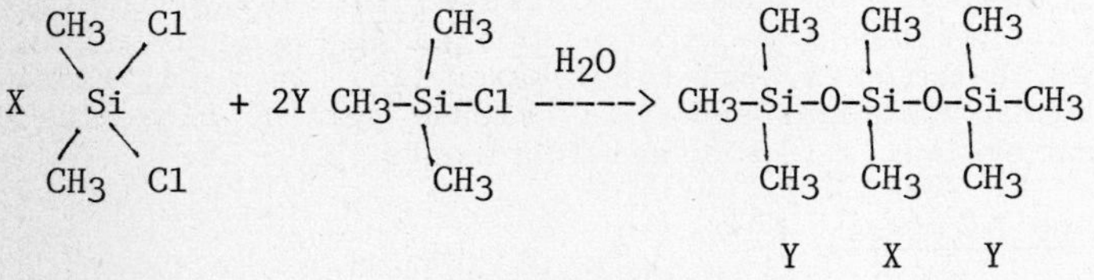

$$
\begin{array}{ccccc}
& CH_3 \quad Cl & & CH_3 & \\
& \diagdown \diagup & & | & H_2O \\
X \quad Si & + 2Y \; CH_3{-}Si{-}Cl & \xrightarrow{\quad\quad} \\
& \diagup \diagdown & & | & \\
& CH_3 \quad Cl & & CH_3 &
\end{array}
$$

$$
\begin{array}{ccccc}
CH_3 & CH_3 & CH_3 \\
| & | & | \\
CH_3{-}Si{-}O{-}Si{-}O{-}Si{-}CH_3 \\
| & | & | \\
CH_3 & CH_3 & CH_3
\end{array}
\qquad (23)
$$

$$
\begin{array}{ccc}
Y & X & Y
\end{array}
$$

Cross-linking of linear polymers can be effected as well as controlled by the judicious use of trichlorosilanes, Equation 24.

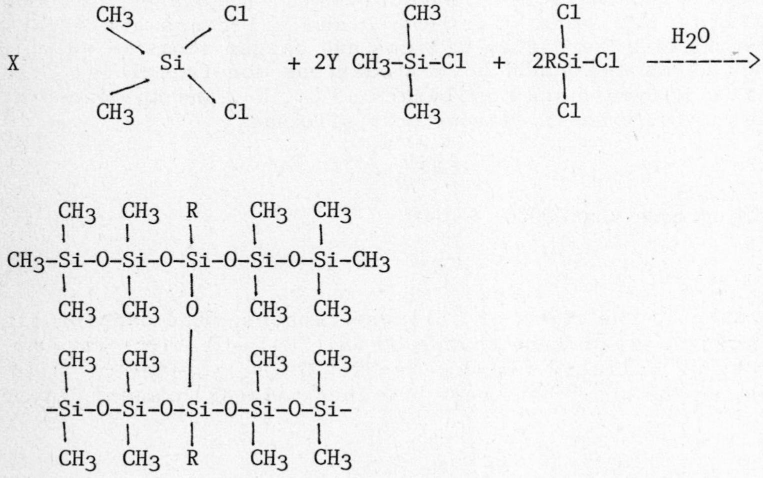

$$
\begin{array}{cccccc}
& CH_3 \qquad Cl & & CH_3 & & Cl \\
& \diagdown \quad \diagup & & | & & | & H_2O \\
X \quad & Si & + 2Y \; CH_3{-}Si{-}Cl & + & 2RSi{-}Cl & \xrightarrow{\quad\quad} \\
& \diagup \quad \diagdown & & | & & | \\
& CH_3 \qquad Cl & & CH_3 & & Cl
\end{array}
$$

$$
\begin{array}{ccccc}
CH_3 & CH_3 & R & CH_3 & CH_3 \\
| & | & | & | & | \\
CH_3{-}Si{-}O{-}Si{-}O{-}Si{-}O{-}Si{-}O{-}Si{-}CH_3 \\
| & | & | & | & | \\
CH_3 & CH_3 & O & CH_3 & CH_3 \\
\\
CH_3 & CH_3 & | & CH_3 & CH_3 \\
| & | & & | & | \\
{-}Si{-}O{-}Si{-}O{-}Si{-}O{-}Si{-}O{-}Si{-} \\
| & | & | & | & | \\
CH_3 & CH_3 & R & CH_3 & CH_3
\end{array}
\qquad (24)
$$

During the reaction sequence cyclic trimers may be produced and are stripped off via distillation or are purified and subsequently polymerized (23) to a high molecular weight polymer via the catalytic action of sulfuric acid, Equation 25.

$$3R_2Si(OH)_2 \longrightarrow$$

$$R = \text{alkyl or aryl}$$

$$(25)$$

In fact, these cyclic trimers may be incorporated into silicone resins of the type employed in surface coatings. In such products, a silanetriol is incorporated to cross-link the resin (this cross-linking reaction has been previously described) and to allow the presence of excess hydroxyl groups after resin preparation. These materials containing excess hydroxyl groups are then applied to a substrate as a surface coating and subsequently cured via a thermal cycle of up to 60 min at 500 °F. When catalyzed, their cure cycle may be reduced to 30–60 min at 400 °F. Some typical catalysts and the levels at which they are usually employed are listed in Table II.

The importance of substituent identity with respect to thermal life of a number of completely condensed polymers of alkyl- or aryltrichlorosilane has been accomplished by Brown (24) as shown in Table III.

Colored enamels based on unmodified silicone resins withstand continuous exposure to 260–370 °C. Black enamels are serviceable to 430 °C, and aluminum combinations are serviceable up to 650 °C. Ceramic frit formulations will operate successfully at 760 °C, and like aluminum finishes, the silicone resin disintegration is followed by –Si–O–frit (aluminum) bond formation (25).

The enhanced thermal stability of the silicones has been postulated by Andrianov and Sokolov (26) to be due to (1) the high energy of the SiOSi bond, (2) the stability of organic radicals to oxidation at the silicon atom, (3) stable SiOSi bond formation within the polymer after partial degradation, and (4) the formation of a layer of silicon–oxygen bonds on the surface of the polymer during thermal oxidation.

A brief summary of the silicones in high-temperature coating applications follows:

Outstanding Advantage	Chief Limitations	Typical Application
1. Air–dry tack free (some formulations)	1. High cost	1. High-temperature stacks
2. Service temperature up to 1000 °F	2. Low film build	2. Boiler and boiler breeder
3. Interior or exterior service	3. Limited solvent resistance	3. Exhaust lines, manifolds, and mufflers
	4. Limited chemical resistance	

Table II. Silicon Curing Catalysts

Metal	Suggested Metal Content Based on Nonvolatile Material (%)	Comments
Zinc	0.3–0.5	Generally considered universal catalyst
Manganese, cobalt	0.1–0.5	Good top hardness, often discolors
Iron	0.1–0.03	Good top hardness, Fast, give poor shelf stability and heat resistance
Lead, tin		Poor shelf life, cause gellation and poor heat resistance

Table III. Thermal Life as a Function of Alkyl or Aryl Substituent on Silicon

Group Bonded to Silicon	Approximate Half-life at 250 °C[a] (h)
Phenyl	>100,000
Methyl	>100,000
Ethyl	6
Propyl	2
Butyl	2
Pentyl	4
Decyl	12
Octadecyl	26
Vinyl	101

[a]Time at which half of the groups are replaced by oxygen.

Trifunctional materials, such as methyltrichlorosilane, generally provide for a hard, sometimes brittle (if CH_3-$SiCl_3$ is employed) resin and enhanced cure rate for organosilicones. High-phenyl-containing resins seem to be more stable to oxidation (see Table III), to have a better shelf life, to have better heat resistance, to be tough, to be more soluble, and to air-dry but to be less thermoplastic than polymers containing a high level of $(CH_3)_2SiO$ repeating units ($\underline{2}$, $\underline{31}$, $\underline{32}$). The resins containing the phenyl-methyl $\ \text{-}CH_3Si(C_6H_5)O\text{-}\ $ unit are tough and flexible with a moderate modulus ($\underline{24}$). Resins containing a large proportion of $\text{-}(CH_3)_2SiO\text{-}$ repeating units, on the other hand, are superior in hardness at high temperatures ($\underline{27}$), flexibility, water repellency, low-temperature properties, chemical resistance, cure rate and ability to withstand thermal shock, gloss retention, arc resistance, and ultraviolet and infrared stability. Although organic groups other than phenyl and methyl have been employed in silicones, their high price limits their marketability (24).

Indeed, silicones are virtually transparent to ultraviolet energy and are thus not subject to the usual degradation and subsequent failure of conventional coatings. Unmodified silicone coatings on exterior exposure for 12 years have been shown to exhibit little loss of gloss, color change, chalking, or any other sign of failure. In fact, silicones are so stable that nonchalking coatings can be made from freely chalking pigments such as titanium dioxide. Silicone coatings are often modified with organic resins in order to achieve the desired hardness, adhesion, and abrasion resistance and to reduce thermoplasticity and produce a faster cure coating. However, modification of coatings results in a compromise or blend of properties, and this should be taken into consideration when formulations are developed. Blending silicones with organic resins can be affected in two ways.

(1) The first is formation of a cold blend in which the silicone resin is added to the organic resin. The usual maximum level of silicones used in cold blends is 50% although maintenance paints typically contain 30%. Cold blending is an effective method for improving heat stability or exterior durability of organic coatings. It is essential, however, that such blends possess good compatibility. There are disadvantages to cold blending in that the types of silicone resins utilized in cold blending are higher in cost than reactive silicones and, of course, physical blending does not chemically combine the silicone and organic resin into one homogeneous polymeric system. Thus, reactive groups on both the silicone resin and organic resin are subject to hydrolysis and contribute little if any to chemical and solvent resistance.

(2) Silicones can also be combined in such a manner that a chemical bond is formed between the silicone reactive intermediate and the organic polymer. The major functionalities providing for bond formation of reactive intermediates are the methoxy ($\equiv$Si–OCH$_3$) and the silanol ($\equiv$SiOH) groups, Equation 26.

$$\equiv\text{Si–O–Me} + \text{HO–organic polymer} \longrightarrow \equiv\text{Si–O–organic polymer} + \text{CH}_3\text{OH}$$

$$\equiv\text{Si–OH} + \text{HO–organic polymer} \longrightarrow \equiv\text{Si–O–organic polymer} + \text{HOH} \qquad (26)$$

The resulting silicone–alkyd copolymers are thus formulated into coatings that have the same general and physical properties as alkyd resins and that can be applied in the usual manner. Silicone–alkyd copolymers generally require no special primer types. While there are numerous silicone intermediates that can be and are employed in the formation of organic–silicone copolymers, a few are shown (19) in Figure 1.

These materials are generally reacted with hydroxyl–terminated organic polymers, with the resultant copolymers possessing dramatically improved weathering resistance, a property whose performance is essentially proportional to the silicone content (24) as shown in Figure 2.

Likewise, silicone incorporation dramatically improves exterior performance over non–silicone–containing counterparts as illustrated in Figure 3 (24).

Further confirmation of the extended weatherability of silicone-containing coatings has been reported by Finzel (25). Coatings containing silicones had lower film erosion rates than control coatings, and the rates were inversely proportional to the silicone content (see Figures 4 and 5).

Recently I have confirmed reports from Dow-Corning Corp. that a Dow-Corning methoxy-terminated intermediate reacts effectively with hydroxyl-terminated acrylic latexes to produce a covalently bonded acrylic-silicone possessing properties expected of silicon-modified polymers. Our work has involved the incorporation of the silicone intermediate onto a hydroxyl-terminated acrylic, which is subsequently combined with a melamine cross-linking agent for the production of a thermosetting silicone-acrylic. This silicone modification has allowed for the production of conventionally pigmented TSA films that retain the base polymer tensile strength while exceeding its elongation, gloss retention, and water resistance.

Silicones as Water Repellents

Silicones are well-known as water repellents. They have been employed on masonry above grade level to repel liquid water not under pressure. They are used above grade level and not under hydrostatic pressure. For instance, they line the masonry pores but do not fill the pores. Thus, they are not a moisture barrier but a moisture repellent.

The clear-colorless nature of silicone repellents allows preservation of the original appearance of the concrete. Their lack of film formation leaves the masonry pores open, thereby allowing transport of moisture outward to the atmosphere. They are highly durable to environmental conditions and will retain their repellent properties for several years. There are two types of water-repellent silicones: (1) one is the resin-solvent type that is characterized as a partially hydrolyzed, partially alcoholized silicone resin dissolved in a hydrocarbon solvent. Generally, such products approach the 5% nonvolatile level. (2) The second is a water-soluble alkaline silicone repellent that is recommended for use on limestone and other nonsiliceous stone and masonry.

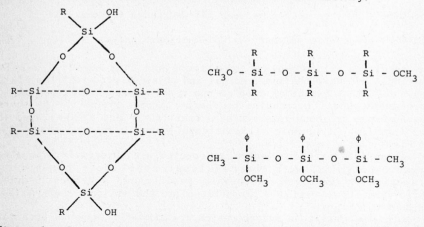

Figure 1.   Structures of some silicon intermediates employed in the formation of silicon copolymers.

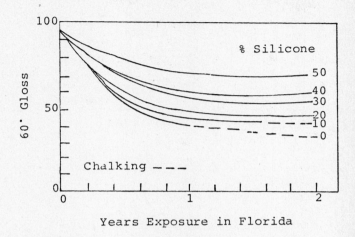

Figure 2. Improvement in gloss stability as the silicon content is increased.

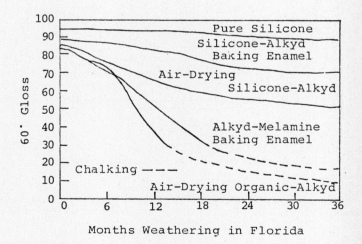

Figure 3. Comparison of 30% and 100% silicon coatings with alkyds.

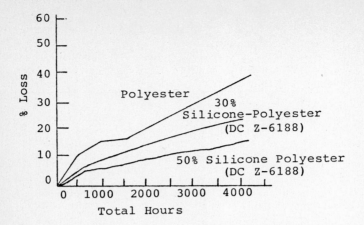

Figure 4.   Film erosion by weight loss in silicon-polyester coil
            coatings.

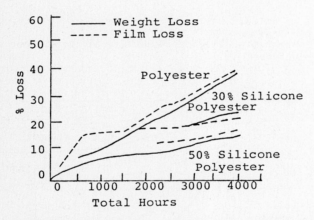

Figure 5.   Correlation of weight loss and film thickness data for
            silicon-copolymer and control coil coatings.

    A summary of outstanding advantages, chief limitations, and
typical applications of silicones as masonry water repellents
follows:

Outstanding Advantages	Chief Limitations	Typical Applications
1. Colorless	1. Not for use on limestone	1. New brick, mortar, sandstone, and poured concrete.
2. Invisible	2. Use only on new masonry	
3. Effective for 10 years		
4. Prevents staining		

Pigmentation

As a general rule, silicone resins are of low viscosity when compared to an organic counterpart, they wet pigments with ease, allowing fine grinds to be achieved in minimum time, and they can be pigmented essentially by conventional methods. Because silicones have extensive utility in heat-resistant coatings, pigmentation must assuredly consider this end use; thus, a 10-15% pigment volume concentration is typical if high gloss and prolonged life during extended high temperatures is a requirement. Although heat life tolerates the use of extender pigments, their use does result in a sharp reduction in gloss, and this decline is somewhat more precipitous than with other coating types. Extenders useful with silicones are magnesium and aluminum silicates, diatomaceous silica, mica, calcium carbonate, and barytes.

The white pigments of choice are led by titanium dioxide, and it too has excellent heat stability. Lithopone, zinc sulfide, zinc oxide, and antimony oxide are additional white pigments of choice.

The low viscosity of silicone resins can encourage settling. However, more often than not, remixing of the soft settled mass yields a smooth uniform enamel. Because low viscosity accompanies the use of silicones, the formulator may be tempted to over pigment, and this should be avoided if poor craze resistance is to be avoided and gloss retention is to be maintained.

The pigmentation of silicones with aluminum in the preparation of a heat-resistant paint is, however, atypical. Course-ground aluminum paste is employed at the rate of 2 lb/gal. unless service temperatures of 500-600 °C are to be maintained. If the more severe, higher temperature conditions are to be experienced, 2.9 to 3.0 lb/gal. of vehicle is an appropriate pigmentation level. This high aluminum level is necessary to provide for the sufficiently high temperature formation of the silicon-oxygen-aluminum protective barrier.

The availability of inorganic colors designed especially for heat resistance now allows for the production of heat-resistant silicones in many colors other than white.

Silicones as Additives

The low surface tension characteristics of silicones make them suitable as additives (additives are defined as any substance added in small quantities to another substance, usually to improve properties—sometimes called modifier) for preventing foaming, flooding, silking, hard settling, floating, and, in some cases, leveling wetting and mar resistance. Furthermore, small amounts of select silane reagents are reported to be effective adhesion promoters (28).

Silane fluids or silicone resins can be utilized as additives to paint formulations. It will suffice to note that the specific end use need will dictate the type and quantity of silane to be employed in any particular application.

Coupling agents are additives employed to augment the uniting of dissimilar materials. Thus, successful coupling agents form a chemical bond with the materials in question. All commercial silane coupling agents are of the general structure $X_3-Si(CH_2)_n-Y$, where n = 0-3, X is a hydrolyzable group on silicon (usually RO-), and Y is

Table IV. Commercial Silane Coupling Agents

No.	Name	Formula	Application
1	Vinyltriethoxysilane	$H_2C=CHSi(OC_2H_5)_3$	Unsaturated polymers
2	Vinyltris(β-methoxyethoxy)silane	$H_2C=CHSi(OCH_2CH_2OCH_3)_3$	Unsaturated polymers
3	Vinyltriacetoxysilane	$H_2C=CHSi(OOCCH_3)_3$	Unsaturated polymers
4	((γ-Methacryloxy)propyl)trimethoxysilane	$H_2C=C(CH_3)COO(CH_2)_3Si(OCH_3)_3$	Unsaturated polymers
5	(γ-Aminopropyl)triethoxysilane	$H_2NCH_2CH_2CH_2Si(OC_2H_5)_3$	Epoxies, phenolics, nylon
6	((γ-(β-Aminoethyl)amino)propyl)trimethoxysilane	$H_2NCH_2CH_2NH(CH_2)_3Si(OCH_3)_3$	Epoxies, phenolics, nylon
7	(γ-Glycidoxypropyl)trimethoxysilane	$H_2C\overset{\displaystyle O}{-\!-\!-}CHCH_2O(CH_3)_3Si(OCH_3)_3$	Almost all resins
8	(γ-Mercaptopropyl)trimethoxysilane	$HSCH_2CH_2CH_2Si(OCH_3)_3$	Almost all resins
9	(β-(3,4-Epoxycyclohexyl)ethyl)trimethoxysilane	$O\!\!<\!\!\bigcirc\!\!-\!-CH_2CH_2Si(OCH_3)_3$	Epoxies
10	(γ-Chloropropyl)trimethoxysilane	$ClCH_2CH_2CH_2Si(CH_3)_3$	Epoxies

an organofunctional group selected for compatibility with a given resin, such as vinyl, amino, mercapto, and epoxy (see Table IV) (28).

It has been reported that the incorporation of silicon into polyamides and polyimides dramatically lowers the glass transition temperature of these polymeric materials, for example

Furthermore, most of the silicon-containing polyimides were soluble in cold, moderately polar solvents such as trichloroethylene and o-dichlorobenzene. By way of comparison, conventional aromatic polyimides soften at temperatures as much as 100 °C higher than their silicon-containing counterparts, while generally being insoluble in almost all cold solvents (29).

Literature Cited

1. Kipping, F. S. J. Chem. Soc. 1927, 130, 104.
2. Bazant, V.; Chvalovski, V.; Rathovsky, J. "Organosilicon Compound I"; Academic: New York, 1956; Vol. 11, pp. 15–24.
3. Ibid., p. 16.
4. Sommer, L. H.; Rockett, J. J. Am. Chem. Soc. 1951, 73, 5130.
5. Bazant, V.; Chvalovski, V.; Rathovsky, J. "Organosilicon Compound I"; Academic: New York, 1965; p. 178.
6. Ibid., pp. 225–33.
7. Gilman, H.; Brook, A. G.; Miller, L. S. J. Am. Chem. Soc. 1953, 75, 4531.
8. Roohow, E. G. U.S. Patent 2 380 995, Aug 1945.
9. Brown, L. H. "Treatise on Surface Coatings"; Myers, R. R.; Long, J. S., Eds.; Dekker: New York, 1972; Vol. I, Part III, Chap. 13, p. 516.
10. Anderson, D. G.; Bradney, M. A. M.; Webster, D. E. J. Chem. Soc. B 1968, 450.
11. Pratt, J. R.; Pinkerton, F. H.; Thames, S. F. J. Hetero. Chem. 1972, 9, 67. Pinkerton, F. H.; Thames, S. F. J. Hetero. Chem. 1971, 8, 257. Pratt, J. R.; Pinkerton, F. H.; Thames, S. F. J. Organometal. Chem. 1972, 38, 29–36.
12. Bazant, V.; Chvalovski, V.; Rathovsky, J. "Organosilicon Compound I"; Academic: New York, 1965; p. 229.
13. Brook, A. C.; Abdesaken, F.; Gutekunst, G.; Kallury, K. Chem. Eng. News 1981, 59(13), 18.
14. West, R.; Fink, J. J.; Michl, J. Chem. Eng. News 1981, 59(51), 89.
15. Kipping, F. S. J. Chem. Soc 1927, 130, 104.
16. Rochow, E. G. "Chemistry of the Silicones"; Wiley: New York, 1946; pp. 222–33.
17. Morozumi, K.; Takenaka, T.; Suzuki, M. Jpn. Kokai 1973, 73(90), 365.
18. Streubel, H. Hermsdorfer Tech. Mitt. 1969, 9(24), 768.
19. Muzovakaya, O. A.; Poleleva, G. S.; Melyakhova, E. I.; Solov'ev, A. A. Khim. Prom. (Moscow) 1969, 45(8), 604.

20. Kristapsons, M.; Rumyantseva, N. P.; Reiznieks, A.; Atrena, A. Virusol. 1970, 15(1), 116.
21. Ladenburg, A. Leikings Ann. 1872, 164, 300.
22. Kipping, F. S. J. Chem. Soc. 1927, 130, 104.
23. Cahn, H. L. In "Technology of Paints, Varnishes, and Lacquers"; Martens, C. R., Ed.; Robert E. Krieger Publishing: Huntington, N.Y., 1968; Chap. 14.
24. Brown, L. H. In "Treatise on Coatings"; Myers, R. R.; Long, J. S., Eds.; Dekker: New York, 1972; Vol. I, Part 3, Chap. 13.
25. Finzel, William A. "A Review of Silicones in Coatings, II"; Dow Corning Corp.: Midland, Mich., 1980; J. Coat. Technol. 1980, 52(660), 55–61.
26. Andrianov, K. A.; Sokolov, N. N. Elektrichestvo 1956, No. 6, 31; Chem. Abstr. 1956, 50, 16171.
27. Clope, R. W.; Glaser, M. A. "Federation Series on Coatings Technology"; FSCT: Philadelphia, Pa., 1970; Unit 14.
28. Plueddemann, E. P. In "Treatise on Coatings"; Myers, R. R.; Long, J. S., Eds.; Dekker: New York, 1972; Vol. I, Part 3, Chap. 9.
29. Pratt, J. R., Ph.D. Dissertation, May 1974, University of Southern Mississippi, Hattiesburg, Miss., p. 89.

Chemistry and Technology of Phenolic Resins and Coatings

J. S. FRY, C. N. MERRIAM, and W. H. BOYD

Union Carbide Corporation, Bound Brook, NJ 08805

Early History
Chemistry
  Raw Materials
  Base—Catalyzed Reactions
  Acid—Catalyzed Reactions
Classification of Phenolic Resins
  Unsubstituted and Heat Reactive
  Unsubstituted and Nonheat Reactive
  Substituted and Heat Reactive
  Substituted and Nonheat Reactive
  Applications
  Waterborne Phenolic Resins
Conclusions

Phenolic resins, being the first truly synthetic resin, already had a highly developed technology by 1924. The phenolic industry had grown to be a large and profitable plastic business closely allied in supplying materials to the other technological industries of that day. Major uses included automotive applications, the telephone, and the new mass communications medium, the radio. Leo H. Baekeland, founder of the phenolic industry, was recognized as a very capable chemist as well as a businessman. While always interested in the basic science of phenolic resins, the business demands kept intruding on his efforts to elucidate the complex nature of phenolic resin chemistry. Baekeland's attitude toward the commercial world is illustrated by the following quote from his presidential address to the September National American Chemical Society (Ithaca) Meeting 51 years ago (1), 1924: "Whether we like to admit it or not, much of the history of science has been shaped by the needs and the outside influences of commerce or industry. brought about by the early manufacture of synthetic dyes.... The very nature of that new industry puts an unprecedented premium on research in organic chemistry. Fame, honors, and monetary rewards were the alluring inducements offered to the best available. I am not one of those who tend to exaggerate the benefits of chemistry in the creation of thousands of new synthetic dyes except for the enormous fund of new chemical knowledge we have gathered thereby and which has helped immensely in other more valuable directions." He goes on to say, "Anyone who looks for nothing in chemistry but a means of getting rich has chosen the wrong career."

Early History

In order to better understand the rapid technical development of phenolic resins, one should go back not 51 years but 76 years to 1909, the year of phenolic resin commercialization, and look at the state of several important industries. The electrical industry was rapidly expanding, though electrical lighting was not yet commonplace. The automotive industry was in its infancy but fast growing. In the communications industry, wireless telegraph was just starting, telephones existed but were not widespread, and radios were unknown. Phonographs were big business. A new material was needed by these industries that was more amenable to mass production methods, had better electrical insulating properties, had more heat stability, and had greater strength. The materials being used and with which phenolics would compete were Celluloid, hard and soft rubber, bituminous binders, and shellac.

The goal of initial phenolic resin efforts was the development of a cheap thermoplastic resin replacement for shellac. The result was the phenol/formaldehyde, nonheat-reactive resins that Baekeland referred to as novolaks. These products did not have the toughness of shellac and, therefore, never became successful substitutes.

During these studies, Baekeland became more interested in the insoluble, infusible masses sometimes obtained. He reasoned that if he could control this reaction, making it occur when, where, and in the form he wanted, he would have a product of commercial value. This effort resulted in the development of the heat-reactive resins and the heat and pressure technique of molding them. The resulting cured phenolic produced hard, strong, infusible, and insoluble objects with good electrical insulating properties. These properties plus the relatively short molding cycles made phenolics just what the aforementioned industries needed. Within 1 year of the announcement of this technology, Bakelite had a dozen customers representing many of the applications still important today. The first commercial use of this heat-reactive resin was found in molding material as a replacement for hard rubber and amber in applications where resistance to heat and pressure was needed. The first customer was the Boonton Rubber Co., who purchased heat-reactive resin, blended it with asbestos fiber, and produced moldings. One of Boonton Rubber's first customers was the Weston Electrical Instrument Co., who, under the leadership of Dr. Edward Weston, was looking for a molding material that could be molded to close tolerances and not distort under heat and pressure. All known molding materials had been found inadequate. Being near the end of his resources, Weston chanced to meet Baekeland, who put him in contact with the Boonton Rubber Co. Through this cooperation, Weston was able to solve his problem, resulting in the first commercial use of phenolic plastic in the electrical industry--the

Weston Insulating Coil Support. Other electrical applications rapidly developed, and by the end of 1911, phenolics were being used in molded push plates and commutators.

The use of phenolics in the electrical industry was quickly followed by its introduction into the automotive industry. This was initiated by Kettering's development of the "Delco" ignition and starting system based on a Bakelite molded distributor head. Kettering found that the Bakelite phenolic, with its dielectric strength and immunity to adverse effects from temperature, acids, oils, and moisture, solved the problems previously associated with the use of hard rubber. The use of phenolics has grown with this industry, and today finds use throughout the transportation industry in such diverse applications as laminated timing gears and bonded brake linings.

Other early customers were Westinghouse, producing laminates, and General Electric, impregnating coils. Use in consumer products was also established early with the molding of transparent pipe stems and cigarette holders.

The use of phenolics in communications developed in the next few years with such objects as telephone handsets. With the development of radios, phenolics found use in laminated panels, molded dials, and coil forms. Phonograph records became a large market with the development by Aylsworth of the use of hexamethylenetetramine as a curing agent for the nonheat-reactive novolak-type phenolics.

Although in 1909 Baekeland obtained patent coverage for the use of phenolics in bonding abrasives, it was not until 1921 that it became commercially useful. Phenolic bonded grinding wheels were found to be capable of being operated safely at higher speeds and with improved quality.

More recent new markets include convenience electrical appliances such as electric frying pans, toasters, and steam irons that require phenolics for both their handles and electrical connections.

The tremendous growth of the electronics and computer industry has also given a boost to phenolic resin sales. Printed circuits are manufactured by bonding copper foil to a laminated phenolic base, printing the area representing the circuit with acid-resisting ink, and then etching away the remainder of the foil. The dimensional stability, acid resistance, and excellent electricals make phenolics a necessity in the application.

The first attempt to use phenolic resins in coatings was the unsuccessful use of nonheat-reactive novolaks dissolved in solvent as a replacement for shellac. The coatings were not good, being very brittle and darkening in color on aging. With the development of the heat-reactive resins, it was found that their alcohol solutions, when applied to objects and baked, gave hard glasslike films with excellent resistance to water, solvents, and most chemicals. These resin solutions were found to be particularly suitable for coating brass beds that were so popular at the time.

Perhaps the story of the first production batch of phenolic resin at Baekeland's new plant in 1911 will illustrate the state of the industry some 70 years ago. This story has some elements in common with present-day scale ups. While the new Perth Amboy, NJ, plant was being installed, Dr. Baekeland hired a new chemist, J. J. Frank, trained him at the Yonkers Laboratory, and set him to work in

the new plant. Present at this first large-scale batch was Dr.
Baekeland. the factory manager, the few new factory workers, some
contractors, and Frank. As the day ran into the evening, lighting
had to be supplied by kerosene lanterns since the electric lighting
had not been completed. The batch proceeded well. The resin was
reacted, dehydrated, and finished, and the alcohol solvent was
added. Dr. Baekeland asked whether the resin was all dissolved or
if there was a lump of "B" (gelled) resin around the agitator.
Frank quickly grabbed a lantern, swung open the manhole door, and
held the lantern to look in. There was a mild "boom," the hot
alcohol vapor ignited, and flames flashed out of the manhole.
Everyone quickly left by the nearest doors and windows, except for
the factory manager and Frank. The manager calmly slammed the still
manhead, extinguishing the flames. Fortunately, Frank was not hurt
badly. He lost his hair, eyebrows, mustache, and his job. The
batch of varnish was saved.

Chemistry

Understanding the chemistry of phenolic resins is particularly
difficult because of the many complex reactions involved and the
insolubility of the final cured products. The initial chemical
efforts were spent in isolating and studying pure, crystallizable
compounds. It was not until 30 years after Baeyer's first report in
1872 of the reaction of phenol with aldehydes (2) that interest
turned to the noncrystalline resinous materials. From 1905 to 1909,
Baekeland studied phenol/formaldehyde reactions and defined the
differences between acid and base catalysts and the effects of the
mole ratio of these reactants. He learned how to control this
reaction and isolate three stages of products: A stage being
soluble and fusible, B stage being insoluble but swellable and
infusible but softenable, and C stage being completely insoluble and
infusible. The results of this work were presented to the New York
Section of the American Chemical Society in 1909 (3). The useful
products known as Bakelite and the novel heat and pressure technique
of curing them were patented (4) and became the basis of the
phenolic industry.
     The years after 1909 were primarily spent on applying this
knowledge. The development of an understanding of the chemistry
involved proceeded slowly. This is understandable when it is
considered that the basis of modern polymer theory, for example,
high molecular weights, functionality, and cross-linking, was not
developed until later. Neither were the powerful tools of chemical
analysis available to define molecular weights, distribution of
products, functional groups, etc. Added to this was the problem
that the final cured products being insoluble, infusible, and
nonreactive could not be analyzed. Studies of model compounds
during the twenties (5, 6), led to an understanding of the initial
reactions. Comprehensive reviews of the development of this
chemistry from the 1920s to the 1950s are available (7, 8), and, for
the purposes of this paper, only a general description of these
reactions will be given.

Raw Materials. Phenolic resins are formed by the condensation of a
phenol with an aldehyde using either base or acid catalysis. Though

there are many phenols that could be used, only a few have gained commercial importance. These can be divided into two classes: unsubstituted and substituted phenols. Phenol, itself, is the most important. Small amounts of other unsubstituted phenols, such as resorcinol, find use. Of the substituted phenols, the most important are the following alkylated phenols: various cresols, *p*-*tert*-butylphenol, *p*-*tert*-amylphenol, and *p*-*tert*-octylphenol. Small amounts of nonyl- and dodecylphenol are also used. The aryl-substituted phenols, *o*- and *p*-phenylphenol, have had a long commercial history, though their high cost and recent scarcity has all but eliminated their use today. Bisphenol A is growing in importance as a raw material in phenolic resins.

Phenol   Orthocresol   Metacresol   Paracresol

(2,2 bis(4-hydroxyphenyl) propane (Bisphenol A)   p-phenyl-phenol   p-tert-butyl-phenol   p-tert-amylphenol   (3-1)

p-tert-octyl-phenol   p-nonylphenol

Though many aldehydes can be made to react with phenol, only formaldehyde, the lowest molecular weight and most reactive aldehyde, is of major commercial importance. Acetaldehyde and butyraldehyde are used only to a limited extent and often in combination with formaldehyde. Furfural is also used, forming phenolic resins that readily oxidize resulting in dark brown to black products.

Since formaldehyde in its pure state is a highly reactive gas, it is commercially used either as a solution in water, known as formalin, or as a solid flake polymeric form known as paraform. Formaldehyde in solution exists not as the pure aldehyde but rather in hydrated forms such as methylene glycol and low molecular weight gylcol ethers:

$$CH_2O + H_2O -----> HOCH_2OH-HOCH_2OCH_2OH + etc.$$

Solid paraform consists of higher molecular weight polymethylene glycol ethers having a degree of polymerization of 6-100. On heating, these polymers readily unzip forming formaldehyde gas again (9).

The initial products formed on reacting phenol with formaldehyde are dependent on the catalyst used and whether it raises or lowers the initial pH of 3.0 for the mixture of phenol and aqueous formaldehyde.

Base-Catalyzed Reactions. When a base catalyst is used, raising the pH above 8, the first products formed are various hydroxybenzyl alcohols. When unsubstituted phenol is used, five different alcohols can be formed, since formaldehyde will react at the two ortho and the one para position to the phenolic hydroxyl. These alcohols, commonly referred to as methylol phenols, are found in varying relative amounts depending on the ratio of formaldehyde to phenol present and various reaction conditions such as time and temperature.

If the reaction is carried further, the methylol groups will condense by two possible paths, forming either dihydroxydiaryl-methanes (methylene links) or dihydroxybenzyl ethers (methylene ether links). These two reactions can occur at either ortho or para positions, and the remaining ortho and para positions can have varying degrees of methylol substitution. This results in a very large number of possible isomers. High ratios of formaldehyde and low reaction temperatures favor the formation of large amounts of methylol groups. Low temperatures and intermediate pH values (4–7) favor methylene ether link formation. High temperatures and high pH values (above 8) favor methylene links.

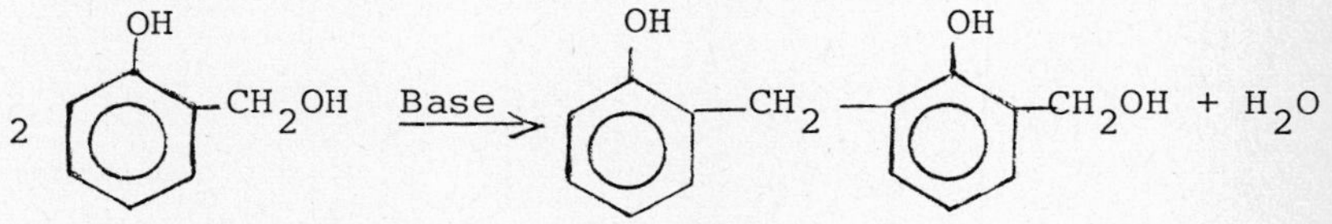

Methylene Link

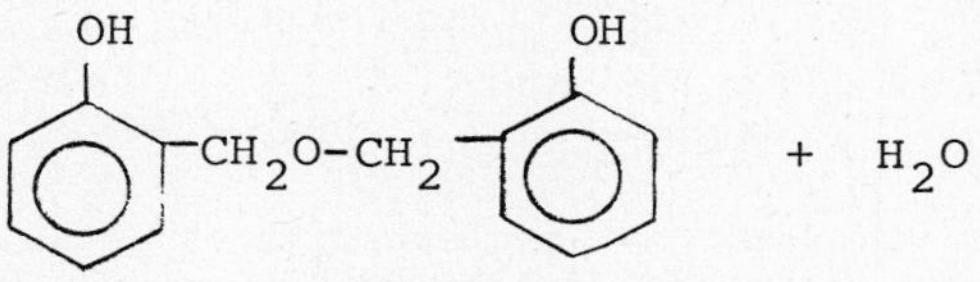

Methylene Ether

Further reaction increases the molecular weight, resulting in highly branched polymers. If an excess of formaldehyde is present, it is possible to form a gelled, cross-linked structure. Further heating increases the cross-link density.

The physical form of these products is dependent on their molecular weight and the amount of reacted formaldehyde. The initial reaction products, mixed methylols of phenol, are low viscosity liquids that are water soluble. This mixture can be separated into pure crystalline compounds. On further condensation, viscous reactive liquids are obtained, which are difficult to isolate in an anhydrous form. These products are soluble in alcohols and other polar solvents. Increasing the molecular weight further with only a slight excess of formaldehyde results in grindable, fusible, resinous solids that are soluble in polar solvents. Further polymerization decreases the solubility and increases the softening point until an infusible, insoluble gel is obtained.

Acid-Catalyzed Reactions. When an acid catalyst is used and the pH of the phenol/formaldehyde mixture is lowered to 0.5–1.5, somewhat less complicated products are formed. The initial reaction of addition of formaldehyde to the aromatic ring results in an unstable intermediate that rapidly condenses to three possible dihydroxydiarylmethanes.

As more formaldehyde is reacted, the molecular weight increases with the formation for tri- and tetranuclear compounds with a methylene link at the various ortho and para positions. As the

molecular weight increases, the number of isomers also increases—
there being 7 possible trinuclear isomer compounds, 27 tetranuclear
isomers, and 99 pentanuclear isomers. This results in a very large
number of different molecules all being present in a resin having a
molecular weight of less than 1000.

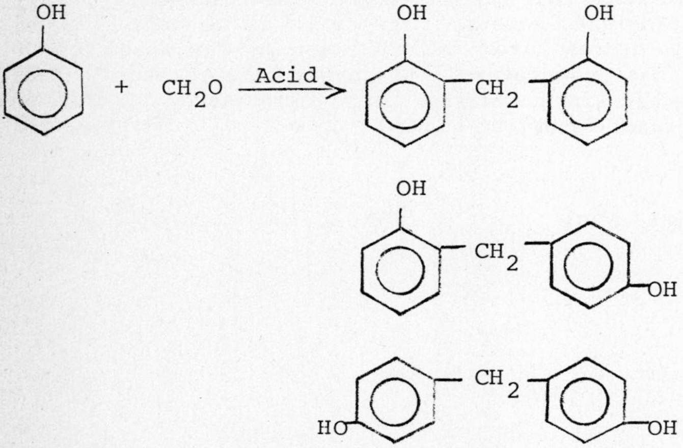

OH

$+$ CH_2O $\xrightarrow{Acid}$ ... CH_2 ...   o,o'-isomer

OH

... CH_2 ... OH   o,p'-isomer

HO ... CH_2 ... OH   p,p'-isomer

     When phenols are used that are substituted ortho or para to the
phenolic hydroxyl, the functionality of the phenol is decreased to
two, decreasing the number of potential structures and forming only
linear polymers.

Classification of Phenolic Resins

Phenolic resins can be divided between heat-reactive and nonheat-
reactive resins and between resins made by using unsubstituted or
substituted phenols. A review of the four resulting classifications
follows.

Unsubstituted and Heat Reactive. The first class, the unsubstituted, heat-reactive resins, are made by using phenol, cresols, and
xylenols. They are multifunctional and thus can be cross-linked to
form films. They are soluble in alcohols, ketones, esters, and
glycol ethers and insoluble in aromatic and aliphatic hydrocarbons.
They will tolerate some water in their solvents and, in some cases,
are completely water soluble. They are compatible with polar resins
such as amino resins, epoxies, polyamides, and poly(vinyl butyral),
though compatibility on curing is dependent on reaction between the
two resins. Less polar resins such as alkyds and drying oils are
incompatible.

     Unsubstituted, heat-reactive phenolic resins are commercially
available as 100% viscous liquids, as water and alcohol solutions,
and as solid resins. The viscous liquids are mixtures of monomers
and dimers of varying methylolation. The water solutions contain
resins of monomers and dimers with the highest degree of
methylolation. Alcohol solutions contain resins that are higher in
molecular weight and are too reactive to isolate as solids.

     Since these resins are so heat reactive, it is often necessary
to store them under refrigerated conditions. Typical alcohol

solutions of the most reactive resins will gel in 3-6 months at ambient temperatures. Even the solid resins, based on phenol, stored at ambient temperatures can polymerize to the point of being insoluble and infusible.

A new, solid, heat-reactive resin based on bisphenol A was introduced recently (10) that is usable after even a year in storage at ambient (70-80 °F) temperatures. This resin polymerizes well at common curing temperatures of 300-400 °F and may be formulated into both solution and powder coatings (11) as will be discussed.

The importance of these resins in high-performance coatings is large. They are usually applied from alcohol solutions by spray, dip, brush, or roller coating techniques. Some solid resins find application as powder coatings.

Once the alcohol solution coating is applied, the coating is baked, first removing the solvents and then heat polymerizing the resin. During this polymerization, water is released by condensing methylol groups, and formaldehyde is released by decomposing methylene ether groups. Some of this formaldehyde reacts, and some is lost with the solvent and water. This release of volatiles during curing limits the film thickness to less than 1 mil. Thicker films will develop blisters and pinholes on baking. For many applications, one-coat, clear films of 0.2-0.7 mils provide all the protection necessary.

These films are baked at temperatures ranging from 135 to 300+ °C. Times may vary from several minutes to several hours. The time/temperature cycle used will have a large effect on the degree and type of cure resulting in different film properties. Increased cure caused by longer times and higher temperatures will result in increased corrosion, chemical and solvent resistance, and decreased flexibility. Short, high-temperature bakes product a more resistant film than long, low-temperature bakes. The color of the coating darkens on baking, going from very light greenish yellows to golden to deep red browns. This color is caused by the formation of various oxidative and unsaturated structures such as quinones, stilbenes, and methides.

Films thicker than 1 mil can be obtained by applying several thin coats. Intermediate coats are given a partial bake of 10-20 min at 135 °C to remove the solvents and most of the reaction volatiles. The solvents in subsequent coats will not lift this partially cured film. After the final coat the film is given a full bake schedule, during which enough flow and curing occurs to fuse the several coats into one film. These thick, glasslike coatings must be cooled slowly in order to minimize stresses and strains.

The resulting cured coatings are less affected by solvents than any other type of organic coatings. They remain unaffected after years of exposure to alcohols, ethers, esters, ketones, aromatic and aliphatic hydrocarbons, and chlorinated solvents. They have excellent resistance to boiling water, aqueous solutions of mild acids, and acidic and neutral salts but poor alkali resistance. A report by the National Association of Corrosion Engineers (12) made recommendations for the use of this type of coating, finding it recommended for over 90% of some 500 chemicals tested.

These coatings are very hard with smooth, dense surfaces and poor flexibility. Their low conductance and low moisture absorption result in good electrical insulating properties, resisting up to 500 V/mil. Resistance to a broad range of temperatures is good, being capable of withstanding dry heat temperatures as high as 370 °C for short periods and having little change in properties on cooling to very low temperatures. Extended exterior exposure of clear and pigmented coatings results in only darkening of color and loss of gloss. These films are completely odorless, tasteless, and nontoxic and find wide use as interior coatings for food cans.

Unsubstituted, heat-reactive coatings find use in a variety of applications on rigid metal substrates where maximum resistance is required. A few examples will suffice to illustrate their use.

Drill Pipe. Iron oxide pigmented coatings are used to protect oil well drill pipe. In this application, phenolics are required for their resistance to abrasion, acids, hydrocarbons, and water at high temperatures and pressures, as encountered in drilling operations.

Printing Plate Backing. Baked phenolics are used to protect the back of copper printing plates that are processed by etching in strong acids. These plates are reetched many times without degrading the backing. Any failure of the backing would cause partial etching of the back and uneven pressure during high-speed printing, resulting in faintly printed areas.

Lining for Solvent Drums. These coatings have been used for many years as linings for drums used to ship solvents. Their excellent resistance to solvents makes them ideal for this application.

Other coating applications are in hardware, such as door handles and hinges; food processing, such as food, fruit, and milk containers; the liquor industry, such as beer and wine tanks and railroad cars; ships, such as propellers and oil tanker interiors; and such miscellaneous areas as belt buckles, munitions cartridge cases, and razor blades.

In the total phenolic market this class is by far the largest, having more than 50% of the total phenolic resin volume. Their noncoating uses include various bonding, laminating, and molding applications. The largest volume usages are the bonding of wood veneer in making plywood and bonding glass and rock wool fiber in making thermal and acoustical insulations. In these applications the low molecular weight resins are used either as 100% liquids or as water solutions. Other large volume bonding applications include the bonding of abrasives for making brake linings, grinding wheels, and sandpaper, the bonding of sand for making foundry molds, and the bonding of wood chips and wafers for making construction board. Molding materials are made by compounding with wood flour, fibers and other fillers. Paper and cloth with liquid or alcohol solutions are used to form many laminated products.

All of these applications are based on the original-type heat-reactive Bakelite resins and the heat and pressure technique for curing them. The phenolic resin brings to these various composites the properties of good adhesion, structural stability, high resistance to most environments, and good electrical properties. The other phase of the composite, be it cellulosics, glass fiber, or chopped chicken feathers, adds mechanical toughness and lowers the cost.

Unsubstituted and Non Heat Reactive. The second class, the unsubstituted, nonheat-reactive resins, are the novolak resins developed by Baekeland and Thurlow by the acid-catalyzed reaction of formaldehyde with an excess of phenol and mixed cresols. They are nonfilm formers, being brittle, permanently fusible, and soluble solids. They are soluble in alcohols, ketones, and esters, though not as polar as their heat-reactive analogues and, therefore, insoluble in water. They are also insoluble in aromatic and aliphatic hydrocarbons. Some uses depend on the fact that heating a novolak resin under pressure with a source of excess formaldehyde will result in a cross-linked, insoluble, infusible, "C"-staged material (13). One way of doing this is to compound with a heat-reactive resin having a large excess of methylol groups that can be used to react with the novolak resin. A second way, practiced first by Aylsworth in 1911, uses hexamethylenetetramine, "hexa," as a source of both formaldehyde and base catalyst. Hexa is a solid product formed by combining formaldehyde and ammonia. This results in the terminology of "two-step" resins. The first reaction or step (acid catalyst with a deficiency of formaldehyde) is brought to completion, resulting in a permanently fusible and soluble resin that, on compounding with hexa, can become reactive again. Pulverized solid resins containing hexa find application in molding materials and in bonding applications where a minimum of penetration is desired. In volume they are second of the four types but coming way behind their heat-reactive counterparts.

Phenolic novolak resins have been also used as coreactants (hardeners) with epoxy resins to produce thermoset systems with high-quality "engineering plastic" properties.

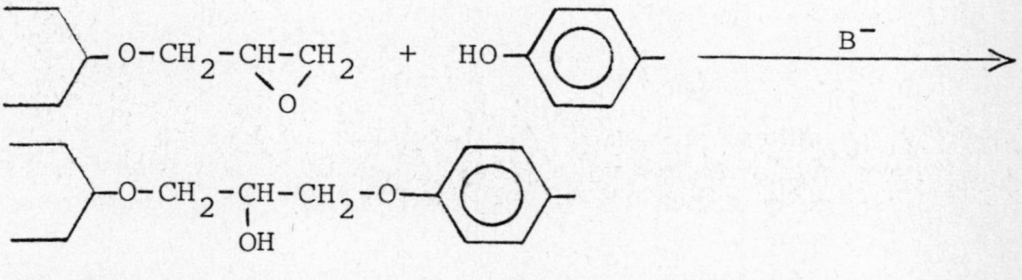

The base-catalyzed reaction of an epoxy resin with the phenolic resin produces a cross-linked polyether structure that is resistant to chemicals and heat and is a good moisture vapor barrier. Since the curing mechanism does not produce byproducts, thick sections may be obtained without voids and low shrinkage. Applications that employ the advantages of epoxy-phenolic formulations include molding materials, laminates, coatings, and adhesives.

Single-package epoxy-phenolic molding materials (14) usually utilize a solid epoxidized novolak and a phenolic novolak resin in a formulation such as that shown below:

Epoxy resin	20%
Phenolic resin	10%
Catalyst	2%
Lubricant (stearate)	1%
Filler--silica	67%

The catalysts used are usually base catalysts such as an alkyl-imidazole or tertiary amines. The catalysts may be blocked or coated in molecular sieves to produce latency. The molding materials are used in transfer molding processes to produce encapsulated semiconductors, discrete devices such as transistors, capacitors, diodes, rectifiers and resistors, and general–purpose components such as coils, bushings, armatures, and potentiometers. Since all the above devices require excellent long-term electrical properties, phenolic resins work well because they can be made low in inorganic ion content. While the older standard resins have given reliable performance, newer, lower free phenol resins will lower the mold staining and cleaning problems and increase production efficiency.

These phenolic resins are also combined with epoxy resins for prepreg laminating. Prepregs consist of a reinforcing fabric, usually glass cloth, impregnated with a partially cured (B–stage) resin system. These materials are used to manufacture electrical circuit boards, corrosion resistant fittings, and recreational equipment such as water skis, tennis rackets, and fishing rods.

In addition to the above applications, this class of phenolic resins may be used with epoxy resins to form either solution or powder coatings for pipes, electrical parts, or metal items that require excellent corrosion resistance. A representative formulation is shown:

	Parts by wt
Phenolic novolak (equiv wt 117)	33
Epoxidized novolak (equiv wt 225)	50.8
Bisphenol A epoxy resin (equiv wt 525)	29.5
Leveling agent	0.4
Base catalyst	0.2
Pigment filler	24.0
Solvent (if not powder coating)	80–100

Phenolic–epoxy adhesive systems for structural bonding use similar formulations but are frequently modified with a linear polymer having reactive end groups (carboxyl, amino) for toughness and vibration resistance.

Substituted and Heat Reactive. The third class, substituted and heat-reactive resins, are made by using para-substituted phenols where the substituent is a four–carbon or higher group such as *tert*-butyl, *tert*-octyl, and phenyl. Small amounts of ortho–substituted phenols and unsubstituted phenols are sometimes coreacted; but, in general, the functionality is 2, and only linear molecules are formed. They are brittle solids that do not form films. The substituent makes the resins less polar and hence they are soluble in ketones, esters, and aromatic hydrocarbons, with limited solubility in alcohols and aliphatic hydrocarbons. The phenolic resins based on longer chain aliphatic phenols are more compatible with drying oils, alkyds, and rubbers.

These resins are less important in coatings than their nonheat-reactive counterparts. They do find use in combination with drying oils in making electrical coil impregnation varnishes. The oils used in this type application are blown with air so that they contain enough oxygen to cross-link but are still fluid. The combination of the heat-reactive resin with the blown oil makes it possible to heat cure very thick coatings in the absence of air.

Combination with neoprene results in rubber coatings used for waterproofing concrete and for general-purpose maintenance (15). In these coatings the phenolic resin is complexed with a tetravalent metal oxide such as magnesium or zinc oxide to form a stable infusible compound. This treatment improves the adhesion and hardness of the neoprene.

The largest use for these resins is in neoprene contact adhesives. A formulation similar to that used in neoprene rubber coatings is used. The phenolic is used to improve adhesion to metals and glass, cohesive strength, elevated temperature strength, and tack of the rubbers. This adhesive finds use in shoe making, automotive upholstery and weatherstripping adherence, laminate joining to wood for tables and counter tops, and trade sales as adhesives for wood, cloth, plastic, rubber, and metal.

Substituted and Non Heat Reactive. The fourth class, substituted and nonheat-reactive resins, are produced with the same substituted phenols as their heat-reactive counterparts. They are brittle, permanently fusible, and soluble solids and thus non film formers. Similar to their nonheat-reactive analogues they have broad solubility and compatibility. Their importance in coatings is large. Being non film formers, they are always used to modify filming resins, such as drying oils and alkyds. Their largest use, from which they receive the oil-soluble phenolic name, is in the preparation of oleoresinous varnishes. Varnishes existed many years before the development of phenolics, being made from naturally occurring oil-soluble resins and drying oils.

The first successful addition of a synthetic resin to a drying oil is reported by Aylsworth in 1914 using an o-cresol phenolic resin (16). This resin and its varnish were not very desirable due to their dark color and poor color stability. Attempts to improve the color and oil solubility of these resins resulted in rosin-modified phenolics. This type resin is made by reacting a heat-reactive phenolic with rosin and then esterifying with glycerine. They were first produced by Kurt Albert of Germany in 1917 (17). They were referred to as Albertols or Albertol acids, depending on whether they were esterified or not (18). Commercial Albertols and Amberols were introduced in this country by the Resinous Products Co. in 1924 and are sold today by the Reichold Corp.

Efforts to further improve the hard resins were fruitless until substituted phenols became available. In the late 1920s, p-phenylphenol became available. It was isolated from the tarry residue of commercial phenol plants. With this new phenol came the development of the first high-performance 100% pure phenolic oil-soluble resin (19), marketed as BR-254 by the Bakelite Corp. in 1929. The para substituent resulted in a light color resin with good light stability. Its combination with drying oils resulted in a more rapid drying varnish film with better exterior durability than had previously been obtainable with any other hard resin. The excellent performance of varnishes containing this resin resulted in its inclusion in many industrial and governmental specifications written during the 1930s. This resin remains today the standard of the industry, but because of diminishing availability and high costs of the monomer, its use in the last several years has decreased to almost nothing.

During the thirties many substituted phenols were screened for applicability in varnishes. Patents on the preparation of oil-soluble resins were issued to Honel [20] using *p-tert*-butylphenol and *p-tert*-amylphenol and to Turkington and Butler [21] using *p-tert*-butylphenol, octylphenol, and others. Turkington and Allen [22] reported the effect of the alkyl substituent on the phenol and the influence on resin and varnish properties. Of some 40 tested, only the acid-catalyzed, nonheat-reactive resins made with *p*-phenylphenol and a few para tertiary alkylated phenols gave good performance.

These premium varnishes can best be described as oligomer solutions of phenolic resins and drying oils. The phenolic resin, in addition to previously discussed properties, has a high glass transition temperature, 80–100 °C, and a molecular weight of 600–1200. The drying oils are any of the naturally occurring triglycerides of mixed C_{10} to C_{22} fatty acids having some degree of unsaturation, low viscosity, low glass transition, −30 °C and lower, and molecular weights of about 900 (for most common C_{18} fatty acids). They are called drying oils because, on standing at room temperature, they will absorb oxygen from the air and cross-link by various reactions, thus being transformed from a low-viscosity liquid to dry solid films. This cross-linking occurs through the unsaturation of the oils. Oils with higher degrees of unsaturation are more reactive and cross-link more tightly. This polymerization is accelerated by heat and transition metal salts commonly called driers. A more complete description can be found in other references [23, 24].

Varnishes have historically been prepared by "cooking" the raw oils and resins at high temperatures, 230–310 °C, until the resin was all dissolved and the oils had polymerized to the desired viscosity. Solvent was then added, the varnish was cooled, and various additives such as driers, UV absorbers, antiskinning agents, and mildewcides were included. With the continuing technological development of phenolics came resins having better solubility in drying oils and higher molecular weights. This made it possible to decrease the amount of cooking to obtain resin solubility and desired varnish viscosity. This work resulted in the development by S. H. Richardson [25] of the cold mix varnish technique, which was introduced at the Paint Industries Show in 1954. Using a new alkylated phenol resin having improved oil solubility and compatibility and higher solution viscosities, it was possible to dissolve resin and oils in varnish-type solvents at room temperature, thus obtaining a stable premium-quality varnish. Hence, it became possible for a paint company having only room-temperature mixing equipment to manufacture phenolic varnishes that, for the majority of applications, performed as well as the best cooked varnishes. The maturity of the varnish industry and its reluctance to change from time-proven procedures resulted in a slow but steady acceptance of this procedure.

The advent of air pollution regulations in the late 1960s greatly increased the use of this technique since no volatiles were given off during manufacture. Producers of cooked varnishes have had to either discontinue making varnishes, make cold mixtures, or invest in the necessary new equipment, such as caustic scrubbers, or remove the 3–5% oily, obnoxious volatiles given off during high-temperature cooking.

In order to better service the cold mixing varnish industry, a convenient 50% solids solution was introduced (26). The small coatings company can produce a complete line of cold-cut varnishes by simply mixing a few liquid raw materials together. Large-volume companies can obtain all raw materials in bulk and continuously prepare cold-cut varnishes with the aid of metering and mixing pumps, thus further reducing manufacturing costs.

In the 1980s, high-solids coatings have become popular due to EPA air pollution requirements and the economics of solvents. Phenolic varnishes were already fairly high-solid coatings (50–70% nonvolatiles), but improvements up to the 75–80% nonvolatile level were made possible by the introduction of a new cold-cut varnish resin (27) that has shown equivalent performance to the traditional ones, particularly in aluminum pigmented maintenance paints. This same resin may also be used to upgrade the corrosion resistance of high-solids alkyd vehicles as noted later under Applications.

Phenolic varnishes, either cooked or cold mixed, offer many excellent coating properties. They brush well, having good flow characteristics, and dry to a high gloss. The drying oils give them good flow and wetting characteristics. They have good adhesion to metal and wood surfaces due to the polarity of the phenolic hydroxyl groups. Intercoat adhesion on recoating is excellent since the varnish solvents soften the surface of the cured film but do not dissolve it. Varnishes can be formulated to air-dry to tack-free films in 4 h. The color of these varnishes is light golden. This color adds a richness to wood, though in pigmented coatings, pure white cannot be made because of this yellowness. On aging this color is stable.

The cured films have fair resistance to dilute acids and alkalis and alcohols and aliphatic hydrocarbon solvents. Though these films will resist short contact with the chemicals, they are not recommended for continuous contact. Strong alkali will dissolve the coatings by hydrolyzing the ester group in the oil and making water-sensitive salts of the phenolic hydroxyls. Solvents such as ketones, esters, and aromatics will dissolve the phenolic resin out of the films and lift the polymerized oil films off the surface as a swollen gel. Resistance to hot and cold water is excellent. The excellent flexibility of these films is well demonstrated by their use in can coatings where metal sheets are first coated and then formed into cans. There is no rupture of the film in the various crimping operations. Abrasion resistance is also good. The excellent exterior durability of these coatings, first recognized in marine spar varnishes, today still makes them one of the most durable clear coatings for the protection of wood.

Applications. Clear phenolic varnishes find use today in industrial coatings as food can linings and maintenance primers and in trade sales coatings as general-purpose clears, floor varnishes, and premium exterior clears. Many of the cooked phenolic varnish can coatings formulations have not changed in the last 40 years, but some cold-cut systems have come into use. The excellent wetting and adhesion to metal, resistance to steam sterilization, flexibility, odorless and tasteless nature, and ability to cure rapidly on baking have kept a place for these coatings in the can industry.

The ease of brushing, 4-h air-dry, and recoatability make these excellent trade sales coatings. The fact that their water and alkali resistance is better than that of most other oil-containing vehicles makes them more durable to cleaning with household detergents. As exterior clears for the protection of wood, phenolic varnishes offer premium performance. Their combination of exterior durability and ability to protect wood from degradation by ultraviolet radiation is unsurpassed by any other coating.

In the mid-1960s, the Baker Castor Oil Co. (now Cas Chem) introduced a new drying oil polymer, Copolymer 186 (28), for use in cold mixing of coatings with improved durability over other drying oils. The use of this product, cold mixed with a phenolic resin, has shown improved exterior durability, being in good condition after 4 years of exterior exposure 45° facing south in New Jersey and 18 months in Florida.

In pigmented coatings, phenolic varnishes find use as vehicles for porch and deck paints where moisture and abrasion resistance are of utmost importance. Their moisture resistance and good metal adhesion make them useful vehicles for metal primers. In concrete paints they outperform most other oil-containing vehicles due to their better alkali resistance.

As a vehicle for aluminum-based maintenance paint used to protect metal, the cold mixture varnish offers maximum exterior durability at a minimum of cost and effort. This easy-to-apply, single-coat finish will maintain film integrity, retard corrosion, and retain good appearance for upward of 4 years, outperforming many other vehicles (29). The durability of this paint is partly due to the "leafing" or floating of the aluminum platelets parallel to the surface, resulting in a good moisture barrier. If the surface treatment of the aluminum flakes is affected by either the vehicle, solvents, or driers on aging, the paint will no longer "leaf." Cold mixture phenolic varnish aluminum paints have been shown to retain excellent leafing after 10 years of aging.

Phenolic oil-soluble resins also find use as additives to other vehicles to increase adhesion, hardness, and alkali and moisture resistance. The addition of 5-20% to an alkyd will significantly improve this type performance.

Another form of phenolic resin/drying oil combination is the phenolic dispersion resin, which is a highly bodied varnish dispersed in fast evaporating aliphatic hydrocarbon solvents. On evaporation of solvents, these dispersions form a dry film. Further oxidative polymerization is not needed. These dispersions are rather brittle when used alone and are, therefore, usually added to varnishes or alkyds to increase the dry rate and solvent resistance. Traffic paints formulated with these dispersions will dry in 3 min. Fast-drying shop primers for metal can be sprayed, air-dried for 5 min, and topcoated with strong solvent-containing lacquers without lifting the primer. The phenolic dispersion/alkyd combination is recommended for fast-drying, corrosion-inhibiting aircraft primer (30).

Waterborne Phenolic Resins. In addition to the unsubstituted, heat-reactive, water-soluble resins mentioned earlier, recent years have seen the introduction of several phenolic dispersions in water. The dispersed resins are higher in molecular weight than the water-soluble resins and have little or even no free phenol content.

The first materials of this class were heat-reactive phenolic resins dispersed in situ during the preparation reaction and stabilized with a mixture of water-soluble gums (polysaccharides). Patented by J. Harding (31), this technology resulted in commercially available products with an average particle size of 2–5 μm (32). These dispersions, supplied at 40–50% total solids, are fully dilutable with water and may also be blended with various latexes to produce compositions with modified properties. Applications involving the use of these dispersions include fiber bonding, pulp molding, paper and fabric impregnation, friction element bonding, coated abrasives, laminates, wood bonding, and adhesives.

A second generation of phenolic dispersions, patented by J. S. Fry (33), involved the post dispersion of phenolic resins in a mixture of water and water-miscible solvents. To conform with air pollution regulations, the solvent was held to 20 volume %, or less, of the volatiles. A heat-reactive phenolic resin dispersion (34) and a phenolic-epoxy codispersion have become commercially available based on the above technology. Supplied at 40–45% solids, these products, which have a small particle size (0.75–1.0 μm), are better film formers than the earlier dispersions. Used alone or in blends with other waterborne materials, corrosion-resistant baking coatings may be formulated for coil coating primers, dip primers, spray primer-surfacers, and chemically resistant one-coat systems. Products of this type are also tackifiers for acrylic latexes, and such systems have been employed as contact, heat seal, and laminating adhesives for diverse substrates.

Conclusions

Phenolics have found many uses over the last 50 years, resulting in a continuing growth. Sales have grown from 100 thousand pounds in 1911 and 1 million pounds in 1914 to 1 billion pounds in 1966. Today's volume is about 1.5 billion pounds and still growing.

Scientific knowledge is also increasing with new information concerning kinetics and molecular structure. The application of various new analytical tools will further increase our knowledge and may assist in improving such properties as color, internal plasticization, molecular weight, and heat stability.

Further growth is expected to occur in screw injection molding, flame retardant applications, phenolic foams, microballoons, and wafer board. Usage in coatings is expected to continue to expand in high-performance applications. Waterborne coatings and adhesives are expanding into new applications also.

Literature Cited

1. Baekeland, L. H. Ind. Eng. Chem. 1924, 16, 10.
2. Baeyer, A. Ber. 1872, 5, 1095.
3. Baekeland, L. H. Ind. Eng. Chem. 1909, 1, 3.
4. Baekeland, L. H. U.S. Patents 939 966, 1909, and 942 852, 1909.
5. Baekeland, L. H.; Bender, H. L. Ind. Eng. Chem. 1925, 17, 3.

6. Megson, N. J. L.; Drummond, A. A. J. Soc. Chem. Ind., London 1930, 49, 251T.
7. Megson, N. J. L. "Phenolic Resin Chemistry"; Academic: New York, 1958.
8. Martin, R. W. "The Chemistry of Phenolic Resins"; Wiley: New York, 1956.
9. Walker, J. Frederic "Formaldehyde"; Reinhold: New York, 1953.
10. Union Carbide Corp. "Product Information, Phenolic Resin BK-5918," F-47489, 1979.
11. Fry, J. S. U.S. Patent 4 182 732, 1980 (to Union Carbide Corp.)
12. Henderson, W. H. A.; Baskett, F. Materials Protect. 1962, Aug.
13. Baekeland, L. H. Ind. Eng. Chem. 1909, 1, 8.
14. Salensky, G. "Epoxy Resin Systems for Electrical/Electronic Encapsulation: Insulation/Circuits"; Lake Publishing Corp.: Libertyville, IL, 1962, May.
15. E. I. Du Pont de Nemours & Co. "Neoprene Maintenance Coatings," PB-15.
16. Aylsworth, J. W. U.S. Patent 1 111 287, 1914.
17. Mann, A. A.; Fonrobert, E. U.S. Patent 1 623 901, 1917.
18. Greth, A. Kunststoffe 1914, 31, 345.
19. Turkington, V. H.; Butler, W. H. U.S. Patent 2 017 877, 1935.
20. Hönel, H. U.S. Patent 1 996 960, 1935.
21. Turkington, V. H.; Butler, W. H. U.S. Patent 2 173 346, 1939.
22. Turkington, V. H.; Allen, I. Ind. Eng. Chem. 1941, 33, 966.
23. Cowan, J. C. "Chemistry and Technology of Drying Oils," chapter in this book.
24. Payne, H. F. "Organic Coating Technology"; Wiley: New York, 1964; Vol. 1.
25. Richardson, S. H. Paint Varnish Prod. 1955, Aug.
26. Union Carbide Corp. "Bakelite Phenolic Resin Solution CKSB-2001," F-4301, 1970.
27. Union Carbide Corp. "UCAR Phenolic Resin CK-2500," Product Information, F-48305, 1982.
28. Cas Chem., Inc. Product Bulletin "Copolymer 186" (85%).
29. Union Carbide Corp. "Comprehensive Exposure Data on Ready-Mixed Aluminum Paints," F-41516, 1967.
30. U.S. Military Specification, TT-P-1757, 1976.
31. Harding, J. U.S. Patent 3 823 103, 1974 (to Union Carbide Corp.).
32. Union Carbide Corp. "Phenolic Bonding Resin BKUA-2260," Product Bulletin F-46526, 1976.
33. Fry, J. S. U.S. Patent 4 124 554, 1978 (to Union Carbide Corp.).
34. Union Carbide Corp. "Phenolic Dispersion BKUA-2370," Product Information, F-47415, 1979.

Tall Oil and Naval Stores

YUN JEN[1] and E. E. McSWEENEY[2]

[1]J. J. Chemicals, Inc., Savannah, GA 31406
[2]Marcam, Inc., Savannah, GA 31411

History of Naval Stores
Production Statistics of Naval Stores
Processes of Naval Stores
  Gum Naval Stores
  Wood Naval Stores
  Tall Oil Naval Stores
Chemistry of Naval Stores
  Rosin
  Turpentine
  Fatty Acid
Major Uses of Naval Stores and Their Trends
  Rosins
  Turpentine
  Fatty Acid
Tree Stimulation

History of Naval Stores

The history of Naval Stores in the United States dates back more
than 400 years when the early explorers needed pitch tar and rosin
for ship caulking (1, 2). As the population gradually moved
southward, the center of the naval stores activities also shifted
(3). In the mid-1600s, the Connecticut River was the region where
most naval store production was concentrated. Around the early
1700s, the activities moved gradually southward through Virginia
into North Carolina. In the late 1800s, the states of Georgia,
Alabama, and Florida became the center of naval store operations.
These southeastern states remain the major producing areas of naval
stores today.
    The varieties of naval store operations grew as the years went
by. Prior to 1910, all of the naval stores were derived from living
coniferous trees. Flows of oleoresin were collected from artificial
wounds in these trees, and this type of naval stores is known as gum
naval stores. In the early days, a box was actually cut into the

0097–6156/85/0285–1159$06.25/0

tree to collect the resin and the collected oleoresin was separated into turpentine and rosin in a still, which incidentally was also used for the purpose of making whiskey. As science and technology developed, numerous improvements were made to facilitate collection and increase yields. A significant impact to the gum naval stores was made when the wooden box was replaced by earthenware or metal cup and a shallow cut rather than a deep gash was found to be sufficient for gum collection, thus preserving the bulk of the wood for lumber stock. In the late 1940s, an important technique of applying a sulfuric acid paste on the wound for the purpose of increasing oleoresin flow was discovered. In spite of these improvements, the operations of gum naval stores are intrinsically labor intensive and not easily adaptable to mechanization. In the past 50 years, there has been a steady downward trend of gum naval store production. Historically, 1908 was the biggest crop year ever in the history of naval stores (4); gum rosin production was well over 1 billion pounds.

During the 1915–25 period, a new type of naval store operation emerged in the South. Plants were built to extract naval stores from the stumps of virgin pines that had been left in the ground after lumbering. The extract is further separated into turpentine, pine oil, and rosin. This, known as the wood naval stores, is sometimes called steam–distilled naval stores, although steam is no longer used. The first plant was built at Gulfport, Miss., in 1909. In 1913 and 1916, Newport Industries built plants at Bay Minette, Ala., and Pensacola, Fla. In 1910, Hercules Powder Co., partly due to its interest in explosives required to excavate stumps and partly due to its interest in the paper–size business, decided to enter this market by establishing extraction plants at Brunswick, Ga., and Hattiesburg, Miss. Several other plants were established shortly thereafter in Louisiana, Mississippi, and Florida. The emergence of large-volume production of wood naval stores in contrast to gum naval stores allowed the former to become the predominant factor in the industry from the early 1910s to the late 1960s. Total rosin production exceeded the billion-pound level in a number of years, the latest having been 1966. Improvements were made in wood naval stores in which the extracted dark rosin was refined to pale grades, equivalent to colors of the gum rosin.

As the wood naval store industry activities expanded, the stumps near the extraction plants rapidly became depleted. For continuing operation, the industry had to search and harvest stumps from distant locations or seek the scattered first-growth stumps that became available as forests were clear cut for pulpwood, resulting in higher raw material cost. Stumps of marginal quality and excessive dirt found their way to the extraction plant, contributing to poor yield and inefficient operation. These problems, combined with the high labor costs, caused many plants to close in the 1960s. This trend continued through the 1970s and 1980s, resulting in there being only one producer operating two plants by 1985.

The rapid growth of sulfate pulping in the South in the 1940s and 1950s led to the birth of the third type of naval stores in this country, the tall oil naval stores. In the tall oil naval store operation, the oleoresinous materials in the trees are recovered in the pulping process in the forms of sulfate turpentine and crude tall oil. The crude tall oil, comprising rosin and fatty acid, can be fractionated into purer forms of these components. Unlike the other two types of naval stores, the tall oil process is highly automated and requires very little manual labor. In the early 1940s, West Virginia Pulp and Paper Co. operated a rudimentary distillation plant to refine tall oil to a crude form of rosin and fatty acid while Union Bag and Paper Co. started the production of acid-refined tall oil. The latter process caused some polymerization as well as color improvement, resulting in a light-colored, noncrystallizing tall oil. In 1949, Arizona Chemical installed a vacuum distillation unit on a commercial scale to separate tall oil into high-quality fatty acids and rosin. Similar units were subsequently placed in operation by Hercules, Union Camp, Newport (now Reichhold), Crosby (now Westvaco), Westvaco, and Glidden (now Sylvachem). By 1970, the tall oil naval stores became the major factor in the total U.S. production of rosin and turpentine. The distillation capacity has outstripped the supply for a number of years and will probably continue to do so. This imbalance was further extended by Georgia Pacific's completion of a new plant in 1980. This plant uses thin film evaporators and packed columns for dry distillation that has been popular in Europe for several years. Relative merits of steam and dry distillation were reviewed in a recent Pulp Chemicals Association Meeting (5, 6).

Production Statistics of Naval Stores (7, 8)

As shown in Figure 1, rosin production peaked at over 500,000 tons in the early 1960s and since has decreased sharply to a low of about 275,000 tons in the recession of 1975. Gum rosin in the United States has been in a steady decline since the early 1930s and now is an insignificant part of the total, although it continues to be a major factor in world production, particularly in China and Portugal (9). Wood rosin production peaked in the early 1950s and has declined steadily since, but it may be leveling out for the near future, especially if the Pinex process is successful. Tall oil rosin climbed from its introduction in the early 1950s until it peaked in 1972 and 1973. Currently it accounts for about two-thirds of the U.S. production and should keep the total about at the current level of 340,000 tons for the foreseeable future with the possibility of some very slight growth (10). Unfortunately, it is quite certain that crude tall oil and the rosin derived therefrom will not grow at the same rate as pine pulp production, 2% per year, for a variety of reasons such as recycling of waste paper, cutting of younger trees, increasing use of pine chips as well as hardwood, and the export of crude tall oil for use per se or for fractionation in Japan and Europe. Better recovery of the potential tall oil in the pinewood consumed is an achievement to be hoped for but realistically cannot be expected to improve much. New pulping processes may also affect both quantity and quality of tall oil, but it is too early to make any predictions.

Turpentine production from the three types of naval stores follows the same pattern as rosin (Figure 2). However, due probably to the simple method of recovery, sulfate turpentine dominated the other two sources as early as 1955. The near-term outlook is for a steady rate of turpentine production. A comprehensive review of turpentine production and use was given recently (11).

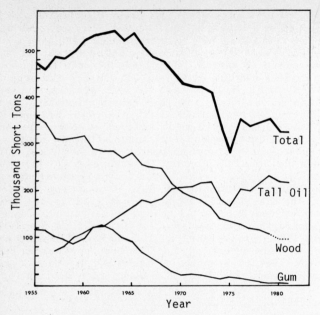

Figure 1.   Production of rosin in the United States.

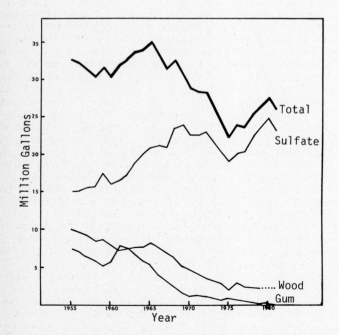

Figure 2.   Production of turpentine in the United States.

It is of interest to note that rosin and turpentine, the two major naval store products, are subject to violent price movement similar to other types of commodities. For example, rosin sold for 24¢/lb in 1919, <2¢/lb in 1932, approximately 8¢/lb in the 50s, 10¢/lb in the 60s, and over 25¢/lb in 1974 and peaked at over 50¢/lb for gum and wood and about 40¢/lb for tall oil in 1981. In 1984, the rosin price was less than 20¢/lb. Even wider fluctuations were seen in the price of turpentine: $2.30/gal. in 1919, $0.22/gal. in 1938, $0.97/gal. in 1946, $0.38/gal. in 1949, $0.55/gal. in 1972, and over $1.00/gal. in 1974. In 1984, the turpentine price was in the range of $0.55/gal.

Processes of Naval Stores

The processing of the three types of naval stores are briefly described below.

Gum Naval Stores (Figure 3). Gum from trees is scraped into a drum and delivered to a gum processing plant. The crude gum is diluted with turpentine and filtered. Turpentine in the filtrate is then separated from the rosin by distillation.

Wood Naval Stores (Figure 4). The stumps are pulled out of the ground now mostly by mechanized equipment and seldom by dynamite, as was practiced in the early days. One advantage of the mechanized equipment is that it can recover stumps in swampy locations, thereby broadening the potential of stump availability.

Aged stumps typically contain 20-25% extractives. After delivery to the plant site, the wood is hogged, shredded, and extracted semicontinuously with petroleum solvents in a series of extraction retorts by using the countercurrent principle. Fresh solvent is used to extract the chips containing the least amount of oleoresin in the retort before the chips are dumped, and the strongest liquor is used to extract fresh chips. The plant is generally operated in such a manner that chips are alternately filled to and dumped from a series of retorts in a rotating schedule. The oleoresin/solvent mixture is fractionally distilled to recover solvent, turpentine, and pine oil. The residue is a dark wood rosin with a color grade of FF on the USDA scale. The FF rosin is upgraded to pale-colored rosin by either an earth-clay adsorption process or furfural extraction process; a very dark rosin is a byproduct that contains high levels of oxidized materials. As noted below, a new source of stumps for extraction is being developed at present.

Tall Oil Naval Stores (Figures 5 and 6). In the sulfate pulping process, as the pine chips are cooked, the volatilized gases are condensed to produce sulfate turpentine. The fatty acid glycerides and rosin acids are converted to their sodium salts, which are separated as soap skimmings after black liquor concentration and settling. The soap skimmings are reacted with a mineral acid to convert the organic salts back to acidic forms as crude tall oil. The crude tall oil is then fractionated in a series of towers to yield fatty acids, rosin acid, distilled tall oil, heads, and pitch residue.

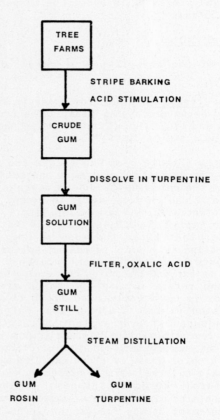

Figure 3.   Gum naval store process.

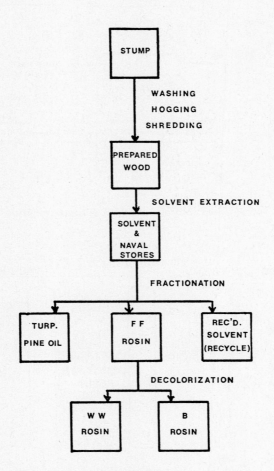

Figure 4.   Wood naval store process.

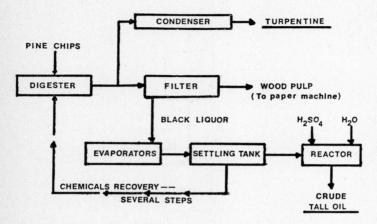

Figure 5.   Recovery of naval store byproducts from Kraft pulping process.

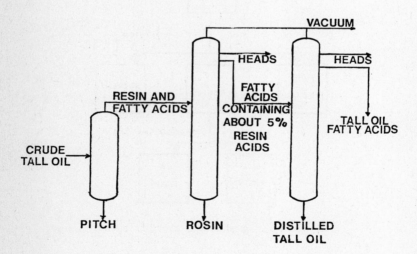

Figure 6.   Fractionation of crude tall oil.

Chemistry of Naval Stores

<u>Rosin</u>. Rosin is primarily a mixture of diterpene resin acids of the abietic and pimaric types. The quality of rosin has traditionally been graded by its color. With the advent of modern analytical tools, scientists are now able to describe the composition of a rosin in terms of different resin acids. Properties of a rosin can usually be predicted by its composition analysis.

Figure 7 shows the structures of commonly found resin acids. It is to be noted that these materials have two major sites for chemical reactions: the carboxylic group and the double bonds. It can be expected that the positions of double bond would play an important role in certain reactions. A benzenenoid structure in dehydroabietic acid would indicate inertness of this molecule toward reactions such as dimerization, oxidation, Diels-Alder, and free radical chain transfer or termination, while resin acids containing conjugated double bonds would undergo these reactions with ease.

The chemical reactivities of gum, wood, and tall oil rosin are significantly different due largely to their respective compositions. Table I shows typical chromatographic analyses of the three types of rosin (<u>12</u>, <u>13</u>). It is interesting to note that the gum rosin contains the least amount of the inert dehydroabietic acid among the three types of rosin, and as expected, it exhibits the greatest adduct potential toward a dieneophile such as maleic anhydride.

The overabundance of one species of resin acid and the presence of neutral materials in a rosin affect its tendency to crystallize. One of the serious problems of tall oil rosin in the early days was its rapid rate of crystallization, making this type of rosin difficult to melt. A commonly used yardstick for keeping rosin and rosin derivatives from serious crystallization problems is to keep any one species of the isomers below 20% of the total. It is generally observed that gum rosin is the least crystallizable and tall oil rosin the most. Heat treatment or minor chemical modification greatly reduces the crystallization tendency of tall oil rosin.

Important reactions of rosin are summarized in Figure 8. Dimerization is generally carried out by a Lewis acid catalyst such as sulfuric acid, boron fluoride, or zinc chloride (<u>14–16</u>). A rosin dimer finds uses in resin preparation because it allows quick buildup of molecular weight through a polyfunctional alcohol or amine.

A very important technique in rendering rosin and its derivatives stable to heat and oxygen is the removal of unsaturation through hydrogenation (<u>17</u>, <u>18</u>). The commonly used catalysts are nickel (<u>19</u>) and palladium on carbon (<u>20</u>), operated under moderately high pressure at elevated temperatures. The preferred feed rosins are gum and wood. Tall oil rosin generally contains small amounts of sulfur compounds from the process, which causes poisoning of the catalyst. For economic reasons, therefore, in the commercial process of tall oil hydrogenation, pretreatment before hydrogenation may be necessary. Where the requirement is the removal of conjugated unsaturation and not total saturation, a disproportionation reaction is carried out. In this reaction, abietic acid, for example, is converted to dehydroabietic and

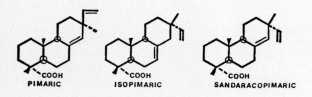

Figure 7.   Common resin acids of pine.

Table I. Typical Rosin Compositions

Components	Gum	Wood	Tall Oil
Pimaric	2	3	3
Palustric	18	10	10
Isopimaric	18	11	7
Abietic	20	45	35
Dehydroabietic	4	8	20
Neoabietic	18	7	4

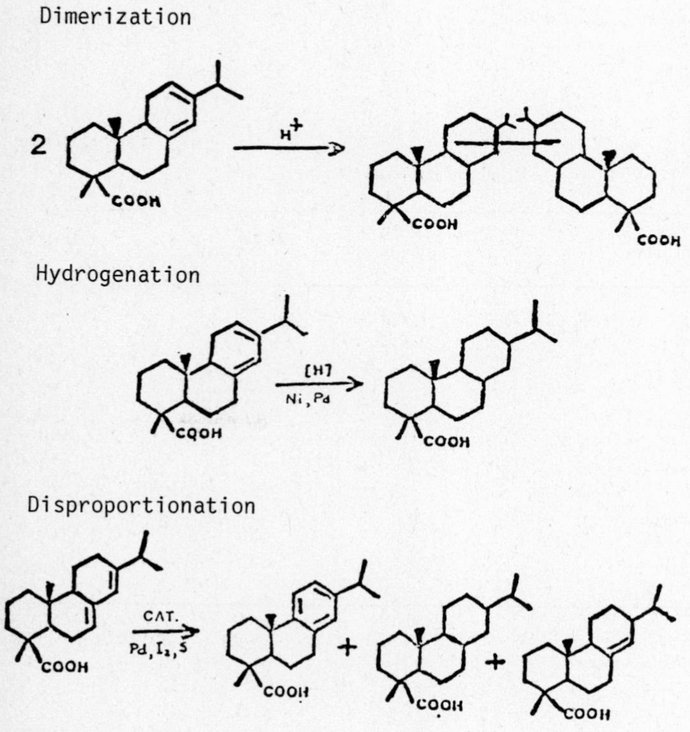
Dimerization
2
H+
COOH
COOH
COOH
Hydrogenation
[H]
Ni, Pd
COOH
COOH
Disproportionation
CAT.
Pd, I2, S
COOH
+
COOH
+
COOH
COOH

Figure 8. Typical reactions of rosin.

dihydroabietic/tetrahydroabietic acids. The removal of conjugation
allows the rosin to perform satisfactorily as an emulsifier in a
free radical emulsion polymerization recipe without undesirable
chain transfer and chain termination side reactions. Commercially
used catalysts for disproportionation are palladium (21), iodine
(22), sulfur dioxide, and certain sulfur compounds (23).

The carboxylic acid group in rosin undergoes the usual ester and
amide formation with alcohols and amines (Figure 9). A large number
of compositions can be prepared from various rosins and modified
rosins.

Perhaps the reaction carried out on rosin in largest volume is
salt formation. Alkali metals form simple salts (24) of rosin that
are water soluble, allowing their use in aqueous systems, such as
paper sizing and emulsion polymerization (25). Alkaline earth metal
salts of rosin are organic solvent soluble and have good pigment
wetting, thus lending themselves well to use as ink vehicles.
Depending on the performance requirements, the rosin part could have
undergone one or more aforementioned modifications before salt
formation (26).

Formaldehyde reactions with rosin have been a subject for many
patent disclosures. It is now generally concluded that two types of
reactions occur. Under mildly acidic, neutral, or alkaline
conditions, a mixture of methylene ether, ethylene bridge, and
methylol derivatives from abietic type acids are formed (27), while
under strong acidic conditions a methyl derivative is formed (28).
The ease of carrying out these reactions makes them attractive means
of modifying rosins for the purpose of crystallization suppression
and introducing an additional functional group into rosin.

Another well-known method of increasing rosin functionality is
by reacting it with a dieneophile such as maleic anhydride, furmaric
acid, and acrylic acid through a Diels-Alder reaction (29). Dibasic
or tribasic acids of rosin are formed readily by this reaction.

In addition to resin acids, rosin contains neutral materials and
oxidized impurities. Tall oil rosin contains about 1-5% fatty acid
and small amounts of sulfur compounds and phenolics. In tall oil
pitch, which is the bottom fraction in the tall oil distillation
process, besides residual amounts of rosin and fatty acids, a
substantial amount of unsaponifiable matters has been identified and
recovered as β-sitosterol (30). Gum rosin contains large amounts of
neutrals that contribute to the typical odor and reduce its tendency
to crystallize. The amount of neutral materials present in a rosin
largely depends upon the location where the raw material originates
and its processing conditions.

<u>Turpentine</u>. Similar to rosins, turpentines differ in content of
monoterpenes depending upon the type and sources. Typical southern
pine derived turpentines have the analyses shown in Table II.
Western U.S. woods contain an appreciable quantity of Δ^3-carene.
There is no gum and wood naval stores production in the west, only
sulfate turpentine.

The most important reaction of monoterpenes is hydration (31).
α-Pinene can be converted to α-terpineol with water in the presence
of a mineral acid. This is the procedure used to make the
commercial product known as pine oil. Further hydration converts it
to terpine hydrate.

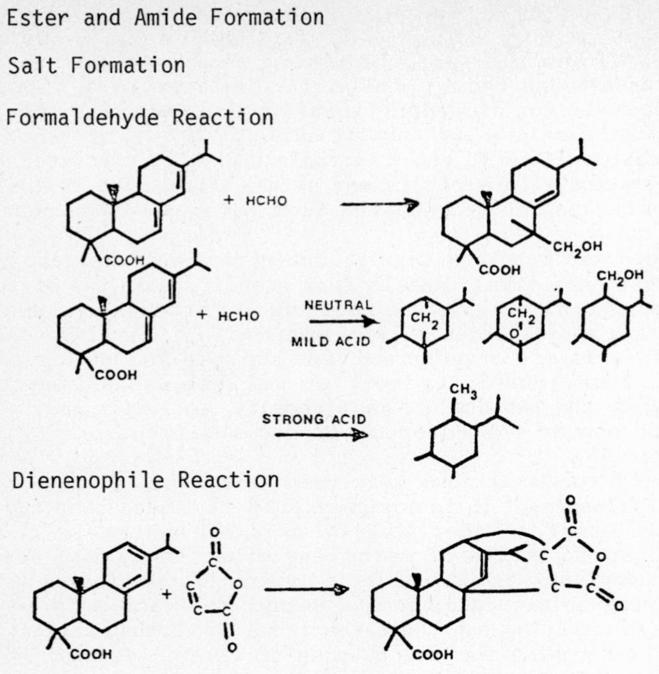

Ester and Amide Formation

Salt Formation

Formaldehyde Reaction

Dienenophile Reaction

Figure 9.   Typical reactions of rosin.

Table II.   Composition of Turpentines

Components	Gum	Wood	Sulfate
α–Pinene	60–65	75–80	60–70
β–Pinene	25–35	0–2	20–25
Camphene		4–8	
Other terpenes	5–8	15–20	6–12

α-PINENE   β-PINENE   CAMPHENE   Δ³-CARENE

Polymerization of terpenes using acidic catalysts such as $AlCl_3$ and BF_3 is another major reaction for terpenes. β-Pinene and dipentene yield almost quantitative amounts of polymer in the 600–2000 MW range, while α-pinene yields only about 60% polymer in this range (32). More recently, it was claimed that an all α-pinene resin can be prepared by using chlorosilane as a catalyst (33). Monoterpenes condense with phenol in the presence of BF_3 to form terpene-phenolic resins.

The quantitative conversion of dipentene and β-pinene to form polymers stirred interest in (1) the pyrolysis of α-pinene to dipentene and allo-ocimene (34) and (2) the synthesis of β-pinene from α-pinene (35). However, due to the availability of citrus limonene as a supplemental source for dipentene, the pyrolysis scheme is no longer economically attractive.

Isomerization of wood turpentine and α-pinene to camphene followed by chlorination is another major chemical reaction of turpentine, for the preparation of an insecticide against boll weevils in cotton, but restrictions in the use of chlorinated insecticides has greatly reduced this use.

Perhaps the most intriguing and fast-growing area of terpene chemistry is in speciality chemicals. A wide range of flavor and fragrance materials can be synthesized from terpenes. Figures 10 and 11 illustrate some of the reaction routes. A major expansion has recently been completed in this area, including a new route to some of these derivatives using α- instead of β-pinene (11).

Fatty Acid. High-grade tall oil fatty acid is comprised essentially of oleic and linoleic acids in equal proportions. Typical carboxylic related reactions such as salt formation, esterification, amidization, reduction, sulfonation, sulfation, and ethylene oxide adduction are well-known. Since these reactions are common to all fatty acids, they will not be discussed further. However, there are a number of other more specific reactions applicable mainly to tall oil fatty acid that are significant industrially; a brief review of these is in order.

Tall oil fatty acid esters can be easily epoxidized in the presence of hydrogen peroxide and acetic acid. For satisfactory results, the tall oil fatty acid should be pretreated to remove trace phenolic impurities, such as dimethoxystilbene, in order to avoid pink coloration.

Oxidative cleavage of the double bond in tall oil derived fatty acid may be carried out with ozone; pelargonic and azelaic acids are the major acids formed (36).

Dimer acid, trimer acid, and small amounts of higher polymers are formed when tall oil fatty acid is treated with an active clay (37). In the same process, part of the fatty acid is isomerized to methyl-branched acids. These can be hydrogenated to produce a mixture of isostearic and stearic acids, which can be separated by a solvent crystallization process. Dimer acids are separated from trimers by thin-film or molecular distillation.

Tall oil fatty acid readily undergoes isomerization and reduction to essentially all monoenoic acid in the presence of a catalyst and a hydrogen donor. The reaction is very similar to the rosin disproportionation discussed earlier. Since tall oil rosin and fatty acid are produced simultaneously, it becomes obvious that

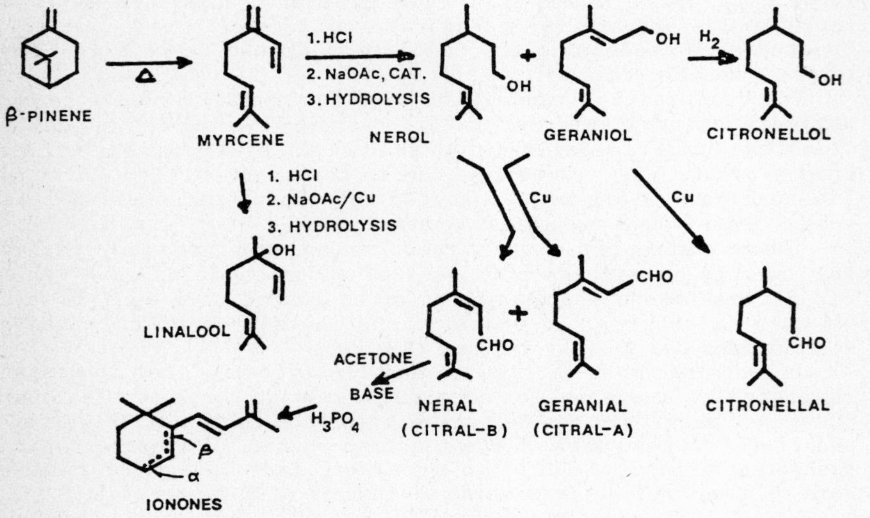

Figure 10.   Synthesis of some fragrance and flavor chemicals from β-pinene.

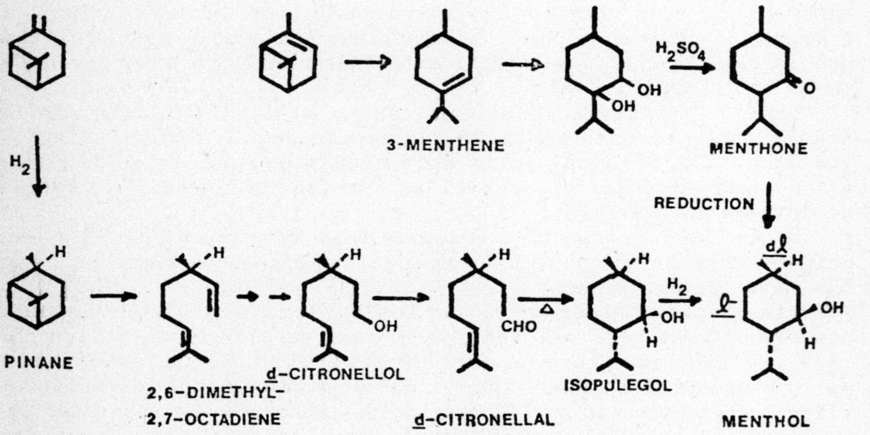

Figure 11.   Synthesis of menthol.

a separation of these two components may not be necessary if both rosin and fatty acid are useful together. Thus, a mixture of rosin and tall oil fatty acid can be treated with iodine to yield a product containing no conjugated unsaturation in either component and useful as a SBR emulsifier.

More recently, acrylic acid has been reacted with tall oil fatty acid to form a C_{21} dibasic acid with oleic and elaidic acids as byproducts (38).

Major Uses of Naval Stores and Their Trends

Rosins. Major uses of rosin in the United States in 1979 (the last year for which such statistics were published), which accounted for 277,000 tons, were as follows: chemical intermediates plus rubber, 33%; paper size, 33%; resins plus ester gums, 25%; coatings, 2%; other, 6%.

The traditional major use for rosin is paper size. Sodium and potassium soaps of rosin or modified rosin are precipitated on paper by alum to serve this purpose. In the late 1940s, maleic anhydride or fumaric acid modified rosins were introduced as fortified size, which cut down the rosin consumption. In the mid-1970s, aqueous rosin emulsions began to replace part of the paste rosin soap market because of better cost performance. Therefore, in spite of growth of the paper industry, there has been no growth in rosin consumption, as the long-term trend of using less rosin per ton of paper produced continues. Two major synthetic sizes, alkyl ketene dimer and alkenylsuccinic anhydride, are estimated to have captured 25% of the dollar volume of the size market in 1984. Therefore, continuing reduction in the use of rosin by the paper industry is expected in the second half of 1980s.

Emulsifiers for styrene-butadiene, styrene-butadiene-acrylonitrile, and neoprene rubbers is the next important area of rosin use. Due to its unique properties of being surface active in aqueous solution and tacking in coagulated rubber, disproportionated rosin finds ready acceptance in this application. However, due to slow growth of SBR, which is by far the most important factor among all rubbers, the consumption of rosin here will rise only very slowly, if at all, in the future. It is significant to note that there appears to be an industry-wide acceptance of a mixed disproportionated rosin-fatty acid emulsifier, which is lower in manufacturing cost, to replace the traditional disproportionated rosin acid and soaps.

Rosin "cooked" with drying oils have been used in varnishes since antiquity. Early in this century these products were greatly improved by "liming" or converting the rosin to "ester gum" by reacting with glycerol. After World War I, greatly improved products were made by reacting rosin with maleic anhydride or fumaric acid before esterifying. Simultaneously phenolic-modified rosins, in which as little as 7-12% of a phenol-formaldehyde condensate greatly improved varnish properties, were developed. These were among the first truly synthetic resins and were introduced in this country by Resinous Products and Chemical Co. and Reichhold Chemicals. This advent resulted in the development of the so-called 4-h varnishes, which revolutionized the coating industry in the 1920s.

Although the advent of the "phenolics" and "maleics" displaced
limed rosin in varnishes, further improvement of resinates by use of
zinc and other metal oxides in addition to calcium led to the
resinates getting a major share of the publication ink production.

The use of rosin in synthetic resins and metal resinates has
kept pace with the growth of the ink, adhesive, and coating
industries. Polymerized and hydrogenated rosin esters have been
introduced to meet the requirements of heat and oxidation stability
in the hot melt coating and adhesive industries. Water-soluble
rosin derivatives for printing inks are being offered to cope with
the problems of air pollution and solvent shortages. Industry now
offers specialty rosin-derived products at premium prices that are
designed to serve special needs of the consumers. In spite of the
inroads of hydrocarbon resins, rosin-based resins maintain a major
proportion of the market in hot melts, pressure-sensitive adhesives,
inks, and various coating formulations. Rosin derivatives for
adhesive and ink industries appear to be the most significant growth
area in the coming years.

In summation, the overall rosin uses appear to be about in
balance with demand.

Turpentine. The major uses of turpentine in 1979 were pine oil,
48%; resins, 16%; insecticides, 16%; other, 11%; and fragrances, 9%.
Statistics on consumption are no longer published. Since 1979, the
use for chlorinated insecticides has decreased markedly because of
environmental restrictions. As a result of this and generally poor
business conditions in 1981 and 1982, significant quantities of
sulfate turpentine have been used as fuel.

The traditional use of turpentine as a paint solvent is no
longer a major factor in turpentine consumption, first due to
replacement with the cheaper mineral spirits and more recently due
to the success of water-based paints and air pollution restrictions.
The only remaining significant solvent use for turpentine is in the
reclaiming of rubber.

The production level of pine oil is related to the population
and standard of living: in this area no sharp increase is expected
in this country. Polyterpene and terpene-phenol resins are finding
good acceptance as ingredients in hot melts and pressure-sensitive
adhesives; their growth will be continuing. Perhaps the greatest
increase in demand for turpentine will be in the flavor and
fragrance chemicals.

Fatty Acid (39, 40). The major uses of tall oil fatty acid in the
United States in 1981 were (40) intermediate chemicals, 47.6%;
protective coatings, 21.2%; soaps detergents, 6.6%; flotation, 4.4%;
and other, 20.1%.

The upgrading of tall oil fatty acid quality in the past two
decades has put this unique class of fatty acid in uses where it
serves special functions. Dimerized tall oil fatty acid is widely
used for resin preparation for use in printing inks, adhesives, and
curing agents for epoxy resins. Isostearic acid, a liquid acid
derived from tall oil fatty acid, finds special applications in
high-performance lubricants and cosmetics. Epoxidized tall oil
esters are used in large quantities as vinyl plasticizers. High-
grade palmitic acid for cosmetics can be recovered from tall oil

heads. Low-temperature crystallization separates tall oil fatty acid into useful oleic-rich and linoleic-rich fractions. In short-oil alkyds, tall oil fatty acid is a preferred ingredient because of its superior overbake stability. Tall oil fatty acid is finding uses in rubber emulsifiers, soaps, detergents, defoamers, ore flotation, asphalt, and many other products. This is not due to its traditional low cost but rather to its desirable qualities. It is significant to note that in 1985 the price of tall oil fatty acid far exceeded that of soybean oil. In the past, these two commodities were in the same comparable price range.

Tree Stimulation

A very significant discovery was reported in 1973 by USDA scientists at Olustee, Fla. (41). Naval store yields from coniferous trees can be dramatically stimulated by a simple spray of paraquat and diquat-quaternary compounds near the bases of the trees a year or two prior to harvesting.

An industry-government laboratory joint task force was organized in late 1973 to conduct further investigations in various apsects of tree stimulation. Included in the study are forestry biology, environmental impact, government regulations, solvent extraction, and sulfate pulping of trees treated with the quaternary compounds. Renewed interest in wood naval stores arose as this scheme would produce inexhaustible supplies of oleoresin-soaked wood.

Results of widespread studies by government, university, and industrial organizations were reported and published annually from 1975 through 1979 (42). Two- to threefold increases in both rosin and turpentine in 18-24 months after treatment were confirmed in many studies, and the technical feasibility of recovering these products in the pulping operation was demonstrated. Opinions regarding financial return from treating pulpwood varied from somewhat pessimistic to quite optimistic. However, at present (mid-1982) there is no real promise for commercialization for pulpwood treatment, possibly because the division of profit between tree grower, harvester, pulp mill, and tall oil recovery and separation leave the process without a champion.

Initial hopes that the lower 5 ft of the treated tree would be sufficiently rich in rosin for extraction also led to disappointment. Instead treating the roots of the pine a year or two before harvesting appears economically viable. Hercules, Inc., is pursuing this vigorously as the Pinex process. First commercial extraction trials were run in the summer of 1982, and 10,000 acres of pulpwood timber were treated for further scale up of the process. Reports on this development and parallel studies at the Olustee Station of the U.S. Forest Service were given early in 1982 (43, 44). At this time it seems probable that this development could stabilize the wood rosin industry but with the additional advantage that the rosin recovered will be similar to gum rosin, which is the preferred form for many uses (45).

One final note: Forestry in the U.S. is one of the few natural resources that is increasing rather than decreasing every year. Naval stores, as byproducts of forestry, will therefore stay with us for many years to come, as an important renewable source of raw materials for the chemical industries.

Literature Cited

1. Zinkel, D. F. "Organic Chemicals from Biomass"; Goldstein, I. S., Ed.; CRC: Boca Raton, Fla., 1981; Chap. 9.
2. Drew, John. Naval Stores Rev. 1981, Nov.–Dec.
3. Cooke, D. W. Thesis, University of Tennessee, Knoxville, 1975.
4. "Naval Stores Statistics, 1914–25 and Previous Years"; Columbia Naval Stores Co.
5. Freese, H. L., et al. Naval Stores Rev. 1982, May–June, 8.
6. Bress, D. F. Naval Stores Rev. 1982, May–June, 13.
7. "Naval Stores"; Crop Reporting Board, Statistical Reporting Service, U.S. Department of Agriculture: Washington, D.C.
8. Naval Stores Rev. 1981 International Yearbook.
9. Stauffer, D. F. Naval Stores Rev. 1981, July–Aug.
10. McSweeney, E. E., paper given at the Pulp Chemicals Association's 8th International Conference, Oct. 6, 1981.
11. Mattson, R. H., paper given at the American Chemical Society Atlanta Meeting, March 31, 1981.
12. Baldwin, D. E., et al. Chem. Eng. Data Ser. 1958, 3, 342–6.
13. Union Camp Corp., unpublished data.
14. Grun, A., et al. Chem. UmschauGebiete Fette, Ole, Wachse Harze 1919, 26, 77.
15. Rummelsburg, A. L. U.S. Patents 2 108 928, 1936; and 2 136 525, 1936 (Hercules, Inc.).
16. Rummelsburg, A. L. U.S. Patent 2 124 675, 1938 (Hercules, Inc.).
17. Byrkit, R. J., Jr. U.S. Patent 2 174 651, 5/17/37 (Hercules, Inc.).
18. Montgomery, J. B., et al. Ind. Eng. Chem. 1958, 50, 313–36.
19. Brykit, R. J., Jr. U.S. Patent 2 094 117, 1937; and U.S. Patent Reissue 21 448, 1940 (Hercules, Inc.)
20. Glasebrook, A. L., et al. U.S. Patent 2 776 276, 1957 (Hercules, Inc.).
21. Fleck, E. E., et al. U.S. Patent 2 239 555, 1941.
22. Hasselstrom, T., et al. J. Am. Chem. Soc. 1941, 63, 1759.
23. Scharrer, R. P. F. U.S. Patent 3 649 612, 1972 (Arizona Chemical Co.).
24. Davidson, R. W. TPPI 1964, 47, 609–16.
25. Hays, J. T., et al. Ind. Eng. Chem. 1947, 39, 1129.
26. Watkins, S. H. U.S. Patent 2 887 475, 1959; Reaville, E. T., et al. U.S. Patent 2 994 635, 1961 (Hercules, Inc.).
27. Minor, J. C., et al. Ind. Eng. Chem. 1954, 46, 1973.
28. Lawrence, R. V. USDA, Olustee, Fla., personal communication.
29. Hovey, A. G., et al. Ind. Eng. Chem. 1940, 32, 272–9.
30. Steiner, C. S., et al. U.S. Patent 2 835 652, 1958 (Swift & Co.).
31. Sheffield, D. H. U.S. Patent 2 060 579, 1936; and 2 178 349, 1939 (Hercules, Inc.).
32. Roberts, W. J., et al. J. Am. Chem. Soc. 1950, 72, 1226.
33. Barkley, L. B., et al. U.S. Patent 3 478 007, 1969 (Picco).
34. Goldblatt, L. A., et al. J. Am. Chem. Soc. 1945, 67, 242; ibid. 1947, 69, 319.
35. Webb, R. L. U.S. Patent 3 264 362, 1966 (Union Camp).
36. Kadesch, R. J. J. Am. Oil Chem. Soc. 1954, 31, 568.

37. Cowan, J. S. <u>J. Am. Oil Chem. Soc</u>. 1962, <u>39</u>, 534–45.
38. Ward, B. F., et al. <u>J. Am. Oil Chem. Soc</u>. 1975, <u>52</u>, 219.
39. McSweeney, E. E., paper presented at the American Oil Chemical Society's New York Meeting, May 1, 1980.
40. Eagleson, D. R., paper presented at the "Symposium on Chemicals and the Pulp and Paper Industry"; American Chemical Society: Atlanta, Georgia, March 1981.
41. <u>Naval Stores Rev</u>. 1974, <u>Jan–Feb</u>.
42. Proceedings, Lightwood Research Coordination Council, Southeast Forest Experimental Station, Asheville, N.C.
43. Propst, M., paper given at the Pulp Chemical Association Meeting, March 5, 1982.
44. Roberts, D. R., paper given at the Pulp Chemical Association Meeting, March 5, 1982.
45. Smith, K., paper presented at the 11th International Naval Stores Meeting; New Orleans, October 1984.

Chemistry and Technology of Alkyd and Saturated Reactive Polyester Resins

H. J. LANSON

LanChem Corporation, East St. Louis, IL 62205

History of Alkyd Resins
Functionality Theory and Synthesis
Processing of Alkyd Resins
Classification and Properties of Alkyd Resins
Formulation and Design of Alkyd Resins
Uses of Alkyd Resins
Saturated Reactive Polyesters
Resin Systems for Reduced Solvent Emissions
Low-Solvent and Solvent-Free Alkyd Resins

Alkyd resins have been defined as the reaction product of a polybasic acid and a polyhydric alcohol. This definition includes polyester resins of which alkyds are a particular type. The specific definition that has gained wide acceptance is that alkyds are polyesters modified with monobasic fatty acids. In recent years, the term nonoil or oil-free alkyd has come into use to describe polyesters formed by the reaction of polybasic acids with polyhydric alcohols in non-stoichiometric amounts. These products are best described as functional saturated polyesters containing unreacted OH and/or COOH groups, and they are finding rapidly increasing uses in organic coatings.

History of Alkyd Resins

Alkyd resins came into commercial use over 50 years ago, and even with the wide array of other polymers for coatings that have appeared in more recent years, they rank as the most important synthetic coating resins and still constitute about 35% of all resins used in organic coatings. A recent survey reported that over 700 million lb of alkyd resins were used in 1980 (Table I) (1).

The formation of alkyd resins is a typical example of condensation polymerization. In 1847, Berzelius reported a resinous product formed by the reaction of tartaric acid and glycerol. In 1901, Watson Smith (England) prepared a brittle resinous polymer by

0097–6156/85/0285–1181$07.00/0
© 1985 American Chemical Society

the reaction of phthalic anhydride and glycerol. In 1910, the search for better electric insulating materials stimulated extensive investigations on the glycerol-phthalic anhydride reaction. The work of Callahan, Arsem, Dawson, and Howell showed that when part of the phthalic anhydride was replaced with monobasic acids, such as butyric and oleic acids, the products were more flexible and had better solubility than the glyceryl phthalate reaction product. However, it was not until 1929 that the chemical reactions involved were understood. In that year, Carothers (2) showed the broad relationship of functionality to polymer formation and extended the idea of polymerization to condensation reactions as well as addition reactions. His early work defined the conditions necessary for intermolecular condensation polymerization of bifunctional monomers to produce linear, soluble, fusible polymers such as the polyesters from glycols and dibasic acids.

In 1929 and 1930, Kienle and Hovey (3) emphasized the fact that, although a bifunctional polymer such as the glycol phthalate polyester is fusible and soluble even at the highest degree of polymerization obtainable, the use of a trifunctional component, such as the glyceryl phthalate polyester, leads to gelation long before completion of the esterification. When glycerol and phthalic anhydride are heated together in equivalent proportions, the acid number drops very rapidly at first because of the dibasic acid reaction with the primary hydroxyls of the glycerol to form rather short, linear polyesters. After sufficient heating, the reaction product is of moderate molecular weight; when cold, it is a clear resinous material. As the reaction proceeds, the secondary hydroxyls of the glycerol come into play; their reaction with molecules of phthalic anhydride connect the short chain to form a complex branched or network type structure. When approximately 80% esterification has occurred, the product becomes infusible and insoluble in common solvents.

Functionality Theory and Synthesis

The conditions for the gelation of such condensation polymers have been treated mathematically by Carothers and others (2); the application of gelation to the glycerol-phthalic anhydride reaction has been studied by Kienle and coworkers. The simplified form of the Carothers equation states that as molecular weight becomes infinite at the gel point, $p = 2/f$ where p is the extent of reaction, and f is the average degree of functionality or the average number of functional groups in the reacting molecules. Only stoichiometric equivalents of interacting functional groups are considered. The Carothers equation, when applied to bifunctional reactants, indicates no gelation even though 100% esterification occurs. When applied to the reaction of equivalent quantities of glycerol and phthalic anhydride, $f = [(2 \times 3)+(3 \times 2)]/5 = 2.4$ and therefore, $p = 2/2.4 = 0.83$ or 83%.

The experiments of Kienle and his coworkers (3) in which equivalent quantities of glycerol and phthalic anhydride were used showed that gelation occurred at 79.5% reaction regardless of reaction temperature. The reactions that did occur are represented by Figure 1. Before gelation occurs, the resin is a pale, transparent, fusible product soluble in strong solvents such as acetone or a

TABLE I.   Consumption of Synthetic Resins
           in Paints and Coatings in 1980

Resin Type	Millions of Pounds
Alkyd	710
Vinyl	390
Acrylic	470
Epoxy	145
Urethane	85
Amino	85
Cellulosic	60
Phenolic	25
Styrene-butadiene	25
TOTAL selected resins	1995

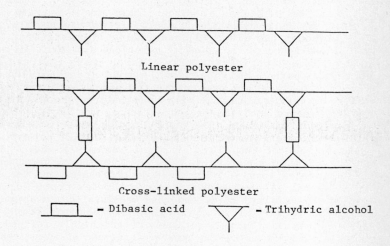

Linear polyester

Cross-linked polyester

▭ – Dibasic acid   ▽ – Trihydric alcohol

Figure 1. Glycerol-phthalic anhydride reactions.

mixture of alcohol and benzene. It is insoluble in petroleum solvents and possesses a high acid number.

In the glycerol-phthalic anhydride reaction, consider the replacement of part of the dibasic acid with a monobasic fatty acid. If the first reaction that takes place is assumed to be between the fatty acid and one hydroxyl of the glycerol, the resulting ester or monoglyceride can be regarded as a glycol because it has only two hydroxyls. If the phthalic group reacts with this modified glycol, a linear polymer should result because the functionality of each reactant is now only two, and cross-linking with its subsequent gelation should not occur. This may be represented as shown in Figure 2.

The monobasic acid modifies the properties of the resin in two ways: first, by its capacity to control functionality and thus allow control of polymer building, and second, by nature of its own inherent physical properties. Although alkyd resins are based on three fundamental building blocks--oils or fatty acids, dibasic acids, and polyhydric alcohols--the permutations and combinations possible with only these three basic components become enormous, and no resin-forming reaction lends itself to greater useful variation and modification than the formation of alkyd resins. Nature has been generous in providing us with a wide array of oils and fatty acids of varying degrees of unsaturation and chain length. Variation in this component alone allows us a wide gradation in film types from soft, colorless, plasticizing films to hard, tough, tack-free films. The mechanism by which alkyds are converted from a liquid to a dry film depends on the alkyd structure and its method of use. Although the alkyd is essentially a plasticizer, as in nitrocellulose lacquers, no chemical reactions are involved in film formation. In a plasticizing alkyd, fatty acids that are fully saturated or contain only one double bond are generally used. Although the fatty acids present in the alkyd are derived from semidrying or drying oils, the alkyd can undergo autooxidation at ambient temperature with the oxygen attacking the unsaturated area of the fatty acid molecule. The general mechanism of film formation is similar to that for glyceride drying oils. However, the molecular weight of the alkyd is significantly higher than the molecular weight of a glyceride oil, and therefore, the number of cross-links needed to give a dry film is decreased, and the drying time is reduced. Also, because relatively few cross-links are required to dry an alkyd film, the less unsaturated semidrying oils can be used to give alkyds with good drying properties. The plentiful supply of low-cost soybean oil and highly refined fatty acids from tall oil spurred the growth in the use of alkyd resins.

Alkyds or oil-modified polyesters are made with drying, semi-drying, and nondrying oils or the fatty acids thereof. The type selected depends on the conditions under which the film will be used, and the color retention and film properties required. The extent and kind of unsaturation in the drying oil fatty acids have a profound effect on the properties of the finished alkyd. By using known mixtures of the various fatty acids present in drying and semidrying oils, scientists have shown a number of interesting relationships between film-forming properties and the fatty acid composition in oil-modified alkyds (4).

Reactants:

—▭—   =   dibasic acid

▽   =   glycerine

⌇⌇⌇⌇⌇⌇⌇⌇⌇   =   monobasic acid

Step I - Conversion of trifunctional polyol to a bifunctional polyol

▽   +   ⌇⌇⌇⌇⌇⌇⌇⌇   ⟶   ▽
                                                    ⌇   monoglyceride

Step II - Conversion to a fatty acid modified linear polymer

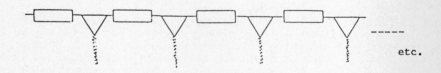

etc.

Figure 2. Modified glycerol-phthalic anhydride reactions.

1. The rate of drying is a function of polyunsaturated or polyenoic acid content. This rate increases rather rapidly up to a polyenoic content of about 50%. Above that figure, a limiting value is gradually approached.
2. The hardness of a film is proportional to the polyunsaturated acid content; at a constant polyunsaturated acid content, a change in the linoleic-linolenic ratio produces no appreciable change in hardness.
3. Color development or after yellowing is proportional to the polyunsaturated acid content; trienoic linolenic acid is five times as potent as dienoic linoleic acid in producing yellowing.
4. The presence of conjugated unsaturation up to a limit of about one-half of the total unsaturation has a beneficial effect on the drying time. However, the ultimate hardness of the alkyd film is not appreciably affected.

From these relationships, it becomes readily apparent that alkyds based on soybean oil give good drying rates and good color retention because soybean oil contains 55-58% polyunsaturated fatty acids, of which only 4-6% are trienoic linolenic acid. One of the interesting facts about oil-modified alkyd resins is that there is not as much difference in the drying rates of soybean-modified alkyds and linseed-modified alkyds as one would expect on the basis of the drying behavior of the oils themselves. This is because the alkyd resin molecules are sufficiently polymerized such that a small amount of cross-linking by way of oxidation causes the film to set to a gellike structure and appear dry even before the final oxidation reactions have been completed.

The advent of lower cost refined fatty acids from tall oil has spurred their use in alkyd resins. These acids contain 43-46% dienoic fatty acids, and although they impart somewhat slower air dry in alkyds than soybean fatty acids, they give slightly better color retention and are especially useful in baking alkyds. Nonoxidizing or plasticizing alkyds are based on coconut or castor oil or short chain-saturated acids such as pelargonic, isononanoic, and isodecanoic acids. Drying and nondrying oils used in the formation of alkyd resins and the fatty acid composition of drying oils are shown in Table II.

For resins of moderate drying rates and good color retention, the standard alkyds are the soybean alkyds. For faster drying and reduced color retention requirements, alkyds based on linseed oil are used. Dehydrated castor oil is used for color-retentive baking alkyds. Tung oil and oiticica are occasionally used with other oils to impart faster drying and earlier hardness. Castor oil and coconut oil are used for color-retentive plasticizing resins because of the nonoxidizing character of the oils. Other oils that are used for special properties are fish, sunflower, walnut seed, and safflower. Safflower oil contains 60-65% linoleic acid with no linolenic acid and is an excellent oil for preparing resins having a combination of excellent drying properties and color retention.

The following compounds are most commonly used for the polyol component of alkyd resins: glycerol, ethylene glycol, pentaerythritol, propylene glycol, dipentaerythritol, neopentyl glycol, sorbitol, diethylene glycol, trimethylolethane, dipropylene glycol, trimethylolpropane, and cyclohexanedimethanol. The structures of

Table II. Typical Composition of Important Drying and Nondrying Oils and Fatty Acids

Property	Soya Oil	Safflower Oil	Dehydrated Castor Oil	Linseed Oil	Coconut Oil	Castor Oil	Segregated Cottonseed Fatty Acids	Tall Oil Fatty Acids (1% Rosin)
Acid number	0.5-6.0	1-4	3-6	2-4	0.5	5-12	195-200*	197-199*
Saponification no.	189-195	188-194	188-194	188-196	250-264	176-187	197-203	199
Iodine no. (Wijs)	120-140	140-150	125-140	155-205	7-10	81-90	140-145	125-130*
Titer (C)*	20-21	15-18	--	19-21	20-24	2-4	5	5
Melting point (C)	-20 to -23	-13 to -18	--	-20	23-26	-10 to -15	--	--
Density (g/L)	923	923	936	930	924	968	899*	902*
Fatty acid distrib.(%)*								
conjugated diene	--	--	22.5	--	--	--	--	9.9
nonconjugated diene	52.0	73.2	61.5	20	2	5	56	36.4
nonconjugated triene	4.5	0.1	--	48	--	--	2	--
monounsaturated	30.5	20.2	8.0	23	7	6	39	52.5
hydroxy monounsaturated	--	--	5.5	--	--	86	--	--
saturated acids	13.0	6.5	2.5	9	91	3	3	1.2

NOTE: All properties were determined on the oil except for those measurements marked with an asterik. These measurements were determined on the fatty acids.

the less commonly used polyols are shown in Figure 3. Of these, glycerol remains the "workhorse" and is closely followed by pentaerythritol. Pentaerythritol, which has four primary hydroxyl groups, forms more complex resins with phthalic anhydride than does glycerol, and its reactivity can be reduced, either by partial replacement with various glycols (functionality of 2), or by the use of larger proportions of fatty acids. The high functionality of pentaerythritol is especially useful in long oil alkyds containing 60% or more of fatty acids, because it imparts faster drying, greater hardness, better gloss and gloss retention, and better water resistance than alkyds based on glycerol of equal fatty acid modification. Sorbitol I is an interesting hexahydric polyol derived from the catalytic hydrogenation of glucose. Under the conditions of esterification, it undergoes intramolecular etherification to form sorbitan II, and to a lesser extent, sorbides. This polyol and other polyols such as trimethylolethane, trimethylolpropane, dipentaerythritol, and cyclohexanedimethanol are used to impart special properties to alkyd resins.

$$
\begin{array}{ccc}
\text{CH2OH} & & \text{CH2} \\
| & & | \\
\text{CHOH} & & \text{CHOH} \\
| & \xrightarrow[\text{heat}]{\text{H+}} & | \\
\text{CHOH} & & \text{CHOH} \quad\text{O} \\
| & & \text{CH} \\
\text{CHOH} & & | \\
| & & \text{CHOH} \\
\text{CHOH} & & | \\
| & & \text{CH2OH} \\
\text{CH2OH} & &
\end{array}
$$

Sorbitol   Sorbitan

I   II

The third basic component of alkyd resins is the polybasic acid. By far, the most important is phthalic anhydride. The increased availability of isophthalic acid at an attractive price has resulted in the increased use of this dibasic acid. This material, unlike phthalic anhydride, will not form intramolecular cyclic structures, and it gives higher molecular weights and higher viscosities. When 1,2,4-trimethylbenzene is oxidized, trimellitic anhydride is obtained (also commercially available). This is an interesting tribasic acid because the anhydride portion of the molecule can be reacted to form a polymeric structure, which leaves one carboxyl available for forming an ammonium or amine salt that imparts water solubility. This reaction will be covered in greater detail later in this chapter. When film flexibility and plasticization by an alkyd are required, dibasic acids such as adipic, azelaic, and sebacic acids and dimerized fatty acids are used. When fire-retardant properties are required in coatings, chlorinated polybasic acids, such as tetrachlorophthalic anhydride or chlorendic anhydride with six chlorine atoms in the molecule, are used. Other polybasic acids frequently used in smaller quantities in alkyd resins are fumaric acid and maleic anhydride. These are dibasic acid reactants, but they also contain a double bond that may react with unsaturated molecules such as unsaturated fatty acids, rosin, and terpenes to form di- and trifunctional reactants as the case may be.

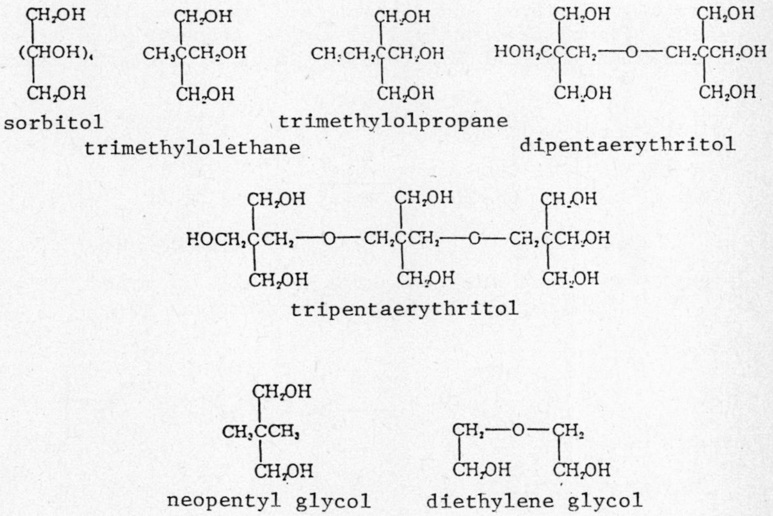

CH₂OH   CH₂OH   CH₂OH   CH₂OH   CH₂OH

(CHOH)₄   CH₃CCH₂OH   CH₂CH₂CCH₂OH   HOH₂CCCH₂—O—CH₂CCH₂OH

CH₂OH   CH₂OH   CH₂OH   CH₂OH   CH₂OH

sorbitol   trimethylolpropane

trimethylolethane   dipentaerythritol

CH₂OH   CH₂OH   CH₂OH

HOCH₂CCH₂—O—CH₂CCH₂—O—CH₂CCH₂OH

CH₂OH   CH₂OH   CH₂OH

tripentaerythritol

CH₂OH

CH₃CCH₃

CH₂OH   CH₂—O—CH₂

neopentyl glycol   CH₂OH   CH₂OH

                          diethylene glycol

Figure 3. Structures of polyols.

Their effect is to increase functionality with a resultant significant increase in viscosity.

Processing of Alkyd Resins

As previously indicated, the usual concept of an alkyd is that of a polyester based on polyhydric alcohols, polybasic acids, and monobasic acids. The most obvious method of producing this polyester is the simultaneous esterification of all ingredients—the "fatty acid method." This method allows greater freedom of formulation because any polyhydric alcohol or polyhydric alcohol blend can be used with fatty acids not available as glycerides, such as the fatty acids from tall oil. On the other hand, the fatty acids derived from glyceride oils are more expensive than the oils and require the use of additional polyol. If a glyceride oil, polyol, and dibasic acid are heated together, a useless heterogeneous mixture is obtained because of the preferential condensation of the polyol and dibasic acid. However, if the oil is treated with glycerol or other polyol in the presence of a catalyst at temperatures of 225–250 °C, the following alcoholysis reaction occurs:

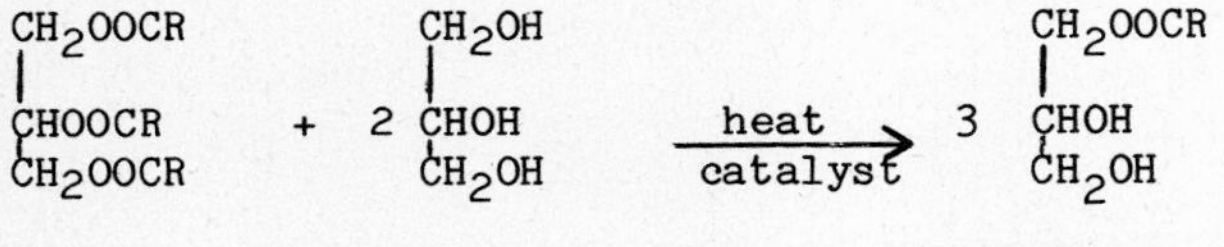

$$
\begin{array}{ccccc}
\mathrm{CH_2OOCR} & & \mathrm{CH_2OH} & & \mathrm{CH_2OOCR} \\
| & & | & \xrightarrow[\text{catalyst}]{\text{heat}} & | \\
\mathrm{CHOOCR} & +\ 2\ \mathrm{CHOH} & & 3 & \mathrm{CHOH} \qquad\qquad (1) \\
\mathrm{CH_2OOCR} & & \mathrm{CH_2OH} & & \mathrm{CH_2OH}
\end{array}
$$

Monoglyceride

When a dibasic acid such as phthalic acid is added, it will esterify the alcoholized oil to form a homogeneous resin:

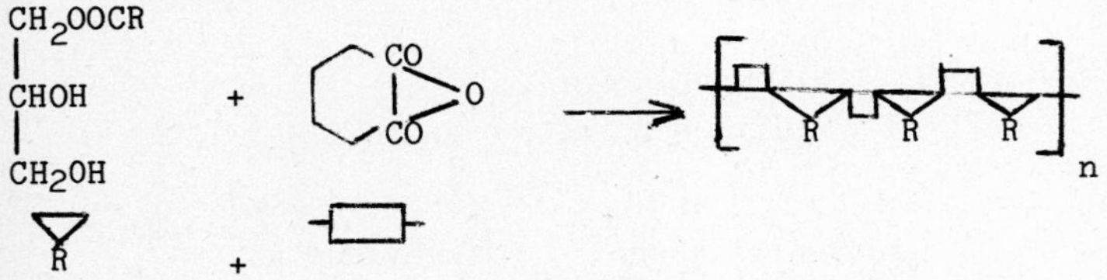

This is the oil–monoglyceride method, which is the most commonly used method at present. In the alcoholysis reaction, the catalysts most frequently used for the reaction are litharge (Pbo), calcium hydroxide, and lithium carbonate or the soaps derived from them. The alcoholysis reaction (Reaction 1) is an oversimplification of a rather complex reaction (5). In reality, the alcoholysis results in an equilibrium mixture that consists of monoglyceride, diglyceride, unchanged triglyceride, and free glycerol when glycerol is the polyol. The degree of alcoholysis has an important bearing on the properties of the resulting resin. During the final reaction with the dibasic acid, esterification of the free hydroxyl groups of the monoglyceride must compete with any unreacted or excess glycerol present. The latter course leads to rapid cross-linking and formation of infusible, gelled glycerol phthalate, which is insoluble in the glyceride oil and oil-modified glycerol phthalate. Therefore, unless sufficient bifunctional monoester is present prior

to the addition of the phthalic anhydride, the ultimate reaction
product will be an insoluble gel of glyceryl phthalate suspended in
oily mixed glycerides. Such products are commercially worthless.
   The alcoholysis reaction is further complicated by the tendency
of the polyhydric alcohols, under the influence of heat and alkaline
catalysts, to undergo interetherification with the resultant loss of
available hydroxyl groups and the formation of higher functional
polyols. In the case of glycerol, the reaction could be written as
follows:

$$
\begin{array}{c}
CH_2OH \\
| \\
CHOH \\
| \\
CH_2OH
\end{array}
\xrightarrow[\text{heat}]{\text{Catalyst}}
\begin{array}{c}
CH_2OH \\
| \\
CHOH \\
| \\
CH_2\text{-----O-----}CH_2
\end{array}
\;+\; H_2O
$$

The tetrafunctional alcohol, pentaerythritol, is more prone to
undergo this type of reaction to form higher functional polypen-
taerythritols:

$$
2\;HOH_2C\!-\!\underset{\substack{|\\CH_2OH}}{\overset{\substack{CH_2OH\\|}}{C}}\!-\!CH_2OH
\xrightarrow[\text{heat}]{\text{Catalyst}}
HOH_2C\!-\!\underset{\substack{|\\CH_2OH}}{\overset{\substack{CH_2OH\\|}}{C}}\!-\!Ch_2\!-\!O\!-\!CH_2\!-\!\underset{\substack{|\\CH_2OH}}{\overset{\substack{CH_2OH\\|}}{C}}\!-\!CH_2OH
$$

$$+\; H_2O$$

Pentaerythritol   Dipentaerythritol

When this reaction occurs, the resulting resins are excessively high
in viscosity and acid number, and these results necessitate careful
control of temperature, agitation, quantity and type of catalyst,
and reaction time. The tendency for a polyol to polymerize through
ether formation increases with the temperature and the presence of
the alkaline catalysts.
   In comparing the fatty acid and alcoholysis processes, it is
evident that the reactions involved are quite different. The
interchange reactions involved in alcoholysis and the subsequent
reaction with the polybasic acid are slow compared to polyesterification. Performance comparisons indicate that performance
differences do exist between alkyds of the same composition that are
prepared by the fatty acid and monoglyceride (alcoholysis) processes
(6).
   Another less used method of forming alkyd resins is by
acidolysis, which might be considered the counterpart of alcoholysis
(7). The reaction that occurs is as follows:

$$
\begin{array}{c}
CH_2OOCR \\
| \\
CHOOCR \\
| \\
CH_2OOCR
\end{array}
\;+\;
\underset{\text{Isophthalic Acid}}{\underset{}{\text{(benzene ring with COOH, COOH)}}}
\longrightarrow
\begin{array}{c}
CH_2\text{-----}OOC\text{-----(benzene ring)-----}COOH \\
| \\
CHOOCR \\
| \\
CH_2OOCR
\end{array}
$$

$$+\; RCOOH$$

This reaction requires temperatures in the range of 275-285 °C, and
its use is limited to polybasic acids such as isophthalic and

terephthalic acids that, unlike phthalic acid or anhydride, do not sublime and are quite insoluble in a monoglyceride mixture until considerable esterification has occurred.   When carefully controlled, this method gives light-colored resins with good drying properties.

Two basic methods are being used to process alkyd resins. In the "fusion" process, the components are charged into the reactor and heated to reaction temperature, usually 225–250 °C, in a stream of inert gas that prevents air oxidation and removes the water of esterification.   The procedure more recently developed on a commercial scale, and for many purposes the best method of alkyd preparation, is the so-called "azeotropic process" or "solvent process." In this process an inert solvent such as xylol is added, and the reaction mixture is heated to the reflux temperature. As water is produced from the reaction, it is distilled over with some of the hydrocarbon solvent. After condensation, it is separated in a gravity separator, and the excess solvent is returned to the reactor. The solvent vapor protects the resin from air oxidation, even more efficiently than the inert gas in the "fusion process"; therefore, lighter colored resins are obtained. The "solvent process" also results in better yields, better process control, faster processing, and greater polymer uniformity. Although the trend is in the direction of solvent processing, a considerable volume of alkyds will still be made by the fusion process because the investment cost is lower on new installations, safety requirements are less stringent than with solvent processing, and certain alkyds are more effectively made by the fusion process.

Classification and Properties of Alkyd Resins

The basic components of an alkyd or oil-modified polyester have been discussed to show the wide difference in properties and performance that can be obtained by variation of the three basic components of a simple alkyd.   Moreover, the degree of fatty acid modifications has a great bearing on the properties of the resin. The components of any particular alkyd resin and the proportions of these components will vary with the intended use of the resin.   By suitable variations, the polymer chemist can achieve a wide range of physical properties and performance characteristics including viscosity, color, drying rate, hardness, toughness, adhesion, solubility, compatibility with other polymers, chemical resistance, color and gloss retention, and durability.

Alkyds are made with oil contents varying from 30 to 80%. It is more accurate to refer to fatty acid content rather than oil content because the oil has ceased to exist as such after it has been reacted into the alkyd. It has been suggested that the fatty acids in 100 parts of theoretically completely esterified alkyd be calculated to triglyceride and used to define (in percent) the "oil length" or extent of oil modification. This method is of questionable value because the advent of new polyols and the increasing use of mixtures of the same to achieve special properties make it quite impossible to calculate the quantity of fatty acid ester.

Alkyds are best classified according to the amount of fatty acids and phthalic anhydride present in the alkyd (Table III). As indicated earlier, the percentage of fatty acid modification greatly influences the properties of alkyd resins (Figure 4). In general, alkyds containing over 48% fatty acids are soluble in petroleum spirits, and those alkyds containing under 48% fatty acids require a more or less aromatic solvent, such as xylene. Solubility and viscosity are directly related to solvent requirement. As the fatty acid modification decreases, "solvent release" becomes more rapid, and the apparent dryness is due to lacquer type drying--the evaporation of solvent from a practically solid resin.   As previously mentioned, the drying of oil-modified alkyds involves the oxidation of the combined fatty acids of these oils. Figure 4 shows that the drying time of alkyds decreases as fatty acid content increases when starting in the 48-50% range. However, below this range, drying time increases as the fatty acid content decreases. As the fatty acid modification decreases, the fatty acid groups become spatially farther apart in the resin molecule. As they become farther apart, the chances of oxidative dimerization of the polyunsaturated fatty acids decrease. At the same time, the rapid evaporation of solvent from the "solid" resin gives the impression of very fast drying. With short oil resins this is only an apparent dryness. The films actually remain thermoplastic for longer periods as shown by print tests and softening when a drop of xylene is placed on the film. Film hardness, as would be expected, is inversely proportional to the degree of fatty acid modification. Rheological characteristics such as brushing, flow, sagging, and ease of dispersion are generally improved as the fatty acid content increases, although other factors such as acid number and pigmentation are also important.

Color gloss and retention of these qualities increase as the proportion of colorless, nonoxidizable "polyol phthalate" increases. Below 47-48% fatty acid modification, alkyds are liable to exhibit failure by cracking, checking, peeling, or other types of film destruction that result from hardness, lack of flexibility, or poor adhesion. Above 50% fatty acid modification, the durability decreases because of the lower content of "polyol phthalate," which is important in regard to durability. This characteristic of durability also depends on several other factors such as the nature of the substrate. For example, for a specific surface like bonderized steel with a proper primer, durability will be at an optimum around 33-38% fatty acid modification. Storage stability depends on a number of factors such as type of oil, degree of polymerization, functionality of the polyols and polybasic acids, excesses of reactants, and solvency of the thinner. However, in general, storage stability increases with an increase in fatty acid content.

Earlier in this chapter, mention was made of the equation derived by Carothers that stated that as the molecular weight becomes infinite at the gel point, $p = 2/f$ where p is the extent of reaction, and f is the average degree of functionality or the average number of functional groups in the reacting molecules, and where only stoichiometric equivalents of interacting functional groups are considered. Deviations from the theory found in the actual synthesis of alkyds and reasons for these deviations have received the attention of Flory, Stockmayer, and other investigators in this field. However, their efforts have not been completely

TABLE III. Classification of Alkyds

Type	Fatty Acid Content (%)	Phthalic Anhydride (%)
Short	30-42	37
Medium	43-54	30-37
Long	55-68	20-30
Very long	68	20

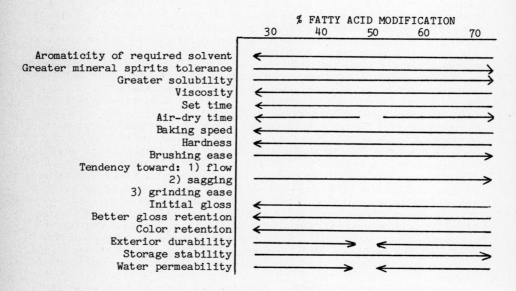

Figure 4. Effect of fatty acid modification on alkyd resin properties.

successful, and a number of modifications and correction factors
have been applied to the Carothers and Flory equations. Even with
these modifications, the theoretical and actual gel points
frequently do not correspond.

Formulation and Design of Alkyd Resins

In alkyd formulations, the polymer chemist generally seeks to
prepare the resin with the highest molecular weight at low acid
numbers without encountering gelation. Patton (8) has derived an
alkyd constant based on an adaptation of Carother's classical
functionality equation. The proposed alkyd constant k, which
represents the ratio mo/ea, is numerically equal to one for all
properly formulated alkyds: k = mo/ea = 1, where mo is the total
number of acid- and hydroxyl-bearing molecules in a given alkyd
composition at the beginning of a condensation reaction between acid
(-COOH) and hydroxyl (-OH) groups, and ea is the total number of
acid equivalents in this composition.

Patton (8) studied a group of alkyd formulations reported in the
literature. The mo/ea ratio was computed for each formulation, and
the mean value for the ratio was 1.022 ± 0.023. Although the value
fails by small fractional amount to coincide with k = 1.00, this
discrepancy can be explained by recalling that the constant 1.00 is
theoretically derived for the ratio mo/ea and pertains to the
conditions of exactly 100% completion of the condensation reaction
at incipient gelation. However, for any practical alkyd synthesis,
formulating this close to gelation is dangerous. Accordingly,
experience dictates that a margin of safety in processing be
provided. Therefore, the difference between the theoretical
constant of 1.00 and the experimentally determined value of 1.022
allows for a safety factor of 0.022. This suggests the use of the
alkyd constant at a practical value of 1.02 rather than at the
theoretical value of 1.00. Patton's analysis revealed that those
alkyds based on phthalic anhydride tend to have low mo/ea values,
whereas those alkyds based on isophthalic acid tend to have higher
values. Experimentally determined alkyd constants were found to be
1.005 ± 0.014 for phthalic anhydride alkyds and 1.046 ± 0.008 for
resins based on isophthalic acid. The use of the alkyd constant
represents a good starting point in the formulation of alkyds and
will reduce the number of experiments necessary for developing and
evaluating alkyd resins. Although the alkyd constant provides
valuable data on reactant proportions, the choice of raw materials
must be carefully evaluated, and with the increasing number of new
polyols, polybasic acids, oils, and monobasic acid modifiers, this
consideration is most important in the design of alkyd resins.

Uses of Alkyd Resins

The wide spectrum of properties of all alkyd resins is broadened by
its use with a variety of reactive chemicals and other resinous
materials. In fact, alkyd resins have the distinction of
participating in more of what we might term "polymeric mergers" than
any other single class of resins. It is of value to examine the
idealized structure of both a long oil III and a short oil IV alkyd
resin. The long oil alkyd illustrated is based on a 1:1:1 mol ratio

of glycerine–phthalic anhydride–fatty acids (~60.5% oil). The short
oil alkyd illustrated is based on a 3:3:1 mol ratio of the same
materials (31.2% oil).

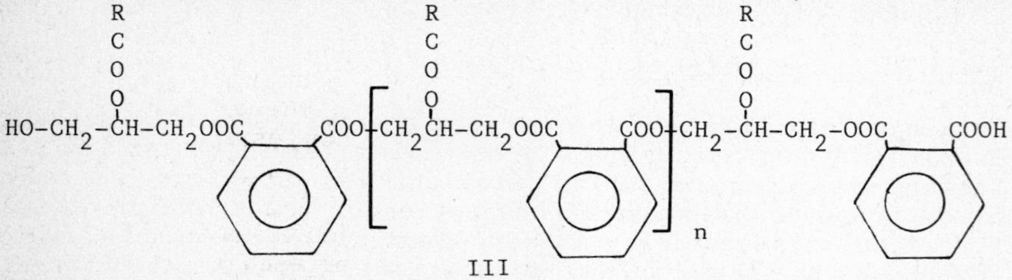

$$
\begin{array}{ccc}
R & R & R \\
C & C & C \\
O & O & O \\
O & O & O
\end{array}
$$

HO–CH$_2$–CH–CH$_2$OOC ⟨C$_6$H$_4$⟩ COO–[CH$_2$CH–CH$_2$OOC ⟨C$_6$H$_4$⟩ COO–CH$_2$–CH–CH$_2$–OOC ⟨C$_6$H$_4$⟩ COOH]$_n$

III

(Idealized structure of a long oil alkyd based on a 1:1:1 mol ratio
of glycerine–phthalic anhydride–fatty acids, ca. 60.5% oil.)

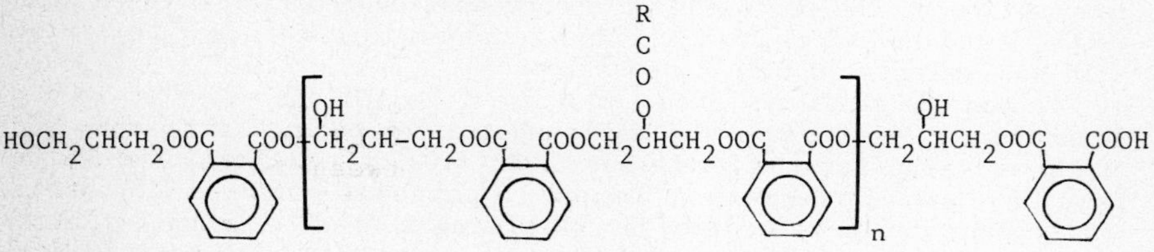

$$
\begin{array}{c}
R \\
C \\
O \\
O
\end{array}
$$

HOCH$_2$CHCH$_2$OOC ⟨C$_6$H$_4$⟩ COO–[CH$_2$CH–CH$_2$OOC ⟨C$_6$H$_4$⟩ COOCH$_2$CHCH$_2$OOC ⟨C$_6$H$_4$⟩ COO–CH$_2$CHCH$_2$OOC ⟨C$_6$H$_4$⟩ COOH]$_n$

(OH groups indicated)

IV

(Idealized structure of a short oil alkyd based on 3:3:1 mol ratio
of glycerine–phthalic anhydride–fatty acids, ca. 31.2% oil.)

In the long oil resin, the large number of long chain fatty acid
groups impart a predominantly nonpolar character to the molecule.
In general, the compatibility of alkyd resins depends upon the
polarity of the molecules and is inversely proportional to the
degree of polymerization. In a short oil alkyd, the high proportion
of hydroxyl groups imparts not only polarity, but centers of
potential reactivity with a host of hydroxyl-reactive materials.
Undoubtedly, the aromatic ring in phthalic anhydride and the ester
linkages contribute polarity. Moreover, the unsaturation in the
fatty acid groups allows interpolymerization with many vinyl
monomers, epoxidation, and the other reactions of the double bond.
Many polymeric materials and reactive functional materials are used
to modify and impart improved film-forming properties to suitably
designed alkyd resins (9):

Nitrocellulose
Urea–formaldehyde resins
Melamine–formaldehyde resins
Vinyl resins
Epoxy resins
Silicones
Cellulose acetobutyrate
Polyamides
Monobasic aromatic acids

Phenolic resins
Chlorinated rubber
Synthetic latices
Chlorinated paraffin
Polyisocyanates
Reactive monomers
Phenolic varnishes
Natural resins

The use of oil-modified alkyds with nitrocellulose in lacquers
is perhaps the first "polymeric merger" that resulted in the
remarkable growth in the use of nitrocellulose lacquers.
Compatibility of alkyds with nitrocellulose extends up to about 50%

fatty acid modification. However, the best compatibility is achieved with shorter oil alkyds that have a higher degree of polarity from ester and excess hydroxyl groups. Alkyds modified with short chain acids such as those from coconut oil are widely used in high-grade furniture lacquers, as are the castor oil-modified alkyds. The latter contain hydroxyl groups in the long fatty acid chain that impart excellent compatibility, flexibility, and adhesion. The upgrading effect of the alkyd in improving the gloss, adhesion, durability, cold check resistance, and build of nitrocellulose lacquers is well known to all in the field of organic coatings. This valuable combination will undoubtedly continue as long as nitrocellulose lacquers are used.

With regard to chlorinated rubber, the linear-type structure and low polarity of the chlorinated rubber molecule favor compatibility with linear, less polar alkyds. In general, alkyds above 50% fatty acid modification, when dissolved in aromatic thinners, are more compatible and combine improved toughness, adhesion, solvent resistance, and durability with the excellent alkali, acid, and water resistance of chlorinated rubber for use in concrete floor paints and swimming pool paints. Today, most of the road-marking paints are based on a combination of chlorinated rubber and a compatible alkyd resin. In higher phthalic alkyds, the high proportion of hydroxyl groups promotes compatibility with urea and melamine-formaldehyde resins and provides sites for reaction with these resins. The short oil alkyds containing about 38-45% phthalic are used primarily in industrial baking finishes, usually in conjunction with urea and melamine resins. Although soya oil or tall oil fatty acids are most often the monobasic modifier, coconut oil or short-chain saturated fatty acids provide the best color retention on baking. This valuable combination still remains the backbone of baking industrial finishes.

The coreaction of phenolic resins with alkyds is an "old timer" in which the excellent gloss retention and durability of alkyds are combined with the water and alkali resistance of phenolics. Alkyd resin can be designed to have compatibility with hydroxyl-modified vinyl chloride-vinyl acetate copolymer resins, and this combination of fast drying, adhesion, excellent water resistance, and durability is specified in a number of coatings for military and marine use.

As noted earlier, oil-modified alkyds contain free hydroxyl groups, and these groups represent sites for reaction with poly-isocyanates. The resultant products possess faster drying, better chemical resistance, and abrasion resistance. Rosin is frequently used in a monobasic acid modifier to replace a portion of the oil fatty acids. As such, it promotes hardness, fast set, and solubility but impairs color, color retention, and durability, depending on the quantity used. Small amounts (up to 5%) improve the solubility without detracting from other properties to any significant degree.

Vinyl- or acrylic-modified alkyds have been prepared by three distinct methods (10): (1) blending of the two polymer entities; (2) copolymerization, or partial copolymerization, involving the unsaturation of the fatty acids; and (3) chemical combination of the polymers by other functional groups, which usually results in an ester linkage between the components.

Alkyd resins modified with compounds such as styrene, vinyl toluene, and methyl methacrylate have become more important in recent years and are widely used where speed of drying approaching that of lacquers is required. Methacrylated alkyds are particularly useful in metal decorating where good flow, gloss, toughness, and durability are required.

Reactive silicone intermediates can be reacted with the hydroxyl groups in long oil air-dry alkyd resins to give copolymers having greatly improved durability, gloss retention, and heat resistance. The use of these silicone-modified air-drying alkyds in maintenance and marine coatings is increasing rapidly. The excellent durability of these coatings compared with other coatings is shown in Figure 5.

Long oil-drying alkyds are combined with latex polymers for use in exterior latex house paints. The early exterior latex house paints were deficient in adhesion to highly chalky surfaces. The replacement of 20-30% of the latex solids with a 100% solid long-oil flexible alkyd has overcome this deficiency and has been a boon to the great increase in the use of exterior latex house paints. At the present time, some improved latex paints are claimed to perform well without the use of these alkyds.

Another important combination is the use of polyamides reacted with air-drying alkyds. Certain alkyd resins can be combined with modified polyamide resins to furnish materials exhibiting a marked degree of thixotropy when dissolved in suitable solvents. This

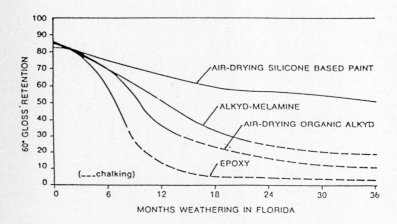

Figure 5. Weathering comparison of various coatings.

valuable property is maintained when such products are used in the manufacture of paints. When the alkyd and polyamide are heated at elevated temperatures, amide and ester interchange reactions occur that split the polyamide chains and affix the fragments onto the alkyd molecule. Further reaction breaks these chains down and distributes the amide groups throughout the molecule. This reaction probably proceeds by way of the free groups on the alkyd and polyamide. The thixotropic behavior of these resins has been ascribed to hydrogen bonding, presumably between the carbonyl and amine groups distributed throughout the alkyd–polyamide complex. The physical properties of thixotropic alkyds are very delicately balanced, and close control of the endpoint of the reaction is vital for the establishment of reproducible properties. These thixotropic alkyds are particularly valuable in architectural and maintenance finishes where they impart nondrip properties, freedom from pigment settling, and ease of brushing, as well as allow the application of thick coats without danger of sagging.

Saturated Reactive Polyesters

For many years, short oil thermosetting alkyd resins with urea– and melamine–formaldehyde resins were the basis of most baketype finishes. The advent of new raw materials and the versatility of the alkyd structure allowed these resin systems to virtually capture the market for baking finishes. However, change is inevitable, and the constant search for superior performance led in more recent years to the development of oil–free alkyds or hydroxylated polyesters that are characterized by the absence of oil or oil–derived fatty acids that were an integral part of conventional baking alkyd. These polyesters are lighter in color than conventional alkyds resins because unsaturated fatty acids that contribute to color and color development are not used. Moreover, the freedom from unsaturation of the hydroxy–functional polyesters eliminates post–embrittlement caused by oxidation of double bonds. The elimination of the long chain fatty acids in these reactive polyesters gave improved chemical resistance. A unique hardness to flexibility relationship made it possible to attain greater hardness with high reverse impact than is possible with current systems.

The rapidly expanding market for precoated sheet metal rolled into coils has been an impetus to the development of superior coatings. The formation of the precoated metal into simple and complex shapes demanded high–speed curing, outstanding adhesion and flexibility, impact resistance, hardness, and other qualities demanded by the different and expanding uses for coated and formed sheet metal. Hydroxyl–functional polyesters can now be designed to provide finishes with a combination of properties unavailable from coil coatings based on vinyls and acrylics.

The preparation of hydroxylated polyesters involves reactions of various polyols and polyol combinations with polybasic acids to give a wide range of properties. In the design of these resins, the polyol source is generally determined by the amount of hydroxyl groups available for further reaction of the polyester chain with cross–linking polymers. The polyols most often used as sources of available hydroxyls are pentaerythritol V, trimethylolethane VI, and trimethylolpropane VII.

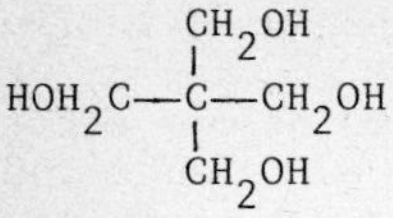

$$
\begin{array}{ccc}
\underset{|}{CH_2OH} & \underset{|}{CH_2OH} & \underset{|}{CH_2OH} \\
HOH_2C{-}\underset{|}{C}{-}CH_2OH & CH_3\underset{|}{C}{-}CH_2OH & CH_3CH_2\underset{|}{C}{-}CH_2OH \\
\underset{}{CH_2OH} & \underset{}{CH_2OH} & \underset{}{CH_2OH} \\
\\
V & VI & VII
\end{array}
$$

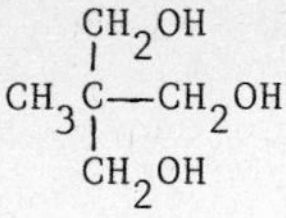

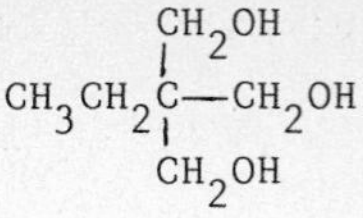

A glycol is included in the polyester to serve as a functionality modifier in preventing premature gelation, to contribute
linearity, and to improve specific performance characteristics such
as resistance to detergents, stains, and heat. Although ethylene
and propylene glycols have been used as functionality modifiers for
many years in conventional alkyds, newer glycols have become
available that impart improved performance to hydroxyl functional
polyesters. Some of the newer glycols used in the design of these
polyesters are 2,2-dimethyl-1-propanol (neopentyl glycol), 2,2,4-
trimethyl-1,3-pentanediol, 1,6-hexanediol, and 1,4-
cyclohexanedimethanol. Among the acids, the phthalic acid series
and adipic acid are most frequently used. In the preparation of
these hydroxylated polyesters, various polyols and polyol
combinations are reacted with polybasic acids to give a wide range
of properties.

A general idealized structure of a hydroxylated polyester based
on 2 mol of trimethylolpropane, 9 mol of 1,6-hexanediol, and 10 mol
of isophthalic acid is the following (11, 12):

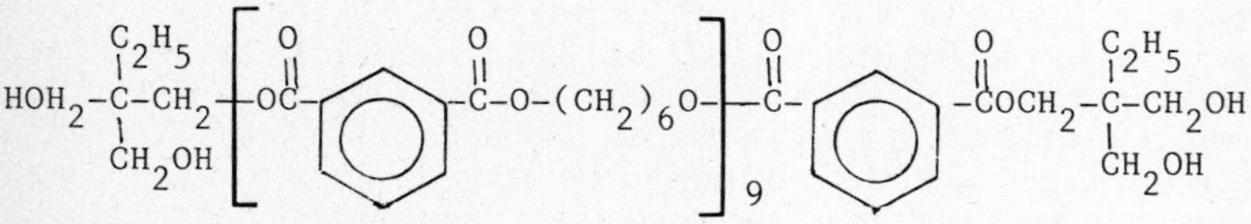

Combinations of isophthalic and adipic acid result in useful
polymers for appliances and automotive topcoats. However, for coil
coatings where extreme flexibility is required, long-chain dibasic
acids such as azelaic and dimerized fatty acids are used. Increased
flexibility can also be effectively achieved by the use of a long-
chain glycol such as 1,6-hexanediol.

The hydroxylated polyesters cross-linked with melamine resins
produce single component baking enamels with the valuable properties
of fast cure, high hardness with good flexibility, and excellent
stain and detergent resistance. These performance qualities have
increased the use of these cross-linked melamine resins for coil
coatings, appliance finishes, and beverage and can coatings (13).
The use of hydroxylated polyesters cross-linked with both aromatic
and aliphatic polyisocyanates is well established and growing,
especially in high-quality maintenance finishes. These 2-component
finishes cure at ambient temperatures, and aliphatic urethane types
are used on aircraft, rail, and other transportation equipment where
excellent gloss retention and abrasion resistance are required.

The hydroxylated polyesters react with special silicone
intermediates to give silicone-polyester copolymers for baked
finishes that provide high heat resistance and outstanding exterior
durability (14). These resins are extensively used in coil coatings
for steel and aluminum building panels where high durability is
required (Figure 6).

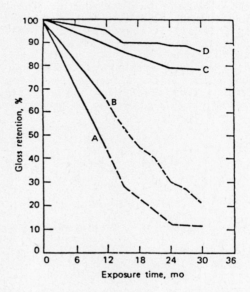

Figure 6. Gloss retention vs. months Florida exposure.
Key: A, acrylic; B, vinyl; C, fluoropolymer; D, silicone
polyester; ---, chalking.

Resin Systems for Reduced Solvent Emissions

The powerful forces of ecology have resulted in antipollution legislation requiring the coatings industry to limit atmospheric emission of photochemically reactive solvents. Research on new application techniques, new polymers, and cross-linking systems has led to new technologies such as powder coatings and UV- and electron-beam-cured coatings. Water-reducible coatings and high solids coatings are also viable approaches to reducing solvent emissions. Alkyd resins and reactive saturated polyesters have been designed to meet the requirements of these new technologies; the use of these resins is rapidly increasing in these areas.

In water-reducible resins, water solubility is developed by the incorporation of carboxyl groups located along the polymer backbone and the end sites of the polymer chain. The acid groups are subsequently neutralized, usually with ammonium hydroxide and/or an organic amine. Trimellitic anhydride VIII (TMA) is particularly effective in designing water-soluble alkyds and polyesters (15, 16). The availability of a wide variety of glycols, polyols, fatty acids, and polybasic acids used with tribasic TMA provides the resin chemist with the opportunity of developing waterborne alkyds of varied properties. Both air-drying and baking types with varying degrees of fatty acid modification can be prepared.

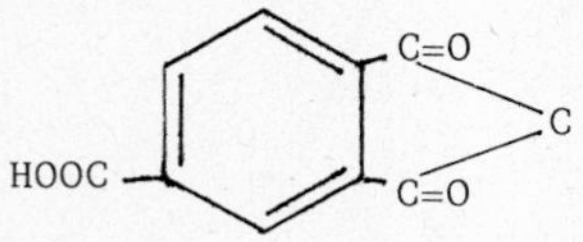

TMA

VIII

The idealized structure of a water-reducible baking-type alkyd based on TMA is

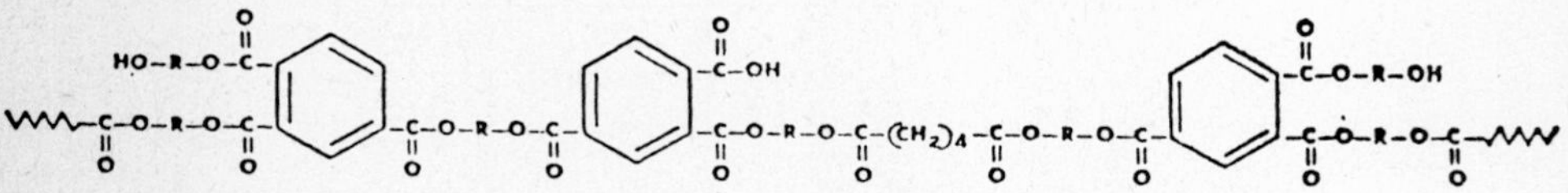

where R is the glycol radical, and the sawtooth line is the unsaturated fatty acid chain. The formulation for this resin is 3 mol of TMA, 7 mol of neopentyl glycol, 1 mol of adipic acid, and 2 mol of tall oil fatty acids. This formulation represents 30.4% fatty acid modification. The resin is processed to an acid number of 50-55 and is solubilized by the addition of dimethylethanolamine. Resins of similar composition are used in electrocoating. In this method of application, an organic coating is deposited on a metal substrate from a water-reducible paint system by means of electrical attraction. This method of application is of particular value for primers and one-coat enamels for geometrically complex metal parts for automotive, appliance, and marine use (17).

Reactive saturated polyester coating resins that are water-soluble can be used to formulate gloss enamels that have excellent application and performance properties. These polyesters have carboxyl and hydroxyl groups and are based on glycols, triols, and polybasic acids. They are used with water-soluble melamine cross-linking agents in baking systems. A less frequently used method of imparting water dispersibility in alkyds is the introduction of a polyoxyethylene glycol in the structure. The resulting polyether group in the alkyd acts as a built-in surfactant that imparts dispersion in water.

Low-Solvent and Solvent-Free Alkyd Resins

High solids coatings offer an alternative solution to the restrictions on the use of organic solvents and the soaring cost of energy. Compared to a conventional industrial enamel applied at 30% volume solids, a 70% volume solids system shows an 81% reduction in total solvent required. These considerations have been an impetus to the development of high-solids alkyds and hydroxylated saturated polyesters. Low viscosity is necessary for good application, and molecular weight build-up must occur in the film applied on the substrate by cross-linking without volatilization of the film-forming materials. Low molecular weight, low-viscosity hydroxylated polyesters can be formulated from glycols, higher polyols, and polybasic acids (18). These high-solids resins can be cross-linked with melamine resins at elevated temperatures or with a polyisocyanate at ambient or slightly elevated temperatures. The melamine cross-linker allows a one-package system, and the use of polyisocyanate requires a two-package system.

Special alkyds and reactive polyesters that exhibit excellent properties have been developed for coatings applied as solvent-free sprayed powders. Polyester powder coatings are available in thermoplastic and thermosetting types (19, 20). The thermoplastic types are essentially linear-saturated, substantially fully reacted polymers such as glycol phthalates or combinations of more complex polyols with various phthalic isomers. The thermosetting polyesters primarily are based on hydroxyl- and carboxyl-rich intermediates cross-linked with hexamethoxymethyl melamines, epoxy resins, or dianhydrides. Polyester-based hybrid systems have experienced the greatest growth in recent years. A typical formulation would be a carboxyl-functional polyester cross-linked with solid epoxy resins of the bisphenol A type: epoxy resin, 35%; carboxyl-functional polyester (acid number 90), 35%; pigment, 27%; and additives, 3%. The use of the carboxyl-functional polyester with an epoxy resin upgrades the weatherability of the epoxy, and the use of this hybrid system is increasing rapidly. Other thermosetting polyester powder coatings are based on hydroxyl-functional polyesters cross-linked with blocked isocyanates.

The polyester structure has been adapted for use in radiation-cured coatings. Low molecular weight, linear polyesters have been made in which the terminal hydroxyl groups have been esterified with acrylic acid. These oligomers have low viscosity and are used increasingly in coatings cured by UV or electron-beam radiation (21).

Thus, beginning with alkyds or monobasic acid-modified poly-esters, the impact of environmental restrictions and economic forces has resulted in the adaptation of the basic polyester structure to new emerging coating technologies and application processes.

Literature Cited

1. "Kline Guide to the Paint Industry," 6th ed.; Susan Rich, Ed.; C. H. Kline & Co.: Fairfield, New Jersey, 1981.
2. Carothers, W. R. J. Am. Chem. Soc. 1929, 51, 2548.
3. Kienle, R. H.; Hovey, A. G. J. Am. Chem. Soc. 1929, 51, 509; ibid., 1930, 52, 3636.
4. Moore, D. T. Ind. Eng. Chem. 1951, 43, 2348.
5. Solomon, D. H.; Hopwood, J. J. J. Appl. Polym. Sci. 1966, 10, 993.
6. Solomon, D. H. J. Oil Colour Chemists Assoc. 1965, 48, 282.
7. Carlston, E. F. J. Am. Oil Chem. Soc. 1959, 36, 28.
8. Patton, T. C. "Alkyd Resin Technology"; Interscience: New York, 1962.
9. Lanson, H. J. Paint Varn. Prod. 1959, 49, 25.
10. "Acrylic-Modified Alkyd Resins--Technical Bulletin"; Celanese Chemical Co.: Dallas.
11. "Polyester Resins for Coatings--Technical Bulletin"; Celanese Chemical Co.: Dallas, 1970.
12. Technical Bulletin CSLR No. 370-M; Eastman Chemical Products, Inc.: Kingsport, Tenn., 1968.
13. Finney, D. C. Paint Varn. Prod. 1969, 59, 27.
14. Spargo, A. Am. Paint & Coatings J. 1968, 80-85.
15. Lerman, M. A. J. Coatings Tech. 1976, 48, 37.
16. Bulletin TMA-120; Amoco Chemicals Corp.: Chicago, 1980.
17. Bulletin TMA-122; Amoco Chemicals Corp.: Chicago, 1977.
18. Gott, L. J. Coatings Tech. 1976, 48, 52.
19. Chandler, R. H. "Polyester Powder Coatings"; Braintree: Essex, England, 1976.
20. Cole, G. E. Modern Paint & Coatings 1982, 47-49.
21. Hoyle, C. E. Am. Paint & Coatings J. 1984, 42-45.

Bibliography

1. Solomon, D. H. "The Chemistry of Organic Film Formers"; Wiley: New York, 1967.
2. Lanson, H. J. "Kirk-Othmer Encyclopedia of Chemical Technology," 3rd ed.; Wiley: New York, 1978; Vol. 2, pp. 18-50.
3. Alkyd Report Bulletins; Hercules Inc.: Wilmington, Del.

Vinyl Resins Used in Coatings

RUSSELL A. PARK

PVC Resins and Compounds Division, Occidental Chemical Corporation, Pottsdown, PA 19464

Historical
Polymerization Techniques
Polymers
Formulation Considerations
Application Techniques
Today and Tomorrow

Historical

Poly(vinyl chloride) is over 100 years old. M. V. Regnault (1) is credited with discovering the basic building block, vinyl chloride, in 1835. He obtained the vinyl chloride by chlorinating ethylene to give 1,2-dichloroethane, which was later treated with an alcoholic caustic potash solution to yield the monomer plus water and potassium chloride. When heated 4 days later, he noted that a white powder was formed under the influence of sunlight. It took 40 years before additional conversion of vinyl chloride was investigated.

In 1912 Zacharias and Klatte carried out the commercial preparation of vinyl chloride by the addition of hydrogen chloride to acetylene in Germany. In 1914 F. Klatte and A. Rollet described the first polymerization of vinyl acetate to poly(vinyl acetate). In 1921 Herman Plausen of Hamburg, Germany, described the polymerization of vinyl chloride. The copolymerization of vinyl chloride and vinyl acetate was disclosed in the United States in 1928.

"Imitation" or substitute products made from poly(vinyl chloride) did not have much commercial acceptance after World War I, probably due to the poor quality of war-manufactured goods.

During 1928-32 compositions were made by dissolving polymers of vinyl halides in a number of high-boiling esters at elevated temperatures to form rubberlike materials.

The development of compounding principles (still felt to be an art by many) and the availability of new raw material sources produced from natural and cracked petroleum gases brought the

0097-6156/85/0285-1205$06.75/0
© 1985 American Chemical Society

industry out of its infancy. World War II pushed it into adoles-
cence when the need for new sources of electrical insulation,
waterproofing, and corrosion protection developed as a military
necessity.

Of the more than 30 billion pounds of ethylene consumed in the
United States in 1980, over 17% was used in the production of
ethylene dichloride/vinyl chloride monomer and vinyl acetate. The
major feedstock for ethylene production is ethane (47%), with other
major feedstocks being gas oil, naphtha, and propane.

In the oxychlorination process, ethylene is first converted to
ethylene dichloride as follows:

$$CH_2=CH_2 + \quad\quad 2HCl \quad + \tfrac{1}{2}O_2 \quad \xrightarrow[\substack{1\ atm \\ vapor\ phase}]{\substack{catalyst \\ 250-315\ ^\circ C}}$$

ethylene   hydrogen chloride   oxygen   vapor phase

$$CH_2ClCH_2Cl \quad + \quad H_2O$$

ethylene dichloride   water

The ethylene dichloride is then dehydrochlorinated by pyrolysis as
follows:

$$CH_2ClCH_2Cl \quad \xrightarrow[3\ atm]{\substack{catalyst \\ 480-500\ ^\circ C}} \quad CH_2=CHCl \quad + \quad HCl$$

ethylene dichloride   vinyl chloride   hydrogen
                                                          chloride

Seventy-five percent of ethylene dichloride consumption is converted
to the vinyl chloride monomer (VCM).

A minor process for VCM production is based on acetylene as
follows:

$$HC\equiv CH \quad + \quad\quad HCl \quad \xrightarrow[\substack{1\ atm \\ vapor\ phase}]{\substack{catalyst \\ 100-210\ ^\circ C}} \quad CH_2=CHCl$$

acetylene   hydrogen chloride   vinyl chloride

Vinyl acetate production is mostly based on acetylene as follows:

$$HC\equiv CH \quad + \quad CH_3COOH \quad \xrightarrow[\substack{over\ 1\ atm \\ vapor\ phase}]{\substack{catalyst \\ 210\ ^\circ C}} \quad CH_3COOCH=CH_2$$

acetylene   acetic acid   vinyl acetate

The newer plants mostly use the following ethylene route:

$$CH_2=CH_2 \quad + \quad CH_3COOH \quad + \quad \tfrac{1}{2}O_2 \quad \text{-----}> \quad CH_3COOCH=CH_2 + H_2O$$

ethylene   acetic acid   oxygen   vinyl acetate   water

Approximately ninety percent of the world's organic chemical production is petroleum based. Approximately 8% of the U.S. production of petroleum products will go to the chemical industry (a significant part of this comes from the natural gas reserves of the United States).

Vinyl chloride is one of many "monomers" that can be made to form a large molecule by successive additions to itself. The vinyl chloride "polymer" thus formed can be represented as shown below. Poly(vinyl chloride) is a linear, that is, straight chain thermoplastic.

$$CH_2=CHCl + CH_2=CH-Cl + CH_2=CH-Cl + \text{etc.} \text{ --->} -[CH_2-CH-Cl]_n$$

When vinyl acetate is introduced into the reactor in the presence of vinyl chloride, a "copolymer" is formed:

$$CH_2=CH-Cl + CH_2=CHO-\underset{\underset{O}{\|}}{C}-CH_3 \text{ ----->} -[CH_2-CHCl-CH_2-CH(O-\underset{\underset{O}{\|}}{C}-CH_3)]_n$$

Four commercial methods are used to polymerize vinyl chloride. These are emulsion polymerization, suspension polymerization, bulk polymerization, and solution polymerization. The first two are the only techniques of significance with respect to fluid vinyl systems.

Polymerization Techniques

Many solution resins used today are produced via suspension polymerization techniques. It is the most widely used process for making plastic resins, in terms of both the number of polymer products and the tonnage production. Practically all of the newer polymers are made by this method. With this polymerization technique vinyl chloride droplets are "suspended" throughout the water phase. Protective colloids or suspending agents are added to prevent agglomeration of the PVC droplets due to the agitation achieved in the reactor. The suspending agents are soluble in water but insoluble in vinyl chloride. As the polymer molecules form, the organic phase increases in viscosity. A polymer phase is produced in the dispersed organic droplets due to the polymer being insoluble in vinyl chloride monomer. As with all polymerization process raw materials, the water and monomer must be in an extremely pure state. Suspension polymerization is considered one of the most economical methods of polymerization since water is generally the suspension medium. This also assists in the removal of exothermic heat of polymerization.

Figure 1 shows the typical steps in the manufacture of suspension (solution type) resins. The vinyl chloride and vinyl acetate monomers are "charged" (added) to a pressure vessel. Water and suspending agent, etc., are added, the actual polymerization being carried out under conditions of controlled pressure and temperature.

"Slurry" from more than one reactor is "dropped" into a "stripper" to remove unreacted monomer and then transferred to a large "blend tank." The suspended polymer particle can easily be separated from the water phase by filtering or centrifuging. The "wet cake" is then sent to a rotary kiln type dryer and bagged. The particle size of polymers obtained in this manner are usually much larger than those obtained with emulsion polymerization. They can be defined by a conventional screen analysis with respect to particle size.

Some solution resins are made via solvent or solution polymerization. This technique of polymerization has been performed in many media that are solvents for monomer but not polymer. Solvents may include methyl alcohol, diethyl ether, dioxane, toluene, benzene, acetone, etc., which may be used singly or as blends. A suitable catalyst for use with methyl alcohol, for example, would be benzoyl peroxide. Hydrogen peroxide may be used with acetone. The resin is separated as a solid by filtering and/or water quenching. This method is usually used for copolymers because of the poor solubility of poly(vinyl chloride).

The cost of the solvents used in the polymerization make this process uneconomical except for some of the more expensive vinyl chloride copolymers and terpolymers used in solution, for example, coating applications.

Emulsion polymerization is considered the generally accepted technique to produce dispersion (plastisol) resins and vinyl latexes today. The monomers are made dispersible by emulsifiers in a continuous water phase. The initiators or catalysts are water soluble while the emulsifying agents stabilize the emulsion formed when the system is agitated. Gellner (2) cites the following advantages of emulsion polymerization over other methods:

"1. Relatively high molecular weight polymers can be produced at a high rate of reaction...
 2. Excellent heat transfer is realized with water as the continuous phase...
 3. Relatively low viscosity at high polymeric solids is of substantial advantage in many applications..."

To these advantages must be added the ability to produce a very small particle size that is essential to preparing an adequate dispersion of resin in plasticizers.

A modification of this technique would be to use a monomer-soluble catalyst and use an homogenization step to form the dispersion.

Figure 2 shows typical steps in the manufacture of an emulsion resin. Basically, the vinyl chloride monomer is added to a pressure vessel (reactor) where in contact with the emulsifier and the initiator the polymerization is carried out under conditions of controlled pressure and temperature. Since it is uneconomical to carry out the reaction to 100% conversion, unreacted monomer is removed in a stripping vessel. If the polymer is to be applied via the latex (water-based coating) technique, the manufacturing process can now be considered complete. If a dispersion resin is to be produced, then latex from the blend tank is transferred to a spray dryer where water is removed. These very small polymer particles

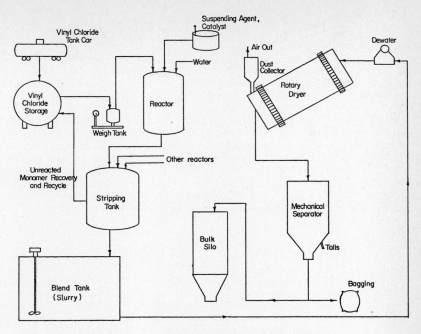

Figure 1.   Suspension polymerization.

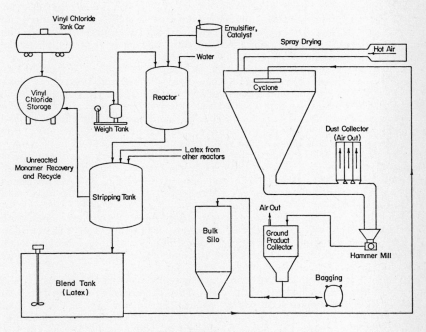

Figure 2.   Emulsion polymerization.

have a tendency to agglomerate during the spray drying operation. This is counteracted by passing the resin through a hammer mill before it is bagged.

The average particle size varies from approximately 2 to 8 μm in diameter with extremes falling below 0.6 and above 20 μm. Note that average opening of a 325-mesh screen is 43 μm. This means that all the dispersion resin particles would show up as 100% through the 325 mesh in a conventional screen analysis.

With bulk polymerization the only materials present in the reactor are monomer and catalyst. The polymer formed from vinyl chloride, as mentioned previously, is not soluble in its own monomer. Very high fluid viscosities are encountered in the reactor with this technique. Until its recent development in Europe, a commercial process for producing bulk polymers did not exist. This European technique has been licensed to a few domestic U.S. resin producers who use it to produce "large particle" type resins.

Polymers

The use of the vinyl polymer in fluid form offers distinct advantages to the manufacturer of vinyl products. The most important obvious advantage is the relatively lower cost of liquid processing equipment and accessories as compared to those required for processing of solid or powder thermoplastics. For example, a spread coating line, for example, knife or roll coater, even with appropriate unwind and takeup accessories, is less expensive than a banbury-mixer-mill calender configuration.

Fortunately or unfortunately (due to the many variables that have to be considered), the liquid vinyls come in a variety of forms. One can use a 100% solid system or a system with varying amounts of a volatile component. The volatile component can be an organic solvent or water, and the solvent may be a fast evaporating or a slow evaporating type.

One of the first uses of vinyl polymers as coatings was in the form of a solution; for example, the resin was dissolved in a pure solvent. Vinyl solution resin systems are defined as lacquers with varying amounts of solid contents. When dissolved in suitable solvents, these resins give coatings with excellent chemical and weathering resistance upon evaporation of the solvent--air dry coatings are obtainable. The solution polymers available today include homopolymers (high physical properties but with very low solids--5–10%), vinyl acetate copolymers (still high physicals but with conventional lacquer solid contents of 10–35%), and terpolymers (with metal adhesion properties that most others do not have and very high solids of 30–40%).

Vinyl chloride has also been reacted with other monomers containing carboxyl, hydroxyl, acetate, maleate, and vinylidene groups. Some of these reactive groups permit cross-linking with other polymers such as alkyds and melamines. Typical solubility characteristics are shown in Figure 3.

With solution vinyls we can obtain very low fluid viscosities and very hard and thin coatings. These properties have to be balanced against dangers of using volatile, flammable solvents and the inability to produce thick coatings with these systems. Solution polymers are the most expensive of the fluid vinyls.

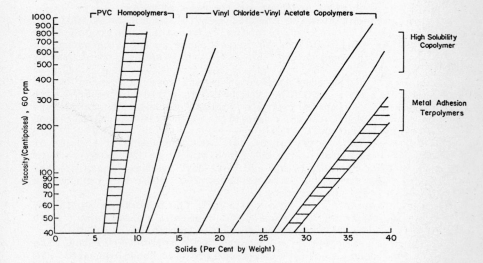

Figure 3. Solubility characteristics of vinyl polymers in MEK. Brookfield LVF viscometer.

Vinyl solution resins are usually made via suspension or solution polymerization techniques. Plastisols are defined as suspensions of homopolymers or copolymers in nonaqueous liquids. The liquids, which are normally vinyl plasticizers, are selected so that they do not solvate the polymer to any extent at room temperature. The suspension is maintained by residual emulsifier left on the particle, and the very small particle size of the polymer itself (all pass through a 200-mesh screen). A finite quantity of plasticizer must be present in order to form the plastisol or "paste": no plasticizer, no plastisol. As with all vinyl systems, consideration may have to be given to plasticization, heat and light stabilization, pigmentation, etc.

The first reference to the term vinyl dispersions is claimed to have occurred in July 1944 (3). Prior to this landmark, the use of these high polymers required either a low-solid solution of the resin in an expensive solvent or the use of heavy-duty, high-pressure equipment to move the polymer in its hot melt (molten) state. Initially, poly(vinyl chloride) resins were ground in an organic media. This required the liquid phase to be polar enough to slightly solvate the resin but not strong enough to actually dissolve it.

In 1947 the first true dispersion resins were introduced in the United States. These emulsion-type polymers could be prepared with 100% solids with relatively simple "stir-in" techniques. Paste-type dispersions are noted to have been developed somewhat earlier and independently in Europe.

Mechanical incorporation of air into the paste or the use of chemical "blowing agents" into the formulations allows foam products to be produced. Plastisols are not "air-dry" systems; that is, they require fusion temperatures of at least 250 °F for the copolymers and 300 °F for the homopolymers.

Plastisols enable us to obtain very low durometer hardness, thick fused sections, and excellent chemical resistance. High durometer hardness values are difficult to obtain.

Organosols are plastisols that contain a volatile diluent. The volatile component is usually selected from true nonsolvents for PVC resins such as the aliphatic hydrocarbons--the main purpose of the volatile ingredient is to lower the viscosity of the paste by contributing more liquid to the formulation. Very low plasticizer levels are then possible, and higher durometer hardness values are attainable (as compared to plastisols).

Generally, the same heat history requirements are required for the fusion or development of physicals with organosols and plastisols. On the other hand, an organosol may require more heat input since some of the applied heat is used only for the volatilization of the diluent. The volatile component prevents fusing thick sections without the danger of bubble formation.

Dispersion or plastisol resins are made exclusively by emulsion polymerization techniques. A vinyl latex is a colloidal suspension (emulsion) of the homo-copolymer particle in water. (The 0.2 μm average particle size is small enough to be in Brownian movement.) Soaps are generally used to form a protective colloid around the particle, and they are surrounded by a negative electrical charge. In some cases air-dry film-forming systems are possible.

Because water is the fluid component, very low viscosities at relatively high solids (50%) are attainable. Latexes are usually cheaper than solution vinyls but similar in cost to dispersion resins. As with organosols and solution vinyls, coating thickness is limited due to the evaporation of the fluid component. Note that more heat is usually required to remove water than organic solvents, and corrosion of equipment must be considered.

One of the latest coating techniques to find commercial acceptance is "powder coating." In this process the coating is in the form of a fine powder that is later fused in place to develop the protective film. Its advantages include elimination of waste in spraying, since overspray can be collected and used again, relative low plasticizer levels, elimination of volatile solvents, and improved long-term storage stability. These advantages must be weighed against the critical coating application parameters, sophisticated coating preparation techniques, and lower molecular weight limitation of the polymers.

Formulation Considerations

Plasticizers have enabled the vinyl compounder to obtain such diverse properties as resistance to extraction by many chemicals to extremely low temperature flexibility. They contribute not only desirable properties in the finished product but also appreciably toward improved processability.

Plasticizers can be classified as either solvent type (primary) or nonsolvent or poor solvent type (secondary). A special category of the nonsolvent type would include polymeric and epoxy polymeric plasticizers.

Plasticizers such as the phthalates, phosphates, etc., are considered the most important. Other monomerics such as adipates, azelates, sebacates, glycolates, citrates, etc., are usually classified as specialty types. The phthalates are considered the most well balanced "property-wise" of all the plasticizers available.

The polymeric plasticizers are used where permanence is of prime importance. The molecular weights of these plasticizers vary quite extensively. Variations from under 850 to 8000 are not uncommon. The cost of polymeric plasticizers is usually in excess of that of the normal monomeric types. Resistance to extraction by soapy water, oils, and migration into nitrocellulose, polystyrene, and rubber are usually superior with polymeric plasticizer systems. Because of their higher viscosities, they are usually more difficult to handle.

Where low-temperature flexibility is of importance, adipates, azelates, sebacates, and the epoxidized stearates and tallates are used. Care must be exercised when using these plasticizers in regard to compatibility limits. The low-temperature plasticizers are usually less compatible than the phthalates because of their lower solvent power for the resin. This condition usually requires more heat and/or higher temperatures to obtain adequate fusion.

The primary reason for utilizing a dispersion resin system is to obtain the benefit of fluid properties. With viscosity characteristics in mind, we must consider this qualification for the plasticizers used in plastisols.

Since they vary quite extensively in oil viscosity and in solvent power, paste viscosity will vary. Because of the influence of viscosity stability upon commercial acceptance, plasticizers such as the long-chain phthalates are common.

With solution vinyl systems the qualifications of plasticizers cited previously apply with the following two additional considerations.

1. The oil viscosity of the plasticizer may now be insignificant due to the high concentration of solvent.
2. Most solution vinyl resins are more soluble than the powder coating or dispersion grade vinyl resins. They will, therefore, require much less plasticizer to reach any specific level of "softening." The compounder has to be extremely careful that he does not over plasticize and obtain a "tacky" or "soft" film.

The basic principles regarding the selection of a plasticizer remain the same with a latex system as with the other polymer systems if only end properties of the final film are to be considered. The vinyl latexes require very little reorientation of thought with respect to compounding philosophy. As the copolymers are introduced, each system will have to be considered on its own merits based on the chemistry of the polymers under study.

With few exceptions, the plasticizer must be added to the latex as an oil-in-water emulsion. This operation requires high-speed, high shearing action equipment.

Some latex systems require a plasticizer to make them film formers at room temperature. One polymer latex system can sometimes be used to plasticize another.

Another modifier for vinyl polymer systems frequently used is a stabilizer.

Polymers of vinyl chloride, under prolonged heat and/or light exposure, may change color and increase in hardness. Heat can cause the splitting out of HCl from the molecule, causing conjugated polymer unsaturation and a color-bearing group. Light can cause a simultaneous oxidation at the point of unsaturation. Light effects, however, are greatly reduced in pigmented systems. Vinyl stabilizers react usually in one of the following ways:

1. They possess reactive dienophilic molecules that break up color-forming polyene systems.
2. They are HCl acceptors that react with the HCl, removing it as an insoluble product.
3. They are selective ultraviolet absorbers, thus reducing total ultraviolet energy absorbed.
4. They are antioxidants and inhibit carbonyl formation from oxygen and polyenes.

The metallic soaps of barium, cadmium, lead, and calcium are commonly used as stabilizers. They are HCl acceptors, but their reaction products often cause cloudiness in clear formulations. Alkyl and aryl phosphates are often used with them to inhibit precipitation of insoluble chlorides. Tin complexes have also been used successfully.

Opaque stabilizers can be used in pigmented and filled compounds.

Many of the metallic compounds and organic auxiliaries used as heat stabilizers also assist in regard to light stability.

There are a number of specific compounds that are used as light stabilizers only. They have been classified chemically as salicylic esters, hydroxybenzophenones, and benzotriazoles.

Because of the requirements for fluid properties in a plastisol, the liquid stabilizers are most often encountered. The chemical nature of the stabilizer is based on the physical properties and processing requirements of the system.

Solid stabilizers are encountered occasionally when they can contribute to increasing the yield or raising the viscosity of the paste, if these properties are required. Solid stabilizers are added, in most cases, after they have been effectively dispersed in plasticizer on a three-roll paint mill.

The UV absorbers in use today are solids thus requiring dispersion via the three-roll paint mill technique.

Stabilizers are not used as frequently with the solution vinyl systems as with the dispersion resin systems. The reasons for this are as follows:

1. Air-dry systems do not require heat stabilizers.
2. When heat is applied to a solvent vinyl system, it is of low magnitude; that is, it is used just to drive off solvent.

When an actual "bake" is required in order to produce a high gloss or due to the treatment of some other part of the item coated, the selection of the stabilizer is very difficult. As mentioned earlier, the copolymer may depolymerize before HCl degradation takes place. This limits the selection of the stabilizer. The sulfur-containing organotin type stabilizers appear to be more effective with respect to improving the heat stability of the solution vinyl resin systems.

The UV absorbers are used to a greater extent with the solution vinyls than with the general-purpose and dispersion resins. This is due to the solution vinyl systems having been used to a greater extent outdoors than any of the other systems. The high color pigment loadings used with solution vinyl often act like UV screeners and protect the vinyl resin from attack.

As with all compounding ingredients used in latex systems, the stabilizer should be emulsified prior to addition to the latex.

Liquid stabilizers containing metallic salts or soaps may be soluble in water-based systems. When solubility occurs, metallic cations may be formed that can injure the stability of the resin emulsion. Various liquid epoxy compounds and organotin stabilizers have been used successfully.

The use of inert materials in vinyls is widely practiced. The filler can be used to lower cost and increase hardness. The most common types of fillers in use today include the calcium carbonates and silicate types. Also available are various silica gels, barytes, gypsum, alums, wood flour, and antimony oxide. Depending upon the oil absorption value of the inert material, a filler will (1) lower tensile properties, (2) increase hardness, (3) lower flexibility, and (4) increase processing temperatures.

Very high filler loadings will cause a whitening effect upon the compound when it is bent.

Since fillers are bought by the pound and many finished goods sold by the yard, it is important to use pound volume figures when using a high-density filler for cost reduction purposes.

The addition of a filler will generally increase the paste viscosity of a plastisol. To what extent is determined by the oil absorption value, density, particle size, shape, and distribution. The high oil absorption fillers increase the yield value in addition to raising the ultimate viscosity of the paste.

Fillers are used frequently with solvent vinyl systems in order to lower cost, increase hiding power of the coating, and increase the total solids of the system. They may detract from ultimate durability due to impurities that are present in the filler. This may manifest itself as poor light stability or film porosity due to water-soluble ingredients present. Basic ingredients may be reactive with carboxyl-containing (metal adhesion type) resins. Iron contamination may cause poor light and heat stability of the system.

The word "filler" is normally associated with vinyl compounding and the word "extender" with latex compounding. They both denote inert materials used to reduce the cost of the coating or compound and provide opacity.

An important feature of general concern to the vinyl plastisol compounder is the oil absorptivity of the inert and its effect on the rheology of the dispersion. A parallel property known as water absorptivity or water demand is of prime importance to the latex compounder.

The water demand of an inert filler is determined by many variables such as particle size, shape and distribution, moisture level of the inert, etc. As a result, each filler must be given a separate identity and treated as such.

The filler should be added to the latex as a water-based dispersion.

The most important properties of a color pigment to be used in a vinyl system are (1) resistance to bleed in solvents, plasticizers, stabilizers, and other additives; (2) heat stability; both high heats for short durations and long-term low heat stability may have to be considered; (3) inertness to the polymer and other formulation ingredients; (4) resistance to migration; (5) resistance to crocking; (6) light stability; (7) chemical resistance (staining, etc.).

Color pigments can best be incorporated into a plastisol formulation by means of a predispersed pigment-plasticizer paste. This paste can be prepared on a three-roll paint mill using a small quantity of a polymeric type plasticizer as the grinding media to give mill tack. The higher viscosities of polymerics are believed necessary in order to increase the shearing action, thus achieving a better dispersion.

Many color pigments can cause alteration of rheological properties far in excess of what their concentration would indicate. Pigments such as phthalocyanine greens and blue, and some blacks, may cause high paste viscosities.

In addition to the pigment characteristics mentioned previously, we have to consider very critically the solubility of the color pigment in solvents when using solution vinyls. Pigments containing iron and zinc have to be evaluated carefully because of their possible adverse effects upon heat stability.

The dispersion of the color pigment becomes more difficult since the quantity of plasticizer required for a three-roll paint mill dispersion may be in excess of that permissible in the final coating. It is frequently desirable to add the color pigment directly to the resin itself. In the case of the nonreactive pigments, it may be possible to mill fuse the resin and add the color pigment directly to the resin upon the mill. Color "chips" are obtained in this manner. The chips can later be incorporated into the solution during mixing.

It may be possible to diperse the pigment in a slowly evaporating solvent such as isophorone or Solvesso 150 by using a differential three-roll mill. This technique is similar to the plasticizer dispersion technique except the plasticizer is replaced by a solvent.

When reactive (carboxyl groups present) resins are used, the pebble mill, or color chip techniques are most frequently encountered. (The color chip is prepared with a nonreactive resin.)

The acidic condition of the resin reduces the compatibility with many pigments, especially those made with basic metals, such as zinc, lead, etc. Some apparently inert pigments contain small amounts of impurities that are basic, thus causing reactivity with the resin. Phthalocyanine blue is an example. This means that only the purest pigments should be used.

The choice of color pigments is extremely broad in the latex systems. Many of the processing considerations cited under fillers with latex systems will apply to color pigments also.

Cellular vinyls can be produced by three basic techniques. These are mechanical, physical, or chemical methods. In the mechanical method, a foam is produced by the vigorous agitation of a fluid producing froth that is later gelled and fused.

In the physical method, an inert gas is incorporated into a fluid by mechanical means. Upon fusion, a foam or cellular product is formed.

Chemical blowing agents are inorganic or organic materials that decomposed under heat to yield at least one gaseous decomposition product. About a dozen or more have commercial significance. The most commonly encountered today is azodicarbonamide (a high-temperature blowing agent).

Volatile ingredients, called diluents, are used to lower the viscosity of a plastisol. By definition, the use of a diluent converts the plastisol to an organosol. The main reason for using a diluent is to lower the viscosity of the paste. The diluent will modify the wetting and solvating characteristics of the plasticizer (dispersant). Since viscosity control is one of the primary reasons for using a diluent, it is usually of an aliphatic or nonsolvating nature. Each dispersant-diluent system should be accurately investigated because there is usually a critical point or composition where a minimum viscosity is obtained.

Organosols, being complex colloidal systems, differ from solutions in that the individual resin particle is dispersed in the suspending liquid, but not dissolved. Formulations low in dispersant tend to agglomerate, are unstable on aging, and may exhibit false body. Resin solvation may be excessive when the dispersant level is high, causing poor paste viscosity stability (i.e., increase in viscosity) on aging.

Liquids with poor solvent action upon the resin make the best diluents. High-solvent power diluents produce sharp, that is, critical viscosity composition curves and may cause a rapid increase of paste viscosity during aging.

The effects of various volatile ingredients upon the rheological properties of a plastisol are shown in Figure 4. The diluents studied include (1) toluene (an aromatic), (2) Solvesso 150 (aromatic but high in C_{10} aromatic and above plus indenes and heavier aromatics), and (3) APCO thinner.

In all cases, an appreciable drop in paste viscosity was observed upon the addition of the volatile ingredients.

Ketones are true solvents for vinyl resins with the solution resin system. The solvent power of the solvent will depend upon many properties of the volatile, including molecular weight and chemical configuration. Depending upon the solubility of the resin, various volatiles such as acetates, nitroparaffins, etc., may be considered primary solvents.

The molecular and chemical composition of the polymer will influence its solubility characteristics. Park (4) has discussed the solvent-resin relationships in detail in "Advances in Chemistry Series 124." They can be summarized as follows: Aromatics such as toluene and xylene are primary solvents for only the most soluble of the vinyl resins. The homopolymers have very slight aromatic tolerances. Aliphatic-type solvents are not considered good solvents for vinyls. As with the aromatics, the extremely soluble resins will tolerate aliphatic solvents if a strong ketone is present. Only fair aliphatic tolerance is obtained with the low molecular weight high vinyl chloride content solution polymers. Aliphatic tolerance of the homopolymers is practically nil. The alcohol tolerance of vinyl resins is very limited. Recent studies with the high solubility type metal adhesion copolymers indicate that appreciable quantities of 2-propanol may be used, if a strong ketone solvent is used.

Other factors, besides solubility, to be considered when selecting a solvent include evaporation rate, density, flash point, boiling range, etc. Much data of this nature are available in the literature along with test methods for additional solvent properties.

Three basic methods are used to manufacture dry powders. There are (1) dry-mix blending, similar to conventional dry blending; (2) hot melt blending, for example, the ingredients are blended (via regular dry blend procedures), fused or melted down and cooled and then ground to final particle size; and (3) wet processing, for example, mixing in a liquid phase followed by spray drying, evaporation, and precipitation.

The resin and resulting particle size of the final compound must be held within vary narrow limits. All liquids must be absorbed uniformly and all solid ingredients thoroughly dispersed.

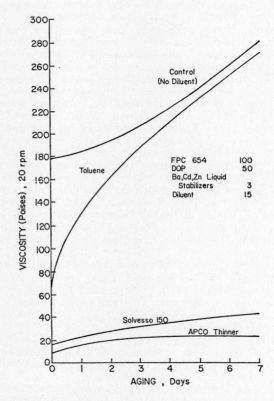

Figure 4. Organosol viscosity stability. Brookfield RVF viscometer.

Application Techniques

The fluid vinyls are extremely versatile with respect to application techniques. Dipping is one of the earliest and simplest methods to apply a coating. Figure 5 illustrates a typical liquid vinyl process. In this case, it is designed to handle plastisols.

All of the liquid vinyl systems are candidates in dipping. Very thick plastisol coatings can and are often applied. Coating thickness control can be achieved first through controlling the flow characteristics of the plastisol and preheating the objects to be coated. Possible limitations to the use of hot dipping include danger of hot forms raising the dip tank paste viscosity, that is, destroying its viscosity stability, and damage to the object being coated by the preheated oven. If conditions leading to these limitations are present, one must rely completely on the flow characteristics of the paste to control the amount of "pickup" in a cold dip tank. Objects commercially available made by plastisol dipping include coated gloves (fabric), disposable surgical gloves, tool handles, seat springs, glass bottles, and lamp sockets.

If an organosol is used, lower plasticizer loadings are possible, but they will limit the thickness of the applied coating. At the same time the volatile solvents force one to select a lower initial preheat temperature of the object to be coated.

It should be noted that both plastisols and organosols have no natural adhesion to metal or glass. If the generic configuration of the item to be coated is inadequate to ensure a locked on coating, a primer coat may have to be applied. Typical primers for this application are of a buna-phenolic or epoxy nature. A small amount of volatile diluent in the coating will help to obtain a tight fit due to the shrinkage of the coating when the solvent is driven off during fusion.

Solution vinyl systems have been used for many years to coat metal parts. Since we have a volatile organic solvent in the system, we have to ensure that (1) the tank remains cool and (2) we have a small liquid surface to prevent premature solvent evaporation. The carboxyl-containing vinyl terpolymers do not need an additional vinyl-to-metal primer.

High-solid vinyl latexes can "skin over" in a dip tank if agitation is not provided. (Note many metal pretreatments make metal surfaces hydrophobic, i.e., it will not be wet out readily by a water-based system).

Powder coatings are used in dipping operations via fluidized bed and electrostatic fluidized beds. It is the oldest form of powder application. It was first used in Germany in 1950 and was introduced to North America in 1955 (5).

Large irregular-shaped objects that are impossible to coat by dipping can be protected by spraying. All the liquid vinyls can be handled in this manner. Low-viscosity systems such as the solution vinyls and latexes can be handled easily on conventional suction feed equipment. Because of the acidic nature of vinyl latexes, the materials of construction of the spray equipment will have to receive special consideration.

Pressure pot type guns and airless spray equipment enable much higher viscosity latexes, vinyl solutions, and plastisols to be sprayed. Hot airless spraying is desirable for very high solids or high viscosity solvent vinyl systems. Equipment of this type with plastisols is not normally considered since heat causes gelation of the plastisol, which is an irreversible process.

Electrostatic spraying of powders is now a commercial reality. Golovoy (6) cites particle size, spherical particle shape, and electrical surface resistivity of the powder blend as important factors in deposition efficiency.

Most of the fluid vinyls find application via spread coating of value. Spread coating has many applications. Typical spread coating equipment would include knife coaters, blade coaters, roll coaters (direct and reverse roll, gravure, transfer roll, cast), bar or rod coaters, and curtain coaters. Figure 6 shows the basic configuration for a spread coating line. Auxiliary trimming and embossing equipment can be included in the line, and all the fluid vinyl systems described above can be used in these systems.

The 100% solid system of plastisols allows very thick coating weights to be applied. Because increased plasticizer lowers viscosity and assists in coating application, very high plasticizer levels (for very soft films) are easily handled. The use of chemical blowing agents and mechanical air incorporation are easily adaptable to plastisol systems.

An organosol will enable the use of a much lower plasticizer level. Low levels of volatile diluents will enable one to produce tough coatings as in the clear wearlayer sheet goods flooring field. Depending upon the diluent levels, fused film thicknesses of approximately 10 mils can be obtained. With both the plastisol and organosol, fusion temperatures in the range of 290–400 °F are required. Other applications of organosols include aluminum siding, tapes, and skin coats on upholstery and handbags. Coil coating (of aluminum siding and steel products for example) is a fast-growing application of this technique.

A solution vinyl system would enable one to deposit and air-dry a spread coat film. For practical production rates, forced solvent evaporation at about 120–180 °F is practiced. Completely unplasticized coatings are attainable, but with a limitation on coating thickness of 0.1–4.0 mils. Less latitude is available with solvent systems than is available with plastisols and organosols with respect to increasing coating viscosity or yield value of the resin. Typical applications of solution vinyls include wallpaper, can coatings, aluminum siding, and tapes. The organic solvents are flammable (as are the organosol diluents); therefore, one needs explosion proof and possibly volatiles collection equipment.

Some of the solution polymers are cross-linkable, enabling thermoset-like properties, such as solvent resistance, to be obtained. The low solution viscosities characteristics of a solvent system generally contribute toward good coating properties. This can be a disadvantage, however, when a porous or loose weave substrate is to be coated.

The vinyl latex systems are similar to the solvent vinyl systems with respect to coating characteristics; that is, they both have low coating viscosity. Removal of at least 50% of the coating weight as water requires ovens with minimum available temperatures of 212 °F. Since organic solvents are not present, fire hazards are diminished. The high emulsifier levels in these systems will limit their applications where water contact or sensitivity are important

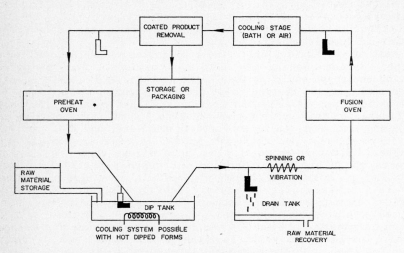

Figure 5. Application of vinyl plastisols by dipping.

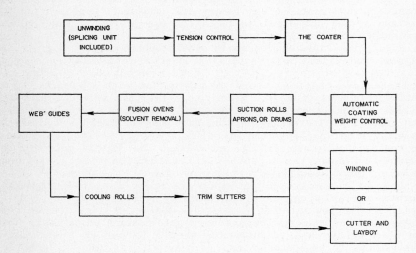

Figure 6. Application of fluid vinyl resins in a spread coating
            line.

factors. They have found applications on wallpaper, window shades, wearlayers in yard goods flooring products, and drapery backing (crushed foam). Fire retardant properties are considered excellent with the vinyl latex polymers. Unplasticized and plasticized films are produced.

Slush molding is used mainly with plastisols. It is usually a continuous operation. The mold (female) is filled above a specific pattern mark and passed through an oven that sets or gels the material in contact with the metal mold surface. These molds are electroformed copper or cast aluminum. The material in the center of the mold that is still a fluid is poured out of the mold with the mold returning to an oven for final fusion. Upon removal from the fusion oven, the mold is cooled and the finished product is stripped from the mold. Vacuum or compressed air can be used with large parts or with those of a complex geometry to assist in product removal. One part of the mold must be open to the atmosphere to permit pouring and product removal.

The low-solid content or, more exactly, the large amount of volatile components that cannot readily be removed—especially in thicker molded parts—completely eliminates solution vinyls and latexes from slush molding operations. It is conceivable that low diluent loading organosols could be used, but commercially this is a rare occurrence. Plastisols, almost by the process of elimination, have the field to themselves. Plastisol objects produced in this manner include rain shoes (transparent), hip boots (foam lined based on slushing different pastes containing blowing agents in series), and doll parts.

With rotational molding, a closed mold is rotated in two or more planes during fusion or solidification of the fluid polymer system. The action of the rotation deposits an even coating of polymer upon the inside of a "clam shell" type mold. After a predetermined time under heat, the molds are removed from the oven and cooled, and the finished product is removed whereupon the cycle is started again. Conventional gas fired ovens or hot salt bath sprays are used for heat transfer, and cooling is usually effected by spraying or immersion in cold water. This technique is well suited to molding completely hollow objects.

Solution vinyls and latexes are restricted from rotational molding for the same basic reason as cited in slush molding operations, that is, the problems connected with removing large volumes of volatile components. Plastisols have the field again nearly to themselves, but low diluent levels in an organosol may be tolerated. It should be noted that rotational molding is adaptable to powder technology also. Dry blend PVC systems, low molecular weight polyethylene, and now other resins have been successfully rotocasted. Plastisol balls, dashboard pads, auto arm rests, traffic cones, hobby horses, and doll parts have been made commercially for many years.

For cavity and inplace molding plastisols are mainly used. Fluids are ideal for the encapsulation of intricate mold inserts. They can be poured into place by gravity or injected at low pressures, thus enabling relatively low cost, lightly constructed molds to be used. Due to the problems inherent in solvent or volatile removal, the plastisols have this field nearly to

Table I. Physical Properties of Fluid Coating Systems

Physical Properties	Plastisols	Organosols
Molecular weight of polymers	Low to very high	Low to very high
Copolymers available	VC/VA	VC/VA
Plasticizer level (range) (phr)	35–100+	25–70
Fusion temperatures (°F)	250–400	300–400
Film thickness (Mil)	0.5–250+	0.5–8
Metal primer required (for adhesion)	Yes[b]	Yes[b]
Gloss range available	Low to high	Low to high
Chemical resistance	Excellent	Excellent
Water resistance	Good to excellent	Good to excellent
Percent solids	100	50–95
Electrical properties	Poor to fair	Poor to fair
Outdoor durability (pigmented systems)	Fair to good	Good
Normal viscosity range Brookfield 20 rpm (cps)	2000–10,000	500–4000
Nature of volatile	None	Aliphatic, naphtha

[a]Plus VC/vinylidene chloride, VC/maleates, terpolymers (COOH, OH types).

[b]Limited self-priming formulas available.

Solutions	Latexes	Powder Coatings
Low to medium	Low to high	Low
VC/VA[a]	VC/acrylates	No
0–25	0–40	0–60
Air–dry	230–400	400–600
0.25–5	0.25–5	1.5–100
No	Yes	Yes
Low to very high	Low to high	Low to high
Good to excellent	Good to excellent	Excellent
Excellent	Poor to fair	Good to excellent
10–50	40–50	100
Good to excellent	Poor	Good to excellent
Good to excellent	Poor to fair	Good to excellent
40–3000	40–3000	Not applicable
Ketone, aliphatic, aromatic, naphtha, esters	Water	None

themselves. Encapsulated electrical and electronic parts, printing plates, automotive air cleaner seals, clay pipe gaskets, etc., are typical products made with these techniques.

Table I cites the physical properties than can be anticipated from fluid coating systems.

Today and Tomorrow

Today's market areas for vinyl chloride dispersion resins, solution resins, and latexes are centered around coating (wearing apparel, upholstery, shoe uppers), molding and dipping (surgical gloves, toys, plating racks, tool handles), bottle closures, coil coating industries, beverage can coatings, protective and decorative lacquers, and nonwoven bindings.

High energy costs encourage polymer compositions that require fewer BTUs for conversion. The use of copolymer dispersion resins offers definite advantages over homopolymers. Figure 7 illustrates this characteristic of the vinyl chloride/vinyl acetate dispersion resins. In this instance, the 5% vinyl acetate copolymer offers distinct tensile superiority to the high molecular weight homopolymer until fusion temperatures of 185 °C are exceeded and then the superiority of the homopolymer is very slight. Reactive dispersion resins (for metal and polymer adhesion) are currently being given commercial evaluations. Reliance upon batch polymerization techniques by polymer manufacturers will also be given close scrutiny. Higher solid latexes will help to lower BTU costs for driving off the water used in latex-based coatings.

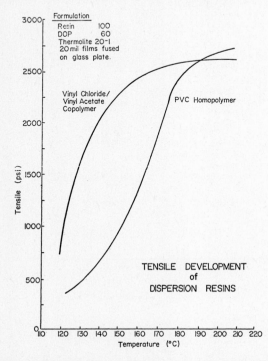

Figure 7.   Tensile development of dispersion resins.

Both polymer manufacturers and polymer converters will continue to investigate new monomers, plasticizers, and additives. Greater use of olefins, maleates, and acrylates as comonomers with vinyl chloride will occur. Regulations and restrictions on solvent use will continue to encourage use of resins that have greater solubility in order to achieve higher solids and/or to tolerate poorer solvents.

Cellular products (foams and froths) will cost less through greater coverage per pound. This approach is currently being used to advantage in the sealing and gasketing area.

The years ahead offer great challenges but also great rewards for those with the vision to take advantage of the vinyl polymers' greatest asset--their versatility and adaptability.

Literature Cited

1. Regnault, M. V. *Ann. Chimi Phys*. 1838, 69(2), 151.
2. Gellner, O. *Chem. Eng*. 1966, 73(16), 74.
3. Whittington, L. R. "A Guide to the Literature and Patents Concerning Polyvinyl Chloride Technology," 2nd ed.; Society of Plastics Engineers, Stamford, Conn. 1963; p. 232.
4. Tess, R. W. "Solvents Theory and Practice"; American Chemical Society: Washington, D.C., 1973; pp. 186–218.
5. Levinson, S. B. *J. Paint Technol*. 1972, 44(570), 38.
6. Golovoy, A. *J. Paint Technol*. 1973, 45(580), 42.

Bituminous Coatings

STEPHEN H. ALEXANDER[1]

Gulf States Asphalt Company, Inc., Houston, TX 77002

Availability of Raw Materials
Manufacturing Processes
    Processes for Coal Tar
    Mining of Gilsonite
Properties
    Chemical Properties
    Physical Properties
Types of Coatings by Use
The Future

The two main raw materials used in the production of bituminous coatings are asphalt and coal tar; a third is a native bitumen, gilsonite.

The petroleum industry, from which asphalt is derived, was born in this country on August 27, 1859, when Colonel Drake drilled 69.5 ft. near Titusville, Pennsylvania, to bring in a producing well of 30 bbl/day (1). Until 1908, the emphasis on products from petroleum was for kerosene used as an illuminating oil. In 1908, Henry Ford began mass-producing his Model "T" Ford, and as automobile production rapidly increased, the emphasis changed from illuminating oil to motor fuel (2).

The coal tar industry began in this country when a horizontal retort was installed in Baltimore in 1816 (3). Not much attention was given to the recovery of coal tar until World War I proved our great dependency on European sources for chemicals available from coal. The shortages resulting from our being cut off from European supplies caused a rapid growth in production of coal tar and related chemicals during and following World War I (4).

[1]Current address: 13826 Kingsride, Houston, TX 77079

0097–6156/85/0285–1229$06.00/0
© 1985 American Chemical Society

Gilsonite was discovered in 1882 by S. H. Gilson of Salt Lake City, who also later commercialized it. It is a natural bitumen found in Utah near the Colorado border. It is a material of good uniformity, occurring in veins from a fraction of an inch to 18 ft. thick. The veins are generally more or less vertical (5).

Availability of Raw Materials

The production of asphalt has increased greatly as the amount of crude oil processed has risen to meet growing demands for petroleum fuels and petrochemicals. From 1925 to 1950, the annual rate of production doubled each decade. After 1950, the rate of growth of asphalt production declined. In 1960, the annual rate was 180% that of 1950, and in 1970, the annual rate was 140% that of 1960.

The factors that have affected the availability of coal tar are quite different from those affecting the availability of asphalt. The interruption of availability of coal tar-derived chemicals from Europe during World War I provided incentive for a rapid increase in production of coke and a greater emphasis on the recovery of the coproduct, coal tar. Production of coal tar doubled in the four years from 1914 to 1918 and doubled again in the nine years from 1918 to 1929. Production rates decreased during the depression years but increased significantly during World War II; production rates peaked in 1957. Since 1957, there have been substantial reductions in the amount of coke needed to produce a ton of iron through increased efficiency by higher blast furnace temperatures, oxygen enrichment, humidification, and injection of supplemental fuel.

In Table I data are given on production of asphalt and coal tar. Data have not been given on volume of coatings produced from these bitumens because of a lack of availability of data. Doerr and Gibson published some information on the use of coal tar for pipe coatings, which is in the range of 200 million lb/year for the years 1950 to 1964 (7). They also indicated that about 30 million lb/year were used for other coating uses (8). These data indicate that approximately 3% (230 million lb) of the coal tar produced is used for coatings manufacture.

Figures show that in 1970 the amount of asphalt used for paving and roofing (6) was 78% and 14.5%, respectively. This leaves 7.5% for all other miscellaneous uses. From knowledge of the industry, it is estimated that at least 10% of this 7.5% goes into coatings uses. For the year 1970, this would be 460 million lb.

Data on the production of gilsonite and the amount used in coatings were not available to the author. Gilsonite was a significant factor when the available grades of petroleum asphalt were limited. Gilsonite's importance for coatings manufacture began to fade in the 1940s, and its use for coatings is now quite limited.

Manufacturing Processes

The first step in the manufacture of asphalt-based coatings consists of the distillation of crude petroleum, which results in distillate fractions (gasoline, naptha, kerosene, diesel fuel, and gas oil). The fraction of crude oil that boils above approximately 300 °C is not distilled, but is withdrawn from the bottom of the distillation

Table I.   Asphalt and Coal Tar Production
           in the United States (6)

Year	Asphalt (million pounds)	Coal Tar
1925	4,800	3,150[a]
1930	6,000	7,800[b]
1940	12,400	6,733
1950	24,000	7,400
1960	43,600	6,880
1970	60,900	7,609
1980	64,100[c]	534(gal.)[d]

[a] For year 1918
[b] For year 1929
[c] From Energy Information Administration,
    U.S. Department of Energy
[d] From Synthetic Organic Chemicals Publication
    (the number shown is million gallons)

tower. This is petroleum asphalt. If the correct crude oil has
been selected and the right processing conditions have been used, it
is possible to get an asphalt for use in coatings from the crude oil
distillation step. If an asphalt different from that produced in
the crude oil distillation step is desired, further modification can
be accomplished by air blowing, by air blowing with a catalyst, by
solvent precipitation, or by a combination of these.

Air blowing is carried out at about 240–270 °C by introducing
small streams of air into the bottom of the asphalt–containing
vessel, and allowing small air bubbles to rise through the asphalt.
This process increases its softening point by dehydrogenation and
condensation polymerization reactions.

The asphalt can be further modified in the air blowing process
by the use of a catalyst, such as phosphorus pentoxide or ferric
chloride. These are usually used in concentrations from 0.5 to 3%,
and they are not catalysts in the true sense because they enter into
the reaction and are at least partially incorporated into the
asphalt. Sometimes these catalysts change reaction rate, but they
are used primarily to modify properties.

Still another method of modifying the properties of asphalt
involves the use of solvent precipitation. This is done by charging
the asphalt into a vessel of propane, butane, or a mixture of these
or similar solvents. This vessel is then held at about 70 °C and
400 psia. The solvents dissolve an oily fraction, and the remaining
asphalt precipitates. Thus, an asphalt of a higher softening point
and greater hardness is produced. Asphalt produced in this manner
can be further modified by air blowing. A block diagram showing
these manufacturing steps appears in Figure 1.

These several processes can be used in series. The composition
of the asphalt contained in the crude oil is an important
consideration. However, the separating and modifying processes
allow the production of asphalts of a fairly wide range of
properties from a given crude oil.

Processes for Coal Tar. Coal tar is produced by the destructive
distillation of coal to give coke and a vapor stream from which the
coal tar is condensed. This crude coal tar can be further distilled
to obtain distillate oils and a coal tar pitch. The pitch can be
used for the manufacture of coatings or can be further modified by
air blowing. Air blowing is not as commonly practiced in modifying
coal tar pitch as it is for modifying asphalt. A block diagram
showing these manufacturing steps appears in Figure 2.

Mining of Gilsonite. Gilsonite is obtained by strip mining. It is
usually blended with other bitumens to produce a coating base. No
prior processing is required.

Properties

Chemical Properties. The chemical makeup and some of the properties
of the three bitumens used in the manufacture of coating are shown
in Table II. The five elements commonly contained in these bitumens
in significant quantities are carbon, hydrogen, oxygen, nitrogen,
and sulfur. Several metals are frequently present in trace amounts;
nickel and vanadium are the most prevalent.

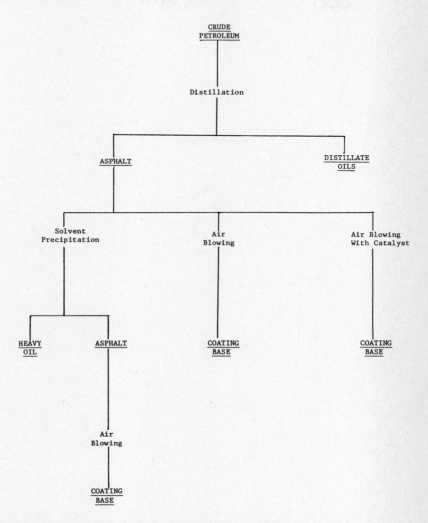

Figure 1.   Manufacturing steps and coproducts in producing asphalt coating base.

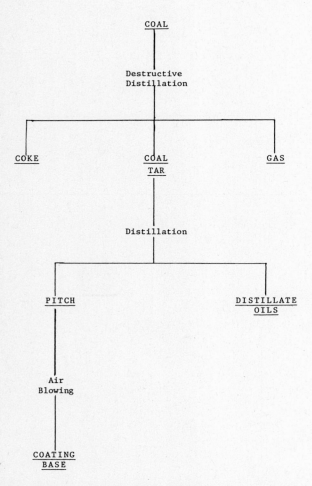

Figure 2. Manufacturing steps and coproducts in producing coal tar pitch coating base.

Table II. Typical Properties of Asphalt, Gilsonite
and Coal Tar Pitch

Property	Asphalt	Gilsonite	Coal Tar
Elemental analysis			
Carbon, wt%	84.5	85.5	89.3
Hydrogen, wt%	9.8	8.4	4.8
C/H ratio	0.72	0.85	1.54
Oxygen, wt%	1.6	1.0	3.2
Nitrogen, wt%	0.9	2.4	1.6
Sulfur, wt%	2.7	0.4	1.1
Molecular weight	450 to 5000	--	220 to 1000
Density, g/mL	1.02	1.06	1.24
Flash point COC, °C.	340	345	245

The carbon/hydrogen ratios are similar for asphalt and gilsonite (0.78 is typical), thereby indicating a similar distribution of paraffinic, naphthenic, and aromatic compounds in these two bitumens. Coal tars have a much higher carbon/hydrogen ratio (1.5 is typical), thereby indicating a high degree of condensed aromatic compounds.

Most of the compounds making up asphalt fall within a molecular-weight range of 450 to 5000. The range for coal tar compounds is somewhat lower: 220 to 1000. The density of asphalt and gilsonite is slightly over 1 g/mL whereas 1.24 g/mL is typical for coal tar. Flash point by the Cleveland Open Cup method is in the range of 340 °C for asphalt and gilsonite and 245 °C for coal tar. Coal tar's lower molecular weight and high aromaticity account for this low flash point.

Physical Properties. Some properties of specific interest in the coating field are: adhesion; weatherability; moisture permeation and absorption; inertness to chemical environments; color; and service temperature range. Bitumens exhibit good adherence to almost any surface. Coatings with excellent weathering properties can be obtained by proper selection of the bitumen and formulation with other materials.

Low permeability and low water absorption are two outstanding features of bitumens for use in coatings manufacture. Table III gives permeability data on four bitumens and uses neoprene and vinyl chloride for comparative purposes. These data demonstrate the relatively low permeability of bitumens.

The use of added fillers in the manufacture of coatings generally increases permeability. For some special uses, high permeability is desirable and can be accomplished by selection of fillers as shown by the example in Table III. Permeation rate in use depends on the permeability constant of the coating and on the coating thickness. Permeation decreases with increasing thickness of the coating. For example, vulcanized neoprene has a permeability constant of 26 which is 6.5 times that of coal tar pitch, which has a permeability constant of 4. Coal tar coatings are typically used 90 mils thick as compared to neoprene film at 34 mils. This is a thickness ratio of 2.6. Combining these two factors (6.5 x 2.6) makes neoprene film 18 times more permeable than a coal tar coating in a typical use situation. More extensive data on permeability and water absorption can be found elsewhere (9).

Bitumens also have good resistance to attack by most inorganic salts and weak inorganic acids. In addition, bitumens are dark brown or black and these colors are very difficult to mask with pigments; therefore, their use should be restricted to applications where these colors are not objectionable. Colored overcoats can be used when a dual coating system can be justified. Bitumens are thermoplastic and have a rather narrow service temperature range unless modified with fibrous fillers and/or synthetic resins.

Types of Coatings by Use

Bituminous coatings can be divided into two major classes by application characteristics: hot applied and cold applied. The hot-applied coatings can be subdivided by composition: nonfilled and filled (usually finely divided mineral). The cold-applied

Table III.   Permeability of Various Organic Films to Water Vapor

Film	Permeability Constant[a] x 10^9	Typical Usage Thickness (in.)	Perms[b]
Asphalt, oxidized	8	0.050	0.023
Coal tar pitch	4	0.090	0.006
Asphalt fibrated coatings (conventional fillers)	20	0.060	0.048
Asphalt fibrated coatings (special filler to increase permeability)	200	0.060	0.475
Neoprene, vulcanized	26	0.034	0.109
Plasticized vinyl chloride	38	0.019	0.286

[a] Permeability constant is based on Fick's law of diffusion, which states that the amount of vapor which diffuses through a membrane is proportional to the area of the membrane, the pressure gradient of the diffusing vapor, and time. It is inversely proportional to the thickness of the membrane. Mathematically it is expressed as $W = k(APT/L)$, where A is the area of the membrane in square centimeters, P is the pressure differential of the vapor in mm Hg, T is the time of diffusion in hours, W is the weight of gas diffused in grams, L is the thickness of the membrane in centimeters, and k is the permeability constant. Thus, the units of k are $g - cm/cm^2 - mm\ Hg - h$.

[b] A perm expresses permeation rate in the units of grains per square foot of membrane, per inch of Hg pressure differential, per hour of time elapsed. (The perm measure does not include the thickness which is part of the permeability constant formula. Thus, a membrane of a given material will have a different "perm" value for each thickness of membrane measured.) The units are $grains/ft.^2 - in.\ Hg - h$.

coatings can also be subdivided by composition: those containing
solvent or those containing water. Both of these subclasses can be
further subdivided into nonfibrated or fibrated types.

A major use of hot-applied, nonfilled asphalt coatings is for
coating metal highway culverts. The largest use of hot-applied,
filled, bituminous coatings is to coat pipe for buried pipelines.
Pipelines so coated are used to transport natural gas, petroleum
products, petroleum crude oil, slurried coal, and carbon dioxide.

Some other major uses, by types, are as follows: Solvent
cutback and water emulsion, nonfibrated coatings are used as
protective coatings for metal and masonry. Solvent cutback and
water emulsion, fibrated coatings are used in roofing, waterproofing, and industrial insulation, and in automobiles for
protection and sound deadening.

Aluminum pigments are compounded with solvent cutback bitumens
to produce a variety of coatings for various uses. After a short
weathering period, they produce brilliant aluminum coatings. Major
uses are in roofing, metal protection and insulation protection. A
block diagram showing the manufacturing steps used to produce these
coatings is shown in Figure 3, and Table IV shows three types of
asphalt roof coatings.

Each coating has its special use. The solvent type has an
extremely low permeability and is especially suited for new
construction. The aluminum type is heat reflective and decreases
the air conditioning load on a building. The emulsion type has a
high permeability and is especially suited as a maintenance coating
where the roofing components have absorbed moisture that needs to be
expelled without causing blistering of the coating. With this type
of coating, the heat of the sun will drive the moisture out, and it
can be expelled through the coating at a sufficiently rapid rate to
prevent pressure buildup and blistering from occurring.

Table V shows properties for two asphalt and two coal tar
coatings used to coat pipe for buried pipelines. Because of its
aromaticity, coal tar is more temperature susceptible (changes more
in viscosity with a given temperature) than asphalt. This
deficiency is overcome to a large degree by compounding with
plasticizing oils and coal dust to produce a coal tar coating
denoted as "wide range". The service temperature range of the
nonmodified coal tar is 33 °C vs. 94 to 100 °C for the two asphalt
and the "wide range" coal tar coatings. Again, because of greater
aromaticity, the coal tar coatings are 20% more dense than the
asphalt coatings.

Table VI shows three types of asphalt-weather and vapor-barrier
coatings used to protect thermal insulation. These examples
illustrate that a wide range in moisture vapor permeability can be
accomplished. The low-permeability material has a rating of .003
perm-in. and is especially useful for protecting insulation on
vessels operating at very low temperature. This condition produces
driving forces that cause moisture to migrate into the thermal
insulation. If this is allowed to happen, ice builds up in the
insulation, and its efficiency is greatly reduced. For
installations operating at higher temperatures, the heat prevents a
moisture buildup problem, but protection from weathering and
blistering resistance is needed. Here a solvent cutback formulated
to have 10 times the permeation (.03 perm-in.) or an emulsion type

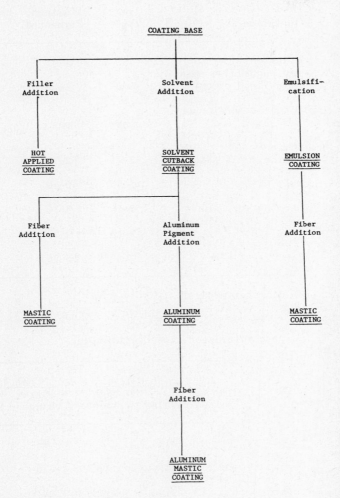

Figure 3.   Manufacturing steps to produce various types of bituminous coatings.

Table IV. Asphalt Roof Coatings

Description	Solvent Type	Aluminum Pigmented	Emulsion Type
Typical properties			
Blistering and sagging on metal at 30° angle in 71 °C oven	none	none	none
Bend test of coated panel over a 1-in. mandrel at 0 °C	pass	pass	pass[a]
Typical composition			
Asphalt softening point (R&B), °C	52–71	52–71	43–54
Asphalt content, wt%	45–55	40–50	30–40
Asbestos fiber content, wt%	8–16	8–16	7–15
Mineral filler, wt%	5–15	0–20	—
Aluminum pigment, wt%	—	15–25	—

[a] Over a 2-inch diameter mandrel

Table V. Typical Properties of Pipe Coatings

| Property | Asphalt | | Coal Tar | |
	Regular	High Temp.	Narrow Range	Wide Range
Softening point, °C	115	132	93	110
Mineral filler content, wt%	25	25	25	30
Service temperature range, °C	–29 to 65	–18 to 82	4 to 37	–29 to 65
Density	1.20	1.20	1.45	1.45
Application Temperature, °C	245 to 260	255 to 275	176 to 204	218 to 246

Table VI. Asphalt Weather and Vapor-Barrier Coatings

Description	Solvent-Type Low-Permeability	Solvent-Type Med-Permeability	Emulsion-Type High-Permeability
Typical properties Solids content, wt%	72	60	58
Water Vapor Permeability (perm-in.a)	0.003	0.03	0.15
Typical composition Asphalt softening point (R&B), °C	71-93	63-88	43-54
Asphalt content, wt%	50-65	30-40	30-40
Asbestos fiber content, wt%	10-20	10-20	7-15
Other fillers usually present	Yes	Yes	Yes

a A perm-in. is the same as perm except that it is expressed per inch of thickness of the membrane. (To convert a perm value to perm-in. value, multiply by the thickness of the membrane in inches.) Units are grain-in./ft.2-in.Hg-h.

of 50 times the permeation (.15 perm-in.) of the conventional coating can be used. [See footnotes for Tables III and VI for definitions of perm and perm-in.)

Table VII shows three asphalt coatings used in thicknesses comparable to industrial paints. Film thicknesses are from 0.7 to 3 mils. These coatings find wide use in industry for the protection of structural steel, for the protection of metal parts in storage (removable with solvents), and as pipe and tank coatings. The aluminum pigmented coating has two main advantages: aluminum appearance is more desirable than black, and the pigment protects the bitumen from the actinic rays, thereby extending the life of the coatings.

Table VIII gives data on a railroad car and three automotive coatings. These are combination protective coatings and sound-deadening materials. A significant property of these compositions is their decibel decay rate at 10 °C. These rates range from 8.2 for the standard automotive undercoating to 10.1 for medium-efficiency sound deadening and 15 for high-efficiency sound deadening. Also, the filler content in the high-efficiency deadener is in the range of 45-60%. It contains only 16-22% asphalt and is a very dense material of about 12.3 lb/gal.

In recent years, the bituminous coatings have been modified with epoxy resins, urethane polymers, and various rubbers. The epoxies and urethanes have made possible the formulation of coatings containing over 50% bitumens. These coatings have low shrinkage during cure and widely varying cure times. Modification with these resins and rubbers allows tailoring of performance characteristics for specific applications.

Table IX gives a typical formulation for coal tar epoxy. It contains approximately 30% epoxy and 25% coal tar. This type of coating will perform better than either the pure coal tar or the pure epoxy coatings in several applications. Because of the relatively low cost of coal tar pitch, the raw material cost is significantly less than for a pure epoxy.

The Future

During the last 10 years, the cost of asphalt and coal tar has increased significantly more than the cost of most other coating raw materials. This is because their cost is so closely related to energy costs, which have increased approximately 10-fold (1000%) in the last 10 years (much faster than general inflation which increased 130% over the same period of time).

The cost of energy now appears to have stabilized such that it will more nearly approximate that of general inflation in the immediate future. Therefore, the cost of bituminous coatings as related to other coating types is also expected to stabilize. In light of these factors, the bituminous coatings and modified bituminous coatings are expected to maintain their market share for the immediate future.

Table VII.   Asphalt Protective Coatings, Thin-Film Type
             for Above-Ground Service

Description	Chassis and Frame Paint	Solvent Type Nonfibrated	Aluminum Pigmented
Typical properties			
Solids content, wt%	45–60	53–67	60–70
Film thickness, mils	0.7	3	2
Service temperature range, °C	–18 to 71	–18 to 71	–18 to 77
Typical composition			
Asphalt softening point (R&B), °C	82–104	82–104	77–104
Asphalt content, wt%	45–60	57–67	17–30
Aluminum pigment, wt%	– –	– –	14–20

Table VIII. Asphalt Coatings and Sound Deadeners for Transportation Vehicles

Description	Railroad Car Coating	Automotive Under- coating	Automotive Under- coating Deadener	Automotive High Efficiency Deadener
Typical properties Solids content, wt%	64	65	78	80
Cold adhesion test, −23 °C, angle	70°	70°	70°	50°
Sound deadening properties, 0.5 lb/ft² , decibels Decay per second at				
−18 °C	--	13.6	15.9	10.4
+10 °C	--	8.2	10.1	15.0
+43 °C	--	3.8	3.9	6.8
Weight, lb/gal	8.0	8.6	10.8	12.3
Typical composition Asphalt softening point (R&B), °C	71–93	71–93	60–82	60–82
Asphalt content, wt%	40–50	28–40	22–32	16–22
Asbestos fiber content, wt%	9–18	10–20	8–16	6–12
Finely divided heavy, nonabrasive mineral filler, wt%	--	0–15	25–50	45–60

Table IX.   Typical Formulation (Coal Tar-
Epoxy Coating)

Component	Wt%
Base component (20 parts by volume)	
Liquid epoxy resin	30
Soft coal tar pitch	25
Xylene	20
Pulverized talc	25
Curing agent (1 part by volume)	
Diethylenetriamine	70
Xylene	30

Literature Cited

1. Oil and Gas Journal, Petroleum Panorama--An Issue to
    Commemorate Oil's First One Hundred Years, 1959, 57, 5, A-3.
2. Ibid., A-20.
3. Rhodes, E. O. Bitum. Mater. 1966, 3, 14.
4. Ibid., 16.
5. Broome, D. C. Bitum. Mater. 1965, 2, 3.
6. Minerals Year Book, U.S. Bureau of Mines.
7. Doerr, J. S.; Gibson, P. B. Bitum. Mater. 1966, 3, 129.
8. Ibid., 133.
9. Alexander, S. H.; Tarver, G. W. Bitum. Mater. 1965, 2, 223.

SCIENCE AND TECHNOLOGY OF PIGMENTS AND PAINT MANUFACTURE

Opaque White Pigments in Coatings

FRED B. STIEG

PigmenTech Consulting, Jekyll Island, GA 31520

Early History
Optical Theories
Optimum Particle Size
Hiding Power
Hiding Power Test Methods
Exterior Durability
Rutile Titanium Dioxide
Advances in Hiding Power Theory
Chloride Process Titanium Pigment and Optimum Particle Size
Pigmentation of Latex Paints
High Dry Hiding of Paints
Light-Scattering Studies
Composite Titanium Pigment
Microvoids
Single-Pigment Concept
Effective PVC and Dilution Efficiency
Plastic Pigments
Optimum Solids

Early History

A little more than 50 years ago the American coating industry could be correctly described as a commercialized art form. Neither science nor advanced technology had been allowed to intrude upon the concepts of "paint making"--some of which were secrets handed down from father to son for a hundred years.

The most widely used opaque white pigment was basic carbonate white lead, manufactured by a process little different from that known by the early Greeks in 400 B.C. Other commonly used pigments were zinc oxide, dating back to 1770; zinc sulfide, first produced in 1783; lithopone, developed in 1847; and basic sulfate white lead, which first appeared in 1855.

0097-6156/85/0285-1249$06.25/0

In the September 1924 issue of <u>Industrial and Engineering Chemistry</u> the following note appeared under the heading "Scandinavian Letter":

> Two Norwegian chemists have developed a process by which the black material ilmenite can be transformed into an intensively white powder consisting of titanium dioxide. A large factory has been built and the new paint is now conquering the world market because of its superiority in covering value and durability, although it costs a little more than lead and zinc white. In the United States a similar process has been worked out independently by American chemists, who have started the production in the States. The Norwegian and American companies have agreed to divide the market between them, the latter covering the two American continents while the former takes care of the rest of the world market.

The American production referred to was of course that of Titanox, the first commercially produced titanium pigment, manufactured by the Titanium Pigment Co. in Niagara Falls, N.Y. This small enterprise was soon taken over by the National Lead Co., representing the American white lead trust, who recognized it as serious competition for their conglomerate of white lead grinders. The major influence of pigmentary titanium dioxide on the future history of the coating industry makes this development a logical starting point for the present review.

Due to the success with which the guild of paint makers had fought off the encroachments of science, practically nothing had been known about the optical properties of any of the commonly used white pigments until only a very few years before the appearance of titanium dioxide, and what had been learned was little used outside of the laboratories of pigment manufacturers.

To paraphrase a comment made by J. C. Gehant in 1932 (1), the entire aspect of the American coatings industry "may be likened to the early operation and appearance when a new building is being erected; to the bystander all is confusion and the ground is covered with the scattered materials and machinery that is to be used. Here and there one sees part of the excavation and the real building is soon to start."

The basic scientific concepts necessary to establish the absolute opacity, particle size distribution, and optimum average particle size of opaque white pigments were among the litter of "machinery" still waiting to be used.

Optical Theories

In 1832 Fresnel (2) had proposed the laws that relate the reflectivity of any white pigment to its refractive index and that of the medium in which it is dispersed:

$$F = \frac{(n_1 - n_2)^2}{(n_1 + n_2)^2}$$

where F = Fresnel reflectivity, n_1 = refractive index of the pigment, and n_2 = refractive index of the medium. Early experiments had confirmed a general relationship between the refractive index of white pigments and their observed hiding power, but the theory proposed by Fresnel provided no explanation for the fact that different samples of the same type of pigment might produce different degrees of white hiding power when dispersed in the same vehicle.

Optimum Particle Size

In 1865 Maxwell (3) had published his famous set of equations from which the Mie (4) theory was derived in 1908. While Maxwell's observations had shown that the light-scattering ability of a solid particle of matter was related not only to its refractive index but also to its particle size--increasing as the particle size was reduced, Mie demonstrated that an optimum particle size existed for light scattering, below which light scattering decreases as the particle size is further reduced. This, of course, adequately explained the deviations in hiding power noted for apparently similar pigments, since little control was exerted over particle size.

While the equations involved in the Mie theory are exceedingly complex, a first approximation of the optimum particle size for any white pigment may be obtained from the equation

$$D = \frac{2}{\pi} \frac{\lambda}{n_1 - n_2}$$

where D = optimum particle diameter for maximum scattering and λ = wavelength of light to be scattered. It will be noted that optimum particle size is also a function of the relative refractive indexes of pigment and binder.

By 1911 Wright (5) had provided methods for experimentally determining the refractive index of pigments. In 1909 Knudson (6) had developed the calculations for obtaining the surface area of pigments from air permeation. In 1910 Thompson (7) produced the first apparatus for the controlled separation of pigments into various particle size groups by the process of elutriation. In 1915 Oden (8) had begun to publish his series of papers on the sedimentation method for determining particle size. In 1921 Green (9) had published a photographic method, and Martin et al. (10) reported on an optical micrometer in 1923.

From the fact that Gehant's "building-site" analogy was made in 1932, it may be correctly inferred that most of the scientific developments referred to in the preceding were still littering the ground about the excavation almost a decade after the date chosen as a starting point for our review. The "builders" of the coating

industry were not yet ready to use them. This was largely due to
the scarcity of scientifically trained personnel; artisans seldom
read foreign technical journals.

What little knowledge existed about the opacity and particle
size of white pigments was comparative rather than absolute in
nature and had been developed primarily by raw-material manufacturers who had devised testing procedures to control the quality of
their products. The builders were not among the paint makers, but
among the pigment makers.

Men such as G. W. Thompson (7) and R. L. Hallet (11) of the
National Lead Co., G. F. A. Stutz (12) and H. Green (9) of the New
Jersey Zinc Co., and J. H. Calbeck (13) of the Eagle-Picher Lead Co.
had cooperated with concerned academists such as Professor A. H.
Pfund (14) of Johns Hopkins University to provide the existing data.

Hiding Power

The most commonly used method for evaluating the comparative opacity
of white pigments was that of color or tinting strength. This was
essentially nothing more than a comparison of the relative amounts
of blue or black colorant required to visually match the depth of
color produced in a standard. The standard was usually basic
carbonate white lead, arbitrarily rated at 100.

The term opacity was reserved for the results of tests made
under standard conditions of composition and/or film thickness and
was felt to be related solely to the opacifying power of the white
pigment. The term hiding power was applied to paints as they were
used. There were considerable differences of opinion, however, as
to the relationship of tinting strength to opacity and of opacity to
hiding power. Hallet (11) was among those who believed that a
definite relationship existed between tinting strength and hiding
power. The data in Table I were obtained from hiding power
measurements made on paints of similar volume composition using
Hallet's "hidemeter" and from tinting strength tests of the same
five pigments.

Many of the difficulties encountered by these early investigators were related to the insistence of the paint makers that
hiding power be determined in "practical" paint systems--that is,
that the amount of pigment present be based upon the amount required
to produce brushing viscosity in a simple linseed oil vehicle.
Since pigments varied in both density and oil absorption, pigment
volume concentrations were seldom comparable. The same paint makers
also insisted upon a "practical" brush application and preferred
visual to instrumental comparisons.

In 1926 Rhodes and Fonda (15) departed somewhat from this
practical format to develop equations for calculating the "ultimate
brightness" (reflectance at infinite film thickness) from photometer
measurements of films of varying thickness applied over a black
substrate. The paints investigated contained equal amounts of
pigment on a weight basis; while Rhodes and Fonda spoke of the
importance of considering the hiding power of a unit volume rather
than a unit weight of pigment, it did not occur to them to formulate
their test paints at a constant pigment volume concentration. Their
derived equation involved the average particle diameter of the

pigment in the first recorded attempt to determine the effect of the size parameter, but their results (shown in Table II) show poor correlation between average particle size and ultimate brightness.

The authors determined pigment particle size using Green's photographic method (9) which could not have been too precise, but it was claimed that good correlation had been obtained with the relative hiding power values of Hallet (11) when calculated for film thicknesses, providing a residual contrast of 2% over their black substrate (see Table III).

The figures for lithopone are identical in Table III because Rhodes and Fonda arbitrarily selected that pigment as the standard for their comparisons and set its value at 80 to conform with Hallet's value. There would seem to have been little significance to the so-called correlation in view of the wide range of reflectances reported for lithopone in Table II.

In the same year (1926) Hallet (16) reappraised his hiding power values and discovered that they could be related to carefully determined tinting strength values by the equation

$$\text{hiding power } (\text{ft}^2/\text{lb}) = 0.1 \times \text{tinting strength} + 3.0$$

Hallet's data, reproduced in Table IV, disclose that the zinc sulfide pigment industry had busied itself since his earlier publication in producing higher hiding pigments to more successfully compete with Titanox. Ordinary lithopone (28% zinc sulfide) had been supplemented with high-strength lithopone (50% zinc sulfide) and titanated lithopone (15% titanium dioxide). Titanium had also been on the move, however, with the Titanox B pigment (25% titanium dioxide) beginning to yield its place to a higher strength calcium sulfate base product, Titanox C (30% titanium dioxide). A pure anatase titanium dioxide had also made its appearance.

The year of 1926 was also the year in which Stutz and Pfund (14) estimated the optimum particle size for zinc oxide in water to be 0.24 μm, based upon relative turbidity measurements for a series of pigments of predetermined particle size. Because of their similar findings in an organic vehicle, they concluded that differences in refractive index did not affect the optimum particle size for light scattering and that it would therefore be the same for all other white pigments. Although their conclusions were erroneous and contrary to the scientifically accepted tenets of both the Fresnel and Mie theories, they did serve to call the attention of paint makers to the importance of pigment particle size. If the same type of pigment could produce differing degrees of opacity depending upon its particle size, then the paint maker could no longer safely assume that all pigments of similar type were interchangeable.

Hiding Power Test Methods

In 1926 Bruce (17) of the National Bureau of Standards reported upon one of the first attempts to improve the precision of hiding power measurements. A thin film was uniformly spread over a black and white glass plate by spinning, and the contrast ratio was determined photometrically. This method was improved upon by Haslam (18), who

Table I. Hiding Power of White Pigments

Pigment	Hiding Power (cm^2/g)	Relative HP (wt basis)	Relative HP (vol basis)	Tinting Strength
Carbonate white lead	40	100	100	100
Basic lead sulfate	30	75	72	85
Titanox	80	200	128	350
Zinc oxide	46	115	96	170
Lithopone	50	125	80	200

Table II. Rhodes and Fonda Data (1926)

Pigment	Refractive Index	Specific Gravity	Diameter (microns)	Ultimate Brightness
Lithopone	2.125	4.35	0.341	0.825
Lithopone	2.125	4.35	0.331	0.935
Lithopone	2.125	4.35	0.375	0.873
Lithopone	2.125	4.35	0.520	0.860
Lithopone	2.125	4.35	0.371	0.875
Basic lead sulfate	1.88	6.4	0.435	0.883
Carbonate white lead	2.01	6.64	0.620	0.860
Carbonate white lead	2.01	6.64	0.385	0.860
Zinc oxide (French)	2.02	5.23	0.350	0.902
Zinc oxide (USP)	2.02	5.23	0.500	0.892
Zinc oxide (American)	2.02	5.45	0.500	0.873

Table III. Relative Hiding Power per Unit Volume of Pigment

Pigment	Rhodes and Fonda	Hallett
Lithopone	80	80
Basic lead sulfate	86	72
Carbonate white lead	99	100
Zinc Oxide	90	96

Table IV. Hiding Power of Common White Pigments

Pigment	Tinting Strength	ft^2/lb
Titanium dioxide (anatase)	1150	115
Zinc sulfide	540	58
Titanox C	450	48
Titanated lithopone	400	44
High-strength lithopone	400	44
Titanox B	480	40
Lithopone	260	27
Zinc oxide	200	20
Carbonate white lead	100	15
Basic lead sulfate	85	13

used a black and white paper chart with off-center spinning to further improve the uniformity of film thickness. In the meantime, Brier and Wagner (19) had developed the drawdown blade.

By this time, the development of new opaque white pigments having substantially more opacity than the traditional white lead "standard" had aroused the interest of the changing paint industry in more practical hiding power test methods. White lead was being replaced in almost all interior finishes by either lithopone or Titanox because of their lower cost, and test methods were needed to assist formulators in matching the hiding power of their modified and original products. In addition, the claims and counterclaims generated by intense competition between the suppliers of zinc sulfide and titanium pigments could not always be accepted at face value.

Since the hiding power requirements of the coatings industry were still at the white lead level, considerably more interest centered around the composite pigments (the several lithopones and Titanox pigments) than in either pure titanium dioxide or zinc sulfide. This was true not only because of the lack of any immediate demand for greater hiding power but also because of the observed fact that these composites produced more white hiding power than might be expected from their relative contents of zinc sulfide and titanium dioxide and were consequently more economical to use than the "pure" pigments. The reasons for this greater optical efficiency were not understood, and the relative strengths indicated for these pigments by the data of Hallet and others were not always realized in practice.

By 1930 the now familiar black and white hiding power chart was in common use. The majority of hiding power comparisons appear to have been made on wet paint films (20), which provided fair correlation with the recently developed black and white Pfund cryptometer (21), but it was generally accepted that observed dry hiding power did not necessarily agree with the results of either of these methods. Stutz and Haslam (22), however, claimed that the cryptometer agreed with "practical" brushouts, providing the dry film brightness was greater than 70%. Pfund (21), however, became the first to call attention to the "dilution effect" as the reason for varying hiding power results if comparisons were not made at the same pigment volume concentration.

Also by 1930 a second producer of anatase titanium dioxide had entered the industry and both suppliers became extremely active in the study of white hiding power—which they of course recognized as the strongest selling point for titanium pigments.

In 1931 two events occurred that were to have a great effect on the future development of opaque white pigment technology. Kubelka and Munk (23) published their famous two-parameter theory of hiding power, and Knoll and Ruska (24) produced the first electron microscope. White hiding power might now be accurately determined, and the ultimate particles of white pigments might now be "seen" for the first time—but neither event attracted any immediate attention among American paint makers since both were reported in foreign scientific journals.

Equally unnoticed, 2 years later in 1933, was the granting of U.S. Patent 1 931 381 to the same P. Kubelka (assigned to the Krebs Pigment & Color Corp.) for the production of pure titanium dioxide from vaporized titanium tetrachloride. Twenty-five years was to pass before a surplus of titanium tetrachloride would result in its commercial application.

In 1934 the two producers of titanium pigments made new contributions to the paint makers' knowledge of white hiding power. Sawyer (25), of Krebs Pigment & Color Corp., working with a gray and white paper chart to better approximate "practical" (that word again!) hiding requirements, produced a series of plotted curves depicting the wet and dry hiding power of basic carbonate white lead, lithopone, and titanium dioxide over a range of PVC levels. He called attention to the progressive loss of hiding power per pound of pigment as the PVC was increased and to the existence of a PVC above which dry hiding power actually decreased for a time before again increasing as the percentage of pigment was still further increased. He correctly attributed wet-to-dry hiding power loss to the difference in pigment volume concentration in the wet and dry films, but erroneously concluded that the "high dry hiding" effect was due to increased pigment packing.

Jacobsen and Reynolds (26) of the National Lead Co., working with gray and black paper charts for the same reason as that conceived by Sawyer, produced hiding power vs. PVC curves for titanium dioxide, zinc sulfide, Titanox C, titanated lithopone, high-strength lithopone, Titanox B, and regular lithopone. While Sawyer had pointed to the data as evidence of the futility of assigning any single hiding power value to a pigment, Jacobsen and Reynolds showed that their method might be used to produce hiding power values for the composite pigments at 28 PVC and for titanium dioxide and zinc sulfide at 15 PVC that were consistent with the long-accepted values of Hallet (see Table V), thus becoming partially responsible for the industry's acceptance of these simple but misleading "hiding power numbers" for another 20 years.

Exterior Durability

It was significant of change, however, that within 10 years time the traditional "standard" of the coating industry, basic carbonate white lead, had disappeared from a tabulation of opaque white pigments.

Up until this time, however, titanium pigments had not seriously competed with white leads in exterior house paints, due to the extremely poor chalk resistance of anatase titanium dioxide. In spite of the common conception of titanium dioxide as an "inert" pigment, due to its lack of chemical reactivity with the acid decomposition products of paint vehicles, the source of soap formation by the basic white leads and zinc oxide, it was quite reactive in a totally different way. Many strange theories were brought forward to account for this activity (27–29, 4), among them the idea that high molecular weight might be required to provide durability. This latter theory might have something to do with the production of the first durable titanium pigment, lead titanate (Titanox L), by the National Lead Co. in 1935.

The true explanation for the poor chalk resistance of titanium dioxide had actually been reported by Renz (30) back in 1921 before commercial titanium pigments were even available, but this of course had not attracted the attention of the paint makers since his report was published in a Swiss technical journal. Renz had found that titanium dioxide, if exposed to ultraviolet light in the presence of an oxidizable medium (such as an organic paint vehicle), was photoreduced to a lower oxide--showing that the poor chalk resistance of titanium dioxide was due to photochemical reactivity.

During the next 5 years (1935-40), a number of anatase titanium dioxide pigments appeared in which inorganic surface treatments-- including, but not limited to, the hydrates of alumina, silica, and antimony--had been used to partially pacify this photochemical reactivity, improving color retention as well as chalk resistance.

Rutile Titanium Dioxide

One year before the end of this period (in 1939), the first chalk- resistant pure titanium dioxide pigment was produced by the National Lead Co. Although never described as such at the time, this was also the first commercial rutile titanium dioxide, which accounted for some of its improved chalk resistance. Since the rutile structure had been attained by overcalcination, its overly large particle size and poor dispersibility prevented it from attaining the increased hiding power that should have accompanied its conversion to rutile, and the pigment was manufactured only until the end of World War II to satisfy the demand of the U.S. Navy for a durable titanium dioxide.

Throughout the same period, lithopones remained strongly competitive with the composite titanium pigments, the latter having strengthened their position only slightly by increasing the titanium dioxide content of Titanox B from 25% to 30%. In 1940, however, the death knell was sounded for the zinc sulfide pigments when National Lead introduced a 30% rutile titanium calcium base pigment. This pigment had a hiding power of 57 ft^2/lb, in terms of Hallet's values (see Table IV), and was essentially as strong as pure zinc sulfide although considerably less expensive.

During the same 5-year period, the Kubelka-Munk theory was finally called to the attention of American paint makers. In 1935 Steele (31) reported on its application to paper in a U.S. paper trade journal. Judd (32, 33) of the National Bureau of Standards, in the same year, devised a graph of the Kubelka-Munk functions showing the relationship of contrast ratio to printing opacity in paper. He followed this 2 years later with a similar graph showing the interrelation of lightness (reflectance over black), reflectivity (reflectance at infinite film thickness), and contrast ratio--and proved that it was applicable to paint films, as well as to paper. The theory was far from achieving general acceptance, however, due to the resistance of those who continued to feel that complete coverage over a black and white substrate was "impractical."

Some investigators (34, 35) who had been using light- transmission measurements to determine the relative opacity of white pigments questioned the validity of the contrast ratio derived from reflectance measurements as an indication of hiding power. The resulting confusion among paint makers was increased by the controversial issue of wet versus dry hiding power (36).

The next 5 years (1940-5) covered most of World War II, and as might be expected, few changes took place in the relative volumes of zinc sulfide and titanium pigments produced. The producers of titanium pigments were unable to increase production because of wartime restrictions and supplied their regular customers on a quota basis, while lithopone continued to be used because of inadequate titanium dioxide stocks.

There were also few technological advances. True enamel-grade rutile titanium dioxide pigments were produced for the first time-- but wartime restrictions delayed their acceptance by the coating industry. Marchese and Zimmerman (37) developed a sophisticated drawdown method for the determination of hiding power using a black and white paper substrate. Spreading rate was plotted against 10 times the log of the contrast ratio, producing a straight-line plot from which hiding power might be read off at any desired contrast ratio.

Between 1945 and 1950, no longer protected by wartime shortages, lithopone in all forms rapidly disappeared from the market, leaving it to the complete domination of titanium pigments, a situation that has prevailed until the present day. The reformulation that was required to effect this changeover was beyond the capabilities of many paint makers and was principally handled by the technical service laboratories of the titanium pigment producers, who were expected to provide equal hiding power at lower cost. A last-minute effort was put forth by one zinc sulfide manufacturer who produced a higher strength pure zinc sulfide (by optimizing particle size) that was slightly superior to titanium dioxide in terms of hiding power per dollar. Apparently, however, the process was too costly for the product to sell at normal zinc oxide prices, and the effort failed. Kubelka (38) transformed the original Kubelka-Munk logarithmic equations to hyperbolic functions, making them more easily handled. In 1949 Jacobsen (39, 40) published two important papers showing the significance of pigment particle size and shape on optical properties and the correlation between the relative photochemical reactivity of a given titanium dioxide pigment and its rate of chalking in an exposed paint film.

The 5 years that followed (1950-5) saw the introduction of a higher strength 50% titanium composite pigment to combat the inroads of pure titanium dioxide and extender for the pigmentation of alkyd flat wall paints. Interest continued in hiding power test methods, and publications were authored by the New York Production Club (41) and by Switzer (42, 43) of ASTM. This progress was uniquely American, as evidenced by the fact that, in England, Arnold (44) was writing about the optical evaluation of pigments for tinting strength in refined linseed oil. Not until 1956 would another Englishman, Tough (45), concede that good agreement existed between contrast-ratio measurements and brushed-out hiding power charts.

Advances in Hiding Power Theory

The 1955-60 period was highly productive in opaque white pigment technology. On the basis of a background of hiding power studies that extended over a 10-year period, Mitton (46-57) and Stieg (58-72) of the Titanium Pigment Corp. (a subsidiary of the National Lead Co.) began a series of papers that were to continue over the next 25 years. During the 5-year period under consideration, Mitton (46-49) concerned himself with a statistical analysis of the precision of hiding power test methods. Stieg (58-61) concentrated upon a geometric analysis of the behavior of opaque white pigments in paint films, particularly in the presence of extenders, explaining for the first time the reasons for wet-to-dry hiding power losses and the differing effects of large- and small-particle extenders on dry hiding power in white paints. Using a rhombohedral model for the distribution of opaque white pigment particles, he derived a simple equation for each type of titanium pigment that would accurately predict its hiding power in any nonporous paint film (see Table VI).

Stieg also identified 9.25 PVC as the pigment concentration at which titanium dioxide should theoretically develop its maximum hiding power per pound, a deduction based upon his equations that was verified 10 years later by Bruehlman and Ross (73).

Chloride Process Titanium Pigment and Optimum Particle Size

During the late forties and early fifties, metallic titanium was being produced from titanium tetrachloride in the United States under heavy government subsidy. The availability of titanium tetrachloride in large quantities prompted Du Pont, as early as 1952, to experiment with the production of chloride-process titanium dioxide, using the old 1933 Kubelka patent that they had acquired along with the Krebs Pigment & Color Corp., but quality was poor. In 1957, however, the government subsidy was withdrawn, leaving Du Pont the owner of a large, newly completed titanium tetrachloride plant. Efforts to salvage this facility, plus the advanced age of the old Krebs titanium pigment plant, led to the opening in 1959 of the first commercial chloride-process titanium pigment plant. By this time quality was acceptable--as a matter of fact, it was superior in both strength and blue-white tone to the older sulfate-process pigments.

The improved quality of this first "chloride" pigment, as it came to be known, was due to a breakthrough in the pigment industry's understanding of optimum particle size that took place during the 1950-60 period. Reconsideration of the Mie theory suggested that its basis upon single-particle scattering might be unrealistic when applied to the pigment populations of paint films. A particle only half the diameter identified as "optimum" for light scattering by Mie theory might produce more hiding power than the larger particle simply because the same volume of pigment would then contain 8 times as many individual particles to do the scattering. Smaller particle size also resulted in bluer tone because of the relationship between particle diameter and the wavelength of light most efficiently scattered. These facts had become much more apparent with the development, by Penndorf (74), of a method for approximating the value of the total Mie scattering coefficient K.

The chloride process was able to take early advantage of this new knowledge, as the desired reduction in particle size required little more than the regulation of a few gas streams. The problem was somewhat more complicated for the sulfate process, being dependent not only upon nucleation in the precipitation from sulfuric acid solution but also upon times and temperatures in the calcination stage. Within 2 years, however, the new blue-toned, higher strength rutile pigments were available from all producers-- by then five in number.

Table V. Hiding Power Values

Pigment	Hiding Power (ft^2/lb)
28% pigment vol concn	
Lithopone	43
Titanox B	58
High-strength lithopone	67
Titanated lithopone	67
Titanox C	74
15% pigment vol concn	
Zinc sulfide	108
Titanium dioxide	180

Table VI. Formulas for the Hiding Powers of Titanium Pigments

Type of Pigment	Hiding Power per lb
Rutile titanium dioxide	370–410 (PVC)$^{1/3}$
Anatase titanium dioxide	298–330 (PVC)$^{1/3}$
50% titanium calcium base	222–213 (PVC)$^{1/3}$
30% titanium calcium base	127–106 (PVC)$^{1/3}$

While the end products of the sulfate and chloride processes were essentially identical, the chloride process possessed the advantages of requiring less manpower for plant operation and of producing fewer waste products due to its use of an essentially iron-free rutile ore, as contrasted to the ilmenite ore employed in the sulfate process, which contained far more iron than titanium, although the least expensive source of the latter element.

The years 1960-5 saw a proliferation of grades of titanium pigment, not only because of increased competition but also because of the rapid growth of the latex paint industry and the assumed need for special pigments designed for water-base coatings. Extra-fine particle size pigments were produced for rubber and plastics, fine particle size pigments for industrial and automotive coatings, larger particle size and high oil absorption pigments for alkyd and latex flat wall paints, and highly durable pigments for exterior finishes and paper laminates.

Pigmentation of Latex Paints

The development of high oil absorption pigments was initiated by the desire of paint makers to increase the dry hiding power of flat latex paints without extensive reformulation of their products. These had originally been formulated at lower PVCs than their solvent-base counterparts because of the lower binding power of the latex polymers, and severe wet-to-dry hiding power losses were a frequent source of customer complaints. The bag-for-bag replacement of a low oil absorption titanium pigment with a high oil absorption "latex grade" was an easy, if wasteful, solution to the problem.

High Dry Hiding of Paints

Stieg (62, 63) turned his attention to an explanation of the "high dry hiding effect," showing that it was directly related to film porosity and that film porosity might be numerically expressed by the equations

$$PI = 1 - \frac{CPVC(1 - PVC)}{PVC(1 - CPVC)}$$

$$LP = 1 - \frac{CPVC(1 - PVC)}{PVC(1 - CPVC)}_x$$

where PI = porosity index for solvent-type flats, LP = latex porosity, CPVC = critical PVC of pigmentation (from oil absorption), PVC = pigment volume concentration in dry film, and x = binding power index for the latex vehicle. He showed that similar high dry hiding effects might be produced at lower cost by increasing formula PVCs and by introducing fine particle size extenders.

Mitton (50-54) applied the Kubelka-Munk equations to the instrumental determination of the tinting strength of white pigments, eliminating the role of the out-moded linseed oil test method and greatly simplifying the computation of scattering coefficients and hiding power by a graphical approach.

By the end of this period the Cabot Corp. had become the second producer of chloride-process titanium dioxide, having purchased an existing titanium tetrachloride plant from National Distillers. It proved more difficult than expected, however, to apply the experience gained in the manufacture of pyrogenic silica--where ultimate particle size was of no great significance--to the production of titanium dioxide, and no commercially acceptable pigment for coatings was ever produced.

In the 5 years that followed (1965-70), three additional producers--American Potash, National Lead, and Pittsburgh Plate Glass--opened chloride-process plants. Sales of composite titanium calcium-base pigments continued to drop, as they had done over the previous 5 years in which Du Pont had ceased to produce them. This was due primarily to their adverse effect upon the stability of latex systems and the rapid replacement of alkyd and oleoresinous flat wall paints by flat latex paints. Two so-called "maximum-durability" pigments were introduced during this period--one by Du Pont and the other by National Lead. These pigments, although produced by different methods, both essentially eliminated the photochemical reactivity of the base titanium dioxide.

Light-Scattering Studies

Volz ($\underline{75}$) questioned the use of filter-type colorimeters rather than spectrophotometers for measuring Kubelka-Munk reflectance (since the latter was originally based upon light of a single wavelength), but his experimental work was not convincing, and Mitton ($\underline{55}$, $\underline{56}$) continued the refinement of his hiding power method in cooperation with ASTM, culminating in 1970 with his publication of a set of tables of Kubelka-Munk K/S values as a function of R_∞. Stieg ($\underline{64}$, $\underline{65}$) combined his porosity index, his hiding power equation for rutile titanium dioxide (see Table VI), and Fresnel reflectivity to successfully predict the relative wet and dry hiding powers of both solvent base and latex coatings.

In 1967 a new field of investigation was opened when Ross ($\underline{76}$) began work on a computer program for the prediction of hiding power using the Kubelka-Munk equations. This was followed in 1969 by Bruehlman and Ross' ($\underline{73}$) paper on the development of a method for determining hiding power from transmission measurements. Both were of value only for exceedingly dilute systems.

In 1970 Slepetys and Sullivan ($\underline{77}$) computed the theoretical light scattering of rutile titanium dioxide as a function of particle size and particle size distribution. All previous work had been done upon monodispersed systems.

Composite Titanium Pigment

In the years between 1970 and the present, possibly the most important single event related to white opaque pigments was the closing of the last plant to manufacture the composite calcium-base titanium pigments, leaving the so-called "pure" rutile titanium dioxide in undisputed possession of the coatings market. The replacement of calcium-base pigments with titanium dioxide and extender was a difficult and expensive task for the coatings industry, since one of the reasons for the former's disappearance had been an unreasonably low pricing policy, and no suitable extender was available at a low enough price to make the blend competitive. Stieg (66) pointed out that the primary advantage of the composite pigment had been a "built-in dilution" that had to be replaced with very fine particle size extender if equivalent film properties and hiding power were to be obtained. Such extenders were both expensive and in short supply.

Microvoids

It had been an important period for several other reasons, however. In 1971 Burrell (78) stirred up considerable interest in the possible use of air bubbles (microvoids) to replace opaque white pigments. Ross (79) computed the theoretical light-scattering power of spherical particles of rutile and spherical air bubbles, using Mie theory, and concluded that foams with thin parallel walls would be required to achieve sufficiently high levels of scattering power to be competitive with titanium dioxide. He pointed out, however, that the computed scattering coefficient for titanium dioxide was 40% above that found in practical coatings. In 1973 the ACS Division of Organic Coatings and Plastic Chemistry conducted a symposium on micrcvoid coatings (80). Allen (81) concluded that the prediction of the optical properties of microvoid coatings was impossible by either Kubelka-Munk or Mie theory because of multiple scattering effects that could not be accounted for. Rosenthal and McBane (82) referred to an unexpected "synergistic response of white pigments and air-filled microvoids." Stieg (67, 68), however, showed that the dry hiding power of air-containing paint films of known composition could be predicted with an accuracy equivalent to the normal hiding power test method. He also devised a nomagraph from which hiding power in square feet per pound of titanium dioxide might be obtained, given a knowledge of formula PVC, effective PVC of titanium dioxide content, and the CPVC of the paint pigmentation.
    Work continued on attempts to predict white hiding power using computers programmed with the Kubelka-Munk functions, despite Allen's earlier conclusions. Billmeyer and Abrams (83) and Billmeyer and Phillips (84) had some limited success at very low PVCs but concluded that "when pigment volume concentration is so high that the pigment particles are at distances apart comparable to their own sizes, the mixing laws...no longer predict the proper $\underline{K}$ and $\underline{S}$ for use in the Kubelka-Munk theory" and, furthermore, "...this is considered too complex for any theoretical approach, although semi-empirical treatments (Stieg's) are quite useful for prediction in this region."
    Cairns (85) publicized a computer program HIDE, which purportedly predicted the hiding power of combinations of both white and colored pigments but which also achieved partial success only at very low pigment volume concentrations.
    Mudgett and Richards (86) returned to the argument of Volz (75), claiming that only one wavelength at a time must be used if the Kubelka-Munk equations are to apply and that the 45° geometry common to most photometers is unsuitable.
    In 1974, Tunstall and Hird (87) published a paper entitled "Effect of Particle Crowding on Scattering Power of Titanium Pigments" in which they agreed with Stieg's 1959 interpretation of

the interference effect. In an apparent effort to avoid Stieg's assumed rhombohedral distribution of particles, they introduced the variable CPVC in place of the constant 0.74. Since particles of pigment are actually separated by an adsorbed binder layer at the CPVC, this innovation somewhat reduces the accuracy of their final equation.

Continued interest in the potential of microvoids to replace increasingly expensive titanium dioxide, particularly on the part of Pittsburgh Plate Glass who were interested in licensing a patented process called Pittment, resulted in a project sponsored by the Paint Research Institute "to demonstrate the synergism between air-filled microvoids and titanium dioxide." In 1975, Kerker et al. (88) published the results of this project, purporting to describe the light-scattering ability of concentric spheres of air and titanium dioxide embedded in a polymeric shell, based upon complex computer models. Due to the use of a grossly inaccurate refractive index for rutile titanium dioxide, however, the conclusions reached by this paper were of questionable value.

During this same period, practical work with microvoids was being done in Australia that resulted in the commercial production and application of polymeric "pigments" by Balm Paint, Ltd. Publications by Kershaw (89) and Lubbock (90) appeared in Australian technical journals, and several U.S. Patents (3 822 224; 3 839 253; 3 933 579) were granted for the production of solid, pigmented solid, and pigmented vesiculated polymeric beads--the latter employing the microvoid concept.

Single-Pigment Concept

The high price of titanium dioxide was also responsible for the start of a movement to replace the high oil absorption, low TiO_2 content latex-grade pigments with blends of high TiO_2 content enamel-grade pigments with fine particle size spacing extenders. Initiated by Stieg (69, 70) in 1976, the single-grade concept was taken up by Tatman (91) and by Ritter and Schelong (92) in 1978. It was also supported by many manufacturers of high oil absorption extenders--particularly calcined clays--who were anxious to have their products employed as "spacers" for titanium dioxide. Generally, paint quality was adversely affected because of the readiness of paint makers to accept the high dry hiding produced by increased film porosity as evidence of a spacing effect.

In 1978 a slurried microvoid pigment under the trade name of Spindrift was introduced in this country. Described by Hislop and McGinley (93) as a pigmented vesiculated polymeric bead, it was relatively large in particle size (12 μm) and hence limited to the pigmentation of latex flat wall paints. The effect of the microvoid in enhancing the optical efficiency of the contained titanium dioxide was partially offset by the effect of particle packing on any free titanium dioxide in the paint system due to this same large particle size.

Effective PVC and Dilution Efficiency

In 1981 Stieg (71) published a test procedure for quantifying the ability of an extender pigment to serve as a "spacer" for titanium dioxide in terms of its dilution efficiency (E_d). This work was based upon an updating of his equation for the effective PVC to be employed in his hiding power equation (Table VI):

$$PVC_{eff} = \frac{\text{vol of TiO}_2}{\text{vol of TiO}_2 + (\text{extender} \times E_d) + \text{binder}}$$

It was shown that dilution efficiencies were not as high as previously assumed for some calcined clays and also that some common extenders possessed dilution efficiencies as high as 30%. As an outgrowth of this work, the first extender to have been deliberately produced as a spacer for titanium dioxide was introduced in the form of an ultrafine, low oil absorption natural calcium carbonate having a dilution efficiency of 99%.

Despite the earlier partial success of the single-pigment concept (69, 70, 91, 92), the market was still saturated with over 90 grades of wet and dry titanium pigment, in a large measure due to the profitability of the highly treated grades to the pigment producers, although some producers were grudgingly admitting that the most highly treated grades (about 80% titanium dioxide) were perhaps less economical to use than originally claimed. In 1981 Stieg (72) demonstrated, however, that a single high-durability grade of rutile pigment might be successfully used to replace all other grades in any type of paint product.

Also in 1981, N L Industries (the new name of the old National Lead Co.), who had closed their chloride plant in this country, announced the development of a continuous-digestion process that would greatly reduce the waste-disposal problem that had plagued all sulfate-process plants. This represented a new stage in the competition between the sulfate and chloride processes, but came too late to greatly affect the outcome. By 1981 Du Pont, completely committed to the chloride process for 20 years, so dominated the titanium pigment industry that it had already been unsuccessfully prosecuted for monopolistic activities, and N L Industries closed its last titanium plant in this country in 1983, importing pigment from Canadian and European subsidiaries. Over a period of little more than 55 years, the coating industry has grown from a commercialized art form practiced by a guildlike group of paint makers, and dominated by reactive white lead and zinc pigments, to a highly technical industry. Its growth during this period closely paralleled the growth of the titanium pigment industry, starting in 1924 with a single anatase composite pigment with extremely poor chalk resistance and little more hiding power than white lead itself and developing by 1970 into as many as 35 grades and subgrades dominated by "pure" rutile titanium dioxides and modified to provide a wide range of oil absorptions, durability, or gloss-producing capability. Hiding power had become completely predictable, at a level almost 20 times that of white lead—by then employed only in a dwindling-few exterior house paints and primers.

Much of the building of coating technology had been done by the pigment makers, but once established as the producers of a stable commodity, titanium pigment suppliers discontinued both application research and technical services to the industry, permitting the mantle of innovation to fall upon the shoulders of the polymer makers and a few extender producers.

Plastic Pigments

Today the consequences are evident. Spurred by the price increases secured by titanium dioxide's apparent domination of the opaque white pigment field, solid and microvoid polymeric "pigments" and highly efficient spacing extenders are being offered—all designed to reduce the industry's consumption of titanium dioxide by increasing its efficiency.

While it is unlikely that the future will see any significant improvements over present-day titanium pigments, it is also unlikely that any new white opaque pigment will ever take its place. Its combination of high refractive index and high specific gravity makes it possible to achieve adequate hiding power in a thin organic coating without exceeding the pigment volumes that limit the gloss, flow, and application characteristics of industrial coatings.

The air interface provided by a microvoid is capable, theoretically, of almost doubling the optical efficiency of the titanium dioxide with which it is used, but the necessary volume of the microvoid multiplies the overall concentration of non-film-forming material in the coating by as much as four times. An ultrafine particle size polymeric bead can perform the function of a titanium dioxide "spacer," but there is little application for such in very low PVC finishes. These considerations essentially restrict the application of plastic pigments and microvoids to more highly pigmented trade sales coatings—and more particularly to flat latex paints, because of their solubility characteristics.

In trade sale coatings, plastic pigments and microvoids will be directly competitive with ultrafine particle size natural calcium carbonates and "thermooptic" calcined clays. It is probable, however, that the future will see greater rather than less emphasis placed upon the fire retardancy of architectural coatings, and the flammability of the plastic products may prove to be a deterrent to their widespread use, as has been true of plastic wall panels.

Optimum Solids

The average amount of titanium dioxide used per gallon of paint has fallen steadily over the past 10–15 years, and it is probable that this trend will continue, not only because of the more widespread use of efficient spacing extenders and microvoids but also because of a basic economic fact pointed out almost 25 years ago by Stieg (58). As the price of titanium dioxide becomes higher, relative to the cost of the film-forming binder, it becomes more economical to formulate paints at higher solids, using less titanium dioxide per gallon. Microcomputers, now available even in battery-powered pocket models, make it possible for the paint formulator to accurately pinpoint the optimum level of solids for any given formulation—the process almost invariably resulting in lower titanium dioxide levels and lower cost, with no reduction in white hiding power. Computer programs (software) are currently being offered in cassette form for paint laboratory use (94).

Literature Cited

1. Gehant, J. C. Paint Oil Chem. Rev. 1932, 94, 8.
2. Fresnel--see any textbook on optics.
3. Maxwell, J. C. Phil. Trans. 1865, 155, 459.
4. Mie, G. Ann. Phys. 1908, 25, 377.
5. Zhukova, A.; Sovalova, A. Lakokrasochnuyu Ind., Za. 1934, 2, 11.
6. Knudson, M. Ann. Phys. 1909, 28, 75.
7. Thompson, G. W. ASTM Proc. 1910, 10, 610.
8. Oden, S. Kolloid-Ztsch. 1916, 18, 33.
9. Green, H. J. Franklin Inst. 1921, 192, 637.
10. Martin, W. H. Proc. Roy. Soc. Can. 1913, 1(III), 219.
11. Hallett, R. L. ASTM Proc. 1920, 20(Part II), 426.
12. Stutz, G. F. A. J. Franklin Inst. 1930, 210, 67.
13. Calbeck, J. H.; Harner, H. R. Ind. Eng. Chem. 1926, 19, 58.
14. Stutz, G. F. A.; Pfund, A. H. Bull., NJ Zinc Co. 1926.
15. Rhodes, F. H.; Fonda, J. S. Ind. Eng. Chem. 1926, 18, 130.
16. Hallett, R. L. ASTM Proc. 1926, 26(Part II), 538.
17. Bruce, H. D. U.S. Bur. Std., Tech. Pap. No. 306, 1926.
18. Haslam, G. S. Ind. Eng. Chem. Anal. Ed. 1930, 2, 69.
19. Brier, J. C.; Wagner, A. M. Ind. Eng. Chem. 1928, 20, 758.
20. Gamble, D. L.; Pfund, A. H. Ind. Eng. Chem. Anal. Ed. 1930, 2, 63.
21. Pfund, A. H. ASTM Proc. 1930, 30(Part II), 882.
22. Stutz, G. F. A.; Haslam, G. S. ASTM Proc. 1930, 30(Part II), 884.
23. Kubelka, P.; Munk, F. Z. Tech. Phys. 1931, 12, 59.
24. Knoll, M.; Ruska, E. Z. Tech. Phys. 1931, 12, 389.
25. Sawyer, R. H. Ind. Chem. Eng. Anal. Ed. 1934, 6, 113.
26. Jacobsen, A. E.; Reynolds, C. E. Ind. Eng. Chem. Anal. Ed. 1934, 6, 393.
27. McGregor, J. R. Paint Oil Chem. Rev. 1938, 100, 37.
28. Wagner, H. Paint Manufact. 1934, 4, 5.
29. Wait, D.; Weber, I. E. J. Oil Colour Chem. Assoc. 1934, 17, 257.
30. Renz, C. Helv. Chim. Acta 1921, 4, 961.
31. Steele, F. A. Paper Trade J. 1935, 100(12), 37.
32. Judd, D. B.; Paper Trade J. 1935, 101(5), 40.
33. Judd, D. B., et al. J. Res. Natl. Bureau Std. 1937, 19, 287.
34. Hanstock, R. F. J. Oil Colour Chem. Assoc. 1937, 20, 5.
35. Mills, H. J. Oil Colour Chem. Assoc. 1940, 23, 245.
36. Baltimore Prod. Club Sci. Sect. Circ. 1938, 568, 261.
37. Marchese, V. J.; Zimmerman, E. K. Natl. Paint Bull. 1944, 8(6), 12.
38. Kubelka, P. J. Opt. Soc. Am. 1948, 38, 448.
39. Jacobsen, A. E. Ind. Eng. Chem. 1949, 41, 523.
40. Jacobsen, A. E. Can. Chem. Proc. Ind. 1949, Feb.
41. N.Y. Prod. Club Off. Dig. 1951, 23, 758.
42. Switzer, M. H. ASTM Bull. 1951, No. 175, 68.
43. Switzer, M. H. ASTM Bull. 1954, No. 197, 60.
44. Arnold, M. H. M. J. Oil Colour Chem. Assoc. 1954, 37, 513.
45. Tough, D. J. Oil Colour Chem. Assoc. 1956, 39, 169.
46. Mitton, P. B. Off. Dig. 1957, 29, 251.

47. Mitton, P. B. Off. Dig. 1958, 30, 156.
48. Mitton, P. B.; White, L. S. Off. Dig. 1958, 30, 1259.
49. Mitton, P. B. Off. Dig. 1959, 31, 318.
50. Mitton, P. B.; Vejnoska, L. J. Off. Dig. 1961, 33, 1264.
51. Mitton, P. B., et al. Off. Dig. 1962, 34, 73.
52. Mitton, P. B.; Jacobsen, A. E. Off. Dig. 1962, 34, 704.
53. Mitton, P. B.; Jacobsen, A. E. Off. Dig. 1963, 35, 871.
54. Mitton, P. B. Off. Dig. 1965, 37, 43.
55. Mitton, P. B.; Madi, A. D. J. Paint Technol. 1966, 38(503), 717.
56. Mitton, P. B., et al. J. Paint Technol. 1967, 39,(512), 536.
57. Mitton, P. B. J. Paint Technol. 1970, 42(542), 159.
58. Stieg, F. B. Off. Dig. 1957, 29, 439.
59. Stieg, F. B. Off. Dig. 1959, 31, 52.
60. Stieg, F. B. Off. Dig. 1959, 31, 736.
61. Stieg, F. B. Paint Ind. 1960, 75, 8.
62. Stieg, F. B. Off. Dig. 1961, 33, 792.
63. Stieg, F. B. Off. Dig. 1962, 34, 1065.
64. Stieg, F. B. J. Paint Technol. 1967, 39(515), 701.
65. Stieg, F. B. J. Oil Colour Chem. Assoc. 1970, 53, 469.
66. Stieg, F. B. J. Paint Technol. 1972, 44(565), 63.
67. Stieg, F. B. J. Paint Technol. 1973, 45(576), 76.
68. Stieg, F. B. Ind. Eng. Chem. Prod. Dev. 1974, 13, 41.
69. Stieg, F. B. J. Coatings Technol. 1976, 48(612), 51.
70. Stieg, F. B. Am. Paint J. 1976, 61(6), 14.
71. Stieg, F. B. J. Coatings Technol. 1981, 53(680), 75.
72. Stieg, F. B. J. Coatings Technol. 1981, 53(682), 65.
73. Bruehlman, R. J.; Ross, W. D. J. Paint Technol. 1969, 41(538), 584.
74. Penndorf, R. B. J. Phys. Chem. 1938, 62, 1537.
75. Volz, H. G. Farbe Lack 1965, 71, 725.
76. Ross, W. D. J. Paint Technol. 1967, 39(511), 515.
77. Slepetys, R. A.; Sullivan, W. F. Ind. Chem. Eng. Prod. Res. Dev. 1970, 9, 266.
78. Burrell, H. J. Paint Technol. 1971, 43(559), 48.
79. Ross, W. D. J. Paint Technol. 1971, 43(563), 49.
80. ACS, papers presented at Chicago Meeting 1973, 33, 258.
81. Allen, E. J. Paint Technol. 1973, 45(584), 65.
82. Rosenthal, W. S.; McBane, B. N. J. Paint Technol. 1973, 45(584), 73.
83. Billmeyer, F. W., Jr.; Abrams, R. L. J. Paint Technol. 1973, 45(579), 23.
84. Billmeyer, F. W., Jr.; Phillips, D. G. J. Paint Technol. 1974, 46(592), 36.
85. Cairns, E. L. J. Paint Technol. 1972, 44(572), 76.
86. Mudgett, P. S.; Richards, L. W. J. Paint Technol. 1973, 45(586), 43.
87. Tunstall, D. F.; Hird, M. J. J. Paint Technol. 1974, 46(588), 33.
88. Kerker, M.; Cooke, D. D.; Ross, W. D. J. Coat. Technol. 1975, 47(603), 33.
89. Kershaw, R. W. Aust. OCCA, Proc. News 1971, 8(8), 4.
90. Lubbock, F. J. Aust. OCCA, Proc. News 1972, 11(5), 12.
91. Tatman, C. C. J. Coat. Technol. 1978, 50(646), 64.
92. Ritter, H. S.; Schelong, M. J. Coat. Technol. 1978, 50(640), 69.
93. Hislop, R. W.; McGinley, P. L. J. Coatings Technol. 1979, 50(642), 69.
94. PigmenTech Consulting, Jekyll Island, GA 31520.

Color Pigments

LAWRENCE R. LERNER[1] and MAX SALTZMAN[2]

[1]Dyes and Pigments Division, Mobay Chemical Corporation, Hawthorne, NJ 07507
[2]Institute of Geophysics and Planetary Physics, University of California, Los Angeles, CA 90024

Inorganic Pigments
  Chrome Pigments
  Iron Oxide Pigments
  Cadmium/Mercury Sulfides
  Colored Titanium Oxides
Organic Pigments
  Phthalocyanines
  Quinacridones
  Vat Pigments
  Isoindoline Pigments
  Azo Pigments
  Miscellaneous

Due to the many developments that have occurred in the pigment field since the 1920s, proper treatment of all important topics is simply not possible in a single chapter. We, therefore, will emphasize the major developments since the end of World War II. Within this period we will cover inorganic pigments somewhat more briefly than the organic pigments due to the greater complexity of the latter field, combined with space limitations and the greater experience of the authors in the organic pigment area.

As only an overview can be provided, the reader is encouraged to seek more complete and detailed information in the papers of Gaertner (1), Heinle (2), Hopmeier (3), Inman (4), Kehrer (5), and Lenoir (6) and books by Zollinger (7) and Lubs (8). We are fortunate that in 1973 an outstanding treatise on the entire field of pigments was published, entitled "Pigment Handbook" by Patton (9), which in its three volumes provides a wealth of information to anyone interested in pigments for any field of application.

The development of new coating resins, plastics, and synthetic fibers, as well as new methods of utilization of all of these materials, has created an unprecedented demand for materials to impart color to these otherwise colorless substances. Not only is it necessary to color these plastics and resins but also it is important that this coloration stand up for some time in spite of exposure to a hostile environment. Many new resins are difficult to pigment and require both high-energy input and fairly high temperatures. Any colorant for such a system must be resistant to both the physical and chemical stresses encountered in the dispersion conditions and in the final application methods.

To begin, let us look at the requirements for a color pigment to satisfy the demands of modern industry. The following is a list of desirable pigment properties: high chroma; fastness to light, weather, migration, and solvents; resistance to processing chemicals and heat.

The first requirement, that of high chroma, is necessary to give bright, clean self colors, both in full tone and in dilution with white pigments or a mixture with nearly colorless plastics. This high chromaticity also permits blending with other colorants to obtain mixtures without substantial loss of cleanliness. Fastness to light, both in full shade and when diluted with white, permits the use of a pigment under conditions of long times of exposure so that refinishing will not be necessary for the effective life of the object being colored or in the case of interior house paint (or even exterior house paint) until the homemaker gets tired of the color. Fastness to weathering is important for automotive, architectural, and engineering coatings, which are exposed to all kinds of weather, including the industrial environment. The multitude of colors obtainable in automobiles today is witness to the success of the pigment industry in supplying the necessary color pigments to provide any color that is desired--including black! Fastness to migration permits the use of a pigment in a plastic film or a coating that is to be in contact with other organic materials, including the seat of one's synthetic fiber clothing or with bare skin. Fastness to solvents is necessary for all paint systems where one color is to be sprayed over another, and it is also important for the stability of any paint formulation containing solvents. Fastness to heat is needed for those materials to be used in today's high-bake finishes as well as those to be incorporated into plastics that are processed at high temperatures such as polypropylene. Resistance to chemicals is almost self-evident in that all pigments are chemical entities that are not truly inert; therefore, they must react in the systems in which they are employed.

For the sake of brevity, we have referred to the fastness of a pigment. It should be understood that the fastness of the pigment powder by itself has no real significance. The only fastness properties that are of any value are those of the complete pigmented system. By this is meant the pigment incorporated in the coating or resin in which it is to be employed under the exact conditions of incorporation and under the exact condition of use. This makes it difficult to pass global judgments as to the usefulness of a pigment, but it has been shown in studies such as those of Smith and Stead (10), Vesce (11, 12), and Levison (13) that this concept of fastness is the only valid one.

Inorganic Pigments

The inorganic pigments used today include many of those known since earliest times, such as iron oxides and mercuric sulfide or

vermillion. The inorganic pigments are the "workhorse" pigments of
the industry and are the basis for the majority of the colored
finishes today. Improvements on these older types accounts for many
of the inorganic pigments in use today.

Chrome Pigments. Chrome pigments based on lead chromate and mixed
lead molybdenum salts have been the basis for yellow and orange
pigments since their discovery in the 19th century.

The lead chromate yellows are based on solid solutions of lead
chromate ($PbCrO_4$) and lead sulfate ($PbSO_4$). The higher the amount
of $PbSO_4$, the lighter the yellow.

The so-called molybdate oranges are solid solutions containing
perhaps 75% lead chromate ($PbCrO_4$) with the rest made up with lead
molybdate ($PbMoO_4$) and lead sulfate ($PbSO_4$). The principal defect
of chrome pigments is their tendency to darken on exposure.
Developments in the 1940s led to the introduction of "predarkened"
pigments in the early 1950s. Treatment of the precipitated pigments
with foreign metals such as antimony, bismuth, or various rare
earths lead to duller colors but improved light fastness. It was
later found that encapsulation of the pigment crystallite with a
"skin" of silica provides a protective barrier that aids chemical
resistance (particularly against sulfur dioxide) and heat
resistance. Various silica coated chrome pigments are available
today.

The most severe blow to the continued use of chrome pigments has
come from the proponents of safety in banning the use of even
insoluble lead pigments in any paint that might be approached by
children. However, even without legislative requirements, many
automotive and other industrial finishes are being formulated
without chrome colors.

Iron Oxide Pigments. The discovery of iron oxide pigments is lost
in antiquity. The cave paintings of early man were made with earth
colors composed of iron oxides. Even today, natural iron oxides are
still in use. The yellows such as the yellow ochers and siennas are
hydrated ferric oxides ($Fe_2O_3 \cdot H_2O$). The nonhydrated ferric oxides
(Fe_2O_3) comprise the reds and browns such as the red ochers and
brown umbers. The blacks are mixed ferrous and ferric oxides (Fe_3O_4
or more precisely $FeO \cdot Fe_2O_3$).

During this century, processes were developed to produce
synthetic iron oxides covering the entire color range of the natural
colors. These synthetic iron oxides had the advantage of greater
purity and better particle size distribution, leading to cleaner,
stronger pigments. In recent years, demand for low-cost transparent
pigments has led to very fine particle sized iron oxides that,
though chemically identical with the previously developed opaque
types, have the high transparency needed, particularly in metallic
automotive finishes.

Cadmium/Mercury Sulfides. One of the major accomplishments of the
past 25 years has been the introduction of a series of red and
maroon pigments based on mercury and cadmium sulfides ($CdS \cdot xHgS$), to
replace the well-known cadmium sulfoselenides. The work was
prompted by the increasing shortage of selenium and its ever
increasing price. In their search for replacement for the standard

cadmium pigments, workers at Imperial Color and Chemical (now a division of Ciba-Geigy) successfully substituted mercury for selenium to produce the mercadium pigments. While they are not as heat resistant as the cadmium sulfoselenides, they are quite satisfactory for many uses. The hue range varies from orange to maroon as the weight percent of CdS decreases from approximately 90% to 74%.

Colored Titanium Oxides. Modification of titanium dioxide in which nickel and antimony are incorporated into the crystal lattice has given a series of weak yellow pigments ranging in hue from a clean almost lemon yellow to a buff. These are quite durable and have the additional advantage of being self-chalking so that on weathering, the original color is not masked by a white chalk which can be the case when the color is based on straight titanium white modified with a tint of organic colorant. Other colored oxides such as those based on cobalt and aluminum have been used to a limited extent, especially for high-temperature applications. An additional interesting use of TiO_2 was developed in the 1960s whereby mica is coated with a controlled thickness of TiO_2 to produce colored pigments that derive their color from light interference phenomena. Introduction of small amounts of a colored material such as Fe_2O_3 allows color both from interference and absorption. These pigments find use in both cosmetic and industrial markets because of their metallic iridescent luster.

Organic Pigments

Phthalocyanines. The modern era of synthetic organic pigments is generally considered to have begun with the discovery and subsequent marketing of copper phthalocyanine pigments in 1935 by Imperial Chemical Industries. Even today, the pigmentary form of this material is considered the standard by which all other pigments are judged. Scheme I shows the synthesis of copper phthalocyanine (14) from phthalic anhydride and urea in the presence of copper. Most of the commercially available pigments of the phthalocyanine class are made from the copper compound or its halogenated derivatives.

The most widely used are the two crystal forms of the same chemical composition; the reddish blue alpha phase and the greenish blue beta phase of copper phthalocyanine, Pigment Blue 15 (I). Metal-free phthalocyanine (II), Pigment Blue 16 (15), is a greenish blue pigment that can be produced by first preparing a phthalocyanine containing a labile metal such as magnesium or calcium and then removing the metal by acid treatment. Although the chemical syntheses of phthalocyanines is relatively simple, the preparation of the pigmentary form requires much ingenuity. It is probable that more work has been done on the "finishing" or "conditioning" of phthalocyanines than on the synthetic methods. Physical and chemical treatment to control particle size and shape, crystal habit, and rheological properties of the pigments includes acid-pasting, salt-grinding, and solvent treatments. The admixture of small amounts of substituted phthalocyanines has also been used to improve the working properties such as resistance to flocculation and stability to solvents.

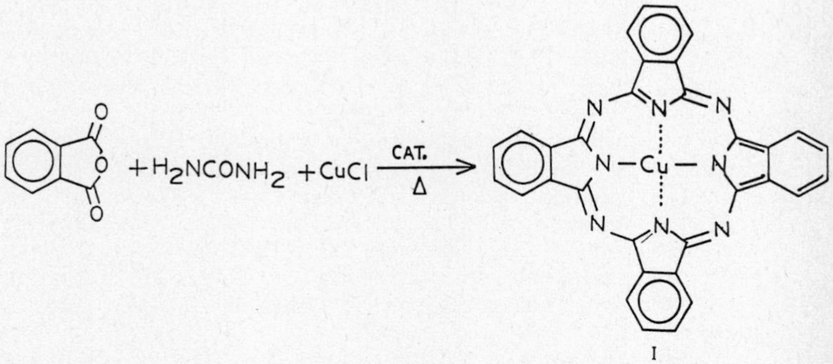

Scheme I. Copper phthalocyanine synthesis.

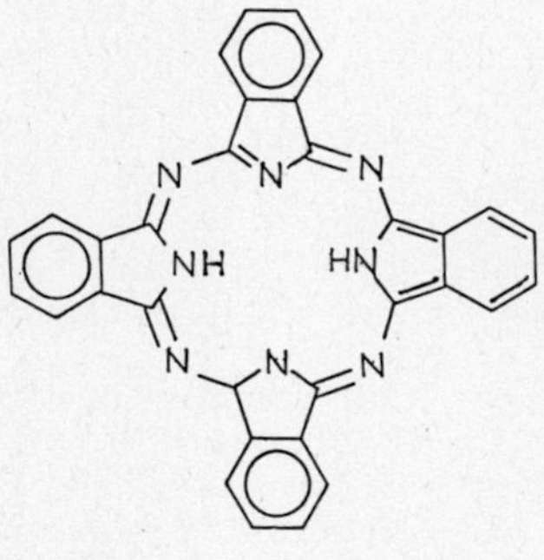

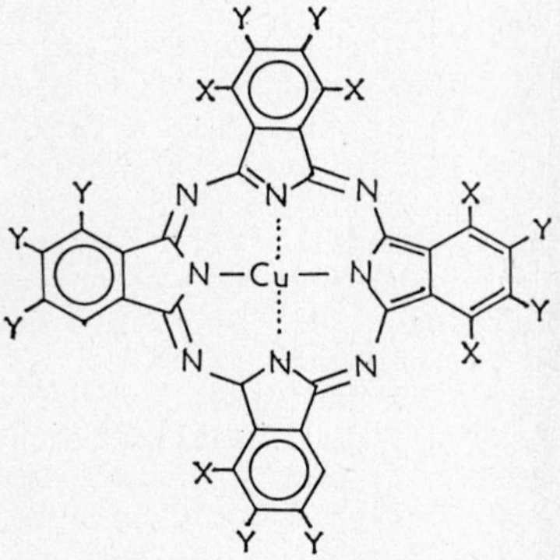

The only other major class of pigmentary phthalocyanines, phthalocyanine greens, are made by the halogenation of the copper compound with either chlorine, bromine, or a mixture of both. Approximately 14 positions are occupied by the halogen. In the case of Pigment Green 7 (III), they are all chlorine, which gives a blue shade of green. Increasing substitution of bromine for chlorine to a total of approximately eight or nine bromine atoms gives increasingly yellow shades of green of which the most yellow is Pigment Green 36 (IV). We pass on from these pigments not because they are unimportant but because they are so superior to pigments of almost any other chemical class that we must search through a very large number of other chemical types to find colors whose pigment properties equal those of the blue and green phthalocyanines.

Quinacridones. Much effort was expended to carry the hue of phthalocyanines in the other direction to produce reds and violets with outstanding fastness properties, but these efforts proved to be unsuccessful. However, in the 1950s a number of developments did take place with other chromophores in this color range to produce new lines of outstanding pigments. The development and commercialization of quinacridone pigments by Du Pont was one of these developments.

Though the synthesis of quinacridone was first reported in 1935 (16, 17), it was originally evaluated as a potential vat dye and found to be of no interest. However, beginning in the late 1950s, patents issued to Du Pont (18) described processes for making three crystal forms of linear trans-quinacridone in pigmentary form. These forms are the red, alpha (α) and gamma (γ), and the violet, beta (β). Scheme II outlines the process described in the Du Pont patents. The process actually starts with dianilinodihydroterephthalic ester (VII), which can be made by the condensation of succinic acid ester to cyclic succinyl succinate (VI) (Step 1), followed by condensation with aniline (Step 2). This intermediate (VII) is then ring closed at high temperature to form dihydroquinacridone (VIII) (Step 3), which is oxidized under controlled conditions to a crude alpha, beta, or gamma crystal form of quinacridone (IX) (Step 4). These crudes are then conditioned to the finished pigment.

Scheme III describes another route that was patented by Harmon Colors (19-21) also starting with dihydroterephthalic ester (VII), but the ester is oxidized and hydrolyzed first to (X) and then ring closed in polyphosphoric acid to IX. Controlled drowning of the acid melt onto various solvents gives the finished pigment directly.

A third route developed by Sandoz (22) shown in Scheme IV allows one to make unsymmetrically substituted quinacridones by stepwise condensation (Steps 1 and 2) of the two amines carried out under carefully controlled conditions (23).

The violet quinacridone (beta form) has found much use in the past in admixture with molybdate orange to make bright "fire engine" reds. With many paint manufacturers removing this lead-based orange from use, a large particle sized opaque -quinacridone red is now being substituted in many cases for the violet, orange combination in industrial finishes. Stronger, more transparent forms of both the alpha and gamma red forms are sold for use in paints and

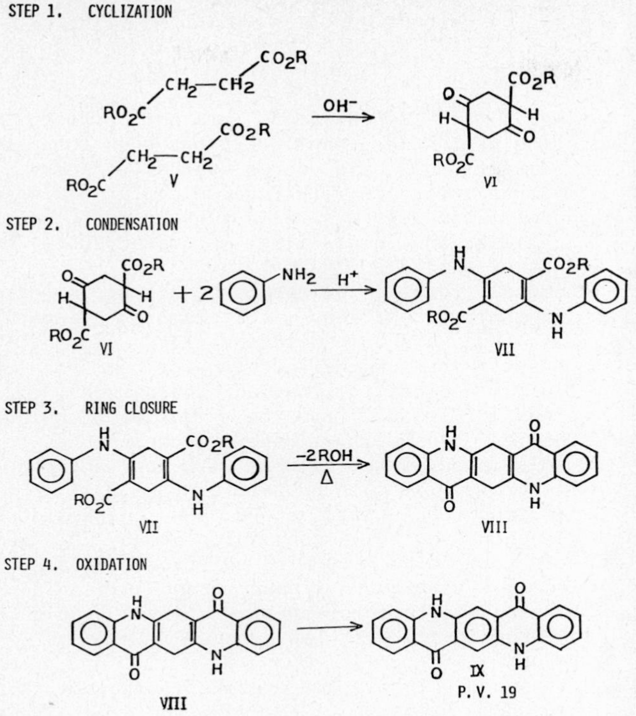
STEP 1. CYCLIZATION
OH⁻
V
VI
STEP 2. CONDENSATION
H⁺
VI
VII
STEP 3. RING CLOSURE
-2ROH
Δ
VII
VIII
STEP 4. OXIDATION
VIII
IX
P. V. 19

Scheme II. Quinacridone via dihydroquinacridone.

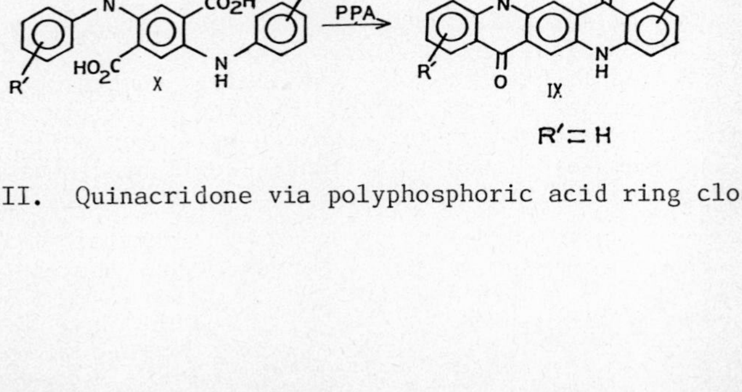
OXIDATION - HYDROLYSIS
VII
X
RING CLOSURE
PPA
X
IX
R′ = H

Scheme III. Quinacridone via polyphosphoric acid ring closure.

plastics. Other crystal forms of quinacridone have been reported but seem to be either special modifications, more crystalline forms, or mixes of the alpha, beta, and gamma types.

Two substituted quinacridones have been introduced commercially and find use: 2,9-dimethylquinacridone, Pigment Red 122 (XI), and 2,9-dichloroquinacridone, Pigment Red 202 (XII). Both are magenta in shade.

Interestingly, the substituents are not ortho to either the carbonyl oxygens or the NH groups. It has been found that ortho-substituted quinacridones have poorer weatherfastness. Apparently ortho substituents interfere with intermolecular hydrogen bonding. It is this hydrogen bonding (Figure 1) that seems to account for the outstanding stability and insolubility of quinacridone, which has a molecular weight of only 312 (24-26).

Other chemical forms of quinacridone are known but are not as outstanding in their properties. For example, linear cis-quina-cridone (XIII) (27) and epindolidione (XIV) (28, 29) are both yellows with insufficient weatherfastness for commercial use.

A third yellow derivative, quinacridonequinone (XV), made as shown in Scheme V or directly from the oxidation of quinacridone, is also not weatherfast by itself. However, chemists at Du Pont found that solid solutions of the quinone with quinacridone leads to pigments with improved fastness properties (30-32).

Today a maroon containing about 60% quinacridone and 40% quinacridonequinone and a gold containing about 75% quinacridone-quinone are sold commercially. Solid solutions can also be prepared between variously substituted linear trans-quinacridones, and a scarlet containing a mixture of approximately 60% quinacridone and 40% 4,11-dichloroquinacridone (an orange) is also on the market.

Vat Pigments. Another chemical class of compounds that has been exploited to produce pigments in the same hue area as quinacridones are the thioindigos. Thioindigos are members of a group of compounds referred to as "vat pigments," which are pigments based on textile vat dyes. These dyes have been known since the beginning of the 20th century, but they were not greatly exploited until the 1950s when Harmon Colors (now part of Mobay Chemical Corp.), under the direction of Vincent Vesce, systematically investigated and introduced an expanded range of vat pigments to the market place (11, 12). A number of thioindigo derivatives were sold initially, but as weatherfastness requirements became more stringent, we are left today with only the tetrachlorothioindigo Pigment Red 88 (XVI) (Scheme VI) as a suitable colorant with color and fastness properties similar to those of quinacridone violet (33). Interestingly, as in the case of quinacridone, this thioindigo did not make a good vat dye. However, the substituents ortho to the carboxyl and sulfur groups are essential for its superior insolubility and lightfastness. Obviously, no hydrogen bonding is possible in the case of thioindigo, and the chlorines may stabilize the molecule either electronically or sterically. Another interesting example of such stabilization will be discussed later in reference to isoindolinone pigments.

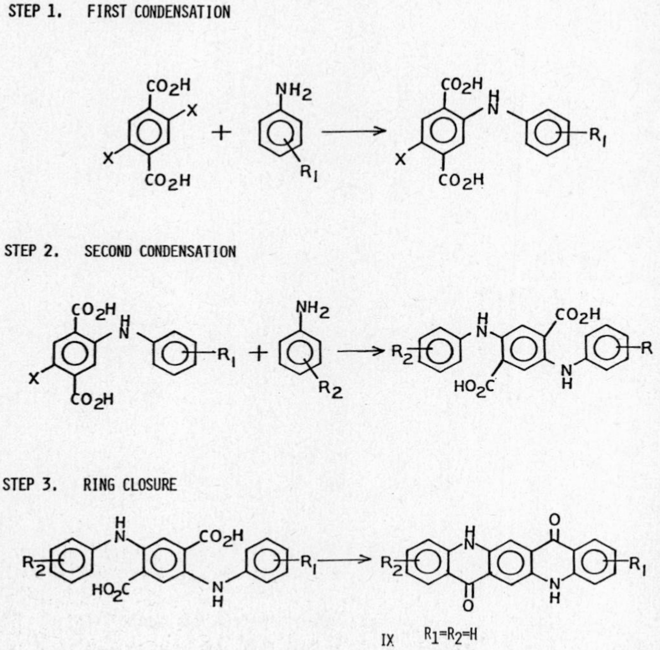

STEP 1.   FIRST CONDENSATION

STEP 2.   SECOND CONDENSATION

STEP 3.   RING CLOSURE

IX $R_1 = R_2 = H$

Scheme IV.   Quinacridone via stepwise condensation.

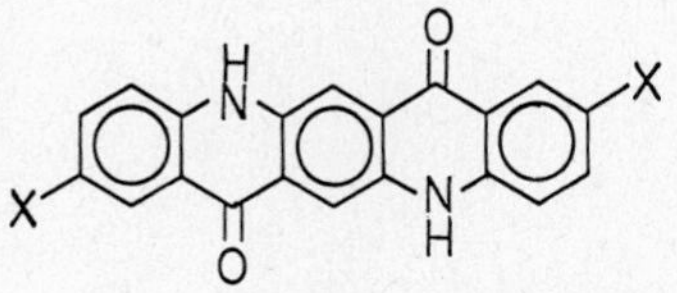

| XI | $X = CH_3$ | P.R. 122 |
| XII | $X = Cl$ | P.R. 202 |

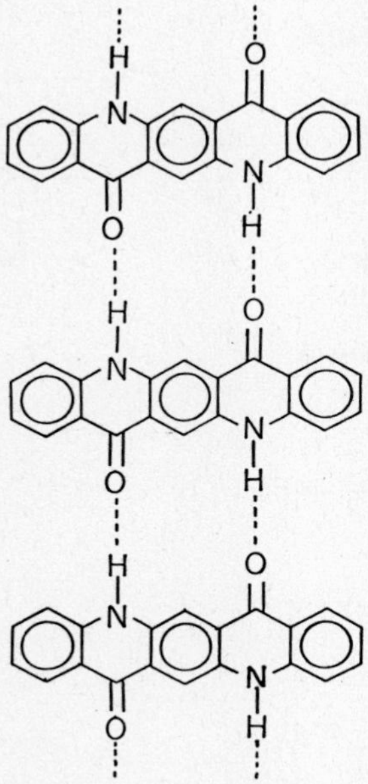

Figure 1.   Quinacridone hydrogen bonding.

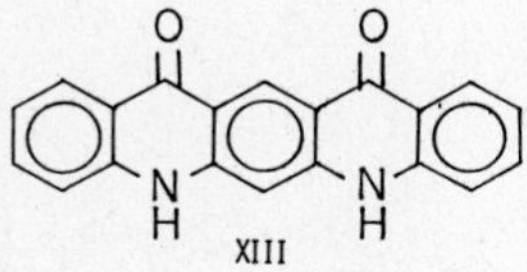

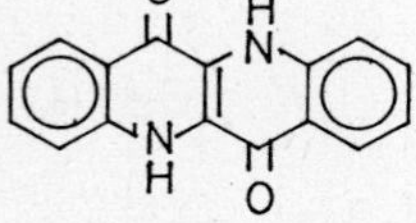

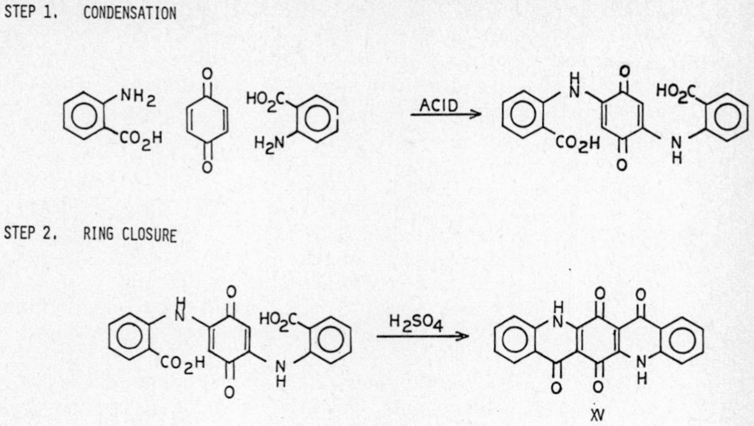

STEP 1.   CONDENSATION

STEP 2.   RING CLOSURE

Scheme V.   Synthesis of quinacridonequinone.

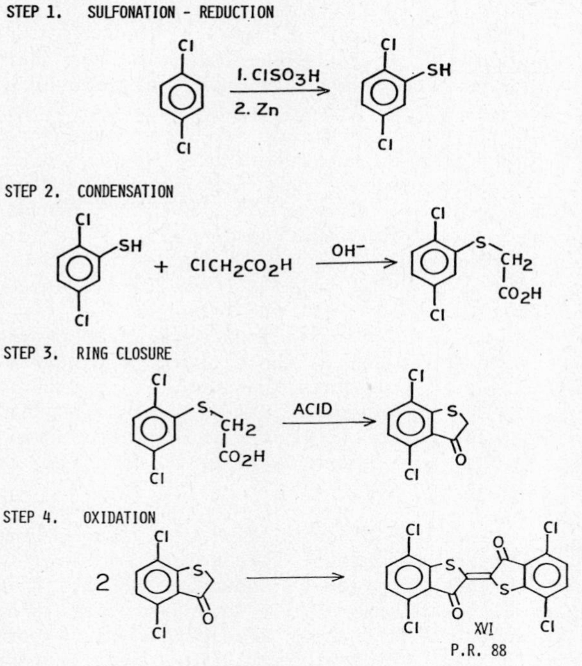

STEP 1.   SULFONATION - REDUCTION

STEP 2.   CONDENSATION

STEP 3.   RING CLOSURE

STEP 4.   OXIDATION

Scheme VI.   Synthesis of thioindigo.

Structures XVII and XVIII show two outstanding yellow vat pigments, flavanthrone, Pigment Yellow 24 (XVII), and anthrapyrimidine, Pigment Yellow 108 (XVIII). Unfortunately, high cost has drastically reduced their usage today.

Another superior vat color is indanthrone blue, Pigment Blue 22 (XIX) (Scheme VII). In certain uses it is even more weatherfast than copper phthalocyanine (34). It is used to redden phthalocyanines, but, again, high cost limits its market.

There are also outstanding vat pigments in the orange range including orange GR, Pigment Orange 43 (XX) (Scheme VIII), brominated anthanthrone, Pigment Red 168 (XXII) (Scheme IX), and brominated pyranthrone, Pigment Red 197 (XXIII) (Scheme X). Cost is also a factor with the pigments, particularly Pigment Orange 43, where the commercial methods of manufacture result only in a 50–55% yield of the desired trans isomer. The much less desirable cis isomer Pigment Red 194 (XXI) has a dull red color and bleeds badly in solvents.

Perhaps the most important class of vat pigments introduced by Harmon Colors is the perylene family with new members being introduced frequently (35). Scheme XI outlines the general synthesis of perylene pigments, and it is of interest that even the dianhydride, Pigment Red 224, is of commercial importance and is sold for use in automotive finishes. The other perylenes find broad use in architectural, plastic, and industrial finishes and vary from red to violet in hue.

Isoindoline Pigments. At about the same time that quinacridone and thioindigo pigments were being developed, research on a new class of yellow pigments was being carried out by Geigy, now a part of Ciba-Geigy (36, 37). The first of these tetrachloroisoindolinine pigments was introduced in the late 1950s and have been important as replacements for the increasingly more expensive flavanthrone and anthrapyrimidine yellows. Scheme XII shows the general structure of these pigments. When the diamine is varied, yellow to dark red pigments can be prepared.

Two important commercial yellows, one from p-phenylenediamine, Pigment Yellow 110 (XXIV), and the other from 2,6-diaminotoluene, Pigment Yellow 109 (XXV), have found use in automotive finishes. Prior to the research at Geigy, the class of isoindolinones had been looked at as potential colorants; however, it was the Geigy chemists who discovered that the chlorine substituents are vital to produce pigments with acceptable weatherfastness and insolubility for commercial use as was shown earlier to be the case for the thioindigo, Pigment Red 88.

Azo Pigments. One of the oldest and most diverse groups of pigments available is the azo pigments. They cover the complete color gamut from yellows to reds to blues as shown in Scheme XIII; basically what is required to synthesize them is a diazotizable aromatic amine that can be coupled either at an electron-rich position on an aromatic ring (XXVI), Pigment Red 1 (Para Red), or to an active methylene position (XXVII), Pigment Yellow 1.

It is clear that almost an infinite number of substituents can be introduced either on the amine or the coupler. A vast number of such azos has been prepared over the last 100 years (38). Among the

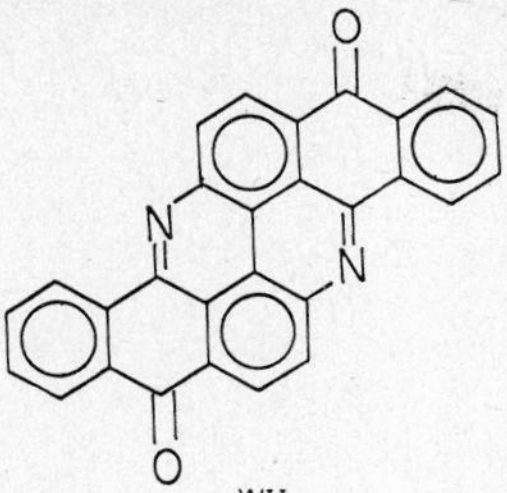

XVII

P.Y. 24

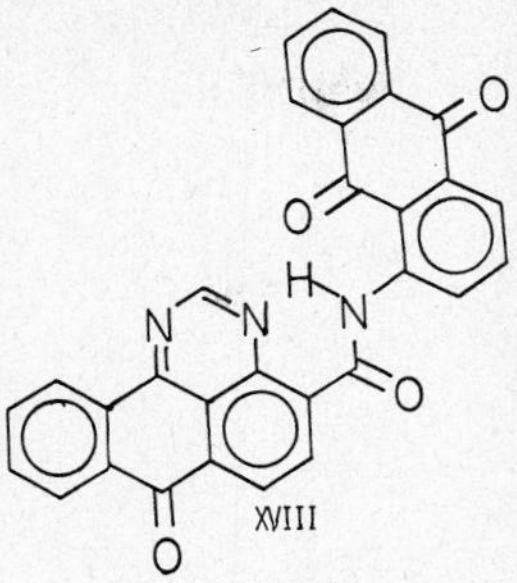

XVIII

P.Y. 108

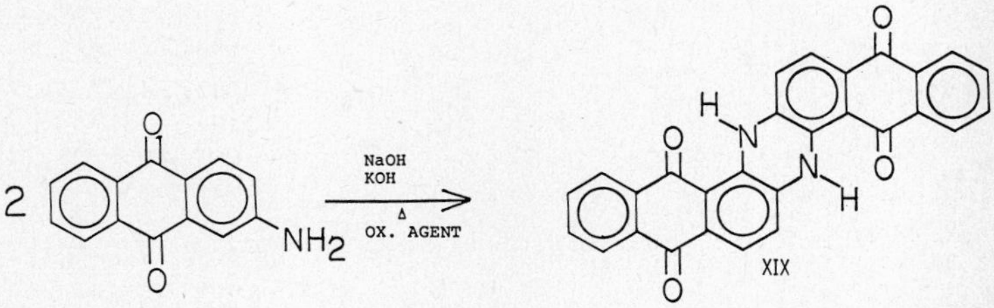

XIX

P.B. 22

Scheme VII. Synthesis of indanthrone blue.

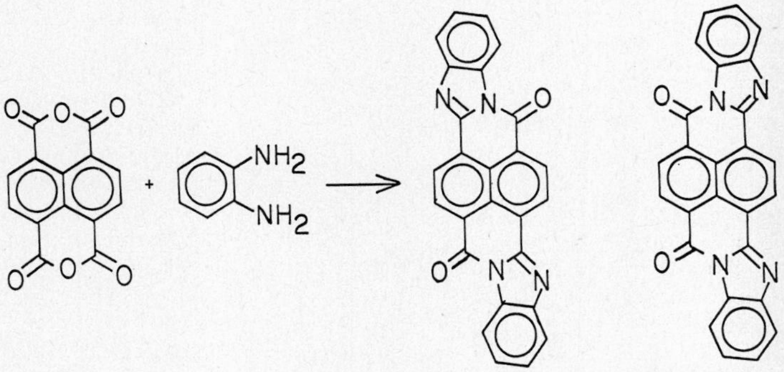

XX

P.O. 43

XXI

P. R. 194

Scheme VIII. Synthesis of orange GR.

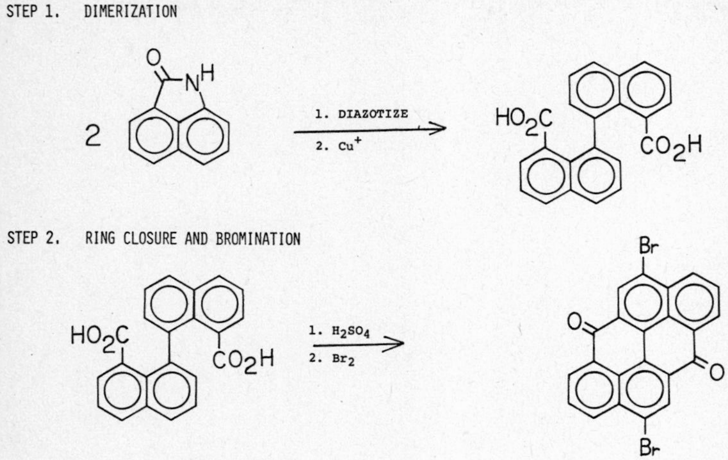
STEP 1. DIMERIZATION
2
1. DIAZOTIZE
2. Cu+
HO2C
CO2H
STEP 2. RING CLOSURE AND BROMINATION
HO2C
CO2H
1. H2SO4
2. Br2
Br
Br

P.R. 168
XXII

Scheme IX. Synthesis of brominated anthanthrone.

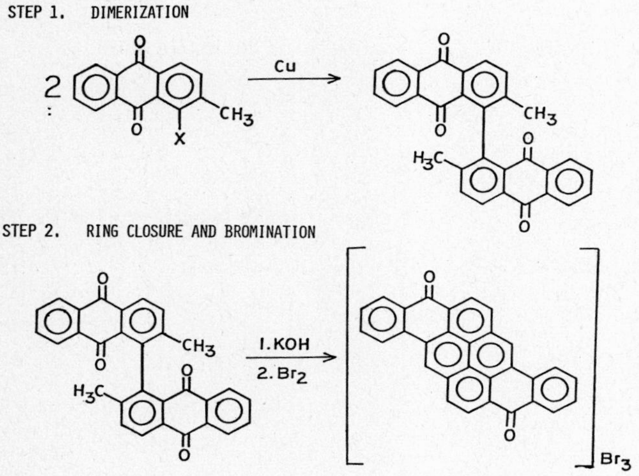
STEP 1. DIMERIZATION
2
CH3
X
Cu
CH3
H3C
STEP 2. RING CLOSURE AND BROMINATION
CH3
H3C
1. KOH
2. Br2
Br3

XXIII

Scheme X. Synthesis of brominated pyrantharone.

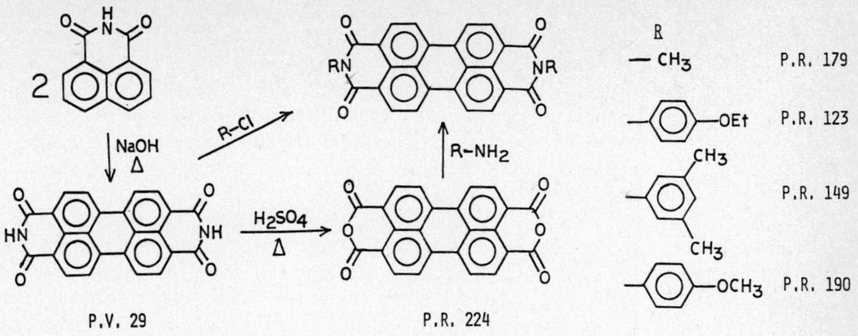

Scheme XI. Synthesis of perylene pigments.

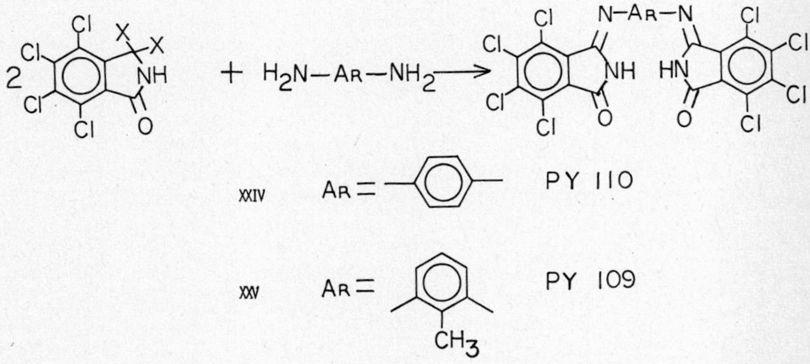

Scheme XII. Tetrachloroisoindolinone synthesis.

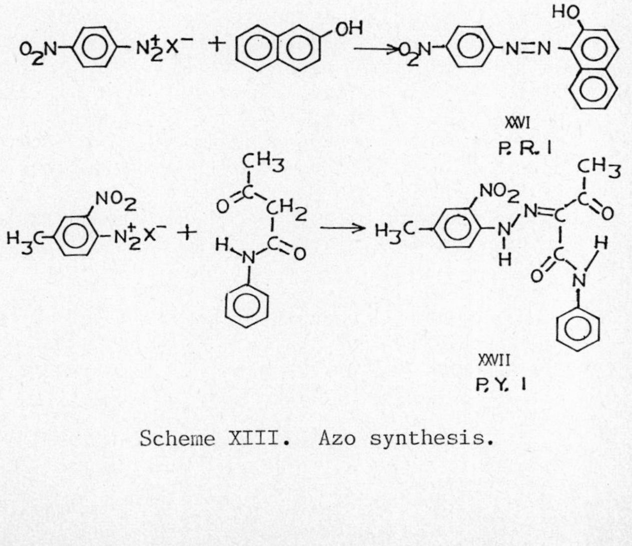

Scheme XIII. Azo synthesis.

most successful have been reds based on deviations of β-hydroxynaphthoic acid (BON) such as Pigment Red 22 (XXVIII), where an amide has been formed from the BON acid group and aniline. Amide formation helps to insolubilize the pigment formed.

Another successful group of BON-derived pigments utilizes various sulfonated derivatives of BON. The resulting dye is insolubilized by forming insoluble metal salts such as seen with Permanent Red 2B, Pigment Red 48 (XXIX).

The banning of lead chromate pigments has resulted in an enormous demand for yellow azo pigments as replacements. Pigment Yellow 74 (XXX) has become important as such a replacement in the past 10 years. The structure is written in the hydrazone form. There has been evidence over recent years that this is the preferred tautomeric form of these types of pigments in the solid state (39-41) versus the more commonly shown azo form (XXXI).

Pigment Yellow 74 shows more color strength than other simple azos of this type, presumably due to a higher extinction coefficient with the nitro group in the para rather than the more common ortho position (42) such as in Pigment Yellow 65 (XXXII).

When Pigment Yellow 74 was first introduced, its weather fastness was only fair. Improvements were made by increasing the particle size of the pigment. The larger crystallite leads to a weaker color but still with strength comparable to other azos and greatly improved weatherfastness and hiding power (43). More careful control of crystallite size and distribution has been used in recent years for many different pigments to optimize various properties (44-47).

Another direction to improve the fastness properties of azos was taken by chemists at Ciba (now Ciba-Geigy). They developed a practical method to form high molecular weight azos (48, 49). Decreased solubility and increased melting point improve light fastness and thermal stability. Scheme XIV shows the general procedure, which calls for carrying out an aqueous coupling with a carboxylic acid containing amine followed by the final condensation step in an organic solvent.

Pigment Red 144 (XXXIII) and Pigment Yellow 93 (XXXIV) are made in this manner. They find extensive use in plastics and fibers.

Another approach to improved azos is to insolubilize them via amide groups. Amines and acetoacetanilides based on aminobenzimidazolones have been exploited by Hoechst (50, 51). Examples are Pigment Red 171 (XXXV) and Pigment Orange 36 (XXXVI). The pigments in this series range from yellow to maroon and generally show very good heat stability and good weatherfastness, especially in masstone, although, in dilution, they are not up to automotive standards.

Other azo pigments have been produced containing various amide or sulfonamide groups such as Pigment Yellow 97 (XXXVII) and Pigment Red 146 (XXXVIII).

Miscellaneous. In this section will be discussed a number of individual pigments that, as of today, are represented by only one successful product, where future developments might be expected, and other areas where there has been much patent activity.

One such pigment is Carbazole Violet, Pigment Violet 23 (XXXIX) (52), synthesized as shown in Scheme XV. It is a brilliant violet

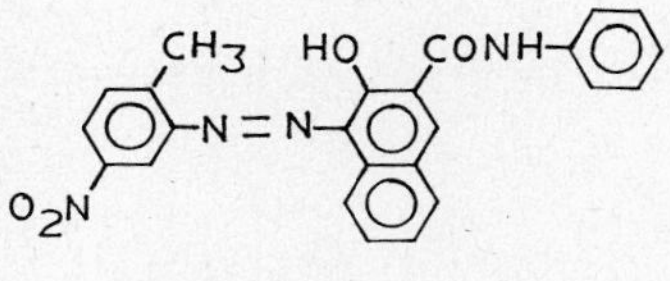

XXVIII

P. R. 22

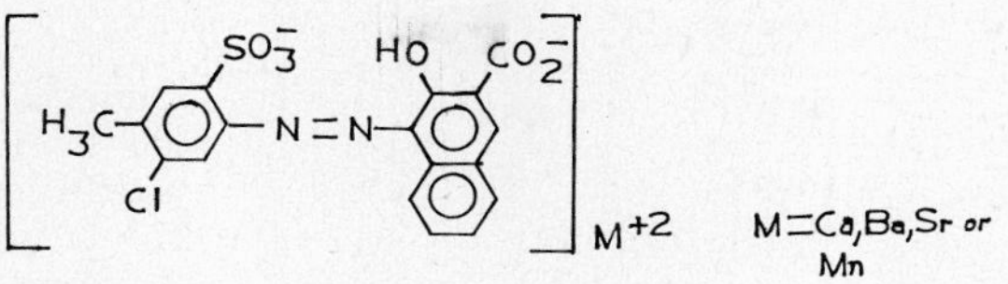

XXIX

P. R. 48

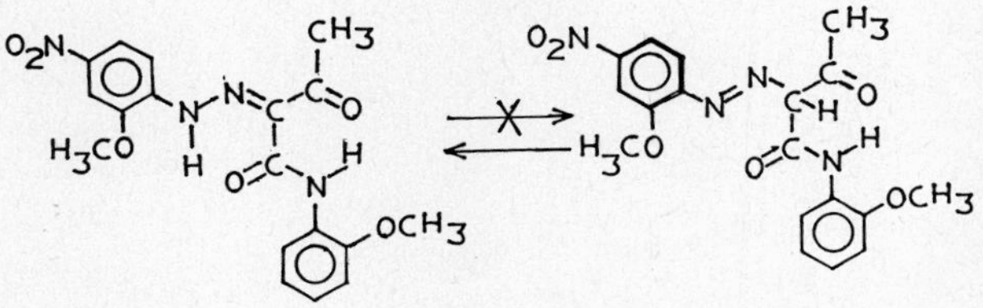

M≡Ca,Ba,Sr or Mn

XXX

P.Y. 74

XXXI

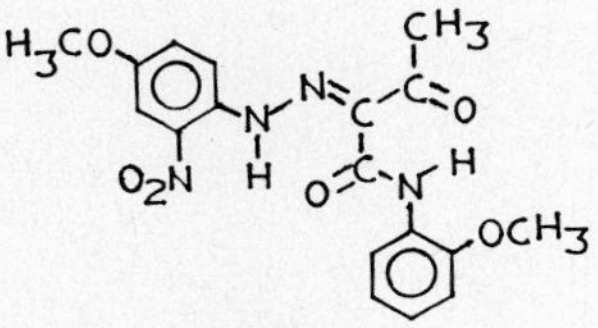

XXXII

P.Y. 65

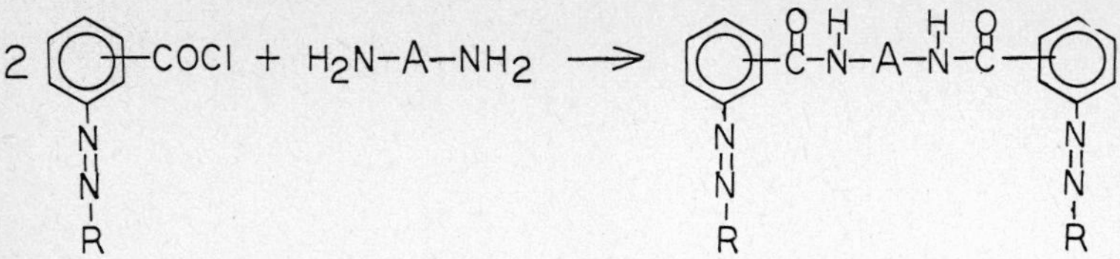

Scheme XIV. Synthesis of condensation azos.

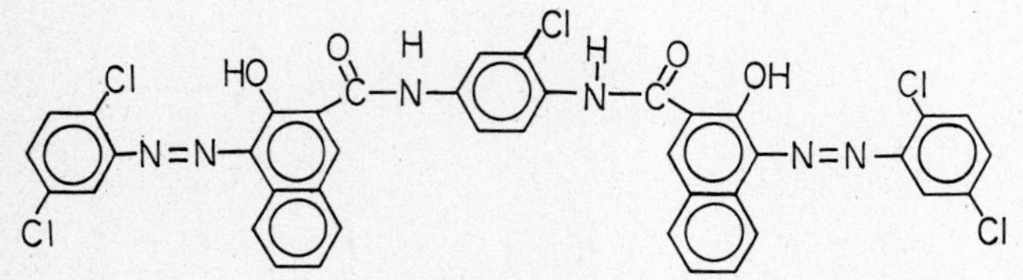

XXXIII

P. R. 144

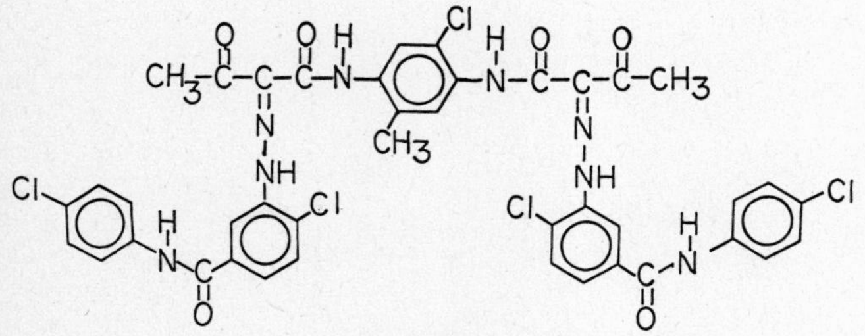

XXXIV

P. Y. 93

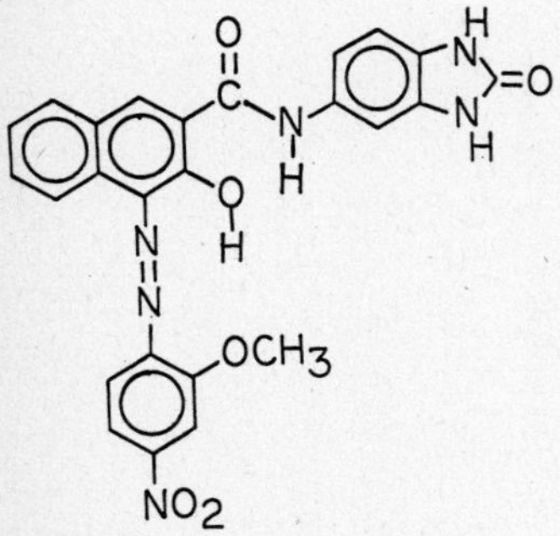

XXXV

P. R. 171

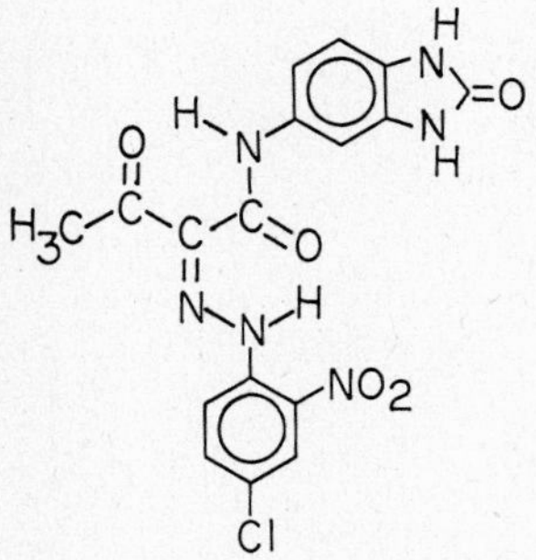

XXXVI

P. O. 36

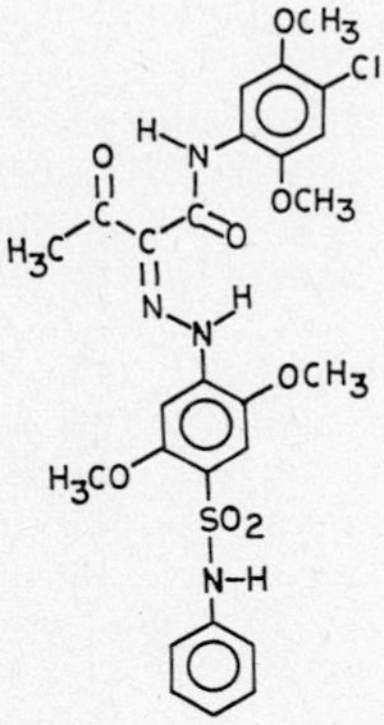

XXXVII
P. Y. 97

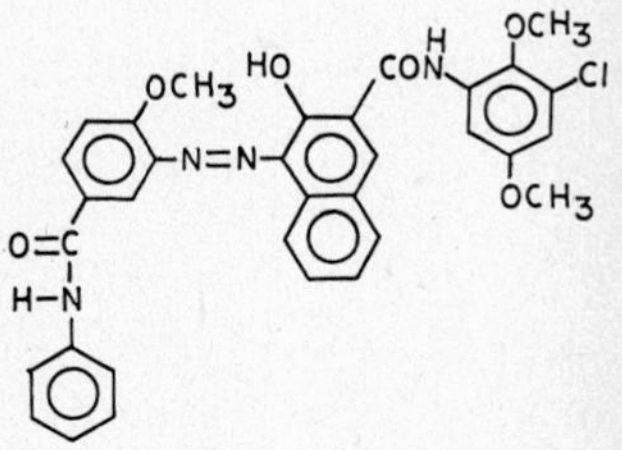

XXXVIII
P. R. 146

STEP 1. CONDENSATION

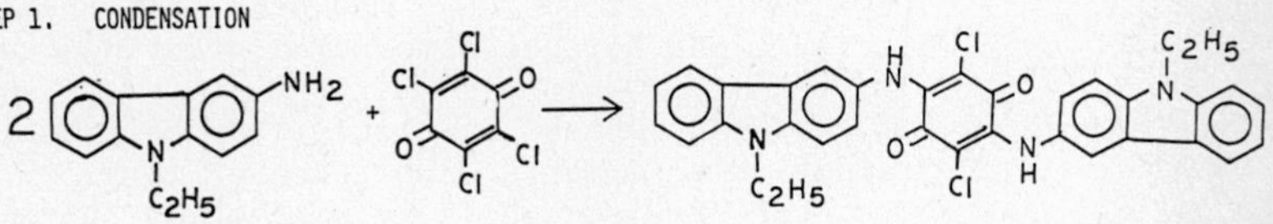

STEP 2. OXIDATION - RING CLOSURE

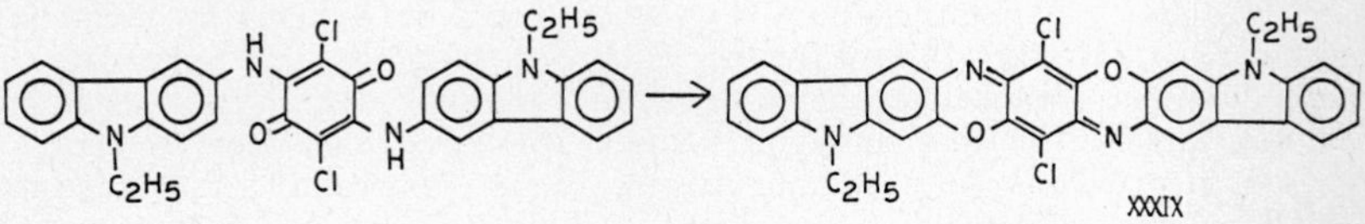

XXXIX
P.V. 23

Scheme XV. Synthesis of carbazole violet.

with very high tinctorial strength and outstanding weatherfastness at all dilutions. It is used for reddening over phthalocyanine blue to give colors closer to ultramarine blue as compared to the greenish blues normally obtained with phthalocyanines. Although this class of chromophores has been known for a number of years, Pigment Violet 23 has thus far been the only commercially successful derivative.

Another unique example is Pigment Green 10 (XL), patented by Du Pont in 1941 (53). It is a nickel chelate of the azo formed by coupling p-chloroaniline to 2,4-dihydroxyquinoline (Scheme XVI). This is the most weatherfast azo-based pigment known and is fast to light even in very great dilution with white. Its major defect is a slight sensitivity to acids. Although it has an unattractive masstone, its very greenish yellow undertone shows to advantage in mixtures with aluminum and titanium dioxide.

Other metal chelate pigments were patented around this period but did not lead to commercial products until recently. A specific example is Pigment Yellow 129 (XLI), which was not exploited for 30 years until Ciba-Geigy developed new processes and made improved products (55, 56). In addition, much work in this area has been carried out by BASF (XLII) (57) and others. XLIII (58) and XLIV (59) are further examples from the patent literature.

A related nonmetallized yellow is Pigment Yellow 139 (XLV), which although known for some time (60-62) has only recently been offered and used commercially. It is a nonbleeding yellow with suitable fastness for industrial finishes.

Another area of interest to pigment researchers both in Germany and Japan, is quinophthalones. XLVI (63) and XLVII (64) are two patented examples. Pigment Yellow 138, marketed by BASF, is believed to be of this type, and further commercial offerings can be expected.

The old area of anthraquinones continues to be actively researched, and recent developments incorporate many of the substituents that have been used successfully with other pigments, such as increased molecular weight and addition of amide groups to increase bleed resistance and weatherfastness. A few older anthraquinone-based pigments were mentioned earlier. Others still available today include Pigment Yellow 123 (XLVIII) and Pigment Red 177 (XLIX). Pigment Red 177 is used for its extremely transparent masstone. It is of particular interest because it contains free amine groups normally considered detrimental to weatherfastness. However, these amines are positioned for good hydrogen bonding with the carbonyl groups, which might help explain its acceptable fastness properties.

Other patented samples are L (65) and LI (66), which are closely related to Pigment Yellow 123 and Pigment Red 177, respectively.

Combining azos and anthraquinones in one molecule has also been carried out and such azoanthraquinones as LII (67) and LIII (68) have been patented and are just recently appearing on the market.

The continued introduction of new resins and more stringent performance requirements necessitates the development both of new and improved older products by pigment manufacturers. To date, this challenge has been met and will continue to be met in the years ahead as present requirements become even more stringent or as new improvements become necessary.

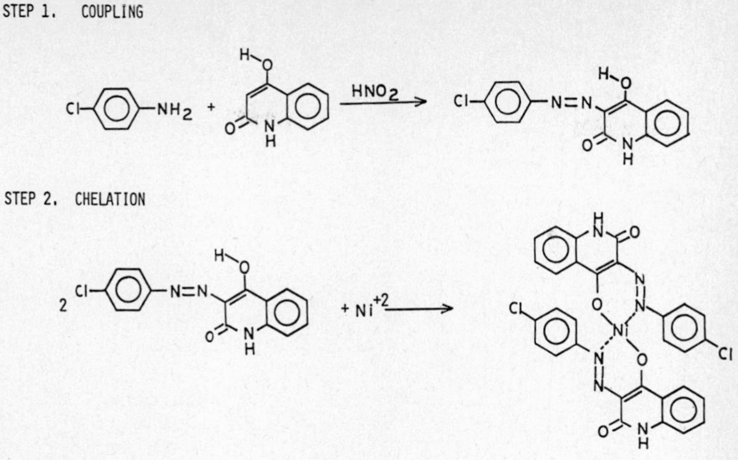
STEP 1. COUPLING
HNO2
STEP 2. CHELATION
+ Ni+2

XL

P. G. 10

Scheme XVI. Synthesis of Pigment Green 10.

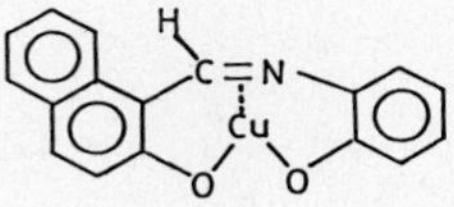

XLI

P. Y. 139

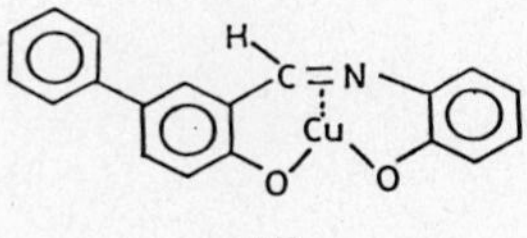

XLII

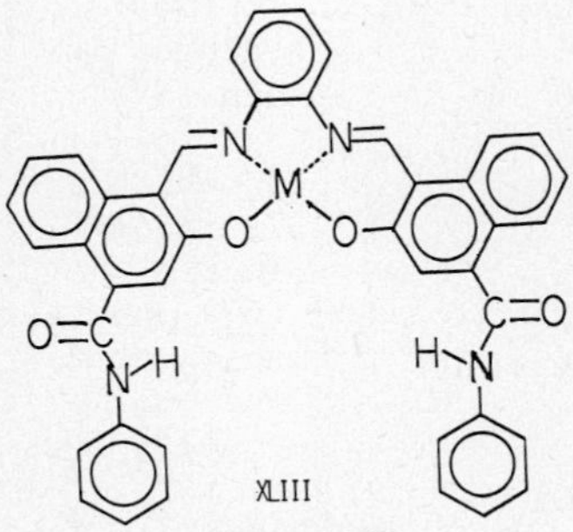

XLIII

ORANGES & REDS
CIBA-GEIGY

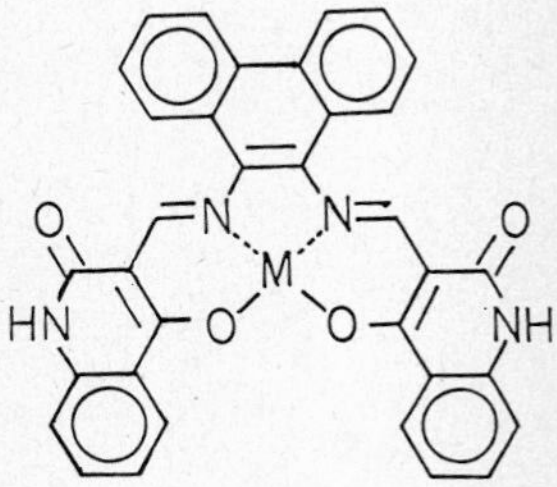

XLIV

YELLOWS
DU PONT

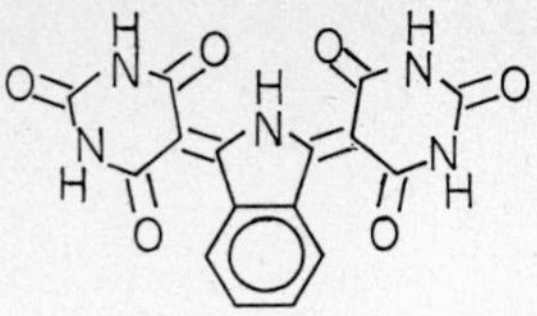

XLV

P. Y. 139

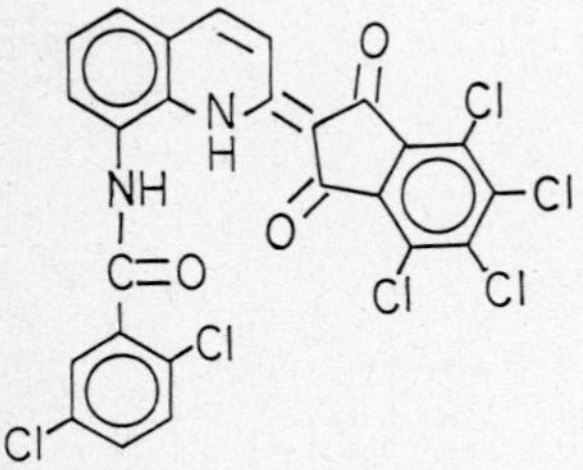

YELLOW
MITSUBISHI
XLVI

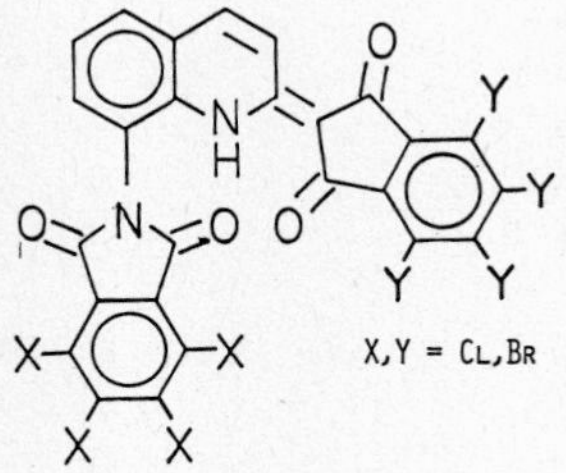

X,Y = CL,BR

YELLOW
BASF
XLVII

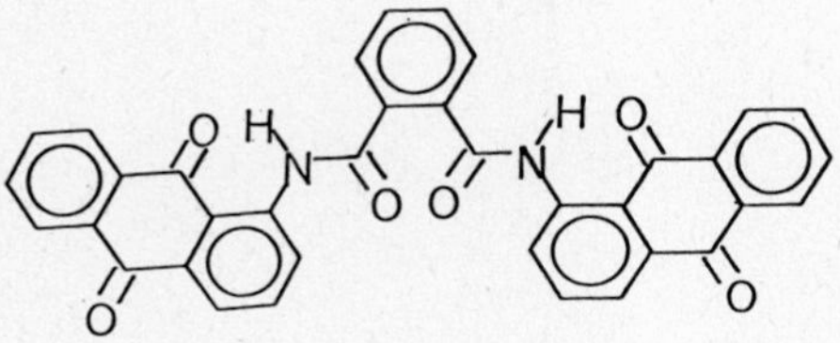

XLVIII
P. Y. 123

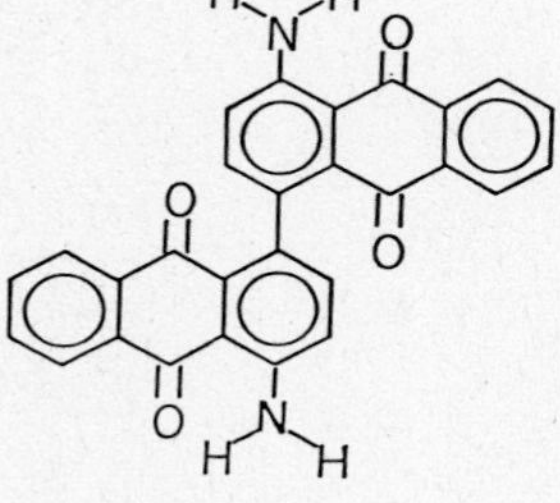

XLIX

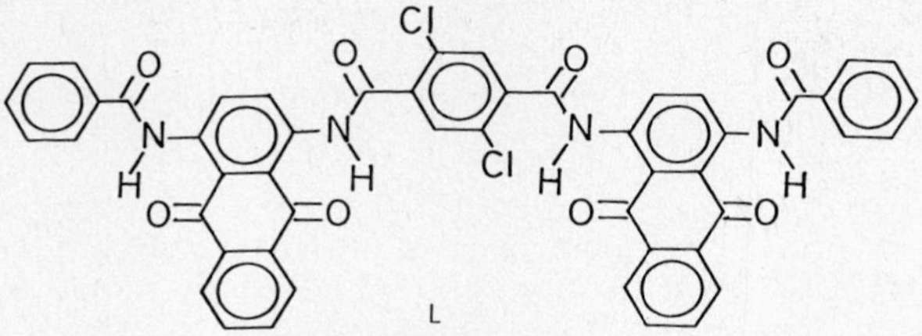

L
SANDOZ

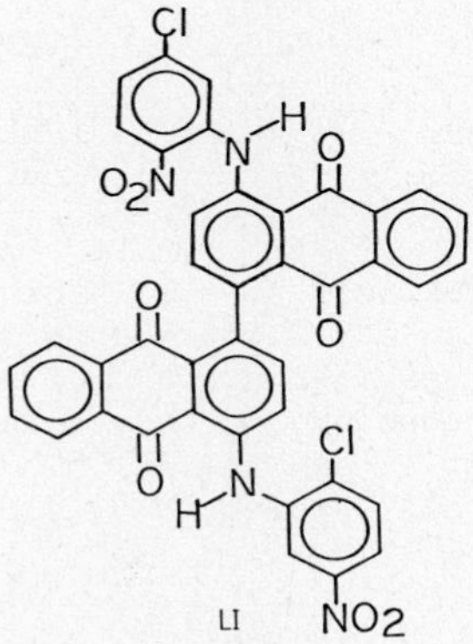

LI
MOBAY

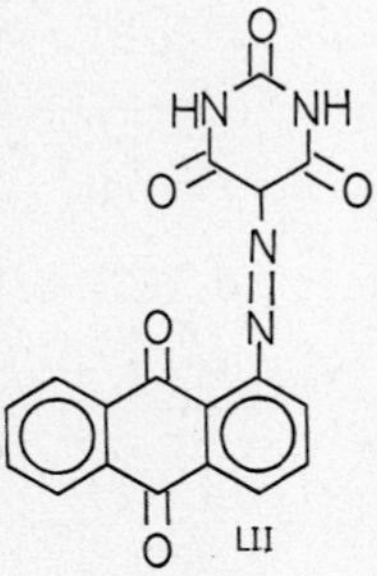

LII
YELLOW
BASF

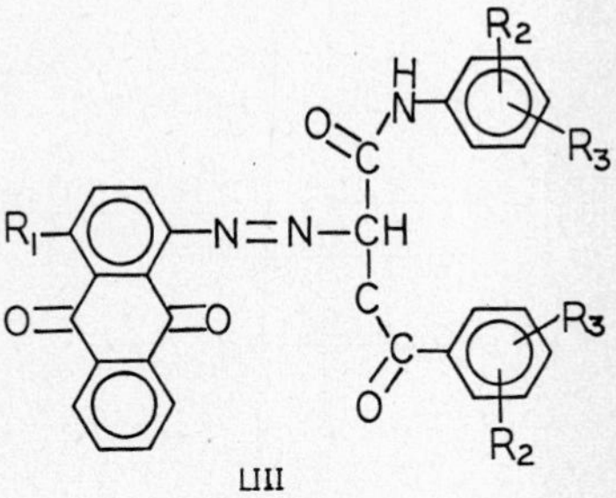

LIII
YELLOWS & ORANGES
BAYER

Literature Cited

1. Gaertner, H. J. Oil Colour Chem. Assoc. 1963, 46, 13.
2. Heinle, K. Farbe Lack 1967, 73, 735.
3. Hopmeier, A. P. In "Encyclopedia of Polymer Science and Technology"; Wiley: New York, 1969; Vol. 10, pp. 157-93.
4. Inman, E. R. In "The Royal Institute of Chemistry Lecture Series 1967"; The Royal Institute of Chemistry: London, England, 1967; No. 1.
5. Kehrer, F. Chimia 1974, 28, 173-83.
6. Lenoir, J. In "The Chemistry of Synthetic Dyes"; Venkataraman, K., Ed.; Academic: New York, 1971; Vol. 5, Chap. 6.
7. Rys, P.; Zollinger, H. "Fundamentals of the Chemistry and Application of Dyes"; Wiley-Interscience: New York, 1972.
8. Lubs, H. A., Ed. "The Chemistry of Synthetic Dyes and Pigments"; Reinhold: New York, 1955.
9. Patton, Temple, Ed. "Pigment Handbook"; Wiley-Interscience: New York, 1973.
10. Smith, F. M.; Stead, D. M. J. Oil Colour Chem. Assoc. 1954, 37, 117-30.
11. Vesce, V. C. Off. Dig. 1959, 28(377)(Part 2), 1-48.
12. Vesce, V. C. Off. Dig. 1959, 31(414)(Part 2), 1-143.
13. Levison, H. W. "Artists' Pigments"; Colorlab: Hallandale, Fla., 1976.
14. Moser, F. H.; Thomas, A. L. "Phthalocyanine Compounds"; American Chemical Society Monograph Series No. 157, American Chemical Society: Washington, D.C., 1963.
15. Dye class number according to "Colour Index," 3rd ed., Society of Dyers and Colourists and the AATCC, 1971.
16. Liebermann, H., et al. Annalen 1935, 518, 245.
17. Anitschkoff, N. Thesis, Berlin, 1934, p. 9.
18. Struve, W. S. U.S. Patent 2,821,529, 1958, and 2,844,485, 1958; Struve, W. S.; Reidinger, A. D. U.S. Patent 2,844,484, 1958.
19. Dien, C. K. U.S. Patent 3,342,823, 1967.
20. Gerson, H.; Santimauro, J. F.; Vesce, V. C. U.S. Patent 3,257,405, 1966.
21. North, R. U.S. Patent 3,940,399, 1976 and 4,100,162, 1978.
22. Sandoz Ltd. British Patent 924 661, 1964; Chem. Abstr. 1964, 60, 699.
23. See also Pollak, P. Prog. Org. Coat. 1977, 5, 254-53.
24. Lincke, F. Farbe Lack 1970, 76, 764-75.
25. Altiparmakian, R. H.; Bohler, H.; Kaul, B. L.; Kehrer, F. Helv. Chim. Acta 1972, 55, 85-100.
26. Zaharia, C. N.; Tarabasanu-Mihailia, C. Rev. Roum Phys. 1976, 24, 335-444.
27. Labana, S. S.; Labana, L. L. Chem. Rev. 1967, 67, 1-18.
28. Jaffe, E. E.; Matrick, H. J. Org. Chem. 1968, 33, 404. Jaffe, E. E.; Matrick, H. U.S. Patent 3,334,102.
29. Kim, C. K.; Maggiulli, C. A. J. Heterocycl. Chem. 1979, 16, 1651-3.
30. Ehrich, F. F. U.S. Patent 3,160,510, 1964.
31. Ehrich, F. F. U.S. Patent 3,148,075, 1964.
32. Ehrich, F. F. U.S. Patent 3,607,336, 1965.

33. Fisher, W. A. "Pigment Handbook"; Wiley-Interscience: New York, 1973; Vol. 1, pp. 673-7.
34. Rys, P.; Zollinger, H. "Fundamentals of the Chemistry and Application of Dyes"; Wiley-Interscience: New York, 1972; pp. 130, 131.
35. Fisher, W. A. "Pigment Handbook"; Wiley-Interscience: New York, 1973; Vol. 1, pp. 667-72.
36. Pugin, A.; von der Crone, J. "New Organic Pigments", Off. Dig. 1965, 37(488), 1071-72.
37. Pugin, A. U.S. Patent 2,973,358, 1961.
38. Review on Azos: Herbst, W.; Hunger, K. Prog. Org. Coat. 1978, 6, 211-70.
39. Pendergrass, D. B., Jr.; Paul, I. C.; Curtin, D. Y. J. Am. Chem. Soc. 1972, 94, 8730, Brown, C. J. J. Chem. Soc. A 1967, 405.
40. Mez, H. C. Ber. Bunsengesellschaft 1968, 72, 389.
41. Griffiths, J. "Colour and Constitution of Organic Molecules"; Academic: New York, 1976; pp. 189-92.
42. Johnson, R. A. U.S. Patent 3,032,546, 1962.
43. Keay, A. M. J. Coat. Technol. 1977, 49, 31-37.
44. Hafner, O. J. Paint Technol. 1975, 147, 65-69.
45. Sappok, R. J. Oil Colour Chem. Assoc. 1978, 61, 299.
46. Chromy, L.; Kaminska, E. Prog. Org. Coat. 1978, 6, 31-48.
47. Vernardakis, T. G. Dyes Pigments 1981, 2, 175.
48. Schmid, M.; Mueller, W. U.S. Patent 2,936,306, 1960.
49. Gaertner, H. J. Oil Colour Chem. Assoc. 1963, 46, 33-39.
50. Lenoir, J. "The Chemistry of Synthetic Dyes"; Reinhold: New York, 1971; Vol. V, p. 372.
51. Schilling, K.; Dietz, E. U.S. Patent 3,109,842, 1963.
52. Lenoir, J. "The Chemistry of Synthetic Dyes"; Reinhold: New York, 1971; Vol. V, p. 421.
53. Kvalnes, D. E.; Woodward, H. E. U.S. Patent 2 396 327, 1946.
54. Schmidt, K.; Wahl, O. U.S. Patent 2 116 913, 1938.
55. Inman, E. R.; MacPherson, I. A.; Stirling, J. A. U.S. Patent 3,700,709, 1972.
56. McCrae, J. M.; Irvine, A. M.; MacPherson, I. A. U.S. Patent 4,143,058, 1979.
57. BASF. British Patent 1,122,938, 1968.
58. Inman, E. R.; McCrae, J. M.; Midcalf, C.; Turner, A. U.S. Patent 3,864,371, 1975.
59. Dhaliwal, P. S. U.S. Patent 3,903,118, 1975.
60. Tartter, A.; Wessbarth, O. German Aus. 1,012,406, 1957.
61. Bock, G.; Elser, W. German Offen. 2,041,999, 1972.
62. Lotsch, W. U.S. Patent 4,166,179, 1979.
63. Imahori, S.; Kaneku, M.; Ono, H. Japanese Kakai 7,650,330, 1976.
64. Fabian, W. German Offen. 2,357,077, 1975.
65. Ehrhardt, K. German Offen. 2,340,951, 1974.
66. Gerson, H. U. S. Patent 3,718,669, 1973.
67. BASF. German Offen 1,544,372 and 1,544,374, 1965.
68. Rolf, M.; Neeff, R.; Mueller, W. German Offen. 2,644,265, 1978; 2,659,676, 1978; and 2,812,635, 1979.

Paint Manufacture

LOUIE F. SANGUINETTI

Manufacturing Committee, Golden Gate Society for Coatings Technology, San Francisco, CA 94107

```
Basic Operations in Paint Manufacture
Oil Absorption and the Milling Process
Bulk Properties of Paints
Types of Mills for Paint Manufacture
   Stone and Colloid Mills
   Roller Mills
   Ball and Pebble Mills
   Sand Mills
   High-Speed Dispersers
   Horizontal Mill
   Kinetic Dispersion Mills
   Other Mills
```

Basic Operations in Paint Manufacture

Paint is a mixture of pigment or combination of pigments and a liquid vehicle that is composed of binders and thinner (1). The formulation of paint is a highly developed combination of science, art, and technology. Since paints are required to meet a wide diversity of end uses, the formulations vary widely. The type and relative quantities of vehicle, pigments, extenders, additives, and volatile thinners determine the final film properties. The paint is usually applied by brush or roller when used in the home; commercially it is also applied by spray, dip, electrical deposition, screens, and curtain coaters. Normally applied in thin films on various surfaces such as wood, metal, concrete or plastic, paint is used for protection, decoration, identification, or some functional application such as an etch resist. Clear coatings, while very important, are not considered as paints since there are no pigments in these coatings.

The manufacture of paint at first glance appears simple. First the pigment is mixed with a portion of the vehicle to form a paste that is then milled to disperse the pigment into the vehicle. After

0097-6156/85/0285-1297$06.00/0

the pigment is dispersed to a predetermined standard and into finely divided particles, the balance of vehicle and additives are added to give the final product.

Other steps are taken in the manufacture of paints. Tinting may be required to obtain the proper color. The paints are clarified in order to strain out any foreign or large particles. After passing quality control, the paints are transferred into containers, labeled, and stored.

The layout of a plant usually takes advantage of gravity flow. The paste or dispersion flows from an elevated floor to letdown tanks where the operations of thinning, tinting, and blending with additives are carried out. Oftentimes, dispersing and letdown are done in the same vessel. Slurries, which are pigments suspended in liquids, are frequently used in paint manufacture. Computers are being used to meter slurries into the tanks.

Oil Absorption and the Milling Process

Various types of mills are used to supply the energy necessary to disperse the pigment in the vehicle. However, the pigment and vehicle must be in the correct ratio for a particular mill in order for the mill to work efficiently.

The pigment manufacturers produce pigments having an optimal particle size (2). These dry pigment particles, however, are attracted to each other and tend to form clusters of pigment known as agglomerates. In order to return the pigment clusters back to their inherent particle size, the pigment clusters are dispersed in a suitable mill to form stable suspended particles in a vehicle.

The paint manufacturer must utilize the milling process so that there is high throughput of mill base or pigment. To do this, the pigment and vehicle must be in the proper ratio so that the liquid fills the voids, each agglomerate in the dry pigment is separated, and each particle is wetted. In the milling process enough energy is necessary to separate the agglomerates and wet the particles.

The oil absorption test is used to measure the amount of vehicle necessary to fill the voids among the pigment particles. The requisite quantity of vehicle includes that which is absorbed on the surface of the pigment. Depending on particle size and shape of the pigment, the void volume usually accounts for the greater portion of the oil absorption. The oil absorption of a pigment or combination of pigments for a given vehicle combination of vehicles is constant. The oil absorption of a pigment is the amount of vehicle by weight that is absorbed by a given amount of dry pigment to form a paste. There are two ways that this property can be measured. One, the "rub out" method, involves the mixing of the pigment with increments of vehicle with a stiff spatula until a stiff paste is formed. This putty like paste must not break or separate as described by ASTM method D-281-31. The second method is like the first method except the end point is brought to a soft paste (ASTM D-1483-60). The tests are subjective but with a little practice consistent results can be attained.

Pigment volume concentration (PVC) of a paint is defined as follows:

$$PVC = \frac{\text{volume of pigment}}{\text{volume of pigment + volume of binder}}$$

In 1949, Asbeck, Laiderman, and Van Lee first defined the "critical pigment volume concentration" (CPVC) (3). CPVC is that level of pigmentation in a dry paint film where just sufficient binder is present to fill all voids among pigment particles. The PVC of a paint affects many properties such as hiding, corrosion resistance, scrub resistance, etc. The critical pigment volume concentration (CPVC) is an important property in that any increase or decrease of binder from the CPVC will affect the film properties; it is more useful to know the volume pigment-binder relationship than to know the weight relationship.

Various mills usually do not "grind" or reduce the pigment particle size; they disperse the agglomerates. The milling equipment must not only wet and disperse but also be able to move this mass. Consequently, the viscosity of the mill base is important.

Bulk Properties of Paints

Viscosity can be defined as the ratio of shear stress to shear rate (4). The shear stress is equal to a force measured in dynes divided over an area upon which it acts in square centimeters. Shear stress has the dimensions of dynes per square centimeter. The shear rate is equal to the velocity of a layer due to shear stress measured in centimeters per second divided by the thickness in centimeters; the unit of shear rate is given in reciprocal seconds. Viscosity often is expressed in poises. The poise has the dimension of dyne seconds per square centimeter. Dyne seconds per square centimeter is the dimension for shear stress divided by shear rate and therefore is equal to viscosity.

Paints are rarely Newtonian in their flow behavior. They are usually found to be thixotropic, pseudoplastic, and sometimes dilatant. Paints are thixotropic if there is initial resistance to viscosity loss by agitation, if there is a reduction in viscosity with continued agitation, and if there is a return to the original viscosity after the paint is no longer agitated. A pseudoplastic paint is similar to a thixotropic paint except there is no initial resistance to viscosity loss or no yield factor. A dilatant paint differs because the viscosity increases or shows resistance to flow as the agitation is increased. Mill bases are usually slightly dilatant and are intended to be that way.

Types of Mills for Paint Manufacture

Mills break up the agglomerates generally by "smashing," "smearing," or a mixture of the two actions called "hybrid" (5). Under these classifications, the "smasher" will use a mill base of low viscosity. The "smearer" will require a mill base of high viscosity.

<u>Stone and Colloid Mills</u>. The original stone mill, used for centuries, consisted of thick large stone disks that rotated slowly (<u>6</u>). These mills were in common use up to the late 1940s. These mills were replaced by the colloid mill and high-speed stone mill.

While these mills are not the same, they differ chiefly in the shape of their rotor/stator configuration (Figure 1). The mills contain two stones of which one rotates (rotor) and the other is a stationary stone (stator). There is also an all metal rotor/stator.

A premix is required for these mills. The mill paste normally has a consistency of 90-140 Kreb units. The mill base should be smooth and free from lumps, since the residence or dwell time in the mill is very short. The vehicle solids for the paste can be as low as 20% for low oil absorption pigments and up to as high as 75% vehicle solids for high absorption pigments. The capacity of the mill will vary with viscosity of the paste.

The paste or mill base is usually fed by gravity to the mill (Figure 2). As the paste reaches the rapidly revolving rotor, it is impelled to the outer edge by centrifugal force. This force pushes the material through the narrow gap in which the clearance between the rotor and stator has been preset. The material is subjected to high stress as it passes through the gap. This stress (smearing) causes the agglomerate particles to disperse.

The gap between the rotor and stator can be adjusted even while the mill is operating. Adjustments are usually necessary since the mill will generate heat and expand the rotor and stator. The adjustment is quick, accurate, and simple. This can be done manually with a calibrated adjustment ring. Care must be taken that enough paste passes through so that the mill will not wear excessively or cause overheating. The gap between the stator and rotor is in a nearly closed position when starting to operate, and the mill is then adjusted immediately. The mill temperature stabilizes rapidly and another adjustment may not be necessary.

The stone or colloid mill, which is easily cleaned, is usually used for architectural paints. Pigments that are difficult to disperse are not usually used on these mills.

<u>Roller Mills</u>. The most popular types of roller mills have been the three- or five-roller mill (Figure 3). The three-roller is mostly used in the paint industry while the five-roller mill is preferred in the ink industry. There are, also, one-, two-, and four-roller mills. All work on the same principle and are of the smearer type.

The rolls on a three-roller mill each rotate in opposite direction to the one or two adjacent rolls. The speeds between the rolls also differ. While different manufacturers may have different speeds for their rolls, typical RPM values are as follows (<u>7</u>):

Roll Number	1	2	3	4	5
Three-roll mill rpm	35	115	345		
Five-roll mill rpm	25	50	100	200	300

In a roller mill, the paste must be tacky so that it adheres to the surface of the rotating rolls. The vehicle portion must be high

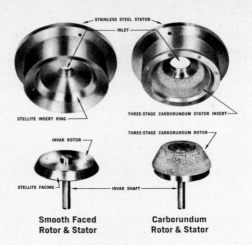

Figure 1. Metal and carborundum rotor and stator for colloid mill. Courtesy Premier Mill Corp.

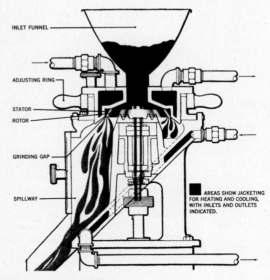

PRINCIPLE OF OPERATION
Material flows by gravity through the mill, where action takes place in the finely set grinding gap between the rotor and stator.

Figure 2. Inside of colloid mill. Courtesy Premier Mill Corp.

in resin solids. The mills have tremendous shearing action that requires that the paste has fluidity and that the vehicle has high viscosity.

Since the premix is a viscous composition, an intensive heavy duty mixer is required. The vehicle and then the pigment are added to the mixer. Sufficient vehicle to wet the entire pigment load must be used, and there must be enough vehicle so that the paste is fluid enough to charge the mill. The pigments must be thoroughly wet and free from lumps.

The paste is fed into the mill in the space between roll one and roll two whereupon the material flows into the narrow gap. Part of the material is rejected and returned to the feed bank. Part of the mill base continues on through where it is split between the feed roll and the center roll. The material on the center roll continues to the apron roll where it is again split between the center roll and the apron roll. The material on the center roll returns back to the feed bank while the material on the apron roll is knifed off onto the apron.

The rolls rotate at different speeds. Clearances between the rolls must be adjusted carefully and are related to the relative speed of each roll. The clearance of the back rolls must be sufficient so that enough material is received by the front roll so that output at the front roll is sustained. The clearance between the back rolls determines the amount of paste that is delivered.

Roller mills have certain advantages. They can handle viscous materials and produce high-quality dispersions of fine particle size; these are necessary in specialty coatings such as inks. Difficult-to-disperse pigments are handled by these mills. However, these mills are costly because the mill base throughput is at best moderate and a premixed paste is required. While the roller mills are used usually for specialty coatings, other mills can handle easy-to-disperse pigments more efficiently.

<u>Ball and Pebble Mills</u>. Ball mills are hardened steel shells with closed ends that use steel balls as the grinding medium (Figure 4). Pebble mills have steel shells and ends but they are lined with burrstone or synthetic stone (porcelain) and they use natural or porcelain balls as the grinding media (<u>8</u>) (Figure 5).

Ball mills disperse faster than pebble mills due to the fact that the steel balls are more dense than the porcelain or natural stone pebbles. Since the steel balls gives a greying effect they are not used in white or light colors. No premixing is needed. Since it is a closed mill, there is no volatile loss. They can be operated with minimal skill requiring little attention during operation. These mills can be used overnight with the batch being ready in the morning; a timer can be used to start and stop the mill. The mill should be vented when necessary or when experience dictates (<u>9</u>).

The efficiency of the mill is affected by the following four main factors:

1. The speed of rotation of the mill.
2. The relative volume of grinding media and mill base.
3. The size and density of the grinding medium.
4. The consistency of viscosity of the mill base.

Figure 3. Three–roll horizontal mill. Courtesy Lehmann/Thropp, Division of Paxson Machine Co.

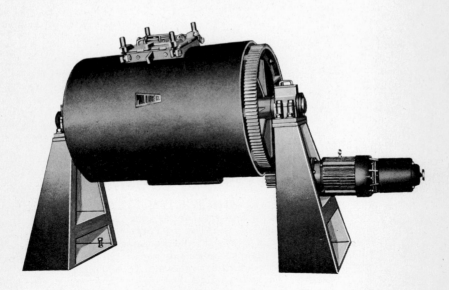

Figure 4. Pebble mill. Courtesy Paul O. Abbe, Inc.

The speed of rotation is in the range of 50–60% of the critical speed (10). The critical speed is the speed at which the centrifugal forces acting on the charge are balanced by the centripetal forces.

The volume of the grinding media as well as the volume of paste in the mill affects the grinding time. The pebble mill should contain about 45–50% pebble charge by volume. From 33% to 40% by mill volume of steel balls should be charged. The mill base to be added should cover the grinding media; this is approximately 20–25% of the mill volume. This volume should fill the void spaces that are left by the grind media. The load of grinding media, the mill paste, and the viscosity may vary, but these variations will change the dispersing time and/or create excessive wear of the mill or grinding media. It is best to stay within the general recommended range of volume of grinding medium, viscosity of mill base, and mill base volume to obtain the best results. Overloading of the mill increases the dispersing time, and the desired results may never be attained since the mill base may continuously ride over the grinding media. Since laboratory mills give good correlation to factory mills, practical results can be obtained in the laboratory.

The consistency of the paste must be adjusted to the type and size of grinding media. The viscosity of the mill base generally will vary with the density of the grinding media; the denser the grinding media, the higher viscosity the mill base. The viscosity of the paste used in milling is in the range of 75–120 Kreb units. Since these are relatively low viscosities, in practice low vehicle solid contents are used with a high ratio of pigment to vehicle.

Various methods have been used to determine the critical speed of the ball mill (11). Coghill and De Vaney calculated the constant in terms of the radius as follows:

$$S = \frac{54.18}{\sqrt{R}} \; ; \text{ a 6-ft diameter mill would then rotate}$$

$$S = \frac{54.18}{\sqrt{3}} = \frac{54.18}{1.732} = 31.3 \text{ critical speed}$$

at 50% of 31.3 = 16 rpm; at 60% of 31.3 = 19 rpm

In the Berliner report, the critical speed in rpm of any mill can be determined by dividing by a constant 76.6 by the square root of the mill diameter in feet (12). In a 6-ft diameter mill, the critical mill speed would be 31.2 rpm. It has been determined that a mill with lifter bars with a 6-ft diameter should have an optimum speed of about 20 rpm.

Mellor's formula gives similar results (13). The formula is rpm = 43.3 $\sqrt{1/(D - d)}$, where D = inside diameter of the mill in feet and d = average diameter of pebbles in feet.

The lifter bars are used for various reasons (Figure 6). The bars are important for lifting the grinding media to the high point of the cascade, producing more effectiveness. In addition, a layer of the grinding media is locked against the wall, thereby reducing slippage against the shell and resulting in less wear of the shell.

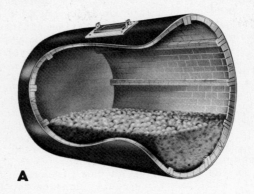

A

Figure 5.   Inside of pebble mill.   Courtesy Epworth Manufacturing
Co., Inc., and Coors Porcelain Co.

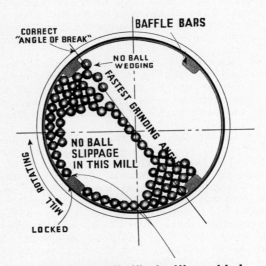

Figure 6.   Control of grinding media with baffle bars in closed
mill.   Courtesy Paul O. Abbe, Inc.

The ball or pebble mill is considered to be a hybrid. The cascading action of the grinding media causes the pigment to be both impacted and intensively sheared by the tumbling balls.

Sand Mills. In a ball mill, if the speed of the mill is increased, the grinding media is immobilized by centrifugal force. In a sand mill, the design is such that the grinding media is not immobilized (14).

The sand mill is in essence a unit that consists of a water-jacketed cylindrical shell containing disks mounted on a centrally located shaft in the shell and a screen at the top of the mill to allow the base to flow through while retaining the media in the mill (Figure 7). The base is pumped through the mill media by a variable pump. The sand mill can be an open or closed mill. The operation of the sand mill is not complicated; detailed instructions are given by the manufacturers.

The consistency of the mill base may vary from 70 to 140 Kreb units. This is in the same range as the viscosity used in ball mills so that ball mill bases are easily applied to the sand mill. However, the vehicle solids in the sand mill base is usually higher than in the ball mill formulation.

The mill, when operating, is in the temperature range of 120–150 °F, but for special cases the temperature may be as high as 300 °F. Care must be taken for heat-sensitive materials. Temperature is controlled by means of the water flowing through the water jacket. The rate of flow of water is usually adjusted so that the water coming out is slightly warmer than the water that is coming into the inlet.

Since the temperature of the mill in operation is high, the consistency of the paste should be formulated for these temperatures. Too thin a base results in excessive wear of the mill. Too thick of a base may keep the media from circulating, give excessive dwell time, and cause the mill to stall or the media to block the screens.

The rate of base entering into the mill is controlled by pumps. Once the pumps are set, the mills practically run by themselves. The pump together with the temperature and viscosity all are factors that influence the efficiency of the mill. In some cases, the paste may not achieve the proper dispersion on the first pass due to the nature of the pigment and vehicle involved. The mill base must then be put through the mill again. However, before doing this, it is necessary to ensure that there is no "kick out" or that the grinding media is not breaking down. Once the mill is put into operation, it should not be allowed to run dry or the mill media will stick to itself and the mill will become a solid mass.

The mill is called a sand mill because of its original use of 20-30-mesh Ottawa sand. In recent years, new media have been developed for the sand mill. Some of the types of media used are as follows:

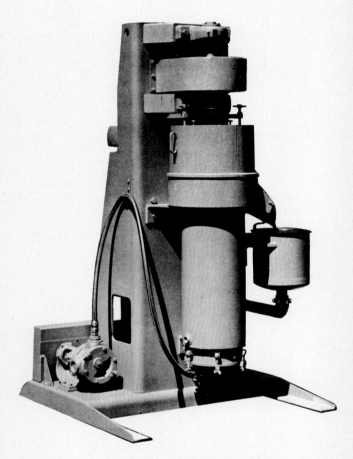

Figure 7. Sand mill. Courtesy Chicago Boiler Co.

Material	Density (g/cm^3)	Media Diameter (in.)
SiO$_2$ (sand)	2.8	0.028
SiO$_2$ (glass beads)	2.8	various size
Ceramic	2.7	0.050
Ceramic	3.6	0.020
Al$_2$O$_3$	3.5	various size
ZRO$_2$	5.4	0.026
Steel	7.1	various size

With all these new types of grinding media, the sand mill is more appropriately called a small media mill.

The ratio of 1:1 of mill base to sand is usually desirable. With different types of grinding media available, the ratio can and does vary. In addition, the size of the media and the density of the media affect the grinding rate (15). Satisfactory results can be obtained with any media by altering the paste formulation. With higher density grinding media, the viscosity of the mill base can be higher with consequent greater throughput of paste. With the higher density media, the size of the media can be smaller because of the larger surface for grinding. While higher density and smaller size media may give better results, a study should be made to include cost, wear of mill, and the degree of dispersion needed.

A relatively good quality premix is required for this mill, but all types of pigments can be used. It is difficult to clean because the paste can dry on the screen and the screen can become clogged with media. These mills disperse by shearing and by some impingement.

High-Speed Dispersers. The high-speed disperser consists of a disk that rotates at high speed in the center of a vertical tank (Figure 8). The disk, mounted on a shaft, rotates at a speed of 4500–5500 ft/min. Although there are stationary shaft dispersers, most high-speed dispersers are designed so that the shaft can go up or down; the shaft in some cases can be moved sideways. The disk is driven either by a variable-speed or a two-speed motor.

The disk (impeller) is the key feature of a high-speed disperser. The original impeller was a sawtooth disk (Figure 9); it and variations of the sawtooth disk are still the most popular. The teeth are formed by bending the edges of the disk alternately up and down. The slant and height of the teeth can vary.

The effectiveness of the high-speed mill is dependent on the viscosity of the mill base. The mill base must be thick enough to avoid turbulence since shearing and attrition efficiency is necessary for the effective use of the mill. There is a need for some dilatancy of the mill base. The mill base must be pulled in from the edge of the tank into the blade. The side of the tank and the shaft should be scraped to make sure that no pigment is caked on the side after all the pigment has been added.

In determination of the mill base, several possible procedures may be used. One of the earliest methods of determining the mill base was given by Taylor (16). The oil demand is determined by the Gardner–Coleman method, and the amount of oil needed is multiplied by 2. This is the amount of vehicle needed for dispersing the pigment in the disperser when the Cowles dissolver is used.

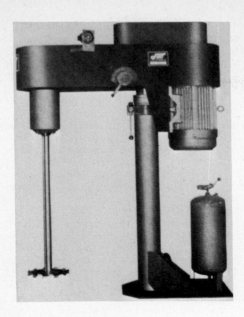

Figure 8. Variable high-speed disperser. Courtesy Jaygo, Inc.

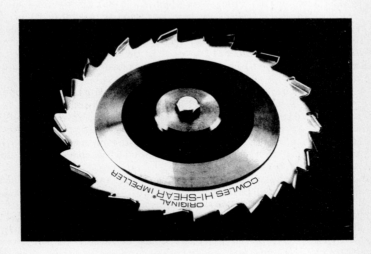

Figure 9. Type of impeller for high-speed disperser. Courtesy
Morehouse-Cowles, Division of Morehouse Industries, Inc.

Another method was presented by Guggenheim (17). The "stirring rod" method of determining the oil absorption was used. The formula as calculated by Guggenheim is the following:

$$F/C = 0.90 + \frac{VS}{145} + \frac{P}{40}$$

where f/c = oil absorption factor for high-speed impeller equipment, VS = percent vehicle solids, and P = viscosity in poises.

Still another method for determining the mill base for the high-speed disperser is the wet point/flow. It was developed in 1946 for ball and pebble mills, but since then it has been successfully used on the sand mill and on the high-speed disperser. The method developed by Daniel is quite simple, but some experience is needed to obtain the correct interpretation (18).

The wet point, the first-stage consistency, is determined by kneading a certain amount of pigment with just enough liquid to form a stiff coherent paste. This end point is similar to the results obtained from the Gardner-Coleman rub out oil absorption procedure. The end point will vary with different combinations of pigments and with different vehicles.

The flow point requires two end points. First, the paste either flows or drops from the spatula. Second, the spatula is dragged over the paste. There should be no strong drag and no permanent ridges. The surface must retain its glossy appearance or immediately regain it.

The dilatancy of the paste is indicated by the percent difference needed to go from the wet point to the flow point. Closely spaced wet and flow points indicate a good base. The gap is usually in the order of 15-30%. The degree of dilatancy can be controlled by the concentration of the solids of the vehicle. In aqueous systems, additives, such as thickeners, can be added to control dilatancy.

The position of the blade within the tank plays an important role (19). The height of the blade from the bottom should be at least half the diameter of the blade. The height can be adjusted for the level of paste up to 2-3 times the diameter of the tank. When the paste is moving in a doughnut shape pattern, about 1/3 of the blade can be seen. The impeller should be moved up or down to obtain the proper vortex.

Other useful equipment on the disperser includes a tachometer, particularly for variable-speed mills (20). Also, an ammeter should be used to measure power transfer.

With the advent of pigments that are consistent in particle size, the high-speed disperser gives high throughput of pigment. The batch can be made and completed in the same tank. However, heat-sensitive pigments and vehicles may cause problems due to the heat generated when milling.

Horizontal Mill. The horizontal mill is a newly designed micromedia mill with a horizontal grinding chamber as shown in Figure 10. The grinding chamber is filled to 70-85% volume with grinding media, which is evenly distributed in the chamber. Suitable types of media are made of glass, ceramic, or steel closely graded in size. The

Figure 10. Horizontal mill. Courtesy Premier Mill Corp.

density of media is not a critical factor. Unlike the sand mill, the horizontal mill does not work against gravity. The premix, adjusted to the type of media used, is pumped into the chamber. In the mill, the premix is subjected to both impact and shear by the grinding media.

A highly efficient cooling system allows the use of heat-sensitive products. This cooling system also allows for longer residence time in the grinding chamber. With increased quantity of grinding media and longer residence time in the mill, the need for multiple passes can be eliminated.

As the paste moves through the mill, it exists through a slotted cylindrical screen that is rotating with the grinding shaft. The centrifugal force prevents the grinding media from clogging and jamming the screen. The screen is self-cleaning, and as a consequence the paste does not cake on the screen, the media does not block the screen, and there is unrestricted product removal from the mill.

Kinetic Dispersion Mills. Another important mill is the Kady mill developed by Charles Kew of the Kinetic Dispersion Corp. (21). By many it is considered another high-speed dissolver because both use a high-speed impeller even though there are basic differences between the two mills.

The Kady mill uses a paste of low viscosity and an impeller rim speed of 8700 ft/min (22). The mill gives little or no shearing action and depends primarily on impact for dispersion. Most of the work is done in the slotted area of the stator. The agglomerates leave the rotor and smash into the stator at high speed where they receive a series of impacts that break up the agglomerates (23). The mill base leaves the stator in a jet stream where some shearing may take place. The mill base should always cover the impeller to achieve the best efficiency and to avoid the introduction of air into the base. The rotor is assisted by two propellers, one of which is above the rotor and the other below. These propellers function as a pump, drawing the batch into the head from the top and the bottom.

No premix is needed in the Kady mill. The vehicle is fed into the tank, the motor is started, and the pigments are added at a relatively fast rate. The mill is allowed to run for about 15–45 min until the required degree of dispersion is attained. The dispersion is rapid, the mill is easily cleaned, but pigment throughput is usually low. Since the mill depends on impact for dispersion, the more easily dispersed pigments are reserved for this mill.

Other Mills. There are many types of mills that have been tried in industry. Two additional types worthy of mention are the Attritor mill and the SWMill.

The Attritor mill (24) has bars that rotate, lift the grinding media, and drop it upon the paste to break up the agglomerates while at the same time the paste is circulated by a pump. The mill is designed for pigments that are difficult to disperse. It is claimed that this mill is more efficient than a ball mill.

In the case of the SWMill (Figure 11), the ingredients are placed directly in the mill because no premixing is required. The

Figure 11. Internal view of SWMill. Courtesy Epworth Manufacturing
            Co., Inc.

grinding media consists of miniature beads of balls. While moving at high speed, rotators develop a deep vortex of paste and grinding media. Grinding occurs very rapidly in the SWMill; the viscosity of the paste is the same as in the sand mill and ball mill. After the paste is properly dispersed, letdown with more vehicle or solvent can be accomplished in the mill or after transfer to another tank. Solvent used to clean the mill is saved to be used in subsequent batches or as a reducer for viscosity control.

Literature Cited

1. Definitions Committee of the Federation of Societies for Coatings Technology. "Paint/Coatings Dictionary"; Federation of Societies for Coatings Technology: Philadelphia, 1978.
2. N L Industries, Titanium Pigment Division. "Pigment Dispersion," No. P-3, p. 1.
3. Ibid., p. 2.
4. Baker Castor Oil Co. "Fundamentals of Viscosity and Thixotropy"; 1971, Technical Bulletin 83c, p. 2.
5. Patton, Temple C. "Paint Flow and Pigment Dispersion," 2nd ed.; Wiley: New York, 1979; p. 380.
6. Payne, H. F. "Organic Coating Technology"; Wiley: New York, 1961; Vol. II, p. 991.
7. Ibid., p. 993.
8. Brown, Harry M. Off. Dig. Fed. Soc. Coat. Technol. 1948, No. 284, 668.
9. Paul O. Abbel, Inc. "Handbook of Ball Mill and Pebble Mill Operation"; Paul O. Abbe, Inc.: Little Falls, NJ; p. 21.
10. Ibid., p. 7.
11. Payne, H. F. "Organic Coating Technology," p. 998.
12. Berliner, J. J., et al. "Pigment Dispersion"; J. J. Berliner Research: New York; No. 4716, p. 10.
13. Baker, Chester, P.; Vozzella, Joseph F. Off. Dig. Fed. Soc. Paint Technol. 1949, No. 294, 435.
14. Payne, H. F. "Organic Coating Technology," p. 1002.
15. Wahl, E. F. J. Paint Technol. 1969, 41(532), 345.
16. Taylor, J. J. "Mathematical Approach Toward Proper Pigment-to-Vehicle Ratios"; Los Angeles, CA, 1956.
17. Guggenheim, Stanford Off. Dig. Fed. Soc. Paint Technol. 1958.
18. Daniel, Frederick, K. J. Paint Technol. 1966, 38(500), 535.
19. Guggenheim, Stanford Off. Dig. Fed. Soc. Paint Technol. 1958.
20. Myers, C. K. "Bud" "Fundamentals of High Speed Dispersion"; Myers Engineering, Inc.: Bell, Calif. 1976; p. 11.
21. Payne, H. F. "Organic Coating Technology," p. 1007.
22. Zimmerman, O. T.; Lavine, Irvin Cost Eng. 1967, 12(1), 4-8.
23. Behrns, L. S. paper presented to the Golden Gate Society for Coatings Technology, Manufacturing Committee Conference, June 1970.
24. Patton, Temple. "Paint Flow and Pigment Dispersion," 2nd ed.; Wiley: New York, 1979; p. 439.

Author Index

Subject Index

A

A-B-A model polymers, synthesis by anionic polymerization, 186
A-B-A thermoplastic elastomers
characteristic features, 183-84
phase arrangement, 184,185f
Acetates
forms, 455
synthesis, 455
Acrylic coating polymers
durability
properties, 1038,1040t,1041
length of the ester side chain vs. film properties, 1035-36
molecular weight effect on film properties, 1034-35
monomer effect on film properties, 1036
presence of functionality in the ester side chain vs. film properties, 1036,1037t
properties vs. composition, 1034-41
solubility, 1036,1038,1039-40t
temperature vs. hardness, 1036,1039f
Acrylic coatings, recent trends, 1042,1045
Acrylic powder coating, properties, 902,903t
Acrylics
fiber formation, 455-56
polymerization, 455
Active site specificity, 82
Addition polymers, definition, 17
Additives, definition, 1137
Adhesion, definition, 30
Adhesive fracture, definition, 300,302
Adhesives
bond strengths from curing agent types, 569,570t
characteristics, 30
development, 567,569
epoxy resins, 569,571
lap shear strengths, 572,575f
Lockheed Vega production line, 567,568f
nitrile-phenolic film adhesive, 571
ultra-high-temperature adhesive, 571-72

Air-quality regulations
definition of photochemically reactive solvents, 686-87
EPA guidelines for solvent content of coatings, 688,689t
Rule 66, 686
transport theory, 688
Air spraying, 807,809
Alkoxycarbenium ions, initiation, 99
Alkyd resin processing
acidolysis, 1191-92
azeotropic process, 1192
fatty acid method, 1190-91
fusion process, 1192
oil-monoglyceride method, 1190-91
Alkyd resins, 1195-97,1198f,1199
classification, 1192-93,1194t
condensation polymerization, 1181-82
consumption in 1980, 1181,1183t
definition, 1181
fatty acid component, 1184,1186,1187t
formulation and design, 1195
history, 1181
low-solvent and solvent-free resins, 1203
modifications, 1184
polybasic acid component, 1188-89
polyol component, 1186,1188,1189f
processing, 1190
properties, 1193,1194f
properties vs. composition, 1184,1186-88,1192-95
reduced solvent emissions, 1202
Alternating copolymers, synthesis, 222
Amino coatings, chemistry, 1109-10
Amino resin chemistry
alkylation, 1107
background, 1104-5
condensation, 1105
methyolation, 1105-6
Amino resins
applications, 1109-12
background, 1101-2
chemistry, 1104-7
formaldehyde toxicity, 1104
future trends, 1113-14
manufacture, 1107-9
trends, 1112
Amorphous domain, definition, 26

1316

Production by Anne Riesberg
Editing and indexing by Karen McCeney and Deborah H. Steiner
Jacket design by Pamela Lewis

Elements typeset by Hot Type Ltd., Washington, D.C.
Printed and bound by Maple Press Co., York, Pa.